Lecture Notes in Computer Science

Commenced Publication in 1973
Founding and Former Series Editors:
Gerhard Goos, Juris Hartmanis, and Jan van Leeuwen

Editorial Board

David Hutchison
   Lancaster University, UK
Takeo Kanade
   Carnegie Mellon University, Pittsburgh, PA, USA
Josef Kittler
   University of Surrey, Guildford, UK
Jon M. Kleinberg
   Cornell University, Ithaca, NY, USA
Alfred Kobsa
   University of California, Irvine, CA, USA
Friedemann Mattern
   ETH Zurich, Switzerland
John C. Mitchell
   Stanford University, CA, USA
Moni Naor
   Weizmann Institute of Science, Rehovot, Israel
Oscar Nierstrasz
   University of Bern, Switzerland
C. Pandu Rangan
   Indian Institute of Technology, Madras, India
Bernhard Steffen
   TU Dortmund University, Germany
Madhu Sudan
   Microsoft Research, Cambridge, MA, USA
Demetri Terzopoulos
   University of California, Los Angeles, CA, USA
Doug Tygar
   University of California, Berkeley, CA, USA
Gerhard Weikum
   Max Planck Institute for Informatics, Saarbruecken, Germany

Demetres D. Kouvatsos (Ed.)

Network Performance Engineering

A Handbook on Convergent
Multi-Service Networks and
Next Generation Internet

Springer

Volume Editor

Demetres D. Kouvatsos
University of Bradford
School of Computing
Informatics and Media
Richmond Road, Bradford, BD7 1DP, UK
E-mail: d.kouvatsos@bradford.ac.uk

ISSN 0302-9743 e-ISSN 1611-3349
ISBN 978-3-642-02741-3 e-ISBN 978-3-642-02742-0
DOI 10.1007/978-3-642-02742-0
Springer Heidelberg Dordrecht London New York

Library of Congress Control Number: 2011926418

CR Subject Classification (1998): C.2, H.3.5-7, H.4.3, H.5.1

LNCS Sublibrary: SL 5 – Computer Communication Networks and Telecommunications

© Springer-Verlag Berlin Heidelberg 2011
This work is subject to copyright. All rights are reserved, whether the whole or part of the material is concerned, specifically the rights of translation, reprinting, re-use of illustrations, recitation, broadcasting, reproduction on microfilms or in any other way, and storage in data banks. Duplication of this publication or parts thereof is permitted only under the provisions of the German Copyright Law of September 9, 1965, in its current version, and permission for use must always be obtained from Springer. Violations are liable to prosecution under the German Copyright Law.
The use of general descriptive names, registered names, trademarks, etc. in this publication does not imply, even in the absence of a specific statement, that such names are exempt from the relevant protective laws and regulations and therefore free for general use.

Typesetting: Camera-ready by author, data conversion by Scientific Publishing Services, Chennai, India

Printed on acid-free paper

Springer is part of Springer Science+Business Media (www.springer.com)

Preface

Over recent years a great deal of progress has been made in the performance modelling and evaluation of the Internet, towards the convergence of multiservice heterogeneous networks, supported by internetworking and the evolution of diverse access and switching technologies. Performance modelling, evaluation and prediction of such networks are of crucial importance in view of their ever-expanding usage and the multiplicity of their component parts and the complexity of their functioning.

However, many important and challenging performance-engineering issues need to be addressed and resolved, such as those involving heterogeneous network architectures and technology integration, traffic modelling and characterization, management, congestion control, routing and quality-of-service (QoS). The ultimate goal is the establishment of a global and wide-scale integrated broadband network infrastructure for the efficient support of multimedia applications with different QoS guarantees. Of particular interest and challenge is the design and engineering of the next- and future-generation Internets, such as those based on the convergence of heterogeneous wireless networks enabled by internetworking and wireless mesh networking technologies. Moreover, of vital interest is the creation of generic evaluation platforms capable of measuring and validating the performance of networks of diverse technology and multi-service interoperability. In this context, robust quantitative methodologies and performance modelling tools are needed, such as those based on queueing network models (QNMs), in order to provide a sound theoretical underpinning of application-driven research leading to credible and cost-effective algorithms for the performance evaluation and prediction of convergent heterogeneous networks under various traffic handling protocols.

The principal objective of the *Network Performance Engineering* handbook is to bring together technical contributions and future research directions in the performance engineering of heterogeneous networks and the Internet by eminent researchers and practitioners from industry and academia worldwide. The handbook consists of 44 extended and revised chapters, which were selected following a rigorous international peer review. These chapters were drawn from selected lectures and tutorials of the recent six HET-NETs International Working Conferences and associated EU PhD courses on the 'Performance Modelling and Evaluation of Heterogeneous Networks'. These events took place in Ilkley, UK (July 2003 – 2005, September 2006), Karlskrona, Sweden (February 2008) and Zakopane, Poland (January 2010) and were staged under the auspices of the EU Networks of Excellence (NoE) *Euro-NGI* and *Euro-FGI* (c.f., *Work-package WP.SEA.6.1)* and with the collaboration of EU academic and industrial consortia and other international organizations. Moreover, this handbook is part of the

Final Deliverables of NoE Euro-NGI and Euro-FGI to the European Commission (c.f., *Deliverable D.SEA.6.1.6c*).

The chapters of the handbook constitute essential introductory material for further research and development in the performance modelling, analysis, design and engineering of heterogeneous networks and of next- and future-generation Internets. They aim to unify relevant material already known but dispersed in the literature, introduce the readers to unfamiliar and unexposed research areas and, generally, illustrate the diversity of research found in the high-growth field of convergent multiservice heterogeneous networks and the Internet. Moreover, the theoretical themes of the handbook, such as those focusing on traffic modelling, quantitative network methodologies and associated performance engineering tools, are also of relevance to the design and development of other types of discrete flow systems such as flexible manufacturing systems and transportation networks.

The chapters of the *Networks Performance Engineering* handbook are broadly classified into 12 parts covering the following topics: 'Measurement Techniques', 'Traffic Modelling and Engineering', 'Queueing Systems and Networks', 'Analytic Methodologies', 'Simulation Techniques', 'Performance Evaluation Studies', 'Mobile, Wireless and Ad Hoc Networks', 'Optical Networks', 'QoS Metrics and Algorithms', 'All IP Convergence and Networking', 'Network Management and Services' and 'Overlay Networks'.

In Part 1, 'Measurement Techniques', Popescu and Constantinescu investigate Kleinrock's independent assumption by carrying out measurements, performance modelling and analysis of end-to-end delay in a chain of IP (Internet Protocol) routers represented by a tandem queueing system with correlated traffic flows. Arlos discusses some of the fundamental aspects of performance measurements with particular focus on the application-level measurements for the estimation of the network's performance properties. Fiedler et al. assess the impact of application-perceived throughput on the performance of networked applications and focus on the process of user-perceived throughput in GPRS (General Packet Radio Service) and UMTS (Universal Mobile Telephone System) systems over small averaging intervals and active measurements of streaming applications. In Part 2, 'Traffic Modelling and Engineering', Markovich and Krieger provide a common methodology for the statistical characterization of peer-to-peer packet traffic arising from passive VoIP (voice over IP) and video measurements, and consider applications using individual Skype flows and the aggregated flow of video packets exchanged with a mobile peer in an overlay network. Nogueira et al. discuss the suitability of MMPP (Markov Modulated Poisson Processes), and evaluate the credibility of parameter-fitting procedures for the characterization of Internet packet traffic flows incorporating self-similarity and long-range dependence over multiple time scales. Iovanna et al. propose an adaptive traffic management system in MPLS (Multi-Protocol Label Switching) networks operating on short timescales, and employ an economics-based figure of merit for the relocation of bandwidth. Fretwell and Kouvatsos present the batch renewal process for modelling both LRD (long range dependent) and SRD (short range dependent) traffic flows in both discrete and continuous time domains, and present applications in the analysis of simple queues and queueing network models. Larijani reviews LAN (Local Area Network) technologies with self-similar and long-range dependent traffic processes, and highlights some modelling methods with particular emphasis on pseudo self-similar models. Liang et al. undertake measurements and analytic studies of IP traffic in a WLAN (Wireless Local Area Network) environment, and carry out an investigation into the characterization on protocol distribution and the modelling of IP packet inter-arrival times. In Part 3, 'Queueing Systems and Networks', Walraevens et al. present analytical techniques for the study of discrete-time two-class traffic queueing systems with priority scheduling disciplines, and determine related performance measures via the probability generating functions approach. Balsamo highlights exact and approximate algorithms for the quantitative evaluation of open and closed queueing networks with finite capacity, and reviews equivalence properties amongst different blocking mechanisms as well as applications into communication networks and distributed computer systems. Anisimov reports on asymptotic investigations and a new methodology for the analysis of queueing systems and networks with heavy traffic, based on the limit theorems of the averaging principle and diffusion approximation types. Levy et al. discuss the fundamental principles and properties related to queue fairness from the perspective of the relevant applications and carry out a comparative study with some emphasis on computer communications networks. In Part 4, 'Analytic Methodologies', Pagano provides a heuristic interpretation of basic concepts and theorems of LDT (Large Deviation Theory) and highlights its applications into the analysis of single queues and network dimensioning as well as rare event simulation. Thomas and Bradley use the Markovian process algebra PEPA to specify and analyze non-product form parallel queues, which are decomposed into their components to obtain, with some degree of confidence, a scalable solution. Harrison and Thomas use the reversed process, based on the RCAT (Reversed Compound Agent Theorem), to derive expressions for the steady state probability distribution of a class of product-form solutions in PEPA for generalized closed, queueing networks with multiple servers, competing services and functional rates within actions. Kouvatsos and Assi undertake an exposition of the 'classical' EME (Extensive Maximum Entropy) formalism and generalised NME (Non-extensive Maximum Entropy) formalism in conjunction with their applicability to the analysis of queues with bursty and/or heavy tails often observed in performance evaluation studies of heterogeneous networks and the Internet exhibiting traffic burstiness, self-similarity and LRD. Ferreira and Pacheco address the usual and level-crossing stochastic ordering of semi-Markov processes, and carry out comparisons against simulation of processes with a given distribution by employing the sample-path approach. Mitrani applies the spectral expansion method to obtain exact solutions for a large class of state-dependent queueing models, and illustrates their applicability in the fields of computing, communication and manufacturing systems. Czachorski and Pekergin present the method of diffusion approximation for the modelling and analysis of single queues and networks of stations with general service times and transient states in the presence of general bursty traffic streams such as multimedia transfers in modern communication networks. In Part 5, 'Simulation Techniques', Dhaou et al. adopt the 'Cross Layer' concept to implement a dynamic simulation tool for the performance optimisation of network models composed of evolving MANETS (Mobile Ad Hoc Networks) and Satellites, which can be often reduced to smaller submodels by decomposition or aggregation methods. M. Vill'en-Altamirano and J. Vill'en-Altamirano present RESTART, an accelerated simulation technique for estimating rare-event probabilities in queueing networks and ultra-reliable systems, based on the choice of the importance function of the system state used for determining when simulation retrials are made. In Part 6, 'Performance Evaluation Studies', Krieger considers a hierarchical algebraic description of a Web graph with host-oriented clustering of pages, and proposes a computation of the stationary distribution of the underlying Markov chain of a random surfer, based on aggregation/disaggregation procedures and algebraic multigrid methods. Pagano and Secchii address the widespread diffusion of TCP (Transmission Control Protocol) over the Internet, and introduce simple approaches to describe the dynamics of an individual source over a simplified network model and detailed techniques for modelling the behaviour of a set of TCP connections over an arbitrary complex network. Mkwawa et al. analyze an open queueing network model (QNM) representing the functional units and application servers of an IMS (IP Multimedia Subsystem)-based testbed architecture implemented by Nokia-Siemens as part of the EU IST (Information Society Technologies) VITAL project and assess the handover process of SIP (Session Initiation Protocol) messages between WLAN and GSM (Global System for Mobile Communications) access networks. Do and Chakka suggest Markovian queueing models, based on generalizations of QBD (Quasi Birth and Death) processes and devise steady-state solutions assessing the impact of burstiness and autocorrelation of traffic flows of packets for the performance evaluation of next-generation networks. Chakka and Do present an analytic methodology for the steady state solution of a complex multi-server Sigma-type queue and its applicability to the performance evaluation of an optical burst switching multiplexer. Wang et al. present a detailed review of various handover schemes proposed in the literature and focus on an analytic model developed for a DGCS (Dynamic Guard Channel Scheme), which manages adaptively the channels reserved for handover calls. Shah et al. employ simulation and analytic methodologies for the performance modelling and optimisation of DOCSIS (Data-over-cable service interface specification) 1.1/2.0 HFC (hybrid fibre coax) networks with particular focus on the contention resolution algorithm, upstream bandwidth allocation strategies, flow-priority scheduling disciplines, QoS provisioning and TCP applications. In Part 7, 'Mobile, Wireless and Ad Hoc Networks', Casares-Giner et al. deal with mobility aspects of wireless mobile telecommunication systems and provide some basic frameworks for mobility models as applied to the performance evaluation of relevant mobility management procedures, such as handover and location update. Remondo reviews some of the main enabling technologies of wireless Ad

Hoc networks, including physical and medium access control layers, networking, transport issues and dynamic routing protocols, and discusses proposals that aim at maintaining service level agreements in ad hoc networks either in isolation or connected to fixed networks. Popescu et al. present ROMA, a new middleware architecture implemented at the application layer, enabling seamless handover in wireless networks with dynamic combinations of services and underlying transport substrates (overlay and underlying networks). Mkwawa and Kouvatsos review some current broadcasting methods in MANETs for the control and routing information of multicast and point-to-point communication protocols in conjunction with recommendations on how to improve the efficiency and performance of tree- and cluster-based methods. Popescu et al. report on some recent developments and challenges focusing on seamless handover, supported by several components such as mobility and connectivity management as well as Internet mobility, under the auspices of the recent EU research projects MOBICOME and PERIMETER, based on IMS technology standards. In Part 8 'Optical Networks', Atmaca and Nguyen review the infrastructure and evolution of MANs (Metropolitan Area Networks) towards OPS (Optical Packet Switching) networks and highlight performance issues in optical networking in metropolitan areas in terms of optical packet format, MAC (medium access control) protocol, QoS and traffic engineering issues. Castel-Taleb et al. focus on packet aggregation mechanisms on the edge router of an optical network and present an efficient aggregation mechanism supporting QoS requirements of IP flows. Moreover, analytical models based on Markov chains are devised in order to assess the packetization efficiency (filling ratio) and determine its mean time for each class. Mouchos et al. review the traffic characteristics of an optical carrier's OC-192 link, based on the IP packet size distribution, traffic burstiness and self-similarity. Under optical burst switching (OBS), a performance evaluation is undertaken involving the dynamic offset control (DOC) and Just Enough Time (JET) allocation protocols. Moreover, parallel generators of optical bursts are implemented and simulated using the Graphics Processing Unit (GPU) and the Compute Unified Device Architecture. In Part 9, 'QoS Metrics and Algorithms', Belzarena and Simon address the estimation of QoS parameters in the Internet using traffic traces, end-to-end active measurements and statistical learning tools, and determine the admission control problem from results of the many sources and small buffer asymptotics. De Vuyst et al. employ the supplementary variables approach in the transform domain to analyze a discrete-time single-server queue at equilibrium with a new type of scheduling mechanism for the control of delay-sensitive and delay-tolerant classes of packet arrivals. In Part 10, 'All IP Convergence and Networking', Sun presents an overview of the fundamental concepts and research issues of IP networking with particular emphasis on the current and next-generation Internets, including some important mechanisms to control and manage diversity and network resources towards the enhancement of network performance and QoS. Popescu et al. report, in the context of the NoE Euro-NGI project ROVER, on some recent advances in the research and development of multimedia distribution over IP and suggest routing in overlay routing as an alternative solution for content distribution. In Part 11, 'Network Management and Services', Akhlaq et al. carry out a performance evaluation study of NIDS (Network Intrusion Detection Systems) on commodity hardware by employing evasive and avoidance strategies on a suitable test bench, and they implement techniques to simulate real-life normal and attack-like traffic flows. In Part 12, 'Overlay Networks', Dragos and Popescu introduce fundamental concepts of unicast QoS in overlay networks and report performance evaluation results by adopting the ORP (Overlay Routing Protocol) developed at BTH (Blekinge Institute of Technology) in Karlskrona, Sweden. Finally, Mkwawa and Kouvatsos review some graph-theoretic-based methods for the selection of a set of topologically diverse routers towards the provision of independent paths, and propose a graph decomposition-based approach for the maximization of path diversity without degrading network performance in terms of latency.

I would like to end this preface by expressing my deepest thanks to the following organizations for their support of the HET-NETs International Working Conferences over the recent years: NoE Euro-NGI and Euro-FGI (EU Commission), INFORMS - The Applied Probability Society (USA), EPSRC – The Engineering and Physical Sciences Research Council (UK), The British Computer Society (UK), IEE – The Institute of Electrical and Electronic Engineers (UK), The ACATS Forum - ATS Network Consortium Proprietary (EU) and the EU IST Consortium 'VITAL' consisting of Telekom Austria AG (Austria), Siemens AG, Solinet GmbH & Alcatel SEL (Germany), Teletel SA, Voiceglobe sprl, Keletron & University of Patras (Greece), Telefonica I+D (Spain) and the University of Bradford (UK). Thanks are also extended to the members of the Advisory Boards and Program Committees of the HET-NETs International Working Conferences, as well as to the expert referees worldwide for their invaluable and timely peer reviews. Thanks are also due to Is-Haka Mkwawa, University of Plymouth (UK), for his expert technical support and kind uploading of the *Network Performance Modelling* handbook to the Springer FTP Server.

January 2011   Demetres D. Kouvatsos

External Reviewers

Samuli Aalto
Ramon Agusti
Sohair Al-Hakeem
Eitan Altman
Jorge Andres
Vladimir Anisimov
Laura Aspirot
Salam Adli Assi
Tulin Atmaca
Zlatka Avramova
Frank Ball
Simonetta Balsamo
Ivano Bartoli
Alejandro Beccera
Monique Becker
Pablo Belzarena
Andre-Luc Beylot
Andreas Binzenhoefer
Jozsef Biro
Pavel Bocharov
Sem Borst
Nizar Bouabdallah
Richard Boucherie
Christos Bouras
Onno Boxma
Chris Blondia
Alexandre Brandwajn
George Bravos
Oliver Brun
Herwig Bruneel
Alberto Cabellos-Aparicio
Patrik Carlsson
Fernando Casadevall
Vicente Casares-Giner
Hind Castel
Llorenc Cerda
Eduardo Cerqueira
Mohamad Chaitou
Ram Chakka

Meng Chen
Stefan Chevul
Tom Coenen
Doru Constantinescu
Marco Conti
Laurie Cuthbert
Tadeusz Czachorski
Koen De Turck
Danny De Vleeschauwer
Stijn De Vuyst
Alexandre Delye Mazieux
Luc Deneire
Felicita Di Giandomenico
Manuel Dinis
Tien Do
Jose Domenech-Benlloch
Rudra Dutta
Joerg Eberspaecher
Antonio Elizondo
Khaled Elsayed
Peder Emstad
David Erman
Melike Erol
Jose Oscar Fajardo Portillo
Fatima Ferreira
Markus Fiedler
Jean-Michel Fourneau
Rod Fretwell
Wilfried Gangsterer
Peixia Gao
Ana Garcia Armada
David Garcia-Roger
Georgios Gardikis
Vincent Gauthier
Alfonso Gazo
Xavier Gelabert Doran
Leonidas Georgiadis
Bart Gijsen
Jose Gil

External Reviewers

Cajigas Gillermo
Stefano Giordano
Jose Gonzales
Ruben Gonzalez Benitez
Annie Gravey
Klaus Hackbarth
Slawomir Hanczewski
Guenter Haring
Peter Harrison
Hassan Hassan
Dan He
Gerard Hebuterne
Bjarne Helvik
Robert Hines
Enrique Hernandez
Helmut Hlavacs
Amine Houyou
Hanen Idoudi
Ilias Iliadis
Dragos Ilie
Paola Iovanna
Andrzej Jajszczyk
Lorand Jakab
Sztrik Janos
Robert Janowski
Terje Jensen
Laszlo Jereb
Mikael Johansson
Hector Julian-Bertomeu
Athanassios Kanatas
Tamas Karasz
Johan Karlsson
Stefan Koehler
Daniel Kofman
Vangellis Kollias
Huifang Kong
Kimon Kontovasilis
Rob Kooij
Goerge Kormentzas
Ivan Kotuliak
Harilaos Koumaras
Demetres Kouvatsos
Tasos Kourtis
Udo Krieger
Koenraad Laevens

Samer Lahoud
Jaakko Lahteenmaki
Juha Leppanen
Amaia Lesta
Hanoch Levy
Wei Li
Yue Li
Fotis Liotopoulos
Renato Lo Cigno
Michael Logothetis
Carlos Lopes
Johann Lopez
Andreas Maeder
Tom Maertens
Thomas Magedanz
Sireen Malik
Lefteris Mamatas
Jose Manuel Gimenez-Guzman
Michel Marot
Alberto Martin
Jim Martin
Simon Martin
Ignacio Martinez Arrue
Jose Martinez-Bauset
Martinecz Matyas
Lewis McKenzie
Madjid Merabti
Bernard Metzler
Geyong Min
Isi Mitrani
Nicholas Mitrou
Is-Haka Mkwawa
Hala Mokhtar
Miklos Boda
Sandor Molnar
Edmundo Monteiro
Ioannis Moscholios
Harry Mouchos
Luis Munoz
Maurizio Naldi
Victor Netes
Pal Nilsson
Simon Oechsner
Sema Oktug
Mohamed Ould-Khaoua

Antonio Pacheco
Michele Pagano
Zsolt Pandi
Panagiotis Papadimitriou
Stylianos Papanastasiou
Nihal Pekergin
Izaskun Pellejero
Roger Peplow
Paulo Pereira
Gonzalo Perera
Jordi Perez-Romero
Rubem Perreira
Guido Petit
Maciej Piechowiak
Michal Pioro
Jonathan Pitts
Vicent Pla
Nineta Polemi
Daniel Popa
Adrian Popescu
Dimitris Primpas
David Remondo-Bueno
David Rincon
Roberto Sabella
Francisco Salguero
Sebastia Sallent
Werner Sandmann
Ana Sanjuan
Lambros Sarakis
Wolfgang Schott
Raffaello Secchi
Maria Simon
Swati Sinha Deb
Charalabos Skianis
Amaro Sousa
Dirk Staehle
Maciej Stasiak
Panagiotis Stathopoulos
Bart Steyaert
Zhili Sun

Kannan Sundaramoorthy
Riikka Susitaival
Janos Sztrik
Yutaka Takahashi
Sotiris Tantos
Leandros Tassiulas
Luca Tavanti
Silvia Terrasa
Geraldine Texier
David Thornley
Florence Touvet
Phuoc Tran-Gia
Chia-Sheng Tsai
Thanasis Tsokanos
Krzysztof Tworus
Rui Valadas
Rob Van der Mei
Vassilios Vassilakis
Vasos Vassiliou
Sandrine Vaton
Tereza Vazao
Speros Velentzas
Dominique Verchere
Pablo Vidales
Nguyen Viet Hung
Manolo Villen-Altamirano
Bart Vinck
Jorma Virtamo
Kostas Vlahodimitropoulos
Joris Walraevens
Xin Gang Wang
Wemke Weij
Sabine Wittevrongel
Mehti Witwit
Michael Woodward
George Xilouris
Mohammad Yaghmaee
Bo Zhou
Stefan Zoels
Piotr Zwierzykowski

Table of Contents

Measurement Techniques

On Kleinrock's Independence Assumption 1
    Adrian Popescu and Doru Constantinescu

Application Level Measurement 14
    Patrik Arlos

Measurements and Analysis of Application-Perceived Throughput via
Mobile Links 37
    Markus Fiedler, Lennart Isaksson, and Peter Lindberg

Traffic Modelling and Engineering

Statistical Analysis and Modeling of Peer-to-Peer Multimedia Traffic ... 70
    Natalia M. Markovich and Udo R. Krieger

Markovian Modelling of Internet Traffic 98
    António Nogueira, Paulo Salvador, Rui Valadas, and
    António Pacheco

Multi-timescale Economics-Driven Traffic Management in MPLS
Networks 125
    Paola Iovanna, Maurizio Naldi, Roberto Sabella, and Cristiano Zema

Modelling LRD and SRD Traffic with the Batch Renewal Process:
Review of Results and Open Issues 141
    Rod J. Fretwell and Demetres D. Kouvatsos

Local Area Networks and Self-similar Traffic 174
    Hadi Larijani

Characterisation of Internet Traffic in Wireless Networks 191
    Lei Liang, Yang Chen, and Zhili Sun

Queueing Systems and Networks

Performance Analysis of Priority Queueing Systems in Discrete Time ... 203
    Joris Walraevens, Dieter Fiems, and Herwig Bruneel

Queueing Networks with Blocking: Analysis, Solution Algorithms and
Properties 233
    Simonetta Balsamo

Switching Queueing Networks 258
  Vladimir V. Anisimov

Principles of Fairness Quantification in Queueing Systems 284
  Hanoch Levy, Benjamin Avi-Itzhak, and David Raz

Analytic Methodologies

Large Deviations Theory: Basic Principles and Applications to
Communication Networks 301
  Michele Pagano

Analysis of Non-product Form Parallel Queues Using Markovian
Process Algebra 331
  Nigel Thomas and Jeremy Bradley

Product-Form Solution in PEPA via the Reversed Process 343
  Peter G. Harrison and Nigel Thomas

Generalised Entropy Maximisation and Queues with Bursty and/or
Heavy Tails 357
  Demetres D. Kouvatsos and Salam A. Assi

Stochastic Ordering of Semi-Markov Processes 393
  Fátima Ferreira and António Pacheco

Spectral Expansion Solutions for Markov-Modulated Queues 423
  Isi Mitrani

Diffusion Approximation as a Modelling Tool 447
  Tadeusz Czachórski and Ferhan Pekergin

Simulation Techniques

Cross Layer Simulation: Application to Performance Modelling of
Networks Composed of MANETs and Satellites 477
  Riadh Dhaou, Vincent Gauthier, M. Issoufou Tiado,
  Monique Becker, and André-Luc Beylot

The Rare Event Simulation Method RESTART: Efficiency Analysis
and Guidelines for Its Application 509
  Manuel Villén-Altamirano and José Villén-Altamirano

Performance Evaluation Studies

An Algebraic Multigrid Solution of Large Hierarchical Markovian
Models Arising in Web Information Retrieval 548
  Udo R. Krieger

An Introduction to Modelling and Performance Evaluation for TCP
Networks 571
    Michele Pagano and Raffaello Secchi

Performance Modelling and Evaluation of a Mobility Management
Mechanism in IMS-Based Networks 594
    Is-Haka M. Mkwawa, Demetres D. Kouvatsos,
    Wolfgang Brandstätter, Gerhard Horak, Alfons Geier, and
    Christoforos Kavadias

Generalized QBD Processes, Spectral Expansion and Performance
Modeling Applications 612
    Tien Van Do and Ram Chakka

Some New Markovian Models for Traffic and Performance Evaluation
of Telecommunication Networks................................ 642
    Ram Chakka and Tien Van Do

Modelling and Analysis of a Dynamic Guard Channel Handover Scheme
with Heterogeneous Call Arrival Processes 665
    Lan Wang, Geyong Min, Demetres D. Kouvatsos, and
    Xiangxiang Zuo

On the Performance Modelling and Optimisation of DOCSIS HFC
Networks 682
    Neelkamal P. Shah, Demetres D. Kouvatsos, Jim Martin, and
    Scott Moser

Mobile, Wireless and Ad Hoc Networks

Mobility Models for Mobility Management 716
    Vicente Casares-Giner, Vicent Pla, and Pablo Escalle-García

Wireless Ad Hoc Networks: An Overview 746
    David Remondo

Broadcasting Methods in MANETS: An Overview 767
    Is-Haka M. Mkwawa and Demetres D. Kouvatsos

ROMA: A Middleware Framework for Seamless Handover 784
    Adrian Popescu, David Erman, Karel de Vogeleer,
    Alexandru Popescu, and Markus Fiedler

Seamless Roaming: Developments and Challenges 795
    Adrian Popescu, David Erman, Dragos Ilie, Markus Fiedler,
    Alexandru Popescu, and Karel de Vogeleer

Optical Networks

Optical Metropolitan Networks: Packet Format, MAC Protocols and
Quality of Service 808
    Tülin Atmaca and Viet-Hung Nguyen

Performance of Multicast Packet Aggregation in All Optical Slotted
Networks 835
    Hind Castel-Taleb, Mohamed Chaitou, and Gerard Hébuterne

Performance Modelling and Traffic Characterisation of Optical
Networks 859
    Harry Mouchos, Athanasios Tsokanos, and Demetres D. Kouvatsos

QoS Metrics and Algorithms

The Search for QoS in Data Networks: A Statistical Approach 891
    Pablo Belzarena and María Simon

Transform-Domain Analysis of Packet Delay in Network Nodes with
QoS-Aware Scheduling 921
    Stijn De Vuyst, Sabine Wittevrongel, and Herwig Bruneel

All IP Convergence and Networking

IP Networking and Future Evolution 951
    Zhili Sun

Content Distribution over IP: Developments and Challenges 979
    Adrian Popescu, Demetres D. Kouvatsos, David Remondo, and
    Stefano Giordano

Network Management and Services

Implementation and Evaluation of Network Intrusion Detection
Systems 988
    Monis Akhlaq, Faeiz Alserhani, Irfan Awan, John Mellor,
    Andrea J. Cullen, and Abdullah Al-Dhelaan

Overlay Networks

Unicast QoS Routing in Overlay Networks 1017
    Dragos Ilie and Adrian Popescu

Overlay Networks and Graph Theoretic Concepts 1039
    Is-Haka M. Mkwawa, Demetres D. Kouvatsos, and Adrian Popescu

Author Index 1055

On Kleinrock's Independence Assumption

Adrian Popescu[1] and Doru Constantinescu[2]

[1] Blekinge Institute of Technology, Sweden
adrian.popescu@bth.se
[2] Telenor Connexion, Sweden
doru.constantinescu@telenor.com

Abstract. This chapter is concerned with the impact of traffic correlations on the end-to-end delay of packets in a chain of IP routers represented by an open tandem queueing system. Particular consideration is given to Kleinrock's independence assumption on the performance impact of traffic correlation on packet switched networks with Poisson arrival processes and exponential packet length distributions. According to this assumption, any traffic correlations can be ignored and the effect on delay performance is negligible, subject to sufficient traffic mixing and moderate-to-heavy traffic loads. In this context, results associated with measurements, traffic modeling and delay analysis of an actual chain of IP routers are reported from experiments conducted at the Blekinge Institute of Technology in Karlskrona, Sweden. It is shown that no experimental evidence was found in support of Kleinrock's independence assumption as traffic correlation has an adverse effect on the end-to-end delay of the tandem queueing system.

1 Introduction

As the Internet has emerged as the backbone of worldwide business and commercial activities, end-to-end (e2e) Quality of Service (QoS) for data transfer becomes a significant factor. In this context, one-way delay is an important QoS parameter. It is a key metric in evaluating the performance of networks as well as the quality of service perceived by end users. This parameter is defined by both the IETF (One Way Delay for IP Performance Metrics) and the International Telecommunications Union - Telecommunications Standardization (IP Packet Transfer Delay).

Today, network capacities are deliberately being overprovisioned in the Internet so that the packet loss rate and the delay are low. However, given the heterogeneity of the network and the fact that the overprovisioning solution is not adopted everywhere, especially not by backbone teleoperators in developing countries, the question arises as to how the delay performance impacts the e2e performance. There are several important parameters that may impact the e2e delay performance in the link, e.g., traffic self-similarity, routing flaps and link utilization [16, 18].

Several publications report on the e2e delay performance, and both Round-Trip Time (RTT) and One-Way Transit Time (OWTT) are considered [4, 6, 18, 19].

Traffic measurements based on both passive measurements and/or active probing are used. As a general comment, it has been observed that RTT and OWTT show large "peak-to-peak" variations, in the sense that maximum delays far exceed minimum delays. Although a range of more than 10:1 in RTTs seems to be common, most of connections have RTTs between 15 and 500 s [12]. Further, it has been observed that OWTT variations (for opposite directions) are generally asymmetric, with different delay distributions. They also seem to be correlated with packet loss rates [19]. Periodic delay spikes and packet losses have been observed, which seem to be a consequence of routing flaps [18].

Typical distributions for OWTT have been observed to have a Gamma-like shape and to possess heavy tail [4,17]. The parameters of the Gamma distribution have been observed to depend upon the path (e.g., regional, backbone) and the time of the day. The heavy tail behavior is due to the presence of self-similarity in Internet packet delay [3]. Typical queueing models like M/M/1, M/G/1 and using Fractional Brownian Models (fBm) for traffic models have been shown to underestimate average queueing delays for link utilization below 70% [18]. Furthermore, another important question is regarding the impact on OWTT performance of diverse correlations existent in a tandem queueing system and whether Kleinrock's independence assumption [14] is valid. Leonard Kleinrock suggested that, under specific conditions (e.g., Poisson arrival processes, packet lengths that are nearly Exponentially distributed, a densely connected network with sufficient traffic mixing and moderate-to-heavy traffic loads), the effects of correlations may become small and therefore completely ignored.

The paper is reporting on some of the results obtained in experiments done at the Blekinge Institute of Technology (BTH) in Karlskrona, Sweden, on measurements, modeling and analysis of delay in a chain of IP routers. Particular focus is given to validating Kleinrock's independence assumption regarding the effect of correlations in a tandem queueing system. Our results show that this assumption is not valid in our experiments.

The paper is organized as follows. In Section 2 we describe the delay components associated with the OWTT in a chain of IP routers. In Section 3, we give an overview on queueing delay in a chain of IP routers, with particular focus on correlations and Kleinrock's independence assumption. In Section 4 we shortly describe the experiments done at BTH. We briefly report in Section 5 on the results obtained in our experiments on measurement, modeling and analysis of delay in a chain of IP routers with particular focus on the validation of Kleinrock's independence assumption. In Section 6 we conclude the paper.

2 One-Way Transit Time Components

One-Way Transit Time (OWTT) is measured by timestamping a specific packet at the sender, sending the packet into the network, and comparing then the timestamp with the timestamp generated at the receiver [1]. Packet timestamping can be done either by software (for the case of delay measurements at the application level) or by hardware (for the case of delay measurements at the network level), and in this case special hardware is used. Clock synchronization between the sender and the receiver nodes is important for the precision of one-way delay measurements [1].

OWTT has several components:

$$OWTT = D_{prop} + \sum_{i=0}^{N} D_{n,i} \qquad (1)$$

where the delay per node i, $D_{n,i}$ is given by:

$$D_{n,i} = D_{tr,i} + D_{proc,i} + D_{q,i} \qquad (2)$$

The components are as follows:

- D_{prop} is the total propagation delay along the physical links that make up the Internet path between the sender and the receiver. This time is solely determined by the properties of the communication channel and the distance. It is independent of traffic conditions on the links.
- N is the number of nodes between the sender and the receiver.
- $D_{tr,i}$ is the transmission time for node i. This is the time it takes for the node i to copy the packet into the first buffer as well as to serialize the packet over the communication link. It depends on the packet length and it is inversely proportional to the link speed.
- $D_{proc,i}$ is the processing delay at node i. This is the time needed to process an incoming packet (e.g., to decode the packet header, to check for bit errors, to lookup routes in a routing table, to recompute the checksum of the IP header) as well as the time needed to prepare the packet for further transmission, on another link. This delay depends on parameters like network protocol, computational power at node i, and efficiency of network interface cards.
- $D_{q,i}$ is the queueing delay in node i. This delay refers to the waiting time in output buffer, and depends upon traffic characteristics, link conditions (e.g., link utilization, interference with other IP packets) as well as implementation details of the node. It is mentioned that we consider routers with best-effort service model, i.e., routers where an output port is modeled as a single output queue.

Statistics like mean, median, maximum, minimum, standard deviation, variance and peakedness are usually used in the calculation of delay for non-corrupted packets. Typical values obtained for OWTT range from tens of μs (between two hosts on the same LAN) to hundreds of ms (in the case of hosts placed in different continents) [5].

For a general discussion, the OWTT delay can be partitioned into two components, a deterministic delay D_d and a stochastic delay D_s:

$$OWTT = D_d + D_s \qquad (3)$$

D_{prop}, D_{tr} and (partly) D_{proc} are contributing to the deterministic delay D_d, whereas the stochastic delay D_s is created by D_q and, at some extent, D_{proc}. The stochastic part of the router processing delay can be observed especially in the case of low and very low link utilization, i.e., when the queueing delays are minor.

3 Queueing Delay in Chained IP Routers

An important delay component in IP networks is the queueing delay in routers and switches and the jitter that may appear in the case of large queueing delays. In a chain of IP routers, there may be many transmission queues (e.g., output ports in routers) that may interact with each other in the sense that a traffic stream leaving a queue enters others queues, likely after merging with other traffic streams coming from other queues (Figure 1).

The direct effect of traffic merging in packet networks is that the character of the arrival process at a downstream queue changes. Since the same packets visit more queues in a tandem queueing system, the service times of each packet at the visited successive queues are typically positively correlated. Furthermore, given that the service times at two queues are dependent, then the packet interarrival times become correlated with packet lengths at the downstream queue. Long packets typically wait less time than short packets at a downstream queue, which is because they need longer time for service at the upstream queue. The consequence is that the downstream queue has more time to empty out.

A similar situation is in the case of a slow truck traveling on a narrow street, with one track only. The truck typically has empty space ahead but more faster cars following behind the truck. Simulation studies have shown that in real situations, when interarrival times and service times are strongly correlated, the average delay per packet at the downstream queue tends to be less than in the case the dependence was not existent. On the other hand, under heavy loads, the average delay tends to be dramatically less. A reverse situation is valid under light traffic conditions as well [11].

Leonard Kleinrock studied the problem of correlations between service and interarrival times in the context of a queueing network model for communication networks [14]. He observed that, if there is sufficient mixing of traffic, then the dependence effect may become small, and therefore it can be completely ignored. Kleinrock suggested consequently that merging several packet streams on a tandem queueing system has an effect similar to restoring the independence of interarrival times and packet lengths. This means that each time a packet is received at a node in a network, an Exponential distribution can be used to generate a new length for the specific packet. This is clearly false since packets

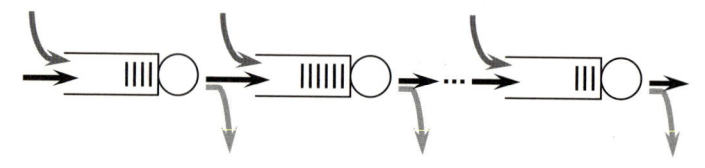

Fig. 1. Tandem queueing

maintain their lengths as they pass through the network, but Kleinrock showed that the effect on delay performance is negligible.

Based on these arguments, it was concluded that it is appropriate to adopt an M/M/1 queueing model for every queue in a tandem queueing system regardless of the interaction of traffic with other traffic flows. This is known as *Kleinrock's independence assumption*, which amounts to ignoring correlations. This assumption seems to be a good approximation for the case of Poisson arrival processes, packet lengths that are nearly Exponentially distributed, a densely connected network with sufficient traffic mixing and moderate-to-heavy traffic loads [2, 14]. It can however significantly overestimate delays in tandem queueing systems with little traffic mixing, where there is strong positive correlation between service and interarrival times.

The process of changing the character of the arrival process at downstream queues is very complex. It is heavily influenced by different aspects, like the presence of different traffic classes (with specific traffic characteristics) sharing the same queue, the presence of Long-Range Dependence (LRD) in traffic, the presence of tandem links with different link utilization and the presence of a large number of traffic sources sharing the network. Today, the situation is such as it is not clear what the arrival processes at downstream queues are, and therefore it is impossible to do a precise analysis like, e.g., in the case of M/M/1 or M/G/1 queueing systems. Delay models based on Poisson assumptions are totally inappropriate for analysis at downstream queues and no analytical solutions are actually known for even a simple tandem queueing system with Poisson arrivals and Exponentially distributed service times [2, 4, 18].

There are several classes of correlations in a queueing system, all of them contributing to the complexity of the process of changing the character of the arrival process at downstream queues [11]. These are:

- autocorrelations in packet interarrival times
- autocorrelations in packet service times
- crosscorrelations among packet interarrival times and packet service times
- crosscorrelations in packet service times for tandem queues

Generally, successive packet interarrival times are often positively correlated [11]. This is valid for successive packet service times as well. Diverse factors like the presence of LRD in traffic and segmentation of large messages into IP packets with maximum 1500 bytes length heavily influence the appearance of correlations. On the other hand, packet interarrival times and packet service times are often negatively correlated with each other. Altogether, the above-mentioned types of correlations tend to make packet delays larger than in the case of independent and identically distributed (iid) packet lengths where there is no dependence [11].

4 Experiments

Measurement of one-way delay relies on time-sensitive parameters and time synchronization of both sender and receiver is required. The BTH research group has done a number of measurement and modeling experiments on OWTT and the associated convolution products in a chain of IP routers, as reported in [7,8,9,20]. A novel measurement system to do delay measurements in IP routers has been developed, which follows specifications of the IETF RFC 2679 [1]. The system uses both passive measurements and active probing. Dedicated application-layer software has been developed to generate UDP traffic with TCP-like characteristics. The generated traffic matches well traffic models observed for the World Wide Web, which is one of the most important contributors to the traffic in Internet [13]. The well-known interactions between TCP sources and network are avoided. UDP is not aware of network congestion, and this means that we could do experiments where the focus was on the network only and not on terminals. The combination of passive measurements and active probing, together with using the DAG monitoring system [10], gave us an unique opportunity to perform precise traffic measurements and also the flexibility needed to compensate for the lack of analytic solutions.

The real value of our work lies in the hop-by-hop instrumentation of the devices involved in the transfer of IP packets. The mixture of passive and active traffic measurements allows us to study changes in traffic patterns relative to specific reference points and to observe different contributing factors to the observed changes. This approach offers us the choice of better understanding of diverse components that may impact on the performance of OWTT as well as to accurately measure queueing delays in operational routers.

In the case of delay measurements through a single router, the one-way delay is given by the time difference between the timestamp reading corresponding to the first bit of packet n approaching DAG interface j (output of the router) and the timestamp reading corresponding to the first bit of the same packet n approaching DAG interface i (input of the same router). In other words, OWTT for a router is defined as the time difference between the moment when the first bit of packet leaves the router and the moment when the first bit of the same packet arrives to router (Figure 2). A similar procedure is used in the case of OWTT measurements through several routers.

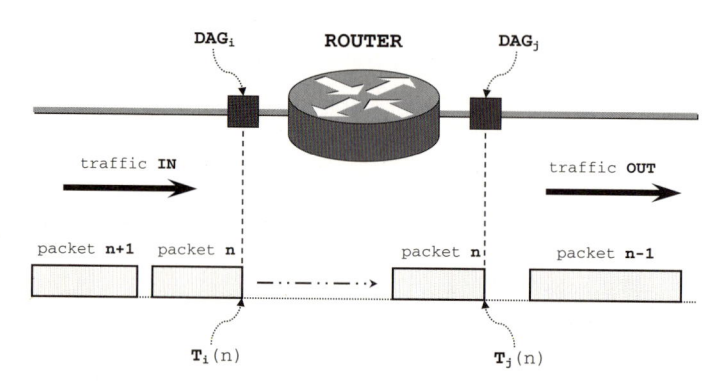

Fig. 2. Timestamping a packet

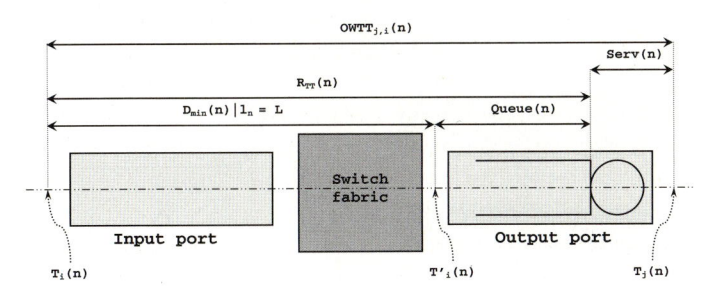

Fig. 3. Router delay model

The consequence therefore is that this measurement methodology allows us to estimate different delay components of OWTT (Figure 3):

$$D_{router} = D_{read} + D_{proc} + D_q \qquad (4)$$

where D_{read} represents the time it takes for router to copy the packet into the input port (including the time to parallelize the packet coming from the communication link), D_{proc} represents the processing delay at a router and D_q represents the queueing delay at a router. Note that the router transmission time D_{tr} contains D_{read} as well as the time to serialize the packet from the output port onto the communication link.

Three classes of experiments have been carried out, which correspond to possible situations existent in IP networks. The experiments cover the delay process for three different scenarios, i.e., delay for a router with a single data flow (no interfering traffic), delay for a router with several traffic flows (crossing traffic) and delay for a chain of three routers (with both crossing and merging traffic).

A number of traces have been generated for every experiment, with specific values for the Hurst parameter H and link utilization L_u.

The reported results are in form of several statistics regarding processing and queueing delays of a router, router delay for a single data flow, router delay for multiple data flows as well as end-to-end delay for a chain of routers [7,8,9,20]. Figure 4 shows an example of results obtained in our experiments for a chain of three routers. The results are in the form of OWTT and the associated histogram measured at the output of the third router. A indicates the source host generating traffic and B, C and D indicate the hosts generating cross traffic. The parameter α is the shape parameter of the generated Pareto traffic model and the parameter ρ is the link utilization. In the upper left corner is the plot for the traffic that enters the chain of IP routers. The associated histogram is plotted at the right of the figure. Below is plotted the traffic at the output of the chain, together with the associated histogram.

Our results confirm results reported earlier that the delay in IP routers is generally influenced by traffic characteristics, link conditions and, to some extent, details in hardware implementation and different Internetwork Operating System (IOS) releases. The delay in IP routers may also occasionally show extreme values, which are due to improper functioning of routers.

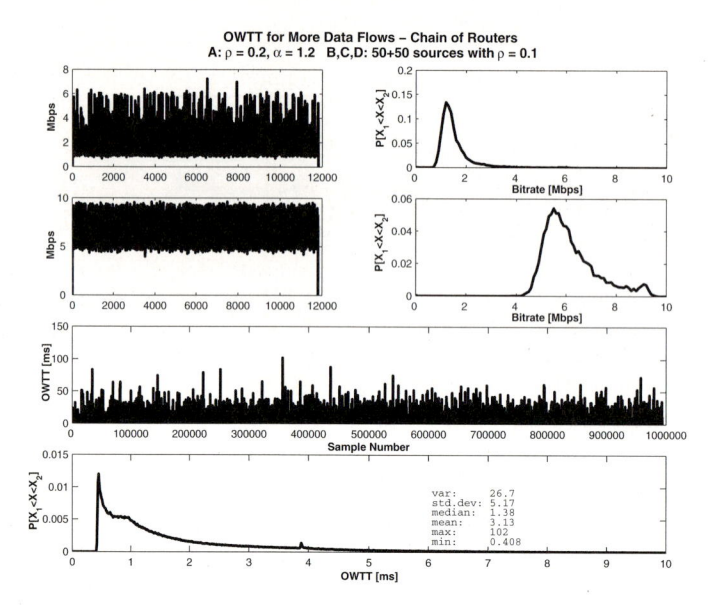

Fig. 4. Example of results on OWTT and the associated histograms

Furthermore, new results have been obtained that indicate that the delay in IP routers shows heavy-tailed characteristics, which can be well modeled with the help of several distributions, either in the form of a single distribution or as a mixture of distributions. There are several components contributing to OWTT in routers, i.e., processing delay, queueing delay and service time. The obtained results have shown that, e.g., the processing delay in a router can be well modeled with the Normal distribution, and the queueing delay is well modeled with a mixture of Normal distribution for the body probability mass and Weibull distribution for the tail probability mass [7,8,9,20]. Furthermore, OWTT has several component delays and it has been observed that the component delay distribution that is most dominant and heavy-tailed has a decisive influence on OWTT. Dual Generalized Pareto distributions are typical examples of distributions that can be used to model OWTT. These distributions correspond to the body probability mass and the tail probability mass, respectively.

A detailed description of the measurement setup, the set of measurements done, the associated modeling methodology as well as the set of results obtained in our experiments are presented in [7,8,9,20].

5 Some Observations

Some of the most important results regarding validation of Kleinrock's independence assumption can be summarized as follows.

5.1 Kleinrock's Independence Assumption

As mentioned above, Kleinrock suggested to adopt an M/M/1 queueing model for each queue in a tandem queueing system regardless of the interaction of these queues and the statistical properties of the traffic streams. Under this assumption, tractable analysis is possible.

The OWTT delay in a chain of IP routers can be partitioned into two components, a deterministic delay D_d and a stochastic delay D_s. D_d is composed of the propagation delay (D_{prop}), transmission delay (D_{tr}) and (partly) processing delay in routers (D_{proc}). On the other hand, D_s is primarily influenced by the router queueing delay D_q. The e2e delay created by the queueing delays in a chain of N routers is thus

$$D_Q = \sum_{i=1}^{N} D_{q,i} \tag{5}$$

where $D_{q,i}$ is the queueing delay in router i.

The Complementary Cumulative Distribution Function (CCDF) of $D_{q,i}$ is

$$P(D_{q,i} > x) = e^{-\varphi_i x}, \qquad \varphi_i = \mu_i(1 - \rho_i) \tag{6}$$

where μ_i is the average service rate on link i and ρ_i is the utilization factor on link i. The CCDF of D_Q is thus [21]

$$P(D_Q > x) = \sum_{i=1}^{N} \theta_i e^{-\varphi_i x} \tag{7}$$

where

$$\theta_i = \prod_{j=1, j \neq i}^{N} \frac{\varphi_i}{\varphi_j - \varphi_i}, \qquad \sum_{i=1}^{N} \theta_i = 1 \tag{8}$$

Equation 7 shows that D_Q has a CCDF that should decay Exponentially. Figure 5 shows however that this is not the case. This figure shows an example of the CCDF of OWTT measured in our experiments and the corresponding Exponential distribution, i.e., an Exponential distribution with the same mean as the one for the measured OWTT. It is observed a clear difference in the sense that OWTT decays more slowly than the Exponential distribution. The conclusion is that Kleinrock's independence assumption is not valid in these experiments.

5.2 End-to-End Delay in a Chain of Routers

The distribution of a sum of independent random variables whose individual distributions are known, is obtained by *convolution*. This operation is equivalent to multiplying individual functions in the frequency domain. Our experiments show however that this is not the case in a chain of IP routers, due to the dependence that may exist among queueing delays in different routers.

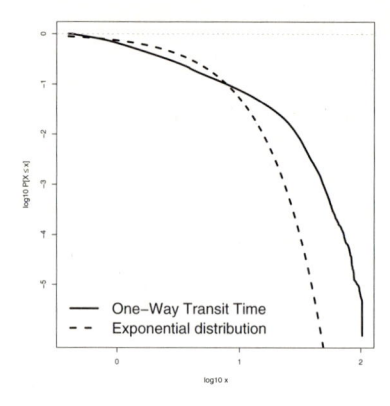

Fig. 5. OWTT and Exponential distribution

Assume a number of N M/G/1 queues with load ρ_i ($i = 1, \ldots, N$) where all queues have i.i.d service times given by a common Probability Density Function (PDF) $f(t)$ and, therefore, identical Laplace-Stieltjes Transform (LST) denoted by $\mathcal{F}_i(s)$. If w_i denotes the waiting time in queue i, $w_i(x, \rho_i)$ the PDF and $W_i(x, \rho_i)$ the Cumulative Distribution Function (CDF) of w_i, then the Pollaczek-Khinchin formula can be used to calculate the LST of the delay in an M/G/1 system [15]

$$\mathcal{W}(s, \rho_i) = \frac{1 - \rho_i}{1 - \rho_i \mathcal{R}_i(s)} \qquad (9)$$

where $\mathcal{R}_i(s)$ is the LST of the remaining service time

$$\mathcal{R}_i(s) = \frac{1 - \mathcal{F}_i(s)}{s f_i} \qquad (10)$$

and f_i is the mean service time, i.e., $f_i = E[\mathcal{F}_i(s)]$. Accordingly, the PDF of the sum of waiting times in N queues, i.e., $w = w_1 + w_2 + \ldots + w_N$, yields the convolution of the waiting times in each queue, provided that they are independent. Then, the LST of the convolution is [22]

$$\mathcal{W}(s, \rho_1, \ldots, \rho_N) = \prod_{i=1}^{N} \frac{1 - \rho_i}{1 - \rho_i \mathcal{R}_i(s)} \qquad (11)$$

This result implies that the e2e delay in a tandem queueing system can be completely described by the convolution product of the individual delay components at each queue. Our experiments show however that this is not the case. For instance, Figure 6 shows an example of CCDF of e2e OWTT measured in our experiments. Figure 6(a) illustrates the CCDF of individual router delays measured for three chained IP routers as well as the convolution product of these delays. Figure 6(b) shows the CCDF of the measured e2e OWTT of the three chained routers and the convolution product of the individual router delays.

(a) CCDF of individual OWTT and convolution product

(b) CCDF of e2e OWTT and convolution product

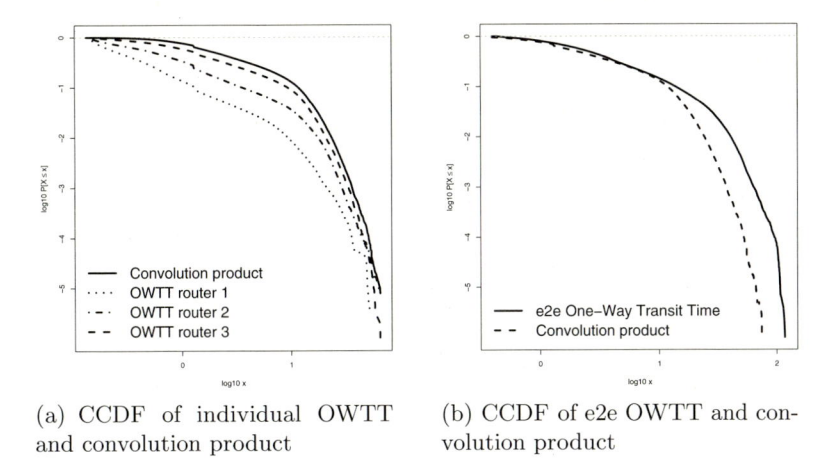

Fig. 6. OWTT and convolution product

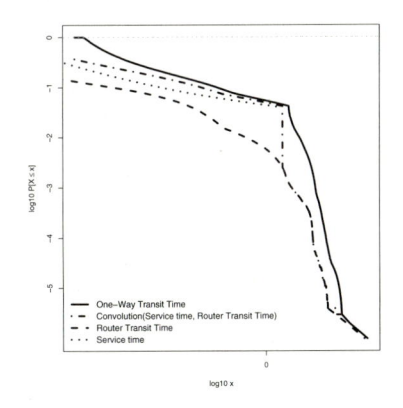

Fig. 7. Router delay and components

It is observed that there is a clear difference between the measured e2e OWTT and the convolution product of the individual router delays and that this difference is most significant in the tail of the distribution. The conclusion therefore is that e2e delay in a chain of routers is not given only by the convolution product of individual router delays. This is because of, e.g., correlations that may exist among queueing delays in different routers.

5.3 Router Delay

Similar to the observations done on e2e delay, we have also observed that the delay in a single IP router is not given only by the convolution product of queueing time and service time. This is due to crosscorrelations that may exist between interarrival times and packet service times within the same router [8].

This observation is sustained, e.g., by the results obtained in our experiments (Figure 7). This figure shows an example of router delays and components.

We observe a clear difference between the measured OWTT (solid line) and the convolution product of the router transit time (R_{TT}) and the service time (dotdashed line). This difference is most significant in the tail of the OWTT.

In other words, the formula

$$F(x) = p \cdot F_1(x) + (1-p) \cdot F_2(x) \tag{12}$$

is not sufficient in this case. Here $F(x)$ represents the CDF of OWTT, $F_1(x)$ the CDF of R_{TT}, $F_2(x)$ the CDF of service time and p is the probability mass for the body of the OWTT distribution. Consequently, $(1-p)$ corresponds to the tail probability mass in the OWTT distribution. There are more parameters contributing to the OWTT like, e.g., crosscorrelations that may exist between interarrival times and packet service times.

6 Conclusions

The paper has reported on results obtained in experiments done at the Blekinge Institute of Technology in Karlskrona, Sweden, on measurements, modeling and analysis of delay in a chain of IP routers. Particular focus has been given to validating Kleinrock's independence assumption regarding the effect of correlations in a tandem queueing system. Our results show that this assumption is not valid in our experiments, and this has been particularly observed in the end-to-end delay distribution.

Planned future work is to further analyze and to model the correlations observed in our measurements as well as to understand their effect on the delay performance in a chain of IP routers.

References

1. Almes, G., Kalidindi, S., Zekauskas, A.: A One-Way Delay Metric for IPPM. IETF, RFC 2679 (1999)
2. Bertsekas, D., Gallager, R.: Data Networks. Prentice-Hall, Englewood Cliffs (1992)
3. Borella, M., Uludaq, S., Sidhu, I.: Self-Similarity of Internet Packet Delay. In: Proceedings of IEEE JCC 1997, Montreal, Quebec, Canada (1997)
4. Bovy, C.J., Mertodimedjo, H.T., Hooghiemstra, G., Uijterwaal, H.: Analysis of End-to-End Delay Measurements in Internet. In: Proceedings of ACM Conference on Passive and Active Leasurements (PAM), Fort Collins, Colorado, USA (2002)
5. Network Measurements Metrics WG (2001),
   http://www.caida.org/outreach/metricswg
6. Claffy, K.C., Polyzos, G.C., Braun, H.W.: Measurement Considerations for Assessing Unidirectional Latencies. Journal of Internetworking 4(3) (1993)
7. Constantinescu, D., Carlsson, P., Popescu, A.: One-Way Transit Time Measurements. Technical report Blekinge Institute of Technology, Karlskrona, Sweden (2004)

8. Constantinescu, D.: Measurements and Models of One-Way Transit Time in Routers, Licentiate Dissertation, Blekinge Institute of Technology, Karlskrona, Sweden (2005)
9. Constantinescu, D., Popescu, A.: Modeling of One-Way Transit Time in IP Routers. In: Proceedings of IEEE Advanced International Conference on Telecommunications (AICT 2006), Guadeloupe, French Caribbean (2006)
10. Endace Measurement System, http://www.endace.com
11. Fendick, K.W., Saksena, V.R., Whitt, W.: Dependence in Packet Queues. IEEE Transactions on Communications 17(11) (1989)
12. Floyd, S.: Building Models for Aggregate Traffic on Congested Links, ICSI Networking Group (2005), http://www.icir.org/models/linkmodel.html
13. Jena, A., Popescu, A., Nilsson, A.: Modeling and Evaluation of Internet Applications. In: Proceedings of the International Teletraffic Congress (ITC18), Berlin, Germany (2003)
14. Kleinrock, L.: Communications Nets: Stochastic Message Flow and Delay (1964)
15. Kleinrock, L.: Queueing Systems. Theory, vol. 1 (1975)
16. Leland, W.E., Taqqu, M.S., Willinger, W., Wilson, D.V.: On the Self-Similar Nature of Ethernet Traffic (Extended Version). IEEE/ACM Transactions on Networking 2(1) (1994)
17. Mukherjee, A.: On the Dynamics and Significance of Low Frequency Components of Internet Load. Internetworking: Research and Experience 5 (1994)
18. Papagiannaki, K., Moon, S., Fraleigh, C., Thiran, P., Diot, C.: Measurement and Analysis of Single-Hop Delay on an IP Backbone Network. IEEE Journal on Selected Areas in Communications 21(5) (2003)
19. Paxson, V.: Measurements and Analysis of End-to-End Internet Dynamics, PhD thesis, University of Berkeley, California, USA (1997)
20. Popescu, A., Constantinescu, D.: Measurement of One-Way Transit Time in IP Routers. In: Proceedings of the 3rd International Working Conference (HET-NETs 2005), Ilkley, UK (2005)
21. Qiong, L.: Delay Characteristics and Performance Control of Wide-Area Networks, Technical Report, University of Delaware, Delaware, USA (2000)
22. Østerbø, O.: Models for Calculating End-to-End Delay in Packet Networks. In: Proceedings of the International Teletraffic Congress (ITC18), Berling, Germany (2003)

Application Level Measurement

Patrik Arlos

Blekinge Institute of Technology,
37179 Karlskrona, Sweden
http://www.bth.se

Abstract. In some cases, application-level measurements can be the only way for an application to get an understanding about the performance offered by the underlying network(s). It can also be that an application-level measurement is the only practical solution to verify the availability of a particular service. Hence, as more and more applications perform measurements of various networks; be that fixed or mobile, it is crucial to understand the context in which the application level measurements operate their capabilities and limitations. To this end in this paper we discuss some of the fundamentals of computer network performance measurements and in particular the key aspects to consider when using application level measurements to estimate network performance properties.

Keywords: Application level measurements, Computer network measurements, Network performance measurments, Accuracy, Quality.

1 Introduction

In recent years computer network measurements (CNM), and in particular application level measurements (ALM), have gained much interest, one reason is the growth, complexity and diversity of network based services. CNM/ALM provide network operations, development and research with information regarding network behaviour. The accuracy and reliability of this information directly affects the quality of these activities, and thus the perception of the network and its services [1],[2].

Measurements are a way of observing events and objects to obtain knowledge. A measurement consists of an observation and a comparison. The observation can be done either by humans or machines. The observation is then compared to a reference. There are two types of references; personal and non-personal. A personal reference is formed by the individual based on his experiences. A non-personal reference has a definition that is known and used by more than one individual, for instance the International System of Units (SI) [7] provides a set of global references.

2 Network Performance Framework

The Network performance framework consists of four modules; generation, measurement, analysis and visualization of analysis results. The framework is depicted

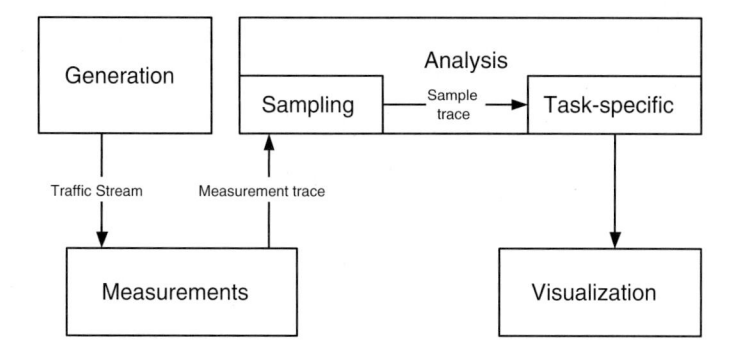

Fig. 1. Network performance measurement framework

in Figure 1. The generation module's task is to generate traffic. The measurement module captures and filters this and other traffic streams at one or multiple points in the network. There are no restrictions on which layers can be used by the generation and measurement modules, both can be done from the physical layer up to the application layer. The measurement module is so named to emphasize that it collects PDUs, measures time and does not perform any analysis. The analysis module processes the data provided from the measurement module, it does this by first sampling the data and then performing a task-specific operation, which is entirely dependent on the type of analysis that is to be performed. The output from the analysis module is then sent to the visualization module, that displays the results from the analysis. Using this framework, it is possible to clarify the semantics, detect and discuss error sources and allow for independent development of the modules. Each module has a specific role. In the following sections an overview will be given on the framework modules.

2.1 Generation

Generation deals with the construction of traffic according to a given set of parameters. It has mainly been used as a part of active measurements, but recently it has also been used together with passive measurements. Traffic generation can be performed at the same level as measurement, and hence it is subject to the same accuracy problems. Instead of detecting events, it generates events. The output from the module is a network traffic stream that is fed into a network and eventually the measurement module. With respect to ALM the generation module is interesting as a lot of ALMs are based on some externally generated data.

2.2 Measurement

The measurement module deals *only* with the collection and filtering of network traffic and associated parameters, i. e. no aggregation or parameter extraction takes place. Filtering is a process that determines if the collected data matches certain criteria, and if it does not the data is discarded. Measurements can be done at various levels in a network, from the physical layer all the way up and into the application layer [3]. In many research publications there is a differentiation between active and passive measurements. Here, no such differentiation will be made since there is no difference between them in the measurement module. Both active and passive measurements use the measurement module to collect PDUs. The output from the measurement module is a *measurement trace*. The trace can be stored in a file or temporarily stored in memory.

2.3 Analysis

The analysis module deals with everything after measurement and prior to visualization, hence this is a very large module. It can be divided into two sub-modules; *sampling* and *task-specific*. The sampling process is used to interpret the measurement trace provided by the measurement module. There are two types of sampling, time-based and event-based, comparable to simulations with fixed-time-increment or event-based-increments [5]. Figure 2 shows the difference. In time-based sampling the PDUs of the measurement trace are arranged on a time line with markers at fixed intervals T_S time units apart. While for event-based sampling, the passing of T_S time units does not need to be the sample criteria, in Figure 2 the criteria is the arrival of a PDU. Time-based sampling can be seen as event-based sampling with the criteria to sample each T_S time unit. The output from the sampling process is a *sample trace*. One or more of these sampling processes can be applied in sequence, the first one operates on the measurement trace and the other operates on intermediate sample traces. One or more of these sample traces are then subject to the task-specific analysis. The task-specific sub-module can be anything from simple averaging to protocol or user behaviour analysis. Furthermore, it is not limited to using only one sample trace. The output from the module is analysis specific and preferably adjusted for the following visualization.

2.4 Visualization

The last module, visualization, presents the analysis results to the user. Since the visualization module is the only visible module, this module will have a profound impact on the interpretation of the results. This module can hide, emphasise or

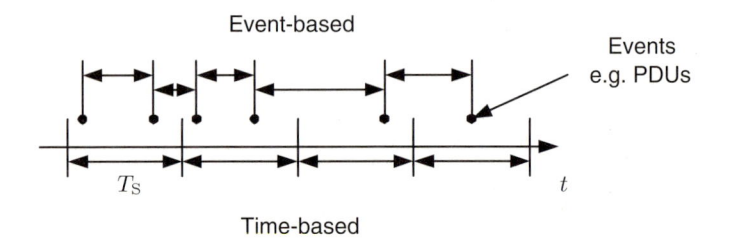

Fig. 2. Difference between time and event based sampling

distort the results obtained by the analysis module. For example, the visualization module can choose whether or not to display confidence intervals (if these are provided by the analysis module). Distortion occurs for instance, if a value is printed with too few digits. Just to mention a couple of examples: the text output from `ping` [6] in a console window and a topology map of a network [4] are both the results of this module.

3 Measurement

The measurement module only deals with the collection and filtering of PDUs and parameters associated with these. The result of the measurement module is called a measurement trace, in other works this is sometimes referred to as a packet trace. The trace can be virtual or physical; a physical trace is stored in a semi-permanent memory like a file, while a virtual or logical trace would only exist in memory. A trace can be as small as one PDU or contain millions of PDUs. The measurement trace is used to reduce the amount of data sent to the analysis module. The content of the measurement trace is in direct relation to the type of analysis that will be performed later on. For instance, some analysis methods are only interested in the PDU arrival times, others in PDU contents and some methods are interested in combinations of these fields.

3.1 Parameters

What parameters should the measurement trace contain? The PDU or at least some parts of it. The trace should also contain the collection location w of the PDU, this includes both where in the network stack (logical location) as well as where in the world (physically location) the PDU was collected. The trace should also include timing information such as when the PDU started to arrive T_A and when the PDU was completely received T_E. If the trace contains both time values, it will be possible to determine behaviour in environments with variable capacities since it is possible to calculate the capacity perceived by the PDU from its length and timing information. In addition to these four, two more values should be included the PDU length L and PDU capture length L_C. These parameters are listed in Table 1. In addition to these parameters a measurement trace should also have a set of meta-data. The meta-data can

Table 1. Measurement trace parameters

Name	Symbol	Description
PDU	p	The PDU, or parts of it
Location	w	Position of collection, logical and physical
Arrival time	T_A	Arrival time of the PDU's first bit/byte [s]
End time	T_E	End time of the PDU's last bit/byte [s]
Length	L	Length of the original PDU [bit]
Capture length	L_C	How much of the PDU is stored here [bit]

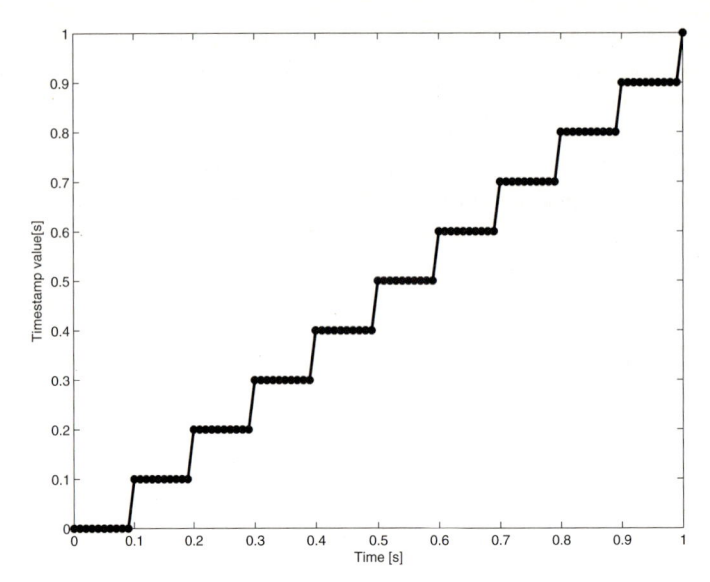

Fig. 3. Timestamp staircase

be filter information, network environment description, capture tools, hardware specifications, software versions etc. This information is however more static than the parameters that possibly change with each PDU. Hence, each measurement trace should be accompanied by its meta-data.

Timestamps. Recall that the PDU arrival time T_A identifies when a PDU started to arrive and PDU end time T_E identifies when the PDU was completely received. These values are usually refered to as timestamps. A timestamp is associated with a timestamp accuracy T_Δ provided by the measurement system including both hardware and software. A timestamp is obtained by reading a counter [8], which is updated at given intervals, and converting it into a time value. The length of these intervals determines the resolution of the timestamp and is the lower bound on the timestamp accuracy. To illustrate this, Figure 3 shows an artificial example of a timestamp sequence. The x-axis shows the true time and the y-axis the timestamp value. The staircase is created because the timestamp counter is not continuously updated; in this example it is updated every 0.1 s. In some cases, the timestamp counter is large enough to keep a smaller value than the update interval, for instance if the counter can support a timestamp with a resolution of 0.001 s. However, a high counter resolution does not increase the timestamp accuracy; it may however give false confidence in the values.

In [9] the author presents the following terminology regarding clocks: *Resolution* is defined as the smallest unit by which a clock is updated, also known as a *tick*. *Offset* specifies the difference between a particular clock and the *true* time as definied by national standards. *Skew* is the frequency difference between the clock and a national standard, or the first derivative of the offset at a particular moment, and *Drift* specifies the second derivative of the offset, or the variation of the skew.

The timestamp accuracy obtained is a combination of all these factors as well as processing delay and scheduling when collecting the timestamp. T_Δ indicates how accurate a timestamp is. The timestamp accuracy is a local value, meaning that unless two clocks are synchronized then their timestamp accuracy cannot be compared. However, if synchronization is applied this will be visible from the timestamp accuracy. Furthermore, the timestamp accuracy does not specify the offset of a timestamp. The true offset value is hard or impossible to include for every PDU, however information about the offset should be included in the meta-data associated with the measurement trace. The timestamp accuracy also includes information about the system that collected the timestamp, in the sense that it reflects the impact of the entire system and not only that of the clock.

Clock Synchronization. Clock synchronization is divided into two tasks; time synchronization and frequency synchronization. Time synchronization is used to give two separate clocks the same value, frequency synchronization is used to make the clocks tick at the same rate. For instance, let two clocks be time-synchronized at time zero. Wait a while, and then read the time values from both clocks simultaneously. Now the first clock might report 120 281 time-units (tu) and clock two reports 120 304 tu. On the other hand, if two clocks are frequency-synchronized but not time-synchronized, the initial time reading will produce two different values for example 1201 tu and 11 029 tu, and when the time is reread after a while the values might be 1450 tu and 11 278 tu. These clocks are frequency synchronized since the same time (249 tu) elapsed on both clocks. By having a common/public reference, it is possible to synchronize many clocks [10], [11], [12], [14].

Depending on how the clocks are synchronized, the timestamp accuracy is affected. Time synchronization involves changing the counter value, this can cause jumps in time, either forward or backward. If such a jump occurs during a measurement the measurement section involving the time correction cannot be used. For this reason, if time-synchronized measurements are required the devices should be synchronized prior to starting the measurement.

Frequency synchronization is on the other hand a continuous process. It usually operates by modifying a variable v which is used to create a synthetic clock frequency S_f. The variable describes the relationship between the crystal frequency C_f and the desired synthetic frequency: $S_f = f(C_f, v)$. The synthetic frequency is then used to update the time counter in a more stable way than if the crystal frequency would be used directly. This is needed since the crystal frequency changes with age and temperature. When a crystal is powered on, its frequency can vary significantly. Thus, before a measurement is started the crystal needs to reach its operating temperature. At this point the synchronization method should be applied, and once the crystal frequency has become stable the measurement can begin. Depending on the equipment and environment, this time can vary significantly but as a rule of thumb, 15–30 minutes should be sufficient to obtain crystal frequency stability [13].

The most common way to synchronize computers on the Internet is the Network Time Protocol, NTP [15], [11]. Since NTP is used so widely, it is interesting

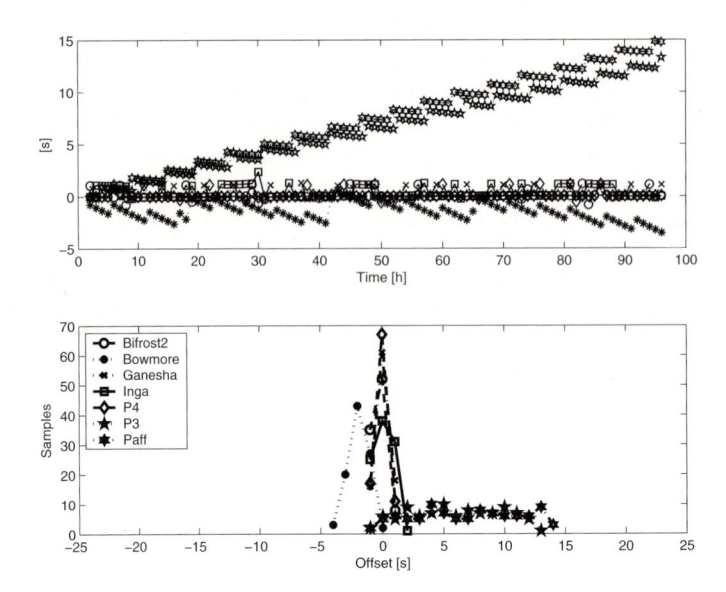

Fig. 4. Clock offset, with or without NTP synchronization

to see how it conditions a computer's clock [16]. Figure 4 shows the time offset of five computers compared to a common reference, each hour the syste time was compared to the refence and logged to a file. The top graph shows the time series and the bottom graph shows the corresponding histogram. Here it is obvious that both P3 and Paff are unsynchronized despite the NTP daemon being started on the P3 and no NTP related errors being detected in any of the logs found on the machine. Bowmore deviates from the others since it shows a negative offset, i.e., it runs slower than the NTP reference. The difference also seems to be growing, Bowmore was synchronized not by running the NTP daemon, but by issuing `ntpdate` once every 24 hours. This command will correctly synchronize the time, but will not correct the frequency. This is visible in the trace, even though the offset is reduced after 18, 42 and 66 hours, there is a drift in the behaviour. A reduction was also expected around hour 90, but this seems to be missing, causing Bowmore to be almost 5 seconds behind the NTP reference. This behaviour is emphasised by the histogram, where the Bowmore's shape has a small tendency to become wider. The remaining four devices: Bifrost2, Inga, Ganesha and P4 are synchronized within -2 to $+2$ seconds.

Timestamping Methods. When collecting timestamps in software there are two primary methods that are used; the Timestamp Counter (TSC) [17] and Get Time of Day (GTOD). The TSC reads the CPUs internal clock counter, usually via an assembler call, while the GTOD uses a system call `gettimeofday` to obtain a time value. The benefit with GTOD is that it reports the time directly, while the TSC reports a counter that represents the number of CPU cycles since the computer was started, with one cycle completed approximately every $1/f_{CPU}$, where f_{CPU} denotes the CPU clock rate. This value has to be divided by f_{CPU} to get a time value. A problem is that the actual cycle time depends on the crystal frequency, hence it is subject to aging, heat and many other sources of errors. Effectively one has to estimate the CPU speed over some interval, preferably determined by some external time source. The TSC method has a clock resolution of $1/f_{CPU}$ which can be quite high, while the resolution for the GTOD method is determined by the operating systems clock, usually in the order of a few μs.

The TSC method enables higher resolutions, < 1 ns for $f_{CPU} > 1$ GHz, and should as such be used. However, the method does come with a set of problems. The conversion from CPU cycles to time is a problem that needs to be addressed and solved. Related to this is synchronization, see [17] the authors they discuss synchronization methods in detail. A third problem that both methods have is the operating system's scheduling. The problem is present for all processes in a multi-tasking system, all user processes can be paused in their execution. If this happens, then regardless of clock method the results will be compromised. An ideal solution is to use the TSC method in combination with code that is executed by the kernel, for instance the network driver [17,18] where scheduling effects can be minimized.

If PDUs are collected in the lower layers of the stack, the impact that both system and stack have on them is minimized. Furthermore, if the TSC approach is used, the PDU timestamps can be quite accurate. However, if measurements are performed at the upper layers, i. e., at the application level, then both stack and system need to be evaluated since the behaviour that is observed is a combination of the network, network stack and system. Hence, conclusions drawn from this data must account for this. Ideally, application level measurements should be backed up with measurements at the physical or link layer to monitor the input to the stack.

As stated before, the location parameter is important and when performing stack or application measurements it is crucial to specify where the PDUs are collected. At first glance it seems obvious that they should only collect the location parameter after the PDU has been obtained, for instance after the `read` command has returned. But by adding a second timestamp before the `read` command, a lot more can be done. It is possible to determine the processing time of the `read` command and indirectly see if there was any buffering in it and the per-PDU processing time can also be evaluated. On the other hand, by adding a second timestamp, more data is created and requires more processing power. It can also be argued that this is evaluation of the system and not of the network.

Another problem is timestamping location. For example, in Linux it is quite easy to access the raw data that is passed from the link layer to the network layer. Now if the application timestamps the PDUs, this is an application layer timestamp, not a link-layer timestamp. In this case to get a link-layer timestamp the kernel has to be modified.

PDU Location. The location parameter is rarely discussed, but it is very important. The location parameter identifies where in the network stack (logical location) and where in the world (physical location) the parameters were collected. One of the reasons that this is rarely mentioned in publications is that the location is usually known to those that perform the measurement and it is not needed to motivate the end results. However it is important to remember and it becomes even more important when comparing measurements at different physical and logical locations. The physical location could for instance be specified using the GPS coordinate system. To determine the logical location can be somewhat more problematic. For example, assume that the logical location identifies a particular layer in the OSI stack. If the logical location is stated as the data link, does this mean the interface towards the physical layer, the network layer or everything in-between? This needs clarification, i.e., saying that the measurements were performed on the data link layer is insufficient. Furthermore, if the timestamping is not performed at the PDU collection location, then there will be a difference between the timing information and the PDU contents. This can at worst cause problems, and atleast confusion.

4 Sampling and Analysis

Once a measurement trace has been obtained, the next step is to analyse it. This is the task of the analysis module, it can range from simple parameter extraction, averaging to modelling of user or application behaviour. Common to all of them is the need to sample the measurement trace. The sampling can be seen as a sub-module of the analysis module. There are two types of sampling; time-based sampling and event-based sampling. The result from the sampling process is a sample trace, which is delivered to the task specific sub-module. The analysis can be performed both in the time domain or in the frequency domain. On top of this, the scaling behaviour can be analysed on different timescales [19].

4.1 Sampling

Sampling describes the process of converting a measurement trace into a format suitable for the subsequent analysis. In its simplest form the sampling process can be a format conversion, i. e., converting a Unix timestamp to a human readable format, or it can involve filtering and simple arithmetics [29]. The sampling process can be done in one operation or a sequence of operations. The two ways of sampling a measurement trace are denoted time-based or event-based, which are comparable to the two approaches that can be used in simulations; fixed-increment time advance or next-event time advance [5].

Sampling differs from the classical signal-processing approach, where the sample instance indicates that a value is to be read from an A/D-converter. The sampling here is more of an evaluation of the current conditions and it can involve simple arithmetics, examples for which will be provided below.

Time-based Sampling. Time-based sampling is the classical procedure for sampling a signal. Given a measurement trace D that contains three parameters; PDU arrival time $T_{A,i}$, PDU length L_i and the PDU p_i. These are then placed

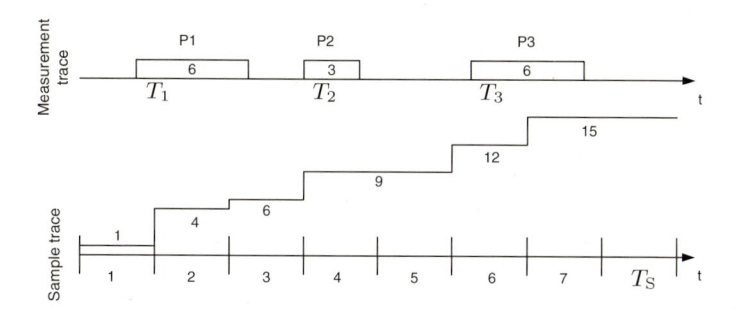

Fig. 5. Time-based sampling, data counter

on a timeline, with markers T_S time units in-between. Within each of these intervals one or more of the parameters are aggregated and the result used in the following analysis. In Figure 5 a simple example is given. The measurement trace is sampled each T_S time unit, at which the total amount of data received up to and including the interval is written to the sample trace. This is quite a simple operation, a slightly more complicated operation would be to sample the amount of data received in the latest interval, since it would involve resetting the counter after each sample interval.

Event-based Sampling. Event-based or adaptive sampling does not necessarily use time as the sample criteria. For instance, the reception of n PDUs can be the criteria for sampling, or that T seconds of silence has passed since the last received frame. However, regardless of what sample criteria is used, the aggregation is done in the same way as in time-based sampling. Using the same measurement trace as before, the resulting event-based sample trace is shown in Figure 6.

Combination and Sequence of Sampling. It is quite common to use a combination of sampling techniques, which are applied in sequence. The first process is applied to the measurement trace, the second to the sample trace produced, the third to the second sample trace and so on [20].

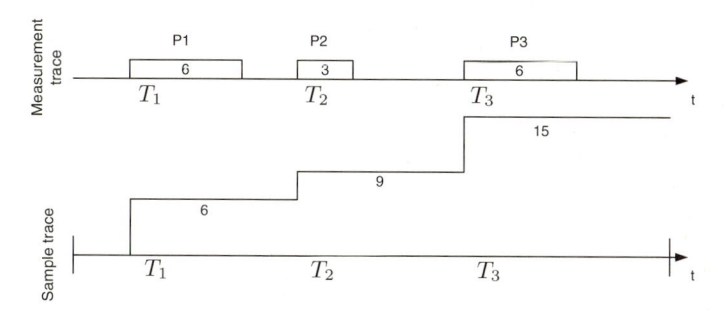

Fig. 6. Event-based sampling, bitrate

Here, a simple notation is introduced, the sampling techniques are listed in the sequence that they are applied. Time-event-based sampling means that the measurement trace was sampled using time-based sampling, and the intermediate trace was then sampled using an event-based criteria. For example, a periodical SNMP query would be a time-time-based sampling, if the SNMP agent internally used time-based sampling of the counters within the device [21]. If the SNMP agent internally used event-based sampling, the correct notation would be event-time-based sampling. An event-event-based sampling could describe a tool that first calculates the PDU inter-arrival time, followed by the inter-arrival time between two PDUs. Time-time-based could be a scaling analysis, like the one performed in [19]. The sample criteria should be supplied in the meta-data associated with a sample trace, this also includes the meta-data from the measurement trace.

4.2 Analyser and Software Impact Numerical Precision

All computers keep time by counting the number of seconds that has passed since a particular time instance, in the majority of systems this is the time since 1970-01-01[1]. At the time of writing the number of seconds that has passed is 1 256 871 240 (2009-10-29 00:00:00). Depending on how this value is represented, it will eventually wrap around and become zero again. These timestamps are stored using a fixed number of bits. For a 32-bit representation the counter will wrap to 0 around 2038-01-19. But this number only holds the seconds, not any fractions of seconds.

For this reason a timestamp is usually divided into two numbers, one for the second and another for the fractional second. When this data is read into an analyser, it might be combined into a single value for simpler processing. Here the problems arise from the limited accuracy in computers. If a large value and a very small value are added together, the new number might drop some of the digits in the smaller number in order to correctly represent the larger number. If the numbers would be kept separate, then the number of operations needed to handle them would (at least) double.

In a computer a float value is represented as two numbers, the exponent and the mantissa, and to further complicate things it is represented in a binary system (base-2) and not with a decimal system (base-10) as we are accustomed to. The mantissa stores the number and the exponent a scaling factor. For example, when storing the value 4049 (base-10) in a system with an 8-bit mantissa and a 4-bit exponent. In a computer this is represented as $N = 0.98828125 \times 2^{12}$, where the mantissa is 0.98828125 and the exponent is 12. Here, the number N does not represent 4049, but 4048 since this is the closest value that can be represented with an 8-bit mantissa. By increasing the mantissa size a better representation of the value can be obtained. Increasing the size to 16 bits, 4049 can be represented correctly.

Should one wish to represent a timestamp as one single value, including both seconds and fractional seconds, then one needs to bear in mind that the analyser's

---
[1] YYYY-MM-DD.

or computer's representation prefer large numbers. That is, even if a timestamp has an accuracy of 100 ns, combining the second and fractional second values may cause a loss of accuracy. To evaluate this for some common analysis software, a small test was created. The test was performed by adding two values, one represented the number of seconds since a given reference and the other represented a number of fractional seconds. The system was then requested to print the new value using its maximum resolution.

In Tables 2 and 3 a comparison between four software systems (Matlab, R, Perl and Python) and three number representations in C++ (`double`, `long double` and `quad double` [22]) is shown. The first column holds the reference date and the second column contains the number of seconds that have elapsed since the reference date. The third column holds the fractional seconds. The values in the second and third columns are added to create a new value x, which is in turn printed by the systems listed in columns four to six. Starting with Table 2, it is clear that if the entire second count since 1970 is kept, one can only be sure that the ten-μs digit is correct. If the reference is changed to 2000-01-01, Matlab and R can be trusted to the one-μs digit. By choosing an even closer reference, 2005-01-01, one may expect to be able to rely on the 100 ns value, this is not the case. But by choosing a reference only a week away, one can trust the value representing 100 ps, and by choosing a reference one day away the smallest number one can rely on is 10 ps. Worth noting here is that if one wishes to have a μs accuracy then one must use 2000-01-01 as the time reference instead of the default 1970 reference. A second comment is that if one performs a measurement that spans a week it is possible to obtain 10 ps if the first day of the measurement is used as a reference. If 1970-01-01 was used as a reference one will only obtain 10 μs.

In Table 3 the output from a C++ program is shown, here if x is stored as a float with `double` precision and 1970 is used as a reference then the representation is quite accurate, and if the fractional is decreased to 0.1 μs then the value is not correctly represented. When x was represented as a `long double` then the value is correctly identified and the same is true when the fractional was only 0.1 μs . The output from the `quad double` representation, when using 1970 as the reference date is not as accurate as the `long double` representation. If one uses a `double` to represent timestamps and these are accurate to one μs , one must use 2000-01-01 as the reference date. If the timestamps are accurate to the nanosecond, then one must use a reference that is less than 24 hours away, and should not compare values that are more than 24 hours apart. For the `long double` representation things look much better, in fact it has the best representation of x for all reference dates and fractional values. The `quad double` representation is almost as accurate as the `long double`, since it seems to be rounding the values differently.

4.3 Task-Specific Analysis

Based on the sample trace, a multitude of different analysis methods are available, however, there are far to many to discuss in the context of this thesis.

Table 2. Matlab, R and Perl accuracy

Reference		Environment		
Seconds	Fractional	Matlab 6.5	R 2.1.0	Perl 5.8.2 (windows) Python 2.5.3 (windows)
1970-01-01				
1116885600	1e-6	1116885600.0000010	1116885600.000001	1116885600.00000095367432
1116885600	1e-7	1116885600	1116885600	1116885600
2000-01-01				
170200800	1e-6	170200800.000001	170200800.00000101	170200800.000001013278961
170200800	1e-7	170200800.000000090	170200800.00000009	170200800.000000089406967
2005-01-01				
12348000	1e-6	12348000.000001	12348000.000001	12348000.0000010002404451
12348000	1e-7	12348000.0000001010	12348000.000000101	12348000.0000001005828381
2005-05-17				
604800	1e-6	604800.00000100001	604800.0000010000	604800.000001000007614493
604800	1e-7	604800.0000001	604800.0000001	604800.000000100000761449
604800	1e-10	604800.00000000012	604800.00000000012	604800.000000000116415322
604800	1e-11	604800	604800.00000000000	604800
2005-05-23				
86400	1e-6	86400.000000999993	86400.000000999993	86400.0000009999930625781
86400	1e-7	86400.0000001	86400.000000100001	86400.0000009999930625781
86400	1e-11	86400.000000000015	86400.000000000015	86400.00000000000145519152
86400	1e-12	86400	86400	86400
2005-05-24				
3600	1e-12	3600.0000000000009	3600.0000000000009	3600.0000000000009094947

Table 3. C++ accuracy

Reference		Environment		
Seconds	Fractional	double	long double	quad double
1970-01-01				
1116885600	1e-6	1116885600.0000009536743..	1116885600.00000100000......	1116885600.00000092........
1116885600	1e-7	1116885600	1116885600.000000100000......	1116885600.00000003........
2000-01-01				
170200800	1e-6	170200800.0000010132789..	170200800.000000999993.....	170200800.000000995........
170200800	1e-7	170200800.0000000894069..	170200800.000000100000......	170200800.000000107........
2005-01-01				
12348000	1e-6	12348000.0000010002404...	12348000.000001000000.....	12348000.0000010003.......
12348000	1e-7	12348000.0000001005828...	12348000.000000099999.....	12348000.0000000988.......
2005-05-17				
604800	1e-6	604800.0000010000076...	604800.000000999999.....	604800.000001000004.....
604800	1e-9	604800.0000000010477...	604800.00000000099998...	604800.000000000981.....
2005-05-23				
86400	1e-6	86400.0000009999930...	86400.000000999999997..	86400.0000010000057....
86400	1e-9	86400.0000000010040...	86400.000000000999996..	86400.0000000010004....
86400	1e-12	86400	86400.000000000001001..	86400.0000000000056....
2005-05-24				
3600	1e-12	3600.0000000000009...	3600.000000000001000..	3600.00000000000097...

What needs to be pointed out is that the analysis might emphasize the errors accumulated in the sample trace. It is easy to believe that the error can be reduced by increasing the amount of data, i. e,. by collecting 100 000 samples instead of 10 000 samples. However, the error in each of these samples is independent on the number of samples collected.

In Figure 7 an example of a measurement trace is shown, consisting of three PDUs 4, 3 and 6 bytes long, it is subject to a time-based non-fractional PDU accounting sampling that calculates the bitrate. The resulting sample trace is denoted as vector $\mathbf{V}_{st}$ and contains realisations of the random variable V_{st}, given in bps. V_{st} contains three non-zero values:

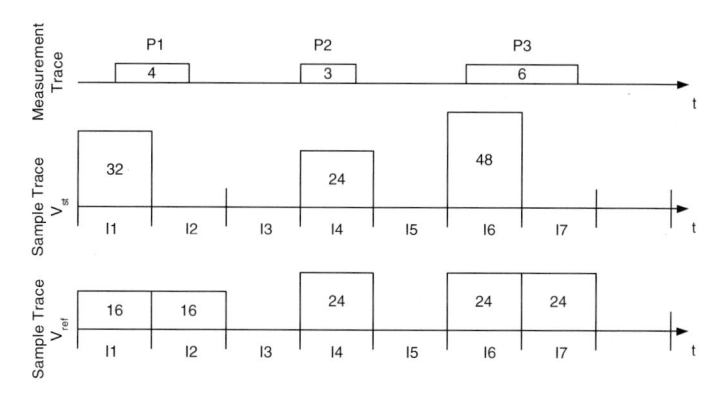

Fig. 7. Analysis problem

$$\mathbf{V}_{\text{st}} = [32, 0, 0, 24, 0, 48, 0] \quad \mathbf{E}[V_{\text{st}}] = 14.86 \quad \mathbf{Var}[V_{\text{st}}] = 393$$

let us compare V_{st} to a sample trace $\mathbf{V}_{\text{ref}}$ that accounted for the fractional PDUs. $\mathbf{V}_{\text{ref}}$ would contain five non-zero samples:

$$\mathbf{V}_{\text{ref}} = [16, 16, 0, 24, 0, 24, 24] \quad \mathbf{E}[V_{\text{ref}}] = 14.86 \quad \mathbf{Var}[V_{\text{ref}}] = 116$$

Now, comparing the mean values of these vectors shows that they are identical, but their higher order statistics differ significantly. However, the last zero sample in $\mathbf{V}_{\text{st}}$ is necessary to achieve the same mean values. For instance, if both traces were such that interval I7 was not included, the statistics would be different.

$$\mathbf{V}_{\text{st}} = [32, 0, 0, 24, 0, 48] \quad \mathbf{E}[V_{\text{st}}] = 17.3 \quad \mathbf{Var}[V_{\text{st}}] = 420$$
$$\mathbf{V}_{\text{ref}} = [16, 16, 0, 24, 0, 24] \quad \mathbf{E}[V_{\text{ref}}] = 13.3 \quad \mathbf{Var}[V_{\text{ref}}] = 119$$

It is therefore important to cover all intervals in which PDUs are supposed to be present. Especially if fractional PDUs are not taken into account, extra sample intervals might need to be added.

To further complicate matters, the time-of-day comes into play, as the network behaviour is influenced by the time-of-day. The same reasoning can be applied for the problems caused by the timestamp accuracy. In general one should assume the worst case scenario, where errors enhance each other after each step of processing. Thus, the best way to go is to perform an error analysis for the entire system and analysis method, in order to determine the quality of the final results.

5 Application Level Measurements

Application level measurements collect PDUs at, or above, the application layer in the network stack [23]. These are usually collected using regular user applications that are executed and scheduled in the user domain by the operating system. These measurements are then used to draw conclusions about the network behaviour. However, the results are not only affected by the network, but also by the hardware, software and in particular the operating system of the computer that performs the measurement. Hence it is necessary to investigate the influence these components have on the measurement results [24].

We will do this by using the timings inbetween packets, i.e. the Inter-Packet Time (IPT), as this should not be changed as long as the packet passes either up or down the stack. However, this holds only if the stack is not congested. If it is then the packets may be delayed, or even buffered before deliverd to the next layer in the stack. This can cause the IPT to either shrink or grow. Many tools have been built indirectly based on this assumption. They usually consist of two parts, a sender and a receiver. The sender is configured to send a packet, pause execution for some time, and then repeat the procedure until a predefined number of packets has been transmitted. The receiving side will then receive the packets and calculate the IPT and compare it to the user defined IPT at the sender. This works fine, if you know that the sender really behaves as desired.

5.1 Setup

We evaluated three different ALM tools; the first (A) was implemented in classical C, the second (B) in Java and the third (C) used C#. Furthermore, the C and C# implementations use UDP for communications, while the Java application uses TCP. The setup is shown in Figure 8. To collect the network data, we use the Distributed Passive Measurement Infrastructure [25]. The PDU are copied by the wiretaps and sent to the MP, were we use DAG3.5E cards [26] synchronized using GPS, to collect them. The hosts (H1 and H2) were identical in terms of hardware for each of the experiments. For tool A it were Pentium-4 2.8 GHz systems with 1 GB of RAM and a built-in 1000Base-TX (configured to operate at 100 Mbps) cards. Tool B and C used Pentium-3 667 MHz, 256 MB RAM and built-in 100Base-TX cards. For Tool A the operating system was a Linux 2.6 system, while for B and C it was Windows XP (SP2). For B the Java version was 1.5.0, and for C the .NET framework was 2.0. The evaluation was done by having host H1 and H2 running the tools and collecting the application-level traces, while the DPMI collected the link layer traces. The data was then analyzed offline using Matlab.

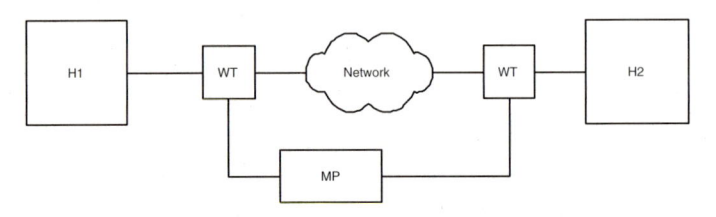

Fig. 8. Setup used to evaluate ALM

Application Level Measurements

All three tools consists of two parts a sender and a receiver, the sender is located on H1, and is configured to transmit its data to H2, were the receiving part resides. The senders are configurable with respect to load into the network, this is done via controlling the inter packet time (IPT), the payload size and the number of packets to send. The receiver applications timestamps the data arrival, and stores these in a static vector that is written to file after the experiment has been completed. Tool A and C used the TSC timestamping method, while B used the GTOD method. For Tool A the sender was configured to send 1472 bytes UDP datagram, corresponding to 1514 bytes at the link layer) once every 1 ms, and for B and C they sent 526 or 538 bytes, corresponding to 576 bytes at the link layer.

5.2 Analysis

To evaluate the quality of the ALM is to estimate the timestamp accuracy error, for details see [28]. Let $T_{x,y}(k)$ be the timestamp obtained at party x at layer y for PDU $k \in (1 \ldots n - 1)$. Party x can either be the sender (s) or the receiver (r), and a layer y can be the application (a) or the link (l) layer. Let $IPT_{x,y}(k, k+1)$ be an IPT for a PDU pair $(k, k+1)$ and $\epsilon_{k,k+1}$ a timestamp accuracy error for this pair, then T_Δ is obtained using:

$$IPT_{x,y}(k, k+1) = T_{x,y}(k+1) - T_{x,y}(k)$$
$$\epsilon_{k,k+1} = IPT_{a,r}(k, k+1) - IPT_{l,r}(k, k+1)$$
$$T_\Delta = |max(\epsilon_{k,k+1})| + |min(\epsilon_{k,k+1})| \quad \forall k$$

Here we'll use a simplified method, were we only compare the statistics (mean and standard deviation) and the minimum and maximum values of the IPT.

5.3 Results

The results are summarised in Table 4. Remember that Tool A has a target of 1 ms, while for B and C the target is 125 ms. Looking at tool A, we observe that the difference between link and application is quite small for all the values. The extreme values are 60-70 μs different than the corresponding link values. Looking on Tool B this is quite different, here the minimum value is 90 ms smaller than the link value, and the maximum is 4 ms larger. For Tool C this looks better,

Table 4. Tools A–C: IPT's statistics at receiver

Param.	Tool A Link [ms]	Tool A App [ms]	Tool B Link [ms]	Tool B App [ms]	Tool C Link [ms]	Tool C Appl. [ms]
min	0.44	0.37	109.96	20.00	65.98	65.97
max	1.56	1.62	236.92	241.00	184.94	184.94
mean	0.99	0.99	125.43	125.43	125.00	125.00
std.dev	0.01	0.02	1.23	5.33	0.75	0.75

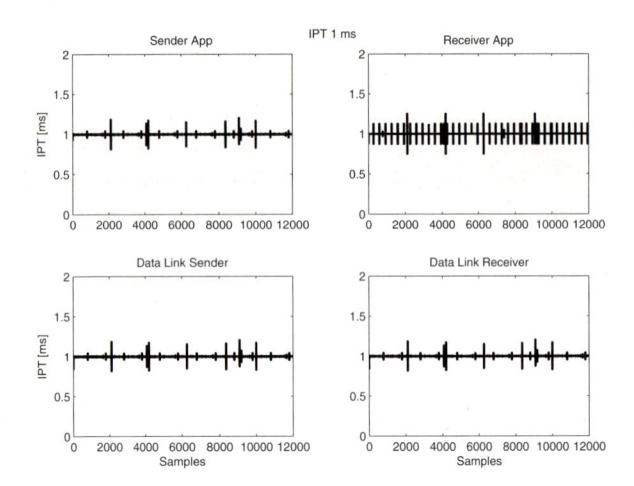

Fig. 9. Tool A: measured IPT at receiver for nominal IPT 1 ms

with only a small 1 μs difference. However, we need to remember that the Tool C had a load of 6 packets/s, while Tool A had a load of 1000 packets/s. Turning our attention to the statistics, all three tools match their target mean quite well. But, again Tool B has difficulties, this time with the standard deviation, its significantly larger than that of the link layer. The main reason for this is that the GOTD of Tool B uses *System.currentTimeMillis()*, which had a resolution of 10 ms. As Tool A and C uses the TSC method they do not suffer from this problem, however they have other problems that we'll come to later.

In Figure 9–11 we show the IPT trace for tools. The top left graph shows the IPT at the sender at the application level, the graph below (bottom, left) shows the IPT at the link layer at the sources. Then the bottom right graph shows the IPT at the receiver, and then the top-right graph shows the IPT at the application layer of the receiver. Looking at Tool A (Figure 9 there is not much variability in the IPT data from sender to the data link receiver, but at the receiver application layer there is a slightly higher variability. For Tool B, shown in Figure 10 the results are quite different, here the IPT has a significant variability at both sender and receiver application, this is coupled to the timestamp resolution offered by Java. However, its interesting to note that at the link layer the IPT is not suffering from this. So, in Java its possible to execute a sleep that is smaller than 10 ms, but the reported time elapsed will be either 0 or 10 ms. For Tool C, the data is similar to Tool A. At both sides the IPT is more or less identical at both application and link layer.

In Figure 12 we see 20 IPT samples for Tool A. For Tool A, first we note that the correlation between the link IPT and the app IPT. Secondly, the app IPT seems to be a lot smoother. Initially, it's tempting to account this smooth behavior to the variability of the CPU frequency, and that the tool used the average CPU frequency that covered the entire experiment. Now as we do not have intermediate timestamps from the GTOD, we cannot recalculate the CPU

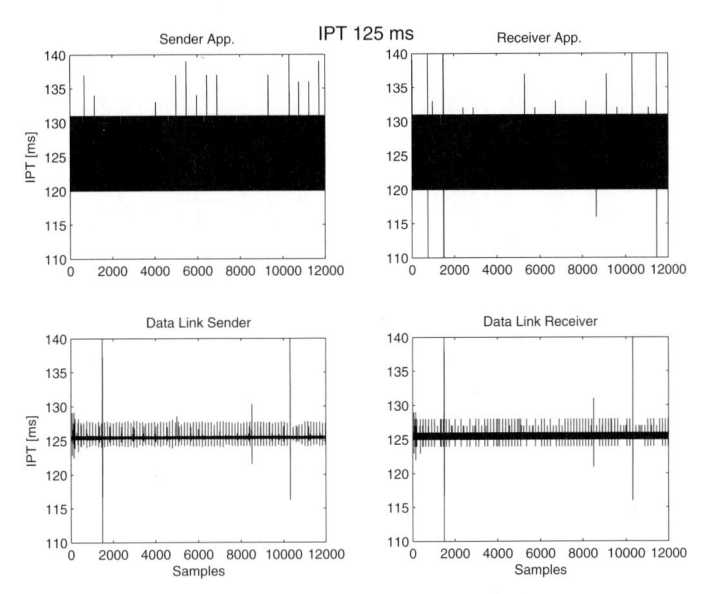

Fig. 10. Tool B: measured IPT at sender and receiver for nominal IPT 125 ms

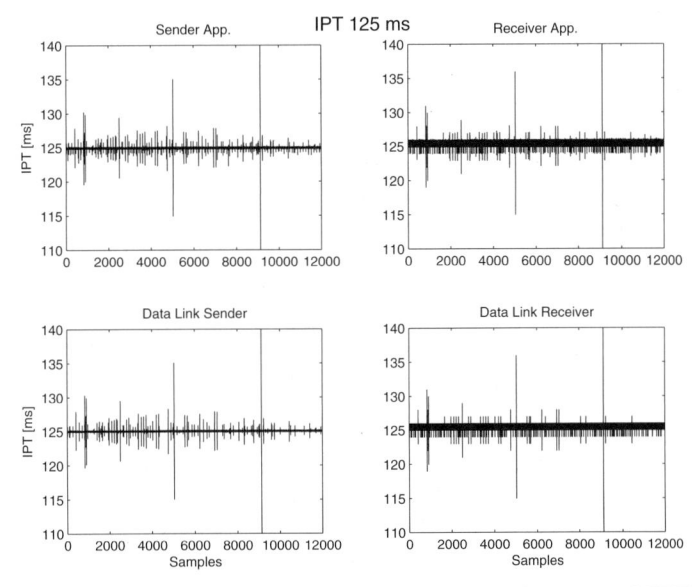

Fig. 11. Tool C: Measured IPT at sender and receiver for nominal IPT 125 ms

frequency for different periods. Thus, we cannot really explain why the application IPT is smoother, nor why the obvious peak and valley, at sample 94 and 95, does not show in the application IPT.

To investigate this further, we used the arrival time of the packets, and created a time trace relative to the arrival time of the first packet, for both link and

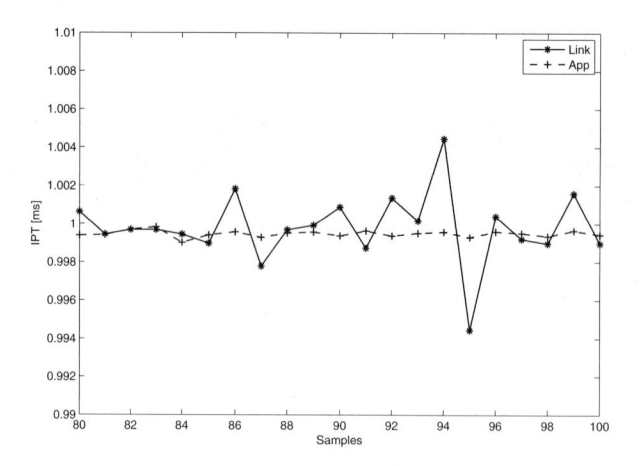

Fig. 12. Tool A: Detailed IPT 20 samples

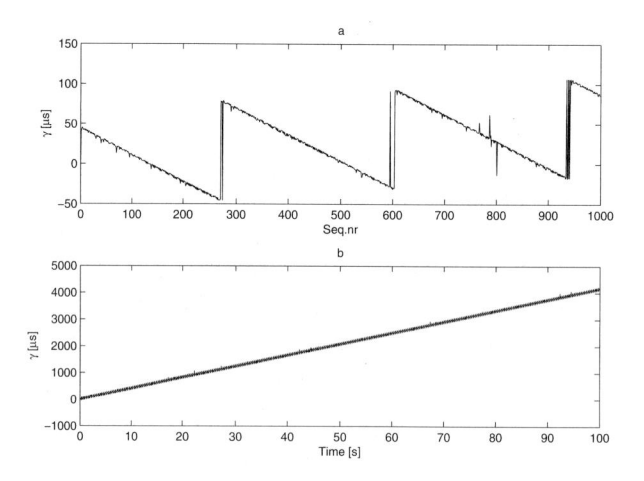

Fig. 13. Tool A: Difference between application packet arrival times and link layer arrival times

application layer. In the next stage we then created the difference between these time traces, like:

$$\gamma(k) = \hat{T}_a(k) - \hat{T}_l(k) = (T_{r,a}(k) - T_{r,a}(0)) - (T_{r,l}(k) - T_{r,l}(0)) \quad (1)$$

As we know that the link layer timestamps are obtained from a properly synchronized source (DAG card + GPS), we can trust these. So, ideally γ should be very very close to zero. As this would indicate that the application layer timestamps are also obtained from a well behaved clock source. The results are shown in Figure 13. The upper graph (a), shows the first 1000 packets, corresonding to the first second. The sawtooth behaviour is obvious, this is a classical view of a clock that drifts. It seems that something tries to correct the clock every

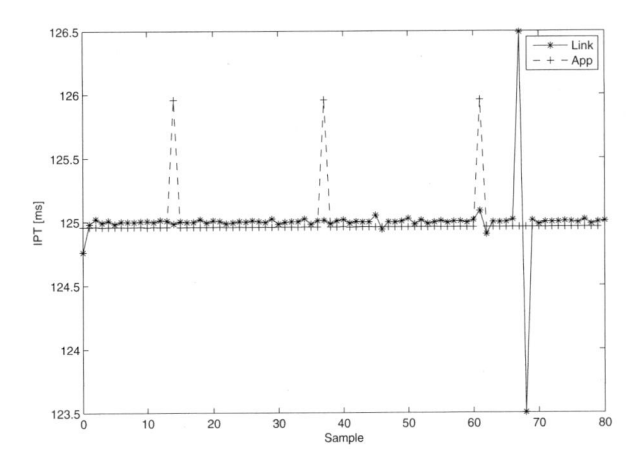

Fig. 14. Tool C: Details IPT for 80 samples

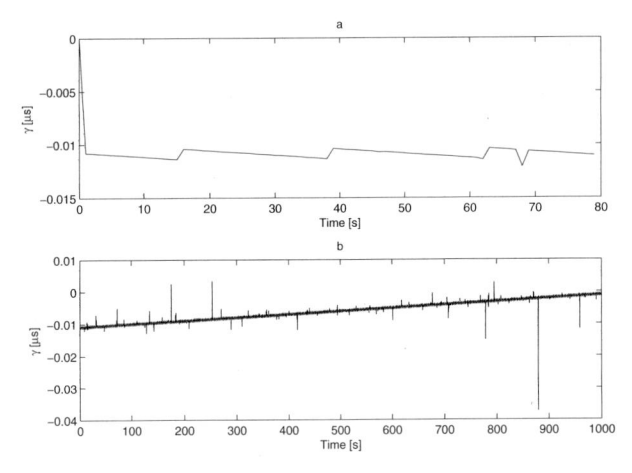

Fig. 15. Tool C: Difference between application packet arrival times and link layer arrival times

300 ms, but it over compensates, hence the clock also deviates even more as time goes. This is clearly visible in the lower graph (b), where we show the difference increases to 4 ms during a 100 second period. This corresponds to a drift of 3.5 s during one day, which is seriouly bad.

Turning our attention to Tool C, we show the a detailed view for 80 IPT samples in Figure 14. First we notice a periodic behaviour, every 23-24 samples. Secondly, the app IPT is consequently lower than the link IPT, not counting the periodic behavior. The increase/decrease indicated by the link at sample 67 passes totally undetected. We repeated the γ evaluation, and the results are shown in Figure 15. Graph a shows the first 80 samples, corresponding to approximately 10 seconds of data. Again we see a sawtooth, however much weaker, but significantly stronger as well, as the scale is in ms, not μs. If when look on the lower graph, the drift is obvious, it goes from a -10 ms to 0 over 1000 s, thus during a day this system drifts 0.864 seconds. This is four times better than that of the four times faster the Pentium-4 system used by tool A.

6 Conclusions

In this paper we described a frame work usefull when discussing network performance. As ALMs are usually conducted with the intention of detecting network performance, the framework is also usefull in this context. Base on the framework we described the associated modules, and the problems associated with each module when it comes to the accuracy of measurements.

We showed how good timestamps can be destroyed by improper use of analysis tools.

We evaluated three ALM tools, one C++, one Java and one C# tool. We showed that all three generated statistical values that lookes similar to those obtained from the link layer. The Java application showed a high standard deviation, due to that the clock that was used to obtain the timestamp seemed to update in steps of 10 ms. Then we investigated the C++/C# tools, that both used the TSC method to obtain timestamps, we detected that both time traces exhibited clock drifts. The drifts seemed to be coupled to the CPU speed of the reciving host, and in this case a 2.8 GHz CPU generated a clock drift of around 4 ms in 100 seconds, while the slower CPU drifted around 1 ms in 100 seconds.

Based on this, if you choose to use the TSC method to obtain timestamps, make sure that you obtain a GTOD timestamp on a regular interval, preferably four times a second (cf. the conditioning behaviour shown in Figure 13 every 300 ms). This will allow you to condition your estimate of the CPU frequency, better, when you make the count to time conversion. The of course, this requires that the system clock is properly conditioned by some other means, either NTP or GPS.

Regardless, where you are measuring you should perform some steps before you can report results obtained from measurements. First you need to identify the parameter(s) and the desired accuracy of these parameters. Then evaluate the accuracy that your HW/SW combination delivers, and if needed replace parts. Do a test where you evaluate the system over the intended measurement period. Evaluate the accuracy that your SW/analysis tool gives, the best way it to use artificial data that gives you full control on the desired output. Perform an error analysis, were you estimate the worst case error obtained in one sample. Then do your measurements, and report that you "measured X, and got $X \pm Y$. Also make sure that your systems are synchronized to a well known reference, and ideally the time should be traceable.

References

1. Schormans, J.A., Timotijevic, T.: Evaluating the Accuracy of Active Measurement of Delay and Loss in Packet Networks. In: MMNS (2003)
2. Chevul, S., Isaksson, L., Fiedler, M., Karlsson, J., Lindberg, P.: Measurement of application-perceived throughput of an E2E VPN connection using a GPRS network. In: 2nd EuroNGI IA.8.3 Workshop (2005)
3. Open Systems Interconnect - Basic Reference Model, Recommendation X.200 (1994)
4. CAIDA, http://www.caida.org/tools/measurement/skitter/
5. Law, A.M., Kelton, W.D.: Simulation, Modelling and Analysis. McGraw-Hill, New York (1991)
6. Muuss, M.: The Story of the PING program, http://ftp.arl.mil/~mike/ping.html
7. Bureau International des Poids et Mesures, http://www.bipm.org/
8. Donnelly, S.: High Precision Timeing in Passive Measurements of Data Networks. Phd Thesis, The University of Waikato (2002)
9. Paxson, V.: On calibrating measurements of packet transit times. SIGMETRICS Perform. Eval. Rev. (1998)
10. Skeie, T., Johannessen, S., Holmeide, Ø.: Highly Accurate Time Synchronization over Switched Ethernet. In: Proceedings of 8th IEEE conference on Emerging Technologies and Factory Automation, ETFA (2001)
11. Mills, D.L.: Improved algorithms for synchronizing computer network clocks. IEEE/ACM Transactions on Networking (1995)
12. Zhang, L., Liu, Z., Honghui Xia, C.: Clock synchronization algorithms for network measurements. In: INFOCOM (2002)
13. Dietz, M.A., Ellis, C.S., Frank Starmer, C.: Clock Instability and Its Effect on Time Intervals in Performance Studies. Technical report DUKE–TR–1995–13 (1995)
14. Wang, J., Zhou, M., Zhou, H.: Clock synchronization for internet measurements: a clustering algorithm. Computer Networks (2004)
15. Mills, D.L.: RFC1305:Network Time Protocol (Version 3): Specification, Implementation and Analysis. IETF (1992)
16. Smotlacha, V.: Experience with precise timekeeping in end-hosts. CESNET Technical Report 18/2004, http://www.ces.net/project/qosip/
17. Veitch, D., Babu, S., Pásztor, A.: Robust Synchronization of Software Clocks Across the Internet. In: Proceedings of the Internet Measurement Conference (2004)
18. Deri, L.: nCap: Wire-speed Packet Capture and Transmission. In: E2EMON (2005)
19. Carlsson, P.: Multi-Timescale Modelling of Ethernet Traffic. Licentiate Thesis, Blekinge Institute of Technology (2003)
20. Claffy, K.C., Polyzos, G.C., Braun, H.-W.: Application of Sampling Methodologies to Network Traffic Characterization. In: SigComm (1993)
21. Carlsson, P., Fiedler, M., Tutschku, K., Chevul, S., Nilsson, A.: Obtaining Reliable Bit Rate Measurements in SNMP-Managed Networks. In: Proceedings of the 15th ITC Specialist Seminar (2002)
22. High-Precision Software Directory, http://crd.lbl.gov/~dhbailey/mpdist/
23. Feng, W.-C., Gardner, M.K., Hay, J.R.: The MAGNeT Toolkit: Design, Implementation and Evaluation. Journal of Supercomputing (2002)
24. Danzig, P.B.: An analytical model of operating system protocol processing including effects of multiprogramming. SIGMETRICS Perform. Eval. Rev. (1991)

25. Arlos, P., Fiedler, M., Nilsson, A.A.: A Distributed Passive Measurement Infrastructure. In: Proceedings of Passive and Active Measurement Workshop (2005)
26. Endace, http://www.endace.com
27. Arlos, P.: On the Quality of Computer Network Measurements. Phd. Thesis, Blekinge Institute of Technology (2005)
28. Arlos, P., Fiedler, M.: A Method to Estimate the Timestamp Accuracy of Measurement Hardware and Software Tools. In: Proceedings of Passive and Active Measurement Workshop (2007)
29. Zseby, T., Molina, M., Duffield, N., Niccolini, S., Raspall, F.: Sampling and Filtering Techniques for IP Packet Selection, http://www.ietf.org/internet-drafts/draft-ietf-psamp-sample-tech-07.txt

Measurements and Analysis of Application-Perceived Throughput via Mobile Links

Markus Fiedler[1], Lennart Isaksson[2,*], and Peter Lindberg[3]

[1] Blekinge Institute of Technology
School of Computing, 371 79 Karlskrona, Sweden
`markus.fiedler@bth.se`
[2] Ericsson India Private Limited
Tamarai Tech Park, Chennai 600032, India
`lennart.isaksson@ericsson.com`
[3] AerotechTelub AB
351 80 Växjö
`peter.lindberg@saabgroup.com`

Abstract. Application-perceived throughput plays a major role for the performance of networked applications and user experience and thus, for network selection decisions. To support the latter, this tutorial paper investigates the process of user-perceived throughput in GPRS and UMTS systems seen over rather small averaging intervals, based on test traffic mimicking the needs of streaming applications, and analyzes the results with aid of summary statistics. These results reveal a clear influence of the network, seen from variations and autocorrelation of application-perceived throughput mostly on the one-second time scale and indicate that applications have to cope with significant jitter when trying to exploit the nominal throughputs. In GPRS, the promised average throughputs are not reached in downlink direction; instead, significant packet loss occurs. Furthermore, with aid of causality arguments for an equivalent bottleneck, bounds for the extra delay of the first packet sent via mobile links is derived from throughput measurements.

Keywords: Throughput, user-perceived Quality of Service, UMTS, GPRS, higher-order statistics, equivalent bottleneck.

1 Introduction

Networked applications, no matter whether connected in a wired or wireless way, rely upon the ability of timely data delivery. The achievable throughput is a quality measure for the very task of a communication system, which is to transport data in time. This is of particular importance for nowadays trendy *streaming applications* such as TV, telephony (as part of Triple Play) and gaming. Especially the higher throughput offered by (beyond-)3G mobile systems as

---

* This work was done when Lennart Isaksson was with Blekinge Institute of Technology.

compared to earlier generations seems to be paving the way for these types of applications into mobile environments.

Throughput is thus one of the most essential enablers for networked applications, if not the most important one. While in general, throughput is defined on network or transport level (*e.g.* for TCP), the *application-perceived throughput* that is investigated in this paper reflects the perspective of the application, *i.e.* captures the behavior of all communication stacks in-between the endpoints. Streaming multimedia applications require some amount of throughput on a regular basis. For an elastic application such as file transfer, the achieved throughput determines the download time. For situation-dependent services, *e.g.* for Intelligent Transport Systems and services (ITS), short response times are of outmost importance. The background of the investigations summarized in this tutorial is the task of automatically choosing the right type of network for ITS applications, comprising streaming, messaging and interactive services. More information on the related PIITSA (Personal Information for Intelligent Transport Systems through Seamless communications and Autonomous decisions) project and the particular tasks to be solved is found in [1].

Looking at the OSI model, the conditions on layer n affect layer $n+1$, $n+2$, *etc.*. In the end, the application and thus the user is affected by any kind of problem in the lower layers. For instance, traffic disturbances on network level are packet reordering, delays and loss. The latter can be compensated *e.g.* by retransmissions, which on the other hand increases the delay. Non-interactive applications usually cope well with (approximately) constant delays, even if they are comparably large. This is however not the case for interactive applications, where long delays disturb the interaction of both communication partners. End-to-end throughput variations reveal loss and delay variations, but not one-way delays [2,1]; the latter can be measured on the network level as described in [3]. On the other hand, comparative throughput measurements on small time scales are capable of revealing existence, nature and severeness of a bottleneck [2]. Reference [1] discusses the concept of application-perceived throughput in detail and provides formulae to calculate application-perceived throughput requirements for streaming, messaging and interactive applications based on user patience, processing times and data volumes.

The current tutorial provides a survey on how to measure throughput as perceived by the user and its applications, combining the appealing possibilities of scaling studies with those of comparative end-to-end measurements. Our active measurements of application-perceived throughput values averaged over some short time interval (typically around one second) during an observation window of duration (typically around one minute) followed by a comparison between key higher-order statistics of sent and received data streams clearly reflect impairments that happen on OSI layers 1 to 7. Initially, we avoid interfering traffic in order to get a clear picture of the best-case throughput properties of a mobile connection. Having several connections sharing the same resources, the user-perceived throughput is likely to be reduced [2]. A specific problem of mobile connectivity consists in the fact that Temporary Block Flows have to be set up [4]. The definition an equivalent bottleneck based on end-to-end throughput measurements and a subsequent application of causality arguments allows deriving approximations for such initial delays beyond the unavoidable minimal one-way transit time.

The remainder of the paper is organized as follows. Section 2 places the method amongst related work, focusing on scaling studies and end-to-end measurements as roots for the throughput modeling used in this work. The statistical parameters used to summarize and compare the perceived throughput processes at server (= sender) and client (= receiver) are presented in Section 3. The setup and implementation of the measurements is described in Section 4, including considerations of the above-mentioned initial delay and of warm-up phases. We then investigate the selected summary statistics and estimations of initial delays obtained by active measurements of application-perceived throughput for UMTS and GPRS in Section 5. There, we also illustrate the impact of the size of the throughput averaging interval. Finally, Section 6 presents conclusions and outlook.

2 Related Work

2.1 Traditional View on Throughput

Traditionally, the notion of throughput (also called bandwidth) is considered as per session, *i.e.* the *averaging interval* ΔT matches the *observation interval* ΔW. This happens with the end of determining file transmission times in rate-shared scenarios such as TCP-based file transfer [5] and the related degree of user satisfaction [6]. In [7,8], average throughputs per session are used for investigating and classifying the performance of different sessions in wireless scenarios. In general, such studies do not consider temporary variations of throughput *during* the sessions.

2.2 Scaling Studies

The consideration of smaller time scales ($\Delta T \ll \Delta W$) and higher-order statistics offers added value for traffic characterization, which will be exemplified in this section.

The use of throughput averaging intervals of

$$\Delta T_i = \Delta T_0 \, k^i, i \in \mathcal{N} \qquad (1)$$

with typical scaling factors $k = 2$ or $k = 10$ became popular in the beginning of the 1990's. The investigation of higher-order statistics (such as variance and autocorrelation) of throughput averaged over ΔT_i allowed the discovery and analysis of scaling phenomena such as self-similarity, multi-fractality and long-range dependence [9,10,11]. Long-range dependence is for instance seen from the lag-k autocorrelation coefficient of a random variable having the form

$$\rho(k) \sim c_k \cdot k^{2H-2}, \qquad (2)$$

where H is the *Hurst factor* that reflects the degree of self-similarity. One typical consequence of long-range dependence is a slowly decaying variance

$$\mathbf{Var}[R_s(\Delta T_i)] \sim c_i \cdot \Delta T_i^{2H-2}, \tag{3}$$

where $R_s(\Delta T)$ is a throughput time series built upon averaging intervals of ΔT. Plotting the variance of the throughput versus ΔT on a double-logarithmic scale allows for the determination of H. The existence of areas with different gradients in such log-log plots might point at multi-fractal properties.

Juva *et al.* [12] analyzed data from the Finnish university network (Funet) using averaging intervals between $\Delta T = 1$ s and 5 min. Their analysis is mainly focused on the mean-variance relationship regarding traffic volumes X_n in consecutive intervals:

$$\mathbf{Var}[X_n] = \phi \cdot \mathbf{E}[X_n]^c. \tag{4}$$

The exponent c turns out to be scale-invariant, which is shown by double-logarithmic scatter plots of throughput variance versus average throughput. Depending on the data measured, type of link, and measured interval, $c \in [0.5, 4.0]$. If the traffic was Poissonian, then both ϕ and c were equal to one.

Since the times of ATM, quantiles of throughput distributions have been used for dimensioning purposes [13,14,15]. The latter publication focuses on the impact of the time scale on the capacity required to keep the probability of capacity overrun below a given level (*e.g.* $\epsilon = 1$ %). Assuming a normal distribution, we obtain the required capacity [15]

$$C(\Delta T, \epsilon) = \mathbf{E}[R_s] + \frac{1}{\Delta T}\sqrt{(-2\log\epsilon - \log2\pi) \cdot \mathbf{Var}[R_s(\Delta T)]}. \tag{5}$$

Another interesting observation in [12] is the fact that even with a considerably large averaging interval ΔT, the autocorrelation does not vanish. Thus, even if long-range dependence might not exist, the observed throughput process seems to have a quite distinct after-effect.

2.3 The Network Management View

Interestingly enough, using $\Delta T \ll \Delta W$ has been common practice in network management for quite a long time, but with the limitations that (i) the corresponding time plots are visually inspected, but not analyzed beyond minimal, maximal and average values, and (ii) typical averaging intervals have been rather long ($\Delta T \geq 5$ min).

The 5 min time scale is also of interest in the context of demand modeling and provisioning. For instance, [16] is using a $\Delta T = 5$ min interval when measuring the point-to-point traffic matrix in the IP backbone. Data are analyzed *e.g.* regarding the mean-variance relationship (4) for different demands in the European and American subnetworks. Recently, shorter averaging intervals have attracted interest. The popular open-source network management tool MRTG [17] that originally employed 5 min intervals comes now with a time resolution of 10 s, which matches the capabilities of SNMP-capable networking equipment

[18]. Modern network management tools such as InfoSim StableNet [19] and the RMON tool NI Observer [20] support $\Delta T = 1$ s. Reference [21] investigates throughput monitoring for $\Delta T \in [50$ ms, 2 s$]$, and [22] is using $\Delta T = 100$ ms interval for studies of web traffic variability as Internet routers appear to have corresponding buffering capabilities. Also, [2] identified $\Delta T = 100$ ms as a useful time scale for describing the throughput of a video conference application. As mentioned above, reference [15] considers link dimensioning as a function of the averaging interval of interest *cf.* (5).

Most studies consider *passive measurements* at one point of reference in a fixed network, *e.g.* on a backbone link, which reflects the typical viewpoint of an operator.

2.4 End-to-End Considerations

In order to reflect the viewpoint of the user, a comparison between a sent and a received packet stream needs to be carried out. *Active measurements* sending probing traffic into a network and deducing the available throughput or bandwidth from these measurements (*cf.* [8]) reveal this view, but in general, merely measurement session averages throughput ($\Delta T = \Delta W$) are reported.

Reference [2] combined the passive observation approach using comparably short averaging intervals with the recently described end-to-end view and used higher-order throughput statistics for the identification and classification of bottlenecks. From throughput histogram difference plots, it can be seen whether the network acts as a shared or shaping bottleneck, *i.e.* increases or decreases the burstiness of a packet stream.

The method can be used for both passive and active measurements. The current tutorial extends the concept presented in [2] as follows:

1. mobile systems are focused;
2. the throughput process is observed on application instead of link level;
3. active measurements on hardly loaded systems are performed in order to reveal basic properties of the systems under study; and
4. additional statistical parameters like standard deviation and autocorrelation are considered.

3 Application-Perceived Throughput Statistics

A typical starting point for traffic characterization purposes are *traces*, *i.e.* lists of packet-related information such as

- T_p: the time when packet p was observed at a point of reference;
- L_p: the length of packet p (payload) at a point of reference;
- any other information such as IP addresses, port numbers, *etc.*

In general, the raw data $\{T_p, L_p\}_{p=0}^{k-1}$ need some kind of post-processing such some further condensation of the information and the calculation of statistical parameters in order to extract and highlight effects of interest. In the following, we perform both steps.

As we are particularly interested the traffic flow properties, we focus on a *discrete-time fluid flow traffic model*. To this aim (in a first step) we collect the contributions of packets observed during short averaging intervals ΔT. We treat the first packet ($p = 0$) of the trace as synchronization packet both at sender and receiver, which is observed at T_0, respectively. This is motivated as the receiving application begins to act upon reception of this packet. Then, we calculate the corresponding throughput time series

$$R_{A,s} = \frac{\sum_{\forall p: T_p \in]T_0+(s-1)\Delta T, T_0+s\Delta T]} L_p}{\Delta T} \quad (6)$$

containing $n = \Delta W/\Delta T$ values. As point of reference, we use the application level (index $_A$). On this level, L_p reflects the payload sent by a server application (index^{in}) or received by a receiver application (index^{out}). The time stamp is taken just before a packet is sent or just upon reception. A detailed description of the corresponding procedure is found in Section 4.

The second step consists in calculating selected summary statistics such as average, standard deviation, throughput histograms and autocorrelation coefficients, which is detailed in the following subsections.

3.1 Average Application-Perceived Throughput

Definition:

$$\bar{R}_A = \frac{1}{n} \sum_{s=1}^{n} R_{A,s} \quad (7)$$

A change of this parameter between server ($\bar{R}_A^{in}$) and client ($\bar{R}_A^{out}$) reflects missing traffic at the end of the observation interval:

$$L = \max\left\{(\bar{R}_A^{in} - \bar{R}_A^{out})\Delta W, 0\right\} \quad (8)$$

That share of traffic might be overdue (*i.e.* appear in the next observation interval) or might have been lost. The use of the max-operator is motivated by the fact that there might be overdue traffic from an earlier interval reaching the receiver in the current observation interval, yielding $\bar{R}_A^{out} > \bar{R}_A^{in}$. The corresponding loss ratio is obtained as

$$\ell = \frac{L}{\bar{R}_A^{in}}. \quad (9)$$

3.2 Standard Deviation of the Application-Perceived Throughput

Definition:

$$\sigma_{R_A} = \sqrt{\frac{1}{n-1} \sum_{s=1}^{n} (R_{A,s} - \bar{R}_A)^2} \quad (10)$$

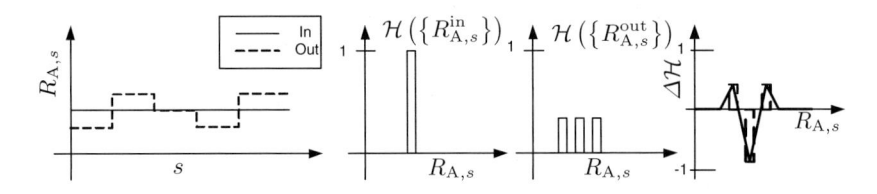

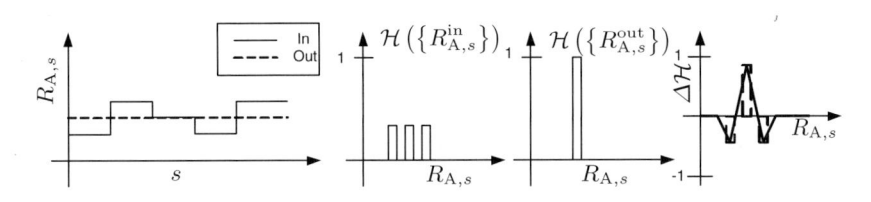

Fig. 1. Anticipated time plot, throughput histograms at input and output and throughput histogram difference plot (from left to right) in case of a shared bottleneck [2]

A rising standard deviation ($\sigma_{R_A}^{out} > \sigma_{R_A}^{in}$) reflects a growing burstiness of the traffic between sender and receiver, while a sinking standard deviation ($\sigma_{R_A}^{out} < \sigma_{R_A}^{in}$) means a reduction of burstiness. The latter case is typical for a shaper [23].

3.3 Application-Perceived Throughput Histogram

Definition:
$$h_{R_A}(i) = \frac{\text{number of } R_{A,s} \in \,](i-1)\Delta R, i\Delta R]}{n} \tag{11}$$

If the throughput histogram at the receiver $\mathcal{H}\left(\{R_{A,s}^{out}\}\right)$ is broader – in terms of non-vanishing values $h_{R_A}(i)$ when plotted versus $i\Delta R$ – than the one at the sender $\mathcal{H}\left(\{R_{A,s}^{in}\}\right)$, the burstiness has increased due to interfering traffic [24]. In the other case, the traffic has been shaped, yielding a more sharp throughput distribution at the receiver $\mathcal{H}\left(\{R_{A,s}^{out}\}\right)$. As shown in figures 1 and 2, the *throughput histogram difference plot* $\Delta\mathcal{H}$ with

$$\Delta h_{R_A}(i) = h_{R_A}^{out}(i) - h_{R_A}^{in}(i) \tag{12}$$

originally defined in [2] and serving as a *bottleneck indicator* helps to visualize these changes perceived by traffic on its way through a network as follows:

– Shared bottleneck ⇔ M shape of the bottleneck indicator;
– Shaping bottleneck ⇔ W shape of the bottleneck indicator.

Compared to standard deviation values, the throughput histograms contain more detailed information about the impact of the bottleneck.

Fig. 2. Anticipated time plot, throughput histograms at input and output and throughput histogram difference plot (from left to right) in case of a shaping bottleneck [2]

3.4 Lag-j Autocorrelation Coefficient of the Application-Perceived Throughput

Definition:
$$\hat{\rho}_{R_A}(j) = \frac{\sum_{s=1}^{n-j}(R_{A,s}-\bar{R}_A)(R_{A,s+j}-\bar{R}_A)}{(n-j)\,\sigma_{R_A}^2} \qquad (13)$$

The autocorrelation coefficients allow to detect and compare degrees of after-effects and periodicities within the throughput processes at the server's ($\hat{\rho}_{R_A}^{in}$) and client's ($\hat{\rho}_{R_A}^{out}$) side. Periodicies are revealed by positive spikes of $\hat{\rho}_{R_A}(j)$ when plotted versus j. Changes of the autocorrelation coefficients (from $\hat{\rho}_{R_A}^{in}$ to $\hat{\rho}_{R_A}^{out}$) reflect changes of after-effects and periodicities within the throughput process imposed by the network.

4 Measurement Setup

4.1 Parameter Settings

According to the prerequisites of the fluid flow model [24,2], several packets should be captured in one averaging interval in order to get a differentiated view on throughput variations. As GPRS allows only for a couple of packets to be sent or received per second, an averaging interval of $\Delta T = 1$ s was chosen. With a typical observation window of duration $\Delta W = 1$ min, the throughput time series to be analyzed consist of $n = 60$ values. In some particular cases, different ΔT values around one second and longer observation windows $\Delta W \leq 5$ min will be applied. Due to the limited resolution in time, time stamping inaccuracies due to the limited exactness of the computer clock should be a minor issue, while the quite short observation interval should damp the potential effect of clock drift.

All throughput histograms in Section 5 will be presented with a throughput resolution of $\Delta R = 1$ kbps and the autocorrelation plots with a confidence interval of 95 %.

4.2 Layer of Interest

When it comes to speed and capacity issues, vendors and providers in general specify the available bit rate on link or physical layer (OSI layers 2 and 1, respectively). However, applications perceive throughput at application layer (OSI layer 7) or even above, which is considered in this work. Due to the overhead introduced by the layers in-between (at least OSI layer 3 and 4 in the Internet context), the average application-perceived throughput is exclusives upper-bounded by the link-layer capacity C_{Link}. Denote Ω_i as overhead introduced by layer i and $\bar{L}$ as average packet length on application layer, we arrive at

$$\frac{\bar{R}_A}{C_{\text{Link}}} = \frac{\bar{L}}{\bar{L}+\Omega_3+\Omega_4}. \qquad (14)$$

For UDP/IP, $\Omega_4 = 8$ B and $\Omega_3 = 20$ B, respectively. For an average packet length of $\bar{L} = 128$ B, the average application-perceived throughput is $\frac{128}{128+20+8} \simeq 82\,\%$ of the link capacity; for $\bar{L} = 480$ B, this share rises to 94 %.

In case the (average) offered traffic on application level $\bar{R}_A^{\text{off}}$ exceeds the throughput indicated by Equation 14, a share of

$$\ell = \frac{\bar{R}_A^{\text{off}} \frac{\bar{L}+\Omega_3+\Omega_4}{\bar{L}} - C_{\text{Link}}}{\bar{R}_A^{\text{off}} \frac{\bar{L}+\Omega_3+\Omega_4}{\bar{L}}} = 1 - \frac{\bar{L}}{\bar{L}+\Omega_3+\Omega_4} \cdot \frac{C_{\text{Link}}}{\bar{R}_A^{\text{off}}} \qquad (15)$$

is lost. In the above mentioned case of $\bar{L} = 480$ B, loss of $\ell \simeq 6$ % is to be expected when the offered traffic on application level matches the link capacity, cf. Table 3.

4.3 Initial Delay

One specific property of mobile networks consists in the fact that the first packet experiences an extraordinarily high delay beyond the usual one-way delay due to the need to set up a so-called Temporary Block Flow [4,25]. As the measurements synchronize on the first observed packet, the receiver starts capturing of the time series $\{R_{A,s}^{\text{out}}\}_{s=1}^n$ too late, yielding quite high throughput values $R_{A,s}^{\text{out}}$ during the first intervals. In the following, we describe how to cope with this problem.

Let us define the network as an equivalent fluid-flow bottleneck as proposed in [2]. The buffer content at the end of interval s is described by

$$X_s = X_{s-1} + (R_{A,s}^{\text{in}} - R_{A,s}^{\text{out}})\Delta T\,;\ X_0 := 0\,. \qquad (16)$$

Obviously, X_s cannot be negative due to reasons of causality: a certain packet cannot be received in an interval prior to when it was sent. Taking this into account and applying (16) in a recursive manner, we arrive at the *causality condition*

$$\sum_{s=1}^n R_{A,s}^{\text{in}} \geq \sum_{s=1}^n R_{A,s}^{\text{out}} \quad \forall n\,. \qquad (17)$$

In case of a late first packet, $R_{A,s}^{\text{out}}$ is typically larger than $R_{A,s}^{\text{in}}$ during the first interval(s), which means that condition (17) is not met. To compensate for this problem, we can delay the time series $\{R_{A,s}^{\text{out}}\}_{s=1}^n$ artificially by $q\Delta T$ by including zero samples in the beginning so that

$$q^* = \min\left\{q : \sum_{s=1}^n R_{A,s}^{\text{in}} \geq \sum_{s=1}^n R_{A,s+q}^{\text{out}} \quad \forall n\right\},\ R_{A,0}^{\text{out}},\ldots,R_{A,q^*-1}^{\text{out}} := 0 \qquad (18)$$

From Equation 18, we derive an estimation of the extra initial delay of the first packet of

$$\tau_0^R = q^* \Delta T\,. \qquad (19)$$

In case the original traces are available, such an estimation can as well be constructed by comparing the cumulative inter-arrival times of the packets in $\{T_p^{\text{in}}, L_p^{\text{in}}\}_{p=0}^{k-1}$ and $\{T_p^{\text{out}}, L_p^{\text{out}}\}_{p=0}^{k-1}$, respectively. The corresponding causality condition reads as follows:

$$T_p^{\text{in}} - T_0^{\text{in}} \leq T_p^{\text{out}} - T_0^{\text{out}} \quad \forall p\,. \qquad (20)$$

We obtain an estimation of the extra initial delay of

$$\tau_0^T = \min\left\{\tau : T_p^{\text{in}} - T_0^{\text{in}} \leq T_p^{\text{out}} - T_0^{\text{out}} + \tau \quad \forall p\right\}. \tag{21}$$

Comparisons of (19) and (21) showed that in case of considerable loss (8) at the beginning of the observation interval, q^* becomes too small. The reason for this is obvious: the loss reduces the output flow $R_{A,s+q}^{\text{out}}$, and thus, (18) is met "too early" when q is increased. For this reason, we construct another bound by taking $\ell > 0$ into account:

$$q^{**} = \min\left\{q : \sum_{s=1}^{n} R_{A,s}^{\text{in}} \geq \sum_{s=1}^{n} R_{A,s+q}^{\text{out}} + \ell \quad \forall n\right\}, R_{A,0}^{\text{out}} \ldots R_{A,q^{**}-1}^{\text{out}} := 0 \tag{22}$$

$$\tau_0^L = q^{**} \Delta T. \tag{23}$$

Please note that ℓ is calculated over the whole observation window in order to ensure that this portion of traffic is very likely to be lost. Estimations of τ_0^R and τ_0^L will be compared with τ_0^T in Section 5.5.

4.4 Warm-Up Phase

A practicable work-around of the problem of the delayed first packet consists in not considering the first k' (say 100) packets of the trace both at sender and receiver, yielding the traces $\{T_p^{\text{in}}, L_p^{\text{in}}\}_{p=k'}^{k-1}$ and $\{T_p^{\text{out}}, L_p^{\text{out}}\}_{p=k'}^{k-1}$ as a basis for analysis, respectively. This way of treating the problem is comparable to the elimination of a warm-up phase in a simulation by not taking the first samples into account. Thus, we observe the some kind of steady-state thoughput process. This work-around was applied except for the study of the initial delay found in Section 5.5.

4.5 UDP Generator

For the measurements of application-perceived throughput, two scenarios were considered: One is called the *downlink scenario*, cf. Figure 3 (a), in which the

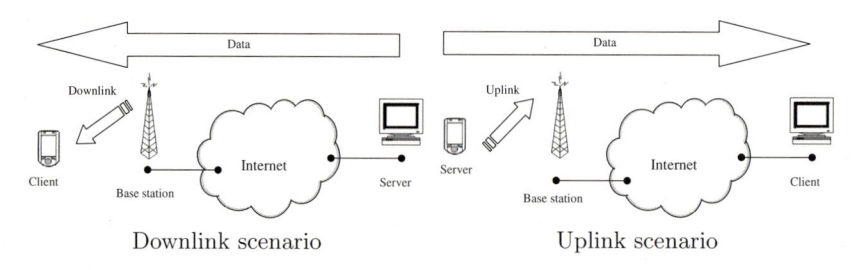

Fig. 3. Mobile scenarios

Table 1. Query Performance Parameters in Server and Client Code

Line	Code
01	[Dllimport("kernel32.dll")]
02	extern static ulong QueryPerformanceCounter(ref ulong x);
03	[Dllimport("kernel32.dll")]
04	extern static ulong QueryPerformanceFrequency(ref ulong x);

client is connected to a base station (BS) and the server to the Internet via 100 Mbps Ethernet. The other one is called the *uplink scenario*, Figure 3 (b), in which the server is connected to an BS and the client to the Internet via 100 Mbps Ethernet. These names are based on the way the data traffic is directed regarding BS.

The measurements are produced by using a User Datagram Protocol (UDP) generator. The generator is trying to send UDP datagrams of constant length as regularly as possible with a sequence number inside each datagram. The datagrams are not sent *back-to-back*, but spaced in order to yield a certain load. The minimal time the packets are spaced is hereafter called *inter-packet delay*. The software tries to keep this value as closely as possible. However, as blocking calls are used, sending packets too fast as compared to the capacity of the link close to the sender can increase the effective inter-packet delay, *cf.* Sections 5.2 and 5.4.

The software was developed in C# running on both server and client, and these are producing and monitoring the datagram stream. The original time stamp resolution is limited to 10 milliseconds, which is not good enough for our purposes. Thus, specific coding was necessary to improve the time stamp resolution to one thousand of a millisecond. This was achieved by using *performance counters* in conjunction with the system time to provide smaller time increments. To this aim the `kernel32.dll` functions `QueryPerformanceCounter`

Table 2. Inter-Packet Delay Algorithm in Server

Line	Code
01	QueryPerformanceCounter(ref ctr0);
	// Perform this line once at start up
...	...
05	QueryPerformanceCounter(ref ctr1);
06	QueryPerformanceCounter(ref ctr2);
07	while (((ctr2 · 1000000 / freq) − (ctr0 · 1000000 / freq))
	< (delayTimeLong · 1000 · (ulong)i))
08	{
09	QueryPerformanceCounter(ref ctr2);
10	}
...	...
15	QueryPerformanceCounter(ref ctr3);
16	s.Send(rndFile,rndFile.Length,ipAddress,remotePort);
17	QueryPerformanceCounter(ref ctr4);

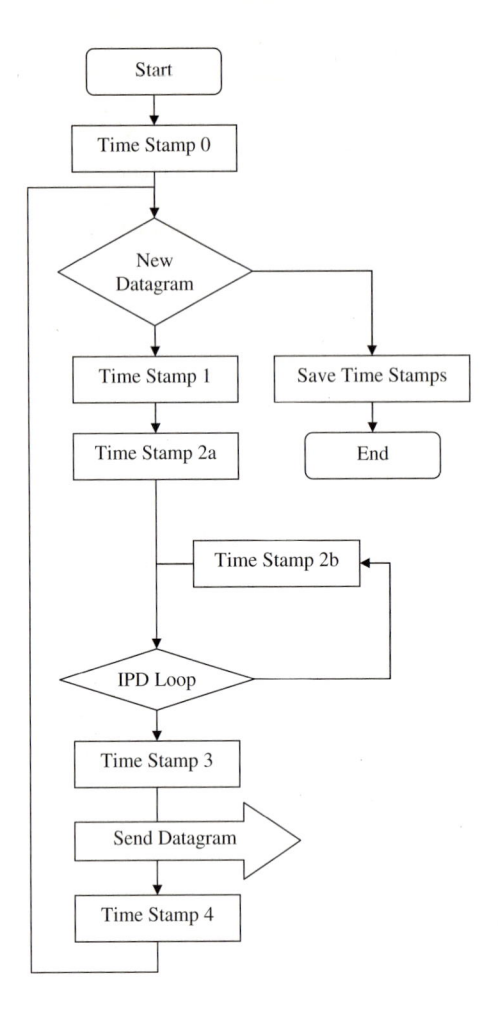

Fig. 4. UDP Generator with Time Stamps

and `QueryPerformanceFrequency` were used, *cf.* Table 1. For each measurement, the process for each run was set to *realtime* in the Microsoft Windows operating system.

All data are saved in a static vector to create the necessary space at the start-up of the program. If with was neglected memory allocations or hard disc access could jeopardize the timestamping, which should be avoided at all cost. At the end of executions all time stamps in the static vector are saved, *cf.* Figure 4 (right-hand side).

The UDP generator starts with saving the starting time in variable $ctr0$ to be used as an absolute value to every other time stamp in the code, *cf.* Figure 4 (TS 0) and Table 2 (line 01), using the `QueryPerformanceCounter`. All parameters are using *ulong* because the amount of numbers represented by the hardware except for i which has to be done by type-casting.

Before and after each the *inter-packet delay* function a time stamp (TS) is executed, *cf.* Figure 4 (TS 0 to 4) and Table 2 (line 05, 06 and 09). To maintain the nominal inter-packet delay, a while-loop containing an absolute measurement is used for all packets to be sent, *cf.* Figure 4 (TS 0, 2a and 2b) and Table 2 ($ctr2 - ctr0 < delayTimeLong \cdot i$). Together, the *while loop* is being hold until the accumulated inter-packet delay time is reached, *cf.* Figure 4 (TS 2a and 2b). Before and after each *send* function a time stamp is executed, *cf.* Figure 4 (TS 3 and 4) and Table 2 (line 15 and 17). The parameters *rndFile*, *rndFile.Length*, *ipAddress*, *remotePort* are used inside each packet. Finally the datagram is sent out. The time stamp 3 ($ctr3$) is used at the sender side.

5 Measurements and Results

5.1 UMTS Downlink

The first UMTS downlink case presented addresses a modestly loaded link with the nominal inter-packet delay time set to 90 ms, *cf.* Figure 5. In this case, the impact of the network is hardly visible; the average throughput is the same on server and client side, while the standard deviation has grown slightly, still, it is small as compared to the average both at sender and receiver. The throughput histograms are almost identical, which is also shown by the throughput histograms different plot that is close to zero, *cf.* Figure 5 (a) (middle). At the server side, the autocorrelation plot, *cf.* Figure 5 (a) (bottom-left), reveals a throughput periodicity of about 9 s, which stems from the fact that the averaging interval is not an integer multiple of the inter-packet delay. However, at the client side, this periodicity is less pronounced, *cf.* Figure 5 (a) (bottom-right). Obviously, the network has destroyed some of the original throughput autocorrelation and its structure. The plots from the time domain, *cf.* Figure 5 (b) time stamp 1, shows some rather small amount of jitter. The jitter indicated from the first time stamp is compensated to sustain the throughput. In this case the true IPD is 89.88 ms (second from bottom). Figure 5 (b) (bottom) show the time plot perceived by the receiver. The mean IPD at the client side is slightly larger than the mean IPD at the server, displaying a distinct burst deviation around sample 450. The client does not indicate any packet loss.

In second case, the nominal inter-packet delay time is set to 30 ms, *cf.* Figure 6. Still the average throughput is the same on both the server and client side, but the standard deviation is much higher at the client side as compared to the server side. The client also starts to perceive more jitter compared to the previous example, *cf.* Figure 6 (b) (bottom). Still, the client does not indicate any packet loss. The autocorrelation structure is again changed significantly by the network.

In the third case, the offered throughput was increased to 384 kbps by reducing the inter-packet delay to 10 ms, *cf.* Figure 6. The server produces a stream at exactly this speed (ΔT is an integer multiple of the nominal inter-packet delay). Consequently, the standard deviation is zero and the autocorrelation undefined. Following the argumentation in Section 4.2, we obviously face an overloaded bottleneck, and data loss amounts to $\ell = (1 - 359.68/384) \simeq 6\,\%$. However, according

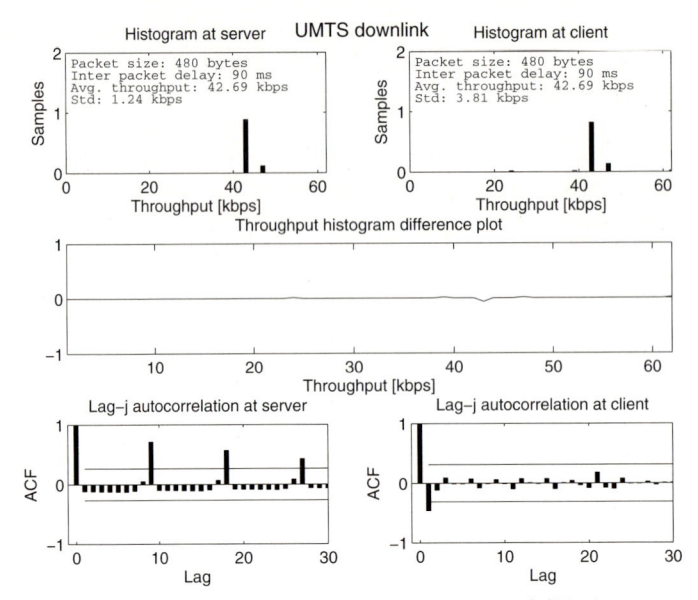

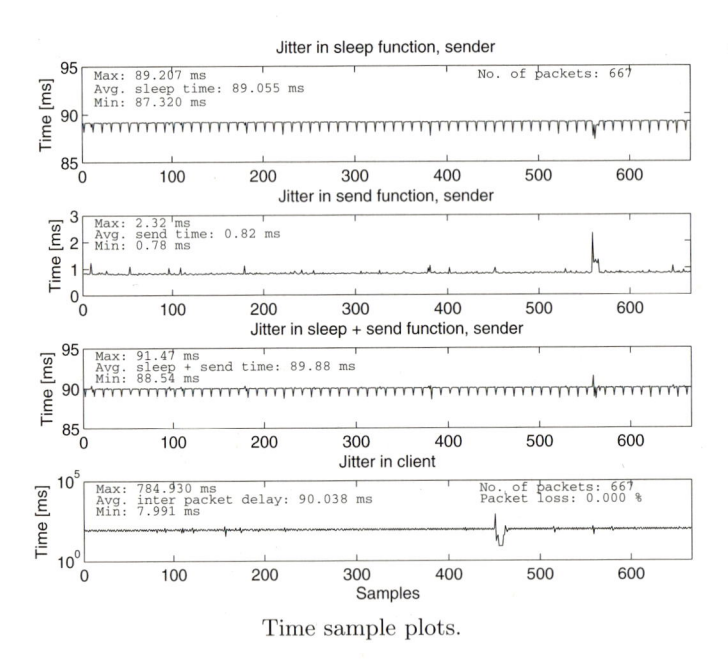

Throughput histograms, difference plot and ACF plots.

Time sample plots.

Fig. 5. UMTS downlink scenario, 90 ms inter-packet delay

to standard deviation and histogram, *cf.* Figure 7 (a) (top-right), the datagrams arrive much more scattered at the client's side, which receives data at rates varying from roughly 175 to 475 kbps. The bottleneck indicator seen in Figure 6 (a)

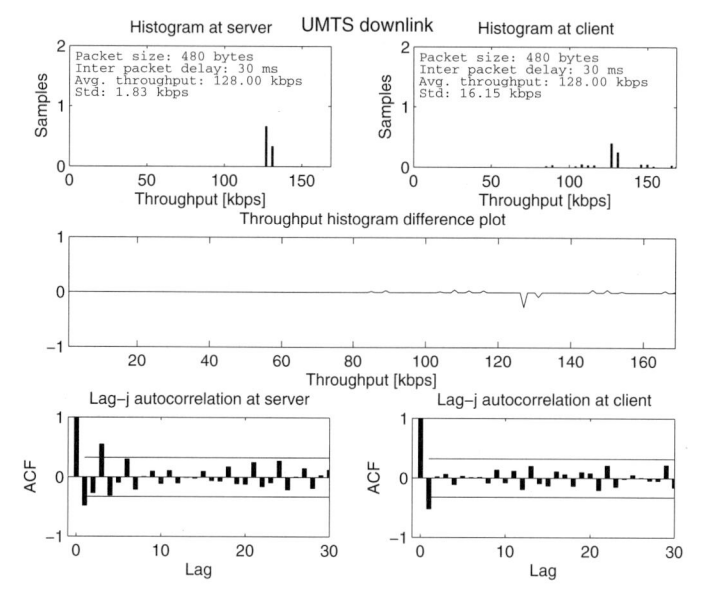

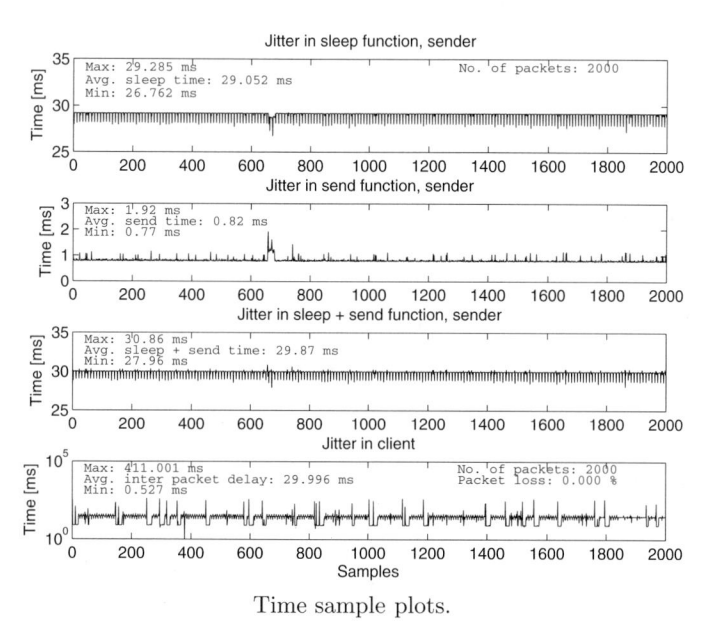

Fig. 6. UMTS downlink scenario, 30 ms inter-packet delay

(middle) reveals a shared bottleneck. The autocorrelation, *cf.* Figure 7 (a) (bottom-right), is rather small but irregular. The jitter perceived by the client is also very high together with some packet loss, *cf.* Figure 7 (b) (bottom).

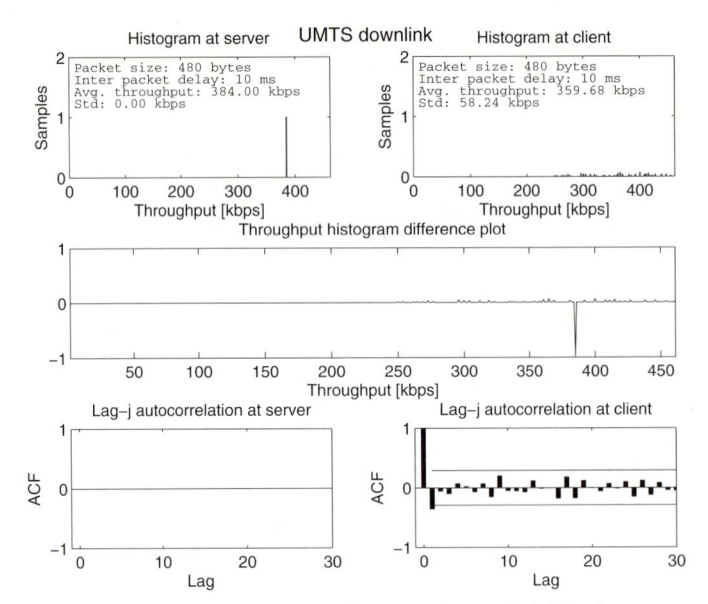

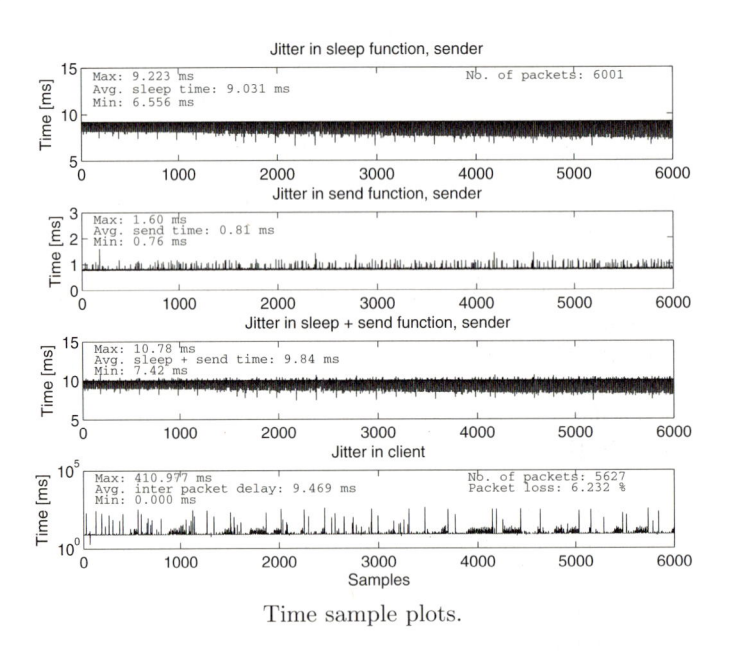

Fig. 7. UMTS downlink scenario, 10 ms inter-packet delay

Table 3 contains an overview of all UMTS downlink measurements that have been carried out. There is a tendency that throughput jitter increases as the offered load approaches the capacity of the downlink (384 kbps).

Table 3. UMTS Downlink with packet size of 480 bytes

Nominal IPD [ms]	Average IPD [ms]	$\bar{R}_A^{in}$ [kbps]	$\bar{R}_A^{out}$ [kbps]	$\sigma_{R_A}^{in}$ [kbps]	$\sigma_{R_A}^{out}$ [kbps]	ℓ [%]
10	9.84	384.00	359.68	0.00	58.24	6.2
11	10.85	349.12	348.42	1.07	68.19	0.2
12	11.85	320.00	319.94	1.83	95.05	0.0
15	14.87	256.00	254.98	1.83	67.30	0.4
20	19.87	192.00	192.00	0.00	55.64	0.0
30	29.87	128.00	128.00	1.83	16.15	0.0
40	39.88	96.06	95.81	2.06	56.80	0.2
50	49.87	76.80	76.80	2.34	4.90	0.0
60	59.89	64.00	63.94	1.83	6.12	0.1
80	79.88	48.00	47.94	1.94	3.71	0.1
90	89.88	42.69	42.69	1.24	3.81	0.0
100	99.89	38.40	38.40	0.00	1.00	0.0

5.2 UMTS Uplink

We now apply the same type of overload simulation to the UMTS uplink. The nominal inter-packet delay was set to 90 ms, cf. Figure 8. In this case, the impact of the network is visible but small; the average throughput is the same on server and client side, while the standard deviation has grown slightly; still, it is small as compared to the average both at sender and receiver. At the server side, the autocorrelation plot, cf. Figure 8 (a) (bottom-left), reveals a throughput periodicity of about 9 s. However, at the client side, this periodicity is less pronounced, cf. Figure 8 (a) (bottom-right). No packet loss is perceived. Figure 8 (b) shows that the sender's jitter is hardly damped by the adaptable sleep function. This indication comes from the burstiness of the data sent thought the link.

In the second case, the offered traffic amounts to 480 × 8 bit / 60 ms = 64 kbps. However, the average throughput at the server's side reached only 59.26 kbps, which means an effective average inter-packet delay of 65 ms. A strange peak at 0 kbps occurs in the throughput histogram at the server's side, cf. Figure 9 (a) (top-left). Obviously, datagrams were buffered and sending was delayed until there is capacity available on the UMTS link. The server was sending at 0 to 112 kbps; the standard deviation is almost as large as the average throughput. At the client side the average throughput also amounts to 59.26 kbps, which means that there is no missing traffic at the end of the observation interval (0 % loss). The throughput histogram is much more compact, displaying throughputs between 35 and 66 kbps. The standard deviation is reduced almost by factor eight, the bottleneck indicator (Figure 9 (a) (middle)) displays a shaping bottleneck. The autocorrelation is quite small, cf. Figure 9 (a) (bottom). Another indication is the time spent in the sleep function is very low, cf. Figure 9 (b) (top), which indicates that sleeping time is already consumed by the send function.

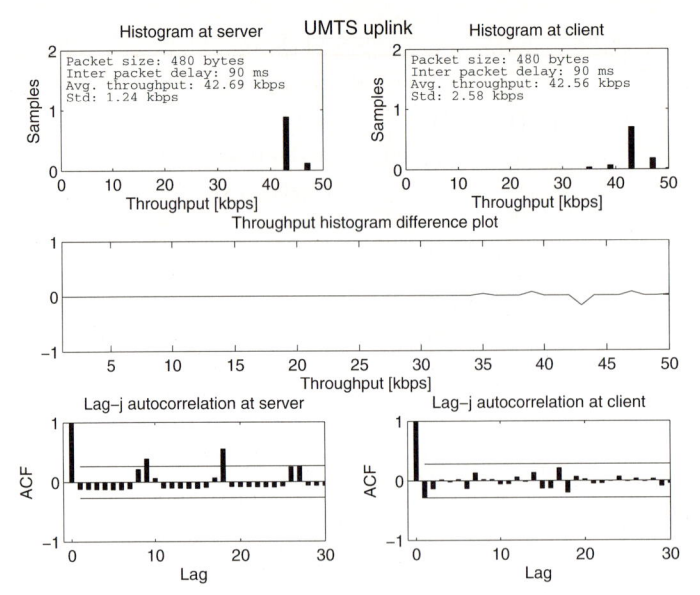

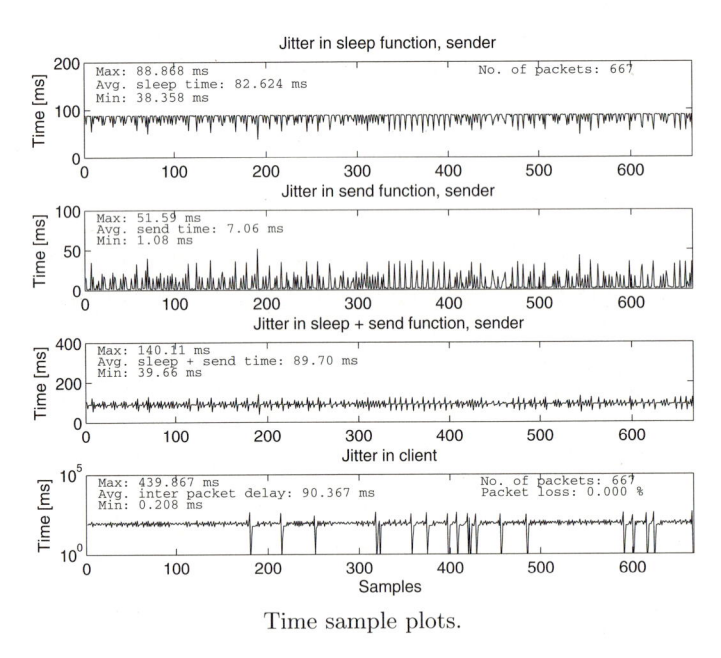

Fig. 8. UMTS uplink scenario, 90 ms inter-packet delay

Table 4 summarizes results from the UMTS uplink measurements. Again, we recognize the trend that the receiver perceives more throughput variation as the sender approaches the nominal capacity of 64 kbps. However, as soon as this

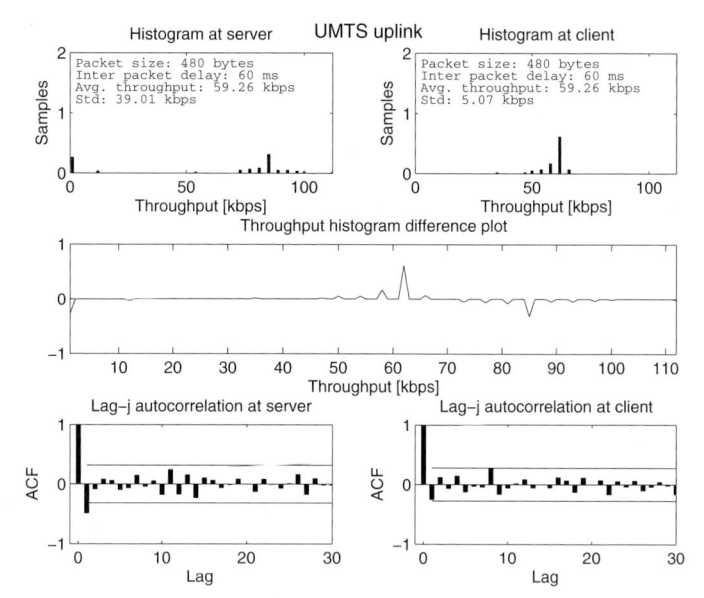

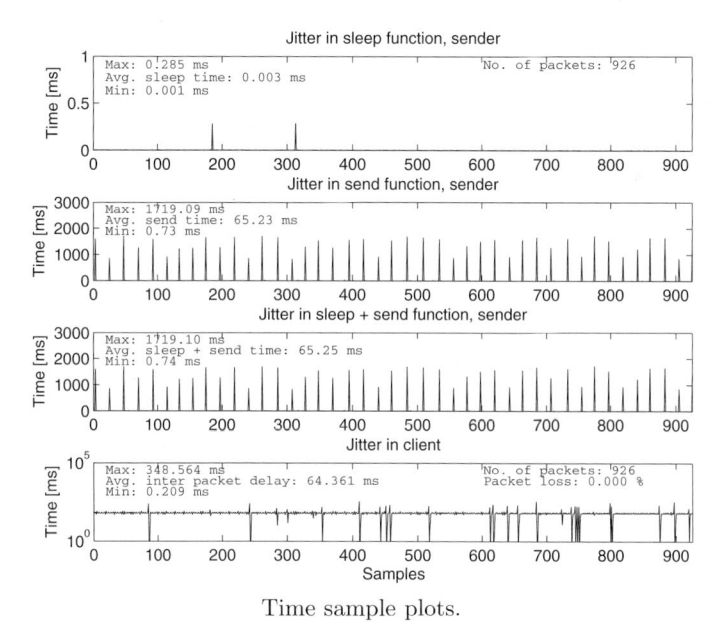

Fig. 9. UMTS downlink scenario, 60 ms inter-packet delay

Table 4. UMTS Uplink with packet size of 480 bytes

Nominal IPD [ms]	Average IPD [ms]	$\bar{R}_A^{in}$ [kbps]	$\bar{R}_A^{out}$ [kbps]	$\sigma_{R_A}^{in}$ [kbps]	$\sigma_{R_A}^{out}$ [kbps]	ℓ [%]
50	65.61	58.62	58.62	41.13	6.92	0
60	65.25	59.26	59.26	39.01	5.07	0
70	69.70	54.91	54.91	1.77	11.71	0
80	79.70	48.00	48.00	1.94	5.27	0
90	89.70	42.69	42.56	1.24	2.58	0
100	99.67	38.40	38.40	0.00	4.30	0

capacity is surpassed, the average inter-packet delay stays below the nominal inter-packet delay, and the throughput at the sender starts to jitter.

5.3 GPRS Downlink

The first GPRS downlink case addresses a modestly loaded link with the nominal inter-packet delay time set to 130 ms, *cf.* Figure 10. The throughput histograms and the standard deviations on server and client side are equal, *cf.* Figure 10 (a) (top). The autocorrelation at the sender, *cf.* Figure 10 (a) (bottom-left), reflects the non-integer ratio of ΔT and the nominal inter-packet delay. It is slightly damped by the network, but the original correlation structure is preserved. According to Figure 10 (b) the packet loss is zero.

The second case addresses a GPRS downlink with the nominal inter-packet delay time set to 70 ms, *cf.* Figure 11, the server sent with an almost constant throughput of 14 to 16 kbps, yielding an average of 14.64 kbps. However, the client received the datagrams in a scattered way at throughputs varying from 1 to 33 kbps which an average of 11.64 kbps. As in the UMTS downlink case, *cf.* Figure 7, the network introduces additional burstiness. From Figure 11 (a) (top), we can see that the standard deviation grows roughly by factor 16. Figure 11 (a) (middle) reveals a shared bottleneck. Data loss amounts to about 20 % with a period of 9 s, *cf.* Figure 11 (b) (bottom). The autocorrelation (Figure 11 (a) (bottom)) at the server side displays a periodic behavior of 3.5 s, while the client perceives a longer period of about 15 s which is probably due to the overload situation with regular throughput breakdowns. Reference [25] reports on abortive file downloads via GPRS, which might have its roots in the here-described problems.

Table 5 provides an overview of different measurements. Quite high losses occur quite frequently. However, no real trend can be seen. The only loss-less case in which the network behaves transparently is the case of a nominal IPD of 130 ms. In all other cases, there is a considerable growth of the standard deviation at the receiver.

5.4 GPRS Uplink

The first GPRS uplink case addresses a modestly loaded link with the nominal inter-packet delay set to 130 ms, *cf.* Figure 12. Already at this load, the packets

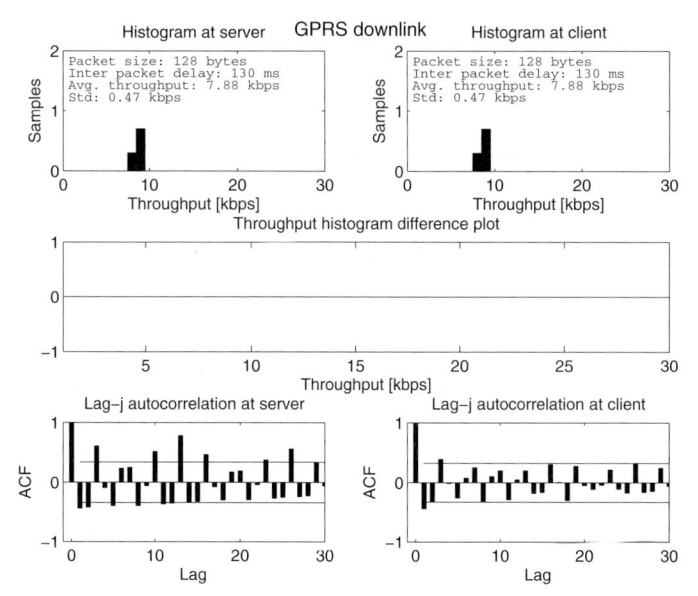

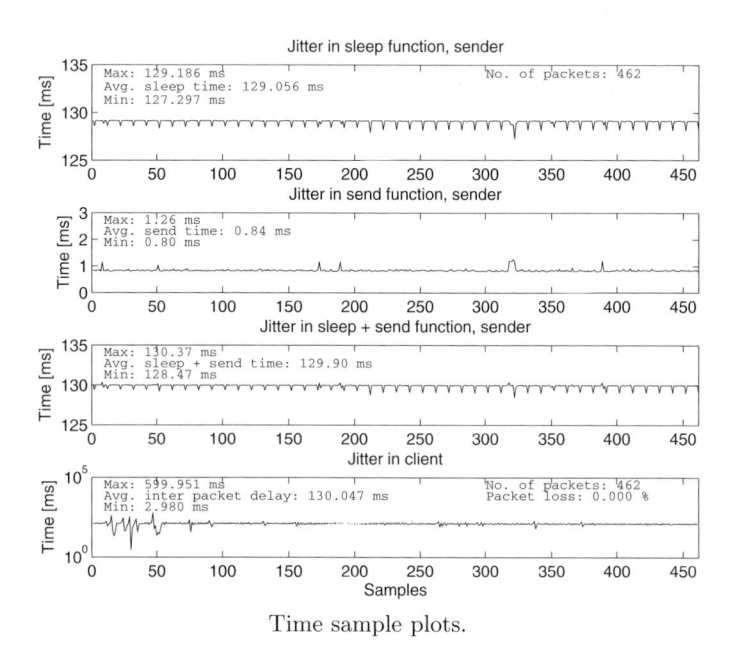

Fig. 10. GPRS downlink scenario, 130 ms inter-packet delay

are received in a bursty way at the client side, *cf.* Figure 12 (a) (top-right); Figure 12 (a) (middle) indicates a shared bottleneck. The autocorrelation at the server side displays a periodic behavior of 3 s with highly correlated values (Figure 12 (a) (bottom)), data loss is zero.

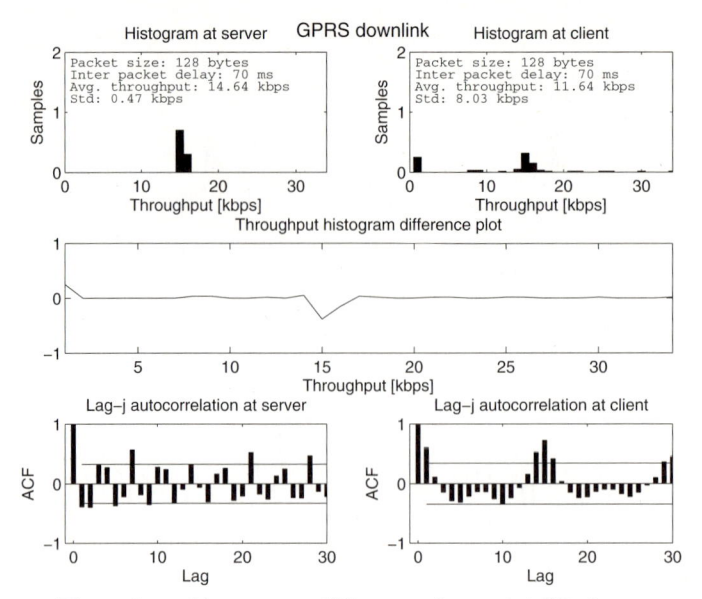

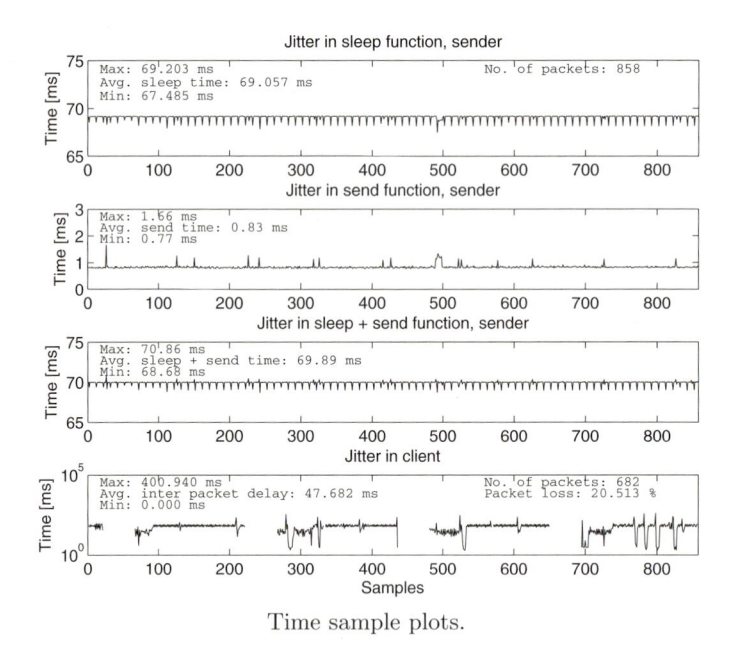

Fig. 11. GPRS downlink scenario, 70 ms inter-packet delay

In the second GPRS uplink case, the nominal inter-packet delay was set to 80 ms, yielding a server transmission rate of 12.82 kbps, *cf.* Figure 13 (a) (top-left). This which is almost matched by the measured averaged throughput, which amounts to 12.77 kbps. At the server and client side datagrams appeared as

Table 5. GPRS Downlink with packet size of 128 bytes

Nominal IPD [ms]	Average IPD [ms]	$\bar{R}_A^{in}$ [kbps]	$\bar{R}_A^{out}$ [kbps]	$\sigma_{R_A}^{in}$ [kbps]	$\sigma_{R_A}^{out}$ [kbps]	ℓ [%]
50	49.89	20.50	16.93	0.58	9.67	15.2
60	59.89	17.07	14.47	0.49	10.37	15.3
70	69.89	14.64	11.64	0.47	8.03	20.5
80	79.89	12.80	10.22	0.52	7.01	20.2
90	89.89	11.38	10.80	0.33	3.19	5.1
100	99.87	10.26	9.73	0.61	4.44	5.0
110	109.89	9.32	8.26	0.31	4.75	11.4
120	119.89	8.53	7.22	0.49	5.89	15.4
130	129.90	7.88	7.88	0.47	0.47	0.0

a scattered stream with throughputs of 1 to 48 kbps at the server and 1 to 20 kbps at the client. The standards deviation is slightly reduced. We observe a shared bottleneck in the histogram difference plot, cf. Figure 13 (a) (middle). The channel destroys the 4 s throughput periodicity but introduces some extra low-term correlation, cf. Figure 13 (a) (bottom). From Figure 13 (b), we see that there exist situations in which sleeping is not an option. Between samples 125–195 and 230–260, such non-sleeping situations are indicated. Packet loss occurs close to sample 150, with a total packet loss ratio of 0.4 %.

Table 6 summarizes the results of the GPRS uplink measurements. From the rise of the standard deviation at the sender $\sigma_{R_A}^{in}$, we can deduct the cases in which an overload situation is given (nominal inter-packet delay between 50 and 80 ms). For longer inter-packet delays, the standard deviation at the receiver $\sigma_{R_A}^{out}$ is similar in size. Loss is neglectible.

5.5 Impact of the Initial Delay

This section visualizes the impact of the initial delay and evaluates the throughput-based criteria presented in Section 4.3.

Figure 14 (a) plots the unshifted throughput time series $R_{A,s}^{in}$ and $R_{A,s}^{out}$ (i.e. $q = 0$) for the UMTS downlink with 90 ms inter-packet delay and $\Delta T = 0.5$ s. Excessive throughput values are observed at the output in the beginning. Figure 15 (b) plots the content of the equivalent bottleneck

$$X_s = X_{s-1} + (R_{A,s}^{in} - R_{A,s+q}^{out})\Delta T\,;\ X_0 := 0, q \in \mathcal{N}^0 \qquad (24)$$

for different values of q. Obviously, $X_s < 0$ for $q < 3$, while a shift by $q^* = 3$ yields causality ($X_s \geq 0\ \forall s$), which leads us to the estimation $\tau_0^R = 3 \cdot 0.5$ s $= 1.5$ s (19).

We now look at a case involving loss. Figure 15 (a) plots the unshifted throughput time series $R_{A,s}^{in}$ and $R_{A,s}^{out}$ (i.e. $q = 0$) for the UMTS downlink with 20 ms inter-packet delay and $\Delta T = 0.5$ s. Again, quite high throughput values are

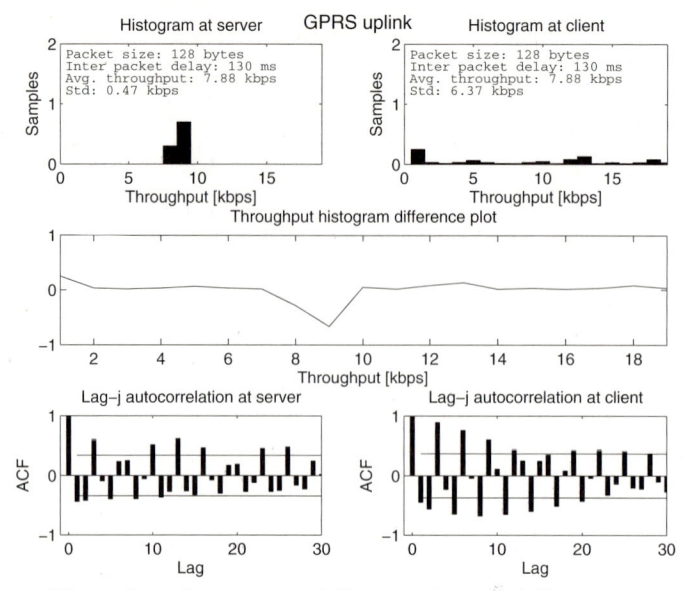

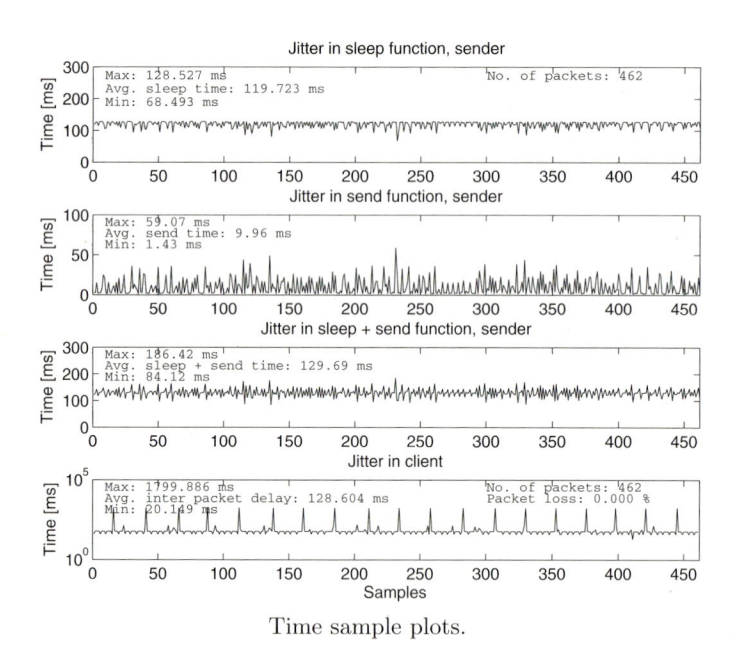

Fig. 12. GPRS uplink scenario, 130 ms inter-packet delay

observed at the client in the beginning, while in the end there is no traffic at all. Figure 15 (b) plots the content of the equivalent bottleneck (24) for different values of q. Finally, a shift by $q^* = 5$ yields causality, which gives $\tau_0^R = 2.5$ s.

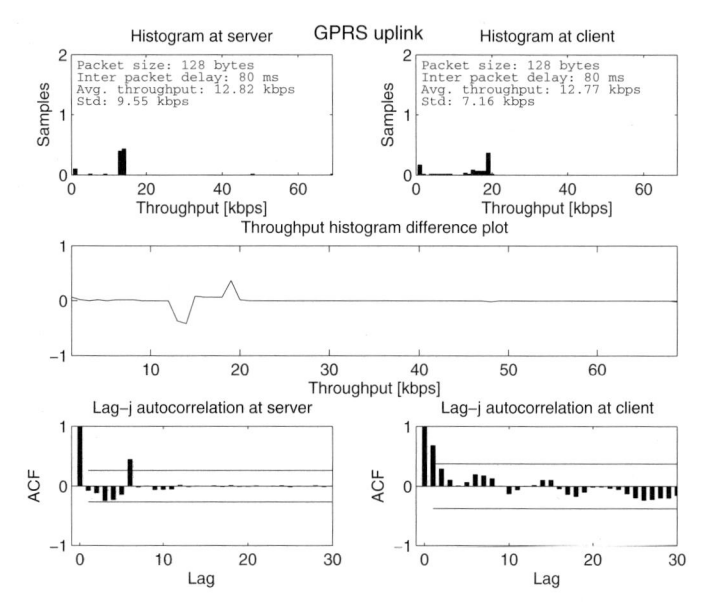

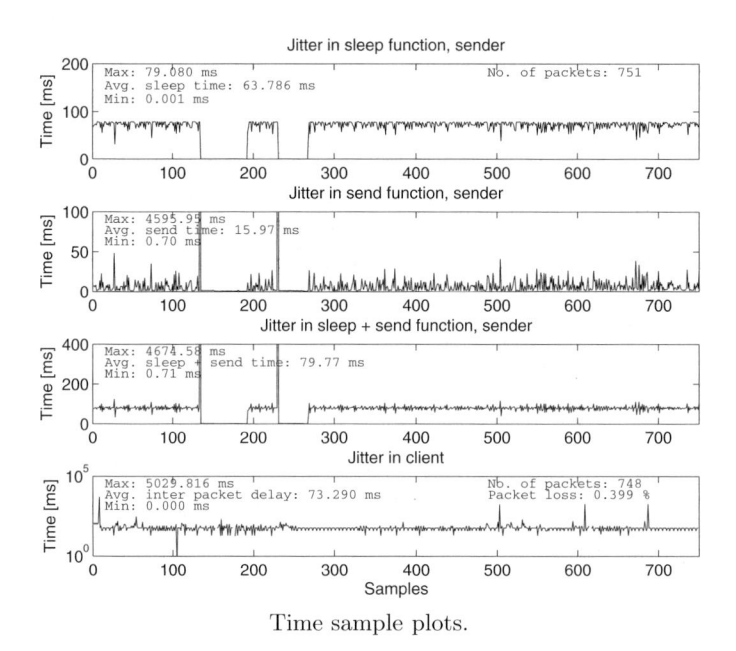

Fig. 13. GPRS uplink scenario, 80 ms inter-packet delay

Please observe the final "staircase" for $q = 5$ visualizing the loss L. Comparing with Table 7, we find that τ_0^R underestimates the time-based initial delay estimation $\tau_0^T \simeq 6.85$ s. This is due to the initial loss. Using the loss-compensated criterion (22), we in fact obtain $\tau_0^L = 7$ s, *i.e.* an upper bound for τ_0^T.

Throughput time series.

Buffer content of the equivalent bottleneck.

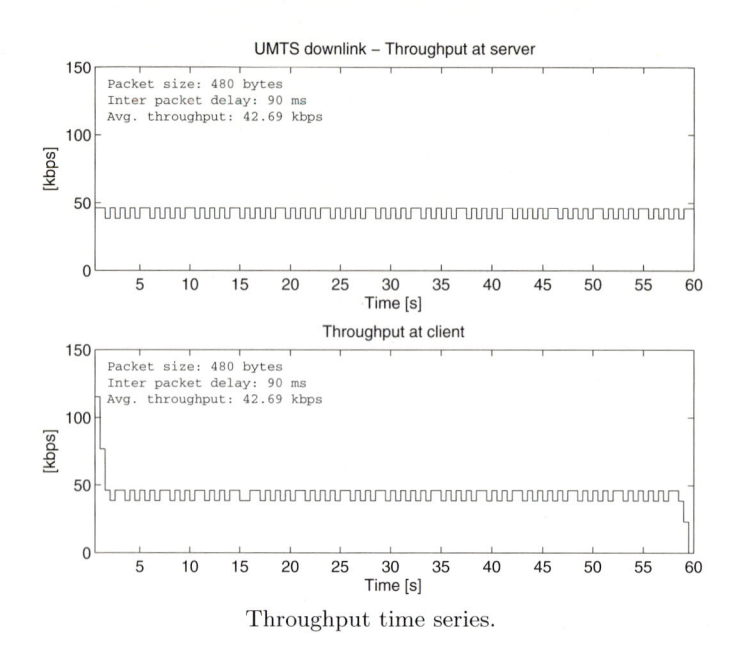

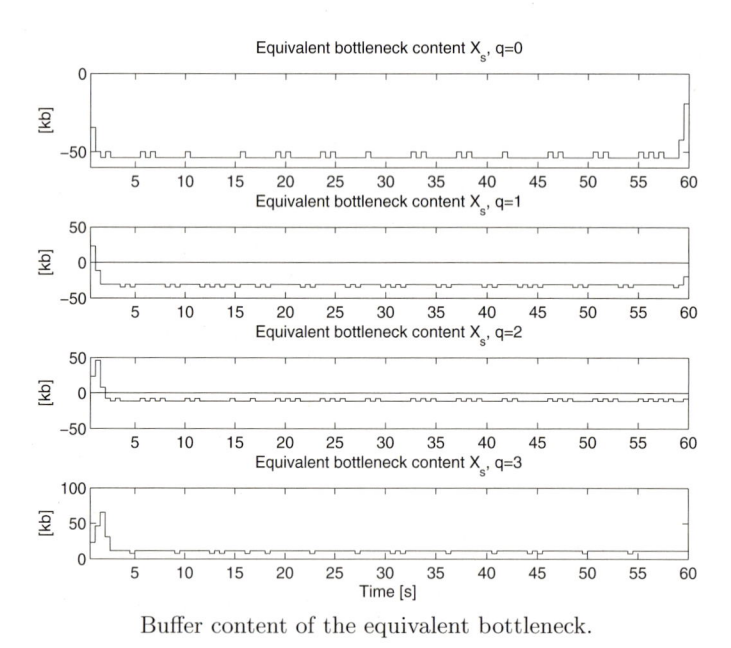

Fig. 14. UMTS downlink scenario, 90 ms inter-packet delay

Table 6. GPRS Uplink with packet size of 128 bytes

Nominal IPD [ms]	Average IPD [ms]	$\tilde{R}_A^{in}$ [kbps]	$\tilde{R}_A^{out}$ [kbps]	$\sigma_{R_A}^{in}$ [kbps]	$\sigma_{R_A}^{out}$ [kbps]	ℓ [%]
50	130.24	8.58	8.58	24.64	2.13	0.0
60	91.24	11.95	11.91	26.46	3.99	0.3
70	104.15	10.02	10.00	24.53	4.55	0.2
80	79.77	12.82	12.77	9.55	7.16	0.4
90	89.67	11.38	11.38	0.33	6.88	0.0
100	99.71	10.24	10.07	0.42	6.40	0.3
105	104.74	9.76	9.52	0.52	6.77	0.2
110	109.72	9.32	9.32	0.31	6.74	0.0
115	114.72	8.91	8.91	0.47	6.62	0.0
120	119.74	8.53	8.38	0.49	6.70	0.0
130	129.69	7.88	7.88	0.47	6.37	0.0

Table 7. UMTS with packet size of 480 bytes

IPD Nominal [ms]	Downlink τ_0^T [ms]	τ_0^R [s]	τ_0^L [s]	Packet Loss [%]	Uplink τ_0^T [ms]	τ_0^R [s]	τ_0^L [s]	Packet Loss [%]
10	N/A	N/A	5.0	8.7				
11	4933	2.0	5.0	5.2				
12	5862	2.0	6.0	6.5				
15	4375	2.0	4.5	4.3				
20	6854	2.5	7.0	7.3				
30	6221	3.0	6.5	6.1				
40	5658	3.5	6.0	4.1				
50	5805	3.5	6.0	4.7	N/A	N/A	N/A	0.0
60	6164	3.0	6.0	5.4	3074	3.0	3.0	0.0
70	1051	1.5	1.5	0.0	4115	4.5	4.5	0.0
80	6558	4.0	7.0	5.1	4288	4.5	4.5	0.0
90	1260	1.5	1.5	0.0	3990	4.5	4.5	0.0
100	5320	4.0	5.5	2.4	4088	4.5	4.5	0.0

Let us now look at the UMTS-related results presented in Table 7, where N/A stands for cases when the minimum in (18), (21) or (22) could not be found due to overload situations. If there is no loss, $\tau_0^R = \tau_0^L$. Except for some few cases, τ_0^L provides a tight upper bound of the extra initial delay. For UMTS, the anticipation of having the losses collected at the beginning of the interval seems to be accurate.

Some results for GPRS are presented in Table 8. Here, we observe that τ_0^R in most cases provides a good bound for τ_0^T, while τ_0^L overestimates the initial delay especially in cases of heavy loss ($\ell \simeq 0.1\ldots0.2$). This has its origin in a loss process rather different from the UMTS case: Loss is not concentrated to the

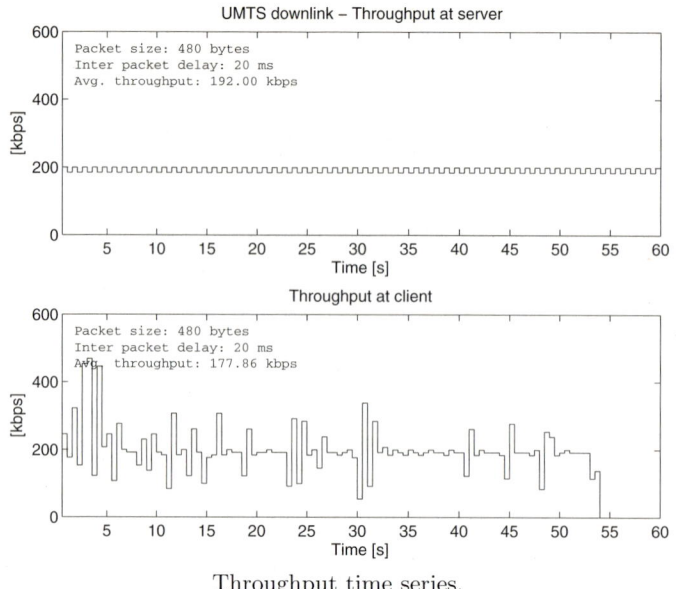

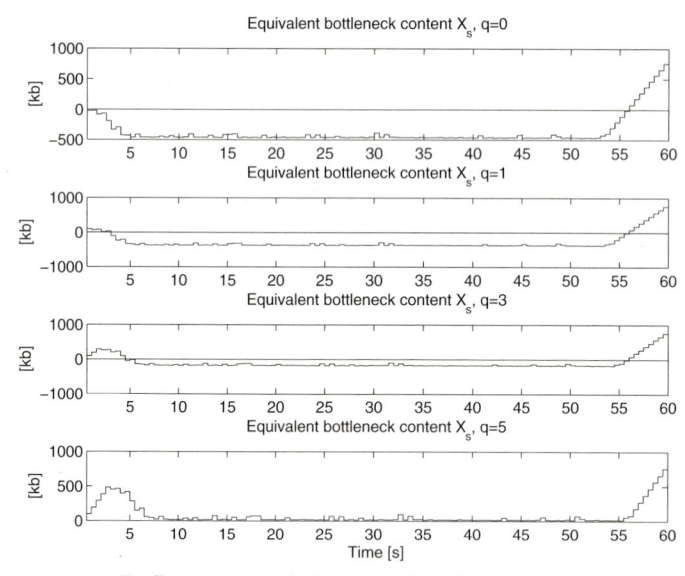

Fig. 15. UMTS downlink scenario, 20 ms inter-packet delay

beginning but appears rather bursty during ΔW, cf. Sections 6.3 and 6.4. Still, τ_0^T can serve as a (conservative) upper bound if no further information on the nature of the loss process is available. Again, in case of negligible loss ($\ell \to 0$), both estimations yield the same value $\tau_0^R = \tau_0^L$.

Table 8. GPRS with packet size of 128 bytes

IPD Nominal [ms]	Downlink				Uplink			
	τ_0^T [ms]	τ_0^R [s]	τ_0^L [s]	Packet Loss [%]	τ_0^T [ms]	τ_0^R [s]	τ_0^L [s]	Packet Loss [%]
80	1526	2.0	11.0	15.1	181	0.5	0.5	0.2
90	808	1.0	1.0	0.1	1493	1.5	1.5	0.1
100	2074	2.5	8.5	9.9	321	0.5	1.0	0.6
110	1870	2.0	8.0	9.8	345	0.5	0.5	0.0
120	1899	0.5	12.5	19.9	305	0.5	0.5	0.0
130	1187	1.5	1.5	0.0	402	0.5	0.5	0.2

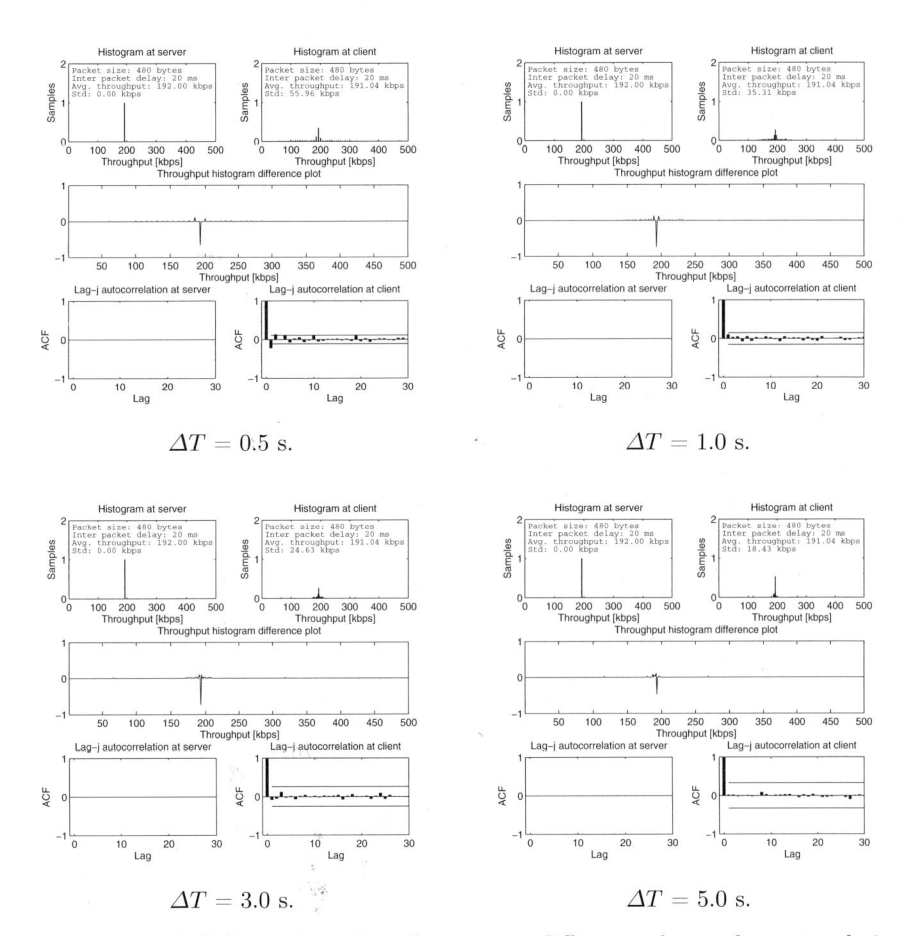

Fig. 16. Impact of ΔT on throughput histograms, difference plots and autocorrelation functions

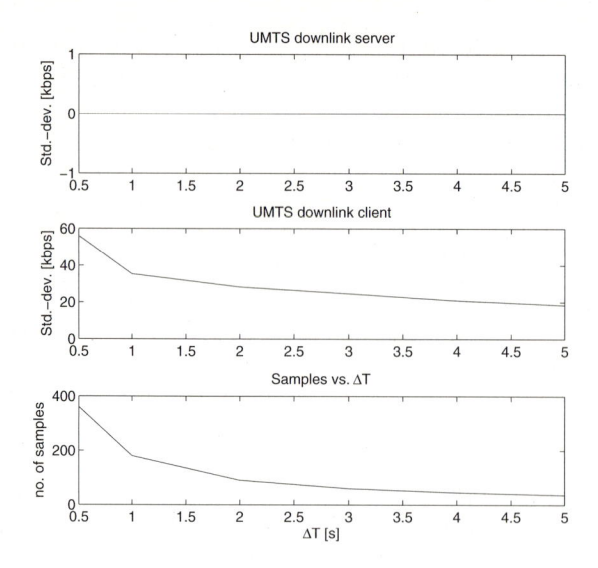

Fig. 17. Impact of the averaging interval ΔT on standard deviation and number of time intervals n

5.6 Impact of the Averaging Interval ΔT

This section highlights the impact of the averaging interval ΔT on the performance measures introduced in Section 3. Figure 16 shows throughput histograms, corresponding difference plots and plots of the lag-j autocorrelation coefficients for the UMTS downlink with an inter-packet delay of 20 ms for different values of ΔT as observed during $\Delta W = 5$ min. Figure 17 illustrates the standard deviations at sender and receiver together with the number of samples as functions of ΔT.

In general, standard deviation and autocorrelation at the receiver get smaller as the averaging interval grows, which means that high-frequent jitter components are averaged out at least to some extent. Also, the throughput difference plots get more narrow as ΔT grows. Interestingly enough, they do not vanish, which means that the network cannot be considered to be transparent on these time scales. In [12], considerable autocorrelation has been observed on a backbone link for $\Delta T = 5$ min. We can conclude that the network can have a distinct long-term after-effect on the throughput of a stream of interest.

6 Conclusions and Outlook

This paper focuses on measurements of application-perceived throughput on rather short averaging intervals. The influence of GPRS, and UMTS networks in both uplink and downlink direction on relevant summary statistics is illustrated and discussed. The throughput-related statistics have shown to be capable of visualizing critical network impacts on application performance. Using the concept of an equivalent bottleneck, it is also possible to derive bounds for the extra delay of the first packet due to the need of setting up a Temporary Block Flow via causality arguments. Thus, we have been able to describe these kind of delays by throughput measurements.

In general, both UMTS and GPRS networks are capable of introducing enormous amounts of jitter, which is seen from throughput deviations based on one-second averages. The impact of these mobile channels on throughput histograms and autocorrelation functions are clearly visible. While UMTS is almost transparent in terms of throughput as long as the application uses only a small share of the nominal capacity, GPRS hardly does without jitter.

In the *downlink* direction, there is a certain risk of data loss even if the nominal capacity of the mobile link is not reached yet. It was observed that the GPRS network in use had considerable problems in delivering packets in downstream direction. Loss ratios of 10 to 20 % were not uncommon, which efficiently jeopardizes the performance of streaming applications.

In the *uplink* direction, the burstiness also rises as long as the nominal capacity of the wireless link is not reached. This effect is reversed as soon as the nominal capacity is surpassed: In case the sending application transmits datagrams too fast, the send function itself acts as a shaper by holding packets until they can be sent, which can be considered as some kind of "force feed-back". Loss is avoided that way, but timing relationships within the stream can differ substantially from the ones imposed by the streaming application.

Summarizing, UMTS seems to be suitable for streaming services as long as the service uses merely a part of the nominal link capacity. GPRS, on the contrary, does not seem to be feasible for streaming services at all and might even fail supporting elastic services such as file transfers because of the large amounts of loss and jitter.

Based on the above analysis and knowledge gained, an application could gain more throughput and being more efficient if the datagrams are sent with some inter-packet delay instead of sending all datagrams as fast as possible. This inter-packet delay, *i.e.* the nominal throughput, should be well-adapted to the throughput supported by the weakest communication link. As the available capacity could be changed during an ongoing connection due to cross traffic, it might be important to adaptively change the inter-packet delay to the new conditions. In general, the application programmer considering the use of mobile channels should be conscious of these throughput variations. Moreover, these results are aimed at optimizing network selection based on requirements provided by users and applications.

Future work will compare the offerings by different providers and will take cross traffic into account. Also, other time scales as well as the relationship between network behavior and user perception needs to be investigated further.

Acknowledgements

The authors would like to thank Stefan Chevul and Johan Karlsson for fruitful discussions and valuable help in the context of this work.

References

1. Fiedler, M., Chevul, S., Isaksson, L., Lindberg, P., Karlsson, J.: Generic Communication Requirements of ITS-Related Mobile Services as Basis for Seamless Communications. In: First EuroNGI Conference on Traffic Engineering, NGI 2005 (2005)
2. Fiedler, M., Tutschku, A., Carlsson, P., Nilsson, A.: Identification of performance degradation in IP networks using throughput statistics. In: 18th International Teletraffic Congress (ITC-18), pp. 399–407. Elsevier, Amsterdam (2003)
3. Carlsson, P., Constantinescu, D., Popescu, A., Fiedler, M., Nilsson, A.: Delay performance in IP routers. In: HET-NETs 2004 Performance Modelling and Evaluation of Heterogeneous Networks (2004)
4. Bettstetter, C., Vogel, H.-J., Eberspächer, J.: GSM Phase 2+, General Packet Radio Service GPRS: architecture, protocols and air interface. IEEE Communications Surveys 2 (1999)
5. Vranken, R., van der Mei, R., Kooij, R., van den Berg, J.: Flow-level performance models for the TCP with QoS differentiation. In: International Seminar on Telecommunication Networks and Teletraffic Theory, pp. 78–87 (2002)
6. Charzinski, J.: Fun Factor Dimensioning for Elastic Traffic. In: 13th ITC Specialist Seminar on Internet Traffic Measurement, Modeling and Management (2000)
7. Ji, Z., Zhou, J., Takai, M., Bagrodia, R.: Scalable simulation of large-scale wireless networks with bounded inaccuracies. In: ACM MSWIM (2004)
8. Yang, C.C.Z., Luo, H.: Bandwidth measurement in wireless mesh networks. Course Project Report (2005), http://www.crhc.uiuc.edu/~cchered2/pubs.html
9. Leland, W.-E., Taqqu, M.S., Willinger, W., Wilson, D.V.: On the self-similar nature of Ethernet traffic (extended version). IEEE/ACM Trans. on Netw. 2(1), 1–15 (1994)
10. Norros, I.: On the use of fractional Brownian motion in the theory of connectionless networks. IEEE JSAC 13(6), 953–962 (1995)
11. Veres, A., Kenesi, Z., Molnár, S., Vattay, G.: TCP's role in the propagation of self-similarity in the Internet. Computer Communications, Special Issue on Performance Evaluation of IP Networks 26(8), 899–913 (2003)
12. Juva, I., Susitaival, R., Peuhkuri, M., Aalto, S.: Traffic characterization for traffic engineering purpose: Analysis of Funet data. In: First EuroNGI Conference on Traffic Engineering, NGI 2005 (2005)
13. Fiedler, M., Addie, R.G.: Verification and application of a second-order scale symmetry for queueing systems. In: 16th International Teletraffic Congress (ITC-16), pp. 807–816. Elsevier, Amsterdam (1999)
14. Haßlinger, G., Hartleb, F., Fiedler, M.: The relevance of the bufferless analysis for traffic management in telecommunication networks. In: IEEE European Conference on Universal Multiservice Networks. IEEE, Los Alamitos (2000)
15. van de Meent, R., Mandjes, M.: Evaluation of 'user-oriented' and 'black-box' traffic models for link provisioning. In: First EuroNGI Conference on Traffic Engineering, NGI 2005 (2005)
16. Gunnar, A., Johansson, M., Telkamp, T.: Traffic Matrix Estimation on a Large IP Backbone - A Comparison on Real Data. In: 2004 ACM SIGCOMM Internet Measurement Conference, pp. 149–160. ACM, New York (2004)
17. Oetiker, T.: The multi router traffic grapher (MRTG), http://www.mrtg.org/
18. Carlsson, P., Fiedler, M., Tutschku, K., Chevul, S., Nilsson, A.A.: Obtaining reliable bit rate measurements in SNMP-managed networks. In: 15th ITC Specialists Seminar on Traffic Engineering and Traffic Management, pp. 114–123 (2002)

19. InfoSim GmbH & Co. KG homepage (2005), http://www.infosim.net/
20. Network Instruments – Protocol Analyzer, Network Protocol Analysis, Monitoring, Management, and Troubleshooting, http://www.networkinstruments.com/
21. Bolliger, J., Gross, T.: Bandwidth monitoring for network-aware applications. In: 10th IEEE International Symposium on High Performance Distributed Computing, HPDC-10 (2001)
22. Morris, R., Lin, D.: Variance of aggregated web traffic. In: IEEE INFOCOM 2000, vol. 1, pp. 360–366 (2000)
23. Fiedler, M., Chevul, S., Radtke, O., Tutschku, K., Binzenhöfer, A.: The network utility function: A practicable concept for assessing network impact on distributed systems. In: 19th International Teletraffic Congress (ITC-19); Technical Report 355, Universität Würzburg (2005)
24. Fiedler, M., Tutschku, K.: Application of the stochastic fluid flow model for bottleneck identification and classification. In: SCS Conference on Design, Analysis, and Simulation of Distributed Systems (DASD 2003) (2003)
25. Hoßfeld, T., Tutschku, K., Andersen, F.-U.: Mapping of file-sharing onto mobile environments: Enhancements by UMTS. Technical Report 343, University of Würzburg (2004)

Statistical Analysis and Modeling of Peer-to-Peer Multimedia Traffic

Natalia M. Markovich[1] and Udo R. Krieger[2]

[1] Institute of Control Sciences, Russian Academy of Sciences
Profsoyuznaya Str. 65, Moscow 117997, Russia
markovic@ipu.rssi.ru
[2] Faculty Information Systems and Applied Computer Science
Otto-Friedrich-Universität, D-96052 Bamberg, Germany
udo.krieger@ieee.org

Abstract. We study peer-to-peer packet traffic arising from passive VoIP and video measurements that are generated by Skype and IPTV clients. We provide a common methodology for the statistical characterization of the packet flows, discuss the user's satisfaction and load estimation. Two main ideas are used in our analysis. Due to the dependence of the data we first partition the observations into independent blocks and deal further with these block-wise independent data. Secondly, loss is generated by packet lengths which exceed the channel capacity in a time unit if the inter-arrival times coincide with this time unit. If the inter-arrival times are random, loss is generated by the lengths of those packets corresponding to transmission rates that exceed the channel capacity. Our methodology is demonstrated by individual Skype flows and the aggregated flow of video packets exchanged with a mobile peer of a SopCast session.

Keywords: Peer-to-peer traffic characterization, multimedia packet traffic, Skype, IPTV, SopCast.

1 Introduction

In recent years, peer-to-peer (P2P) multimedia applications like Skype, IPTV and on-line games have become a powerful service platform for the generation and transport of voice and video over IP. Due to its free access Skype, for instance, is now a real competitor of the traditional telephony services and has millions of customers.

Due to randomly appearing peers the main feature of a P2P application is determined by the random structure of its overlay network. The peers are both receivers and senders of chunks of information at the same time. They determine random transmission processes of this information to a cloud of receivers. The management and control of the traffic dynamics is complicated since it depends on this randomness in the overlay network and the transmission processes. To improve the understanding of the transport mechanisms in P2P networks, we investigate the statistical properties of P2P multimedia traffic at the interaction level and time scale of the packet layer.

Regarding Skype traffic different types of media like voice, video, and text messages are transferred by a client. Considering IPTV life sessions, video and voice streams are transmitted by the P2P network. Currently, there are several important P2P TV applications, including PPlive, PPStream, SopCast and TVAnts. Many authors have already tried to classify the VoIP traffic of Skype sessions and the packet flows of IPTV sessions regarding the used applications, applied encoding schemes etc., see for example [6], [12], [22]. These studies concern both the packet and flow level characterization of the monitored traffic streams.

In our study we do not intend to classify the gathered P2P traffic. But we study passive measurements of VoIP flows arising from a Skype client and video traffic generated by an IPTV session at the Ethernet packet layer. We focus on the inter-arrival time (IAT) and packet length (PL) processes and provide a common methodology for the statistical characterization, the analysis of the user's satisfaction and the load estimation of the packet transmission. This approach is possible since the analysis of the packet transmission over IP has a common foundation irrespective of the P2P video or voice transfer.

The characterization implies that we have to investigate whether the traffic is stationary, long- or short-range dependent or independent, self-similar (i.e. scale invariant), and heavy-tail distributed. The analysis can be done by a common and rigorous mathematical methodology. It is illustrated by examples of an aggregated IPTV traffic flow to a mobile peer and the VoIP traffic in a WLAN environment between two Skype users.

Since the data are mostly dependent, we have to partition it into independent blocks and to deal with representatives of these blocks like maxima, minima and averages just like with independent data. Particularly, this procedure allows us to fit the distribution of the maximum of the IATs between packets.

Another important question of our study concerns the user's satisfaction. It is determined at the packet layer by the loss and delay of transmitted packets. Both impact on the quality of service (QoS) and the user's quality of experience (QoE). The extremes (i.e. maximal and minimal values) of the IATs between delivered packets and of the PLs influence on the speech and image perception more than the non-extremal values of these indices. Hence, we model the distribution of the maximal IAT between packets and find its quantiles as indices of the quality. Besides, we propose the mean byte loss, the mean delivery time variation of packets per cluster and the quantiles of lossless periods as new indices of the quality.

The rest of the paper is organized as follows. In Section 2 the used data sets of the P2P multimedia flows are described. In Section 3 the common methodology to detect stationarity, long-range dependence (LRD), self-similarity and the heaviness of tails is presented. In Section 4 the partitioning of the observations into independent blocks as preliminary tool for the analysis of dependent data is explained. In Section 5 we present indices of the user's satisfaction during a packet transmission and illustrate them by means of Skype traffic data. In Section 6 the estimation of the offered traffic load in a finite time interval is presented. In Section 7 the Extreme Value Distribution and its quantiles describing the maximum of the IATs between transmitted packets are evaluated for IPTV data. Finally, some conclusions are presented.

2 Description of the Multimedia Packet Traffic

Actually, we use in our study the inter-arrival times (IATs) between packets and the packet lengths (PLs) as the main source of information. In the following the sequence of n IATs between the packets of a multimedia traffic flow is denoted by $X_1, X_2..., X_n$ and $Y_1, Y_2..., Y_n$ are the associated PLs. Our proposed methodology is demonstrated by means of two illustrative data sets containing peer-to-peer video and voice-over-IP (VoIP) traffic.

2.1 Description of the VoIP Data

To reveal the features of our statistical techniques for multimedia traffic characterization, we have used P2P VoIP traffic generated by Skype clients, cf. [26]. Due to the peer-to-peer character of Skype and the random nature of its overlay network relaying the generated packet flows, it is in general difficult to monitor the traffic between two communicating hosts along a path in the overlay network. Hence, one must gather the Skype packet traffic related to a particular site, cf. [8].

We have also followed this approach. Communication sessions between two Skype clients within a LAN test bed and their encoded voice samples have been gathered by means of Wireshark at Otto-Friedrich University Bamberg in 2006 (see Fig. 1 taken from [15], [16, Fig. 1]). The collected PLs and IATs between Ethernet packets of a representative single Skype flow will be used as first illustrative data set in our study. The latter flow has been generated based on a mixture of short representative sessions of monologues, dialogs and music clips with German and English male and female speakers and lasts 135 seconds. The resulting unidirectional VoIP packet stream has been isolated in a pre-processing phase. The variable bit rate wideband Internet Speech Audio Codec (iSAC) has been applied by the clients as basic voice encoding scheme with a sampling frequency of 16 kHz. It is able to respond to varying network conditions and generates variable data rates ranging from 10 to 32 kbps.

This data set illustrates the typical features of Skype flows in current home environments. It has been arising from a transmission path with several wired and two wireless links. Here both the sending mobile client (mobile host 2 at 192.168.182.22) and the receiving mobile client (mobile host 1 at 192.168.1.4) are first traversing private IEEE802.11 WLAN segments 1 and 5, respectively, with DSL attachments to the public Internet and then an Internet path 2,3,4 including

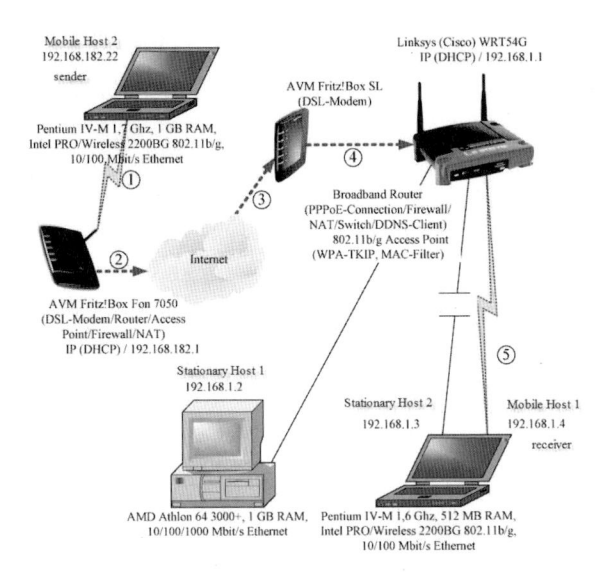

Fig. 1. LAN test bed for voice over IP communication by Skype clients (see [16, Fig. 1])

a tier-1 carrier exchange point between Telefonica's and Deutsche Telekom's ISP networks (see Fig. 1, cf. [16, Fig. 1]). Therefore, this network path can be considered as typical VoIP over WLAN environment that a majority of Skype users traverse today between two private homes.

To evaluate the load and delivery variation profile, we also need the extremes of the PLs and IATs within independent subsets (called blocks) of data, see Section 4. All descriptive statistics of these random variables (r.v.s) arising from our representative VoIP data set are stated in Table 1.

2.2 Description of the IPTV Data

The second illustrative data set contains P2PTV traces generated by the P2P IPTV system SopCast [27]. A comprehensive measurement study of a typical IPTV home scenario including a wireless access to the Internet has been performed during the second quarter of 2009 by the Computer Networks Laboratory of Otto-Friedrich University Bamberg, Germany.

In this wireless scenario the SopCast client is running on a desktop IBM Thinkcentre with 2.8 GHz Intel Pentium 4 processor, 512 MB RAM, and Windows XP Home. It is attached by a Netgear WG111 NIC operating the IEEE802.11g MAC protocol over a wireless link to the corresponding ADSL router acting as gateway to the Internet.

Watching a popular sport channel, representative traces arising from sessions of 30 minutes have been gathered by Wireshark at the mobile host. The descriptive statistics of a representative aggregated flow to the observed SopCast client are stated in Tab. 2.

Table 1. Description of the VoIP data arising from a Skype packet flow

R.V.	Sample Size	Min	Max	Mean	StDev	Skewness	Kurtosis
Inter-arrival times (sec)	4605	$1.9 \cdot 10^{-5}$	$2.01 \cdot 10^{-1}$	$3.1 \cdot 10^{-2}$	$8.635 \cdot 10^{-3}$	5.183	79.75
Packet lengths (bytes)	4605	45	284	160.27	25.808	−0.921	2.32
Maxima of inter-arrival times (sec)	72	$5.8 \cdot 10^{-2}$	$2.01 \cdot 10^{-1}$	$7.3 \cdot 10^{-2}$	$2.6 \cdot 10^{-2}$	3.622	14.253
Minima of inter-arrival times (sec)	72	$1.9 \cdot 10^{-5}$	$9.1 \cdot 10^{-2}$	$3.3 \cdot 10^{-2}$	$6.816 \cdot 10^{-4}$	0.028	−1.56
Maxima of packet lengths (bytes)	72	85	284	197.69	1688	−1.405	6.857

Table 2. Description of the IATs between packets in seconds and the block maxima corresponding to IAT blocks of size 400 (IAT400) arising from the aggregated flow to the observed peer

R.V.	Sample Size	Min	Max	Median	Mean	StDev	Skewness	Kurtosis
IAT	$6.553 \cdot 10^4$	$2.1 \cdot 10^{-5}$	0.625	$5.58 \cdot 10^{-4}$	$4.934 \cdot 10^{-3}$	0.016	13.56	313.087
IAT400	163	0.021	0.625	0.105	0.138	0.102	1.773	4.193

3 Statistical Characterization of P2P Packet Flows

3.1 Detection of Stationarity

Before applying any statistical analysis method to time series it is the first step to check whether the data are stationary. In practice it is not realistic to observe a pure stationary process. In this case we can partition the observations at our disposal into homogeneous sequences of data which are approximately stationary.

The weak stationarity of a stochastic process $\{X_t, t \geq 0\}$ requires that the first two moments and the autocorrelation function (ACF) do not change in time, i.e. $\mu = \mathbb{E}(X_t)$, $\sigma^2 = \text{Var}(X_t)$, and the ACF between X_t and X_s only depends on the difference $|t - s|$. Non-stationarity is equivalent to the presence of a deterministic or stochastic trend in the data.

All tests on stationarity, e.g., Cochran's test [9], the runs test [3], the R/S method [23] (see also Section 3.3), are based on a partitioning of the data into independent blocks and the comparison of averages, deviations from averages, standard deviations or other statistical characteristics calculated by means of these blocks.

The runs test, for instance, recommends to divide the time series into equal-sized time intervals, to compute a mean value for each interval and to count the number of runs of the mean values above and below the median value of the series. Then one has to compare the calculated number of counts with the value that one would expect if the observations were independent of each other.

In [5] the R/S test is applied to detect the presence of deterministic trends in the data. A survey of recent methods is provided by [10]. However, all these tests have own constraints and drawbacks. An important constraint is determined by the independence of data blocks that is difficult to achieve. Therefore, we consider here some rough tools to check whether the mean and variance do not change in time.

Let $\{X_i, i = 1, 2, ..., n\}$ denote the original time series. Regarding the partitioned data we then calculate the averages and variances within each block of size m numbered by the integer $k = 1, 2, ..., [n/m]$

$$X^{(m)}(k) = \frac{1}{m} \sum_{i=(k-1)m+1}^{km} X_i,$$

$$V^{(m)}(k) = \frac{1}{m-1} \sum_{i=(k-1)m+1}^{km} \left(X_i - X^{(m)}(k)\right)^2$$

and the sample variance of $X^{(m)}(k)$

$$\widehat{\text{Var}} X^{(m)} = \left[\frac{m}{n}\right] \sum_{k=1}^{[n/m]} \left(X^{(m)}(k)\right)^2 - \left(\left[\frac{m}{n}\right] \sum_{k=1}^{[n/m]} X^{(m)}(k)\right)^2. \quad (1)$$

To test the stationarity with regard to the homogeneity of the mean, we check the difference of the sample variances

$$D(m) = \widehat{\text{Var}} X^{(m)} - \widehat{\text{Var}} X^{(m-1)}$$

for successive values of m, cf. [23].

To check the homogeneity of the variances, one can apply Cochran's test. The idea of this test is simply to calculate the ratios

$$G_k = V^{(m)}(k) / \sum_{i=1}^{[n/m]} V^{(m)}(i), \quad k = 1, 2, ..., [n/m].$$

Then it is the objective to select the maximal value $G_{max} = \max_k G_k$ among them and to compare it with the quantiles of the distribution of $G_{\max}$ for a required level. Choran's test requires the independence and normality of the block data. In Section 4 we discuss a method to partition our data into independent blocks. However, the normality of the data over the blocks is not always fulfilled. According to the Central Limit Theorem the average over the blocks can be asymptotically normal distributed if the representatives of the blocks are independent and their second moment is finite. In Section 3.2 we discuss a way to detect the presence of heavy tails in the data and to understand how many moments of the distribution are finite. Note, that the normal distribution is light-tailed.

Moreover, it is known that it is difficult to distinguish between stationary processes with a long memory and non-stationary processes (cf. [4, Chap. 7.4, p. 141f]). We show in Section 3.3 that both illustrative data sets exhibit a long-range dependent (LRD) behavior.

Example 1: We check first the stationarity of the Skype packet data, namely, the IATs and PLs (regarding their description see Section 2.1).

The means of the IATs and PLs do not change much (see Fig. 2(b), 2(d), cf. also [15]). One cannot conclude definitely from the visual analysis that the variances do not change much in order to expect the non-stationarity of the IATs and PLs (see Fig. 2(a), 2(b), 2(e), 2(f), cf. also [15]).

Applying Cochran's test, we get $G_{\max} = 0.19$ regarding the IATs and $G_{\max} = 0.114$ for the PLs, cf. [15]. For $m = 145$ and $[n/m] = 31$ the 5% quantile of $G_{\max}$ is equal to 0.0457. The null hypothesis regarding the equality of the variances should be rejected since the calculated values $G_{\max}$ of the IATs and PLs are larger than the bound 0.0457. Nevertheless, this conclusion may be unreliable due to the deviation of the data from normality. The latter assumption constitutes the main constraint of this test. Due to a positive kurtosis the IATs and PLs are not normal distributed. Their non-zero skewness' indicate that the distributions of both characteristics are asymmetric, see Tab.1.

To proceed in a mathematically rigorous way, we partition our data into independent blocks and check the stationarity of their representatives by an inversion test, cf. [3]. In Section 4.2 we consider the techniques to select such blocks and apply it to the Skype data of the example.

The null hypothesis states that the underlying sequence contains independent stationary random observations, i.e., a trend does not exist. For this purpose one calculates the statistic

$$A = \sum_{i=1}^{n-1} \sum_{j=i+1}^{n} \mathbb{1}(X_i > X_j).$$

The hypothesis is accepted at level $\alpha = 0.05$ if $A_{n,1-\alpha/2} < A \leq A_{n,\alpha/2}$ holds, where $A_{n,1-\alpha/2}$ and $A_{n,\alpha/2}$ are quantiles of the distribution function (DF) of A. The bounds of A are given by $[1014, 1400]$.

We investigate the stationarity and independence of the maxima of the PLs and the increments $X_t - X_{t-1}$ of the maxima and minima of the IATs between Skype packets in the blocks. Since the values of A fall into the mentioned interval (see Tab. 3), the null hypothesis should be accepted for the maxima of PLs as well as the maxima and minima of the IATs.

3.2 Detecting the Heaviness of Tails of the Distributions

Let $F(x)$ denote the distribution function (DF) of the underlying r.v. X, e.g., the IAT. Roughly speaking, heavy-tailed distributions are those long-tailed distributions whose tails decay to zero slower than an exponential tail. The tail is determined by the function $1 - F(x)$. Regularly varying distributions with

$$1 - F(x) = x^{-\alpha} \ell(x), \qquad (2)$$

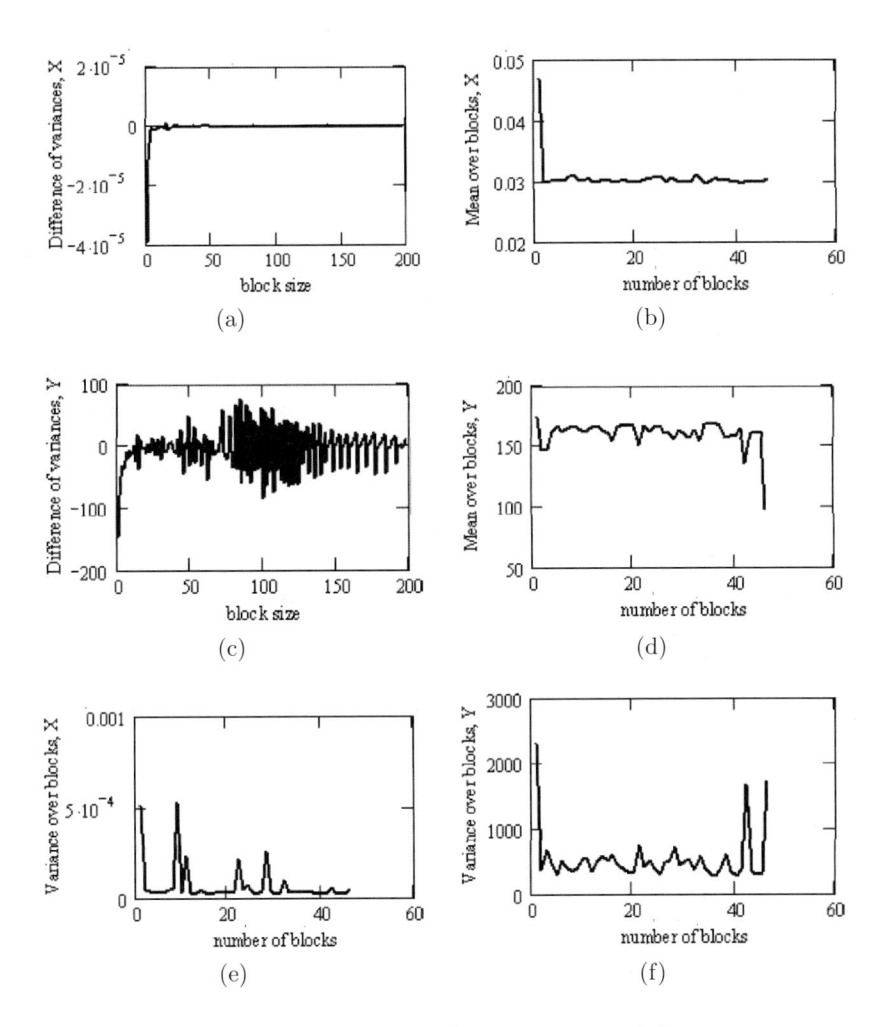

Fig. 2. The differences of variances $D(m)$ and the means $X^{(m)}(k)$ within each block of the inter-arrival times between Skype packets (a) and (b) and the lengths of Skype packets (c) and (d); the size of blocks $m = 100$ was chosen for (b) and (d); the variance $V^{(m)}(k)$ within each block of inter-arrival times (e) and packet lengths (f)

where $\ell(x)$ is a slowly varying function with the property $\lim_{x \to \infty} \ell(xt)/\ell(x) = 1$ for any $t > 0$, constitute the widest class of heavy-tailed distributions.

The tail index α or its reciprocal $\gamma = 1/\alpha$ called the extreme value index (EVI) show the shape of the tail of the distribution F. A positive sign of α implies that the distribution is heavy-tailed. The smaller α the heavier is the tail. Further, α indicates the number of finite moments, namely, $\mathbb{E}x^\beta < \infty$ holds for $\beta < \alpha$ if the distribution is regularly varying. In contrast to light-tailed distributions, not all moments of heavy-tailed ones are finite. All moments are infinite for super-heavy-tailed distributions.

Table 3. Inversion test results (cf. [16, Table V])

	Maxima of Packet Length	Increments of Inter-arrival Times	
		Maxima	Minima
A	1358	1294	1249

There are several rough tools that allow us to distinguish between light and heavy tails, see, e.g., [14] for a survey. Here we describe the most evident methods, namely, the estimation of the tail index and the mean excess function.

To estimate the EVI γ, we use the popular Hill's estimator

$$\widehat{\gamma}^H(n,k) = \frac{1}{k}\sum_{i=1}^{k} \ln X_{(n-i+1)} - \ln X_{(n-k)}.$$

Here $X_{(1)} \leq X_{(2)} \leq \ldots \leq X_{(n)}$ denote the order statistics of the sample $\{X_i, i = 1, \ldots, n\}$ and k is a smoothing parameter. One can select k corresponding to the stability interval of the Hill's plot $\{(k, \widehat{\gamma}^H(n,k)), k = 1, \ldots, n-1\}$.

One can estimate γ by a bootstrap procedure as an automatic method. This scheme implies the averaging of the Hill's estimates constructed over bootstrap re-samples that are taken from the underlying sample with repetitions, see [14, pp. 22–25].

There are numerous estimators of the tail index, but many of them are very sensitive to dependence in the data. The Hill's estimator, however, can be applied to dependent data subject to specific mixing conditions, cf. [19].

The mean excess function

$$e(u) = \mathbb{E}(X - u | X > u) \qquad (3)$$

provides another method that can indicate a heavy tail. The increase (or decrease) of $e(u)$ implies a heavy-tailed (or light-tailed) distribution. Its constant value corresponds to an exponential distribution. Its linear increase indicates a Pareto-like distribution. Usually, the sample mean excess function

$$e_n(u) = \sum_{i=1}^{n}(X_i - u)\mathbb{1}(X_i > u) / \sum_{i=1}^{n}\mathbb{1}(X_i > u),$$

is calculated, where $\mathbb{1}(A)$ is the indicator function of the event A. Relatively large values of u are usually not considered due to the few observations exceeding these thresholds. It leads to unreliable estimates $e_n(u)$.

Example 2: Let us consider the IPTV IAT data as a typical example. Regarding the stability interval of the Hill's plot, one can find the corresponding value of the tail index $\hat{\alpha} \approx 2$, see Fig. 3(a). The bootstrap method over 200 bootstrap re-samples taken from the IPTV IAT sample with repetitions yields a similar value $\hat{\alpha} = 1.883$, cf. [18]. Assuming a regular varying DF, this value implies that only the first moment of the IAT distribution is finite and, hence, the distribution is heavy-tailed.

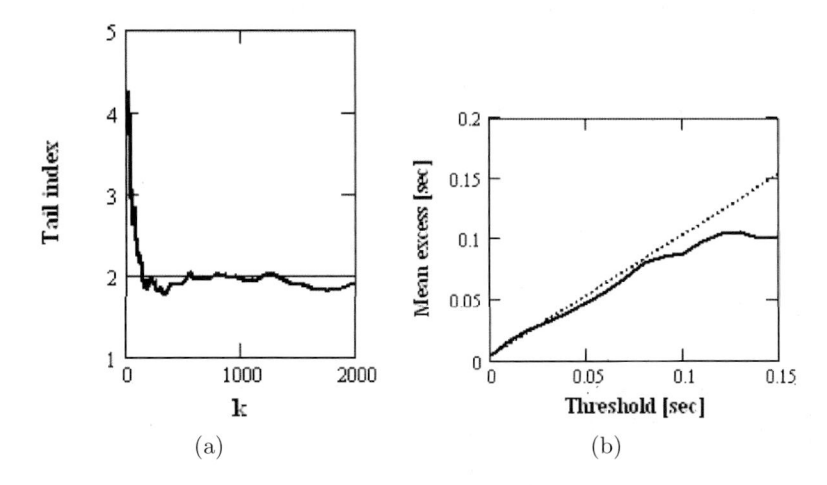

Fig. 3. Estimation of the tail index of the IPTV IATs by the reciprocal of Hill's estimator against the number of the largest order statistics k (a) and the mean excess function of the IPTV IATs against the threshold u (b)

Since the Hill's estimate may be corrupted by dependence if necessary mixing conditions are not fulfilled, we also estimate the bootstrapped Hill's estimate of the independent block maxima of the IPTV IATs calculated over 500 bootstrap re-samples. $m = 400$ has been selected as block size to provide independent blocks, see [18]. Then we have obtained $\hat{\alpha} = 3.39$. This implies that the first three moments of this distribution are finite. The distribution of the IAT block maxima is heavy-tailed.

By Fig. 3(b) one can conclude that the IPTV IAT distribution is a mixture of a Pareto-like and an exponential distributions since $e_n(u)$ increases almost linear up to the threshold $u = 0.12$ and is almost constant beyond 0.12. The latter is the 99.8% empirical quantile of the IATs.

3.3 Detection of Long-Range Dependence

Long-range dependence (LRD) of a time series $\{X_t, t = 1, \ldots, n\}$ means that the ACF $\rho_X(h) = \mathbb{E}\left((X_t - \mu)(X_{t+h} - \mu)\right)/\sigma^2$, $\mu = \mathbb{E}(X_t), \sigma^2 = \mathrm{Var}(X_t)$, remains sufficiently large in magnitude over a long period of time. The values of the ACF can be small, but in case of LRD their cumulative effect is significant, i.e., $\sum_{h=0}^{\infty} |\rho_X(h)| = \infty$.

The dependence structure may be derived by calculating the sample ACF and the extremal index and executing Portmanteau tests. An LRD property may be detected by an estimation of the Hurst parameter.

Estimation of the Autocorrelation Function. The classical sample ACF is calculated by the formula

$$\widehat{\rho}(h) = \frac{\sum_{t=1}^{n-h}(X_t - \overline{X}_n)(X_{t+h} - \overline{X}_n)}{\sum_{t=1}^{n}(X_t - \overline{X}_n)^2}$$

with $\overline{X}_n = 1/n \sum_{i=1}^{n} X_i$. For normal distributed characteristics one may conclude that independence or short-range dependence occurs if the ACF is dying after a few lags h inside the 95% Gaussian confidence window $\pm 1.96/\sqrt{n}$. For non-Gaussian r.v.s this conclusion may be wrong.

Note that the ACF does not exist when the variance is infinite. In this case one can use the modified estimate without a centering by the sample mean,

$$\widehat{\rho}_n(h) = \sum_{t=1}^{n-h} X_t X_{t+h} / \sum_{t=1}^{n} X_t^2,$$

instead of $\rho_X(h)$, cf. [20]. This estimate may be unreliable for non-linear processes.

In practice one can use the classical sample ACF even if the distribution is heavy-tailed, cf. [20, p. 349]. However, it is difficult to check the null hypothesis that the data stems from an independent sample because the confidence intervals of the ACF cannot be defined easily for heavy-tailed distributions in contrast to the Gaussian bounds of the $\widehat{\rho}(h)$, see [7]. The conclusion is that one has to apply additional tests to check the dependence.

Example 3: We consider the sample ACFs of our illustrative Skype and IPTV data sets. The ACFs of both the IATs and PLs of a Skype flow are small but do not decrease at large lags, see Fig. 4(a), 4(b) (cf. also [15]). This property implies that both the IATs and PLs of a Skype packet flow may be LRD.

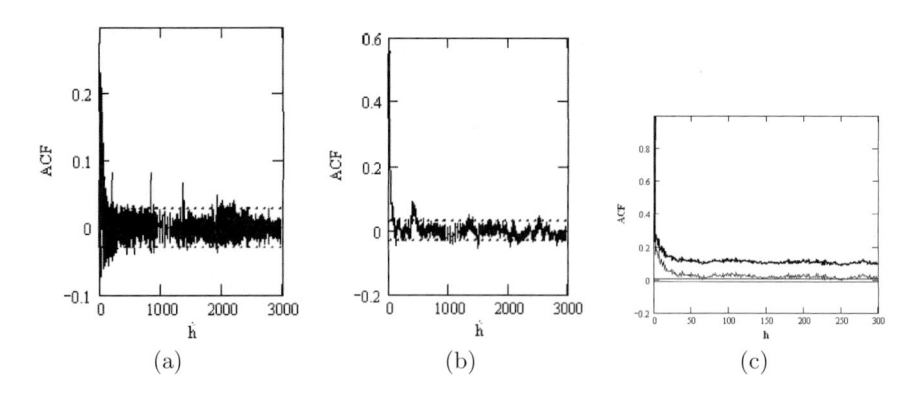

Fig. 4. The sample ACF $\widehat{\rho}(h)$ of the IATs between Skype packets (a) and of the lengths of Skype packets (b). The sample ACFs of the IATs between IPTV packets $\widehat{\rho}(h)$ (thin line) and $\widehat{\rho}_n(h)$ (solid line) (c). All estimates are shown with Gaussian 95% confidence intervals with the bounds $\pm 1.96/\sqrt{n}$.

Since the variance of the IATs of IPTV video data is infinite (see example 2 in Section 3.2) we consider both sample ACFs $\widehat{\rho}(h)$ and $\widehat{\rho}_n(h)$. They do not decrease as the lag h increases and do not remain within the Gaussian 95% confidence interval, see Fig.4(c) (cf. also [18]). Both facts may confirm the LRD of the IPTV IATs.

Portmanteau Tests. We consider two Portmanteau tests, namely, the Ljung-Box test and Runde's test. The first one is appropriate if the variance of the underlying r.v. is finite whereas the second is valid for time series with infinite variance. These tests check the null hypothesis regarding the independence of the underlying data.

According to the Ljung-Box test the test statistic

$$Q = n(n+2)\sum_{j=1}^{h} \widehat{\rho}^2(j)/(n-j)$$

has approximately a chi-square distribution with h degrees of freedom, cf. [7], [13]. The i.i.d. hypothesis should be rejected at level η if $Q > \chi_\eta^2(h)$ holds, where $\chi_\eta^2(h)$ is the ηth quantile of the chi-square distribution with h degrees of freedom, i.e., $Pr\{\chi^2 > \chi_\eta^2(h)\} = \eta$, $0 < \eta < 1$.

Given the tail index $1 < \alpha < 2$, Runde's test [21] uses the test statistic

$$Q_R = (n/\ln n)^{2/\alpha} \sum_{j=1}^{h} \widehat{\rho}^2(j).$$

The condition $1 < \alpha < 2$ implies that the second moment of the distribution is infinite, see Section 3.2. The quantiles of the limiting stable distribution of Q_R can be found in [21]. Some of them are given in Tab. 4, cf. [18]. To use Runde's test we have to estimate first the tail index (see Section 3.2 for corresponding methods). Since this test is valid for symmetric r.v.s, one has to construct a new symmetric r.v. based on the underlying r.v.s $\{X_i\}$. For this purpose we consider $Y_i = s_i X_i$, where s_i is a discrete r.v. that takes the values $+1$ and -1 with the probabilities 0.5. Y_i has the same tail index α as X_i since the DF of Y_i is determined by

$$\mathbb{P}\{Y_i \leq x\} = 1/2\mathbb{P}\{|s_i X_i| \leq x\} = 1/2\mathbb{P}\{X_i \leq x\}.$$

Now we can check the independence of Y_i. If $\{Y_i\}$ are independent then $\{X_i\}$ are independent, too, since

$$\begin{aligned}1/2^n \; \mathbb{P}\{X_1 \leq x_1\}...\mathbb{P}\{X_n \leq x_n\} &= \mathbb{P}\{Y_1 \leq x\}...\mathbb{P}\{Y_n \leq x\} \\ &= \mathbb{P}\{Y_1 \leq x_1,...,Y_n \leq x_n\} = \mathbb{P}\{s_1 X_1 \leq x_1,...,s_n X_n \leq x_n\} \\ &= 1/2^n \mathbb{P}\{|s_1 X_1| \leq x_1,...,|s_n X_n| \leq x_n\} = 1/2^n \mathbb{P}\{X_1 \leq x_1,...,X_n \leq x_n\},\end{aligned}$$

holds, cf. [16].

Table 4. Results of the Ljung-Box and Runde's test for IPTV data

Data	Lags	Q_R	$Q_h(0.05)$	Data	Lags	Q	$\chi^2_{0.05}(h)$
IAT	2	$4.01 \cdot 10^3$	13.53	IAT block	10	13.273	18.3
	3	$5.917 \cdot 10^3$	16.32	maxima	20	25.515	31.4
	4	$7.82 \cdot 10^3$	18.28	$\{X_i^{400}\}$	30	40.696	43.8
	5	$9.344 \cdot 10^3$	19.17				

Example 4: We apply Runde's test to the IATs between IPTV packets, see Section 2.2. The latter test is appropriate since the variance of the IAT distribution is infinite, see example 2 in Section 3.2. Since the values of Runde's statistic Q_R with the estimated $\hat{a} = 1.883$ exceed the critical values $Q_h(0.05)$ of the limiting distribution of Q_R for the $0.05-$level given in [21], the null hypothesis regarding the independence of IPTV IATs should be rejected, see Tab. 4.

In contrast to that, the block maxima $\{X_i^{400}\}$ of the IPTV IATs calculated over equal-sized blocks of size 400 may be independent. We can apply the Ljung-Box test to this data set since the estimate of its tail index is equal to $\hat{a} = 3.39$ which is larger than 2, see example 2 in Section 3.2. Hence, the variance is finite. Since the values Q do not exceed $\chi^2_{0.05}(h)$, the null hypothesis regarding the independence of these block maxima should be accepted, see Tab. 4.

Estimation of the Extremal Index. To check the dependence additionally, a constant $\theta \in [0, 1]$ of the process known as its extremal index is calculated. θ shows the change in the limiting distribution of the sample maximum due to dependence in the process. According to the theory of extremes typically

$$\mathbb{P}\{M_n \leq u\} \approx \mathbb{P}^{\theta}\{\widetilde{M}_n \leq u\} = F^{n\theta}(u) \qquad (4)$$

holds for sufficiently large n and u. Here M_n and $\widetilde{M}_n$ are the maximum of the sequence of dependent r.v.s $\{X_1, ..., X_n\}$ and the sequence of associated independent r.v.s $\{\widetilde{X}_1, ..., \widetilde{X}_n\}$ with the same DF $F(x)$, cf. [2]. For independent identically distributed (i.i.d.) sequences $\theta = 1$ holds.

The dependence leads to clusters in the data. The value $\theta < 1$ gives some indication on the clustering behavior and, hence, the dependence in the underlying sequence. $1/\theta$ determines the mean number of exceedances over some threshold per cluster, cf. [2]. Hence, estimates of $1/\theta$ are equal to the ratio of the number of exceedances over the threshold to the number of clusters. Its estimates are only distinguished by different definitions of a cluster in the data.

Regarding the blocks estimate

$$\overline{\theta}^B(u) = \frac{n \sum_{j=1}^{k} \mathbb{1}(M_{(j-1)r, jr} > u)}{rk \sum_{i=1}^{n} \mathbb{1}(X_i > u)} \qquad (5)$$

of θ, the cluster is a block of data with at least one exceedance over a threshold u. $M_{i,j} = \max(X_{i+1}, ..., X_j)$, k is the number of blocks, $r = [n/k]$ is the number of observations in each block, and $[\cdot]$ denotes the integer part of a number, cf. [2]. The estimate $\hat{\theta}$ is selected in a u-region where the plot of the extremal index against u does not change much.

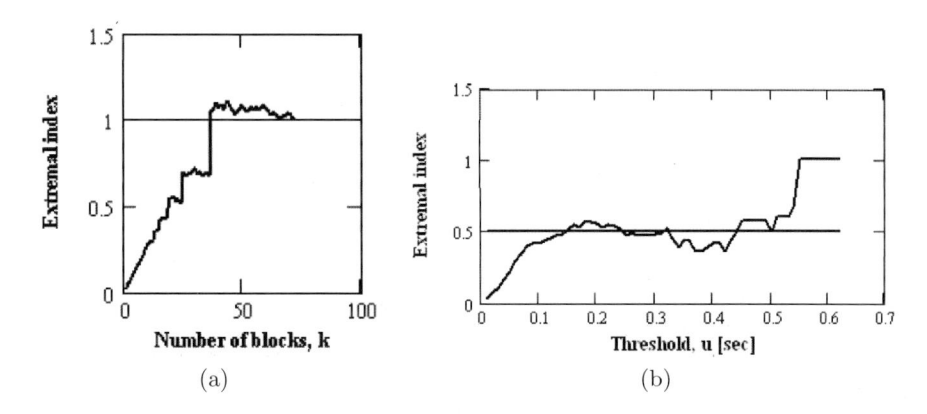

Fig. 5. Blocks estimates of the extremal index of the maximal Skype PLs within blocks against the number of blocks (a) (cf. [16, Fig. 4a]) and of the IPTV IATs against the threshold u (b)

Example 5: We consider the maximal PLs of a Skype flow within blocks that are separated by those IATs arising from the 98.4% empirical quantile of the IATs, see example 9 in Section 4.2. For the number of blocks equal to 72, $\bar{\theta}^B(u^*)$ is equal to 1, see Fig. 5(a) taken from [16, Fig. 4a]. This implies that the maxima of the PLs corresponding to these 72 blocks are independent. Here $u^* = 197.69$ has been selected which is equal to the mean of the PL maxima.

Let us consider the IPTV IATs. By Fig. 5(b) one can find that $\hat{\theta} \approx 0.5$ corresponds to the interval of the approximate stability of the plot. It implies the dependence of the IATs.

Estimation of the Hurst Parameter. Fractional Gaussian noise and fractional ARIMA are often used as ideal models of LRD time series. For such models the ACF has the common property $\rho_X(h) \sim c_\rho h^{2(H-1)}$ as $h \to \infty$, and c_ρ is a constant. The value $H = 1/2$ implies $\rho_X(h) = 0$ due to $c_\rho = 0$ and corresponds to independence. The closer the value $1/2 < H < 1$ is to 1 the longer reaches the dependence.

To estimate the Hurst parameter H, we can use the R/S and aggregated variance methods as well as Abry-Veitch's wavelet technique, cf. [1], [23]. These methods assume the self-similarity of the underlying time series. The detection of self-similarity is considered in Section 3.4.

According to the aggregated variance method one plots the logarithm of $\widehat{\mathrm{Var} X}^{(m)}$ (see (1)) versus $\log m$. A straight regression line approximating the points has the slope $\beta = 2H - 2$, $-1 \leq \beta < 0$.

According to the R/S method the estimate of the Hurst parameter H is given by a slope of the plot $\log(R(l_i, r)/S(l_i, r))$ against $\log(r)$, where $i = 1, ...K$, and r denotes a range. For this purpose one has to divide the time series into K intervals of length $[n/K]$. $R(l_i, r)/S(l_i, r)$ is computed by the formula

Table 5. Estimation of the Hurst parameter by the data of a Skype flow

r.v.	Estimation methods		
	R/S	Aggregated Variance	Abry-Veitch
Inter-arrival times (sec)	0.6	0.6	0.301 ± 0.023
Packet lengths (bytes)	0.7	0.675	0.729 ± 0.034

$$\frac{R(l,r)}{S(l,r)} = \frac{1}{S(l,r)} \left(\max_{0 \le i \le l} \mu_i(l,r) - \min_{0 \le i \le l} \mu_i(l,r) \right),$$

where

$$S(l,r) = \left(\frac{1}{r} \sum_{i=l+1}^{l+r} \left(X_i - \overline{X_{l,r}} \right)^2 \right)^{1/2},$$

$$\mu_i(l,r) = \sum_{j=1}^{i} \left(X_{l+j} - \overline{X_{l,r}} \right), \qquad \overline{X_{l,r}} = \frac{1}{r} \sum_{i=l+1}^{l+r} X_i,$$

holds for each lag r, starting at points $l_i = i[n/K] + 1$ such that $l_i + r \le n$ holds. We may further take the average values of the R/S statistics, $\overline{R(l_i,r)}/S(l_i,r)$, $i = 1,...,K$.

Following the approach of Abry and Veitch [1] and applying their Matlab code 'LDestimate' [1], one can furthermore compute the wavelet estimate of the Hurst parameter H by means of the slope of a regression line in the logscale diagram, i.e. the log-log plot of the relation between scale $a_j = 2^j$ and the variance estimate $\mu_j = 1/n_j \sum_{k=1}^{n_j} |d_X(j,k)|^2$ of the wavelet details determined by the discrete wavelet-transform coefficients $d_X(j,k)$ of the process X_t.

Example 6: We first consider the Skype packet data. The values of the Hurst parameter obtained by the described methods (see Tab. 5, cf. [15]) imply the possibility that no strong LRD occurs regarding the IAT and PL processes of a Skype flow.

Regarding the IATs of an IPTV flow we have obtained $\widehat{H} \approx 0.56$ by the R/S method. It implies a weak long-range dependence of the IATs of an IPTV packet stream, see Fig.6(a), cf. [18].

3.4 Detection of Self-similarity

A time series $\{X_t, t \ge 0\}$ is self-similar or scale invariant with Hurst parameter $H \in (0,1)$ if for any real $a > 0$ and $t \ge 0$ $X_t \stackrel{d}{=} a^{-H} X_{at}$ holds, i.e. the statistical properties of both sides of this equation are identical.

To check the self-similarity, we apply Higuchi's method, cf. [11]. The latter works as follows. Using a given time series $X_1, X_2, ..., X_n$, one first constructs a new time series X_k^m defined by $X_m, X_{m+k}, X_{m+2k},..., X_{m+[(n-m)/k]k}$, $m = 1, 2, ..., k$. Then one computes

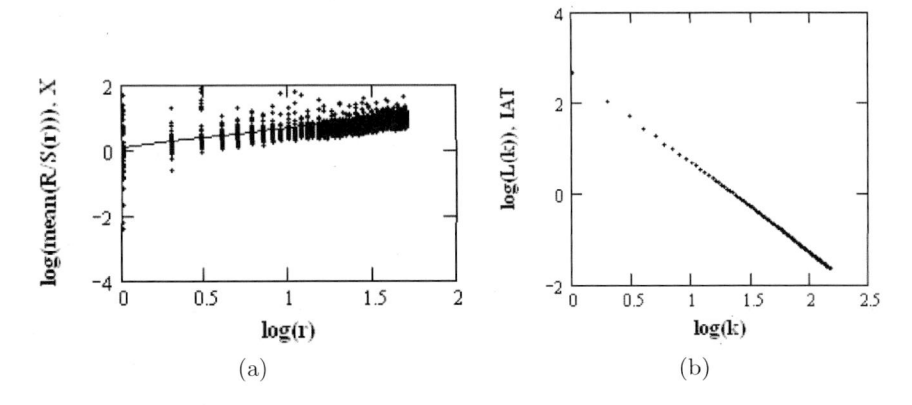

Fig. 6. Estimation of the Hurst parameter of IPTV IATs by the R/S method (a). Testing the self-similarity by Higuchi's method using $\log \overline{L(k)}$ versus $\log k$ for IPTV IATs (b).

$$L_m(k) = \frac{n-1}{k^2[(n-m)/k]} \sum_{i=1}^{[(n-m)/k]} |X_{m+ik} - X_{m+(i-1)k}|,$$

and draws a log-log plot of the statistic $\overline{L(k)}$ (that is the average value over k sets of $L_m(k)$) versus k. A constant slope D in $\overline{L(k)} \propto k^{-D}$ indicates self-similarity.

Example 7: Considering the IPTV data Fig. 6(b) (see also [18]) shows that the IPTV IATs behave like a self-similar process since the slope of the corresponding plot does not change.

3.5 Application to Packet Data of Skype and IPTV Flows

The results of the statistical characterization of our illustrative data sets obtained in [15] to [18] are summarized in Tab. 6. Apart of the dependent IAT and PL

Table 6. Characterization of Skype and IPTV data (yes = '+', no = '-')

Features	Skype Data			IPTV Data	
	IAT	PL	Block duration	IAT	IAT Block maxima
Independence	−	−	+	−	+
LRD	+	+	−	+	−
Self-similarity	+	+	+	+	+
Heavy-tailed with finite variance	+	−	−	−	+
Heavy-tailed with infinite variance	−	−	+	+	−
Light-tailed	−	+	−	−	−

series, the independent block representatives, namely, the time duration of Skype blocks and the block maxima of IPTV IATs are presented.

4 Principles of Data Segmentation

The statistical analysis of dependent data often requires to partition the data into independent blocks beforehand. Then one can deal with representatives of these blocks in the same manner like with independent data.

For instance, such popular methods like an empirical DF, the maximum likelihood method and many other techniques require i.i.d. data. Following this line of reasoning, our further analysis in Sections 5-7 requires independent blocks, too.

4.1 Equal-Sized Data Blocking

One can simply partition data into equal-sized blocks and find an appropriate minimal block size such that the r.v.s in the blocks are independent and the number of such blocks is large enough. A small number of blocks leads to a small sample size of the representatives of the blocks at our disposal and thus, to a large variance of an estimation by these representatives.

Example 8: We have partitioned the IPTV IATs into equal-sized blocks of the sizes 30, 40, 50, 100, 200, 300, 400, 500, 700, 1000. We have found that 400 is the minimal block size such that the maxima over such blocks are independent. The independence follows from the Ljung-Box test since the test statistic Q does not exceed the 5% quantiles $\chi^2_{0.05}(h)$ of the χ^2 distribution for different lags h, see Tab. 4, cf. also [18].

4.2 Non Equal-Sized Data Blocking

One can partition the time series of packets into non equal-sized blocks that are separated by long time intervals. Then one can expect that the data in the blocks may be independent. More exactly, it is proposed to partition the IATs $X_1, ..., X_n$ into blocks by the exceedances arising from their sufficiently high empirical quantile, see Fig. 7(a), cf. [16]. The exceedances indicate the boundaries of the blocks where the left or the right bound is included in the block. The blocks of the IATs determine the partitioning of the packet sequence and their PLs $Y_1, ..., Y_n$, see Fig. 7(b), cf. [16]. If a partition of the PLs were provided by their own quantile these subsets could be different. The cumulative inter-arrival time lengths within these blocks called block durations are determined by

$$L_j = \sum_{i=k_j}^{k_j-1+N_j} X_i, \qquad j = 1, ..., N_s, \tag{6}$$

cf. [16]. Here N_j is the random size of the jth block depending on the quantile of X_i, $N_0 = 0$, and $k_j = \sum_{m=0}^{j-1} N_m + 1$ is the number of first IAT in the jth block. N_s is the number of blocks.

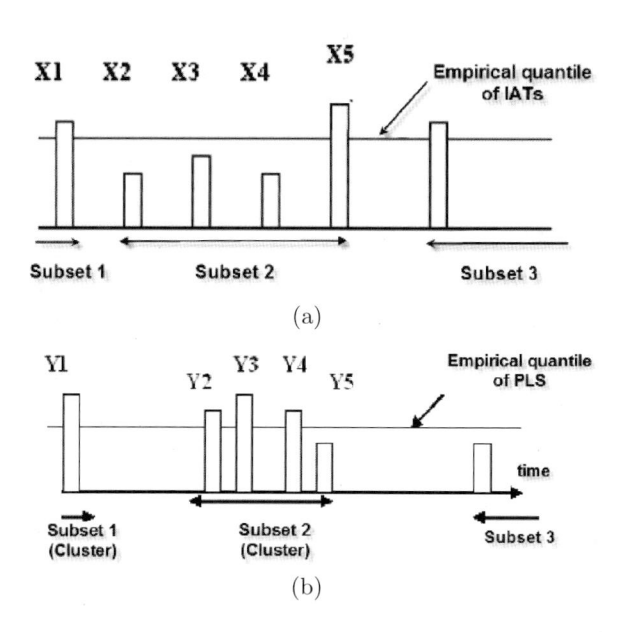

Fig. 7. Partition of the IATs between packets into subsets (blocks) by exceedances over the empirical quantile of the IATs (a). Partition of the PLs into subsets that are separated by long IATs (b).

Example 9: To generate a sufficient number $N_s = 72$ of independent blocks of Skype packets (the description of extremes of these blocks is given in Tab. 1), we use the minimal possible 98.4% empirical quantile of the Skype IATs. It is equal to 0.057 sec. The independence of representatives of such blocks can be easily checked by methods described in Section 3.3. The moderate number of blocks is the price of independence. The characterization of the block durations $\{L_j\}$ is stated in Table 6.

5 Characteristics of Skype User's Satisfaction and Their Estimation

In [8] the bitrate, jitter and round-trip time are selected as factors that influence on the call duration of a Skype session and, hence, on the user's satisfaction. However, the call duration may also reflect the user behavior and cannot be considered as a direct indicator of the user's satisfaction. Here we investigate the IATs between packets received by a Skype client and their PLs which reflect the user's activity and interrelate with the bitrate and delay variation. They influence on the loss and thus on the quality of service (QoS) and quality of experience (QoE) aspects. We propose the mean byte loss, the mean delivery time variation per cluster and the quantiles of lossless periods as indicators of the user's satisfaction.

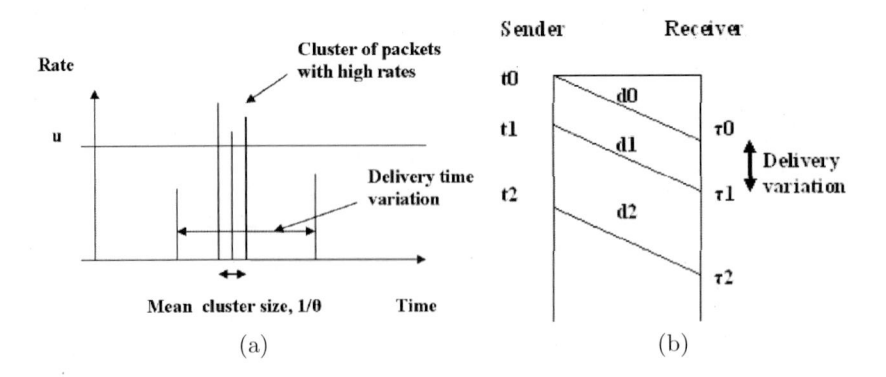

Fig. 8. Clusters of packets are the source of loss (a). Sender-receiver relation of the delivered packets (b).

To analyze all these aspects, we use a bufferless fluid model and assume that the packet stream is approximated by a continuous flow. Its rate is determined by the ratio of the PL Y_i per IAT X_i, i.e. $R_i = Y_i/X_i$, $i = 1, \ldots, n$. It is supposed to be constant between arrivals and only changing at arrival epochs. It is assumed that the IATs are not caused by a silence period of the user but are integral part of one flow.

In case that the IATs between packets are constant, the exceedances of PLs over a threshold u cause loss and delay since the corresponding packets are not delivered. The threshold u is equal to the channel capacity in a time unit. The latter is equal to the IAT. However, in case that the IATs between packets are random, the packets corresponding to exceedances of the required rates $\{R_i\}$ over a capacity u cause loss and delivery delay. Since the rate is defined as ratio of the PL to the IAT, large rates may be generated by frequent packets which arise in the clusters of data, see Fig. 8(a), or by rare large packets. We shall focus on this situation.

The delivery time variation of completely transmitted and correctly received packets is determined by

$$y_i = \tau_i - \tau_{i-1} = t_i + d_i - (t_{i-1} + d_{i-1}) = \Delta t_i + \Delta d_i.$$

Here Δd_i is the delay jitter and Δt_i are the IATs at the sender, see Fig. 8(b). The clusters are generated by packets corresponding to the rates that exceed the channel capacity u. The delivery time variation of packets is equal to the time between two consecutively transmitted and correctly received packets, see Fig. 8(a). Hence, one can estimate the mean delivery time variation of packets per cluster, i.e., the mean IAT between successfully transmitted packets by

$$d = (1 + 1/\theta)\mathbb{E}X.$$

Here $\mathbb{E}X$ is the mean IAT and $1/\theta$ is the mean cluster size, i.e. the mean number of rates exceeding the capacity u per cluster. $1 + 1/\theta$ forms the mean number of

IATs between successfully transmitted packets. θ is calculated by exceedances of the rates $\{R_i\}$ beyond the channel capacity u using the blocks estimator (5). d may be estimated by

$$\widehat{d} = (1 + 1/\overline{\theta}^B(u))\overline{X}$$

where $\overline{X}$ estimates the mean IAT.

Example 10: We consider our Skype flow data where the IATs are random. The estimate of the mean delivery time variation of packets per cluster $\widehat{d} = 0.111$ sec arises for the average IAT $\overline{X} = 0.031$ sec. The extremal index of the rate is given by $\overline{\theta}^B \approx 0.38$ (see Fig. 9) for $k = 150$ equal-sized Skype blocks.

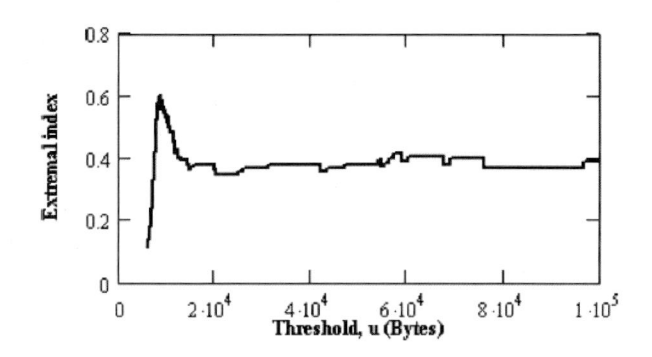

Fig. 9. Blocks estimate of the extremal index of the transmission rate of a Skype flow. The approximate value 0.38 of the extremal index corresponds to the stable u-region and is accepted as an estimate $\overline{\theta}^B$.

Applying a bufferless fluid model, we now estimate the loss, the mean loss and the corresponding channel capacity. First, we consider the case when the IATs are equal. Then the overall byte loss for an observation time

$$E_n(u) = \sum_{i=1}^{n} Y_i \mathbb{1}(Y_i > u)$$

is generated by the PLs $\{Y_i\}$ that exceed a threshold u given in bytes. The latter is equal to the channel capacity in a time unit. The mean byte loss is determined by

$$e_n(u) = \sum_{i=1}^{n} Y_i \mathbb{1}(Y_i > u) / \sum_{i=1}^{n} \mathbb{1}(Y_i > u).$$

Now let us assume that the IATs between packets in the P2P stream are random. Then the loss is generated by packets corresponding to high rates which exceed the channel capacity. Thus, the overall byte loss is given by

$$\widehat{E}(u) = \sum_{i=1}^{n} Y_i \mathbb{1}(R_i > u).$$

The mean byte loss is determined by

$$\hat{e}(u) = \sum_{i=1}^{n} Y_i \mathbb{1}(R_i > u) / \sum_{i=1}^{n} \mathbb{1}(R_i > u), \qquad (7)$$

where u denotes the channel capacity.

The time between packets corresponding to rate exceedances beyond the capacity determines a lossless period. The lossless periods may coincide with periods without packet transmission when the rate exceedances correspond to consecutive packets, see 8(a). We calculate the empirical quantiles of lossless periods in the following example.

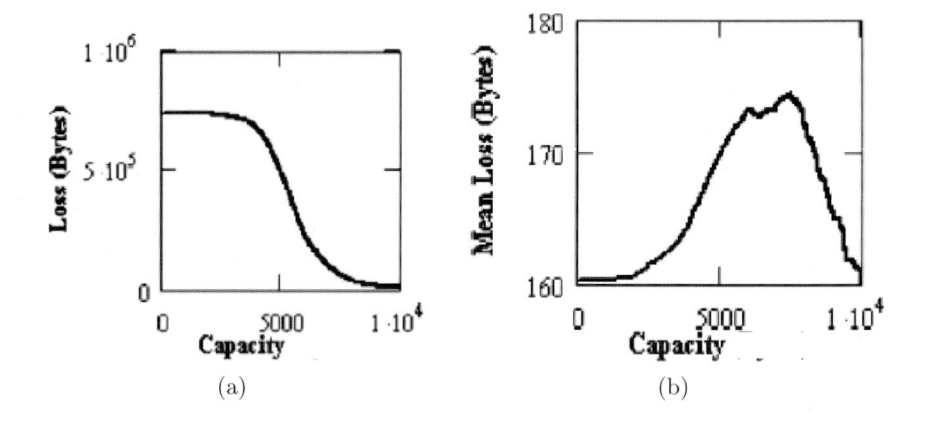

Fig. 10. Estimation of the overall byte loss $\hat{E}(c)$ (a) and the mean byte loss $\hat{e}(c)$ (b) against the channel capacity

Example 11: We consider our Skype data with random IATs and calculate $\hat{E}(c)$ and $\hat{e}(c)$, see Fig. 10. The channel capacity corresponding to the 3% overall byte loss is equal to $c^* = 8.534$ kbps. Note that the Internet speech audio codec (iSAC) dynamically adjusts the transmission rate from 10 to 32 kbps. The mean byte loss $\hat{e}(c^*)$ is equal to 168 bytes. $\hat{e}(c)$ increases up to $c_m \approx 7.4$ kbps and decreases beyond c_m. Indeed, the number of packets corresponding to the rates exceeding u (the denominator in (7)) decreases as u increases. However, the numerator of (7) may behave not predictable since large rates can correspond to frequent small packets and frequent (or rare) large packets. The overall byte loss decreases as the capacity increases.

The empirical $50, 75, 80, 85, 95, 97.5, 99.9\%$ quantiles of lossless periods arising from exceedances of the rates $\{R_i\}$ beyond c^* are equal to $30, 34, 35, 36, 39, 44, 121$ ms, respectively. It implies that the loss-free time may exceed these values with the probabilities $50, 25, 20, 15, 5, 2.5, 0.1\%$ and the corresponding overall byte loss is equal to 3%.

6 Estimating the Offered Load

In Section 4 we have considered the partitioning of the packet flow into independent blocks of durations $\{L_j\}$, see (6). Then one can evaluate the overall volume of packets transmitted during a fixed time interval $[0, t]$ by the formula

$$V^*(t) = \sum_{j=1}^{N_t} V_j = \sum_{j=1}^{N_t} \sum_{i=1}^{N_j} Y_i.$$

Here N_t denotes the number of packet blocks arriving before time t, $N_t = \max\{n : t_n < t\}$, and $t_n = \sum_{i=1}^{n} L_i$ is the cumulative time interval corresponding to the n blocks. V_j is the packet volume of the jth block, cf. [17]. For simplicity and without loss of generality, we re-enumerate the packets within blocks from 1 to N_j, where N_j denotes the random number of packets in the jth block.

In [17] it has been shown that the appearances of the volumes $\{V_j\}$ can be considered as a renewal process if we assume that the volume V_j of the jth subset is concentrated at some point of the time interval L_j, e.g. at the beginning. All properties of the renewal process are fulfilled if we assume that the durations of blocks $\{L_j\}$, $j = 1, 2, \ldots$, (or IATs between the cumulative volumes $\{V_j\}$) are i.i.d r.v.s.

To find the expectation of the overall volume at time t, one can use Wald's equation

$$\mathbb{E}(V^*(t)) = \mathbb{E}\left(\sum_{j=1}^{N_t} V_j\right) = \mathbb{E}(N_t)\mathbb{E}(V_j) \tag{8}$$

since the volumes of blocks V_j and their number N_t at time t are independent. Then the variance of $V^*(t)$ is determined by

$$\text{Var}(V^*(t)) = \text{Var}\left(\sum_{j=1}^{N_t} V_j\right) = \text{Var}(V_j)\mathbb{E}(N_t) + (\mathbb{E}(V_j))^2 \text{Var}(N_t) \tag{9}$$

if the second moment of the volume exists, i.e. $\mathbb{E}V_j^2 < \infty$ cf. [14], [25].

Furthermore,

$$H(t) = \mathbb{E}(N_t) = \sum_{n=1}^{\infty} \mathbb{P}\{t_n < t\}$$

denotes the renewal function, cf. [14]. It exhibits simple analytic forms just for a few distributions of the IATs like the uniform, normal and exponential distributions, e.g., it is linear $H(t) = \lambda t$ for an exponential distribution with intensity λ. Usually, the distribution of the IATs $\{L_i\}$ of packet volumes V_j is unknown. Thus, one has to apply nonparametric estimators of $H(t)$, e.g., a histogram-type estimator recommended in [14]. Using the IATs $\{\tau_j, j = 1, \ldots, N\}$, it is determined by

$$\widetilde{H}(t, k, N) = \sum_{n=1}^{k} \frac{1}{N_n} \sum_{i=1}^{N_n} \mathbb{1}(t \geq t_n^i).$$

Here $t_n^i = \sum_{q=1+n(i-1)}^{n \cdot i} \tau_q$, $i = 1, \ldots, N_n$, and $N_n = \left[\frac{N}{n}\right]$, $n = 1, \ldots, k$, are observations of the r.v. t_n. In our consideration we take $\{L_j\}$ and N_S instead of $\{\tau_j\}$ and N. N_S denotes the number of blocks or the number of $\{L_j\}$, respectively.

To select the parameter k for a fixed t one can use the formula

$$k^* = \arg\min\{k : \widetilde{H}(t, k, N) = \widetilde{H}(t, k+1, N), k = 1, \ldots, N-1\}.$$

More details about these methods can be found in [14, Chapter 8].

Due to the limited number of IATs $\{L_i\}$ that are at our disposal the histogram-type estimate becomes constant after a sufficiently large time t,[1] i.e., $\widetilde{H}(t, k, N_S) = k$ holds for $t \in [t_{\max}(k), \infty)$, where

$$t_{\max}(k) = \max_{1 \le n \le k} \max_{1 \le i \le [N_S/n]} t_n^i \le \sum_{i=1}^{N_S} L_i$$

and $k \le N_S$ is some fixed number.

For large t one can use the well-known linear approximation of the renewal function

$$H(t) = \frac{t}{\mu} + \frac{\sigma^2}{2\mu^2} - \frac{1}{2} + o(1),$$

if the mean μ and variance σ^2 of $\{L_j\}$ are finite[2]. Another approximation

$$H(t) = \frac{t}{\mu} + \frac{t^2(1-F(t))}{\mu^2(\alpha-1)(2-\alpha)} + o(1) \qquad (10)$$

for $t \to \infty$ (cf. [24]), where $F(t)$ is the DF of L_j, is valid for regularly varying distributions (2) and $1 < \alpha < 2$. This is the case if the variance of $\{L_j\}$ is infinite. One can rewrite (10) using the estimate $t^{-\alpha}$ instead of $1-F(t)$ and replacing μ and α by their estimates, e.g., by $\widehat{\mu}$ which is the average of $\{L_j\}$ and by Hill's estimate $\widehat{\alpha} = 1/\widehat{\gamma}^H(n, k_0)$, respectively. Then we get

$$\widehat{H}(t) = \frac{t}{\widehat{\mu}} + \frac{t^{2-\widehat{\alpha}}}{\widehat{\mu}^2(\widehat{\alpha}-1)(2-\widehat{\alpha})} \qquad (11)$$

for $t > t_{\max}(k)$.

According to [17] one can estimate the mean overall volume $\mathbb{E}(V^*(t))$ at time t by the formula

$$\overline{V^*}(t) = H^*(t)\overline{V}. \qquad (12)$$

Here $\overline{V}$ is the sample average of $\{V_j\}$ which can be used instead of $\mathbb{E}(V_i)$ in (8),

$$H^*(t) = \begin{cases} \widetilde{H}(t, k, N_S), & t \le t_{max}, \\ t/\widehat{\mu} + \widehat{\sigma}^2/(2\widehat{\mu}^2) - 1/2, & t > t_{max}, \sigma^2 < \infty, \\ \widehat{H}(t), & t > t_{max}, \sigma^2 = \infty, \end{cases} \qquad (13)$$

and $\widehat{\sigma}$ is the standard deviation of $\{L_j\}$.

---

[1] This is a typical feature of all histogram-type estimates.
[2] The notation $o(1)$ means that the approximation is valid up to an arbitrary constant.

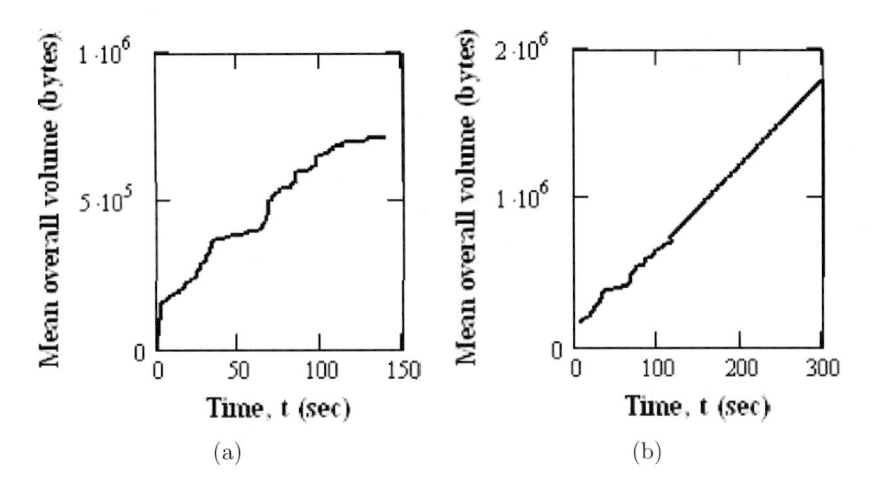

Fig. 11. The estimate of the mean overall volume $\overline{V^*}(t)$ in the time intervals $[0, t_{max}] = [0, 138.914]$ (a) and $t \in [5, 300]$ (b). For $t > 138.914$ $\widetilde{H}(t, k, N) = k^* = 72$ holds and it is replaced by a linear model (cf. [17, Fig. 7])

Example 12: We consider again the Skype data. In this case the variance of the block volume $\mathrm{Var}(V_j)$ is infinite, since the tail index of V_j is about 1.5 as shown in [17]. Then the variance $\mathrm{Var}(V^*(t))$ is infinite which follows from (9). Thus, we can estimate $\mathbb{E}(V^*(t))$ by formulae (12) and (13) only. The sample average of $\{V_j\}$ is given by $\overline{V} = 10.18$ Kbytes. The Hill's estimate of the tail index α of $\{L_j\}$ falls into the interval $[1.468, 1.56]$ and one can expect that the distribution of $\{L_j\}$ is regularly varying, cf. [17]. Hence, we can apply (11) and take $\alpha = 1.5$. Figures 11(a), 11(b) (see also [17, Fig. 7]) depict the mean offered load of the packets corresponding to a pre-defined time t. In Figure 11(b) the estimate of $H(t)$ coincides with $\widetilde{H}(t, k, N_S)$ in $t \in [5, 138.914]$ and with (11), where $\alpha = 1.5$ holds for $t \in (138.914, 300]$. The mean overall volume increase evidently as time increases. One can calculate how much traffic load arises at time t by means of $\overline{V^*}(t)$. The considered Skype traffic is non-Poissonian, since $H^*(t)$ and $\overline{V^*}(t)$ are not linear at relatively small times t. For Poisson traffic $H^*(t) = \lambda t$ holds and $\overline{V^*}(t)$ is linear at any t.

7 Distribution of the Maximum of Inter-Arrival Times

In Section 4 we have considered the partitioning of the packet flow into independent blocks. Knowing these blocks, one can fit the distribution of representatives of these blocks accurately.

We consider here the IPTV data and their equal-sized blocks of size 400 described in the example 8 of Section 4.1. Then we can fit the distribution of the block maxima $\{X_i^{400}\}$. It is well known that the Generalized Extreme Value (GEV) distribution with DF

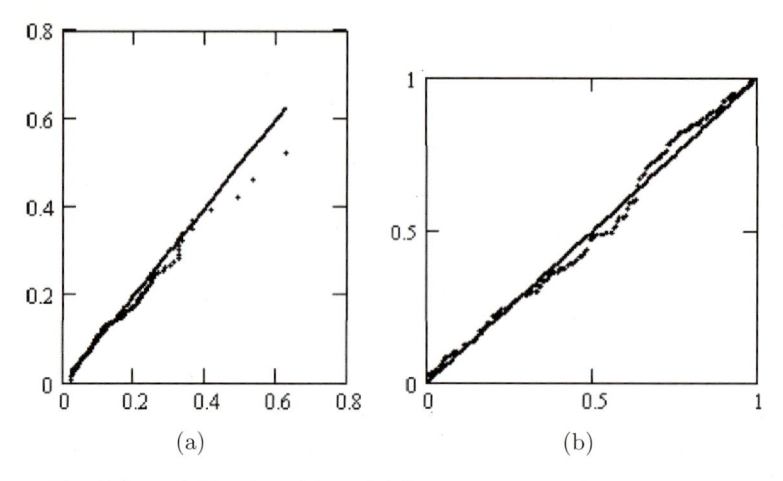

Fig. 12. The QQ- and PP-plots (a) and (b), respectively, of the GEV distribution of the IPTV IAT block maxima with parameters $\gamma = 0.21666$, $\sigma = 0.05887$, $\mu = 0.0881$.

$$F(x) = \exp\left(-\left(1 + \gamma \frac{x - \mu}{\sigma}\right)^{-1/\gamma}\right)$$

is an appropriate model to fit the maximum. Since the block maxima $\{X_i^{400}\}$ are independent, we can apply the maximum likelihood method to find the parameters of a GEV. By different goodness-of-fit tests with a 5% confidence level the following values $\gamma = 0.21666$, $\sigma = 0.05887$ and $\mu = 0.0881$ were found to be the best ones, see Tab. 7. They provide QQ- and PP-plots which are close to the empirical data, see Fig. 12 (cf. also [18]).

Table 7. Test results of the GEV distribution fitted to IAT block maxima $\{X_i^{400}\}$

Parameters	Goodness-of-Fit Tests	
	Kolmogorov-Smirnov	Anderson-Darling
$\gamma = 0.21666$, $\sigma = 0.05887$, $\mu = 0.0881$	0.07075	0.8341

Using (4) now and a GEV approximation of $\mathbb{P}\{\widetilde{M}_n \leq u\} = \mathbb{P}\{X_1^{400} \leq u\}$, we can approximate the distribution of the maximum of the IPTV IATs by the formula

$$\mathbb{P}\{M_n \leq x\} \approx \exp\left(-\left(1 + \gamma \frac{x - \mu^*}{\sigma^*}\right)^{-1/\gamma}\right). \tag{14}$$

Table 8. The GEV distribution fitted to the IAT maximum and its high quantiles

Parameters	High quantiles		
	95%	97.5%	99%
$\gamma = 0.21666$, $\sigma^* = 0.05$, $\mu^* = 0.051$	0.263	0.337	0.452

Here $\mu^* = \mu - \sigma(1-\theta^\gamma)/\gamma$, and $\sigma^* = \sigma\theta^\gamma$ hold, cf. [2, p. 377]. We can also find the quantiles of this maximum. In Fig. 5(b) we have shown that $\widehat{\theta} \approx 0.5$ arises for the IPTV IATs. Then one can calculate the new parameters μ^* and σ^* by this estimate $\widehat{\theta}$. Regarding the IPTV IAT the quantiles of the maximum M_n can be obtained by the formula

$$q_p = \frac{(-\ln(1-p))^{-\gamma} - 1}{\gamma} \sigma^* + \mu^*,$$

see Table 8, cf. [18]. The high $95, 97.5, 99, 99.9\%$ quantiles imply a delay between packets which can only be exceeded with the small probabilities $5, 2.5, 1, 0.1\%$, respectively.

8 Conclusions

In recent years peer-to-peer (P2P) multimedia applications have became a powerful service platform for the generation and transport of voice and video streams over IP. The application of variable bitrate encoding schemes and the packet based voice and video transfer raises a large variety of new questions regarding traffic characterization.

In our study we have developed a general methodology concerning the statistical analysis of P2P packet flows using two types of information. These characteristics are given by the inter-arrival times between packets and the lengths of the transported packets. We have focussed on the case when the inter-arrival times are random entities. Such a situation arises, for instance, when a wireless access to the Internet by Skype or IPTV clients is considered.

We deal with observations which are time series, i.e. they are dependent. Thus, we have presented principles of data blocking to partition the observations into independent blocks and to deal with independent data instead of dependent ones.

Our methodology includes the statistical characterization of a P2P packet stream regarding the stationarity, long-range dependence, self-similarity and heaviness of tail of the associated distributions. In addition to that, we have presented some important characteristics of the Skype user's satisfaction such as the overall byte loss, the mean byte loss, the mean delivery time variation of packets per cluster and the quantiles of lossless periods. These characteristics extend the list of known indices like the bitrate, jitter and round-trip time.

The proposed statistical methodology is accompanied by several examples which illustrate its application to packet flows of representative Skype and IPTV sessions.

Further, we have considered the problem to evaluate the channel capacity which is required to guarantee an appropriate overall byte loss. We have always assumed that the loss is caused by packets corresponding to exceedances of the transmission rate beyond the channel capacity of a bufferless fluid model and that the rate is equal to the ratio of the packet length to the adjacent inter-arrival time. This is a natural assumption for random inter-arrival times between packets.

Moreover, we have estimated the offered traffic load in a finite observation period. For this purpose we have observed that the cumulative traffic volumes arising from independent blocks of packets create a renewal process.

In summary, the proposed statistical methodology provides a powerful and versatile approach for a traffic characterization in the Internet. It can be applied to any correlated data arising from monitored packet streams and is not limited to P2P data used here as illustration.

Acknowledgment. The authors acknowledge the partial support by research grants of the EU FP6-NoE project "EuroFGI" under contract 028022, the COST Action IC0703 "Data Traffic Monitoring and Analysis (TMA)", and the German Ministry of Education and Reseach (BMBF) under contract MDA 08/015.

Regarding the measurement studies of P2P traffic the authors are also grateful to Mr. Schweßinger and Mr. Eittenberger for their support.

References

1. Abry, P., Veitch, D.: Wavelet Analysis of Long-Range Dependence Traffic. IEEE Transactions on Information Theory 44(1), 2–15 (1998)
2. Beirlant, J., Goegebeur, Y., Teugels, J., Segers, J.: Statistics of Extremes: Theory and Applications. Wiley, Chichester (2004)
3. Bendat, J.S., Piersol, A.G.: Random Data: Analysis and Measurement Procedures. J. Wiley & Sons, New York (1986)
4. Beran, J.: Statistics for Long-Memory Processes. Chapman & Hall, New York (1994)
5. Bhattacharya, R.N., Gupta, V.K., Waymire, E.: The Hurst effect under trends. J. Appl. Probab. 20, 649–662 (1983)
6. Bonfiglio, D., Mellia, M., Meo, M., Rossi, D., Tofanelli, P.: Revealing Skype Traffic: When Randomness Plays with you. In: Proceedings of ACM SIGCOMM 2007, Kyoto, August 27–31 (2007)
7. Brockwell, P.J., Davis, R.A.: Introduction to Time Series and Forecasting, 2nd edn. Springer Texts in Statistics, New York (2002)
8. Chen, K.-T., Huang, C.-Y., Huang, P., Lei, C.-L.: Quantifying Skype user satisfaction. In: Proceedings ACM SIGCOMM 2006, Pisa, Italy, September 11-15 (2006)
9. Cochran, W.G.: The distribution of the largest of a set of estimated variances as a fraction of their total. Ann. of Eugenics 11, 47–52 (1941)
10. Giraitis, L., Leipus, R., Philippe, A.: A test for stationarity versus trends and unit roots for a wide class of dependent errors. Econometric Theory 22(6), 989–1029 (2006)

11. Higuchi, T.: Approach to an irregular time series on the basis of the fractal theory. Physica D 31, 277–283 (1988)
12. Liu, F., Li, Z.: A Measurement and Modeling Study of P2P IPTV Applications. In: Proceedings of the 2008 International Conference on Computational Intelligence and Security, vol. 1, pp. 114–119 (2008)
13. Ljung, G.M., Box, G.E.P.: On a Measure of Lack of Fit in Time Series Models. Biometrika 65, 297–303 (1978)
14. Markovich, N.M.: Nonparametric Estimation of Univariate Heavy-Tailed Data. J. Wiley & Sons, Chichester (2007)
15. Markovich, N.M., Krieger, U.R.: Statistical Analysis of VoIP Flows Generated by Skype Users. In: Proceedings of IEEE International Workshop on Traffic Management and Traffic Engineering for the Future Internet, FITraMEn, Porto, Portugal, December 11-12 (2008)
16. Markovich, N.M., Krieger, U.R.: Statistical Characterization of QoS Aspects Arising From the Transport of Skype VoIP Flows. In: Proceedings of The First International Conference on Evolving Internet (INTERNET 2009), IARA, Cannes/La Bocca, August 23-29, pp. 9–14 (2009)
17. Markovich, N.M., Krieger, U.R.: Statistical Analysis and Modeling of Skype VoIP Flows. Computer Communications 33, 11–21 (2010)
18. Markovich, N.M., Krieger, U.R.: Characterizing Packet Traffic in Peer-to-Peer Video Applications. Technical Report, Otto-Friedrich University Bamberg (2009) (submitted)
19. Novak, S.Y.: Inference of heavy tails from dependent data. Siberian Advances in Mathematics 12(2), 73–96 (2002)
20. Resnick, S.I.: Heavy-Tail Phenomena. Probabilistic and Statistical Modeling. Springer, New York (2006)
21. Runde, R.: The asymptotic null distribution of the Box-Pierce Q-statistic for random variables with infinite variance. J. of Econometrics 78, 205–216 (1997)
22. Silverston, T., Fourmaux, O., Botta, A., Dainotti, A., Pescapé, A., Ventre, G., Salamatian, K.: Traffic analysis of peer-to-peer IPTV communities. Computer Networks 53, 470–484 (2009)
23. Taqqu, M.S., Teverovsky, V., Willinger, W.: Estimators for long-range dependence: an empirical study. Fractals 3, 785–798 (1995)
24. Teugels, J.L.: Renewal theorems when the first or the second moment is infinite. Annals of Statistics 39, 1210–1219 (1968)
25. Trivedi, K.S.: Probability & Statistics with Reliability, Queuing, and Computer Science Applications. Prentice Hall of India, New Delhi (1997)
26. Skype, http://www.skype.com
27. SopCast, http://www.sopcast.com

Markovian Modelling of Internet Traffic

António Nogueira[1], Paulo Salvador[1], Rui Valadas[2], and António Pacheco[3]

[1] University of Aveiro / Institute of Telecommunications Aveiro
Campus de Santiago, 3810-193 Aveiro, Portugal
{nogueira,salvador}@ua.pt
[2] Instituto Superior Técnico - UTL
Department of Electrical and Computing Engineering
Av. Rovisco Pais, 1049-001 Lisboa, Portugal
{rui.valadas}@ist.utl.pt
[3] Instituto Superior Técnico - UTL
Department of Mathematics and CEMAT
Av. Rovisco Pais, 1049-001 Lisboa, Portugal
{apacheco}@math.ist.utl.pt

Abstract. This tutorial discusses the suitability of Markovian models to describe IP network traffic that exhibits peculiar scale invariance properties, such as self-similarity and long range dependence. Three Markov Modulated Poisson Processes (MMPP), and their associated parameter fitting procedures, are proposed to describe the packet arrival process by incorporating these peculiar behaviors in their mathematical structure and parameter inference procedures. Since an accurate modeling of certain types of IP traffic requires matching closely not only the packet arrival process but also the packet size distribution, we also discuss a discrete-time batch Markovian arrival process that jointly characterizes the packet arrival process and the packet size distribution. The accuracy of the fitting procedures is evaluated by comparing the long range dependence properties, the probability mass function at each time scale and the queuing behavior corresponding to measured and synthetic traces generated from the inferred models.

Keywords: Long range dependence, self-similarity, time scale, Markov Modulated Poisson Process, packet arrival (size) process.

1 Introduction

The growing diversity of services and applications for IP networks has been driving a strong requirement to make frequent measurements of packet flows and to describe them through appropriate traffic models. Several studies have already shown that IP traffic may exhibit properties of self-similarity and/or long-range dependence (LRD) [1, 2, 3, 4, 5], peculiar behaviors that have a significant impact on network performance. However, matching LRD is only required within the time-scales of interest to the system under study [6, 7]: for example, in order to analyze queuing behavior the selected traffic model only needs to capture the correlation structure of the source up to the so-called critical time-scale or correlation horizon, which is directly related to the maximum buffer size [5,8,9]. One of the consequences of this result is that more traditional traffic models such as Markov Modulated Poisson Processes (MMPPs) can still be used to model traffic exhibiting LRD [10, 11, 12, 13]. However, providing a good match of the LRD characteristics through an accurate fitting of the autocovariance tail is not enough for accurate prediction of the queuing behavior [14]. In general, an accurate prediction of the queuing behavior requires detailed modeling of the first-order statistics, not just the mean, and for certain types of network traffic it demands the incorporation of time-dependent scaling laws [15, 16, 17].

This tutorial discusses the suitability of Markovian models, based on MMPPs, for modeling IP traffic. Traffic modeling is usually concerned with the packet arrival process, aiming to fit its main characteristics. In order to describe the packet arrival process, we will present three traffic models that were designed to capture self-similar behavior over multiple time scales. The first model is based on a parameter fitting procedure that matches both the autocovariance and marginal distribution of the counting process [18]. The MMPP is constructed as a superposition of L two-state MMPPs (2-MMPPs), designed to match the autocovariance function, and one M-MMPP designed to match the marginal distribution. Each 2-MMPP models a specific time scale of the data. The second model is a superposition of MMPPs, where each MMPP describes a different time scale [19,20]. The third model is obtained as the equivalent to an hierarchical construction process that, starting at the coarsest time scale, successively decomposes MMPP states into new MMPPs to incorporate the characteristics offered by finner time scales [21]. Both models are constructed by fitting the distribution of packet counts in a given number of time scales. For all three traffic models, the number of states is not fixed *a priori* but is determined as part of the inference procedure. The accuracy of the different models will be evaluated by comparing the probability mass function (PMF) at each time scale, as well as the packet loss ratio (PLR) corresponding to measured traces (exhibiting LRD and self-similar behavior) and traces synthesized according to the proposed models.

It is known that the accurate modeling of certain types of IP traffic requires matching closely not only the packet arrival process but also the packet size distribution [22,23]. In this way, we also present a discrete-time batch Markovian arrival process (dBMAP) [24, 25, 26] that jointly characterizes the packet arrival process and the packet size distribution, while achieving accurate prediction of queuing behavior for IP traffic exhibiting LRD behavior. In this dBMAP, packet arrivals occur according to a dMMPP and each arrival is further characterized by a packet size with a general distribution that may depend on the phase of the dMMPP. This allows having a packet size distribution closely related to the packet arrival process, which is in contrast with other approaches [22, 23] where the packet size distribution is fitted prior to the matching of the packet arrival rates.

2 Notions of Self-similarity and Long-Range Dependence

Consider the continuous-time process $Y(t)$ representing the traffic volume (e.g. in bytes) from time 0 up to time t and let $X(t) = Y(t) - Y(t-1)$ be the corresponding increment process (e.g. in bytes/second). Consider also the sequence $X^{(m)}(k)$ that is obtained by averaging $X(t)$ over non-overlapping blocks of length m, that is

$$X^{(m)}(k) = \frac{1}{m} \sum_{i=1}^{m} X((k-1)m + i), k = 1, 2, \ldots \quad (1)$$

The fitting procedure that are presented in this work are based on the aggregated processes $X^{(m)}(k)$.

$Y(t)$ is exactly self-similar when it is equivalent, in the sense of finite-dimensional distributions, to $a^{-H}Y(at)$, for all $t > 0$ and $a > 0$, where H ($0 < H < 1$) is the Hurst parameter. Clearly, the process $Y(t)$ can not be stationary. However, if $Y(t)$ has stationary increments then again $X(k) = X^{(1)}(k)$ is equivalent, in the sense of finite-dimensional distributions, to $m^{1-H}X^{(m)}(k)$. This illustrates that a traffic model developed for fitting self-similar behavior must preferably enable the matching of the distribution on several time scales.

Long-range dependence is associated with stationary processes. Consider now that $X(k)$ is second-order stationary with variance σ^2 and autocorrelation function $r(k)$. Note that, in this case, $X^{(m)}(k)$ is also second-order stationary. The process $X(k)$ has long-range dependence (LRD) if its autocorrelation function is non-summable, that is, $\sum_n r(n) = \infty$. Intuitively, this means that the process exhibits similar fluctuations over a wide range of time scales. Taking for instance the October Bellcore trace, that is publicly available [1], it can be seen from Figure 1 that the fluctuations over the 0.01, 0.1 and 1s time scales are indeed similar.

Equivalently, one can say that a stationary process is LRD if its spectrum diverges at the origin, that is $f(v) \sim c_f |v|^{-\alpha}$, $v \to 0$. Here, α is a dimensionless scaling exponent, that takes values in $(0, 1)$; c_f takes positive real values and has dimensions of variance. On the other hand, a short range dependent (SRD) process is simply a stationary process which is not LRD. Such a process has $\alpha = 0$ at large scales, corresponding to white noise at scales beyond the so-called characteristic scale or correlation horizon. The Hurst parameter H is related with α by $H = (\alpha + 1)/2$.

There are several estimators of LRD. In this paper we use the semi-parametric estimator developed in [27], which is based on wavelets. Here, one looks for alignment in the so-called Logscale Diagram (LD), which is a log-log plot of the variance estimates (y_j) of the discrete wavelet transform coefficients representing the traffic

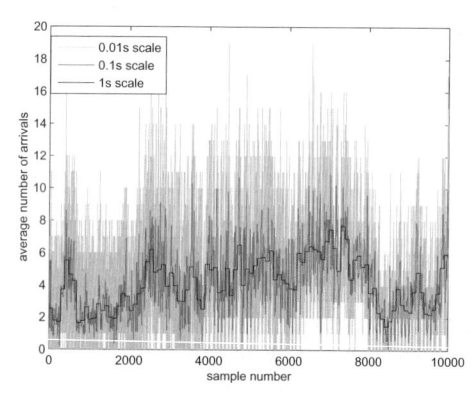

Fig. 1. LRD processes exhibit fluctuations over a wide range of time scales (Example: trace pOct)

process, against scale (j), completed with confidence intervals about these estimates at each scale. It can be thought of as a spectral estimator where large scale corresponds to low frequency. The main properties explored in this estimator are the stationarity and short-term correlations exhibited by the process of discrete wavelet transform coefficients and the power-law dependence in scale of the variance of this process. Traffic is said to be LRD if, within the limits of the confidence intervals, the log of the variance estimates fall on a straight line, in a range of scales from some initial value j_1 up to the largest one present in data and the slope of the straight line, which is an estimate of the scaling exponent α, lies in $(0,1)$.

There is a close relationship between long-range dependent and self-similar processes. In fact, if $Y(t)$ is self-similar with stationary increments and finite variance then $X(k)$ is long-range dependent, as long as $\frac{1}{2} < H < 1$. The process $X(k)$ is said to be exactly second-order self-similar ($\frac{1}{2} < H < 1$) if

$$r(n) = 1/2 \left[(n+1)^{2H} - 2n^{2H} + (n-1)^{2H}\right] \tag{2}$$

for all $n \geq 1$, or is asymptotically self-similar if

$$r(n) \sim n^{-(2-2H)} L(n) \tag{3}$$

as $n \to \infty$, where $L(n)$ is a slowly varying function at infinity. In both cases the autocovariance decays hyperbolically, which indicates LRD. Any asymptotically second-order self-similar process is LRD, and vice-versa.

3 Background on Markovian Models

The dBMAP stochastic process may be regarded as an Markov random walk whose additive component takes values on the nonnegative integers, $I\!N_0$. Thus, we say that a Markov chain $(Y,J) = \{(Y_k, J_k), k \in I\!N_0\}$ on the state space $I\!N_0 \times S$ is a dBMAP if

$$P(Y_{k+1} = m, J_{k+1} = j | Y_k = n, J_k = i) = \begin{cases} 0 & m < n \\ p_{ij}\, q_{ij}(m-n) & m \geq n \end{cases} \tag{4}$$

where $\mathbf{P} = (p_{ij})_{i,j \in S}$ is a stochastic matrix and, for each pair $(i,j) \in S^2$, $q_{ij} = \{q_{ij}(n), n \in I\!N_0\}$ is a probability function over $I\!N_0$, and we let $\mathbf{Q}(n) = (q_{ij}(n))_{i,j \in S}$. This implies, in particular that J is a Markov chain, called the *Markov component* or *phase* of (Y,J) and S is the set of modulating states or the phase set. When the dBMAP (Y,J) is used to model an arrival process, Y_k may be interpreted as the total number of arrivals until instant k. (X,J) is also a dBMAP, where X_n represents the total number of packets that arrive until instant n.

An important particular case of the dBMAP is the dMMPP. We say that the process (Y,J) on the state space $I\!N_0 \times S$ is a dMMPP with parameters $(\mathbf{P}, \Lambda)$, where $\mathbf{P} = (p_{ij})_{i,j \in S}$ is a stochastic matrix and $\Lambda = (\lambda_{ij})_{i,j \in S} = (\lambda_i 1_{\{i=j\}})_{i,j \in S}$ is a diagonal matrix with nonnegative entries (i.e., $\lambda_i \geq 0$, $i \in S$), if it is a dBMAP with parametrization $(\mathbf{P}, \{\mathbf{Q}(n), n \in I\!N\})$, where

$$q_{ij}(n) = e^{-\lambda_j} \frac{\lambda_j^n}{n!} \tag{5}$$

for $i, j \in S$ and $n \in \mathbb{N}$; i.e., $q_{ij} = \{q_{ij}(n), n \in \mathbb{N}_0\}$ is the probability function of a Poisson random variable with mean λ_j. Thus a dMMPP is a dBMAP for which the number of arrivals in a given instant of time is only a function of the current phase of the dBMAP and when the process is in phase j the number of arrivals at an instant has a Poisson distribution with mean λ_j; the parameter λ_j may be null, in which case no arrivals occur in phase j. So, (Y, J) is a dMMPP with set of modulating states S and parameter (matrices) $\mathbf{P}$ and $\mathbf{\Lambda}$, and write

$$(Y, J) \sim \text{dMMPP}_S(\mathbf{P}, \mathbf{\Lambda}) \tag{6}$$

where $\mathbf{\Lambda} = (\lambda_{ij}) = (\lambda_i \delta_{ij})$. The matrix $\mathbf{P}$ is the transition probability matrix of the modulating Markov chain J, whereas $\mathbf{\Lambda}$ is the matrix of Poisson arrival rates. If S has cardinality r, we say that (Y, J) is a dMMPP of order r (dMMPP$_r$). The stationary distribution of J is denoted by $\pi = [\pi_1 \, \pi_2, \ldots \, \pi_r]$.

The superposition of independent dMMPPs is still an dMMPP. More precisely, if $(Y^{(l)}, J^{(l)}) \sim \text{dMMPP}_{r_l}(\mathbf{P}^{(l)}, \Lambda^{(l)})$, $l = 1, 2, \ldots, L$, are independent, then their superposition $(Y, J) = (\sum_{l=1}^{L} Y^{(l)}, (J^{(1)}, J^{(2)}, \ldots, J^{(L)}))$ is a dMMPP$_S(\mathbf{P}, \Lambda)$, where $S = \{1, 2, \ldots, r_1\} \times \ldots \times \{1, 2, \ldots, r_L\}$,

$$\mathbf{P} = \mathbf{P}^{(1)} \otimes \mathbf{P}^{(2)} \otimes \ldots \otimes \mathbf{P}^{(L)} \tag{7}$$

and

$$\Lambda = \Lambda^{(1)} \oplus \Lambda^{(2)} \oplus \ldots \oplus \Lambda^{(L)} \tag{8}$$

with $\oplus$ and $\otimes$ denoting the Kronecker sum and product, respectively.

4 M2^L-MMPP - A Second-Order Self-similar Model

This section describes a parameter fitting procedure, based on MMPPs, that matches both the autocovariance and the marginal distribution of the counting process, leading to accurate estimates of queuing behavior for network traffic exhibiting LRD behavior. This work was firstly published in [18] and was also motivated by the need to keep the number of states of the MMPP at a minimum in order to reduce the complexity associated with the calculation of the performance metrics of interest.

Matching simultaneously the autocovariance and marginal distribution of the counting process is a difficult task since every MMPP parameter influences both characteristics. With the purpose of achieving some degree of decoupling when matching

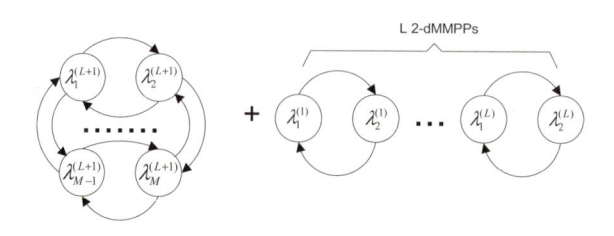

Fig. 2. Superposition of an M-dMMPP and L 2-dMMPP models

these two statistics, the proposed MMPP, $(X, J) \sim \text{M2}^L\text{-dMMPP}$, is constructed as a superposition of L independent 2-dMMPPs, $(X^{(l)}, J^{(l)}) \sim \text{dMMPP}_2(\mathbf{P}^{(l)}, \Lambda^{(l)}), l = 1, 2, \ldots, L$, that capture the autocovariance function of the increments of the arrival process and one M-dMMPP, $(X^{(L+1)}, J^{(L+1)}) \sim \text{dMMPP}_M(\mathbf{P}^{(L+1)}, \Lambda^{(L+1)})$, that approximates the distribution of the increments of the arrival process. This superposition step is graphically illustrated in Figure 2. In this approach L and M are not fixed *a priori* but instead are computed as part of the fitting procedure.

Let us define the increment processes $Y^{(1)}, Y^{(2)}, \ldots, Y^{(L+1)}$ and Y associated to $X^{(1)}, X^{(2)}, \ldots, X^{(L+1)}$, and X, respectively:

$$Y_k^{(l)} = X_{k+1}^{(l)} - X_k^{(l)}, \; l = 1, 2, \ldots, L+1 \tag{9}$$

and

$$Y_k = X_{k+1} - X_k \tag{10}$$

for $k = 0, 1, \ldots$. Note that Y_k is the (total) number of arrivals at sampling interval k and $Y_k^{(l)}$ is the number of arrivals that are due to the l-th arrival process, so that, in particular,

$$Y_k = \sum_{l=1}^{L+1} Y_k^{(l)}, \quad k = 0, 1, 2, \ldots. \tag{11}$$

Moreover, $Y^{(1)}, Y^{(2)}, \ldots, Y^{(L+1)}$, and Y, are stationary sequences.

In order to characterize the marginal distributions of the L 2-dMMPPs processes, $Y^{(1)}, Y^{(2)}, \ldots, Y^{(L)}$, the M-dMMPP, $Y^{(L+1)}$, and the resulting process, Y, we denote by $\{f_l(k), k = 0, 1, 2, \ldots\}, l = 1, 2, \ldots, L+1$, and $\{f(k), k = 0, 1, 2, \ldots\}$, their (marginal) probability functions, respectively. As the univariate distributions of $Y^{(1)}$, $Y^{(2)}, \ldots, Y^{(L+1)}$ are mixtures of Poisson distributions, we denote the probability function of a Poisson random variable with mean μ by $\{g_\mu(k), k = 0, 1, 2, \ldots\}$, for $\mu \in [0, +\infty)$, so that $g_\mu(k) = e^{-\mu} \frac{\mu^k}{k!}, k = 0, 1, 2, \ldots$. For $l = 1, 2 \ldots, L$, the marginal distribution of $Y^{(l)}$ (that is, the distribution of $Y_k^{(l)}$, for $k = 0, 1, \ldots$) is a mixture of two Poisson distributions with means $\lambda_1^{(l)}$ and $\lambda_2^{(l)}$ and weights $\pi_1^{(l)}$ and $\pi_2^{(l)}$, respectively. Thus the probability functions of $Y^{(l)}, l = 1, 2, \ldots, L$, are given by

$$f_l(k) = \pi_1^{(l)} g_{\lambda_1^{(l)}}(k) + \pi_2^{(l)} g_{\lambda_2^{(l)}}(k), k = 0, 1, 2, \ldots \tag{12}$$

and their autocovariance functions are

$$\gamma_k^{(l)} = \text{Cov}(Y_0^{(l)}, Y_k^{(l)}) = \pi_1^{(l)} \pi_2^{(l)} |\lambda_2^{(l)} - \lambda_1^{(l)}|^2 e^{kc_l}, k = 0, 1, 2, \ldots \tag{13}$$

where $c_l = \ln(1 - p_{12}^{(l)} - p_{21}^{(l)})$. Note that, in particular, the autocovariance functions of $Y^{(1)}, Y^{(2)}, \ldots, Y^{(L)}$ exhibit an exponential decay to zero.

As we want the M-dMMPP to approximate the distribution of the increments of the arrival process but to have no contribution to the autocovariance function of the increments of the M2^L-dMMPP, we choose to make $J^{(L+1)}$ a Markov chain with no memory whatsoever. This is accomplished by choosing

$$\mathbf{P}^{(L+1)} = \begin{bmatrix} \pi_1^{(L+1)} & \pi_2^{(L+1)} & \cdots & \pi_M^{(L+1)} \\ \pi_1^{(L+1)} & \pi_2^{(L+1)} & \cdots & \pi_M^{(L+1)} \\ \cdots & \cdots & \cdots & \cdots \\ \pi_1^{(L+1)} & \pi_2^{(L+1)} & \cdots & \pi_M^{(L+1)} \end{bmatrix} \quad (14)$$

The probability function of $Y^{(L+1)}$ is given by

$$f_{L+1}(k) = \sum_{j=1}^{M} \pi_j^{(L+1)} g_{\lambda_j^{(L+1)}}(k), \ k = 0, 1, 2, \ldots \quad (15)$$

and the autocovariance function of $Y^{(L+1)}$ is null for all positive lags; i.e.,

$$\gamma_k^{(L+1)} = \text{Cov}\left(Y_0^{(L+1)}, Y_k^{(L+1)}\right) = 0, \ k \geq 1. \quad (16)$$

Taking into account (11), it follows that the probability function of Y is given by:

$$f(k) = (f_1 * f_2 * \ldots * f_{L+1})(k) = \\ = \sum_{j_1=1}^{2} \sum_{j_2=1}^{2} \cdots \sum_{j_L=1}^{2} \sum_{j_{L+1}=1}^{M} \left(\prod_{l=1}^{L+1} \pi_{j_l}^{(l)} \right) g_{\sum_{l=1}^{L+1} \lambda_{j_l}^{(l)}}(k) \quad (17)$$

where $*$ denotes the convolution of probability functions and the autocovariance function is given by

$$\gamma_k = \text{Cov}(Y_0, Y_k) = \sum_{l=1}^{L+1} \text{Cov}\left(Y_0^{(l)}, Y_k^{(l)}\right) \\ = \sum_{l=1}^{L} \pi_1^{(l)} \pi_2^{(l)} |\lambda_2^{(l)} - \lambda_1^{(l)}|^2 e^{kc_l}, k = 1, 2, \ldots \quad (18)$$

The inference procedure is illustrated in the flow diagram of Figure 3 and can be divided in four major steps.

A. Approximation of the empirical autocovariance by a weighted sum of exponentials and identification of the time scales

Our approach approximates the autocovariance by a large number of exponentials and then aggregates exponentials with a similar decay into the same time-scale, which is close to the approaches considered in [10, 13, 28] (Figure 4). As a first step, we approximate the empirical autocovariance by a sum of K exponentials with real positive weights and negative real time constants. We chose K as $\sqrt{k_{max}}$, where k_{max} represents the number of points of the empirical autocovariance. This is accomplished through a modified Prony algorithm [29]. The Prony algorithm returns two vectors, $\mathbf{a} = [a_1, \ldots, a_K]$ and $\mathbf{b} = [b_1, \ldots, b_K,]$, which correspond to the approximating function

$$C_K(\mathbf{a}, \mathbf{b}) = \sum_{i=1}^{K} a_i e^{-b_i k}, \quad k = 1, 2, 3, \ldots \quad (19)$$

At this point we identify the components of the autocovariance that characterize the different time-scales by defining L different time-scales in which the autocovariance decays, $b_i, i = 1, \ldots, K$, fall in the same logarithmic scale. The components of the

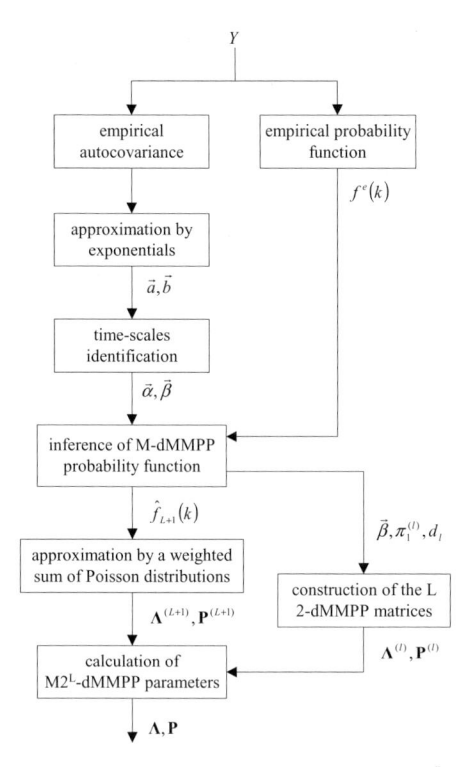

Fig. 3. Flow diagram of the inference procedure of the $M2^L$-MMPP model

decays that belong to the same traffic scale are aggregated in one component with the following parameters:

$$\alpha_l = \sum_{k=i_l}^{i_{l+1}-1} a_k \quad \text{and} \quad \beta_l = -\frac{\sum_{k=i_l}^{i_{l+1}-1} a_k b_k}{\alpha_l}. \tag{20}$$

where b_{i_l} and $b_{i_{l+1}-1}$ correspond to the first and last decay of the time scale. These parameters are used to fit the autocovariance function of the 2-dMMPP $Y^{(l)}$, since

$$\alpha_l = d_l^2 \pi_1^{(l)} \pi_2^{(l)} \quad \text{and} \quad \beta_l = c_l \tag{21}$$

where $\pi_i^{(l)}, i = 1, 2$ corresponds to the steady-state probabilities of $Y^{(l)}$, $d_l = |\lambda_2^{(l)} - \lambda_1^{(l)}|$ and $\beta_l = ln(1 - p_1^{(l)} 2 - p_2^{(l)} 1$, i.e., the fitted autocovariance function of $Y_1 + Y_2 + \ldots + Y_L$ is

$$\sum_{l=1}^{L} \alpha_l e^{k\beta_l}, \quad k = 1, 2, \ldots. \tag{22}$$

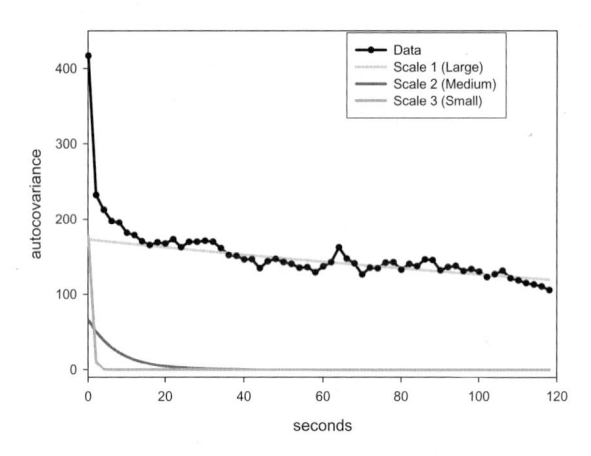

Fig. 4. Approximation of the autocovariance function

B. Inference of the M-dMMPP probability function and of the L 2-dMMPP parameters

The relation between the probability functions of the 2-dMMPPs, the M-dMMPP and the M2^L-dMMPP is defined by (17). In order to simplify the deconvolution of $f_{L+1}(k)$ and $f_l(k), l = 1, ..., L$, we consider that the Poisson arrival rate is zero in one state of each 2-dMMPP source; that is, $\lambda_1^{(l)} = 0$ and $\lambda_2^{(l)} = d_l$, for $l = 1, ..., L$. From (21), it follows that $d_l = \sqrt{\frac{\alpha_l}{\pi_1^{(l)} \pi_2^{(l)}}}$, $l = 1, 2, ..., L$. The probability function of the M-dMMPP, f_{L+1}, is fitted jointly with the parameters $\pi_1^{(l)}$, $l = 1, ..., L$, through the following constrained minimization process:

$$\min_{\{\pi_1^{(l)}, l=1,...,L\}, \{f_{L+1}(k), k=0,1,...\}} \sum_k |o^e(k)| \qquad (23)$$

where

$$o^e(k) = f^e(k) - \left(\hat{f}_1 \oplus ... \oplus \hat{f}_L \oplus f_{L+1}\right)(k) \qquad (24)$$

subject to (21) and

$$0 < \pi_1^{(l)} < 1, l = 1, 2, ..., L, \quad f_{L+1}(k) > 0, k = 0, 1, ...,$$
$$\text{and} \quad \sum_{k=0}^{+\infty} f_{L+1}(k) = 1. \qquad (25)$$

with f^e denoting the empirical probability function of the data. We denote by $\hat{f}_{L+1}$ the fitted probability function of the M-dMMPP. Note that $\pi_1^{(l)}$ is not allowed to be 0 or 1 because, in both cases, the lth 2-dMMPP would degenerate into a Poisson process. The constrained minimization process given by (23)–(25) is a non-linear programming problem and in general, it is computationally demanding to obtain the global optimal solution. Accordingly, to solve this problem we consider two approximations: (i) we make $\pi_1^{(l)} = \pi_1^{(l+1)}, l = 1, ..., L-1$ and (ii) restrict the range of possible $\pi_1^{(l)}$ solutions to be discrete and such that $\pi_1^{(l)} = 0.001k$, $k = 1, \ldots, 999$. Then a search process is used to find the minimum value of the objective function.

At this point all parameters of the 2-dMMPPs, $Y^{(1)}, Y^{(2)}, \ldots, Y^{(L)}$, have been determined and their corresponding 2-dMMPP matrices can be constructed in the following way:

$$\mathbf{P}^{(l)} = \begin{bmatrix} 1 - \pi_2^{(l)}(1 - e^{\beta_l}) & \pi_2^{(l)}(1 - e^{\beta_l}) \\ \pi_1^{(l)}(1 - e^{\beta_l}) & 1 - \pi_1^{(l)}(1 - e^{\beta_l}) \end{bmatrix} \quad (26)$$

$$\Lambda^{(l)} = \begin{bmatrix} 0 & 0 \\ 0 & d_l \end{bmatrix} \quad (27)$$

C. Inference of the M-dMMPP parameters

The next step is the inference of the number of states and Poisson arrival rates of the M-dMMPP from $\hat{f}_{L+1}$. To do this, we infer $\hat{f}_{L+1}$ as a weighted sum of Poisson probability functions, i.e., as the probability function of a finite Poisson mixture with an unknown number of components. The matching is carried out through an algorithm that progressively subtracts a Poisson probability function from $\hat{f}_{L+1}$, which is described in the flowchart of Figure 5. We represent the ith Poisson probability function, with mean φ_i, by $g_{\varphi_i}(k)$. We define $h^{(i)}(k)$ as the difference between $\hat{f}_{L+1}(k)$ and the weighted sum of Poisson probability functions at the ith iteration. Initially, we set $h^{(1)}(k) = \hat{f}_{L+1}(k)$. In each step, we first detect the maximum of $h^{(i)}(k)$. The corresponding k-value, $\varphi_i = [h^{(i)}]^{-1} (\max h^{(i)}(k))$, will be considered the ith Poisson rate of the M-dMMPP. We then calculate the weights of each Poisson probability function, $\mathbf{w}_i = [w_{1i}, w_{2i}, \ldots, w_{ii}]$, through the following set of linear equations:

$$\hat{f}_{L+1}(\varphi_l) = \sum_{j=1}^{i} w_{ji} g_{\varphi_j}(\varphi_l), \qquad l = 1, \ldots, i. \quad (28)$$

This assures that the fitting between $\hat{f}_{L+1}(k)$ and the weighted sum of Poisson probability functions is exact at φ_l points, for $l = 1, 2, \ldots, i$. The final step in each iteration is the calculation of the new difference function

$$h^{(i)}(k) = \hat{f}_{L+1}(k) - \sum_{j=1}^{i} w_{ji} g_{\varphi_j}(k). \quad (29)$$

The algorithm stops when the maximum of $h^{(i)}(k)$ is lower than a pre-defined percentage of the maximum of $\hat{f}_{L+1}(k)$ and M is made equal to i. After M has been determined, the parameters of the M-dMMPP, $\{(\pi_j^{(L+1)}, \lambda_j^{(L+1)}), j = 1, 2, \ldots, M\}$, are then set equal to

$$\pi_j^{(L+1)} = w_{jM} \quad \text{and} \quad \lambda_j^{(L+1)} = \varphi_j. \quad (30)$$

D. M2^L-dMMPP model construction

Finally, the M2^L-dMMPP process can be constructed using equations (7) and (8), where $\Lambda^{(L+1)}$, $\mathbf{P}^{(L+1)}$, $\Lambda^{(i)}$ and $\mathbf{P}^{(i)}$, $i = 1, \ldots, L$, were calculated in the last two steps.

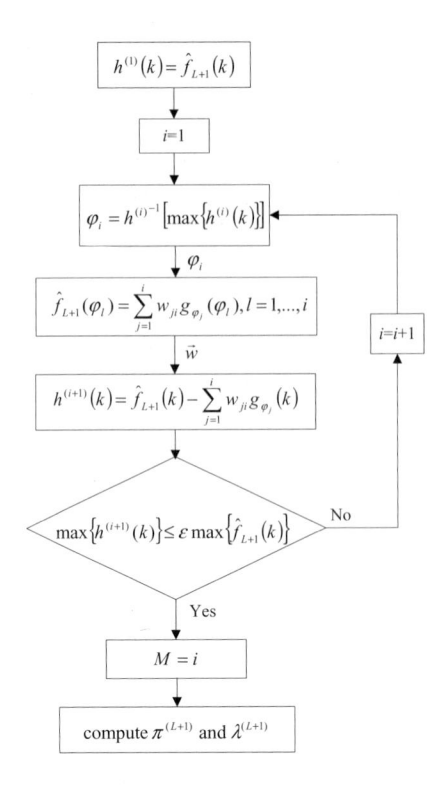

```
                    h^{(1)}(k) = \hat{f}_{L+1}(k)

                              i=1
                               ↓
              φ_i = h^{(i)-1}[max{h^{(i)}(k)}]  ←
                               ↓ φ_i                    │
         \hat{f}_{L+1}(φ_i) = Σ_{j=1}^{i} w_{ji} g_{φ_j}(φ_i), l=1,...,i
                               ↓ \vec{w}              i=i+1
         h^{(i+1)}(k) = \hat{f}_{L+1}(k) - Σ_{j=1}^{i} w_{ji} g_{φ_j}(k)
                               ↓                        │
                                                       No
              max{h^{(i+1)}(k)} ≤ ε max{\hat{f}_{L+1}(k)} ─┘
                               ↓ Yes
                            M = i
                               ↓
                   compute π^{(L+1)} and λ^{(L+1)}
```

Fig. 5. Algorithm for calculation of the number of states and Poisson arrival rates of the M-dMMPP

4.1 Efficency Results

In [18] the efficiency of this fitting procedure was evaluated by applying it to trace UA, a trace of IP traffic measured at University of Aveiro (UA) that is representative of Internet access traffic produced within a University campus environment. The UA trace was measured on July 10th, 2001, between 10.15am and 3.08pm, and comprises 20 millions packets with a mean rate of 1138 packets/s and a mean packet size of 557 bytes. This trace exhibits LRD behavior, which was confirmed by applying the method described in [27] (Figure 7). The sampling interval of the counting process was considered as 0.1 seconds, so octave j corresponds to 0.1×2^j seconds.

The performance of this fitting procedure (and all the others that will be described in the next sections) was evaluated using several evaluation criteria: (i) comparing both the probability and autocovariance functions of the packet arrival counts obtained with the fitted dMMPPs (theoretical) and with the original data traces; (ii) analyzing queuing behavior by comparing the PLR obtained, through trace-driven simulation, with the original data traces and simulated traces generated from the fitted dMMPPs (Figure 6). The results of trace driven simulation for the fitted traces were based on 10 replicas.

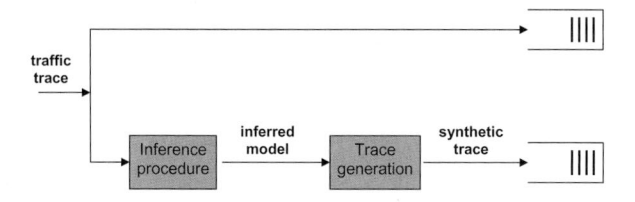

Fig. 6. Methodology for testing queueing behavior

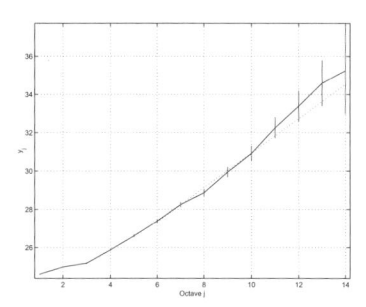

Fig. 7. Scaling analysis, UA

Fig. 8. Probability function, UA

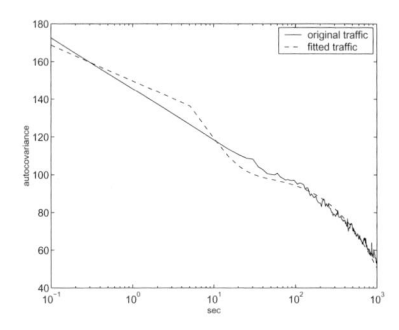

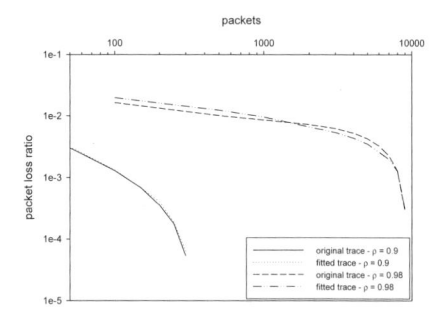

Fig. 9. Autocovariance function, UA

Fig. 10. Packet loss ratio versus buffer size, UA

The empirical autocovariance function was fitted by two exponentials with parameters $\boldsymbol{\alpha} = [1.00 \times 10^2 \; 6.87 \times 10^1]$ and $\boldsymbol{\beta} = [-6.91 \times 10^{-5} \; -1.28 \times 10^{-2}]$. With a resulting 12-states dMMPP, the fitting of the probability and autocovariance functions is very good (Figures 8 and 9, respectively), which reveals itself sufficient to get a very close matching of the PLR curve (Figure 10). The considered service rates are 685 KBytes/s and 629 KBytes/s, corresponding to link utilizations of $\rho = 0.9$ and $\rho = 0.98$, respectively. Both the original and the fitted traces exhibit LRD, with estimated Hurst parameters of $\widehat{H} = 0.952$ and $\widehat{H} = 0.935$, respectively.

5 Distributional Self-similar Models

This section proposes two traffic models, based on dMMPPs, designed to capture self-similar behavior over multiple time scales by fitting the empirical distribution of packet counts at each time scale. The number of time scales, L, is fixed *a priori* and the time scales are numbered in an increasing way, from $l = 1$ (corresponding to the largest time scale) to $l = L$ (corresponding to the smallest time scale).

5.1 Superposition Model

This model was firstly proposed in [19] and is based on the superposition of dMMPPs, where each dMMPP represents a specific time scale. Figure 11 illustrates the construction methodology of the dMMPP for the simple case of three time scales and two-state dMMPPs in each time scale. The dMMPP associated with time scale l is denoted by dMMPP$^{(l)}$ and the corresponding number of states by $N_{(l)}$. The flowchart of the inference procedure is represented in Figure 12 where, basically, the following three steps can be identified.

A. Computation of the data vectors (corresponding to the average number of arrivals per time interval) at each time scale

Having defined the time interval at the smallest time scale, Δt, the number of time scales, L, and the level of aggregation, a, the aggregation process starts by computing the data sequence corresponding to the average number of arrivals in the smallest time scale, $D^{(L)}(k), k = 1, 2, \ldots, N$. Then, it calculates the data sequences of the remaining time scales, $D^{(l)}(k), l = L - 1, \ldots, 1$, corresponding to the average number of arrivals in intervals of length $\Delta t a^{(L-l)}$. This is given by

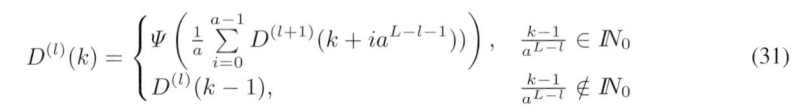

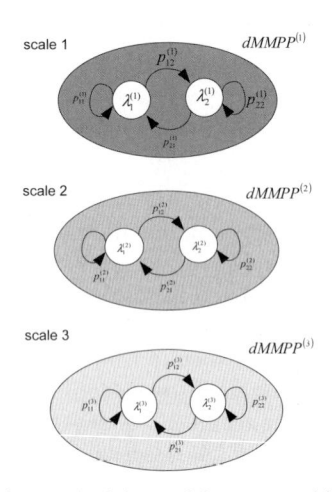

$$D^{(l)}(k) = \begin{cases} \Psi\left(\frac{1}{a}\sum_{i=0}^{a-1} D^{(l+1)}(k + ia^{L-l-1})\right), & \frac{k-1}{a^{L-l}} \in \mathbb{N}_0 \\ D^{(l)}(k-1), & \frac{k-1}{a^{L-l}} \notin \mathbb{N}_0 \end{cases} \quad (31)$$

Fig. 11. Construction methodology of the superposition dMMPP model

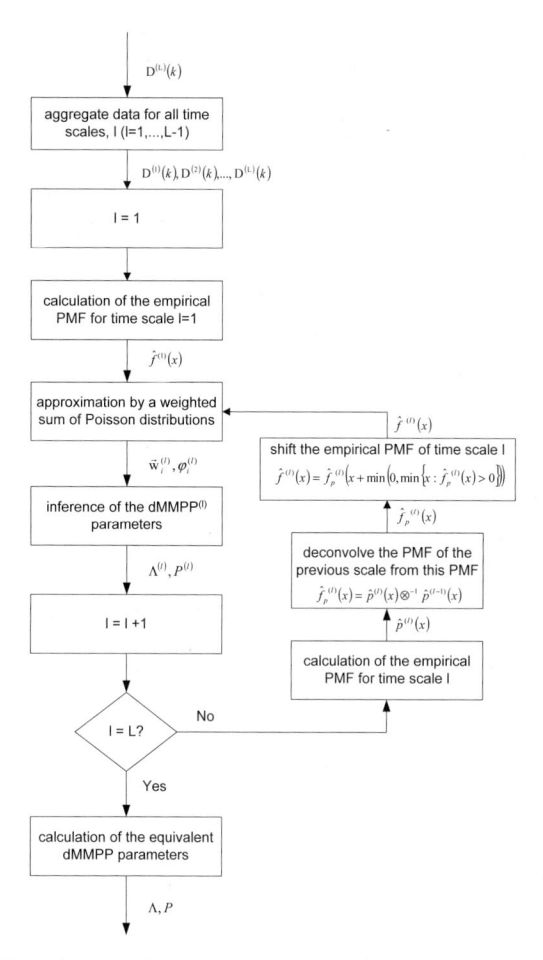

Fig. 12. Flow diagram of the inference procedure of the superposition model

where $\Psi(x)$ represents round toward the integer nearest x. Note that all data sequences have the same length N and that $D^{(l)}(k)$ is formed by sub-sequences of a^{L-l} successive equal values; these sub-sequences will be called *l-sequences*. The empirical distribution of $D^{(l)}(k)$ will be denoted by $\hat{p}^{(l)}(x)$.

Figure 13 illustrates the aggregation process for the particular case of considering only three time scales and an aggregation level of $a = 2$. The top part of the picture corresponds to the finest time scale (scale 3) and represents the number of arrivals per sampling interval. At time scale 2, the Figure represents the average number of arrivals per time interval of length $2\Delta t$, while at time scale 1 it represents the average number of arrivals per time interval of length $4\Delta t$, since the aggregation level is equal to 2.

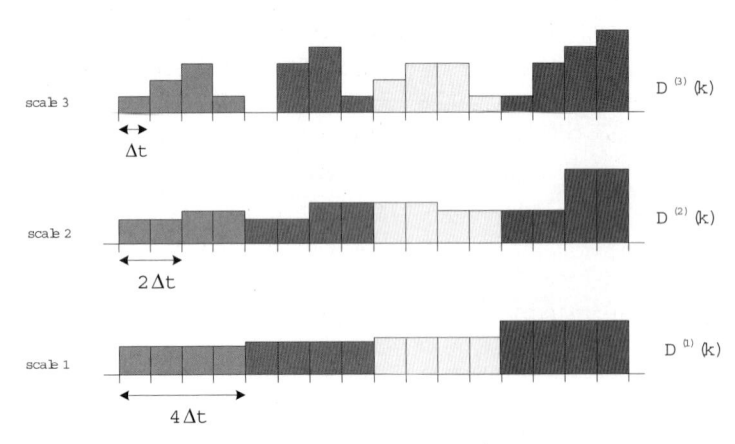

Fig. 13. Illustration of the aggregation process

B. For all time scales (going from the largest to the smallest one), calculation of the corresponding empirical PMF and inference of a dMMPP that matches the resulting PMF

Each dMMPP will be inferred from a PMF that represents its contribution to a particular time scale. For the largest time scale, this PMF is simply the empirical one. The traffic components due to time scale l, $l = 2, ..., L$, are obtained through deconvolution of the empirical PMFs of this and the previous time scales, i.e., $\hat{f}_p^{(l)}(x) = [\hat{p}^{(l)} \otimes^{-1} \hat{p}^{(l-1)}](x)$. However, this may result in probability mass at negative arrival rates for the dMMPP$^{(l)}$, which will occur whenever $\min\{x : \hat{p}^{(l-1)}(x) > 0\} < \min\{x : \hat{p}^{(l)}(x) > 0\}$. To correct these results, the dMMPP$^{(l)}$ will be fitted to

$$\hat{f}^{(l)}(x) = \hat{f}_p^{(l)}(x + e^{(l)}) \tag{32}$$

where $e^{(l)} = \min\left(0, \min\{x : \hat{f}_p^{(l)}(x) > 0\}\right)$, which assures $\hat{f}^{(l)}(x) = 0$, $x < 0$. The additional factors that are now introduced will be removed in the final step of the inference procedure.

The number of states, $N_{(l)}$, and the parameters of the dMMPP$^{(l)}$, $\{(\pi_j^{(1)}, \lambda_j^{(1)}), j = 1, 2, ..., N_{(1)}\}$, that adjusts the empirical PMF $\hat{f}^{(l)}(x)$ are calculated using the same procedure described in step 3 of the M2^L-dMMPP inference procedure.

The next step consists of associating one of the dMMPP$^{(l)}$ states with each time interval of the arriving process. Recall that the data sequences aggregated at time scale l have a^{L-l} successive equal values called l-sequences. The state assignment process considers only the first time interval of each l-sequence, defined by $i = a^{L-l}(k-1) + 1, k \in \mathbb{N}, i \in E^{(l)}$, where $E^{(l)}$ represents the set of time intervals associated with dMMPP$^{(l)}$. The state that is assigned to l-sequence i is calculated randomly according to the probability vector $\boldsymbol{\theta}^{(l)}(i) = \left\{\theta_1^{(l)}(i), ..., \theta_{N_{(l)}}^{(l)}(i)\right\}$, with

$$\theta_n^{(l)}(i) = \frac{g_{\lambda_n^{(l)}}(D^{(l)}(i))}{\sum_{j=1}^{N_{(l)}} g_{\lambda_j^{(l)}}(D^{(l)}(i))}. \tag{33}$$

for $n = 1, ..., N_{(l)}$, where $\lambda_j^{(l)}$ represents the Poisson arrival rate of the jth state of dMMPP$^{(l)}$, and $g_\lambda(y)$ represents a Poisson probability distribution function with mean λ. The elements of this vector represent the probability that the state j had originated the number of arrivals $D^{(l)}(k)$ at time interval k from time scale l.

After this step, we infer the dMMPP$^{(l)}$ transition probabilities, $p_{od}^{(l)}$, with $o, d = 1, ..., N_{(l)}$, by counting the number of transitions between each pair of states. If $n_{od}^{(l)}$ represents the number of transitions from state o to state d of the dMPPP$^{(l)}$, then

$$p_{od}^{(l)} = \frac{n_{od}^{(l)}}{\sum_{m=1}^{N_{(l)}} n_{om}^{(l)}}, o, d = 1, ..., N_{(l)} \tag{34}$$

The transition probability and the Poisson arrival rate matrices of the dMMPP$^{(l)}$ are then given by

$$\mathbf{P}^{(l)} = \begin{bmatrix} p_{11}^{(l)} & p_{12}^{(l)} & \cdots & p_{1N_{(l)}}^{(l)} \\ p_{21}^{(l)} & p_{22}^{(l)} & \cdots & p_{2N_{(l)}}^{(l)} \\ \cdots & \cdots & \cdots & \cdots \\ p_{N_{(l)}1}^{(l)} & p_{N_{(l)}2}^{(l)} & \cdots & p_{N_{(l)}N_{(l)}}^{(l)} \end{bmatrix} \tag{35}$$

$$\Lambda^{(l)} = \begin{bmatrix} \lambda_1^{(l)} & 0 & \cdots & 0 \\ 0 & \lambda_2^{(l)} & \cdots & 0 \\ \cdots & \cdots & \cdots & \cdots \\ 0 & 0 & \cdots & \lambda_{N_{(l)}}^{(l)} \end{bmatrix} + e^{(l)}\mathbf{I} \tag{36}$$

The diagonal matrix of the steady-state probabilities is designated by $\Pi^{(l)}$.

Figure 14 schematically illustrates the main steps of the construction process for the superposition model, considering only the first two time scales. As was previously said, the empirical PMF corresponding to the largest time scale (scale 1) is estimated and the dMMPP that best adjusts it is inferred. For the next immediate scale (scale 2), the empirical PMF is estimated and then it is deconvolved from the PMF corresponding to time scale 1. The dMMPP that describes the contribution of time scale 2 for the arrival process is calculated based on the PMF that results from this deconvolution operation.

C. Calculation of the final dMMPP through the superposition of the dMMPPs inferred for each time scale

The equivalent dMMPP process is constructed using equations (7) and (8), where matrices $\mathbf{P}^{(l)}$ and $\Lambda^{(l)}$, $l = 1, ..., L$, were calculated in the last subsection. Besides, the additional factors introduced in 32 must be removed. Thus, the final $\Lambda^{(l)}$ will be given by

$$\Lambda = \Lambda - \sum_{l=2}^{L} e^{(l)} \cdot \mathbf{I} \tag{37}$$

where $\mathbf{I}$ is the identity matrix.

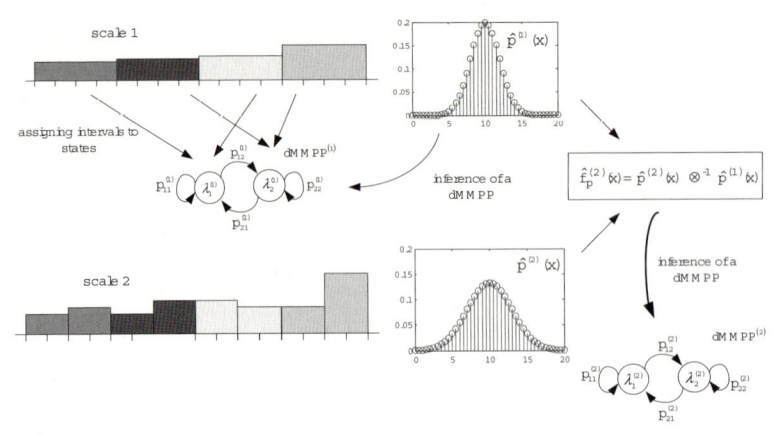

Fig. 14. Procedure for calculating the empirical PMFs and inferring the partial dMMPPs of the superposition model

5.2 Hierarchical Model

This model was firstly presented in [21] and is constructed using an hierarchical procedure, that successively decomposes dMMPP states into new dMMPPs, thus refining the traffic process by incorporating the characteristics offered by finer time scales (Figure 15). The procedure starts at the largest time scale by inferring a dMMPP that matches the empirical PMF corresponding to this time scale. As part of the parameter fitting procedure, each time interval of the data sequence is assigned to a dMMPP state; in this way, a new PMF can be associated with each dMMPP state. At the next finer time scale, each dMMPP state is decomposed into a new dMMPP that matches the contribution of this time scale to the PMF of the state it descends from. In this way, a child dMMPP gives a more detailed description of its parent state PMF. This refinement process is iterated until a pre-defined number of time scales is integrated. Finally, a dMMPP incorporating this hierarchical structure is derived.

The construction process of the hierarchical model can be described through a tree where, except for the root node, each tree node corresponds to a dMMPP state and each tree level to a time scale. A dMMPP state will be represented by a vector indicating the path in the tree from its higher level ancestor (i.e. the state it descends from at the largest scale, $l = 1$) to itself. Thus, a state at time scale l will be represented by some vector $s = (s_1, s_2, ..., s_l)$, $s_i \in \mathbb{N}$. Each dMMPP will be represented by the state that generated it (i.e. its parent state), that is, dMMPP^s will represent the dMMPP generated by state s. The root node of the tree corresponds to a virtual state, denoted by $s = \emptyset$, that is used to represent the dMMPP of the largest time scale, $l = 1$. This dMMPP will be called the root dMMPP. Thus, the dMMPP states in the tree are characterized by $s = (s_1, s_2, ..., s_l)$, $l \in \mathbb{N}$, with $s_{i+1} \in \{1, 2, ..., N_{s_{i]}}\}$, $i = 0, 1, ..., l-1$; here, $s_{j]}$ denotes the sub-vector of s given by $(s_1, s_2, ..., s_j)$, with $j < |s|$, and $s_{0]} = \emptyset$, where $|s|$ denotes the length of vector s. Note that, using this notation, a vector s can either represent state s or the dMMPP generated by s. Besides, the time scale of dMMPP^s is $|s| + 1$.

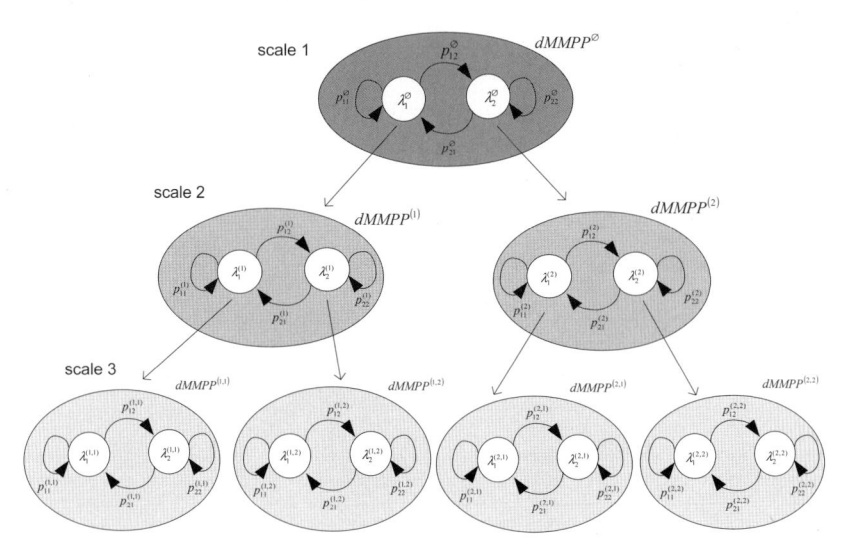

Fig. 15. Construction methodology of the hierarchical dMMPP model

Finally, let E^s denote the set of time intervals associated with state s, i.e., with dMMPP^s. Using this notation, the set associated with dMMPP$^\emptyset$ will be $E^\emptyset = \{1, 2, ..., N\}$, where N is the number of time intervals at the smallest time scale. Starting from $E^\emptyset$, the sets E^s are successively partitioned at each time scale in a hierarchical fashion. Thus, if states s and t are such that $|s| = |t| = l$ and $s \neq t$, then $E^s \cap E^t = \emptyset$ and $\bigcup_{s:|s|=l} E^s = E^\emptyset$. Moreover, if state s is a parent of state t, that is $t = (s, j)$, then $E^t \subseteq E^s$ and $\bigcup_{j=1,...,N_s} E^{(s,j)} = E^s$.

The inference procedure is represented schematically in the flowchart of Figure 16, where the following three main steps can be identified.

A. Computation of the data vectors (corresponding to the average number of arrivals per time interval) for each time scale

This step is equal to the one described for the superposition model.

B. For all time scales (going from the largest to the smallest one), calculation of the corresponding empirical PMF and inference of a dMMPP that matches the resulting PMF

Each dMMPP will be inferred from a PMF that represents its contribution to a particular time scale. For the largest time scale, this PMF is simply the empirical one, but for all other time scales l, $l = 2, ..., L$, the PMF represents the contribution of the time scale to the PMF of its parent state. The contribution of a dMMPP at time scale l generated from state s corresponds also to the deconvolution of empirical PMFs, but now calculated over the set of time intervals E^s, at this time scale $l = |s| + 1$ and the previous time scale $l - 1 = |s|$, i.e., $\hat{f}_p^s(x) = \left[\hat{p}^{s,|s|+1} \otimes^{-1} \hat{p}^{s,|s|}\right](x)$, where $\hat{p}^{s,l}$ represents the PMF obtained from the data sequence $D^l(k), k \in E^s$. Note that the two empirical PMFs are obtained from the same set of time intervals but aggregated at

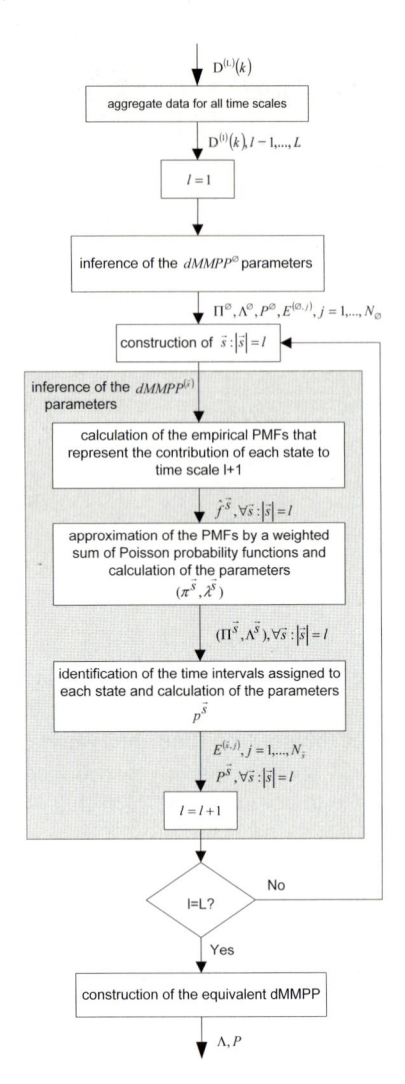

Fig. 16. Flow diagram of the inference procedure of the hierarchical model

different levels. Once again, these operations may result in probability mass at negative arrival rates for the dMMPP^s, which will occur whenever $\min \{x : \hat{p}^{s,|s|}(x) > 0\} < \min \{x : \hat{p}^{s,|s|+1}(x) > 0\}$. These results must be corrected using equation 32, with (l) replaced by s.

The number of states, N_s, and the parameters of the dMMPP^s, $\{(\pi_j^s, \lambda_j^s), j = 1, 2, \ldots, N_s\}$, that adjusts the empirical PMF $\hat{f}^s(x)$ are calculated using the same procedure described in step 3 of the M2^L-dMMPP inference procedure.

The next step consists of associating one of the dMMPP^s states with each time interval of the arriving process. The set of time intervals associated with dMMPP^s is E^s and the goal here is to partition E^s into subsets $E^{(s,j)}, j = 1, \ldots, N_s$. Now, the state assignment process considers only the first time interval of each l-sequence, defined by

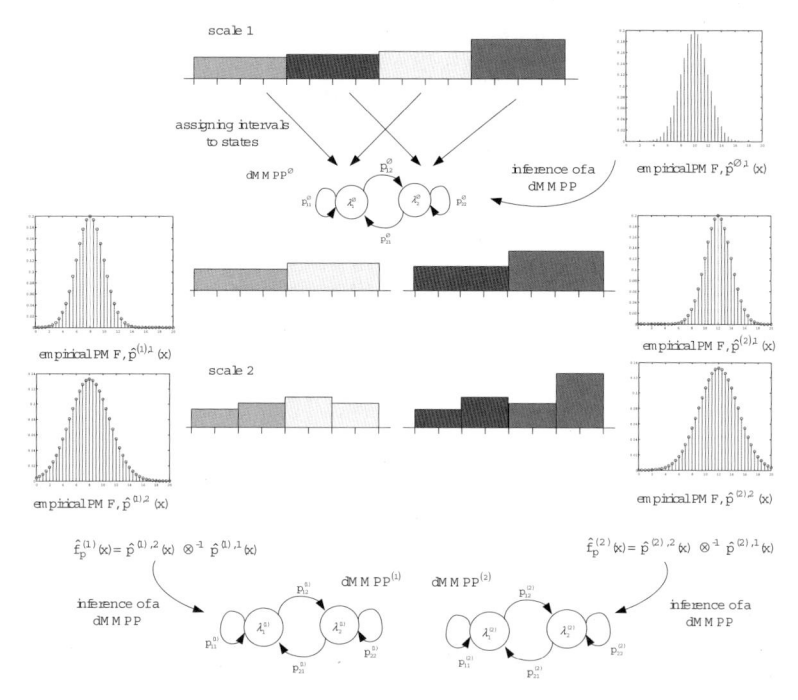

Fig. 17. Procedure for calculating the empirical PMFs and inferring the partial dMMPPs of the decomposition model

$i = a^{L-(|s|+1)}(k-1) + 1, k \in \mathbb{N}, i \in E^s$. The state that is assigned to l-sequence i is calculated randomly according to the probability vector $\boldsymbol{\theta}^s(i) = \{\theta_1^s(i), \ldots, \theta_{N_s}^s(i)\}$, with

$$\theta_n^s(i) = \frac{g_{\lambda_n^s}\left(D^{|s|+1}(i)\right)}{\sum_{j=1}^{N_s} g_{\lambda_j^s}\left(D^{|s|+1}(i)\right)} \tag{38}$$

for $n = 1, \ldots, N_s$.

The dMMPP^s transition probabilities, p_{od}^s, $o, d = 1, \ldots, N_s$, are calculated through equation 34 with (l) replaced by s. In this way, the transition probability and the Poisson arrival rate matrices are also given by equations 35 and 36, respectively.

Figure 17 schematically illustrates the main steps of the construction process for the decomposition model, considering only the first two time scales. For the largest time scale (scale 1), the empirical PMF is estimated and the dMMPP that best adjusts it is inferred (dMMPP^0). Each time interval of the data sequence is then assigned to each dMMPP state and the next step consists on estimating the empirical PMFs associated to each state. For the next immediate scale (scale 2), the empirical PMFs associated to each state will also be estimated and then they are deconvolved from the PMFs corresponding to time scale 1 and to the same states. The dMMPPs that describe the contribution of time scale 2 for the arrival process are calculated based on the PMFs that result from these deconvolution operations.

C. Calculation of matrices Λ and P of the dMMPP that incorporates the hierarchical structure

In this step we have to construct a dMMPP equivalent to the tree structure of dMMPPs derived in previous steps. The goal is to incorporate in the model the level of detail given by the finer time scale, so the equivalent dMMPP will have a number of states equal to the number of states in smallest time scale of the tree structure, L. These can be identified by $\boldsymbol{s} = (s_1, s_2, ..., s_L)$; each state is associated with its ancestor states $\boldsymbol{s}_{i+1]} = (s_1, s_2, ..., s_{i+1}), i = 0, 1, ..., L-1$ of the dMMPP$^{\boldsymbol{s}_{i]}}$.

Thus, the states of the equivalent dMMPP will have Poisson rates which are the sum of the Poisson rates of its ancestors in the tree structure, i.e.,

$$\lambda_{\boldsymbol{s}} = \sum_{j=0}^{L-1} \lambda_{s_{j+1}}^{\boldsymbol{s}_{j]}} \tag{39}$$

The transition between each pair of states is determined by the shortest path in the tree structure, going through the root dMMPP, that joins the two states. Any pair of states descend from one or more common dMMPPs. The first one, at the time scale with higher l, will be denoted by $\boldsymbol{s} \wedge \boldsymbol{t} = (s_1, s_2, ..., s_k)$ where $k = \max\{i : s_j = t_j, j = 1, 2, ..., i\}$.

We first consider the case of $\boldsymbol{s} \neq \boldsymbol{t}$. The probability of transition from $\boldsymbol{s}$ to $\boldsymbol{t}$, $p_{\boldsymbol{s},\boldsymbol{t}}$, is given by the product of three factors. The first factor accounts for the time scales where $\boldsymbol{s}$ and $\boldsymbol{t}$ have the same associated states and is given by

$$\phi_{\boldsymbol{s},\boldsymbol{t}} = \begin{cases} \prod_{j=0}^{|\boldsymbol{s} \wedge \boldsymbol{t}|-1} p_{s_{j+1},s_{j+1}}^{\boldsymbol{s}_{j]}}, & |\boldsymbol{s} \wedge \boldsymbol{t}| \neq 0 \\ 1, & |\boldsymbol{s} \wedge \boldsymbol{t}| = 0 \end{cases} \tag{40}$$

The second factor accounts for the transition in the time scale where $\boldsymbol{s}$ and $\boldsymbol{t}$ are associated to different states of the same dMMPP, which corresponds to $p_{s_{|\boldsymbol{s} \wedge \boldsymbol{t}|+1}, t_{|\boldsymbol{s} \wedge \boldsymbol{t}|+1}}^{\boldsymbol{s} \wedge \boldsymbol{t}}$. The third factor accounts for the steady-state probabilities of states associated to $\boldsymbol{t}$ in the time scales that are not common to $\boldsymbol{s}$ and is given by

$$\psi_{\boldsymbol{s},\boldsymbol{t}} = \prod_{j=|\boldsymbol{s} \wedge \boldsymbol{t}|+1}^{L-1} \pi_{t_{j+1}}^{\boldsymbol{t}_{j]}} \tag{41}$$

where an empty product is equal to one.
Finally, for $\boldsymbol{s} \neq \boldsymbol{t}$,

$$p_{\boldsymbol{s},\boldsymbol{t}} = \phi_{\boldsymbol{s},\boldsymbol{t}} p_{s_{|\boldsymbol{s} \wedge \boldsymbol{t}|+1}, t_{\boldsymbol{s} \wedge \boldsymbol{t}+1}}^{\boldsymbol{s} \wedge \boldsymbol{t}} \psi_{\boldsymbol{s},\boldsymbol{t}} \tag{42}$$

In case $\boldsymbol{s} = \boldsymbol{t}$, it is simply

$$p_{\boldsymbol{s},\boldsymbol{t}} = \phi_{\boldsymbol{s},\boldsymbol{t}} \tag{43}$$

5.3 Efficiency Results

These fitting procedures were applied to the Kazaa trace, a trace measured at the backbone of a Portuguese ISP network characterizing the downstream traffic from 10

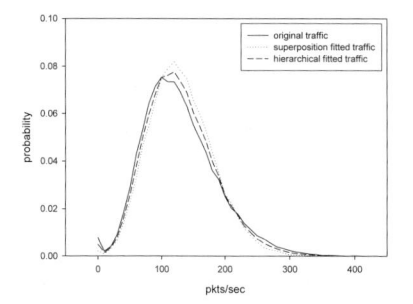

Fig. 18. PMF at the smallest time scale, Kazaa

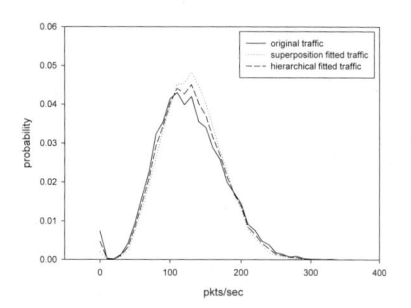

Fig. 19. PMF at the intermediate time scale, Kazaa

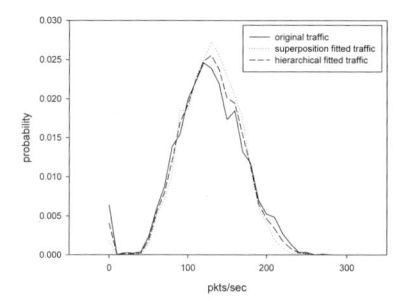

Fig. 20. PMF at the largest time scale, Kazaa

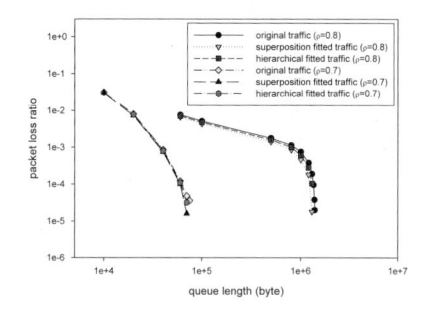

Fig. 21. Packet loss ratio versus buffer size, Kazaa

users of the file sharing application Kazaa. The Kazaa trace was measured on October 18th 2001, between 10.26pm and 11.31pm, and comprises 1 million packets with a mean rate of 131140 packets/s and a mean packet size of 1029 bytes. This trace exhibits self-similar characteristics and three different time scales were considered: 0.1s, 0.2s and 0.4s. Larger aggregation levels were also considered, with good fitting results. Both fitting approaches were able to capture the traffic LRD behavior and the agreement between the PMFs corresponding to the original and dMMPP fitted traces, for the smallest, intermediate and largest time scales, was very good, as can be seen from figures 18, 19 and 20. These results were achieved with resulting dMMPPs having about 288 states in the superposition model and 38 states in the hierarchical model.

Considering queuing performance, Figure 21 shows that PLR behavior is very well approximated by the equivalent dMMPPs for both utilization ratios ($\rho = 0.7$ and $\rho = 0.8$). However, as the utilization ratio increases the deviation slightly increases, because the sensitivity of the metrics variation to a slight difference in the compared traces is higher. Thus, the proposed fitting approaches provide a close match of the PMFs at each time scale and this agreement reveals itself sufficient to drive a good queuing performance in terms of packet loss ratio.

The computational complexity of both fitting methods is small. This complexity, as well as the number of states of the resulting dMMPPs, is directly related to the level of accuracy used to approximate the empirical PMFs at each time scale by weighted sums of Poisson probability functions. The performance of both inference procedures is very similar. Thus, it is not easy to recommend one of approaches over the other based solely on their associated performances. One argument that clearly favors the hierarchical approach is that the numbers of states of the resulting dMMPPs are smaller than the corresponding numbers for the superposition approach. This may be due to the fact that in the hierarchical approach and as the time scale increases, dMMPPs are fitted to successively smaller sets of intervals whose arrivals characteristics tend to increase in homogeneity and, thus, tend to have associated a smaller number of states than the dMMPP fitted through the superposition approach for the same time scale. However, the contribution of each time scale for the characterization of the aggregate traffic characteristics is interpreted in an easier and more natural way through the superposition approach. Note also that, for the same number of states, a smaller number of dMMPPs and corresponding parameters tend to be needed to compute the final dMMPP using the superposition approach than using the hierarchical approach.

6 Joint Characterization of Packet Arrivals and Packet Sizes - dBMAP

The dBMAP jointly characterizes the packet arrival process and the packet size distribution, being able to achieve an accurate prediction of the queuing behavior for IP traffic exhibiting LRD behavior. In this process, that was firstly presented in [30], packet arrivals occur according to a dMMPP (that can be any one of the previously described models) and each arrival is further characterized by a packet size with a general distribution that may depend on the phase of the dMMPP (Figure 22). This construction process allows having a packet size distribution closely related to the packet arrival process, and is in contrast with the approach followed by [22] where the packet size distribution is fitted prior to the matching of the packet arrival rates.

Lets consider that the packets have independent sizes, with the size of packets arriving in phase i having probability function $q_i = \{q_i(n), n \in \mathbb{N}\}$. If we let (X, J)

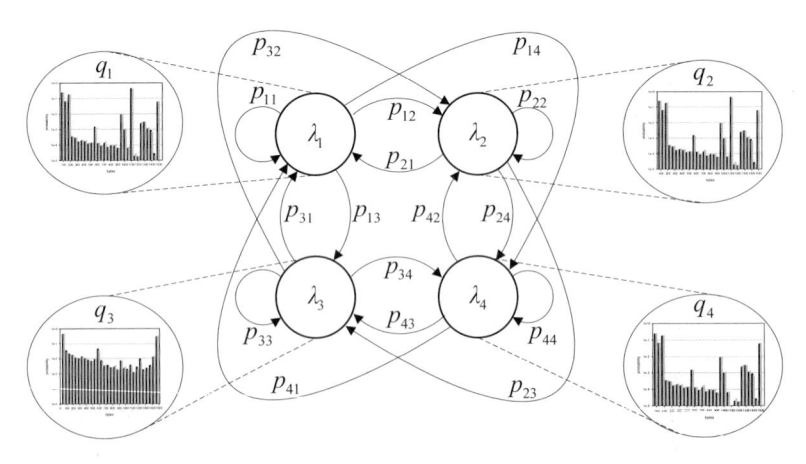

Fig. 22. Construction methodology of the BMAP model

denote the dMMPP, on the state space $I\!N_0 \times S$ and having parametrization $(\mathbf{P}, \Lambda)$, that models the packet arrival process, then the byte arrival process (Y, J) is a dBMAP, on the state space $I\!N_0 \times S$, satisfying equation 4 with

$$q_{ij}(n) = \sum_{l=0}^{+\infty} e^{-\lambda_j} \frac{\lambda_j^l}{l!} q_j^{(l)}(n) \qquad (44)$$

for $i, j \in S$ and $n \in I\!N_0$, where $q_j^{(l)}$ denotes de convolution of order l of q_j. Thus, (Y, J) is a dBMAP on the state space $I\!N_0 \times S$, such that, for $n, m \in I\!N_0$,

$$P(Y_{k+1} = m+n, J_{k+1} = j | Y_k = m, J_k = i) = p_{ij} \sum_{l=0}^{+\infty} e^{-\lambda_j} \frac{\lambda_j^l}{l!} q_j^{(l)}(n) \qquad (45)$$

which we express by saying that (Y, J) has *type-II parametrization* $(\mathbf{P}, \Lambda, \{q_i, i \in S\})$. S is the phase set of the (Y, J) dBMAP.

The packet size characterization is carried out in an independent way for each state of the inferred dMMPP and involves two steps: (i) association of each time slot to one of the dMMPP states and (ii) inference of a packet size distribution for each state of the dMMPP. In the first step, we scan all time slots of the empirical data. A time slot in which k packet arrivals were observed is randomly assigned to a state, according to the probability vector $\boldsymbol{\theta}(k) = \{\theta_1(k), \ldots, \theta_{N_B}(k)\}$, where $\theta_i(k)$ represents the probability that the observed k packet arrivals were originated in state i and N_B is the number of states of the dMMPP. This is given by

$$\theta_i(k) = \frac{\pi_i g_{\lambda_i}(k)}{\sum_{j=1}^{N_B} \pi_j g_{\lambda_j}(k)} \qquad (46)$$

where λ_j represents the Poisson packet arrival rate of the dMMPP and π_j the corresponding steady-state probability (as stated before, $g_\lambda(y)$ represents a Poisson probability distribution function with mean λ).

The inference of the packet size distribution in each state resorts to histograms. The inference of each histogram uses only the packets that arrived during the time slots previously associated with the state for which we are inferring the packet size distribution. Note that some low-probability states may have no packets associated with them, making impossible the packet characterization specifically for these states. We associate a packet size distribution to these states that considers all data packets, i.e., the packet size distribution unconditioned on the dMMPP states. The histograms result in the packet size distributions $q_i = \{q_i(n), n \in I\!N\}$, for $i = 1, 2, \ldots, N_B$.

6.1 Efficiency Results

Reference [30] evaluated the efficiency of a dBMAP where the packet arrival process was modelled using the M2^L-dMMPP described in section 4 and the packet size process was modelled using the procedure described in section 6. The UA trace was also used to assess the efficiency of this traffic model, so the results of applying the M2^L-dMMPP

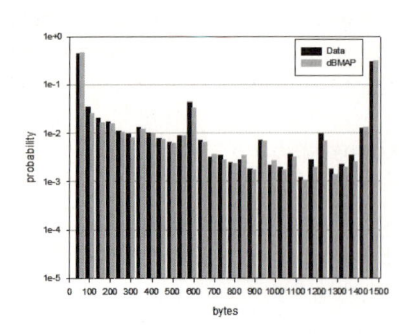

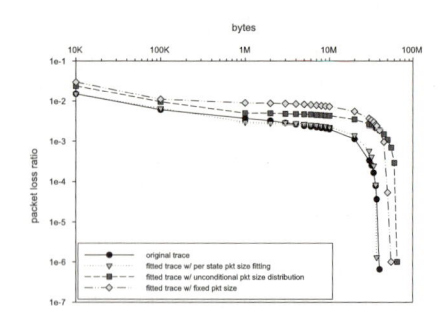

Fig. 23. Packet size distribution, UA

Fig. 24. Packet loss ratio versus buffer size, UA

fitting procedure to the packet arrival process were already presented in section 4. The packet size distribution is essentially bimodal with two pronounced peaks around 40 and 1500 bytes, presenting also non negligible values at 576 and 885 bytes. There was an excellent agreement between the original and fitted packet size distributions (Figure 23), leading to a good match between the original and fitted distributions of the bytes/s processes.

For the dBMAP, four types of input traffic are considered in the trace-driven simulation: (i) the original trace, (ii) a trace generated according to the fitted dBMAP, (iii) a trace where the arrival instants were generated according to the fitted dMMPP arrival process and the packet size according to the unconditional packet size distribution of the fitted dBMAP and (iv) a trace where the arrival instants were also generated according to the fitted dMMPP arrival process but the packet size is fixed and equal to the average packet size of the original trace. In order to analyze queuing behavior, we considered a queue with a service rate of 700 Kbytes/s, corresponding to a link utilization of $\rho = 0.90$, and varied the buffer size from 10 Kbytes to 60 Mbytes. As it can be observed in Figure 24, there is a close agreement between the curves corresponding to the original trace and to the trace generated according to the fitted 12-dBMAP, for all buffer size values. In contrast, for the other two curves corresponding to traces where the packet size is fitted independently of the packet arrival process, significant deviations are obtained. Thus, detailed modeling of the packet size and of the correlations with the packet arrivals is clearly required.

7 Conclusion

Accurate modeling of certain types of IP traffic involves the description of the packet arrival process and the packet size distribution. This tutorial discussed the suitability of Markovian models to describe traffic that exhibits self-similarity and long range dependence behaviours. Three traffic models, based on MMPPs, were designed to describe the packet arrival process by capturing the self-similar behavior over multiple time scales: the first model is based on a parameter fitting procedure that matches both the autocovariance and marginal distribution of the counting process and the

MMPP is constructed as a superposition of L two-state MMPPs, designed to match the autocovariance function, and one M-MMPP designed to match the marginal distribution. The second model is a superposition of MMPPs, where each MMPP describes a different time scale of the packet arrival process. The third model is obtained as the equivalent to an hierarchical construction process that, starting at the coarsest time scale, successively decomposes MMPP states into new MMPPs to incorporate the characteristics offered by finner time scales. For all three traffic models, the number of states is not fixed *a priori* but is determined as part of the inference procedure. In order to closely match not only the packet arrival process but also the packet size distribution a dBMAP was also presented and discussed: packet arrivals occur according to a dMMPP and each arrival is further characterized by a packet size with a general distribution that may depend on the phase of the dMMPP. This allows having a packet size distribution closely related to the packet arrival process. The accuracy of the proposed models was evaluated by comparing the probability mass function at each time scale, as well as the packet loss ratio corresponding to measured traces and to traces synthesized according to the proposed models. The accuracy analysis was based on traffic traces exhibiting LRD and self-similar behaviors.

References

1. Leland, W., Taqqu, M., Willinger, W., Wilson, D.: On the self-similar nature of Ethernet traffic (extended version). IEEE/ACM Transactions on Networking 2(1), 1–15 (1994)
2. Beran, J., Sherman, R., Taqqu, M., Willinger, W.: Long-range dependence in variable-bit rate video traffic. IEEE Transactions on Communications 43(2/3/4), 1566–1579 (1995)
3. Crovella, M., Bestavros, A.: Self-similarity in World Wide Web traffic: Evidence and possible causes. IEEE/ACM Transactions on Networking 5(6), 835–846 (1997)
4. Paxson, V., Floyd, S.: Wide-area traffic: The failure of Poisson modeling. IEEE/ACM Transactions on Networking 3(3), 226–244 (1995)
5. Ryu, B., Elwalid, A.: The importance of long-range dependence of VBR video traffic in ATM traffic engineering: Myths and realities. ACM Computer Communication Review 26, 3–14 (1996)
6. Grossglauser, M., Bolot, J.C.: On the relevance of long-range dependence in network traffic. IEEE/ACM Transactions on Networking 7(5), 629–640 (1999)
7. Nogueira, A., Valadas, R.: Analyzing the relevant time scales in a network of queues. In: Proceedings of SPIE's International Symposium ITCOM 2001 (August 2001)
8. Heyman, D., Lakshman, T.: What are the implications of long range dependence for VBR video traffic engineering? IEEE/ACM Transactions on Networking 4(3), 301–317 (1996)
9. Neidhardt, A., Wang, J.: The concept of relevant time scales and its application to queuing analysis of self-similar traffic. In: Proceedings of SIGMETRICS 1998/PERFORMANCE 1998, pp. 222–232 (1998)
10. Yoshihara, T., Kasahara, S., Takahashi, Y.: Practical time-scale fitting of self-similar traffic with Markov-modulated Poisson process. Telecommunication Systems 17(1-2), 185–211 (2001)
11. Salvador, P., Valadas, R.: Framework based on markov modulated poisson processes for modeling traffic with long-range dependence. In: van der Mei, R.D., de Bucs, F.H.S. (eds.) Internet Performance and Control of Network Systems II, August 2001. Proceedings SPIE, vol. 4523, pp. 221–232 (2001)

12. Salvador, P., Valadas, R.: A fitting procedure for Markov modulated Poisson processes with an adaptive number of states. In: Proceedings of the 9th IFIP Working Conference on Performance Modelling and Evaluation of ATM & IP Networks (June 2001)
13. Andersen, A., Nielsen, B.: A Markovian approach for modeling packet traffic with long-range dependence. IEEE Journal on Selected Areas in Communications 16(5), 719–732 (1998)
14. Hajek, B., He, L.: On variations of queue response for inputs with the same mean and autocorrelation function. IEEE/ACM Transactions on Networking 6(5), 588–598 (1998)
15. Feldmann, A., Gilbert, A., Willinger, W.: Data networks as cascades: Investigating the multifractal nature of internet WAN traffic. In: Proceedings of SIGCOMM, pp. 42–55 (1998)
16. Feldmann, A., Gilbert, A.C., Huang, P., Willinger, W.: Dynamics of IP traffic: A study of the role of variability and the impact of control. In: SIGCOMM, pp. 301–313 (1999)
17. Riedi, R., Véhel, J.: Multifractal properties of TCP traffic: a numerical study. Technical Report No 3129, INRIA Rocquencourt, France (February 1997), www.dsp.rice.edu/~riedi
18. Salvador, P., Valadas, R., Pacheco, A.: Multiscale fitting procedure using Markov modulated Poisson processes. Telecommunications Systems 23(1-2), 123–148 (2003)
19. Nogueira, A., Salvador, P., Valadas, R., Pacheco, A.: Fitting self-similar traffic by a superposition of mmpps modeling the distribution at multiple time scales. IEICE Transactions on Communications E84-B(8), 2134–2141 (2003)
20. Nogueira, A., Salvador, P., Valadas, R., Pacheco, A.: Modeling self-similar traffic through markov modulated poisson processes over multiple time scales. In: Proceedings of the 6th IEEE International Conference on High Speed Networks and Multimedia Communications (July 2003)
21. Nogueira, A., Salvador, P., Valadas, R., Pacheco, A.: Hierarchical approach based on mmpps for modeling self-similar traffic over multiple time scales. In: Proceedings of the First International Working Conference on Performance Modeling and Evaluation of Heterogeneuous Networks (HET-NETs 2003) (July 2003)
22. Klemm, A., Lindemann, C., Lohmann, M.: Traffic modeling of IP networks using the batch Markovian arrival process. Performance Evaluation 54(2), 149–173 (2003)
23. Gao, J., Rubin, I.: Multifractal analysis and modeling of long-range-dependent traffic. In: Proceedings of International Conference on Communications ICC 1999, June 1999, pp. 382–386 (1999)
24. Lucantoni, D.M.: New results on the single server queue with a batch Markovian arrival process. Stochastic Models 7(1), 1–46 (1991)
25. Lucantoni, D.M.: The BMAP/G/1 queue: A tutorial. In: Donatiello, L., Nelson, R. (eds.) Models and Techniques for Performance Evaluation of Computer and Communication Systems, pp. 330–358. Springer, Heidelberg (1993)
26. Pacheco, A., Prabhu, N.U.: Markov-additive processes of arrivals. In: Dshalalow, J.H. (ed.) Advances in Queueing: Theory and Methods, ch. 6, pp. 167–194. CRC, Boca Raton (1995)
27. Veitch, D., Abry, P.: A wavelet based joint estimator for the parameters of LRD. IEEE Transactions on Information Theory 45(3) (April 1999)
28. Feldmann, A., Whitt, W.: Fitting mixtures of exponentials to long-tail distributions to analyze network performance models. Performance Evaluation 31(3-4), 245–279 (1997)
29. Osborne, M., Smyth, G.: A modified prony algorithm for fitting sums of exponential functions. SIAM J. Sci. Statist. Comput. 16, 119–138 (1995)
30. Salvador, P., Pacheco, A., Valadas, R.: Modeling IP traffic: Joint characterization of packet arrivals and packet sizes using BMAPs. Computer Networks Journal 44, 335–352 (2004)

Multi-timescale Economics-Driven Traffic Management in MPLS Networks

Paola Iovanna[1], Maurizio Naldi[2], Roberto Sabella[1], and Cristiano Zema[3]

[1] Ericsson Telecomunicazioni S.p.a.
Via Moruzzi 1, 56124 Pisa, Italy
{paola.iovanna,roberto.sabella}@ericsson.com
[2] Dipartimento di Informatica, Sistemi e Produzione
Università di Roma "Tor Vergata"
Via del Politecnico 1, 00133 Rome, Italy
naldi@disp.uniroma2.it
[3] CoRiTeL c/o Ericsson Telecomunicazioni S.p.a.
Via Anagnina 203 00118 Roma, Italy
cristiano.zema@ericsson.com

Abstract. Today's networking environment is characterized by significant traffic variability and squeezing profit margins. An adaptive and economics-aware traffic management approach is needed to cope with such environment. An adaptive traffic management system is proposed that acts on short timescales (from minutes to hours) and employs an economics-based figure of merit to rellocate bandwidth. The tool works in an MPLS context. Both underload and overload deviations from the optimal bandwidth allocation are sanctioned through the economical evaluation of the consequences of such non-optimality. A description of the traffic management system is provided together with some simulation results to show its operations.

1 Introduction

Traffic on the Internet is more and more subject to extensive variability, which is reflected both in its patterns and in its statistical characteristics. This is due to the variety of services supported by the TCP/IP suite and to the appearance of new consumer styles that accompany those services. While the telephone network (relying on a circuit-switched infrastructure) essentially provided a single service, i.e. the conversational voice service, new services appear now and again on the Internet (of which the most disruptive, as to sheer traffic volume, is the peer-to-peer file exchange, a.k.a. P2P [1] [2]). For example, the Internet is now used to transfer larger and larger files (e.g. movies), with the ensuing hours-long transfer sessions, as well as to enable engaging interactive activities (e.g. online games, or jam sessions). According to established classifications, the variety of the traffic streams on the Internet can be characterized either by their nature or by their size or by their lifetime. In the first domain we may have streaming traffic, which is characterized by bandwidth and whose support is driven by real-time requirements, and elastic traffic, which is instead characterized by the file volume and whose support is driven by integral file transfer requirements. As to the stream size an established terminology considers *mice* and *elephant* streams, where the mice are small transfers (e.g. downloading a simple Web page) and the elephants are the large ones (e.g. downloading a video file). In addition to these two dimensions we may consider the stream lifetime with *dragonflies* streams lasting less than 2 seconds and *tortoise* streams lasting more than 15 minutes [3]. The presence of traffic streams living at various timescales coupled with certain characteristics of the TCP control protocol is also deemed responsible for the radical change in the statistical characteristics of traffic streams, namely the presence of long-range dependence [4] [5] [6]. In addition, traffic patterns have also changed, for a number of factors, among which:

- Mobile services have extended the range of time usable for communications purposes;
- Asynchronous services (e.g. e-mail) or downloading service (e.g. the Web) don't require the presence of two parties;
- Downloading services (e.g. P2P) don't require the presence of humans if not to trigger the communication session, and can give rise to very long traffic exchanges.

As a result hourly traffic profiles are less and less predictable, and are often flatter than in the past, so that the concept of peak hour, traditionally used in dimensioning procedures, is fading (as shown in [3] or [7] heavy downloading and back-up services typically have their peak in the night).

Modern networks must be able to cope with such traffic variability: they must be adaptable. In turn that basically means that traffic management solutions should be dynamic and rely on online traffic monitoring. Cognitive packet networks (CPN) can be considered as a pioneer example of self-aware networks [8], in that they adaptively select paths so as to offer a best-effort QoS to the end-users. That concept has been further advanced in the proposition of self-adaptive networks, where a wider set of QoS requirements (including strict QoS guarantees) is satisfied by the introduction of a traffic management system acting on two timescales within an MPLS infrastructure [9]. As in the established approach to QoS, constraints are imposed on a number of parameters, such as blocking probability for services offered over a connection-oriented network and packet-loss, average delay and jitter for the services provided by connectionless networks [10] [11].

However, network design and management procedures can't be based on QoS considerations alone, since the economical issue is of paramount importance and is the ultimate goal of the activities of any company. The quality of service delivered to the customers is itself evaluated in economical terms, since the QoS constraints are typically embodied in a Service Level Agreement (SLA), where precise QoS obligations are taken by the service provider and an economical value is associated to those obligations, under the form of penalties or compensations. SLA's are now the established way to incorporate QoS guarantees in the provisioning of communications services, e.g. in leasing of transmission capacity

[7], Internet services spanning multiple domains [12], MPLS-based VPNs [13], or wireless access [14].

In addition, it is to be considered that QoS constraints could be easily met by extensive overprovisioning, though this solution could make network operations unaffordable in the long run. Even if the practice of overprovisioning is limited in extent, the amount of bandwidth that is currently unused and unnecessarily left to the customer's disposal could be assigned otherwise, providing additional revenues: its less than careful management represents therefore an opportunity cost and a source of potential economical losses.

An efficient traffic management system should implement a trade-off between the contrasting goals of delivering the required QoS (driving towards overprovisioning) and exploiting the available bandwidth as much as possible (driving towards underprovisioning). Deviations in either way are amenable to an economical evaluation, so that traffic management economics appears as the natural common framework to manage network operations.

In this paper we propose a novel engine for the traffic management system envisaged for self-adaptive networks in [9], using economics as the single driver, so to cater both for QoS violations and for bandwidth wastage. In the new formulation the traffic management system is driven by a newly defined cost function, which accounts for both overprovisioning and underprovisioning occurrences, and practical suggestions are provided to link the parameters of such cost function to relevant economical parameters associated to network operations. The new traffic management system is described in Section 2, while its two major components, i.e., the forecasting blocks and the cost computation blocks are described in Sections 3 and 4 respectively. In Section 6 we report the results of extensive simulations to show its behaviour for a complete set of network services under different traffic patterns.

2   The Traffic Management System: Overview

We consider a traffic management system acting in an MPLS context, where the traffic is channelled on LSPs (Label Switched Path), in turn accomodated on traffic tunnels (though there is typically a one-to-one association between LSPs and traffic tunnels, as we assume in the following). The main goal of MPLS traffic engineering is the correct allocation of bandwidth to LSPs so to achieve an effective use of the network resources. For this purpose we resume the proposal of a traffic management system acting on two timescales put forward in [9]. In this section we describe in detail the system.

A schematic diagram of the traffic management system is reported in Fig. 1, with the components defined in Table 1. The system is composed of two macroblocks, representing respectively the functions intervening for short term operations (the Short Term Management Subsystem, or STMS, for short) and for long term ones (the Long Term Management Subsystem, LTMS). In addition, we use two blocks (blocks A and E in Fig. 1), that are common to both kinds of operations. Block A is responsible for collecting traffic data on both transmission links and LSPs. These data are then fed to the forecasting engines on the

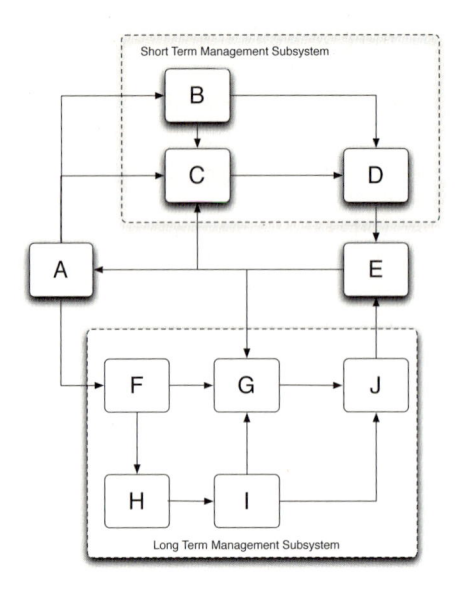

Fig. 1. Traffic Management System

Table 1. Composition of the traffic management system

Block	Function
A	Traffic measurement
B	Traffic nowcasting engine
C	Cost Computation
D	LSP Adjustment
E	Network structure infobase
F	Long term Traffic Forecasting
G	Long term Cost Computation and Comparison
H	Traffic Matrix Estimation
I	Global Path Design
J	Long term LSP Adjustment

two timescales (respectively blocks B and F). Block E is instead responsible for keeping the overall network picture up-to-date, i.e., the network topology, the transmission capacity of each transmission links, the set of active LSPs, and the bandwidth allocated to each LSP. This information is updated on the basis of the decisions taken by the two management subsystems (namely, blocks D and J), and is then supplied to the traffic measurement block to drive the measurement process (i.e., to indicate what are the network entities - transmission links and LSPs - for which traffic data are to be collected).

The STMS (Short Term Management Subsystem) relies on the the traffic measurements block, which monitors each traffic tunnel and uses its output to forecast the evolution of traffic for the next time interval (the domain of the SMTS is on timescales of the order of magnitude of hours, so that the forecasting engine is more aptly named nowcasting). The nowcasting engine (block B in Fig. 1) employs the Exponential Smoothing technique in the versions proposed and analysed in [15] to build a time series of traffic. This time series is in turn fed as an input to the cost computation block (block C in Fig. 1), which evaluates the cost associated to the current combination of traffic and allocated capacity. In order to do so, that block has to receive information on the bandwidth currently allocated to each LSP (provided by the network infobase block), so to be able to compare allocation and occupation. Rather than minimizing deviations from the QoS objectives (which is the common approach to bandwidth management, e.g., the one also adopted in [15]), in the STMS here proposed bandwidth allocation is instead driven by the willingness to maximize the provider's revenues. This is accomplished by taking into account the economics of bandwidth use, and is described in detail in Section 4. The cost computation block gathers information both on the current occupation state (provided by block A) and on the future use (provided by block B), since this allows to evaluate the trend of costs. On the basis of the trend observed for the cost the STMS may take some correcting actions, e.g., the following ones:

- Modification of LSP attributes (e.g., their bandwidth);
- Rerouting of LSPs;
- Termination of LSPs, in particular of the lower priority ones (pre-emption);
- Dynamic routing of new unprecedented requests.

This action are decided in block D, whose actual decision criteria and scope of intervention may be left to the operator and are not dealt with in detail in this paper. A possible strategy could be to limit short time scale actions to LSP bandwidth adjustments, leaving LSP termination and re-routing to long term management.

While the STMS leads to small changes in LSPs, the aim of the Long Term Management Subsystem is to assess if the traffic picture is so distant from that adopted during the network design process to warrant design a new routing plan and a new set of LSPs (including in the latter term also the simple rearrangement of the existing flows on the current set of LSPs). The decision to go for a radical change in the network structure is taken on the basis of the comparison between the costs associated to the current set of LSPs and those incurred if the set of LSPs is redesigned (with a hysteresis allowance to cater for switching costs and avoid too frequent redesign operations). In LTMS the traffic measurement are fed to a forecasting block (block F in Fig. 1), which again adopts the Exponential Smoothing technique but with larger smoothing factors. The output of the forecasting block gives us the future occupation of the current set of LSPs. In order to build the alternative set of LSPs, as deriving from the complete redesign, the future traffic matrix has to be estimated from the measurements on LSPs and links. The resulting traffic matrix is fed to the design engine (embodied in the Global Path Design block, indicated as block I in Fig. 1). We can now compare the two scenarios:

1. Scenario A, represented by future origin-destination traffic flowing on the current set of LSPs;

2. Scenario B, represented by future origin-destination traffic flowing on the set of LSPs indicated by the Global Path Design block.

The cost computation and comparison block (block G in Fig. 1) receives the sets of LSPs and the pertaining occupation level in both scenarios and can compute the costs pertaining to the two scenario. The resulting comparison is fed to the decision block J, which has to decide whether to stay with the current set of LSPs or proceed with the redesign. Again, the actual decision criteria may be left to the network operator.

3 Traffic Measurement and Forecasting

Any traffic management decision has to be driven first by traffic data. For this purpose our system includes a traffic measurement subsystem (labelled as Block A in Fig. 1), which in turn feeds two traffic prediction subsystems, respectively on short timescales (named nowcasting) and on longer timscales (labelled as blocks B and F in the same picture). Prediction is needed to match the timeframes of traffic data and of the intervention of the traffic management system: the decisions taken by LSP adjustment blocks are accomplished in the future (though near in the case of the STMS), i.e., when the traffic has changed with respect to present. In this section we review the characteristics of the measurement and prediction subsystems.

The aim of the traffic measurement subsystem is to provide the traffic data to feed the nowcasting algorithm. Such measurements are conducted on each LSP (and on each transmission link) currently set up in the network. Namely for each LSP a counter is defined that measures the cumulative number of bytes being transferred on that LSP during a given period of time. Typically we can consider period of 5 minutes (in agreement with the time resolution of measurements provided by SNMP-based devices); at the end of each period the byte count is transferred to the nowcasting block and the counter is reset. The byte count can be divided by the period length to obtain the average bandwidth employed during that period. The choice of the period duration can be chosen as the result of a trade-off between readiness of reaction (by reducing that duration under 5 minutes, down e.g. to 60 or 30 seconds) and accuracy of measurement and of the subsequent forecasting (each measurement represents in itself an estimation of the average bandwith of the underlying traffic stochastic process).

The traffic nowcasting subsystem subsystem gets the latest traffic measurements from block A and provides a forecast for the next time interval. Two forecasting methods are considered, both based on the Exponential Smoothing (ES) approach:

1. ES with linear extrapolation (ESLE);
2. ES with predicted increments (ESPI).

Both methods are not new, having been proposed and analysed in [15]. We now proceed to describe them. In the following we indicate by M_j the traffic measurement performed at time j and by F_j the traffic forecast for the same time.

In both methods the classic Exponential Smoothing recursive formula is adopted unless when both underestimation ($F_j < M_j$) and a growing trend ($M_j > M_{j-1}$) are observed at the same time. In that case different forecasting algorithms are used in the two methods. A complete definition of the two methods follows.

ESLE method. If both underestimation and a growing trend take place the forecast is equal to the latest measurement (M_j) plus the latest measured increase ($M_j - M_{j-1}$). The complete algorithm reads therefore as follows:

Algorithm 1. (ESLE)
---
if $M_j > M_{j-1}$ AND $F_j < M_j$ then
$\quad F_{j+1} = 2M_j - M_{j-1}$
else
$\quad F_{j+1} = \alpha F_j + (1-\alpha)M_j$
end if

ESPI method. In the predicted increments method, when both underestimation and a growing trend take place the forecast is equal to the latest forecast plus a specified increment. This increment is equal to: a) a fixed fraction of the latest measured increment $z > \alpha(M_j - M_{j-1})$ on the first interval the mentioned conditions apply; b) the estimated increase Δ_{j+1} on following time intervals as long as those conditions apply. In case b) the estimate of the increase is obtained by a parallel basic ES approach, i.e. $\Delta_{j+1} = \alpha \Delta_j + (1-\alpha)(M_j - M_{j-1})$. The complete algorithm reads therefore as follows: For the purpose of estimating traffic on longer horizon we can rely as well on the exponential smoothing techniques presented so far. We have, however, to smooth out the short term fluctuation we may instead be interested when acting on shorter timescales. For this purpose we can follow either of two approaches:

1. Aggregate then Forecast (AF);
2. Ultra-smoothing (US).

In the former case we abandon the short time window (e.g., 5 minutes) adopted in the STMS and consider a larger time window, e.g. one day. We aggregate then the traffic measurements collected during each day in a single reference value for the whole day. Aggregating over a whole day allows us to remove the short term fluctuations. The daily reference values represent the input for the long term forecast system, where the smoothing factor α can take values in the same range as in the nowcasting use. This approach is that adopted, e.g., in the ITU-T Recommendation E.500 [16].

In the latter approach the traffic data are fed to the forecasting engine with the same granularity adopted in the nowcasting case, but the smoothing factor is much larger. Though its optimal value can be determined empirically, e.g., by a least square fitting with respect to an observed time series track, it can be guessed that its value may be even larger than 0.95.

Algorithm 2. (ESPI)

if $M_j > M_{j-1}$ AND $F_j < M_j$ then
   if $M_{j-1} < M_{j-2}$ OR $F_{j-1} > M_{j-1}$ then
      $F_{j+1} = M_j + z$
   else
      $\Delta_{j+1} = \alpha \Delta_j + (1-\alpha)(M_j - M_{j-1})$
      $F_{j+1} = F_j + \Delta_{j+1}$
   end if
else
   $F_{j+1} = \alpha F_j + (1-\alpha) M_j$
end if

4 An Economic Figure of Merit

In order to manage traffic properly we have to know the current state of traffic management (i.e., its value and the bandwidth allocation) and a measure of adequacy of bandwidth allocation. In the past the latter was chosen so to achieve specific targets on QoS, embodied by bounds on loss and delay figures [17]. However, offering QoS is not a goal in itself, but rather a means to conduct a rewarding business. The offer of differentiated QoS is since long a reality and is associated to differentiated prices. In a QoS-based approach the measure of adequacy is typically the efficiency in the usage of transmission resources subject to constraints on the QoS achieved. However, such approach may be unrelated to the overall economic goal of the provider, since it fails to consider the economic figures associated to the usage of bandwidth. In fact, the cost is in that case associated to the capital cost incurred in building the transmission infrastructure, that has to be used as much as possible; no care is taken for the costs associated to alternative uses of the same bandwidth. We need to introduce a measure of adequacy capable of taking into account a wider view of costs associated to bandwidth allocation decisions. In this section we propose a new figure of merit for traffic management, that takes into account the monetary value of bandwidth allocation decisions.

An improper bandwidth allocation may impact on the provider's economics (i.e., lower revenues or highers costs) in basically two opposite ways. If the LSP is overused, congestion takes place, leading to failed delivery of packets and possible SLA (Service Level Agreement) violations. On the other hand, when the LSP is underused (by allocating too much bandwidth to a given user) chunks of bandwidth are wasted that could be sold to other users (and the provider incurs an opportunity cost). Common approaches to bandwidth management either focus on just the first issue, overlooking bandwidth waste, or anyway lack to provide an economics-related metric valid for both phenomena. A first attempt to take into account both phenomena has been made by Tran and Ziegler [15] through the introduction of the Goodness Factor (GF). In the GF definition the relevant parameter is the load factor X on the transmission link (the LSP in our case), i.e., the ratio between the expected traffic and the allocated bandwidth.

The optimal value for such parameter, i.e., the maximum value that meets QoS constraints, is X_{opt}. The GF is then defined as

$$GF = \begin{cases} X/X_{opt} & \text{if } X < X_{opt} \\ X_{opt}/X & \text{if } X_{opt} \leq X < 1 \\ (1/X - 1)/X_{opt} & \text{if } X \geq 1 \end{cases} \quad (1)$$

The curve showing the relationship between the GF and the load factor is shown in Fig. 2 (dotted curve) when the optimal load factor is 0.7. It can be seen that over- and under-utilization are associated to different signs and can therefore be distinguished from each other. The value of the GF associated to the optimal situation is 1, so that less-than-optimal situations are marked by deviations of the GF from 1. The most remarkable pro of the GF is that it takes into account both underloading and overloading. However, it fails to put them on a common scale, since it doesn't take into account the relative monetary losses associated to the two kinds of phenomena: the worst case due to under-utilization bears $GF = 0$, while the worst case due to over-utilization leads to the asymptotic value $GF = -1/X_{opt}$. In addition, the Goodness Factor function as defined by expr. 1 is discontinuous when going to severe congestion ($X > 1$). In addition to the economic figure of merit we describe in the following, we have also developed a continuous version of the Goodness Factor, where the function behaviour when the load factor falls in the $X_{opt} \leq X \leq 1$ range is described by a quadratic function; the modified version of the Goodness Factor is given by expr. (2) and shown in Fig. 2 (solid curve). This modified version of the GF will be used for the simulation analysis reported in Section 6.

$$GF = \begin{cases} X/X_{opt} & \text{if } X < X_{opt} \\ 1 - \left(\frac{X - X_{opt}}{1 - X_{opt}}\right)^2 & \text{if } X_{opt} \leq X < 1 \\ (1/X - 1)/X_{opt} & \text{if } X \geq 1 \end{cases} \quad (2)$$

The resulting GF value, as measured during the monitoring period, changes more smoothly than what would appear from Fig. 2. In Fig. 3 we report the observed GF (in the original Tran-Ziegler formulation) in a simulation where the load factor X follows a Gaussian distribution with a standard deviation equal to 0.1 ($X_{opt} = 0.6$ in this instance). As can be seen the transition to negative values is quite gradual and takes place when the load factor is 110%.

In our approach we introduce a cost function whose value depends on the current level of LSP utilization, putting on a common ground both under- and over-utilization. The minimum of the cost function is set by default to zero when the LSP utilization is equal to a predefined optimal level, set according to QoS requirements. As we deviate from the optimal utilization level the cost function grows. The exact shape of the function can be defined by the provider, since it depends on its commercial commitments. However we can set some general principles and provide a simple instance. If a SLA is violated due to insufficient bandwidth allocation, the provider faces a cost due to the penalty defined in the SLA itself. On the other hand an opportunity cost may be associated to the

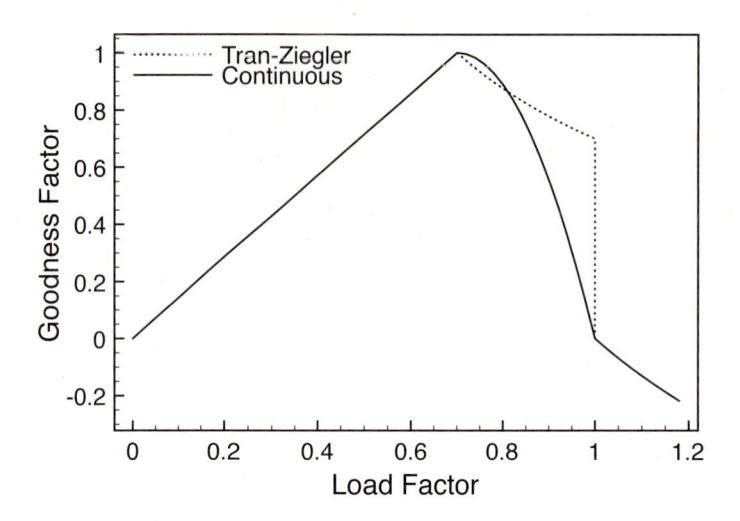

Fig. 2. Goodness Factor

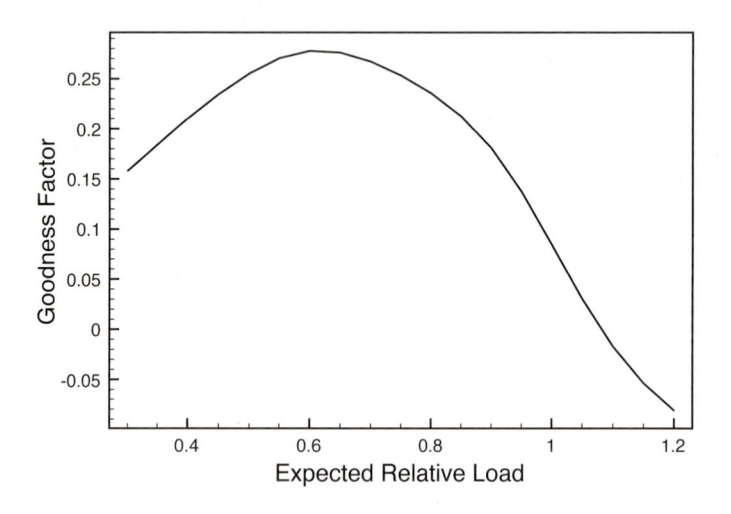

Fig. 3. Observed Goodness Factor

bandwidth unused on an LSP; the exact value of the cost may be obtained by considering, e.g., the market price of leased lines. A very simple example of the resulting cost function is shown in Fig. 4. The under-utilization portion takes into account that leased bandwidth is typically sold in chunks (hence the function is piecewise constant), e.g., we can consider the typical steps of 64 kbit/s, 2 Mbit/s, 34 Mbit/s, and so on. The over-utilization portion instead follows a logistic curve, that asymptotically leads to the complete violation of all SLAs acting on that LSP, and therefore to the payment of all the associated penalties.

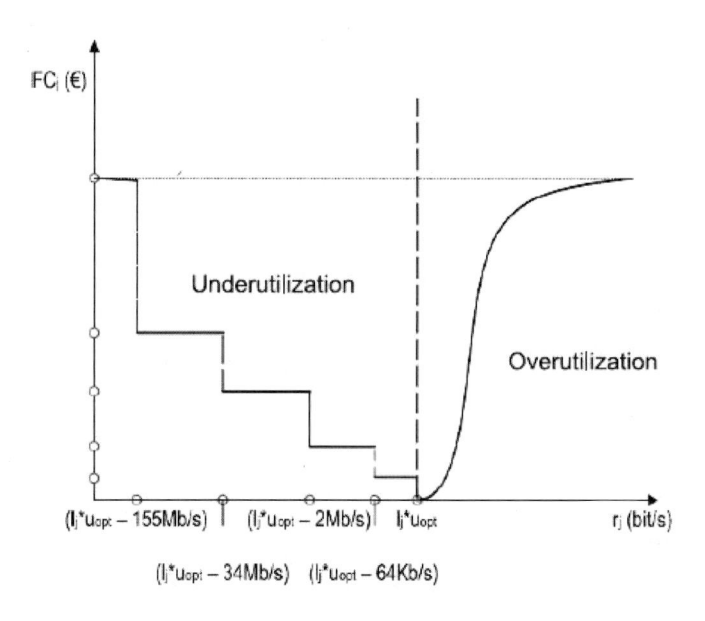

Fig. 4. Cost function of STMS

5 Long Term Traffic Generation

In order to test the capabilities of the traffic management system we have to adopt a set of models to simulate the traffic flowing on the network. Since we act on two timescales we need models capable of coping with the nonstationarity occurring on such long timescales. We have opted for a separable model, where the average value is supposed to vary during the day and modulates the stochastic models adopted for shorter timescales. In this section we focus on the model adopted for the variation of the average traffic intensity along the day and over a set of days.

We consider each day to be subdivided into a number N of intervals: for example, we could consider a subdivision into 15 minutes intervals (so to have $N = 60$). In a very simple fashion, we assume the day-to-day variation to be linear, while there is an underlying intra-day variation. The latter is modelled through the subdivision of the day into three hourly ranges:

1. Low traffic range from 0.00 to x.00 hours;
2. High traffic range from x.00 hours to y.00 hours;
3. Low traffic range from y.00 hours to 24.00 hours.

This assumption is justified by the traffic observations reported in Fig. 5 [18]. We could, e.g. assume the second hourly range to start at 10.00 hours and end at 20.00 hours. The overall expression for the average traffic intensity in day i and in the j-th intra-daily period ($j \in \{1, 2, \ldots, N\}$) is

$$X_{ij} = A \cdot (1 + \lambda i) \cdot \beta_j, \tag{3}$$

where

$$\beta_j = \begin{cases} \gamma & j < \frac{x}{24} N \\ \theta & \frac{x}{24} N < j < \frac{y}{24} N \\ \gamma & \frac{y}{24} N < j \leq N \end{cases} \tag{4}$$

The parameter A is the average traffic intensity in the first day of the period. The parameter λ is the variation of the average traffic intensity over a day. For example, if we suppose the traffic to increase by 6% over 30 days, we may set $\lambda = 0.06/30 = 0.02$. As to the parameters γ and θ, they have to meet the constraint due to the average daily value:

$$\frac{\gamma \cdot x + \theta \cdot (y - x) + \gamma \cdot (24 - y)}{24} = 1. \tag{5}$$

If $x = 10$ and $y = 20$, the previous constraint is

$$14\gamma + 10\theta = 24. \tag{6}$$

We can consider also the constraint on θ/γ, e.g., the ratio between the intensities in the high- and low-traffic hourly ranges. A suitable value, after observing Fig. 5, is $\theta/\gamma = 2.5$. We end up with the following pair of equations

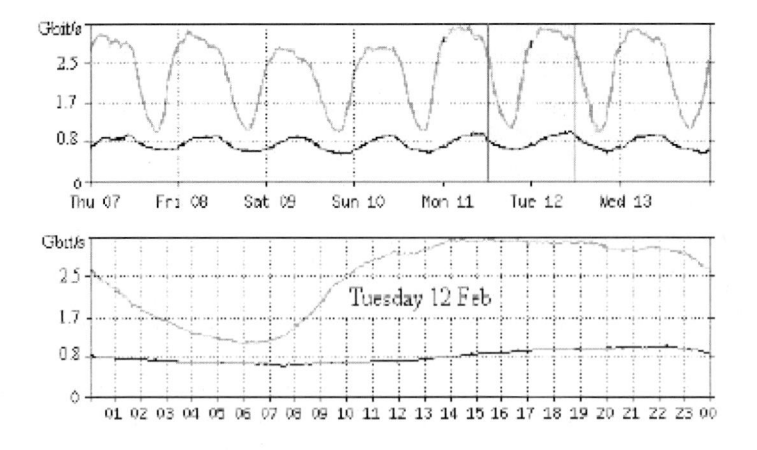

Fig. 5. Traffic profiles on a real network [18]

$$14\gamma + 10\theta = 24,$$
$$\theta = 2.5\gamma, \tag{7}$$

whose solution gives us the values

$$\gamma = 24/39 \sim 0.615,$$
$$\theta = 2.5\gamma \sim 1.538. \tag{8}$$

6 Simulation Analysis

In Section 4 we have introduced the cost function as a new performance metric and have described its qualities that justify the replacement of the Goodness Factor. In this section we show how the two metrics behave in a simulated context. For this purpose we have set up a simulator through the use of the Network Simulator (ns2) [19].

The simulation scenario considers a single LSP on which we have generated traffic over an interval of the overall duration of 6 hours with a sampling window size of 5 minutes. The traffic was a mix resembling the UMTS service composition, including the following services (the figures within parentheses are the percentages on the overall volume):

- Voice (50%);
- SMS (17.7%);
- WAP (10.9%);
- HTTP (7.8%);
- MMS (5.7%);
- Streaming (4.1%);
- E-mail (3.8%).

This traffic mix was simulated at the application layer by employing the most established model for each service as reported in [20].

In this context we have accomplished the following operations;

1. Monitoring the rate;
2. Applying the nowcasting engine;
3. Computing the Goodness Factor and the Cost Function;
4. Readjusting the LSP bandwidth according to the value of the load factor and of the Cost Function.

As to the last issue, the value of the load factor provides the direction to follow in the readjustment of the LSP bandwidth. The optimal load factor was set at 0.82, so that whenever this threshold is exceeded the bandwidth is increased (the reverse action takes place when the load factor falls below 0.82). The value of the Cost Function provides an measure of the adequacy of bandwidth readjustments.

In Fig. 6 the observed rate and the load factor are shown together during the 6 hours interval. Though the rate exhibits significant peaks, the load factor is kept tightly around the optimal value by the bandwidth readjustment operations.

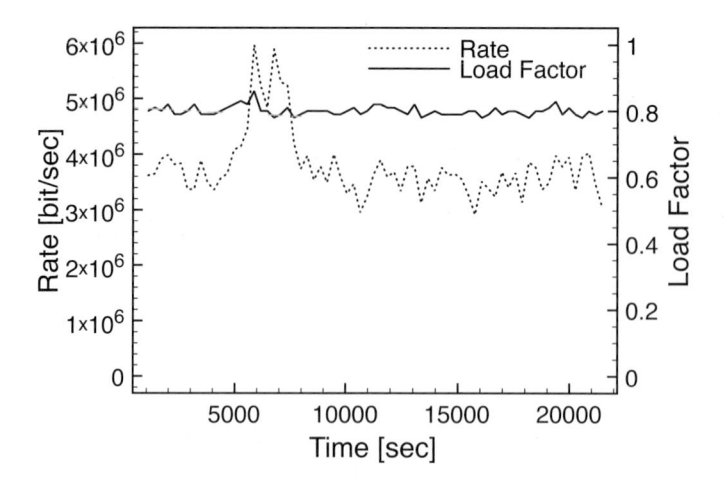

Fig. 6. Load on LSP

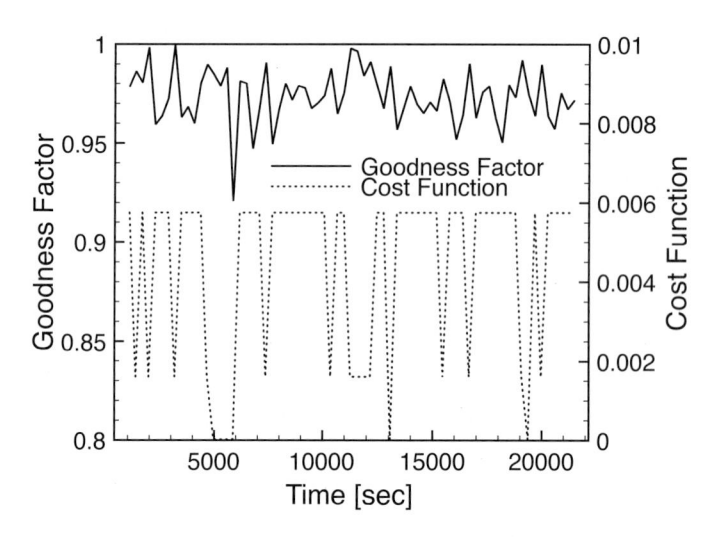

Fig. 7. Performance indicators

The performance indicators are both shown in Fig. 7. Here the optimal value is represented by the null line for both indicators. The line representing the Cost Function exhibits an oscillation between two values since for most of the time the LSP is slightly underloaded due to the continuous bandwidth readjustments, so that the load factor falls in the under-utilization area, where the cost function has a stair-wise appearance. This is due to the granularity by which bandwidth is sold, which may make small changes in the load factor not relevant for the opportunity cost. On the other hand, the continuous changes of the Goodness Factor would induce readjustments when there's nothing to gain by reallocating bandwidth.

7 Conclusions

A traffic management system acting on short timescales and employing an economics-based figure of merit has been introduced to base traffic management on the consequences of bandwidth mis-allocation. Such figure of merit marks the deviations from the optimal allocation due to under- and over-utilization, and improves a previously defined Goodness Factor proposed by Tran and Ziegler. The traffic management system allows to adjust bandwidth allocation so to achieve an economically efficient use of the network resources.

References

1. Liotta, A., Lin, L.: The Operator's Response to P2P Service Demand. IEEE Comm. Mag. 45(7), 76–83 (2007)
2. Sen, S., Wang, J.: Analyzing Peer-To-Peer Traffic Across Large Networks. IEEE/ACM Trans. Networking 12(2), 219–232 (2004)
3. Brownlee, N., Claffy, K.C.: Understanding internet traffic streams: dragonflies and tortoises. IEEE Comm. Magazine 40(10), 110–117 (2002)
4. Karagiannis, T., Molle, M., Faloutsos, M.: Long-range dependence: Ten years of internet traffic modeling. IEEE Internet Computing 8(5), 57–64 (2004)
5. Park, C., Hernández-Campos, F., Marron, J.S., Smith, F.D.: Long-range dependence in a changing internet traffic mix. Comput. Netw. 48(3), 401–422 (2005)
6. Gong, W.B., Liu, Y., Misra, V., Towsley, D.: Self-similarity and long range dependence on the internet: a second look at the evidence, origins and implications. Comput. Netw. 48(3), 377–399 (2005)
7. Jajszczyk, A.: Automatically Switched Optical Netwoks: Benefits and Requirements. IEEE Comm. Mag. 453(72), S10–S15 (2005)
8. Gelembe, E., Lent, R., Nunez, A.: Self-aware networks and QoS. Proc. IEEE 92(9), 1478–1489 (2004)
9. Sabella, R., Iovanna, P.: Self-Adaptation in Next-Generation Internet Networks: How to React to Traffic Changes While Respecting QoS? IEEE Trans. Syst., Man, and Cybernetics - Part B: Cybernetics 36(6), 1218–1229 (2006)
10. Xiao, X., Ni, L.M.: Internet QoS: A Big Picture. IEEE Network 13(2), 8–18 (1999)
11. Giacomazzi, P., Musumeci, L., Saddemi, G., Verticale, G.: Two different approaches for providing qos in the internet backbone. Comp. Comm. 29(18), 3957–3969 (2006)
12. Bhoj, P., Singhal, S., Chutani, S.: SLA management in federated environments. Comp. Netw. 35(1), 5–24 (2001)
13. Ash, J., Chung, L., D'Souza, K., Lai, W.S., Van der Linde, H., Yu, Y.: AT&T's MPLS OAM Architecture, Experience, and Evolution. IEEE Comm. Mag. 42(10), 100–111 (2004)
14. Das, S.K., Lin, H., Chatterjee, M.: An econometric model for resource management in competitive wireless data networks. IEEE Netw. Mag. 18(6), 20–26 (2004)
15. Tran, H.T., Ziegler, T.: Adaptive bandwidth provisioning with explicit respect to QoS requirements. Comp. Comm. 28(16), 1862–1876 (2005)
16. International Telecommunications Union ITU-T. Recommendation E.500 - Traffic intensity measurement principles (1998)
17. Carter, S.F.: Quality of service in BT's MPLS-VPN platform. BT Tech. J. 23(2), 61–72 (2005)

18. Benameur, N., Roberts, J.W.: Traffic Matrix Inference in IP Networks. Netw. Spat. Econ. 4(1), 103–114 (2004)
19. Issariyakul, T., Hossain, E.: Introduction to Network Simulator NS2. Springer, Heidelberg (2009)
20. Iovanna, P., Naldi, M., Sabella, R.: Models for services and related traffic in Ethernet-based mobile infrastructure. In: HET-NETs 2005 Performance Modelling and Evaluation of Heterogeneous Networks, Ilkley, UK (2005)

Modelling LRD and SRD Traffic with the Batch Renewal Process: Review of Results and Open Issues

Rod J. Fretwell and Demetres D. Kouvatsos

Networks and Performance Engineering Research Group
Informatics Research Institute, University of Bradford,
Bradford BD7 1DP, United Kingdom
{R.J.Fretwell,D.Kouvatsos}@Bradford.ac.uk

Abstract. The batch renewal process is the least biased choice of a process given only the measures of count and interval correlations at all lags. This article reviews the batch renewal process for modelling both LRD (long range dependent) and SRD (short range dependent) traffic flows. The exposition focuses mainly in the discrete-space discrete-time domain and in the wider context of general traffic in that domain. However, corresponding results in the continuous-time domain are also presented. Moreover, some applications of the batch renewal process in simple queues and in queueing network models are undertaken and associated analytic performance results are devised. The article concludes with open research problems and issues relating to the batch renewal process.

1 Introduction

Over the past two decades there has been great interest in (auto)correlated traffic because of its adverse impact upon performance of high speed telecommunications systems by buffer congestion and blocking or packet loss, transmission delay and jitter (delay variability).

In 1986 Sriram and Whitt [19] considered the effect on a multiplexer of superposition of a number of identical renewal processes (modelling telephony talkspurts). They observed that "the aggregate arrival process possesses exceptional long-term positive dependence" and reported the adverse effect on performance of "dependence among interarrival times" in terms of congestion in the queue, delay and blocking probability, provided that the buffer were sufficiently large that "...many interarrivals times interact in the queue." However, when the buffer was small the impact of traffic correlation was restrained: the queue behaved more like one fed by a renewal process.

Gusella [8] collected traces of traffic in a large Ethernet and in 1991 proposed that traffic correlation be characterized by the indices of dispersion. He illustrated his proposal by computing the sample IDC's (indices of dispersion for counts) and IDI's (indices of dispersion for intervals) for measurements of traffic generated by each of six workstations and gave a procedure for fitting the indices of dispersion to the parameters of a 2-phase MMPP (Markov modulated Poisson process). It is of interest to note that Sriram and Whitt [19] used an approximation based on fitting a 2-phase MMPP as also did Heffes and Lucantoni [9] (for the same model as in [19]) and that the recommended fitting procedure is different in all three papers.

Generally the models used most commonly for early investigations into the impact of correlated traffic were simple forms of the Neuts process [18], predominantly MMPP's with small numbers of phases. This class of models can address only short-range dependent (SRD) traffic.

By using indices of dispersion, Gusella implicitly assumed SRD traffic and one of his concerns was for the possible non-stationarity in the traffic over the longer periods of time. However, the lengths of his traces were short relative to the extensive, precise traces of Bellcore LAN traffic which were collected subsequently. From analysis of those data first Fowler and Leland [4] (1991) reported LAN traffic with unbounded IDC and, in 1994, Leland, Taqqu, Willinger and Wilson discerned "the self-similar nature of Ethernet traffic" [15].

Similar effects have been reported subsequently by many researchers, from simulation studies and analysis of a variety of models, and have led to the present consensus that, in general terms, traffic correlation adversely affects queue congestion, waiting times and blocking probabilities and that long term positive correlation can have significant impact, even when the magnitude of the correlation is relatively low. Consequently there has been more interest in models, such as fractional Brownian motion (fBM), and in Pareto distributions of interarrival times [7], which can capture the asymptotically 'hyperbolic' decline in covariances for long-range dependent (LRD) processes.

The popular models for SRD traffic can be fitted tolerably well to covariances in measure traffic at small lags but are limited necessarily to geometrically declining covariances at long lags. Contrarily, the popular models of LRD traffic can be fitted precisely to the (asymptotic) decline in covariances with increasing long lags for measured traffic but do not provide for matching covariances at shorter lags. However, the batch renewal process can match both correlation of counts and correlation of intervals at all lags.

The observation of that property of the batch renewal process (first reported in [11]), derived from consideration of the duality implicit in Gusella's argument (in [8]) for equality $I_\infty = J_\infty$ of the limits I_∞ for the IDC and J_∞ for the IDI as lags tend to infinity in a wide sense stationary process. The duality is most readily apparent in the discrete space discrete time domain. Section 2 of this article addresses different views of general discrete-space discrete-time traffic and shows that one of those views leads naturally to introducing the batch renewal process. An essentially similar argument for the continuous-time domain is given by Li [16].

The next three sections focus upon the batch renewal process itself. Section 3 shows how to construct a batch renewal process which matches measured correlation, whether LRD or SRD and to arbitrary accuracy. Section 4 presents solution methods for simple single-server queues fed by general batch renewal processes and closed-form results for the sGGeo [12] in particular. The sGGeo is a batch renewal process which has proved useful as an investigative tool (a role in which the sGGeo features in Section 8). Section 5 shows how burst structure is induced in the deparures from a finite-buffer queue fed by correlated traffic.

Before illustrating other applications of the batch renewal process, Section 6 gives consideration to a more general class of traffic processes and to the ways such processes might be modelled. Then, because the batch renewal process is *the least biased choice of process given only measures of correlation* [12], it has application as the standard for reference in comparison with other correlated traffic processes. An example of such usage is provided in Section 7 which reports an investigation into the effect of bias consequent upon chosing some other process to capture traffic correlation. Section 8 shows an application in which the sGGeo is used in examining the impact of SRD traffic correlation upon the accuracy of a fast algorithm for approximate analysis of networks.

The article concludes with a review of some open problems and research topics in Section 9.

2 External Views of Traffic and Traffic Processes

In classic queueing theory, traffic is described as the sequence of instants at which customers arrive to the queue system or, usually, as the sequence of interarrival times (the intervals between successive arrivals). In this view of traffic, we number the customers consecutively, in order of arrival instant, and then define the nth interarrival time x_n to be the time between the instant of the $(n-1)^{\text{th}}$ arrival and that of the nth arrival.

In digital computer systems and telecommunications systems there is usually a *natural* unit of time. For example, in an output buffer of an ATM switch, the output port transmits an ATM cell at regular intervals at at rate determined by the output line transmission speed; the (fixed) time to transmit one cell is the natural unit of time in this case. As far as buffer performance is concerned, those cells that arrived during one transmission period might just as well have arrived all together at the start of that period.

Each time period is called a *slot* and the instant that marks the end of one slot (and the beginning of the next) is a called an *epoch*. In discrete time models, events are deemed to occur at epochs only.

So, in digital systems, there is another natural way of viewing traffic: that is, in terms of the numbers of arrivals at successive epochs.

Usually, when discussing models of traffic processes, we are concerned with the *internal* representation of the process. For example, we may describe a DBMAP (discrete time batch Markovian arrival process) as a traffic process in which there is an underlying Markov chain over a countable space J such that whenever the process be in phase $i \in J$ there are n arrivals generated and a transition to phase $j \in J$ with probability $d_{ij}(n)$, $n \in \mathbb{N}_0$. That description gives an internal representation of the process because it describes how the traffic is generated, not what the traffic is.

On the other hand, when we are dealing with observed traffic we are concerned with the *external* view of the traffic process, without necessarily knowing what internal representation might have generated the traffic.

When we record, for example, the size c_t of a message segment detected at time t or the number c_t of individual ATM cells that arrive at an output port buffer during the tth transmission slot, we are, in effect, viewing the traffic as a sequence $\{c_0, c_1, \ldots, c_T\}$ of counts at the successive epochs numbered $0, 1, \ldots, T$. We may then choose to regard that sequence $\{c_0, c_1, \ldots, c_T\}$ as if it were a finite subsequence of the infinite sequence $\{c_t : t \in \mathbb{Z}, c_t \in \mathbb{N}_0\}$ which, in turn, we may choose to regard as being a possible realization of a count process $\{c(t) : t \in \mathbb{Z}, c(t) \in \mathbb{N}_0\}$.

Alternatively, we may choose to regard the traffic as a sequence of interarrival times. We may number the customers consecutively, in order of arrival (applying arbitrary ordering on simultaneous arrivals), and then define the nth interarrival time x_n to be the number of slots between the epoch of the $(n-1)^{th}$ arrival and that of the nth arrival. Just as for the counts view of traffic, we may choose to regard a sequence of observed interarrival times as comprising some finite subsequence from a realization of a persistent interarrival time process $\{x(n) : n \in \mathbb{Z}, x(n) \in \mathbb{N}_0\}$, i.e. the random function $x(n)$ is the duration of the interval between the nth individual arrival and the $(n+1)^{th}$ arrival.

Count processes and interarrival processes are equivalent in the sense that for every realization of a count process we can construct an equivalent realization of a corresponding interarrival process and *vice versa*.

There is some symmetry in the duality between these two views of traffic.

- Interarrival times greater than zero correspond to intervals between successive points (in the count process) at which the counts are greater than zero.
- Counts greater than zero correspond to intervals between successive points (in the interarrival process) at which the interarrival times are greater than zero.
- When interarrival times are *iid* (independent and identically distributed), as from a renewal process, the corresponding count process is covariance stationary.
- When counts are *iid*, as from a batch Bernoulli process, the corresponding interarrival process is covariance stationary.

If the two views of traffic, as a counts process or as an interarrivals time process, which are presented above, are perceived as being in opposition to each other then there is an intermediate, more symmetric view of traffic. In this view the traffic is described as an alternating process of (non-empty) batches and intervals (at least one slot long) between batches. An equivalent description is a 2-dimensional process $\{\xi(s), \kappa(s) : s \in \mathbb{Z}, \xi(s), \kappa(s) \in \mathbb{N}_0\}$ in which the component $\kappa(s)$ represents the number of simultaneous arrivals in the sth (non-empty) batch and $\xi(s)$ represents the interval between the $(s-1)^{th}$ batch and the sth batch. Figure 1 illustrates the relation between this 2-dimensional process and the counts process and the interarrival time process.

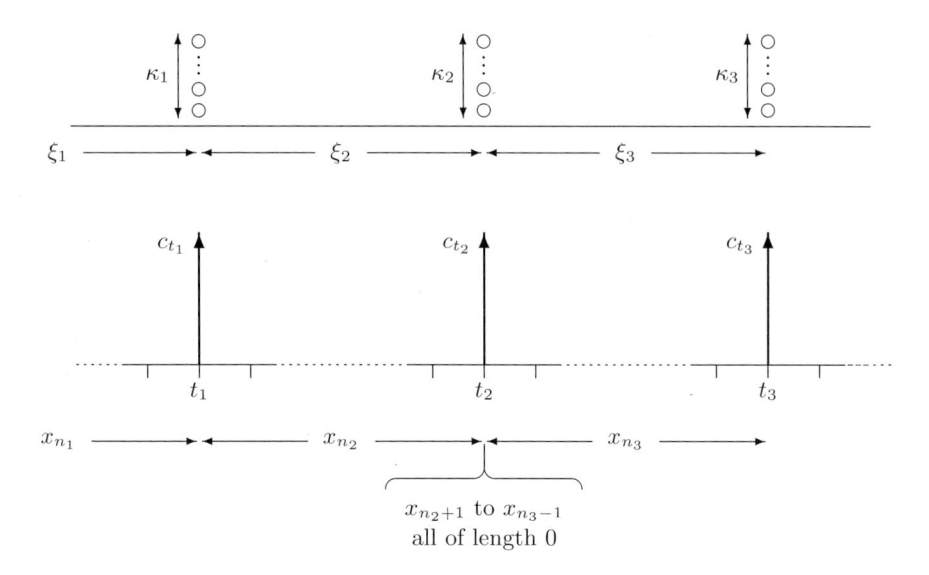

Fig. 1. Relationships between a realization $\{\ldots,(\xi_1,\kappa_1),(\xi_2,\kappa_2),(\xi_3,\kappa_3),\ldots\}$ of the process $\{\xi(s),\kappa(s)\}$ and realizations of the counts process $\{c(t)\}$ and of the interarrival times process $\{x(n)\}$. In this illustration $n_2 = n_1 + c_{t_1}, \ldots, t_2 = t_1 + x_{n_2}, \ldots$

The features of the duality between the two previous views of traffic extend to this 2-dimensional view.

- When the batch sizes $\kappa(\cdot)$ are *iid* and the intervals $\xi(\cdot)$ between batches are *iid* the 2-dimensional process is a batch renewal process. Then the corresponding count process is covariance stationary and the corresponding interarrival process is covariance stationary.

To describe more precisely the relationship between the sequence $\{\xi(s),\kappa(s)\}$ and the sequence $\{x(n)\}$, let n_s be the number of the interval between the last individual arrival of batch s and the first of batch $s+1$: equivalently, let the individual arrivals be numbered such that arrival n_s be the last member of batch s, arrivals $n_s+1, n_s+2, \ldots, n_{s+1}$ be the the first, second, $\ldots$, last member (respectively) of batch $s+1$. Then $x(n_s) = \xi(s)$ and $n_{s+1} = n_s + \kappa(s+1)$. The $\kappa(s+1)$ members of batch $s+1$ arrive simultaneously: the intervals between them are each of zero duration; so $x(n) = 0$ for $n_s < n < n_{s+1}$.

Obviously, each of the sequences $\{\xi(s),\kappa(s)\}$, $\{c(t)\}$ and $\{x(n)\}$ contains the same information (although in a different form) about the traffic process. Each of the sequences can be derived from either of the other two.

The usual measures of traffic correlation assume that the sequences $\{c(t)\}$ and $\{x(n)\}$ be wide sense stationary.

Definition. A random sequence $\{x(n) : n = \ldots, -2, -1, 0, 1, 2, \ldots\}$ is stationary in the wide sense (equivalently, stationary in Khinchin's sense) if

- the random function $x(n)$ has finite mean $\mathsf{E}\left[x(n)\right] = x$ which is constant (independent of n) and
- the correlation function $\mathsf{Cov}\left[x(n), x(m)\right] \stackrel{\Delta}{=} \mathsf{E}\left[(x(n) - x)(x(m) - x)\right]$ is finite and depends on the lag $n-m$ only.

Observe that $\mathsf{Cov}\left[x(n), x(n+\ell)\right] = \mathsf{Cov}\left[x(n+\ell), x(n)\right]$, by symmetry of the definition, and that $\mathsf{Cov}\left[x(n+\ell), x(n)\right] = \mathsf{Cov}\left[x(n), x(n-\ell)\right]$, by change of variable n to $n-\ell$. Consequently, $\mathsf{Cov}\left[x(n), x(n+\ell)\right] = \mathsf{Cov}\left[x(n), x(n-\ell)\right]$. Only the magnitude of the lag is significant and it is therefore necessary to consider positive lags only.

Traffic correlation is customarily expressed either as the correlation functions on $\{c(t)\}$ and $\{x(n)\}$ or as the indices of dispersion. The index of dispersion for counts is defined to be the sequence $\{I_t : t = 1, 2, \ldots\}$ where

$$I_t = \frac{\mathsf{Var}\left[c(i+1) + \cdots + c(i+t)\right]}{\mathsf{E}\left[c(i+1) + \cdots + c(i+t)\right]} = \frac{\mathsf{Var}\left[c(i+1) + \cdots + c(i+t)\right]}{t\,\mathsf{E}\left[c(i)\right]}. \tag{1}$$

The index of dispersion for intervals is defined to be the sequence $\{J_n : n = 1, 2, \ldots\}$ where

$$J_n = \frac{\mathsf{Var}\left[x(i+1) + \cdots + x(i+n)\right]}{\mathsf{E}\left[x(i+1) + \cdots + x(i+n)\right]^2 / n} = \frac{\mathsf{Var}\left[x(i+1) + \cdots + x(i+n)\right]}{n\,\mathsf{E}\left[x(i)\right]^2}. \tag{2}$$

Observe that, if λ be the intensity of the traffic, $\mathsf{E}\left[c(t)\right] = c = \lambda$ and $\mathsf{E}\left[x(n)\right] = x = 1/\lambda$.

The indices of dispersion and the correlation functions contain exactly the same information and are related in a simple way.

$$t\,I_t = \sum_{i=1}^{t} i\,K_{t-i} \quad \text{and} \quad n\,J_n = \sum_{j=1}^{n} j\,L_{n-j} \tag{3}$$

where

$$K_\ell = \begin{cases} \dfrac{1}{\lambda}\,\mathsf{Var}\left[c(t)\right] & \ell = 0 \\ 2\dfrac{1}{\lambda}\,\mathsf{Cov}\left[c(t), c(t+\ell)\right] & \ell = 1, 2, \ldots \end{cases} \tag{4}$$

and

$$L_\ell = \begin{cases} \lambda^2\,\mathsf{Var}\left[x(n)\right] & \ell = 0 \\ 2\lambda^2\,\mathsf{Cov}\left[x(n), x(n+\ell)\right] & \ell = 1, 2, \ldots \end{cases}. \tag{5}$$

In particular, J_1 is the square coefficient of variation C_x^2 of the intervals $x(n)$ between successive individual arrivals and, for bounded indices of dispersion,

$$J_\infty = I_\infty. \tag{6}$$

3 The Batch Renewal Process That Matches Measured Correlation

The batch renewal process is the *least biased choice* of process given only the count covariances $\{\text{Cov}\left[c(t), c(t+\ell)\right] : t, \ell \in \mathbb{Z}\}$ and the interarrival covariances $\{\text{Cov}\left[x(n), x(n+\ell)\right] : n, \ell \in \mathbb{Z}\}$ [12]. This section shows how to identify the batch renewal process that matches exactly the given covariances. The approach is first to derive covariance generating functions for a general batch renewal process and then to solve the corresponding equations to express the batch renewal process probability generating functions in terms of the covariance generating functions.

The exposition is in two parts. The first is applicable when both the count covariances are summable and also the interval covariances are summable (Short Range Dependent processes). The second subsection applies to cases in which either the count covariances are not summable or the interval covariances are not summable (Long Range Dependent processes).

The following notation is used.

$\{a(t) = \mathbf{P}\left[\xi(s) = t\right] : t = 1, 2, \ldots\}$ the *pmf* (probability mass function) of the interval between successive batches

$\{b(n) = \mathbf{P}\left[\kappa(s) = n\right] : n = 1, 2, \ldots\}$ the *pmf* of the batch size

$a = \mathbf{E}\left[\xi(s)\right]$, C_a^2 the mean and SCV (squared coefficient of variation) of the interval between successive batches

$b = \mathbf{E}\left[\kappa(s)\right]$, C_b^2 the mean and SCV of the batch size

$\lambda = b/a$ the mean arrival rate

$A(\omega) = \sum_{t=1}^{\infty} a(t)\,\omega^t$ the *pgf* (probability generating function) of $\{a(t)\}$

$B(z) = \sum_{n=1}^{\infty} b(n) z^n$ the *pgf* of $\{b(n)\}$

3.1 Short Range Dependent Processes

We shall say that a process is short range dependent if both the count covariances are summable and the interval covariances are summable and shall see that condition, in the case of the batch renewal process, is equivalent to the variances of counts and of intervals both being finite. Finite variances are assumed in this sub-section.

Calculation of the various expectations (mean, variance and covariances) is greatly facilitated by exploiting conditional independence.

- Independence of batch size $\{\kappa(s)\}$ implies conditional independence of the count $c(t)$ at epoch $t \in \mathbb{Z}$ given only that $c(t) > 0$.
- Independence of intervals $\{\xi(s)\}$ between batches implies conditional independence of the interval $x(n)$ between individual arrivals ($n \in \mathbb{Z}$) given only that the interval $x(n) > 0$.

But only random variables with values greater than zero contribute to expectations.

$$\lambda = \mathsf{E}\left[c(t)\right] = \mathsf{P}\left[c(t) > 0\right]\mathsf{E}\left[c(t) \mid c(t) > 0\right] + \mathsf{P}\left[c(t) = 0\right]\mathsf{E}\left[c(t) \mid c(t) = 0\right]$$

$$= \frac{1}{\mathsf{E}\left[\xi(s)\right]}\mathsf{E}\left[\kappa(s)\right] + 0$$

$$= \frac{1}{a}b = b/a \qquad (7)$$

$$\mathsf{Var}\left[c(t)\right] = \mathsf{E}\left[c(t)^2\right] - \mathsf{E}\left[c(t)\right]^2 = \mathsf{P}\left[c(t) > 0\right]\mathsf{E}\left[c(t)^2 \mid c(t) > 0\right] - \mathsf{E}\left[c(t)\right]^2$$

$$= \frac{1}{\mathsf{E}\left[\xi(s)\right]}\mathsf{E}\left[\kappa(s)^2\right] - \mathsf{E}\left[c(t)\right]^2$$

$$= \frac{1}{\mathsf{E}\left[\xi(s)\right]}\left(\mathsf{Var}\left[\kappa(s)\right] + \mathsf{E}\left[\kappa(s)\right]^2\right) - \mathsf{E}\left[c(t)\right]^2$$

$$= \frac{1}{a}b^2(C_b^2 + 1) - \left(\frac{b}{a}\right)^2$$

$$= \frac{b^2}{a}\left(C_b^2 + 1 - \frac{1}{a}\right) \qquad (8)$$

and, for $\ell = 1, 2, \ldots$,

$$\mathsf{P}\left[c(t) = n, c(t+\ell) = k, n > 0, k > 0\right]$$
$$= \mathsf{P}\left[c(t) > 0\right]$$
$$\times \mathsf{P}\left[c(t) = n \mid c(t) > 0\right]$$
$$\times \mathsf{P}\left[c(t+\ell) > 0 \mid c(t) = n, c(t) > 0\right]$$
$$\times \mathsf{P}\left[c(t+\ell) = k \mid c(t+\ell) > 0, c(t) = n, c(t) > 0\right]$$
$$= \mathsf{P}\left[c(t) > 0\right]$$
$$\times \mathsf{P}\left[c(t) = n \mid c(t) > 0\right]$$
$$\times \mathsf{P}\left[c(t+\ell) > 0 \mid c(t) > 0\right]$$
$$\times \mathsf{P}\left[c(t+\ell) = k \mid c(t+\ell) > 0\right]$$
$$= \frac{1}{a}b(n)\phi_\ell b(k) \qquad (9)$$

so that

$$\mathsf{E}\left[c(t)c(t+\ell)\right] = \frac{b^2}{a}\phi_\ell \qquad (10)$$

where $\phi_\ell = \mathsf{P}\left[c(t+\ell) > 0 \mid c(t) > 0\right]$ is the probability that some integral number of intervals between batches be exactly ℓ slots long. Clearly ϕ_ℓ must satisfy

$$\phi_\ell = \begin{cases} 1 & \ell = 0 \\ \sum_{t=1}^{\ell} a(t)\,\phi_{\ell-t} & \ell = 1, 2, \ldots \end{cases} \tag{11}$$

and so is generated by

$$\sum_{\ell=0}^{\infty} \phi_\ell\, w^\ell = 1 + \sum_{\ell=1}^{\infty} \sum_{t=1}^{\ell} a(t)\,\phi_{\ell-t} w^\ell$$

$$= 1 + \sum_{t=1}^{\infty} \sum_{\ell=t}^{\infty} a(t)\,\phi_{\ell-t} w^\ell$$

$$= 1 + A(w) \sum_{\ell=0}^{\infty} \phi_\ell\, w^\ell$$

$$= \frac{1}{1 - A(w)} \tag{12}$$

The structure of the batch renewal process makes it relatively simple to derive the form of covariances by exploiting conditional independence. For example, the distribution of count $c(t)$ at epoch t depends only upon the condition $c(t) > 0$. Also, calculation of the covariances makes obvious that they are stationary. Then, assuming that the variances $\mathrm{Var}\left[\xi(s)\right]$ and $\mathrm{Var}\left[\kappa(s)\right]$ are finite, it can be seen that the covariances are given by the generating functions

$$K(w) = \sum_{\ell=0}^{\infty} K_\ell w^\ell = 1/\lambda \left(\mathrm{Var}\left[c(t)\right] + 2 \sum_{\ell=1}^{\infty} \mathrm{Cov}\left[c(t), c(t+\ell)\right] w^\ell \right)$$

$$= b\left(C_b^2 + \frac{1 + A(w)}{1 - A(w)} - \frac{1}{a}\frac{1 + w}{1 - w} \right) \tag{13}$$

and

$$L(z) = \sum_{\ell=0}^{\infty} L_\ell z^\ell = \lambda^2 \left(\mathrm{Var}\left[x(n)\right] + 2 \sum_{\ell=1}^{\infty} \mathrm{Cov}\left[x(n), x(n+\ell)\right] z^\ell \right)$$

$$= b\left(C_a^2 + \frac{1 + B(z)}{1 - B(z)} - \frac{1}{b}\frac{1 + z}{1 - z} \right) \tag{14}$$

Under the assumption that the analyst has been able to express the covariances in the form of the generating functions $K(w)$ and $L(z)$, construction of the corresponding batch renewal process reduces to solving equations (13) and (14) for $A(w)$ and $B(z)$, as follows.

Setting $w = 0$ in (13) and $z = 0$ in (14) immediately yields

$$K(0) = b\left(C_b^2 + 1 - 1/a\right) \tag{15}$$
$$L(0) = b\left(C_a^2 + 1 - 1/b\right) \tag{16}$$

and, by considering limits as $\omega \to 1-$ in (13) and $z \to 1-$ in (14),

$$K(1-) = b\left(C_a^2 + C_b^2\right) \quad \text{and} \quad L(1-) = b\left(C_a^2 + C_b^2\right). \tag{17}$$

It may be observed that the existence of those limits is equivalent to saying that the covariances be summable by the method of Abel. Furthermore, $K(1-) = I_\infty$ and $L(1-) = J_\infty$, where I_∞ and J_∞ are the limits for the IDC and IDI as lags tend to infinity.

Then, from equations (15), (16) and (17), b must satisfy

$$K(0) + L(0) - K(1-) = 2b - 1 - \lambda \tag{18}$$

and, using (18), equations (13) and (14) can be manipulated to give

$$A(\omega) = 1 - \frac{K(0) + L(0) - K(1-) + 1 + \lambda}{K(\omega) + L(0) - K(1-) + 1 + \lambda \dfrac{1+\omega}{1-\omega}} \tag{19}$$

and

$$B(z) = 1 - \frac{K(0) + L(0) - K(1-) + 1 + \lambda}{K(0) + L(z) - K(1-) + \dfrac{1+z}{1-z} + \lambda}. \tag{20}$$

Constructing the Covariance Generating Functions $K(\omega)$ and $L(z)$. For the analyst to have decided that the process be short range dependent, it is likely that the graph of log covariance against lag is (approximately) piece-wise linear — which is equivalent to saying that the correlation function is (approximated by) the weighted sum of geometric terms or that the generating function ($K(\omega)$ or $L(z)$) is a rational function (of ω or of z, respectively).

In that case, the geometric components may be extracted progressively, begining with line segment for the longest lags, until adequate fit with the data be obtained.

Direct Numerical Solution. The main objection to direct calculation of the component distributions $a(t)$ and $b(n)$ is that, for fixed precision arithmetic, rounding errors accumulate and are likely to become significant when dealing with covariances at the longer lags.

Where the analyst has algebraic expressions for the measures of correlation (such that $L(1) \equiv I_\infty = K(1) \equiv J_\infty$) equations (19) and (20) can be employed directly to produce the *pgf*'s of the component distributions of the appropriate batch renewal process.

In considering the case of measurements of the correlation of actual traffic it is apparent that there are fundamental problems to construction of a general numerical algorithm to determine the corresponding batch renewal process. For example, equation (19) gives the recurrence relationship

$$a(t) = \frac{K_t + 2\lambda - \sum_{\ell=1}^{t-1} a(\ell)(K_{t-\ell} + 2\lambda)}{K_0 + L_0 - K(1-) + 1 + \lambda} \quad \text{for } t = 2, 3, \ldots,$$

which calculation requires the difference between numbers of similar magnitude.

Firstly there is the (lesser) difficulty of estimating $K(1) = L(1)$, which is equivalent to estimating geometric tails to complement the truncated sets of measurements. Secondly the form of the recurrence relationship suggests that the effect of rounding errors might accumulate rapidly. This difficulty is inherent. By defining ϕ_ℓ by its generating function

$$\sum_{\ell=0}^{\infty} \phi_\ell \omega^\ell \triangleq \frac{1}{1 - A(\omega)}$$

the essence of the recurrence relation is seen to be

$$a(\ell) = \phi_\ell - \sum_{t=1}^{\ell} a(t)\phi_{\ell-t} \quad \ell = 1, 2, \cdots$$

where

$$\phi_0 = 1 \quad \text{and} \quad \phi_\ell = \frac{K_\ell + 2\lambda}{K_0 + L_0 - K(1) + 1 + \lambda} \quad \ell = 1, 2, \cdots.$$

Consequently, actual traffic measurements should be converted to an algebraic representation and then the algebraic method be used.

Generally, when the logarithm of the measured correlation be plotted (with error bars) against the corresponding lag, the resulting graph may be (or may be approximated by) a series of straight line segments — which is equivalent to saying that the correlation function is (approximated by) the weighted sum of geometric terms or that the generating function ($K(\omega)$ or $L(z)$) is a rational function (of ω or of z, respectively).

The simplest form of the graphs of $\log K_\ell$ against ℓ and $\log L_\ell$ against ℓ is when both are straight line graphs. This case may arise naturally, because of the characteristics of the traffic source, or may arise from the practicalities in actual traffic measurements. The size of the data sets may be limited by the time period for which the traffic process may be regarded as being wide sense stationary. Then the practical recourse is to fit a straight line to the data points. When $\log K_\ell$ and $\log L_\ell$ are linear in ℓ the corresponding batch renewal process is of the simplest non-trivial form. It is the form which is used for the arrival process to the queue in sections 4.3.

SRD Batch Renewal process in Continuous Time. The results for the batch renewal process in continuous time are similar to those for the discrete-time domain. Corresponding to (13) and (14) we have[16]

$$K(\theta) = b\left(C_b^2 + \frac{1+A(\theta)}{1-A(\theta)} - \frac{1}{a}\frac{2}{\theta}\right) \tag{21}$$

and

$$L(z) = b\left(C_a^2 + \frac{1+B(z)}{1-B(z)} - \frac{1}{b}\frac{1+z}{1-z}\right) \tag{22}$$

where $A(\theta)$ is now the Laplace transform of the density of intervals between batches and $K(\theta)$ generates the count covariances.

By considering the limits

$$K(0) = \lim_{\theta \to 0} K(\theta) = b\left(C_a^2 + C_b^2\right)$$

$$L(1-) = \lim_{z \to 1-} L(z) = b\left(C_a^2 + C_b^2\right)$$

$$K(\infty) = \lim_{\theta \to \infty} = b\left(C_a^2 + 1\right)$$

$$L(0) = \lim_{z \to 0} L(z) = b\left(C_a^2 + 1 - \frac{1}{b}\right)$$

we obtain[16]

$$A(\theta) = 1 - \frac{K(\infty) + L(0) - K(0) + 1}{K(\theta) + L(0) - K(0) + 1 + \lambda\frac{2}{\omega}} \tag{23}$$

and

$$B(z) = 1 - \frac{K(\infty) + L(0) - K(0) + 1}{K(\infty) + L(z) - K(0) + \frac{1+z}{1-z}}. \tag{24}$$

3.2 Long Range Dependent Processes

We shall say that a process is long range dependent if either the count covariances are not summable or the interval covariances are not summable.

This subsection addresses the case in which either the count covariances are not summable or the interval covariances are not summable. For illustration, consider the case when the interval covariances are not summable but the count covariances are summable. In this case, by taking limits as $z \to 1-$ in equation (14), $L(1-) = \infty$ and the scv C_b^2 of batch size is infinite — an instance of "the infinite variance syndrome". Consequently, even though the sample variance of counts is finite (necessarily), we have to treat the counts as arising from a process with infinite variance. Thus, the generating function $K(\omega)$ cannot be used unmodified. Instead, define $K_+(\omega)$ by

$$K_+(\omega) = \sum_{\ell=1}^{\infty} K_\ell \omega^\ell = 2/\lambda \sum_{\ell=1}^{\infty} \text{Cov}\left[c(t), c(t+\ell)\right] \omega^\ell \qquad (25)$$

and then the analysis of the preceding sub-section can be adapted to yield

$$A(\omega) = 1 - \frac{L(0) - K_+(1-) + 1 + \lambda}{K_+(\omega) + L(0) - K_+(1-) + 1 + \lambda \frac{1+\omega}{1-\omega}} \qquad (26)$$

and

$$B(z) = 1 - \frac{L(0) - K_+(1-) + 1 + \lambda}{L(z) - K_+(1-) + \frac{1+z}{1-z} + \lambda} . \qquad (27)$$

For the analyst to have decided that the process be long range dependent, it is likely that the graph of log covariance against log lag would be asymptotically linear. If the asymptotic slope be $-s$ then $K(\omega)$ can be represented as the sum of two terms, with one term having the form $C\bigl((1-\omega)^{-1-s} - 1\bigr)$. Components may be extracted progressively until adequate fit with the data be obtained.

3.3 Improper Batch Renewal Processes

For some correlation structures the corresponding batch renewal process is improper, i.e. the constituent distributions contain negative probabilities. The cause can be seen by considering equation (18) for SRD processes or the corresponding relation (such as)

$$L(0) - K_+(1-) = 2b - 1 - \lambda$$

for LRD processes. In each case, the left hand side of the equation may be so small that the mean batch size b does not exceed 1: indeed b may be negative. Whereas, from the formulation of the batch renewal process, b is the expected size of a non-empty batch.

The question then arises as to whether such improper batch renewal process may be used for performance prediction of (for example) a buffer fed by the traffic. Possible approaches are discussed in Section 9.

4 Simple Queues Fed by Batch Renewal Process Traffic

4.1 $GI^G/D/1/N$

Consider the discrete-time $GI^G/D/1/N$ censored queue under DF.

- The arrivals are from the batch renewal process.
- The service time is fixed at one slot.
- The queue capacity is N, including the customer in service.
- At an epoch, an arrival may take the place released by a departure.
- When the system becomes full, other customers in the arriving batch are lost.

Events (arrivals and departures) occur at discrete points in time (epochs) only. The intervals between epochs are called *slots* and, without loss of generality, may be regarded as being of constant duration. At an epoch at which both arrivals and departures occur, the departing customers release the places, which they had been occupying, to be available to arriving customers (*departures first* memory management policy). The service time for a customer is one slot and the first customer arriving to an empty system (after any departures) receives service and departs at the end of the slot in which it arrived (*immediate service* policy). By GI^G arrivals process is meant the intervals between batches are independent and of general distribution and the batch size distribution is general (batch renewal process).

Let the state of the queue be the number of customers in the queue (buffered or receiving service). Because the transitions at epochs are deemed to be instantaneous the *pmf* for the stationary distribution of queue length is simply the time average probability for each state observed during slots only.

The solution described in this section is based upon the observation that, between arrival epochs, the state in each slot is determined completely by the state in the previous slot. (By 'arrival epoch' is meant an epoch at which there is a batch of arrivals). The number of customers in the queue is reduced by one departure at each epoch until either the queue becomes empty or an arrival epoch is reached. Thus, given the state in the slot immediately following an arrivals epoch, the evolution of the queue until the next arrivals epoch depends only upon the interval between the two batches of arrivals. But, at the arrival epoch, the change in state (after accounting for any departure at that epoch) depends only upon the size of the arriving batch.

The steady state behaviour of the queue may be solved by considering the state at points immediately before and immediately after each batch of arrivals. It is apparent that each point is an embedding point for a Markov chain.

Consider two (related) Markov chains embedded at arrival epochs.

- For the first chain (chain 'A'), the state is the number of customers in the queue after allowing for any departure at that epoch but discounting the new arrivals at that epoch. Let $p_N^A(n)$ be the probability that the state be n, $n = 0, \ldots, N-1$ (where N is the capacity of the queue).
- For the second chain (chain 'D'), the state is the number of customers in the queue after allowing for any departure at that epoch but including the new arrivals. Let $p_N^D(n)$ be the probability that the state be n, $n = 1, \ldots, N$.

Equivalently, one might treat the departures as *actually* occuring before arrivals and focus on the two points 1) at which the departures have already gone and the arrivals have not yet come, 2) immediately after the arrivals.

To see the relation between the two Markov chains, first consider the state of each chain at an arrival epoch. Chain 'D' may be in state n, $n = 1, \ldots, N-1$, when chain 'A' is in state k, $k = 0, \ldots, n-1$, and there be just $n-k$ arrivals in the batch. Alternatively, chain 'D' may be in state N when chain 'A' is in state k, $k = 0, \ldots, N-1$, and there be at least $N-k$ arrivals in the batch. Therefore

$$p_N^D(n) = \begin{cases} \sum_{k=0}^{n-1} p_N^A(k)\, b(n-k) & n = 1, \ldots, N-1 \\ \sum_{k=0}^{N-1} p_N^A(k) \sum_{r=N-k}^{\infty} b(r) & n = N \end{cases} \qquad (28)$$

Next, consider the state of each chain at successive arrival epochs. At the later epoch the chain 'A' may be in state n, $n = 1, \ldots, N-1$, when chain 'D' is in state k, $k = n+1, \ldots, N$, at the earlier arrival epoch and there be just $k-n$ departures in the interval between the two arrival epochs, i.e. the interval is $k-n$ slots long. Alternatively, at the later epoch the chain 'A' may be in state 0 when chain 'D' is in state k, $k = 1, \ldots, N$, at the earlier arrival epoch and the interval is at least k slots long. Therefore

$$p_N^A(n) = \begin{cases} \sum_{k=1}^{N} p_N^D(k) \sum_{t=k}^{\infty} a(t) & n = 0 \\ \sum_{k=n+1}^{N} p_N^D(k)\, a(k-n) & n = 1, \ldots, N-1 \end{cases} \qquad (29)$$

Performance statistics and measures of interest are obtainable, in obvious ways, in terms of the solutions to equations (28) and (29) for the two Markov chains.

Queue Length Distribution. If the second Markov chain (chain 'D') be in state k at an arrival epoch then, in each successive slot of the interval to the next arrival epoch, the queue will be in state k, $k-1$, etc. until either the queue becomes empty or the next batch arrives.

If the interval to the next arrivals epoch be t slots then, if $t \leq k$, the queue visits states $k, \ldots, k-t+1$ for one slot each but, if $t > k$, the queue visits states $k, \ldots, 1$ for one slot each and then remains in state 0 for the remaining $t-k$ slots (see Figure 2).

Thus, the time average probability $p_N(n)$ that the queue be in state n is given by

$$p_N(n) = \begin{cases} \dfrac{1}{a} \sum_{k=1}^{N} p_N^D(k) \sum_{t=k+1}^{\infty} (t-k)\, a(t) & n = 0 \\ \dfrac{1}{a} \sum_{k=n}^{N} p_N^D(k) \sum_{t=k-n+1}^{\infty} a(t) & n = 1, \ldots, N \end{cases} \qquad (30)$$

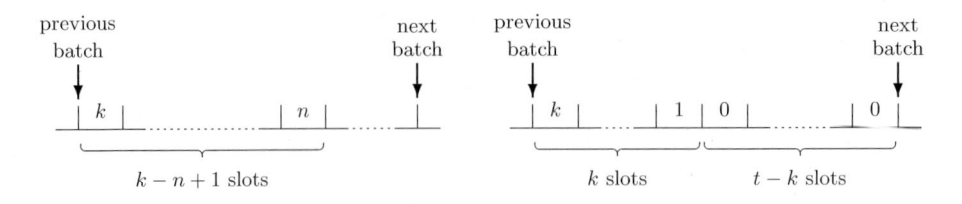

```
previous                      next    previous                           next
 batch                        batch    batch                             batch
   ↓                            ↓        ↓                                 ↓
   | k |                | n |   |       | k |           | 1 | 0 |         | 0 |
       _____ _____/              \____ ____/ \____ ____/
                     v                          v           v
              k − n + 1 slots                k slots    t − k slots
```

Fig. 2. Ways in which queue length n ($n > 0$) and queue length 0 may be reached during an interval between batches

in which $a(t)/a$ is the probability that an arbitrary slot be at any given position within an interval of t slots between batches.

Blocking Probability. If chain 'A' be in state k at an arrival epoch there are $N-k$ places available in the queue to the arriving batch. So, if the batch contain $N-k+r$ arrivals, r of those arrivals are blocked.

For an arbitrary arrival, the probability that it be in a batch of size n is $nb(n)/b$ and the probability that it be in any given position in the batch is $1/n$. Therefore the marginal probability π_N^B that any individual arrival be turned away is

$$\pi_N^B = \sum_{k=0}^{N-1} p_N^A(k) \sum_{r=1}^{\infty} \frac{r}{b} b(N-k+r) \tag{31}$$

which, by reference to (28–30), can be seen to satisfy the flow balance equation

$$\lambda(1 - \pi_N^B) = 1 - p_N(0). \tag{32}$$

Waiting Time. Because service time is one slot per customer the waiting time of an arrival is given by its position in the queue at the instant of its arrival, given that the arrival enter the queue and not be blocked. If there be k in the queue (i.e. Markov chain 'A' be in state k at that arrival epoch) then the arrival in position $t-k$ of the batch will enter the queue provided that $t \leq N$ and will then remain in the queue for t slots. Thus,

$$P\left[\text{waiting time} = t \mid k \text{ in queue, arrival not blocked}\right]$$
$$= \frac{P\left[\text{customer in position } t-k \text{ of batch}, k < t \leq N\right]}{P\left[\text{arrival not blocked}\right]}$$
$$= \frac{\sum_{r=t-k}^{\infty} \frac{rb(r)}{b} \frac{1}{r}}{1 - \pi_N^B} = \frac{\sum_{n=t}^{\infty} b(n-k)}{b(1 - \pi_N^B)}$$

Therefore the conditional probability $w_N(t)$ that an arbitrary arrival spend t slots in the queue, given that the arrival not be blocked, is given by

$$w_N(t) = \frac{\sum_{n=t}^{\infty}\sum_{k=0}^{t-1} p_N^A(k) b(n-k)}{b(1-\pi_N^B)} = \frac{\sum_{n=t}^{\infty}\sum_{k=0}^{t-1} p_N^A(k) b(n-k)}{\sum_{t=1}^{N}\sum_{n=t}^{\infty}\sum_{k=0}^{t-1} p_N^A(k) b(n-k)} \quad (33a)$$

which may be manipulated, using equations (28–30) to show that

$$w_N(t) = \frac{p_N(t)}{1 - p_N(0)} \quad \text{for } t = 1, \ldots, N. \quad (33b)$$

It may be observed that the relation (33b), between waiting time and stationary queue length, is what should be expected when the service time is deterministic at one customer per slot [21].

Example. This section shows some numerical results for the batch renewal process with LRD counts. In the chosen case, the inter-batch *pgf* is of the form

$$A(\omega) = 1 - a + a\omega + (a-1)(1-\omega)^{2-s} \quad (34)$$

where $0 < s < 1$ and, for a proper *pmf*, $1 < a < \dfrac{2-s}{1-s}$, and the batch size *pmf* is of the form

$$b(n) = \begin{cases} 1 - \eta & n = 1 \\ \eta\nu(1-\nu)^{n-2} & n = 2, 3, \ldots \end{cases} \quad (35)$$

For the graphs in Figures 3 and 4 the following parameter values were used for three values of s.

- $a = 2$.
- $\eta = 1/8$, $\nu = 3/8$ (giving $b = 4/3$, $\lambda = 2/3$).
- Queue capacity $N = 100$.

For Figure 3 the calculation is the recursive relation derived by inverting equation 25 of Section 3.2. The graphs show the positive count covariances; at small lags some covariances are zero or negative in the cases shown. It is clear that asymptotic slopes of the graphs approach $-s$ rapidly.

For Figure 4 the algorithm is the general method given in Section 4.1.

4.2 Continuous-Time $GI^G/G/1/N$ Queues

The approach, taken in the previous section, of considering two Markov chains embedded immediately before and immediately after each batch of arrivals can

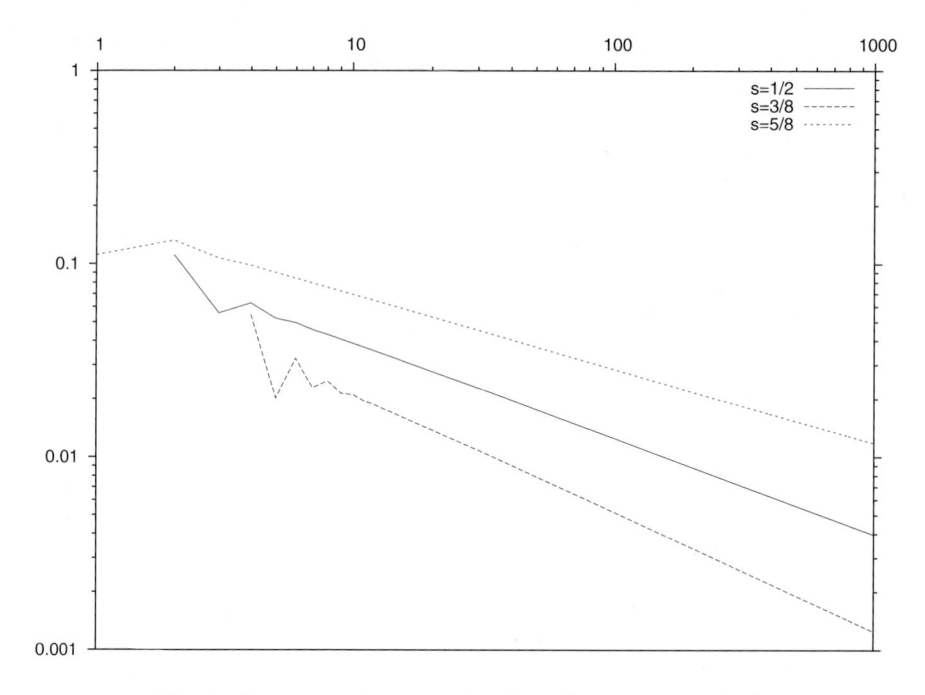

Fig. 3. Count covariances against lags (logarithmic scales)

be extended to a queue with general service time distribution. Li obtained the following general results in the continuous-time domain[16] in terms of the density $g(k,t)$ of the probability that k customers can complete service in time t (i.e. k depart provided that there are at least k in the system, otherwise all customers depart).

Relationship betwwen the chains. c.f. (28) and (29),

$$p_N^D(n) = \begin{cases} \sum_{k=0}^{n-1} p_N^A(k)\, b(n-k) & n = 1, \ldots, N-1 \\ \sum_{k=0}^{N-1} p_N^A(k) \sum_{r=N-k}^{\infty} b(r) & n = N \end{cases} \quad (36)$$

$$p_N^A(n) = \begin{cases} \sum_{k=1}^{N} p_N^D(k) \int_0^{\infty} \sum_{r=k}^{\infty} g(r,t)\, a(t)\, dt & n = 0 \\ \sum_{k=n}^{N} p_N^D(k) \int_0^{\infty} g(k-n, t)\, a(t)\, dt & n = 1, \ldots, N \end{cases} \quad (37)$$

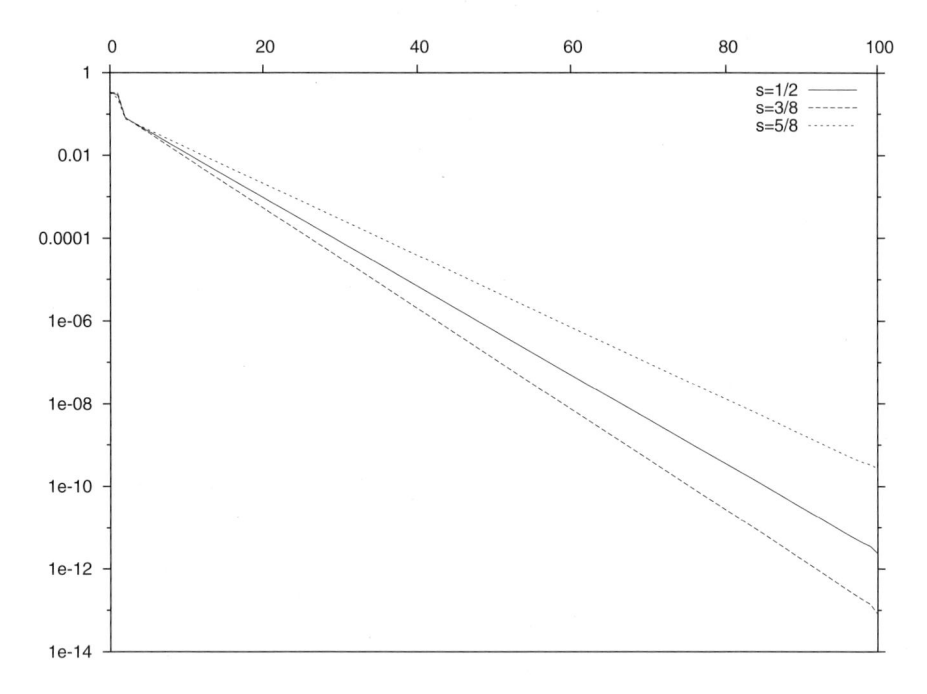

Fig. 4. Queue length distribution (logarithmic scale)

Queue Length Distribution. c.f. (30)

$$p_N(n) = \begin{cases} \dfrac{1}{a} \sum_{k=1}^{N} p_N^D(k) \int_0^\infty \int_0^t \sum_{r=k}^{\infty} g(r,s)\, ds\, a(t)\, dt & n=0 \\ \dfrac{1}{a} \sum_{k=n}^{N} p_N^D(k) \int_0^\infty \int_0^t g(k-n,s)\, ds\, a(t)\, dt & n=1,\ldots,N \end{cases} \qquad (38)$$

Blocking Probability. c.f. (31)

$$\pi_N^B = \sum_{k=0}^{N} p_N^A(k) \sum_{r=1}^{\infty} \frac{r}{b} b(N-k+r) \qquad (39)$$

Waiting Time Density. c.f. (33a)

$$w_N(t) = \frac{\sum_{k=0}^{N-1} p_N^A(k) \sum_{i=1}^{N-k} g(k+i,t) \sum_{r=i}^{\infty} b(r)}{b(1-\pi_N^B)} \qquad (40)$$

4.3 The sGGeo/D/1/N Queue

The sGGeo[1] is the simplest batch renewal process in which there is both count correlation and interval correlation.

$$a(t) = \begin{cases} 1-\sigma & t=1 \\ \sigma\tau(1-\tau)^{t-2} & t=2,3,\ldots \end{cases} \qquad b(n) = \begin{cases} 1-\eta & n=1 \\ \eta\nu(1-\nu)^{n-2} & n=2,3,\ldots \end{cases} \qquad (41)$$

For the sGGeo, the covariances of counts and the covariances of intervals (between individual arrivals) both decline geometrically, viz.

$$\begin{aligned} \text{Cov}\left[c(t), c(t+\ell)\right] &= \lambda^2(a-1)\beta_a^{\ell} & \text{where } \beta_a = 1-\sigma-\tau \\ \text{Cov}\left[x(n), x(n+\ell)\right] &= \frac{1}{\lambda^2}(b-1)\beta_b^{\ell} & \text{where } \beta_b = 1-\eta-\nu \end{aligned} \qquad (42)$$

Remarks. The sGGeo may be appropriate to model a traffic source for which only the first two moments of message size and of intervals between messages are known. It is also the appropriate model of measured traffic when either the decline in covariances is geometric (c.f. equation 42) or there be so few measurements that the best procedure is to fit a straight line to the plot of the logarithms of measured covariances against lags.

The sGGeo/D/1/N Queue Length Distribution. By using the particular forms (41) in application of the general methods of Section 4.1 it is seen that the sGGeo/D/1/N queue length distribution has the form

$$p_N(n) = \begin{cases} \frac{1}{Z_N}(1-\lambda) & n=0 \\ \frac{1}{Z_N}\lambda(1-y) & n=1 \\ \frac{1}{Z_N}\lambda y(1-x)x^{n-2} & n=2,\ldots,N-1 \\ \frac{1}{Z_N}\lambda y(1-x)x^{N-2}\frac{1}{1-\beta_a x} & n=N \end{cases} \qquad (43)$$

where $\beta_a = 1-\sigma-\tau$ and $\beta_b = 1-\eta-\nu$ are as defined at (42) and x and y are given by

$$1-y = \frac{1-x}{1-\beta_b}, \qquad x = \frac{\sigma(1-\eta-\nu)+\eta}{\sigma+(1-\sigma-\tau)\eta}$$

and the normalizing constant Z_N may be written

$$Z_N = 1 - \lambda y \frac{1-\beta_a}{1-\beta_a x} x^{N-1} \qquad (44)$$

---

[1] The sGGeo process is so named because both the constituent distributions (i.e. of the batch sizes and of the intervals between batches) have the form of a Generalized Geometric (GGeo) shifted by one.

The sGGeo/D/1/N Mean Queue Length

$$L_N = \frac{1}{Z_N}\left(\lambda + \lambda\frac{b-1}{1-\lambda}\frac{1-\beta_a\beta_b}{(1-\beta_a)(1-\beta_b)}(1-x^{N-1})\right.$$
$$\left. + N\frac{b-1}{a-1}\frac{\beta_a}{1-\beta_a}x^{N-1}\right) \quad (45)$$

Figure 5 shows the effect of correlation on mean queue length and that the effect is constrained for small values of buffer capacity N.

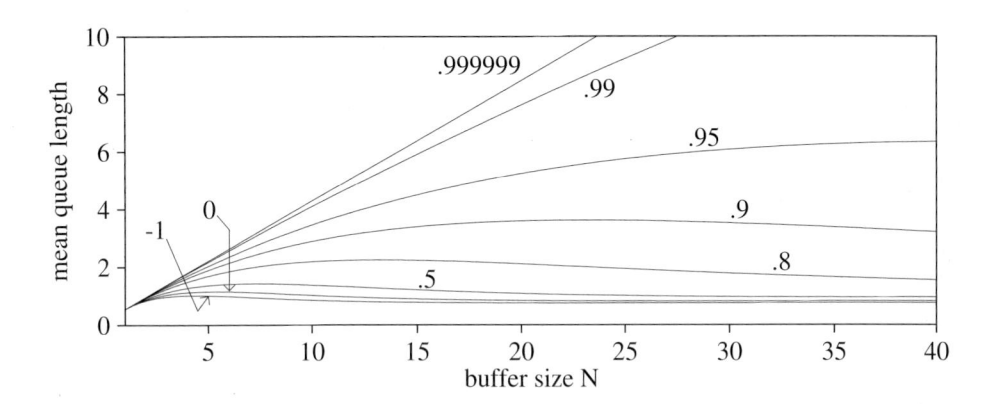

Fig. 5. Mean queue length against buffer size N for mean batch size $b = 1.5$, mean interval $a = 7.5$ slots between batches, intensity $\lambda = 0.2$, $\beta_a = 0.8$ and various values of β_b

The sGGeo/D/1/N Blocking Probability. Because the probability π_N^B, that any individual arrival be turned away, satisfies the flow balance equation $\lambda(1 - \pi_N^B) = 1 - p_N(0)$, it follows from equation (43) that

$$\pi_N^B = \frac{1-\lambda}{\lambda}\frac{1-Z_N}{Z_N} = \frac{(1-\lambda)y\frac{1-\beta_a}{1-\beta_a x}x^{N-1}}{1-\lambda y\frac{1-\beta_a}{1-\beta_a x}x^{N-1}}. \quad (46)$$

This relation shows that the asymptotic behaviour of π_N^B with increasing buffer size N is log-linear:

$$\frac{\pi_{N+1}^B}{\pi_N^B} \longrightarrow x \quad \text{as } N \longrightarrow \infty \quad (47)$$

Indeed π_N^B may approach its asymptote for relatively small values of N, as is illustrated by the graphs in Figure 6. Expressions (47) and (44) also show that

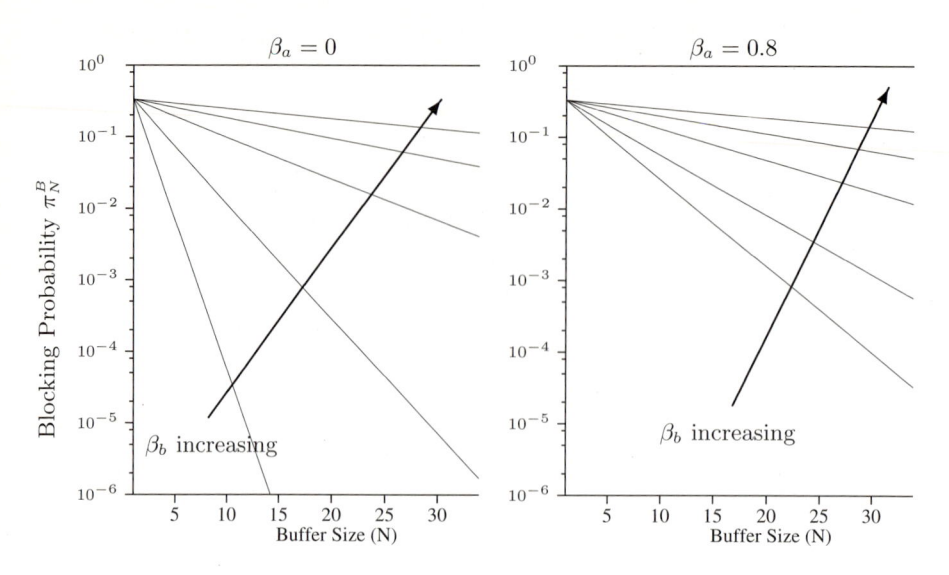

Fig. 6. Blocking probability against buffer size for mean batch size $b = 1.5$, mean interval $a = 7.5$ slots between batches, intensity $\lambda = 0.2$ with $\beta_b = 0, 0.5, 0.8, 0.9, 0.95$

$$\pi_N^B \to \pi_1^B = 1 - \frac{1}{b} \quad \text{as } x \to 1 \tag{48}$$

i.e. as $\beta_a \to 1$ or as $\beta_b \to 1$.

5 Effect of a Queue on Correlation—Creation of Burst Structure

The departure process from a $\text{GI}^G/\text{D}/1/\text{N}$ queue is determined by the cycle of busy period followed by idle period. For each slot the server is busy there is a departure. Consecutive departures constitute a burst. Because the intervals between batches are independent each cycle of busy period followed by idle period is independent of other busy/idle cycles. The distribution of one burst length (busy period) and successive silence (idle) period is governed by the following relationships.

$$busy(n, i) = \sum_{k=1}^{\min(N,n)} busy(n, i; k) \, b_N(k) \tag{49}$$

$$busy(n, i; k) = \begin{cases} a(n+i) & n = k \\ \sum_{\ell=0}^{k-1} \sum_{q=\ell+1}^{\min(N,n-k+\ell)} a(k-\ell) \, b_{N-\ell}(q-\ell) \, busy(n-k+\ell, i; q) & n > k \end{cases} \tag{50}$$

where $busy(n,i)$ is the marginal probability that the server be busy for n slots and idle for i slots, $busy(n,i;k)$ is the conditional probability that the server be busy for n slots and idle for i slots given that the busy period begin with k customers in the queue and where $b_k(n)$ is the probability that just n arrivals join the queue from a batch when there be k spaces in the queue.

Observe that, for $n < N$, both $busy(n,i;k)$ and $busy(n,i)$ take the same values in the finite buffer system as they do in the infinite buffer system. Observe further that

$$busy(n+1,i;1) = \sum_{q=1}^{\min(N,n)} a(1)\, b_N(q)\, busy(n,i;q) = a(1)\, busy(n,i) \qquad (51)$$

and that, when the idle period is independent of the busy period, the probability that the idle period be i slots is

$$\frac{a(i+!1)}{1-a(1)} \qquad (52)$$

Example

When the batch renewal process has both batch sizes and intervals between batches distributed as shifted Generalized Geometric (as in the example of Section 4.3) the idle periods are independent of the busy periods and are distributed geometrically. Thus only the busy period distribution needs to be considered. A typical form is shown in figure 7.

In departures from an infinite buffer the burst length is distributed as the sum of two geometrics. For moderate values of β_a (correlation of counts) there is a marked knee in the graph.

For finite buffers the form of the burst length distribution is more complex. Two features are obvious in figure 7.

- First, there is a 'hump' or accumulation of mass at burst lengths just longer than the buffer size N. The reason is intuitively obvious because, on the one hand, the probability of any busy periods less than N slots is the same for both finite and infinite buffer queues but, on the other hand, in comparison to the infinite buffer the finite buffer reduces the probability of longer busy periods.
- Secondly, the tail of the distribution depends upon the location of the knee. This is most readily explicable in terms of the limited 'memory' of the finite buffer queue: at any time the state of a queue of capacity N and deterministic service time of one slot is independent of its state at any time which is more than N slots earlier. If the knee occurs after the finite buffer distribution separates from the infinite buffer distribution (at burst length N) then the queue 'memory' includes the knee, which appears as waviness in the tail. Whereas, if the finite buffer distribution does not include the knee the tail is relatively straight.

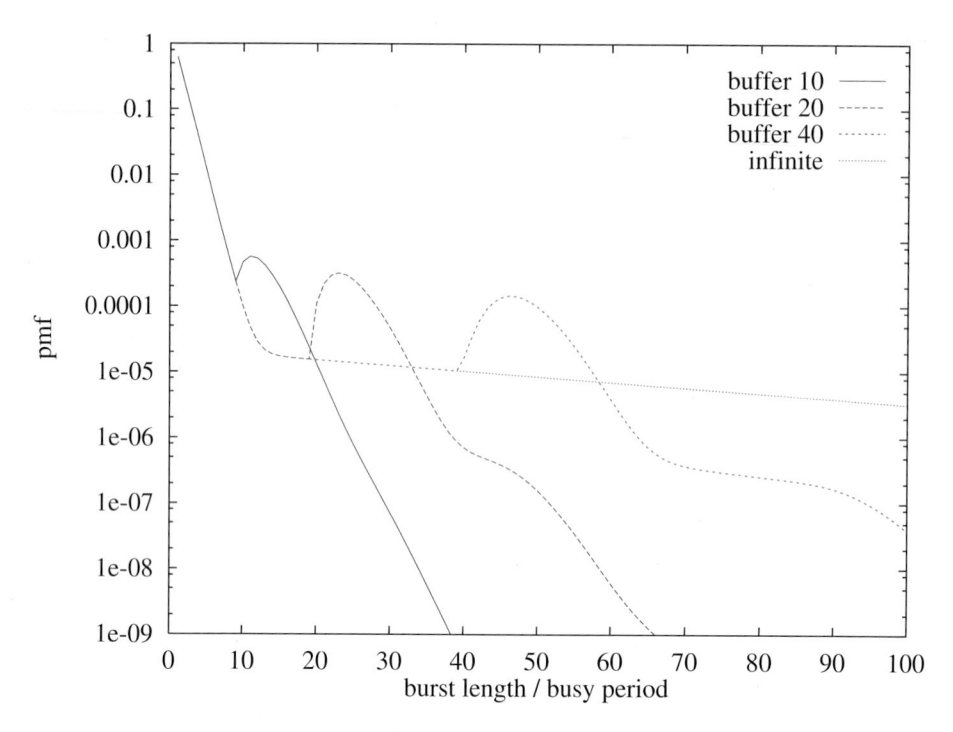

Fig. 7. Pmf of departure process burst length (busy period) for mean batch size $b = 1.25$, mean interval $a = 6.25$ slots between batches, intensity $\lambda = 0.25$, $\beta_a = 0.25$ and $\beta_b = 0.99$ for finite buffers of size 10, 20 and 40 and for infinite buffer

6 Equivalence in Discrete Space Discrete Time Processes

In this section we consider a large class of internal models of discrete time traffic processes. The discussion has relevance to the design of the experiment that is described in section 7 and also to points raised in Section 9. The batch renewal process is always representable in the class, e.g. as a trivial batch Markov renewal process that has one phase only.

Correspondences between some representations of processes are well known, for example between the semi-Markov and Markov renewal processes [2] and between the MAP and Neuts process [17]. In the context of discrete time processes, correspondence between other representations can be seen and this observation leads to the notion of a class of processes in which each member admits a variety of representations, so that the class might equally well be defined in terms of any of the representations. The class that is considered here is that of *all processes that admit representation as MMBBP's over countable phase spaces*. Figure 8 shows some relationships between three representations of that class: batch Markov renewal process or SMP; DBMAP; MMBBP.

1. An arbitrary MMBBP may be described as a batch Markov renewal process in which the sojourn (in a phase between two points of the batch Markov

```
                    2
   MMBBP  ⇄  DBMAP
             4
     |      ↗
    1|    ↗3
     ↓  ↙
  batch Markov
  renewal process
```

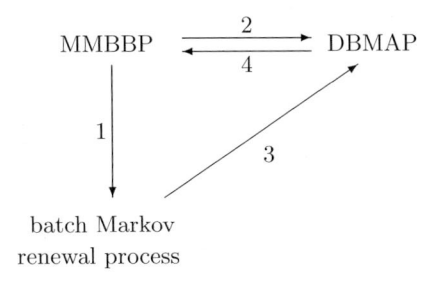

Fig. 8. Equivalence of discrete time models. The arrows show subset relations, e.g. arrow 1 shows that the Markov modulated processes are special cases of the batch Markov renewal process.

renewal process) is always one slot long. Therefore the set of MMBBP's is a subset of the set of the batch Markov renewal processes.

2. An arbitrary MMBBP may be described as a DBMAP in which the number of arrivals generated at an epoch is conditionally independent of the phase in the next slot given the current phase of the DBMAP. Therefore the set of MMBBP's is a subset of the set of the DBMAP's.

Then, by virtue of the transitivity of the subset relation, it is sufficient to show two further subset relations, such as those labelled 3 and 4 in Figure 8, to establish equivalence between all three representations.

3. For an arbitrary discrete time batch Markov renewal process there may be constructed a DBMAP that is equivalent to the MMBBP in the sense that, at every point in the evolution of the batch Markov renewal process, the same behaviour of the constructed DBMAP is exactly the same as that of the batch Markov renewal process. So, each batch Markov renewal process is representable as a DBMAP. Figure 9 illustrates the essential feature in the construction, which is that the DBMAP should contain a phase j_t for each phase j of the Markov renewal process and each possible sojourn t slots in phase j of the Markov renewal process. Whenever, in the Markov renewal process, there are k arrivals at point n together with a transition from phase i to phase j for a sojourn of t slots there should be correspondingly at epoch τ_n in the DBMAP k arrivals together with a transition from phase i_1 to

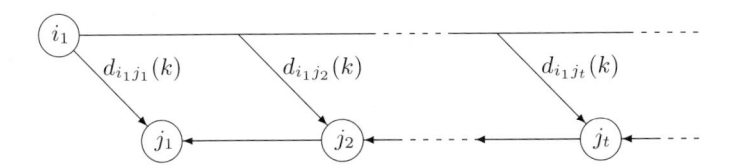

```
  (i₁)
   ╲  d_{i₁j₁}(k)   ╲ d_{i₁j₂}(k)        ╲ d_{i₁jₜ}(k)
    ╲                ╲          - - - -    ╲
    (j₁) ←          (j₂) ←  - - - -      (jₜ) ← - - -
```

Fig. 9. DBMAP phases and transitions corresponding to Markov renewal process phase transition $i \to j$

DBMAP phase ϕ	i	j	k
MMBBP phase ϕ'	(i,j)	(j,k)	(k,ℓ)
slot number	t	$t+1$	$t+2$
epoch number	$t-1$	t	$t+1$	$t+2$

Fig. 10. Relation between phase $\phi(\cdot)$ of DBMAP and phase $\phi'(\cdot)$ of equivalent MMBBP: $\phi'(t) = (i,j)$ whenever $\phi(t) = i$ and $\phi(t+1) = j$

phase j_t followed by successive transitions to phases $j_{t-1}, \ldots, j_1$ at the $t-1$ successive epochs $\tau_n+1, \ldots, \tau_n+t-1 = \tau_{n+1}-1$.
4. For an arbitrary DBMAP an equivalent MMBBP may be constructed and so, in that sense, each DBMAP is representable as a MMBBP. Figure 10 illustrates the essential feature in the construction, which is that the MMBBP should contain a phase (i,j) for each phase transition $i \to j$ of the DBMAP. Whenever in the DBMAP there be n arrivals generated at an epoch t together with a transition from phase i to phase j and followed (at epoch $t+1$) by a transition to phase k, there should be correspondingly at epoch t in the constructed process n arrivals generated and transition from phase (i,j) to phase (j,k).

The equivalence between the three representations of traffic models may suffice to show why the MMBBP may be used for the experiment that is described in Section 7. However, restricting consideration to just three types of internal model does seem arbitrary. That the equivalence might be more general provokes the conjecture that the class (of *all processes that admit representation as MMBBP's over countable phase spaces*) might properly suffice for internal models of all realizable (discrete time) traffic.

7 Biased Results from Other Models

The batch renewal process is, in information theoretic terms, the least biased choice of traffic process given only the customary measures of correlation (e.g. indices of dispersion) [12]. In the batch renewal process there is no semblance of burst structure or, indeed, of any other feature other than correlation. The batch renewal process may be described fairly as "pure correlation". The question then arises as to what is the effect upon (say) queue performance caused by the bias of chosing some other model for traffic that is characterised by correlation? This section describes an experiment designed to provide some insight.

The reference model chosen was the MMBBP/D/1 queue. Because nothing was known about the impact of chosing some process other than the batch renewal process, it was desirable to choose a form of arrivals process that readily permitted extremes of behaviour. The arrivals process was chosen to be a

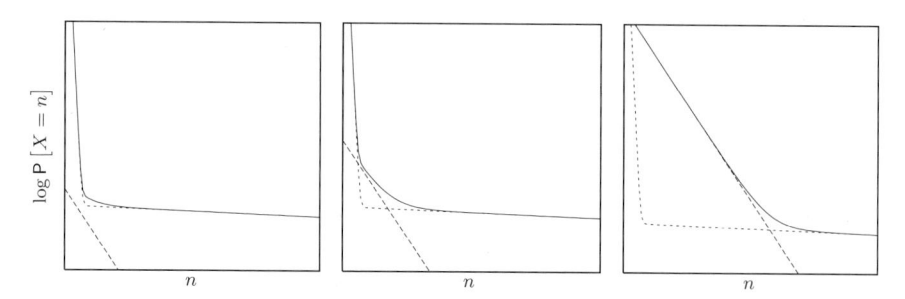

Fig. 11. Effect of adding a third phase to a 2-phase distribution
$\mathsf{P}\left[X=n\right] = (1-\alpha)\bigl(A(1-x_1)x_1{}^n + (1-A)(1-x_2)x_2{}^n\bigr) + \alpha(1-x_3)x_3{}^n$
with $A = 0.998$, $x_1 = 0.01$, $x_2 = 0.99$, $x_3 = 0.7$ and $\alpha = 10^{-4}, 10^{-2}, 0.5$.
Solid line: 3-phase distribution $\mathsf{P}\left[X=n\right]$; dotted line: 2-phase distribution, as with $\alpha = 0$; dashed line: contribution $\alpha(1-x_3)x_3{}^n$ of third phase.

2-phase MMBBP in which the distributions of counts in each phase were GGeo (an extremal 2-phase distribution). Figure 11 is intended to show why it was thought that two phases might provide extreme behaviour by illustrating how often, in a distribution composed of the weighted sum of geometric terms, either two phases dominate or the distribution is 'smoothed'. The other choices — deterministic service, with one customer served per slot, and infinite capacity — were made to avoid effects not directly related to the arrivals process.

The first part of the experiment was conducted with high intensity traffic ($\lambda = 0.9$ customer/slot), mean intensity 0.8 in one phase and 1.1 in the other but while varying mean sojourn in the phases and variance of counts in each phase. For each set of MMBBP parameters, the corresponding batch renewal process (i.e. that with counts and intervals covariances identical to those of the MMBBP) was determined as described in Section 3, the queue length distributions were computed for both the MMBBP/D/1 and the ·/D/1 queue fed by the corresponding batch renewal process. Figure 12 shows a typical result. The smaller geometric rate (steeper first segment) in the MMBBP/D/1 queue length distribution clearly implies lower waiting time, less jitter and, extrapolating to the finite buffer case, lower cell loss rate as compared with the distribution attributable to the correlation alone. In other words, the MMBBP yields optimistic results in the cases considered.

That observation from the first part was formulated as *the proposition*

> the smallest geometric rate (steepest segment) in the MMBBP/D/1 queue length distribution is less than that of the queue fed by the corresponding batch renewal process

The experiment was then extended to randomly generated MMBBP's. For each of 2–, 3– and 6–phase models, 4000 MMBBP's were generated randomly: for each MMBBP the phase transition matrix entries were taken randomly from a uniform distribution and then each row of the matrix was normalised; for each MMBBP the mean intensity for each phase was taken randomly from a uniform

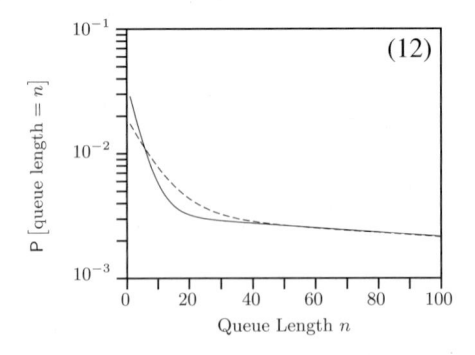

[Graph showing P[queue length = n] vs Queue Length n, with solid and dashed curves, labeled (12)]

Fig. 12. Typical result from first part of experiment: solid line for MMBBP; dashed line for batch renewal process

distribution and scaled (5 times) to give overall mean rates (i.e. with respect to the stationary phase distribution for the particular MMBBP) of 0.1, 0.3, 0.5, 0.7 and 0.9. For each MMBBP with each value of mean rate, the largest poles of the MMBBP/D/1 and corresponding batch renewal process queue length distribution generating functions were computed and compared. Then all the (60 000) computations and comparisons were repeated after biasing each transition matrix to ensure diagonal dominance: the bias was applied to each row by halving each off-diagonal entry and increasing the diagonal entry in compensation.

Table 1 shows the proportion of cases in which *the proposition* was true. The difference between the "unbiased" and "biased" columns shows the significance of diagonal dominance in the phase transition matrix, especially for the 6-phase models, and that the effect is increased somewhat by high intensity. However

Table 1. Proportion of cases supporting *the proposition*

degree	λ	unbiased	biased
2	0.1	81%	>99%
	0.3	81%	>99%
	0.5	81%	>99%
	0.7	81%	>99%
	0.9	81%	>99%
3	0.1	80%	>99%
	0.3	80%	>99%
	0.5	80%	>99%
	0.7	80%	>99%
	0.9	80%	>99%
6	0.1	46%	88%
	0.3	46%	88%
	0.5	46%	89%
	0.7	46%	93%
	0.9	45%	94%

diagonal dominance alone is not sufficient (as shown by the "biased" column) nor is it necessary: in the "unbiased" column the proportion of results that support *the proposition* is far greater than that of diagonally dominant matrices among randomly generated matrices. The results strongly suggest that the MMBBP is likely to yield optimistic results, under conditions of positive traffic correlation of counts and intervals arising from long sojourn in each phase, c.f. the "biased" column of Table 1.

8 Cost Effective Approximation for Queueing Network Analysis

Exact analysis of queueing network models of communications systems can be intractable. So it is important to know in what circumstances a simpler process can approximate to tolerable accuracy the behaviour of the more complex process.

The formulation of the batch renewal process makes clear (c.f. equations 13 and 14) why correlation in SRD processes has much the same impact on measures of queue performance as does variability of interarrival time in renewal processes and variability of counts from batch Bernoulli processes. This observation leads to the notion that the effect of both counts and interarrival correlation in SRD traffic input to a queue might be captured (to some tolerable approximation) by variability in either counts alone (i.e. by an 'equivalent' batch Bernoulli process) or interarrival times (i.e. by an 'equivalent' renewal process). The accuracy on measures of queue performance of substituting some 'equivalent' processes for SRD traffic processes has been investigated in [1,3].

There had previously been some indication that such 'equivalent' processes provided tolerable accuracy in analysis of queueing networks. Typically, at each queue in the network the arrivals traffic is a superposition of departures from other queues and of external traffic. Clearly such traffic is correlated. However, in a number of fast algorithms based upon entropy maximisation and queue-by-queue decomposition (such as given in [10]), the input to each queue is treated *as if* it were completely free of correlation (c.f. Jackson networks). In [10] and similar algorithms, the input to each queue is, in effect, replaced by an 'equivalent' process with the same mean and variance of interarrival time. Nevertheless, the algorithms typically give good accuracy, in comparison with simulation results, when applied to networks of arbitrary topology and complexity. They had previously been used for networks for which input traffic was uncorrelated.

In a particular application of queue-by-queue decomposition, Kouvatsos et al.[13] considered networks fed by SRD traffic represented by the sGGeo process. In this context, the existing algorithm devised in [10], based upon the principle of maximum entropy, was extended to treat input sGGeo traffic as if substituted, with a tolerable accuracy, by uncorrelated traffic represented by an ordinary GGeo [10,12] for which the count distribution had the same mean and variance as those of the sGGeo.

Another queue, that might be used as a building block in analysis of an open queueing network of nodes with multiple servers and with correlated traffic is the GI^G/Geo/c, analyzed by Writtevrongel, Bruneel and Vinck [20]. This queue has a general batch renewal arrival process, infinite buffer and c servers with independent geometrically distributed service times. The analysis of this queue was based on the use of generating functions in conjunction with complex analysis and contour integration. Consequently, new analytic expressions for the generating functions of the system contents during an arrival slot were obtained as well as at an arbitrary slot. Moreover, the delay analysis the queue under the first-come-first-served discipline was presented.

An alternate approach, for each queue in a network with correlated input, is that followed by Laevens [14] to relate the output process of each queue to its input.

9 Open Problems and Research Topics

1. For given correlation, the batch renewal process is known to be the least biased choice of all possible processes which exhibit that correlation. However there are some patterns of correlation for which the corresponding batch renewal process is improper. So it might be as well to add to the previous statement the rider "... provided that the batch renewal process be proper". The question then arises as to what might be the least biased choice of process
   (a) when the correlation be such that the batch renewal process not be proper,
   (b) when other constraints (in addition to correlation) be applicable; perhaps the most interesting is that of the least biased choice of process for given count and intervals correlation given also that the traffic is on a line i.e. no simultaneous arrivals.
2. Given that there are some patterns of correlation for which the corresponding batch renewal process is improper, the question then arises as to whether such improper batch renewal process may be used for performance prediction of (for example) a buffer fed by the traffic.
   (a) For finite buffer queues, numeric methods may be derived, possibly based upon the general relations given in Section 4. In effect, that would be to attempt to find the stationary vector for a transition matrix which has some entries negative. Clearly, a naïve implementation would be unstable.
   (b) For batch renewal processes having especially simple forms (such as the sGGeo example used in Section 4) it may be possible to derive explicit closed form solutions (assuming that the parameters corresponded to a proper batch renewal process) and then apply the explicit form (with the parameters of the improper batch renewal process). However, except for the simplest forms, this approach is unworkable.
   (c) A potentially rewarding approach is to view the observed traffic as having resulted from a stream from which customers have been removed.

That is equivalent to saying that the traffic contains 'negative customers' [6]. Then the observed traffic might be modelled as the merge of traffic from a (proper) batch renewal process plus randomly inserted negative customers. This approach appears feasible because, for any traffic stream, injecting (positive) customers randomly affects the generating functions such that $K(0) + L(0)$ increases faster than $K(1-)$ and $L(0)$ increases faster than $K_+(1-)$.

3. The demonstration in Section 6 between some representations of traffic processes provokes further questions. For example, the constructions used to demonstrate relations 3 and 4 of Figure 8 depend upon the phase space being discrete but none of the relations depend in any way upon the form of distributions of counts at each point of the process nor, indeed, upon the such distibutions being discrete.

    The equivalence argument works just as well when the counts have continuous distribution. So, could the duality between counts and interarrival times be exploited to show equivalence between various forms of (discrete space) continuous time models?

4. It would be very useful to know if the conjecture (that is offered at the end of Section 6) were true. Then, when attempting to match some measured traffic, there would be no inherent bias in seeking an appropriate model only amongst the most convenient representation.

    However, the conjecture might be not testable. A conjecture of equal practical value might be expressed as *there is no test that in finite time discriminates between traffic generated by process X and that from some MMBBP over a countable phase space* for every process X that is not known to be a member of the class.

    Again, that second conjecture might not be testable. But it might be possible to make some progress towards proving or disproving the second conjecture for some characteristics of traffic or for some interesting subclasses of traffic.

5. It is clearly important to have fast algorithms, such as those mentioned in Section 8, that provide reliably accurate approximations for network performance. Our confidence in their results is based upon the empirical evidence that the results have been good for all cases – so far.

    (a) More experiments are needed on networks with correlated inputs.

    (b) The bases of the implicit assumptions and approximations inherent in the algorithms need to be examined – both to discover simple expressions for the accuracy of the results (or bounds on the errors) and also to see whether the accuracy could be improved without degrading the speed of the algorithms.

References

1. Dimakopoulos, G.A.: On the Approximation of Complex Traffic Models on ATM Networks, M.Phil. Dissertation. Postgraduate School of Computing and Mathematics, University of Bradford (2000)

2. Disney, R.L., Kiessler, P.C.: Traffic Processes in Queueing Networks: A Markov Renewal Approach. The John Hopkins University Press, Baltimore (1987)
3. Fretwell, R.J., Dimakopoulos, G.A., Kouvatsos, D.D.: Ignoring Count Correlation in SRD Traffic. In: Bradley, J.T., Davies, N.J. (eds.) Proc. 15th UK Perf. Eng. Workshop, pp. 285–294. UK Performance Engineering Workshop Publishers (1999)
4. Fowler, H.J., Leland, W.E.: Local Area Network Traffic Characteristics, with Implications for Broadband Network Congestion Management. IEEE JSAC 9(7), 1139–1149 (1991)
5. Andrade, J., Martinez-Pascua, M.J.: Use of the IDC to Characterize LAN Traffic. In: Kouvatsos, D. (ed.) Proc. 2nd. Workshop on Performance Modelling and Evaluation of ATM Networks, pp. 15/1–15/12 (1994)
6. Gelenbe, E.: Random Neural Networks with Positive and Negative Signals and Product Form Solution. Neural Computation 1(4), 502–510 (1989)
7. Gordon, J.J.: Pareto Process as a Model of Self-Similar Packet Traffic. In: Proc. Globecom 1995, Singapore, pp. 2232–2236 (1995)
8. Gusella, R.: Characterizing the Variability of Arrival Processes with Indexes of Dispersion. IEEE JSAC 9(2), 203–211 (1991)
9. Heffes, H., Lucantoni, D.M.: A Markov Modulated Characterization of Packetized Voice and Data Traffic and Related Statistical Multiplexer Performance. IEEE JSAC 4(6), 856–868 (1986)
10. Kouvatsos, D.D., Tabel-Aouel, N.M., Denazis, S.G.: Approximate Analysis of Discrete-time Networks with or without Blocking. In: Perros, H.G., Viniotis, Y. (eds.) High Speed Networks and their Performance, vol. C-21, pp. 399–424. North-Holland, Amsterdam (1994)
11. Kouvatsos, D.D., Fretwell, R.: Discrete Time Batch Renewal Processes with Application to ATM Switch Performance. In: Hillston, J., et al. (eds.) Proc. 10th. UK Computer and Telecomms. Performance Eng. Workshop, September 1994, pp. 187–192. Edinburgh University Press, Edinburgh (1994)
12. Kouvatsos, D., Fretwell, R.: Closed Form Performance Distributions of a Discrete Time $GI^G/D/1/N$ Queue with Correlated Traffic. In: Fdida, S., Onvural, R.O. (eds.) Enabling High Speed Networks, October 1995, pp. 141–163. IFIP Publication, Chapman and Hall (1995)
13. Kouvatsos, D.D., Awan, I.U., Fretwell, R., Dimakopoulos, G.: A Cost-effective Approximation for SRD Traffic in Arbitrary Multi-buffered Networks. Computer Networks 34, 97–113 (2000)
14. Laevens, K.: The Output Process of a Discrete Time $GI^G/D/1$ Queue. In: Proc. 6th IFIP Workshop on Performance Modelling and Evaluation of ATM Networks, Research Papers, pp. 20/1–20/10 (July 1998)
15. Leland, W.E., Taqqu, M.S., Willinger, W., Wilson, D.V.: On the Self-Similar Nature of Ethernet Traffic (Extended Version). IEEE/ACM Transactions on Networking 2(1), 1–14 (1994)
16. Li, W.: Performance Analysis of Queues with Correlated Traffic, PhD Thesis (University of Bradford) (2007)
17. Lucantoni, D.M.: The BMAP/G/1 Queue: A Tutorial. In: Donatiello, L., Nelson, R. (eds.) SIGMETRICS 1993 and Performance 1993. LNCS, vol. 729, pp. 330–358. Springer, Heidelberg (1993)
18. Neuts, M.F.: A Versatile Markovian Point Process. J. Appl. Prob. 16, 764–779 (1979)

19. Sriram, K., Whitt, W.: Characterizing Superposition Arrival Processes in Packet Multiplexers for Voice and Data. IEEE JSAC 4(6), 833–846 (1986)
20. Wittevrongel, S., Bruneel, H., Vinck, B.: Analysis of the discrete-time $G^{(G)}$/Geom/c queueing model. In: Gregori, E., Conti, M., Campbell, A.T., Omidyar, G., Zukerman, M. (eds.) NETWORKING 2002. LNCS, vol. 2345, pp. 757–768. Springer, Heidelberg (2002)
21. Xiong, Y., Bruneel, H.: Buffer Contents and Delay for Statistical Multiplexers with Fixed Length Packet Train Arrivals. Performance Evaluation 17(1), 31–42 (1993)

Local Area Networks and Self-similar Traffic

Hadi Larijani

Dept. of Communication, Network & Elec. Eng.
School of Eng. and Computing
Glasgow Caledonian University, UK
hla@gcal.ac.uk

Abstract. Ethernet is one of the most popular LAN technologies. The capacities of Ethernet have steadily increased to Gbps and it is also being studied for MAN implementation. With the discovery that real network traffic is self-similar and long-range dependent, new models are needed for performance evaluation of these networks. One of the most important methods of modelling self-similar traffic is Pseudo self-similar processes. The foundations are based on the theory of decomposability, which was developed approximately 20 years ago. Many researchers have revisited this theory recently and it is one of the building blocks for self-similar models derived from short-range dependent processes. In this paper we will review LANs, self-similarity, several modelling methods applied to LAN modelling, and focus on pseudo self-similar models.

Keywords: Ethernet, self-similarity, decomposability, pseudo self-similar processes.

1 Introduction

Performance modeling of computer networks is essential to predict the effects of increases in traffic and to allow network managers to plan the size of upgrades to equipment. Historically the assumption has been made that traffic followed Poisson assumptions. That is, the probability of a packet arriving for onward transmission during any short interval is independent of the arrivals in any of the intervals, and depends only on the mean arrival rate and the length of the interval considered. Similarly, the length of packets is usually considered to be exponentially distributed. Both these assumptions are false, but models using them have been successfully validated, mostly in the wide area network field.

Recent measurements of network teletraffic have revealed properties which may have significant consequences to the modelling of computer networks, especially Broadband ISDN and ATM networks. Although self-similarity is not a new concept to the teletraffic community and its origins date back to Mandelbrot's paper in 1965 [1], its consequences were not fully appreciated till the 1990's. One of the most highly acclaimed papers which might be considered as the spark to an explosion of research into this area and its wide scope of applications is the fascinating paper written by Leland, Willinger, Taqqu and Wilson [2]. As Stallings [3] very well stated, this paper rocked the field of network performance modelling and it is arguably the most important networking paper of the decade. The main finding of this paper is that real traffic does not obey the Poisson assumptions that have been used for years for analytical modelling. 'Real' network traffic is more bursty and exhibits greater variability than previously suspected. The paper reported the results of a massive study of Ethernet traffic and demonstrated that it has self-similar statistical properties at a range of time scales: milliseconds, seconds, minutes, hours, even days and weeks. What this means is that the network looks the same when measured over time intervals ranging from milliseconds to minutes to hours. It was later found that self-similarity also occurs in ATM traffic, compressed digital video streams, World Wide Web (WWW) traffic, Wide Area Networks (WAN) traffic, etc., [4]. Self – similar traffic is very different from both conventional telephone traffic and from the currently accepted norm for models of packet traffic. Conclusions of recent empirical studies regarding the nature of network traffic have all concurred on one issue: the data is self – similar in nature. What are the ramifications of this discovery? Consequently, the following needs to be addressed:

1. What are the performance implications of self – similar data traffic upon telecommunication systems?
2. How can researchers utilize queuing models to study this behavior?

Note that some researchers do not believe that queuing models are sufficient and suggest research on new tools for this end [5]. Self – similarity has immense impact on a wide variety of fields such as: traffic modelling, source characterization, performance evaluation, analytical modelling, buffer sizing, control mechanisms, etc. For example, Partridge [6] foresaw the implications of this in congestion control and stated 'anyone interested in congestion control should read the paper' [2]. A.A. Kherani [7] has looked into the effect of adaptive window control in LRD network traffic. His study indicates that the buffer behaviour in the Internet may not be as poor as predicted from an open loop analysis of a queue fed with LRD traffic; and it shows that the buffer behaviour (and hence the throughput performance for finite buffers) is sensitive to the distribution of file sizes.

A. Veres et al. [8] analyzed how TCP congestion control can propagate self-similarity between distant areas of the Internet. This property of TCP is due to its congestion control algorithm, which adapts to self-similar fluctuations on several timescales. The mechanisms and limitations of this propagation were investigated. It was demonstrated that if a TCP connection shared a bottleneck link with a self-similar background traffic flow, it propagates the correlation structure of the background traffic flow asymptotically, above a characteristic timescale. The cut-off timescale depends on the end-to-end path properties, e.g. round-trip time and average window size, and the receiver window size in case of high-speed connections. It was also shown that even short TCP connections can propagate long-range correlations effectively. In case when TCP encounters several bottleneck hops, the end-user perceived end-to-end traffic was also long-range dependent and it was characterized by the largest Hurst exponent. Through simple examples, it was shown that self-similarity of one TCP stream can be passed on to other TCP streams that it was multiplexed with.

1.1 What Is Self-similarity?

Self-similarity and fractals are notions pioneered by B.B. Mandelbrot [9]. He describes the phenomenon where a certain property of an object – for example, a natural image, the convergent subdomain of certain dynamical systems, a time series (the mathematical object of interest) – is preserved with respect to scaling in space and time. If an object is self-similar or fractal, its parts, when magnified, resemble – in a suitable sense – the shape of the whole object.

Stochastic self-similarity is of more importance to our study of network traffic. Analysis and modelling of. computer network traffic is a daunting task considering the amount of available data. This is quite obvious when considering the spatial dimension of the problem, since the number of interacting computers, gateways and switches can easily reach several thousands, even in a local area network (LAN) setting. This is also true on the time dimension: Willinger and Paxson in [10] cite the figures of 439 million packets and 89 gigabytes of data for a single week record of the activity of a university gateway in 1995. The complexity of the problem further increases when considering wide area network (WAN) data [11]. In light of the above, it is clear that a notion of importance for modern network engineering is that of invariants, i.e., characteristics that are observed with some reproducibility and independently of the precise settings of the network under consideration. In this study we focus on one such invariant related to the time dimension of the problem, namely, long-range dependence or self-similarity. A striking feature, which collaborate the conjecture that self-similarity, long-range dependence, and heavy-tailness are really meaningful traffic invariants, is that they can be observed, to some extent, without using any specific experimental protocol. Accordingly, the traffic data in Figure 1 corresponds to actual 100 Mb/s Ethernet traffic, which was measured on a server in Drexel University [12]. To generate this trace, all packets of private connections with this server, broadcasting, and multicasting were captured and time-stamped during several hours. Cappe et. al. [12] only consider byte counts (size of the transferred data) measured on 10 ms intervals, which is the data represented in the top plot of Figure 1. The overall length of the record is about three hours (exactly, 10^4 s). The three other plots in Figure 1 correspond to the "aggregated" data obtained by accumulating the data counts on increasing time intervals. The striking feature in it is that the aggregation is not really successful in smoothing out the data. The aggregated traffic still appears bursty in the bottom plot despite the fact that each point in it is obtained as the sum of one thousand successful values of the series displayed in the top plot of Figure 1.

Similar characteristics have been observed in many different experimental setups, including both LAN and WAN data (e.g., [13], [14], [15], and the references therein).

Unlike deterministic fractals, the objects corresponding to Figure 1 do not possess exact resemblance of their parts with the whole at finer details. Here, we assume that the measure of "resemblance" is the shape of a graph with the magnitude suitably normalized. Indeed, for measured traffic traces, it would be too much to expect to observe exact, deterministic self-similarity given the stochastic nature of many network events (e.g., source arrival behaviour) that collectively influence actual network traffic. If we adopt the view that traffic series are sample paths of stochastic processes and relax the measure of resemblance, say, by focusing on certain statistics of the rescaled time series, then it may be possible to expect exact self-similarity of the mathematical objects and approximate similarity of their specific realizations with respect to these relaxed measures. Second - order statistics are statistical properties that capture burstiness or variability, and the autocorrelation function is a yardstick with respect to which scale invariance can be fruitfully defined. The shape of the autocorrelation function – above and beyond its preservation across rescaled time series – will play an important role. In particular, correlation, as a function of time lag, is assumed to decrease polynomially as opposed to exponentially. The existence on nontrivial correlation "at a distance" is referred to as *long-range dependence*.

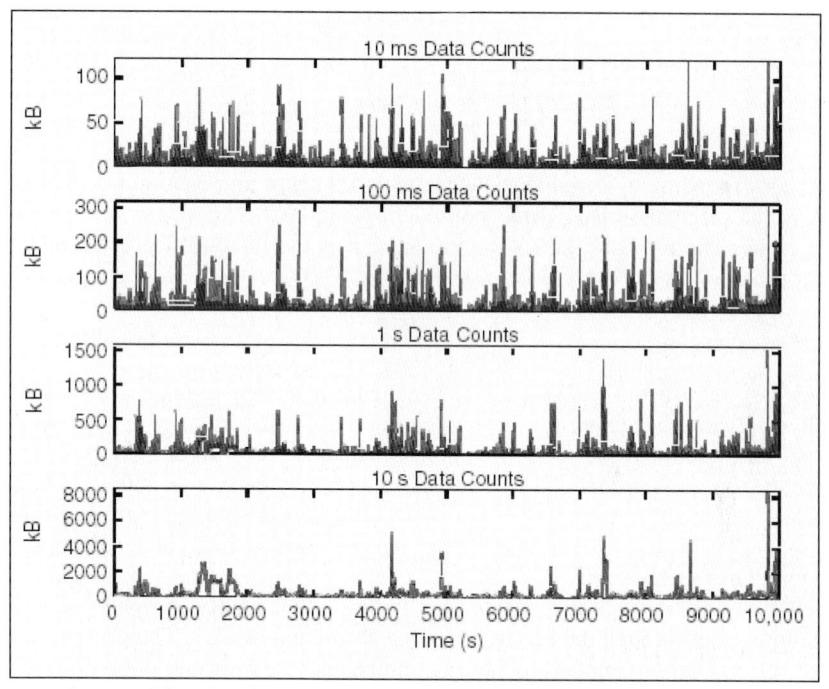

▲ *1. The Drexel data (10,000 s in total) viewed through four different aggregation intervals: from top to bottom, 10 ms, 100 ms, 1 s, and 10 s.*

Fig. 1. Example of Self-Similar Traffic Trace

1.2 Self-similar Processes: Basic Definitions

A self-similar process is invariant in distribution under scaling of time. Intuitively, if we look at several pictures of a self-similar process at different time scales they will all look *similar*. Figure 1 shows the visual differences between distributions of traditional models and self-similar processes in a few scales. There are a number of different, non-equivalent definitions of self-similarity. The standard one states that a continuous time process $Y = \{Y(t), t \in T\}$ is *self-similar* (with self-similarity parameter *H*) if it satisfies the condition:

$$Y(t) \stackrel{d}{=} \alpha^{-H} Y(\alpha t), t \in T, \forall \alpha > 0, 0 \leq H < 1 \quad (1)$$

where the equality is in the sense of finite-dimensional distributions. *H* is known as the *Hurst* parameter, in honour of an early pioneer of the study of self – similarity [16]. While a process *Y* satisfying (1) can never be stationary, it is typically assumed to have stationary increments.

A second definition of self-similarity, more appropriate in the context of standard time series theory, involves a stationary sequence $X = \{X(t), i \geq 1\}$.

Let

$$X^{(m)}(k) = \frac{1}{m} \sum_{i=(k-1)m+1}^{km} X(i) \quad k = 1, 2, \ldots, \quad (2)$$

be the corresponding aggregated sequence with a level of aggregation *m*, obtained by dividing the original series *X*, into non-overlapping blocks of size *m* and averaging every block. The index, *k*, labels the block. If *X* is the increment process of a self-similar process *Y* defined in (1) i.e., (*X(i)* = *Y(i+1)* - *Y(i)*), then for all integers *m*,

$$X \stackrel{d}{=} m^{1-H} X^{(m)} \quad (3)$$

A stationary sequence *X* = *{X(i), i ≥ 1}* is called *exactly self- similar* if it satisfies (3) for all aggregation levels *m*. The second definition of self-similarity is closely related (but not equivalent) to the first. A stationary sequence *X={X(i) ,i ≥ 1}* is said to be *asymptotically self-similar* if (3) holds as $m \to \infty$. Similarly, we call a covariance-stationary sequence *X* = *{X(i), i ≥ 1} exactly second-order self-similar* or *asymptotically second-order self-similar* if $m^{1-H} X^{(m)}$ has the same variance and autocorrelation as *X*, for all *m*, or as $m \to \infty$.

Self-similarity is often investigated not through the equality of finite-dimensional distributions, but through the behaviour of the absolute moments. Thus, a third definition of self-similarity (implied by but not equivalent to the second definition) is simply that the moments must scale. Thus consider:

$$\mu^{(m)}(q) := E|X^{(m)}|^q = \left| \frac{1}{m} \sum_{i=1}^{m} X(i) \right|^q \quad (4)$$

If *X* is self-similar, then $\mu^{(m)}(q)$ is proportional to $m^{\beta(q)}$, i.e., $\log \mu^{(m)}(q)$ is linear to $\log m$.

For a fixed *q*:

$$\log \mu^{(m)}(q) = \beta(q) \log m + C(q) \quad (5)$$

In addition, the exponent $\beta(q)$ is linear with respect to *q*. In fact, since $X^{(m)}(i) \stackrel{d}{=} m^{H-1} X(i)$, we have: $\beta(q) = q(H-1)$.

1.3 Long-Range Dependence

A stochastic process satisfying the following relation is said to exhibit *long-range dependence* (see [17] or [18]):

Let $X = (X_t : t = 0,1,2,...)$ be a *covariance stationary* (sometimes called *wide-sense stationary*) stochastic process with mean μ, variance σ^2 and auto correlation function

$$r(\kappa) \sim \kappa^{-\beta} L_1(\kappa) \qquad \text{as } \kappa \to \infty, \qquad (6)$$

where $0 < \beta < 1$ and L_1 is slowly varying at infinity, that is $\lim_{t \to \infty} L_{1(tx)} / L_{1(x)} = 1$ for all $x > 0$.

1.4 Ethernet

Ethernet is the most widely used local area (LAN) technology. It was used to fill the middle ground between long – distance, low – speed networks and specialized, computer – room networks carrying data at high speeds for very limited distances. Ethernet is well suited to applications where a local communication medium must carry sporadic, occasionally heavy traffic at high peak data rates.

Ethernet network architecture has its origins in the 1960's at the University of Hawaii, where the earliest and simplest random access scheme was developed, (pure-ALOHA) [19]. Another more efficient random access scheme called CSMA (carrier sense multiple access) was developed by Kleinrock's [20] team at UCLA. Ethernet uses an access method called carrier sense multiple access/ collision detection (CSMA/ CD), which was developed at Xerox corporation's Palo Alto Research Center (PARC) in the early 1970's. This was used as the basis for the Institute of Electrical and Electronic Engineers (IEEE) 802.3 specification released in 1980.Shortly after the 1980 IEEE 802.3 specification, Digital Equipment Corporation, Intel Corporation, and Xerox Corporation jointly developed and released Ethernet specification version 2.0, that was substantially compatible with IEEE 802.3.

Together, Ethernet and IEEE 802.3 currently maintain the greatest market share of any LAN protocol. Today, the term Ethernet is often used to refer to all carrier sense multiple access/ collision detection (CSMA/CD) LAN's that generally conform to Ethernet specifications, IEEE 802.3.

1.5 Self-similar Traffic Modeling

It has been known for a long time that network traffic is self – similar in nature. In fact, Mandelbrot was the first to apply the self- similarity concept to the analysis of communication systems [1]. As a consequence of Leland's et al. paper [13] much work has recently appeared addressing various aspects of self-similarity [21]. This research can be classified into the following three categories:

<u>Network Traffic Trace Analysis:</u> This research typically analyses traffic traces from production networks so that statistical tools can be employed to identify the presence of self – similarity. This genre of papers indicates that Long Range Dependence (LRD) is an omnipresent phenomenon encompassing both local area and wide area network

traffic. Furthermore, work in this area has consistently demonstrated that sources such as the WWW and VBR video services exhibit self – similar properties [21].

Simulation and Analytical Models: Research in this area attempts to investigate the effect of self – similar data traffic upon telecommunication systems using either simulation or asymptotic analytical models. These papers conclude that data traffic with properties of self – similarity and LRD significantly degrade system performance. One important result in this area has been the development of Fractional Brownian Noise (FBN) models. FBN models not only capture properties of self – similarity but LRD in the counting process. Consequently, many researchers have utilized computer simulations of FBN models to study the impact of LRD upon queuing behaviour. Typical analytical approaches attempt to find asymptotic bounds concerning selected performance characteristics of the queue (e.g. buffer overflow probability) [22].

Conceptual Analysis: The third category of papers attempts to physically understand how self - similarity arises in production networks. One model proposed by Willinger, Taqqu, and Sherman which attempts to address this issue utilizes an ON/OFF source model (also commonly known as "packet train model") [23]. Their ON/OFF model purports that self – similarity arises as a consequence of independent contributions from power – tail sources. The reason is that mathematical analysis indicates that the superposition of many power – tail ON/OFF sources with alternating ON and OFF periods produces aggregate network traffic which exhibits properties of self-similarity and LRD. The authors conclude that their ON/OFF model is successful in describing characteristics of the measured traffic.

Various methods have been presented for modelling self-similar processes. The two major families of self-similar time series models are fractional Gaussian noise (i.e. increment processes of fractional Brownian motion) and fractional ARIMA processes (auto-regressive integrated moving-average), (a generation of the popular ARIMA time series models). Other stochastic approaches to modelling self-similar features that have been presented are:

— Shot-noise processes
— Linear models with long-range dependence
— Renewal reward processes and their superposition
— Renewal processes or ``zero-rate" processes
— Aggregation of simple short-range-dependent models
— Wavelet analysis
— Approaches based on the theory of chaos and fractals
— Batch Renewal Process

Willinger et al have provided an excellent comprehensive review in [24]. R. Fretwell and D. Kouvatsos [25] use the batch renewal process for both LRD and SRD traffic. They also show some applications of the batch renewal process in simple queues and in queuing network models.

Generally we can group the different approaches researchers take in modeling self-similarity into two distinct 'camps':

- One group tries to develop approaches that attempt mimicking LRD with the help of short-range dependent models (e.g. [26]).
- The second group tries to advent a new set of tools for modeling self-similar processes (e.g. [27]).

K. Maulik and S. Resnick [28] attempt to model this phenomenon, by a model that connects the small time scale behavior with behavior observed at large time scales of bigger than a few hundred milliseconds. There have been separate analyses of models for high speed data transmissions, which show that appropriate approximations to large time scale behavior of cumulative traffic are either fractional Brownian motion or stable Lévy motion, depending on the input rates assumed. Their paper tries to bridge this gap and develops and analyzes a model offering an explanation of both the small and large time scale behavior of a network traffic model based on the infinite source Poisson model. Previous studies of this model have usually assumed that transmission rates are constant and deterministic.

They considered a nonconstant, multifractal, random transmission rate at the user level which results in cumulative traffic exhibiting multifractal behaviour on small time scales and self-similar behaviour on large time scales.

We follow the first approach, and a Markov chain model which shows self-similarity is developed, based on ideas presented by Robert and LeBoudec [29].

There are two levels of modeling: application and aggregate level. Although it is true that network traffic is governed by many physical factors a 'good' model should incorporate those features which are relevant for the problem under consideration. Some of the many factors affecting network traffic flows are:

- user behavior
- data generation, organisation, and retrieval
- traffic aggregation
- network controls
- network evolution

In the Section 2 Pseudo Self-Similar models will be discussed, and a numerical solution suggested for the queuing behavior of a Self-Similar LAN. Experiments on a live Ethernet network will be presented in Section 3 with validation by Opnet simulation application. Finally conclusions will be presented in Section 4.

2 Pseudo Self-similar Models

2.1 Foundations of the Model

Courtois's [30] theory of decomposability is based on the important observation that large computing systems can effectively be regarded as nearly decomposable systems. Systems are arranged in a hierarchy of components and subcomponents with strong interactions within components at the same level and weaker interactions between other components. Near decomposability has been observed in domains such as: in economics, biology, genetics and social sciences. The pioneers of this theory are

Simon and Ando [31]. What they stated is that aggregation of variables in a nearly decomposable system; we must separate the analysis of the short - term and the long - term dynamics. They proved two major theorems. The first says that a nearly decomposable system can be analyzed by a completely decomposable system if the intergroup dependences are sufficiently weak compared to the intragroup ones. The second theorem says that even in the long - term, the results obtained in the short - term will remain approximately valid in the long - term, as far as the relative behavior of the variables of the same group is concerned. Robert and Le Boudec [29] state that LAN traffic is composed of different timescales. The Markov chain proposed is in fact decomposable at several levels. In a first step, the development is done for only one level of decomposability. The Markov chain to consider is presented in section 2.3 and it is characterized by its transition matrix $(n*n)A$ and its state probabilities, $\pi(\pi_{t+1} = \pi_t A)$, A is nearly decomposable. Let A^* be completely decomposable, then A^* is composed of squared submatrices placed on the diagonal:

$$A^* = \begin{pmatrix} A_1^* & \cdots & 0 & 0 \\ 0 & A_2^* & \cdots & 0 \\ 0 & \cdots & \cdots & 0 \\ 0 & 0 & \cdots & A_N^* \end{pmatrix}$$

The remaining elements are zero. If we apply the first theorem of Simon and Ando [31] we can develop the general form of matrices that are nearly completely decomposable with the form described in section 2.3. A is defined in section 2.3 and A^* is defined as below:

$$A^* = \begin{cases} 1-1/a-1/a^2-\cdots-1/a^{n-1} & 1/a & 1/a^2 & \cdots & 0 \\ q/a & 1-q/a & 0 & \cdots & 0 \\ (q/a)^2 & 0 & 1-(q/a)^2 & \cdots & 0 \\ \cdots & \cdots & \cdots & \cdots & 0 \\ 0 & 0 & 0 & \cdots & 1 \end{cases}$$

with $q \langle a$, A^* is a non-ergodic matrix.

2.2 Pseudo Long-Range Dependent Process

Mathematically, the difference between short – range and long – range dependencies is clear, for a short - range dependent process:

$$\sum_{\tau=0}^{\infty} Cov(X_t, X_{t+\tau}) \text{ is convergent}$$

Spectrum at 0 is finite

$Var(X^{(m)})$ is for large m asymptotically of the form $VarX / m$

The averaged process $X_k^{(m)}$ tends to second-order pure noise as $m \to \infty$

For a long-range dependent process:

$$\sum_{\tau=0}^{\infty} Cov(X_t, X_{t+\tau}) \text{ is divergent}$$

Spectrum at 0 is singular

$Var(X^{(m)})$ is of the form $m^{-\beta}$ (for large m asymptotically)

The averaged process $X_k^{(m)}$ does not tend to second-order pure noise as $m \to \infty$.

All stationary autoregressive-moving average processes of finite order an all Markov chains (including semi – Markov processes) are included in the first category. In the second category, we have the fractional Brownian motion, ARIMA processes, and chaotic maps which have long – range dependencies. If we look more closely to these definitions we see that a process having "long – term dependences", but which is limited, is considered as a short – term dependent process. This is exactly the case with Ethernet measurements at Bellcore. If we consider the number of Ethernet packets arriving at a time interval 1 s to be our process, then over 4-5 orders of magnitude we observe long-term dependences. So our process looks the same (distribution wise) for 10, 100, 1000, 10000 s. However, in the order of days, researchers at Bellcore have observed a stabilization of the index of dispersion indicating a lack of self-similarity. So according to our previous definitions, a short - term dependent process would be sufficient to model LAN traffic. The difference with the other processes (Poisson, ON-OFF, etc.) is striking and that is why they should be categorized differently.

Therefore Robert and Le Boudec proposed to name them Pseudo long – range dependent processes:

"A pseudo long – range dependent process is able to model (as well as an (exactly) long – range process) aggregated traffic over several timescales".

This reflects the fact that in practice, we have always a finite set of data, and asymptotic conditions are never met.

2.3 Suggested Process

A discrete time Markov modulated model for representing self-similar data traffic was proposed by Robert and Le Boudec [29]. The cell arrivals on a slotted link is considered: call X_t the random variable representing the number of cells (assumed to be 0 or 1) during the tth time slot, namely during time interval $[t-1, t)$. Let $Y_t = i$ be the modulator's state i, $i \in 1, 2, 3, ..., n$, at time t. The arrivals of the cells are modulated by a n-stated discrete time Markov chain with transition probabilities α_{ij} (t_1, t_2) = $Pr(Y_{t2} = j \mid Y_{t1} = i)$. Let ϕ_{ij} denote the probability of having j cells in one time slot, given that the modulator's state is i; more specifically $\phi_{ij} = Pr(Y_t = j \mid Y_t = i)$.

The modulated chain state probabilities are noted as $\pi_{it} = \Pr(Y_t = i)$, i is referred to the modulator's stated and t to the time. The Markov modulated chain is assumed stationary and homogeneous.

The Markov chains that they suggested using is the following:

$$A = \begin{cases} 1-1/a-1/a^2-\cdots-1/a^{n-1} & 1/a & 1/a^2 & \cdots & 1/a^{n-1} \\ q/a & 1-q/a & 0 & \cdots & 0 \\ (q/a)^2 & 0 & 1-(q/a)^2 & \cdots & 0 \\ \cdots & \cdots & \cdots & \cdots & \cdots \\ (q/a)^{n-1} & 0 & 0 & \cdots & 1-(q/a)^{n-1} \end{cases}$$

$$\Lambda = \begin{pmatrix} 1 & 0 & 0 & \cdots & 0 \\ 0 & 0 & 0 & \cdots & 0 \\ 0 & 0 & 0 & \cdots & 0 \\ \cdots & \cdots & \cdots & \cdots & \cdots \\ 0 & 0 & 0 & \cdots & 0 \end{pmatrix}$$

So, the Markov chain has only three parameters: a, q, plus the number of states in the Markov chain n.

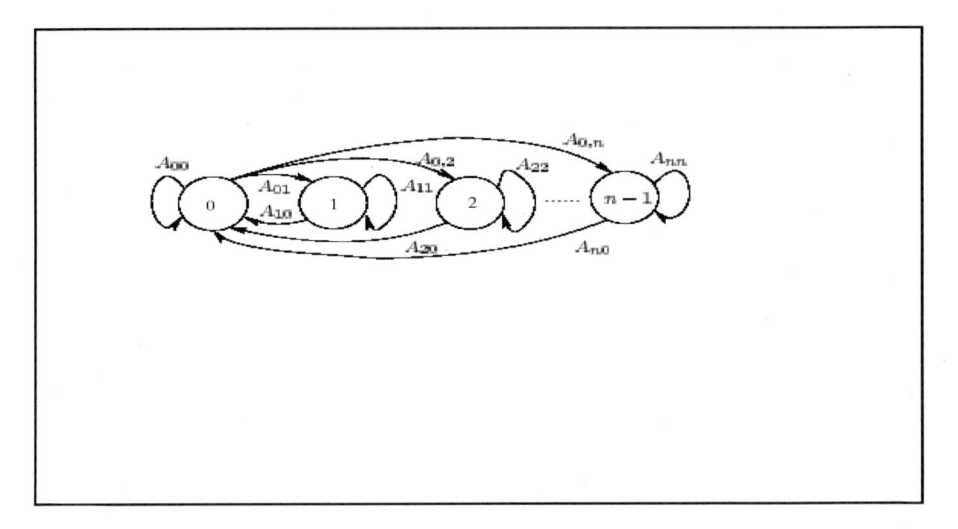

Fig. 2. The state-transition diagram of the modulating Markov chain of the Pseudo Self-similar Traffic (PSST) model

Notice that the parameters α and q need to fulfill certain conditions so that A is indeed a stochastic matrix describing a discrete-time Markov chain; q, α >0, $q < $ α such that $0 \leq A_{0,0} \leq 1$.

In sequel, we denote with $A^k_{i,j}$ the entry in row i and column j of A^k. We furthermore define $N = (N_t, t \in \mathbb{N})$ as the discrete-time stochastic process describing the number of arrivals over time, as described by the Markov-Modulated Bernoulli process (MMBP).

3 Experiments

3.1 Overview of Experiments

As shown in the Figure 3 diagram traces of packets are used as input to a S-Plus script which will evaluate the Hurst parameter using the aggregate variance method. Then another S-Plus script will evaluate parameters: n, q, and a in Robert's Pseudo-Self

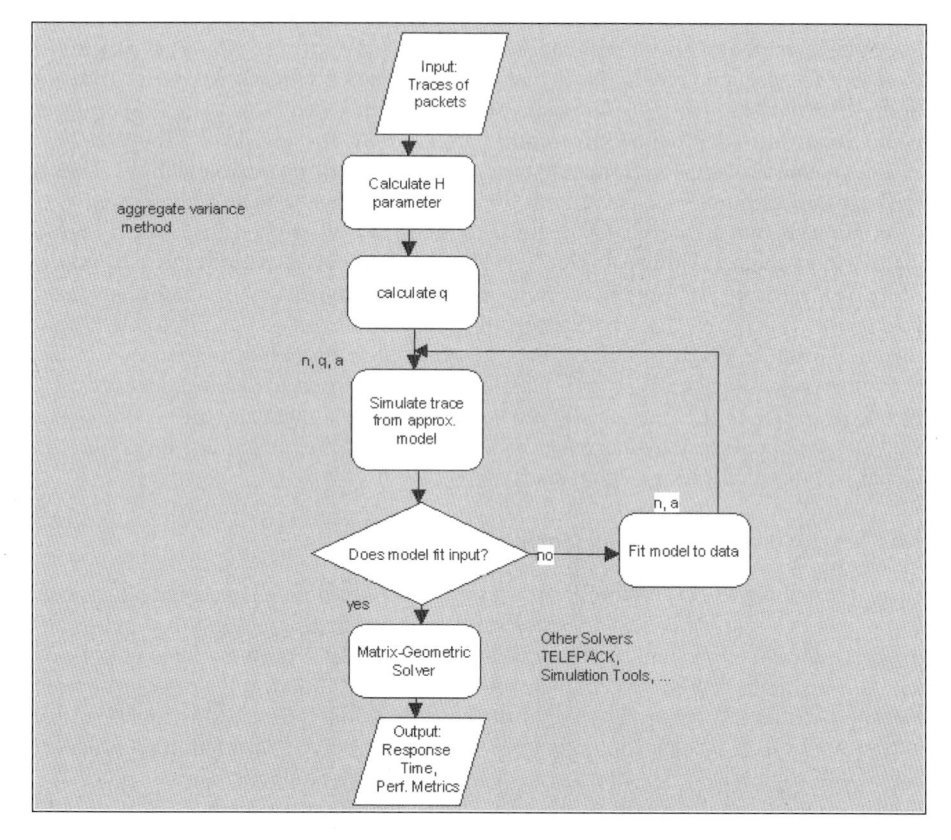

Fig. 3. Overview of Experiments

similar Markov Chain described in chapter 2.3. A close fit to the calculated Hurst parameter from the traffic traces is evaluated iteratively. Then we used the Matrix-Geometric method to calculate the mean queue length and response time.

3.2 Packet Traces

Packet traces were generated by a custom made self-similar packet generator. Traces using Vern Paxson's [32] fast approximation algorithm proved more suitable. Traces were generated with *Hurst parameter* varying from 0.55 to 0.95. Traces sufficient for an hour simulation were prepared. Two different metrics were considered:

- Self-similar packet length
- Self-similar inter-arrival times

3.3 Experiment Results

In the first results we obtained from the experiment and which were published in [33], we showed the effect of different parameters related to Stephen Roberts pseudo self-similar models (in section 2) to mean queue length. More specifically we showed that increases in the *a* parameter (see section 2.3) consequently increase the mean queue length. This is in agreement with many studies that have proven self-similarity traffic increases mean queue length. Note that increased values of the *a* parameter translate to higher *Hurst* parameters. The results were obtained by solving a single user queue with arrivals from the pseudo self –similar process with varying *Hurst* parameters.

The obvious conclusion from our study was that modelling queues with self-similar traffic, using traditional Poison arrivals greatly underestimates mean queue length.

In the next two experimental studies [34] and [35] we studied the effect of artificially generating self-similar traffic in an Ethernet LAN on response time. The studies were conducted in an Ethernet where a custom-made packet generator generated packets with specified Hurst parameters. The response time was monitored by a Ping function using a high-resolution clock based on the processor frequency. Results show that response time is in most cases self-similar when the arrival process has a Hurst parameter in the interval (0.55 0.95). In another simulation study using Opnet we validated the three previous experimental studies. Packet length was based on a bi-modal with probabilities based on distributions in [36].

3.4 Novel Numerical Solution Queuing Model

Using the input pseudo self-similar process (PSST) (section 2.3) in a queuing model is possible in a number of ways. First, a discrete time queue could be constructed, and the process used as the input to that queue. Since at most 1 customer arrives per slot, and one customer can be served, this would not be very interesting. A more interesting approach would be to use several of the self similar processes as input, so that queues might have a chance to build up. This would involve construction of a Markov chain with each state representing the states of the individual self similar processes. Construction of such a chain is straightforward, and then the analysis of a discrete time queue with that input process could be conducted.

A second approach, which we followed, is to (incorrectly) assume that the process is continuous, and to assume that when the input process is in state 1 arrivals form a Poisson process with rate 1, and that the server gives exponentially distributed service times with mean 1 whatever the state of the input. This means that we are analyzing an M/M/1 queue in a Markovian environment, which has been the subject of many studies. We solve this queuing model using the matrix-geometric technique. More details can be found in [33].

The state of the system can be denoted by a pair of integers, (I; J), with I representing the number of customers in the system, and J, $1 \leq J \leq n$, representing the state of the input process. When the input process is in state 1, jobs arrive in a Poisson stream at rate 1, and in all other states of the input process there are no arrivals. Whatever the state of the input process, service takes place at rate 1.

The steady state probabilities can be denoted as $\pi_{ij} = \Pr(I = i \wedge J = j)$ and can be related using the balance equations, and if the vector $\pi_i = (\pi_{i1}, \pi_{i2}, \pi_{i3}, \ldots, \pi_{in})$ is defined, then the balance equations can be expressed as:

$$A_2 \pi_{i+1} + A_1 \pi_I + A_0 \pi_{i-1} \quad (7)$$

The matrices A_2 and A_0 are diagonal matrices, with entries consisting of the arrival and service rates, respectively, in the corresponding state of the input process.

A_1 is $P - I - A_2 - A_0$, where P is where P is the transition matrix of the input process.

Neuts [37] shows that: $\pi_i = R^i \pi_0$.

where R is the unique solution of:

$$A_2 R^2 + A_1 R + A_0 = 0 \quad (8)$$

And

$$\pi_0 = (1-R)\alpha \quad (9)$$

where α is the steady state distribution of the Markov chain representing the input process.

4 Conclusions

4.1 Effect of Packet Size and Hurst Parameter

In [34] and [35] we showed results of two major experiments based on a novel packet generator and high resolution ping function. The fundamental findings were that response time in experiments with random packet length proved to be self-similar, and experiments with bi-modal proved that response time was non-self-similar. Simulations on similar network with the same traces as the experiments showed that the Hurst parameter has a declining effect on the delay, and that the bi-modal packet length has a similar effect.

4.2 Comparison of Traffic Generator (Measurement Approach) and Simulation Model

The difference in the results of the measurement study and simulation study can be attributed to the following factors:

- Measurement approach was conducted in an open environment with traffic from other sources than the self-similar traffic generator
- Ping function also contributed to traffic, this had a high overhead in lower time resolutions.
- Number of traffic sources were not identical, even though packet traces were identical
- Opnet simulation Kernel had a different effect than the self-similar generator and measurement approach

In summary the measurement study showed the effect of self-similarity can under conditions be passed down to delay, and the simulation study showed how the bimodal packet distribution can affect the delay, this factor is more than the *Hurst* parameter.

References

[1] Mandelbrot, B.B.: Self-similar error clusters in communication systems and the concept of conditional stationarity. IEEE Transactions on Communications Technology Com 13, 71–90 (1965)
[2] Leland, W.E., Taqqu, M.S., Willinger, W., Wilson, D.V.: On the self-similar nature of Ethernet traffic. IEEE/ACM Transactions in Networking 2, 1–15 (1994)
[3] Stallings, W.: Viewpoint: Self-similarity upsets data traffic assumptions. IEEE Spectrum 34, 28–29 (1997)
[4] Willinger, W., Taqqu, M.S., Erramili, A.: A bibliographical guide to self-similar traffic and performance modeling for modern high-speed networks. In: Stochastic Networks: Theory and Applications, pp. 339–366 (1996)
[5] Erramilli, A., Narayan, O., Willinger, W.: Experimental Queueing Analysis with Long-Range Dependent Packet Traffic. IEEE/ACM Transactions in Networking 4, 209–223 (1996)
[6] Partridge, C.: The end of simple traffic models. IEEE Network 7, 3 (1993)
[7] Kharani, A.A., Kumar, A.: Long range dependence in network traffic and the closed loop behaviour of buffers under adaptive window control. Performance Evaluation 61, 95–127 (2005)
[8] Veresa, A., Kenesic, Z., Molnarc, S., Vattayd, G.: TCP's role in the propagation of self-similarity in the Internet. Computer Communications 26, 899–913 (2003)
[9] Mandelbrot, B.B.: The Fractal Geometry of Nature. W.H. Freeman, New York (1982)
[10] Willinger, W., Paxson, V.: Discussion of the paper by S.I. Resnick. Annals in Statistics 25(5), 1856–1866 (1997)
[11] Liu, Z., Niclausse, N., Jalpa-Villanueva, C., Barbier, S.: Traffic model and performance evaluation of web servers. INRIA, Sophia Antipolis, France, Tech. Rep. RR-3840 (1999)
[12] Cappe, O., Moulines, E., Pesquet, J.-C., Petropulu, A., Yang, X.: Long-Range Dependence and Heavy-Tail Modeling for Teletraffic Data. IEEE Signal Processing Magazine 19, 14–27 (2002)

[13] Leland, W.E., Taqqu, M.S., Willinger, W., Wilson, D.V.: On the self-similar nature of Ethernet traffic. Computer Communication Review 23, 183–193 (1993)
[14] Paxson, V., Floyd, S.: Wide-area traffic: The failure of Poisson modeling. IEEE/ACM Trans. Networking 3, 226–244 (1995)
[15] Beran, J., Sherman, R., Taqqu, M.S., Willinger, W.: Long-range dependence in variable-bit-rate video traffic. IEEE Trans. Commun. 43, 1566–1579 (1995)
[16] Hurst, H.E.: Long-term storage capacity of reservoirs. Transactions of the American Society of Civil Engineers 116, 770–799 (1951)
[17] Beran, J.: Statistical methods for data with long-range dependence. Statistical Science 7, 404–427 (1992)
[18] Cox, D.R.: Long-range dependence: a review. Statistics: An Appraisal. In: David, H.A., David, H.T. (eds.) pp. 55–74. Iowa State Univ. Press (1984)
[19] Abramson, N.: The ALOHA System- Another alternative for computer. In: Fall Joint Comput. Conf., pp. 281–285. AFIPS Press, Monvale (1970)
[20] Kleinrock, L., Tobagi, F.A.: Packet switching in radio channels: Part 1-Carrier Sense multiple-access modes and their throughput- delay characteristics. IEEE Trans. Commun. Com. 23, 1400–1416 (1975)
[21] Crovella, M.E., Bestavros, A.: Self - similarity in world wide web traffic:evidence and possible causes. IEEE/ACM Transactions on Networking 5(6), 835–846 (1997)
[22] Likhanov, N., Tsybokov, B., Georganas, N.D.: Analysis of an ATM buffer with self - similar (fractal) input traffic. In: Proc IEEE INFOCOM 1995, Boston, pp. 985–992 (1995)
[23] Willinger, W., Taqqu, M.S., Erramili, A.: Self-Similarity in High-Speed Packet Traffic: Analysis and Modeling of Ethernet Traffic Measurements. Statistical Sciences 10, 67–85 (1995)
[24] Willinger, W., Taqqu, M.S., Erramili, A.: A bibliographical guide to self- similar traffic and performance modeling for modern high-speed networks. In: Kelly, F.P., Zacharay, S., Zeidin, I. (eds.) Stochastic Networks: Theory and Applications, pp. 339–366. Oxford University Press, Oxford (1996)
[25] Fretwell, R., Kouvatsos, D.: LRD and SRD traffic: review of results and open issues for the batch renewal process. Performance Evaluation 48, 267–284 (2002)
[26] Jagerman, D.L., Melamed, B.: The transition and autocorrelation structure of tes processes part i: General Theory. Stochastic Models 8, 193–219 (1992)
[27] Erramilli, A., Singh, R.P., Pruth, P.: Chaotic maps as models of packet traffic. In: Labetoulle, J., Roberts, J.W. (eds.) The Fundamental Role of Teletraffic in the Evolution of telecommunications Networks, Proc. ITC-14, Antibes, Juan-les-Pins, France, June 1994, pp. 329–338. Eleseveir, Amstrerdam (1994)
[28] Maulik, K., Resnick, S.: Small and Large Time Scale Analysis of a Network Traffic Model. Queueing Systems 43, 221–250 (2003)
[29] Robert, S., Le Boudec, J.-Y.: New models for pseudo self-similar traffic. Performance Evaluation 30(1), 57–68 (1997)
[30] Courtios, P.J.: Decomposability, Queue and Computer Applications. Academic Press, New York (1977)
[31] Simon, H., Ando, A.: Aggregation of variables in dynamic systems. Econometrica 29, 111–139 (1961)
[32] Paxson, V.: Fast, Approximate Synthesis of Fractional Gaussian Noise for Generating Self-Similar Network Traffic. Computer Communications Review 27, 226–244 (1997)
[33] King, P.J.B., Larijani, H.A.: Queueing Consequences of Self-similar Traffic. In: Proc. Fourteenth UK Computer and Telecommunications Performance Engineering Workshop, Edinburgh, pp. 182–186 (1998)

[34] Larijani, H.A., King, P.J.B.: Effect of Packet Size on the Response Time in an Ethernet LAN with Self-Similar Traffic. In: Proc. IFIP ATM & IP 2000, Ilkley, West Yorkshire, U.K (2000)
[35] Larijani, H.A., King, P.J.B.: Response Time in an Ethernet LAN with Self-Similar Traffic. In: Proc. PGNET 2000, Liverpool (2000)
[36] Christensen, K.J., Molle, M.L., Yeger, B.: The design of a station-centric network model for evaluating changes to the IEEE 802.3 Ethernet standard. Simulation 72(1), 33–47 (1999)
[37] Neuts, M.F.: Matrix Geometric solutions in Stochastic Models: An Algorithmic Approach. John Hopkins University Press, Baltimore (1981)

Characterisation of Internet Traffic in Wireless Networks

Lei Liang, Yang Chen[*], and Zhili Sun

Centre for Communication System Research (CCSR),
University of Surrey, Guildford, Surrey, United Kingdom
{L.Liang,Z.Sun}@Surrey.ac.uk

Abstract. Wireless connection technologies provide users (Internet Protocol) IP network access without the physical hardware connection of the wired networks. One of the applications of these technologies is the Wireless Local Area Network (WLAN), which is based on the IEEE802.11x Wireless Fidelity (WiFi) standard and is widely deployed as a flexible extension to data network or an alternative for the wired Local Area Network (LAN). In this context, the design, control and performance analysis of future wireless networks requires the study and credible characterisation of WLAN traffic. This tutorial presents measurements and analytic studies of IP traffic in a WLAN environment. Moreover, an investigation is reported into the characterisation on protocol distribution and modelling of IP packet inter-arrival times.

Keywords: WLAN, IP traffic measurement, traffic modelling and packet inter-arrival time.

1 Introduction

Traffic measurement is crucial to traffic engineering (TE) functions [1]. It provides insight of network traffic, network operation state and problem anticipation. It is also crucial for optimising the network resources to meet the requirements of traffic condition and quality of service (QoS) requirements. It can also provide the feedback data for the engineer to adaptively optimise network performance in response to events and stimuli originating within and outside the network. It is essential to determine the QoS in the network and to evaluate the effectiveness of traffic engineering policies. And experience indicates that measurement is most effective when acquired and applied systematically.

Measurement in support of the TE functions can occur at different levels of abstraction. For example, measurement can be used to derive packet level characteristics, flow level characteristics, user or customer level characteristics, traffic aggregate characteristics, component level characteristics, and network wide characteristics [1]. The measurement presented in this paper has been carried out to study the packet level characteristics of aggregate WLAN traffic.

---

[*] Yang Chen has left the University of Surrey.

The history of wireless networking can stretch back over fifty years ago, during World War II, when the United States Army first used radio signals for data transmission. They developed a radio data transmission technology, which was heavily encrypted. A group of researchers in 1971 at the University of Hawaii created the first packet based radio communications network named ALOHNET with inspiration from that army system. It is essentially the very first WLAN.

In the last few years, the wireless connection technology is improving, making it easier and cheaper from companies to set up WiFi access point for access to the Internet. That resulted in a rise of WLAN usage worldwide to set up hotspots, including university campuses, airports, hospitals, companies, and warehouses. The fast development of the WLAN brings new challenges to researchers. One of them is the need to understand the characteristics of the WLAN traffic. This paper presented a measurement experiment in WLAN using packet monitor software.

The paper is organized as following: the second section gives background knowledge of WiFi including the relevant standards, and the third section presents the measurement methodologies, parameters and environment, measurement results and analysis will be presented for traffic characterisation in the forth section, and finally the last section draws conclusions based on the measurement and analytical results.

2 Background of WiFi

WLANs are currently the most common form of wireless networking for day-to-day business operations. An industry standard has to be developed to enable the WLAN widely accepted and ensure the compatibility and reliability among all manufacturers of the devices. The first standard for WLAN IEEE 802.11 [2] was defined in 1997 by the Institute of Electrical and Electronics Engineers (IEEE) as a part of a family of IEEE 802.x standards for local and metropolitan area networks. It operates at a radio frequency (RF) band between 2.4GHz to 2.5GHz with data rates of 1Mbps and 2Mbps. It also provided a set of fundamental signalling methods and services. It addressed the difference between wireless LAN and wired LAN and provided the method to integrate both LANs together. There are new main amendments defined in IEEE following IEEE 802.11. They are IEEE 802.11a, IEEE 802.11b, IEEE 802.11d and IEEE 802.11g. All of them have commercial products in the market now.

The IEEE 802.11a [3], defined in September 1999, gives method to provide a WLAN with data payload communication capabilities up to 54 Mbps in a frequency band around 5 GHz. IEEE 802.11b [4] in 1999 and IEEE 802.11g [5] in 2003 provide methods to increase the data payload rate up to 11Mbps and 33 Mbps respectively in the 2.4GHz frequency band defined in IEEE 802.11. IEEE 802.11d [6] in 2001 provided specifications for conformant operation beyond the original six regulatory domains of IEEE 802.11 and enabled an IEEE 802.11 mobile station to roam between regulatory domains. 802.11n is a recent amendment which improves upon the previous 802.11 standards by adding multiple-input multiple-output (MIMO) and many other newer features. The maximum bit rate can reach 600 Mbps. Table 1 shows the list of IEEE 802.11 families.

Table 1. List of IEEE 802.11 families

Name	Time of Definition	Document Type
IEEE 802.11	1997	Original Standard
IEEE 802.11a	1999	Amendment
IEEE 802.11b	1999	Amendment
IEEE 802.11d	2001	Amendment
IEEE 802.11g	2003	Amendment
IEEE 802.11g	2009	Amendment

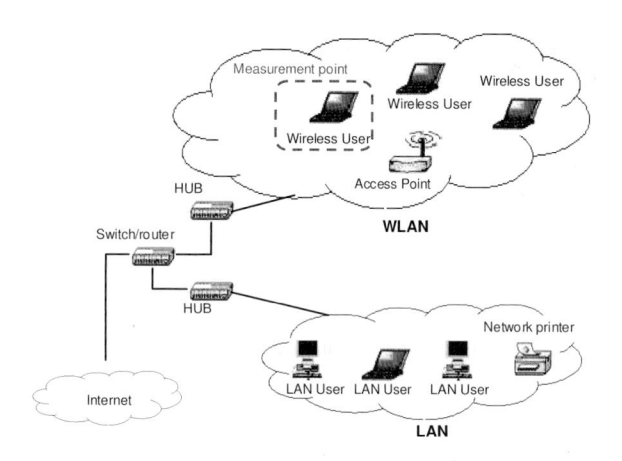

Fig. 1. Basic WLAN layout example

WLAN is normally deployed within a single building or a campus area. Fig. 1 illustrates the basic layout of a WLAN with connection to wired LAN, where several laptops are connected with the core wired network through an Access Point (AP), and the wired network has server, switch, hub, workstation, network printer, and so on. It can be seen from the Fig. 1 that the WLAN is an extension to the wired LAN.

A WLAN may consist of three network components. They are wireless network card, wireless access point and wireless bridge.

The wireless network cards include PCI cards for workstations and PC cards for laptops and other mobile devices. The card can work in an ad-hoc mode, as in client-to-client scenario (ad hoc), or in a pure client-to-AP mode (infrastructure).

The wireless network cards connect to an AP. An AP is essentially a hub that gives wireless clients the ability to attach to the wired LAN backbone. With the use of the cell structures, more than one AP can be set up in a given area. This is similar to the cell phone coverage in mobile communication systems. Wireless bridges can provide high speed longer range outdoor links between buildings.

The IEEE 802.11 standard gives two different ways to configure a WLAN using these network components: ad-hoc and infrastructure. In the ad-hoc WLAN, computers are brought together to form a network "on the fly." As shown in Fig. 2, there is no structure to the network; there are no fixed points; and usually every node is able to communicate with every other node. A good example of this is the aforementioned meeting where employees bring laptop computers together to communicate and share design or financial information. Although it seems that order would be difficult to maintain in this type of network, algorithms such as the spokesman election algorithm (SEA) have been designed to "elect" one machine as the base station (master) of the network with the rest being slaves. Another algorithm in ad-hoc network architectures uses a broadcast and flooding method to all other nodes to establish their identifications and connections.

The second type of network structure used in WLANs is the infrastructure as shown in Fig. 3. This architecture uses fixed network AP with which mobile nodes can communicate. These AP are sometime connected to landlines to widen the LAN's capability by bridging wireless nodes to other wired nodes. If service areas overlap, handovers can occur. This structure is very similar to the present day cellular networks around the world.

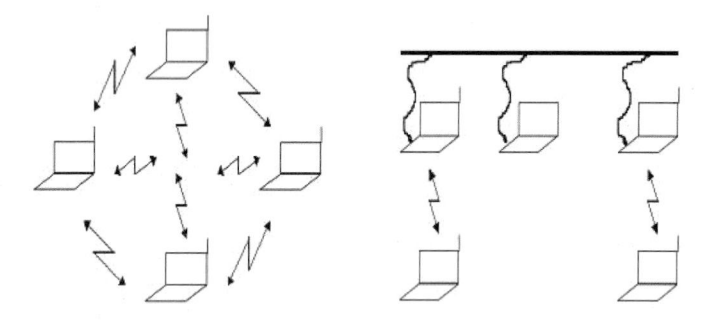

Fig. 2. Ad-hoc WLAN architecture

Fig. 3. Infrastructure WLAN architecture

3 Measurement Environment and Methodologies

A series of WLAN measurements have been carried out in an academic building in the University of Surrey. The network is a typical infrastructure WLAN having around 20 to 50 users in different time of a day. The users have laptop computers with wireless cards and using Microsoft Windows XP operation system. Fig. 1 shows the measurement environment. The laptops are connected to the wired core network through an AP. The Measurement Point (MP) is one of the laptops.

To measure WLAN traffic, one has to address an appropriate measurement methodology. Many measurement methodologies have been used for traffic measurement in the history, i.e. using LOG files and capturing packets form the networks using some software and hardware. The measurement methodologies can be divided into two main groups: passive approach and active approach. Both have their values and should be regarded as complementary, in fact they can be used in conjunction with one another. The MP can be used as a normal laptop while it measures the traffic situation of the whole WLAN. In principal, this is a passive measurement.

Then we have to decide what tools we can use to measure the WLAN traffic. There are so many traffic monitoring and measurement software in the market and on the Internet. Some of them are very powerful with hardware equipments. However, commercial software can be very expensive. Much open source software is distributed on the Internet free of charge. But most of them are programmed for common use or some other special purposes that don't exactly fit in a particular measurement of our interests. As a Windows version of the famous UNIX packet capture tool TCPDump, WinDump was chosen as the measurement tool in our measurement for its powerful functions and easy-handling output.

Measurement parameters are the other important factor to be considered. Some performance and QoS parameters are very important for traffic engineering such as delay, jitter and packet loss. The IETF IP Performance Metrics (IPPM) working group has carried out studies and defined metrics of these parameters. They are very useful to evaluate the performance of a network. There is a group of parameters that can reflect the network traffic status. These parameters at packet level include throughput, packet length, packet inter-arrival time, packet burstness and so on. They are useful for capacity management, queue management and traffic prediction. This paper is going only present the packet inter-arrival time, which is a key factor in our measurement and analysis. It can be calculated using the following formula:

$$IA = AT(i) - AT(i-1)` \qquad (1)$$

where IA stands for inter-arrival time, $AT(i)$ and $AT(i-1)$ are the arrival time of the ith packet and its previous packet respectively.

During the measurement, the MP kept running WinDump software for thirty minutes as one measurement interval to capture WLAN packets from all the active users. Totally there were more than ten measurements taken at different time periods of a day. The IP packets were monitored and captured. These packets include different applications of WWW surfing, FTP, real-time online video and online games, etc.

4 Measurement Results and Analysis

Over ten sets of measurement results are stored for analysis. Results including packet length, protocol type, packet arrival time, number and volume of captured packets, and so on. All results are stored in pure text files. Every measurement interval of thirty minutes generated a text file with size ranged from 7Mb to 40Mb corresponding to the network usage at different times of a day. The number of packets captured during each measurement interval varies from 100,000 to 600,000.

4.1 Packet Type Analysis

Software was developed using Borland C++ Builder. It handled the raw data of the measurement output and generate summary in the Table 2.

Ten sets of data are shown in Table 2. Each of them presents results of one measurement interval. For each of the measurement interval, the number of the captured packets ranged from 93,744 to 598,819. This variance is mainly due to the difference

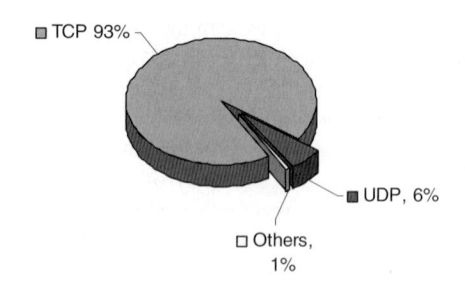

Fig. 4. Protocol percentage pie for WLAN traffic

Table 2. Summary of WLAN Measurement Output

Interval	No. of packets captured	No. of TCP packets	No. of UDP packets	Percentage of TCP packets (%)	Percentage of UDP packets (%)	TCP packets total length (Bytes)	UDP packets total length (Bytes)
1	164,280	154,549	6,888	94.08	4.19	76,400,952	579,987
2	598,819	593,464	2,514	99.11	0.4	354,190,737	623,560
3	547,050	541,934	2,636	99.06	0.5	329,917,033	307,035
4	468,306	463,013	2,861	98.87	0.7	252,739,983	290,285
5	238,616	234,931	1,589	98.46	0.7	127,104,315	164,625
6	93,744	89,455	1,897	95.42	2.02	60,215,251	184,117
7	520,495	465,212	51,526	89.38	9.9	271,855,736	9,759,117
8	576,802	522,682	50,407	90.62	8.74	350,224,666	9,719,882
9	546,570	427,324	110,400	78.18	20.2	225,362,090	21,303,531
10	205,778	199,625	3,441	97.01	1.07	79,533,200	351,498
Total	3,960,460	3,692,189	234,159	93.23	5.91	2,127,543,963	43,283,637

of the number of users in the WLAN and the applications running during each interval. Taking the interval 6 as an example, it was taken in a midnight when there were only less than ten users in the whole WLAN, which leading to the least number of packets captured among all measurement intervals.

Applications also affected the number of packets generated in each measurement interval. For instance, web surfing users generate much less packets than online video watching users because they did not generate packets continuously when they read web pages Totally, 3,960,460 packets were captured during these ten measurement intervals, which include 3,692,189 TCP packets and 234,159 UDP packets. That means TCP protocol is still counted as the majority used by Internet applications, i.e. around 93.23% of the captured packets are TCP packets as shown in Fig. 4.

The results showed that all of the online video services captured in the measurement used TCP to transport data rather using UDP. This is conflicting with the basic understanding that real-time applications are very sensitive to network latency and should use UDP as transport protocol to avoid retransmission and latency caused by congestion control function of TCP.

However, it is also understandable that online video service prefers TCP in the current best-effort Internet because it can provide reliable delivery with less packet loss and less distortion of the video services. It is also because online video service is relevantly immune to latency because it is not an interactive and real time service and, thus, network latency can be compensated by local playback buffer.

4.2 Packet Inter-arrival Time Analysis

The self-developed software also calculated the packet inter-arrival times for each measurement interval using formula (1). The objective was to find out what distribution the packet inter-arrival times of WLAN traffic follows. To establish the most suitable statistic distribution to model the measured curves, many distributions were tested by varying relevant parameters. These distributions include Chi-squared distribution, exponential distribution, inverse Gaussian (Wald) distribution, lognormal distribution, Pareto distribution and Rayleigh distribution [7]. Both Probability Density Functions (PDFs) and Cumulative Distribution Functions (CDFs) of these distributions were used to compare with the measurement plots of each of the traces. However, we found the CDF is a better way to present all of the theoretical curves and the measurement curves as well as being easier to use.

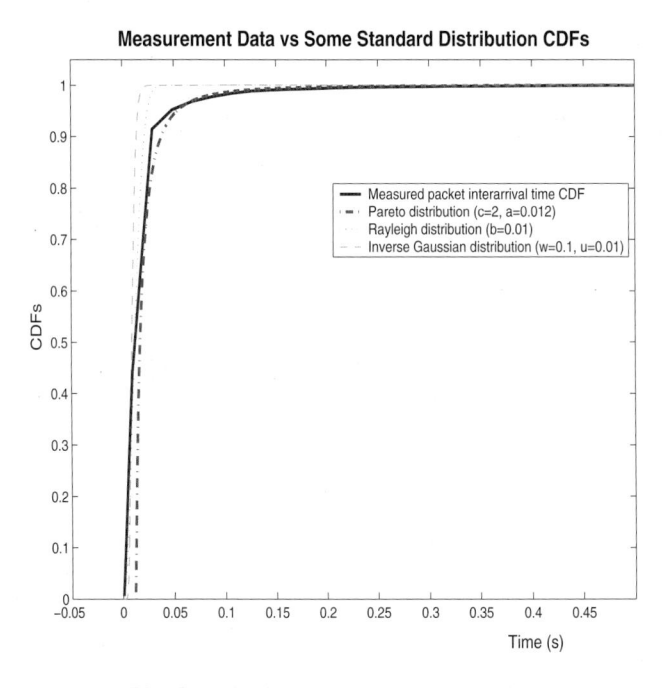

Fig. 5. Packet inter-arrival time CDF fitting

The fitting results show that we can easily tell which distribution is the best fit by the CDF of the data for the very clear fitting difference of all the experimented standard distributions. Fig. 5 shows the Inverse Gaussian (green dashed line), Rayleigh (magenta dotted line), Pareto (red dash-dot line) and one measured packet interarrival time (blue solid line) CDFs. It's clear that the different distributions have notably different CDF that can be distinguished by human eyes.

From all of these distributions, it was seen that the Pareto distribution is the best fit to nine measurement results. One measurement result can be fitted by using Inverse Gaussian distribution. Table 3 shows the fitting result.

Table 3. Packet Inter-arrival Time Fitting Results

Measurement	Max. Inter-arrival Time	Min. Inter-arrival Time	Best fit distribution
1	1.9402	5×10^{-6}	Pareto
2	0.4004	5×10^{-6}	Pareto
3	0.3642	3×10^{-6}	Pareto
4	0.1736	3×10^{-6}	Pareto
5	0.3925	5×10^{-6}	Pareto
6	0.5594	5×10^{-6}	Inverse Gaussian
7	0.1644	4×10^{-6}	Pareto
8	0.2081	1.2×10^{-5}	Pareto
9	0.3057	4×10^{-6}	Pareto with cut-off
10	2.9733	3×10^{-6}	Pareto

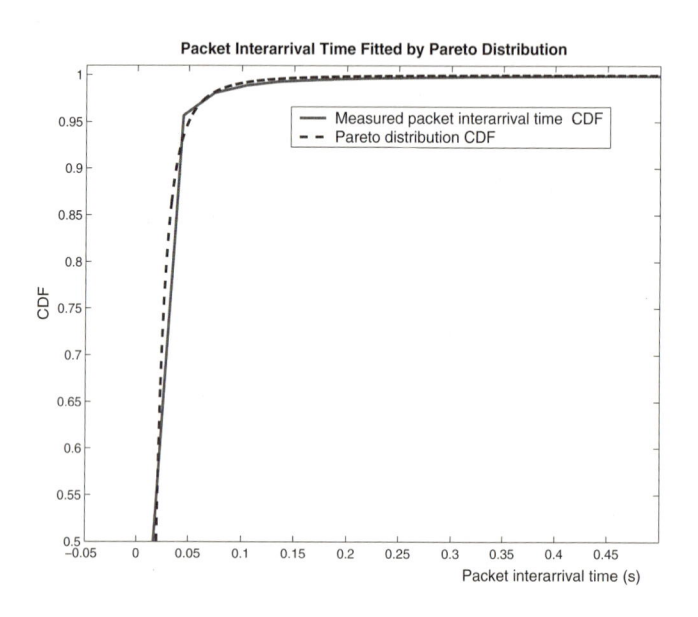

Fig. 6. Packet inter-arrival time fitted by Pareto distribution

For those measurements that can be fitted by Pareto distribution, Fig. 6 shows one of them as an example in this paper. The solid line is the measured packet inter-arrival time CDF and the dashed line is the CDF of a Pareto distribution.

The mathematic presentation of the Pareto distribution PDF is:

$$f_T(t) = \frac{ca^c}{t^{c+1}} \qquad (2)$$

where $c > 0$ is the shape parameters of Pareto distribution and $a > 0$ is the location parameter. Fig. 7 shows the fitting of the measured packet inter-arrival time in measurement 6 using Inverse Gaussian distribution.

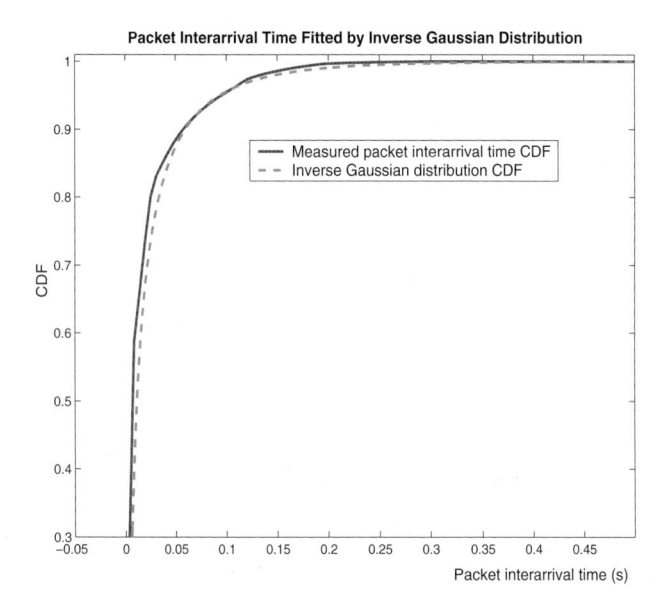

Fig. 7. Packet inter-arrival time fitted by Pareto distribution

The mathematic presentation of the Inverse Gaussian distribution PDF is:

$$f_T(t) = \left[\frac{\lambda}{2\pi t^3}\right]^{\frac{1}{2}} \exp\left\{\frac{-\lambda(t-\mu)^2}{2\mu^2 t}\right\} \qquad (3)$$

where the parameters $\lambda > 0$ and $\mu > 0$ are the scale parameter and location parameter respectively for the Inverse Gaussian distribution. The mean of the distribution is μ and the variance μ^3/λ. For different measurements, we found slightly various values of λ and μ. The characters w and u in the legend of Fig. 5 correspond to the variables λ and μ in equation (3).

The main reason that this measurement is fitted with a different distribution from the other measurements is the less packets number. The measurement was taken in the midnight and the number of users in the WLAN was far less than the peak day hours, leading to much less packet captured in the measurement interval. Statistically, the time between each packet carried in the WLAN became larger and it finally resulted in an Inverse Gaussian distributed packet inter-arrival time.

Therefore, we can use Pareto distribution to model the packet inter-arrival time for WLAN traffic during the day when many users are active and Inverse Gaussian distribution to model it at late night with few active users.

There is an exception among all the nine Pareto distribution fitted measurements. In the measurement 9, we found a sharp rise cut off the measured packet inter-arrival time CDF around the point of 0.09 second. The sharp cut-off in the CDF plotting makes it much less accurate if we still trying to use one Pareto distribution to fit the measured curves. To model this cut-off distribution, we have to consider the distribution and the cut-off separately. For example, if a distribution has a PDF function without cut-off $Y = f_X(x)$ where $x \geq k$, the function with cut-off occurring at $x=m$ can be expressed as:

$$Y' = \begin{cases} f_X(x), & k \leq x < m \\ \beta, & x = m \\ 0, & x > m \end{cases} \quad (4)$$

where $\beta = 1 - \int_k^m f_X(x)dx$.

This method was introduced by [8]. Thus the mathematic expression of the Pareto distribution for situation happened in measurement 9 should be modified to:

$$Y' = \begin{cases} \dfrac{cb^c}{x^{c+1}}, & T_{min} \leq x < T_{cut} \\ \beta, & x = T_{cut} \\ 0, & x > T_{cut} \end{cases} \quad (5)$$

where $\beta = 1 - \int_{T_{min}}^{T_{cut}} \dfrac{cb^c}{x^{c+1}} dx$, T_{min} is the minimum packet inter-arrival time and T_{cut} is the cut-off point.

5 Conclusions

A series of measurements have been presented for the study of IP traffic characteristics in a WLAN environment. The software called WinDump was chosen to be the measurement tool and the whole measurement campaign was scheduled into a set of 30 minutes intervals covering different time of a day in different user behavioural conditions. The packets captured include applications of web surfing, FTP, online gaming and video streaming, etc. Totally, the results of ten measurement intervals were studied and presented.

The results show that TCP is still the main transport protocol for network services. A total of 3,692,189 TCP packets were captured in the ten measurement intervals out of 3,960,460 IP packets, i.e. 93.23% packets are TCP packet. The UDP protocol was mainly used by online card game application. Online video streaming also used TCP protocol although it is a real-time application. This is because TCP can provide reliable packet delivery without packet loss over the best-effort Internet, where the network latency can be compensated using local buffering technologies.

The Pareto distribution was found to be the best fitting to the packet inter-arrival times in nine (out of ten) measurement intervals. The much less usage of the WLAN at midnight, made the Inverse Gaussian distribution to fit better the packet inter-arrival times in one measurement interval. Thus, it is better to model the traffic of a WLAN using the Pareto distribution during the busy day time and using Inverse Gaussian distribution when few users are active during the night. The cut-off phenomenon was observed during measurements and modelled using method expressed in equation (5).

Future works may focus on the study of particular applications using either TCP or UDP protocol and may include web surfing, FTP, online gaming and online video streaming and so on. In this context, new results will contribute towards a better understanding of the network requirements for different applications and provide input to the network management and design configuration.

References

1. Awduche, D., Chiu, A., Elwalid, A., Widjaja, I., Xiao, X.: Overview and Principles of Internet Traffic Engineering. IETF RFC 3272, pp. 4–21 (2002)
2. IEEE Std 802.11-1997 Information Technology - Telecommunications and Information Exchange between Systems - Local and Metropolitan Area Networks - Specific Requirements, Part 11: Wireless LAN Medium Access Control (MAC) and Physical Layer (PHY) Specifications. IEEE (1997)
3. Supplement to IEEE Standard for Information Technology - Telecommunications and Information Exchange between Systems - Local and Metropolitan Area Networks - Specific Requirements, Part 11: Wireless LAN Medium Access Control (MAC) and Physical Layer (PHY) Specifications: High-speed Physical Layer in the 5 GHz band. IEEE (1999)
4. Supplement to IEEE Standard For Information Technology - Telecommunications and Information Exchange between Systems - Local and Metropolitan Area Networks - Specific Requirements, Part 11: Wireless LAN Medium Access Control (MAC) and Physical Layer (PHY) Specifications: High-speed Physical Layer Extension in the 2.4 GHz Band. IEEE (1999)
5. IEEE Standard for Information Technology - Telecommunications and Information Exchange between Systems - Local and Metropolitan Area Networks - Specific Requirements, Part II: Wireless LAN Medium Access Control (MAC) and Physical Layer (PHY) Specifications. IEEE (2003).
6. IEEE Standard for Information Technology - Telecommunications and Information Exchange between Systems - Local and Metropolitan Area Networks - Specific Requirements, Part 11: Wireless LAN Medium Access Control (MAC) and Physical Layer (PHY) Specifications, Amendment 3: Specifications for Operation in Additional Regulatory Domains. IEEE (2001)

7. Evans, M., Hastings, N., Peacock, B.: Statistical Distributions, pp. 45–136. John Wiley & Sons, Inc., Chichester (1993)
8. Universal Mobile Telecommunications System (UMTS) - Selection Procedures for the Choice of Radio Transmission Technologies of the UMTS (UMTS 30.03 version 3.2.0). TR 101 112, v3.2.0, Universal Mobile Telecommunications Systems, European Telecommunications Standards Institute, pp. 34–35 (April 1998)

Performance Analysis of Priority Queueing Systems in Discrete Time

Joris Walraevens, Dieter Fiems, and Herwig Bruneel

SMACS Research Group
Department for Telecommunications and Information Processing (IR07)
Ghent University - UGent
Sint-Pietersnieuwstraat 41, B-9000 Gent, Belgium
{jw,df,hb}@telin.UGent.be
http://telin.UGent.be/smacs

Abstract. The integration of different types of traffic in packet-based networks spawns the need for traffic differentiation. In this tutorial paper, we present some analytical techniques to tackle discrete-time queueing systems with priority scheduling. We investigate both preemptive (resume and repeat) and non-preemptive priority scheduling disciplines. Two classes of traffic are considered, high-priority and low-priority traffic, which both generate variable-length packets. A probability generating functions approach leads to performance measures such as moments of system contents and packet delays of both classes.

1 Introduction

In recent years, there has been much research devoted to the incorporation of multimedia applications in packet-based networks. Different types of traffic need different Quality of Service (QoS) standards. For real-time applications, it is important that mean delay and delay-jitter are bounded, while for non real-time applications, the throughput and loss ratio are the restrictive quantities. In order to guarantee acceptable delay boundaries to delay-sensitive traffic (such as voice/video), several scheduling schemes – for switches, routers, ... – have been proposed and analyzed, each with their own specific algorithmic and computational complexity. The most drastic in this respect is the strict priority scheduling. With this scheduling, as long as delay-sensitive (or high-priority) packets are present in the queueing system, this type of traffic is served. Delay-insensitive packets can thus only be transmitted when no delay-sensitive traffic is present in the system. As already mentioned, this is the most drastic way to meet the QoS constraints of delay-sensitive traffic, but also the easiest to implement.

Within this tutorial paper, we focus on the analysis of queues with this priority scheduling discipline. We give an overview of the different types of priority scheduling disciplines and, for the most part, we show and explain some techniques to analytically analyze discrete-time queues with a priority scheduling discipline. Assume packets arriving to a buffer (located in a switch, router, multiplexer, ...) being categorized in two distinct classes, the high-priority class and the low-priority class. In a queue with a priority scheduling discipline, the high-priority packets are transmitted ahead of the low-priority packets, i.e., when a server becomes available, a high-priority packet is always scheduled for transmission. Only, when there are no high-priority packets in the buffer at that time, a low-priority packet is selected for transmission. Priority scheduling disciplines come in two basic flavors, i.e., *non-preemptive* and *preemptive*. In the former, transmission of a packet is never interrupted once it is in service. So, if new high-priority packets arrive while a low-priority packet is served, they have to wait until the low-priority packet leaves the server. In a queue with a preemptive priority scheduling discipline on the other hand, those newly arriving high-priority packets interrupt transmission of the low-priority packet in service. Within the latter type of priority scheduling, two different strategies can be distinguished, depending on what happens when an interrupted low-priority packet re-enters the server. If the packet can resume its service where it was interrupted, i.e., when the part that was already transmitted before the interruption does not have to be retransmitted again, it is called a preemptive *resume* priority scheduling discipline. In a preemptive *repeat* priority scheduling on the other hand, the packet has to be retransmitted completely after the interruption.

In the literature, there have been a number of contributions with respect to priority scheduling. An overview of some basic priority queueing models in continuous time can be found in [1–3] and references therein. Discrete-time priority queues with deterministic service times equal to one slot have been studied in [4–16]. Khamisy and Sidi [4] analyze the system contents of the different classes, for a queue fed by a two-state Markov-modulated arrival process. Takine et al. [5] present the system content and delay for Markov-modulated high-priority arrivals and geometrically distributed low-priority arrivals. Laevens and Bruneel [6] analyze the system content and delay in the case of a multi-server queue. Choi et al. [7] and Walraevens et al. [12, 15] analyze a priority queue with train arrivals with resp. fixed, geometrically distributed and generally distributed train lengths. Walraevens et al. [8] study the system content and packet delay, in the special case of an output queueing switch with Bernoulli arrivals. Mehmet Ali and Song [9] examine a priority queue with on-off sources. Van Velthoven et al. [10] and Demoor et al. [13] tackle priority queues with finite (high-priority) buffer space. Kamoun [11] analyzes a priority queue with service interruptions. Finally, Walraevens et al. [14, 16] study the transient behavior and the output process resp. of a priority queue. All these papers have a single-slot service time in common; as a result no distinction has to be made between preemptive and non-preemptive priority scheduling.

Continuous-time *non-preemptive* priority queues have been considered in [17–26]. Discrete-time non-preemptive queues are the subject of [27–35]. Rubin and Tsai [27] study the mean waiting time, for a discrete-time queue fed by an i.i.d. arrival process. Hashida and Takahashi [28] analyze the packet delay by means of a delay-cycle analysis. Takine et al. [29] and Takine [30] study a discrete-time MAP/G/1 queue, using matrix-analytic techniques. Walraevens et al. examine the system content [31] and the packet delay [32] in a two-class non-preemptive priority queue with i.i.d. number of per-slot arrivals and general service times using generating functions. The results presented in section 3 are largely based on the latter two papers. This analysis is furthermore extended to a general number of classes in [33]. Maertens and al. [34] investigate the tail behavior of the total content in a priority buffer. Finally, Demoor et al. [35] analyze a priority queue with finite capacity for high-priority customers.

Continuous-time *preemptive resume* priority queues have been analyzed in [36–47]. Discrete-time preemptive resume priority queues are the subject of [48–53]. Walraevens et al. [49] and Ndreca and Scoppola [52] analyze a two-class preemptive priority queue with geometric service times. Walraevens et al. [50] study a priority queue with general high-priority service times and geometric low-priority service times, while Lee [48] and Walraevens et al. [53] handle a two-class priority queue with generally distributed service times for both classes. Van Houdt and Blondia [51] analyze a three-class priority queue. Queues with a *preemptive repeat* priority scheduling discipline are studied less frequently than their non-preemptive and preemptive resume counterparts. Continuous-time models are studied in [54, 55]. Discrete-time preemptive repeat priority queues are the subject of [56–58]. Mukherjee et al. [56] study a preemptive repeat with resampling scheduling of voice traffic over data traffic in a ring-based LAN. Resampling, in this context, means that the length of a repeated service time is not necessarily equal to the length of the first (interrupted) service time. It is a new sample (with the same distribution). Walraevens et al. [57, 58] analyze resp. the preemptive repeat priority queue with resampling and without resampling. Queues with resampling and without resampling resp. are also known as preemptive repeat *different* and preemptive repeat *identical* priority queues.

Finally, Hong and Takagi [59] and Kim and Chae [60] analyze priority models which are combinations of non-preemptive and preemptive priority.

In this tutorial paper, we show some analytic techniques for analyzing the performance of queues with a preemptive or non-preemptive priority scheduling discipline. The analysis is largely based on the probability generating functions (pgfs) approach. We discuss two main methods to analyze priority queues. In the first method, a non-preemptive priority queue with two classes is analyzed. The joint pgf of the system contents of both classes and the pgfs of the delays of packets of both classes are calculated. Starting from these pgfs, it is shown how moments and approximate tail probabilities are calculated. In the second method, performance of low- and high-priority traffic is assessed separately in the case of a preemptive priority scheduling discipline. Here, a single-class model can be used to assess performance of the high-priority traffic as the preemptive priority discipline implies that high-priority traffic perceives the system as one without low-priority traffic. Low-priority traffic performance, on the other hand, is assessed with a single-class model with service interruptions. From the point of view of the low-priority class, the server is interrupted whenever high-priority packets are served and is available otherwise. We obtain a stochastic model for the perceived interruption process and present the analysis of the corresponding queueing model with interruptions.

So, queueing models with service interruptions are highly applicable for modeling the low-priority class in priority queues. To end this introduction, we will make a (short) literature overview of queueing models with service interruptions. Continuous-time queues with service interruptions are the subject of (a.o.) two recent papers [61, 62]. Research on discrete-time queues with service interruptions dates back to the 70's. Early papers include those by Hsu [63] and Heines [64]. Both authors treat the single-server system with Bernoulli server interruptions and a Poisson arrival process. The former considers queue contents at random slot boundaries whereas the latter considers queue contents at service completion times. A single-server system with an i.i.d. arrival and a correlated on/off server interruption process is treated in [65–67]. Woodside and Ho [66] and Yang and Mark [67] model the on- and off-periods as a series of i.i.d. shifted geometric random variables, whereas Bruneel [65] assumes that the series of consecutive on- and off-periods share a common general distribution. The only restriction in the latter contribution is that the common probability generating function of the on-periods must be rational. Alternatively, correlation in the interruption process is captured by means of a Markovian process by Lee [68]. In a more general setting – that is, no assumptions are made regarding the nature of the interruption process – relationships between queue contents at different time epochs are derived by Bruneel [69].

Georganas [70] and Bruneel [71] treat multi-server systems with i.i.d. customer arrival and server interruption processes. The latter extends the former in the sense that it does not assume that all outputs are either available or not. The delay analysis of the latter system is presented by Laevens and Bruneel [72]. A multi-server system with a correlated interruption process is considered by Bruneel [73]. The interruption process is modeled as an on/off process (geometrical on-periods). The number of available servers during the consecutive on-slots, are modeled by means of an i.i.d. series of non-negative random variables whereas no servers are available during off-periods.

Some contributions also allow a certain degree of correlation in the arrival process. Bruneel [74] assumes that both arrival and interruption processes are on/off processes with geometric on- and off-periods. A stochastic number of customers (an i.i.d. series) enters the system during arrival-on periods, whereas no customers arrive in the system during arrival-off periods. The interruption process is similar as the one analyzed by Yang and Mark [67] in the case of uncorrelated arrivals. This interruption process is also considered by Ali et al. [75] and by Kamoun [76]. The former authors assume that customer arrivals stem from a superposition of two-state Markovian on-off sources, while the latter author considers a so-called train-arrival process.

All the former discrete-time queueing models with service interruptions have a fixed customer service time of a single slot in common. A queueing system where customers have a fixed multiple-slot service-time, is considered by Inghelbrecht et al. [77]. The interruption process is again similar as the one treated by Yang and Mark [67]. The presence of multiple-slot service times and interruptions implies that a packet's transmission may be interrupted. The contribution considers both the case that the packet transmission is continued after the interruption (CAI) and the case that transmission is repeated after the interruption (RAI). These modes correspond to preemptive resume and preemptive repeat priority scheduling, discussed above, respectively. Interruption models with generally distributed service times and a Bernoulli interruption process are considered by Fiems et al. [78, 79]. In [78], results for the CAI and RAI transmission modes are presented whereas some variants are considered in [79]. In particular we mention the repeat after interruption with resampling mode (in which the service time of an interrupted packet is resampled upon retransmission) and the partial repeat after interruption mode (in which only part of the packet has to be retransmitted after an interruption). The same authors consider CAI and RAI modes in the case of a Markovian interruption process [80] and in the case of a renewal-type interruption process [81]. The results presented in section 4 are based on the latter contribution.

The remainder of this paper is outlined as follows. In the next section we provide a more detailed description of the queueing model under consideration. In sections 3 and 4, we analyze the priority system in the case of a non-preemptive priority discipline and in the case of a preemptive priority discipline respectively. Some conclusions are drawn in section 5.

2 Mathematical Model

We consider a discrete-time single-server queueing system with infinite buffer space. Time is assumed to be slotted. There are two types of traffic arriving in the system, namely packets of class 1 and packets of class 2. We denote the number of arrivals of class j during slot k by $E_j^{(k)}$ ($j = 1, 2$). Both types of packet arrivals are assumed to be i.i.d. from slot-to-slot and are characterized by the joint probability mass function $e(m,n) \triangleq \Pr[E_1^{(k)} = m, E_2^{(k)} = n]$, and joint probability generating function (pgf) $E(z_1, z_2) \triangleq \mathrm{E}[z_1^{E_1^{(k)}} z_2^{E_2^{(k)}}]$. Notice that the number of packet arrivals from different classes (within a slot) can be dependent. If necessary for the analysis though, we will loosen this condition and assume the number of arrivals of both classes in a slot mutually independent. Further, we define the marginal pgfs of the number of arrivals of class 1 and class 2 during a slot by $E_1(z) \triangleq \mathrm{E}[z^{E_1^{(k)}}] = E(z,1)$ and $E_2(z) \triangleq \mathrm{E}[z^{E_2^{(k)}}] = E(1,z)$ respectively. We furthermore denote the arrival rate of class j ($j = 1, 2$) by $\overline{E}_j = E_j'(1)$. The variance of the number of per-slot arrivals of class-j is given by $\sigma_{E_j}^2 = E_j''(1) - (E_j'(1))^2 + E_j'(1)$.

The service times of the class-j packets are assumed to be i.i.d. and are characterized by the probability mass function $s_j(m) \triangleq \Pr[\text{service of a class-}j \text{ packet takes } m \text{ slots}]$, $m \geq 1$, and pgf $S_j(z) = \sum_{m=1}^{\infty} s_j(m) z^m$, with $j = 1, 2$. We furthermore denote the mean and variance of the service time of a class-j packet by $\overline{S}_j = S_j'(1)$ and $\sigma_{S_j}^2 = S_j''(1) - (S_j'(1))^2 + S_j'(1)$. We define the arrival load offered by class-j packets as $\rho_j \triangleq \overline{E}_j \overline{S}_j$ ($j = 1, 2$). The total arrival load is then given by $\rho_T \triangleq \rho_1 + \rho_2$.

The system has one server that provides the transmission of packets. Class-1 packets are assumed to have priority over class-2 packets, and within one class the service discipline is First Come First Served (FCFS). So, if there are any class-1 packets in the queue when the server becomes empty, the one with the longest waiting time will be served next. If, on the other hand, no class-1 packets are present in the queue at that moment, the class-2 packet with the longest waiting time, if any, will be served next.

3 Non-preemptive Priority Queues

In this section, we analyze non-preemptive priority queues. We derive the joint pgf of the system contents of both priority classes and calculate the pgfs of the packet delays of both classes. From these pgfs, we show how to calculate moments and (approximate) tail probabilities of the respective stochastic variables.

3.1 System Content at Service Initiation Epochs

To be able to analyze the system content at random slot boundaries and the packet delays of both classes, we first analyze the system content at the beginning of so-called start slots. These are slots at the beginning of which a packet (if available) can enter the server. Note that every slot during which the system is empty, is also a start slot. We denote the system content of class-j packets at the beginning of the l-th start slot by $U_{s,j}^{(l)}$ ($j = 1, 2$). Clearly, the set $\{U_{s,1}^{(l)}, U_{s,2}^{(l)}\}$ forms a Markov chain, since the arrival process is i.i.d. and the buffer solely contains entire messages at the beginning of start slots. If $S^{(l)}$ indicates the service time of the packet that enters service at the beginning of start slot l (which is - by definition - regular slot k) the following system equations can be established:

1. If $U_{s,1}^{(l)} = U_{s,2}^{(l)} = 0$:

$$U_{s,1}^{(l+1)} = E_1^{(k)}, \ U_{s,2}^{(l+1)} = E_2^{(k)}.$$

   The only packets present in the system at the beginning of start slot $l + 1$ are the packets that arrived during the previous slot, i.e., start slot l.

2. If $U_{s,1}^{(l)} = 0$ and $U_{s,2}^{(l)} > 0$:

$$U_{s,1}^{(l+1)} = \sum_{i=0}^{S^{(l)}-1} E_1^{(k+i)}, \ U_{s,2}^{(l+1)} = U_{s,2}^{(l)} + \sum_{i=0}^{S^{(l)}-1} E_2^{(k+i)} - 1.$$

   The class-2 packet in service leaves the system just before start slot $l + 1$. $S^{(l)}$ is characterized by probability mass function $s_2(m)$.

3. If $U_{s,1}^{(l)} > 0$:

$$U_{s,1}^{(l+1)} = U_{s,1}^{(l)} + \sum_{i=0}^{S^{(l)}-1} E_1^{(k+i)} - 1, \ U_{s,2}^{(l+1)} = U_{s,2}^{(l)} + \sum_{i=0}^{S^{(l)}-1} E_2^{(k+i)}.$$

$S^{(l)}$ is characterized by probability mass function $s_1(m)$.

We assume that the system is stable, implying that the equilibrium condition $\rho_T < 1$ is met. We define $U_s(z_1, z_2) \triangleq \lim_{l \to \infty} \mathrm{E}\left[z_1^{U_{s,1}^{(l)}} z_2^{U_{s,2}^{(l)}}\right]$. Using the system equations, we derive a functional equation for U_s:

$$[z_1 - S_1(E(z_1, z_2))] U_s(z_1, z_2) = \frac{z_1 S_2(E(z_1, z_2)) - z_2 S_1(E(z_1, z_2))}{z_2} U_s(0, z_2)$$
$$+ z_1 \frac{z_2 E(z_1, z_2) - S_2(E(z_1, z_2))}{z_2} U_s(0, 0). \quad (1)$$

It now remains for us to determine the unknown function $U_s(0, z_2)$ and the unknown parameter $U_s(0,0)$. This can be done in two steps. First, we notice that $U_s(z_1, z_2)$ must be analytic for all values of z_1 and z_2 such that $|z_1| < 1$ and $|z_2| < 1$. In particular, this should be true for $z_1 = Y(z_2)$, with $Y(z_2) \triangleq S_1(E(Y(z_2), z_2))$ and $|z_2| < 1$, since it follows from (an extension of) Rouché's theorem [82] that $z_1 = S_1(E(z_1, z_2))$ has exactly one solution $|Y(z_2)| < 1$ for all such z_2. Notice that $Y(1)$ equals 1. The above implies that if we insert $z_1 = Y(z_2)$ in equation (1), where $|z_2| < 1$, the left hand side of this equation vanishes. The same must then be true for the right hand side, yielding

$$U_s(0, z_2) = U_s(0, 0) \frac{z_2 E(Y(z_2), z_2) - S_2(E(Y(z_2), z_2))}{z_2 - S_2(E(Y(z_2), z_2))}. \quad (2)$$

The following expression for $U_s(z_1, z_2)$ can now be derived by combining equations (1) and (2):

$$U_s(z_1, z_2) = U_s(0, 0) \left[\frac{z_1(z_2 E(z_1, z_2) - S_2(E(z_1, z_2)))}{(z_1 - S_1(E(z_1, z_2)))(z_2 - S_2(E(Y(z_2), z_2)))} \right.$$
$$+ \frac{S_2(E(Y(z_2), z_2))(S_1(E(z_1, z_2)) - z_1 E(z_1, z_2))}{(z_1 - S_1(E(z_1, z_2)))(z_2 - S_2(E(Y(z_2), z_2)))}$$
$$\left. + \frac{E(Y(z_2), z_2)(z_1 S_2(E(z_1, z_2)) - z_2 S_1(E(z_1, z_2)))}{(z_1 - S_1(E(z_1, z_2)))(z_2 - S_2(E(Y(z_2), z_2)))} \right]. \quad (3)$$

Finally, in order to find an expression for $U_s(0,0)$, we put $z_1 = z_2 = 1$ and use de l'Hôpital's rule in equation (3). Therefore, we need the first derivative of $Y(z)$ for $z = 1$ and this follows from its definition

$$Y'(1) = \overline{S}_1(\overline{E}_1 Y'(1) + \overline{E}_2) = \frac{\overline{E}_2 \overline{S}_1}{1 - \rho_1}. \quad (4)$$

We then obtain $U_s(0,0)$:

$$U_s(0,0) = \frac{1 - \rho_T}{1 - \rho_T + \overline{E}_1 + \overline{E}_2}. \quad (5)$$

Substituting the expression for $U_s(0,0)$ in (3) gives a fully determined version of $U_s(z_1, z_2)$.

3.2 System Content at the Beginning of Arbitrary Slots

The system content of priority class j at the beginning of a slot k in steady state is denoted by $U_{r,j}^{(k)}$ ($j = 1, 2$). Define the steady-state joint pgf $U_r(z_1, z_2) \triangleq \mathrm{E}[z_1^{U_{r,1}^{(k)}} z_2^{U_{r,2}^{(k)}}]$. In order to derive an expression for $U_r(z_1, z_2)$, we condition on the status of the server during slot k. There are three possibilities: the server can be idle, a low-priority or a high-priority packet can be in service during slot k. The server is idle during a slot if and only if the system was empty at the beginning of the slot. On the other hand, if the server is busy during slot k, a class-j packet is being served with probability ρ_j/ρ_T ($j = 1, 2$). We relate the system content at the beginning of a random slot to the system content at the beginning of the preceding start slot. The elapsed service time of the packet in service (if any) during slot k is given by $\tilde{S}$. The system content at the beginning of slot k is a superposition of the system content at the beginning of the last preceding start slot and the arrivals during $\tilde{S}$, yielding

$$U_r(z_1, z_2) = U_r(0,0) + (1 - U_r(0,0)) \left\{ \frac{\rho_2}{\rho_T} \frac{U_s(0, z_2) - U_s(0,0)}{U_s(0,1) - U_s(0,0)} \tilde{S}_2(E(z_1, z_2)) \right.$$
$$\left. + \frac{\rho_1}{\rho_T} \frac{U_s(z_1, z_2) - U_s(0, z_2)}{1 - U_s(0,1)} \tilde{S}_1(E(z_1, z_2)) \right\}. \quad (6)$$

Hereby is $\tilde{S}_j(z)$ ($j = 1, 2$) defined as the pgf of the elapsed service time of the class-j packet in service at the beginning of slot k. It is shown in e.g. [83] that

$$\tilde{S}_j(z) = \frac{S_j(z) - 1}{\overline{S}_j(z - 1)}, \quad (7)$$

for $j = 1, 2$. It now remains for us to determine the unknown parameter $U_r(0, 0)$. Keeping in mind that, if the server is idle during slot k, slot k is a start slot, $U_r(0, 0)$ can easily be found as follows:

$$U_r(0,0) = \Pr[U_{r,1}^{(k)} = U_{r,2}^{(k)} = 0]$$
$$= \Pr[U_{s,1}^{(l)} = U_{s,2}^{(l)} = 0 | \text{slot } k \text{ is a start slot}] \Pr[\text{slot } k \text{ is a start slot}],$$

with start slot l the start slot directly preceding slot k. Conditioning on the possibilities of a slot being a start slot, we find

$$U_r(0,0) = U_s(0,0) \left[U_r(0,0) + \frac{1 - U_r(0,0)}{\overline{S}_1} \frac{\rho_1}{\rho_T} + \frac{1 - U_r(0,0)}{\overline{S}_2} \frac{\rho_2}{\rho_T} \right] = 1 - \rho_T. \quad (8)$$

Using equations (3), (5), (7) and (8) in (6), we derive a fully determined version for $U_r(z_1, z_2)$:

$$U_r(z_1, z_2) = (1 - \rho_T) \left\{ \frac{S_1(E(z_1, z_2))(z_1 - 1)}{z_1 - S_1(E(z_1, z_2))} + \frac{E(Y(z_2), z_2) - 1}{E(z_1, z_2) - 1} \right.$$

$$\times \left[\frac{z_1 S_2(E(z_1,z_2))(S_1(E(z_1,z_2))-1)}{(z_1 - S_1(E(z_1,z_2)))(z_2 - S_2(E(Y(z_2),z_2)))} \right.$$
$$+ \frac{z_1 z_2 (S_2(E(z_1,z_2)) - S_1(E(z_1,z_2)))}{(z_1 - S_1(E(z_1,z_2)))(z_2 - S_2(E(Y(z_2),z_2)))}$$
$$\left. + \frac{z_2 S_1(E(z_1,z_2))(1 - S_2(E(z_1,z_2)))}{(z_1 - S_1(E(z_1,z_2)))(z_2 - S_2(E(Y(z_2),z_2)))} \right] \right\}. \quad (9)$$

From the two-dimensional pgf $U_r(z_1, z_2)$, we can easily derive expressions for the pgfs of the system contents of high- and low-priority packets at the beginning of an arbitrary slot - denoted by $U_{r,1}(z)$ and $U_{r,2}(z)$ respectively - yielding

$$U_{r,1}(z) \triangleq \lim_{k \to \infty} \mathrm{E}\left[z^{U^{(k)}_{r,1}}\right] = U_r(z,1)$$
$$= \frac{S_1(E_1(z))(z-1)}{z - S_1(E_1(z))} \left[1 - \rho_T + \overline{E}_2 \frac{S_2(E_1(z)) - 1}{E_1(z) - 1}\right], \quad (10)$$

$$U_{r,2}(z) \triangleq \lim_{k \to \infty} \mathrm{E}\left[z^{U^{(k)}_{r,2}}\right] = U_r(1,z)$$
$$= (1 - \rho_T) \frac{S_2(E_2(z))(z-1)}{z - S_2(E(Y(z),z))} \frac{E(Y(z),z) - 1}{E_2(z) - 1}. \quad (11)$$

3.3 Packet Delay

The packet delay is defined as the total time period a tagged packet spends in the system, i.e., the number of slots between the end of the packet's arrival slot and the end of its departure slot. We denote the steady-state delay of a tagged class-j packet by D_j and its pgf by $D_j(z)$ ($j = 1, 2$). Before deriving expressions for $D_1(z)$ and $D_2(z)$, we first define some notions and stochastic variables we will frequently use in this subsection. We denote the arrival slot of the tagged packet by slot k. If slot k is a start slot, it is assumed to be start slot l. If slot k is not a start slot on the other hand, the last start slot preceding slot k is assumed to be start slot l. We denote the number of class-j packets that arrive during slot k, but which are served before the tagged packet by $\tilde{E}_j^{(k)}$ ($j = 1, 2$). Since we only analyze the integer part of the delay, the precise time instant within the slot at which the tagged packet arrives, is not important. Only the order of service of all packets arriving in the same slot has to be specified. The class-1 packets will be serviced before the class-2 packets, and within a class the order of service is FCFS. We furthermore denote the service time of the tagged class-j packet by $\hat{S}_j$ ($j = 1, 2$). We finally denote the service time and the elapsed service time of the packet in service (if any) during the arrival slot of the tagged packet by S and $\tilde{S}$ respectively. The latter random variable is the amount of service that the packet being served has already received at the beginning of the tagged packet's arrival slot. Assume S and $\tilde{S}$ equal to 0 if no service is ongoing.

Delay of High-Priority Packets. We have that the delay of a tagged class-1 packet - arriving during slot k - is given by

$$D_1 = (S - \tilde{S} - 1)^+ + \sum_{m=1}^{U_{s,1}^{(l)}-1} \check{S}_{1,m} + \sum_{i=1}^{\tilde{S}} \sum_{m=1}^{E_1^{(k-i)}} S_{1,m}^{(k-i)} + \sum_{m=1}^{\tilde{E}_1^{(k)}} S_{1,m}^{(k)} + \hat{S}_1 ,$$

with $(x)^+ = \max(x, 0)$, the $S_{1,m}^{(k)}$'s the service times of the class-1 packets that arrived during slot k, but that are served before the tagged class-1 packet, the $S_{1,m}^{(k-i)}$'s ($0 \leq i \leq \tilde{S}$) the service times of the class-1 packets that arrived during slot $k - i$, and with $\check{S}_{1,m}$ the service times of the class-1 packets already in the queue at the beginning of the ongoing service (thus without the possible packet in service during slot k). We make the convention that a sum $\sum_{m=l}^{k}$ is 0 if $k < l$. Using this equation and conditioning on the type of the packet that is in service (no service, class 1 or class 2, we can derive an expression for $D_1(z)$:

$$D_1(z) = \tilde{E}_1(S_1(z)) S_1(z) \left\{ 1 - \rho_T + \rho_2 \frac{S_2^*\left(\frac{E_1(S_1(z))}{z}, z\right)}{z} \right.$$

$$\left. + \rho_1 \frac{U_s(S_1(z), 1) - U_s(0, 1)}{(1 - U_s(0, 1)) S_1(z)} \frac{S_1^*\left(\frac{E_1(S_1(z))}{z}, z\right)}{z} \right\}, \quad (12)$$

with $\tilde{E}_1(z) \triangleq \mathrm{E}[z^{\tilde{E}_1^{(k)}}]$, $S_2^*(x, z) \triangleq \mathrm{E}[x^{\tilde{S}} z^S | U_{s,1}^{(l)} = 0, U_{s,2}^{(l)} > 0]$ and $S_1^*(x, z) \triangleq \mathrm{E}[x^{\tilde{S}} z^S | U_{s,1}^{(l)} > 0]$. The random variable $\tilde{E}_1^{(k)}$ can be shown to have the following pgf (see e.g. [83]):

$$\tilde{E}_1(z) = \frac{E_1(z) - 1}{\overline{E}_1(z - 1)} . \quad (13)$$

If a class-j packet is in service during slot k, S is characterized by the probability mass function $s_j(m)$ ($j = 1, 2$). The conditional joint pgf of $\tilde{S}$ and S when a class-j packet is in service has the following form:

$$S_j^*(x, z) = \frac{S_j(xz) - S_j(z)}{\overline{S}_j(x - 1)} , \quad (14)$$

with $j = 1, 2$. We now obtain the following expression for $D_1(z)$ from equation (12) together with equations (3), (13) and (14):

$$D_1(z) = \frac{1}{\overline{E}_1} \frac{S_1(z)(z-1)}{z - E_1(S_1(z))} \frac{E_1(S_1(z)) - 1}{S_1(z) - 1} \left(1 - \rho_T + \rho_2 \frac{S_2(z) - 1}{\overline{S}_2(z-1)} \right) . \quad (15)$$

Delay of Low-Priority Packets. An expression for $D_2(z)$ is a bit more involved. We tag a class-2 packet that enters the buffer during slot k (in steady state). Let us refer to the packets in the system at the end of slot k, but that have to be served before the tagged packet as the "primary packets". So, basically, the tagged class-2 packet can enter the server, when all primary packets and all class-1 packets that arrived after slot k (i.e., while the tagged packet is waiting in the queue) are transmitted. In order to analyze the delay of the tagged class-2 packet, the number of class-1 packets and class-2 packets that are served between the arrival slot of the tagged class-2 packet and its departure slot is important, not their precise service order. Therefore, we consider an equivalent virtual system with an altered service discipline. We assume that, from slot k on, the order of service for class-1 packets (those in the queue at the end of slot k and newly arriving ones) is Last Come First Served instead of FCFS in the equivalent system (the transmission of class-2 packets remains FCFS). So, a primary packet can enter the server, when the system becomes free (for the first time) of class-1 packets that arrived during and after the service time of the primary packet that precedes it in the queue according to the new service discipline. Let $V_{1,m}^{(i)}$ denote the length of the time period during which the server is occupied by the m-th class-1 packet that arrives during slot i and its class-1 "successors", i.e., the time period starting at the beginning of the service of that packet and terminating when the system becomes free (for the first time) of class-1 packets which arrived during and after its service time. Analogously, let $V_{2,m}^{(i)}$ denote the length of the time period during which the server is occupied by the m-th class-2 packet that arrives during slot i and its class-1 "successors". The $V_{j,m}^{(i)}$'s ($j = 1, 2$) are called sub-busy periods, initiated by the m-th class-j packet that arrived during slot i. We have the following general expression for D_2:

$$D_2 = (S - \tilde{S} - 1)^+ + \sum_{i=1}^{S-\tilde{S}-1} \sum_{m=1}^{\tilde{E}_1^{(k+i)}} V_{1,m}^{(k+i)} + \sum_{j=1}^{2} \sum_{m=1}^{\tilde{E}_j^{(k)}} V_{j,m}^{(k)}$$

$$+ \sum_{j=1}^{2} \sum_{i=1}^{\tilde{S}} \sum_{m=1}^{E_j^{(k-i)}} V_{j,m}^{(k-i)} + \sum_{m=1}^{U_{s,1}^{(l)}-1} \check{V}_{1,m} + \sum_{m=1}^{U_2^{(l)}-1} \check{V}_{2,m} + \hat{S}_2,$$

with the $\check{V}_{j,m}$'s the sub-busy periods, initiated by the m-th class-1 packet already in the queue at the beginning of start slot l and 1_X the indicator function of X. It is clear that the length of the sub-busy periods initiated by class-1 packets are i.i.d. and thus have the same pgf $V_1(z)$. Also the length of the sub-busy periods initiated by class-2 packets are i.i.d., and their pgf is denoted by $V_2(z)$. Using the equation for D_2 and conditioning on which class is being served, we derive an expression for $D_2(z)$:

$$D_2(z) = \tilde{E}(V_1(z), V_2(z)) S_2(z) \left\{ 1 - \rho_T + \rho_2 \frac{U_s(0, V_2(z)) - U_s(0,0)}{(U_s(0,1) - U_s(0,0))V_2(z)} \right.$$

$$\times \quad \frac{S_2^*\left(\frac{E(V_1(z),V_2(z))}{zE_1(V_1(z))}, zE_1(V_1(z))\right)}{zE_1(V_1(z))} + \frac{U_s(V_1(z),V_2(z)) - U_s(0,V_2(z))}{(1 - U_s(0,1))V_1(z)}$$

$$\times \rho_1 \frac{S_1^*\left(\frac{E(V_1(z),V_2(z))}{zE_1(V_1(z))}, zE_1(V_1(z))\right)}{zE_1(V_1(z))} \bigg\}, \quad (16)$$

with pgfs $\tilde{E}(z_1,z_2) \triangleq \mathrm{E}[z_1^{\tilde{E}_1^{(k)}} z_2^{\tilde{E}_2^{(k)}}]$, $S_2^*(x,z) \triangleq E[x^{\tilde{S}} z^S | U_{s,1}^{(l)} = 0, U_{s,2}^{(l)} > 0]$ and $S_1^*(x,z) \triangleq E[x^{\tilde{S}} z^S | U_{s,1}^{(l)} > 0]$. The random variables $\tilde{E}_1^{(k)}$ and $\tilde{E}_2^{(k)}$ have the following joint pgf (extension of a technique used in e.g. [83]):

$$\tilde{E}(z_1,z_2) = \frac{E(z_1,z_2) - E_1(z_1)}{\overline{E}_2(z_2 - 1)}. \quad (17)$$

The $S_j^*(x,z)$'s ($j = 1,2$) are again given by equation (14). Finally, we have to find expressions for $V_1(z)$ and $V_2(z)$. These pgfs satisfy the following relations:

$$V_j(z) = S_j(zE_1(V_1(z))), \quad (18)$$

with $j = 1,2$. This can be understood as follows: when the m-th class-j packet that arrived during slot i enters service, $v_{j,m}^{(i)}$ consists of two parts: the service time of that packet itself, and the service times of the class-1 packets that arrive during its service time and of their class-1 successors. This leads to equation (18). Equation (16) together with equations (3), (14) and (17) leads to:

$$D_2(z) = \frac{1 - \rho_T}{\overline{E}_2} \frac{S_2(z)(E(V_1(z),V_2(z)) - E_1(V_1(z)))}{zE_1(V_1(z)) - E(V_1(z),V_2(z))} \frac{1 - zE_1(V_1(z))}{1 - V_2(z)}, \quad (19)$$

with $V_j(z)$ ($j = 1,2$) implicitly given by equation (18).

3.4 Calculation of Moments

The functions $Y(z)$, $V_1(z)$ and $V_2(z)$ can only be explicitly found in case of some simple arrival and service processes. Their derivatives for $z = 1$, necessary to calculate the moments of the system content and the packet delay, on the contrary, can be calculated in closed form. For example, $Y'(1)$ is given by equation (4) and the first derivatives of $V_j(z)$ for $z = 1$ are given by $V_j'(1) = \overline{S}_j/(1-\rho_1)$, $j = 1,2$. Now, we can calculate the mean values of the system contents and packet delays of both classes by taking the first derivatives of the respective pgfs for $z = 1$. We find

$$\overline{D}_1 = \frac{\overline{S}_1}{2} + \frac{(\sigma_{E_1}^2 \overline{S}_1 + \overline{E}_1^2 \sigma_{S_1}^2)}{2(1-\rho_1)\overline{E}_1} + \frac{\overline{E}_2(\sigma_{S_2}^2 + \overline{S}_2(\overline{S}_2 - 1))}{2(1-\rho_1)}, \quad (20)$$

$$\overline{D}_2 = \frac{\overline{S}_2}{2} + \frac{\sigma_{E_2}^2 \overline{S}_2}{2(1-\rho_T)\overline{E}_2} + \frac{\overline{E}_2 \sigma_{S_2}^2}{2(1-\rho_T)(1-\rho_1)} + \frac{\sigma_{E_1}^2 \overline{S}_1^2 + \overline{E}_1 \sigma_{S_1}^2}{2(1-\rho_T)(1-\rho_1)}$$

$$-\frac{\rho_1(\overline{S}_2-1)}{2(1-\rho_1)}+\frac{\overline{S}_1\sigma_{E_1E_2}}{(1-\rho_T)\overline{E}_2}. \tag{21}$$

$\sigma_{E_1E_2}$ is the covariance of E_1 and E_2. We only showed the expressions for the mean packet delay (as we will do throughout this paper), but the mean system content can be found in a similar way. Alternatively, one can always use the discretized version of Little's law [84] to calculate the mean system content from the mean packet delay. In a similar way, expressions for higher order moments can be calculated by taking the appropriate derivatives of the respective generating functions as well.

3.5 Tail Behavior

The tail distributions of system content and packet delay are often used to impose statistical bounds on the guaranteed QoS for both classes, and are therefore important performance measures. From the pgfs of the system contents and packet delays of class-1 and class-2 packets derived in subsections 3.2 and 3.3, approximations of the tail probabilities can be derived using complex contour integration and residue theory. In order to determine the asymptotic behavior of the tail distribution, the dominant singularity of the respective generating function is important. We concentrate on the packet delay when no long-tail behavior is encountered in numbers of per-slot arrivals or service times.

First, we concentrate on the class-1 packet delay. The dominant singularity z_H of $D_1(z)$ is a zero of $z - E_1(S_1(z))$ (see equation (15)) and this singularity is a single pole. In the neighborhood of this pole, we can approximate $D_1(z)$ by

$$D_1(z) \approx \frac{K_1}{z_H - z}, \tag{22}$$

where K_1 is found by taking the limit $z \to z_H$ in (22). Using residue theory, we find, for large enough n,

$$\Pr[D_1 = n] \approx \frac{1}{\overline{E}_1}\frac{S_1(z_H)(z_H-1)[(1-\rho_T)(z_H-1)+\overline{E}_2(S_2(z_H)-1)]}{z_H(S_1(z_H)-1)(E_1'(S_1(z_H)))S_1'(z_H)-1)}z_H^{-n}. \tag{23}$$

The tail behavior of the class-2 delay is a bit more involved, since it is not a priori clear what the dominant singularity is of $D_2(z)$. This is due to the occurrence of the function $V_1(z)$ in (19), which is only implicitly defined. First we take a closer look at this function $V_1(z)$. The first derivative of $V_1(z)$ is given by

$$V_1'(z) = \frac{S_1'(zE_1(V_1(z)))E_1(V_1(z))}{1-zS_1'(zE_1(V_1(z)))E_1'(V_1(z))}. \tag{24}$$

Consequently, $V_1(z)$ has a singularity z_B, where the denominator of $V_1'(z)$ becomes 0. Thus $z_B S_1'(z_B E_1(V_1(z_B)))E_1'(V_1(z_B)) = 1$. Since $V_1(z)$ remains finite in the neighborhood of z_B, this singularity is not a simple pole. Application of the results from [85] $V_1(z)$ is, in the neighborhood of z_B, approximately given by

$$V_1(z) \approx V_1(z_B) - K_V\sqrt{z_B - z}, \qquad (25)$$

with $K_V = \sqrt{\dfrac{2E_1(V_1(z_B))}{z_B[z_B^2(E_1'(V_1(z_B)))^3 S_1''(z_B E_1(V_1(z_B))) + E_1''(V_1(z_B))]}}$, which can be found by taking the limit $z \to z_B$ of (25) and using (18). From equation (25), it becomes obvious that z_B is a square-root branch point of $V_1(z)$. $V_1(z)$ has thus two real solutions when $z < z_B$ (the solution we are interested in is the one where $V_1(z) < 1$, if $z < 1$), which coincide at z_B, and has no real solution when $z > z_B$. z_B is also a branch point of $D_2(z)$. A second potential singularity z_L of $D_2(z)$ on the real axis is given by the positive zero of the denominator which is a zero of $zE_1(V_1(z)) - E(V_1(z), V_2(z))$. The tail behavior of the class-2 packet delay is thus characterized by z_L or z_B, depending on which is the dominant (i.e., smallest) singularity. It depends on the number of arrivals and service time distributions which singularity dominates. Three types of tail behavior may thus occur, namely when $z_L < z_B$, when $z_L = z_B$ and when z_L does not exist. In those three cases, $D_2(z)$ can be approximated in the neighborhood of its dominant singularity by:

$$D_2(z) \approx \begin{cases} \dfrac{K_2^{(1)}}{z_L - z} & \text{if } z_L < z_B \\[2mm] \dfrac{K_2^{(2)}}{\sqrt{z_B - z}} & \text{if } z_L = z_B \\[2mm] D_2(z_B) - K_2^{(3)}\sqrt{z_B - z} & \text{if } z_L \text{ does not exist}, \end{cases}$$

where the constants $K_2^{(i)}$ ($i = 1, 2, 3$) can be found by investigation of the behavior of $D_2(z)$ in the neighborhood of this dominant singularity. By using residue theory once again (see [86] for more details), the asymptotic behavior of D_2 is given by

$$\Pr[D_2 = n] \approx \begin{cases} \dfrac{K_2^{(1)}}{z_L} z_L^{-n} & \text{if } z_L < z_B \\[2mm] \dfrac{K_2^{(2)}}{\sqrt{z_B \pi}} n^{-1/2} z_B^{-n} & \text{if } z_L = z_B \\[2mm] \dfrac{K_2^{(3)}}{2}\sqrt{\dfrac{z_B}{\pi}}\, n^{-3/2} z_B^{-n} & \text{if } z_L \text{ does not exist}. \end{cases}$$

The first expression shows geometric tail behavior, while the second and third expressions show non-geometric tail behavior.

4 Preemptive Priority Queues

In this section, we consider the preemptive resume and preemptive repeat priority scheduling disciplines. For ease of analysis, we here additionally assume that there is no correlation between the number of class-1 and class-2 packets arriving during the same slot, that is, $E(z_1, z_2) = E_1(z_1)E_2(z_2)$. This assumption allows us to study high-priority and low-priority performance separately by use of a single-class queueing system. The influence of class-1 traffic on class-2 traffic can be incorporated with interruptions.

The following subsection considers performance of class-1 traffic. The other sections then focus on performance of class-2 traffic. In subsection 4.2, we deduce an appropriate description of the interruption process perceived by class-2 traffic. The analysis of this queueing system with interruptions is then presented in subsections 4.3 to 4.5.

4.1 High-Priority Traffic

Preemptive priority implies that high-priority class-1 traffic is not influenced by low-priority class-2 traffic. That is, a class-1 packet receives service as if there is no low-priority traffic at all. Therefore, performance of the class-1 traffic can be assessed by means of a standard queueing model without priorities. In particular, the assumed nature of arrival and service processes yields that class-1 traffic can be assessed by the $Geo^X/G/1$ queueing model. This model is investigated by amongst others, Bruneel and Kim [83], by Takagi [87] and also by Hunter [88]. Alternatively, we may also retrieve our results from the results in the previous section by assuming that there is no class-2 traffic. That is, we assume: $E(z_1, z_2) = E_1(z_1)$. One easily verifies that the non-preemptive system then reduces to a single-class system. Substitution of the former expression in equations (10) and (15), then yields the pgf $U_{r,1}(z)$ of the class-1 system content at random slot boundaries,

$$U_{r,1}(z) = (1-\rho_1)\frac{(z-1)S_1(E_1(z))}{z - S_1(E_1(z))},$$

and the pgf $D_1(z)$ of the class-1 delay,

$$D_1(z) = \frac{1-\rho_1}{\overline{E}_1}\frac{E_1(S_1(z))-1}{z-E_1(S_1(z))}\frac{(z-1)S_1(z)}{1-S_1(z)},$$

respectively. The moment generating property of pgfs then yields e.g. following expression for mean class-1 packet delay $\overline{D}_1$,

$$\overline{D}_1 = \frac{\rho_1(1-\rho_1) + \sigma_{S_1}^2 \overline{E}_1^2 + \overline{S}_1 \sigma_{E_1}^2}{2(1-\rho_1)\overline{E}_1}. \tag{26}$$

4.2 Interruption Process

Consider low-priority class-2 traffic. Low-priority traffic is only served whenever there are no high-priority packets in the system. That is, a low-priority packet perceives the server as one that alternates between an available state and a blocked state. Slots during which no class-1 packets are served are called available slots or A-slots. Similarly, slots during which a class-1 receives service are called blocked slots or B-slots. Contiguous periods of A-slots (B-slots) are referred to as A-periods (B-periods). One may verify that due to the nature of the class-1 arrival process, the consecutive A-periods as well as the consecutive B-periods constitute series of i.i.d. random variables.

If the high-priority queue is empty at the beginning of a slot, it remains empty during the next slot if there are no arrivals. That is, an A-period continues during the next slot with probability $\alpha = E_1(0)$. This implies that the consecutive A-periods share a common geometrical distribution. Let $A(z)$ denote the corresponding pgf, then we get, $A(z) = (1-\alpha)z/(1-\alpha z)$. Let the sub-busy period of a packet denote the number of slots between the first service slot of this packet and the beginning of the slot where for the first time the number of packets in the system is one less. Note that this definition of sub-busy period is essentially the same as in the preceding section. Clearly, a sub-busy period consists of the time the packet occupies the server (i.e., the packet length) and the sub-busy periods of all class-1 arrivals during this time. That is,

$$V_1 = S_1 + \sum_{i=1}^{S_1} \sum_{j=1}^{E_1^{(i)}} V_{ij}.$$

Here V_1 denotes a random class-1 packet's sub-busy period, S_1 denotes this packet's length, $E_1^{(i)}$ denotes the number of arrivals during the i-th service slot of this packet and V_{ij} denotes the sub-busy period of the j-th arrival during the i-th service slot of the packet. Due to the nature of the arrival process, the sub-busy periods V_{ij}'s are independent random variables sharing the same pgf of the sub-busy period V_1. Some standard z-transform manipulations transform the former equation into

$$V_1(z) = S_1(z E_1(V_1(z))). \tag{27}$$

The busy period of class-1 traffic – that is, the B-period for class-2 traffic – then equals the sum of the sub-busy periods of all arrivals during a slot, given that there is at least one arrival,

$$B(z) = \frac{E_1(V_1(z)) - E_1(0)}{1 - E_1(0)}.$$

The latter follows from the fact that a busy period starts with a non-empty batch of packets arriving in an empty system.

Although equation (27) only provides an implicit expression for $V_1(z)$, it allows to retrieve various moments by evaluation of the appropriate derivatives for $z = 1$ (as discussed in subsection 3.4). Therefore, one may retrieve moments of A- and B-periods as well. In particular, mean lengths of A- and B-periods are given by,

$$\overline{A} = \frac{1}{1-\alpha}, \quad \overline{B} = \frac{\rho_1}{1-\rho_1} \frac{1}{1-\alpha}.$$

For preemptive resume priority scheduling, the transmission of the packet is resumed after interruptions. We will therefore further refer to this scheduling discipline as the *continue after interruption mode* (CAI). Similarly, as transmission is repeated in case of the preemptive repeat priority scheduling, we will further refer to this mode as the *repeat after interruption mode* (RAI). Note that the interruption process under investigation may find other applications as well. B-periods are an abstraction for some kind of server unavailability which does not necessary have to be linked with priority queueing models.

4.3 Effective Service Times

In a first step, we derive expressions for the pgfs of the effective service times of packets. First of all, for ease of explanation, we assume that a packet exists of a number of cells, where each cell needs 1 slot service time (so basically the number of cells in a packet is equal to the number of slots in that packet's service time). The effective service time of an arbitrary packet is defined as the time period elapsed (expressed in slots) between the beginning of the slot during which the first cell of a packet enters the service unit, and the end of the slot during which the last cell of a packet is served. In other words, the effective service time of a packet includes the slots during which the server is interrupted, and in case of RAI (preemptive repeat), the slots required for repeating service of certain cells. Due to the nature of the output process and the packet length distributions, the effective service times of consecutive packets also constitute a series of independent positive random variables, with distributions only depending on the state of the server – described by the availability of the server (A or B) together with the number of remaining B-slots in case the server is unavailable – during the slot preceding the start of the effective service time of the packet and on the operation mode under consideration. This implies that once we know the pgfs of the effective service times for the different operation modes, the evaluation of the system under consideration reduces to the evaluation of an equivalent system without server interruptions but with (state-dependent) service times given by the effective service times.

Continue after Interruption. Recall that the continue after interruption mode corresponds to the preemptive resume priority scheduling discipline. Let $t_{k,A}^{CAI}(n)$ denote the probability that the effective service time of a packet of length k (in cells) equals n slots given that the slot preceding the effective service time is an A-slot. The continue after interruption mode is a memoryless operation mode, in the sense that from a system point of view, once the first cell of a packet of length k has been served, there is no difference between serving the remaining $k-1$ cells of this packet and servicing a new packet of length $k-1$. Therefore, conditioning on the state of the server during the first slot of the effective service time yields,

$$t_{k,A}^{CAI}(n) = \alpha t_{k-1,A}^{CAI}(n-1) + (1-\alpha)\sum_{j=1}^{\infty} b(j) t_{k-1,A}^{CAI}(n-j-1), \qquad (28)$$

for $n \geq k$ and for $k > 1$ whereas for $n < k$ and $k > 1$ this probability equals 0. Let $T_{k,A}^{CAI}(z)$ denote the conditional pgf corresponding to $t_{k,A}^{CAI}(n)$, then, using standard z-transform manipulations, equation (28) easily transforms into,

$$T_{k,A}^{CAI}(z) = (\alpha z + (1-\alpha)zB(z))\, T_{k-1,A}^{CAI}(z)\,, \tag{29}$$

for $k > 1$. Clearly, equation (29) is also valid for $k = 1$ if one defines $T_{0,A}^{CAI}(z) = 1$, i.e., a zero-length packet requires no service time. Equation (29) then easily yields explicit expressions for the effective service time of a packet conditioned on the packet length and given that the server was available during the slot preceding the effective service time. Summation over all possible packet lengths with respect to their probabilities, then yields following expression for the pgf of the effective service time of a random packet given that the server was available during the slot preceding the effective service time,

$$T_{A}^{CAI}(z) = S\left(\alpha z + (1-\alpha)zB(z)\right)\,. \tag{30}$$

Finally, taking the appropriate derivatives of (30) yields expressions for the various moments of the corresponding random variable.

Repeat after Interruption. The memoryless property that was used in the previous section is not valid in case of RAI (preemptive repeat) as the server has to completely repeat transmission of the packet after an interruption. Consider an arbitrary slot that is part of a packet's effective service time. We define the remaining service time of a packet as the number of slots that are necessary to complete transmission of a packet in case there would be no interruptions. It is clear that in case of RAI (as opposed to CAI), the remaining service time for a particular packet is not a decreasing function in time, as after an interruption this value equals the packet length (in slots) again. Analogously, the remaining effective service time is defined as the number of slots it will effectively take to complete service (including interruptions and repetitions) at a certain point in time during a packet's effective service time.

Let $t_{k,l,A}^{RAI}(n)$ denote the probability that the remaining effective service time of a packet of length k equals n slots given that the remaining service time equals l slots and that the slot preceding the remaining effective service time is an A-slot. Conditioning on the state of the server during the first slot of the remaining effective service time then yields,

$$t_{k,l,A}^{RAI}(n) = \alpha t_{k,l-1,A}^{RAI}(n-1) + (1-\alpha)\sum_{j=1}^{\infty} b(j) t_{k,k-1,A}^{RAI}(n-j-1)\,,$$

for $k,l > 1$ and for $n \geq l$, whereas the latter probability equals 0 for $k,l > 1$ and $n < l$. Let $T_{k,l,A}^{RAI}(z)$ denote the corresponding conditional pgf, then

$$T_{k,l,A}^{RAI}(z) = \alpha z T_{k,l-1,A}^{RAI}(z) + (1-\alpha)zB(z) T_{k,k-1,A}^{RAI}(z)\,, \tag{31}$$

for $k,l > 1$. It is easy to verify that the latter equation remains valid for $l = 1$ by defining $T^{RAI}_{k,0,A}(z) = 1$, i.e., if there are no more cells to send, the service ends in the current slot with probability 1. The former equation is a first order linear recursive equation and therefore easily solved. Substitution of $l = k-1$ then determines the unknown function $T^{RAI}_{k,k-1,A}(z)$. In particular the pgf of the complete effective service time conditioned on the length of the packet and the state of the server during the slot preceding the effective service is then given by,

$$T^{RAI}_{k,k,A}(z) = \frac{(\alpha z)^{k-1}(1-\alpha z)(\alpha z + (1-\alpha)zB(z))}{1 - \alpha z - (1 - \alpha^{k-1}z^{k-1})(1-\alpha)zB(z)}, \qquad (32)$$

for $k > 1$. One can easily verify that this expression remains valid for the trivial case of single slot service times ($k = 1$). Summation over all possible packet lengths (in slots) with respect to the packet length probabilities then yields the pgf of the effective service time given that the server is available during the preceding slot,

$$T^{RAI}_A(z) = \sum_{k=1}^{\infty} s_2(k) T^{RAI}_{k,k,A}(z). \qquad (33)$$

Note that this expression is in general not explicit due to the infinite sum. The moment-generating property of pgfs however, allows to determine the various moments of the effective service time explicitly by evaluation of the appropriate derivatives of the pgf for $z = 1$.

Remarks. Clearly, the server is not always available during the slot that precedes the effective service time. Therefore, let $T_{B,m}(z)$ denote the pgf of the effective service time of a random packet given the server is blocked during the slot preceding the effective service and given that the server remains blocked for another m slots after this slot (the server operates in one of the modes under consideration).

Consider now the decomposition of the effective service time of a packet in two components: the number of slots up to the first non-interrupted slot (i.e., the effective service of the first cell of the packet) and the remaining effective service time. Both components are independent random variables. It is clear that the first component (and its pgf) does not depend on the operation mode whereas the second component does not depend on the state of the server during the slot preceding the effective service as the last slot of the first component is by definition an A-slot. Let $X_A(z)$ and $X_{B,m}(z)$ denote the pgfs of the first component given the state during the slot preceding the effective service and let $Y_{(mode)}(z)$ denote the pgf of the second component only depending on the operation mode, then

$$T_A(z) = X_A(z)Y_{(mode)}(z), \quad T_{B,m}(z) = X_{B,m}(z)Y_{(mode)}(z),$$

with

$$X_A(z) = \alpha z + (1-\alpha)zB(z), \quad X_{B,m}(z) = z^{m+1}.$$

The former expression follows from the fact that the first cell is either transmitted directly (with probability α) or immediately after a interruption (with probability $(1-\alpha)$) in the case that the preceding slot is an A-slot. The latter expression follows from the fact that the first cell of a packet is transmitted immediately after the interruption in case the slot preceding the packet's effective service time is a B-slot followed by another m B-slots. Elimination of $Y_{(mode)}(z)$ in the equations above then yields,

$$T_{B,m}(z) = \frac{z^m}{\alpha + (1-\alpha)B(z)} T_A(z). \tag{34}$$

Equations (30) and (33) also imply that whereas for CAI the n-th moment of the effective service time depends on the moments of the underlying packet length distribution up to and including order n, this is not the case for RAI. For the latter operation modes, the n-th moment depends on the complete packet length distribution. In particular the first moments of the effective service time given that the slot preceding this effective service time is an A-slot, are given by,

$$\overline{T}_A^{CAI} = \frac{\overline{S_2}}{\sigma}, \tag{35}$$

$$\overline{T}_A^{RAI} = \frac{1}{\sigma}\frac{\alpha}{1-\alpha}\left(S_2\left(\frac{1}{\alpha}\right) - 1\right), \tag{36}$$

for CAI and RAI respectively. Here σ denotes the fraction of slots that the server is available, that is,

$$\sigma = \frac{\overline{A}}{\overline{A}+\overline{B}} = \frac{1}{1+(1-\alpha)\overline{B}}. \tag{37}$$

Let us now assume the existence of all moments of the B-periods and assume that α is nonzero. For CAI, the existence of the n-th moment of the packet length in cells then implies the existence of the n-th moment of the effective service time, whereas this is not the case for the RAI operation mode. Let R_{S_2} denote the radius of convergence of the pgf $S_2(z)$, then, one can verify that for RAI the n-th moment exists if $\alpha^{-n} < R_{S_2}$ and does not exist if $\alpha^{-n} > R_{S_2}$. For $\alpha^{-n} = R_{S_2}$, the existence depends on the behavior of $S_2(z)$ and its derivatives on their common radius of convergence. The additional condition for RAI also implies, that for finite radii of convergence and given α, only a finite number of moments exist. In particular, one can easily verify that for $\alpha \in (R_{S_2}^{-1}, R_{S_2}^{-1/2})$ the respective effective service time distributions are heavy-tailed in case of RAI.

4.4 System Content

We now use the results of the preceding section to establish expressions for the pgf of the class-2 system content – i.e., the number of packcts present in the system – at packet departure times and at random slot boundaries. Since the effective service time of a packet includes interruptions and possible service repetitions of packets in case of RAI, results of the previous section allows a unified analysis for both operation modes.

At Packet Departure Times. Let $U_{d,2}^{(n)}$ denote the class-2 system content at the beginning of the slot following the departure slot of the n-th class-2 packet, i.e., at the departure time of the n-th class-2 packet. For positive $U_d^{(n)}$, service of the $(n+1)$-th class-2 packet can start immediately as this packet is already present in the system. Therefore, as the previous slot was an A-slot since there was a class-2 packet departure, it will take T_A slots to the next departure, where T_A denotes the random variable representing the effective service time of a class-2 packet given its effective service is preceded by an A-slot, and whose pgf is given by (30) or (33) depending on the operation mode under consideration. The system content $U_{d,2}^{(n+1)}$ is then given by

$$U_{d,2}^{(n+1)} = U_{d,2}^{(n)} - 1 + \sum_{j=1}^{T_A} E_2^{(j)}, \qquad \text{for } U_{d,2}^{(n)} > 0, \qquad (38)$$

with $E_2^{(j)}$ the number of class-2 packets arriving in the system during the j-th slot of the effective service time of the $(n+1)$-th class-2 packet. If, on the other hand, the class-2 buffer is empty after the departure of the n-th class-2 packet, service of the next class-2 packet cannot start immediately. Let w denote the first slot following the departure slot during which one or more packets arrive in the system, and let $E_{2,w}$ and Θ_w denote the number of class-2 arrivals and the state of the server during this slot respectively. As service of the $(n+1)$-th class-2 packet starts in the slot following slot w and its effective service time is described by the random variable T_{Θ_w}, $U_{d,2}^{(n+1)}$ is given by,

$$U_{d,2}^{(n+1)} = E_{2,w} - 1 + \sum_{j=1}^{T_{\Theta_w}} E_2^{(j)}, \qquad \text{for } U_{d,2}^{(n)} = 0, \qquad (39)$$

with $E_2^{(j)}$ the number of packets arriving in the system during the j-th slot of the effective service time of the $(n+1)$-th packet. As the numbers of packets arriving during consecutive slots constitute a series of i.i.d. random variables, the common pgf of the $E_2^{(j)}$'s in (38) and (39) equals $E_2(z)$. Furthermore, as the only distinction regarding the number of arrivals between a random slot and the slot w is that we are certain there arrives at least one packet in the system during slot w, the pgf of $E_{2,w}$ is given by

$$E_{2,w}(z) = \frac{E_2(z) - E_2(0)}{1 - E_2(0)}. \qquad (40)$$

Now, let $q_{k,B,n}$ denote the probability that the k-th slot following an A-slot is a B-slot followed by another n B-slots, and let $Q_B(x,z) = \sum_{k=1}^{\infty} \sum_{n=0}^{\infty} q_{k,B,n} x^k z^n$

denote the corresponding z-transform (note that this is not a pgf). Analogously, let $q_{k,A}$ denote the probability that the k-th slot following an A-slot is an A-slot and let $Q_A(x) = \sum_{k=1}^{\infty} q_{k,A} x^k$ denote the corresponding z-transform. Then, conditioning on the number of slots since the last preceding A-slot yields,

$$q_{k,A} = \alpha q_{k-1,A} + \sum_{j=1}^{k-1} b(j) q_{k-j-1,A},$$

$$q_{k,B,n} = (1-\alpha) \sum_{j=n+1}^{n+k} b(j) q_{k+n-j,A}.$$

for $k \geq 1$ and for $n \geq 0$, whereas $q_{0,A} = 1$ and $q_{0,B,n} = 0$ for all $n \geq 0$. Standard z-transform manipulations then yield,

$$Q_A(x) = \frac{\alpha x + (1-\alpha) x B(x)}{1 - \alpha x - (1-\alpha) x B(x)},$$

$$Q_B(x,z) = (1-\alpha) x (Q_A(x) + 1) \frac{B(x) - B(z)}{x - z}.$$

Due to the nature of the arrival process, slot w (i.e., the first slot with at least one class-2 packet arrival after the departure of the n-th packet) is the k-th slot ($k \geq 1$) after the last departure slot with probability $g(k)$,

$$g(k) = E_2(0)^{k-1}(1 - E_2(0)).$$

Furthermore, this slot is an A-slot (B-slot followed by n B-slots) with probability $q_{k,A}$ ($q_{k,B,n}$) as the server is available during the last slot of the effective service time of the preceding packet. Summation over all possible values of k with respect to the probabilities $g(k)$ yields the probabilities γ_A and $\gamma_{B,n}$ that the server is available during slot w or remains unavailable for another n slots following slot w respectively. Let $\Gamma_B(z)$ denote the z-transform of $\gamma_{B,n}$ then,

$$\begin{cases} \gamma_A = \dfrac{1 - E_2(0)}{E_2(0)} Q_A(E_2(0)), \\ \Gamma_B(z) = \dfrac{1 - E_2(0)}{E_2(0)} Q_B(E_2(0), z). \end{cases} \qquad (41)$$

Now, assume the existence of a stationary distribution of the system contents, i.e., $U_{d,2}(z) = U_{d,2}^{(k+1)}(z) = U_{d,2}^{(k)}(z)$. From (34), (38) and (39), it then follows that the pgf of the class-2 system content at departure times is given by,

$$U_{d,2}(z) = \frac{U_{d,2}(0) T_A(E_2(z))}{z - T_A(E_2(z))} \left\{ \gamma_A E_{2,w}(z) + \frac{\Gamma_B(E(z)) E_{2,w}(z)}{\alpha + (1-\alpha) B(E_2(z))} - 1 \right\}, \quad (42)$$

where $T_A(z)$ is given by (30) or (33) depending on the operation mode. The unknown parameter $U_{d,2}(0)$ in (42) can then be determined by applying the normalization condition $U_{d,2}(1) = 1$, leading to

$$U_{d,2}(0) = \frac{\sigma}{\overline{E_2} \gamma_A}(1 - E_2(0))\left(1 - \overline{E_2}\,\overline{T}_A\right), \qquad (43)$$

with $\overline{T}_A$ given by (35) or (36) for CAI and RAI operation modes respectively and with σ given by expression (37). Substitution of equations (41) and (43) into (42) then yields the pgf of the steady-state class-2 system content at departure times,

$$U_{d,2}(z) = \frac{\sigma(1-\overline{T}_A\overline{E}_2)}{\overline{E}_2} \frac{E_2(z)}{Q_A(E_2(z))} \frac{T_A(E_2(z))}{T_A(E_2(z)) - z}.$$

Random Slot Boundaries. Let $U_{r,2}(z)$ denote the pgf of the (stationary) system content at random slot boundaries and assume that there are no bulk arrivals (all arrivals occur at distinct epochs within slots). According to Bruneel [89], the pgf of the system content at random slot boundaries then relates to the pgf of the system content at arrival times $U_{a,2}(z)$ as,

$$U_{r,2}(z) = \frac{U_{a,2}(z)(z-1)\overline{E}_2}{E_2(z) - 1}. \tag{44}$$

Again under the assumption that there are no bulk arrivals, the system content at arrival and departure times have the same distribution (see e.g. Kleinrock [90] or Takagi [3]), or equivalently, $U_{a,2}(z) = U_{d,2}(z)$, yielding,

$$U_{r,2}(z) = \frac{U_{d,2}(z)(z-1)\overline{E}_2}{E_2(z) - 1}. \tag{45}$$

As both system content at random slot boundaries and system content at packet departure times do not depend on the exact arrival epochs within the consecutive slots, the former expression remains valid for systems with possible bulk arrivals.

Remarks. We assumed that the system under consideration reaches equilibrium. This is only the case if the buffer empties infinitely often during time, i.e., $U_r(0) > 0$, or equivalently, if the effective system load $\rho_{eff} = \overline{T}_A \overline{E}_2$ is less then the number of servers,

$$\rho_{eff} < 1. \tag{46}$$

Substitution of (35) or (36) then yields explicit conditions for the existence of the stationary distribution of the buffer contents for CAI and RAI respectively. Note that for CAI $\rho_{eff} = \rho_T$, as there are no retransmissions. For RAI, we get $\rho_{eff} \geq \rho_T$ as the effective load includes possible retransmissions.

The existence of a stationary distribution however does not imply the existence of moments of this distribution. Let us assume that all moments of the given distributions (number of arrivals in a slot of both classes, length of the packets of both classes) exist. Taking the first derivative of (42) or (45) reveals that the mean system content in both cases depends on both mean and variance of the effective service time, or in general, taking the appropriate derivatives reveals that the n-th moment of the stationary system content is a function of the moments of the effective service times up to order $(n+1)$. This implies that where for CAI – due to our initial assumptions – the equilibrium condition guarantees a finite mean system content, this is not the case for RAI. In the latter case, the n-th moment of the system content distribution is finite as long as both the equilibrium condition and the condition for having a finite $(n+1)$th moment of the effective service time for RAI are satisfied (cfr section 4.3).

4.5 Unfinished Work and Packet Delay

Let $W_2^{(k)}$ denote the unfinished class-2 work at the beginning of slot k, i.e., the number of slots it would take to empty the class-2 buffer under the assumption that there are no new class-2 packet arrivals. Note that this definition implies that the unfinished work takes the interruptions and possible service repetitions into account. Consider now the unfinished work $W_2^{(k+1)}$ at the beginning of slot $(k+1)$. These random variables are related as,

$$W_2^{(k+1)} = (W_2^{(k)} - 1)^+ + \sum_{j=1}^{E_2^{(k)}} T^{(j)}, \qquad (47)$$

where $E_2^{(k)}$ denotes the number of arriving class-2 packets in slot k and $T^{(j)}$ denotes the effective service time of the j-th class-2 packet arriving in slot k. The unfinished work at the beginning of slot $(k+1)$ equals the unfinished work at slot k, diminished with the work done in slot k (if there is any) and augmented with the additional work arriving in slot k. For each class-2 packet arriving in slot k, an additional number of slots, equal to its effective service time is necessary to completely empty the class-2 buffer.

If the class-2 buffer is not empty at the beginning of slot k, the effective service times of all packets entering the system in slot k are preceded by an A-slot as the server was available during the last slot of the preceding class-2 packet's effective service time. This is also the case for all but the first packet entering the system during slot k if the system is empty at the beginning of slot k. The state of the server preceding the first packet's effective service time is an A-slot with probability γ_A or a B-slot followed by another m B-slots with probability $\gamma_{B,m}$ as was shown in the previous section. Let $W_2^{(k)}(z)$ denote the pgf corresponding with $W_2^{(k)}$, then, from equation (47),

$$W_2^{(k+1)}(z) = W_2^{(k)}(0)\left(H(z) - \frac{E_2(T_A(z))}{z}\right) + W_2^{(k)}(z)\frac{E_2(T_A(z))}{z},$$

with

$$H(z) = E_2(0) + (E_2(T_A(z)) - E_2(0))\left(\gamma_A + \sum_{m=0}^{\infty} \gamma_{B,m} \frac{T_{B,m}(z)}{T_A(z)}\right).$$

Now, assume that the system reaches equilibrium – i.e., the equilibrium condition (46) is satisfied – and let $W_2(z)$ denote the pgf of the stationary distribution, i.e.,

$W_2(z) = W_2^{(k)}(z) = W_2^{(k+1)}(z)$. As an empty buffer implies zero unfinished work and vice versa, i.e., $W(0) = U_r(0)$, the pgf of the unfinished work in equilibrium is given by,

$$W_2(z) = \frac{\sigma}{\gamma_A}(1 - \rho_{eff})\frac{zH(z) - E_2(T_A(z))}{z - E_2(T_A(z))} . \tag{48}$$

Consider a particular (tagged) class-2 packet arrival. The packet delay D_2 is defined as the number of slots between the end of the arrival slot and the end of the departure slot of this packet. Let $W_{2,t}$ denote the unfinished work at the beginning of this packet's arrival slot and let $\tilde{E}_2$ denote the numbers of packets arriving in the same slot but before the tagged packet, then,

$$D_2 = (W_{2,t} - 1)^+ + \sum_{j=1}^{\tilde{E}_2+1} T^{(j)}, \tag{49}$$

with $T^{(j)}$ the effective service time of the j-th packet arriving in the system in the tagged packet's arrival slot.

The pgf of the unfinished work at the beginning of the tagged packet's arrival slot is given by (48) due to the i.i.d. nature of the arrival process. Furthermore – similar as equation (13) in the preceding section – the pgf $\tilde{E}_2(z)$ corresponding to $\tilde{E}_2$ is given by,

$$\tilde{E}_2(z) = \frac{E_2(z) - 1}{\overline{E_2}(z - 1)} . \tag{50}$$

If the unfinished work $W_{2,t}$ is nonzero, all effective service times $T^{(j)}$ are preceded by an A-slot as service of these packets starts immediately after service of the preceding packet. This is also the case for all $T^{(j)}$ but $T^{(1)}$ when the queue is empty at the beginning of the tagged packet's arrival slot. For the latter, the preceding slot is again either an A-slot or a B-slot followed by another m B-slots with probability γ_A and $\gamma_{B,m}$ respectively. Let $D_2(z)$ denote the pgf corresponding to D_2, from (48) to (50) then follows,

$$D_2(z) = \frac{\sigma}{\overline{E_2}}(1 - \rho_{eff})\frac{z}{Q_A(z)}\frac{E_2(T_A(z)) - 1}{E_2(T_A(z)) - z}\frac{T_A(z)}{T_A(z) - 1} . \tag{51}$$

The moment-generating property of pgfs then allows the calculation of explicit expressions for the moments of the class-2 packet delay. In particular mean class-2 packet delay is given by,

$$\overline{D}_2 = \frac{\overline{T}_A \sigma_{E_2}^2}{2\overline{E}_2(1 - \rho_{eff})} + \frac{\overline{E}_2 \sigma_{T_A}^2}{2(1 - \rho_{eff})} + (1 - \alpha)\frac{\sigma \sigma_B^2}{2} + \frac{\overline{T}_A}{2}$$
$$- (1 - \alpha)(1 - \alpha\sigma\overline{B})\frac{\overline{B}}{2} , \tag{52}$$

Here $\sigma_{T_A}^2$ and σ_B^2 are the variances of T_A and a B-period respectively.

5 Conclusions

In this paper, we analyzed the high- and low-priority system content and packet delay in a queueing system with a two-class priority scheduling discipline. Two basic types of priority scheduling are analyzed, namely, non-preemptive and preemptive priority scheduling. For each queueing system, a different analysis method was used. A generating-functions-approach was adopted in both, which led to closed-form expressions for some of the relevant performance measures. The results could be used to analyze performance of buffers in a packet-based networking context. Several extensions of the models and analyses are possible, such as a general number of priority classes, correlation in the arrival process, ...

Acknowledgments. The first two authors are Postdoctoral Fellows with the Research Foundation, Flanders (F.W.O.-Vlaanderen), Belgium.

References

1. Miller, R.: Priority queues. Annals of Mathematical Statistics 31, 86–103 (1960)
2. Kleinrock, L.: Queueing systems. Computer applications, vol. II. John Wiley & Sons, New York (1976)
3. Takagi, H.: Queueing analysis: a foundation of performance evaluation, vacation and priority systems, part 1, vol. 1. North-Holland, Amsterdam (1991)
4. Khamisy, A., Sidi, M.: Discrete-time priority queues with two-state Markov Modulated arrivals. Stochastic Models 8(2), 337–357 (1992)
5. Takine, T., Sengupta, B., Hasegawa, T.: An analysis of a discrete-time queue for broadband ISDN with priorities among traffic classes. IEEE Transactions on Communications 42(2-4), 1837–1845 (1994)
6. Laevens, K., Bruneel, H.: Discrete-time multiserver queues with priorities. Performance Evaluation 33(4), 249–275 (1998)
7. Choi, B., Choi, D., Lee, Y., Sung, D.: Priority queueing system with fixed-length packet-train arrivals. IEE Proceedings-Communications 145(5), 331–336 (1998)
8. Walraevens, J., Steyaert, B., Bruneel, H.: Performance analysis of a single-server ATM queue with a priority scheduling. Computers & Operations Research 30(12), 1807–1829 (2003)
9. Mehmet Ali, M., Song, X.: A performance analysis of a discrete-time priority queueing system with correlated arrivals. Performance Evaluation 57(3), 307–339 (2004)
10. Van Velthoven, J., Van Houdt, B., Blondia, C.: The impact of buffer finiteness on the loss rate in a priority queueing system. In: Horváth, A., Telek, M. (eds.) EPEW 2006. LNCS, vol. 4054, pp. 211–225. Springer, Heidelberg (2006)
11. Kamoun, F.: Performance analysis of a discrete-time queuing system with a correlated train arrival process. Performance Evaluation 63(4-5), 315–340 (2006)
12. Walraevens, J., Wittevrongel, S., Bruneel, H.: A discrete-time priority queue with train arrivals. Stochastic Models 23(3), 489–512 (2007)
13. Demoor, T., Walraevens, J., Fiems, D., Bruneel, H.: Mixed finite-/infinite-capacity priority queue with interclass correlation. In: Al-Begain, K., Heindl, A., Telek, M. (eds.) ASMTA 2008. LNCS, vol. 5055, pp. 61–74. Springer, Heidelberg (2008)

14. Walraevens, J., Fiems, D., Bruneel, H.: Time-dependent performance analysis of a discrete-time priority queue. Performance Evaluation 65(9), 641–652 (2008)
15. Walraevens, J., Wittevrongel, S., Bruneel, H.: Performance analysis of a priority queue with session-based arrivals and its application to E-commerce web servers. International Journal On Advances in Internet Technology 2(1), 46–57 (2009)
16. Walraevens, J., Fiems, D., Wittevrongel, S., Bruneel, H.: Calculation of output characteristics of a priority queue through a busy period analysis. European Journal of Operational Research 198(3), 891–898 (2009)
17. Stanford, D.: Interdeparture-time distributions in the non-preemptive priority $\Sigma M_i/G_i/1$ queue. Performance Evaluation 12(1), 43–60 (1991)
18. Sugahara, A., Takine, T., Takahashi, Y., Hasegawa, T.: Analysis of a nonpreemptive priority queue with SPP arrivals of high class. Performance Evaluation 21(3), 215–238 (1995)
19. Abate, J., Whitt, W.: Asymptotics for M/G/1 low-priority waiting-time tail probabilities. Queueing Systems 25(1-4), 173–233 (1997)
20. Takine, T.: The nonpreemptive priority MAP/G/1 queue. Operations Research 47(6), 917–927 (1999)
21. Isotupa, K., Stanford, D.: An infinite-phase quasi-birth-and-death model for the non-preemptive priority M/PH/1 queue. Stochastic Models 18(3), 387–424 (2002)
22. Drekic, S., Stafford, J.: Symbolic computation of moments in priority queues. INFORMS Journal on Computing 14(3), 261–277 (2002)
23. Bouallouche-Medjkoune, L., Aissani, D.: Quantitative estimates in an $M_2/G_2/1$ priority queue with non-preemptive priority: the method of strong stability. Stochastic Models 24, 626–646 (2008)
24. Iftikhar, M., Singh, T., Landfeldt, B., Caglar, M.: Multiclass G/M/1 queuing system with self-similar input and non-preemptive priority. Computer Communications 31, 1012–1027 (2008)
25. Al-Begain, K., Dudin, A., Kazimirsky, A., Yerima, S.: Investigation of the $M_2/G_2/1/\infty, N$ queue with restricted admission of priority customers and its application to HSDPA mobile systems. Computer Networks 53, 1186–1201 (2009)
26. Chen, Y., Chen, C.: Performance analysis of non-preemptive GE/G/1 priority queueing of LER system with bulk arrivals. Computers and Electrical Engineering 35, 764–789 (2009)
27. Rubin, I., Tsai, Z.: Message delay analysis of multiclass priority TDMA, FDMA, and discrete-time queueing systems. IEEE Transactions on Information Theory 35(3), 637–647 (1989)
28. Hashida, O., Takahashi, Y.: A discrete-time priority queue with switched batch Bernoulli process inputs and constant service time. In: Proceedings of ITC 13, Copenhagen, pp. 521–526 (1991)
29. Takine, T., Matsumoto, Y., Suda, T., Hasegawa, T.: Mean waiting times in non-preemptive priority queues with Markovian arrival and i.i.d. service processes. Performance Evaluation 20, 131–149 (1994)
30. Takine, T.: A nonpreemptive priority MAP/G/1 queue with two classes of customers. Journal of Operations Research Society of Japan 39(2), 266–290 (1996)
31. Walraevens, J., Steyaert, B., Bruneel, H.: Performance analysis of the system contents in a discrete-time non-preemptive priority queue with general service times. Belgian Journal of Operations Research, Statistics and Computer Science (JORBEL) 40(1-2), 91–103 (2000)
32. Walraevens, J., Steyaert, B., Bruneel, H.: Delay characteristics in discrete-time GI-G-1 queues with non-preemptive priority queueing discipline. Performance Evaluation 50(1), 53–75 (2002)

33. Walraevens, J., Steyaert, B., Moeneclaey, M., Bruneel, H.: Delay analysis of a HOL priority queue. Telecommunication Systems 30(1-3), 81–98 (2005)
34. Maertens, T., Walraevens, J., Bruneel, H.: Priority queueing systems: from probability generating functions to tail probabilities. Queueing Systems 55(1), 27–39 (2007)
35. Demoor, T., Walraevens, J., Fiems, D., De Vuyst, S., Bruneel, H.: Analysis of a non-preemptive priority queue with finite high-priority capacity and general service times. In: Proceedings of the 4th International Conference on Queueing Theory and Applications (QTNA 2009), Singapore, ID12 (2009)
36. Miller, D.: Computation of steady-state probabilities for M/M/1 priority queues. Operations Research 29(5), 945–958 (1981)
37. Sandhu, D., Posner, M.: A priority M/G/1 queue with application to voice/data communication. European Journal of Operational Research 40(1), 99–108 (1989)
38. Takine, T., Hasegawa, T.: The workload in the MAP/G/1 queue with state-dependent services: its application to a queue with preemptive resume priority. Communications in Statistics - Stochastic Models 10(1), 183–204 (1994)
39. Takahashi, Y., Miyazawa, M.: Relationship between queue-length and waiting time distributions in a priority queue with batch arrivals. Journal of the Operations Research Society of Japan 37(1), 48–63 (1994)
40. Boxma, O., Cohen, J., Deng, Q.: Heavy-traffic analysis of the M/G/1 queue with priority classes. In: Proceedings of ITC 16, Edinburgh, pp. 1157–1167 (1999)
41. Sharma, V., Virtamo, J.: A finite buffer queue with priorities. Performance Evaluation 47(1), 1–22 (2002)
42. Takada, H., Miyazawa, M.: A Markov Modulated fluid queue with batch arrivals and preemptions. Stochastic Models 18(4), 529–652 (2002)
43. Liu, Y., Gong, W.: On fluid queueing systems with strict priority. IEEE Transactions on Automatic Control 48(12), 2079–2088 (2003)
44. Jin, X., Min, G.: Performance analysis of priority scheduling mechanisms under heterogeneous network traffic. Journal of Computer and System Sciences 73, 1207–1220 (2007)
45. Tarabia, A.: Two-class priority queueing system with restricted number of priority customers. AEÜ-International Journal of Electronics and Communications 61(8), 534–539 (2007)
46. Tzenova, E., Adan, I., Kulkarni, V.: Output analysis of multiclass fluid models with static priorities. Performance Evaluation 65(1), 71–81 (2008)
47. Horvath, A., Horvath, G., Telek, M.: A traffic based decomposition of two-class queueing networks with priority service. Computer Networks 53, 1235–1248 (2009)
48. Lee, Y.: Discrete-time $Geo^x/G/1$ queue with preemptive resume priority. Mathematical and Computer Modelling 34(3-4), 243–250 (2001)
49. Walraevens, J., Steyaert, B., Bruneel, H.: Performance analysis of a GI-Geo-1 buffer with a preemptive resume priority scheduling discipline. European Journal of Operational Research 157(1), 130–151 (2004)
50. Walraevens, J., Steyaert, B., Bruneel, H.: A packet switch with a priority scheduling discipline: Performance analysis. Telecommunication Systems 28(1), 53–77 (2005)
51. Van Houdt, B., Blondia, C.: Analyzing priority queues with 3 classes using tree-like processes. Queueing Systems 54 (2), 99–109 (2006)
52. Ndreca, S., Scoppola, B.: Discrete-time GI/Geom/1 queueing system with priority. European Journal of Operational Research 189, 1403–1408 (2008)
53. Walraevens, J., Steyaert, B., Bruneel, H.: Analysis of a discrete-time preemptive resume priority buffer. European Journal of Operational Research 186(1), 182–201 (2008)

54. Sumita, U., Sheng, O.: Analysis of query processing in distributed database systems with fully replicated files: a hierarchical approach. Performance Evaluation 8(3), 223–238 (1988)
55. Yoon, C., Un, C.: Unslotted 1- and p_i-persistent CSMA-CD protocols for fiber optic bus networks. IEEE Transactions on Communications 42(2-4), 158–465 (1994)
56. Mukherjee, S., Saha, D., Tripathi, S.: A preemptive protocol for voice-data integration in ring-based LAN: performance analysis and comparison. Performance Evaluation 11(3), 339–354 (1995)
57. Walraevens, J., Steyaert, B., Bruneel, H.: A preemptive repeat priority queue with resampling: performance analysis. Annals of Operations Research 146(1), 189–202 (2006)
58. Walraevens, J., Fiems, D., Bruneel, H.: The discrete-time preemptive repeat identical queue. Queueing Systems 53(4), 231–243 (2006)
59. Hong, S., Takagi, H.: Analysis of transmission delay for a structured-priority packet-switching system. Computer Networks and ISDN Systems 29(6), 701–715 (1997)
60. Kim, K., Chae, K.: Discrete-time queues with discretionary priorities. European Journal of Operational Research 200(2), 473–485 (2010)
61. Fidler, M., Persaud, R.: M/G/1 priority scheduling with discrete pre-emption points: on the impacts of fragmentation on IP QoS. Computer Communications 27(12), 1183–1196 (2004)
62. Fiems, D., Maertens, T., Bruneel, H.: Queueing systems with different types of server interruptions. European Journal of Operational Research 188(3), 838–845 (2008)
63. Hsu, J.: Buffer behavior with Poisson arrival and geometric output processes. IEEE Transactions on Communications 22, 1940–1941 (1974)
64. Heines, T.: Buffer behavior in computer communication systems. IEEE Transactions on Communications 28, 573–576 (1979)
65. Bruneel, H.: A general treatment of discrete-time buffers with one randomly interrupted output line. European Journal of Operational Research 27(1), 67–81 (1986)
66. Woodside, C., Ho, E.: Engineering calculation of overflow probabilities in buffers with Markov-interrupted service. IEEE Transactions on Communications 35(12), 1272–1277 (1987)
67. Yang, O., Mark, J.: Performance analysis of integrated services on a single server system. Performance Evaluation 11, 79–92 (1990)
68. Lee, D.: Analysis of a single server queue with semi-Markovian service interruption. Queueing Systems 27(1–2), 153–178 (1997)
69. Bruneel, H.: Buffers with stochastic output interruptions. Electronics Letters 19(18), 735–737 (1983)
70. Georganas, N.: Buffer behavior with Poisson arrivals and bulk geometric output processes. IEEE Transactions on Communications 24(8), 938–940 (1976)
71. Bruneel, H.: A general model for the behaviour of infinite buffers with periodic service opportunities. European Journal of Operational Research 16, 98–106 (1984)
72. Laevens, K., Bruneel, H.: Delay analysis for discrete-time queueing systems with multiple randomly interrupted servers. European Journal of Operational Research 85, 161–177 (1995)
73. Bruneel, H.: A discrete-time queueing system with a stochastic number of servers subjected to random interruptions. Opsearch 22(4), 215–231 (1985)
74. Bruneel, H.: On buffers with stochastic input and output interruptions. International Journal of Electronics and Communications (AEU) 38(4), 265–271 (1984)

75. Ali, M., Zhang, X., Hayes, J.: A discrete-time queueing analysis of the wireless ATM multiplexing system. In: Lorenz, P. (ed.) ICN 2001. LNCS, vol. 2093, pp. 429–438. Springer, Heidelberg (2001)
76. Kamoun, F.: Performance evaluation of a queuing system with correlated packet-trains and server interruption. Telecommunication Systems 41(4), 267–277 (2009)
77. Inghelbrecht, V., Laevens, K., Bruneel, H., Steyaert, B.: Queueing of fixed-length messages in the presence of server interruptions. In: Proceedings Symposium on Performance Evaluation of Computer and Telecommunication Systems, SPECTS 2k, Vancouver, Canada (July 2000)
78. Fiems, D., Steyaert, B., Bruneel, H.: Performance evaluation of CAI and RAI transmission modes in a GI-G-1 queue. Computers and Operations Research 28(13), 1299–1313 (2001)
79. Fiems, D., Steyaert, B., Bruneel, H.: Randomly interrupted GI-G-1 queues, service strategies and stability issues. Annals of Operations Research 112, 171–183 (2002)
80. Fiems, D., Steyaert, B., Bruneel, H.: Analysis of a discrete-time GI-G-1 queueing model subjected to bursty interruptions. Computers and Operations Research 30(1), 139–153 (2002)
81. Fiems, D., Steyaert, B., Bruneel, H.: Discrete-time queues with generally distributed service times and renewal-type server interruptions. Performance Evaluation 55(3-4), 277–298 (2004)
82. Adan, I., Van Leeuwaarden, J., Winands, E.: On the application of Rouché's theorem in queueing theory. Operations Research Letters 34(3), 355–360 (2006)
83. Bruneel, H., Kim, B.: Discrete-time models for communication systems including ATM. Kluwer Academic Publisher, Boston (1993)
84. Fiems, D., Bruneel, H.: A note on the discretization of Little's result. Operations Research Letters 30(1), 17–18 (2002)
85. Drmota, M.: Systems of functional equations. Random Structures & Algorithms 10(1-2), 103–124 (1997)
86. Flajolet, P., Odlyzko, A.: Singularity analysis of generating functions. SIAM Journal on discrete mathematics 3(2), 216–240 (1990)
87. Takagi, H.: Queueing Analysis; A foundation of performance evaluation. Discrete-time systems, vol. 3. Elsevier Science Publishers, Amsterdam (1993)
88. Hunter, J.J.: Mathematical Techniques of Applied Probability. Operations Research and Industrial Engineering, vol. 2. Academic Press, New York (1983)
89. Bruneel, H.: Performance of discrete-time queuing systems. Computers and Operations Research 20, 303–320 (1993)
90. Kleinrock, L.: Queueing systems. Theory, vol. I. John Wiley & Sons, New York (1975)

Queueing Networks with Blocking: Analysis, Solution Algorithms and Properties

Simonetta Balsamo

Dipartimento di Informatica
Università Ca' Foscari di Venezia
via Torino, 155 Mestre-Venezia, Italy

Abstract. Queueing network models with finite capacity queues and blocking are used for modeling and performance evaluation of systems with finite resources and population constraints, such as communication and computer systems, traffic, production and manufacturing systems. Various blocking types can be defined to represent different system behaviors, network protocols and technologies. Queueing networks with blocking are difficult to analyze, except for the special class of product-form networks. Most of the analytical methods proposed in literature provide an approximate solution with a limited computational cost. We introduce queueing networks with finite capacity queues and blocking, the main solution techniques for their analysis, both exact and approximate algorithms, and some network properties. We discuss the conditions under which exact solutions can be derived, and criteria for the appropriate selection of approximate methods. We present equivalence properties among different types of blocking types, the analysis of heterogeneous networks, and some application examples.

Keywords: Queueing Networks, Blocking, Product-form models, Equivalence properties.

1 Introduction

Performance analysis of various systems, including communication and computer systems, as well as production and manufacturing systems can be carried out through queueing network models. System performance analysis consists of the derivation of a set of figures of merit, that typically includes queue length distribution and some average performance indices such as mean response time, throughput, and utilization. Queueing networks with finite capacity queues and blocking have been introduced to represent systems with finite capacity resources and population constraints. When a queue reaches its maximum capacity then the flow of customers into the service center is stopped, both from other service centers and from external sources in open networks, and the blocking phenomenon arises. Various blocking mechanisms have been defined and analyzed in the literature to represent distinct behaviors of real systems with limited resources [42, 47, 53, 55, 57, 58]. Some comparisons and equivalences among blocking types have been presented for queueing networks with various topologies in [9, 10, 42, 45, 47, 57, 58]. Performance analysis of queueing networks with blocking can be exact or approximate. Exact solution algorithms have been proposed to evaluate both average performance indices, queue length distribution [10, 42, 47], and passage time distribution [7, 10, 11]. Under exponential assumption one can define and analyze the continuous-time Markov chain underlying the queueing network. In some special cases queueing networks with blocking show a product-form solution, under particular constraints, for various blocking types; a survey is presented in [10]. Some solution algorithms for product-form networks with finite capacities have been defined [8, 19, 50].

However, except for this special class of models, queueing networks with blocking do not have a product-from solution and a numerical solution of the associated Markov chain is seriously limited by the space and time computational complexity that grows exponentially with the model number of components. Hence recourse to approximate analytical methods or simulation is necessary. Several approximate solution methods for queueing networks with blocking have been proposed in literature both for open and closed models and surveys of some methods have been presented in [6, 42, 47]. Most of these methods provide an approximate solution with a limited computational cost. However, they do not provide any bound on the introduced approximation error. They are usually validated by comparing numerical results with either simulation results or exact solutions. Many approximation methods are heuristics based on the decomposition principle applied to the underlying Markov process or, more often, to the network itself. Some methods consider a forced solution of a product-form network and some approximations are based on the maximum entropy principle.

In this paper we focus on queueing networks with finite capacity queues and blocking, their exact and approximate analysis, their properties and applications. We consider various blocking mechanisms that represent different system behaviors. We review the main solution methods and algorithms to analyze queueing networks with blocking to evaluate a set performance indices, the conditions and some criteria for the appropriate selection of the solution method. We consider exact and approximate analytical methods for open and closed queueing networks with blocking. We recall some equivalence properties that allow the solution of heterogeneous models and some application examples.

The paper is structured as follows. Section 2 introduces the model definition of queueing networks with finite capacity queues, the various blocking mechanisms and the main performance indices. Section 3 describes the exact analysis of queueing networks with blocking based on analytical methods, that includes the approach based on the Markov chain definition and analysis, and the special cases of networks with product-form solutions. Section 4 recalls the main principles and approaches proposed for the approximate analytical solution of networks with blocking. Section 5 compare some approximation methods for closed and open networks with blocking, and for different network topologies. In Section 6 we recall some equivalence properties of network models with different blocking types, and we illustrate an application example.

2 Model Definition and Blocking Mechanisms

A queueing network consists of a set of service centers, each formed by a queue and a set of identical servers that provide service to a set of customers. Let us consider a network with N queues with finite capacity, one class of customers, and probabilistic routing. The network may be open or closed. For a closed network let K denote the number of customers. For an open network let λ_i be the external arrival rate (from outside the network) to station i, $1 \leq i \leq N$. For the sake of simplicity we usually assume exponential service time distribution and Poisson arrivals. The service rate of station i is denoted by μ_i, and K_i is the number of servers, usually just single servers, $1 \leq i \leq N$. The finite capacity of node i is denoted by B_i, $1 \leq i \leq N$. Let $\mathbf{P} = [p_{ij}]$ denote the routing probability matrix, where p_{ij} is the probability for a customer to go to station j after being served by station i, $1 \leq i, j \leq N$, and p_{i0} is the probability that a customer leaves the network after being served by station i. Let $\mathbf{e} = (e_1, \ldots, e_N)$ denote the solution of the traffic equations, defined as follows:

$$e_i = \lambda_i + \sum_{j=1}^{N} e_j p_{ji}, \quad 1 \leq i \leq N \tag{1}$$

If the routing probability matrix is irreducible, this system has a unique solution for open networks and an infinite number of solutions for a closed network, unique up to a multiplicative constant.

In queueing networks with finite capacities when a customer attempts to enter a finite capacity queue that is full, it can be blocked. We shall now introduce the definition of some blocking mechanism that describe the blocked customers behavior.

2.1 Blocking Types

Various blocking types have been defined to represent different system behaviors. We now recall three of the most commonly used blocking types defied for computer, communication, networks, and production systems [10, 42, 53].

- Blocking After Service (BAS): if a job attempts to enter a full capacity queue j upon completion of a service at node i, it is forced to wait in node i server, until the destination node j can be entered. The server of source node i stops processing jobs (it is blocked) until destination node j releases a job, and its service will be resumed as soon as a departure occurs from node j. At that time the job waiting in node i immediately moves to node j. If more than one node is blocked by the same node j, then a scheduling discipline must be considered to define the unblocking order of the blocked nodes when a departure occurs from node j.
- Blocking Before Service (BBS): a job declares its destination node j before it starts receiving service at node i. If at that time node j is full, the service at node i does not start and the server is blocked. If a destination node j

becomes full during the service of a job at node i whose destination is j, node i service is interrupted and the server is blocked. The service of node i will be resumed as soon as a departure occurs from node j. The destination node of a blocked customer does not change. Two subcategories distinguish whether the server can be used as a buffer when the node is blocked: BBS-SO (server occupied) and BBS-SNO (server not occupied). Hereafter we consider BBS-SO blocking, which is simply called BBS.
- Repetitive Service Blocking (RS): if, upon completion of its service at ode i, a job attempts to enter a destination queue j, which is full, the job is looped back into the sending queue i, whereupon it receives a new, independent and identically distributed service according to the service discipline. Two subcategories distinguish whether the job, after receiving a new service, chooses a new destination node independently of the one that it had selected previously: RS-RD (random destination) and RS-FD (fixed destination).

Another kind of blocking, called *generalized blocking* or *kanban blocking*, is defined when the server continues processing customers in the queue even if the destination node is full [17, 18, 38, 39]. The customers that have completed service at node i, but cannot be sent to the next node, continue to share the buffer space of node i along with the other customers that are either waiting for service or being served upon. The customers arriving at a node when the queue is full are lost. This blocking is defined to model manufacturing systems. For particular values of the parameters that define this blocking type, it reduces to other blocking types, including BBS-SO and BAS.

Other types of blocking mechanisms model population constraints in a network, by assuming that the number of customers are in the range $[L, U]$, i.e., L and U are the minimum and maximum populations admitted, respectively. Let $a(n)$ denote a load dependent arrival rate function and $d(n)$ a non-negative departure blocking function, where $n \geq 0$ is the overall network population. Then we set $a(n) = 0$ for $n \geq U$ and $d(n) = 0$ for $n \leq L$. The blocking types defined for population constraints include the following types [36, 57, 58].

- Stop Blocking: the service rate at each node depends on the number n of customers in the network, according to function $d(n)$. When $d(n) = 0$ the service at each node is stopped. Service at a node is resumed upon arrival of a new customer to the network.
- Recirculate Blocking: a job upon completion of its service at node i leaves the network with probability $p_{i0}d(n)$, when n is the total network population and it is forced to stay in the network with probability $p_{i0}[1 - d(n)]$, where p_{i0} is the routing probability. Hence, a job completing the service at node i enters node j with state dependent routing probability $p_{ij} + p_{i0}[1 - d(n)]p_{0j}$, $1 \leq i, j \leq N$, $n \geq 0$.

Closed queueing networks with finite capacity queues and blocking can deadlock, depending on the blocking type. Deadlock prevention or detection and resolving techniques must be applied. Deadlock prevention for blocking types BAS, BBS and RS-FD requires that the overall network population K is less than the total buffer capacity of the nodes in each possible cycle in the network, whereas for RS-RD blocking it is sufficient that routing matrix **P** is irreducible and K is less than the total buffer capacity of the nodes in the network [10, 42]. Moreover, to avoid deadlocks for BAS and BBS blocking types we assume $p_{ii} = 0$, $1 \leq i \leq N$. In the following we shall consider deadlock-free queueing networks in steady-state conditions.

2.2 Performance Indices

Queueing networks with blocking are used to model real life systems with finite capacities and to estimate various performance indices. These performance metrics may be defined for each node, and, in multiclass networks, for each chain and/or class. Performance indices are defined in terms of distributions of various random variables, or in terms of average or mean rate of a performance measure. The most commonly used average performance indices are, for each node i: the utilization U_i, the throughput X_i, the average queue length L_i and the mean response time T_i. Performance indices evaluated in terms of random variable distribution are the queue length n_i, i.e., the number of customers at node i, and the number of active servers at node i, that is servers that are neither empty nor blocked. More complex analysis can be carried out to derive more detailed performance indices, such as the customer passage time distribution through the node, and the cycle time distribution for closed network [7, 11].

The definition of the performance indices, both probabilities and average values, depends on the blocking type. Let $PB_i(n_i)$ denote the probability that node i is not empty and blocked when there are n_i customers in node i, and let $PB_i = \sum_{n_i} PB_i(n_i)$ the overall blocking probability. They depend on the blocking type [10].

Then the performance indices for each node i can be defined as follows:

- queue length distribution $\pi_i(n_i)$, $max(0, K - \sum_{j \neq i} B_j) \leq n_i \leq B_i$
- utilization $U_i = 1 - \pi_i(0) - PB_i$
- throughput $X_i = \sum_{n_i} [\pi_i(n_i) - PB_i(n_i)]\mu_i(n_i)$, for load dependent service rate and $X_i = U_i \mu_i$ for constant service rate
- mean queue length $L_i = \sum_{n_i} n_i \pi_i(n_i)$
- mean response time $T_i = L_i/X_i$
- mean cycle time $\sum_j x_j T_j / x_i$.

In networks with blocking we can further define a specific performance index called *effective utilization*. It is defined as the fraction of time that the node is neither empty nor blocked, that is a measure of the useful work of the node. Similarly, for BBS blocking where the interrupted service is repeated and for RS blocking where the job can be looped back, we can also define the *effective throughput* as a measure of the useful work of the node. It is given by the fraction of throughput that is not due to the service repetition because of blocking.

In this paper we mainly focus on the evaluation of the queue length distribution and the average performance indices. We now recall the main method used for exact analysis of queuing networks with blocking.

3 Exact Analysis of Networks with Finite Capacities

Exact analysis of queueing networks with finite capacities and blocking can be obtained by representing the model with a stochastic Markov process, and specifically with a continuous-time Markov chain. By using exact analysis of queueing networks with blocking one can evaluate of a set of average performance indices, the joint queue length distribution at arbitrary times and at arrival times, and possibly the passage time and cycle time distributions. We shall now recall the exact solution based on the Markov process associated to queueing networks with blocking to evaluate the queue length distribution and average performance indices. Then we present the special class of product-form networks with blocking that can be solved by more efficient techniques.

3.1 Markov Process Analysis

Under the assumptions of exponential delays and independence between service times and inter-arrival times, the network can be represented by a continuous-time, homogeneous Markov chain. Let $\mathbf{S} = (S_1, \ldots, S_N)$ denote the state of the network with N nodes, and let E be the state space, i.e., the set of all feasible states. The network model evolution can be represented by a continuous-time ergodic Markov chain with discrete state space E and transition rate matrix $\mathbf{Q}$. The stationary and transient behaviour of the network can be analyzed by the underlying Markov process. Under the hypothesis of an irreducible network routing matrix $\mathbf{P}$, there exists a unique steady-state queue length probability distribution, denoted by $\pi = [\pi(\mathbf{S})], \forall \mathbf{S} \in E$. It can be obtained by solving the homogeneous linear system of the global balance equations

$$\pi \mathbf{Q} = \mathbf{0}, \qquad (2)$$

subject to the normalising condition $\sum_{\mathbf{S} \in E} \pi(\mathbf{S}) = 1$ and where $\mathbf{0}$ is the all zero vector.

The definition of state $\mathbf{S}$, state space E and the transition rate matrix $\mathbf{Q}$ depends on the network characteristics and on the blocking type of each node [9, 10, 42, 47, 57, 58]. Each process state transition corresponds to a particular set of events on the network model, such as a job service completion at a node and the simultaneous transition towards another node or an external arrival at a node. This correspondence depends on the blocking type.

For example for RS-RD blocking, by definition the servers cannot be blocked. That is the server is always active and servicing a customer, if $n_i \geq 0$. Therefore, under exponential assumptions, node i state definition is simply $S_i = n_i$. For BAS blocking we have to consider the server activity and the scheduling of the nodes that are blocked by a full destination node. Then, under exponential assumptions, the process state of node i can be defined as $S_i = (n_i, s_i, \mathbf{m}_i)$, where n_i is the number of jobs in node i, s_i is the number of servers of node i blocked by a full destination node and therefore containing a served job, $0 \leq s_i \leq min(n_i, K_i)$, for K_i servers of node i, and $\mathbf{m}_i$ is the list of nodes blocked by node i. For BBS blocking, since a job declares its destination node j before it starts receiving service, and it can be blocked when node j is full, then the state can be defined as $S_i = (n_i, \mathbf{NS}_i)$, where n_i is the number of jobs in node i, and $\mathbf{NS}_i$ is a vector defined only for nodes that can be blocked. The j-th component $NS_i(j)$ denotes the number of node i servers that are servicing jobs destined to node j, and for an open network $NS_i(0)$ denotes the number of node i servers with jobs that will leave the network.

For each blocking type one can define the corresponding transition rate matrix $\mathbf{Q}$ and solve the liner system (2) to derive the steady-state distribution π. From vector π one can derive the queue length distribution of node i, π_i, and other average performance indices of node i, such as throughput (X_i), utilization (U_i), average queue length (L_i) and mean response time (T_i).

By summarizing, exact analysis of queueing network model with finite capacities based on a continuous-time Markov process requires:

1. Definition of system state $\mathbf{S}$ and state space E according to the blocking type.
2. Definition of transition rate matrix $\mathbf{Q}$ according to the blocking type and the network topology.
3. Solution of global balance equations (2) to derive the steady-state distribution $\pi(\mathbf{S}), \forall \mathbf{S} \in E$.
4. Computation, from the steady-state distribution π, of the average performance indices for each node of the network.

The numerical solution based of the Markov chain analysis is seriously limited by the space and time computational complexity that grows exponentially with the model number of components. For open network the Markov chain is infinite and, unless a special regular structure of matrix $\mathbf{Q}$ allows to derive closed form expression of the solution π, one has to approximate the solution on a truncated state space. For closed networks the time computational complexity of liner system (2) is determined by the space state E cardinality that grows exponentially with the buffer sizes $(B_i \leq K, 1 \leq i \leq N)$ and N. Although the state space cardinality of the process can be much smaller than that of the process underlying the same network with infinite capacity queues (which is exponential in K and N), it still remains numerically untractable as the number of model components grows.

When special constraints are satisfied we can apply exact analysis based on product-form, that we now introduce, or, in many practical cases, it is necessary to apply approximate solution methods.

3.2 Product-Form Networks

In some special cases, queueing networks with blocking have a product-form solution, under certain constraints on network parameters and for various blocking types. Various product-form networks with finite capacities have been defined [1, 3, 9, 10, 25, 27, 40, 41, 58, 59]. A detailed description of product-form solutions of networks with blocking and equivalence properties among different blocking network models is presented in [9] and in [10, 59]. Some efficient algorithms for some closed product-form networks with blocking have been defined [7, 19, 50] and can be applied to derive the performance indices, under some constraints.

Product-form solutions for the joint queue length distribution π for single class open or closed networks under given constraints, depending both on the network topology and the blocking mechanism, can be defined as follows:

$$\pi(\mathbf{S}) = \frac{1}{G} V(n) \prod_{i=i}^{N} g_i^{n_i}, \quad \forall \mathbf{S} \in E \tag{3}$$

where G is a normalising constant and $n = \sum_{i=1}^{N} n_i$ is the total network population, n_i is the number of customers in node i, defined in the node state S_i. The definition of functions V and g_i depends on some network parameters, which include the solution e_i of the traffic balance equations (1) and the service rates $\mu_i, 1 \leq i \leq N$, on the blocking type and some additional constraints.

We shall now recall the main product-form results. For the sake of simplicity we provide the product-form definition for single class networks.

Consider the following five network topologies: two-node networks, cyclic topology, central server (or star topology), reversible routing networks, and arbitrary topology. The first three are special cases of closed networks, the last two apply to closed and open networks.

Reversible routing. A routing matrix $\mathbf{P}$ is said to be reversible if the following conditions hold:

$$e_i p_{ij} = e_j p_{ji}, \quad \lambda_i = e_i p_{i0} \quad \forall 1 \leq i, j \leq N \tag{4}$$

where $\mathbf{e} = [e_1, \ldots, e_N]$ is the solution of system (1).

Note that for closed networks, only the first condition of this definition has to be verified. Two-node networks are a special case of reversible routing.

In order to define some cases of product-form we introduce the following definitions.

Condition 1. (Non-empty condition). The non-empty condition for closed networks requires that at most one node can be empty, i.e., $K \geq B - B_{min}$, where $B = \sum_{1 \leq i \leq N} B_i$ and $B_{min} = \min_{1 \leq i \leq N} B_i$.

Condition 2. (Strictly non-empty condition). This condition is said to hold strictly when each node can never be empty, i.e., the inequality is strict: $K > B - B_{min}$.

Condition 3. (Single destination node). Each node i with finite capacity is the only destination node for each upstream node, i.e., if $B_i < K$ and $p_{ji} > 0$ then $p_{ji} = 1$, $1 \leq i, j \leq N$.

Condition 4. (Only one node blocked). At most one node can be blocked, i.e., if $K = B_{min} + 1$.

Definition: A-type node. An A-type node has arbitrary service time distribution, symmetric scheduling discipline or exponential service time, identical for each class at the same node, when the scheduling is arbitrary [3].

Product-form solution (3) has been derived for networks with different blocking types and with different topologies. Some product-forms hold for both homogeneous networks, that is where each node operates with the same blocking mechanism, and non-homogeneous ones, where different nodes in the networks work under different blocking mechanisms. Table 1 summarizes the main cases of allowed combination of blocking types for each network topology, under some additional constraints, i.e., conditions 1 through 4 defined above, and where product-form (3) is defined by formulas $F1$ through $F5$ as follows.

Table 1. Product-form heterogeneous networks with blocking

Network topology	Blocking types	Product-form formulas
Two nodes	BAS, BBS, RS	F1
Cyclic topology	BBS, RS	F2 and Condition 1
Central server (star)	BBS, RS (central node with RS)	F3
Reversible routing	RS-RD, Stop	F4
Arbitrary routing	BBS, RS-FD	F2 and Conditions 2 and 3
Arbitrary routing	BAS	F5 and Condition 4

Let us define $\rho_i = e_i/\mu_i$, where μ_i is the service rate of node i, and e_i the solution of the system of traffic equations (1).

Product-form F1. For multiclass networks with BCMP-type nodes [13] and class independent capacities, formula $F1$ defines: $V(n) = 1$ and $g_i(n_i) = \rho_i^{n_i}$.

Product-form F2. For single class network and nodes with exponential service time distribution, and load independent service rates μ_i, formula $F2$ defines: $V(n) = 1$ and $g_i(n_i) = 1/y_i$, where $\mathbf{y} = (y_1, \ldots, y_N)$ is the solution of the equations $\mathbf{y} = \mathbf{yP'}$, and matrix $\mathbf{P'} = [p'_{ij}]$ is defined in terms of the routing probability matrix $\mathbf{P}$ and the service rates as follows: $p'_{ij} = \mu_j p_{ji}, p'_{ii} = 1 - \sum_{j \neq i} p'_{ji}, 1 \leq i,j \leq N$.

Product-form F3. It applies to multiclass central server networks with A-type nodes, the class type of a job fixed in the system, state-dependent routing depending on the class type, and blocking functions dependent on node. Let 1 denote the central node. Let $b_i(n_i)$ denote the blocking function of node i, that is the probability that a job arriving at node i, is accepted when there are n_i customers. For single class exponential networks, load dependent service rates $\mu_i(n_i) = \mu_i f_i(n_i)$, and the state-dependent routing defined as $p_{1j}(n_j) = w_j(n_j)w(K - n_1), \forall n_j, p_{j1} = 1$, and $2 \leq j \leq N$, formula $F3$ defines:

$$V(n) = \prod_{l=i}^{K-n_1} w(l-1) \prod_{j=2}^{N} \prod_{l=i}^{n_j} w_j(l-1), \quad g_i(n_i) = \prod_{l=i}^{n_i} \frac{1}{\mu_i} \frac{b_i(l-1)}{f_i(l)}, \forall i. \quad (5)$$

For the definition of formula $F3$ for multiclass central server networks expression refer to [3].

Product-form $F4$. It applies to single class networks with A-type nodes. For the case load dependent service rates $\mu_i(n_i) = \mu_i f_i(n_i)$, and blocking function $b_i(n_i)$ for each node i, formula $F4$ defines: $V(n) = 1$ and $g_i(n_i) = \rho_i^{n_i} \prod_{l=i}^{n_i} \frac{b_i(l-1)}{f_i(l)}$.

Product-form $F5$. For multiclass networks and nodes with FCFS service discipline, exponential service time, and class independent capacities. Formula $F5$, like $F1$, defines: $V(n) = 1$ and $g_i(n_i) = \rho_i^{n_i}$.

Note that product-form formula (3) generalizes the closed-form expression for BCMP networks [13], and in certain cases, corresponds to the same solution as for queueing networks with infinite capacity queues computed on the truncated state space defined by the network with finite capacities. Product-forms for queueing networks with finite capacities are proved mostly by applying two approaches: i) reversibility of the underlying Markov process, ii) duality.

The former approach applies to reversible routing networks with finite capacity, whose underlying Markov process is shown to be obtained by truncating the reversible Markov process of the network with infinite capacity. This allows us to immediately derive a product-form solution from the theorem for truncated Markov processes of reversible Markov processes. This theorem states that the truncated process shows the same equilibrium distribution as the whole process normalised on the truncated sub-space. For example networks with RS blocking, BCMP-type networks with finite capacity and reversible routing **P** have product-form steady-state distribution given by formula $F1$ defined above [3, 27, 41]. Note that this solution is the BCMP product-form, renormalised over the reduced state space.

The latter approach, duality, applies to networks with arbitrary topology (possibly non-reversible) routing, for which the product-form solution is derived by the definition of a dual network that has the same equilibrium probability distribution. The dual network is proved to be in product-form under the *non-empty condition* (condition 1). For example, consider a cyclic closed network with single class, load independent exponential service rates and BBS or RS blocking. We can define a dual network which has the same steady-state joint queue length distribution [25]. It is obtained from the original one by reversing the connections between the nodes. It is formed by N nodes and $(B-K)$ customers, which correspond to the 'holes' of the original (primal) network, where $B = \sum_{i=1}^{N} B_i$ is the total capacity of the network. When a customer moves from node i in the original network, a hole moves backward to node i in the dual one. The state of n_i customers in node i of the original network, corresponds to $B_i - n_i$ holes in node i of the dual one contains. The underlying Markov process that describes the evolution of customers in the network is equivalent to the one that describes the evolution of the holes in the dual network. Hence, when the non-empty condition is satisfied, the total number of holes in the dual network cannot exceed the minimum capacity, i.e., $(B-K) \leq B_{min}$, and the dual network has a product-form solution like a network without blocking. Then the product-form solution for the primal network is given by equation (3) with formula $F2$ defined above [25]. This solution can be extended to arbitrary topology networks with load independent service rates for RS blocking, as proved in [27]. Another remarkable example of duality is for closed cyclic networks with phase-type (general) service distributions and BBS blocking for which the throughput of the network is shown to be symmetric with respect to its population, i.e., $X(B - K) = X(B)$ [22].

3.3 Algorithm for Closed Networks with Blocking

Product-form closed networks with blocking can be analyzed by some efficient algorithms [8, 19, 50]. They can be applied if some additional constraints are verified. They provide the model solution with a time computational complexity linear in the number of network components, i.e., they require $O(NK)$ operations, for a network with N service centers and K customers. There are two types of algorithms for product-form closed networks with blocking: Convolution and MVA (Mean Value Analysis). Note that we cannot directly apply the algorithms already known for BCMP networks, such as convolution algorithm and MVA [49], because of the different state space definition. However, the main idea of the two algorithms is similar to the non blocking case. Convolution algorithm aims to evaluating the normalizing constant G in formula (3) and average performance indices. MVA provides a direct computation of a set of average performance indices (mean response time, throughput, and mean queue length).

Convolution algorithm. We shall now briefly recall a Convolution algorithm for product-form queueing networks with blocking, whose computational complexity has a linear time computational complexity in the number of network components. With respect to the algorithm for BCMP networks, a Convolution algorithm for queuing networks with finite capacities takes into account the set of constraints on the queue lengths. This corresponds to a state space limitation that leads to a new definition of recursive equations to compute the normalizing constant.

The Convolution algorithm applies to networks with RS and BBS blocking, arbitrary topology, load independent service rates, and product-form solution given by formula $F1$ or $F2$. The algorithm computes the normalizing constant G in formula (3). This is obtained by on a set of recursive equations to evaluate functions $G_j(n)$, that can be interpreted as the normalizing constant of the network with finite capacity queues and with the first j nodes and n customers, $1 \leq j \leq N$, $\forall$ feasible $n \leq K$. The algorithm eventually computes $G = G_N(K)$. It defines a set of different recursive equations depending on the network population and the finite capacities. Once the last function $G_N(n)$, for each feasible n, has been computed, one can derive for each node i, the marginal queue length distribution $\pi_i(n_i)$, $\forall n_i$, and the average performance indices, i.e., the mean queue length L_i, the mean response time T_i, the node throughput X_i and utilization U_i, the mean busy period, and the blocking probabilities.

The time computational complexity of the algorithm depends on the network parameters and is $O(NC)$, where $C = max_{1 \leq i \leq N}(B_i - a_i)$, and $a_i = max(0, K - \sum_{j \neq i} B_j)$ is minimum feasible queue length of node i. A detailed description of the algorithm is given in [8].

MVA algorithm. The MVA algorithm directly computes a set of average performance indices, without evaluating the normalizing constant. The algorithm recursively evaluates the mean queue length, mean response time, and throughput. Other performance indices that can be derived are utilization, mean busy period and blocking probabilities for each node.

An MVA algorithm is defined for the class of product-form networks with cyclic topology and with BBS-SO and RS blocking [19]. In this case product-form $F2$ holds when the non-empty condition is satisfied, and we can define a dual network without blocking with identical product-form state distribution. Hence, by duality, this algorithm simply applies the standard MVA algorithm for networks without blocking to the dual network (see [19] for details). Note that such a MVA algorithm is not a direct application of the arrival theorem, as we have in MVA for queueing networks without blocking [49], since it is based on the dual network that is without blocking. The arrival theorem for network with blocking is discussed in [7, 10, 14].

Another MVA algorithm has been extended to a class of product-form networks with RS blocking, load independent service rates, and with $F2$ or $F3$ product-form [50]. The MVA algorithm has a time computational complexity of $O(B_{max}NK)$ operations, where $B_{max} = \max_{1 \leq i \leq N} B_i$.

4 Approximate Analysis of Networks with Finite Capacities

General queueing networks with blocking that have not a product-from solution can be analyzed by approximate analytical methods or by simulation. Several approximate techniques for open or closed queueing networks with finite capacity queues have been proposed to evaluate average performance indices and queue length distributions [10, 47, 54]. Most of the methods provide an approximate solution with a limited computational cost, but they do not give any bound on the introduced approximation error. The accuracy of the methods is usually validated by comparing numerical results with either simulation results or exact solutions.

Various heuristics have been defined by taking into account both the network model characteristics and the blocking type [4, 15, 16, 20, 23–26, 28–35, 37, 42, 48, 53–56, 60, 61]. Approximate methods for queuing networks with finite capacities are defined on the basis of the following principles:

- decomposition applied to the Markov process or to the network,
- forced product-form solution,
- structural properties for special cases,
- maximum entropy.

The various approaches can be applied under some constraints and for some blocking type, and they show different accuracy and time computational complexity. The methods based on forced product-form solution try to apply the product-form results to networks that do not satisfy the required constraints, possibly making some iterative check to appropriately select the approximation parameters. They usually are quite efficient from the computational viewpoint, but with unknown approximation error. Networks with particular topologies can be solved by special approximation methods that take advantage of their structure. Maximum entropy approximations apply the maximum entropy method (ME) to match the performance indices, which leads to a closed-form solution of the queue length distribution. ME approximation can be applied under quite general conditions and provide a good accuracy [29–33, 35]. We now discuss the decomposition approach that is widely used.

Network and process decomposition. Many approximate methods are heuristics based on the *decomposition* principle applied to the underlying Markov process or directly to the network.

Decomposing a Markov process consists in identifying a state space E partition of into H subsets E_h, $1 \leq h \leq H$, which leads to a decomposition of the rate matrix $\mathbf{Q}$ into H^2 submatrices. Hence the solution of the entire system of global balance equations (2) is reduced to the solution of H subsystems of smaller dimension. Each subsystem is related to a subset E_h, so obtaining the conditioned state probability denoted by $Prob(\mathbf{S} \mid E_h)$, $\forall$ state $\mathbf{S} \in E_h$, $\forall h$. Then these solutions are combined to obtain the overall process solution, i.e., the state distribution as

$$\pi(\mathbf{S}) = Prob(\mathbf{S} \mid E_h) Prob(E_h) \qquad (6)$$

where $Prob(E_h)$ is the aggregated probability of subset E_h, $\forall h$. Then the decomposition technique substitutes the direct computation of $\pi(\mathbf{S})$ with the computation of probabilities $Prob(\mathbf{S} \mid E_h)$ and $Prob(E_h)$, $\forall S, \forall E_h$. Exact process decomposition in general cannot be efficiently applied, except for special cases.

Approximate methods based on the decomposition of the Markov process provide an approximate evaluation of these probabilities. They require to:

- identify a partition of E into H subsets, so decomposing state space E and transition rate matrix $\mathbf{Q}$,
- compute the conditional state probabilities $Prob(\mathbf{S} \mid E_h)$ and the aggregate probabilities $Prob(E_h)$ for each subset E_h, $\forall h$, and compute state probability π by formula (6).

A critical issue is the definition of the state space partition that affects both the accuracy and the time computational complexity of the approximate algorithm. If the partition of E corresponds to a network partition into subnetworks then subsystems are (possibly modified) subnetworks.

The decomposition principle applied to the queueing network is based on the aggregation theorem for queueing networks. It performs in three steps: 1) network decomposition into a set of subnetworks, 2) analysis of each subnetwork in isolation to define an aggregate component, 3) definition and analysis the new aggregated network. Step 1 is a NP-complete problem, so it is the most critical issue. One should then choose simple subnetworks to apply efficient solution methods at step 2. At step 3 aggregation can be exactly applied only for product-form networks, and it is approximated otherwise, in general with unknown error. Network decomposition can be very efficient when the isolated subnetworks at step 2 and the aggregated network at step 3 are simple to analyze. The various approaches determine the subnetwork parameters. Many approximate methods use iterative aggregation-disaggregation procedures, for which conditions and speed of convergence should also be considered. Few approximations have known accuracy. An open issue is the definition of approximate methods with known error, such as bound solutions.

Approximate method comparison. We now present a review and comparison of some approximate methods by considering their accuracy, efficiency and the class of models to which they can be applied. Specifically, we consider the algorithm rationale and the model assumptions, i.e., constraints on the network parameters such as topology, types of service distributions, queue capacities, and blocking type. The approximation accuracy is evaluated by comparing numerical results with either simulation or exact results [6, 10].

We shall now consider some significant approximations for the two classes of closed and open networks. Table 2 summarizes the conditions under which the methods for closed and open networks can be applied, i.e., the constraints on network topology, service centers (service time distribution, number of servers and queue capacity, and blocking type).

Table 2. Approximate methods for queuing networks with blocking

Methods for closed networks	Network costraints topology - node type - blocking types	
Throughput Approximation (TA)	cyclic - G/M/1/B	BAS - BBS
Network Decomposition (ND)	cyclic - G/M/1/B	BBS
Variable Queue Capacity Decomp. (VQD)	cyclic[1] - G/M/1/B	BBS
Matching State Space (MSS)	general - G/M/1/B	BAS
Approximate MVA (AMVA)	general - G/M/1/B	BAS
Maximum Entropy Algorithm (ME)	general - G/GE/1/B	RS-RD
Methods for open networks		
Tandem Exponential Network Decomp. (TED)	tandem - G/M/1/B	BAS
Tandem Phase-Type Network Decomp. (TPD)	tandem - G/M/1/B	BAS
Acyclic Network Decomposition (AND)	acyclic - G/M/1/B	BAS
Maximum Entropy Algorithm (ME-O)	general - G/GE/1/B	RS-RD

4.1 Approximate Methods for Closed Networks with Finite Capacities

We consider the following six algorithms for closed queuing networks, based on various principles:

- Throughput Approximation (TA) [46]
- Network Decomposition (ND) [23]
- Variable Queue Capacity Decomposition (VQD) [56]
- Matching State Space (MSS) [1]
- Approximate MVA (AMVA) [2]
- Maximum Entropy Algorithm (ME) [32, 35]

They applied to homogeneous networks, i.e., each node has the same blocking type. We assume FCFS service discipline at each node. Table 3 reports the key idea of each approximation method.

Cyclic networks. The first three methods (TA, ND and VCD) evaluate the throughput of cyclic networks with exponential service time distribution. TA and VCD algorithm compute the network throughput $X(K)$ as a function of network population K.

Throughput Approximation (TA) applies to cyclic networks with BBS or BAS blocking and exponential service times [46]. It evaluates the network throughput, assuming that it is a symmetrical function, that is $X(K) = X(B - K)$, where $B = \sum_i B_i$. This property holds for BBS blocking as proved under the more general assumption of phase-type service distribution in [22], and it reaches its maximum value for $K = K^* = \lfloor \frac{B}{2} \rfloor$. The algorithm directly computes few values of function $X(K)$ with exact analytical methods and computes the other values by fitting the curve through those known points. For BAS blocking the symmetry property of the throughput does not hold, but a similar shape of the curve as for BBS blocking is conjectured, supported by experimental results, where K^* is approximated by an iterative scheme that depends on the queue capacities and the service rates. The main drawback of this method is the cumbersome computational complexity required to evaluate the exact throughput. Hence, it can be used for parametric analysis of the throughput by varying the network population and only for networks with a limited number of nodes and customers.

Network Decomposition (ND) approximates the throughput of the cyclic network with BBS blocking by a network decomposition method [23]. At step 1 the network is partitioned into N one-node subnetworks. At step 2 each subnetwork is analyzed in isolation as an $M/M/1/B_i$ network with arrival rate λ_i^* and load dependent service rate $\mu_i^*(n)$, $\forall n$, to derive the marginal queue length distribution $\pi_i^*(n)$, $\forall n$, $\forall i$. Parameters λ_i^* and $\mu_i^*(n)$ are defined by a set of equations and are approximated for each subnetwork. The isolated queue is approximated by taking into account the blocking of customers due to the finite capacity of the downstream nodes. The authors consider two cases depending on whether all the nodes have finite capacity or there is one infinite capacity node.

Table 3. Approximate methods for closed networks with blocking: main idea

Method	Key idea
TA	Exact model analysis for some network population and throughput interpolation by varying network population K.
ND	Network decomposition into nodes analyzed in isolation as M/M/1/B.
VCD	Network decomposition and aggregation into a single composite node with state dependent service rate and variable buffer size.
MSS	Analysis of the QN without blocking by choosing the network population to approximately match the state space cardinality.
AMVA	Modified and forced MVA algorithm to consider blocking.
ME	Approximate product-form for the queue length distribution based on maximum entropy.

They define the parameters by a fixed-point equation for λ_i^* and an iterative algorithm. It starts with a throughput approximate interval $[X_{min}(0), X_{max}(0)]$, computes new parameters λ_i^* and $\mu_i^*(n)$ at each step and appropriately updates the k-th throughput approximation $[X_{min}(k), X_{max}(k)]$, until a convergence condition is satisfied. Such conditions check the approximate throughput interval width, and some consistency control on the network. If all nodes have finite capacity an additional iteration cycle is required to compute probabilities $\pi_i^*(B_i)$ (see [23] for details). Convergence has not been proved, but it has been observed. The time computational complexity is of $O(kN^4 B_{max}^3)$ operations, for k iteration steps.

Variable Queue Capacity Decomposition (VQD) method can be applied to cyclic[1] networks with BBS blocking [56], and we assume that node 1 has infinite capacity ($B_1 = \infty$). The algorithm is based on the network decomposition principle applied to nested subnetworks. The key idea is that given a node i, all the downstream nodes $(i+1, \ldots, N)$ are aggregated in a single composite node C_{i+1} with load dependent service rate and a variable queue capacity. The approximation evaluates the composite node C_{i+1} parameters (load dependent service rate, and the fraction of time in which the queue capacity is n, given the network population). The algorithm starts with the analysis of the two-node subnetwork formed by the last two nodes $(N-1, N)$ to define the composite aggregate node C_{N-1}, that is seen by node $N-2$. Then the algorithm goes backward from node $i = N-2$ to node 1 eventually to the two-node network formed by $(1, C_2)$ that represents the entire aggregated network, and from which one obtains the approximated throughput. The analysis of each two-node network where the composite node has variable queue capacity (or variable buffer) is carried out by considering two corresponding two-node networks where a composite node has fixed buffer and infinite buffer, respectively (see [56] for details). The algorithm is very simple, non-iterative and its time computational complexity is of $O(NK^3)$ operations.

---

[1] With a node with unlimited capacity.

Arbitrary topology networks. The three methods (MSS, AMVA and ME) apply to arbitrary topology networks. MSS and AMVA methods assume networks with BAS blocking, exponential service time, and evaluate the network throughput. ME algorithm assumes RS-RD blocking, generalized exponential service time and evaluates the queue length distribution and average performance indices.

The basic idea of Matching State Space (MSS) method [1] is to approximate the behavior of the network with blocking with that of a network without blocking by choosing the population to approximately match the state space cardinality of the underlying Markov chain. The assumption is that the two networks with nearly the same state space cardinality should have similar throughputs. The algorithm defines a new network with infinite capacity queues and K' customers so that the state space cardinality of the underlying Markov process, say $C'(K')$, is nearly equal to that of the Markov process of the original network with K customers, $C(K)$. The algorithm determines K' to approximate the state space matching, that is to minimize the difference function $|C'(K') - C(K)|$. Then the network without blocking is analysed (see [1] for details). The algorithm implementation is simple and the time computational complexity is of $O(N^3 + NK^2)$ operations.

Approximate MVA (AMVA) [2] analyzes networks with BAS blocking and exponential service times by a modification of the MVA algorithm originally defined for product-form networks with unlimited queue capacities [49]. The MVA algorithm is based on Little's theorem and the arrival theorem. Note that the arrival theorem and the MVA algorithm, as defined for networks without blocking, cannot be immediately applied to networks with blocking. Let $T_i(n)$, $L_i(n)$ and $X_i(n)$ denote the average response time, mean queue length and throughput of node i when there are n customers in the network. For load independent service center the MVA is based on the following recursive scheme, for $1 \leq n \leq K$:

- $T_i(n) = \frac{1}{\mu_i}[1 + L_i(n-1)], \forall i$
- $X_i(n) = ne_i/[\sum_j e_j T_j(n)], \forall i$
- $L_i(n) = X_i(n)T_i(n), \forall i.$

The approximation algorithm modifies the first equation trying to take into account blocking. In particular if node i is full, it cannot accept new customers and there is at least one node j blocked by node i, then approximation defines:

- $T_i(n) = \frac{1}{\mu_i}L_i(n-1)$
- $T_j(n) = \frac{1}{\mu_j}[1 + L_j(n-1)] + \frac{1}{\mu_i}(e_j p_{ji})/e_i$

For node i only the customers already in the node contribute to the average response time, while for the blocked node j the response time increases of a blocking time due to node i (see [2] for further details). The algorithm can be simply implemented and the time computational complexity is of $O(N^3 + kNK)$ operations where k is the number of iterations of the approximate iterative computation at step n.

Maximum Entropy Algorithm (ME) [32, 35] evaluates the queue length distribution and average performance indices of a network with RS-RD blocking and generalized exponential service time. The approximation is based on the principle of maximum entropy and is an extension of the algorithm defined for open networks and more general cases, such as multiclass networks and priorities [29–31, 33]. Let $a_i = max(0, K - \sum_{j \neq i} B_j)$ be the minimum number of customers in node i. The algorithm approximates the joint queue length distribution $\pi(\mathbf{S})$ for each network state S by maximizing the entropy function

$H(\pi) = -\sum_\mathbf{S} \pi(\mathbf{S}) log(\pi(\mathbf{S}))$

subject to the following constraints

- normalization: $\sum_\mathbf{S} \pi(\mathbf{S}) = 1$
- u_i is the probability of more than a_i customers in i: $\sum_{n_i > a_i} \pi(n_i) = u_i$
- L_i is the mean queue length: $\sum_{n_i=a_i}^{B_i} h_i(n_i)\pi_i(n_i) = L_i$
- Φ_i is the probability that node i is full: $\sum_{n_i=a_i}^{B_i} f_i(n_i)\pi_i(n_i) = \Phi_i$

where $h_i(n_i)=min(0, n_i - a_i - 1)$ and $f(n_i) = max(0, n_i - B_i + 1)$. By the Lagrange's method of undetermined multipliers the algorithm determines an approximation of $\pi(\mathbf{S})$ that has the following product-form expression:

$$\pi(\mathbf{S}) = \frac{1}{Z} \prod_{i=1}^{N} x_i(n_i) y_i^{h_i(n_i)} z_i^{f_i} \quad (7)$$

where Z is a normalizing constant, $x_i(n_i) = 1$ if $n_i = a_i$, and $x_i(n_i) = x_i$ if $a_i < n_i \leq B_i$, and x_i, y_i and z_i are the Lagrangian coefficients corresponding to constraints above. The network cannot be decomposed into single nodes and the coefficients do not have a closed form expression. The algorithm approximates the closed network with a pseudo open network without exogenous departures and arrivals. This open network is analysed by the approximation based on the same principle applied to open networks, introducing an additional constraint on the average queue lengths $K = \sum_i L_i$ and slight modifications to derive the coefficients of formula (7). Then the coefficients are iteratively approximated. The algorithm details are given in [32, 35]. The time computational complexity of the algorithm depends on the algorithm for open networks and for the iterative approximation, with k iteration, is of $O(kN^2K^2)$ operations.

4.2 Approximate Methods for Open Networks with Finite Capacities

We consider the following algorithms for open queuing networks, as reported in Table 2 that shows the corresponding constraints on the network topology, the type of service centers and the blocking type:

- Tandem Exponential Network Decomposition (TED) [20]
- Tandem Phase-Type Network Decomposition (TPD) [48]
- Acyclic Network Decomposition (AND) [37]
- Maximum Entropy Algorithm for Open networks (ME-O) [35, 51]

All the algorithms are based on network decomposition and ME-O method on the maximum entropy. Decomposition define one-node subnetworks as $M/M/1/B$ queues by TED and AND, $M/Cox/1/B$ queue by the other algorithms.

Tandem networks. The TED [20] and TPD [48] algorithms approximate the throughput of the tandem network with BAS blocking by network decomposition. The network is partitioned into N one-node subnetworks $T(i)$, $1 \leq i \leq N$. Subnetwork $T(i)$ represents the isolated node i and is analyzed as an $M/M/1/B_i$ queue by TED and as an $M/PH_n/1/B_i$ queue by TPD (with phase-type service distribution). The method define appropriate parameters to derive marginal probability $\pi_i(n)$, $\forall n$ of each subnetwork $T(i)$. Since $T(1)$ and $T(N)$ correspond to the first and last node of the tandem network the first has arrival rate λ (exogenous arrival rate) and the last service rate μ_N. The remaining $2(N-1)$ unknowns have to be determined. The approximation is based on an iterative scheme to approximate the subnetworks unknown parameters(see [20] for details). TED algorithm requires $O(kNB_{max}^2)$ operations, where k is the number of iterations. The authors proved the algorithm convergence, and numerical results show that it is fast. TPD method solve subsystems $T(i)$ with a matrix-geometric technique and distinguish two cases depending on whether the first node has finite capacity. When all the nodes have finite capacity it has an additional iterative cycle to estimate the effective arrival rates (see [48]). Convergence has not been proved. the algorithm requires $O(k_1 \sum_{2 \leq i \leq N} k_i(N - i + 1)^3 B_i^2)$ operations where k_i is the number of iterations to compute the arrival rate of system $T(i)$.

Acyclic and arbitrary topology networks. The last two methods AND and ME-O apply to more general topology networks and evaluate the queue length distribution and are respectively based on network decomposition and the maximum entropy principle.

The Acyclic Network Decomposition (AND) method [37] extends TED approximation to acyclic networks with exponential service time distribution and BAS blocking. Like TED, the approximation is based on a network decomposition into N single node subsystems $T(i)$. Each subsystem is analyzed as an $M/M/1/B_i$ system, but AND method defines a new set of equations to determine the subsystems parameters (service and arrival rates). If node i has U_i predecessor nodes, then each subsystem $T(i)$ receives arrivals from U_i exponential sources with unknown rates, one source from each predecessor j (i.e., any node j such that $p_{ij} > 0$). These rates are approximated by an iterative procedure. To this aim AND algorithm evaluates the probability that at arrival time at $T(i)$ from the j-th source there are n other nodes blocked by node i, and the probability that at the end of a service system $T(i)$ is empty. These probabilities appear in the new formulas defined for the unknown rates (see [37] for details). The $T(i)$ subsystems are eventually analyzed to derive marginal probabilities $\pi_i(n)$, $\forall n$, $\forall i$. The time computational complexity of AND is bounded by $O(kN[(U + B_{max})^2 + U^3 + 2^{U+1}])$, where k is the iteration number and $U = max_i U_i$.

Maximum Entropy Algorithm for Open networks (ME-O) approximation [35, 51] analyses a more general classes of networks with arbitrary topology, generalized exponential service time distribution, and RS-RD blocking. It is similar to the ME method by the same authors for closed networks, and the approximation is based on the maximum entropy. The open networks is decomposed into N subsystems $T(i)$, each analysed as $GE/GE/1/B$ nodes with appropriate parameters by considering blocking. The analysis of the i-th $GE/GE/1/B$ systems is based on an iterative scheme that computes: 1) the arrival rate by the traffic equations, 2) the probability that, at service completion time at i, node j is full, $\forall j$, 3) the queue length probability π_i defined by a product-form whose coefficient are the Lagrange multipliers corresponding to the constraints of the maximum entropy problem, and 4) the coefficient of variation of the interarrival time at $T(i)$. The iterative scheme is repeated until convergence of the coefficient of variations at step 4. The probability computation at step 2 requires the solution of non-linear system that can lead to numerical instability and problems of convergence, which, however, is rarely observed. There is no proof of convergence and uniqueness of the solution. See [35, 51] for details. The time computational complexity is of $O(\Omega^3)$, where Ω is the cardinality of the set of probabilities computed at step 2.

5 Algorithms Comparison

Table 4 shows a comparison of approximate methods for closed and open networks. It shows the performance indices evaluated by every method, their accuracy and efficiency.

For closed networks, approximation ND is more accurate than VCD and the difference increases with the number of network nodes. TA is more accurate than ND and its accuracy is more stable than that of ND as the number of network nodes increases. However, ND is more efficient than TA, which is limited to small networks. If $K < NB_{max}$ then VCD approximation is better than ND, while the opposite is true otherwise. VCD approximation is less efficient than ND for large network population K. Note that VCD and TA provide the throughput for all the network population from 1 to K. ND is based on a fixed-point iteration and can show some numerical instability. ND and AT apply to a more general class than VCD approximation.

By comparing methods MSS and AMVA, we observe that the former is more accurate and more efficient. The approximations are quite different, since their rationales are not related. They are stable and their accuracy seems to be independent of network parameters (N, μ_i and B_i), but dependent on the topology. Specifically they provide better results for central server networks and worse results for cyclic networks.

For open tandem exponential networks with BAS blocking the two approximation algorithms TED and TPD have nearly the same accuracy, with quite similar approximations, for sign and value. Their accuracy increases for small blocking probabilities, i.e., for networks with large B_i or large μ_i with respect to the arrival rate. The approximation accuracy of TPD is influenced by capacity queue unbalancing, while that of TED is affected by service rate unbalancing. TPD is slightly better than TED for high blocking probabilities. TED is certainly more efficient and has a simpler implementation than TPD, which can show numerical instability that can affect the algorithm convergence.

Table 4. Comparison of approximate methods for networks with blocking

Method	Index	Accuracy	Efficiency
TA	$X(K)$	Very good	Poor for $N > 5$
ND	X	Good	Good
VQD	$X(K)$	Good for $N \leq 4$	Fair
MSS	X_i	Fair	Good
AMVA	L_i, X_i, T_i	Fair for X	Very good
ME	L_i, X_i, T_i	Fair	Fair
TED	L_i, X_i, T_i, π_i Very good		Very good
TPD	L_i, X_i, T_i, π_i Very good		Slow for networks with all finite capacity nodes, fair otherwise.
AND	L_i, X_i, T_i, π_i Very good		Very good
ME-O	L_i, X_i, T_i, π_i Good		Fair

Finally, the maximum entropy methods, ME and ME-O, for closed open networks apply to the more general class of networks with arbitrary topology, generalized exponential service time distribution, and RS-RD blocking. The throughput accuracy of ME is not affected by the topology and the symmetry of network parameters (μ_i and B_i, $\forall i$), but it depends on the coefficient of variation of the service distributions. The approximation error grows with these coefficients of variation. The accuracy of the ME-O method decreases with the presence of cycles in the networks.

6 Application Examples of Networks with Blocking

Some equivalence, insensitivity and monotonicity properties of queueing networks with finite capacities have been proved [10, 12, 21, 22, 43, 44, 52, 57, 58].

Insensitivity properties lead to the identification of the factors that affect system performance. Monotonicity provides insights in the system behavior. It can be applied in parametric analysis to study the impact of various parameters (e.g., system load, buffer dimension) on system performance, to solve optimization problems or for bounding analysis. Equivalencies are defined in terms of state probability distribution π, average performance indices, or passage time distribution. Most of the equivalencies derive from the identity of the network processes. However, even if two networks have identical Markov processes, the meaning of corresponding states may be different. Then performance measures may be not equivalent, because the equivalence in terms of π does not necessarily lead to equivalence in terms of average performance indices.

Examples of equivalences are between networks with and without blocking that immediately leads to the extension of efficient computational solution algorithms defined for BCMP networks. Such equivalences hold for exponential networks with RS-RD blocking with reversible routing and product-form $F4$, and with arbitrary routing and product-form $F2$, for which an equivalent product-form network without blocking can be defined (see [10, 12]). Several equivalences can be defined between networks with different blocking types, and between homogeneous and non-homogeneous networks. Some examples are:

- BBS and RS types are equivalent for multiclass two-node networks with BCMP type nodes and class independent capacities,
- BAS is reducible to BBS for cyclic networks provided that node capacities are augmented by 1.
- for central server topology networks, BAS is reducible to BBS with node capacities B_i, $2 \leq i \leq M$, augmented by 1.
- BBS and RS-FD types are equivalent for networks with arbitrary routing, single class, with exponential nodes, load independent service rates and if condition 3 holds (single destination node) defined in Section. 3.2.
- Stop and Recirculate blocking are equivalent for multiclass open Jackson networks with class type fixed.

These results can be applied, for example, to define more efficient methods or to extend solution algorithms to more general classes of models of networks with different blocking types or network parameters. A detailed description of equivalence properties can e found in [10, 12, 21, 43].

A simple application. A simple application example is a store-and-forward packet switching network with virtual circuits modeled at level 3 in OSI reference model. Under independence assumptions, the window flow control can be represented by a closed cyclic queueing network with finite capacities and RS blocking. Under exponential assumptions and if the non empty conditions (condition 1) is satisfied, then product form solution (3) with formula $F2$ holds and we can apply Convolution algorithm or MVA to derive the performance indices, such as network throughput, delay, and buffer occupancy.

Another simple application of finite capacity networks to model communications and computer systems is shown in Figure 1 that represents an heterogeneous network with blocking modelling two computer systems connected through a communication link. We assume that nodes $C1$ and $C2$ represent computer CPU subsystem with RS-RD blocking, nodes $D1$ and $D2$ are computer disk subsystem with BAS blocking, nodes $N1$ and $N2$ are computer network access with BAS blocking, and nodes $N2$ and $N4$ communication links with BBS

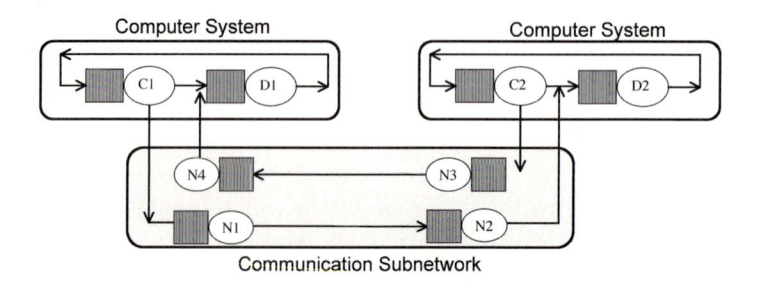

Fig. 1. Example

blocking. The customers of the model represent jobs (in computer systems) and packets (in communication subnetwork). Under exponential assumptions a heterogeneous network reducible to a homogeneous queuing network with RS-RD blocking can represent the system. The network has arbitrary topology and we can can apply the ME approximate solution algorithm to derive the performance indices, such as the average response time and system throughput.

Moreover if nodes $D1$ and $D2$ have RS-RD blocking, then the network has product-form solution $F2$ and we can apply the convolution algorithm to evaluate the system performance.

References

[1] Akyildiz, I.F.: On the Exact and Approximate Throughput Analysis of Closed Queueing Networks with Blocking. IEEE Trans. Soft. Eng. 14, 62–71 (1988)
[2] Akyildiz, I.F.: Mean value analysis of blocking queueing networks. IEEE Trans. Soft. Eng 14, 418–429 (1988)
[3] Akyildiz, I.F., Von Brand, H.: Exact solutions for open, closed and mixed queueing networks with rejection blocking. J. Theor. Comp. Sci. 64, 203–219 (1989)
[4] Altiok, T., Perros, H.G.: Approximate analysis of arbitrary configurations of queueing networks with blocking. Ann. Oper. Res. 9, 481–509 (1987)
[5] Awan, I.U., Kouvatsos, D.D.: Approximate analysis of QNMs with space and service priorities. In: Kouvatsos, D.D. (ed.) Performance Analysis of ATM Networks, ch. 25, pp. 497–521. Kluwer, IFIP Publication (1999)
[6] Balsamo, S.: Closed Queueing Networks with Finite Capacity Queues: Approximate analysis. In: Proc. ESM 2000, SCS, Europ. Sim. Multiconf. Ghent, May 23-26 (2000)
[7] Balsamo, S., Clo', C., Donatiello, L.: Cycle Time Distribution of Cyclic Queueing Network with Blocking. Performance Evaluation 14(3) (1993)
[8] Balsamo, S., Clo', C.: A Convolution Algorithm for Product Form Queueing Networks with Blocking. Annals of Operations Research 79, 97–117 (1998)
[9] Balsamo, S., De Nitto, V.: A survey of Product-form Queueing Networks with Blocking and their Equivalences. Annals of Operations Research 48 (1994)
[10] Balsamo, S., De Nitto, V., Onvural, R.: Analysis of Queueing Networks with Blocking. Kluwer Academic Publishers, Dordrecht (2001)
[11] Balsamo, S., Donatiello, L.: On the Cycle Time Distribution in a Two-stage Queueing Network with Blocking. IEEE Trans. on Soft. Eng. 13, 1206–1216 (1989)
[12] Balsamo, S., Iazeolla, G.: Some Equivalence Properties for Queueing Networks with and without Blocking. In: Agrawala, Tripathi (eds.) Performance 1983. North-Holland, Amsterdam (1983)
[13] Baskett, F., Chandy, K.M., Muntz, R.R., Palacios, G.: Open, closed, and mixed networks of queues with different classes of customers. J. of ACM 22, 248–260 (1975)
[14] Boucherie, R., Van Dijk, N.: On the arrival theorem for product form queueing networks with blocking. Performance Evaluation 29, 155–176 (1997)
[15] Boxma, O., Konheim, A.G.: Approximate analysis of exponential queueing systems with blocking. Acta Informatica 15, 19–66 (1981)
[16] Brandwajn, A., Jow, Y.L.: An approximation method for tandem queueing systems with blocking. Operations Research 1, 73–83 (1988)

[17] Buzacott, J.A., Shanthikumar, J.G.: Design of Manufacturing Systems using Queueing Models. Queueing Systems: Theory and Applications (1992)
[18] Cheng, D.W.: Analysis of a tandem queue with state dependent general blocking: a GSMP perspective. Performance Evaluation 17, 169–173 (1993)
[19] Clo', C.: MVA for Product-Form Cyclic Queueing Networks with RS Blocking. Annals of Operations Research 79 (1998)
[20] Dallery, Y., Frein, Y.: On decomposition methods for tandem queueing networks with blocking. Operations Research 14, 386–399 (1993)
[21] Dallery, Y., Liu, Z., Towsley, D.F.: Equivalence, reversibility, symmetry and concavity properties in fork/join queueing networks with blocking. J. of the ACM 41, 903–942 (1994)
[22] Dallery, Y., Towsley, D.F.: Symmetry property of the throughput in closed tandem queueing networks with finite buffers. Op. Res. Letters 10, 541–547 (1991)
[23] Frein, Y., Dallery, Y.: Analysis of Cyclic Queueing Networks with Finite Buffers and Blocking Before Service. Performance Evaluation 10, 197–210 (1989)
[24] Gershwin, S.B.: An efficient decomposition method for the approximate evaluation of tandem queues with finite storage space and blocking. Oper. Res. 35, 291–305 (1987)
[25] Gordon, W.J., Newell, G.F.: Cyclic queueing systems with restricted queues. Oper. Res. 15, 286–302 (1967)
[26] Hillier, F.S., Boling, W.: Finite queues in series with exponential or Erlang service times - a numerical approach. Oper. Res. 15, 286–303 (1967)
[27] Hordijk, A., Van Dijk, N.: Networks of queues with blocking. In: Kylstra, K.J. (ed.) Performance 1981, pp. 51–65. North Holland, Amsterdam (1981)
[28] Jun, K.P., Perros, H.G.: An approximate analysis of open tandem queueing networks with blocking and general service times. Europ. Journal of Operations Research 46, 123–135 (1990)
[29] Kouvatsos, D.D.: Maximum Entropy Methods for General Queueing Networks. In: Potier (ed.) Proc. Modeling Tech. and Tools for Perf. Analysis, pp. 589–608. North-Holland, Amsterdam (1983)
[30] Kouvatsos, D.D.: A Universal Maximum Entropy Solution for Complex Queueing Systems and Networks. In: Karmeshu (ed.) Entropy Measures, maximum Entropy Principles and Emerging Applications, pp. 137–162. Springer, Heidelberg (2003)
[31] Kouvatsos, D., Awan, I.U.: Arbitrary closed queueing networks with blocking and multiple job classes. In: Proc. Third Int. Work. on Queueing Networks with Finite Capacity, Bradford, UK, July 6-7 (1995)
[32] Kouvatsos, D.D., Awan, I.U.: MEM for arbitrary closed queueing networks with RS blocking and multiple job classes. Annals of Oper. Res. 79, 231–269 (1998)
[33] Kouvatsos, D., Awan, I.U.: Entropy maximization and open queueing networks with priorities and blocking. Performance Evaluation 51, 191–227 (2003)
[34] Kouvatsos, D., Denazis, S.G.: Entropy maximized queueing networks with blocking and multiple job classes. Performance Evaluation 17, 189–205 (1993)
[35] Kouvatsos, D.D., Xenios, N.P.: MEM for arbitrary queueing networks with multiple general servers and repetitive-service blocking. Perf. Ev. 10, 106–195 (1989)
[36] Lam, S.S.: Queueing networks with capacity constraints. IBM J. Res. Develop. 21, 370–378 (1977)
[37] Lee, H.S., Bouhchouch, A., Dallery, Y., Frein, Y.: Performance Evaluation of open queueing networks with arbitrary configurations and finite buffers. In: Proc. Third Int. Work. on Queueing Networks with Finite Capacity, Bradford, UK, July 6-7 (1995)

[38] Mishra, S., Fang, S.C.: A maximum entropy optimization approach to tandem queues with generalized blocking. Perf. Evaluation 30, 217–241 (1997)
[39] Mitra, D., Mitrani, I.: Analysis of a Kanban discipline for cell coordination in production lines I. Management Science 36, 1548–1566 (1990)
[40] Onvural, R.O.: Some Product Form Solutions of Multi-Class Queueing Networks with Blocking. Perf. Evaluation 10(3) (1989)
[41] Onvural, R.O.: A Note on the Product Form Solutions of Multiclass Closed Queueing Networks with Blocking. Performance Evaluation 10, 247–253 (1989)
[42] Onvural, R.O.: Survey of Closed Queueing Networks with Blocking. ACM Computing Surveys 22(2), 83–121 (1990)
[43] Onvural, R.O., Perros, H.G.: On Equivalencies of Blocking Mechanisms in Queueing Networks with Blocking. Oper. Res. Letters 5, 293–298 (1986)
[44] Onvural, R.O., Perros, H.G.: Equivalencies Between Open and Closed Queueing Networks with Finite Buffers. Performance Evaluation (1988)
[45] Onvural, R.O., Perros, H.G.: Some equivalencies on closed exponential queueing networks with blocking. Performance Evaluation 9, 111–118 (1989)
[46] Onvural, R.O., Perros, H.G.: Throughput Analysis in Cyclic Queueing Networks with Blocking. IEEE Trans. Software Engineering 15, 800–808 (1989)
[47] Perros, H.G.: Queueing networks with blocking. Oxford University Press, Oxford (1994)
[48] Perros, H.G., Altiok, T.: Approximate analysis of open networks of queues with blocking: tandem configurations. IEEE Trans. Soft. Eng. 12, 450–461 (1986)
[49] Raiser, M., Lavenberg, S.S.: Mean Value Analysis of closed multi-chain queueing networks. Journal of ACM 27, 217–224 (1989)
[50] Sereno, M.: Mean Value Analysis of product form solution queueing networks with repetitive service blocking. Performance Evaluation 36-37, 19–33 (1999)
[51] Skianis, C.A., Kouvatsos, D.D.: Arbitrary open queueing networks with service vacation periods and blocking. Annals of Operations Research 79, 143–180 (1998)
[52] Shanthikumar, G.J., Yao, D.D.: Monotonicity Properties in Cyclic Queueing Networks with Finite Buffers. In: Perros, Altiok (eds.) First Int. Work. on Queueing Networks with Blocking. North Holland, Amsterdam (1989)
[53] Akyildiz, Perros (eds.): Special Issue on Queueing Networks with Finite Capacity Queues. Performance Evaluation, vol. 10(3). North Holland, Amsterdam (1989)
[54] Onvural, R.O. (ed.): Special Issue on Queueing Networks with Finite Capacity. Performance Evaluation, vol. 17(3). North-Holland, Amsterdam (1993)
[55] Balsamo, S., Kouvatsos, D.: Special Issue on Queueing Networks with Blocking Performance Evaluation Journal, vol. 51(2-4). North Holland, Amsterdam (2003)
[56] Suri, R., Diehl, G.W.: A variable buffer size model and its use in analytical closed queueing networks with blocking. Management Sci. 32(2), 206–225 (1986)
[57] van Dijk, N.: On stop = repeat servicing for non-exponential queueing networks with blocking. J. Appl. Prob. 28, 159–173 (1991)
[58] van Dijk, N.: Stop = recirculate for exponential product form queueing networks with departure blocking. Oper. Res. Lett. 10, 343–351 (1991)
[59] Van Dijk, N.: Queueing networks and product form. John Wiley, Chichester (1993)
[60] Yao, D.D., Buzacott, J.A.: Modeling a Class of State Dependent Routing in Flexible Manufacturing Systems. Annals of Oper. Research 3, 153–167 (1985)
[61] Yao, D.D., Buzacott, J.A.: Modeling a class of flexible manufacturing systems with reversible routing. Oper. Res. 35, 87–93 (1987)

Switching Queueing Networks

Vladimir V. Anisimov

GlaxoSmithKline, Research Statistics Unit
NFSP - South, Third Avenue, Harlow, Essex, CM19 5AW, United Kingdom
Vladimir.V.Anisimov@gsk.com
http://biometrics.com

Abstract. The paper is devoted to the asymptotic investigation of switching queueing systems and networks. The method of analysis uses the limit theorems of averaging principle and diffusion approximation types for the class of "Switching Processes" developed by the author. This class can be used to describe hierarchic stochastic systems with random switching due to internal and external factors.

Different classes of overloaded switching queueing models (heavy traffic conditions) operating under the influence of internal and external random environment, e.g. state-dependent models with Markov and semi-Markov switching, are investigated. The approximation of models with fast switching by simpler models with aggregated state space and averaged transition rates are also considered.

These results form a new methodology for the investigation of transient phenomena for queueing models in heavy-traffic conditions and provide an analytic approach to performance evaluation of queueing networks of a complex structure. Different examples are considered.

Keywords: Queueing models, networks, switching process, averaging principle, diffusion approximation, heavy-traffic, asymptotic aggregation.

1 Introduction

The real communication and computer networks have as usual rather complex structure and operate under the influence of various internal and external factors that may change (switch) the behaviour of the system. The main features of these systems are the stochasticity, presence of different time scales for different subsystems (inner fast computer time and slow user interaction time, etc), and the hierarchic structure. Wide classes of such systems can be adequately described with the help of so-called "Switching Processes" (SP's).

The class of SP's has been introduced and studied in the author papers [2,3] and book [4]. SP's are an adequate mathematical tool for describing stochastic systems that can switch their behaviour at some random epochs of time which may depend on the previous trajectory.

A SP can be described as a two-component process $(x(t), \zeta(t))$, $t \geq 0$, with the property that there exists a sequence of Markov epochs of times $t_1 < t_2 < \cdots$ such that in each interval $[t_k, t_{k+1})$, $x(t) = x(t_k)$, and the behaviour of the process $\zeta(t)$ in this interval depends only on the values $(x(t_k), \zeta(t_k))$. Here $x(t)$ is a discrete switching component and the epochs $\{ t_k \}$ are called switching times. SP's can be described in terms of constructive characteristics [2,3] and are very suitable in analyzing and asymptotic investigating of stochastic systems with "rare" and "fast" switching [3,4,5,7,10,14,17,18,19,20].

Wide classes of queueing models can be described in terms of SP's. The class of switching queueing models includes, as examples, state-dependent queueing systems and networks in a Markov and semi-Markov environment of the type $(M_{M,Q}/M_{M,Q}/m/k)^r$, $(SM_{M,Q}/M_{M,Q}/m/k)^r$ (so-called Markov or semi-Markov modulated models [35]), models under the influence of flows of external events or internal perturbations, unreliable systems, retrial queues, hierarchic queueing models, etc.

Therefore, the developed asymptotic theory of SP's can be effectively applied to the investigation of wide classes of queueing systems and networks.

There are several directions in the asymptotic investigation of queueing models. The first direction is dealing with the analysis of overloaded queueing models (heavy traffic conditions). For this class of models, the convergence of queueing processes to the solutions of differential equations (averaging principle – AP) and to the diffusion processes (diffusion approximation – DA) can be proved using the corresponding asymptotic results for SP's [6,7,8,9,10,20]. These results are applied to different classes of queueing models [11,12,13,14,16,17,19,20,22].

Note that a different analytic approach based on the asymptotic results for semi-groups of linear perturbed operators associated with corresponding Markov processes was developed by V.S. Korolyuk et al. and different applications to queueing models [29,30,31,32,33,34] are studied.

Another direction is dealing with the asymptotic approximation of Markov and semi-Markov state-dependent queueing models with fast switching by simpler Markov models with averaged transition rates. These results are related to the asymptotic decreasing of dimension and aggregation of states and are based on the asymptotic results on the convergence in the class of SP's with slow switches [1,3,4,5] and the conditions when a SP of rather complicated structure can be approximated by a SP of a simpler structure, in particular, by a Markov or a semi-Markov process. Different applications to reliability and queueing models and dynamic systems are considered in [4,13,14,15,18,19,20,21].

These results form the basis of the methodology for the investigation of transient phenomena in switching queueing systems and networks. Numerous examples are considered.

2 Switching Processes

Consider a general definition of the class of switching processes (SP). Let

$$\mathcal{F}_k = \{(\zeta_k(t, x, \alpha), \tau_k(x, \alpha), \beta_k(x, \alpha)), t \geq 0, x \in X, \alpha \in \mathcal{R}^r\}, k \geq 0,$$

be jointly independent in index k parametric families, where $(X, \mathcal{B}_X)$ is a measurable space, $\zeta_k(t, x, \alpha)$ for each fixed k, x, α is a random process with trajectories belonging to the Skorokhod space D_∞^r (the space of right-continuous functions with left-side limits which are also called cadlag functions), and $\tau_k(x,\alpha), \beta_k(x,\alpha)$, are possibly dependent on $\zeta_k(\cdot,x,a)$ random variables, $\tau_k(\cdot) \geq 0, \beta_k(\cdot) \in X$. We assume that the vectors from $\mathcal{R}^r$ are column vectors and the variables introduced are measurable in the ordinary way in the pair (x,a) concerning σ-algebra $\mathcal{B}_X \times \mathcal{B}_{\mathcal{R}^r}$. Let also (x_0, S_0) be the independent of $\mathcal{F}_k, k \geq 0$, initial vector in $X \times \mathcal{R}^r$. We introduce the following recurrent sequences:

$$t_0 = 0, \ t_{k+1} = t_k + \tau_k(x_k, S_k),$$
$$S_{k+1} = S_k + \xi_k(x_k, S_k), \ x_{k+1} = \beta_k(x_k, S_k), \ k \geq 0, \quad (1)$$

where $\xi_k(x,\alpha) = \zeta_k(\tau_k(x,\alpha), x, \alpha)$, and set

$$\zeta(t) = S_k + \zeta_k(t - t_k, x_k, S_k),$$
$$x(t) = x_k, \text{ as } t_k \leq t < t_{k+1}, \ t \geq 0. \quad (2)$$

Then a two-component process $(x(t), \zeta(t)), t \geq 0$, is called a SP [2,3]. We also introduce an embedded process

$$S(t) = S_k \text{ as } t_k \leq t < t_{k+1}, \ t \geq 0, \quad (3)$$

and call it a recurrent process of a semi-Markov type (RPSM) [6].

It is worth to notice that the general definition of SP allows feedback between the discrete switching component $x(\cdot)$ and the switched component $\zeta(\cdot)$ (case of feedback). Consider a particular case when there is no component $x(\cdot)$ and the process $\zeta(\cdot)$ is switching at some random times t_k. Let $\mathcal{F}_k = \{(\zeta_k(t,\alpha), \tau_k(\alpha)), t \geq 0, \alpha \in \mathcal{R}^r\}, k \geq 0$, be jointly independent in index k families of random processes $\zeta_k(t,\alpha)$ with trajectories belonging to the Skorokhod space $\mathcal{D}_\infty^r$ and random variables $\tau_k(\alpha)$ measurable in the natural way. Introduce the following recurrent sequences:

$$t_0 = 0, \ t_{k+1} = t_k + \tau_k(S_k), \ S_{k+1} = S_k + \xi_k(S_k), \ k \geq 0, \quad (4)$$

where $\xi_k(\alpha) = \zeta_k(\tau_k(\alpha), \alpha)$, and set

$$\zeta(t) = S_k + \zeta_k(t - t_k, S_k) \text{ as } t_k \leq t < t_{k+1}, \ t \geq 0. \quad (5)$$

Then the process $\zeta(t), t \geq 0$, is a special case of a SP which is switched at the times t_k that are constructed recurrently using the values of $\zeta(\cdot)$ at switching times. This definition is used to describe some classes of queueing models.

In what follows we assume that SP is regular, i.e. the component $x(\cdot)$ with probability one has a finite number of jumps in each finite interval.

Now consider as the illustration some special subclasses of SP's.

Assume that the characteristics of the families $\mathcal{F}_k, k \geq 0$, do not depend on the parameters a and k. Then $x_k, \ k \geq 0$, is a homogeneous Markov process (MP) and $x(t), t \geq 0$, is a semi-Markov process (SMP). Assume also that the variables $\tau_k(x)$ at each $x \in X$ are independent of the processes $\zeta_k(t,x), t \geq 0$. In that case, if the variables $\tau_k(x)$ have the exponential distribution, then $x(t), t \geq 0$, is a MP. If in addition at each $x \in X$, $\zeta_k(t,x), t \geq 0$, is the process with independent increments, then the two-component process $(x(t), \zeta(t)), t \geq 0$, forms a MP which is homogeneous in the second component [24]. If the variables $\tau_k(x)$ have arbitrary distributions, then the process $\zeta(t)$ is a process with independent increments and semi-Markov switching introduced in [1].

Suppose now that the process $\zeta_k(t,x)$ at each $x \in X$ is a MP. Then the process $\zeta(t), t \geq 0$, in the book of [23] is called a piecewise Markov aggregate and in the book [26] it is called a MP with semi-Markov interference of chance. If the processes $\zeta_k(t,x)$ take the values in a Banach space and described by a semigroup of operators, then $\zeta(t)$ in [27,28] is called a random evolution.

Consider separately a class of recurrent processes of semi-Markov type (RPSM) which is a special subclass of SP introduced above. This process is a step-wise process. It has a simpler structure than a general SP and is a convenient tool for description of wide classed of queueing systems and networks. Below we consider some special models of RPSM which will be used at the description of stochastic queues and the problems of asymptotic analysis.

2.1 Recurrent Processes of a Semi-Markov Type

Let
$$\mathcal{F}_k = \{(\xi_k(\alpha), \tau_k(\alpha)), a \in \mathcal{R}^r\}, k \geq 0,$$
be jointly independent families of random variables with values in the space $\mathcal{R}^r \times [0, \infty)$, and S_0 be an independent of $\mathcal{F}_k, k \geq 0$ random variable in $\mathcal{R}^r$. Note that the variables $\xi_k(\alpha)$ and $\tau_k(\alpha)$ can be dependent. Introduce the following recurrent sequences:
$$t_0 = 0, \ t_{k+1} = t_k + \tau_k(S_k), \ S_{k+1} = S_k + \xi_k(S_k), \ k \geq 0 \qquad (6)$$
and set
$$S(t) = S_k \text{ as } t_k \leq t < t_{k+1}, \ t \geq 0. \qquad (7)$$
Then the process $S(t)$ forms a simple recurrent process of a semi-Markov type (RPSM) (in this case a discrete switching component $x(t)$ is absent).

If the distributions of the families $\mathcal{F}_k$ do not depend on the parameter k, the process $S(t)$ is a homogeneous SMP. If the distributions of the families $\mathcal{F}_k$ do not depend on both parameters α and k, then the times $t_0 \leq t_1 \leq \ldots \leq t_k \ldots$, form a recurrent flow and $S(t)$ can be interpreted as a reward renewal process.

Consider now a RPSM with additional Markov switching. Let
$$\mathcal{F}_k = \{(\xi_k(x, \alpha), \tau_k(x, \alpha)), x \in X, \alpha \in \mathcal{R}^r\}, k \geq 0$$
be jointly independent families of random variables taking values in $\mathcal{R}^r \times [0, \infty)$, and let $x_l, l \geq 0$, be a MP independent of $\mathcal{F}_k, k \geq 0$, with values in some space X, (x_0, S_0) be the initial value. We put
$$t_0 = 0, \ t_{k+1} = t_k + \tau_k(x_k, S_k), \ S_{k+1} = S_k + \xi_k(x_k, S_k), k \geq 0, \qquad (8)$$

and set
$$S(t) = S_k, \ x(t) = x_k, \ t_k \le t < t_{k+1}, \ t \ge 0. \tag{9}$$

Then the two-component process $(x(t), S(t))$ forms a RPSM with additional Markov switching. If the distributions of variables $\tau_k(x,\alpha)$ do not depend on the parameters α and k, then $x(t)$ is a SMP.

Consider now a general RPSM with feedback between both components. Let
$$\mathcal{F}_k = \{(\xi_k(x,\alpha), \tau_k(x,\alpha), \beta_k(x,\alpha)), x \in X, \alpha \in \mathcal{R}^r\}, k \ge 0,$$
be jointly independent families of random variables with values in the space $\mathcal{R}^r \times [0,\infty) \times X, (x_0, S_0)$ be the initial value. We put
$$t_0 = 0, \ t_{k+1} = t_k + \tau_k(x_k, S_k),$$
$$S_{k+1} = S_k + \xi_k(x_k, S_k), \ x_{k+1} = \beta_k(x_k, S_k), \ k \ge 0, \tag{10}$$
and set
$$S(t) = S_k, \ x(t) = x_k \ \text{as} \ t_k \le t < t_{k+1}, \ t \ge 0. \tag{11}$$

Then the two-component process $(x(t), S(t)), t \ge 0$ forms a RPSM with feedback between both components at the switching times.

2.2 Processes with Markov and Semi-Markov Switching

Consider a special case when some random process is switched by the external Markov or semi-Markov environment. Let $\mathcal{F}_k = \{\zeta_k(t,x,\alpha), t \ge 0, x \in X, \alpha \in \mathcal{R}^r\}$, $k \ge 0$, be the jointly independent parametric families of random processes with trajectories in Skorokhod space $D_\infty{}^r$, where $(X, \mathcal{B}_X)$ is a measurable space. Let also $x(t), t \ge 0$, be the independent of $\mathcal{F}_k$, $k \ge 0$, right-continuous SMP with values in X, S_0 be the initial value. We suppose that the variables introduced are measurable in the ordinary way in the pair (x,a) concerning σ-algebra $\mathcal{B}_X \times \mathcal{B}_{\mathcal{R}^r}$.

Denote by $0 = t_0 < t_1 < \cdots$ the times of sequential jumps of $x(\cdot)$ and put $x_k = x(t_k), k \ge 0$. We construct the process with Markov (or semi-Markov) switching in the following way. Put $S_{k+1} = S_k + \xi_k$, where $\xi_k = \zeta_k(\tau_k, x_k, S_k)$, $\tau_k = t_{k+1} - t_k$, and set
$$\zeta(t) = S_k + \zeta_k(t-t_k, x_k, S_k) \ \text{as} \ t_k \le t < t_{k+1}, \ t \ge 0. \tag{12}$$

Then the two-component process $(x(t), \zeta(t)), t \ge 0$, is a process with Markov switching (PMS) if $x(t)$ is a MP, or a process with semi-Markov switching (PSMS), if $x(t)$ is a SMP. The component $x(t)$ stands for a switching environment. Let us introduce also the imbedded process
$$S(t) = S_k \ \text{as} \ t_k \le t < t_{k+1}. \tag{13}$$

Then $(x(t), S(t))$ is a RPSM with independent Markov switching.

Consider a special case where $\{\zeta(t,x), t \ge 0,\}$ is the family of MP's and denote by $\zeta(t,x,\alpha)$ the process $\zeta(t,x)$ with the initial value α. Then the process $(x(t), \zeta(t))$ forms a Markov random evolution (when $x(t)$ is a MP), or a semi-Markov random evolution (when $x(t)$ is a SMP).

3 Switching Queueing Models

Let us consider as the illustration different classes of switching queueing models.

3.1 Markov Models

A state-dependent system $M_Q/M_Q/1/\infty$. A system consists of one server with infinite buffer (infinitely many places for waiting). The calls arrive one at a time and are served according to FIFO discipline. Let nonnegative functions $\{\lambda(q), \mu(q), q \geq 0\}$ be given. Denote by $Q(t)$ the total number of calls in the system at time t. The system operates as follows. If at time t, $Q(t) = q$, then the local arrival rate is $\lambda(q)$. This means that the probability that a new call arrives in a small interval $[t, t+h]$ is $\lambda(q)h + o(h)$. Correspondingly, the local service rate is $\mu(q)$ (the probability that a call in service completes service in a small interval $[t, t+h]$ is $\mu(q)h + o(h)$). After service completion the call leaves the system.

It is well known, that the process $Q(t), t \geq 0$, is a Birth-and-Death process. Let us represent it in a recurrent form. Denote by $t_1 < t_2 < \ldots$ the times of any change in the system (arrival of a call or service completion), and put $Q_k = Q(t_k + 0), k \geq 0$. Suppose that $t_0 = 0$, $Q(0+) = Q_0$.

Let us define the family of jointly independent random variables $\{\tau_k(q), \xi_k(q), q \geq 0\}$, $k \geq 0$, where $\tau_k(q)$ has an exponential distribution with parameter $\Lambda(q) = \lambda(q) + \mu(q)\chi(q > 0)$, $\xi_k(q)$ is an independent of $\tau_k(q)$ variable such that

$$\xi_k(q) = \begin{cases} +1, & \text{with prob. } \lambda(q)\Lambda(q)^{-1}, \\ -1, & \text{with prob. } \mu(q)\chi(q > 0)\Lambda(q)^{-1}, \end{cases}$$

and $\chi(A)$ is the indicator of the set A. Define the following recurrent sequences:

$$\tilde{t}_0 = 0, \ \tilde{Q}_0 = Q_0, \ \tilde{Q}_{k+1} = \tilde{Q}_k + \xi_k(\tilde{Q}_k), \ \tilde{t}_{k+1} = \tilde{t}_k + \tau_k(\tilde{Q}_k), \ k \geq 0, \quad (14)$$

and put

$$\tilde{Q}(t) = \tilde{Q}_k, \quad \text{as } \tilde{t}_k \leq t < \tilde{t}_{k+1}, \ t \geq 0. \quad (15)$$

As one can see, the process $\tilde{Q}(t)$ is a RPSM (see Section 2.1) and by definition the finite dimensional distributions of the process $\tilde{Q}(t)$ coincide with corresponding distributions of the queueing process $Q(t)$.

This representation provides also an idea for how to study the limiting behavior of $Q(t)$. If we can prove that the appropriately scaled two component process $(\tilde{t}_k, \tilde{Q}_k)$ weakly converges to the process $(y(u), q(u))$, $u \geq 0$, where the components $y(u)$ and $q(u)$ are possibly dependent, then under some regular assumptions we can expect that the appropriately scaled process $\tilde{Q}(t)$ weakly converges to the superposition of $y(u)$ and $q(u)$ in the form $q(y^{-1}(t))$, where $y^{-1}(t)$ is the inverse function.

Similar representation can be written for systems with many servers. For example, consider a system $M_Q/M_Q/r/\infty$, and assume that given $Q(t) = q$, the local rate of incoming calls is $\lambda(q)$ and service rate for each busy server is

$\mu(q)$. Then $Q(t)$ is a Birth-and-Death process with birth and death rates $\lambda(q)$ and $min(q,r)\mu(q)$, respectively, and in the expressions above, $\Lambda(q) = \lambda(q) + min(q,r)\mu(q)$,

$$\xi_k(q) = \begin{cases} +1, & \text{with prob. } \lambda(q)\Lambda(q)^{-1}, \\ -1, & \text{with prob. } min(q,r)\mu(q)\Lambda(q)^{-1}. \end{cases}$$

The representation (14),(15) has a similar form for Markov networks and also for batch arrivals and service. In these cases the variables $\xi_k(q)$ may take vector values, and the variables $\tau_k(q)$ have the exponential distributions. By analogy, we can write similar representations for more general systems with non-Markov arrival process and non-exponential service. For these cases we need to choose in the appropriate way the switching times $\widetilde{t}_k$ and construct corresponding processes reflecting the behavior of queueing processes in the intervals $[\widetilde{t}_k, \widetilde{t}_{k+1})$.

For further exploration notice that in fact the exponentiality of $\tau_k(q)$ is not essential for the asymptotic analysis. This means, if we can prove quite general theorems on the convergence of the recurrent processes, constructed according to the relations (14),(15), then these theorems can be used for the analysis of more general queueing models, for which the queueing processes have representations similar to (14),(15).

In this way we can analyze rather general switching queueing models. For these models the queueing processes can be represented in terms of SP's in the form similar to (14),(15).

Queueing system $M_{M.Q}/M_{M.Q}/1/m$. Consider as a more complicated example a state-dependent queueing system in a Markov environment. Let $z(t)$, $t \geq 0$, be a homogeneous MP with finite state space $I = \{1, ..., d\}$ and transition rates $a(i,l)$, $i,l = \overline{1,d}$, $i \neq l$. $z(t)$ stands for the external Markov environment. Let also the family of nonnegative functions $\lambda(i,j)$, $\mu(i,j)$, $i = \overline{1,d}$, $j = \overline{0, m+1}$, be given. The system consists of one server with m places for waiting. The calls enter the system one at a time. Denote by $Q(t)$ the number of calls in the system at time t, $0 \leq Q(t) \leq m+1$. If $z(t) = i$ and $Q(t) = j$, then the local input rate for incoming calls is $\lambda(i,j)$ and the local service rate is $\mu(i,j)$ ($\mu(i,0) \equiv 0$). The call upon arrival at the empty system is immediately taken for service. When the server is busy, the call joins the queue. After completion service the call leaves the system and the next call from the queue, if any, is immediately taken for service. If a call enters the system and at that time $Q(t) = m+1$, then this call is lost.

To describe this system as a switching queueing model consider a two-component MP $x(t) = (z(t), Q(t))$, $t \geq 0$, with state space $I \times \{0, ..., m+1\}$ and transition rates $a((i,j),(l,q))$, $i,l = \overline{1,d}$, $j,q = \overline{0, m+1}$, where

$$a((i,j),(l,j)) = a(i,l), \; i,l = \overline{1,d}, \; j = \overline{0, m+1};$$
$$a((i,j),(i,j+1)) = \lambda(i,j), \; i = \overline{1,d}, \; j = \overline{0, m};$$
$$a((i,j),(i,j-1)) = \mu(i,j), \; i = \overline{1,d}, \; j = \overline{1, m+1},$$

(other rates are zeros). Then the two-component process $x(t)$ stands for the switching component (or environment). Notice that the process $x(t)$ belongs to the class of so called quasi-Birth-and-Death processes introduced in [35].

Let also $\zeta_k(t,(i,m+1))$ be a Poisson process with the rate $\lambda(i,m+1)$, $i = \overline{1,d}$, and $\zeta_k(t,(i,j)) \equiv 0$ as $j < m+1$.

We construct a SP $(x(t), \zeta(t))$, $t \geq 0$, using the Markov component $x(t)$ and processes $\zeta_k(\cdot)$ according to formula (12) where $x(t) = (z(t), Q(t))$ and $S_0 = 0$. Then this process is a process with independent increments and Markov switching, see section 2.2, the component $Q(t)$ is the value of the queue and $\zeta(t)$ is the number of calls lost in the interval $[0,t]$.

Observing the process $(x(t), \zeta(t))$ we can also calculate other characteristics of the system. Let $\nu^+(t)$ ($\nu^-(t)$) be the number of jumps up ($+1$) (and down (-1) correspondingly) of the process $Q(t)$ in interval $[0,t]$. Then $\nu^+(t)$ is the number of calls entered the system in interval $[0,t]$ and $\nu^-(t)$ is the number of calls served in this time interval.

Notice that if the rate of input process $\lambda(i,j) \equiv \lambda(i)$ (depends only on the state of Markov environment), then the input process is usually called a Markov modulated input process [35] and in fact this is a Poisson process with random rate $\lambda(z(t))$ or a doubly-stochastic Poisson process.

In a similar way we can describe Markov model $\overline{M}_{Q,B}/\overline{M}_{Q,B}/1/\infty$ which includes state-dependent systems with batch arrivals and service, systems with different types of calls, impatient calls, etc. [20].

3.2 Non-Markov Systems

Semi-Markov system $SM/M_{SM,Q}/1$. Consider a queueing system which is described in the following way. Let $x(t), t \geq 0$, be a right continuous SMP with values in some measurable space (X, B_X) and let the functions $\mu(x,m), x \in X$, $m = 0, 1, 2, ...$ be given ($\mu(x,m)$ are measurable relatively σ-algebra B_X and stand for the local transition rates). Let also $t_1 < t_2 < ...$ be a sequence of the times of jumps of $x(t)$. We say that the input flow is a semi-Markov one if the calls enter the system one at a time at the times t_k. The system has one server and the service rate at time t is $\mu(x(t), Q(t))$, where $Q(t)$ is the number of calls in the system at time t. After completion service the calls leave the system.

To describe the process $(x(t), Q(t))$ as a SP we introduce the jointly independent families of stepwise decreasing MP's $\{\eta_k(t,x,m), t \geq 0, x \in X, m = 1, 2, ...\}$, $k \geq 0$, with values in $\{0, 1, 2, ...\}$ such that $\eta_k(0,x,m) = m$ for any x, m, k, the transition from state j is possible only to state $j - 1$, and at small h,

$$\Pr\{\eta(t+h,x,m) = j - 1 \mid \eta(t,x,m) = j\} = \mu(x,j)h + o(h),$$

where we assume that $\mu(x,0) \equiv 0$. Let also $\{(\tau_k(x), \beta_k(x), x \in X\}, k \geq 0$, be the independent of the introduced processes jointly independent families of random variables defining the transitions of a SMP $x(t)$ in the following way: $\tau_k(x) > 0$, $\beta_k(x) \in X$, and

$$P(t, x, A) = \mathbf{P}\{\tau_k(x) < t, \beta_k(x) \in A\}$$
$$= \mathbf{P}\{t_{k+1} - t_k < t, x(t_{k+1}) \in A \mid x(t_k) = x\},$$
$$t > 0, x \in X, A \in B_X, k \geq 0, \tag{16}$$

where $P(t, x, A)$ is a transition probability for SMP $x(t)$. Then we introduce the families of processes $\zeta_k(t, x, m)$ in the following way:

$$\zeta_k(t, x, m) = \eta(t, x, m), \ t < \tau_k(x),$$
$$\zeta_k(\tau_k(x), x, m) = \eta_k(\tau_k(x), x, m) + 1.$$

By definition the process $(x(t), Q(t))$ is a SP which is defined by the families

$$\{(\zeta_k(t, x, m), \tau_k(x), \beta_k(x)), t \geq 0, x \in X, m = 0, 1, ...\}, k \geq 0,$$

according to relations (1),(2). This process belongs to the class of Markov processes with semi-Markov switching.

By analogy we can describe more complicated queueing system and networks of the type $SM_Q/M_{SM,Q}/1/\infty$ where the input process depends on the values of the queue in the system and is constructed using independent families of random variables $\{\tau_k(x, m), x \in X, m \geq 0\}, k \geq 0$, and a MP $x_k, k \geq 0$, with values in X as follows: the calls enter the system one at a time. If a call enters the system at time t_k and the total number of calls in the system becomes Q, then the next call enters at time $t_{k+1} = t_k + \tau_k(x_k, Q)$. The service in the system is provided in the same way as in the system $SM/M_{SM,Q}/1/\infty$.

In this case the two-component process $(x(t), Q(t))$ is a SP, but the component $x(t)$ itself is not a semi-Markov process and there is a feedback between the input flow and the values of the queue.

More example of switching queueing models including system and networks with semi-Markov type switching (the input flow is a Poisson process modulated by the external semi-Markov environment and by the values of the queue) of the type $M_{SM,Q}/M_{SM,Q}/1/\infty$ are considered in the author's book [20].

There are also examples of the systems $G_Q/M_Q/1/\infty$ with dependent arrival flows, polling systems, different classes of Markov and semi-Markov queueing systems and networks with unreliable servers and also some classes of retrial queues (see also [12,14,16]).

4 Averaging Principle and Diffusion Approximation for Switching Processes

In this section we consider the limit theorems for SP's in the case of "fast" switching. Consider a sequence of SP's $(x_n(t), \zeta_n(t)), t \geq 0$, depending on some scaling parameter n on the expanding interval $[0, nT]$, where $n \to \infty$. Suppose that SP depends on n in such a way that the number of switches on each interval $[na, nb], 0 < a < b < T$, tends in probability to infinity. In this case we can expect that under some natural assumptions a normalized trajectory of $\zeta_n(nt)$ uniformly converges in probability to a deterministic function which is a solution of some differential equation (Averaging Principle - AP), and a normalized difference between trajectory of $\zeta_n(nt)$ and this solution weakly converges in Skorokhod space $\mathcal{D}_T$ to some diffusion process (Diffusion Approximation - DA). As sample trajectories of a limiting process are continuous, this convergence implies a weak convergence of functionals, which are continuous with respect to the uniform convergence [25,36]. We call this convergence a J-convergence according to [36]. Note that after re-scaling of time we can consider the process in the interval $[0,T]$ in the scale of time nt and in this case the number of switches in each interval $[a,b]$ tends to infinity (switches happen rapidly).

A new approach based on the investigation of the asymptotic properties of recurrent processes of a semi-Markov type (RPSM), theorems about the convergence of recurrent sequences to the solutions of stochastic differential equations and the convergence of superposition of random functions is developed.

4.1 Averaging Principle and Diffusion Approximation for RPSM

Consider the limit theorems for RPSM in the case of fast switching. Let at each n=1,2..., $\mathcal{F}_{nk} = \{(\xi_{nk}(z), \tau_{nk}(z)), z \in \mathcal{R}^r\}, k \geq 0$, be jointly independent at different k families of random variables with values in $\mathcal{R}^r \times [0,\infty)$ and distributions not depending on k, and let S_{n0} be the independent of $\mathcal{F}_{nk}, k \geq 0$, initial value in $\mathcal{R}^r$. According to Section 2.1 we introduce recurrent sequences

$$t_{n0} = 0, \ t_{nk+1} = t_{nk} + \tau_{nk}(S_{nk}), \ S_{nk+1} = S_{nk} + \xi_{nk}(S_{nk}), \ k \geq 0, \quad (17)$$

and define RPSM as follows:

$$S_n(t) = S_{nk} \ \text{as} \ t_{nk} \leq t < t_{nk+1}, \ t > 0. \quad (18)$$

As under natural assumptions the normalized trajectory of RPSM after n switches is of the order n, we consider the dependence of the argument in recurrent equations on the re-scaled trajectory S_{nk}/n with the purpose to obtain a state-dependence property in the limiting equations.

Assume that there exist the functions $m_n(\alpha) = \mathbf{E}\tau_{n1}(n\alpha), b_n(\alpha) = \mathbf{E}\xi_{n1}(n\alpha)$. In the following by symbol $\xrightarrow{P}$ we denote the convergence in probability and by symbol $\stackrel{W}{\Rightarrow}$, a weak convergence.

Theorem 1. *(Averaging principle). Suppose that for any $N > 0$,*

$$\lim_{L\to\infty} \limsup_{n\to\infty} \sup_{|\alpha|\leq N} \left\{ \mathbf{E}\tau_{n1}(n\alpha)\chi(\tau_{n1}(n\alpha) > L) \right.$$
$$\left. + \mathbf{E}|\xi_{n1}(n\alpha)|\chi(|\xi_{n1}(n\alpha)|) > L) \right\} = 0, \quad (19)$$

and as $\max(|\alpha_1|, |\alpha_2|) \leq N$,

$$|m_n(\alpha_1) - m_n(\alpha_2)| + |b_n(\alpha_1) - b_n(\alpha_2)| \leq C_N|\alpha_1 - \alpha_2| + \alpha_n(N), \quad (20)$$

where C_N are some constants, $\alpha_n(N) \to 0$ uniformly in $|\alpha_1| \leq N, |\alpha_2| \leq N$, there exist functions $m(a) > 0$ and $b(a)$ such that for any $\alpha \in \mathcal{R}^r$ as $n \to \infty$,

$$m_n(\alpha) \to m(\alpha), \; b_n(\alpha) \to b(\alpha), \tag{21}$$

and

$$n^{-1} S_{n0} \xrightarrow{P} s_0. \tag{22}$$

Then

$$\sup_{0 \leq t \leq T} |n^{-1} S_n(nt) - s(t)| \xrightarrow{P} 0, \tag{23}$$

where the function $s(t)$ satisfies the following ordinary differential equation

$$ds(t) = m(s(t))^{-1} b(s(t)) dt, \tag{24}$$

and T is any positive number such that $y(+\infty) > T$ with probability one, where

$$y(t) = \int_0^t m(\eta(u)) du, \tag{25}$$

and $\eta(t)$ is a solution of the differential equation

$$d\eta(u) = b(\eta(u)) du, \; \eta(0) = S_0, \tag{26}$$

(it is supposed that a unique solution of equation (26) exists in each interval).

Now we consider the convergence of the normalized process

$$\gamma_n(t) = \frac{1}{\sqrt{n}} (S_n(nt) - ns(t)), t \in [0, T],$$

to a diffusion process. Denote

$$\tilde{b}_n(\alpha) = m_n(\alpha)^{-1} b_n(\alpha), \; \tilde{b}(\alpha) = m(\alpha)^{-1} b(\alpha),$$
$$\rho_n(\alpha) = \xi_{n1}(n\alpha) - b_n(\alpha) - \tilde{b}(\alpha)(\tau_{n1}(n\alpha) - m_n(\alpha)),$$
$$D_n^2(\alpha) = \mathbf{E} \rho_n(\alpha) \rho_n(\alpha)^* \tag{27}$$

(here and in what follows we denote the conjugate vector by symbol *), and put

$$q_n(\alpha, z) = \sqrt{n} \Big(\tilde{b}_n(\alpha + \frac{1}{\sqrt{n}} z) - \tilde{b}(\alpha) \Big). \tag{28}$$

Theorem 2. *(Diffusion approximation) Let conditions (20)-(22) hold where in (20) a condition $\alpha_n(N) \to 0$ is replaced by $\sqrt{n} \alpha_n(N) \to 0$, there exist continuous matrix-valued functions $D^2(\alpha)$ and $Q(\alpha)$ and a vector-valued function $g(a)$ such that as $n \to \infty$, uniformly in each bounded region,*

$$D_n^2(\alpha) \to D^2(\alpha), \tag{29}$$

$$q_n(\alpha, z) \to Q(\alpha) z + g(a), \tag{30}$$

for any $z \in \mathcal{R}^r$,
$$\gamma_n(0) \stackrel{w}{\Rightarrow} \gamma_0, \qquad (31)$$
where γ_0 is a proper random variable, and for any $N > 0$,
$$\lim_{L \to \infty} \limsup_{n \to \infty} \sup_{|\alpha| < N} \Big\{ \mathbf{E} \tau_{n1}^2(n\alpha) \chi(\tau_{n1}(n\alpha) > L)$$
$$+ \mathbf{E} |\xi_{n1}(n\alpha)|^2 \chi(|\xi_{n1}(n\alpha)|) > L) \Big\} = 0. \qquad (32)$$

Then for any T such that $y(+\infty) > T$, the sequence of processes $\gamma_n(t)$ J-converges in the interval $[0, T]$ to the diffusion process $\gamma(t)$ satisfying the following stochastic differential equation:
$$d\gamma(t) = \Big(Q(s(t))\gamma(t) + g(s(t)) \Big) dt + D(s(t)) m(s(t))^{-1/2} dw(t), \qquad (33)$$
where $\gamma(0) = \gamma_0$, and $s(\cdot)$ satisfies the equation (24).

The proof of theorems can be found in [10,20]. Note that J-convergence means the weak convergence of measures in Skorokhod space $\mathcal{D}_T$.

4.2 Averaging Principle and Diffusion Approximation for RPSM with Markov Switching

Consider the next level of complexity when RPSM is switched by some external Markov process with fast switching. Let at each $n \geq 0$,
$$F_{nk} = \{(\xi_{nk}(x,z), \tau_{nk}(x,z)), x \in X, z \in \mathcal{R}^r\}, \; k \geq 0, \qquad (34)$$
be jointly independent families of random variables with values in the space $\mathcal{R}^r \times [0, \infty)$ and distributions not depending on $k \geq 0$, and let $x_{ni}, i \geq 0$, be an independent of $F_{nk}, k \geq 0$, homogeneous MP with values in some space X, S_{n0} be the initial value. Notice that the variables $\xi_{nk}(x,z)$ and $\tau_{nk}(x,z)$ can be dependent. We construct RPSM $(x_n(t), S_n(t)), t \geq 0$, according to Section 2.1. Put $t_{n0} = 0$ and denote
$$S_{nk+1} = S_{nk} + \xi_{nk}(x_{nk}, S_{nk}), t_{nk+1} = t_{nk} + \tau_{nk}(x_{nk}, S_{nk}), \; k \geq 0. \qquad (35)$$
Let
$$S_n(t) = S_{nk}, \; x_n(t) = x_{nk} \text{ as } t_{nk} \leq t < t_{nk+1}. \qquad (36)$$
The process $x_n(\cdot)$ stands for some external environment and in general it is not a MP or SMP as it depends on the values of a switching component S_{nk}.

Suppose that MP $x_{nk}, k \geq 0$, at each $n > 0$ has a stationary measure $\pi_n(A), A \in \mathcal{B}_X$. Assuming that corresponding integrals exist, denote
$$m_n(x, \alpha) = \mathbf{E}\tau_{n1}(x, n\alpha), \; b_n(x, \alpha) = \mathbf{E}\xi_{n1}(x, n\alpha), \qquad (37)$$
$$m_n(\alpha) = \int_X m_n(x, \alpha) \pi_n(dx), \; b_n(\alpha) = \int_X b_n(x, \alpha) \pi_n(dx).$$

Let us introduce a strong mixing coefficient

$$\alpha_n(k) = \sup\{|P\{x_{ni} \in A, x_{n,i+k} \in B\} \\ - P\{x_{ni} \in A\}P\{x_{n,i+k} \in B\}| : A, B \in \mathcal{B}_X, i \geq 0\}. \quad (38)$$

Theorem 3. *(Averaging principle)* Suppose that there exist a sequence of integers r_n such that

$$n^{-1}r_n \to 0, \; \sup_{k \geq r_n} \alpha_n(k) \to 0, \quad (39)$$

for any $N > 0$,

$$\lim_{L \to \infty} \limsup_{n \to \infty} \sup_{|\alpha| \leq N} \sup_x \{\mathbf{E}\tau_{n1}(x, n\alpha)\chi(\tau_{n1}(x, n\alpha) > L) \\ + \mathbf{E}|\xi_{n1}(x, n\alpha)|\chi(|\xi_{n1}(x, n\alpha)| > L)\} = 0, \quad (40)$$

and for any x as $\max(|\alpha_1|, |\alpha_2|) \leq N$,

$$|m_n(x, \alpha_1) - m_n(x, \alpha_2)| + |b_n(x, \alpha_1) - b_n(x, \alpha_2)| \\ \leq C_N|\alpha_1 - \alpha_2| + a_n(N), \quad (41)$$

where C_N are some bounded constants, $a_n(N) \to 0$ uniformly in $|\alpha_1| \leq N$, $|\alpha_2| \leq N$, there exist functions $m(\alpha) > 0$ and $b(\alpha)$ such that for any $\alpha \in \mathcal{R}^r$,

$$m_n(\alpha) \to m(\alpha), \; b_n(\alpha) \to b(\alpha), \quad (42)$$

and

$$n^{-1}S_{n0} \xrightarrow{P} s_0, \quad (43)$$

where s_0 is some (possibly random) value. Then in the interval $[0, T]$,

$$\sup_{0 \leq t \leq T} |n^{-1}S_n(nt) - s(t)| \xrightarrow{P} 0, \quad (44)$$

where a function $s(t)$ is a solution of an ordinary differential equation

$$ds(t) = m(s(t))^{-1}b(s(t))dt, \; s(0) = s_0, \quad (45)$$

and T should satisfy the relation $y(+\infty) > T$ with probability one, where

$$y(t) = \int_0^t m(\eta(u))du,$$

and $\eta(t)$ is a solution of an ordinary differential equation

$$d\eta(u) = b(\eta(u))du, \; \eta(0) = s_0 \quad (46)$$

(it is supposed that a unique solution of equation (46) exists in each interval).

Now we study the conditions of the convergence of the sequence of processes

$$\kappa_n(t) = \frac{1}{\sqrt{n}}(S_n(nt) - ns(t))$$

to some diffusion process. Let us introduce the uniformly strong mixing coefficient for the process x_{nk}:

$$\varphi_n(r) = \sup_{x,y,A} |P\{x_{nr} \in A \mid x_{n0} = x\} - P\{x_{nr} \in A \mid x_{n0} = y\}|.$$

Put

$$\begin{aligned}
&\tilde{b}_n(\alpha) = b_n(\alpha)m_n(\alpha)^{-1}, \; \tilde{b}(\alpha) = b(\alpha)m(\alpha)^{-1},\\
&\rho_{nk}(x,\alpha) = \xi_{nk}(x,n\alpha) - b_n(x,\alpha) - \tilde{b}(\alpha)(\tau_{nk}(x,n\alpha) - m_n(x,\alpha)),\\
&D_n^2(x,\alpha) = \mathbf{E}\rho_{n1}(x,\alpha)\rho_{n1}(x,\alpha)^*, \qquad (47)\\
&\gamma_n(x,\alpha) = b_n(x,\alpha) - b_n(\alpha) - \tilde{b}(\alpha)(m_n(x,\alpha) - m_n(\alpha)).
\end{aligned}$$

Theorem 4. *(Diffusion approximation) Suppose that for some fixed $r > 0$ and $q \in [0,1)$,*

$$\varphi_n(r) \le q, \; n > 0, \qquad (48)$$

the condition (41) with the relation $\sqrt{n}\alpha_n(N) \to 0$ holds, conditions (42), (43) are true, and for any $N > 0$ the following conditions are satisfied:

1) $$\lim_{L \to \infty} \lim_{n \to \infty} \sup_{|\alpha| \le N} \sup_x \{\mathbf{E}\tau_{n1}(x,n\alpha)^2 \chi(\tau_{n1}(x,n\alpha) > L)$$
$$+ \mathbf{E}|\xi_{n1}(x,n\alpha)|^2 \chi(|\xi_{n1}(x,n\alpha)| > L)\} = 0; \qquad (49)$$

2) as $\max(|\alpha_1|, |\alpha_2|) \le N$,

$$|D_n(x,\alpha_1)^2 - D_n(x,\alpha_2)^2| \le C_N |\alpha_1 - \alpha_2| + \alpha_n(N), \qquad (50)$$

where $\alpha_n(N) \to 0$ uniformly in $|\alpha_1| \le N, |\alpha_2| \le N$;

3) there exists a function $q(\alpha,z)$ such that for any N in the region $|\alpha| \le N$,

$$|q(\alpha,z)| \le C_N(1 + |z|),$$

uniformly in $|\alpha| \le N$ at each fixed z,

$$\sqrt{n}(\tilde{b}_n(\alpha + \frac{1}{\sqrt{n}}z) - \tilde{b}(\alpha)) \to q(\alpha,z), \qquad (51)$$

and there exist functions $D(\alpha)$ and $B(\alpha)$ such that for any $\alpha \in \mathcal{R}^m$,

$$D_n^2(\alpha) = \int_X D_n^2(x,\alpha)\pi_n(dx) \to D^2(\alpha), \qquad (52)$$

$$B_n^{(1)}(\alpha)^2 + B_n^{(2)}(\alpha)^2 \to B(\alpha)^2, \qquad (53)$$

where

$$B_n^{(1)}(\alpha)^2 = \int_X \gamma_n(x,\alpha)\gamma_n(x,\alpha)^*\pi_n(dx),$$
$$B_n^{(2)}(\alpha)^2 = \sum_{k\geq 1} E\gamma_n(x_{n0},\alpha)\gamma_n(x_{nk},\alpha)^*,$$

with $P\{x_{n0} \in A\} = \pi_n(A), A \in B_X$, and also

$$\kappa_n(0) \stackrel{W}{\Rightarrow} \kappa_0, \qquad (54)$$

where κ_0 is some proper random variable.

Then for any $T > 0$ satisfying the conditions of Theorem 3 the sequence of processes $\kappa_n(t)$ J-converges in the space $\mathcal{D}_T^r$ to the diffusion process $\kappa(t)$ satisfying the following stochastic differential equation: $\kappa(0) = \kappa_0$,

$$d\kappa(t) = q(s(t), \kappa(t))dt + m(s(t))^{-\frac{1}{2}}(D(s(t))^2 + B(s(t))^2)^{\frac{1}{2}}dw(t), \qquad (55)$$

where $w(t)$ is the standard Wiener process in $\mathcal{R}^r$, and a solution of (55) exists and is unique.

The proof of theorems can be found in [8,9,10,20]. Note that AP and DA type theorems for general SP's can be found in [10,20].

These results equip us with the technique for study the limit theorems of AP and DA type for wide classes of queueing systems and networks.

5 AP and DA in Overloaded Switching Queueing Models

Consider state-dependent Markov and semi-Markov queueing models at the presence of the ergodic Markov or semi-Markov environment, as well. We assume that characteristics of the system depend on some parameter n, $n \to \infty$, and the arrival and service processes as well as the routing matrix may depend on the current value of the queueing process $Q_n(t)$ (a vector of queues or a workload process) and possibly some random environment $x_n(t)$. In general, the environment may also depend on the queueing process and in this case it will not be a MP or SMP (case of feedback). We suppose also that a number of calls (or a value of a workload process) in the system is asymptotically large, which may be caused by a high load or by a large initial value of the queueing process. These results are partially published in [17] and also can be found in the book [20].

5.1 Markov Queueing Models

Consider first for the illustration of a general approach some classes of overloaded state-dependent Markov queueing systems and networks.

A system $M_Q/M_Q/1/\infty$. Consider a system described in section 3.1. There is one server with infinitely many places for waiting. We study AP and DA for the queueing process in transient conditions and assume that the initial number of calls is of the order n. In this case assume that the input and service rates depend on the normalized number of the calls in the system in the following way. If at time t there are Q calls in the system, then the input rate is $\lambda(Q/n)$ and the service rate is $\mu(Q/n)$ where $\lambda(q)$ and $\mu(q)$ are given functions. Denote by $Q_n(t)$ the number of calls in the system at time t. Suppose that as $n \to \infty$,

$$Q_n(0)/n \xrightarrow{P} s_0. \tag{56}$$

Denote by $s(t)$ a solution of the differential equation:

$$ds(t) = b(s(t))dt, \ s(0) = s_0, \tag{57}$$

where $b(q) = \lambda(q) - \mu(q)$. The following result follows from Theorems 1, 2.

Theorem 5. *1) Suppose that (56) is true, $s_0 > 0$, the functions $\lambda(q), \mu(q)$ satisfy the local Lipschitz condition, $\lambda(q) + \mu(q) > 0$ as $q \in (0, \infty)$, and for some fixed $T > 0$ there exists an interval $[0, A]$ such that the equation*

$$d\eta(t) = b(\eta(t))(\lambda(\eta(u)) + \mu(\eta(u)))^{-1}dt, \ \eta(0) = s_0, \tag{58}$$

has a unique solution $\eta(t) > 0$, $t \in (0, A)$, and in addition $y(A) > T$, where

$$y(t) = \int_0^t (\lambda(\eta(u)) + \mu(\eta(u)))^{-1}du. \tag{59}$$

Then

$$\sup_{0 \le t \le T} |n^{-1}Q_n(nt) - s(t)| \xrightarrow{P} 0, \tag{60}$$

where $s(t)$ is a unique solution of (57).

2) Suppose in addition that the functions $\lambda(q), \mu(q)$ are continuously differentiable in $(0, \infty)$ and

$$n^{-1/2}(Q_n(0) - ns_0) \xRightarrow{w} \zeta_0, \tag{61}$$

where ζ_0 is a proper random variable.

Then the sequence of processes $\zeta_n(t) = n^{-1/2}(Q_n(nt) - ns(t))$ J-converges in $\mathcal{D}_T$ to the diffusion process $\zeta(t)$ satisfying the following stochastic differential equation: $\zeta(0) = \zeta_0$,

$$d\zeta(t) = (\lambda'(s(t)) - \mu'(s(t)))\zeta(t)dt + (\lambda(s(t)) + \mu(s(t)))^{1/2}dw(t), \tag{62}$$

Consider as an illustration of AP a simulation of the system $M_Q/M_Q/1/\infty$ with the rates: $\lambda(s) \equiv \lambda$, $\mu(s) = \mu s$. This case corresponds to a system $M/M/\infty$ and equation (57) implies: $s(t) = \lambda/\mu + (s_0 - \lambda/\mu)e^{-\mu t}$, $t \ge 0$.

Figure 1 illustrates the convergence of the trajectory $Q_n(nt)/n$ to the function $s(t)$ for two values of n, $n = 5$ and $n = 100$. At large n, the trajectory is close to $s(\cdot)$.

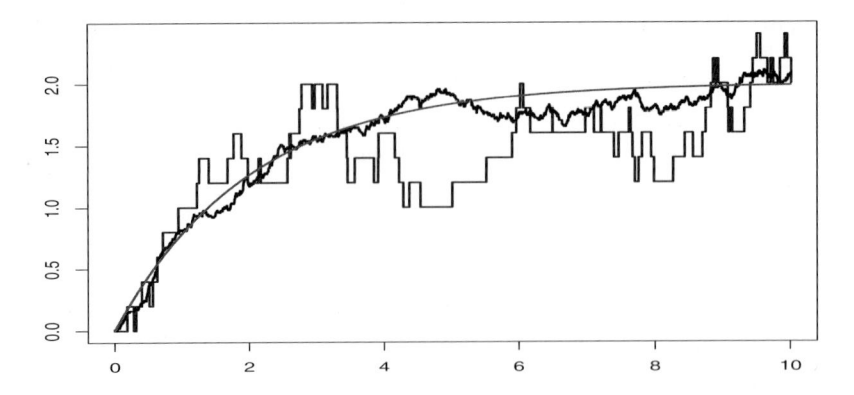

Fig. 1. Two sample paths of $Q_n(nt)$ at $s_0 = 5$, $\lambda = 1, \mu = 0.5$. Step-wise line corresponds to $n = 5$, wavy line – $n = 100$. Graph of $s(t)$ is shown by a continuous solid line.

5.2 Non-Markov Queueing Models

System $GI/M_Q/1/\infty$. For the illustration of the approach we consider first rather simple queueing system $GI/M_Q/1/\infty$ with recurrent input and exponential service with rate depending on the state of the queue in the system and study AP and DA in the overloaded case.

Assume that the calls enter the system one at a time at the times $t_1 < t_2 < ...$ of the events of the renewal flow (the variables $t_{k+1} - t_k, k = 1, 2, ..$, are independent identically distributed variables). Suppose that the distribution of inter-arrival times $t_{k+1} - t_k$ coincides with the distribution of some variable τ. Let the non-negative function $\mu(\alpha)$, $\alpha \geq 0$, be given. There is one server and infinitely many places for waiting. If a call enters the system at time t_k and the number of calls becomes equal to Q, then the service rate in the interval $[t_k, t_{k+1})$ is $\mu(n^{-1}Q)$. After service completion the call leaves the system. Let Q_{n0} be the initial number of calls, and $Q_n(t)$ be the total number of calls in the system at time t. Assuming that corresponding expressions exist, denote

$$m = \mathbf{E}\tau, \; b(\alpha) = (1 - \mu(\alpha)m), \; d^2 = \mathbf{Var}\,\tau, \; D^2(\alpha) = m\mu(\alpha) + d^2/m^2.$$

Theorem 6. *Suppose that $m > 0$, the function $\mu(\alpha)$ is locally Lipschitz and has no more then linear growth and $n^{-1}Q_n(0) \xrightarrow{P} s_0 > 0$. Then the relation (60) holds where*

$$ds(t) = (m^{-1} - \mu(s(t)))dt, \; s(0) = s_0,$$

and T is any positive value such that $s(t) > 0$, $t \in [0, T]$.

Suppose in addition that the function $\mu(\alpha)$ is continuously differentiable and

$$n^{-1/2}(Q_n(0) - s_0) \xRightarrow{W} \gamma_0.$$

Then the sequence of processes $\gamma_n(t) = n^{-1/2}(Q_n(nt) - ns(t))$ J-converges in the interval $[0,T]$ to the diffusion process $\gamma(t)$: $\gamma(0) = \gamma_0$,

$$d\gamma(t) = -\mu'(s(t))\gamma(t)dt + \sqrt{\mu(s(t)) + d^2/m^3}\, dw(t).$$

Semi-Markov system $SM/M_{SM,Q}/1/\infty$. Consider more general overloaded queueing system with semi-Markov input and Markov-type service where the service rate depends on the state of the system and the state of some SMP. Let $x(t)$, $t \geq 0$, be a SMP with values in X which stands for some external random environment. Denote by $\tau(x)$ a sojourn time in the state x. Let the non-negative function $\mu(x,\alpha)$, $x \in X$, $\alpha \geq 0$, be given. There is one server and infinitely many places for waiting. Assume that the calls enter the system one at a time at the times $t_1 < t_2 < \ldots$ of the jumps of the process $x(t)$. Denote $x_k = x(t_k + 0)$. If a call enters the system at time t_k and the number of calls in the system becomes equal to Q, then the service rate in the interval $[t_k, t_{k+1})$ is $\mu(x_k, n^{-1}Q)$. After service completion the call leaves the system. Let Q_{n0} be the initial number of calls, and $Q_n(t)$ be the total number of calls in the system at time t.

Consider the case when the embedded MP x_k, $k \geq 0$, does not depend on parameter n and is uniformly ergodic with stationary measure $\pi(A)$, $A \in \mathcal{B}_X$. Assuming that corresponding expressions exist, denote

$$m(x) = \mathbf{E}\tau(x),\ m = \int_X m(x)\pi(dx),\ c(\alpha) = \int_X \mu(x,\alpha)m(x)\pi(dx),$$
$$b(\alpha) = (1 - c(\alpha))m^{-1},\ G(\alpha) = c'(\alpha),$$
$$g(x,\alpha) = 1 - m(x)(1 - c(\alpha) + \mu(x,\alpha)m)m^{-1},$$
$$d^2(x) = \mathbf{Var}\,\tau(x),\ d^2 = \int_X d^2(x)\pi(dx),$$
$$e_1(\alpha) = \int_X \mu^2(x,\alpha)d^2(x)\pi(dx),\ e_2(\alpha) = \int_X \mu(x,\alpha)d^2(x)\pi(dx),$$
$$D^2(\alpha) = c(\alpha) + e_1(\alpha) + 2(1 - c(\alpha))e_2(\alpha)m^{-1} + (1 - c(\alpha))^2 d^2 m^{-2}.$$

Theorem 7. *Suppose that $m > 0$, the function $\mu(x,\alpha)$ is locally Lipschitz with respect to α uniformly in $x \in X$, the function $c(\alpha)$ has no more then linear growth and $n^{-1}Q_n(0) \xrightarrow{P} s_0 > 0$. Then the relation (60) holds where*

$$ds(t) = m^{-1}(1 - c(s(t))dt,\ s(0) = s_0,$$

and T is any positive value such that $s(t) > 0$, $t \in [0,T]$.

Suppose in addition that variables $\tau(x)^2$ are uniformly integrable, the function $c(\alpha)$ is continuously differentiable, and $n^{-1/2}(Q_n(0) - s_0) \xRightarrow{w} \gamma_0$.

Then the sequence of processes $\gamma_n(t) = n^{-1/2}(Q_n(nt) - ns(t))$ J-converges in the interval $[0,T]$ to the diffusion process $\gamma(t)$: $\gamma(0) = \gamma_0$,

$$d\gamma(t) = -m^{-1}G(s(t))\gamma(t)dt + m^{-1/2}\left(D^2(s(t)) + B^2(s(t))\right)^{1/2}dw(t),$$

where

$$B^2(a) = \mathbf{E}\Big(g(x_0,\alpha)^2 + 2\sum_{k=1}^{\infty} g(x_0,\alpha)g(x_k,\alpha)\Big),$$

and $P\{x_0 \in A\} = \pi(A), A \in \mathcal{B}_X$.

Similar results can be proved for the system $M_{SM,Q}/M_{SM,Q}/1/\infty$ where the input and service rates depend on the state of some external SMP and the value of the queue, for the system $SM_Q/M_{SM,Q}/1/\infty$ where an input forms some process of a semi-Markov type with occupation times depending on the value of the queue, some systems with unreliable servers, polling systems and retrial models. More examples can be found in the author's book [20].

6 Queueing Networks

6.1 Markov Networks

Consider a state-dependent queueing network $(M_Q/M_Q/1/\infty)^r$ consisting of r nodes with one server in each node and infinitely many places for waiting. Denote by $Q_n(i,t)$ the number of calls in the i-th node at time t and let $\overline{Q}_n(t) = (Q_n(i,t), i=\overline{1,r})$ be the column vector.

We assume that the time goes to infinity in the scale nt and consider the AP and DA for the normalized vector-valued queueing process $\overline{Q}_n(nt)$. Let the functions $\{\lambda_i(\overline{q}), \mu_i(\overline{q}), p_{ij}(\overline{q}), i=\overline{1,r}, j=\overline{0,r}, \overline{q} \in [0,\infty)^r\}$ be given. The network is operating in the following way. If at some time-point u, $\overline{Q}_n(u) = \overline{Q}$, then the local input rate in the i-th node is $\lambda_i(\overline{Q}/n)$ and the local service rate is $\mu_i(\overline{Q}/n)$. If at this time a call has completed service in node i, then either with probability $p_{ij}(\overline{Q}/n)$ it goes from i-th to j-th node, $j=\overline{1,r}$ or with probability $p_{i0}(\overline{Q}/n)$ it leaves the network. This network belongs to Jackson's type networks.

Let $\overline{\lambda}(\overline{q}) = (\lambda_1(\overline{q}),...,\lambda_r(\overline{q}))$, $\overline{\mu}(\overline{q}) = (\mu_1(\overline{q}),...,\mu_r(\overline{q}))$ be the column vector-valued functions,

$$P(\overline{q}) = ||p_{ij}(\overline{q})||_{i,j=\overline{1,r}}, \quad a(\overline{q}) = \sum_{i=1}^{r}(\lambda_i(\overline{q}) + \mu_i(\overline{q})),$$

and for a vector-valued function $\overline{f}(\overline{q}) = (f_1(\overline{q}),...f_r(\overline{q}))$ with $\overline{q} = (q_1,...,q_r)$ we denote by $f'(\overline{q})$ the matrix derivative: $f'(\overline{q}) = ||\partial f_i(\overline{q})/\partial q_j||_{i,j=\overline{1,r}}$.

Theorem 8. *Assume that as $n \to \infty$,*

$$n^{-1}\overline{Q}_n(0) \xrightarrow{\mathrm{P}} \overline{s}_0, \tag{63}$$

the functions $\overline{\lambda}(\overline{q}), \overline{\mu}(\overline{q}), P(\overline{q})$, satisfy a local Lipschitz condition, and for some fixed $T > 0$ there exists $A > 0$ such that the system of differential equations

$$d\overline{\eta}(t) = (\overline{\lambda}(\overline{\eta}(t)) + (P^*(\overline{\eta}(t)) - I)\overline{\mu}(\overline{\eta}(t)))a(\overline{\eta}(t))^{-1}dt, \quad \overline{\eta}(0) = \overline{s}_0,$$

has a unique solution $\overline{\eta}(t)$ such that $\overline{\eta}(t) > 0$ in each component, $t \in (0, A)$, and $\int_0^A a(\overline{\eta}(t))^{-1} dt > T$.
Then
$$\sup_{0 \le t \le T} |n^{-1} \overline{Q}_n(nt) - \overline{s}(t)| \xrightarrow{P} 0, \qquad (64)$$

where the vector-valued function $\overline{s}(t)$ satisfies the equation:

$$d\overline{s}(t) = (\overline{\lambda}(\overline{s}(t)) + (P^*(\overline{s}(t)) - I)\overline{\mu}(s(t)))dt, \quad \overline{s}(0) = \overline{s}_0. \qquad (65)$$

If in addition
$$n^{-1/2}(\overline{Q}_n(0) - n\overline{s}_0) \xrightarrow{W} \overline{\gamma}_0, \qquad (66)$$

and the functions $\overline{\lambda}(\overline{q}), \overline{\mu}(\overline{q}), P(\overline{q})$ are continuously differentiable, then the sequence of processes $\overline{\gamma}_n(t) = n^{-1/2}(\overline{Q}_n(nt) - n\overline{s}(t))$ J-converges in the interval $[0,T]$ to a multidimensional diffusion process $\overline{\gamma}(t)$ satisfying the following stochastic differential equation

$$d\overline{\gamma}(t) = G(\overline{s}(t))\overline{\gamma}(t)dt + B(\overline{s}(t))d\overline{w}(t), \quad \overline{\gamma}(0) = \overline{\gamma}_0. \qquad (67)$$

Here

$$G(\overline{q}) = (\lambda(\overline{q}) + ((P^*(\overline{q}) - I)\mu(\overline{q}))', \quad B(\overline{q})^2 = ||b_{ij}(\overline{q})||_{i,j=\overline{1,r}},$$
$$b_{ij}(\overline{q}) = -\mu_i(\overline{q})p_{ij}(\overline{q}) - \mu_j(\overline{q})p_{ji}(\overline{q}), i \ne j,$$
$$b_{ii}(\overline{q}) = -2\mu_i(\overline{q})p_{ii}(\overline{q}) + \lambda_i(\overline{q}) + \mu_i(\overline{q}) + \sum_k \mu_k(\overline{q})p_{ki}(\overline{q}), \ i = \overline{1,r},$$

and P^* is a transposed matrix.

6.2 Non-Markov Queueing Networks

Consider now fluid limit (AP) and DA for some classes of non-Markov networks. The method of analysis is based on the representation of a queueing process as a corresponding SP by choosing in the appropriate way switching times and constructing corresponding processes in switching intervals.

Network $(M_{SM,Q}/M_{SM,Q}/1/\infty)^r$ with semi-Markov switching. Suppose that characteristics of the network depend on parameter n in the following way. A SMP $x(t)$ and other variables introduced below do not depend on n. But if at the time t, $x(t) = x$ and $\overline{Q}_n(t)/n = \overline{q}$, then the local arrival and service rates and transition probabilities as well as random sizes of batches $\eta(x,\overline{q})$, $\kappa_i(x,\overline{q})$ depend on the pair $(x,\overline{q})$. Denote by $t_1 < t_2 < \ldots$ the times of sequential jumps of $x(t)$. Suppose that the imbedded MP $x_k = x(t_k)$, $k \ge 0$, is ergodic with stationary distribution π_x, $x \in X = \{1, 2, .., d\}$. Let $Q_n^{(i)}(t)$ be the size of the queue in node i at time t, and $\overline{Q}_{n0}$ be the initial value.

Denote $\overline{Q}_n(t) = (Q_n^{(1)}(t), ..., Q_n^{(r)}(t))$, $t \geq 0$, and for any $x \in X, i = \overline{1,r}, \overline{q} \in \mathcal{R}^r$, introduce the following variables:

$$m(x) = \mathbf{E}\tau(x), \quad P_0(x,\overline{q}) = ||p_{ij}(x,\overline{q})||_{i,j=\overline{1,r}}, \quad \overline{a}(x,\overline{q}) = \mathbf{E}\overline{\eta}(x,\overline{q}),$$

$$g_i(x,\overline{q}) = \mathbf{E}\kappa_i(x,\overline{q}), \quad \overline{g}(x,\overline{q}) = (\mu_1(x,\overline{q})g_1(x,\overline{q}), ..., \mu_r(x,\overline{q})g_r(x,\overline{q})),$$

$$m = \sum_{x \in X} m(x)\pi_x, \quad \overline{c}(x,\overline{q}) = \lambda(x,\overline{q})\overline{a}(x,\overline{q}) + (P_0(x,\overline{q})^* - I)\overline{g}(x,\overline{q}),$$

$$\overline{b}(\overline{q}) = \sum_{x \in X} m(x)\overline{c}(x,\overline{q})\pi_x, \quad d^2(x) = \mathbf{Var}\,\tau(x), \quad d_i^2(x,\overline{q}) = \mathbf{E}\kappa_i^2(x,\overline{q}),$$

$$J^2(x,\overline{q}) = \lambda(x,\overline{q})\mathbf{E}\eta(x,\overline{q})\eta(x,\overline{q})^*.$$

Let $F^2(x,\overline{q}) = ||f_{ij}(x,\overline{q})||_{i,j=\overline{1,r}}$ be the matrix with the following entries:

$$f_{ij}(x,\overline{q}) = -\mu_i(x,\overline{q})p_{ij}(x,\overline{q})d_i^2(x,\overline{q}) - \mu_j(x,\overline{q})p_{ji}(x,\overline{q})d_j^2(x,\overline{q}), i,j = \overline{1,r}, i \neq j;$$

$$f_{ii}(x,\overline{q}) = \mu_i(x,\overline{q})(1 - 2p_{ii}(x,\overline{q}))d_i^2(x,\overline{q}) + \sum_{k=1}^{r} \mu_k(x,\overline{q})p_{ki}(x,\overline{q})d_k^2(x,\overline{q}).$$

Denote

$$D^2(x,\overline{q}) = d^2(x)\Big(\overline{c}(x,\overline{q}) - m^{-1}\overline{b}(\overline{q})\Big)\Big(\overline{c}(x,\overline{q}) - m^{-1}\overline{b}(\overline{q})\Big)^*$$
$$+ m(x)\Big(F^2(x,\overline{q}) + J^2(x,\overline{q})\Big),$$

$$D^2(\overline{q}) = \sum_{x \in X} D^2(x,\overline{q})\pi_x, \tag{68}$$

$$\overline{\gamma}(x,\overline{q}) = m(x)(\overline{c}(x,\overline{q}) - m^{-1}\overline{b}(\overline{q})).$$

Let the matrix $B^2(\overline{q})$ be calculated using the variables $\overline{\gamma}(x,\overline{q})$ with the help of MP x_k according to (52), (53) in Theorem 4. We put $H^2(\overline{q}) = D^2(\overline{q}) + B^2(\overline{q})$. Define $H(\overline{q})$ according to the relation $H(\overline{q})H(\overline{q})^* = H^2(\overline{q})$. Let $\overline{s}(t)$ be a solution of the equation

$$d\overline{s}(t) = m^{-1}\overline{b}(\overline{s}(t))dt, \quad \overline{s}(0) = \overline{s}_0. \tag{69}$$

Theorem 9. *1) Assume that the functions* $\lambda(x,\overline{q}), \mu_i(x,\overline{q}), \overline{a}(x,\overline{q}), g_i(x,\overline{q})$, $p_{ij}(x,\overline{q})$ *for any* $x \in X, i = \overline{1,r}, j = \overline{1,r+1}$, *are locally Lipschitz with respect to* $\overline{q} \in int\{\mathcal{R}_+^m\}$, *and* $\mathbf{E}\tau(x)^2 < \infty, x \in X$. *Let also* $m > 0$, *for any bounded and closed domain* $G \in int\{\mathcal{R}_+^m\}$,

$$\mathbf{E}\kappa_i(x,\overline{q})^2 \leq C_G, \quad \mathbf{E}|\eta(x,\overline{q})|^2 \leq C_G, \quad i = \overline{1,r}, x \in X, \overline{q} \in G, \tag{70}$$

where $C_G < \infty$, *the function* $\overline{b}(\overline{q})$ *has no more than linear growth,* $n^{-1}\overline{Q}_n(0) \xrightarrow{P} \overline{s}_0 > \overline{0}$, *and there exists* $T > 0$ *such that* $\overline{s}(t) > \overline{0}$, $t \in [0,T]$, *in each component.*
Then

$$\sup_{0 \leq t \leq T} |n^{-1}\overline{Q}_n(nt) - \overline{s}(t)| \xrightarrow{P} 0. \tag{71}$$

2) Assume in addition that there exists a continuous matrix derivative $G(\overline{q}) = \overline{b}'(\overline{q})$, $\overline{q} \in int\{\mathcal{R}_+^m\}$, $\mathbf{E}\tau(x)^3 < \infty$, $x \in X$, and for any bounded and closed domain $G \in int\{\mathcal{R}_+^m\}$,

$$\mathbf{E}\kappa_i(x,\overline{q})^3 \leq C_G, \ \mathbf{E}|\eta(x,\overline{q})|^3 \leq C_G, \ i = \overline{1,r}, x \in X, \overline{q} \in G. \tag{72}$$

Let also
$$n^{-1/2}(\overline{Q}_n(0) - n\overline{s}(0)) \stackrel{W}{\Rightarrow} \overline{\gamma}_0, \tag{73}$$

and the function $H^2(\overline{q})$ is continuous.

Then the sequence $\overline{\gamma}_n(t) = n^{-1/2}(\overline{Q}_n(nt) - n\overline{s}(t))$ J-converges in D_T^r to the diffusion process $\overline{\gamma}(t)$:

$$d\overline{\gamma}(t) = G(\overline{s}(t))\overline{\gamma}(t)dt + m^{-1/2}H(\overline{s}(t))d\overline{w}(t), \ \overline{\gamma}(0) = \overline{\gamma}_0. \tag{74}$$

The proof is given in the book [20].

7 Aggregation in Markov Models with Fast Switching

In this section we consider the approximation of Markov type queueing models with fast Markov switching by Markov models with averaged transition rates. Consider first some general setting. Let $(x_n(t), \zeta_n(t))$ be a two-component MP. The following result is proved in [18,20].

Assume that a component $x_n(\cdot)$ has fast switching and satisfy the asymptotic mixing condition. Assume also that the transition rates of $\zeta_n(t)$ depend on the component $x_n(\cdot)$ and the process $\zeta_n(t)$ has rather slow transitions, that means, the number of transitions in any finite interval is bounded in probability.

Then under rather general conditions the component $\zeta_n(\cdot)$ J-converges in Skorokhod space to a MP with transition rates averaged by some quasi-stationary measures constructed by $x_n(\cdot)$. The convergence of a stationary distribution of $(x_n(\cdot), \zeta_n(\cdot))$ is studied as well.

Example 1: $\zeta_n(t)$ is a MP with local rates $\lambda(i, j, x_n(t))$, where $x_n(t)$ is a fast Markov environment.

Example 2: $\zeta_n(t)$ is a quasi-Birth-and-Death process, that has fast transitions in each level and slow transitions between levels.

Consider now as an example an approximation of a queueing model with fast Markov switching by a simpler queueing model with averaged transition rates.

7.1 System $M_{M,Q}/M_{M,Q}/1/N$ with Fast Markov Switching

Consider a queueing system in a fast Markov environment. A system consists of one server and N waiting places. Assume that the calls arrive according to a state-dependent Poisson process with Markov switching and also at the times of jumps of a switching MP. If the system is full, an arriving call is lost. Let $x_0(t), t \geq 0$, be an ergodic MP with values in $X = \{x_1, ..., x_r\}$. Denote by $\rho(x), x \in X$, its stationary distribution. Let $b(x)$ be the exit rate from state

$x, x \in X$. We define a Markov environment with fast switching as follows: $x_n(t) = x_0(V_n t), t \geq 0$, where V_n is some scaling factor, $V_n \to \infty$.

Denote by $\varphi_0(\cdot)$ and $\varphi_n(\cdot)$ uniformly strong mixing coefficients for processes $x_0(\cdot)$ and $x_n(\cdot)$, respectively. According to ergodicity of $x_0(\cdot)$, there exist $q < 1$ and $L > 0$ such that $\varphi_0(L) \leq q$. Then $\varphi_n(L/V_n) = \varphi_0(L) \leq q$, and therefore $x_n(\cdot)$ is asymptotically mixing in any fixed interval.

Let $\{\lambda(x,i), \mu(x,i), \alpha_A(x,i), \alpha_S(x,i), x \in X, i \geq 0\}$ be the non-negative functions. Denote by $Q_n(t)), t \geq 0$, the total number of calls in the system at time t. The system is switched by the process $(x_n(t), Q_n(t))$ as follows: if $(x_n(t), Q_n(t)) = (x,i)$, then the local arrival rate is $\lambda(x,i)$, and the local service rate is $\mu(x,i)$. Moreover, if at the time t_{nk} of k-th jump of $x_n(t)$, $(x_n(t_{nk}-0), Q_n(t_{nk}-0)) = (x,i)$, then either an additional call may enter the system with probability $V_n^{-1}\alpha_A(x,i)$, or a call on service may complete service with probability $V_n^{-1}\alpha_S(x,i)$ (no changes in the system with probability $1 - V_n^{-1}(\alpha_A(x,i) + \alpha_S(x,i))$. Put

$$\widehat{\lambda}(i) = \sum_{x \in X} \lambda(x,i)\rho(x), \quad \widehat{\mu}(i) = \sum_{x \in X} \mu(x,i)\rho(x),$$

$$\widehat{\alpha}_A(i) = \sum_{x \in X} \alpha_A(x,i)b(x)\rho(x), \quad \widehat{\alpha}_S(i) = \sum_{x \in X} \alpha_S(x,i)b(x)\rho(x), \quad (75)$$

$$A(i) = \widehat{\lambda}(i) + \widehat{\alpha}_A(i), \quad \Gamma(i) = \widehat{\mu}(i) + \widehat{\alpha}_S(i), \quad i = 0,..,N+1,$$

where we set $\Gamma(0) = 0, A(N+1) = 0$. Let $M_Q/M_Q/1/N$ be an averaged state-dependent queueing system (system with averaged rates and probabilities) operating as follows: as $Q(t) = i$, the arrival rate is $A(i)$ and the service rate is $\Gamma(i)$, where $Q(t)$ is a number of calls in the system at time t.

Theorem 10. *If MP $Q(\cdot)$ is regular and $Q_n(0) = q_0$, then for any $N \leq \infty$, $Q_n(\cdot)$ J-converges in each finite interval $[0,T]$ to $Q(\cdot)$, where $Q(0) = q_0$.*

The approximation of a stationary distribution can be proved as well. Denote by $\rho_n(x,i), x \in X, i = 0,..,N+1$, the stationary distribution (if it exists) of the process $(x_n(t), Q_n(t))$. Assume that $N < \infty$ and keep notation (75).

Theorem 11. *Suppose that $A(i) > 0, \Gamma(i) > 0, \lambda(x,i) + \mu(x,i) > 0, x \in X, i = 0,..,N+1$. Then $Q(\cdot)$ is ergodic, $\rho_n(x,i)$ exists at large n and as $n \to \infty$,*

$$\rho_n(x,i) \to \rho(x)\Pi(i), \quad x \in X, \quad i = 0,..,N+1,$$

where $\Pi(i), i = 0,..,N+1$, is the stationary distribution of $Q(\cdot)$.

Similar result holds when $N = \infty$.

Analogous results can be proved for batch systems $BM_{M,Q}/BM_{M,Q}/1/N$, priority models, models with unreliable servers with slow failures/repairs, models with fast semi-Markov switching [18,20].

Another interesting direction of the investigation of transient phenomena is the class of so-called retrial queues and queues with negative arrivals. Some results in this direction are obtained in [12,14,16,20].

7.2 Aggregation in Heavy Traffic Conditions

A new direction of the investigation is the class of queueing models in heavy traffic conditions with fast switching, where the switching component itself allows an asymptotic aggregation of states. A more formal setting is the following. Consider a switching queueing model which is described by a two component process $(x_n(t), Q_n(t))$, where $x_n(t)$ is a MP with state space satisfying the conditions of the asymptotic aggregation of state space [1,3,4] (see also [20]). This means that the state space can be subdivided on the aggregated regions: fast transitions within each region and slow transitions between regions. The transition rates of the queueing process $Q_n(t)$ may depend on the states of $(x_n(t), Q_n(t))$. Assume also that for the process $Q_n(t)$ the number of transitions in any finite interval is going in probability to infinity and the values of $Q_n(t)$ are also asymptotically unbounded, e.g. $Q_n(t)$ is the queue in heavy traffic conditions.

Basing on the results of Section 5 we can see that in the case when the component $x_n(t)$ satisfies an asymptotic mixing condition, an AP and DA type theorems for $Q_n(t)$ hold. However, in this case the process $x_n(t)$ satisfy the conditions of the asymptotic aggregation of states and does not satisfies asymptotic mixing conditions.

Thus, we can expect that under rather general conditions the process $\widehat{x}_n(t)$ weakly converges to $y(t)$, where $\widehat{x}_n(t)$ is an aggregated process (the state space is the space of regions) and $y(t)$ is a MP with averaged transition rates [1,3,4,20]. Correspondingly, under heavy-traffic conditions, the process $Q_n(t)/n$ in each region converges to a solution of differential equation.

Therefore, the two-component process $(\widehat{x}_n(t), Q_n(t)/n)$ should converge to a MP $(y(t), Q_0(t))$, where $Q_0(t)$ satisfied a differential equation with Markov switching of the form $dQ_0(t) = A(y(t), Q_0(t))dt$, and $A(i, q)$ are the aggregated rates in each region. The results of this type are still not fully investigated. Some results in this direction are published in [13,33].

8 Conclusions

The asymptotic results described in Sections 4-6 allow to approximate the behaviour of queueing processes for rather complicated switching queueing systems and networks under heavy traffic conditions by much simpler processes, in particular, by solutions of differential equations or by diffusion processes. Section 7 deals with the approximation of complex queueing models by simpler ones with the rates averaged by some stationary measures.

These results are also valid for the approximation of various functionals defined on the trajectory of the queuing process, e.g., the cumulative reward or losses, time to reach a critical region, time spend by a queueing process in a given region, different stationary characteristics, etc. In particular, various performance measures of operating of complex queueing models can be approximated by corresponding measures of much simpler processes.

Therefore, these results provide us with the new analytic technique for the performance evaluation of complex queueing systems and networks.

Acknowledgments. The author is thankful to Professor Demetres Kouvatsos for a long term fruitful collaboration.

References

1. Anisimov, V.V.: Asymptotic Consolidation of the States of Random Processes. Cybernetics 9(3), 494–504 (1973)
2. Anisimov, V.V.: Switching Processes. Cybernetics 13(4), 590–595 (1977)
3. Anisimov, V.V.: Limit Theorems for Switching Processes and their Applications. Cybernetics 14(6), 917–929 (1978)
4. Anisimov, V.V.: Random Processes with Discrete Component. Limit Theorems. Publ. Kiev Univ. (1988) (Russian)
5. Anisimov, V.V.: Limit Theorems for Switching Processes. Theory Probab. and Math. Statist. 37, 1–5 (1988)
6. Anisimov, V.V., Aliev, A.O.: Limit Theorems for Recurrent Processes of Semi-Markov Type. Theory Probab. and Math. Statist. 41, 7–13 (1990)
7. Anisimov, V.V.: Averaging Principle for Switching Processes. Theory Probab. and Math. Statist. 46, 1–10 (1992)
8. Anisimov, V.V.: Averaging Principle for the Processes with Fast Switching. Random Oper. & Stoch. Eqv. 1(2), 151–160 (1993)
9. Anisimov, V.V.: Limit Theorems for Processes with Semi-Markov Switching and their Applications. Random Oper. & Stoch. Eqv. 2(4), 333–352 (1994)
10. Anisimov, V.V.: Switching Processes: Averaging Principle, Diffusion Approximation and Applications. Acta Applicandae Mathematicae 40, 95–141 (1995)
11. Anisimov, V.V.: Asymptotic Analysis of Switching Queueing Systems in Conditions of Low and Heavy Loading. In: Alfa, A.S., Chakravarthy, S.R. (eds.) Matrix-Analytic Methods in Stochastic Models. Lect. Notes in Pure and Appl. Mathem. Series, vol. 183, pp. 241–260. Marcel Dekker, Inc., New York (1996)
12. Anisimov, V.V.: Averaging Methods for Transient Regimes in Overloading Retrial Queuing Systems. Mathematical and Computing Modelling 30(3/4), 65–78 (1999)
13. Anisimov, V.V.: Diffusion Approximation for Processes with Semi-Markov Switches and Applications in Queueing Models. In: Janssen, J., Limnios, N. (eds.) Semi-Markov Models and Applications, pp. 77–101. Kluwer Academic Publishers, Dordrecht (1999)
14. Anisimov, V.V.: Switching Stochastic Models and Applications in Retrial Queues. Top 7(2), 169–186 (1999)
15. Anisimov, V.V.: Asymptotic Analysis of Reliability for Switching Systems in Light and Heavy Traffic Conditions. In: Limnios, N., Nikulin, M. (eds.) Recent Advances in Reliability Theory: Methodology, Practice and Inference, pp. 119–133. Birkhäuser, Boston (2000)
16. Anisimov, V.V., Artalejo, J.R.: Analysis of Markov Multiserver Retrial Queues with Negative Arrivals. Queueing Systems 39(2/3), 157–182 (2001)
17. Anisimov, V.V.: Diffusion Approximation in Overloaded Switching Queueing Models. Queueing Systems 40(2), 141–180 (2002)
18. Anisimov, V.V.: Averaging in Markov Models with Fast Markov Switches and Applications to Queueing Models. Annals of Operations Research 112(1), 63–82 (2002)
19. Anisimov, V.V.: Averaging in Markov Models with Fast Semi-Markov Switches and Applications. Communications in Statistics - Theory and Methods 33(3), 517–531 (2004)

20. Anisimov, V.V.: Switching Processes in Queueing Models. Wiley & Sons, ISTE, London (2008)
21. Anisimov, V.V., Zakusilo, O.K., Dontchenko, V.S.: The Elements of Queueing Theory and Asymptotic Analysis of Systems. Publ. "Visca Scola", Kiev (1987) (in Russian)
22. Anisimov, V.V., Lebedev, E.A.: Stochastic Queueing Networks. Markov Models. Kiev Univ., Kiev (1992) (in Russian)
23. Buslenko, N.P., Kalashnikov, V.V., Kovalenko, I.N.: Lectures on the Theory of Complex Systems. Sov. Radio, Moscow (1973) (in Russian)
24. Ežov, I.I., Skorokhod, A.V.: Markov Processes which are Homogeneous in the Second Component. Theor. Probab. Appl. 14, 679–692 (1969)
25. Ethier, S.N., Kurtz, T.G.: Markov Processes, Characterization and Convergence. J. Wiley & Sons, New York (1986)
26. Gikhman, I.I., Skorokhod, A.V.: Theory of Random Processes, vol. II. Springer, Heidelberg (1975)
27. Griego, R., Hersh, R.: Random Evolutions, Markov Chains, and Systems of Partial Differential Equations. Proc. Nat. Acad. Sci. 62, 305–308 (1969)
28. Hersh, R.: Random Evolutions: Survey of Results and Problems. Rocky Mount. J. Math. 4(3), 443–475 (1974)
29. Korolyuk, V.S., Swishchuk, A.V.: Random Evolutions. Kluwer Acad. Publ., Dordrecht (1994)
30. Korolyuk, V.S., Korolyuk, V.V.: Stochastic Models of Systems. Kluwer, Dordrecht (1999)
31. Korolyuk, V.S., Limnios, N.: Evolutionary Systems in an Asymptotic Split State Space. In: Limnios, N., Nikulin, M. (eds.) Recent Advances in Reliability Theory: Methodology, Practice and Inference, pp. 145–161. Birkhäuser, Boston (2000)
32. Korolyuk, V.S., Limnios, N.: Average and Diffusion Approximation for Evolutionary Systems in an Asymptotic Split Phase State. Ann. Appl. Prob. 14(1), 489–516 (2004)
33. Korolyuk, V.S., Limnios, N.: Stochastic Systems in Merging Phase Space. World Scientific, Singapore (2005)
34. Korolyuk, V.S., Korolyuk, V.V., Limnios, N.: Queueing Systems with Semi-Markov Flow in Average and Diffusion Approximation Schemes. Methodol. Comput. Appl. Probab. 11, 201–209 (2009)
35. Neuts, M.: Structured Stochastic Matrices of M/G/1 Type and Their Applications. Marcel Dekker, New York (1989)
36. Skorokhod, A.V.: Limit Theorems for Random Processes. Theory Probab. Appl. 1, 289–319 (1956)

Principles of Fairness Quantification in Queueing Systems

Hanoch Levy[1], Benjamin Avi-Itzhak[2], and David Raz[3]

[1] School of Computer Science, Tel-Aviv University, Tel-Aviv, Israel
hanoch@post.tau.ac.il
[2] RUTCOR, Rutgers University, New Brunswick, NJ, USA
aviitzha@business.rutgers.edu
[3] HIT, Holon Institute of Technology, Holon, Israel
davidra@hit.ac.il

Abstract. Queues serve as a major scheduling device in computer networks, both at the network level and at the application level. A fundamental and important property of a queue service discipline is its fairness. Recent empirical studies show fairness in queues to be highly important to queueing customers in practical scenarios. The objective of this tutorial is to discuss the issue of queue fairness and its dilemmas, and to review the research conducted on this subject. We discuss the fundamental principles related to queue fairness in the perspective of the relevant applications, with some emphasis on computer communications networks. This is conducted in the context of the recent research in this area and the queueing related fairness measures which have been proposed in recent years. We describe, discuss and compare their properties, and evaluate their relevance to the various practical applications.

Keywords: Queueing, Fairness.

1 Introduction

Queues serve as a major building block in computer networks and are used to schedule and prioritize tasks both at the network level and at the application level. With the advances of the Internet more and more services move from the "physical world" into the "network controlled" world and require the use of computer and communications controlled queues. Examples include file servers used for the download of music, video, games and other applications, and call-centers.

Why do we use queues in these applications as well as in other real life applications, such as banks, supermarkets, airports, computer systems, Web services and numerous other systems? What purpose do ordered-queues serve? Perhaps the major reason for using a queue at all is to provide *fair service* to the customers. Furthermore, experimental psychology studies show that fair scheduling in queueing systems is indeed highly important to humans. Nonetheless, *Queueing Theory*, the theory that deals with analyzing queues and their efficient operation, has hardly dealt with the questions of what is a fair queue and *how fair* is a queueing policy.

The fairness factor associated to waiting in queues has been recognized in many works and applications. Larson in his discussion paper on the disutility of waiting,[20] recognizes the central role played by 'Social Justice', (which is another name for fairness), and its perception by customers. This is also addressed in Rothkopf and Rech in their paper discussing perceptions in queues.[30] Aspects of fairness in queues were discussed earlier by quite a number of authors: Palm [22] deals with judging the annoyance caused by congestion, Mann [21] discusses the queue as a social system and Whitt [32] addresses overtaking in queues, to mention just three.

Empirical evidence of the importance of fairness of queues was recently provided in Rafaeli et. al. ([26,27]) who studied, using an experimental psychology approach, the reaction of humans to waiting in queues and to various queueing and scheduling policies. The studies revealed that for humans waiting in queues the issue of fairness is highly important, perhaps some times more important than the duration of the wait.

This tutorial aims at addressing the subject of queue fairness, discuss the issues and dilemmas related to it, present the fundamental underlying assumptions and tie them to the real-life applications. Since we deal with the introduction of a new measure for a quantity that is somewhat abstract and not very tangible, several questions should be brought up and discussed. What is the *physical entity*, or *performance objective* that should be dealt with? At what *level of detail* should the system be measured? What are the *physical properties* that affect the measure? How *intuitive* and *appealing* is the measure? And, how does the *measure relate* to the *relevant applications*? These questions are discussed and examined in this tutorial in the context of a few fairness measures proposed recently in the literature. We find it constructive to present many of the examples used in this tutorial in the context of "physical queues" (such as a supermarket queue).

To start the discussion one may ask herself what would be a fair service order in a call center or in a supermarket queue? Most people would instinctively respond that First-Come-First-Served (FCFS) is the fairest order, that is, serving jobs in increasing order of seniority is most fair. In fact, Kingman [17] pronounces this in viewing FIFO (First-In-First-Out) as the 'fairest' queue discipline. This brings up the first factor playing a role in queue scheduling fairness, namely, *queue seniority*.

Aiming at understanding the problem better, one may pose the following more elaborate scenario, which some readers may associate with their own personal experience. Mr. Short arrives at the supermarket counter holding only one item. Waiting at the queue he finds ahead of him Mrs. Long carrying a fully loaded cart of items. Would it be fair to have Mr. Long served ahead of Short and Short waiting for the full processing of Mrs. Long's loaded cart? Or, would it be more fair to advance Short in the queue and serve him ahead of Long?

This dilemma may cause some to "relax" their strong belief in the absolute fairness of FCFS. In fact, the dilemma brings to the discussion a new physical factor, that of *service requirement*. The basic intuition thus suggests that prioritizing short jobs over long jobs may also be fair. It is the trade-off between these two physical factors, *seniority* (prioritize Mrs. Long) and *service requirement* (prioritize Mr. Short), that creates the dilemma in this case. This tradeoff, as well as the "Long vs. Short" scenario, will accompany us in this paper in attempting to understand fairness in queues.

What bothers one in a queue? What is the *performance objective* one aims for while staying in a queue? Understanding this issue should form the basis for proposing a proper fairness measure, since the measure must be built around the *performance objective* of interest to the queue customers. This question is discussed in Section 3 and Section 4, following the presentation of the model in Section 2 First, in Section 3, we note that this article focuses on *job-based* systems. We describe what job based systems (as opposed to *flow-based* systems) are, and review the real-life applications that are associated with these models For the sake of completeness, we note also that significant literature (in the context of fairness) has been devoted to *flow-based* systems. In Section 4 we briefly review this subject and some of the literature that dealt with it.

Next, in Section 5, we discuss the performance objectives associated with job-driven systems. We claim that in addition to the natural candidate, namely that of *waiting times* (*delays*), one may have to consider, in some cases, the performance objective of service (that is whether a service is granted to the job or not) which might be important alongside the waiting times. It should be noted that queueing theory has been mainly occupied with the performance metrics of waiting time and dealt less with the metrics of service (see e.g. text books on the subject, [7,8,12,13,18,19]).

Having dealt with the performance objective, we then (in Section 6) ask at what granularity level the performance metrics should be dealt with. Our conclusion is that it is desirable to deal with the performance metrics at three granularity levels, *individual discrimination, scenario fairness* and *system fairness*. The addressing of fairness at all three levels is similar to the addressing of the waiting time measure that can also be evaluated at these three levels. More importantly, this allows theoreticians and practitioners, as well as queue users, to develop good feel and intuition towards the measure, which will assist in getting used to the measure and to using it.

The physical properties which are at the heart of queue scheduling in general and fairness in particular, namely *seniority* and *service requirement* are posed and discussed in Section 7. These are best illuminated via the "Short vs. Long" example presented above. The tradeoff between serving Short first and serving Long first reflects the tradeoff between seniority and service requirement. This translates into the modeling dilemma (and difficulty) of how to account for both seniority and service requirement in quantifying fairness, a dilemma that seems to be in the heart of fairness quantification models.

Having discussed what fairness measures should quantify, and the seniority and service requirement factors, we then (Section 8) review the approaches for quantifying fairness in (job based) queueing systems. Analytic treatment and quantification of queue fairness have been quite limited in the literature, and been addressed only very recently. Three references that propose measures or a criterion for fairness of queues are Avi-Itzhak and Levy in [1], Raz, Levy and Avi-Itzhak in [28], who propose measures, and Wierman and Harchol-Balter in [33], who propose a criterion. The modeling dilemma of seniority versus service requirement seems to be at the heart of these queue fairness modeling attempts: The approach proposed in [1] centralizes on the *seniority* factor. In contrast, the approach proposed by [33], focuses on the *service-requirement* factor. Lastly, in attempting to give *seniority and service requirement* even treatment, the approach of [28] focuses on neither of them and chooses to focus on a third factor, that of *resource allocation*.

Lastly (Section 9), after discussing the properties of the proposed fairness measures, we focus on the real-life applications and examine how (and whether) the various queueing fairness measures apply to the various applications. This issue, too, seems to strongly tie to the *seniority* versus *service-requirement* dilemma: The selection of a proper quantification approach to an application depends on how strong are the roles *seniority and service requirement* play in the application.

2 Model

We consider a general queueing system consisting of a single server (in some cases we will consider multiple servers). Jobs, denoted $J_1, J_2, \ldots$ arrive at the system at arbitrary arrival epochs, denoted $a_1, a_2, dots$, respectively. In many queueing models, jobs are associated in a one-to-one manner with customers (to be denoted $C_1, C_2, \ldots$). For convenience of notation we assume that $a_i \leq a_{i+1}$. Job J_i requests some service at the server, the amount of which we denote by s_i; For simplicity, we will measure the service requirement in units of time. The server grants service to the customers according to some scheduling policy. Once J_i receives its full amount of service s_i (which does not have to be given continuously or at full rate) it leaves the system, and the epoch when it leaves is called its *departure epoch*, denoted d_i. The duration J_i stays in the system is called *system time* and is denoted $t_i = d_i - a_i$. The duration J_i waits and does not get service is the *waiting time* of J_i, denoted by w_i and is given by $w_i = t_i - s_i = (d_i - a_i) - s_i$, except for processor sharing disciplines, where this conventional definition of *waiting time is not applicable*. These notations a_i, s_i, d_i, t_i, w_i are used to denote the actual values, attributed to J_i in a specific sample path of the system. The same letters capitalized are used to denote the corresponding random variables.

3 Job-Based Systems, Flow-Based Systems and Their Corresponding Applications

Queueing model applications can be classified into 1) *Flow-based systems*, and 2) *Job-based systems*. In the former, customer C_i is associated with a stream (or flow) of jobs $J_1^i, J_2^i, \ldots$ arriving at epochs $a_1^i, a_2^i, \ldots$ respectively. Of interest is the performance experienced by the whole flow. In the latter, each customer, say C_i, is associated with a single job J_i. Of interest is therefore the performance experienced by that job.

The applications associated with flow-based systems are:

1. **Network level communications network devices**, such as routers or load balancers. These devices normally do not view the applications since they operate at a lower layer. Rather, they view flows of packets where each flow can be associated with a customer (or a group of customers), typically via the session details appearing on the packet...

The applications associated with job-based systems are:

1. **Application level communications network devices**, such as Web or FTP servers. These devices are concerned with the application (and have a control on its scheduling) and thus may view the whole application (e.g. a file or a Web page) as a single job. This view is due to the fact that from the customer point of view the objective is the completion of the whole job.
2. **Computer systems**, in which a customer (or a customer's computer application) submits a job to the system and the customer gets satisfied when the service of the job is completed.
3. **Call centers**, in which customers call into a call center to receive service, possibly wait in a virtual queue (while listening to some music) until being answered by "the next available agent". Call center queueing systems are conceptually identical to physical queueing facilities, such as banks, except that the service is done over the phone, the customer waits on the telephone line (instead of physically waiting in a physical line). Other differences are that the system's operator has much more control over the scheduling process and that, unless special technology is applied, other customers are not seen by the customer.
4. **"Physical queueing systems"**, like banks, supermarkets, public offices and the like, in which customers physically enter a queue where they wait for their service and then get served. In the supermarket, for instance, the job is the processing of the whole customer's cart including the payment process. Once being served the customer leaves the system.

4 A Short Review of Flow-Based Fairness

Much work has been conducted in the context of communications networks where the concern is with *flows* traversing a communications node and in allocating the *bandwidth fairly* among the *flows*. As the focus of our work is on job-based fairness, we only briefly review the literature on flow-based fairness.

In flow-based systems each customer is associated with a flow of objects (normally packets) that need be processed/forwarded by the system. The rate of the flow may vary over time. One may deal with these flows either in the context of a single communications device (e.g. a router) or over a network (or a networks path) where a number of network devices may be traversed by a single flow.

Within the context of a single device, one of the earliest attempts to define fairness is the *Q-factor* (Wang and Morris [31]). Two measures that have been used quite widely, mainly in the Weighted-Fair-Queueing related literature, are 1) *The Relative Fairness Bound (RFB)* (used by [10] and others) and 2) The *Absolute Fairness Bound (AFB)* (see, e.g. [11,16]). Note that some of the references use AFB measure without using the term AFB. To define these measures let $S(j, \tau, t)$ be the amount of service provided by a service unit to session j, during the time interval (τ, t), and let $g(j)$ be the service rate allocated by the unit to session j (which could be viewed as the weight of the session). Let $S(j, \tau, t)$ be the amount of service that an idealized *Generalized Processor Sharing (GPS)* server grants to session j in the interval (τ, t). Given that sessions i and j always have packets to send, the *Relative Fairness Bound (RFB)* is given by:

$$RFB = \left| \frac{S(i,\tau,t)}{g(i)} - \frac{S(j,\tau,t)}{g(j)} \right| \qquad (1)$$

Taking the maximum of this value over all values of i, j, and t, yields a bound for the system. The number derived from the maximum operation can be viewed as the unfairness of the system: The higher the number the more it is unfair.

The *Absolute Fairness Bound (AFB)* is given by:

$$RFB = \left| \frac{S(i,\tau,t)}{g(i)} - \frac{G(j,\tau,t)}{g(i)} \right| \qquad (2)$$

Taking the maximum of this value over all values of i, j, and t, yields a bound for the system, which can be viewed as a measure of system unfairness.

Note that both measures evaluate how tightly close the service policy to GPS. It is easy to see that when evaluated for the Processor Sharing The AFB measure seems to be more complicated to calculate than the RFB, and thus RFB is used more frequently in the literature. The relationship between these measures is studied in [34].

Much of the literature in this area has focused on devising efficient schedules that are "fair"; the emphasis on these studies has been on their algorithmic side, where the major algorithm is Weighted Fair Queueing (WFQ). References to some of these works include [5,9,10,11,23,24,25,29].

Within the context of a whole network or a network path, one may consider 1) The Max-Min fairness allocation [14], 2) Proportional fairness [15], and 3) Balanced Fairness [6].

5 The Performance Issue: Delay vs. Service

What performance measure should be accounted for when quantifying queue fairness (for job-driven system)? The immediate and most natural candidate is the job delay, which is either the waiting time or the waiting plus service time experienced by the job. This has been the main quantity (perhaps almost the sole quantity) used in queueing theory to evaluate queueing systems (see, e.g.

text books on the subject, [7,8,13,18,19]) and is frequently being looked at via the expected delay or its variance in steady state. Under this quantity, customer satisfaction decreases with the delay experienced by the job and thus customer's objective is to minimize delay. The use of this quantity seems to be appropriate when the major performance issue associated with job queueing is indeed the delay experienced in the system. As an example, consider the waiting line in a supermarket where the annoyance of arriving late to a queue is merely due to the waiting in queue, since the service itself (purchasing the products) is guaranteed.

While delay has been the main performance metrics used by queueing theory, one should also consider job service, which received much less attention. By job service we refer to the actual service (*not* service time) given to the job by the system. To understand this performance objective, one should consider systems where the service is not guaranteed, e.g. systems where the service includes the selling of a finite-quantity product, or systems where the server is shut down at a predetermined time. In these systems, customers whose processing by the server is delayed (e.g. due to the prioritization of other customers) may encounter the situation in which the product is not available, or the server is shut down before their turn to be served comes, and thus they experience service degradation. In a simplistic representation, this can be a zero-one variable where zero means that no service is given and one means that a full service is given. The use of this variable seems to be appropriate when the major performance issue is that of service, for example a queueing line for scarce concert tickets.

Note that both quantities, delay and service, are affected by scheduling decisions and thus can be the subject of a fairness measure.

6 At What Granularity Level Should Fairness Be Quantified and Measured

In quantifying a physical property of a queueing system we identify three granularity classifications of the measure: The *job-individual measure*, the *scenario measure*, and the *system measure*. All three play a role in traditional queueing analysis, e.g. in dealing with system *delays*. In the context of delays, these are, respectively, the delay experienced by a specific job, the average delay when computed over a finite set of jobs under a specific scenario and the expected delay in the system under steady state. We may define these measures, in the context of queue fairness as:

1. **Job-Individual Discrimination:** This is a quantity attributed to the individual job (customer). It represents the performance experienced by a specific job (customer) under a specific scenario (or a sample path). For example, consider the discrimination experienced either by Short or by Long in the Long vs. Short case.
2. **Scenario (Sample path) fairness:** A summary-statistics that summarizes the performance as experienced by a (finite) set of jobs under a particular scenario (a sample path). For example, consider some averaging of the

discriminations experienced by Long and Short and the other jobs present at the system at that time.
3. **System fairness:** A summary statistics of a probabilistic measure (e.g. expected value or variance) of the performance as experienced by an arbitrary job, when the system is in steady state. This can be extended to a similar measure under transient behavior of the system.

It should be noted that queueing theory has dealt explicitly mainly with the third type of quantity (expected delay or its variance), as the other quantities are somewhat trivial in the context of customer delay. In the context of fairness, it is nonetheless important to make explicit use of the job-individual and scenario quantities as well, since humans can feel them better and associate with them better than with the third quantity. This is important to building confidence in the fairness measure, which is somewhat abstract, non-tangible and difficult to feel.

7 The Physical Entities Playing Role in Queue Fairness: Seniority and Service Requirement

What are the basic physical quantities playing a role in queue fairness? Two fundamental quantities determine the queueing process and the job scheduling decisions. These are the arrival epochs and service times, a_i and s_i. As our goal is to focus on the pure queueing process and neutralize other external parameters, we will deal with these variables only. To this end, we are not accounting for external parameters, such as payments made by customers or a gold/silver/bronze classification of customers.

Since these quantities are the only ones determining the queueing and scheduling process, they also serve as the fundamental variables for determining scheduling fairness. For convenience of presentation, we will translate the arrival time epoch and get the following two basic physical quantities: 1) *Seniority*, and 2) *Service requirement*. The seniority of J_i at epoch t is given by $t - a_i$. The service requirement of J_i is s_i. One may recall that *seniority* and *service-requirement* were in the heart of the dilemma in the Short vs. Long scenario.

It is natural to expect that a "fair" scheduling discipline will give preferential service to highly senior jobs, and to low service-requirement jobs. This can be stated formally in the following two fundamental principles:

1. *(Weak) Service-requirement Preference Principle*: If all jobs in the system have the same arrival time, then for jobs J_i and J_j, arriving at the same time and residing concurrently in the system, if $s_i < s_j$ then it will be more fair to complete service of J_i ahead of J_j than vice versa.
2. *(Weak) Seniority Preference Principle*: If all jobs in the system have the same service times, then for jobs J_i and J_j, residing concurrently in the system, if $a_i < a_j$ then it will be more fair to complete service of J_i ahead of J_j than vice versa.

A stronger form of the preference principles is as follows:

1. *Strong Service-requirement Preference Principle*: For jobs J_i and J_j, arriving at the same time and residing concurrently in the system, if $s_i < s_j$ then it will be more fair to complete service of J_i ahead of J_j than vice versa.
2. *Strong Seniority Preference Principle*: For jobs J_i and J_j, residing concurrently in the system and requiring equal service times, if $a_i < a_j$ then it will be more fair to complete service of J_i ahead of J_j than vice versa.

The seniority preference principle is rooted in the common belief that jobs arriving at the system earlier "deserve" to leave it earlier. The service-requirement preference principle is rooted in the belief that it is "less fair" to have short jobs wait for long ones. It should be noted that when $a_i < a_j$ and $s_i > s_j$ (the Short vs. Long case) the two principles conflict with each other, and thus the relative fairness of the possible scheduling of J_i and J_j is likely to depend on the relative values of the parameters.

One may view these two preference principles as two axioms expressing one's basic belief in queue fairness. As such, one may expect that a fairness measure will follow these principles. A fairness measure is said to follow a preference principle if it associates higher fairness values with schedules that are more fair. A formal definition is given next:

Definition 1. *Consider jobs J_i and J_j, requiring equal service times and obeying $a_i < a_j$. Let π be a scheduling policy where the service of J_i is completed before that of J_j and π' be identical to π, except for exchanging the service schedule of J_i and J_j. A fairness measure is said to adhere to the strong seniority preference principle if the fairness value it associates with π is higher than that it associates with π'.*

Similar definitions can be given to the service-time preference principle and to the weak-versions of the preference principles.

It is easy to see that if a fairness measure adheres to the *strong preference principle* (either Service-requirement or Seniority) then it must adhere to the corresponding *weak preference principle*.

7.1 Scheduling Policies and the Preference Principles

To illustrate the preference principles in the context of scheduling policies we review several common policies and examine whether they follow the preference principles. A formal definition is:

Definition 2. *A scheduling policy π is said to follow the strong seniority preference principle if for every two jobs J_i and J_j, requiring equal service times and obeying $a_i < a_j$, π completes the service of J_i ahead of that of J_j.*

A similar definition can be given for the strong service-time preference principle and for the two weak preference principles.

Using these definitions, one can classify common scheduling policies as follows:

1. **FCFS**: The *First-Come-First-Served* scheduling follows the strong seniority preference principle. On the other hand, since it gives no special consideration to shorter jobs, it does not follow the service-time preference principle (weak or strong).
2. **LCFS** and **ROS**: The *Last-Come-First-Served* and *Random order of Service* policies do not follow the seniority preference principle (either strong or weak). Further, neither do they follow the service-time preference principle.
3. **SJF** and **LJF**: The *Shortest Job First* policy follows the strong service-time preference principle. Nonetheless - it does not follow the seniority preference principle (both strong and weak). The *Longest Job First* policy follows none of the principles.
4. **PS**: The *Processor Sharing* policy follows both the strong seniority preference principle and the strong service-time preference principle.
5. **FQ**: *Fair Queueing*, which is the non-weighted version of Weighted Fair Queueing ([23,24]), serves the jobs in the order they complete service under Processor Sharing (unless some of the jobs are not present at the server at the time that the service decisions must be taken). This property and the fact that PS follows both of the strong preference principles, imply that FQ follows both the strong seniority preference principle and the strong service-time preference principle.

8 Fairness Measures: A Review of Proposed Measures and Their Properties

Having discussed the performance issues associated with queue fairness, we next turn to review recent measures proposed in the literature. We will examine how these measures treat the basic performance issues and how they fit with the various applications.

8.1 Order Fairness

The order fairness measure was studied in [1]. The basic underlying model used in that study assumes that all service times are identical. In that context the major factor of interest is that of job-seniority. The study deals with a specific sample path of the system, and examines a realization π of the service order (that is, a feasible sequence of job indices reflecting the order of service), and with a fairness measure $F(\pi)$ defined on the service order. The paper assumes several elementary axioms on the properties of . The major axiom is:

Axiom 1. *Monotonicity of $F()$ **under neighbor jobs interchange:** If two neighboring jobs are interchanged to modify π and yield a new service order π' then $F()$ increases if the interchange yields advancing the more senior of the two jobs ahead of the less senior job, and it decreases if the interchange advances the less senior job ahead of the more senior job. If the seniority of the interchanged jobs is the same - $F()$ is not affected by the interchange.*

The additional axioms deal with 2) *reversibility of the interchange*, 3) *independence on position and time*, and 4) *fairness change is unaffected by jobs not interchanged*.

The reader may recognize that the core axiom of this approach, Axiom 1, is simply a mathematical form to express the *seniority preference principle* presented in Section 7.

The results derived in [1] show that for a specific sample path the quantity $c\sum_i a_i \Delta_i + \alpha$, where Δ_i is the *order displacement* of J_i (number of positions J_i is pushed ahead or backwards on the schedule), and $c > 0$ and α are arbitrary constants satisfying the basic axioms. This quantity is the unique form satisfying the axioms applied to any feasible interchange (not necessarily of neighbors). Under steady state this quantity is equivalent to the *variance of the waiting time* (with a negative sign). Thus, when all *service times* are *identical* the *waiting time variance* can serve as a surrogate for the *system's unfairness measure*.

Properties. The main properties of this measure are:

1. The measure adheres to the strong *Seniority Preference Principle* (Section 7). This can be verified by recalling that the unfairness function for a sample path is given by $\sum_i a_i \Delta_i$ and by examining the change of this function due to the exchange of J_i and J_j.
2. When all service times are identical, the fairest policy in the family of work conserving and uninterrupted service policies is FCFS. The most unfair policy under these conditions is LCFS.
3. The measure does not adhere[1] to the *Service-requirement Preference Principle* (Section 7): If one uses the variance of waiting time as a measure of unfairness, then there are cases where it is more fair to serve a long job ahead of a short job. For example consider a system with two jobs only, J_1 and J_2 whose service times are $s_1 = 1, s_2 = \epsilon \to 0$. Serving the longer job J_1 first leads to a waiting time variance that approaches 0 while serving the shorter job J_2 first leads to a waiting time variance that approximately equals 1/4.

8.2 Normalized-Delay Based Fairness

Normalized-delay based fairness was presented in [33] in which a fairness criterion (as opposed to a fairness measure) was proposed. The aim of this criterion is to address the differences in service times between different jobs and to follow a principle under which short jobs should be given some preferential service over long jobs (similar to the Service-time preference principle).

Under the criterion proposed in [33] each job is characterized by its service time only. The measure of interest is the "slow down" ("normalized response time") $S(x) \stackrel{def}{=} T(x)/x$ where x is the service time and $T(x)$ is a random variable denoting the delay (response time) experienced by a customer whose service time

---

[1] In fact, it might not be appropriate to examine this principle as the measure is built for equal service-time situations.

is x. The expected slowdown for a job of size x is $E[S(x)]$, and a scheduling policy is said to be fair for given load and service distribution if $E[S(x)] < 1/(1-\rho)$ for all values of x, where ρ is the system's load. A service policy is *always fair* if it is fair under all loads and all service distributions. A service policy is *always unfair* if it is not fair under all loads and all service distributions. Other policies are *sometimes unfair*. The "slow down" was used as a means for evaluating fairness in queues earlier in [4] and in [3].

Properties. The properties of this criterion are:
1. The criterion classifies a large class of service disciplines, common in computer systems, into "always fair" "always unfair" and "sometimes fair". A few examples are:
   (a) **Always fair:** Processor Sharing (PS) and Preemptive LCFS.
   (b) **Always unfair:** All non-size based non-preemptive policies, in particular FCFS. Also age-based polices are always unfair, in particular Feedback Scheduling (FB).
   (c) **Sometimes Unfair:** Shortest Remaining Processing Time (SRPT).
2. The criterion is relatively easy to apply for general service time distributions (M/G/1 type systems) as the measure of interest is $E[T(x)/x]$.
3. The *Seniority Preference Principle* does not hold. Specifically, the classification stated above implies that under this criterion FCFS is "always unfair" while LCFS preemptive is "always fair"; these predictions contradict the *Seniority Preference Principle*.
4. Intuitively speaking, this criterion may possibly adhere to the *Service time Preference Principle* if the criterion is extended to be a measure and after some adaptations. We make this intuitive statement based on recognizing that the criterion favors prioritization of short jobs over long jobs. However it is an open question whether this preference principle always holds and under what formulation.

8.3 Resource Allocation Based Fairness

A Resource Allocation Queueing Fairness Measure (RAQFM) was introduced in [28]. The measure aims at accounting both for seniority and service-requirements, and does it by focusing on the fair sharing of the system resources. The method can apply to multiple servers, but for the sake of presentation will be described for a single server system.

The basic philosophy behind the method is that at every epoch t at which there are $N(t)$ jobs present in the system, they all are entitled to an equal share of the server's time. Thus, the temporal *warranted service rate* to be given to a job at that epoch, is given by: $1/N(t)$. The overall warranted service of job i is given by integrating this value over the duration that J_i stays in the system. Subtracting this warranted service from the granted service (which is the service granted to J_i, namely its service time, s_i) yields the *discrimination* of J_i, denoted $\delta_i = \int_{a_i}^{d_i} \frac{1}{N(t)} dt$. Note that the discrimination may be positive or negative.

Taking summary statistics over all discriminations experienced by the customers yields an unfairness measure for the system. This measure can apply to a specific scenario (sample path), to yield the unfairness of that path. Similarly, taking expectations of this measure over all sample paths yields the system unfairness. One of the basic properties of the discrimination function is that it is a zero-sum function (namely the total discrimination in the system, at every epoch, is 0). Thus, the expected value of discrimination is meaningless, and the proper summary statistics is the second moment (or variance) or expected absolute value of discrimination.

Properties. The main properties of this measure are:

1. The measure adheres to the *Strong Seniority Preference principle*), for work conserving and uninterrupted service policies. This is proven in [2].
2. When all service times are identical, the fairest policy in the family of work-conserving and uninterrupted-service policies is FCFS. The most unfair policy under these conditions is LCFS.
3. The measure adheres to the weak *Service-requirement Preference Principle*, for work conserving and uninterrupted service policies. Nonetheless, it does not adhere to the strong version of this principle, as there exist some counter examples. These properties are proven in [2].
4. The measure yields to analysis for the family of Markovian (M/M type) queues. It is an open subject of research whether (and how) it yields to analysis for general service time (e.g. M/G/1) type systems.

8.4 Summary of Major Properties

The two measures and the criterion can be roughly classified according to their treatment of the seniority and service time physical properties. This summary is depicted in Figure 1 where, in the first column, we indicate the major focus of the measures: Order fairness accounts mainly for seniority, normalized-delay fairness accounts mainly for the service times, and resource allocation accounts for both. The second column of the figure indicates what type of applications would fit to these measures: Order fairness fits service sensitive applications while normalized-delay fairness and resource-allocation fairness fit delay sensitive applications.

9 Application Perspective

How do the measures outlined above fit with the various queueing applications? This, as we discuss next, depends on the characteristics of the applications. The mapping of applications to fairness measures is given (partially) in Figure 2.

Some applications are characterized by high sensitivity to service and lower sensitivity to *delay*; these are outlined in the first row of applications in Figure 2. Perhaps most of the "historical queueing systems", e.g. the waiting line for bread,

Measure/ Criterion	Physical quantities accounted for		Can treat Applications which are sensitive to	
	Seniority	Service time	Service delivery	Delay
Order fairness measure (8.1)	●		○	
Normalized delay criterion (8.2)		●		○
Resource allocation measure (8.3)	●	●		○

Fig. 1. Summary of major properties of fairness measures

Application characteristics	Applications	Fairness sensitive to		Performance Objectives	Applicable Fairness Measure
		Seniority	Service time		
Service sensitive	Airline reservation Call center, "Line for bread"	●		Service	Order fairness (8.1)
Delay sensitive, identical service times	Airport immigration lines	●		Delay	Order fairness (8.1) + Resource allocation (8.3)
Delay sensitive, variable service time	Call centers, Supermarkets, Banks, Computer systems	●	●	Delay	Order fairness (8.3)
Delay sensitive, seniority-blind customers	Computer systems		●	Delay	Normalized-delay (8.2) + resource allocation (8.3)

Fig. 2. Fairness-related Properties of Applications and the applicable measures

can be categorized under this category (this "historical" queueing experience is perhaps the reason for having many people strongly believing in the notion of servicing customers by order of arrival). Today's applications are waiting lines for limited-supply products, such as the queue for highly demanded concert-tickets. Another application is the waiting line in a call center specializing in selling airline tickets, in which often some of the tickets (e.g. special price or special date tickets) are at very low supply. The fairness of these applications is very sensitive to job seniority and is less sensitive to service times (third and fourth) For these service sensitive (seniority sensitive) applications order fairness (Section 8.1) fits well, as it focuses on job seniority. Recall, however, that that measure assumes identical service times. Thus, its use in the case of non-identical service times, though sounds reasonable, still needs to be studied and understood.

Other applications are characterized by high sensitivity to delay and low sensitivity to service (as the service is more or less guaranteed). These, in fact, form the majority of today's applications.

Within this class we first distinguish applications where the service times of all customers are more or less identical (nearly deterministic); these are denoted on the second row of Figure 2. These include, for example, waiting lines for (unmarked) theater tickets. In this case, even if the supply is large and thus the performance is sensitive mainly to delay, one can apply *order fairness* (Section 8.1), since service times are more or less identical. One can also apply in this case the *resource allocation measure* (Section 8.3) as it can handle this equally well.

Second, within this class of delay sensitive applications, we distinguish applications where the service time varies across customers (third line in Figure 2). These include waiting lines in supermarkets, airlines counters, public offices, and call centers with unlimited products. Here customers will be sensitive both to seniority and to service times. In these applications neither order-fairness (does not account for service time differences) nor normalized-delay fairness (does not account for seniority) can be used; the *resource allocation fairness measure* (Section 8.3), which accounts both for service times and seniority, is the most appropriate

Within the class of delay-sensitive applications one may recognize some applications where the customers are *not aware* of the relative seniority of the jobs in the system (fourth line of Figure 2). These may include jobs performed in a computer system where the customers who submit the jobs cannot know the relative seniority/status of their jobs. In such applications, the blindness of customers to relative seniority may allow one to use the *normalized-delay fairness* approach (Section 8.2). Similarly, the "blindness" of customers to service requirements of other customers may allow the use of the *order fairness* measure.

Note however, that even under the blindness conditions, it is likely that customers will require the system not to be seniority-blind or service requirement-blind, namely not to use the normalized-delay fairness. That is, even if justice cannot be seen, customers may want it to be done.

10 Concluding Remarks

We argued that fairness is a fundamental property of queueing systems and that it is highly important to customers. Little work has been done on this subject in the past; an increase in research in this area occurred in recent years, which contributed to better understanding of the subject. Nevertheless, more research must be conducted to have a good understanding of the issue. For example, there exist a huge number of queueing systems and queueing scheduling policies, which were studied in the past and where the focus has been on the delay of the individual customer. Fairness evaluation of these systems will contribute greatly to the understanding of the relative benefits of these systems.

References

1. Avi-Itzhak, B., Levy, H.: On measuring fairness in queues. Advances in Applied Probability 36(3), 919–936 (2004)
2. Avi-Itzhak, B., Levy, H., Raz, D.: A resource allocation queueing fairness measure: Properties and bounds. Queueing Systems Theory and Application 56(2), 65–71 (2007)
3. Bansal, N., Harchol-Balter, M.: Analysis of SRPT scheduling: Investigating unfairness. In: Proceedings of ACM Sigmetrics 2001 Conference on Measurement and Modeling of Computer Systems, pp. 279–290 (2001)
4. Bender, M., Chakrabarti, S., Muthukrishnan, S.: Flow and stretch metrics for scheduling continuous job streams. In: Proceedings of the 9th Annual ACM-SIAM Symposium on Discrete Algorithms, San Francisco, CA, pp. 270–279 (1998)
5. Bennet, J.C.R., Zhang, H.: WF^2Q: Worst-case fair weighted fair queueing. In: Proceedings of IEEE INFOCOM 1996, San Francisco, March 1996, pp. 120–128 (1996)
6. Bonald, T., Proutière, A.: Insensitive bandwidth sharing in data networks. Queueing Systems 44(1), 69–100 (2003)
7. Cooper, R.B.: Introduction to Queueing Theory, 2nd edn. North-Holland (Elsevier), Amsterdam (1981)
8. Daigle, J.D.: Queueing Theory for Telecommunications. Addison-Wesley, Reading (1991)
9. Demers, A., Keshav, S., Shenker, S.: Analysis and simulation of a fair queueing algorithm. Internetworking Research and Experience 1, 3–26 (1990)
10. Golestani, S.J.: A self-clocked fair queueing scheme for broadband application. In: Proceedings of IEEE INFOCOM 1994, Toronto, Canada, June 1994, pp. 636–646 (1994)
11. Greenberg, A.G., Madras, N.: How fair is fair queueing? Journal of the ACM 3(39), 568–598 (1992)
12. Gross, D., Harris, C.L.: Fundamentals of Queueing Theory. Wiley & Sons, New York (1974)
13. Hall, R.W.: Queueing Methods for Services and Manufacturing. Prentice-Hall, Englewood Cliffs (1991)
14. Jaffe, J.M.: Bottleneck flow control. IEEE Transactions on Communications 29(7), 954–962 (1981)
15. Kelly, F.P.: Charging and rate control for elastic traffic. European Transactions on Telecommunications 8, 33–37 (1997)

16. Keshav, S.: An Engineering Approach to Computer Networking: ATM Networks, the Internet, and the Telephone Network. Addison Wesley Professional, Reading (1997)
17. Kingman, J.F.C.: The effect of queue discipline on waiting time variance. Proceedings of the Cambridge Philosophical Society 58, 163–164 (1962)
18. Kleinrock, L.: Queueing Systems. Theory, vol. 1. Wiley, Chichester (1975)
19. Kleinrock, L.: Queueing Systems. Computer Applications, vol. 2. Wiley, Chichester (1976)
20. Larson, R.C.: Perspective on queues: Social justice and the psychology of queueing. Operations Research 35, 895–905 (1987)
21. Mann, I.: Queue culture: The waiting line as a social system. Am. J. Sociol. 75, 340–354 (1969)
22. Palm, C.: Methods of judging the annoyance caused by congestion. Tele (English Ed.) 2, 1–20 (1953)
23. Parekh, A.: A Generalized Processor Sharing Approach to Flow Control in Integrated Services Networks. Ph.D. thesis, MIT (February 1992)
24. Parekh, A., Gallager, R.G.: A generalized processor sharing approach to flow control in integrated services networks: The single node case. IEEE/ACM Trans. Networking 1, 344–357 (1993)
25. Parekh, A., Gallager, R.G.: A generalized processor sharing approach to flow control in integrated services networks: The multiple node case. IEEE/ACM Trans. Networking 2, 137–150 (1994)
26. Rafaeli, A., Barron, G., Haber, K.: The effects of queue structure on attitudes. Journal of Service Research 5(2), 125–139 (2002)
27. Rafaeli, A., Kedmi, E., Vashdi, D., Barron, G.: Queues and fairness: A multiple study experimental investigation. Tech. rep., Faculty of Industrial Engineering and Management, Technion. Haifa, Israel (2003) (under review), http://iew3.technion.ac.il/Home/Users/anatr/JAP-Fairness-Submission.pdf
28. Raz, D., Levy, H., Avi-Itzhak, B.: A resource-allocation queueing fairness measure. In: Proceedings of Sigmetrics 2004/Performance 2004 Joint Conference on Measurement and Modeling of Computer Systems, New York, NY, June 2004, pp. 130–141 (2004); Performance Evaluation Review, 32(1), 130–141
29. Rexford, J., Greenberg, A., Bonomi, F.: Hardware-efficient fair queueing architectures for high-speed networks. In: Proceedings of IEEE INFOCOM 1996, March 1996, pp. 638–646 (1996)
30. Rothkopf, M.H., Rech, P.: Perspectives on queues: Combining queues is not always beneficial. Operations Research 35, 906–909 (1987)
31. Wang, Y.T., Morris, R.J.T.: Load sharing in distributed systems. IEEE Trans. on Computers C 34(3), 204–217 (1985)
32. Whitt, W.: The amount of overtaking in a network of queues. Networks 14(3), 411–426 (1984)
33. Wierman, A., Harchol-Balter, M.: Classifying scheduling policies with respect to unfairness in an M/GI/1. In: Proceedings of ACM Sigmetrics 2003 Conference on Measurement and Modeling of Computer Systems, San Diego, CA, June 2003, pp. 238–249 (2003)
34. Zhou, Y., Sethu, H.: On the relationship between absolute and relative fairness bounds. IEEE Communication Letters 6(1), 37–39 (2002)

Large Deviations Theory: Basic Principles and Applications to Communication Networks

Michele Pagano

Dipartimento di Ingegneria dell'Informazione
Università di Pisa, Via Caruso, I-56122 Pisa, Italy
`michele.pagano@iet.unipi.it`

Abstract. The *theory* of large deviations refers to a collection of techniques for estimating properties of rare events such as their frequency and most likely manner of occurrence. Loosely speaking, LDT can be seen as a refinement of the classical limit theorems of probability theory and it is useful when simulation or numerical techniques become increasingly difficult as a parameter of interest tends to its limit.

The first part of this tutorial deals with the behaviour of the empirical mean of IID RVs, the most natural framework to introduce the basic concepts and theorems of LDT and to highlight their heuristic interpretation.

Then, the large deviation principle for the single server queue is presented and its implications on network dimensioning are discussed. Finally, the tutorial overviews the application of LDT to rare event simulation, for the choice of the optimal change of measure in Importance Sampling.

Keywords: LDT, Rare Events, Contraction Principle, Queues, LRD.

1 Introduction

In the framework of teletraffic engineering, many challenging issues have arisen in the last two decades as a consequence of the fast growth of network service demand. The search for *global* network architectures, which should handle heterogeneous applications and different quality of service (QoS) guarantees [1], has determined a widespread interest for novel performance evaluation techniques, able to cope with the increasing size (and complexity) of telecommunication systems. The need for new mathematical approaches is also related to the adoption of more sophisticated traffic models, the so-called Long Range Dependent (LRD) processes, able to take into account the long memory features of real traffic [2,3].

In case of stringent QoS requirements, network performance are determined by events with a small probability of occurring, but with severe consequence when they occur. Since these events are linked to *large deviations* from the normal behaviour of the system, the so-called theory of large deviations (LDT) represents a natural candidate for analysing *rare events in large systems*.

In a nutshell, LDT studies the tails of distributions of certain random variables. Since, by definitions, probabilities of rare events are involved, it is also known as the theory of rare events. As a matter of fact, LDT only applies to certain types of rare events, caused by a large number of unlikely things occurring together (*conspiracy*), rather then a single event of small probability. For instance, winning a lottery is not a large deviations event, since it is determined by a single trial that cannot be broken into more than one sub-event [4].

Unlike classical limit theorems, LDT also provides a nice qualitative theory to *understand* rare events and the typical way they occur (*most likely path*). Indeed, the probability of a rare event is often reduced to a deterministic optimisation problem. If a cost is assigned to each sample path that would cause the rare event, its probability only depends on the *cheapest path*, i.e., on the cheapest way the event can happen. This concept is described in [4] as the *strong law of rare events*: if there is a unique cheapest path, then as the *asymptotic parameter* gets large, conditioned on the occurrence of the rare event, with overwhelming probability the system followed the cheapest path for any bounded interval of time before the rare event occurred.

This deeper insight into the system behaviour can be successfully employed, for instance, to design proper control systems and to speed-up the simulation of rare events. Indeed, a control affects the probability of the rare event iff it affects the cheapest path; as a consequence, the time scale on which the control should operate is implicitly determined by the *most likely time* of occurrence of the rare event. In the framework of simulation, unlike crude Monte Carlo, the application of speed-up techniques generally requires some additional information about the behaviour of the system, such as the one provided (although in an asymptotic and eventually approximate form) by the LDT.

As pointed out by many authors [5], there is no *real* theory of large deviations and often the same result may be reached in different (and apparently unrelated) ways. Hence, as a whole LDT refers to a set of basic definitions, that by now are standard, and a variety of tools for the analysis of small probability events in completely different frameworks (such as statistical mechanics, information theory, parameter estimation and traffic engineering, just to name a few application fields).

The aim of this tutorial, which is heavily based on [6], is to introduce the basic LDT concepts, highlighting their heuristic interpretation from an engineering perspective and focusing on their applications (or, at least, on some of them) in the framework of queueing systems and computer networks. In more detail, the rest of the paper is organised as follows. Section 2 describes the key LDT principles, starting from simple practical examples and generalising the results to more abstract frameworks. Then Section 3 deals with the application of LDT to the single server queue, focusing on two well-known asymptotic regimes: large-buffer and many-sources asymptotics, while a few more advanced topics (queueing performance in presence of LRD traffic and LDT-based changes of measures) are sketched in Section 4. Finally, hints on further readings conclude the tutorial.

2 Basic LDT Results

The theory of large deviations is concerned with the asymptotic estimation of probabilities of rare events. In its basic form, the theory considers the limit of normalisations of $\log \mathbb{P}(A_n)$ for a sequence of events with asymptotically vanishing probability. Although the topic may be traced back to the early 1900s (see [5] for more detailed historical notes, interpretations and references), its general abstract characterisation by means of a *large deviation principle* was formalised only in 1966 by Varadhan [7], who is considered one of the founders of the *modern* theory of large deviations, together with Donsker (in the West) as well as Freidlin and Wentzell (in the East).

The following subsections will review the basic concepts that by now are standard, starting from the case of independent, identically distributed (IID) random variables (RVs) and introducing some more advanced tools (such as the large deviation principle and the contraction principle), which will be applied in the following to queueing systems.

2.1 Large Deviations of IID RVs

Let us consider the most classical topic of probability theory, namely the behaviour of the empirical mean of IID RVs. Before stating the general result (Cramér's theorem), let us consider some simple examples (see Chapter 2 in [6] for further details).

Sums of Standard RVs. Let $X_i \in \mathcal{N}(0,1)$[1] and consider the empirical mean

$$M_n = \frac{1}{n} S_n \quad \text{where} \quad S_n = \sum_{i=1}^{n} X_i . \tag{1}$$

Since $M_n \in \mathcal{N}(0, 1/n)$, it is easy to show that:

1. for any $a > 0$

$$\lim_{n \to \infty} \mathbb{P}(M_n \geq a) = 0 \quad \text{(Weak Law of Large Numbers)} \tag{2}$$

2. for any interval A

$$\lim_{n \to \infty} \mathbb{P}(\sqrt{n} M_n \in A) = \frac{1}{\sqrt{2\pi}} \int_A e^{-\frac{1}{2}x^2} dx \quad \text{(Central Limit Theorem)} \tag{3}$$

3. for any $a > 0$

$$\mathbb{P}(M_n \geq a) = \frac{1}{\sqrt{2\pi}} \int_{a\sqrt{n}}^{\infty} e^{-\frac{1}{2}x^2} dx \tag{4}$$

and therefore

$$\lim_{n \to \infty} \frac{1}{n} \log \mathbb{P}(M_n \geq a) = -\frac{a^2}{2}, \tag{5}$$

which is a typical large deviations result.

Roughly speaking, according to (3), the *typical* value of M_n is of the order of $1/\sqrt{n}$, but with small probability (of the order of $e^{-na^2/2}$, as suggested by (5)), M_n takes relatively large values.

It is well known from elementary probability theory that (2) and (3) remain valid as long as $\{X_i\}$ are IID RVs of zero mean and unit variance and can be easily modified

---
[1] As usual, $\mathcal{N}(\mu, \sigma^2)$ will denote a Gaussian RV with mean μ and variance σ^2.

in case of IID RVs with mean μ and variance σ^2. Instead, as far as (5) is concerned, the limit still exists (under quite general assumptions), but its value depends on the specific distribution of X_i. This is precisely the content of Cramér's theorem. In order to understand the kind of approximations involved in LDT, it is useful to derive a result similar to (5) in a slightly less trivial framework.

Sums of Bernoulli RVs. Let $X_i \in \mathcal{B}(p)$, i.e., $\mathbb{P}(X_i = 1) = p = 1 - \mathbb{P}(X_i = 0)$; in this case M_n can be seen as the proportion of heads in n independent tosses of a biased coin, which has probability p of coming up heads. Suppose that n is large and consider the probability that M_n exceeds a, for some $a > p$. Through direct calculation[2] (for notational convenience, suppose that $na < n$ is an integer):

$$\begin{aligned}
\mathbb{P}(M_n \geq a) &= \sum_{j=na}^{n} \binom{n}{j} p^j (1-p)^{n-j} \quad (S_n \text{ has a Binomial distribution}) \\
&\approx \binom{n}{na} p^{na} (1-p)^{n(1-a)} \quad \text{(Principle of the largest term)} \\
&= \frac{n!}{(na)!(n-na)!} p^{na} (1-p)^{n(1-a)} \\
&\approx \frac{1}{\sqrt{2\pi n(1-a)a}} a^{-na} (1-a)^{-n(1-a)} p^{na} (1-p)^{n(1-a)} \\
&\approx \left(\frac{a}{p}\right)^{-na} \left(\frac{1-a}{1-p}\right)^{-n(1-a)} \\
&= e^{-n\left(a \log \frac{a}{p} + (1-a) \log \frac{1-a}{1-p}\right)},
\end{aligned}$$

where the Stirling's formula was used to approximate the binomial coefficient:

$$n! \approx \sqrt{2\pi n}\, n^n e^{-n}.$$

Hence, an expression similar to (5) can be written also in case of Bernoulli RVs:

$$\lim_{n \to \infty} \frac{1}{n} \log \mathbb{P}(M_n \geq a) = a \log \frac{a}{p} + (1-a) \log \frac{1-a}{1-p} \triangleq H(a;p) \quad (6)$$

and $H(a;p)$ is known as the relative entropy, or Kullback-Leibler divergence, of the probability distribution $(a, 1-a)$ with respect to the probability distribution $(p, 1-p)$.

It is worth noticing that a *single term* in the sum is sufficient to determine its correct exponential decay rate (in n). It turns out that this feature is characteristic of many situations where LDT is applicable and is known as *principle of the largest term*, which is often expressed in the probability context by the phrase "*rare events occur in the most likely way*".

---

[2] It is straightforward to verify that the largest term in the sum corresponds to $j = na$.

LDT Rate Function. The limit (5) depends on the specific distribution of X_i through the so-called *rate function* Λ^*, which is defined as the Fenchel-Legendre transform of the Cumulant Generating Function. Before stating the general LDT result for sums of IID RVs, it is worth introducing the definition of rate function and its main properties.

Let $\Lambda(\theta)$ denote the Logarithmic Moment Generating Function or Cumulant Generating Function[3] of a real-valued RV X, i.e.,

$$\Lambda(\theta) \triangleq \log M(\theta) = \log \mathbb{E}\left(e^{\theta X}\right) \tag{7}$$

where

$$M(\theta) \triangleq \mathbb{E}\left(e^{\theta X}\right) \tag{8}$$

is the Moment Generating Function of X.

Let $\Lambda^*(x)$ be the *convex dual* or *Fenchel-Legendre transform* of $\Lambda(\theta)$:

$$\Lambda^*(x) \triangleq -\log\left(\inf_\theta e^{-\theta x} M(\theta)\right) = \sup_\theta \left(\theta x - \log M(\theta)\right) = \sup_\theta \left(\theta x - \Lambda(\theta)\right) \tag{9}$$

Figure 1 gives a graphical interpretation of the previous definition: $\Lambda^*(x)$ is the smallest amount by which the straight line $x\,\theta$ (with slope x) has to be pushed down so as to lie below the graph of $\Lambda(\theta)$ $\forall \theta \in \mathbb{R}$.

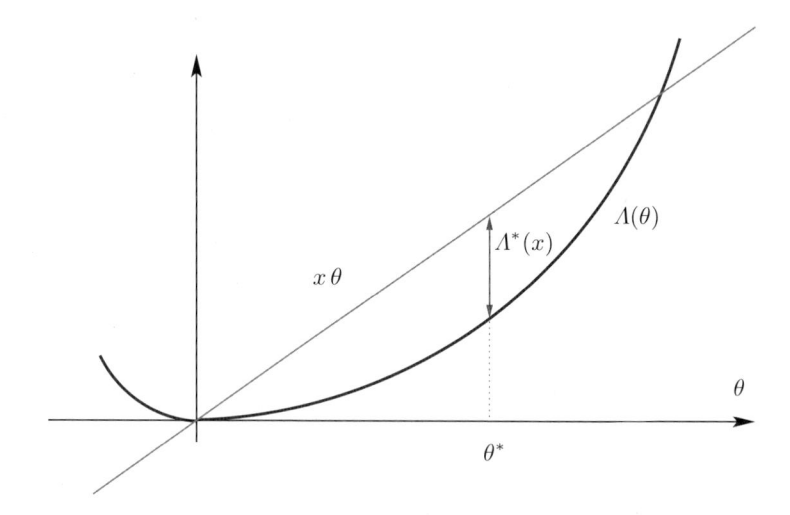

Fig. 1. Graphical interpretation [8] of the Fenchel-Legendre transform

The most relevant properties of Λ^* are recalled (the corresponding proofs are given, for example, in [5]) in the following:

1. $\Lambda^*(x)$ is convex, i.e., $\forall \lambda \in [0, 1]$:

$$\Lambda^*\left(\lambda x_1 + (1-\lambda)x_2\right) \leq \lambda \Lambda^*(x_1) + (1-\lambda)\Lambda^*(x_2)$$

---

[3] Indeed, the cumulants of X are just the derivatives of $\Lambda(\theta)$ evaluated at $\theta = 0$.

2. $\Lambda^*(x)$ is non-negative
3. $\Lambda^*(x)$ has its minimum for $x = \mu \stackrel{\Delta}{=} \mathbb{E}(X)$ and $\Lambda^*(\mu) = 0$
4. $\Lambda^*(x)$ is lower semicontinuous, i.e., the level sets $\{x : \Lambda^*(x) \leq \alpha\}$ are all closed for $\alpha \in \mathbb{R}$
5. If the supremum in (9) is attained at a point θ^* in the interior of the interval where $M(\theta)$ is finite, then $M(\theta)$ is differentiable at θ^*, so that

$$\Lambda^*(x) = -\log \mathbb{E}\left(e^{\theta^*(X-x)}\right) = \theta^* x - \Lambda(\theta^*)$$

6. Let c be the greatest lower bound for a RV X, i.e.,

$$\mathbb{P}(X < c) = 0 \quad \text{and} \quad \mathbb{P}(X \leq c+\epsilon) > 0 \quad \forall \epsilon > 0 \ .$$

Then
(a) $\Lambda^*(x) = \infty$ for $x < c$
(b) $\Lambda^*(c) < \infty \iff \mathbb{P}(X = c) > 0$

Table 1 gives the expressions of Λ and Λ^* for some common distributions, highlighting the similarities with the preliminary examples reported in this section. For instance, in the case of Bernoulli RVs, $\Lambda^*(x) = H(x, p)$ and (6) justifies the name of *rate function* given to Λ^* in the LDT framework: indeed it is the function that specifies the rate of convergence for the Weak Law of Large Numbers.

Cramér's Theorem. Cramér's theorem (1938) is the most general result for IID RVs, stated in generic large deviations form.

Table 1. Examples of rate functions

X	$\Lambda(\theta) = \log \mathbb{E}\left(e^{\theta X}\right)$	$\Lambda^*(x) = \sup_{\theta \in \mathbb{R}}(\theta x - \Lambda(\theta))$
$\mathcal{N}(\mu, \sigma^2)$	$\theta\mu + \frac{1}{2}\theta^2\sigma^2$	$\frac{1}{2\sigma^2}(x - \mu)^2$
$\mathcal{B}(p)$	$\log\left(1 - p + pe^\theta\right)$	$\begin{cases} x \log \frac{x}{p} + (1-x) \log \frac{1-x}{1-p} & 0 \leq x \leq 1 \\ \infty & \text{otherwise} \end{cases}$
$\text{Exp}(\lambda)$	$\log \frac{\lambda}{\lambda - \theta}$	$\begin{cases} x\lambda - 1 - \log(x\lambda) & x > 0 \\ \infty & \text{otherwise} \end{cases}$
$\text{Poisson}(\lambda)$	$\lambda\left(e^\theta - 1\right)$	$\begin{cases} \lambda + x\left(\log \frac{x}{\lambda} - 1\right) & x > 0 \\ \lambda & x = 0 \\ \infty & \text{otherwise} \end{cases}$

Theorem 1 (Cramér's theorem). *Let X_i be IID (real valued) RVs and define*

$$S_n = \sum_{i=1}^{n} X_i \ .$$

Let $\Lambda(\theta)$ denote the Logarithmic Moment Generating Function of X_i, i.e.,

$$\Lambda(\theta) = \log \mathbb{E}\left(e^{\theta X_i}\right)$$

and let Λ^* be the convex conjugate of Λ:

$$\Lambda^*(x) \stackrel{\Delta}{=} \sup_{\theta} \left(\theta x - \Lambda(\theta)\right) \ . \qquad (10)$$

For all closed sets F,

$$\limsup_{n \to \infty} \frac{1}{n} \log \mathbb{P}\left(\frac{S_n}{n} \in F\right) \leq - \inf_{x \in F} \Lambda^*(x) \qquad \textit{Upper Bound for closed sets} \quad (11)$$

and, for all open sets G,

$$\liminf_{n \to \infty} \frac{1}{n} \log \mathbb{P}\left(\frac{S_n}{n} \in G\right) \geq - \inf_{x \in G} \Lambda^*(x) \qquad \textit{Lower Bound for open sets} \quad (12)$$

i.e., for any set $B \subset \mathbb{R}$:

$$\begin{aligned}
- \inf_{x \in B^o} \Lambda^*(x) &\leq \liminf_{n \to \infty} \frac{1}{n} \log \mathbb{P}\left(\frac{S_n}{n} \in B\right) \\
&\leq \limsup_{n \to \infty} \frac{1}{n} \log \mathbb{P}\left(\frac{S_n}{n} \in B\right) \leq - \inf_{x \in \bar{B}} \Lambda^*(x)
\end{aligned} \qquad (13)$$

where B^o denotes the interior of B and $\bar{B}$ its closure.

A complete proof of the theorem goes beyond the scope of this tutorial and can be found, for instance, in [6] or, in a more general form, in [5]. However, it is quite useful to draw here some general remarks:

1. In the theorem, no conditions, not even existence of the mean, are required for the RVs X_i.
2. The **Lower Bound** (12) is *local* (the bound for open balls implies the bound for all open sets) and its proof uses an *exponential change of measure* [9] argument, as in Importance Sampling (more on Importance Sampling and LDT-based changes of measures in section 4.2)

$$\frac{d\mu_\theta}{d\mu}(x) = e^{\theta x - \Lambda(\theta)} = \frac{1}{M(\theta)} e^{\theta x} \qquad (14)$$

where μ and μ_θ denote the law of the original and tilted RVs respectively.

In order to derive a bound on the probability that the sample mean S_n/n lies in $(x - \delta, x + \delta)$ we seek a tilt parameter θ^* that makes the mean of the tilted distribution equal to x. From a heuristic point of view, this tilted RV captures the idea of being close in distribution to X_i, conditional on having a value close to x.

Indeed, the tilted measure μ_θ identifies the *most likely way* by which the mean of a large sample turns out to be close to x. More precisely, conditional on the sample mean S_n/n being in $(x-\delta, x+\delta)$, the empirical distribution of $X_1, \ldots, X_n$ approaches μ_θ as $n \to \infty$.

3. The **Upper Bound (Chernoff's Bound)** holds for all closed sets $F \subset \mathbb{R}$ and *all* n, not just on a logarithmic scale in the limit as $n \to \infty$. This means that (11), presented as a *classical* LDT upper bound, in case of sums of IID RVs can be strengthened as follows:

$$\frac{1}{n}\log \mathbb{P}\left(\frac{S_n}{n} \in F\right) \leq -\inf_{x \in F} \Lambda^*(x) \ . \tag{15}$$

4. When **the limit exists** (i.e., limsup and liminf are equal), the Cramér's Theorem implies that

$$\mathbb{P}\left(\frac{S_n}{n} \in B\right) \approx e^{-n \inf_{x \in B} \Lambda^*(x)} \ . \tag{16}$$

The last approximation highlights three important features of LDT:

(a) The asymptotic probability that the sample mean lies in B tends to zero *exponentially fast (in n)*.

(b) Λ^* gives the *exact* (ignoring terms that are subexponential in n) decay rate of the family of probabilities $\mathbb{P}(M_n \in B)$ and is commonly known as *rate function*.

(c) The speed of convergence essentially depends on *one point*, denoted in the following as $\hat{x}$, the so-called *dominating point* of the set B, i.e., the point where the rate function $\Lambda^*(x)$ attains its infimum (*principle of the largest term*). For instance, the three sets in fig. 2 (which refers to the sums of exponential RVs with mean 1) have the same probability in the large deviations limit.

5. Since Cramér's theorem *only* gives *logarithmic asymptotics*, (16) implies that

$$\mathbb{P}\left(\frac{S_n}{n} \in B\right) = \phi(n)\, e^{-n\Lambda^*(\hat{x})}$$

for some subexponential (at ∞) function $\phi(\cdot)$

$$n^{-1}\log \phi(n) \to 0 \qquad \text{as } n \to \infty \ .$$

For instance, $\phi(n)$ can be any polynomial function n^α or even $\exp(n^{1-\epsilon})$; this means that the LDT approximation may be very inaccurate and better results are sometimes available (for instance, the Bahadur-Rao exact asymptotics [10] for Normal RVs). On the other hand, in many cases LDT represents the only available analytical tool for the analysis of complex systems.

6. Cramér's theorem has a multivariate counterpart [5] dealing with the large deviations of the empirical mean of IID random vectors X_i in $\mathbb{R}^d$. In that case, the

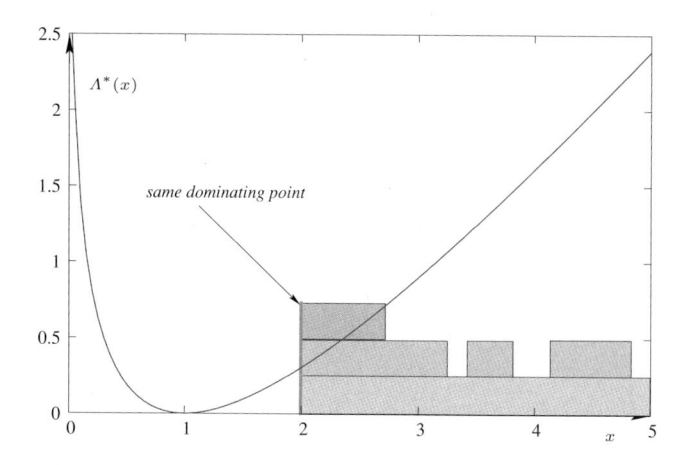

Fig. 2. Application of Cramér's theorem to sums of Exp(1) RVs

definition of the logarithmic cumulant generating function is the straightforward generalisation of (7):

$$\Lambda(\theta) \triangleq \log M(\theta) = \log \mathbb{E}\left(e^{\langle \theta, X_i \rangle}\right) , \tag{17}$$

where

$$\langle \theta, x \rangle = \sum_{j=1}^{d} \theta_j x_j$$

is the usual scalar product in $\mathbb{R}^d$ and x_j denotes the jth component of x.

2.2 General Principles of Large Deviation Theory

The general theory of large deviations has a beautiful and powerful formulation due to Varadhan [7], based on the so-called *Large Deviation Principle* (LDP), which leads to asymptotic results similar to (13), but under more general conditions.

Let S_n be any sequence of RVs, not necessarily the partial sum of IID RVs. In general, Cramér's theorem cannot be invoked *as is* (for instance if the RVs are correlated, as it often happens in computer networks); however, the "scaled" sequence[4] S_n/n may happen to show the *same asymptotic behaviour* proved for the partial sums of IID RVs. In LDT terms, this means that the sequence S_n/n satisfies an LDP.

The following subsections introduce the definition of LDP (at first in $\mathbb{R}^d$ and then its generalisation in Hausdorff spaces) and the main tools that can be used *to build* an LDP.

---

[4] In some cases (see section 4.1 for a relevant application in the field of network performance) it will be necessary to change the scaling factor and consider the sequences S_n/v_n for an adequate choice of the deterministic scaling factors v_n.

Large Deviations Principle in $\mathbb{R}^d$. In its abstract formulation [5], the *large deviation principle* characterises the limiting behaviour of a family of Borel probability measures on a Hausdorff space in terms of a *rate function*.

As in [6], to make the concept more intuitive to non-specialists, preliminary definitions of rate function (not simply the convex conjugate of the Logarithmic Moment Generating Function) and LDP are given in the framework of $\mathbb{R}^d$-valued RVs.

In the following $\mathbb{R}^*$ will denote the extended real numbers, $\mathbb{R} \bigcup \{\infty\}$.

Definition 1 (Rate function). *A function $I : \mathbb{R}^d \to \mathbb{R}^*$, is a* rate function *if*

1. $I(x) \geq 0$ *for all $x \in \mathbb{R}^d$*
2. *I is lower semicontinuous, i.e., the level sets $\{x : I(x) \leq \alpha\}$ are all closed, for $\alpha \in \mathbb{R}$*
3. *It is called a* good rate function *if in addition the level sets are all compact*

The definition of lower semicontinuity implies that I is allowed to jump down, but not to jump up; indeed, a function I is lower semicontinuous (according to a definition equivalent to the previous one) iff

$$\text{whenever } x_n \to x \quad \liminf_{n \to \infty} I(x_n) \geq I(x) \ .$$

It is easy to verify, for instance, that Λ^* in Cramér's theorem is a *good rate function*, where the term "good" has been introduced to highlight that some LDT results (such as the widely used contraction principle) only hold if the rate function has this additional property (i.e., if the level sets are not only closed, but also compact).

Definition 2 (Large Deviations Principle). *Let $(X_n, n \in \mathbb{N})$ be a sequence of RVs taking values in $\mathbb{R}^d$. X_n satisfies a* large deviations principle *in $\mathbb{R}^d$ with rate function I if I is a rate function and, for any measurable set $B \subset \mathbb{R}^d$*

$$\begin{aligned} - \inf_{x \in B^\circ} I(x) &\leq \liminf_{n \to \infty} \frac{1}{n} \log \mathbb{P}\left(X_n \in B\right) \\ &\leq \limsup_{n \to \infty} \frac{1}{n} \log \mathbb{P}\left(X_n \in B\right) \leq - \inf_{x \in \bar{B}} I(x) \end{aligned} \quad (18)$$

where B° denotes the interior of B and $\bar{B}$ its closure.

For example, Cramér's theorem states that the empirical mean S_n/n of IID RVs satisfies an LDP with good rate function Λ^* given by (10). Further examples of LDP will be given in the following sections.

Gärtner-Ellis Theorem. The *Gärtner-Ellis theorem* defines under which hypotheses the sequence S_n/n satisfies an LDP (for instance, an LDP can be derived for dependent random processes, such as Markov chains and autoregressive processes) and says how to calculate the corresponding rate function.

Roughly speaking, the generalisation of Cramér's theorem to any sequence of RVs mainly relies on the existence of a sufficiently "well-behaved" non trivial *limiting scaled cumulant generating function*

$$\Lambda(\theta) = \lim_{n \to \infty} \frac{1}{n} \log \mathbb{E} e^{\theta S_n} \quad (19)$$

and, given the existence of the exponential moments of S_n, this essentially requires that the autocorrelation of the increments of S_n decays sufficiently fast. For instance, this result can be used to prove an LDP for a queue with a weakly dependent input flow [6]. To state the theorem properly (for sake of generality, for random vectors in $\mathbb{R}^d$), it is useful to recall the following definition:

Definition 3 (Essential smoothness). *A convex function $\Lambda : \mathbb{R}^d \to \mathbb{R}^*$ is essentially smooth if*

1. $(\mathcal{D}_\Lambda)^o$ in non-empty
2. $\Lambda(\cdot)$ is differentiable throughout $(\mathcal{D}_\Lambda)^o$
3. $\Lambda(\cdot)$ is steep, i.e., for any sequence θ_n in $(\mathcal{D}_\Lambda)^o$ which converges to a boundary point of $\mathcal{D}_\Lambda$

$$\lim_{n \to \infty} |\nabla \Lambda(\theta_n)| = +\infty$$

where $\mathcal{D}_\Lambda$ denotes the effective domain of $\Lambda(\cdot)$, i.e.,

$$\mathcal{D}_\Lambda = \{\theta : \Lambda(\theta) < \infty\}$$

Theorem 2 (Gärtner-Ellis Theorem). *Let S_n be a sequence of random vectors in $\mathbb{R}^d$ with cumulant generating functions:*

$$\Lambda_n(\theta) = \log \mathbb{E}\left(e^{\langle \theta, S_n \rangle}\right) . \tag{20}$$

Assume that:

1. *The limiting scaled cumulant generating function*

$$\Lambda(\theta) = \lim_{n \to \infty} \frac{1}{n} \Lambda_n(\theta) \tag{21}$$

exists in $\mathbb{R}^*$ for each $\theta \in \mathbb{R}^d$
2. *$\Lambda(\theta)$ is finite in a neighbourhood of $\theta = 0$, i.e., $0 \in (\mathcal{D}_\Lambda)^o$*
3. *Λ is essentially smooth and lower-semicontinuous.*

Then, the sequence S_n/n satisfies an LDP in $\mathbb{R}^d$ with good rate function Λ^*

$$\Lambda^*(x) \triangleq \sup_{\theta \in \mathbb{R}^d} (\langle \theta, x \rangle - \Lambda(\theta)) . \tag{22}$$

To illustrate the meaning of the Gärtner-Ellis theorem, it is useful to consider the empirical mean S_n/n of real-valued RV X_i, where

$$S_n = X_1 + X_2 + \cdots + X_n ,$$

under different correlation structures (calculations may be found in [6]):

- **IID RVs**: it is trivial to prove that

$$\Lambda_n(\theta) \triangleq \log \mathbb{E} e^{\theta S_n} = n \log \mathbb{E} e^{\theta X_i}$$

and hence the rate function given by (22) coincides with (10). This explains why the Gärtner-Ellis theorem is sometimes (for instance, in [6]) referred to as the *generalised Cramér's theorem*.

- **Additive functionals of Markov chains:** let $(\xi_n, n \in \mathbb{N})$ be an irreducible Markov chain, taking values in a finite set E with transition matrix $P = \{p_{ij}\}$. Let f be a function from E to $\mathbb{R}$ and define $X_n = f(\xi_n)$; finally, let $Q(\theta)$ denote the (non-negative irreducible) $E \times E$ matrix whose $\{ij\}$-element is

$$q_{ij}(\theta) = e^{\theta f(i)} p_{ij}$$

and let $\rho(\theta)$ denote its spectral radius (Perron-Frobenius eigenvalue). Then, S_n/n satisfies an LDP with

$$\Lambda(\theta) = \log \rho(\theta) \ .$$

In queueing applications, this result is quite relevant, since it permits to identify the rate function for Markov-modulated fluid sources (S_n/n represents the average data rate over n time slots).

- **Gaussian autoregressive processes:** the samples X_i are defined by the (stable) recursion

$$X_i = \sum_{k=1}^{r} a_k X_{i-k} + \epsilon_i \qquad i \in \mathbb{Z}$$

where the ϵ_i are independent standard normal RVs. The covariance structure of $(X_i, i \in \mathbb{Z})$ is usually described through its Fourier transform

$$\mathcal{S}_X(\omega) = \sum_{k=-\infty}^{\infty} \mathbb{E}(X_0 X_k) \, e^{i\omega k}$$

which is called the power spectral density of the process. It is easy to show that $\mathcal{S}_X(\omega) = |A(\omega)|^2$, where

$$A(\omega) \stackrel{\Delta}{=} 1 - \sum_{j=1}^{r} a_j e^{i\omega j} \ .$$

Then, S_n/n satisfies an LDP with rate function

$$I(x) = \frac{x^2}{2\mathcal{S}_X(0)} \ .$$

It is worth mentioning that different Gaussian processes having the same power spectral density at zero have the same rate function. The underlying assumption is that $\mathcal{S}_X(\omega)$ is finite and differentiable on $[-\pi, \pi]$. This basically requires that the correlations decay sufficiently fast; for LRD processes the spectrum has a singularity at zero and, to use LDT, it will require a different scaling in n (see section 4.1).

Large Deviations Principle in a Hausdorff space. In the study of queueing systems, it is sometimes useful to consider the large deviations of the sample mean of random processes (i.e., infinitely dimensional objects); for instance, Schilder's theorem gives an expression for the probability of the sample mean (which is now a *path*, i.e., a function of time) of n IID Gaussian processes being in some set $\mathcal{S}$. To include such results in the general theory, it is necessary to rephrase the large deviation principle in a more powerful way, making use of the classical abstract language of LDT.

Definition 4 (Large Deviations Principle). *Let $(\mu_n, n \in \mathbb{N})$ be a sequence of Borel probability measures on a Hausdorff space $\mathcal{X}$ and let $\mathcal{B}$ be the Borel σ-algebra. μ_n satisfies a* large deviations principle *on $\mathcal{X}$ with rate function I if I is a rate function and, for all $B \in \mathcal{B}$*

$$-\inf_{x \in B^\circ} I(x) \leq \liminf_{n \to \infty} \frac{1}{n} \log \mu_n(B) \\ \leq \limsup_{n \to \infty} \frac{1}{n} \log \mu_n(B) \leq -\inf_{x \in \bar{B}} I(x) \quad (23)$$

A few comments permit to better clarify the LDP concept in its general form:

1. If X_n is a sequence of RVs with distribution μ_n, then we may equivalently say that the sequence X_n satisfies the LDP.
2. If $\mathcal{X}$ is a space of functions indexed by $\mathbb{R}$ or $\mathbb{N}$, the LDP is usually called *sample path LDP*.
3. If X_n satisfies an LDP in a regular Hausdorff space $\mathcal{X}$ with rate function I, and with rate function J, then $I = J$ (uniqueness of the rate function).
4. A set $A \subset \mathcal{X}$ is called an I-continuity set if

$$\inf_{x \in A^\circ} I(x) = \inf_{x \in \bar{A}} I(x) \ .$$

For such a set (for instance, if $\mathcal{X} = \mathbb{R}$ and I is continuous, then all intervals are I-continuity sets), if it is measurable, then (23) becomes

$$\lim_{n \to \infty} \frac{1}{n} \log \mu_n(A) = -\inf_{x \in A} I(x) \ . \quad (24)$$

Starting from the existence of an LDP, it is possible to give a precise definition (see [6] for the proof) of one of the *most* famous LDT results, the *principle of the largest term*.

Indeed, if I is a good rate function and $A \subset \mathcal{X}$ is closed, then the infimum is attained at some $\hat{x} \in A$. This $\hat{x}$ is the most likely way for an event A to occur, since $I(\hat{x})$ dominates in $\mathbb{P}(X_n \in A)$.

Theorem 3 (Rare events occur in the most likely way). *Suppose X_n satisfies an LDP with good rate function I, and C is a closed set with*

$$\inf_{x \in C} I(x) = k < \infty \ .$$

This infimum must be attained; suppose it is attained in C° and let B be a neighbourhood of $\{x \in C : I(x) = k\}$. Then

$$\mathbb{P}(X_n \notin B \mid X_n \in C) \to 0 \ . \quad (25)$$

The Contraction Principle. The contraction principle is one of the most useful tools in LDT; indeed, once we have an LDP for one sequence of RVs, we can *effortlessly*[5]

---

[5] At least in principle; in practise it might be quite difficult to establish the continuity of a given function, and to compute the resulting rate function.

establish LDPs for a whole other class of random sequences, obtained via continuous transformations.

For example, in queueing applications, starting from the LPD for the arrival process, if a quantity of interest can be written as a continuous function (in some Hausdorff space) of the arrivals, then it will be possible to deduce an LDP for that quantity.

Theorem 4 (Contraction Principle). *Let $\mathcal{X}$ be a Hausdorff space and suppose that X_n satisfies an LDP in $\mathcal{X}$ with good rate function I, and that $f : \mathcal{X} \to \mathcal{Y}$ is a continuous map to another Hausdorff space $\mathcal{Y}$.*
Then $f(X_n)$ satisfies an LDP in $\mathcal{Y}$, with good rate function

$$J(y) = \inf_{x \in \mathcal{X}: f(x) = y} I(x) \ . \tag{26}$$

Although the proof of the theorem is rather technical (mainly to prove that J is a good rate function), it is easy to give a heuristic justification taking into account the basic idea behind the LDT limits in the spirit of (16):

$$\mathbb{P}\left(f(X_n) \approx y\right) \approx \mathbb{P}\left(X_n \approx f^{-1}(\{y\})\right) \approx$$

$$\approx e^{-n \inf_{x \in f^{-1}(\{y\})} I(x)} = e^{-n \inf_{x: f(x) = y} I(x)}$$

Unfortunately, the hypotheses of the contraction principle are too restrictive for its application in the framework of *many flows scaling*; hence in [6] a generalisation is given, in which Y_n is only exponentially equivalent to $f(X_n)$ (i.e., the probability they differ even by ϵ decays superexponentially for all $\epsilon > 0$) and f is continuous only on the subspace where the rate function is finite.

2.3 Sample Path Large Deviations

In many applications, interest lies in the probability that a *path* of a random process hits a particular set; the LDT tools are to be developed in an infinitely dimensional framework and quite often are rather abstract. For sake of brevity, only Gaussian processes are considered in this section, since they represent a widely used model for aggregated traffics [3].

More in detail, as an example of Sample Path LDT, the Schilder's theorem (for Brownian motion) is deeply discussed and then the result is extended to a wider class of Gaussian processes.

Schilder's Theorem. Schilder's theorem analyses the most likely paths of a standard Brownian motion $B(t)$, while a sample path LDP for a generic random walk is given by the Mogulskij's theorem [5].

Before stating the theorem, it might be useful to recall the main properties of the Brownian motion and the definition of absolutely continuous function.

Definition 5 (standard Brownian motion). *A standard Brownian motion is characterised by the following properties:*

- $B(\cdot)$ is Gaussian, i.e., its finite-dimensional distributions are multivariate normal
- $B(t) \in \mathcal{N}(0,t)$
- $B(0) = 0$
- $B(t)$ has independent increments, i.e., $(B(t+u) - B(u))$ is independent of $B(u)$ and
$$B(t+u) - B(u) \in \mathcal{N}(0,t)$$
- $B(\cdot)$ has continuous sample paths

Definition 6 (**Absolutely continuous function**). *A function $f : [0,1] \to \mathbb{R}$ is absolutely continuous if for all $\epsilon > 0$ there exists a $\delta > 0$ such that for every finite collection of non-overlapping intervals $\{[s_i, t_i], 1 \leq i \leq N\}$*

$$\sum_{1 \leq i \leq N} (t_i - s_i) < \delta \quad \Rightarrow \quad \sum_{1 \leq i \leq N} |f(t_i) - f(s_i)| < \epsilon$$

Theorem 5 (**Schilder's Theorem**). *Let $(B(t), t \in [0,1])$ be a standard Brownian motion, taking values in $\mathcal{C}[0,1]$, the space of continuous functions $f : [0,1] \to \mathbb{R}$ equipped with the supremum norm:*

$$\|f\| = \sup_{0 \leq t \leq 1} |f(t)| .$$

Then $\left(B^n(t) \stackrel{\Delta}{=} \frac{1}{\sqrt{n}} B(t), n \in \mathbb{R}^+\right)$ satisfies a sample path LDP in $\mathcal{C}[0,1]$ with good rate function

$$I(f) = \begin{cases} \frac{1}{2} \int_0^1 \dot{f}(t)^2 dt & \text{if } f \text{ is absolutely continuous and } f(0) = 0 \\ \infty & \text{otherwise} \end{cases} \quad (27)$$

A heuristic argument, based on the Cramér's theorem for Gaussian RVs, can lead to a simple (and instructive) justification of (27). A rigorous proof of the theorem and its extension to $[0,T]$ (for any $T < \infty$) can be found in [5].

Let $\Pi_K f$ be the polygonalised version of f, i.e., the piecewise linear approximation of f at n/K, $(0 \leq n \leq K)$; then, assuming $f(0) = 0$ (otherwise the probability is 0 since, by definition, $B(0) = 0$; hence if $f(0) \neq 0$, the corresponding rate function should be $I(f) = \infty$)

$$\mathbb{P}(B^n(\cdot) \approx f(\cdot)) \approx \mathbb{P}(\Pi_K B^n(\cdot) \approx \Pi_K f(\cdot)) .$$

Since $B^n(\cdot)$ has independent increments, the latter can be written as

$$\prod_{i=0}^{K-1} \mathbb{P}\left(B^n\left(\frac{i+1}{K}\right) - B^n\left(\frac{i}{K}\right) \approx f\left(\frac{i+1}{K}\right) - f\left(\frac{i}{K}\right)\right)$$

and, taking into account that $B^n(t) \in \mathcal{N}\left(0, \frac{t}{n}\right)$,

$$\prod_{i=0}^{K-1} \mathbb{P}\left(\mathcal{N}\left(0, \frac{1}{nK}\right) \approx f\left(\frac{i+1}{K}\right) - f\left(\frac{i}{K}\right)\right) .$$

Since in the LDT limit
$$\mathbb{P}\left(\mathcal{N}\left(0, \frac{\sigma^2}{L}\right)\right) \approx e^{-L \frac{1}{2\sigma^2} x^2} ,$$
it is easy to show that
$$\frac{1}{n} \log \mathbb{P}\left(B^n(\cdot) \approx f(\cdot)\right) \approx -\frac{K}{2} \sum_{i=0}^{K-1} \left(f\left(\frac{i+1}{K}\right) - f\left(\frac{i}{K}\right)\right)^2$$
$$= -\frac{1}{2} \sum_{i=0}^{K-1} \frac{1}{K} \left(\frac{f\left(\frac{i+1}{K}\right) - f\left(\frac{i}{K}\right)}{1/K}\right)^2$$
$$\xrightarrow{K \to \infty} -\frac{1}{2} \int_0^1 \dot{f}(t)^2 \, dt ,$$
which gives the expression of rate function $I(f)$ when $f(0) = 0$:
$$I(f) = \frac{1}{2} \int_0^1 \dot{f}(t)^2 dt .$$

The previous computations highlight that, at least informally, multivariate Cramér's theorem can be seen as a special (finite-dimensional) case of Schilder's theorem. Moreover, it is worth noticing that the cost of a path f is exclusively determined by the derivative along the path.

Generalised Schilder's Theorem. In [11], Schilder's theorem is extended to the general case of a (non-trivial) centred Gaussian process $A(t)$, with $A(0) = 0$ and stationary increments. The variance function of $A(\cdot)$ is denoted by $v(t)$; the standard Brownian motion $B(\cdot)$ is only a special case, with $v(t) = t$, for which the resulting expressions are relatively transparent (as shown in the previous section). A key role in the following will be played by the covariance function of $A(\cdot)$:

$$\Gamma(s, t) = \text{Cov}(A(s), A(t)) = \frac{1}{2}(v(t) + v(s) - v(|t - s|)) . \qquad (28)$$

An intrinsic difficulty of the generalisation of Schilder's theorem is that the rate function $I(\cdot)$ cannot be given explicitly. The case of Brownian motion is an exception: indeed, due to the *independence* of the increments, it was possible to derive an explicit formula for $I(f)$ (see the heuristic justification of the theorem). To state the general theorem, it is necessary to introduce a *path state* Ω and a *reproducing kernel Hilbert space* R, equipped with inner product $\langle \cdot, \cdot \rangle_R$ and norm $\|\cdot\|_R$.

The *path space* Ω is defined as
$$\Omega = \left\{\omega : \mathbb{R} \to \mathbb{R}, \text{ continuous}, \omega(0) = 0, \lim_{t \to \pm\infty} \frac{\omega(t)}{1 + |t|} = 0\right\}$$
equipped with the norm
$$\|\omega\|_\Omega = \sup_{t \in \mathbb{R}} \frac{|\omega(t)|}{1 + |t|} .$$

$A(\cdot)$ can be realised on Ω under the assumption that $v(\cdot)$ increases slower than quadratically. For example, this is the case for fractional Brownian motion, one of the most relevant LRD traffic models, which is characterised by $v(t) = t^{2H}$, where $1/2 < H < 1$ (see section 4.1 for a discussion on LRD and its implications on traffic engineering).

In addition to Ω, a central role is played by a linear subspace of Ω, which consists of smoother functions than the typical paths of $A(\cdot)$ and which can be given a Hilbert space structure. This space, the *reproducing kernel Hilbert space* R, is defined starting from the set of functions $\{\Gamma(s, \cdot)\}$, equipped with the inner product

$$\langle \Gamma(s, \cdot), \Gamma(\cdot, t) \rangle_R = \Gamma(s, t) .$$

The space R is obtained by closing this set of functions with linear combinations, and completing with respect to the norm

$$\|w\|_R^2 = \langle w, w \rangle_R .$$

The inner product definition generalises to the *reproducing kernel property*:

$$\langle w, \Gamma(s, \cdot) \rangle_R = w(s) \qquad w \in R . \tag{29}$$

To give a heuristic understanding of the space R, let us consider a centred Gaussian distribution on $\mathbb{R}^d$. In this case, the space R is $\mathbb{R}^d$ itself, but equipped with an inner product such that the density of the distribution can be written as

$$f(x) = \text{const} \cdot \exp\left(-\frac{1}{2}\|x\|_R^2\right) .$$

Thus, minimising $\|\cdot\|_R$ corresponds to maximising the density.

Theorem 6 (Generalized Schilder's Theorem). *Let $A(\cdot) \in \Omega$ be a (non trivial) centred Gaussian process, with variance function $v(t)$. Then $\left(\frac{1}{\sqrt{n}} A(\cdot), n \in \mathbb{R}^+\right)$ satisfies a sample path LDP in Ω with good rate function*

$$I(f) = \begin{cases} \frac{1}{2}\|f\|_R^2 & \text{if } f \in R \\ \infty & \text{otherwise} \end{cases} \tag{30}$$

In [10], the generalised Schilder's theorem is written in terms of the *sample-mean path*

$$\frac{1}{n}\sum_{i=1}^{l} A_n(\cdot)$$

of a sequence of IID centred Gaussian processes with variance function $v(t)$. Informally, the theorem gives an expression for the probability of the *sample-mean path* being in some set $\mathcal{S}$ (that represents a collection of paths):

$$\mathbb{P}\left(\frac{1}{n}\sum_{i=1}^{n} A_i(\cdot) \in \mathcal{S}\right) \approx \exp\left(-n \inf_{f \in \mathcal{S}} I(f)\right) = \exp\left(-\frac{n}{2} \inf_{f \in \mathcal{S}} \|f\|_R^2\right) .$$

Hence, the probability decays exponentially in n and the corresponding exponential decay rate equals the minimum of $I(f)$ over all $f \in \mathcal{S}$.

The minimising $\hat{f}(\cdot)$ corresponds to the *most likely path* in $\mathcal{S}$. Conditional on the sample-mean path being in the set $\mathcal{S}$, with overwhelming probability this happens via a path that is *close to* $\hat{f}$. In other words, as $n \to \infty$

$$\frac{1}{n} \log \mathbb{P}\left(\frac{1}{n} \sum_{i=1}^{n} A_i(\cdot) \in \mathcal{S}\right) \to -I\left(\hat{f}\right)$$

and the decay rate is fully dominated by the likelihood of the most likely element in $\mathcal{S}$.

Unfortunately, finding the minimum of $I(f)$ over all $f \in \mathcal{S}$ is, in general, a hard variational problem. Indeed, the optimisation should be done over all paths in $\mathcal{S}$ (which are infinitely dimensional objects), and, according to (29), the objective function $I(f)$ is only explicitly given if f can be written as a linear combination of covariance functions $\Gamma(s, \cdot)$.

3 Large Deviations for Queues

One of the primary issues in queueing theory is to analyse the (steady-state) buffer content distribution. This problem can be solved explicitly only in a few special cases, and the goal of LDT is to get approximate estimations of the parameters of interest that are sufficiently close to the actual values at least in some asymptotic conditions. In more detail, two asymptotic scalings are usually considered: the *large buffer* regime and the *many-sources* regime, and for both of them LDT permits to obtain logarithmic asymptotics. Although in the following only the overflow probability in stationary condition will be considered, it is worth mentioning that LDT may be applied to estimate other quantities such as the most likely way a queue became big, the exit probability, the distribution of busy periods and even the way steady state is reached (see, as a comprehensive illustrative example, the analysis of the M/M/1 queue in [12]).

The main result in this section is the LDP for the single server queue, which, at least under some restrictive hypotheses, can be established through direct calculation. More general results are achieved making use of abstract LDT tools, such as the contraction principle (in an adequately chosen Hausdorff space), once an LDP for the arrival process is known.

3.1 The Single Server Queue

The evolution of a (FIFO) single server queue is described, in both continuous and discrete time settings, by the Lindley's recursion

$$Q_{n+1} = (Q_n + X_{n+1})^+ \tag{31}$$

where $x^+ = \max(0, x)$ and the interpretation of the different entities depends on the specific settings.

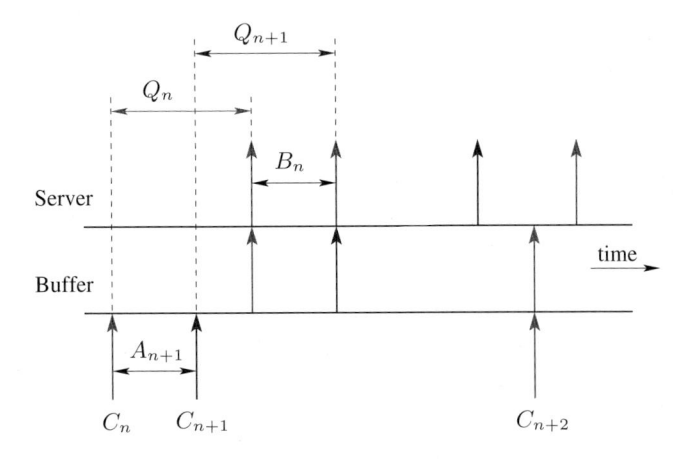

Fig. 3. Continuous time version of Lindley's recursion

In continuous time (see fig. 3), customers are labelled by the integers (C_n denotes the nth arrival) and X_{n+1} is the difference between the service time B_n of C_n and the interarrival time A_{n+1} between C_n and C_{n+1}; in this case Q_n gives the waiting time of C_n, i.e., the time spent in the queue before commencing service.

In discrete time (slotted time model) the interpretation of (31) is straightforward: X_n is the difference between the amount of work A_n that arrives at the queue at time n (or, more in general, during the interval $(n-1, n)$) and the amount of work C_n that the server can process at time n; hence, Q_n represents the amount of work remaining in the queue just after time n. In the following we shall adopt the latter interpretation, but most of the results can be easily adapted to waiting time in the continuous time framework.

In the rather common case of constant rate server (which can be used, for instance, to model a transmission line), Lindley's recursion becomes

$$Q_{n+1} = (Q_n + A_{n+1} - C)^+ \qquad (32)$$

Equation (32) may have different solutions, depending on boundary conditions (see [6,10] and references therein for a detailed analysis). For example, let us consider the queue size at time $n = 0$, subject to the boundary condition that the queue was empty at $n = -\infty$. It is easy to prove that

$$Q_0^{-\infty} = \sup_{n \geq 0} S_n - Cn \qquad (33)$$

where S_n is the cumulative arrival process, i.e.,

$$S_n \stackrel{\Delta}{=} A_{-n+1} + A_{-n+2} + \cdots + A_{-1} + A_0$$

and, by convention, $S_0 = 0$. If the arrival process $(A_n, n \in \mathbb{Z})$ is stationary, then $Q_0^{-\infty}$ has the same distribution as $Q_n^{-\infty}$ for any $n \in \mathbb{Z}$ and this distribution is called the *steady state distribution* of queue size. Moreover, if $(A_n, n \in \mathbb{Z})$ is also ergodic and $\mathbb{E}A_0 < C$ (i.e., $\mathbb{E}X_0 < 0$), then the limit does not depend on the initial condition.

In the following, $(X_n, n \in \mathbb{Z})$ will be a stationary ergodic sequence of RVs with $\mathbb{E}X_0 < 0$ (stability condition) and Q will denote the unique equilibrium distribution, i.e., in case of constant rate server:

$$Q = \sup_{n \geq 0} S_n - Cn \ . \tag{34}$$

In other word, the steady-state buffer content (a reflected additive recursion, which can only assume non negative values) is distributionally equal to the supremum of a free (i.e., nonreflected) process with negative drift (the supremum is non-negative!).

To conclude this overview on Lindley's recursion, it is worth noticing that this framework, although it has been developed for a slotted system, can be directly extended (Reich's theorem) to the *steady-state queue length in continuous time*

$$Q = \sup_{t \geq 0} A(-t, 0) - Ct \ ,$$

where $A(s,t)$ denotes the amount of traffic offered to the system in $[s,t)$. Moreover, if the arrival process is time reversible, then

$$Q = \sup_{t \geq 0} A(t) - Ct \ . \tag{35}$$

3.2 LDT Asymptotics

Since LDT is an asymptotic theory, the solution of Lindley's recursion is analysed in some asymptotic conditions. In particular, two different regimes (large buffer and many sources) are discussed in the next subsections, presenting the key results and highlighting the ideas behind the proofs, which can be found in [6] together with illustrative examples.

Large-buffer regime. In the large-buffer regime, traditionally the most investigated limit (and not only in the field of LDT), the objective is to find asymptotic expansions of the queue size complementary probability $\mathbb{P}(Q > q)$ for $q \to \infty$.

In this section we will consider a (single server FIFO) queue with constant service rate C and arrival process $(A_t, t \in \mathbb{Z})$, A_t being the amount of work arriving at time t. In [6] an LDP for queue size is derived, at first, under the assumption that the A_t were IID and then weakening this assumption as in the Gärtner-Ellis theorem (see section 2.2). For sake of brevity, here only the more general statement is given, followed by some remarks on the proof and on the *interpretation* of the LDP.

Theorem 7 (LDP for queue size). *Let $(A_t, t \in \mathbb{Z})$ be a stationary random process, with $\mathbb{E}A_0 < C$ and let*

$$\Lambda_t(\theta) = \log \mathbb{E} e^{\theta S_t} \ .$$

Suppose that

1. the limit

$$\Lambda(\theta) = \lim_{t \to \infty} \frac{1}{t} \Lambda_t(\theta) \tag{36}$$

exists in $\mathbb{R}^*$ for each $\theta \in \mathbb{R}$

2. $\Lambda(\theta)$ is essentially smooth, and finite in a neighbourhood of $\theta = 0$
3. $\Lambda_t(\theta)$ is finite for all t whenever $\Lambda(\theta) < \theta C$

Then, for $q > 0$:
$$\lim_{l \to \infty} \frac{1}{l} \log \mathbb{P}\left(\frac{Q}{l} > q\right) = -I(q) \tag{37}$$

where
$$\begin{aligned} I(q) &= \inf_{t \in \mathbb{R}^+} t \Lambda^*(C + q/t) \\ &= \inf_{t \in \mathbb{R}^+} \sup_{\theta \geq 0} \theta(q + Ct) - t \Lambda(\theta) \\ &= q \sup\{\theta > 0 : \Lambda(\theta) < \theta C\} \end{aligned} \tag{38}$$

Some remarks may be useful to better understand the theorem.

1. Equation (37) is usually written in a visually simpler equivalent form
$$\lim_{q \to \infty} \frac{1}{q} \log \mathbb{P}(Q > q) = -I(1) \tag{39}$$

   The notation used in the theorem is "closer" to the standard formulation of an LDP. In any case there are two differences:
   (a) Upper and lower bounds happen to agree, so the theorem proves a limit.
   (b) It is a *restricted sort* of LDP, since the theorem only concerns intervals $[q, +\infty)$ and not general events.

2. The assumption that $\Lambda(\theta)$ is finite in a neighbourhood of the origin is necessary to guarantee the exponential decay of the queue size complementary probability. Completely different behaviours are associated to LRD traffic flows (section 4.1) as well as to heavy-tailed distributions [13].

3. The lower bound is proved by estimating the probability that the queue overflows over some fixed timescale. In other words, the approximation
$$\mathbb{P}\left(\sup_t S_t - Ct \geq q\right) \approx \sup_t \mathbb{P}(S_t - Ct \geq q) \tag{40}$$
   is justified (for large q) on a logarithmic scale.

4. The *most likely time* for the queue to fill up to some high level q is $l\hat{t}$, where $\hat{t}$ is the optimising parameter for $I(1)$ according to its definition in (38). Thus the most likely rate for the queue to build up is $1/\hat{t}$ and does not depend on q.

5. Another interpretation of the theorem is that $S_t - Ct$ is effectively a simple random walk with negative drift, in that
$$\mathbb{P}\left(\sup_t S_t - Ct \geq q_1 + q_2\right) \approx \mathbb{P}\left(\sup_t S_t - Ct \geq q_1\right) \mathbb{P}\left(\sup_t S_t - Ct \geq q_2\right)$$
   for large q_1 and q_2. Thus, the (weak) dependence of the A_t is invisible at the macroscopic scale (although it does contribute to the value of $I(1)$ through $\Lambda(\theta)$).

6. If the service is a RV C_t, it is possible to apply the theorem to the random process $A_t - C_t$ (rather than to A_t) and set $C = 0$. Under the usual assumption of independence between service and arrival processes, it is easy to show that (if the limits exist)
$$\Lambda(\theta) = \Lambda_A(\theta) + \Lambda_C(-\theta) .$$

Many-sources regime. An important limitation of large-buffer regime is that it does not give reasonably accurate results about overflow probability for small buffers, which can be desirable in case of applications with stringent delay constraints. Moreover, it might be useful to take into account that the input traffic can be often seen as the superposition of many IID streams.

These thoughts has led to the interest for the so-called many-sources regime. In this setting it is assumed that the number of source N grows large and, at the same time, the queueing resources (buffer and bandwidth) are scaled accordingly. In more detail, the buffer threshold is replaced by Nq and the service capacity by NC. Despite the fact that the load remains constant, it is clear that the overflow probability decays to 0.

Let $A_t^{(i)}$ denote the amount of work arriving from source i at time t. Assume that

1. for each i, $\left(A_t^{(i)}, t \in \mathbb{Z}\right)$ is a stationary sequence of RVs
2. these sequences are independent of each other and identically distributed.

If the total amount of work arriving at the queue in the interval $(-t, 0]$ is denoted by S_t^N, the queue length at time 0 is given by

$$Q^N = \sup_{t \geq 0} S_t^N - NCt$$

and, in the spirit of LRD, we will consider the behaviour of $\mathbb{P}\left(Q^N \geq Nq\right)$ as the number of sources becomes large.

Theorem 8 (LDP for queue size with many sources). *Let S_t^1 be the amount of work produced by a typical source in the interval $(-t, 0]$ with $\mathbb{E}S_1^1 < C$ and let*

$$\Lambda_t(\theta) = \log \mathbb{E}e^{\theta S_t^1}.$$

Suppose that

1. the limit

$$\Lambda(\theta) = \lim_{t \to \infty} \frac{1}{t} \Lambda_t(\theta) \tag{41}$$

exists, and is finite and differentiable in a neighbourhood of the origin
2. for all t, $\Lambda_t(\theta)$ is finite for θ in a neighbourhood of the origin

Then

$$-I(q+) \leq \liminf_{N \to \infty} \log \mathbb{P}\left(Q^N > Nq\right)$$
$$\leq \limsup_{N \to \infty} \log \mathbb{P}\left(Q^N > Nq\right) \leq -I(q) \tag{42}$$

where

$$I(q) = \inf_{t \in \mathbb{N}} \Lambda_t^*(q + Ct) = \inf_{t \in \mathbb{N}} \sup_{\theta \in \mathbb{R}} \theta(q + Ct) - \Lambda_t(\theta) \tag{43}$$

It is interesting to point out a few differences with respect to the previous theorem:

1. Expression (42) involves both an upper and a lower bound, as in classical LDT statements. If $\Lambda_t(\cdot)$ is continuous for each t, then the two bounds agree and we obtain a straightforward limit.
2. Once again the proof makes use of the principle of the largest term and of the most likely way in which the rare event may occur. However, in the many-sources limit, the optimising $\hat{t}$ is simply the most likely time to overflow and typically depends on q in a non-linear way.
3. The assumption (41) is a way to control the distribution of S_t^1 for large t and is needed to prove the upper bound (but not the lower bound), although it does not appear in the result ($I(q)$ depends only on Λ_t).
4. The previous condition does not allow for LRD sources; however, even in that case the probability of large queues still decays exponentially in the number of sources (but, as shown in section 4.1, a different scaling will be required in the definition of the limit expression for Λ).

3.3 Continuous Mapping Approach

It is quite complicated to apply the previous approach (based on direct calculation of the rate function for the overflow probability) to other queue parameters and to more complex network scenarios. An interesting alternative is represented by the use of the contraction principle, once the quantity of interest is expressed as a continuous function of all random inputs. In this way it is possible to analyse different service disciplines (priority queue, processor sharing), consider finite buffers and transient behaviours.

More in detail, let A denote any random influence on the network (in general it can be seen as a vector of arrival and service processes). Many relevant quantities (such as the queue size or the departure process at some queue) can be written as functions $f(A)$. In a nutshell, the continuous mapping approach consists of the following steps:

1. Consider a sequence of queueing networks indexed by L, in which the Lth network has a vector of inputs A^L, a version of A which is speeded up in time and scaled down in space (the exact scaling depends on the specific framework).
2. Prove a sample path LDP for A^L *in some topological space*.
3. Show that f is continuous *on that space*.
4. Use the contraction principle to derive an LDP for $f(A^L)$.
5. Simplify the resulting rate function (typically, the rate function for this LDP will be given as the solution to a variational problem).

A big advantage of this procedure is that, once a sample path LDP is proved for A^L, we can obtain LDPs for different quantities, under the assumption that they can be written as a continuous function of the inputs.

Another useful consequence of the application of the contraction principle is that we can not only estimate the probability of a rare event, but also find the most likely path to that event.

Large-buffers revisited. In spite of its apparent simplicity, the continuous mapping approach requires some technical work to identify the proper space in which $f(A)$ is actually continuous as well as to simplify the rate function. All these issues are deeply analysed in [6], at first identifying proper continuous queueing maps and then analysing, in two separate chapters, the two classical scaling regimes. Since the main goal of this tutorial is to give an introduction to LDT for non specialists, we will simply focus on the application of the contraction principle to derive an LDP for queues with large buffers.

In a single server queue with deterministic service rate C, the queue size at time 0 can be written as a function of the cumulative arrival process

$$Q_0 = \sup_{t \geq 0} S_t - C \cdot t \triangleq f(A)$$

where A denotes the entire input process $(S_t, t \geq 0)$. Since it will be more convenient to work in continuous time, we introduce its polygonalised version (with step 1), $\tilde{A} = \Pi_1 A$, defined for $t \in \mathbb{R}^+$. At this point it is meaningful to define the scaled processes:

$$\tilde{A}^L(t) = \frac{1}{L}\tilde{A}(Lt) \qquad (44)$$

and the continuous-time version (Reich's theorem) of the queue size function:

$$\tilde{f}(\tilde{A}) = \sup_{t \in \mathbb{R}^+} \tilde{A}(t) - C \cdot t \ . \qquad (45)$$

It is easy to verify through direct substitution that

$$\tilde{f}(\tilde{A}^L) = L^{-1} f(A) = L^{-1} Q_0$$

and this means that

$$\mathbb{P}\left(\tilde{f}(\tilde{A}) > b\right) = \mathbb{P}(Q_0 > Lb) \ . \qquad (46)$$

Let us assume that $\tilde{A}^L(t)$ satisfies an LDP in *some topological space* with *good rate function* I, i.e.,

$$\frac{1}{L} \log \mathbb{P}\left(\tilde{A}^L \in B\right) \approx -\inf_{a \in B} I(a) \ . \qquad (47)$$

If $\tilde{f}$ is continuous on *that space*, then the contraction principle gives estimates for the left hand side of (46) and hence for the overflow probability:

$$\frac{1}{L} \log \mathbb{P}(Q_0/L > b) \approx -J(b) \qquad (48)$$

where

$$J(b) = \inf_{a: f(a) > b} I(a) \ . \qquad (49)$$

The expression of the rate function J justifies that the probability of a rare event can be estimated by considering only the *optimal manner* for that event to occur. In other words, the most likely way for the rare event $\{Q_0 > Lb\}$ to occur is when the input process $\tilde{A}^L$ is close to the optimising a, which represents the *most likely path to overflow*.

In the previous discussion, we assumed the existence of a topological space in which A^L satisfies an LDP and on which f is continuous. As stated in [6] (where some instructive counterexamples are also shown), this involves a trade-off and, in general, the selection of the proper topological space depends on the application. It turns out that a suitable choice for the single server queue (and for many other systems) is the space C_μ, defined as the set of continuous functions $x : \mathbb{R}^+ \to \mathbb{R}$, for which $x(0) = 0$ and

$$\lim_{t \to \infty} \frac{x(t)}{t+1} = \mu , \qquad (50)$$

equipped with the topology induced by the *scaled uniform norm*

$$\|x\| = \sup_{t \in \mathbb{R}^+} \left| \frac{x(t)}{t+1} \right| . \qquad (51)$$

Indeed, the queue size function (45) is continuous on C_μ, where μ represents the mean arrival rate (the polygonalised version of the cumulative arrival process corresponds to $x(t)$ in the definition (50)).

It is worth mentioning that in the many-sources regime it is necessary to work in a *larger space* and use the extended version of the contraction principle. The problem is related to the definition of the mean arrival rate in (50); without going into details (see [6] for the definition of a "proper" space), a simple example is enough to highlight the trouble. Indeed, let us consider N constant rate flows, where each rate is drawn independently from $\mathcal{N}\left(\mu, \sigma^2\right)$; then the limit

$$\lim_{t \to \infty} \frac{S_t^N}{t+1}$$

is not necessarily μ, since it is a RV $\in \mathcal{N}\left(\mu, \sigma^2/N\right)$.

4 Applications of LDT to Networks

The results of the previous sections can be applied to more complex scenarios and to general problems related to network dimensioning and planning. Just to show the heterogeneous capabilities of LDT, two completely different issues will be addressed in this section: the analysis of LRD traffic flows and the use of LDT to speed-up the simulation of rare events through Importance Sampling, which is based on a change of measure argument similar to the one employed in the proof of Cramér's theorem.

4.1 Long Range Dependence and Large Deviations

In the early 1990s, researchers at AT&T [2] claimed, on the basis of a huge collection of high-quality traffic measurements, that Internet traffic presents Long Range Dependence. The main consequences were the search for new traffic models, able to take into account this feature in a parsimonious way, and the analysis of queueing performance under the new traffic paradigm. The first issue has led to the introduction of self-similar (or, more in general, asymptotically self-similar) traffic models, among which the most popular is fractional Brownian motion. The main drawback of these models is the lack of analytical results for queueing performance; indeed, even in the case of a single server queue, only asymptotic results are available.

Basic definitions. For sake of completeness, we recall here the main definitions related to Long Range Dependence and Self-similarity (see, for example, [14] for a complete overview).

Definition 7 (Long Range Dependence). *Let $(X_n, n \in \mathbb{Z})$ be a second order stationary process with autocorrelation function $\rho(k)$ and power spectral density $S_X(\omega)$. X_n is Long Range Dependent (LRD) iff (the following properties are all equivalent):*

- *$\rho(k)$ decreases as a non summable power law when k tends to infinity*
$$\rho(k) \sim k^{-\alpha} \quad \text{as } k \to \infty \quad \text{where } 0 < \alpha < 1$$
- *$S_X(\omega)$ diverges as an integrable power law near the origin*
$$S_X(\omega) \sim \omega^{-\beta} \quad \text{as } \omega \to 0 \quad \text{where } 0 < \beta < 1 \text{ and } \beta = 1 - \alpha$$
- *The variance of the aggregated process decays more slowly than the sample size*
$$\operatorname{Var}\left(\frac{1}{n}\sum_{i=0}^{n-1} X_i\right) \sim n^{-\alpha} \quad \text{as } n \to \infty$$

One related phenomenon is self-similarity: roughly speaking, a dilated portion of the sample path of a self-similar process cannot be (statistically) distinguished from the whole. Indeed, self-similar processes have fluctuation at every time-scale, and the Hurst parameter relates the size of fluctuations to their time-scale according to (52).

Definition 8 (Self-similarity for continuous time processes). *Let $(Y_t, t \in \mathbb{R})$ be a continuous time process. Y_t is self-similar with self-similarity parameter H (Hurst parameter) iff*

$$c^{-H} Y_{ct} \stackrel{(d)}{=} Y_t \quad \forall c > 0 \tag{52}$$

i.e., if for any $k \geq 1$, for any $t_1, t_2, \ldots, t_k \in \mathbb{R}$ and for any $a > 0$

$$\left(Y_{at_1}, Y_{at_2}, \ldots, Y_{at_k}\right) \quad \text{and} \quad \left(a^H Y_{t_1}, a^H Y_{t_2}, \ldots, a^H Y_{t_k}\right)$$

have the same distribution

Typically self-similar processes are used to characterise the cumulated workload over a given time interval; in this framework the most popular and well-known self-similar model, widely adopted [3] for its parsimonious structure, is fractional Brownian motion (fBm).

Definition 9 (fractional Brownian motion). *A standard fractional Brownian motion $(Z_H(t), t \in \mathbb{R})$ with Hurst parameter H is characterised by the following properties:*

- *$Z_H(\cdot)$ is Gaussian, i.e., its finite-dimensional distributions are multivariate normal*
- *$Z_H(t) \in \mathcal{N}\left(0, |t|^{2H}\right)$*
- *$Z_H(\cdot)$ has stationary increments, i.e., $Z_H(u+t) - Z_H(u) \sim Z_H(t)$*
- *$Z_H(0) = 0$*
- *$Z_H(\cdot)$ has continuous sample paths*

From the above definition, it follows that Brownian motion is only a special case (for $H = 1/2$) of fBm; in that case, the analysis was much simpler since the increments were not only stationary, but also independent. Instead, for $H \neq 1/2$ the increments of $Z_H(t)$ are correlated and, if $1/2 < H < 1$, they exhibit Long Range Dependence.

Implications of LRD for Queues. Let $X(s,t] = X(t) - X(s)$ denote the amount of work arriving at a single server queue (with deterministic service rate C) in the time interval $(s,t]$, where $(X(t), t \in \mathbb{R})$ is a LRD process with drift μ (where $\mu < C$ to assure the stability of the queue) and $\mathsf{Var} X(-t, 0] \sim \sigma^2 t^{2H}$.

In the large-buffer regime (section 3.2), the LDP for the queue size basically states that

$$\lim_{q \to \infty} \frac{1}{q} \log \mathbb{P}(Q > q) = -\delta \tag{53}$$

with the underlying assumption of the existence of a *sufficiently well-behaved* limiting cumulant generating function

$$\Lambda(\theta) = \lim_{t \to \infty} \frac{1}{t} \Lambda_t(\theta) = \lim_{t \to \infty} \frac{1}{t} \log \mathbb{E} e^{\theta X(-t, 0]}.$$

If the limit exists, then the Taylor-Maclaurin expansion implies that

$$\mathsf{Var} X(-t, 0] \sim t \Lambda''(0). \tag{54}$$

This is not the case for LRD processes since $\mathsf{Var} X(-t, 0] \sim \sigma^2 t^{2H}$. However, a variant of (53) still holds when there is some sequence $(v_t, t \in \mathbb{N})$ taking values in $\mathbb{R}^+$, with $v_t/\log t \to \infty$, such that the limit

$$\Lambda(\theta) = \lim_{t \to \infty} \frac{\Lambda_t(\theta v_t/t)}{v_t} = \lim_{t \to \infty} \frac{1}{v_t} \log \mathbb{E} e^{\theta X(-t, 0] v_t/t} \tag{55}$$

exists, and is finite and differentiable in a neighbourhood of the origin. In that case, (54) becomes

$$\mathsf{Var} X(-t, 0] \sim \frac{t^2}{v_t} \Lambda''(0) \tag{56}$$

and a natural choice for LRD traffic is to put $v_t = t^{2(1-H)}$. If, under this scaling, the limit $\Lambda(\theta)$, defined by (55), exists and is well-behaved, then the queue size does not decay exponentially; instead

$$\lim_{q \to \infty} \frac{1}{q^{2(1-H)}} \log \mathbb{P}(Q > q) = -\delta \tag{57}$$

where

$$\delta = \inf_{t > 0} t^{2(1-H)} \Lambda^*(C + 1/t). \tag{58}$$

The special case of Gaussian processes has been deeply investigated for the analytical tractability and for the relevance in traffic modelling (the *physical* motivations and all the underlying difficulties are discussed, for instance, in [10,15]) of such processes. In that framework, logarithmic as well as exact asymptotics are known, although the latter (which go beyond the LDT set-up) are much harder to obtain [16,17]. For instance, if $X(t)$ is an fBm with drift μ, variance parameter σ and Hurst parameter H, i.e.,

$$X(t) = \mu t + \sigma Z_H(t),$$

the LDP (57) can be rewritten as follows:

$$\lim_{q \to \infty} \frac{1}{q^{2(1-H)}} \log \mathbb{P}(Q > q) = -\gamma^2/2 \qquad (59)$$

where

$$\gamma = \frac{(C-\mu)^H}{\sigma} \kappa \quad \text{and} \quad \kappa = \frac{1}{H^H (1-H)^{1-H}} \,. \qquad (60)$$

Hence, as a function of q, $\mathbb{P}(Q > q)$ decays in a Weibullian way[6] i.e., roughly as $\exp\left(-q^{2-2H}\right)$ and if $H \in (1/2, 1)$, the decay is slower than exponential.

It is worth mentioning that the same asymptotic expression (called *basic approximation* in [11]) for the overflow probability can be obtained directly from the solution of the Lindley's recursion, taking into account the principle of the largest term (in the spirit of approximation (40)) and the Chernoff bound (the optimising $\hat{t}$ represents the *most likely time-scale of overflow*). The application of the generalised Schilder's theorem (that gives the sample path LDP for a general Gaussian process and hence permits to identify the most likely path to overflow) has been extended to heterogeneous traffic flows (for Gaussian processes superposition means just adding the variance functions) as well as to more complex queueing systems, such as priority queues, generalised processor sharing schedulers [15] and tandem queues [18]. Such results, which could be justified, at least in principle, invoking the contraction principle applied to the proper continuous function $f(A)$, are indeed quite accurate over the full range of buffer sizes and even for quite high traffic levels [11].

4.2 LDT and Rare Event Simulation by Means of Importance Sampling

Importance Sampling (IS) is a popular technique devised to build unbiased estimators not suffering from the smallness of the probability of interest. This is achieved by changing the law of the process so that to favour the occurrence of the target rare event and taking this change into account by reweighting the estimation according to the *likelihood ratio*, which, in measure-theoretic terms, is the Radon-Nikodym derivative of the original law with respect to the new one [9].

The efficiency of an IS-based algorithm depends on the choice of a "proper" *change of measure* to reduce the variance of the estimate. It is well known that the optimal change of measure (zero-variance pdf) involves the knowledge of the probability we want to estimate and therefore cannot be practically adopted. The issue is commonly tackled by restricting potential IS measures to a parametric class and determining the optimal change of measure within this restricted class[7]. The most common approach is represented by the use of a so-called exponential change of measure (ECM), already introduced in section 2.1 as a technique to prove the lower bound in Cramér's theorem.

Roughly speaking, LDT states that a target rare set is most likely to be reached by following the path $\hat{f}$ that minimises the corresponding rate function. Thus, simulating

---

[6] The exact asymptotics of $\mathbb{P}(Q > q)$, in which a crucial role is played by the so-called Pickands constant, factorise into the *same* Weibullian term and a hyperbolic prefunction [10].

[7] Since the topic requires by itself a complete tutorial, in the following only the basic ideas are sketched; see [9] for all the relevant definitions.

the system under the change of measure that favours that path is the quickest way to reach the rare set. For random walks (and hence for the G/G/1 framework), the previous heuristic idea is formally justified by the following theorem:

Theorem 9 (Siegmund, Lehtonen/Nyrhinen). *An IS estimator for the probability that a random walk with negative drift exceeds some positive level x is* asymptotically optimal, *iff it is built according to the ECM, where the twisting parameter $\theta^* > 0$ is chosen such that $\Lambda(\theta^*) = 0$, where $\Lambda(\cdot)$ denotes the cumulant generating function of the increments.*

In conclusion, the most likely way in which the random walk can cross level x is by moving linearly at rate $\Lambda'(\theta^*)$, which is exactly the new drift associated to the ECM (14). The previous theorem has a very nice interpretation for an M/M/1 queue: under the optimal ECM, the arrival and service rates are simply twisted. Unfortunately a similar result cannot be extended to Jackson queueing networks and, in general, state-dependent heuristics are required [19].

Finally, it is worth noticing that, when the input traffic is fBm, a change of measure based on the most likely path is *not asymptotically efficient* [20] even for the single server queue. An intuitive explanation is that the main contribution to the asymptotics of the second order moment of the IS estimator is determined by paths which give a very small contribution to the overflow probability, but for which the likelihood ratio is very large. Asymptotic optimality can be achieved by the use of more refined IS techniques [21,22], but at the cost of a higher computational complexity. An alternative approach, which retains the simplicity of ECM-based IS with a lower variance of the estimates (although the algorithm is not asymptotically efficient), is the so-called Bridge Monte-Carlo method, based on the idea of expressing the overflow probability in terms of the *bridge* of the input process [23].

5 Conclusions

The theory of large deviations is a powerful tool for the analysis and simulation of rare events. For lack of space and for its specific target, this tutorial could only introduce some basic principles and show how the general ideas may be used in the framework of computer networks. The interested reader (see also [4] for a more detailed review of the literature) can find in [5] a general introduction to the theory and in [7] a condensed and rigorous overview of the main results. A good compromise between general theory (in the first part of the book) and applications (to performance evaluation in communication and computer architectures) is represented by [12], while [6], the key reference for this tutorial, focuses on queueing systems. The book, starting from the elementary case of IID arrivals, shows how abstract LDT theorems permit to extend the results to very general scenarios and finally deals with more tangible concepts, such as effective bandwidths, scaling properties (which are used, for instance in [24], to analyse the effect of TCP on network stability) and hurstiness, an LDP-oriented characterisation of Long Range Dependence. Finally, [10] is not a book on large deviations, but the analysis of Gaussian queues represents a natural framework to derive and heuristically justify some of the main LDT results.

References

1. Xiao, X., Ni, L.M.: Internet QoS: A Big Picture. IEEE Network 13(2), 8–18 (1999)
2. Leland, W.E., Taqqu, M.S., Willinger, W., Wilson, D.V.: On the self-similar nature of Ethernet traffic (extended version). IEEE/ACM Trans. Netw. 2(1), 1–15 (1994)
3. Norros, I.: On the use of fractional Brownian motion in the theory of connectionless networks. IEEE Journal of Selected Areas in Communications 13(6), 953–962 (1995)
4. Weiss, A.: An introduction to large deviations for communication networks. IEEE Journal on Selected Areas in Communications 13(6), 938–952 (1995)
5. Dembo, A., Zeitouni, O.: Large deviations techniques and applications, 2nd edn. Applications of Mathematics, vol. 38. Springer, Heidelberg (1998)
6. Ganesh, A., O'Connell, N., Wischik, D.: Big Queues. Lecture Notes in Mathematics. Springer, Heidelberg (2004)
7. Varadhan, S.R.S.: Large Deviations and Applications. SIAM, Philadelphia (1984)
8. Bucklew, J.A.: Large Deviation Techniques in Decision, Simulation and Estimation. Wiley, Chichester (1990)
9. Heidelberger, P.: Fast simulation of rare events in queueing and reliability models. ACM Trans. Model. Comput. Simul. 5(1), 43–85 (1995)
10. Mandjes, M.: Large deviations for Gaussian queues. Wiley, Chichester (2007)
11. Addie, R., Mannersalo, P., Norros, I.: Most probable paths and performance formulae for buffers with Gaussian input traffic. European Transactions on Telecommunications 13(3), 183–196 (2002)
12. Shwartz, A., Weiss, A.: Large Deviations for Performance Analysis. Chapman & Hall, Boca Raton (1995)
13. Zwart, A.P.: Queueing Systems with Heavy Tails. PhD thesis, Eindhoven University of Technology (2001)
14. Beran, J.: Statistics for Long-Memory Processes. Chapman & Hall/CRC (1994)
15. Mannersalo, P., Norros, I.: A most probable path approach to queueing systems with general Gaussian input. Computer Networks 40(3), 399–411 (2002)
16. Narayan, O.: Exact asymptotic queue length distribution for fractional Brownian traffic. Advances in Performance Analysis 1(1), 39–63 (1998)
17. Dębicki, K.: A note on LDP for supremum of Gaussian processes over infinite horizon. Statistics & Probability Letters 44(3), 211–220 (1999)
18. Mandjes, M., Mannersalo, P., Norros, I.: Gaussian tandem queues with an application to dimensioning of switch fabrics. Computer Networks 51(3), 781–797 (2007)
19. Zaburnenko, T.S.: Efficient heuristics for simulating rare events in queuing networks. PhD thesis, University of Twente (2008)
20. Baldi, P., Pacchiarotti, B.: Importance Sampling for the Ruin Problem for General Gaussian Processes. Technical report, Universités de Paris 6 & Paris 7 (2004)
21. Dieker, A.B., Mandjes, M.R.H.: Fast simulation of overflow probabilities in a queue with Gaussian input. ACM Transactions on Modeling and Computer Simulation 16(2), 119–151 (2006)
22. Dupuis, P., Wang, H.: Importance Sampling, Large Deviations and differential games. Technical report, Lefschetz Center for Dynamical Systems, Brown University (2002)
23. Giordano, S., Gubinelli, M., Pagano, M.: Bridge Monte-Carlo: a novel approach to rare events of Gaussian processes. In: Proc. of the 5th St.Petersburg Workshop on Simulation, St. Petersburg, Russia, pp. 281–286 (2005)
24. Raina, G., Wischik, D.: Buffer sizes for large multiplexers: TCP queueing theory and instability analysis. In: EuroNGI Conference on Next Generation Internet Networks, Rome (April 2005)

Analysis of Non-product Form Parallel Queues Using Markovian Process Algebra

Nigel Thomas[1] and Jeremy Bradley[2]

[1] School of Computing Science, Newcastle University, UK
nigel.thomas@ncl.ac.uk
[2] Department of Computing, Imperial College London, UK
jb@doc.ic.ac.uk

Abstract. In this paper we use the Markovian process algebra PEPA to specify and analyse a class of queueing models which, in general, do not give rise to a product form solution but can nevertheless be decomposed into their components to obtain a scalable solution. Such a decomposition gives rise to expressions for marginal probabilities which may be used to derive potentially interesting system performance measures, such as the average number of jobs in the system. It is very important that some degree of confidence in such measures can also be given; however, we show here that it is not generally possible to calculate the variance exactly from the marginal probabilities. Hence, two approximations for the variance of the total population are presented and compared numerically.

1 Introduction

Systems of Markovian queues which give rise to product form solutions have been widely studied in the past. In this paper an alternative (non-product form) method of model decomposition is considered that can be found in the queueing network literature, *quasi-separability*. Quasi-separability was developed in the study of queueing systems which suffer breakdowns [4,8], and generalised by Thomas et al [7,5,6] using the Markovian process algebra PEPA [3]. Decompositions of this kind are extremely useful when tackling models with large state spaces, especially when the state space grows exponentially with the addition of further components.

Quasi-separability can be applied to a range of models to derive numerical results very efficiently. While it does not generally give rise to expressions for joint probability distributions it does provide exact results for many performance measures, possibly negating the need for more complex numerical analysis. As such it is a very useful means of reducing the state space of large models. Not all performance measures of interest can be derived exactly from this decomposition. In particular, whilst the average number of jobs in the system may be calculated exactly, in general its variance cannot. It is clearly advantageous however, to gain some confidence in the calculated mean as a useful performance measure without having to solve a much more complicated model. Our proposed solution to this problem is to approximate the variance of the system state. Variance is an extremely important performance measure, knowing how much a system can vary from its mean performance is an essential practical consideration. If the variation of behaviour is large then having only the mean figure for a sojourn is probably not much use for evaluation. Furthermore it has been suggested that, it in certain situations, it is more desirable for a system to be reliably predictable (more deterministic), i.e. have a low variance, rather than fast, as might be indicted by a low mean [1]. In this paper we consider a class of models consisting of a number of nodes in parallel which share a source of jobs. Each node consists of a finite length queue and one or more servers. Jobs are shared amongst the nodes on an a priori basis according to a routing vector which is dependent on the state of a scheduler. The scheduler state may change independently or in response to changes in the behaviour of the nodes. We show that if the scheduler state is not dependent on the number of jobs in the queues, then the system may be decomposed such that each node may be studied in isolation.

There are some advantages in using a process algebraic approach to tackle this problem. Firstly, the formal specification provided by the process algebra facilitates an automatic derivation of the decomposed models and therefore allows such solutions to be applied by non-experts. Secondly, we are able to explore such decompositions in a general setting in order to understand more about the properties of such models and the relationship with other possible solution methods.

In Section 2 we introduce the Markovian process algebra PEPA. In Section 3 the model is presented, followed by a PEPA representation of the model. In Section 4 discuss the decomposition and we show how mean and variance can be calculated from the marginal queue size probabilities derived. Some numerical results are presented in Section 5 for a specific example and some concluding remarks are made in Section 6.

2 PEPA

A formal presentation of PEPA is given in [3], in this section a brief informal summary is presented. PEPA, being a Markovian Process Algebra, only supports actions that occur with rates that are negative exponentially distributed. Specifications written in PEPA represent Markov processes and can be mapped to a continuous time Markov chain (CTMC). Systems are specified in PEPA in terms of *activities* and *components*. An activity (α, r) is described by the type of the activity, α, and the rate of the associated negative exponential distribution, r. This rate may be any positive real number, or given as unspecified using the symbol $\top$.

The syntax for describing components is given as:

$$P ::= (\alpha, r).P | P + Q | P/L | P \bowtie_L Q | A$$

The component $(\alpha, r).P$ performs the activity of type a at rate r and then behaves like P. The component $P + Q$ behaves either like P or like Q, the resultant behaviour being given by the first activity to complete.

The component P/L behaves exactly like P except that the activities in the set L are concealed, their type is not visible and instead appears as the unknown type τ.

Concurrent components can be synchronised, $P \bowtie_L Q$, such that activities in the cooperation set L involve the participation of both components. In PEPA the shared activity occurs at the slowest of the rates of the participants and if a rate is unspecified in a component, the component is passive with respect to the activities of that type. $A \stackrel{def}{=} P$ gives the constant A the behaviour of the component P. The shorthand $P||Q$ is used to denote synchronisation over no actions, i.e. $P \bowtie_\emptyset Q$. We employ some further shorthand that has been commonly used in the study of large parallel systems. We denote $\prod_{i=1}^{N} A_i$ to be the parallel composition of indexed components, $A_1||\ldots||A_N$.

In this paper we consider only models which are cyclic, that is, every derivative of components P and Q are reachable in the model description $P \bowtie_L Q$. Necessary conditions for a cyclic model may be defined on the component and model definitions without recourse to the entire state space of the model.

3 The Model

Jobs arrive into the system in a Poisson stream with rate λ. There are N nodes, each consisting of one or more servers with an associated bounded queue. All jobs arrive at a scheduler which directs jobs to a particular node according to its current state. Jobs sent to a queue which is full are lost. The system model is illustrated in Figure 1.

If, at the time of arrival, a new job finds the scheduler in configuration i, then it is directed to node k with probability $q_k(i)$. These decisions are independent of each other, of past history and of the sizes of the various queues. Thus, a routing policy is defined by specifying 2^N vectors,

$$\mathbf{q}(i) = [q_1(i), q_2(i), \ldots, q_N(i)] \ , \ i \subset \Omega_N \ , \tag{1}$$

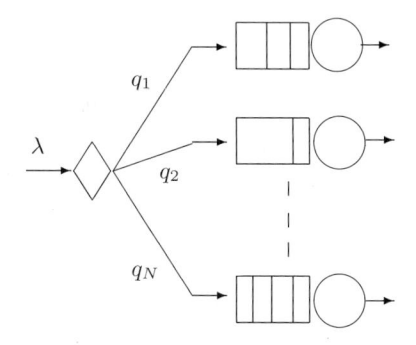

Fig. 1. A single source split among N nodes

such that for every i,

$$\sum_{k=1}^{N} q_k(i) = 1 \ .$$

The system state at time t is specified by the pair $[I(t), \mathbf{J}(t)]$, where $I(t)$ indicates the current scheduler configuration and $\mathbf{J}(t)$ is an integer vector whose k'th element, $J_k(t)$, is the number of jobs in queue k ($k = 1, 2, \ldots, N$). Under the assumptions that have been made, $X = \{[I(t), \mathbf{J}(t)]\, ,\, t \geq 0\}$ is an irreducible Markov process.

We now use PEPA to specify this class of queueing system.

$$Queue_{k,0} \stackrel{def}{=} (arrive_k, \top).Queue_{k,1}$$
$$Queue_{k,j} \stackrel{def}{=} (arrive_k, \top).Queue_{k,j+1}$$
$$+ (service_k, \top).Queue_{k,j-1} \, , \, 0 < j < K$$
$$Queue_{k,K} \stackrel{def}{=} (service_k, \top).Queue_{k,K-1}$$

$$Scheduler_i \stackrel{def}{=} \sum_{k=1}^{N} (service_k, \mu_{k,i}).Scheduler_i$$
$$+ \sum_{k=1}^{N} (arrive_k, q_k(i)\lambda).Scheduler_i$$
$$+ \sum_{\forall h \neq i} (switch, \alpha_{i,h}).Scheduler_h$$

$$\left(\prod_{k=1}^{N} Queue_{k,0}\right) \bowtie_{\mathcal{L}} Scheduler_1$$

Where $\mathcal{L} = \bigcup_{k=1}^{N} \{service_k, arrive_k\}$.

Clearly for this model to be irreducible we must restrict the rates of the variables $\alpha_{i,h}$ which control the switching of scheduler states, such that for each i there is at least one h such that $\alpha_{i,h} > 0$. Furthermore, for each i there must exist paths such that $\alpha_{i,a1}\alpha_{a1,a2}\ldots\alpha_{aX,1} > 0$ and $\alpha_{1,b1}\alpha_{b1,b2}\ldots\alpha_{bN,i} > 0$. That is, every scheduler state must be reachable from every other.

As we have seen, when the routing probabilities depend on the system configuration, the process is not separable (i.e., it does not have a product-form solution). As the capacity of the system becomes large, i.e. each queue has a large bound and N is also large, the direct solution becomes increasingly costly. Hence it is practically relevant to explore more efficient means of solving this class of model.

4 Quasi-Separability

A decomposition based on quasi-separability allows expressions to be derived for marginal distributions just as with a product form solution, however unlike product form these marginal distributions cannot, in general, be combined to form the joint distribution for the whole model. Despite the lack of a solution for the joint distribution, many performance measures of interest can still be derived exactly. Clearly, since exact expressions for marginal probabilities can be found, it is possible to derive any performance measure that depends on a single component. In addition it is possible to obtain certain whole system performance measures in the form of long run averages, such as the average state of the system and average response time in a queueing network.

A system that is amenable to a quasi-separable solution can be considered informally in the following way. The entire system operates within a single environment, which may be made up of several sub-environments. Several components operate within this environment such that their behaviour is affected by the state of the environment. The state of each component does not alter the state transitions of either the environment or the other components. The behaviour of such components can clearly be studied in isolation from the other components as long as the state of the environment is considered also. The restriction on the behaviour of the components imposed here is unnecessarily strong. We can also consider models where the state space of the components can be separated into that part which does have an impact on state transitions in the environment or other components and that part which has no external influence, not even on the other part of that component. Such a separation requires that the part of a component that influences the state of the environment is considered to be part of the environment for the purposes of model decomposition.

Models such as these have appeared in the literature of the study of queueing systems with breakdowns and rerouting of jobs [4,8]. In such models the environment is generally made up of the operational state of servers in the system. For instance each server might be either working or broken, so for a system of N servers the environment has 2^N states. The routing of jobs to queues is dependent on the operational state of the system i.e. the state of the environment. Such models can generally be decomposed into single queue systems with Markov-modulated arrivals and breakdowns. This type of model is conceptually quite simple; there are only two aspects to the state of the components, but in general there may be many aspects of state that must be considered.

Consider an irreducible Markov process, $X(t)$, which consists of N separate components. The state of each component i can be described by a set of K_i separate variables. Denote by $\mathcal{V}_i$ the set of K_i variables which describe the state of component i. If it is possible to analyse the behaviour of each component, i, of the system exactly by only considering those variables that describe it, i.e. $\mathcal{V}_i$, then the system is said to be *separable*. In this case all the components are statistically independent and a product form solution exists.

For the system to be *quasi-separable* it is necessary only that it is possible to analyse the behaviour of each component, i, of the system exactly by only considering those variables that describe it, $\mathcal{V}_i$, and a subset of the variables from all the other components. Thus the elements of $\mathcal{V}_i$ can be classified into the subsets of either system state variables, $\mathcal{S}_i$ or component state variables $\mathcal{C}_i$, such that:

- the state of $c(t) \in \mathcal{C}_i$ changes at a rate which is independent of the state of any variable $v(t) \in \mathcal{C}_j$, $\forall\, j$ such that $j \neq i$.
- the state of $s(t) \in \mathcal{S}_i$ changes at a rate which is independent of the state of any variable $v(t) \in \mathcal{C}_j$, $1 \leq j \leq N$.

If $\mathcal{C}_i \neq \emptyset$, $\forall\, i$, the system can be decomposed into N submodels such that the submodel of the system with respect to the behaviour of component i specifies the changes in the system state variables $\mathcal{S} = \bigcup_{i=1}^{N} \mathcal{S}_i$ and the component state variables $\mathcal{C}_i$. In general the analysis of these submodels gives rise to expressions for their steady-state marginal probabilities if the submodels have stationary distributions with state spaces which are infinite in at most one dimension. As stated above, these marginal probabilities do not, in general, give rise to expressions for the joint probability of the whole system, i.e. no product form solution exists. For quasi-separability to be useful the state space of the submodels should be significantly smaller than the state space of the entire model.

4.1 Deriving Mean and Variance from Marginal Probabilities

If the state space of a model is being reduced then the available information is also reduced unless a product form solution exists. The submodels consist of the system state variables $\mathcal{S} = \bigcup_{i=1}^{N} \mathcal{S}_i$ and the component state variables $\mathcal{C}_i$, hence the steady state solution of such a system gives probabilities of the form $p(\mathbf{S}, \mathbf{c}) = p(\mathcal{S} = \mathbf{S}, \mathcal{C}_i = \mathbf{c})$. A solution of the entire model would give rise to probabilities of the form $p(\mathbf{S}, \mathbf{C}) = p(\mathcal{S} = \mathbf{S}, \mathcal{C} = \mathbf{C})$, where $\mathcal{C} = \{\mathcal{C}_1, \ldots, \mathcal{C}_N\}$ and $\mathbf{C} = \{\mathbf{C}_1, \ldots, \mathbf{C}_N\}$. These probabilities are related in the following way for the submodel involving component i subject to the quasi-separability condition,

$$p(\mathcal{S} = \mathbf{S}, \mathcal{C}_i = \mathbf{c}) = \sum_{\forall \mathbf{C}\, s.t.\, \mathbf{C}_i = \mathbf{c}} p(\mathcal{S} = \mathbf{S}, \mathcal{C} = \mathbf{C})$$

If it is possible to associate a value, x_{ij} with each state of a component i then the average state of the component can easily be found. In addition the average of the sum of all components can be found exactly. Thus,

$$E[x_i] = \sum_{\forall j} \sum_{\forall \mathbf{S}} x_{ij} p(\mathcal{S} = \mathbf{S}, \mathcal{C}_i \equiv x_{ij})$$

Gives the average state of the component, which can be used to derive the average sum,

$$E[x] = \sum_{\forall i} E[x_i]$$

Consider, for example, the following case involving just two values:

$$E[x,y] = \sum_{i=1}^{n}\sum_{j=1}^{m}(i+j)p(i,j) = \sum_{i=1}^{n}\sum_{j=1}^{m}ip(i,j) + \sum_{i=1}^{n}\sum_{j=1}^{m}jp(i,j)$$

$$= \sum_{i=1}^{n}i\sum_{j=1}^{m}p(i,j) + \sum_{j=1}^{m}j\sum_{i=1}^{n}p(i,j)$$

$$= \sum_{i=1}^{n}ip(i,.) + \sum_{j=1}^{m}jp(.,j)$$

$$= E[x] + E[y]$$

Clearly it is an advantageous property to be able to derive system performance measures from marginal probabilities when they can be found. However, the mean is a special case as the sum of the values is trivially separated. If we consider the same example on variance the problem is evident.

$$V[x,y] = \sum_{i=1}^{n}\sum_{j=1}^{m}(i+j)^2 p(i,j) - E^2(x,y)$$

$$= \sum_{i=1}^{n}\sum_{j=1}^{m}(i^2 + 2ij + j^2)p(i,j) - E^2(x,y)$$

$$= \sum_{i=1}^{n}\sum_{j=1}^{m}i^2 p(i,j) + \sum_{i=1}^{n}\sum_{j=1}^{m}j^2 p(i,j) + \sum_{i=1}^{n}\sum_{j=1}^{m}2ij p(i,j) - E^2(x,y)$$

$$= \sum_{i=1}^{n}i^2 p(i,.) + \sum_{j=1}^{m}j^2 p(.,j) + \sum_{i=1}^{n}\sum_{j=1}^{m}2ij p(i,j) - E^2(x,y)$$

In this case there is one term involving $p(i,j)$ which cannot be broken down to the marginal probabilities, $p(i,.)$ and $p(.,j)$. In the more general case where there are N components, there will be N terms involving just the marginal probabilities, but $(N-1)!$ terms involving the joint distribution. Clearly then it is not possible to calculate the variance exactly except when a product form solution exists.

The obvious (traditional) solution to this problem is to generate an approximate solution to variance by substituting $p(i,j)$ with $p(i,.)p(.,j)$, i.e. a product based approximation. In the case of quasi-separability the situation is slightly complicated since the submodels give rise to marginal probabilities involving not only component variables (as in the simple example used here), but also system state variables. The simplest solution (henceforth referred to as the *component state approximation*) would be to eliminate the system state variables by summing over all possible values:

$$p(\mathbf{c}) \approx \prod_{i=1}^{N}\sum_{\forall \mathbf{S}} p(\mathbf{S},\mathbf{c}_i) \qquad (2)$$

where $\mathbf{c} = \{c_1, \ldots, c_N\}$. An alternative approach (henceforth referred to as the *system state approximation*) is to attempt to derive approximations for every possible system state:

$$p(\mathbf{S}, \mathbf{c}) \approx \frac{\prod_{i=1}^{N} p(\mathbf{S}, c_i)}{p(\mathbf{S})^{N-1}} \quad (3)$$

In the following section we will compare these two methods through a numerical example.

5 Example: Multiple Queues with Unreliable Servers

Now consider the following three queue example expressed in PEPA.

$Queue_{k,0} \stackrel{def}{=} (arrive_k, \top).Queue_{k,1}$

$Queue_{k,j} \stackrel{def}{=} (arrive_k, \top).Queue_{k,j+1}$
$\phantom{Queue_{k,j} \stackrel{def}{=}} + (service_k, \top).Queue_{k,j-1} \ , \ 0 < j < K$

$Queue_{k,K} \stackrel{def}{=} (service_k, \top).Queue_{k,K-1}$

$Scheduler_0 \stackrel{def}{=} (repair, \eta).Scheduler_3 + (arrive_1, q_1\lambda).Scheduler_0$
$\phantom{Scheduler_0 \stackrel{def}{=}} + (arrive_2, q_2\lambda).Scheduler_0 + (arrive_3, q_3\lambda).Scheduler_0$

$Scheduler_1 \stackrel{def}{=} (arrive_1, \lambda).Scheduler_1 + (service_1, \mu_1).Scheduler_1$
$\phantom{Scheduler_1 \stackrel{def}{=}} + (fail_1, \xi_1).Scheduler_0$

$Scheduler_2 \stackrel{def}{=} (arrive_2, \lambda).Scheduler_2 + (service_2, \mu_2).Scheduler_2$
$\phantom{Scheduler_2 \stackrel{def}{=}} + (fail_2, \xi_2).Scheduler_0$

$Scheduler_3 \stackrel{def}{=} (arrive_3, \lambda).Scheduler_3 + (service_3, \mu_3).Scheduler_3$
$\phantom{Scheduler_3 \stackrel{def}{=}} + (fail_3, \xi_3).Scheduler_0$

$Scheduler_4 \stackrel{def}{=} (fail_1, \xi_1).Scheduler_2 + (fail_2, \xi_2).Scheduler_1$
$\phantom{Scheduler_4 \stackrel{def}{=}} + (arrive_1, \frac{q_1\lambda}{q_1+q_2}).Scheduler_4 + (arrive_2, \frac{q_2\lambda}{q_1+q_2}).Scheduler_4$
$\phantom{Scheduler_4 \stackrel{def}{=}} + (service_1, \mu_1).Scheduler_4 + (service_2, \mu_2).Scheduler_4$

$Scheduler_5 \stackrel{def}{=} (fail_1, \xi_1).Scheduler_3 + (fail_3, \xi_3).Scheduler_1$
$\phantom{Scheduler_5 \stackrel{def}{=}} + (arrive_1, \frac{q_1\lambda}{q_1+q_3}).Scheduler_5 + (arrive_2, \frac{q_3\lambda}{q_1+q_3}).Scheduler_5$
$\phantom{Scheduler_5 \stackrel{def}{=}} + (service_1, \mu_1).Scheduler_5 + (service_3, \mu_3).Scheduler_5$

$Scheduler_6 \stackrel{def}{=} (fail_2, \xi_2).Scheduler_3 + (fail_3, \xi_3).Scheduler_2$
$\phantom{Scheduler_6 \stackrel{def}{=}} + (arrive_2, \frac{q_2\lambda}{q_2+q_3}).Scheduler_6 + (arrive_3, \frac{q_3\lambda}{q_2+q_3}).Scheduler_6$
$\phantom{Scheduler_6 \stackrel{def}{=}} + (service_2, \mu_2).Scheduler_6 + (service_3, \mu_3).Scheduler_6$

$Scheduler_7 \stackrel{def}{=} (fail_1, \xi_1).Scheduler_6 + (fail_2, \xi_2).Scheduler_5$

$+(fail_3, \xi_3).Scheduler_4 + (arrive_1, q_1\lambda).Scheduler_7$
$+(arrive_2, q_2\lambda).Scheduler_7 + (arrive_3, q_3\lambda).Scheduler_7$
$+(service_1, mu_1).Scheduler_7 + (service_2, \mu_2).Scheduler_6$
$+(service_3, \mu_3).Scheduler_6$

$$(Queue_{1,0} \| Queue_{2,0} \| Queue_{3,0}) \bowtie_{\{arrive_1, service_1, arrive_2, service_2, arrive_3, service_3\}} Scheduler_7$$

This model represents three queues whose servers suffer independent failures and subsequent repairs. A repair will repair the entire system, but will only be triggered once all the servers have failed. The scheduler attempts to route jobs to active servers, if any exist. In the case of all the servers being broken ($Scheduler_0$) the scheduler routes jobs to all queues in the same proportion as if all were working.

Clearly this model fits the decomposition class introduced in the previous section. The number of jobs in each queue is not dependent on the number in the other queues, however, all queue lengths are dependent on the behaviour of the scheduler component. Thus we can decompose this model into three smaller ones, defined by the following system equations.

$$Queue_1 \bowtie_{\{arrive_1, service_1\}} Scheduler_7$$

$$Queue_2 \bowtie_{\{arrive_2, service_2\}} Scheduler_7$$

$$Queue_3 \bowtie_{\{arrive_3, service_3\}} Scheduler_7$$

Each of these models has $8(K+1)$ states in the underlying CTMC, whereas the original model has $8(K+1)^3$ states. Obviously if K is large, then this is a considerable saving, possibly meaning that the decomposed models are numerically tractable when the full model is not.

The immediate advantage of this decomposition is that we can quickly find global average metrics as described above, which can then be used to optimise parameters. In the case of this example we can numerically optimise the routing probabilities q_k to minimise the average number of jobs in the system or the average response time. Performing optimisations of this kind on the whole model would be extremely costly.

5.1 Numerical Results

We now turn our attention to the problem of estimating the variance of the total number of jobs in the system. For this exercise we will assume that the three servers are identical, meaning of course that the optimal (static) routing probabilities will be equal, i.e. $q_k = \frac{1}{3}$. In all cases the queues were bounded at

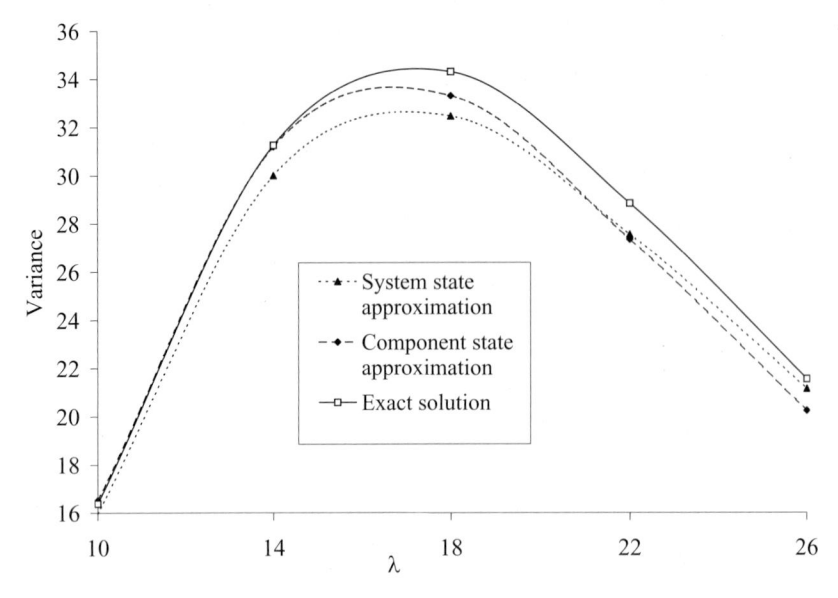

Fig. 2. Variance of the total number of jobs against arrival rate
$\mu_k = 10$, $\eta = 10$, $\xi = 1$, $q_k = \frac{1}{3}$

$K = 10$. We will then investigate how the two approximations perform as we vary the load and the duration of repair periods.

Figure 2 shows the relationship between variance and load. At low load the variance is low, as the number of jobs in any queue rarely grows very large. As the load increases, so does the variance, until leveling off and then decreasing due to the effect of the bound. The variance decreases at high load as the queues become full most of the time. As can be seen, with these parameters, both approximations work well.

Figure 3 shows the variance as a function of the failure rate, ξ_k. The repair rate is also varied in direct proportion to the failure rate, so that the probability of being in any given scheduler state is the same for each value of ξ_k. When the failure (and repair) rate is relatively large the interruptions to service are relatively brief and so few arrivals occur when all the servers are broken. However, when the repair rate is decreased, the duration period for which all servers are broken increases and so the queue will fill up. Thus, when the repair rate is small, there becomes a big difference between the queue lengths in $Scheduler_7$ and the queue lengths in $Scheduler_0$. As the repair rate continues to decrease, this difference does not increase any more as the queues cannot exceed their bound, hence the variance levels off.

In Figure 3 there is much less correspondence between the approximations and the exact result. Apart from $\xi_k = 0.1$, there is fairly good correlation between the system state approximation and the exact result. In other examples we have observed that when the approximations closely agree, they are accurate, however, that does not hold here when $\xi_k = 0.1$.

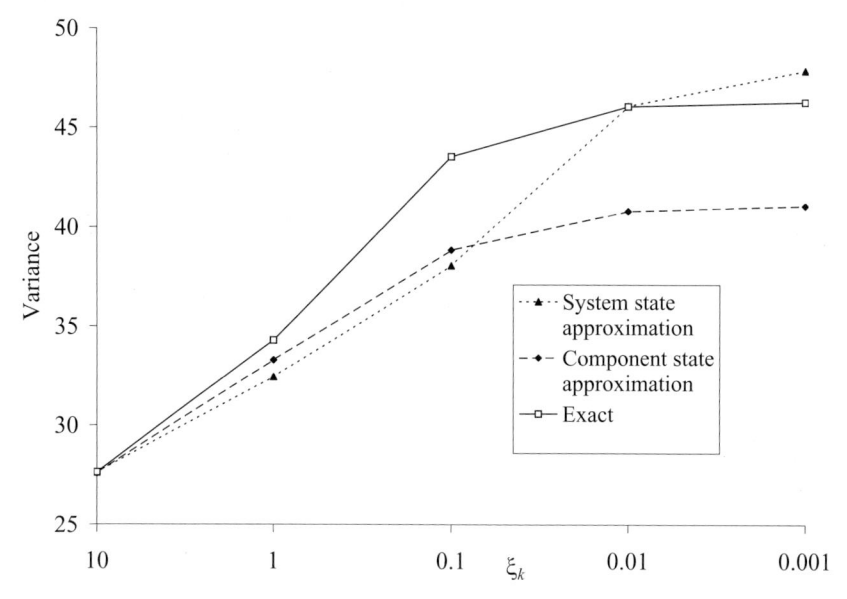

Fig. 3. Variance of the total number of jobs varied with failure rate
$\lambda = 18$, $\eta = 10\xi$, $\mu_k = 10$, $q_k = \frac{1}{3}$

6 Conclusions

In this paper we have shown how a class of queueing model can be specified using PEPA and formally decomposed into a number of submodels. These submodels are easier to solve numerically, but have the weakness that it is not possible to derive the joint queue length probabilities exactly. As a consequence we have investigated two approximations for the joint queue length probability which can be used to predict the variance of the total population.

The approach has been illustrated through a significant example. This has shown that there is a huge potential saving in computational effort through this method. However, computing the approximations is not trivial and their accuracy is not universal. Therefore, the main lesson is that this decomposition is mainly useful as a way of obtaining metrics which are based entirely on the marginal queue length probabilities. Estimates of other metrics are clearly useful and particularly so if there is a very high cost of obtaining an exact solution.

Acknowledgements

Both authors are supported by the EPSRC funded AMPS project, which is held jointly at Imperial College London (EP/G011737/1) and Newcastle University (EP/G011389/1).

References

1. Bradley, J., Davis, N.: Measuring improved reliability in stochastic systems. In: Proceedings of 15th UK Performance Engineering Workshop, University of Bristol, pp. 121–130 (1999)
2. Clark, G., Gilmore, S., Hillston, J., Thomas, N.: Experiences with the PEPA performance modelling tools. IEE Proceedings - Software 146(1) (1999)
3. Hillston, J.: A Compositional Approach to Performance Modelling. Cambridge University Press, Cambridge (1996)
4. Mitrani, I., Wright, P.E.: Routing in the Presence of Breakdowns. Performance Evaluation 20, 151–164 (1994)
5. Thomas, N.: Extending Quasi-separability. In: Proceedings of 15th UK Performance Engineering Workshop, University of Bristol, pp. 131–140 (1999)
6. Thomas, N., Bradley, J.: Approximating variance in non-product form decomposed models. In: Proceedings of the 8th International Workshop on Process Algebra and Performance Modelling. Carleton Scientific Publishers (2000)
7. Thomas, N., Gilmore, S.: Applying Quasi-Separability to Markovian Process Algebra. In: Proceedings of 6th International Workshop on Process Algebra and Performance Modelling (1998)
8. Thomas, N., Mitrani, I.: Routing Among Different Nodes Where Servers Break Down Without Losing Jobs. In: Quantitative Methods in Parallel Systems, pp. 248–261. Springer, Heidelberg (1995)

Product-Form Solution in PEPA via the Reversed Process

Peter G. Harrison[1] and Nigel Thomas[2]

[1] Department of Computing, Imperial College London, UK
pgh@doc.ic.ac.uk
[2] School of Computing Science, Newcastle University, UK
nigel.thomas@ncl.ac.uk

Abstract. In this paper we use the reversed process to derive expressions for the steady state probability distribution of a class of product-form PEPA models. In doing so we exploit the *Reversed Compound Agent Theorem* (RCAT) to compute the rates within reversed components of a model. The class of model is, in essence, a generalised, closed, queueing network that might also be solved by mean value analysis, if full distributions are not needed, or approximated using a fluid flow approximation. A general formulation of RCAT is given and the process is illustrated with a running example, including several new variations that consider effects such as multiple servers, competing services and functional rates within actions.

1 Introduction

Quantitative methods are vital for the design of efficient systems in ICT, communication networks and other logistical areas such as business processes and healthcare systems. However, the resulting models need to be both accessible to the designer, rather than only to the performance specialist, and efficient. A sufficiently expressive formalism is needed that can specify models at a high level of description and also facilitate separable and hence efficient mathematical solutions. Stochastic process algebra (SPA) is a formalism that has the potential to meet these requirements.

One approach to tackling the state space explosion problem common to all compositional modelling techniques is through the exploitation of, so called, *product-form solutions*. Essentially, a product-form is a decomposed solution where the steady state distribution of a whole system can be found by multiplying the marginal distributions of its components. The quest for product-form solutions in stochastic networks has been a major research area in performance modelling for over 30 years. Most attention has been given to queueing networks and their variants such as G-networks [12], but there have also been other significant examples, e.g. [1,2,7]. However, these have typically been derived in a rather ad-hoc way: guessing that such a solution exists, and then verifying that the Kolmogorov equations of the defining Markov chain are satisfied and appealing to uniqueness.

The Reversed Compound Agent Theorem (RCAT) [3] for MPA is a compositional result that finds the reversed stationary Markov process of a cooperation between two interacting components, under syntactically checkable conditions [3,4,5]. From this a product-form follows simply. RCAT thereby provides an alternative methodology that unifies many product-forms, far beyond those for queueing networks. At the time, the original study of product-form G-networks was significantly different from previous product-form analyses since the property of *local balance* did not hold and the traffic equations were non-linear. In contrast, the RCAT-based approach goes through unchanged – the only difference is that there are co-operations between one departure transition type and another, as well as between departure transitions and arrival transitions as in conventional queueing networks.

In this paper we use the RCAT approach to find product-form solutions to a class of PEPA models. This class has previously been shown to be amenable to solution by mean value analysis [10] (for expected values only, of course) and fluid flow approximation [11]. A similar model, expressed quite differently, was included in [4].

The context of the present work is as follows:

- Models are expressed explicitly using the full PEPA syntax.
- Reversed components and models are also fully expressed using full PEPA.
- The class of model considered uses active-active co-operation, not currently supported under RCAT or its extensions.
- The class includes the 'counting' approach to specifying (large) groups of identical components in parallel.

2 General RCAT Algorithm

Whilst dependent on reversed processes for its original derivation, an *application* of RCAT can be done purely mechanically if the steady state probabilities are known for the component-processes; from these the reversed processes could be computed if desired, as discussed in the next section. For simplicity, we consider the cooperation $P_1 \bowtie_L P_2$; the treatment is similar for n-way cooperations.

1. From P_k construct R_k by setting the rate of every instance of action $a \in L$ that is passive in P_k to x_a, for $k = 1, 2$ (note that each a will be passive for only one k);
2. For each active action type a in $R_k, k = 1, 2$, check that a certain quantity $\overline{r_a}$ is the same for all of its instances, i.e. for all transitions $i \to j$ that a denotes in the state transition graph of R_k. This quantity is computed as

$$\overline{r_a} = \pi_k(i) r_a^i / \pi_k(j)$$

   where r_a^i is the specified forward rate of the (any, if more then one) instance of action type a going out of state i (must be the same value for all i);
3. Noting that the symbolic reversed rate $\overline{r_a}$ will in general be a function of the x_b ($b \in L$), solve the equations $x_a = \overline{r_a}$ for each $a \in L$ and substitute the solutions for the variables x_a in each R_k;

4. Check the enabling conditions (detailed in [4], but not part of the specific focus of this paper) for each co-operating action in each process P_k. For queueing networks, these are as in the original RCAT, namely that all passive actions be enabled in all states and that all states also have an incoming instance of every active action;
5. The required product-form for state $\underline{s} = (s_1, s_2)$ is now $\pi(\underline{s}) \propto \pi_1(s_1)\pi_2(s_2)$ where $\pi_k(s_k)$ is the equilibrium probability (which may be unnormalised) of state s_k in R_k.

For irreducible closed networks, this product-form can always be normalised to give the required steady state probabilities. For irreducible open networks, a separate analysis of ergodicity is required. Notice that all synchronisations in RCAT are between active and passive pairs of actions. In what follows we will relax this condition somewhat.

3 Reversed Processes and Product-Form

RCAT depends on properties of the reversed process. Essentially a reversed process is one that would be observed if time were reversed. For every stationary Markov process, there is a reversed process with the same state space and the same steady state probability distribution, i.e. $\pi_i = \pi'_i$, where π_i and π'_i are the steady state probabilities of being in state i in the forward and reversed process respectively. Furthermore, the forward and reversed processes are related by the transitions between states; there will be a non-zero transition rate between states j and i in the reversed process, $q'_{j,i}$ iff there is a non-zero transition rate between states i and j in the forward process, $q_{i,j}$. A special case is the *reversible* process, where the reversed process is stochastically identical to the forward process, so that $q'_{j,i} = q_{i,j}$; an example is the M/M/1 queue.

The reversed process is easily found if we already know the steady state probability distribution (see Kelly [9] for example). The forward and reversed probability fluxes balance at equilibrium, i.e.

$$\pi'_i q'_{i,j} = \pi_j q_{j,i}$$

and so, since $\pi_i = \pi'_i$, we find:

$$q'_{i,j} = \frac{\pi_j q_{j,i}}{\pi_i}$$

In practical cases we do not already know the steady state distribution for the forward process, and if we did we'd have little need to find the reversed process. However, we can use this result directly to find reversed components within a model (if the components are relatively small), or we could guess the possible reversed rates and use this result as a check.

RCAT uses a simpler methodology to derive the reversed components at the syntactic level, based on finding cycles within a component description. Essentially every choice operator in PEPA introduces a new cycle if the successor behaviours on each side are different. For a given sequential component S, we can compute the following:

1. Define q_i to be the total outgoing rate from a state i (component behaviour). The conservation of outgoing rate (the first of Kolmogorov's criteria, [9]) gives $q'_i = q_i$, for all behaviours i.
2. Find a covering set of cycles.
3. For each cycle apply the second of Kolmogorov's criteria to give a system of non-linear equations that uniquely define the set of rates in the reversed sequential component $\overline{S}$.

For example, consider the following sequential PEPA component.

$$Task_1 \stackrel{def}{=} (read, \xi).Task_2$$
$$Task_2 \stackrel{def}{=} (compute, (1-p)\mu).Task_1 + (compute, p\mu).Task_3$$
$$Task_3 \stackrel{def}{=} (write, \eta).Task_1$$

There are two cycles.

1. from $Task_1$ to $Task_2$ and back to $Task_1$
2. from $Task_1$ to $Task_2$ to $Task_3$ and back to $Task_1$

Thus, using Kolmogorov's criteria we can compute the following.

$$q'_{1,2}q'_{2,1} = q_{1,2}q_{2,1} = \xi(1-p)\mu$$
$$q'_{1,3}q'_{3,2}q'_{2,1} = q_{1,2}q_{2,3}q_{3,1} = \xi p\mu\eta$$

Furthermore, we know that

$$q'_1 = q'_{1,2} + q'_{1,3} = q_1 = q_{1,2} = \xi$$
$$q'_2 = q'_{2,1} = q_2 = q_{2,1} + q_{2,3} = \mu$$
$$q'_3 = q'_{3,2} = q_3 = q_{3,1} = \eta$$

Hence, the reversed component $\overline{Task_i}$ is easily computed.

$$\overline{Task_1} \stackrel{def}{=} (read, (1-p)\xi).\overline{Task_2} + (read, p\xi).\overline{Task_3}$$
$$\overline{Task_2} \stackrel{def}{=} (compute, \mu).\overline{Task_1}$$
$$\overline{Task_3} \stackrel{def}{=} (write, \eta).\overline{Task_2}$$

For clarity, we illustrate the states of the forward and reversed components in Figure 1.

Once we know the rates in both the forward and reversed processes, we can compute the steady state probabilities using a simple chain rule:

$$\pi_j = \frac{\pi_j}{\pi_{j-1}} \frac{\pi_{j-1}}{\pi_{j-2}} \cdots \frac{\pi_1}{\pi_0} \pi_0$$

Given a sequence of actions from some source state we can therefore compute the steady state probability of being in the resultant target state j in relation to the source state 0 as follows:

$$\pi_j = \frac{q_{0,1}q_{1,2}\cdots q_{j-1,j}}{q'_{j,j-1}\cdots q'_{2,1}q'_{1,0}}\pi_0$$

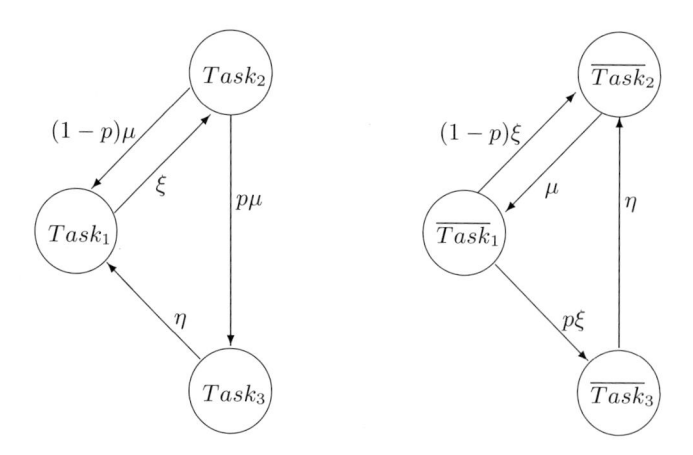

Fig. 1. States of the forward and reversed components

4 A Closed Queueing Model

Now consider a model of a closed queueing network of N jobs circulating around a network of M service stations, denoted $1, \ldots, M$; each station is either a queueing station or an infinite server station, where the number of queueing stations is M_q. Let $\mathcal{M}_q$ be the set of all queueing stations and at each one there is an associated queue (bounded at N) operating a FCFS policy and one server. The servers are able to serve jobs of only one type; each job type, j, is served at rate r_j by queueing station j. At each infinite server station, i, jobs of type i experience a random delay with mean $1/r_i$. All services are negative exponentially distributed.

There are M job types. For each specified job type, there is exactly one station (either queueing station or infinite server station) which may serve it. When a job of type j completes a service at a given station, it will proceed to service at a station (possibly the same station) as a job of type k according to some routing probability p_{jk}.

In PEPA a queueing station, i, can be modelled as follows.

$$QStation_i \stackrel{def}{=} (service_i, r_i).QStation_i$$

Note that r_i is always specified as finite, and not $\top$. This is because passive actions are subject to the *apparent rate* in PEPA. In addition we impose the restriction that the action enabled at each queueing station is unique to that queueing station, i.e. $\forall i, j \in \mathcal{M}_q$,

$$\mathcal{A}ct(QStation_i) \cap \mathcal{A}ct(QStation_j) = \emptyset$$

The infinite server stations are not represented explicitly.

Each job will receive service from a sequence of stations determined by a set of routing probabilities,

$$Job_i \stackrel{def}{=} \sum_{j=1}^{M} (service_i, p_{ij}r_i).Job_j \ , \ 1 \leq i \leq M$$

where, $0 \leq p_{ij} \leq 1$ and
$$\sum_{j=1}^{M} p_{ij} = 1 , \ 1 \leq i \leq M$$

Let $\mathcal{S}_i$ be the set of all job types which perform $service_i$ actions, i.e. $\mathcal{S}_i = j$ if $service_i \in \mathcal{A}(Job_j)$.

The entire system can then be represented as follows:

$$\left(\prod_{\forall i \in \mathcal{M}_{\text{II}}} (QStation_i) \right) \bowtie_{\mathcal{L}} Job_1[N] \tag{1}$$

where
$$\mathcal{L} = \bigcup_{\forall i \in \mathcal{M}_{\text{II}}} \{service_i\}$$

It is important to note that in this classification the actions which cause the queue station to change mode are not shared actions. The notation $P[N]$ denotes that there are N copies of the component Job_1, but we do not represent each individual in the underlying state space description as would be the case in $Job_1 || \ldots || Job_1$. Instead we are only concerned with the number of components behaving as Job_1, Job_2, etc, at any time. Thus, this representation can be considered to be an explicitly 'lumped' version of $Job_1 || \ldots || Job_1$.

4.1 Reversed PEPA Process

The model presented in the previous section does not meet the existing criteria for RCAT, principally because RCAT is only defined over active-passive cooperation[1]. However, obtaining the reversed process is relatively straightforward since the server components are static. Hence it is only necessary to reverse the single (sequential) job component, which can be done using the most basic result on reversed processes (see Section 3).

This yields a reversed component with the following structure:

$$\overline{Job_i} \stackrel{def}{=} \sum_{k=1}^{M} (service_j, q_{jk} r_j).\overline{Job_k} , \ 1 \leq i,j \leq M$$

Where, $0 \leq q_{jk} \leq 1$ and
$$\sum_{k=1}^{M} q_{jk} = 1 , \ 1 \leq j \leq M$$

The new routing probabilities, q_{ij}, can be found by applying Kolmogorov's criteria. That is, equate the product of rates around a given cycle in the Job component with the product of the rates around the corresponding reversed cycle

---
[1] An alternative approach does use RCAT but requires the problem to be mapped into an equivalent problem with functional rates, whereupon the result of [6] can be used

in $\overline{Job}$. Note that $q_{ij} = 0$ in the reversed component iff $p_{ji} = 0$ in the forward component.

The full reversed model can then be specified as

$$\left(\prod_{\forall i \in \mathcal{M}} (QStation_i) \right) \underset{\mathcal{L}}{\bowtie} \overline{Job}_1[N] \tag{2}$$

Where

$$\mathcal{L} = \bigcup_{\forall i \in \mathcal{M}_{\Pi}} \{service_i\}$$

The forward and reversed processes can be used to find a product-form solution (if one exists) by applying Kolmogorov's generalised criteria.

5 Example: A Simple Information Processing System

We now explore the derivation of the reversed process and its use in finding a product-form solution through a specific simple example. Consider the following PEPA specification of a simple information processing system.

$$Channel_1 \stackrel{def}{=} (read, \xi).Channel_1$$
$$Process \stackrel{def}{=} (compute, \mu).Process$$
$$Channel_2 \stackrel{def}{=} (write, \eta).Channel_2$$

$$Task_1 \stackrel{def}{=} (read, \xi).Task_2$$
$$Task_2 \stackrel{def}{=} (compute, (1-p)\mu).Task_1 + (compute, p\mu).Task_3$$
$$Task_3 \stackrel{def}{=} (write, \eta).Task_1$$

The entire system is then specified as

$$(Channel_1 || Process || Channel_2) \underset{\mathcal{L}}{\bowtie} Task_1[N]$$

Items are read from an input channel and processed. If the item meets certain criteria, then it is sent to the output channel before the next item is read, otherwise it is discarded and the next item read.

The reversed component $\overline{Task}_i$ is easily computed as we have already observed.

$$\overline{Task}_1 \stackrel{def}{=} (read, (1-p)\xi).\overline{Task}_2 + (read, p\xi).\overline{Task}_3$$
$$\overline{Task}_2 \stackrel{def}{=} (compute, \mu).\overline{Task}_1$$
$$\overline{Task}_3 \stackrel{def}{=} (write, \eta).\overline{Task}_2$$

The system state is described by the triple, $\{i, j, k\}$, where i is the number of components behaving as $Task_1$, j is the number of components behaving as

$Task_2$ and k is the number of components behaving as $Task_3$. Clearly, $i+j+k = N$. We choose $\{N,0,0\}$ as a reference state and consider transitions from that state to an arbitrary state $\{i,j,k\}$. Hence we can derive specifications for the steady state probabilities depending on the co-operation set $\mathcal{L}$.

- **All actions are resource limited:** $\mathcal{L} = \{read, compute, write\}$

  To reach state $\{i,j,k\}$ there must be $j+k$ read actions, each one at rate ξ, followed by k compute actions (leading to $Task_3$) at rate $p\mu$. In the reverse process, going from state $\{i,j,k\}$ to state $\{N,0,0\}$, there must be k write actions at rate η, followed by $j+k$ compute actions at rate μ. Hence,

  $$\pi_{\{i,j,k\}} = \left(\frac{\xi}{\mu}\right)^{j+k} \left(\frac{p\mu}{\eta}\right)^k \pi_{\{N,0,0\}}$$
  $$= \left(\frac{\xi}{\mu}\right)^j \left(\frac{p\xi}{\eta}\right)^k \pi_{\{N,0,0\}}$$

- **No actions are resource limited:** $\mathcal{L} = \emptyset$

  If actions are not shared they will occur at a rate proportional to the sum of the number of participants. Thus, when the state is $\{N,0,0\}$, the *read* action will occur at rate $N\xi$. Hence, the product of the rates of $j+k$ read actions will be $N!\xi^{j+k}/i!$. Similarly, from state $i,j+k,0$ the product of the rates of k compute actions will be $(j+k)!(p\mu)^k/j!$. In the reverse process, starting in state $\{i,j,k\}$, there must be k write actions, the product of whose rates is $k!\eta^k$, followed by $j+k$ compute actions with rate product $(j+k)!\mu^{j+k}$. Thus,

  $$\pi_{\{i,j,k\}} = \frac{N!}{i!j!k!} \left(\frac{\xi}{\mu}\right)^j \left(\frac{p\xi}{\eta}\right)^k \pi_{\{N,0,0\}}$$

- ***compute* and *write* are resource limited:** $\mathcal{L} = \{compute, write\}$

  When the cooperation set is a subset of the total action set, we have a situation which lies between the two previous cases. In this case only *read* is a mass action, and in the sequence we consider this is only a factor in the forward transitions. Hence, the $j+k$ read actions occur with rate product $N!\xi^{j+k}/i!$, whereas the k compute actions (leading to $Task_3$) each occur at rate $p\mu$. In the reverse process, going from state $\{i,j,k\}$ to state $\{N,0,0\}$, there must be k write actions at rate η, followed by $j+k$ compute actions at rate μ. Hence,

  $$\pi_{\{i,j,k\}} = \frac{N!}{i!} \left(\frac{\xi}{\mu}\right)^j \left(\frac{p\xi}{\eta}\right)^k \pi_{\{N,0,0\}}$$

– **Only** *compute* **is resource limited:** $\mathcal{L} = \{compute\}$

This case is very similar to the previous one, with the exception that the *write* action will be mass action. Hence,

$$\pi_{\{i,j,k\}} = \frac{N!}{i!k!} \left(\frac{\xi}{\mu}\right)^j \left(\frac{p\xi}{\eta}\right)^k \pi_{\{N,0,0\}}$$

5.1 Multiple Service Stations

We now consider the same model components, but with multiple instances of the process component, representing the case where the node has multiple servers.

$$(Input||Process[K]||Output) \bowtie_{\mathcal{L}} Task_1[N]$$

All components are as defined above. Clearly this extension does not affect the structure of the reversed components, or the solution when $compute \notin \mathcal{L}$. Thus, the reversed model is given by the description,

$$(Input||Process[K]||Output) \bowtie_{\mathcal{L}} \overline{Task_1}[N]$$

– **All actions are resource limited:** $\mathcal{L} = \{read, compute, write\}$

As in the previous subsection, in going from $\{N,0,0\}$ to $\{i,j,k\}$, we will see $j+k$ *read* actions followed by k *compute* actions. As $read \in \mathcal{L}$ each *read* action will occur at rate ξ, However, the rate of *compute* actions will depend on the number of servers and the number of participants. Clearly if $j+k \leq K$ then the system will behave as if $compute \notin \mathcal{L}$, i.e.

$$\pi_{\{i,j,k\}} = \frac{1}{j!} \left(\frac{\xi}{\mu}\right)^j \left(\frac{p\xi}{\eta}\right)^k \pi_{\{N,0,0\}}, \quad j+k \leq K$$

If $j > K$ then each forward *compute* action will occur at rate $Kp\mu$. Each reverse *compute* action will occur at rate $K\mu$ until there are fewer than K $\overline{Task_2}$ components.

$$\pi_{\{i,j,k\}} = \frac{1}{K^{j-K}K!} \left(\frac{\xi}{\mu}\right)^j \left(\frac{p\xi}{\eta}\right)^k \pi_{\{N,0,0\}}, \quad j > K$$

Finally, if $j+k > K$ and $j \leq K$ then initially *compute* actions will occur at rate $Kp\mu$ (forward) or $K\mu$ (reverse), but once the volume of participants falls below K, then the (cumulative) rate will also fall. In the forward case this means the product of the rates of *compute* actions in the forward process is $K!K^{j+k-K}(p\mu)^k/j!$ and in the reverse it is $K!K^{j+k-K}$. Hence,

$$\pi_{\{i,j,k\}} = \frac{1}{j!} \left(\frac{\xi}{\mu}\right)^j \left(\frac{p\xi}{\eta}\right)^k \pi_{\{N,0,0\}}, \quad j+k \leq K$$

- **Only *compute* is resource limited:** $\mathcal{L} = \{compute\}$

In the case where the other actions are not shared, the equations are similar as it merely a case of incorporating the cumulative action rates for *read* and *write*.

$$\pi_{\{i,j,k\}} = \frac{N!}{i!j!k!} \left(\frac{\xi}{\mu}\right)^j \left(\frac{p\xi}{\eta}\right)^k \pi_{\{N,0,0\}} \; j+k \le K$$

$$\pi_{\{i,j,k\}} = \frac{N!}{i!k!K^{j-K}K!} \left(\frac{\xi}{\mu}\right)^j \left(\frac{p\xi}{\eta}\right)^k \pi_{\{N,0,0\}}, \; j > K$$

5.2 Multiple Services from a Station

We consider a variation of this model whereby a job may request service from the same server more than once in a cycle. To do this we will use the same action type and rate in more than one derivative of the job. We do this to avoid a race condition at the server, which would distort the service rate. An alternative approach would be to use a functional rate, which we consider in the next subsection.

Now consider the following model where a shared input/output channel is used.

$$Channel \stackrel{def}{=} (io, \xi).Channel$$
$$Process \stackrel{def}{=} (compute, \mu).Process$$

$$Task_1 \stackrel{def}{=} (io, \xi).Task_2$$
$$Task_2 \stackrel{def}{=} (compute, (1-p)\mu).Task_1 + (compute, p\mu).Task_3$$
$$Task_3 \stackrel{def}{=} (io, \xi).Task_1$$

$$(IO || Process) \underset{\mathcal{L}}{\bowtie} Task_1[N]$$

Clearly the renaming of the *read* and *write* actions (to *io*) has no effect on the structure of the reversed process.

$$\overline{Task_1} \stackrel{def}{=} (io, (1-p)\xi).\overline{Task_2} + (read, p\xi).\overline{Task_3}$$
$$\overline{Task_2} \stackrel{def}{=} (compute, \mu).\overline{Task_1}$$
$$\overline{Task_3} \stackrel{def}{=} (io, \xi).\overline{Task_2}$$

$$(IO || Process) \underset{\mathcal{L}}{\bowtie} \overline{Task_1}[N]$$

However, the solution of the model will clearly be different, since the rates have been changed and there is potentially more competition for the channel (if *io* is in $\mathcal{L}$).

- **All actions are resource limited:** $\mathcal{L} = \{io, compute\}$

  In the forward process, the sequence of actions we consider is not affected by this change, since we are not concerned with io actions from $Task_3$ and there are no components behaving as $Task_3$ when io actions are performed from $Task_1$. However, in the reverse process, we perform k io actions starting in state $\{i, j, k\}$. From the processor sharing-like semantics of PEPA, the rate at which each reverse io action occurs will be dependent on the total number of components behaving as $Task_3$ and $Task_1$. Initially this will be $k\xi/(i+k)$, but will alter as the number of $Task_3$'s decreases. Hence the product of the rate of the k reversed io actions will be $i!k!\xi^k/(i+k)!$. Thus,

  $$\pi_{\{i,j,k\}} = \frac{(i+k)!}{i!k!} \left(\frac{\xi}{\mu}\right)^j p^k \pi_{\{N,0,0\}}$$

- **No actions are resource limited:** $\mathcal{L} = \emptyset$

  Clearly this case is the same as the initial model with $\mathcal{L} = \emptyset$, with the small modification that the rate η has been replaced by ξ. Thus,

  $$\pi_{\{i,j,k\}} = \frac{N!}{i!j!k!} \left(\frac{\xi}{\mu}\right)^j p^k \pi_{\{N,0,0\}}$$

  Similarly, the case when only *compute* is resource limited is also the same as previously, since there is no competition on the io action.

- **Only io is resource limited:** $\mathcal{L} = \{io\}$

  In this case the *compute* action will have a cumulative rate in both the forward and reversed processes, as we have previously observed. Hence,

  $$\pi_{\{i,j,k\}} = \frac{(i+k)!}{i!j!k!} \left(\frac{\xi}{\mu}\right)^j p^k \pi_{\{N,0,0\}}$$

5.3 Service Rate Dependent on a Functional Rate

Finally, we consider the case where a station may offer more than one kind of service but the rate at which each service type occurs is governed by a *function* – f_r and f_w below.

$$Channel \stackrel{def}{=} (read, f_r\xi).Channel + (write, f_w\eta).Channel$$
$$Process \stackrel{def}{=} (compute, \mu).Process$$

$$Task_1 \stackrel{def}{=} (read, f_r\xi).Task_2$$
$$Task_2 \stackrel{def}{=} (compute, (1-p)\mu).Task_1 + (compute, p\mu).Task_3$$
$$Task_3 \stackrel{def}{=} (write, f_w\eta).Task_1$$

The entire system is then specified as

$$(Channel || Process) \underset{\mathcal{L}}{\bowtie} Task_1[N]$$

where, $\mathcal{L} = \{read, compute, write\}$.

In general, a (so called) *functional rate* can be dependent on properties of the evolution of a model. This does not alter the derivation of the reversed model. In particular, note that the *Channel* component is still considered to be static, despite the choice between actions, and hence the reversed *Channel* component is identical to the forward *Channel* component.

The reversed model is thus given by the system equation,

$$(\overline{Channel || Process}) \underset{\mathcal{L}}{\bowtie} \overline{Task_1}[N]$$

where, $\mathcal{L} = \{read, compute, write\}$ and $\overline{Task_1}$ is the reversed component, similar to that derived earlier.

$$\overline{Task_1} \stackrel{def}{=} (read, (1-p)f_r\xi).\overline{Task_2} + (read, pf_r\xi).\overline{Task_3}$$
$$\overline{Task_2} \stackrel{def}{=} (compute, \mu).\overline{Task_1}$$
$$\overline{Task_3} \stackrel{def}{=} (write, f_w\eta).\overline{Task_2}$$

A longer discussion of functional rates in the reversed process was given in [6]. In this instance f_r and f_w can be any functions of i, j and k, but we will restrict ourselves to three simple pairs of functions involving i and k.

- $f_r = \frac{i}{i+k}$, $f_w = \frac{k}{i+k}$

  This is the simple processor sharing function pair and is exactly the same as the previous model, with the exception that there are different rates for *read* and *write*. Thus, if $\mathcal{L} = \{read, compute, write\}$,

  $$\pi_{\{i,j,k\}} = \frac{(i+k)!}{i!k!} \left(\frac{\xi}{\mu}\right)^j \left(\frac{p\xi}{\eta}\right)^k \pi_{\{N,0,0\}}$$

- $f_r = ci$, $f_w = ck$, where c is a constant.

  If $c = 1$ this function pair gives the same behaviour as the initial model when $read, write \notin \mathcal{L}$. If $c \neq 1$ there is only a small change.

  $$\pi_{\{i,j,k\}} = \frac{N!}{i!k!} \left(\frac{c\xi}{\mu}\right)^j \left(\frac{p\xi}{\eta}\right)^k \pi_{\{N,0,0\}}$$

- $f_r = \frac{i\xi}{i\xi+k\eta}$, $f_w = \frac{k\eta}{i\xi+k\eta}$

  This function pair allocates resource in proportion to service demand. That is, the faster an action is and the more participants it has, relative to the other, the greater the resource that would be allocated. Once more, in the forward process, the function does not affect the behaviour of the *read* action in the sequence we are interested in (as there are no $Task_3$'s). However, in the reverse process, the rate product is rather more complex:

$$\frac{k!\eta^{2k}}{\prod_{x=1}^{k}(x\eta + i\xi)}$$

Hence,

$$\pi_{\{i,j,k\}} = \frac{\prod_{x=1}^{k}(x\eta + i\xi)}{k!\eta^{2k}} \left(\frac{\xi}{\mu}\right)^{j} \left(\frac{p\xi}{\eta}\right)^{k} \pi_{\{N,0,0\}}$$

6 Conclusions and Further Work

We have used properties of reversed processes to find the steady state probabilities in a class of PEPA models with active-active cooperation. This can be seen as an application of the *Reversed Compound Agent Theorem*, similarly extended to such rate synchronisation. Despite this class lying outside the existing classification of RCAT, the explicit derivation of the reversed process and its use in deriving expressions for the equilibrium distribution is shown to be relatively straight forward. In fact, for reasons of completeness and clarity we have computed the reversed process completely, whereas we have only used a subset of the reversed actions in deriving each product-form solution. Therefore it should be apparent that, in general, it is not necessary to compute all the reversed rates, and in many instances a solution may be derived with only those rates which are most easily found. Indeed, this is the approach adopted for the practical application of the existing RCAT in Section 2.

Of course, we should not be surprised that such results exist for what is, essentially, a closed queueing network. However, through the running example we were able to show the robustness of the solution under a variety of different conditions. Such robustness would clearly be desirable if we were considering a range of different deployment options, for example. There are obviously other extensions to the basic example that we could have considered and some of these could break the product-form solution. For example, we could consider that the server components have alternative modes of operation, which in general would give rise to models without product-form.

Acknowledgements

The authors are supported by the EPSRC funded CAMPA project, which has enabled an extended research visit by Dr Thomas to Imperial College London.

References

1. Balbo, G., Bruell, S., Sereno, M.: Embedded processes in generalized stochastic Petri net. In: Proc. 9th Intl. Workshop on Petri Nets and Performance Models, pp. 71–80 (2001)

2. Boucherie, R.J.: A Characterisation of Independence for Competing Markov Chains with Applications to Stochastic Petri Nets. IEEE Trans. on Software Eng. 20(7), 536–544 (1994)
3. Harrison, P.G.: Turning back time in Markovian process algebra. Theoretical Computer Science (January 2003)
4. Harrison, P.G.: Reversed processes, product-forms and a non-product-form. Linear Algebra and Its Applications (July 2004)
5. Harrison, P.G.: Compositional reversed Markov processes, with applications to G-networks. Performance Evaluation (2004)
6. Harrison, P.G.: Product-forms and functional rates. Performance Evaluation 66, 660–663 (2009)
7. Henderson, W., Taylor, P.G.: Embedded Processes in Stochastic Petri Nets. IEEE Trans. on Software Eng. 17(2), 108–116 (1991)
8. Hillston, J.: A Compositional Approach to Performance Modelling. Cambridge University Press, Cambridge (1996)
9. Kelly, F.P.: Reversibility and stochastic networks. Wiley, Chichester (1979)
10. Thomas, N., Zhao, Y.: Mean value analysis for a class of PEPA models. In: Bradley, J.T. (ed.) EPEW 2009. LNCS, vol. 5652, pp. 59–72. Springer, Heidelberg (2009)
11. Thomas, N.: Using ODEs from PEPA models to derive asymptotic solutions for a class of closed queueing networks. In: Proceedings 8th Workshop on Process Algebra and Stochastically Timed Activities. University of Edinburgh, Edinburgh (2009)

Generalised Entropy Maximisation and Queues with Bursty and/or Heavy Tails

Demetres D. Kouvatsos and Salam A. Assi

Networks and Performance Engineering Research Group (NetPEn),
Informatics Research Institute (IRI), University of Bradford, Bradford, BD7 1DP, UK
D.Kouvatsos@Bradford.ac.uk, S.A.Assi@leeds.ac.uk

Abstract. An exposition of the 'extensive' (EME) and 'non-extensive' (NME) maximum entropy formalisms is undertaken in conjunction with their applicability into the analysis of queues with bursty and/or heavy tails that are often observed in performance evaluation studies of heterogeneous networks and Internet exhibiting traffic burstiness, self-similarity and long-range dependence (LRD). The credibility of these formalisms, as methods of inductive inference, for the study of physical systems with both short-range and long-range interactions is explored in terms of four potential consistency axioms. Focusing on stable single server queues, it is shown that the EME and NME state probabilities are characterized by generalised types of modified geometric and Zipf-Mandelbrot distributions depicting, respectively, bursty generalized exponential and/or heavy tails with asymptotic power law behaviour. Numerical experiments are included to highlight the credibility of the maximum entropy solutions and assess the combined impact of traffic burstiness and self-similarity on the performance of the queue.

Keywords: Extensive maximum entropy (EME) formalism, non-extensive maximum entropy formalism (NME), heterogeneous networks, traffic characterisation, burstiness, self-similarity, short-range dependence (SRD), long-range dependence (LRD), queueing systems, performance evaluation, fractional Brownian motion (fBm), generalised exponential (GE) distribution, generalised (modified) geometric (GGeo) distribution, generalised Zipf-Mandelbrot (G-Z-M) distribution.

1 Introduction

Empirical traffic characterisation studies in networks of diverse technology and the Internet have shown that traffic flows often exhibit burstiness, self-similarity and/or long-range dependence (LRD). These properties can be attributed to the heavy-tailedness of the traffic distributions involved, causing performance degradation and queues with bursty and/or heavy tails with asymptotic power law behaviour (c.f., [8,14,19,55,58,61,62,66,71]).

Traffic distributions with heavy tails, which decay much more slowly than those of an exponential distribution, are often employed to generate workloads in simulation studies on the performance modelling and engineering of high speed telecommunication systems. These, however, tend to be rather inflexible, computationally expensive and may display unusual characteristics [15,65]. Analytic mechanisms for estimating the tail index of Internet traffic with heavy tails can be seen in [65], based on the Pareto distribution [12,21].

This tutorial presents an exposition of the formalisms (or, principles) of the extensive (EME) maximum entropy and non-extensive (NME) maximum entropy, a generalisation, in conjunction with their applications into the analysis of stable single server queues with bursty generalised exponential (GE-type) (e.g., [17,35,40]) and/or heavy tails with asymptotic power law behaviour (e.g., [4,30,49,52,53]), as appropriate. These methodologies are based on the maximisation of the i) classical Boltzmann-Gibbs-Shannon extensive entropy function devised in the fields of Information Theory [67] and Thermodynamics [11] and ii) generalised Havrda-Charvat-Tsallis entropy function proposed in the fields of Quantification Theory [22], Information Theory [64] and Statistical Physics [73]. The credibility of the EME and NME formalisms, as methods of inductive inference, is explored in terms of four consistency axioms proposed in [68] concerning the application of the principles of EME and minimum relative (or, cross) extensive entropy (EMRE) [47,54]. Focusing on stable single server queues, it is shown, subject to appropriate mean value constraints, that the EME and NME solutions for the state probability distributions are characterised, respectively, by i) modified geometric (Geo)(c.f., [70]) and generalised Geo (GGeo) (c.f., [37,38,45]) type distributions and ii) the Zipf-Mandelbrot (Z-M) (c.f., [30,57,73]) and generalised Z-M (G-Z-M)(c.f., [4,5,49,52,53]) type distributions. Moreover, efficient analytic algorithms, based on the Newton-Raphson numerical method, are described and typical numerical experiments are included to highlight the credibility of the EME and NME solutions and the adverse impact of the corresponding bursty GE-type and heavy queue length tails on queue performance.

The concepts of the classical Boltzmann-Gibbs-Shannon extensive entropy and the Havrda-Charvat-Tsallis non-extensive entropy are explored in Section 2. The Z-M and GE type distributions are introduced in Section 3. The EME and NME formalisms are reviewed in Section 4. The Tsallis [73] NME solution in Statistical Physics is presented in Section 5. The consistency of the EME and NME formalisms, as methods of inductive inference, for the analysis of physical systems with short-range and long-range interactions is addressed in Section 6. The EME and NME probability distributions of stable single server queues with bursty GE-type and/or heavy tails, respectively, are devised, in conjunction with associated numerical algorithms, in Section 7. Typical numerical experiments are illustrated in Section 8. Concluding remarks follow in Section 9.

2 On the Interpretation of the Classical and Generalised Entropies

2.1 The Classical Extensive Entropy Function

For a general physical system Q with an integer number of possible (microscopic) configurations or states $N(N > 0)$ and 'short-range interactions', the classical Boltzmann-Gibbs entropy in Statistical Physics [11] or, equivalently, Shannon's information theoretic entropy [67], $H^*(p_N)$ is defined by

$$H^*(p_N) = -c \sum_{n=1}^{N} p_N(n) \log p_N(n) \qquad (1)$$

where $c(c > 0)$ is a positive constant, $N(N > 0)$ is the integer number of possible (microscopic) configurations or states and $\{p_N(n), n = 1, 2, ..., N\}$ are the associated event or state probabilities. The entropy function $H^*(p_N)$ can be interpreted as a measure of uncertainty or information content that is implied by p_N about the physical system Q with short-range interactions (i.e., 'short memory'). The principle of ME is a probability method of inductive inference originally proposed by Jaynes [24,25] in Statistical Physics and it is based on the maximisation of the extensive entropy function $H^*(p_N)$, subject to suitable mean value constraints. For short-range interactions, such as and "holding matter together" in Statistical Physics, quantities such as "entropy and energy" are considered as 'extensive' variables in the sense that the total entropy and energy of the system are both "proportional to the system size" (c.f., [16]).

By applying the method of Lagrange undetermined multipliers to maximise the Boltzmann-Gibbs-Shannon extensive entropy, subject to the normalisation and the first moment constraint, it can be shown that at the limit, as $N \to +\infty$ (e.g., [70]), the ME state probability distribution $\{p_N(n), n = 1, 2, ...\}$ is characterised by the Geo distribution (c.f., [34,70]). Note that in a more general context, an investigation into the credibility of the principles of EME and EMRE was undertaken by Shore and Johnson [26,68,69] in terms of four axioms of inductive inference. Expositions of the ME principle and generalisations, as applied in various fields of Science and Engineering, can be seen in [28,29].

By analogy, traffic flows in queues exhibiting short-range dependence (SRD), such as those represented by Poisson (regular), compound Poisson (bursty) (c.f., [45]) and batch renewal (BR) (bursty and correlated) (c.f., [48]) processes, influence the creation of stable queueing systems with short-range interactions, as appropriate, where the state and entropy variables are extensive (c.f., [53]). In this context, the EME formalism is based on the maximisation of the extensive entropy, $H^*(p_N)$, subject to normalisation and suitable mean value constraints (e.g., the utilisation and the fist moment). To this end, the EME solution is determined by applying the method of Lagrange's undetermined multipliers leading to the characterisation, for example, of modified geometric (Geo), generalised Geo (GGeo) (e.g., [45,47]) and shifted GGeo (c.f., [48]) types of state probabilities for single server queues.

2.2 A Generalised Non-extensive Entropy Function

For a general system Q with an integer number of possible (microscopic) configurations or states $N(N > 0)$ and 'long-range interactions', such as gravity in Statistical Physics, "energy and entropy are no longer extensive quantities" [16].This increases the complexity of the physical system Q for which the state probability distribution associated say, with energy, has heavy tails and power law behaviour and thus, it can no longer be determined by maximizing the classical extensive entropy, $H^*(p_N)$.

To address this problem, Tsallis [73] proposed an generalisation of Boltzmann-Gibbs-Shannon EE function $H^*(p_N)$ to a 'non-extensive' entropy function, $H^*(p_{q,N})$, namely

$$H^*(p_{q,N}) = c(1 - \sum_{n=1}^{N} p_{q,N}(n)^q)/(q-1) \qquad (2)$$

where $c(c > 0)$ is a positive constant, q is a real number known as the 'non-extensivity' parameter measuring the degree of the system's 'long-range interactions', $N(N > 0)$ is the integer number of possible (microscopic) configurations or states and $\{p_{q,N}(n), n = 1, 2, ..., N\}$ are the associated state probabilities of system, Q. As $q \to 1$, $H^*(p_{q,N})$ reduces to the Boltzmann-Gibbs-Shannon entropy function, $H^*(p_N)$.

The non-extensive entropy, $H^*(p_{q,N})$, devised by Tsallis [73] turned out to be identical to the one proposed earlier by Havrda-Charvat [22] in the field of Quantification Theory of classification processes. In the context of the canonical ensemble in Statistical Physics, applying the method of Lagrange undetermined multipliers to maximise $H^*(p_{q,N})$, subject to the normalization and mean (generalised internal) energy constraint, the form of the GME state probability distribution of energy can be determined (c.f.,[73]). As it was observed in [30], this NME solution follows the Z-M distribution [57], which has heavy tails and asymptotic power law behaviour.

The role of constraints towards the characterisation of the NME metrics was investigated by Tsallis et al [74]. This NME solution was formulated in [63] in accordance with the information theoretic approach advocated by Jaynes [24,25]. Reviews of the NME formalism can be found in [75,76] whilst related applications and extensions into the study of power law behaviour in interdisciplinary applications were reported in [18]. Power-law distributions were also devised in [20] for the citation index of scientific publications and scientists. Moreover, the principle of maximum likehood was employed in [72] to show that the Tsallis NME distribution estimate is a non-extensive generalisation of the Gaussian distribution.

By analogy, traffic processes exhibiting self-similarity and LRD, such as fractional Brownian (fBm) [9], influence the formation of queues with long-range interactions. In this case, the state and entropy variables are non-extensive leading to NME state probabilities at equilibrium, which follow the form of the Z-M (c.f., [30,73] and G-Z-M type distributions (c.f., [4,49,52,53]).

Note that NME solutions for single server queues with or without finite capacity at equilibrium were first established in Assi [4] and Kouvatsos and Assi [49] by maximising the non-extensive Havrda-Charvat entropy function [22] and other generalised entropy measures reported in [28], [29], subject to GE-type queueing theoretic mean value constraints[1].

3 The Zipf-Mandelbrot (Z-M) and the Generalised Exponential (GE) Distributions

3.1 The Zipf-Mandelbrot (Z-M) Probability Distribution

The Z-M probability distribution [57] is a power-law discrete time distribution on ranked data. It is a generalisation in the discrete time domain of the Zipf (Z) distribution (c.f., [1,27,77]). The Z-M probability mass function is of the form

---

[1] The work of Tsallis [73] in Statistical Physics was not known at the time to the authors of [4,49], who instead optimised the non-extensive entropy measure of Havrda-Charvat [22], which is identical to that devised by Tsallis [73].

$$p(n, u, s) = \frac{(n+u)^{-s}}{\sum_{i=1}^{N}(i+u)^{-s}} \qquad (3)$$

where N is the number of elements, n is a real number representing their rank, u is a real number and $s(s > 1)$ is the value of the exponent characterising the distribution and is given by $s = \frac{1}{1-q}$, where q is the non-extensivity parameter. Note that the NME solution devised by Tsallis [73], based on the maximisation of the extended Havrda-Charvat-Tsallis entropy function, subject to the normalisation and the first moment constraint, is of the Z-M type.

At the limit as $N \to +\infty$, the sum $\sum_{n=1}^{+\infty}(n+u)^{-s}$ is the Hurwitz-Zeta function (c.f.,[3,7]). For finite $N(N < +\infty)$ and $u = 0$, the Z-M distribution becomes the Zipf (Z) distribution [1] and, moreover, for $W \to +\infty$ and $u = 0$, it is known as the Zeta distribution. A discussion of the Z and Z-M distributions with reference to Tsallis [73] EME solution can be seen in Aksenov et al [2].

The Zipf and Z-M distributions can be viewed as the discrete counterparts of Pareto [59] and generalised Pareto [10] continuous time distributions, respectively. These distributions, due to their self-similar properties, have strong implications towards the convergence of multiservice heterogeneous networks and the design and functionality of the next generation Internet (NGI).

3.2 The Generalised Exponential (GE) Distribution

The GE distribution is a mixed interevent time distribution of the form (c.f., Fig. 1)

$$F(t) = P(A \leq t) = 1 - \tau e^{-\sigma t}, \quad t \geq 0, \qquad (4)$$

$$\tau = \frac{2}{C^2 + 1} \qquad (5)$$

$$\sigma = \tau \nu \qquad (6)$$

where A is a mixed-time random variable of the interevent-time, whilst $(1/\nu, C^2)$ are the mean and squared coefficient of variation (SCV) of random variable A. The GE-type distribution is versatile, possessing pseudo-memoryless properties, which make the derivation of exact and approximate solutions of many GE-type queueing systems and networks analytically tractable.

For $C^2 > 1$, the GE is an extremal case of the family of Hyperexponential-2 (H_2) distributions with the same (ν, C^2) having a corresponding counting process equivalent to a compound Poisson process (CPP) with parameter $2\nu/(C^2 + 1)$ and geometrically distributed bulk sizes with mean, $(1 + C^2)/2$ and SCV, $(C^2 - 1)/(C^2 + 1)$. The CPP is expressed by

$$P(N_{cp} = n) = \begin{cases} \sum_{i=1}^{n} \frac{\sigma^i}{i!} e^{-\sigma} \binom{n-1}{i-1} \tau^i (1-\tau)^{n-i}, & n \geq 1 \\ e^{-\sigma}, & n = 0 \end{cases} \qquad (7)$$

where N_{cp} is a CPP random variable of the number of events per unit time corresponding to a stationary GE-type random variable of interevent time [45].

The choice of the GE distribution is further motivated by the fact that measurements of actual interarrival or service times may be generally limited and so only few parameters can be computed reliably. Typically, only the mean and variance may be relied upon and thus, a choice of a distribution that implies least bias (i.e., introduction of arbitrary and therefore, false assumptions) is that of GE-type distribution [45]. For example, the GE distribution is applicable when smoothing schemes are introduced at the adaptation level (e.g., for a stored video source in ATM networks) with the objective of minimising or even eliminating traffic flow correlation (c.f., [6]). Moreover, under renewality assumptions, the GE distribution is most appropriate to model simultaneous packet arrivals at input and output port queues of a network generated by different bursty sources (e.g., voice or high resolution video) with known first two moments. In this context, the burstiness of the arrival process is characterised by the SCV of the interarrival time or, equivalently, the mean size of the incoming bulk.

$$1 - \tau = \frac{C^2 - 1}{C^2 + 1}$$

$$\tau = \frac{2}{C^2 + 1}$$

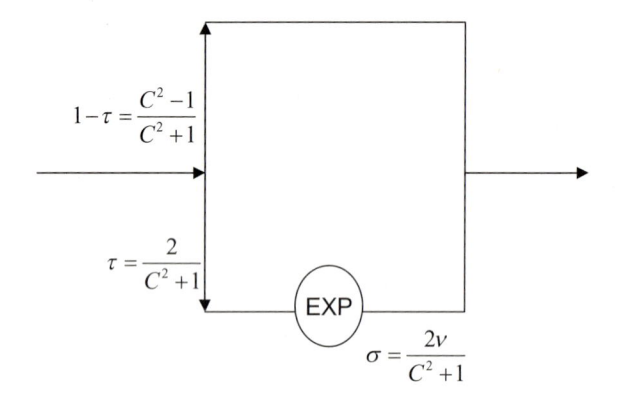

$$\sigma = \frac{2v}{C^2 + 1}$$

Fig. 1. The GE distribution with parameters τ and σ ($0 \leq \tau \leq 1$)

The GE distribution may also be employed to model traffic flows with SRD in the continuous time domain with small error. For example, an SRD process may be approximated by an ordinary GE distribution whose first two moments of the count distribution match the corresponding first two SRD moments (c.f., [56]). Similarly, a traffic process with SRD may be approximated in a discrete time domain with a small error by an ordinary GGeo distribution (c.f., [47]). This approximation of a correlated arrival process by an uncorrelated traffic process may facilitate (under certain conditions) problem tractability with tolerable accuracy and, thus, the understanding of the performance behaviour of external SRD traffic in the interior of the network. It can be further argued that, for a given buffer size, the shape of the autocorrelation curve, from a certain point on wards, does not influence system behaviour. Thus, in the context of system performance evaluation, an SRD model may be used under certain conditions to approximate accurately real traffic with LRD.

4 The EME and NME Formalisms

Consider a general physical system Q that has a set S of possible discrete states $\{S_0, S_1, S_2, ...\}$, which may be finite or countable infinite and state $\{S_n, n = 0, 1, ...\}$ may be specified arbitrarily. Suppose the available information about Q places a number of constraints on $\{p(S_n), S_n \in S\}$ or, $\{p_q(S_n), S_n \in S\}$, the probability distribution that the either an extensive or non-extensive system Q is at state S_n, where q indicates the non-extensivity parameter. Without loss of generality, it is assumed that these take the form of mean values of several suitable functions $\{f_1(S_n), f_2(S_n), ..., f_m(S_n)\}$ or, $\{f_{1,q}(S_n), f_{2,q}(S_n), ..., f_{m,q}(S_n)\}$, for an extensive or non-extensive system Q, respectively, where m is less than the number of feasible states.

4.1 The Extensive Maximum Entropy (EME) Formalism

For an extensive system Q with short-range interactions, the principle of EME (c.f., [24,25]) states that, of all distributions satisfying the constraints supplied by the given information, the minimally prejudiced distribution state probability distribution $p(S_n)$ is the one that maximises the system's Boltzmann-Gibbs-Shannon extensive entropy function, $H^*(p)$ (c.f., [11,67]), namely

$$H^*(p) = -c \sum_{S_n \in S} p(S_n) \ln\{p(S_n)\} \tag{8}$$

subject to the constraints

$$\sum_{S_n \in S} p(S_n) = 1 \tag{9}$$

$$\sum_{S_n \in S} f_k(S_n) p(S_n) = F_k \tag{10}$$

where c ($c > 0$) is a positive constant, $\{F_k, k = 1, 2, ..., m\}$ are the prescribed mean values defined on the set of functions $\{f_k(S_n), k = 1, 2, ..., m\}$.

The maximisation of the extensive entropy $H^*(p)$, subject to constraints (9) - (10), can be carried out using Lagrange method of undetermined multipliers leading to the solution

$$p(S_n) = \frac{1}{Z} \exp\left\{-\sum_{k=1}^{m} \beta_k f_k(S_n)\right\} \tag{11}$$

where $\{\beta_k, k = 1, 2, ..., m\}$ are the Lagrangian multipliers determined from the set of constraints (10) and Z, know in statistical physics as the 'partition function' is given by

$$Z = \exp\{\beta_0\} = \sum_{S_n \in S} \exp\left\{-\sum_{k=1}^{m} \beta_k f_k(S_n)\right\} \tag{12}$$

where β_0 is a Lagrangian multiplier determined by the normalisation constraint (9). Note that maximising the entropy $H^*(p)$ for an extensive system Q with a finite number of states, subject only to normalisation constraint, the EME solution of the state probability distribution is characterised by the uniform distribution which, in information theoretic terms, as the least biased distribution estimate.

Jaynes [24,25] showed for extensive systems that "if the prior information includes all constraints actually operative in a random experiment, then the distribution predicted by the maximum entropy can be overwhelmingly realised in more ways than any other distribution". Moreover, Shore and Johnson [68,69] established thyat the NME principle is "a uniquely correct self-consistent method of inference for estimating a probability distributions based on the available information".

The EME formalism has been utilised in the performance analysis of queueing systems since expected values of various distributions of interest are usually known in terms of moments of the interarrival and service time distributions. Applications of the EME formalism towards the exact and approximate analysis of single queues and arbitrary queueing network models (QNMs) with or without blocking and multiple classes under different scheduling rules and buffer management schemes can be seen, for example, in [17,35,36,37,38,39,40,41,42,43,44,45,46,50,51,56,70,78].

4.2 A Non-extensive Maximum Entropy (NME) Formalism

For a non-extensive physical system Q with long-range interactions, a NME framework can be established to determine the form of state probability distribution, $p_q(S_n)$ that maximises the Havrda-Charvat-Tsallis non-extensive entropy function $H^*(p_q)$ (c.f., [22,73]), namely

$$H^*(p_q) = c \frac{1 - \sum_{k=1}^{m} p_q(S_n)^q}{q - 1} \quad (13)$$

subject to the constraints

$$\sum_{S_n \in S} p_q(S_n) = 1 \quad (14)$$

$$\sum_{S_n \in S} f_{k,q}(S_n) p_q(S_n) = F_{k,q} \quad (15)$$

where $c (c > 0)$ is a positive constant, q the non-extensitivity parameter, $\{F_{k,q}, k = 1, 2, ..., m\}$ are the prescribed mean values defined on the set of extended functions $\{f_{k,q}(S_n), k = 1, 2, ..., m\}$.

By employing the Lagrange method of undetermined multipliers, the maximisation of non-extensive entropy, $H(p_q)$, subject to mean value constraints (14) - (15), leads to a least biased G-Z-M type solution for the NME state probability distribution, namely

$$p_q(S_n) = \frac{1}{Z_q} \left[1 + \sum_{k=1}^{m} \beta_k (1 - q) f_{k,q}(S_n) \right]^{\frac{1}{q-1}} \quad (16)$$

where $\{\beta_k\}, k = 1, 2, ..., m\}$ are the Lagrangian multipliers corresponding to the constraints (15) and Z_q is the normalizing constant expressed by

$$Z_q = \exp\{\beta_0\} = \sum_{S_n \in S} \left[1 + \sum_{k=1}^{m} \beta_k (1-q) f_{k,q}(S_n)\right]^{\frac{1}{q-1}} \quad (17)$$

where β_0 is the Lagrangian multiplier determined by the normalisation constraint (14). Note that the derivation of the m-fold G-Z-M type distribution (17) is attributed to the employment of the m mean value constraints (15) in the context of the NME formalism.

Maximising the entropy function, $H^*(p_q)$ for a non-extensive system Q with a finite number of states, subject only to the normalisation constraint, a uniform state probability distribution, as in the case of the Boltzmann-Gibbs-Shannon entropy function $H^*(p)$, is obtained. The $H^*(p_q)$ entropy function can be described as a low-order truncation of Renyi's entropy [64], which has a well known information theoretic interpretation [16].

Applications of the NME formalism towards the analysis of single server queues and QNMs with or without finite capacity can be seen in [4,5,30,31,32,33,49,52,53]. More specifically, NME state probability distributions of G-Z-M type were devised for stable single server generalised exponential (GE) type queues with or without finite capacity, subject to normalization, utilisation (i.e., the probability of a non-empty queue), mean queue length (MQL) and full buffer state probability GE-type constraints, as appropriate[4,49]. The original NME solution derived by Tsallis [73] in Statistical Physics was adopted by Karmeshu and Sharma [30] for the analysis of a single server queue with infinite capacity, subject to normalization and a MQL constraint based on a formula proposed by Norros [60]. The NME formalism in [73] was also applied for the analysis of i) a single server queue with infinite capacity, subject to normalization and a fractional moment constraint [31] and ii) queueing networks with applications to broadband networks exhibiting LRD traffic [32,33]. Moreover, NME solutions of G-Z-M type were determined in [5,52] for stable single server queues with infinite and/or finite capacities. These NME solutions are based on the ones reported in [4,37,38,40,49] and employ as a MQL constraint a heuristic generalisation of Norros formula [60]. More recently, a critique of the NME formalism, as a method of inductive inference for non-extensive systems and the characterisation of NME state probability of a finite capacity queue as a generalisation of the Z-M type distribution can be seen in [53].

5 The Tsallis EME Solution in Statistical Physics

In the context of the canonical ensemble in Statistical Physics, let Q be a non-extensive physical system with a finite number of states $\{S_n = n, n = 1, 2, ..., N\}$, where $N(N > 0)$ is the integer number of possible (microscopic) configurations. Suppose the available information about Q imposes the normalisation and mean generalised internal energy, $E_{q,N}$ constraints on the state probability distribution $\{p_{q,N}(n), n = 1, 2, .., N\}$.

Tsallis [73] maximised the generalised entropy function, $H^*(p_{q,N})$ (c.f., (2), namely

$$H^*(p_{q,N}) = c \frac{1 - \sum_{n=1}^{N} p_{q,N}^q}{q-1} \quad (18)$$

subject to the constraints of normalisation and mean generalised internal energy, $E_{q,N}$, namely

$$\sum_{n=1}^{N} p_{q,N}(n) = 1 \qquad (19)$$

$$\sum_{n=1}^{N} \varepsilon_n p_{q,N}(n) = E_{q,N} \qquad (20)$$

where $c(c > 0)$ is a constant, and $\{\varepsilon_n, n = 1, 2, ..., N\}$ are real numbers referred to as generalised spectrum.

By applying the method of Lagrange undetermined multipliers, the following Z-M distribution is obtained (c.f., [73]):

$$p_{q,N}(n) = \frac{[1 + \beta(1-q)\varepsilon_n]^{\frac{1}{q-1}}}{Z_q} \qquad (21)$$

where beta is the Lagrangian multiplier corresponding to the mean generalised internal energy, $E_{q,N}$ constraint and

$$Z_q = \sum_{n=1}^{N} [1 - \beta(q-1)\varepsilon_n]^{\frac{1}{q-1}} \qquad (22)$$

The NME solution (21) of the state probability $\{p_{q,N}(n)\}$ reduces at the limit $q \to 1$ to the classical Boltzmann-Gibbs-Shannon EME solution (11), namely

$$p_N(n) = \frac{e^{-\beta\varepsilon_n}}{Z_1} \qquad (23)$$

where

$$Z_1 = \sum_{n=1}^{N} e^{-\beta\varepsilon_n} \qquad (24)$$

6 The EME and NME Formalisms as Methods of Inductive Inference and Consistency Axioms

The principles of EME [24,25] and EMRE [54] were shown by Johnson [26], Shore and Johnson [68,69] to be uniquely correct methods of inductive inference for extensive systems, subject to a prior probability estimate, as appropriate and new information given in the form of mean values. Clearly, in the context of ME formalism, the prior distribution for an extensive system Q with a finite number of states is the uniform.

The approach adopted in [68] was based on the fundamental assumption that the use of the EME and EMRE principles as methods of inductive inference, should lead to consistent results when there are different ways to solve a problem by taking into account the same information. This fundamental requirement was formalised in terms of four consistency axioms [68], namely uniqueness, invariance, system independence and subset independence. It was shown that optimizing any function other than extensive entropy and relative entropy will lead to inconsistencies unless the function in question and the extensive entropies share, respectively, identical maxima or minima. In other words, given new constraint information, there is only one distribution satisfying these constraints that can be chosen by a procedure based on EME and EMRE formalisms satisfying the consistency axioms.

The credibility of some non-extensive entropy principles (c.f., [28,29]), as applied into the special case of single server queues of GE-type, was originally addressed in [4,49] by employing the four consistency axioms for extensive systems proposed in [68]. More recently, a formal study into the relevance of these axioms on the credibility of the NME principle, as a method of inductive inference, for the analysis of generic extensive systems was undertaken in Kouvatsos and Assi [53]. An exposition of this investigation is highlighted below.

An overview of this investigation is highlighted below by focusing on the Havrda-Charvat-Tsallis non-extensive entropy, $H^*p_{q,N}$, where the Boltzmann-Gibbs-Shannon EE function, H^*p_N is a special case for $q \to 1$.

6.1 Uniqueness

According to this axiom, "If the same problem is solved twice in exactly the same way, the same answer is expected in both cases i.e., the solution should be unique" (c.f., [68]).

Focusing on the Havrda-Charvat-Tsallis non-extensive entropy function (c.f., 18) [22,73], let Q be a general non-extensive system having without loss of generality (wlog) a finite set S of N, $N > 0$ possible discrete states $\{S_n, n = 1, 2, ..., N\}$ and Ω be a closed convex set of all probability distributions $\{p_{q,N}(S_n), S_n \in S\}$ such that $p_{q,N}(S_n) > 0$ for $S_n \in S, n = 1, 2, ..., N$ and $\sum_{n=1}^{N} p_{q,N}(S_n) = 1$. Let $f_{q,N}, h_{q,N} \in \Omega$ be two probability distributions defined on S having the same extended entropy functions, namely $H^*(f_{q,N}) = H^*(h_{q,N})$. Moreover, let $u_n = f_{q,N}(S_n)$, $v_n = h_{q,N}(S_n)$ and $r(x_n) = x_n^q$, where $x_n = u_n, v_n$.

In this context, the non-extensive entropy (c.f., 18) can be rewritten as

$$H^*(x_n) = \frac{c}{q-1}1 - \sum_{n=1}^{W} r(x_n) \qquad (25)$$

Let wlog that $0 < q < 1$. Since the second derivative of $r(x_n)$ with respect to x_n is given by $r(x_n)'' = q(q-1)x_n^{q-2} < 0$, it follows that the function $r(x_n)$ is strictly convex for $x_n = u_n, v_n$. This leads to the condition

$$\alpha r(u_n) + (1-\alpha) r(v_n) < r(\alpha u_n + (1-\alpha)v_n) \qquad (26)$$

where $\alpha \in [0, 1]$ and $u_n \neq v_n$. Multiplying by $(\frac{c}{q-1})$ and subtracting 1 from both sides and, moreover, summing both sides of (26) over i, then the following expression is obtained

$$H^*(f_{q,N}) = H^*(h_{q,N}) = \\ \alpha H^*(f_{q,N}) + (1-\alpha)H^*(h_{q,N}) \leq H^*(\alpha f_{q,N} + (1-\alpha)h_{q,N}) \quad (27)$$

The inequality is strict unless $f_{q,N} = h_{q,N}$. If $f_{q,N} \neq h_{q,N}$ and $H^*(f_{q,N}) = H^*(h_{q,N})$ then, since Ω is convex, there is a distribution (i.e., weighted average) given by $\alpha f_{q,N} + (1-\alpha)h_{q,N}$, which belongs to Ω and has a non-extensive entropy greater than $H^*(f_{q,N}) = H^*(h_{q,N})$. Therefore, "there cannot be two distinct probability distributions $f_{q,N}, h_{q,N} \in \Omega$ having the same maximum non-extensive entropy in Ω" (c.f., [53]).

Thus, the NME formalism satisfies the axiom of uniqueness [68].

6.2 Invariance

The invariance axiom states that "The same solution should be obtained if the same inference problem is solved twice in two different coordinate systems" (c.f., [68]).

Following the analytic methodology in [68] and adopting the notation of subsection 6.1, let Γ be a coordinate transformation from state $S_n \in S(n = 1, 2, ..., N)$ to state $R_n \in R(n = 1, 2, ..., N)$, where R be a transformed set of N possible discrete states $\{R_n, n = 1, 2, ..., N\}$ with $(\Gamma p_{q,N})(R_n) = J^{-1}p_{q,N}(S_n)$, where J is the Jacobian $J = \partial(R_n)/\partial(S_n)$.

Moreover, let $\Gamma\Omega$ be the closed convex set of all probability distributions $\Gamma p_{q,N}$ defined on R such that $\Gamma p_{q,N}(R_n) > 0$ for all $R_n \in R, n = 1, 2, ..., N$ and $\sum_{n=1}^{N} \Gamma p_{q,N}(R_n) = 1$.

It can be clearly seen that, transforming of variables from $S_n \in S$ into R_n in R, the Havrda-Charvat-Tsallis extended entropy function (c.f., (18)) is transformation invariant, namely

$$H^*(p_{q,N}) = H^*(\Gamma p_{q,N}) \quad (28)$$

Thus, "the NME formalism satisfies the axiom of invariance [68] since the minimum in $\Gamma\Omega$ corresponds to the minimum in Ω" (c.f., [53]).

6.3 System Independence

This axiom of system independence, which is also referred as the additivity property, states that "It should not matter whether one accounts for independent information about independent systems separately in terms of different probabilities or together in term of joint probability" (c.f., [68]).

Consider two general non-extensive systems Q and R each of which having wlog a finite set of $N, N > 0$ possible discrete states $\{x_n, n = 1, 2, ..., N\}$ and $\{y_n, n = 1, 2, ..., N\}$, respectively. Moreover, let X and Y be the random variables describing the state of the systems Q and R with corresponding state probabilities $f_{q,N}(x_n) = Pr\{X = x_n\}$ and $g_{q,N}(y_n) = Pr\{Y = y_n\}$, respectively.

Assuming that Q and R are independent systems, then the joint probability, $h_{q,N}(x_k, y_n) = Pr(x_k, y_n), k, n = 1, 2, ..., N$ is clearly given by

$$h_{q,N}(x_n, y_n) = Pr(X = x_k, Y = y_n) = f_{q,N}(x_k)g_{q,N}(y_n) \quad (29)$$

For the joint probability $h_{q,N}(x_n, y_n)$, the Havrda-Charvat-Tsallis non-extensive entropy function (c.f., (18)) can be written as

$$H^*[(h_{q,N}] = c\frac{1 - \sum_k \sum_n h_{q,N}^q}{q - 1} \qquad (30)$$

From the definition of (18), it clearly follows that

$$H^*[h_{q,N}] \neq H^*(f_{q,N}) + H^*(g(Y_{q,N})) \qquad (31)$$

The inequality (31) implies that, in information theoretic terms, "the joint NME state probability distribution of two independent non-extensive systems Q and R defies, due to the presence of long-range interactions, the axiom of system independence (c.f., [68]). Thus, this attribute of the NME formalism, as a method of inductive inference, is clearly most suitable for the quantitative studies of non-extensive dynamic systems with heavy queue tails and asymptotic power law behaviour" (c.f., [53]).

Note that in the case of $q \to 1$ limit, equation (31) becomes

$$H^*[h_{q,N}] = H^*(f_{q,N}) + H^*(g(Y_{q,N})) \qquad (32)$$

This result verifies that the joint EME state probability distribution, as expected, satisfies the axiom of system independence (c.f., [68]). This is "an appropriate property of the EME formalism, as a method of inductive inference, for the study of extensive systems with short-range interactions "(c.f., [53]).

6.4 Subset Independence

The axiom of subset independence states that "It does not matter whether one treats an independent subset of system states in terms of a separate conditional density or in terms of the full system density" (c.f., [68]).

Consider a general non-extensive system Q that has wlog a finite number, $L, L > 0$, of disjoint sets of discrete states $\{S_i^*, i = 1, 2, ..., L\}$, whose union is S. Let $\{x_{ij}, i = 1, 2, ..., L; j = 1, 2, ..., L_i\}$ be a conditional state belonging to the set $\{S_i^*, i = 1, 2, ..., L_i\}$, where L_i is the finite number of possible conditional states in S_i^*. Moreover, let ξ_i be the probability that a state of the system Q is in the set $\{S_i^*, i = 1, 2, ..., L_i\}$ such that $\sum_i \xi_i = 1$. Moreover, let probability $f_{q,i}(x_{ij}) \in \Omega_i$, where Ω_i, is the closed convex set of all probability distributions on S_i^* i.e., $\{fq, i(x_{ij}) = Pr\{X_i = x_{ij}\}$, where X_i is the state conditional rv of the system $S_i^*, i = 1, 2, ..., L$. Moreover, let x be an aggregate state of system Q and probability $f_q(x) \in \Omega$, where Ω is the closed convex set of all probability distributions on S i.e., $f_q(x) = Pr\{X = x\}$, where X is the rv describing the aggregate state of the system S.

Clearly, $\xi_i, i = 1, 2, ..., L$ can be expressed by

$$\sum_{S_i^*} f_{q,i}(x_{ij}) = \xi_i \qquad (33)$$

The overall non-extensive entropy function of system Q, $H^*(f_q)$, defined on the total number of states in the union S of states $\{S_i^*, i = 1, 2, ..., L\}$ can be written as

$$H^*(f_q) = \frac{c}{q-1}(1 - \sum_i \sum_{S_i} \xi_i f_{q,i}(x_{ij})^q) \tag{34}$$

where $f_q \in \Omega$. Equation (34) can be rewritten in the form

$$H^*(f_q) = \sum_i \xi_i \frac{c}{q-1}(1 - \sum_{S_i^*} f_{q,i}(x_{ij})^q) \tag{35}$$

However, the conditional non-extensive entropy, $H_i^*(f_{q,i})$, defined on the set of states $S_i, i = 1, 2, ..., L$, is expressed by

$$H_i^*(f_{q,i}) = \frac{c}{q-1}(1 - \sum_{S_i^*} f_{q,i}(x_{ij})^q) \tag{36}$$

Hence, it follows from equations (35) and (36) that

$$H^*(f_q) = \sum_i \xi_i H_i^*(f_{q,i}) \tag{37}$$

Therefore, "maximising the generalised aggregate entropy function, $H^*(f_q)$, subject to an aggregate set of available constraints, it is equivalent to maximising each generalised conditional entropy function, $H_i^*(f_{q,i})$, individually, subject to a conditional set of available constraints. Thus, the Havrda-Charvat-Tsallis NME formalism satisfies the axiom of subset independence [68]"(c.f., [53]).

7 Queues with Bursty GE-type and Heavy Length Tails

This section reviews the applications of the EME and NME formalisms into the analysis of stable single server queues with a single class of jobs and infinite/finite capacity, exhibiting, respectively, bursty GE-type tails (c.f., Section 7.1)[17,37,38,40,41,45,50]) and heavy queue tails (c.f., Section 7.2) [4,5,30,49,52,53]).

The EME and NME solutions for the state probability distributions of these queues are presented subject to two sets of constraints, namely Set_1 : { normalisation, server utilisation, mean queue length} for stable infinite capacity queues and Set_2 : $\{Set_1$, full buffer state probability} for finite capacity queues. The selection of the constraints is motivated by the fact that they can, generally, capture a great deal of the queueing system's dynamic behaviour and, moreover, they can be expressed in terms of usually known input system parameters, such as the mean arrival rate λ, the SCV of the interarrival time, (Ca^2), the mean service rate μ, the SCV of the service times, (Cs^2) and the non-extensivity parameter, q (c.f., [22], [73]), or, equivalently, the Hurst self-similarity parameter, H (c.f., [9,23]). Thus, analytic closed form EME and NME expressions for the state probabilities of these queues can be devised, as useful information theoretic approximations, leading to closed form and computationally efficient EME and NME state probability distributions for the cost-effective assessment of the impact of bursty and self-similar traffic flows on the performance of the queues.

7.1 Queues with Bursty GE-Type Length Tails

7.1.1 An ME Solution for a GE-Type Tailed Queue with Infinite Capacity

Consider a stable single server GE/GE/1 queue with infinite capacity and GE-type interarrival and service times. Let $\{(1/\lambda, Ca^2), (1/\mu, Cs^2)\}$ be the means and SCVs of the interarrival and service times, respectively and, at any given time, $p(n), n = 0, 1, ...,$ be the state probability of having n messages in the system.

Suppose that prior information about the state probability is expressed in terms of the following mean value constraints:

- Normalization (NORM),

$$\sum_{n=0}^{\infty} p(n) = 1 \qquad (38)$$

- Server Utilisation (UTIL), ρ

$$\sum_{n=0}^{\infty} h(n)p(n) = 1 - p(0) = \rho, \quad 0 < \rho < 1 \qquad (39)$$

where $h(n)$ is an auxiliary function defined by $h(n) = \begin{cases} 0, n = 0 \\ 1, n \neq 0 \end{cases}$

- Mean Queue Length (MQL), L

$$\sum_{n=0}^{\infty} np(n) = L \qquad (40)$$

The form of the EME state probability distribution $\{p(n), n = 0, 1, ...\}$ can be completely specified (c.f., [17,40]) by maximizing the system's Boltzmann-Gibbs-Shannon extensive entropy (c.f., [11,67]), namely

$$H^*(p) = -c \sum_{i=1}^{\infty} p(n) \log p(n) \qquad (41)$$

subject to the prior information expressed by the constraints (38), (39) and 40). By applying the method of Lagrange undetermined multipliers, the EME solution for the state probability distribution of a stable GE/GE/1 queue is determined by the following GGeo-type distribution (c.f., [17,40]):

$$p(n) = \frac{1}{Z} g^{h(n)} x^n, \qquad n = 0, 1, 2, ... \qquad (42)$$

where $\rho = \lambda/\mu$, Z is the normalizing constant given by

$$Z = \exp\{-\beta\} = \frac{1}{p(0)} = \frac{1}{1-\rho} \qquad (43)$$

and $g = \exp\{-\beta_1\}$ and $x = \exp\{-\beta_2\}$ are the Lagrangian coefficients and $\{\beta, \beta_1$ and $\beta_2\}$ are the Lagrangian multipliers corresponding to the constraints NORM (38),

UTIL (39) and MQL (40), respectively. By substituting the EME expression (41) into the constraints (40) and (39), the following expressions are obtained

$$x = \frac{L - \rho}{L} \tag{44}$$

$$g = \frac{\rho(1 - x)}{(1 - \rho)x} \tag{45}$$

where the MQL, L is given by (c.f., [40,45])

$$L = \frac{\rho}{2}\left\{1 + \frac{Ca^2 + \rho Cs^2}{1 - \rho}\right\} \tag{46}$$

Substituting (46) into (44) and (45), the following expressions are obtained for the Lagrange coefficients x and g:

$$x = \frac{Ca^2 + \rho Cs^2 + \rho - 1}{Ca^2 + \rho Cs^2 - \rho + 1} \tag{47}$$

$$g = \frac{2\rho}{Ca^2 + \rho Cs^2 + \rho - 1} \tag{48}$$

7.1.2 An ME Solution for a GE-Type Tailed Queue with Finite Capacity

Consider a single server GE/GE/1/N queue with GE-type interarrival and service times and finite capacity, N. Let $\{(1/\lambda, Ca^2), (1/\mu, Cs^2)\}$ be the means and SCVs of the interarrival and service times, respectively and, at any given time, $p_N(n), n = 1, 2, ..., N$ be the state probability of having n messages in the system.

Suppose that the known prior information is expressed in terms of the following mean value constraints (c.f., [37,38,45]):

- NORM,

$$\sum_{n=0}^{N} p_N(n) = 1 \tag{49}$$

- UTIL, U_N

$$\sum_{n=0}^{N} h_N(n) p_N(n) = U_N, \quad 0 < U_N < 1 \tag{50}$$

where $h_N(n)$ is an auxiliary function defined by $h_N(n) = \begin{cases} 1, n = 0 \\ 1, n \neq 0 \end{cases}$

- MQL, L_N

$$\sum_{n=0}^{N} n p_N(n) = L_N \quad U_N \leq L_N < N \tag{51}$$

- Full buffer state probability (FBUF-SP), ϕ_N

$$\sum_{n=0}^{N} s_N(n) p_N(n) = \phi_N, \quad 0 < \phi_N < 1, \tag{52}$$

where $s_N(n)$ is an auxiliary function defined by $s_N(n) = \begin{cases} 0, n < N \\ 1, n = N \end{cases}$
and ϕ_N satisfies the flow balance equation, namely

$$\lambda(1 - \pi_N) = \mu U_N \tag{53}$$

where π_N is the blocking probability that an arriving job to find the queue at full capacity and λ and μ are the mean arrival and service rates, respectively.

The form of the EME queue length distribution, $p_N(n)$, can be characterised by maximising the Boltzmann-Gibbs-Shannon extensive entropy[11,67]

$$H^*(p_N) = -c \sum_{n=1}^{N} p_N(n) \log p_N(n) \tag{54}$$

subject to the constraints (49) - (52). By applying the method of Lagrange undetermined multipliers, the EME solution for the state probability distribution of a stable G/G/1/N queue is characterised by the following GGeo-type distribution (c.f., (c.f., [38]):

$$p_N(n) = \frac{1}{Z_N} g_N^{h_N(n)} x_N^n y_N^{s_N(n)}, \quad n = 1, 2, \ldots, N \tag{55}$$

where g_N, x_N and y_N are the Lagrangian coefficients corresponding to constraints UTIL (50), MQL (51) and FBUF-SP (52), respectively and Z_N is the normalizing constant (49), namely

$$Z_N = \frac{1}{p_N(0)} = \sum_{n \in S} g_N^{h_N(n)} x_N^{l_N(n)} y_N^{s_N(n)} \tag{56}$$

The analytic forms of Lagrangian coefficients g_N and x_N can be established by replacing the EME solution (55) into expressions (50) and (51). Note that, however, for a GE/GE/1/N queue, it has been shown that, for $(0 < N < +\infty)$, the Lagrangian coefficients g_N and x_N are asymptotically invariant irrespective of the buffer capacity, N (c.f.,[37,38,40]) i.e., the ME state probabilities of stable GE/GE/1 and GE/GE/1/N queues for different buffer sizes N are exactly 'parallel' distributions [47] i.e., $p_{N+1}(n) = c_N p_N(n)$, $N \in [1, +\infty)$ for some proportionality constant c_N. Thus, the Lagrangian coefficients g_N and x_N have identical expressions with those of a stable infinite capacity GE/GE/1 queue, given by (47) and (48), respectively i.e., $g_N = g$ and $x_N = x$. Note that, more generally, this parallel property holds exactly for the state probabilities of the stable M^G/G/1 and M^G/G/1/N queues with batch Poisson M^G arrival process and general (G) service times. Moreover, it also holds, under certain conditions, for those of the stable GI^G/D/1 and GI^G/D/1/N queues with GI^G and D denoting, respectively, a general batch renewal arrival process and a constant service time distribution (c.f., [47]).

Moreover, an analytic expression for the Lagrangian coefficient y_N can be devised by focusing on the flow balance condition (53), which involves the server utilisation, $U_N = 1 - P_N(0)$ (c.f., UTIL (50)) and the blocking probability, π. The latter is determined by (c.f., [44,45])

$$\pi_N = \sum_{n=0}^{N} \delta(n)(1-\sigma)^{N-n} p_N(n) \qquad (57)$$

where $\sigma = \frac{2}{Ca^2+1}$, $r = \frac{2}{Cs^2+1}$ and

$$\delta(n) = \begin{cases} \frac{r}{r(1-\sigma)+\sigma} & n = 0 \\ 1 & n \neq 0 \end{cases}$$

Substituting the expression for the EME solution for $p_N(0) = Z_N^{-1}$ (c.f., (55)), the blocking probability, π_N (c.f., (57)), together with those of Lagrangian coefficients x and g (c.f., (47) and (48)) into the global balance equation (53) and solving with respect to the Lagrangian coefficient y, the following expression is obtained

$$y_N = \frac{1-\rho}{1-x} \cdot \frac{\sigma}{r(1-\sigma)+\sigma} \qquad (58)$$

i.e., the Lagrangian coefficient y_N is also asymptotically invariant with respect to N i.e., $y_N = y$, where y is a constant.

7.2 Queues with Heavy Length Tails and Power Laws

G-Z-M type solutions for the EME state probability distributions of stable non-extensive single server GE-type queues with or without finite capacity and heavy length tails, based on the optimisation of the Havrda-Charvat-Tsallis non-extensive entropy, subject to NORM, UTIL, MQL and FBUF-SP constraints of (strictly) GE-type, as appropriate, can be found in [4,49].

An EME solution of the Z-M type (21), originally devised by Tsallis [73] in Statistical Physics, was employed as a state probability distribution for the case of $N \to +\infty$ by Karmeshu and Sharma [30] to study the long tail behaviour of queue lengths of a single server queue with infinite capacity, subject to normalization and a MQL constraint[2]. Moreover, the EME solution suggested in [30] does not incorporate into the EME formalism the fundamental server utilisation constraint associated with a stable single server queue with infinite capacity, namely $\rho = 1 - p(0)$ and, thus, it is incorrect in queueing theoretic terms as it violates Little's Law (c.f., [34]) at the service facility. Note that the utilisation constraint was used explicitly in many earlier works in the field of entropy maximisation and queueing systems (e.g.,[37,38,45,4,49]).

The expression of the MQL adopted in [30] can be described as a reinterpretation of the formula devised by Norros [60] in the context of Local Area Networks (LANs) for the estimation of a threshold, w representing "the size of the storage requirement" for a stationary storage model. The latter was represented by a stochastic process

---

[2] The work in [30] made insufficient attribution to reuse therein of the EME solution of Tsallis [73]. Consequently, [30] was not included at the time in the references and reviewing material of [5,52].

$V(t), t \in (-\infty, +\infty)$ with fractional Brownian net input process $A(t), t \in (-\infty, +\infty)$ described by

$$A(t) = \lambda t + (\alpha \lambda)^{1/2} Z(t), t \in (-\infty, +\infty) \quad (59)$$

where λ is the mean input rate, $\mu, \mu > \lambda$ is the mean service rate, α is a variance coefficient, $Z(t)$ is a normalised fractional Brownian motion (fBm) with Hurst self-similarity parameter, $H, H \in (1/2, 1)$ (c.f., (c.f., [9,23])). More specifically, Norros [60] described as a typical requirement for LANs that "the probability that the amount of work in the system" exceeding a storage requirement, w should be equal to a "Quality-of-Service (QoS) parameter, ε". This was expressed by a probabilistic relationship at the "maximum allowed load", namely

$$\varepsilon = P(V(t) > w), t \in (-\infty, +\infty) \quad (60)$$

w is expressed for $H \in (1/2, 1)$ by

$$w = const \frac{\rho^{\frac{1}{2(1-H)}}}{(1-\rho)^{\frac{H}{1-H}}} \quad (61)$$

$\rho = \frac{\lambda}{\mu}$ and the $const$ depends on H, α and ε but not on ρ, μ and w.

Note that, by analogy, different values of H in formula (61) with $const \sim 1$ may correspond to exact MQL expressions of various single server queues at equilibrium such as those of a stable S-S/M/1 queue with a suitable self-similar (S-S) arrival process having mean arrival rate λ and Hurst parameter H and exponential service time distribution (M) with mean service rate μ (c.f., [53]). For $H = 1/2$, Norros formula (61) reduces to the MQL formula of a stable M/M/1 queue. The latter, in the context of the EME formalism, corresponds to the case of the non-extensivity parameter $q = 1$. Moreover, for $1/2 < q, H < 1$, as the non-extensive queue has the strongest long range interactions for $q \to 1/2$ or, equivalently, $H \to 1$, it is customary in traffic modelling and characterisation studies to employ a simple relation between the non-extensivity parameter, q and Hurst parameter, H, namely $q = 1.5 - H$ (e.g., [30,52,65,53]).

In order to assess the combined impact of traffic burstiness and correlation on queueing system performance, a new heuristic generalization of Norros formula (c.f., [60,30]) was conjectured in [52] and employed as a MQL constraint, L_H in the context of the EME framework for the analysis of a stable single server gS-S/GE/1 queue with infinite capacity under the implicit assumptions of a a general S-S (gS-S) arrival process and a GE-type service time distribution with parameters λ, Ca^2, H and μ, Cs^2, respectively. This heuristic expression is given by

$$L_H = \frac{\rho^{\frac{1}{2(1-H)}}}{2^{\frac{1}{2(1-H)}}} \left(\frac{[1 - \rho + Ca^2 + \rho Cs^2]^{\frac{1}{2(1-H)}}}{(1-\rho)^{\frac{H}{1-H}}} \right), \quad \frac{1}{2} < H < 1 \quad (62)$$

where Ca^2 and Cs^2 are the SCVs of the interarrival and service times, respectively and H is the Hurst parameter taking values in the interval $\frac{1}{2} < H < 1$. Note that the heuristic formula (62) takes explicitly into account the adverse impact of self-similarity as well as burstiness of the traffic process via parameters H and Ca^2 and Cs^2 respectively, on queueing system performance. The formula (62) reduces correctly to the Norros formula (61) when $Ca^2 = Cs^2 = 1$ and $const \sim 1$. Moreover, for $H = \frac{1}{2}$ (i.e., $q \to 1$) equation (62) yields the mean queue length formula 46 of a stable GE/GE/1 queue (c.f., [37,38,40]).

EME solutions of G-Z-M type for stable single server gS-S/GE/1 queues with or without finite capacities, subject to NORM, UTIL, MQL and FBUF-SP constraints, as appropriate, satisfying the flow balance condition (c.f., [5,52,53]) are highlighted in Subsections 7.2.1 and 7.2.1, respectively. Related references are made to the works in [4,30,49] as appropriate.

7.2.1 A NME Solution of a Heavy Tailed Queue with Infinite Capacity

Consider a stable single server gS-S/GE/1 queue with an infinite capacity, a gS-S arrival process with mean arrival rate λ, interarrival time SCV, Ca^2 and Hurst self-similarity parameter, H and a GE-type service time distribution with mean service rate, μ and service time SCV, Cs^2. Moreover, at any given time, let $p_q(n), n = 0, 1, ..$, be the state probability of having n messages in the system.

Suppose that prior information about the state probability can be expressed in terms of the following mean value constraints (c.f., [4,49,52]):

- NORM,

$$\sum_{n=0}^{\infty} p_q(n) = 1 \qquad (63)$$

- UTIL, ρ

$$\sum_{n=0}^{\infty} h_q(n) p_q(n) = 1 - p_q(0) = \rho \qquad (64)$$

where ρ is the server utilisation i.e., $rho = \frac{\lambda}{\mu}$, $0 < \rho < 1$ and $h_q(n)$ is an auxiliary function defined by $h_q(n) = \begin{cases} 0, n = 0 \\ 1, n \neq 0 \end{cases}$

- MQL, L_q

$$\sum_{n=0}^{\infty} n p_q(n) = L_q \qquad (65)$$

where L_q is determined by expression (62) for $q \leftarrow 1.5 - H$ (i.e.,. $L_q = L_{(1.5-H)}$).

The form of the EME queue length distribution, $\{p_q(n), n = 0, 1, ...\}$, can be determined by maximising the Havrda-Charvat-Tsallis non-extensive entropy, $H^*(p_q)$, namely

$$H^*(p_q) = c \frac{1 - \sum_{n=0}^{\infty} p_q(n)^q}{q - 1} \qquad (66)$$

subject to constraints (63)-(64). By applying the method of Lagrange undetermined multipliers, the NME solution for the state probability distribution of a gS-S/G/1/N queue is characterised by the following G-Z-M-type distribution (c.f., (c.f., [52]):

$$p_q(n) = \frac{1}{Z_q} [1 + \alpha(1-q)n + \beta(1-q)h(n)]^{\frac{1}{q-1}}, \quad n = 0, 1, ... \qquad (67)$$

where β and α are Lagrange's multipliers corresponding to the constraints (64) and (65), respectively and Z_q is the normalising constant expressed by

$$Z_q = \sum_{n=0}^{\infty} [1 + \alpha(1-q)n + \beta(1-q)h(n)]^{\frac{1}{q-1}} = \zeta \left[\frac{1}{1-q}, \frac{1+\beta(1-q)h(n)}{\alpha(1-q)} \right] \qquad (68)$$

where $\zeta \left[\frac{1}{1-q}, \frac{1+\beta(1-q)h(n)}{\alpha(1-q)} \right] = \sum_{n=0}^{\infty} \left[n + \frac{1+\beta(1-q)h(n)}{\alpha(1-q)} \right]$ denotes the Hurwitz-Zeta function (c.f., [7,3]). Moreover, the MQL, L_q can be expressed in terms of the Lagrange multipliers and the Hurwitz-Zeta function (c.f.,[7,3]), namely

$$L_q = \sum_{n=0}^{\infty} \frac{n \left[1 + \alpha(1-q)n + \beta(1-q)h_q(n)\right]^{\frac{1}{q-1}}}{\sum_{n=0}^{\infty} \left[1 + \alpha(1-q)n + \beta(1-q)h_q(n)\right]^{\frac{1}{q-1}}} \qquad (69)$$

or

$$L_q = \frac{\zeta \left[\frac{q}{1-q}, \frac{1+\beta(1-q)h(n)}{\alpha(1-q)} \right]}{\zeta \left[\frac{1}{1-q}, \frac{1+\beta(1-q)h(n)}{\alpha(1-q)} \right]} - \frac{1+\beta(1-q)h(n)}{\alpha(1-q)} \alpha(1-q) \qquad (70)$$

where $\frac{q}{1-q} > 1, q > \frac{1}{2}$.

By substituting the NME state probability (67) into constraints (65) and (64), the Lagrangian multipliers, α and β, can be obtained numerically via Newton-Raphson method. For $L_q = L$, where the L is given by (46), the EME solution $p_q(n), n = 0, 1, ..$ (67) becomes identical to the EME in [4,49]. In the absence of the UTIL constraint (64) and when $Ca^2 = Cs^2 = 1$, the EME solution (67) reduces to Tsallis EME solution for $N \to +\infty$) (c.f., [30], [73]). Moreover, the expressions for L_q (c.f., (69), (70) reduce to those in [30].

At the limit $q \to 1$, the EME solution (67) becomes the EME state probability distribution (42) of a stable single server GE/GE/1 queue, namely

$$p(n) = \frac{e^{-\alpha n - \beta h(n)}}{\sum_{n=0}^{\infty} e^{-\alpha n - \beta h(n)}} = \frac{x^n g^{h(n)}}{Z}, \quad n = 0, 1, ... \qquad (71)$$

where Lagrangian coefficients x and g correspond to the mean queue length and server utilisation constraints, namely $x = e^{-\alpha}, g = e^{-\beta}, Z = \sum_{n=0}^{\infty} x^n g^{h(n)}$ (c.f., [38,40,45]).

For $q < 1$ and for large number of messages n, $p_q(n)$ follows asymptotically a power law given by

$$p_q(n) \sim n^{\frac{1}{q-1}}, \quad \frac{1}{2} < q < 1 \qquad (72)$$

where larger values of $q \in [\frac{1}{2}, 1]$ indicate traffic flows with smaller degree of self-similarity ([52]). This power law turns out to be identical to the one devised in [30].

7.2.2 An NME Solution of a Heavy Tailed Queue with Finite Capacity

Consider a single server gS-S/GE/1/N queue with finite capacity, N, a general gS-S arrival process with mean arrival rate λ, interarrival time SCV, Ca^2 and Hurst self-similarity parameter, H and a GE-type service time distribution with mean service rate, μ and service time SCV, Cs^2. Moreover, at any given time, let $p_{q,N}(n), n = 0, 1, .., N$, be the state probability of having n messages in the queue.

Suppose that the prior information about the NME state probability can be expressed in terms of the following mean value constraints (c.f.,[4,49,52,53]):

- NORM,

$$\sum_{n=0}^{N} p_{q,N}(n) = 1 \qquad (73)$$

- UTIL, $U_{q,N}$

$$\sum_{n=0}^{N} h_{q,N}(n) p_{q,N}(n) = 1 - p_{q,N}(0) = U_{q,N}, \quad 0 < U_{q,N} < 1 \qquad (74)$$

where $h_{q,N}(n)$ is an auxiliary function defined by $h_{q,N}(n) = \begin{cases} 0, n = 0 \\ 1, n \neq 0 \end{cases}$

- MQL, $L_{q,N}$

$$\sum_{n=0}^{N} n p_{q,N}(n) = L_{q,N} \qquad (75)$$

- FBUF-SP, $\phi_{q,N}$

$$\sum_{n=0}^{N} s_{q,N}(n) p_{q,N}(n) = \phi_{q,N}, \quad 0 < \phi_{q,N} < 1, \qquad (76)$$

where the auxiliary function, $s_{q,N}(n)$, is defined by $s_{q,N}(n) = \begin{cases} 0, n < N \\ 1, n = N \end{cases}$ and $\phi_{q,N}$ satisfies the flow balance equation, namely

$$\lambda(1 - \pi_{q,N}) = \mu U_{q,N} \qquad (77)$$

where $\pi_{q,N}$ is the blocking probability that an arrival message find a full capacity queue and λ and μ are the arrival and service rates, respectively.

The form of the NME queue length distribution, $p_{q,N}(n), n = 0, 1, ..., N$, can be characterised by maximising the Havrda-Charvat-Tsallis non-extensive entropy (18), namely

$$H^*(p_{q,N}) = c \frac{1 - \sum_{n=0}^{N} p_{q,N}(n)^q}{q - 1} \qquad (78)$$

subject to constraints (73)-(76). By employing the method of Lagrange undetermined multipliers, the NME solution for the state probability distribution of a gS-S/G/1/N queue is characterised by the following G-Z-M-type distribution (c.f., (c.f., [52,53]):

$$p_{q,N}(n) = \frac{1}{Z_{q,N}} [1 + \alpha_N(1-q)n + \beta_N(1-q)h_{q,N}(n) + \gamma_N(1-q)s_{q,N}(n)]^{\frac{1}{q-1}} \qquad (79)$$

where β_N, α_N and γ_N are the Lagrangian multipliers corresponding to the constraints (74), (75) and (76), respectively and Z_{q_N} is the normalizing constant expressed by

$$Z_{q,N} = \sum_{n=0}^{N} [1 + \alpha_N(1-q)n + \beta_N(1-q)h_{q,N}(n) + \gamma_N(1-q)s_{q,N}(n)]^{\frac{1}{q-1}} \quad (80)$$

where $\frac{1}{1-q} > 1, q > 0$. Note that to simplify the computation implementation of the NME solution of the gS-S/GE/1/N queue and related numerical experiments, it is assumed wlog that $rho < 1$ and the Lagrange multipliers α_N and β_N are asymptotically invariant to the buffer size, N i.e., for the stable gS-S/GE/1 and gS-S/GE/1/N queues, $\alpha_N = \alpha$ and $\beta_N = \beta$).

Moreover, the mean queue length, $L_{q,N}(n)$ can be expressed in terms of the Lagrange multipliers and the Hurwitz-Zeta function [7], namely

$$L_{q,N} = \sum_{n=0}^{N} \frac{n\left[1 + \alpha(1-q)n + \beta(1-q)h_{q,N}(n) + \gamma_N(1-q)s_{q,N}(n)\right]^{\frac{1}{q-1}}}{\sum_{n=0}^{N}\left[1 + \alpha(1-q)n + \beta(1-q)h_{q,N}(n) + \gamma_N(1-q)s_{q,N}(n)\right]^{\frac{1}{q-1}}} \quad (81)$$

or

$$L_{q,N} = \frac{\zeta\left[\frac{q}{1-q}, \frac{1+\beta(1-q)h_{q,N}(n)+\gamma_N(1-q)s_{q,N}(n)}{\alpha(1-q)}\right]}{\zeta\left[\frac{1}{1-q}, \frac{1+\beta(1-q)h_{q,N}(n)+\gamma_N(1-q)s_{q,N}(n)}{\alpha(1-q)}\right]} - \frac{1+\beta(1-q)h_{q,N}(n)+\gamma_N(1-q)s_{q,N}(n)}{\alpha(1-q)}\alpha(1-q) \quad (82)$$

where $\frac{q}{1-q} > 1, q > \frac{1}{2}$ and the two $zeta$ functions are of the Hurwitz-Zeta type (c.f.,[7,3]).

The Newton-Raphson method can be applied on the normalisation (73) and flow balance condition (77) to determine numerically the Lagrangian multiplier, γ_N. Moreover, the blocking probability, π, can be computed by using the flow balance condition (77).

Clearly, as $N \to \infty$, the NME solution, $p_{q,N}(n)$ (79) reduces to that of $p_N(n)$(67). Moreover, at the limit $q \to 1$, $p_{q,N}(n)$ (79) reduces to

$$p_N(n) = \frac{e^{-\alpha n - \beta h_N(n) - \gamma_N s_N(n)}}{\sum_{n=0}^{N} e^{-\alpha n - \beta h_N(n) \gamma_N s_N(n)}} = \frac{x^n g^{h_N(n)} y^{s_N(n)}}{\sum_{n=0}^{N} x^n g^{h_N(n)} y^{s_N(n)}}, \quad n = 0, 1, ..., N \quad (83)$$

which is the ME solution for the state probability $p_{q,N}(n)$ (55) of the GE/GE/1/N queue with $x = e^{-\alpha}, g = e^{-\beta}$ and $y = e^{-\gamma}$.

For $q < 1, \rho < 1$ and for large number of messages n, $p_{q,N}(n)$ follows asymptotically, as expected, a power law, which is identical to the one obtained for a stable gS-S/GE/1 (c.f., 72), namely

$$p_N(n) \sim n^{\frac{-1}{1-q}}, \quad \frac{1}{2} < q < 1 \quad (84)$$

7.2.3 'Quality-of-Service' Parameter, ε

The 'quality-of-service' (QoS) parameter ε defined by Norros [60] as the probability that the "maximum amount of work is greater than a certain threshold level, w can be reinterpreted as a buffer overflow probability in the context of single server queues with infinite [30] and finite capacities[52]. This can be clearly expressed by the probabilistic relationship (c.f., [30,52,5])

$$\varepsilon = Pr(N_0 > w) = 1 - Pr(N_0 \leq w) \qquad (85)$$

where N_0 is at any given time the random variable of the state of (or, equivalently, the number of messages of) a single server queue and $0 < N_0 \leq N$ or $+\infty$, as appropriate.

The buffer overflow probability (85) and power laws for asymptotically large w were determined in [30] in the context of the NME analysis of an S-S/M/1 queue, subject to the normalisation and MQL constraint. Generalisations of these results, as applied in the NME analysis of stable gS-S/GE/1 and gS-S/GE/1/N queues, have been established in [52,53] and are highlighted below.

The probability distribution of the NME queue length distribution, $\{p_{q,N}(n), n = 0, 1, ..., N\}$, given in (79) can be rewritten as,

$$p_{q,N}(n) = \frac{\left[\frac{1+\beta(1-q)h_{q,N}(n)+\gamma_N(1-q)s_{q,N}(n)}{\alpha(1-q)} + n\right]^{\frac{1}{q-1}}}{\zeta\left[\frac{1}{1-q}, \frac{1+\beta(1-q)h_{q,N}(n)+\gamma_N(1-q)s_{q,N}(n)}{\alpha(1-q)}\right]}, \qquad q > 0, n = 0, 1, ..., N \qquad (86)$$

where $\zeta\left[\frac{1}{1-q}, \frac{1+\beta(1-q)h_{q,N}(n)+\gamma_N(1-q)s_{q,N}(n)}{\alpha(1-q)}\right]$ denotes a Hurwitz-Zeta function (c.f.,[3,7]).

Using (86), the overflow probability, $P(N_0 > x)$, is expressed by

$$Pr(N_0 > w) = 1 - \sum_{n=0}^{w} \frac{\left[\frac{1+\beta(1-q)h_{q,N}(n)+\gamma_N(1-q)s_{q,N}(n)}{\alpha(1-q)} + n\right]^{\frac{1}{q-1}}}{\zeta\left[\frac{1}{1-q}, \frac{1+\beta(1-q)h_{q,N}(n)+\gamma_N(1-q)s_{q,N}(n)}{\alpha(1-q)}\right]} \qquad (87)$$

Equation (87) can be rewritten, using Hurwitz-Zeta function properties [3] as,

$$Pr(N_0 > w) = - \frac{\left(1/\zeta\left[\frac{1}{1-q}, \frac{1+\beta(1-q)h_{q,N}(n)+\gamma_N(1-q)s_{q,N}(n)}{\alpha(1-q)}\right]\right)\left(\frac{1-q}{q}\right)\left[x + \frac{1+\beta(1-q)h_{q,N}(n)+\gamma_N(1-q)s_{q,N}(n)}{\alpha(1-q)}\right]^{\frac{-q}{1-q}}}{\left(1/\zeta\left[\frac{1}{1-q}, \frac{1+\beta(1-q)h_{q,N}(n)+\gamma_N(1-q)s_{q,N}(n)}{\alpha(1-q)}\right]\right)\left(\frac{1-q}{q}\right)} \sum_{n=w}^{N} \int_0^1 u\left(u + n + \frac{1+\beta(1-q)h_{q,N}(n)+\gamma_N(1-q)s_{q,N}(n)}{\alpha(1-q)}\right)^{\frac{-(2-q)}{1-q}} du \qquad (88)$$

Allowing $N \to \infty$ for $rho < 1$, the overflow probability expressions (87) and (88) reduce to those associated with the NME solution (67) for the state probability $p_q(n)$ of a stable gS-S/GE/1 queue.

For large values of threshold w, asymptotic power laws can be established for the overflow probability (85) (c.f., [30,52]). In the context of an gS-S/GE/1/N queue, the power law is expressed by [52]

$$Pr(N_0 > w) \sim A_{q,N} w^{\frac{-q}{1-q}} \qquad (89)$$

where

$$A_{q,N} = \frac{\left(1/\zeta\left[\frac{1}{1-q}, \frac{1+\beta(1-q)h_{q,N}(n)+\gamma(1-q)s_{q,N}(n)}{\alpha(1-q)}\right]\right)}{\left(\frac{1-q}{q}\right)} = \left(\frac{1}{Z_{q,N}}\right)\left(\frac{1-q}{q}\right) \qquad (90)$$

The above asymptotic analysis indicates that the NME queue length distribution (79) (c.f., [52]) possesses heavy tails, pointing, as it was also observed in [13,30], to the requirement for large-scale buffer dimensioning in order to accommodate bursty and self-similar traffic flows.

In the limiting case $q \to 1$, the expression (88) can be rewritten as,

$$P(N_0 > w) = \frac{1}{Z_{1,N}}\left(\frac{1}{\alpha q}\right)$$
$$[1+\beta(1-q)h_{1,N}(n)+\gamma_N(1-q)s_{1,N}(n)+\alpha(1-q)w]^{\frac{-q}{1-q}}$$
$$-\frac{1}{Z_{1,N}}\left(\frac{1-q}{q}\right)\sum_{n=w}^{N}\int_0^1 u$$
$$(1+\beta(1-q)h_{1,N}(n)+\gamma_N(1-q)s_{1,N}(n)+\alpha(1-q)(u+n))^{\frac{-(2-q)}{1-q}} du$$
$$(91)$$

where

$$Z_{1,N} = \sum_{n=0}^{N}[1+\alpha(1-q)n+\beta(1-q)h_{1,N}(n)+\gamma_N(1-q)s_{1,N}(n)] \qquad (92)$$

(c.f., (80)). At the limit $q \to 1$ equation (88) reduces to an exponentially decaying probability

$$P(N_0 > X) \sim e^{-\alpha x - \beta h_{q,N}(n) - \gamma s_{q,N}(n)} \qquad (93)$$

which is the asymptotic result corresponding to the ME solution (55) of a GE/GE/1/N queue (c.f., [38,45]).

7.2.4 Server Utilisation and Blocking Probability

The probability that the server is busy (i.e., the server utilization), $U_{q,N}$ is determined, after some manipulation, by

$$U_{q,N} = 1 - p_{q,N}(0) =$$
$$1 - \frac{[1+\alpha(1-q)n+\beta(1-q)h_{q,N}(n)+\gamma_N(1-q)s_{q,N}(n)]^{\frac{1}{q-1}}}{\sum_{n=0}^{N}[1+\alpha(1-q)n+\beta(1-q)h_{q,N}(n)+\gamma_N(1-q)s_{q,N}(n)]^{\frac{1}{q-1}}} \qquad (94)$$

or

$$U_{q,N} = 1 - \frac{[\alpha(1-q)+\beta(1-q)h_{q,N}(n)+\gamma_N(1-q)s_{q,N}(n)]^{\frac{1}{q-1}}}{\zeta\left[\frac{1}{1-q}, \frac{1+\beta(1-q)h_{q,N}(n)+\gamma_N(1-q)s_{q,N}(n)}{\alpha(1-q)}\right]} \qquad (95)$$

Note that the analytic expression (94) for the server utilisation $U_{q,N}$ is a generalisation to the one devised in [30].

Moreover, the blocking probability, $\pi_{q,N}$ can be determined by using the flow balance condition (77) and it is given by

$$\pi = 1 - \frac{U_{q,N}}{\rho} \tag{96}$$

7.3 NME Analytic Algorithms

NME analytic algorithms for the analysis of stable single server queues with or without d finite capacity, respectively, can be based on the well known Newton-Raphson numerical method. Outlines of these algorithms, which are extensions of those devised in [30], are highlighted below.

7.3.1 Algorithm I: The Stable gS-S/GE/1 Queue with Heavy Length Tails

Input Data
λ, μ Mean arrival and service rates, respectively;
Ca^2, Cs^2 SCVs for interarrival and service times, respectively;
q Non-extensivity parameter;
Begin
Step 1: Calculate $H = 1.5 - q$ and the MQL, L (c.f., (69));
Step 2: Initialise Lagrangian multipliers $\{\beta, \alpha\}$ corresponding to the UTIL (64) and MQL (c.f., (65)) constraints;
Step 3: Substitute the NME solution for the equilibrium state probability $\{p(n), n = 0, 1, 2, ...\}$ (c.f., (67)) into the UTIL and MQL constraints (64) and (65), respectively and solve the resulting system of non-linear equations using the Newton-Raphson numerical method;
Step 4: Obtain new values for the Lagrangian multipliers $\{\alpha, \beta\}$;
Step 5: Return to Step 3 until convergence of α and β;
Step 6: Compute the NME state probabilities $\{p_q, (n), n = 0, 1, ...\}$ given by (67);
End.

Output Statistics
The numerical values of Lagrangian multipliers $\{\alpha, \beta\}$ and the state probabilities, $\{p_q(n), n = 0, 1, ...\}$.

7.3.2 Algorithm II: The gS-S/GE/1/N Queue with Heavy Length Tails

Input Data
λ, μ Mean arrival and service rates, respectively;
Ca^2, Cs^2 SCVs for interarrival and service rates, respectively;
N Finite buffer capacity;
q Non-extensivity parameter;

$\{\alpha, \beta\}$ Lagrange multipliers corresponding to the MQL and UTIL constraints, respectively;

Comment: Due to the asymptotic invariance assumption of the Lagrange multipliers with respect to the buffer size, N (c.f., Section 7.2), the estimated multipliers, $\{\alpha, \beta\}$ via Algorithm I are now used as input data for Algorithm II;

Begin
Step 1: Initialise the Lagrange multiplier γ_N corresponding to the FBUF-SP constraint $pq, N(N) = \phi_{q,N}$ (c.f., (76));
Step 2: Apply the Newton-Raphson numerical method on the flow balance equation (77) to determine numerically the Lagrange multiplier γ_N;
Step 3: Obtain a new value for the Lagrange multiplier γ_N;
Step 4: Return to Step 2 until convergence of γ_N;
Step 5: Compute the NME state probabilities $\{p_{q,N}(n), n = 0, 1, ..., N\}$ given by (79);
Step 6: Compute the blocking probability, $\pi_{q,N}$ using expression (96);
End.

Output Statistics
The numerical values of Lagrange multiplier γ_N, state probabilities, $\{p_{q,N}(n), n = 0, 1, ..., N\}$ and blocking probability, $\pi_{q,N}$.

8 Numerical Results

This section reviews a subset of typical numerical experiments (c.f., Figures 2-6), which are, generally, extended versions to those first appeared in [52]. Moreover, they are enhancements to those carried out in [4,30,37,38]. They illustrate the credibility of the GME solutions and related algorithms and also assess the adverse impact of combined bursty and self-similar traffic flows on the performance of the queue.

A plot of the queue length distribution $p_{q,N}(n)$ of a finite capacity gS-S/GE/1/N queue versus state n for different values of non-extensivity parameter $q = 1.5 - H$ is shown in Fig. 2. It can be seen that for decreasing values of q from $q = 0.9$ to $q = 0.6$ or, equivalently, increasing self-similarity parameter H from $H = 0.6$ to $H = 0.9$ (n.b., $H = 0.5$ for a GE/GE/1/N queue), imposes gradually, as expected, heavier long tail behaviour on the state probabilities, $p_{q,N}(n)$.

The queue length distribution $p_{q,N}(n)$ of a gS-S/GE/1/N queue versus n for $q = 0.6$ or, $H = 0.9$ and different values of Ca^2 is shown in Fig. 3. It can be observed that, for small states n, higher input traffic burstiness (i.e., variability) corresponding to larger values of Ca^2 has no much influence, as expected, on the tails of the state probabilities. However, as the values of state n increase beyond a small threshold value and in the presence of high self-similarity (c.f., $H = 0.9$), the traffic burstiness imposes progressively, heavier tail behaviour on the state probabilities.

Similarly, under high self-similarity with $q = 0.6$, the relation between the utilisation, $U_{q,N}$ and $\rho = \lambda/\mu$ of a finite capacity gS-S/GE/1/N queue for different values of

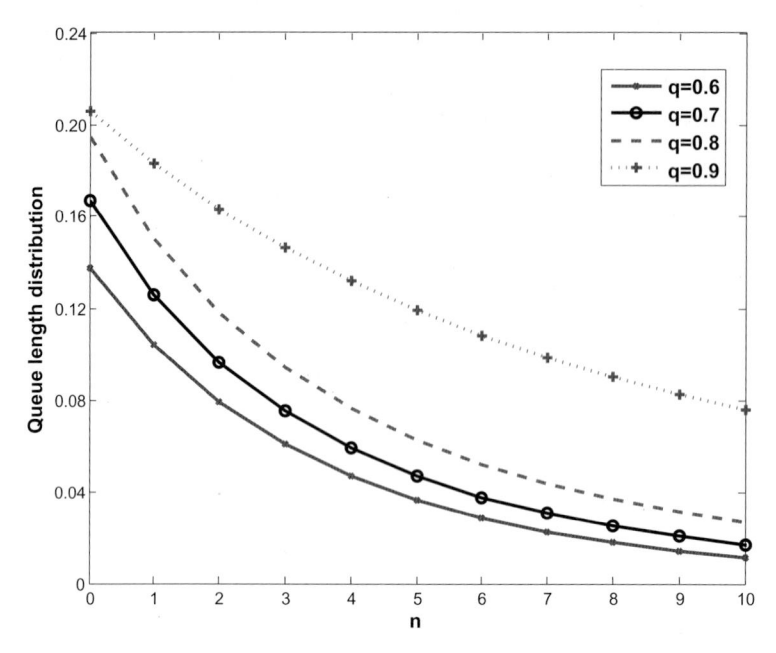

Fig. 2. The relation between $p_{q,N}(n)$ and n for a gS-S/GE/1/N queue with $Ca^2 = 4, Cs^2 = 1$, $N = 30$, $\lambda = 0.8$, $\mu = 1.0$ and $\{q = 0.6, 0.7, 0.8, 0.9\}$

Ca^2 is plotted in Fig. 4. It can be seen that increasing values of Ca^2 (i.e., traffic burstiness) correspond to larger utilisation values whilst all curves display, as anticipated, heavier tail behaviour with very high utilisations even for lower values of rho.

The relationship between the utilisation, $U_{q,N}$ and ρ for $Ca^2 = 3, 20$ and different values of q is plotted in Figs. 5(a) and 5(b). It can be observed in Fig. 5(a) that for smaller values of ρ under moderate traffic burstiness (variability) at $Ca^2 = 3$, the utilisation $U_{q,N}$ for smaller values of non-extensivity parameter $q < 1$ (i.e., larger values of self-similarity parameter, H) is decreasing. This indicates that increasing self-similarity in traffic flows does not have an adverse effect on queue performance when the server is underutilised. However, this 'utilisation anomaly' does no longer persists when the utilisation for smaller values of q is increasing sharply after some threshold value of ρ and, thus, from that point onwards self-similar traffic has an increasing hostile effect on performance, as expected, with a worst case scenario at the smallest value of $q = 0.6$ (or highest value of $H = 0.9$).

Note that, as it was observed earlier in [13], the acute transition from low to high utilisation at lower values of q, as the parameter ρ is increasing, is a typical attribute of a LRD network traffic with moderate variability as $\{Ca^2$ approaching 1 (c.f., [30,52]). However, this 'utilisation anomaly' of the plot $U_{q,N}$ versus ρ in Fig. 5(a) is no longer valid as the traffic displays 'burstier' characteristics with Ca^2 attaining increasing values much greater than one. It is seen in Fig. 5(b) that the relationship between $U_{q,N}$ and ρ reaches a more distinguished pattern as the adverse impact of higher traffic burstiness on utilisation $U_{q,N}$, even for small values of ρ, is quite evident. More specifically,

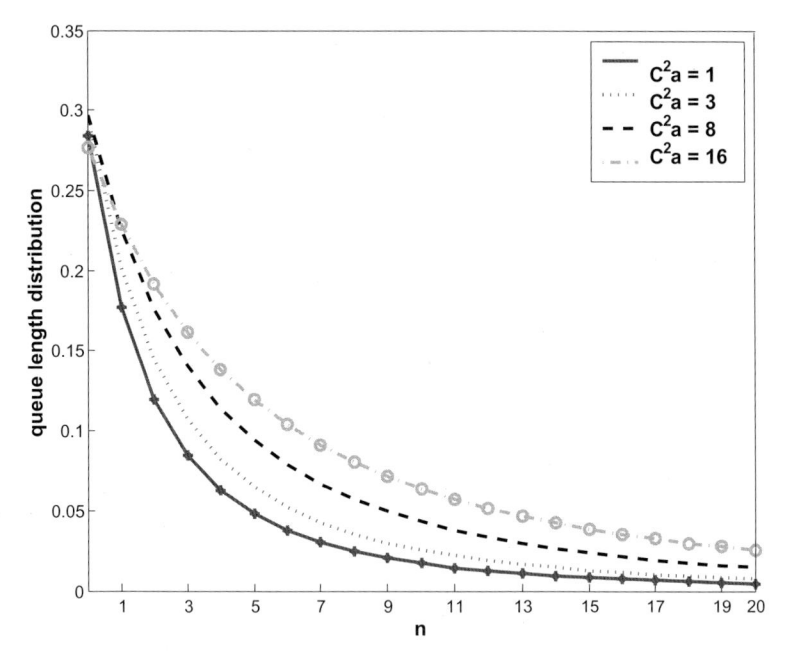

Fig. 3. The relation between $p_{q,N}(n)$ and n for a gS-S/GE/1/N queue with $\lambda = 0.8$, $\mu = 1.0$, $Cs^2 = 9$, $N = 20$, $q = 0.6$ and $\{Ca^2 = 1, 3, 8, 16\}$

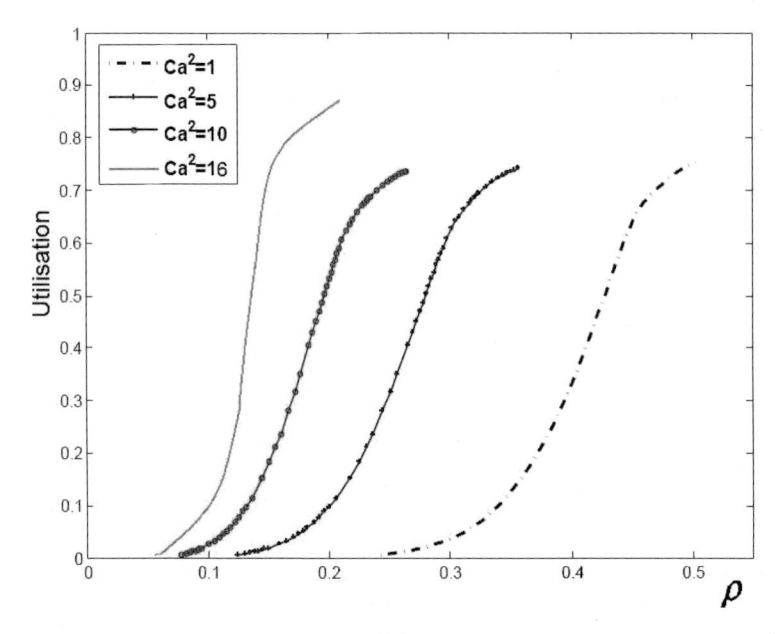

Fig. 4. The relation between $U = 1 - p_{q,N}(0)$ and ρ for a gS-S/GE/1/N queue for $\{Ca^2 = 1, 5, 10, 16\}$ with $Cs^2 = 3$, $N = 20$ and $q = 0.6$

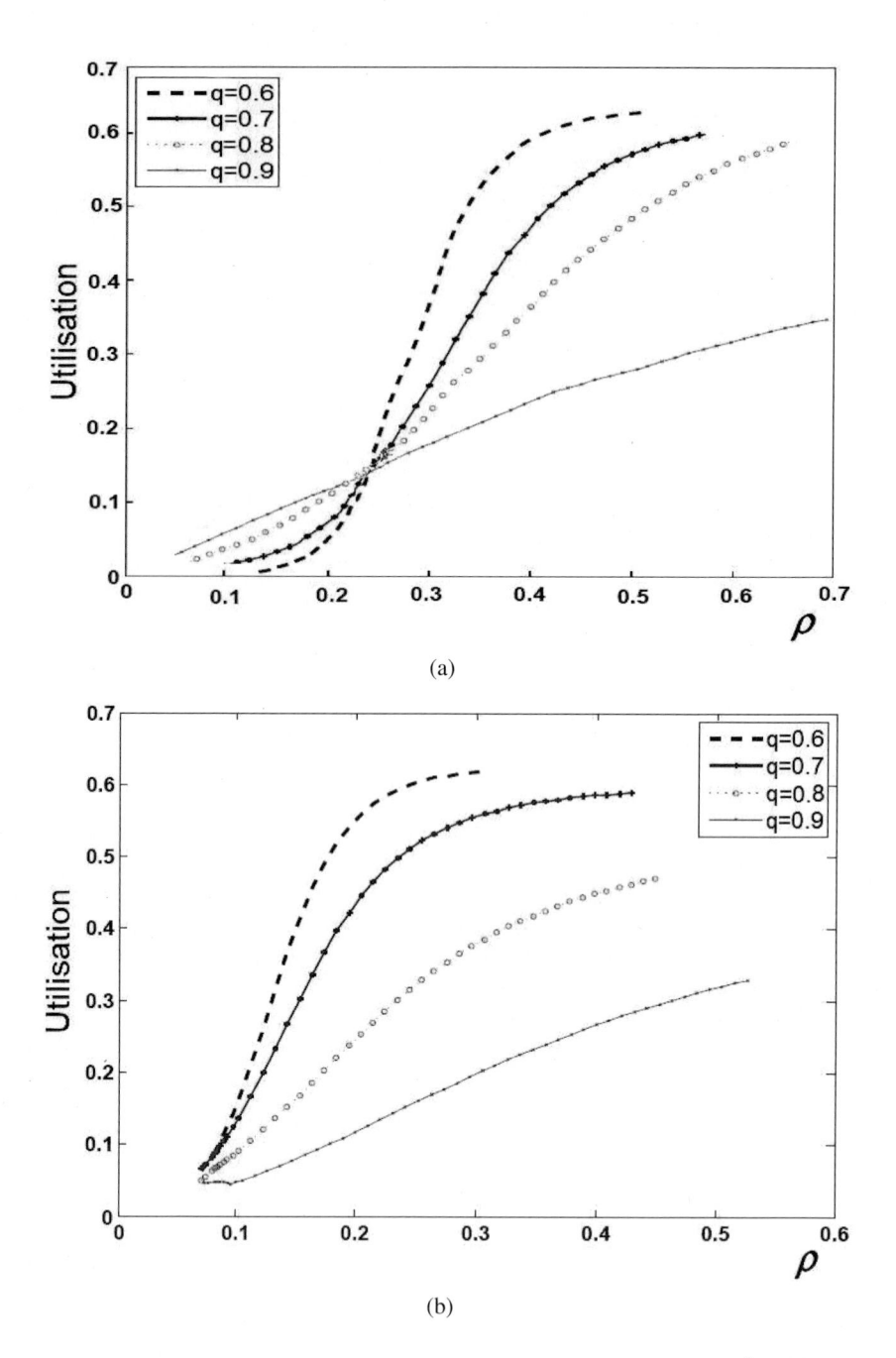

Fig. 5. The relation between $U = 1 - p_{q,N}(0)$ and ρ for a gS-S/GE/1/N with $Cs^2 = 4$, $N = 20$ and $\{q = 0.6, 0.7, 0.8, 0.9\}$ and (a) $Ca^2 = 3$ or (b) $Ca^2 = 20$

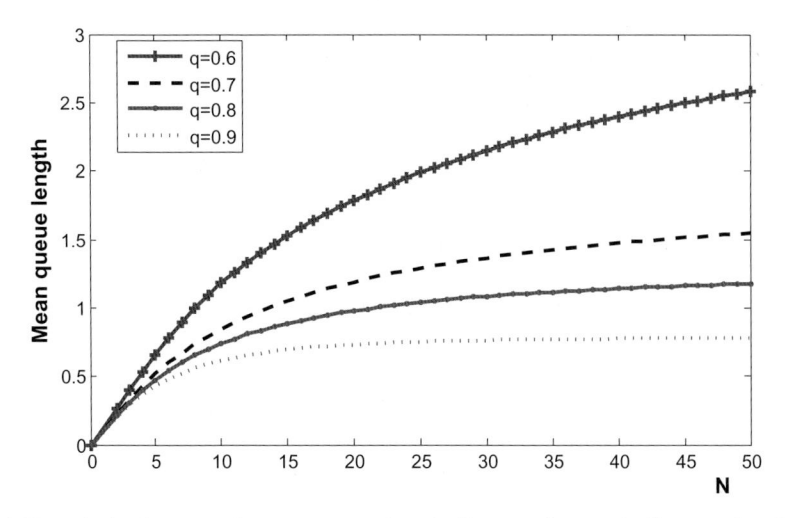

Fig. 6. The relation between the mean queue length, $L_{q,N}$, and queue buffer capacity, N, for a gS-S/GE/1/N queue for $\{q = 0.6, 0.7, 0.8, 0.9\}$ with $\lambda = 0.45$, $\mu = 1.0$, $Ca^2 = 4$ and $Cs^2 = 9$

when $Ca^2 = 20$, the curves of Fig. 5(b) have a much lower intersection point towards ρ approaching 0 whilst those in Fig. 5(a) are quite independent from each other up to a higher threshold value of ρ. This more extremal type of behaviour displayed in Fig. 5(b) illustrates the adverse impact of traffic flows with high levels of combined burstiness and self-similarity on queue performance.

Finally, a plot of the MQL $L_{q,N}$ of a gS-S/GE/1/N queue versus finite buffer capacity, N for different values of q is shown in Fig. 6. For all buffer capacities N, under the presence of even moderate traffic burstiness, it can be seen that as $q \to 1$ or $H \to 0.5$, the MQL L tends to an optimistic bound at $q = 0.9$ approaching the mean queue length of a GE/GE/1/N queue at $q = 1$). For smaller values of q and, thus, stronger influence of self-similar traffic, the MQLs are gradually increasing and, eventually, attain a pessimistic bound as $q \to 0.5$ or $H \to 1$.

The numerical experiments of Figures (2 - 6) demonstrate the credibility and robustness of the NME s power state probabilities with heavy queue length tails and, moreover, assess effectively the adverse combined impact of traffic variability and self-similarity on queue performance.

9 Conclusions

Empirical traffic characterization studies in networks of diverse technology and the Internet have shown that traffic flows often exhibit burstiness, self-similarity and/or LRD causing performance degradation and the formation of queues with bursty and/or heavy length tails. In this context, a review of the EME and NME formalisms was undertaken for the study of general physical systems with short-range and long range interactions, respectively. The exposition was based on the maximization of the classical Gibbs-Boltzmann-Shannon and generalised Havrda-Charvat-Tsallis entropy functions, subject to new information in the form of suitable mean value constraints. The credibility of the EME and NME formalisms, as methods of inductive inference, was explored in terms of four consistency axioms, namely uniqueness, invariance, system independence and subset independence. It was verified that the classical EME formalism satisfies all four axioms and thus, it is a most appropriate methodology for the analysis of queueing systems with short range interactions. Moreover, it was established that the NME formalism does not comply with the axiom of system independence, even though it does satisfy the other three axioms. Thus, the NME formalism is a most suitable methodology for the study queueing systems with long range interactions. Furthermore, it was shown that the state probability distribution of a stable single server queue with or without and finite capacity is characterized by a GGeo and G-Z-M type distributions depicting, respectively, GE-type tails and heavy tails with asymptotic power law behaviour. Typical numerical experiments were included to highlight the credibility and robustness of the EME and NME solutions and verify the adverse impact of combined traffic burstiness and correlation on the performance of the queue.

The NME solutions for single server queues provide simple and cost-effective analytic building blocks for the establishment of further theoretical insights leading to NME product-form approximations of G-Z-M type distributions and queue-by-queue decomposition algorithms for the analysis of complex QNMs with bursty, self-similar and/or LRD traffic flows (c.f.,[53]).

References

1. Adamic, L.A., Huberman, B.A.: Zipf's Law and the Internet. Glottometrics 3, 143–150 (2002)
2. Aksenov, S.V., Savageau, M.A., Jentschura, U.D., Becher, J., Soff, G., Mohr, P.J.: Application of the Combined Nonlinear-condensation Transformation to the Problems in Statistical Analysis and Theoretical Physics. Computer Physics Communication 150, 1–20 (2003)
3. Apostol, T.M.: Introduction to Analytic Number Theory. Springer, New York (1976)
4. Assi, S.A.: An Investigation into Generalised Entropy Optimisation with Queueing Systems Applications. MSc Dissertation. Dept. of Computing, School of Informatics, University of Bradford (2000)
5. Assi, S.A.: Performance Analysis of Interconnection Networks with Blocking and Wormhole Routing. PhD Thesis, Dept. of Computing, School of Informatics, University of Bradford (2008)
6. Ball, F., Hutchinson, D., Kouvatsos, D.D.: VBR Video Traffic Smoothing at the AAL SAR Level. In: Proceedings of the 4th IFIP Workshop on Performance Modelling and Evaluation of ATM Networks Ilkley, pp. 28/1–28/10 (1996)
7. Bateman, H.: Higher Transcendental Functions, Vol. 1. McGraw-Hill, New York (1953)
8. Benny, B.: Broadband Wireless Access. Kluwer Academic Publishers, Dordrecht (2000)
9. Beran, J.: Statistics for Long-Memory Processes. Chapman & Hall, Boca Raton (1994), ISBN 0-412-04901-5
10. Castilloa, J., Daoudib, J.: Estimation of the Generalized Pareto Distribution. Statistics and Probability Letters 79(5), 684–688 (2009)
11. Chakrabarti, C.G., Kajal De: Boltzmann-Gibbs Entropy: Axiomatic Characterisation and Application. Internat. J. Math. & Math. Sci. 23(4), 243–251 (2000)

12. Chlebus, E., Ohri, R.: Estimating Parameters of the Pareto Distribution by Means of Zipf's Law: Application to Internet Research. In: Global Telecommunications Conference, vol. 2(5), pp. 5–28 (2005)
13. Choudhury, G.L., Whitt, W.: Long-tail Buffer-content Distributions in Broadband Networks. Performance Evaluation 30, 177–190 (1997)
14. Crovella, M.E., Bestavros, A.: Self-Similarity in World Wide Web Traffic: Evidence and Possible Causes. IEEE/ACM Transaction Networking 5(6), 835–846 (1997)
15. Crovella, M.E., Lipsky, L.: Long-lasting Transient Conditions in Simulations with Heavy-Tailed Workloads. In: Proc. of Winter Simulation Conference, pp. 1005–1012 (1997)
16. Tsallis Statistics, Statistical Mechanics for Non-extensive Systems and Long-Range Interactions. Notebooks, http://www.cscs.umich.edu/~crshalizi/notabene/tsallis.html (January 23:22 29, 2007)
17. El-Affendi, M.A., Kouvatsos, D.D.: Maximum Entropy Analysis of the M/G/1 and G/M/1 Queueing Systems at Equilibrium. Acta Informatica 19, 339–355 (1983)
18. Gell-Mann, M., Tsallis, C.: Non-Extensive Entropy Interdisciplinary Applications. Oxford University Press, Oxford (2004)
19. Gong, W., Liu, Y., Misra, V., Towsley, D.: Self-similarity and Long Range Dependence on the Internet: A Second Look at the Evidence, Origins and Implications. Computer Networks: The International Journal of Computer and Telecommunications Networking Archive 48(3), 377–399 (2005)
20. Gupta, H.M., Campanha, J.R., Pesce, R.A.G.: Power-Law Distributions for the Citation Index of Scientific Publications and Scientists. Brazilian Journal of Physics 35(4a) (2005)
21. Harris, C.M.: The Pareto Distribution as a Queue Service Discipline. Operations Research 16(2), 307–313 (1968)
22. Havrda, J.H., Charvat, F.: Quantification Methods of Classificatory Processes: Concept of Structural Entropy. Kybernatica 3, 30–35 (1967)
23. Hurst, H.E.: Long-term Storage Capacity of Reservoirs. Transactions of the American Society of Civil Engineers 116, 770–808 (1951)
24. Jaynes, E.T.: Information Theory and Statistical Mechanics. Physical Review 106, 620–630 (1957)
25. Jaynes, E.T.: Information Theory and Statistical Mechanics II. Physical Review 108, 171–190 (1957)
26. Johnson, R.W.: Comments on and Correction to 'Axiomatic Derivation of the Principle of Maximum Entropy and Minimum Cross-Entropy. IEEE transactions on Information Theory IT-29(29) (1983)
27. Johnson, N.L., Kotz, S., Kemp, A.W.: Univariate Discrete Distributions. Series in Probability and Mathematical Statistics. J. Wiley & Sons, New York (1992)
28. Kapur, J.N.: Maximum-Entropy Models in Science and Engineering. Wiley Eastern Limited, Chichester (1989)
29. Kapur, J.N., Kesavan, H.K.: Entropy Optimisation Principles with Applications. Academic Press, London (1992)
30. Karmeshu, Sharma, S.: Long Tail Behaviour of Queue Lengths in Broadband Networks: Tsallis Entropy Framework. Technical Report, School of Computing and System Sciences. J. Nehru University, New Delhi, India (August 2005)
31. Karmeshu, Sharma, S.: Queue Length Distribution of Network Packet Traffic: Tsallis Entropy Maximization with Fractional Moments. IEEE Communications Letters 10(1), 34–36 (2006)

32. Karmeshu, Sharma, S.: Power Law and Tsallis Entropy: Network Traffic and Applications. Stud. Fuzz. 206, 162–178 (2006)
33. Karmeshu, Sharma, S.: q-Exponential Product-Form Solution of Packet Distribution in Queueing Networks: Maximisation of Tsallis Entropy. IEEE Communication Letters 10(8) (2006)
34. Kleinorck, L.: Queueing Systems. John Wiley and Sons, New York (1975)
35. Kouvatsos, D.D.: Maximum Entropy Methods for General Queueing Networks. In: Proc. of Int. Conf. on Modelling Techniques and Tools for Performance Analysis, May 1984, INRIA, Paris (1984)
36. Kouvatsos, D.D.: Maximum Entropy Methods for General Queueing Networks. In: Potier, D. (ed.) Modelling Techniques and Tools for Performance Analysis, pp. 589–608. North-Holland, Amsterdam (1985)
37. Kouvatsos, D.D.: Maximum Entropy and the G/G/1/N Queue. Acta Informatica 23, 545–565 (1986)
38. Kouvatsos, D.D.: Maximum Entropy Queue Length Distribution for G/G/1 Finite Capacity Queue. Performance Evaluation (ACM, IFIP WG 7.3) 14(1), 224–236 (1986)
39. Kouvatsos, D.D.: A Universal Maximum Entropy Algorithm for the Analysis of General Closed Networks. In: Hasegawa, T., et al. (eds.) Computer Networking and Performance Evaluation, pp. 113–124. North Holland, Amsterdam (1986)
40. Kouvatsos, D.D.: A Maximum Entropy Analysis of the G/G/1 Queue at Equilibrium. J. Opl. Res. Soc. 39(2), 183–200 (1988)
41. Kouvatsos, D.D., Almond, J.: Maximum Entropy and Two-Station Cyclic Queues with Multiple General Servers. Acta Informatica 26, 241–267 (1988)
42. Kouvatsos, D.D., Georgatsos, P., Tabet-Aouel, N.: A Universal Maximum Entropy Algorithm for General Multiple Class Open Networks with Mixed Service Disciplines. In: Potier, D., Puigjaner, R. (eds.) Modelling Techniques and Tools for Computer Performance Evaluation, pp. 397–419. North-Holland, Amsterdam (1989)
43. Kouvatsos, D.D., Tabet-Aouel, N.: Product-Form Approximation for an Extended Class of General Closed Queueing Network. In: Kingetal, P. (ed.) Performance 1990, IFIP WG 7.3 and BCS, pp. 301–315. North-Holland, Amsterdam (1990)
44. Kouvatsos, D.D., Denazis, S.G.: Entropy Maximised Queueing Networks with Blocking and Multiple Job Classes. Performed Evaluation 17, 189–205 (1993)
45. Kouvatsos, D.D.: Entropy Maximization and Queueing Network Models. Annals of Operation Research 48, 63–126 (1994)
46. Kouvatsos, D.D., Awan, I.: MEM for Arbitrary Closed Queueing Networks with RS-blocking and Multiple Job Classes. Annals of Operations Research, Special Issue on Queueing Networks with Blocking, Baltzer Science 79, 231–269 (1998)
47. Kouvatsos, D.D., Fretwell, R.F., Skianis, C.A.: MRE: A Robust Method of Inference for Finite Capacity Queues. In: Gelenbe, E. (ed.) System Performance Evaluation: Methodologies and Applications, pp. 249–260. CRC Press LLC (2000)
48. Kouvatsos, D.D., Awan, I., Fretwell, R., Dimakopoulos, G.: A Cost-Effective Approximation for SRD Traffic in Arbitrary Multi-Buffered Networks. Computer Networks 34, 97–113 (2000)
49. Kouvatsos, D.D., Assi, S.A.: An Investigation into Generalised Entropy Optimisation with Queueing Systems Applications. In: Merabti, M. (ed.) The Proceedings of the 3rd Annual Postgraduate Symposium on the Convergence of Telecommunications, Networking and Broadcasting (PGNet 2002), pp. 409–414. Liverpool John Moores University Publisher (2002)
50. Kouvatsos, D.D., Awan, I.: Entropy Maximization and Open Queueing Networks with Priorities and Blocking. Special Issue on Queueing Networks with Blocking 51, 191–227 (2003)

51. Kouvatsos, D.D.: A Universal Maximum Entropy Solution for Complex Queueing Systems and Networks. Invited Paper, Studies in Fuzziness and Soft Computing, Special Issue on Entropy Measures, Maximum Entropy and Emerging Applications 150, 137–162 (2003)
52. Kouvatsos, D.D., Assi, S.A.: On the Analysis of Queues with Long Range Dependent Traffic: An Extended Maximum Entropy Approach. In: Proceedings of the 3rd Euro-NGI Conference on Next Generation Internet Networks - Design and Engineering for Heterogeneity, Trodheim, Norway, May 2007, pp. 226–233 (2007)
53. Kouvatsos, D.D., Assi, S.A.: On the Analysis of Queues with Heavy Tails: A Non-Extensive Maximum Entropy Formalism and a Generalisation of the Zipf-Mandelbrot Distribution. Special IFIP LNCS issue in Honour of Guenter Haring, University of Vienna, Vienna, Austria (to appear, 2011)
54. Kullback, S.: Information Theory and Statistics. Wiley, New York (1959)
55. Leland, W.E., Taqqu, M.S., Willinger, W., Wilson, D.V.: On the Self-Similar Nature of Ethernet Traffic (Extended Version). IEEE/ACM Transaction on Networking 2(1), 1–15 (1994)
56. Li, W.: Performance Analysis of Queues with Correlated Traffic: An Investigation into Batch Renewal Process and Batch Markovian Arrival Process and their Performance Impact on Queueing Models. PhD Thesis. NetPEn - Networks and Performance Engineering Research Group, Informatics Research Institute (IRI), University of Bradford, UK (2007)
57. Mandelbrot, B.B.: The Fractal Geometry of Nature. W.H. Freeman, New York (1982)
58. Mouchos, C.: Traffic and Performance Evaluation for Optical Networks. PhD Thesis, University of Bradford, Bradford, UK (2009)
59. Newman, M.E.J.: Power Laws, Pareto Distributions and Zipf's Law. Contemporary Physics 46, 323–351 (2005)
60. Norros, I.: A Storage Model with Self-similar Input. Queueing Systems and their Applications 16, 387–396 (1994)
61. Paxson, V., Floyd, S.: Wide Area Traffic: The Failure of Poisson Modelling. IEEE/ACM Transactions on Networking 3, 236–244 (1995)
62. Paxson, V.: End-to-End Internet Packet Dynamics. IEEE/ACM Transactions on Networking 7, 277–292 (1999)
63. Plastino, A., Plastino, A.R.: Tsallis Entropy and Jaynes Information Theory formalism. Brazilian Journal of Physics 29(1) (1999)
64. Renyi, A.: On Measures of Entropy and Information. In: Proceedings of the 4th Berkely Symposium Math. Stat. And Probability, vol. 1, pp. 547–561 (1961)
65. Rezaul, K.M., Grout, V.: A Comparison of Methods for Estimating the Tail Index of Heavy-tailed Internet Traffic. In: 'Innovative Algorithms and Techniques in Automation, Industrial Electronics and Telecommunications, pp. 219–222. Springer, Dordrecht (2007)
66. Sahinoglu, Z., Tekinay, S.: On Multimedia Networks: Self-similar Traffic and Network Performance. IEEE Communication Magazine 37, 48–52 (1999)
67. Shannon, C.E.: A Mathematical Theory of Communication. Bell Syst. Tech. J. 27, 379–423, 623–656 (1948)
68. Shore, J.E., Johnson, R.W.: Axiomatic Derivation of the Principle of ME and the Principle of Minimum Cross-Entropy. IEEE Transaction on Information Theory IT-26, 26–37 (1980)
69. Shore, J.E., Johnson, R.W.: Properties of Cross-Entropy Minimization. IEEE Tran., Information Theory IT-27, 472–482 (1981)
70. Shore, J.E.: Information Theoretic Approximations for M/G/1 and G/G/1 Queueing Systems. Acta Inf. 17, 43–61 (1982)
71. Stalling, W.: High Speed Networks and Internets: Performance and Quality of Service, 2nd edn. Pearson, London (2002)

72. Suyari, H.: Law of Error in Tsallis Statistics. IEEE Transaction on Information Theory 51(12) (2005)
73. Tsallis, C.: Possible Generalisation of Boltzmann-Gibbs Statistics. Journal of Statistical Physics 52(1-2), 479–487 (1988)
74. Tsallis, C., Mendes, R.S., Plastino, A.R.: The Role of Constraints within Generalised Nonextensive Statistics. Physica A 261, 534–554 (1998)
75. Tsallis, C., Baldovin, F., Cerbino, R., Pieribon, P.: Introduction to Nonextensive Statistical Mechanics and Thermodynamics, arXCiv: cond-mat/0309093 v1 (Septmeber 04, 2003)
76. Tsallis, C., Brigatti, E.: Nonextensive Statistical Mechanics: a Brief Introduction. Continuum Mechanics and Thermodynamics 16, 223–235 (2004)
77. Zipf, G.K.: Human Behaviour and the Principle of Least Effort. Addison-Wesley, Cambridge (1949)
78. Walstra, R.: Iterative Analysis of Networks of Queues. PhD Thesis, Tech. Rep. CSTRI-166, Toronto University, Toronto University, Canada (December 1984)

Stochastic Ordering of Semi-Markov Processes

Fátima Ferreira[1] and António Pacheco[2]

[1] University of Trás-os-Montes and Alto Douro
Mathematics Department and CM-UTAD, Vila Real, Portugal
[2] Instituto Superior Tcnico - Technical University of Lisbon
CEMAT and Mathematics Department, Lisbon, Portugal

Abstract. In this tutorial we address the stochastic ordering of semi-Markov processes in the usual and level-crossing stochastic ordering senses. We highlight the sample-path approach for the comparison of semi-Markov processes and for the simulation of processes with a given distribution.

Keywords: Markov renewal processes, sample-path approach, semi-Markov processes, stochastic ordering.

1 Introduction

The desire to confront random quantities is probably as old as probability theory itself; in this line of reasoning, Bawa [4] traces the origins of stochastic ordering or dominance in the works of J. Bernoulli in 1713, *Ars Conjectardi*. However, it has been mainly in the last decades that stochastic ordering has progressively became to be recognized as an important tool in the area of applied stochastic processes, as illustrated, e.g, in the bibliographies of Bawa [4], Levy [31], and Mosler and Scarsini [39]. The most popular approaches used to establish stochastic ordering results are: coupling constructions (Lindvall [35]; and Thorisson [50]), sample-path approaches (El-Taha and Stidham [13]; and Stoyan [48]) and some pure analytic results (Kijima [26]; and Shaked and Shanthikumar [44]).

The rich history of applications of stochastic ordering is also made clear, e.g., in Shaked and Shanthikumar [44], which specifically expands on the applications of stochastic ordering in the areas of statistical inference, risk theory, economics, biology, scheduling, operations research, queueing theory, and reliability theory, and reinforced by Arnold [1], Cabral Morais [38], van Doorn [52], Joe [23], Kijima and Ohnishi [27], Lindvall [35], Marshall and Olkin [36], Müller and Stoyan [40], Stoyan [48], Szekli [49], Thorisson [50], and Tong [51]. A reflection of the relevance of stochastic ordering in applications is the significative number of fairly recent books on stochastic processes that have included as chapters or parts of the book stochastic ordering concepts and results, e.g., Baccelli and Brémaud [3], Kijima [26], Kulkarni [29], Last and Brandt [30], and Ross [43].

Semi-Markov processes (SMPs) and the related Markov renewal processes (MRPs) have a well established theory (see, e.g., Çinlar [9], Kulkarni [29], and Limnios and Oprişan [33]) and have many applications (see, e.g., Asmussen [2];

Disney and Kiessler [11]; and Janssen and Limnios [22]). The latter fact is easily understood if we note that SMPs and MRPs are basically in a one-to-one correspondence and, moreover, SMPs generalize both discrete-time Markov chains (DTMCs) and continuous-time Markov chains (CTMCs), whereas MRPs generalize renewal processes and the so called Markovian arrival processes. Thus, the stochastic ordering of SMPs has broad impact in applied stochastic processes.

The analysis of SMPs and MRPs started to be developed in the 1950's by important probabilists, namely: Levy [32], Smith [45,46], Pyke [41,42], Feller [14], and Çinlar [7,8]. However, explicit references to the stochastic comparison of SMPs appeared only several years later, with the work of Sonderman [47] in the usual (in distribution) stochastic ordering sense.

In the tutorial we will review the literature on the stochastic ordering of SMPs in the usual stochastic ordering sense, as well as in the level-crossing stochastic ordering sense, recently proposed by A. Irle and J. Gani [21] and investigated by its proponents and the authors of this tutorial. A process X is said to be smaller than Y in the usual sense if there are copies $\hat{X}$ and $\hat{Y}$ (i.e., processes with the same distributions as the original ones) of the processes X and Y defined on a common probability space such that their trajectories are ordered in the almost sure sense. Similarly, a process X is said to be smaller in level-crossing than Y if it takes X stochastically longer than Y to exceed any given level. As illustrated by Irle and Gani [21], the level-crossing ordering of stochastic processes in the usual sense is (strictly) weaker than the usual stochastic ordering of the processes.

We will start in Section 2 with the presentation of the definition of MRPs and SMPs and present in Section 3 procedures to simulate such processes. Then, in sections 4 and 5, we briefly review the main results on the stochastic comparison of SMPs in the usual and in the level-crossing stochastic ordering senses, respectively. We will follow the sample-path approach, which is useful for simulating pairs of stochastically ordered processes, and will address only SMPs with totally ordered state spaces.

2 Markov Renewal and Semi-Markov Processes

In this section, we provide the definitions of a Markov Renewal process (MRP) and of a semi-Markov process (SMP). We relate these two types of processes and give their characterizations in terms of their associated: (transition distribution) kernel, embedded (transition) kernel, and (failure) rate kernel. We start with the definition of MRP [cf., e.g., [9] or [29]].

2.1 Markov Renewal Processes

Definition 1. *We say that a bivariate process $(Z, S) = (Z_n, S_n)_{n \in \mathbb{N}}$ is a MRP with (countable) phase space I and kernel $Q = (Q_t)_{t \in \mathbb{R}_+}$, where $Q_t = (Q_{ij}(t))_{i,j \in I}$ is a family of sub-distribution functions such that $\sum_{j \in I} Q_{ij}(t)$ is a distribution function, for all $i \in I$, if it is a Markov process on $I \times \mathbb{R}_+$ such that $S_0 = 0$ and*

$$Q_{ij}(t) = \mathbf{P}\left(Z_{n+1} = j,\, S_{n+1} - S_n \leq t \mid Z_n = i, S_n = s\right)$$

for all $n \in \mathbb{N}$, $i, j \in I$ and $s, t \in \mathbb{R}_+$.

MRPs are used, e.g., in the modelling of arrival processes to queueing networks, where S models the network arrival epochs and Z models the influence of environmental factors in the structure of the interarrival times. In this context, $Q_{ij}(t)$ denotes the probability that, given that after an arrival the process is in phase i, the next arrival will put the process in phase j and will take place within t time units. From the definition of MRP it follows that this last event does not depend on the last arrival epoch, i.e.,

$$Q_{ij}(t) = \mathbf{P}\left(Z_{n+1} = j,\, S_{n+1} - S_n \leq t \mid Z_n = i\right).$$

Embedded kernel characterization. Another natural characterization of a MRP is through its embedded kernel, which separates the embedded transition probabilities from the distributions of the holding times in states between transitions. From the definition of MRP, it follows that if (Z, S) is a MRP with kernel Q, then Z is a discrete time Markov chain (DTMC) with one-step transition probability matrix $P = Q(\infty)$ with

$$p_{ij} = Q_{ij}(\infty) = \mathbf{P}\left(Z_{n+1} = j \mid Z_n = i\right)$$

denoting the probability that if the previous phase transition leads the process to phase i the phase process will next move to phase j.

On the other hand, conditional to the next phase being j, i.e., given that the process makes a transition from phase i to phase j, then the amount of time the process stays in phase i before moving to phase j has distribution function

$$F_{(i,j)}(t) = \mathbf{P}\left(S_{n+1} - S_n \leq t \mid Z_n = i, Z_{n+1} = j\right) = \frac{Q_{ij}(t)}{Q_{ij}(\infty)}$$

where, by convention, we let $F_{(i,j)}(t) = 1$, for all $t \in \mathbb{R}_+$, whenever $p_{ij} = 0$. It thus follows that

$$Q_{ij}(t) = p_{ij}\, F_{(i,j)}(t),\quad t \in \mathbb{R}_+$$

for all $i, j \in I$, and we say that the MRP (Z, S) has embedded kernel (P, F), where $P = (p_{ij})_{i,j \in I}$ is a stochastic transition probability matrix and $F = (F_{(i,j)})_{i,j \in I}$ is a matrix of distribution functions of nonnegative random variables such that if $p_{ij} = 0$, then $F_{(i,j)}(t) = 1$, $t \in \mathbb{R}_+$. Thus, a MRP is completely characterized by its embedded kernel.

Rate kernel characterization. Alternatively to the previous characterizations, a MRP may also be characterized via its failure rate kernel. Consider a MRP W with phase space I and kernel Q such that the sub-distributions $Q_{ij}(t)$ are absolutely continuous. Then, $q = (q_t)_{t \in \mathbb{R}_+}$ with $q_t = (q_{ij}(t))_{i,j \in I}$ such that

$$q_{ij}(t) = \frac{\mathrm{d}Q_{ij}(t)}{\mathrm{d}t}$$

is called the density kernel of W, and $q_i(t) = \sum_{j \in I} q_{ij}(t)$ denotes the density of the time needed for a transition from phase i to take place. Moreover, letting

$$r_{ij}(t) = \frac{q_{ij}(t)}{1 - \sum_{l \in I} Q_{il}(t)}$$

then $R = (R_t)_{t \in \mathbb{R}_+}$, with $R_t = (r_{ij}(t))_{i,j \in I}$, is called the failure rate kernel of W and

$$r_i(t) = \sum_{j \in I} r_{ij}(t) = \frac{q_i(t)}{1 - \sum_{l \in I} Q_{il}(t)}$$

denotes the failure rate at time t of the time needed for a transition from phase i to take place.

In this case, as $r_i(t)$ characterizes $q_i(t)$ through $q_i(t) = r_i(t) \exp\left\{-\int_0^t r_i(s) \, ds\right\}$ [cf., e.g., [24]], it immediately follows that $r_{ij}(t)$ characterizes $q_{ij}(t)$ through

$$q_{ij}(t) = r_{ij}(t) \exp\left\{-\int_0^t r_i(s) \, ds\right\}, \quad i, j \in I, \ t \in \mathbb{R}_+. \tag{1}$$

Thus, the MRP W is completely characterized by its failure rate kernel.

2.2 Semi-Markov Processes

We now introduce the definition of a SMP in terms of its usual characterizations.

Definition 2. *A process $W = (W_t)_{t \in \mathbb{R}_+}$ is a SMP with countable state space I and (admitting) kernel Q (embedded kernel (P, F); failure rate kernel R) if*

$$W_t = Z_n, \quad S_n \leq t < S_{n+1} \tag{2}$$

for some MRP (Z, S) with phase space I and kernel Q (embedded kernel (P, F); failure rate kernel R).

The most common description of the evolution of an SMP is through its embedded kernel. An SMP with embedded kernel (P, F) and initial probability distribution vector p evolves as follows. The process starts in phase i with probability p_i, and afterwards changes from phase to phase according to the transition probability matrix P. It moves to phase k after entering phase j, with probability p_{jk}, independently of previous phase changes. After deciding the next phase to visit, say k, from phase j, the process stays in phase j before making the transition to phase k a random holding time, independent of previous holding times in phases and phase transitions, having distribution function $F_{(j,k)}(t)$. If the SMP has kernel Q, then $p_{jk} = Q_{jk}(\infty)$ and $F_{(j,k)}(t) = Q_{jk}(t)/p_{jk}$ case $p_{jk} > 0$.

We end the section noting that given a SMP W with state space I, the process $(Z_n, S_n)_{n \in \mathbb{N}}$ with

$$(Z_0, S_0) = (W_0, 0) \quad \text{and} \quad \begin{cases} S_{n+1} = \inf\{t \geq S_n : W_t \neq W_{t-}\} \\ Z_{n+1} = W_{S_{n+1}} \end{cases}$$

for $n \in \mathbb{N}$, is a MRP with phase space I. In particular, if $Q((P, F); R)$ denotes the kernel (embedded kernel; failure rate kernel) of $(Z_n, S_n)_{n \in \mathbb{N}}$, then W admits the kernel (embedded kernel; failure rate kernel) $Q((P, F); R)$, called the natural kernel of W.

Conversely, if (Z, S) is a MRP with phase space I and kernel Q (embedded kernel (P, F); failure rate kernel R), then the process W with

$$W_t = Z_n, \quad S_n \leq t < S_{n+1}$$

is a SMP with state space I and kernel Q (embedded kernel (P, F); failure rate kernel R).

3 Simulation of Semi-Markov Processes

Having introduced the usual characterizations of a SMP, we proceed to describe a procedure to simulate (generate) a SMP with countable totally ordered state space I, order isomorphic to a subset of integers, and a given parametrization. For that, let F^{-1} denote the generalized inverse function of a distribution function F, i.e.,

$$F^{-1}(u) = \inf\{t : F(t) \geq u\}, \quad \text{for } u \in [0, 1]$$

with the convention that $\inf \emptyset = +\infty$. Moreover, to simplify the writing and avoid extra notation for the associated distribution functions, if p denotes a probability vector and Z denotes a random variable or distribution, like the exponential distribution with rate λ, Exp (λ), then we let p^{-1} and Z^{-1} denote the generalized inverse functions of the distribution function associated to p and Z, respectively.

The generalized inverse function is in the base of the standard method to simulate copies of random variables with prescribed distributions. In fact, if U is a uniform random variable on $(0, 1)$, Unif$(0, 1)$, and F is an arbitrary distribution function, then $F^{-1}(U)$ is a random variable with distribution function F, i.e.,

$$U \sim \text{Unif}(0, 1) \Longrightarrow F^{-1}(U) \sim F. \tag{3}$$

At this point, it is important to note that, given two distribution functions F and G, then

$$F(t) \geq G(t), \quad t \in \mathbb{R} \Longrightarrow F^{-1}(u) \leq G^{-1}(u), \quad u \in (0, 1). \tag{4}$$

This fact ([49], Lemma C) is of paramount importance in the simulation of stochastic ordered random variables and processes.

We proceed to address the simulation of a SMP from its embedded kernel.

3.1 Simulation via Embedded Kernel

To simulate a SMP W with state space I and embedded kernel (P, F) it suffices to simulate a MRP (Z, S) with phase space I and embedded kernel (P, F) and then obtain $W = (W_t)_{t \in \mathbb{R}}$ from

$$W_t = Z_n, \quad S_n \le t < S_{n+1}.$$

In turn, to simulate a MRP (Z, S) with state space I and embedded kernel (P, F), it suffices to simulate a DTMC Z with state space I and a sequence $S = (S_n)_{n \in \mathbb{N}}$ in such a way that Z has associated transition probability matrix P and, conditional to $Z_n = i$ and $Z_{n+1} = j$, the time interval between the n-th and the $n + 1$-th phase transitions $H_{n+1} = S_{n+1} - S_n$ has distribution $F_{(i,j)}(\cdot)$ and is independent of $(Z_k, S_k)_{k<n}$. In fact, this procedure leads to a sequence $(Z, S) = (Z_n, S_n)_{n \in \mathbb{N}}$ for which

$$\begin{aligned} Q_{ij}(t) &= \mathbf{P}\left(Z_{n+1} = j, S_{n+1} - S_n \le t \mid Z_n = i\right) \\ &= \mathbf{P}\left(Z_{n+1} = j \mid Z_n = i\right) \mathbf{P}\left(S_{n+1} - S_n \le t \mid Z_n = i, Z_{n+1} = j\right) \\ &= p_{ij} F_{(i,j)}(t) \end{aligned}$$

that is, to a MRP with embedded kernel (P, F).

For that, let $(U_n)_{n \in \mathbb{N}}$ and $(V_n)_{n \in \mathbb{N}}$ be two sequences of independent uniform random variables on $(0, 1)$, defined on independent probability spaces $\Lambda_1 = (\Omega_1, \mathcal{F}_1, \mathbf{P}_1)$ and $\Lambda_2 = (\Omega_2, \mathcal{F}_2, \mathbf{P}_2)$, respectively, and construct (Z, S) on the product space $\Lambda = \Lambda_1 \times \Lambda_2$ as follows.

For $\omega = (\omega_1, \omega_2) \in \Omega$, use $(U_n(\omega_1))_{n \in \mathbb{N}}$ to construct $Z(\omega_1)$ on Λ_1 from

$$Z_0(\omega_1) = p^{-1}(U_0(\omega_1))$$
$$Z_{n+1}(\omega_1) = [p_{Z_n(\omega_1) \cdot}]^{-1}(U_{n+1}(\omega_1)), \quad n \in \mathbb{N}$$

where p denotes the initial phase probability vector. At the same time, use the sequences $(Z_n(\omega_1))_{n \in \mathbb{N}}$ and $(V_n(\omega_2))_{n \in \mathbb{N}_+}$ to generate the time intervals between state transitions $H(\omega) = (H_n(\omega))_{n \in \mathbb{N}_+}$, making

$$H_{n+1}(\omega) = \left[F_{(Z_n, Z_{n+1})(\omega_1)}\right]^{-1}(V_{n+1}(\omega_2)), \quad n \in \mathbb{N}. \tag{5}$$

Finally, obtain the renewal sequence $S(\omega)$ by setting $S_0(\omega) = 0$ and

$$S_{n+1}(\omega) = S_n(\omega) + H_{n+1}(\omega), \quad n \in \mathbb{N}.$$

By construction and in view of (3):

- Z_0 has probability vector p.
- $Z_{n+1} \mid Z_n = i$ has probability vector $p_{i \cdot}$, for $n \in \mathbb{N}$.
- $[H_{n+1} \mid Z_n = i, Z_{n+1} = j]$ has distribution function $F_{(i,j)}(\cdot)$, for $n \in \mathbb{N}$.

As in addition, for each $n \in \mathbb{N}$, (U_{n+1}, V_{n+1}) is independent of $\{U_0, (U_m, V_m)_{m \le n}\}$ it follows that, given Z_n, $(Z_{n+1}, S_{n+1} - S_n)$ is independent of $(Z_k, S_k)_{k<n}$ and, thus, the process (Z, S) is a MRP with embedded kernel (P, F).

The previous procedure leads to the algorithm presented in Fig. 1 for the simulation of an SMP with initial phase probability vector p and embedded kernel (P, F).

Input: Independent sequences of independent Unif(0, 1) random variables $(U_n)_{n\in\mathbb{N}}$ and $(V_n)_{n\in\mathbb{N}}$ and $N \in \mathbb{N}_+$
$Z_0 := p^{-1}(U_0), \quad S_0 := 0$
for $n = 0, 1, \ldots, N-1$ **do**
$\quad Z_{n+1} := [p_{Z_n \cdot}]^{-1}(U_{n+1})$
$\quad S_{n+1} = S_n + [F_{(Z_n, Z_{n+1})}]^{-1}(V_{n+1})$
end for
Output: $W_t := Z_n$ for $S_n \le t < S_{n+1}$, $0 \le n < N$

Fig. 1. Simulation of an SMP with initial phase probability vector p and embedded kernel (P, F)

3.2 Simulation via Rate Kernel

In a similar manner, to simulate a SMP W with failure rate kernel R it suffices to simulate a MRP (Z, S) with failure rate kernel R and then obtain W from

$$W_t = Z_n, \quad S_n \le t < S_{n+1}.$$

The simulation of a MRP (Z, S) with failure rate kernel R is fairly different from the simulation based on the embedded kernel. In this case, it is generated a Poisson process with rate modulated by the state of the process and, conditional to the fact that after the last Markov renewal epoch the MRP moved or stayed in phase i, it is taken into account the failure rate in phase i to decide if the next Poisson arrival epoch will make part or not of the random sequence S. Then, if so, it is decided what the next phase will be with a procedure which assures that, if an event takes place t units of time after the transition instant to phase i then the next phase will be j with probability $\frac{r_{ij}(t)}{\lambda_i}$, where λ_i denotes the Poisson uniformization rate in phase i. The following lemma will be useful for such a construction.

Lemma 1. *Let J be an ordered set, order-isomorphic to some bounded or unbounded interval of $\mathbb{Z}$, i be an element of J, $\beta = (\beta_j)_{j\in J}$ be a sub-stochastic vector, and U be a uniform random variable on $(0, 1)$. Then, the random variable*

$$\text{Failure}(i, \beta, U) = \begin{cases} 1 & U \notin \left]\sum_{k \le i} \beta_k, 1 - \sum_{k > i} \beta_k\right] \\ 0 & \text{otherwise} \end{cases} \quad (6)$$

is a Bernoulli random variable with parameter $\sum_{k \in J} \beta_k$. In addition, if we let $\beta^{(i)}$ denote the probability vector obtained from β making

$$\beta_j^{(i)} = \begin{cases} \beta_j & j \ne i \\ 1 - \sum_{l \ne i} \beta_l & j = i \end{cases}, \quad j \in J \quad (7)$$

and let $F_{\beta^{(i)}}(\cdot)$ be its associated distribution function, then the random variable

$$\mathrm{NewState}(i,\beta,U) = F^{-1}_{\beta^{(i)}}(U) \qquad (8)$$

takes values on J and has probability function $\beta^{(i)}$.

To simulate a MRP (Z, S) with initial phase distribution p and failure rate kernel R, with bounded failure transition rate from each phase of the MRP, as before, let $(U_n)_{n\in\mathbb{N}}$ and $(V_n)_{n\in\mathbb{N}}$ be two sequences of independent uniform random variables on $(0, 1)$, defined on independent probability spaces $\Lambda_1 = (\Omega_1, \mathcal{F}_1, \mathbf{P}_1)$ and $\Lambda_2 = (\Omega_2, \mathcal{F}_2, \mathbf{P}_2)$, and construct (Z, S) on the product space $\Lambda = \Lambda_1 \times \Lambda_2$ as next described.

The sequence $(V_n)_{n\in\mathbb{N}}$ along with the consecutive phases of the phase process Z are used to simulate on Λ a sequence of arrival epochs of a modulated Poisson process $(T_m)_{m\in\mathbb{N}}$ with rate vector $\lambda = (\lambda_i)_{i\in I}$ such that

$$\lambda_i \geq \sup_t r_i(t)$$

where λ_i denotes the modulated Poisson rate in phase i, so that the MRP and the associated Poisson process are inter-dependent. On the other hand, the sequence $(U_n)_{n\in\mathbb{N}}$ along with the Failure procedure, defined in Lemma 1, are used to decide whether or not these potential renewal epochs correspond to effective failure time instants and should be included as real renewal epochs. In case the answer is affirmative, the NewState procedure, defined in Lemma 1, is used to generate the phases associated to the Poisson arrival epochs.

Specifically, for $\omega = (\omega_1, \omega_2) \in \Omega$, generate the initial phase and time from

$$Z_0(\omega) = p^{-1}(U_0(\omega_1)) \quad \text{and} \quad S_0(\omega) = 0$$

and, let $\hat{Z}_0(\omega) = Z_0(\omega)$ and $T_0(\omega) = 0$. Then, starting with $n = 0$, proceed recursively for $m \in \mathbb{N}_+$ as follows, where at the end of the cycle m will denote the index of the epoch of the uniformizing Poisson process corresponding to S_{n+1}. Generate new arrival epochs of the uniformizing Poisson process, letting

$$T_m(\omega) = T_{m-1}(\omega) + \left[\mathrm{Exp}(\lambda_{Z_n(\omega)})\right]^{-1}(V_m(\omega_2))$$

until

$$\mathrm{Failure}(Z_n(\omega), r_{Z_n(\omega)}\cdot(T_m(\omega) - S_n(\omega))/\lambda_{Z_n(\omega)}, U_m(\omega_1)) = 1,$$

in which case we consider that a new phase change takes place $T_m(\omega) - S_n(\omega)$ instants after the previous renewal epoch $S_n(\omega)$.

In this case, add the time $T_m(\omega)$ to the Markov renewal time sequence, and determine the new phase $Z_{n+1}(\omega)$ of the Markov renewal phase sequence, i.e.,

$$S_{n+1}(\omega) = T_m(\omega)$$
$$Z_{n+1}(\omega) = \mathrm{NewState}(Z_n(\omega), r_{Z_n(\omega)}\cdot(T_m(\omega) - S_n(\omega))/\lambda_{Z_n(\omega)}, U_m(\omega_1))$$

and, finally, increment n by one unit.

As shown in [15], the previous procedure guarantees that the generated process is a MRP with failure rate kernel R. This follows since, in view of Lemma 1,

> **Input:** Independent sequences $(U_n)_{n\in\mathbb{N}}$ and $(V_n)_{n\in\mathbb{N}}$ of independent Unif(0,1) random variables, a set of nonnegative numbers $\lambda = (\lambda_i)$ such that $\lambda_i \geq \sup_t r_i(t)$, and $N \in \mathbb{N}_+$
> $Z_0 := p^{-1}(U_0)$
> $S_0 := 0, \; T_0 := 0$
> $n := 0, \; m := 0$
> **while** $(n < N)$ **do**
>   **do**
> $m := m + 1$
> $T_m := T_{m-1} + [\text{Exp}(\lambda_{Z_n})]^{-1}(V_m)$
>   **until** $(\text{Failure}(Z_n, r_{Z_n}.(T_m - S_n)/\lambda_{Z_n}, U_m) = 1)$
> $S_{n+1} := T_m$
> $Z_{n+1} := \text{NewState}(Z_n, r_{Z_n}.(S_{n+1} - S_n)/\lambda_{Z_n}, U_m)$
> $n := n + 1$
> **end while**
> **Output:** $W_t := Z_n$ for $S_n \leq t < S_{n+1}, \; 0 \leq n < N$

Fig. 2. Simulation of an SMP with initial phase probability vector p and rate kernel R

conditional to the fact that an event of the Poisson process takes place $t = T_m - S_n$ units of time after the last transition epoch, S_n, at which the MRP is in phase $Z_n = i$, then a failure occurs at that instant with probability $r_i(t)/\lambda_i$, in which case $(Z_{n+1}, S_{n+1}) = (j, T_m) = (j, S_n + t)$ with probability $r_{ij}(t)/r_i(t)$.

The presented procedure leads to the algorithm presented in Fig. 2 for the simulation of a MRP with initial phase probability vector p and rate kernel R.

4 Usual Stochastic Ordering of Semi-Markov Processes

In this section we present the main results for the comparability of two SMPs in the usual stochastic ordering sense. For this purpose, we start with the definition of stochastic ordering of random vectors and stochastic processes, in the usual stochastic ordering sense (c.f., e.g., [44] or [40]).

Definition 3. *Given two real-valued random vectors $X = (X_1, X_2, \ldots, X_n)$ and $Y = (Y_1, Y_2, \ldots, Y_n)$ whose components take values on an ordered set J, we say that X is stochastically smaller than Y in the usual sense, written $X \leq_{st} Y$, if and only if*

$$\mathbf{P}(X \in U) \leq \mathbf{P}(Y \in U), \quad \text{for all increasing sets}^1 \; U \text{ in } J^n.$$

Roughly speaking, we say that a random vector X is stochastically smaller than a random vector Y, in the usual sense, if X is less likely than Y to take large values, where by large we mean values in any increasing set.

---
[1] Given an ordered set J, $U \subseteq J^n$ is called an increasing set if $x \in U$ and $x \leq y$ implies that $y \in U$, with $\leq$ denoting the componentwise ordering for vectors.

For the particular case of two real-valued random variables X and Y, the previous definition specializes into

$$X \leq_{st} Y \iff \mathbf{P}(X \geq x) \leq \mathbf{P}(Y \geq x), \quad x \in \mathbb{R}$$

as the upper sets of $\mathbb{R}$ are the intervals of the form $[u, \infty)$ or (u, ∞), $u \in \mathbb{R}$.

For the sake of simplicity, throughout the paper, order relation symbols will be applied indistinctively to compare random variables or their associated distribution functions, i.e., $X \leq_{st} Y$ is equivalent to $F^X \leq_{st} F^Y$, for random variables X and Y, with F^X and F^Y denoting the distribution functions of X and Y, respectively.

An important property of the usual stochastic order is that it is closed under convolutions [40]. That is, given sequences of independent random variables $\{X_m, 1 \leq m \leq n\}$ and $\{Y_m, 1 \leq m \leq n\}$, for a positive integer n, then

$$[X_m \leq_{st} Y_m, 1 \leq m \leq n] \implies \sum_{m=1}^{n} X_m \leq_{st} \sum_{m=1}^{n} Y_m. \tag{9}$$

For finite measure vectors, i.e., nonnegative vectors with finite sum of their entries, we have the following definition.

Definition 4. *Given two finite measure vectors $a = (a_i)_{i \in I}$ and $b = (b_i)_{i \in I}$ with indices on a countable ordered set I, then we say that a is smaller than b in the usual ordering sense, written $a \leq_{st} b$, if*

$$\sum_{j \geq k} a_j \leq \sum_{j \geq k} b_j, \quad k \in I. \tag{10}$$

In this case, if a and b are probability vectors we say that a is stochastically smaller than b, in the usual sense.

If X and Y are discrete random variables with support in the same ordered set I, with respective probability vectors p^X and p^Y, then $X \leq_{st} Y \iff p^X \leq_{st} p^Y$.

The usual stochastic ordering of two stochastic processes establishes the stochastic ordering of all their finite dimensional distributions, as follows.

Definition 5. *Given two stochastic processes $X = (X(t))_{t \in \mathbb{R}_+}$ and $Y = (Y(t))_{t \in \mathbb{R}_+}$ with common partially ordered state space $(I, \leq)$, then the process X is said to be stochastically smaller than Y in the usual stochastic ordering sense, written $X \leq_{st} Y$, if and only if*

$$(X(t_1), X(t_2), \ldots, X(t_n)) \leq_{st} (Y(t_1), Y(t_2), \ldots, Y(t_n))$$

for all $n \in \mathbb{N}_+$ and $t_1, t_2, \ldots, t_n \in \mathbb{R}_+$.

Alternative characterizations of the usual stochastic ordering, useful to establish stochastic ordering results in various applications, have been proposed in the literature [cf., e.g., [25,26,40]].

Theorem 1. *Given two stochastic processes X and Y with common partially ordered state space $(I, \leq)$, the following conditions are equivalent to $X \leq_{st} Y$:*

(i) *For all $n \geq 1$, $t_1, t_2, \ldots, t_n \in \mathbb{R}_+$, and non-decreasing real function f,*

$$E[f(X(t_1), X(t_2), \ldots, X(t_n))] \leq E[f(Y(t_1), Y(t_2), \ldots, Y(t_n))].$$

(ii) *There exists $\hat{X} =_{st} X$ and $\hat{Y} =_{st} Y$ defined on a common probability space such that*
$$\mathbf{P}(\hat{X}(t) \leq \hat{Y}(t), \text{ for all } t \geq 0) = 1.$$

(iii) *There exists a coupling $(\hat{X}, \hat{Y})$ of X and Y with support on $\{(x, y) \in E \times E : x \leq y\}$.*

As will be seen further, characterization (ii), establishing that the usual stochastic order of two stochastic processes is equivalent to the pathwise comparability of some equivalent versions of these processes, was a key tool in the derivation of the main results presented throughout this tutorial paper.

Sufficient conditions for the stochastic ordering in the usual sense of two SMPs were established by Sonderman [47].

Theorem 2 (Sonderman [47, Theorem 3.2]). *For $W = X, Y$, let W be a SMP with ordered phase space I, order-isomorphic to some subset of $\mathbb{Z}$, initial probability vector p^W and failure rate kernel $R^W = (R^W(t))_{t \in \mathbb{R}_+}$, with $R^W(t) = (r_{ij}^W(t))_{i,j \in I}$. Then, $X \leq_{st} Y$ if the initial phase distributions and the failure rates satisfy*

$$p^X \leq_{st} p^Y$$

and, for all $s, t \in \mathbb{R}_+$ and $i \leq j$,

$$\sum_{k \leq n} r_{ik}^X(s) \geq \sum_{k \leq n} r_{jk}^Y(t), \quad n < i, \tag{11}$$

and

$$\sum_{k \geq n} r_{ik}^X(s) \leq \sum_{k \geq n} r_{jk}^Y(t), \quad n > j. \tag{12}$$

A proof of the previous result, specially tailored for the simulation of pathwise ordered equivalent versions of the involved SMPs, can be found in [15]. There, under the conditions of Theorem 2, copies of the SMPs to be compared with ordered sample-paths are constructed in a common probability space. Namely, a coupling $(X^\star, Y^\star)$ of (X, Y) such that $X^\star \leq Y^\star$ is constructed in the following way.

Let $(U_n)_{n \in \mathbb{N}_+}$ and $(V_n)_{n \in \mathbb{N}_+}$ denote two sequences of independent uniform random variables on $(0, 1)$, defined on independent probability spaces $\Lambda_1 = (\Omega_1, \mathcal{F}_1, \mathbf{P}_1)$ and $\Lambda_2 = (\Omega_2, \mathcal{F}_2, \mathbf{P}_2)$, respectively. Construct the processes $X^\star$ and $Y^\star$ on the product probability space $(\Omega, \mathcal{F}, \mathbf{P}) = \Lambda_1 \times \Lambda_2$, simulating two MRPs (Z^X, S^X) and (Z^Y, S^Y) with failure rate kernel R^X and R^Y in such a way that the construction of $X^\star$ and $Y^\star$ from

$$W_t^\star = Z_n^W, \quad S_n^W \le t < S_{n+1}^W, \quad W = X, Y \qquad (13)$$

leads to $X_t^\star(\omega) \le Y_t^\star(\omega)$, for all $t \in \mathbb{R}_+$ and $\omega \in \Omega$.

For that, generate the potential transitions epochs on both MRPs through a common doubly stochastic Poisson process with rate modulated by the phases of the two processes. Namely, whenever $X^\star$ is in phase i and $Y^\star$ is in phase j, use for uniformization rate a value λ_{ij} such that

$$\lambda_{ij} \ge 2 \sup_t \max\{r_i^X(t), r_j^Y(t)\}. \qquad (14)$$

Specifically, for $\omega = (\omega_1, \omega_2) \in \Omega$, generate the initial phases and times from

$$Z_0^W(\omega) = [p^W]^{-1}(U_0(\omega_1)) \quad \text{and} \quad S_0^W(\omega) = 0, \quad W = X, Y. \qquad (15)$$

Then, starting with $T_0(\omega) = 0$, $\hat{Z}_0^W(\omega) = Z_0^W(\omega)$ and $n_W = 0$, $W = X, Y$, proceed recursively for $m \in \mathbb{N}_+$ as follows. First let

$$\lambda^\star = \lambda_{Z_{n_X}^X(\omega)\, Z_{n_Y}^Y(\omega)}$$

denote the uniformization rate for the phase vector $(Z_{n_X}^X(\omega), Z_{n_Y}^Y(\omega))$ and let

$$T_m(\omega) = T_{m-1}(\omega) + [\operatorname{Exp}(\lambda^\star)]^{-1}(V_m(\omega_2))$$
$$\hat{Z}_m^X(\omega) = \operatorname{NewState}(Z_{n_X}^X(\omega), r_{Z_{n_X}^X(\omega)}^X.(T_m(\omega) - S_{n_X}^X(\omega))/\lambda^\star, U_m(\omega_1))$$
$$\hat{Z}_m^Y(\omega) = \operatorname{NewState}(Z_{n_Y}^Y(\omega), r_{Z_{n_Y}^Y(\omega)}^Y.(T_m(\omega) - S_{n_Y}^Y(\omega))/\lambda^\star, U_m(\omega_1)).$$

In sequence, for each $W = X, Y$ for which

$$\operatorname{Failure}(Z_{n_W}^W(\omega), r_{Z_{n_W}^W(\omega)}^W.(T_m(\omega) - S_{n_W}^W(\omega))/\lambda^\star, U_m(\omega_1)) = 1 \qquad (16)$$

i.e., for which the time $T_m(\omega)$ is a transition epoch of Z^W, include the time $T_m(\omega)$ and the phase $\hat{Z}_m^W(\omega)$ as a new pair of the sequence (Z^W, S^W) by making

$$S_{n_W+1}^W(\omega) = T_m(\omega) \quad \text{and} \quad Z_{n_W+1}^W(\omega) = \hat{Z}_m^W(\omega)$$

and let $n_W = n_W + 1$.

Finally, for $W = X, Y$, construct the SMP $W^\star$ from

$$W_t^\star(\omega) = \hat{Z}_m^W(\omega), \quad T_m(\omega) \le t < T_{m+1}(\omega) \qquad (17)$$

for $m \in \mathbb{N}$, which is equivalent to generate $W^\star$ from

$$W_t^\star(\omega) = Z_n^W(\omega), \quad S_n^W(\omega) \le t < S_{n+1}^W(\omega)$$

for $n \in \mathbb{N}$.

By construction, for $W = X, Y$ the MRP (Z^W, S^W) has failure rate kernel R^W and thus the generated SMP $W^\star$ is such that $W^\star =_{st} W$. In addition, as for each pair of states (i,j) such taht $\hat{Z}_m^X(\omega) = i \le j = \hat{Z}_m^Y(\omega)$ the uniformization rates are selected with the guaranty that $\sum_{k \ne i} \frac{r_{ik}^X(s)}{\lambda_{ij}} \le 0.5$ and $\sum_{k \ne j} \frac{r_{jk}^Y(t)}{\lambda_{ij}} \le 0.5$, from conditions (11)-(12) it follows that

$$\text{NewState}(i, r_{i\cdot}^X(s)/\lambda_{ij}, u) \le \text{NewState}(j, r_{j\cdot}^Y(t)/\lambda_{ij}, u)$$

for $s, t \in \mathbb{R}_+$ and $u \in (0,1)$. Thus, the generation of the next transitions from a common uniform generator guarantee that $\hat{Z}_{m+1}^X(\omega) \le \hat{Z}_{m+1}^Y(\omega)$, for all $\omega \in \Omega$. As such, by induction on m and in view of (17), the proposed construction leads to two SMPs $X^\star$ and $Y^\star$ such that $X_t^\star(\omega) \le Y_t^\star(\omega)$, for all $\omega \in \Omega$ and $t \in \mathbb{R}_+$.

The decribed procedure leads to the algorithm of Fig. 3 to simulate, in a common probability space, two st-ordered SMPs X and Y with initial probability vectors p^X and p^Y, and failure rate kernels R^X and R^Y, respectively, under the conditions of Theorem 2.

Input: Independent sequences of independent Unif$(0,1)$ random variables $(U_n)_{n \in \mathbb{N}}$ and $(V_n)_{n \in \mathbb{N}}$, a nonnegative matrix $\lambda = (\lambda_{ij})_{i,j \in I}$ such that $\lambda_{ij} \ge 2 \sup_t \max\{r_i^X(t), r_j^Y(t)\}$, and a positive value TMAX

$n_X := n_Y := 0; \quad T_0 := 0, \ m := 0$
$Z_0^X := [p^X]^{-1}(U_0); \quad Z_0^Y := [p^Y]^{-1}(U_0)$
while $(\min\{S_{n_X}^X, S_{n_Y}^Y\} < \text{TMAX})$ **do**
  $\lambda^\star = \lambda_{Z_{n_X}^X Z_{n_Y}^Y}$
  do
    $m := m+1$
    $T_m := T_{m-1} + [\text{Exp}(\lambda^\star)]^{-1}(V_m)$
  until $(\text{Failure}(Z_{n_X}^X, r_{Z_{n_X}^X}^X(T_m - S_{n_X}^X)/\lambda^\star, U_m) = 1$ or
      $\text{Failure}(Z_{n_Y}^Y, r_{Z_{n_Y}^Y}^Y(T_m - S_{n_Y}^Y)/\lambda^\star, U_m) = 1)$
  for $W = X, Y$ **do**
    if $(\text{Failure}(Z_{n_W}^W, r_{Z_{n_W}^W}^W(T_m - S_{n_W}^W)/\lambda^\star, U_m) = 1)$ **then**
      $Z_{n_W+1}^W := \text{NewState}(Z_{n_W}^W, r_{Z_{n_W}^W}^W(T_m - S_{n_W}^W)/\lambda^\star, U_m)$
      $S_{n_W+1}^W := T_m$
      $n_W := n_W + 1$
    end if
  end for
end while
Output: $X_t := Z_l^X$ for $S_l^X \le t < S_{l+1}^X, \ 0 \le l < n_X$
    $Y_t := Z_l^Y$ for $S_l^Y \le t < S_{l+1}^Y, \ 0 \le l < n_Y$

Fig. 3. Simulation of two st-ordered CTMCs, under the conditions of Theorem 2

4.1 Usual Stochastic Ordering of CTMCs

For the particular case of two CTMCs, Sonderman's result specialize into the earlier derived Kirstein's [28] sufficient conditions for the stochastic ordering in the usual sense of two CTMCs.

Corollary 1 (Kirstein [28]). *Let X and Y be CTMCs with ordered state space I, order isomorphic to a subset of $\mathbb{Z}$, initial probability vectors p^X and p^Y, and (infinitesimal) generator matrices Q^X and Q^Y, respectively. Then $X \leq_{st} Y$ provided that*

$$p^X \leq_{st} p^Y \tag{18}$$

and

$$\sum_{m \geq n} q^X_{im} \leq \sum_{m \geq n} q^Y_{jm}, \quad \text{for all } i \leq j \text{ and } (n \leq i \text{ or } n > j). \tag{19}$$

In fact, for CTMCs, the failure rate at time t of the time needed for a transition from phase i to phase j to take place does not depend on t, i.e., $r_{ij}(t) = r_{ij}$, for all $t \in \mathbb{R}_+$. As a consequence, Sonderman's conditions (11)-(12) are equivalent to conditions (19) since

$$q^W_{ij} = \begin{cases} r^W_{ij} & j \neq i \\ -\sum_{l \neq i} r^W_{il} & j = i \end{cases}, \quad W = X, Y.$$

Under the conditions of Corollary 1, the simulation of st-ordered uniformizable CTMCs can be done in a simpler manner. In fact, it suffice to uniformize both

---

Input: Independent sequences of independent Unif$(0, 1)$ random variables $(U_n)_{n \in \mathbb{N}}$ and $(A_n)_{n \in \mathbb{N}_+}$, and a positive value TMAX

$\lambda := \sup_{i \in I} 2\{q^X_i, q^Y_i, 1\}$

$P^{\bar{X}} := \mathbf{I} + \dfrac{Q^X}{\lambda}; \qquad P^{\bar{Y}} := \mathbf{I} + \dfrac{Q^Y}{\lambda}$

$\bar{X}_0 := [p^X]^{-1}(U_0); \qquad \bar{Y}_0 := [p^Y]^{-1}(U_0)$

$T_0 := 0; \qquad n := 0$

while $(T_n < \text{TMAX})$ **do**

$\quad \bar{X}_{n+1} := [p^{\bar{X}}_{\bar{X}_n}]^{-1}(U_{n+1})$

$\quad \bar{Y}_{n+1} := [p^{\bar{Y}}_{\bar{Y}_n}]^{-1}(U_{n+1})$

$\quad T_{n+1} := T_n + [\text{Exp}(\lambda)]^{-1}(A_{n+1})$

$\quad n := n + 1$

end while

Output: $X^*_t := \bar{X}_l$ for $T_l \leq t < T_{l+1}, 0 \leq l < n$

$\qquad\quad Y^*_t := \bar{Y}_l$ for $T_l \leq t < T_{l+1}, 0 \leq l < n$

Fig. 4. Simulation of two st-ordered CTMCs, under the conditions of Corollary 1

chains at a common uniformization rate, and then simulate transitions in both chains from a common sequence of independent uniform generators and, at same time, use another independent sequence of independent uniform generators to simulate the holding times in states before transitions on both chains from another common generator, as presented in the algorithm of Figure 4, where **I** denotes de identity matrix of an appropriate dimension.

The strongest generalization of such result was achieved by [37], who provides the characterization of the usual ordering of CTMCs with partially ordered state spaces in terms of conditions on their infinitesimal transition rates to upper sets.

5 Level-Crossing Ordering of Semi-Markov Processes

In this section we will focus on the comparability of SMPs in the level-crossing ordering sense, which compares stochastic processes in terms of the times they take to reach or exceed high levels. Specifically, a process X is said to be stochastic smaller in level-crossing than Y if it takes X stochastically longer to reach or exceed any given level than it does Y.

The analysis of this stochastic ordering for processes with common ordered state spaces, order isomorphic to a subset of integers, was pioneered by A. Irle and J. Gani motivated by problems of comparing random times for the detection of words. As remarked by these authors, the usual stochastic ordering was too strong to be used in the envisaged context. As such, in their pioneering work [21], times for detection of words were modelled as first passage times to up-cross levels in skip-free to the right DTMCs[2] and were directly compared in the usual stochastic ordering sense. Specifically, [21, Theorem 4.1] shows that, for two skip-free to the right DTMCs with common ordered state space, the ordering in distribution of their transition probabilities for any common initial state (which does not guarantee the usual stochastic ordering of the respective DTMCs) implies the level-crossing ordering of the DTMCs.

Imposing extra stochastic ordering conditions on the holding times in states before transitions Irle [20] established sufficient conditions for the level-crossing ordering of skip-free to the right SMPs, paying particular attention to the ordering of uniformizable CTMCs and birth-and-death processes, along with Wiener processes. These results were later improved by Ferreira and Pacheco [16,17] for skip-free to the right DTMCs, SMPs and CTMCs with common ordered state spaces. The lc-ordering analysis had further developments in [18] for general (i.e., non-skip-free to the right) DTMCs, SMPs and CTMCs with totally ordered state spaces.

Hereafter we give an overview of the main results derived in these papers for the level-crossing of two SMPs. For the sake of simplicity, the results will be presented only in terms of the level-crossing ordering of stochastic processes in

---

[2] We recall that a trajectory of a stochastic process with ordered state space I, order-isomorphic to some bounded or unbounded interval of $\mathbb{Z}$, is said to be *skip-free to the right* if it does not have jumps up more than one level and the stochastic process itself is *skip-free to the right* if its trajectories are almost surely skip-free to the right.

the usual sense. As such, in the following, we just refer to level-crossing ordering instead of level-crossing ordering in the usual sense. Nevertheless, we note that the results derived in [18] are valid in a general framework in which the comparison of the passage times to up-cross levels may be made using, aside the usual stochastic ordering, any integral stochastic order relation for positive variables closed for convolution, which includes many important cases, such as the Laplace transform and the increasing concave order [40,44].

5.1 Preliminaries

Let Γ be either the set of natural numbers $\mathbb{N}$, positive integers $\mathbb{N}_+$ or real nonnegative numbers $\mathbb{R}_+$. Given a set I, order isomorphic to a bounded or unbounded interval of $\mathbb{Z}$, and $y \in I$, we let $\bar{I} = I \setminus \{\sup I\}$, where $\sup I$ is the supremum of set I, $I^A = I \cap A$ denote the restriction of I to states in A, and $I^{\leq y} = I^{(-\infty, y]}$ denote the restriction of I to states smaller or equal to y.

Moreover, if $W = (W_t)_{t \in \Gamma}$ is a stochastic process with state space I, we let S_y^W denote the hitting time of the set of values greater or equal to y, i.e.,

$$S_y^W = \inf\{t \in \Gamma : W_t \geq y\} = \inf\{t \in \Gamma : W_t \in I^{\geq y}\}$$

where $\inf \emptyset = +\infty$. Finally, to introduce the definition of level-crossing ordering of stochastic processes, we let $S_{x,y}^W$ denote the hitting time of the set of values greater or equal to y when departing from state x, i.e.,

$$S_{x,y}^W = [\inf\{t \in \Gamma : W_t \geq y\}|W_0 = x].$$

Definition 6. *Let $X = (X_t)_{t \in \Gamma}$ and $Y = (Y_t)_{t \in \Gamma}$ be stochastic processes with ordered state space I. Then, the process X is said to be smaller in level-crossing than Y, denoted $X \leq_{lc} Y$, if, for any common initial state x, $S_{x,y}^Y \leq_{st} S_{x,y}^X$, for all $y \in I$, i.e.,*

$$X \leq_{lc} Y \iff S_{x,y}^Y \leq_{st} S_{x,y}^X, \text{ for all } x, y \in I.$$

The next result, provided in [17, Theorem 1], asserts that stochastic processes are stochastically monotone increasing in the level-crossing ordering sense with respect to time-clock speed, i.e., if the time-clock speed of a process is increased, then the resulting process is faster in level-crossing than the original process.

Theorem 3 (Ferreira and Pacheco [17, Theorem 1]). *Given a stochastic process $X = (X_t)_{t \in \mathbb{R}_+}$ with ordered state space, the α-parameterized family of processes $\{X^{(\alpha)}, \alpha > 0\}$ where $X_t^{(\alpha)} = X_{\alpha t}$, for $t \in \mathbb{R}_+$, denote the time-clock speed change of X by factor α, is stochastically increasing in the level-crossing ordering sense, i.e., $X^{(\alpha_1)} \leq_{lc} X^{(\alpha_2)}$, for all $\alpha_1 \leq \alpha_2$.*

Given a stochastic process $W = (W_t)_{t \in \Gamma}$ with ordered state space I, we let $W^{\leq y}$, $y \in I$, denote the process W restricted to the state space $I^{\leq y}$ in such a way that state y is made absorbing and all states of W greater or equal to y are collapsed into state y, namely,

$$W_t^{\leq y} = \begin{cases} W_t & t < S_y^W \\ y & t \geq S_y^W \end{cases}.$$

Note that if W is a SMP with ordered state space I and $y \in I$, then $W^{\leq y}$ is also a SMP whose parameters are easily derived from the parameters of the original process. With this notation, we consider the following definition.

Definition 7. Let $\triangle$ denote a property and W be a stochastic process with ordered state space I. Then, W has the lower-$\triangle$ property if and only if the process $W^{\leq x}$ has the $\triangle$ property, for all $x \in I$.

In the next two subsections, we present sufficient conditions and algorithms to simulate level-crossing ordered SMPs, treating separately the cases where we impose the condition of one of the processes involved in the comparison being skip-free to the right and the case where we do not.

5.2 Level-Crossing Ordering of Skip-Free to the Right SMPs

As mentioned in the introduction, the pioneering result for the level-crossing ordering of SMPs was provided in [20] and established sufficient conditions for the level-crossing ordering of skip-free to the right SMPs, with the random times to up-cross levels being compared in the usual stochastic ordering sense. Specifically, using the characterization of an SMP via its embedded kernel, which separates the embedded transition probabilities from the distributions of the holding times in states between transitions, ([20], Theorem 2.1) establishes that the level-crossing ordering of two skip-free to the right SMPs follows from the ordering in distribution of their transition probabilities from common states, and from the reversed order of the holding times in common states before the processes make transitions.

Theorem 4 (Irle [20, Theorem 2.1]). Let $X = (X_t)_{t \in \mathbb{R}_+}$ and $Y = (Y_t)_{t \in \mathbb{R}_+}$ be two lower-regular skip-free to the right SMPs with ordered state space I, order-isomorphic to some bounded or unbounded interval of $\mathbb{Z}$, and embedded kernel (P^X, F^X) and (P^Y, F^Y), respectively. Then $X \leq_{lc} Y$ provided that

$$p_{i \cdot}^X \leq_{st} p_{i \cdot}^Y, \quad i \in I$$

and

$$F_{(a,b)}^X \geq_{st} F_{(c,d)}^Y, \quad a, b, c, d \in I.$$

By means of a sample-path based coupling approach [34], this result was later improved in [16] by removing the stochastic ordering conditions involving the transition probabilities from the highest state (if it exists), removing the lower-regularity and the skip-free to the right properties of the faster of the two processes (in level-crossing), and relaxing the conditions on the comparison of the times between transitions in X and Y (namely that $F^X_{(a,b)} \geq_{st} F^Y_{(c,d)}$ for all $a, b, c, d \in I$) to $F^X_{(a,b)} \geq_{st} F^Y_{(a,c)}$ for all $a \in \bar{I}$ and $b, c \in I$, such that $b \leq c$.

Theorem 5 (Ferreira and Pacheco [16, Theorem 4.1]). Let $X = (X_t)_{t \in \mathbb{R}_+}$ and $Y = (Y_t)_{t \in \mathbb{R}_+}$ be two SMPs with ordered state space I, order-isomorphic to some bounded or unbounded interval of $\mathbb{Z}$, and embedded kernels (P^X, F^X) and (P^Y, F^Y), respectively, such that

$$p^X_{i\cdot} \leq_{st} p^Y_{i\cdot}, \quad i \in \bar{I} \tag{20}$$

and

$$F^X_{(a,b)} \geq_{st} F^Y_{(a,c)} \tag{21}$$

holds simultaneously for all $a \in \bar{I}$ and $b, c \in I$, with $b \leq c$, such that $p^X_{ab} p^Y_{ac} > 0$. If the processes X is skip-free to the right and lower-regular, then $X \leq_{lc} Y$.

To prove this result, the authors showed that, under the conditions of Theorem 5, we may construct, in a common probability space, copies of the SMPs to be compared with level-crossing ordered sample-paths, i.e., to construct

$$X^\star =_{st} X, \quad Y^\star =_{st} Y, \quad \text{such that} \quad X^\star \leq_{lc} Y^\star.$$

For that, using two independent sequences of independent uniform random variables, $(U_n)_{n \in \mathbb{N}}$ and $(V_n)_{n \in \mathbb{N}}$, the authors begin to simulate two DTMCs, $\hat{X}$ and $\hat{Y}$, (with common initial state) with transition probability matrices P^X and P^Y, respectively, such that $\hat{X} \leq_{lc} \hat{Y}$. The main idea of the proof consists in simulating $\hat{Y}$, the faster of the two DTMCs in level-crossing, in advance and to simulate transitions in both chains from a common uniform generator only when the slower of the two DTMCs, $\hat{X}$, reaches successively each one of the states visited by $\hat{Y}$. Accordingly, they propose to first generate $\hat{Y}$ using the standard procedure to simulate a DTMC from a sequence of independent uniform random variables, $(U_n)_{n \in \mathbb{N}}$. Then, to simulate $\hat{X}$ based on the generated sample path of $\hat{Y}$ and the skip-free to the right property of X: (a) whenever $\hat{X}$ reaches the next state on the sample path of $\hat{Y}$, the next transition in $\hat{X}$ is simulated using the generator previously used to simulate the transition from the next state in $\hat{Y}$; and (b) any other transition in $\hat{X}$ is simulated from an independent sequence of independent uniform random variables, $(V_n)_{n \in \mathbb{N}}$.

From the construction, as the embedded DTMCs at transition epochs of X and Y satisfy the conditions (20), it readily follows that, whenever $\hat{X}$ reaches the next state on the sample path of $\hat{Y}$, the next transition will put $\hat{X}$ in a smaller state than the next state visited by $\hat{Y}$

$$\hat{X}_n = \hat{Y}_m \implies \hat{X}_{n+1} = \left[p^X_{\hat{X}_n \cdot}\right]^{-1}(p) \leq \hat{Y}_{m+1} = \left[p^Y_{\hat{Y}_m \cdot}\right]^{-1}(p), \quad p \in (0, 1).$$

As a consequence, $\hat{X}$ will need at least as many transitions as $\hat{Y}$ to reach a state greater or equal to any given desired state, so that $\hat{X} \leq_{lc} \hat{Y}$.

At the same time, they propose to use two other independent sequences of independent uniform random variables, $(A_n)_{n\in\mathbb{N}}$ and $(B_n)_{n\in\mathbb{N}}$, to simulate the holding times in states before transitions on both processes (say $(H_n^{X^\star})_{n\in\mathbb{N}_+}$ and $(H_n^{Y^\star})_{n\in\mathbb{N}_+}$) in the following way: (a) the holding times in states given the next state visited for the two SMPs are computed from a common generator, the sequence $(A_n)_{n\in\mathbb{N}}$, whenever the corresponding transitions in $\hat{X}$ and $\hat{Y}$ are simulated from a common uniform generator, the sequence $(U_n)_{n\in\mathbb{N}}$; and (b) the holding times in any other state (of $X^\star$) are simulated from the sequence $(B_n)_{n\in\mathbb{N}}$, i.e., the holding times in states given the next state visited in $X^\star$ are computed from the sequence $(B_n)_{n\in\mathbb{N}}$ whenever the corresponding transitions in $\hat{X}$ are simulated from the sequence $(V_n)_{n\in\mathbb{N}}$. Finally, the SMPs $X^\star$ and $Y^\star$ are obtained by letting, for $W = X, Y$,

$$W_t^\star = \hat{W}_n, \text{ for } S_n^{W^\star} \leq t < S_{n+1}^{W^\star},$$

where $S_0^{W^\star} = 0$ and $S_{n+1}^{W^\star} = S_n^{W^\star} + H_{n+1}^{W^\star}$.

By construction, $X^\star =_{st} X$, $Y^\star =_{st} Y$ and the lc-ordered DTMCs $\hat{X}$ and $\hat{Y}$ are constructed in such a way that the sequence of states visited by $\hat{X}$ until it reaches a state greater or equal to any desired given state includes the sequence of states visited by $\hat{Y}$ to reach the same set of states. As the holding times of $X^\star$ and $Y^\star$ in states when $X^\star$ reaches successively the states visited by $Y^\star$ are simulated from a common uniform generator, from (21), we conclude that $X^\star$ spends at least as much time as $Y^\star$ in each of the states visited by $Y^\star$ before reaching the desired set of states. As the usual stochastic order is closed for convolution (9), it follows that $S_l^{X^\star} \geq S_l^{Y^\star}$, for all $l \in I$, i.e., $X^\star \leq_{lc} Y^\star$.

Based in this procedure, the algorithm of Figure 5 simulates, under the conditions of Theorem 5, two lc-ordered SMPs, $X^\star$ and $Y^\star$, with common initial probability vector p, and embedded kernels (P^X, F^X) and (P^Y, F^Y) respectively.

As noted by the authors ([16], Theorem 4.2), the conditions (21) in Theorem 5, on the stochastic ordering of the holding times between transitions, can be further relaxed to

$$F^X_{(a,b)} \oplus F^X_{(b,b+1)} \oplus F^X_{(b+1,b+2)} \oplus \cdots \oplus F^X_{(\min(c,a+1)-1,\min(c,a+1))} \geq_{st} F^Y_{(a,c)}$$

for all $a \in \bar{I}$ and $b, c \in I$, with $b \leq c$, such that $P^X_{ab}\left(\prod_{k=b+1}^{\min(c,a+1)} P^X_{k-1,k}\right) P^Y_{ac} > 0$, with $\oplus$ denoting convolution. However, such a relaxation is paid at the cost of obtaining conditions that are much more difficult to check.

5.3 Level-Crossing Ordering of General SMPs

The level-crossing ordering of general (i.e., non-skip-free to the right) SMPs was addressed in [18]. Under stronger stochastic ordering conditions on the transition probabilities departing from certain states and on the holding times in states between transitions, the authors asserted the following result.

```
Input: Independent sequences of independent Unif(0,1) random variables
       $(U_n)_{n \in \mathbb{N}}$, $(V_n)_{n \in \mathbb{N}}$, $(A_n)_{n \in \mathbb{N}_+}$ and $(B_n)_{n \in \mathbb{N}_+}$, and a positive value TMAX
$\hat{X}_0 := \hat{Y}_0 := p^{-1}(U_0)$
$S_0^{X^\star} := S_0^{Y^\star} := 0$
$k := 0$
while $(S_k^{Y^\star} < \text{TMAX})$ do
    $k := k+1$
    $\hat{Y}_k := [p_{\hat{Y}_{k-1},\cdot}^Y]^{-1}(U_k)$
    $S_k^{Y^\star} := S_{k-1}^{Y^\star} + [F_{(\hat{Y}_{k-1}, \hat{Y}_k)}^Y]^{-1}(A_k)$
end while
$n := 0$
$m := 0$
while $(S_n^{X^\star} \leq \text{TMAX})$ do
    $m := m+1$
    $n := n+1$
    $\hat{X}_n := [p_{\hat{X}_{n-1},\cdot}^X]^{-1}(U_m)$
    $S_n^{X^\star} := S_{n-1}^{X^\star} + [F_{(\hat{X}_{n-1}, \hat{X}_n)}^X]^{-1}(A_m)$
    while $\left(\hat{X}_n < \hat{Y}_m \ \& \ S_n^{X^\star} < \text{TMAX}\right)$ do
        $n := n+1$
        $\hat{X}_n := [p_{\hat{X}_{n-1},\cdot}^X]^{-1}(V_n)$
        $S_n^{X^\star} := S_{n-1}^{X^\star} + [F_{(\hat{X}_{n-1}, \hat{X}_n)}^X]^{-1}(B_n)$
    end while
end while
Output: $X_t^\star := \hat{X}_l$ for $S_l^{X^\star} \leq t < S_{l+1}^{X^\star}$, $0 \leq l < n$
        $Y_t^\star := \hat{Y}_l$ for $S_l^{Y^\star} \leq t < S_{l+1}^{Y^\star}$, $0 \leq l < k$
```

Fig. 5. Simulation of two *lc*-ordered SMPs under the conditions of Theorem 5

Theorem 6. *Let $X = (X_t)_{t \in \mathbb{R}_+}$ and $Y = (Y_t)_{t \in \mathbb{R}_+}$ be two SMPs with ordered state space I, order-isomorphic to some bounded or unbounded interval of $\mathbb{Z}$, and embedded kernel (P^X, F^X) and (P^Y, F^Y), respectively, such that X is lower-regular. Then $X \leq_{lc} Y$, if*

$$p_{x,\cdot}^X \leq_{st} p_{y,\cdot}^Y, \quad \text{for all } x, y \in \bar{I}, \ x \leq y \tag{22}$$

and

$$F_{(a,b)}^X \geq_{st} F_{(c,d)}^Y \tag{23}$$

for all $a, c \in \bar{I}$ and $b, d \in I$, with $a \leq c$, $b \leq d$, and $p_{ab}^X p_{cd}^Y > 0$.

This result was proved using a sample-path based coupling approach. The authors showed how to simulate, under conditions (22)-(23) of Theorem 6, SMPs $X^\star$ and $Y^\star$, departing from the same state, such that $X^\star =_{st} X$, $Y^\star =_{st} Y$, and $X^\star \leq_{lc} Y^\star$.

The simulation in this case is quite simple as the DTMCs embedded at transition epochs and the holding times in states before transitions are generated in a synchronized manner. The copies of the DTMCs embedded at transition epochs, $\hat{X}$ and $\hat{Y}$, are simulated from an independent sequence of independent uniform random variables, $(U_n)_{n \in \mathbb{N}_+}$, making $\hat{W}_n = [P^W_{\hat{W}_n}]^{-1}(U_n)$ for $W = X, Y$. Simulating the transitions on both processes from a common uniform generator, the conditions (22) guarantee that, before $\hat{Y}$ reaches the highest level: transitions on $\hat{X}$ departing from states smaller or equal to the ones from which $\hat{Y}$ departs on the same instant always put $\hat{X}$ in states smaller or equal than the ones for which $\hat{Y}$ makes the transition, i.e.,

$$\hat{X}_n \leq \hat{Y}_n \Longrightarrow \hat{X}_{n+1} = [P^X_{\hat{X}_n}]^{-1}(U_n) \leq [P^Y_{\hat{Y}_n}]^{-1}(U_n) = \hat{Y}_{n+1}.$$

Thus, when the DTMCs start from a common level, this procedure leads to $\hat{X}_n \leq \hat{Y}_n$, for all $n \leq \inf\{m \in \mathbb{N} : \hat{Y}_m = \sup I\}$, and consequently to $\hat{X} \leq_{lc} \hat{Y}$.

At the same time, the sequences of holding times in states between transitions for both processes, say $(H_n^{X^\star})_{n \in \mathbb{N}_+}$ and $(H_n^{Y^\star})_{n \in \mathbb{N}_+}$, are simulated from an independent sequence of independent uniform random variables, $(A_n)_{n \in \mathbb{N}_+}$, making $H_n^{W^\star} = [F^W_{(\hat{W}_{n-1}, \hat{W}_n)}]^{-1}(A_n)$, for $W = X, Y$. Finally, the SMPs $X^\star$ and $Y^\star$ are obtained by letting, for $W = X, Y$,

$$W_t^\star = \hat{W}_n, \text{ for } S_n^{W^\star} \leq t < S_{n+1}^{W^\star},$$

where $S_0^{W^\star} = 0$ and $S_{n+1}^{W^\star} = S_n^{W^\star} + H_{n+1}^{W^\star}$.

By construction, $X^\star =_{st} X$ and $Y^\star =_{st} Y$. In addition, before $\hat{Y}$ reaches the higher state, the embedded DTMCs $\hat{X}_n$ and $\hat{Y}_n$ are strictly ordered, then $\hat{X}$ will need at least as many transitions as $\hat{Y}$ to reach a state greater or equal to any given state. Since, in addition, the holding times in states before transitions are simulated in both processes from a common uniform generator, then, from (4) and (23), before $\hat{Y}$ reaches the highest state: $X^\star$ spends in each sucessive state at least as much time as $Y^\star$. Thus, as the usual stochastic order is closed under convolution (9), we necessarily have $S_l^{X^\star} \geq_{st} S_l^{Y^\star}$, for all $l \in I$, i.e., $X^\star \leq_{lc} Y^\star$. Based on the described procedure, the algorithm of Figure 6 simulates two lc-ordered SMPs $X^\star$ and $Y^\star$, with common initial probability vector p, under the conditions of Theorem 6.

For SMPs with equal transitions probabilities other than the supremum of the state space, we can relax the conditions (23) of Theorem 6, on the times between state transitions, to conditions involving transitions between the same state in both processes, establishing that an increase of the times between state transitions of an SMP in the usual stochastic ordering sense gives rise to an increase of the associated (upper) level-crossing times in the same sense, as next stated.

Corollary 2 (Ferreira and Pacheco [18, Corollary 3]). Let $X = (X_t)_{t \in \mathbb{R}_+}$ and $Y = (Y_t)_{t \in \mathbb{R}_+}$ be two SMPs with ordered state space I, order-isomorphic to some bounded or unbounded interval of $\mathbb{Z}$, and embedded kernels (P^X, F^X) and (P^Y, F^Y), respectively, such that X is lower-regular. Then $X \leq_{lc} Y$ if

$$p^X_{i.} =_{st} p^Y_{i.} \quad \text{and} \quad F^X_{(i,j)} \geq_{st} F^Y_{(i,j)} \tag{24}$$

for all $i \in \bar{I}$, and for all $i \in \bar{I}$ and $j \in I$ such that $p^X_{ij} > 0$, respectively.

This result extends, to non-skip-free to the right SMPs, results for skip-free to the right SMPs with common embedded transition probability matrices, with respect to: the Laplace transform and the mean value order, in [10], and the usual stochastic order, in [20].

Input: Independent sequences of independent Unif(0, 1) random variables $(U_n)_{n \in \mathbb{N}}$ and $(A_n)_{n \in \mathbb{N}}$, and a positive value TMAX
$\hat{X}_0 := \hat{Y}_0 := p^{-1}(U_0)$
$S^{X^\star}_0 := S^{Y^\star}_0 := 0$
$n := 0$
while $\left(\min\{S^{X^\star}_n, S^{Y^\star}_n\} < \text{TMAX}\right)$ **do**
$\quad n := n + 1$
$\quad \hat{X}_n := [p^X_{\hat{X}_{n-1}.}]^{-1}(U_n)$
$\quad \hat{Y}_n := [p^Y_{\hat{Y}_{n-1}.}]^{-1}(U_n)$
$\quad S^{X^\star}_n := S^{X^\star}_{n-1} + [F^X_{(\hat{X}_{n-1},\hat{X}_n)}]^{-1}(A_n)$
$\quad S^{Y^\star}_n := S^{Y^\star}_{n-1} + [F^Y_{(\hat{Y}_{n-1},\hat{Y}_n)}]^{-1}(A_n)$
end while
Output: $X^\star_t := \hat{X}_k$ for $S^{X^\star}_k \leq t < S^{X^\star}_{k+1}$, $0 \leq k < n$
$\qquad\quad Y^\star_t := \hat{Y}_k$ for $S^{Y^\star}_k \leq t < S^{Y^\star}_{k+1}$, $0 \leq k < n$

Fig. 6. Simulation of two lc-ordered SMPs under the conditions of Theorem 6

5.4 Level-Crossing Ordering of CTMCs

A CTMC W with ordered state space I and generator matrix $Q^W = (q^W_{ij})_{i,j \in I}$, whose corresponding transition rate from state i is $q^W_i = -q^W_{ii} = \sum_{j \neq i} q^W_{ij}$, may be interpreted as an SMP with one-step embedded transition probability matrix $P^W = (p^W_{ij})_{i,j \in I}$, where

$$p^W_{ij} = \begin{cases} (1-\delta_{ij})\frac{q^W_{ij}}{q^W_i} & q^W_i > 0 \\ \delta_{ij} & q^W_i = 0 \end{cases}$$

with δ denoting the Kronecker delta function, i.e., $\delta_{ij} = 1$ if $i = j$ and $\delta_{ij} = 0$ if $i \neq j$, and holding times in state i exponentially distributed with rate q_i, regardless of the state visited at the next transition.

In view of the previous, the translation of Theorem 5 and Theorem 6 for the level-crossing ordering of two CTMCs goes as follows.

Corollary 3. *Let X and Y be CTMCs with state space I, order-isomorphic to some bounded or unbounded interval of $\mathbb{Z}$, vectors q^X and q^Y of transition rates from states, and embedded transition probability matrices P^X and P^Y, respectively. Then $X \leq_{lc} Y$ if either*

(i) X is lower-regular, and
$$q_i^X \leq q_j^Y \quad \text{and} \quad p_{i\cdot}^X \leq_{st} p_{j\cdot}^Y, \quad \text{for all } i,j \in \bar{I} \text{ with } i \leq j. \tag{25}$$

(ii) X is skip-free to the right and lower-regular, and
$$q_i^X \leq q_i^Y \quad \text{and} \quad p_{i\cdot}^X \leq_{st} p_{i\cdot}^Y, \quad \text{for all } i \in \bar{I}. \tag{26}$$

As next stated, these results were further improved in [16,18] by means of an adequate modulated adaptive uniformization of the CTMCs.

Theorem 7 (Ferreira and Pacheco ([16, Theorem 5.1], [18, Theorem 5])). *Let X and Y be CTMCs with state space I, order-isomorphic to some bounded or unbounded interval of $\mathbb{Z}$, and generator matrices Q^X and Q^Y, respectively. Then $X \leq_{lc} Y$ if either*

(i) *X and Y are lower-regular and there exists a matrix $\bar{\beta} = (\beta_{i,j})_{i,j \in \bar{I}}$, with entries in $(0, 1]$, such that*
$$\sum_{n \geq m} q_{in}^X \leq \beta_{i,j} \sum_{n \geq m} q_{jn}^Y, \text{ for all } i,j \in \bar{I} \text{ such that } i \leq j \text{ and } (m \leq i \text{ or } m > j). \tag{27}$$

(ii) *X and Y are lower-regular, X is skip-free to the right, and there exists a vector $\bar{\alpha} = (\alpha_i)_{i \in \bar{I}}$, with entries in $(0, 1]$, such that*
$$\sum_{n \geq m} q_{in}^X \leq \alpha_i \sum_{n \geq m} q_{in}^Y, \quad \text{for all } i \in \bar{I} \text{ and } m \in I. \tag{28}$$

Figure 7 presents two algorithms for the simulation of level crossing ordered CTMCs: one for general CTMCs satisfying (i), and the other for the case in which the slower CTMC is skip-free to the right and the CTMCs satisfy (ii). Specifically, the algorithm presented on the left-hand side [right-hand side] of Figure 7 simulates, under the conditions (i) [(ii)], two CTMCs X and Y such that $X \leq_{lc} Y$. Once again, these algorithms are proposed based on a sample-path based coupling proofs of these results, provided in [16,18], of which we next give a brief sketch.

The first construction uses two dependent modulated Poisson uniformization processes with rates modulated by the states of the two processes at appropriately chosen times, and generates independently the transitions in both processes from a common generator sequence. Namely, if at the time of occurrence of the nth event of the modulated Poisson uniformization process associated to X (Y) the process X (Y) goes to state i (j), then, the amounts of time X and Y stay in states i and j until the next events take place in the modulated Poisson uniformization processes associated to X and Y are generated from a common generator and have exponential distributions with rates λ_{ij}^X and λ_{ij}^Y, respectively, such that λ_{ij}^Y is greater or equal to $2\max\{q_j^Y, q_i^X/\beta_{ij}, 1\}$ and $\lambda_{ij}^X = \beta_{ij}\lambda_{ij}^Y$, so that
$$0 < \lambda_{ij}^X = \beta_{ij}\lambda_{ij}^Y \leq \lambda_{ij}^Y, \quad \text{for all } i \in \bar{I} \wedge j \in I \text{ such that } i \leq j. \tag{29}$$

Moreover, the probability vectors of the state the processes X and Y go to after those events occur are

$$\hat{p}_{i\cdot}^{(X,j)} = \mathbf{e}_i + \frac{q_{i\cdot}^X}{\lambda_{ij}^X} \quad \text{and} \quad \hat{p}_{j\cdot}^{(Y,i)} = \mathbf{e}_j + \frac{q_{j\cdot}^Y}{\lambda_{ij}^Y}$$

respectively. Since, under conditions (i), we have $\lambda_{ij}^X \geq \lambda_{ij}^Y$ and $\hat{p}_{i\cdot}^{(X,j)} \leq_{st} \hat{p}_{j\cdot}^{(Y,i)}$, whenever $i \leq j < \sup I$, the conditions of Theorem 6 are satisfied, and thus the procedure described guarantees that $X \leq_{lc} Y$.

If, in addition, the slower CTMC is skip-free to the right, then instead of comparing the upper sums of different rows of the generator matrices of the two CTMCs, as (27) imposes, we may compare only the upper sums for the same rows of the generator matrices of the two CTMCs, as stated in (28). In this case, the procedure to simulate lc-ordered CTMCs is presented on the right-hand side of Figure 7 and is based on two uniformizing Poisson processes with state dependent rates modulated by the states of the process itself, such that, whenever X (Y) is in state i (j) the uniformizing Poisson process of X (Y) has rate $\beta_i \lambda_i$ (λ_j), where

$$\lambda_k = \max\{q_k^Y, q_k^X/\beta_k, 1\}. \tag{30}$$

The amounts of time X and Y stay in states i and j until the next events of the corresponding modulated Poisson uniformization processes take place are generated from a common generator and have exponential distributions with rates $\beta_i \lambda_i$ and λ_j, respectively, and the probability vectors of the state the processes X and Y go to after those events occur are

$$\hat{p}_{i\cdot}^X = \mathbf{e}_i + \frac{q_{i\cdot}^X}{\beta_i \lambda_i} \quad \text{and} \quad \hat{p}_{j\cdot}^Y = \mathbf{e}_j + \frac{q_{j\cdot}^Y}{\lambda_j}.$$

Since, under conditions (ii), we have $\beta_i \lambda_i \leq \lambda_i$ and $\hat{p}_{i\cdot}^X \leq_{st} \hat{p}_{i\cdot}^Y$, for all $i \in \bar{I}$, i.e., the conditions of Theorem 5 are satisfied, it follows that $X \leq_{lc} Y$.

We end this section noting that, uniformizing the CTMCs X and Y with two possibly different constant (i.e., non state-dependent) uniformization rates, say α^X and α^Y, respectively, such that $\alpha = \alpha^X/\alpha^Y \leq 1$, then conditions (28) specialize into

$$\exists \alpha \in (0,1]: \sum_{m \geq n} q_{im}^X \leq \alpha \sum_{m \geq n} q_{im}^Y, \quad \text{for all } i \in \bar{I} \text{ and } n \in I.$$

This result was first achieved in ([17], Theorem 3.1) by observing that the level-crossing ordering is stochastically monotone increasing with respect to time clock speed-ups, and constitutes itself a first generalization of ([20], Corollary 2.1) which establishes the same conclusion for two lower-uniformizable skip-free to the right CTMCs with the constant α taking the value one.

It is important to note also that the sufficient conditions for the level-crossing ordering of two general CTMCs are weaker but related to Kirstein's (19) sufficient conditions for the stochastic ordering of CTMCs in the usual sense. In fact, if such conditions are valid for $\beta_{i,j} = 1$, then the CTMCs, departing from a common state, will be also ordered in the usual sense.

Input: independent sequences of independent Unif$(0,1)$ random variables $(U_n)_{n\in\mathbb{N}}$ and $(A_n)_{n\in\mathbb{N}_+}$, the matrix $(\beta_{ij})_{i,j\in I}$, and a positive value TMAX

for $(i,j \in I)$ **do**

$$\lambda_{ij} := 2\max\{q_i^Y, q_i^X/\beta_{ij}, 1\}$$
$$\hat{p}_{i\cdot}^{(X,j)} := \mathbf{e}_i + \frac{q_i^X}{\beta_{ij}\lambda_{ij}}$$
$$\hat{p}_{j\cdot}^{(i,Y)} := \mathbf{e}_j + \frac{q_j^Y}{\lambda_{ij}}$$

end for

$Z_0^X := Z_0^Y := \mathbf{p}^{-1}(U_0)$
$S_0^X := S_0^Y := 0$
$n := 0$

while $\left(\min\{S_n^X, S_n^Y\} < \text{TMAX}\right)$

  $n := n+1$
  $Z_n^X := \left[\hat{P}_{Z_{n-1}^X\cdot}^{(X,Z_{n-1}^Y)}\right]^{-1}(U_n)$
  $Z_n^Y := \left[\hat{P}_{Z_{n-1}^Y\cdot}^{(Y,Z_{n-1}^X)}\right]^{-1}(U_n)$
  $S_n^X := S_{n-1}^X + \left[\mathrm{Exp}(\beta_{Z_{n-1}^X Z_{n-1}^Y}\lambda_{Z_{n-1}^X Z_{n-1}^Y})\right]^{-1}(A_n)$
  $S_n^Y := S_{n-1}^Y + \left[\mathrm{Exp}(\lambda_{Z_{n-1}^X Z_{n-1}^Y})\right]^{-1}(A_n)$

end while

Output:
$X_t := Z_k^X$ for $S_k^X \leq t < S_{k+1}^X$, $0 \leq k < n$
$Y_t := Z_k^Y$ for $S_k^Y \leq t < S_{k+1}^Y$, $0 \leq k < n$

Input: independent sequences of independent Unif$(0,1)$ random variables $(U_n)_{n\in\mathbb{N}}$, $(V_n)_{n\in\mathbb{N}_+}$, $(A_n)_{n\in\mathbb{N}_+}$ and $(B_n)_{n\in\mathbb{N}_+}$, the vector $(\beta_i)_{i\in I}$, and a value TMAX

for $(i \in I)$ **do**

$$\lambda_i := \max\{q_i^Y, q_i^X/\beta_i\}$$
$$\hat{p}_{i\cdot}^X := \mathbf{e}_i + \frac{q_i^X}{\beta_i\lambda_i}$$
$$\hat{p}_{i\cdot}^Y := \mathbf{e}_i + \frac{q_i^Y}{\lambda_i}$$

end for

$Z_0^X := Z_0^Y := \mathbf{p}^{-1}(U_0)$
$S_0^X := S_0^Y := 0$
$k := 0$

while $(S_k^Y < \text{TMAX})$

  $k := k+1$
  $Z_k^Y := \left[\hat{P}_{Z_{k-1}^Y\cdot}^Y\right]^{-1}(U_k)$
  $S_k^Y := S_{k-1}^Y + \left[\mathrm{Exp}(\lambda_{Z_{k-1}^Y})\right]^{-1}(A_k)$

end while

$S_0^X := 0$, $m := 0$, $n := 0$

while $(S_n^X < \text{TMAX})$

  $m := m+1$
  $n := n+1$
  $Z_n^X := \left[\hat{P}_{Z_{n-1}^X\cdot}^X\right]^{-1}(U_m)$
  $S_n^X := S_{n-1}^X + \left[\mathrm{Exp}(\beta_{Z_{n-1}^X}\lambda_{Z_{n-1}^X})\right]^{-1}(A_m)$

  while $(Z_n^X < Z_m^Y \ \& \ S_n^X < \text{TMAX})$

    $n := n+1$
    $Z_n^X := \left[\hat{P}_{Z_{n-1}^X\cdot}^X\right]^{-1}(V_n)$
    $S_n^X := S_{n-1}^X + \left[\mathrm{Exp}(\beta_{Z_{n-1}^X}\lambda_{Z_{n-1}^X})\right]^{-1}(B_n)$

  end while
end while

Output:
$X_t := Z_l^X$ for $S_l^X \leq t < S_{l+1}^X$, $0 \leq l < n$
$Y_t := Z_l^Y$ for $S_l^Y \leq t < S_{l+1}^Y$, $0 \leq l < k$

Fig. 7. Algorithm for the simulation of two lc-ordered CTMCs with initial probability vector $\mathbf{p}$, under conditions: (27), on the left-hand side; and, (28), on the right-hand side

5.5 Some Applications

As an illustration of the applicability of the results presented in the previous section, we apply those results to Poisson shock models and birth-and-death processes (with possible catastrophes) to derive sets of sufficient conditions for the level-crossing ordering of such processes.

Level-crossing ordering of two birth-and-death processes with catastrophes. Let I be a subset of $\mathbb{N}$, $\lambda = (\lambda_i)_{i \in I}$, $\mu = (\mu_i)_{i \in I}$ and $\beta = (\beta_i)_{i \in I}$ be nonnegative vectors such that $\lambda_{\sup I} = 0$ if I is bounded above and $\mu_{\inf I} = \beta_{\inf I} = 0$, and $C = (c_{ij})_{i,j \in I}$ be a lower-triangular stochastic matrix.

A $(I, \lambda, \mu, \beta, C)$ birth-and-death process with catastrophes (BDC process) is a skip-free to the right CTMC with state space I and generator matrix Q, where

$$q_{ij} = \beta_i c_{ij} + \mu_i \delta_{j,i-1} + \lambda_i \delta_{j,i+1}, \quad i \neq j. \tag{31}$$

In such processes, the nonnegative parameters λ_i, μ_i and β_i are interpreted as the birth, death and catastrophe rates of the process in state i. In addition, the matrix C is seen as the catastrophe probability matrix with c_{ij} denoting the probability that the state resulting from a catastrophe taking place in state i is j.

A direct application of Theorem 7 to BDC processes leads to the following set of sufficient conditions for their level-crossing ordering.

Theorem 8. *For $W = X, Y$, let W be an $(I, \lambda^W, \mu^W, \beta^W, C^W)$ BDC process. Then:*

(i) $X \leq_{lc} Y$ *provided that, for some vector $\bar{\alpha} = (\alpha_i)_{i \in \bar{I}}$ with entries in $(0,1]$, the following conditions hold*

$$\lambda_i^X \leq \alpha_i \lambda_i^Y \wedge \mu_i^X \geq \alpha_i \mu_i^Y \wedge \beta_i^X \geq \alpha_i \beta_i^Y, \quad \text{for all } i \in \bar{I} \tag{32}$$

$$c_{i\cdot}^X \leq_{st} c_{i\cdot}^Y, \quad \text{for all } i \in \bar{I}. \tag{33}$$

(ii) $X \leq_{st} Y$ *provided that the following conditions hold*

$$\lambda_j^X \leq \lambda_j^Y \wedge \mu_i^X \geq \mu_m^Y \wedge \beta_i^X \geq \beta_m^Y, \quad \text{for all } j \text{ and } i \leq m \tag{34}$$

$$c_{i\cdot}^X \leq_{st} c_{j\cdot}^Y, \quad \text{for all } i, j \in \bar{I} \text{ such that } i \leq j. \tag{35}$$

Important types of catastrophe families are described, e.g., in [5] and [12]. These include: Binomial(p), $0 \leq p \leq 1$; Geometric(p), $0 \leq p \leq 1$; Uniform; Deterministic(f), where $f = (f_i)_{i \in I}$ is a vector such that $f_i \leq i$, for all $i \in I$; and Total. Some details on the catastrophe probability matrices associated to each of these types of catastrophe families are given in Table 1. In the following we use the denotation of the type of catastrophe indistinctly of the associated catastrophe probability matrix; thus we write, e.g., $C = \text{Binomial}(p)$ whenever C is the catastrophe probability matrix of Binomial(p) catastrophes.

Table 1. Important types of catastrophe families

Type of catastrophe	c_{ij} $(0 \leq j \leq i)$	$\sum_{j=0}^{k} c_{ij}$ $(0 \leq k \leq i)$
Binomial (p), $p \in [0,1]$	$\binom{i}{j} p^{i-j}(1-p)^j$	$\sum_{j=0}^{k} \binom{i}{j} p^{i-j}(1-p)^j$
Geometric (p), $p \in [0,1]$	$p^i \delta_{j0} + (1-p)p^{i-j} \mathbf{1}_{\{j>0\}}$	p^{i-k}
Uniform	$1/(i+1)$	$(k+1)/(i+1)$
Deterministic(f), $0 \leq f_i \leq i$	δ_{jf_i}	$\mathbf{1}_{\{k \geq f_i\}}$
Total	δ_{j0}	1

Table 2. Some ordering relations associated to catastrophe probability matrices

C^X	C^Y	$c_{i\cdot}^X \leq_{st} c_{i\cdot}^Y$	$C^X \leq_K C^Y$
Binomial (p_1)	Binomial (p_2)	$p_1 \geq p_2$	$p_1 \geq p_2$
Geometric (p_1)	Geometric (p_2)	$p_1 \geq p_2$	$p_1 \geq p_2$
Binomial (p_1)	Geometric (p_2)	$p_1 \geq p_2$	$p_1 \geq p_2$
Uniform	Uniform	yes	yes
Deterministic(f^X)	Deterministic(f^Y)	$f_i^X \leq f_i^Y$	$f^X \leq f^Y$ and $f^X \uparrow$
Total	arbitrary	yes	yes

Table 2 presents some situations where the ordering relations (33) and (35) involving catastrophe probability matrices hold, which are relevant for the use of Theorem 8. Note that, in particular, binomial and geometric catastrophes [[5] and [12]] are stochastically decreasing in the parameter and total catastrophes [6] are the smallest catastrophes, in sense of both (33) and (35).

Nothe that, Theorem 8 (i) implies that BDC processes stochastically increase in the level-crossing ordering sense as the catastrophe distribution in each state increases stochastically in the usual sense. That is: BDC processes, X and Y, that share the birth, death and catastrophe rates but have different catastrophe probability matrices, C^X and C^Y, satisfy $X \leq_{lc} Y$ provided that $c_{i\cdot}^X \leq_{st} c_{i\cdot}^Y$ for all $i \in \bar{I}$. Thus, e.g., BDC processes with binomial catastrophes stochastically decrease in level-crossing with the catastrophe probability. Note also that if two BDC processes, X and Y, have the same catastrophe probability matrix, it suffices to show that (32) holds to conclude that $X \leq_{lc} Y$.

Level-crossing ordering of two birth-and-death processes. Particular consequences of Theorem 8 follow for the level-crossing ordering of birth-and-death processes (BD processes), i.e., BDC processes $(I, \lambda, \mu, \beta, C)$ with null vector β (which we denote by (I, λ, μ, C) BD processes), thus turning irrelevant the form of the catastrophe probability matrix C.

Corollary 4. *For $W = X, Y$, let W be an (I, λ^W, μ^W) BD process. If*

$$\lambda_i^X \leq \alpha_i \lambda_i^Y \quad \wedge \quad \mu_i^X \geq \alpha_i \mu_i^Y, \quad \text{for all } i \in \bar{I} \tag{36}$$

for some vector $\bar{\alpha} = (\alpha_i)_{i \in \bar{I}}$ with entries in $(0,1]$, then $X \leq_{lc} Y$. Moreover, if X and Y are irreducible, with conditions (36) holding for $\alpha_i = \alpha$, $i \in \bar{I}$, for some constant $\alpha \in (0,1)$, then the same conclusion is obtained if (36) is replaced by

$$\sup_{i \neq \sup I} \frac{\lambda_i^X}{\lambda_i^Y} \leq \alpha \leq \inf_{i \neq \inf I, \sup I} \frac{\mu_i^X}{\mu_i^Y}. \tag{37}$$

We note that for irreducible BD processes both Kirstein's conditions (19), for the st-ordering, and Irle's conditions, for the lc-ordering and derived in [20], given respectively by

$$\lambda_i^X \leq \lambda_i^Y \quad \text{and} \quad \mu_i^X \geq \mu_i^Y, \quad i \in \bar{I}$$

and

$$\lambda_i^X + \mu_i^X \leq \lambda_i^Y + \mu_i^Y \quad \text{and} \quad \frac{\mu_i^X}{\lambda_i^X} \geq \frac{\mu_i^Y}{\lambda_i^Y}, \quad i \in \bar{I},$$

are not equivalent to (36).

As the number of customers in a $M/M/s/c$ system [see, e.g., [19]] with arrival rate η and death rate γ, where the system capacity c may be either finite or infinite, can be seen as a BD processes on $\mathbb{N}$ having birth rates $\lambda_i = \eta \mathbf{1}_{\{0 \leq i \leq c-1\}}$ and death rates $\mu_i = \gamma \min(i, s)$, Corollary 4 applies directly to derive sufficient conditions for the level-crossing ordering of two $M/M/s/c$ systems, as follows.

Corollary 5 (Ferreira and Pacheco ([17, Corollary 4.2])). *For $W = X, Y$, let W denote the number of customers in an $M/M/s^W/c^W$ system with arrival rate λ^W and service rate μ^W. If $c^X \leq c^Y$ and*

$$\frac{\lambda^X}{\lambda^Y} \leq \alpha \leq \frac{\mu^X}{\mu^Y} \min\left(1, \frac{s^X}{s^Y}\right) \tag{38}$$

for some $\alpha \in (0,1]$, then $X \leq_{lc} Y$.

References

1. Arnold, B.C.: Majorization and the Lorenz Order: A Brief Introduction. Springer, Berlin (1987)
2. Asmussen, S.: Applied Probability and Queues, 2nd edn. Springer, New York (2003)
3. Baccelli, F., Brémaud, P.: Elements of Queueing Theory: Palm-Martingale Calculus and Stochastic Recurrences, 2nd edn. Springer, Berlin (2003)
4. Bawa, V.S.: Stochastic dominance: a research bibliography. Management Science 28(6), 698–712 (1982)
5. Brockwell, P.J., Gani, J., Resnick, S.I.: Birth, immigration and catastrophe processes. Advances in Applied Probability 14(4), 709–731 (1982)
6. Chao, X., Zheng, Y.: Transient analysis of immigration birth-death processes with total catastrophes. Probability in the Engineering and Informational Sciences 17(1), 83–106 (2003)
7. Çinlar, E.: Markov renewal theory. Advances in Applied Probability 1, 123–187 (1969)

8. Çinlar, E.: On semi-Markov processes on arbitrary spaces. Proc. Cambridge Philos. Soc. 66, 381–392 (1969)
9. Çinlar, E.: Introduction to Stochastic Processes. Prentice-Hall, Englewood Cliffs (1975)
10. Di Crescenzo, A., Ricciardi, L.M.: Comparing first-passage times for semi-Markov skip-free processes. Statistics and Probability Letters 30(3), 247–256 (1996)
11. Disney, R.L., Kiessler, P.C.: Traffic Processes in Queueing Networks: A Markov Renewal Approach. Johns Hopkins, Baltimore (1987)
12. Economou, A., Fakinos, D.: A continuous-time Markov chain under the influence of a regulating point process and applications in stochastic models with catastrophes. European Journal of Operational Research 149(3), 625–640 (2003)
13. El-Taha, M., Stidham Jr., S.: Sample-Path Analysis of Queueing Systems. Kluwer, Boston (1999)
14. Feller, W.: On semi-Markov processes. Proc. Nat. Acad. Sci. U.S.A. 51, 653–659 (1964)
15. Ferreira, F.: Embedding, Uniformization and Stochastic Ordering in the Analysis of Level-Crossing Times and $GI^X/M(n)//c$ Systems. Ph.D thesis, Instituto Superior Técnico - Technical University of Lisbon, Lisbon, Portugal (2007)
16. Ferreira, F., Pacheco, A.: Level-crossing ordering of semi-Markov processes and Markov chains. Journal of Applied Probability 42(4), 989–1002 (2005)
17. Ferreira, F., Pacheco, A.: Level-crossing ordering of skip-free to the right continuous time Markov chains. Journal of Applied Probability 42(1), 52–60 (2005)
18. Ferreira, F., Pacheco, A.: Comparison of level crossing times for Markov and semi-Markov processes. Statistics and Probability Letters 77(2), 151–157 (2007)
19. Gross, D., Harris, C.M.: Fundamentals of Queueing Theory, 3rd edn. Wiley, Chichester (1998)
20. Irle, A.: Stochastic ordering for continuous-time processes. Journal of Applied Probability 40(2), 361–375 (2003)
21. Irle, A., Gani, J.: The detection of words and an ordering for Markov chains. Journal of Applied Probability 38A, 66–77 (2001)
22. Janssen, J., Limnios, N. (eds.): Semi-Markov Models and Applications. Kluwer, Dordrecht (1999)
23. Joe, H.: Multivariate Models and Dependence Concepts. Chapman and Hall, London (1997)
24. Kalbfleisch, J.D., Prentice, R.L.: The Statistical Analysis of Failure Time Data. Wiley, New York (1980)
25. Kamae, T., Krengel, U., O'Brien, G.L.: Stochastic inequalities on partially ordered spaces. The Annals of Probability 5(6), 899–912 (1977)
26. Kijima, M.: Markov Processes for Stochastic Modeling. Chapman and Hall, London (1997)
27. Kijima, M., Ohnishi, M.: Stochastic orders and their applications in financial optimization. Mathematical Methods of Operations Research 50(2), 351–372 (1999)
28. Kirstein, B.M.: Monotonicity and comparability of time-homogeneous Markov processes with discrete state space. Mathematische Operations Forschung und Statistik 7, 151–168 (1976)
29. Kulkarni, V.G.: Modeling and Analysis of Stochastic Systems. Chapman and Hall, London (1995)
30. Last, G., Brandt, A.: Marked Point Processes on the Real Line: The Dynamic Approach. Springer, New York (1995)
31. Levy, H.: Stochastic dominance and expected utility: survey and analysis. Management Science 38(5), 555–593 (1992)

32. Levy, P.: Processus semi-markoviens. In: Proceedings of the International Congress of Mathematicians 1954, Amsterdam, vol. III, pp. 416–426. Erven P. Noordhoff N.V., Groningen (1956)
33. Limnios, N., Oprişan, G.: Semi-Markov Processes and Reliability. Birkhäuser, Boston (2001)
34. Lindvall, T.: Comparisons between certain queuing systems. Probability in the Engineering and Informational Sciences 16(1), 1–17 (2002)
35. Lindvall, T.: Lectures on the Coupling Method. Dover Publications, Mineola (2002)
36. Marshall, A.W., Olkin, I.: Inequalities: Theory of Majorization and its Applications. Academic Press, New York (1979)
37. Massey, W.A.: Stochastic orderings for Markov processes on partially ordered spaces. Mathematics of Operations Research 12(2), 350–367 (1987)
38. Morais, M.J.C.: Stochastic Ordering in the Performance Analysis of Quality Control Schemes. PhD thesis, Instituto Superior Técnico, Technical University of Lisbon, Portugal (2002)
39. Mosler, K., Scarsini, M.: Stochastic Orders and Applications: A Classified Bibliography. Springer, Heidelberg (1993)
40. Müller, A., Stoyan, D.: Comparison Methods for Stochastic Models and Risks. Wiley, Chichester (2002)
41. Pyke, R.: Markov renewal processes: definitions and preliminary properties. The Annals of Mathematical Statistics 32, 1231–1242 (1961)
42. Pyke, R.: Markov renewal processes with finitely many states. The Annals of Mathematical Statistics 32, 1243–1259 (1961)
43. Ross, S.M.: Stochastic Processes, 2nd edn. Wiley, New York (1996)
44. Shaked, M., Shanthikumar, J.G.: Stochastic Orders and Their Applications. Academic Press, San Diego (1994)
45. Smith, W.L.: Regenerative stochastic processes. Proc. Roy. Soc. London. Ser. A. 232, 6–31 (1955)
46. Smith, W.L.: Renewal theory and its ramifications. J. Roy. Statist. Soc. Ser. B 20, 243–302 (1958)
47. Sonderman, D.: Comparing semi-Markov processes. Mathematics of Operations Research 5(1), 110–119 (1980)
48. Stoyan, D.: Comparison Methods for Queues and Other Stochastic Models. Wiley, Chichester (1983)
49. Szekli, R.: Stochastic Ordering and Dependence in Applied Probability. Springer, New York (1995)
50. Thorisson, H.: Coupling, Stationarity, and Regeneration. Springer, New York (2000)
51. Tong, Y.L.: Probability Inequalities in Multivariate Distributions. Academic Press, New York (1980)
52. van Doorn, E.A.: Stochastic Monotonicity and Queueing Applications of Birth-Death Processes. Springer, New York (1981)

Spectral Expansion Solutions for Markov-Modulated Queues

Isi Mitrani

School of Computing Science, Newcastle University, NE1 7RU, UK

Abstract. This tutorial deals with the solution of a large class of models where the behaviour of an unbounded queue is influenced by the evolution of a Markovian environment. The latter, in turn, may be affected by the state of the queue. Several examples of such models, with applications in the fields of computing, communication and manufacturing, are given. The spectral expansion method for obtaining exact solutions is described. A simple and easily computable approximation which is asymptotically exact in heavy traffic is also presented. Some illustrative examples are included.

Keywords: Queueing models, Markovian environment, Exact solutions, Approximations.

1 General Model

There are many computer, communication and manufacturing systems which give rise to queueing models where the arrival and/or service mechanisms are influenced by some external processes. In such models, a single unbounded queue evolves in an environment which changes state from time to time. The instantaneous arrival and service rates may depend on the state of the environment and also, to a limited extent, on the number of jobs present.

The system state at time t is described by a pair of integer random variables, (I_t, J_t), where I_t represents the state of the environment and J_t is the number of jobs present. The variable I_t takes a finite number of values, numbered $0, 1, \ldots, N$; these are also called the environmental *phases*. The possible values of J_t are $0, 1, \ldots$. Thus, the system is in state (i, j) when the environment is in phase i and there are j jobs waiting and/or being served.

The two-dimensional process $X = \{(I_t, J_t) \,;\, t \geq 0\}$ is assumed to have the Markov property, i.e. given the current phase and number of jobs, the future behaviour of X is independent of its past history. Such a model is referred to as a *Markov-modulated queue*. The corresponding state space, $\{0, 1, \ldots, N\} \times \{0, 1, \ldots\}$ is known as a *lattice strip*.

A fully general Markov-modulated queue, with arbitrary state-dependent transitions, is not tractable. However, one can consider a sub-class of models which are sufficiently general to be useful, and yet can be solved efficiently. Those models satisfy the following restrictions:

D. Kouvatsos (Ed.): Next Generation Internet, LNCS 5233, pp. 423–446, 2011.
© Springer-Verlag Berlin Heidelberg 2011

(i) There is a threshold M, such that the instantaneous transition rates out of state (i,j) do not depend on j when $j \geq M$.
(ii) the jumps of the random variable J are bounded.

When the jumps of the random variable J are of size 1, i.e. when jobs arrive and depart one at a time, the process is said to be of the *Quasi-Birth-and-Death* type, or QBD (the term *skip-free* is also used (Latouche et al., [10]).

The requirement that all transition rates cease to depend on the size of the job queue beyond a certain threshold is not too restrictive. Note that there is no limit on the magnitude of the threshold M, although it must be pointed out that the larger M is, the greater the complexity of the solution. Similarly, although jobs may arrive and/or depart in fixed or variable (but bounded) batches, the larger the batch size, the more complex the solution.

The object of the analysis of a Markov-modulated queue is to determine the joint steady-state distribution of the environmental phase and the number of jobs in the system:

$$p_{i,j} = \lim_{t \to \infty} P(I_t = i, J_t = j) \; ; \; i = 0, 1, \ldots, N \; ; \; j = 0, 1, \ldots . \tag{1}$$

That distribution exists for an irreducible Markov process if, and only if, the corresponding set of balance equations has a positive solution that can be normalized.

The marginal distributions of the number of jobs in the system, and of the phase, can be obtained from the joint distribution:

$$p_{\cdot,j} = \sum_{i=0}^{N} p_{i,j} . \tag{2}$$

$$p_{i,\cdot} = \sum_{j=0}^{\infty} p_{i,j} . \tag{3}$$

Various performance measures can then be computed in terms of these joint and marginal distributions.

The following are some examples of systems that are modelled as Markov-modulated queues.

1.1 A Multiserver Queue with Breakdowns and Repairs

A single, unbounded queue is served by N identical parallel servers (Mitrani and Avi-Itzhak, [12], Neuts and Lucantoni, [16]). Each server goes through alternating periods of being operative and inoperative, independently of the others and of the number of jobs in the system. The operative and inoperative periods are distributed exponentially with parameters ξ and η, respectively. Thus, the number of operative servers at time t, I_t, is a Markov process on the state space $\{0, 1, \ldots, N\}$. This is the environment in which the queue evolves: it is in phase i when there are i operative servers.

Jobs arrive according to a Poisson process, with a rate which may depend on the state of the environment, I_t. That is, when there are i operative servers, the instantaneous arrival rate is λ_i. Jobs are taken for service from the front of the queue, one at a time, by available operative servers. The required service times are distributed exponentially with parameter μ. An operative server cannot be idle if there are jobs waiting to be served. A job whose service is interrupted by a server breakdown is returned to the front of the queue. When an operative server becomes available, the service is resumed from the point of interruption, without any switching overheads.

The process $X = \{(I_t, J_t)\,;\, t \geq 0\}$ is of the Quasi-Birth-and-Death type. The transitions out of state (i, j) are:

(a) to state $(i-1, j)$ $(i > 0)$, with rate $i\xi$;
(b) to state $(i+1, j)$ $(i < N)$, with rate $(N-i)\eta$;
(c) to state $(i, j+1)$ with rate λ_i;
(d) to state $(i, j-1)$ with rate $\min(i, j)\mu$.

Note that only transition (d) has a rate which depends on j, and that dependency vanishes when $j \geq N$.

Remark. The breakdown and repair processes could be generalized without destroying the QBD nature of the process. For example, the servers could break down and be repaired in batches, or a server breakdown could trigger a job departure. The environmental state transitions can be arbitrary, as long as the queue changes in steps of size 1.

In this example, as in all models where the environment state transitions do not depend on the number of jobs present, the marginal distribution of the number of operative servers can be determined without finding the joint distribution first. Moreover, since the servers break down and are repaired independently of each other, that distribution is binomial:

$$p_{i,\cdot} = \binom{N}{i} \left(\frac{\eta}{\xi+\eta}\right)^i \left(\frac{\xi}{\xi+\eta}\right)^{N-i} \;;\; i = 0, 1, \ldots, N\,. \qquad (4)$$

Hence, the steady-state average number of operative servers is equal to

$$E(X_t) = \frac{N\eta}{\xi+\eta}\,. \qquad (5)$$

The overall average arrival rate is equal to

$$\lambda = \sum_{i=0}^{N} p_{i,\cdot} \lambda_i\,. \qquad (6)$$

This gives us an explicit condition for stability. The offered load must be less than the processing capacity:

$$\frac{\lambda}{\mu} < \frac{N\eta}{\xi+\eta}\,. \qquad (7)$$

1.2 Manufacturing Blocking

Consider a network of two nodes in tandem, such as the one in figure 1 (Buzacott and Shanthikumar, [1], Konheim and Reiser, [9]). Jobs arrive into the first node in a Poisson stream with rate λ, and join an unbounded queue. After completing service at node 1 (exponentially distributed with parameter μ), they attempt to go to node 2, where there is a finite buffer with room for a maximum of $N-1$ jobs (including the one in service). If that transfer is impossible because the buffer is full, the job remains at node 1, preventing its server from starting a new service, until the completion of the current service at node 2 (exponentially distributed with parameter ξ). In this last case, server 1 is said to be 'blocked'. Transfers from node 1 to node 2 are instantaneous.

Fig. 1. Two nodes with a finite intermediate buffer

The above type of blocking is referred to as 'manufacturing blocking'. (An alternative model, which also gives rise to a Markov-modulated queue, is the 'communication blocking'. There node 1 *does not start* a service if the node 2 buffer is full.)

In this system, the unbounded queue at node 1 is modulated by a finite-state environment defined by node 2. We say that the environment, I_t, is in state i if there are i jobs at node 2 and server 1 is not blocked ($i = 0, 1, \ldots, N-1$). An extra state, $I_t = N$, is needed to describe the situation where there are $N-1$ jobs at node 2 and server 1 is blocked.

The above assumptions imply that the pair $X = \{(I_t, J_t)\,;\, t \geq 0\}$, where J_t is the number of jobs at node 1, is a QBD process. Note that the state $(N, 0)$ does not exist: node 1 may be blocked only if there are jobs present.

The transitions out of state (i, j) are:

(a) to state $(i-1, j)$ $(0 < i < N)$, with rate ξ;
(b) to state $(N-1, j-1)$ $(i = N,\ j > 0)$, with rate ξ;
(c) to state $(i+1, j-1)$ $(0 \leq i < N-1,\ j > 0)$, with rate μ;
(d) to state (N, j) $(i = N-1,\ j > 0)$, with rate μ;
(e) to state $(i, j+1)$ with rate λ.

The only dependency on j comes from the fact that transitions (b), (c) and (d) are not available when $j = 0$. In this example, the j-independency threshold is $M = 1$. Note that the state $(N, 0)$ is not reachable: node 1 may be blocked only if there are jobs present.

1.3 Phase-Type Distributions

There is a large and useful family of distributions that can be incorporated into queueing models by means of Markovian environments (Neuts, [15]). Those distributions are 'almost' general, in the sense that any distribution function either belongs to this family or can be approximated as closely as desired by functions from it.

Let I_t be a Markov process with state space $\{0, 1, \ldots, N\}$ and generator matrix $\tilde{A}$. States $0, 1, \ldots, N-1$ are transient, while state N, reachable from any of the other states, is absorbing (the last row of $\tilde{A}$ is 0). At time 0, the process starts in state i with probability α_i ($i = 0, 1, \ldots, N-1$; $\alpha_1 + \alpha_2 + \ldots + \alpha_{N-1} = 1$). Eventually, after an interval of length T, it is absorbed in state N. The random variable T is said to have a 'phase-type' (PH) distribution with parameters $\tilde{A}$ and α_i.

The exponential distribution is obviously phase-type ($N = 1$). So is the Erlang distribution—the convolution of N exponentials. The corresponding generator matrix is

$$\tilde{A} = \begin{bmatrix} -\mu & \mu & & & \\ & -\mu & \mu & & \\ & & \ddots & \ddots & \\ & & & -\mu & \mu \\ & & & & 0 \end{bmatrix},$$

and the initial probabilities are $\alpha_0 = 1$, $\alpha_1 = \ldots = \alpha_{N-1} = 0$.

Another common PH distribution is the 'hyperexponential', where $I_0 = i$ with probability α_i, and absorbtion occurs at the first transition. The generator matrix of the hyperexponential distribution is

$$\tilde{A} = \begin{bmatrix} -\mu_0 & & & \mu_0 \\ & -\mu_1 & & \mu_1 \\ & & \ddots & \vdots \\ & & -\mu_{N-1} & \mu_{N-1} \\ & & & 0 \end{bmatrix}.$$

The corresponding probability distribution function, $F(x)$, is a mixture of exponentials:

$$F(x) = 1 - \sum_{i=0}^{N-1} \alpha_i e^{-\mu_i x}.$$

The PH family is very versatile. It contains distributions with both low and high coefficients of variation. It is closed with respect to mixing and convolution: if X_1 and X_2 are two independent PH random variables with N_1 and N_2 (non-absorbing) phases respectively, and c_1 and c_2 are constants, then $c_1 X_1 + c_2 X_2$ has a PH distribution with $N_1 + N_2$ phases.

A model with a single unbounded queue, where either the interarrival intervals, or the service times, or both, have PH distributions, is easily cast in the framework of a queue in Markovian environment. Consider, for instance, the M/PH/1 queue. Its state at time t can be represented as a pair (I_t, J_t), where J_t is the number of jobs present and I_t is the phase of the current service (if $J_t > 0$). When I_t has a transition into the absorbing state, the current service completes and (if the queue is not empty) a new service starts immediately, entering phase i with probability α_i.

The PH/PH/n queue can also be represented as a QBD process. However, the state of the environmental variable, I_t, now has to indicate the phase of the current interarrival interval and the phases of the current services at all busy servers. If the interarrival interval has N_1 phases and the service has N_2 phases, the state space of I_t would be of size $N_1 N_2^n$.

1.4 Checkpointing and Recovery in the Presence of Faults

The last example is not a QBD process. Consider a system where transactions, arriving according to a Poisson process with rate λ, are served in FIFO order by a single server. The service times are i.i.d. random variables distributed exponentially with parameter μ. After N consecutive transactions have been completed, the system performs a checkpoint operation whose duration is an i.i.d. random variable distributed exponentially with parameter β. Once a checkpoint is established, the N completed transactions are deemed to have departed. However, both transaction processing and checkpointing may be interrupted by the occurrence of a fault. The latter arrive according to an independent Poisson process with rate ξ. When a fault occurs, the system instantaneously rolls back to the last established checkpoint; all transactions which arrived since that moment either remain in the queue, if they have not been processed, or return to it, in order to be processed again (it is assumed that repeated service times are resampled independently).

This system can be modelled as an unbounded queue of (uncompleted) transactions, which is modulated by an environment consisting of completed transactions and checkpoints. More precisely, the two state variables, $I(t)$ and $J(t)$, are the number of transactions that have completed service since the last checkpoint, and the number of transactions present that have not completed service (including those requiring re-processing), respectively.

The Markov-modulated queueing process $X = \{[I(t), J(t)] \,;\, t \geq 0\}$, has the following transitions out of state (i, j):

(a) to state $(0, j + i)$, with rate ξ;
(b) to state $(0, j)$ $(i = N)$, with rate β;
(c) to state $(i, j + 1)$, with rate λ;
(d) to state $(i + 1, j - 1)$ $(0 \leq i < N,\ j > 0)$, with rate μ;

Because transitions (a), resulting from arrivals of faults, cause the queue size to jump by more than 1, this is not a QBD process.

2 Spectral Expansion Solution

Let us now turn to the problem of determining the steady-state joint distribution of the environmental phase and the number of jobs present, for a Markov-modulated queue. The solution method that we shall present is called 'Spectral Expansion', for reasons that will become apparent.

We shall start with the most commonly encountered case, namely the QBD process, where jobs arrive and depart singly. The starting point is of course the set of balance equations which the probabilities $p_{i,j}$, defined in 1, must satisfy. In order to write them in general terms, the following notation for the instantaneous transition rates will be used.

(a) Phase transitions leaving the queue unchanged: from state (i,j) to state (k,j) $(0 \leq i, k \leq N\,;\, i \neq k)$, with rate $a_j(i,k)$;
(b) Transitions incrementing the queue: from state (i,j) to state $(k, j+1)$ $(0 \leq i, k \leq N)$, with rate $b_j(i,k)$;
(c) Transitions decrementing the queue: from state (i,j) to state $(k, j-1)$ $(0 \leq i, k \leq N\,;\, j > 0)$, with rate $c_j(i,k)$.

It is convenient to introduce the $(N+1) \times (N+1)$ matrices containing the rates of type (a), (b) and (c): $A_j = [a_j(i,k)]$, $B_j = [b_j(i,k)]$ and $C_j = [c_j(i,k)]$, respectively (the main diagonal of A_j is zero by definition; also, $C_0 = 0$ by definition). According to the assumptions of the Markov-modulated queue, there is a threshold, M ($M \geq 1$), such that those matrices do not depend on j when $j \geq M$. In other words,

$$A_j = A\,;\ B_j = B\,;\ C_j = C\,,\ j \geq M\,. \tag{8}$$

Note that transitions (b) may represent a job arrival coinciding with a change of phase. If arrivals are not accompanied by such changes, then the matrices B_j and B are diagonal. Similarly, a transition of type (c) may represent a job departure coinciding with a change of phase. Again, if such coincidences do not occur, then the matrices C_j and C are diagonal.

By way of illustration, here are the transition rate matrices for the model of the multiserver queue with breakdowns and repairs. In this case the phase transitions are independent of the queue size, so the matrices A_j are all equal:

$$A_j = A = \begin{bmatrix} 0 & N\eta & & & & \\ \xi & 0 & (N-1)\eta & & & \\ & 2\xi & 0 & & \ddots & \\ & & \ddots & & \ddots & \eta \\ & & & & N\xi & 0 \end{bmatrix}.$$

Similarly, the matrices B_j do not depend on j:

$$B = \begin{bmatrix} \lambda_0 & & & \\ & \lambda_1 & & \\ & & \ddots & \\ & & & \lambda_N \end{bmatrix}.$$

Denoting
$$\mu_{i,j} = \min(i,j)\mu \; ; \; i = 0, 1, \ldots, N \; ; \; j = 1, 2, \ldots,$$
the departure rate matrices, C_j, can thus be written as
$$C_j = \begin{bmatrix} 0 & & & \\ & \mu_{1,j} & & \\ & & \ddots & \\ & & & \mu_{N,j} \end{bmatrix} \; ; \; j = 1, 2, \ldots,$$

These matrices cease to depend on j when $j \geq N$. Thus, the threshold M is now equal to N, and
$$C = \begin{bmatrix} 0 & & & \\ & \mu & & \\ & & \ddots & \\ & & & N\mu \end{bmatrix}.$$

2.1 Balance Equations

Using the instantaneous transition rates introduced above, the balance equations of a general QBD process can be written as

$$p_{i,j} \sum_{k=0}^{N} [a_j(i,k) + b_j(i,k) + c_j(i,k)]$$

$$= \sum_{k=0}^{N} [p_{k,j} a_j(k,i) + p_{k,j-1} b_{j-1}(k,i) + p_{k,j+1} c_{j+1}(k,i)], \quad (9)$$

where $p_{i,-1} = b_{-1}(k,i) = c_0(i,k) = 0$ by definition. The left-hand side of (9) gives the total average number of transitions out of state (i,j) per unit time (due to changes of phase, arrivals and departures), while the right-hand side expresses the total average number of transitions into state (i,j) (again due to changes of phase, arrivals and departures). These balance equations can be written more compactly by using vectors and matrices. Define the row vectors of probabilities corresponding to states with j jobs in the system:

$$\mathbf{v}_j = (p_{0,j}, p_{1,j}, \ldots, p_{N,j}) \; ; \; j = 0, 1, \ldots. \quad (10)$$

Also, let D_j^A, D_j^B and D_j^C be the diagonal matrices whose ith diagonal element is equal to the ith row sum of A_j, B_j and C_j, respectively. Then equations (9), for $j = 0, 1, \ldots$, can be written as:

$$\mathbf{v}_j [D_j^A + D_j^B + D_j^C] = \mathbf{v}_{j-1} B_{j-1} + \mathbf{v}_j A_j + \mathbf{v}_{j+1} C_{j+1}, \quad (11)$$

where $\mathbf{v}_{-1} = \mathbf{0}$ and $D_0^C = B_{-1} = 0$ by definition.

When j is greater than the threshold M, the coefficients in (11) cease to depend on j:

$$\mathbf{v}_j[D^A + D^B + D^C] = \mathbf{v}_{j-1}B + \mathbf{v}_j A + \mathbf{v}_{j+1}C, \qquad (12)$$

for $j = M+1, M+2, \ldots$.

In addition, all probabilities must sum up to 1:

$$\sum_{j=0}^{\infty} \mathbf{v}_j \mathbf{e} = 1, \qquad (13)$$

where $\mathbf{e}$ is a column vector with $N+1$ elements, all of which are equal to 1.

The first step is to find the general solution of the infinite set of balance equations with constant coefficients, (12). The latter are normally written in the form of a homogeneous vector difference equation of order 2:

$$\mathbf{v}_j Q_0 + \mathbf{v}_{j+1} Q_1 + \mathbf{v}_{j+2} Q_2 = \mathbf{0}; \quad j = M, M+1, \ldots, \qquad (14)$$

where $Q_0 = B$, $Q_1 = A - D^A - D^B - D^C$ and $Q_2 = C$.

Associated with equation (14) is the so-called 'characteristic matrix polynomial', $Q(x)$, defined as

$$Q(x) = Q_0 + Q_1 x + Q_2 x^2. \qquad (15)$$

Denote by x_k and $\mathbf{u}_k$ the 'generalized eigenvalues', and corresponding 'generalized left eigenvectors', of $Q(x)$. In other words, these are quantities which satisfy

$$det[Q(x_k)] = 0,$$

$$\mathbf{u}_k Q(x_k) = \mathbf{0}; \quad k = 1, 2, \ldots, d, \qquad (16)$$

where $det[Q(x)]$ is the determinant of $Q(x)$ and d is its degree. In what follows, the qualification *generalized* will be omitted.

The above eigenvalues do not have to be simple, but it is assumed that if one of them has multiplicity m, then it also has m linearly independent left eigenvectors. This tends to be the case in practice. So, the numbering in (16) is such that each eigenvalue is counted according to its multiplicity.

It is readily seen that if x_k and $\mathbf{u}_k$ are any eigenvalue and corresponding left eigenvector, then the sequence

$$\mathbf{v}_{k,j} = \mathbf{u}_k x_k^j; \quad j = M, M+1, \ldots, \qquad (17)$$

is a solution of equation (14). Indeed, substituting (17) into (14) we get

$$\mathbf{v}_{k,j} Q_0 + \mathbf{v}_{k,j+1} Q_1 + \mathbf{v}_{k,j+2} Q_2 = x_k^j \mathbf{u}_k [Q_0 + Q_1 x_k + Q_2 x_k^2] = \mathbf{0}.$$

By combining any multiple eigenvalues with each of their independent eigenvectors, we thus obtain d linearly independent solutions of (14). On the other hand, it is known that there cannot be more than d linearly independent solutions (Gohberg et al., [7]). Therefore, any solution of (14) can be expressed as a linear combination of the d solutions (17):

$$\mathbf{v}_j = \sum_{k=1}^{d} \alpha_k \mathbf{u}_k x_k^j \; ; \; j = M, M+1, \ldots , \qquad (18)$$

where α_k ($k = 1, 2, \ldots, d$), are arbitrary (complex) constants.

However, the only solutions that are of interest in the present context are those which can be normalized to become probability distributions. Hence, it is necessary to select from the set (18), those sequences for which the series $\sum \mathbf{v}_j \mathbf{e}$ converges. This requirement implies that if $|x_k| \geq 1$ for some k, then the corresponding coefficient α_k must be 0.

So, suppose that c of the eigenvalues of $Q(x)$ are strictly inside the unit disk (each counted according to its multiplicity), while the others are on the circumference or outside. Order them so that $|x_k| < 1$ for $k = 1, 2, \ldots, c$. The corresponding independent eigenvectors are $\mathbf{u}_1$, $\mathbf{u}_2$, ..., $\mathbf{u}_c$. Then any normalizable solution of equation (14) can be expressed as

$$\mathbf{v}_j = \sum_{k=1}^{c} \alpha_k \mathbf{u}_k x_k^j \; ; \; j = M, M+1, \ldots , \qquad (19)$$

where α_k ($k = 1, 2, \ldots, c$), are some constants.

The set of eigenvalues of the matrix polynomial $Q(x)$ is called its 'spectrum'. Hence, expression (19) is referred to as the 'spectral expansion' of the vectors $\mathbf{v}_j$. The coefficients of that expansion, α_k, are yet to be determined.

Note that if there are non-real eigenvalues in the unit disk, then they appear in complex-conjugate pairs. The corresponding eigenvectors are also complex-conjugate. The same must be true for the appropriate pairs of constants α_k, in order that the right-hand side of (19) be real. To ensure that it is also positive, the real parts of x_k, $\mathbf{u}_k$ and α_k should be positive.

So far, expressions have been obtained for the vectors $\mathbf{v}_M$, $\mathbf{v}_{M+1}$, ...; these contain c unknown constants. Now it is time to consider the balance equations (11), for $j = 0, 1, \ldots, M$. This is a set of $(M+1)(N+1)$ linear equations with $M(N+1)$ unknown probabilities (the vectors $\mathbf{v}_j$ for $j = 0, 1, \ldots, M-1$), plus the c constants α_k. However, only $(M+1)(N+1) - 1$ of these equations are linearly independent, since the generator matrix of the Markov process is singular. On the other hand, an additional independent equation is provided by (13).

In order that this set of linearly independent equations has a unique solution, the number of unknowns must be equal to the number of equations, i.e. $(M+1)(N+1) = M(N+1) + c$, or $c = N + 1$. This observation implies the following rather general result.

Proposition 1. *The QBD process has a steady-state distribution if, and only if, the number of eigenvalues of $Q(x)$ strictly inside the unit disk, each counted according to its multiplicity, is equal to the number of states of the Markovian* environment, $N+1$. Then, assuming that the eigenvectors of multiple eigenvalues are linearly independent, the spectral expansion solution of (12) has the form

$$\mathbf{v}_j = \sum_{k=1}^{N+1} \alpha_k \mathbf{u}_k x_k^j \ ; \ j = M, M+1, \ldots \ . \tag{20}$$

In summary, the spectral expansion solution procedure consists of the following steps:

1. Compute the eigenvalues of $Q(x)$, x_k, inside the unit disk, and the corresponding left eigenvectors $\mathbf{u}_k$. If their number is other than $N+1$, stop; a steady-state distribution does not exist.
2. Solve the finite set of linear equations (11), for $j = 0, 1, \ldots, M$, and (13), with $\mathbf{v}_M$ and $\mathbf{v}_{M+1}$ given by (20), to determine the constants α_k and the vectors $\mathbf{v}_j$ for $j < M$.
3. Use the obtained solution in order to determine various moments, marginal probabilities, percentiles and other system performance measures that may be of interest.

Careful attention should be paid to step 1. The 'brute force' approach which relies on first evaluating the scalar polynomial $det[Q(x)]$, then finding its roots, may be very inefficient for large N. An alternative which is preferable in most cases is to reduce the quadratic eigenvalue-eigenvector problem

$$\mathbf{u}[Q_0 + Q_1 x + Q_2 x^2] = \mathbf{0} \ , \tag{21}$$

to a linear one of the form $\mathbf{u}Q = x\mathbf{u}$, where Q is a matrix whose dimensions are twice as large as those of Q_0, Q_1 and Q_2. The latter problem is normally solved by applying various transformation techniques. Efficient routines for that purpose are available in most numerical packages.

This linearization can be achieved quite easily if the matrix $C = Q_2$ is non-singular (Jennings, [8]). Indeed, after multiplying (21) on the right by Q_2^{-1}, it becomes

$$\mathbf{u}[H_0 + H_1 x + I x^2] = \mathbf{0} \ , \tag{22}$$

where $H_0 = Q_0 C^{-1}$, $H_1 = Q_1 C^{-1}$, and I is the identity matrix. By introducing the vector $\mathbf{y} = x\mathbf{u}$, equation (22) can be rewritten in the equivalent linear form

$$[\mathbf{u}, \mathbf{y}] \begin{bmatrix} 0 & -H_0 \\ I & -H_1 \end{bmatrix} = x[\mathbf{u}, \mathbf{y}] \ . \tag{23}$$

If C is singular but B is not, a similar linearization is achieved by multiplying (21) on the right by B^{-1} and making a change of variable $x \to 1/x$. Then the relevant eigenvalues are those outside the unit disk.

If both B and C are singular, then the desired result is achieved by first making a change of variable, $x \to (\gamma + x)/(\gamma - x)$, where the value of γ is chosen so that the matrix $S = \gamma^2 Q_2 + \gamma Q_1 + Q_0$ is non-singular. In other words, γ can have any value which is not an eigenvalue of $Q(x)$. Having made that change of variable, multiplying the resulting equation by S^{-1} on the right reduces it to the form (22).

The computational demands of step 2 may be high if the threshold M is large. However, if the matrices B_j ($j = 0, 1, \ldots, M-1$) are non-singular (which is often the case in practice), then the vectors $\mathbf{v}_{M-1}, \mathbf{v}_{M-2}, \ldots, \mathbf{v}_0$ can be expressed in terms of $\mathbf{v}_M$ and $\mathbf{v}_{M+1}$, with the aid of equations (11) for $j = M, M-1, \ldots, 1$. One is then left with equations (11) for $j = 0$, plus (13) (a total of $N+1$ independent linear equations), for the $N+1$ unknowns x_k.

Having determined the coefficients in the expansion (20) and the probabilities $p_{i,j}$ for $j < N$, it is easy to compute performance measures. The steady-state probability that the environment is in state i is given by

$$p_{i,\cdot} = \sum_{j=0}^{M-1} p_{i,j} + \sum_{k=1}^{N+1} \alpha_k u_{k,i} \frac{x_k^M}{1 - x_k}, \qquad (24)$$

where $u_{k,i}$ is the ith element of $\mathbf{u}_k$.

The conditional average number of jobs in the system, L_i, given that the environment is in state i, is obtained from

$$L_i = \frac{1}{p_{i,\cdot}} \left[\sum_{j=1}^{M-1} j p_{i,j} + \sum_{k=1}^{N+1} \alpha_k u_{k,i} \frac{x_k^M (M - M x_k + x_k)}{(1 - x_k)^2} \right]. \qquad (25)$$

The overall average number of jobs in the system, L, is equal to

$$L = \sum_{i=0}^{N} p_{i,\cdot} L_i . \qquad (26)$$

2.2 Batch Arrivals and/or Departures

Consider now a Markov-modulated queue which is not a QBD process, i.e. one where the queue size jumps may be bigger than 1. As before, the state of the process at time t is described by the pair (I_t, J_t), where I_t is the state of the environment (the operational mode) and J_t is the number of jobs in the system. The state space is the lattice strip $\{0, 1, \ldots, N\} \times \{0, 1, \ldots\}$. The variable J_t may jump by arbitrary, but bounded amounts in either direction. In other words, the allowable transitions are:

(a) Phase transitions leaving the queue unchanged: from state (i, j) to state (k, j) ($0 \leq i, k \leq N$; $i \neq k$), with rate $a_j(i, k)$;
(b) Transitions incrementing the queue by s: from state (i, j) to state $(k, j+s)$ ($0 \leq i, k \leq N$; $1 \leq s \leq r_1$; $r_1 \geq 1$), with rate $b_{j,s}(i, k)$;
(c) Transitions decrementing the queue by s: from state (i, j) to state $(k, j-s)$ ($0 \leq i, k \leq N$; $1 \leq s \leq r_2$; $r_2 \geq 1$), with rate $c_{j,s}(i, k)$,

provided of course that the source and destination states are valid.

Obviously, if $r_1 = r_2 = 1$ then this is a Quasi-Birth-and-Death process.

Denote by $A_j = [a_j(i,k)]$, $B_{j,s} = [b_{j,s}(i,k)]$ and $C_{j,s} = [c_{j,s}(i,k)]$, the transition rate matrices associated with (a), (b) and (c), respectively. There is a threshold M, such that

$$A_j = A \; ; \; B_{j,s} = B_s \; ; \; C_{j,s} = C_s \; ; \; j \geq M \; . \tag{27}$$

Defining again the diagonal matrices D^A, D^{B_s} and D^{C_s}, whose ith diagonal element is equal to the ith row sum of A, B_s and C_s, respectively, the balance equations for $j > M + r_1$ can be written in a form analogous to (12):

$$\mathbf{v}_j[D^A + \sum_{s=1}^{r_1} D^{B_s} + \sum_{s=1}^{r_2} D^{C_s}] = \sum_{s=1}^{r_1} \mathbf{v}_{j-s} B_s + \mathbf{v}_j A + \sum_{s=1}^{r_2} \mathbf{v}_{j+s} C_s \; . \tag{28}$$

Similar equations, involving A_j, $B_{j,s}$ and $C_{j,s}$, together with the corresponding diagonal matrices, can be written for $j \leq M + r_1$.

As before, (28) can be rewritten as a vector difference equation, this time of order $r = r_1 + r_2$, with constant coefficients:

$$\sum_{\ell=0}^{r} \mathbf{v}_{j+\ell} Q_\ell = \mathbf{0} \; ; \; j \geq M \; . \tag{29}$$

Here, $Q_\ell = B_{r_1-\ell}$ for $\ell = 0, 1, \ldots r_1 - 1$,

$$Q_{r_1} = A - D^A - \sum_{s=1}^{r_1} D^{B_s} - \sum_{s=1}^{r_2} D^{C_s} \; ,$$

and $Q_\ell = C_{\ell - r_1}$ for $\ell = r_1 + 1, r_1 + 2, \ldots r_1 + r_2$.

The spectral expansion solution of this equation is obtained from the characteristic matrix polynomial

$$Q(x) = \sum_{\ell=0}^{r} Q_\ell x^\ell \; . \tag{30}$$

The solution is of the form

$$\mathbf{v}_j = \sum_{k=1}^{c} \alpha_k \mathbf{u}_k x_k^j \; ; \; j = M, M+1, \ldots \; , \tag{31}$$

where x_k are the eigenvalues of $Q(x)$ in the interior of the unit disk, $\mathbf{u}_k$ are the corresponding left eigenvectors, and α_k are constants ($k = 1, 2, \ldots, c$). These constants, together with the the probability vectors $\mathbf{v}_j$ for $j < M$, are determined with the aid of the state-dependent balance equations and the normalizing equation.

There are now $(M + r_1)(N + 1)$ so-far-unused balance equations (the ones where $j < M + r_1$), of which $(M + r_1)(N + 1) - 1$ are linearly independent, plus one normalizing equation. The number of unknowns is $M(N+1) + c$ (the vectors $\mathbf{v}_j$ for $j = 0, 1, \ldots, M - 1$), plus the c constants α_k. Hence, there is a unique solution when $c = r_1(N + 1)$.

Proposition 2. *The Markov-modulated queue has a steady-state distribution if, and only if, the number of eigenvalues of $Q(x)$ strictly inside the unit disk, each counted according to its multiplicity, is equal to the number of states of the Markovian environment, $N+1$, multiplied by the largest arrival batch, r_1. Then, assuming that the eigenvectors of multiple eigenvalues are linearly independent, the spectral expansion solution of (28) has the form*

$$\mathbf{v}_j = \sum_{k=1}^{r_1*(N+1)} \alpha_k \mathbf{u}_k x_k^j \; ; \; j = M, M+1, \ldots . \tag{32}$$

For computational purposes, the polynomial eigenvalue-eigenvector problem of degree r can be transformed into a linear one. For example, suppose that Q_r is non-singular and multiply (29) on the right by Q_r^{-1}. This leads to the problem

$$\mathbf{u}\left[\sum_{\ell=0}^{r-1} H_\ell x^\ell + I x^r\right] = \mathbf{0}, \tag{33}$$

where $H_\ell = Q_\ell Q_r^{-1}$. Introducing the vectors $\mathbf{y}_\ell = x^\ell \mathbf{u}$, $\ell = 1, 2, \ldots, r-1$, one obtains the equivalent linear form

$$[\mathbf{u}, \mathbf{y}_1, \ldots, \mathbf{y}_{r-1}] \begin{bmatrix} 0 & & & -H_0 \\ I & 0 & & -H_1 \\ & \ddots & \ddots & \\ & & I & -H_{r-1} \end{bmatrix} = x[\mathbf{u}, \mathbf{y}_1, \ldots, \mathbf{y}_{r-1}].$$

As in the quadratic case, if Q_r is singular then the linear form can be achieved by an appropriate change of variable.

3 A Simple Approximation

The spectral expansion solution can be computationally expensive. Its numerical complexity depends crucially on the number of environmental phases: that number determines the number of eigenvalues and eigenvectors that have to be evaluated, and influences the size of the set of simultaneous linear equations that have to be solved. Moreover, when N is large, there may be numerical problems concerned with ill-conditioned matrices. In some cases, both the complexity and the numerical stability of the solution are adversely affected when the system is heavily loaded.

For these reasons, it may be worth abandoning the exact solution, if one can develop a reasonable approximation which is simple, easy to implement, robust and computationally cheap. Such an approximation can be extracted from the spectral expansion solution. The idea is to use a 'restricted' expansion, based on a single eigenvalue and its associated eigenvector. The eigenvalue provides a geometric approximation for the queue size distribution, while the eigenvector approximates the distribution of the environmental phase.

An attractive feature of the geometric approximation is that its accuracy improves when the offered load increases. In the heavy-traffic limit, i.e. when the system approaches saturation, the approximation becomes asymptotically exact.

In order to keep the presentation simple, the discussion will be restricted to QBD Markov-modulated queues whose solution is given by Proposition 1, with simple eigenvalues. However, the applicability of the proposed approximation is much more general.

A central role in the approximation is played by the largest eigenvalue that appears in (20), and its left eigenvector. Assume, without loss of generality, that the eigenvalues are numbered in increasing order of modulus, so that the largest is x_{N+1}. When the queue is stable, x_{N+1} is real and positive. Moreover, it has a positive eigenvector. From now on, x_{N+1} will be referred to as the 'dominant eigenvalue', and will be denoted by γ.

The expression (20) implies that *the tail* of the joint distribution of the queue size and the environmental phase is approximately geometrically distributed, with parameter equal to the dominant eigenvalue, γ. To see that, divide both sides of (20) by γ^j and let $j \to \infty$. Since γ is strictly greater in modulus than all other eigenvalues, all terms in the summation vanish, except one:

$$\lim_{j \to \infty} \frac{\mathbf{v}_j}{\gamma^j} = \alpha_{N+1} \mathbf{u}_{N+1} . \qquad (34)$$

In other words, when j is large,

$$\mathbf{v}_j \approx \alpha_{N+1} \mathbf{u}_{N+1} \gamma^j . \qquad (35)$$

This product form implies that when the queue is large, its size is approximately independent of the environmental phase. The tail of the marginal distribution of the queue size is approximately geometric:

$$p_{.,j} \approx \alpha_{N+1} (\mathbf{u}_{N+1} \cdot \mathbf{1}) \gamma^j , \qquad (36)$$

where $\mathbf{1}$ is the column vector defined in (13).

These results suggest seeking an approximation of the form

$$\mathbf{v}_j = \alpha \mathbf{u}_{N+1} \gamma^j , \qquad (37)$$

where α is some constant.

Note that γ and $\mathbf{u}_{N+1}$ can be computed without having to find *all* eigenvalues and eigenvectors. There are techniques for determining the eigenvalues that are near a given number. Here we are dealing with the eigenvalue that is nearest to but strictly less than 1.

If (37) is applied to all $\mathbf{v}_j$, for $j = 0, 1, \ldots$, then the approximation depends on just one unknown constant, α. Its value is determined by (13) alone, and the expressions for $\mathbf{v}_j$ become

$$\mathbf{v}_j = \frac{\mathbf{u}_{N+1}}{(\mathbf{u}_{N+1} \cdot \mathbf{1})} (1 - \gamma) \gamma^j \ ; \ j = 0, 1, \ldots . \qquad (38)$$

This last approximation avoids completely the need to solve a set of linear equations. Hence, it also avoids all problems associated with ill-conditioned matrices. Moreover, it scales well. The complexity of computing γ and $\mathbf{u}_{N+1}$ grows roughly linearly with N when the matrices A, B and C are sparse. The price paid for that convenience is that the balance equations for $j \leq M$ are no longer satisfied.

Despite its apparent over-simplicity, the geometric approximation (38) can be shown to be asymptotically exact when the offered load increases.

3.1 The Heavy Traffic Limit

Consider the case where a parameter associated with arrivals or services changes so that system becomes heavily loaded and approaches saturation. The parameters governing the evolution of the environment are assumed to remain fixed. Then the dominant eigenvalue, γ, is known to approach 1 (Gail et al., [6]). When $\gamma = 1$ (i.e., there is a double eigenvalue at 1), the process $X = \{(I, J)\}$ is recurrent-null; when γ leaves the unit disc, the process is transient. Hence, instead of taking a limit involving a particular parameter, e.g., $\lambda \to \lambda_{max}$ (where λ_{max} is the arrival rate that would saturate the system), we can equivalently treat the heavy-traffic regime in terms of the limit $\gamma \to 1$.

Since there is no equilibrium distribution when X is recurrent-null, we must have
$$\lim_{\gamma \to 1} \mathbf{v}_j = \mathbf{0} \; ; \; j = 0, 1, \ldots \; . \tag{39}$$

Hence, in order to talk sensibly about the 'limiting distribution', some kind of normalization must be applied. Multiply the queue size by $1 - \gamma$ and consider the process $Y = \{[I, J(1 - \gamma)]\}$. The limiting joint distribution of Y will be determined by means of the vector Laplace transform
$$\mathbf{h}(s) = [h_0(s), h_1(s), \ldots, h_N(s)] \; , \tag{40}$$

where
$$h_i(s) = \lim_{\gamma \to 1} E[\delta(I = i) e^{-s(1-\gamma)J}] \; ; \; i = 0, 1, \ldots N \; , \tag{41}$$

and $\delta(B)$ is the indicator of the boolean B: it is equal to 1 if B is true, 0 otherwise. In terms of the vectors $\mathbf{v}_j$, (40) is expressed as
$$\mathbf{h}(s) = \lim_{\gamma \to 1} \sum_{j=0}^{\infty} \mathbf{v}_j e^{-s(1-\gamma)j} \; . \tag{42}$$

The objective will be to show that both the exact distribution, where the vectors $\mathbf{v}_j$ are given by (20), and the geometric approximation, where they are given by (38), have the same limiting distribution.

Consider first the exact distribution. When all eigenvalues are simple, the equations (20) and (39) imply that
$$\lim_{\gamma \to 1} \alpha_k \mathbf{u}_k = \mathbf{0} \; ; \; k = 1, 2, \ldots N + 1 \; . \tag{43}$$

This can be seen by taking $N+1$ consecutive equations (20) and setting their left-hand sides to 0; the Vandermonde matrix involving powers of different eigenvalues is non-singular, and so the only solution is $\alpha_k \mathbf{u}_k = \mathbf{0}$.

On the other hand, since the environmental process has a finite number of states, and since the corresponding transition rates are fixed, the stationary marginal distribution of the environmental phase always exists and has a non-zero limit when $\gamma \to 1$. Denote that limit by the vector $\mathbf{q}$. This is the limiting eigenvector corresponding to the eigenvalue 1; it satisfies the equations

$$\mathbf{q}G = \mathbf{0} \; ; \; (\mathbf{q} \cdot \mathbf{1}) = 1 , \tag{44}$$

where G is the generator matrix of the environmental process. In terms of the matrix polynomial (15), G is the limiting matrix $Q(1) = Q_0 + Q_1 + Q_2$, obtained by replacing the changing traffic parameter with its limit. In particular, if the matrices B and C are diagonal, then $G = A - D^A$.

Hence, we can write

$$\lim_{\gamma \to 1} \sum_{j=0}^{\infty} \mathbf{v}_j = \mathbf{q} . \tag{45}$$

Moreover, in view of (39), equation (45) holds if the lower index of the summation is $j = M$ (or any other non-negative integer), instead of $j = 0$.

Substituting (20) into (45) and changing the lower summation index to $j = M$ yields

$$\lim_{\gamma \to 1} \sum_{k=1}^{N+1} \alpha_k \mathbf{u}_k \frac{x_k^M}{1 - x_k} = \mathbf{q} . \tag{46}$$

However, the first N eigenvalues do not approach 1, while the last one, $x_{N+1} = \gamma$, does. Hence, according to (43), the first N terms in (46) vanish and leave

$$\lim_{\gamma \to 1} \frac{\alpha_{N+1} \mathbf{u}_{N+1}}{1 - \gamma} = \mathbf{q} . \tag{47}$$

Now, substituting (20) into (42), and arguing as for (47), we see that only the term involving the dominant eigenvalue survives:

$$\mathbf{h}(s) = \lim_{\gamma \to 1} \sum_{j=M}^{\infty} e^{-s(1-\gamma)j} \sum_{k=1}^{N+1} \alpha_k \mathbf{u}_k x_k^j$$

$$= \lim_{\gamma \to 1} \sum_{k=1}^{N+1} \alpha_k \mathbf{u}_k \sum_{j=M}^{\infty} x_k^j e^{-s(1-\gamma)j}$$

$$= \lim_{\gamma \to 1} \sum_{k=1}^{N+1} \alpha_k \mathbf{u}_k \frac{x_k^M e^{-s(1-\gamma)M}}{1 - x_k e^{-s(1-\gamma)}}$$

$$= \lim_{\gamma \to 1} \frac{\alpha_{N+1} \mathbf{u}_{N+1}}{1 - \gamma e^{-s(1-\gamma)}} . \tag{48}$$

Combining this with (47) leads to

$$\mathbf{h}(s) = \mathbf{q} \lim_{\gamma \to 1} \frac{1-\gamma}{1 - \gamma e^{-s(1-\gamma)}} = \mathbf{q} \frac{1}{1+s} \,. \tag{49}$$

The last limit follows from L'Hospital's rule. The Laplace transform appearing in the right-hand side of (49) is that of the exponential distribution with mean 1. Thus we have established the following rather general result:

Proposition 3. *In any Markov-modulated queue, in the heavy-traffic limit $\gamma \to 1$, the environmental state I and the normalized queue size $(1-\gamma)J$ are independent of each other. The first has distribution $\mathbf{q}$, while the second is distributed exponentially with mean 1.*

It now remains to compare the limit (49) with the corresponding one for the geometric approximation, (38). Denote the approximate limiting vector Laplace transform by $\hat{\mathbf{h}}(s)$; it is given by (42), with $\mathbf{v}_j$ replaced by the approximations (38):

$$\hat{\mathbf{h}}(s) = \lim_{\gamma \to 1} \frac{\mathbf{u}_{N+1}}{(\mathbf{u}_{N+1} \cdot \mathbf{1})} \sum_{j=0}^{\infty} (1-\gamma)\gamma^j e^{-s(1-\gamma)j}$$

$$= \lim_{\gamma \to 1} \frac{\mathbf{u}_{N+1}}{(\mathbf{u}_{N+1} \cdot \mathbf{1})} \lim_{\gamma \to 1} \frac{1-\gamma}{1 - \gamma e^{-s(1-\gamma)}} = \frac{1}{1+s} \lim_{\gamma \to 1} \frac{\mathbf{u}_{N+1}}{(\mathbf{u}_{N+1} \cdot \mathbf{1})} \,, \tag{50}$$

again using L'Hospital's rule.

The last limit in the right-hand side of (50) is simply the vector $\mathbf{q}$. This can be seen by arguing that the normalized left eigenvector of the eigenvalue γ must approach the normalized left eigenvector of the eigenvalue 1. Alternatively, multiply both sides of (47) by the column vector $\mathbf{1}$:

$$\lim_{\gamma \to 1} \frac{\alpha_{N+1}(\mathbf{u}_{N+1} \cdot \mathbf{1})}{1 - \gamma} = 1 \,. \tag{51}$$

Hence rewrite (47) as

$$\lim_{\gamma \to 1} \frac{\mathbf{u}_{N+1}}{(\mathbf{u}_{N+1} \cdot \mathbf{1})} = \mathbf{q} \,. \tag{52}$$

Thus we have

$$\hat{\mathbf{h}}(s) = \mathbf{q} \frac{1}{1+s} = \mathbf{h}(s) \,. \tag{53}$$

So, in heavy traffic, the geometric approximation is asymptotically exact, in the sense that it yields the same limiting normalized distribution of environmental phase and queue size as the exact solution.

4   Numerical Examples

It is instructive to present some numerical experiments aimed at evaluating the accuracy of the geometric approximation in the context of two different models of Markov-modulated queues. In all cases, the exact values of the performance measures are computed by applying the full spectral expansion solution (20).

The first system examined is the network of two nodes in tandem, with manufacturing blocking at node 1. The model is illustrated in figure 2. The parameters are λ (external arrival rate), μ (service rate at node 1), ξ (service rate at node 2) and N (the storage capacity at node 2 is $N-1$).

In this system, the unbounded queue at node 1 is modulated by a finite-state environment defined by node 2. The environment, I, is in state i if there are i jobs at node 2 and server 1 is not blocked ($i = 0, 1, \ldots, N-1$). An extra state, $I = N$, is needed to describe the situation where there are $N-1$ jobs at node 2 and server 1 is blocked.

The pair $X = \{(I, J)\}$, where J is the number of jobs at node 1, is a QBD process. The transitions out of state (i, j) were given earlier.

Because the environmental process is coupled with the queueing process, the marginal distribution of the former (i.e., the number of jobs at node 2), cannot be determined without finding the joint distribution of I and J. There is no simple expression for the stability condition.

Figure 2 illustrates the close agreement between the exact solution of this model and the geometric approximation (38), when the system is heavily loaded. The performance measure is the average size of the unbounded queue; it is plotted against the arrival rate, λ. The service rates at nodes 1 and 2 are the same. Hence, the busier node 1, the higher the likelihood that the buffer will fill

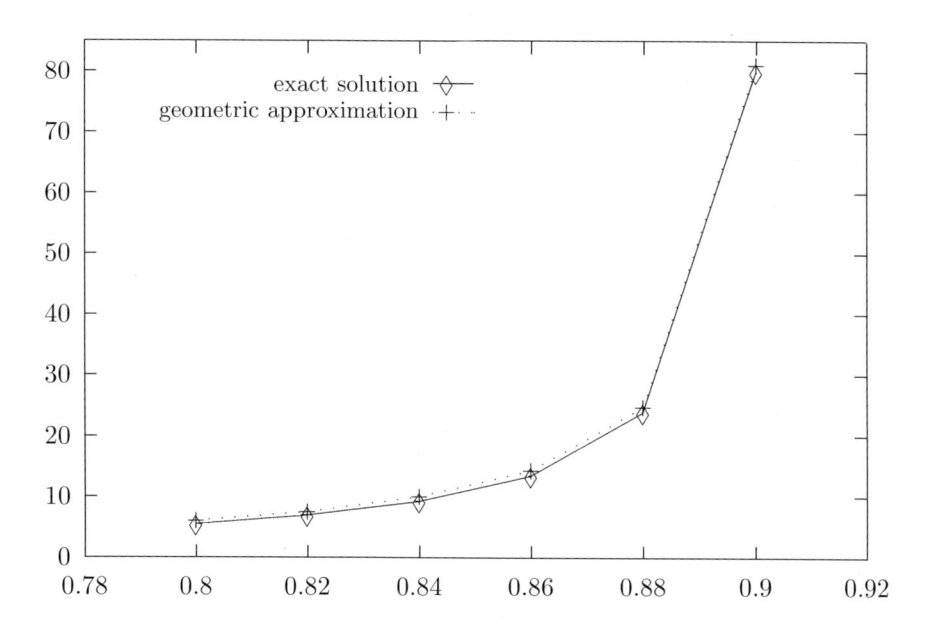

Fig. 2. Manufacturing blocking: Average node 1 queue size against arrival rate
$N = 10, \mu = 1, \xi = 1$

up and cause blocking. Because of that, the saturation point is not at $\lambda = 1$ (as it would be if node 1 was isolated), but at approximately $\lambda = 0.909$.

The geometric approximation for the marginal distribution of the environmental variable, I, indicating the number of jobs at node 2 and whether or not node 1 is blocked, is given by (38) as $\mathbf{q} \approx \mathbf{u}_{N+1}/(\mathbf{u}_{N+1} \cdot \mathbf{1})$. Since there are two environmental states, $I = N-1$ and $I = N$, representing $N-1$ jobs at node 2, the average length of the node 2 queue, L_2, is given by

$$L_2 = \sum_{i=1}^{N-1} i q_i + (N-1) q_N ,$$

where q_i is the $i+1$st element of the vector $\mathbf{q}$.

In figure 3, the average unbounded queue size is plotted against N. Increasing the size of the finite buffer enlarges the environmental state space. Consequently, the exact solution needs to compute more eigenvalues and eigenvectors, and solve larger sets of linear equations.

The accuracy of the geometric approximation is seen to increase with N. This is not really surprising, because enlarging the intermediate buffer reduces the coupling between the two nodes, making them behave more like independent queues. Nevertheless, the exact solution begins to experience numerical difficulties when $N > 35$. The software (Matlab) starts issuing warnings to the effect that the matrix is ill-conditioned, and the results may not be reliable

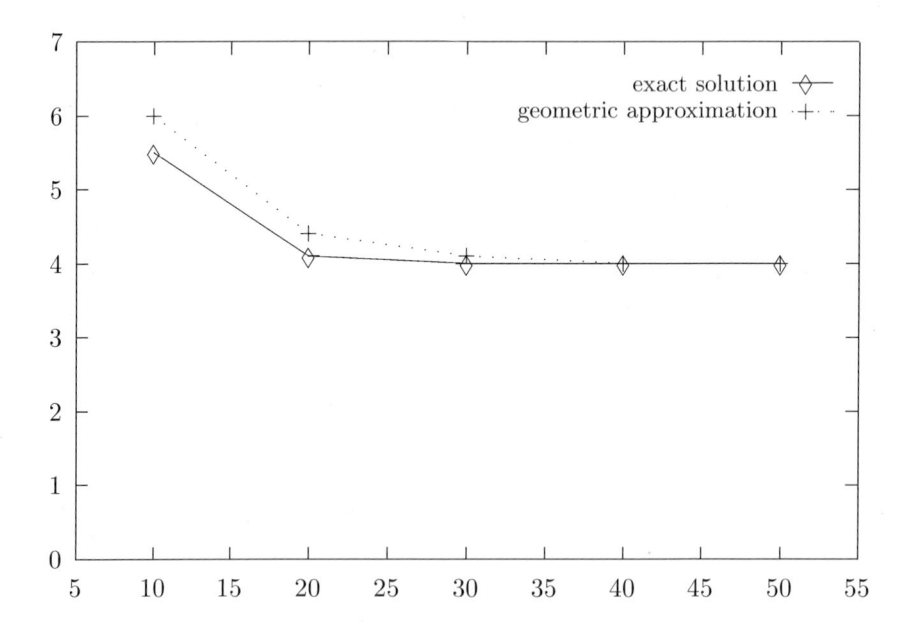

Fig. 3. Manufacturing blocking: Average node 1 queue size against N
$\lambda = 0.8$, $\mu = 1$, $\xi = 1$

(as it happens, the results returned seem fine). Of course the approximation displays no such symptoms, since it has no equations to solve.

The second model to be evaluated is that of the multiserver queue with breakdowns and repairs, described in section 1.1. The parameters are λ (arrival rate; it will be assumed independent of the operative state of the servers), μ (service rate), ξ (breakdown rate), η (repair rate) and N (number of servers. The queue evolves in a Markovian environment which is in phase i ($i = 0, 1, \ldots, N$) when there are i operative servers.

In applying the geometric approximation to this model, there is a choice of approaches. One could use (37) for $j \geq N$, together with the balance equations for $j < N$. This will be referred to as the 'partial geometric' approximation. Alternatively, the geometric approximation (38) can be used for all $j \geq 0$.

Intuitively, the partial geometric approximation can be expected to be more accurate, since it satisfies more of the balance equations. In fact, the results in figure 4 suggest that the opposite is true. The average queue size is plotted against the arrival rate, with parameters chosen so that the system is heavily loaded (the saturation point is $\lambda = 6.666...$). It turns out that the simple geometric approximation is more accurate than the more complex partial geometric one. There seem to be two opposing effects here. On the one hand, relying only on the dominant eigenvalue tends to overestimate the average queue size; on the other hand, the additional approximation introduced by ignoring the boundary balance equations reduces that overestimation.

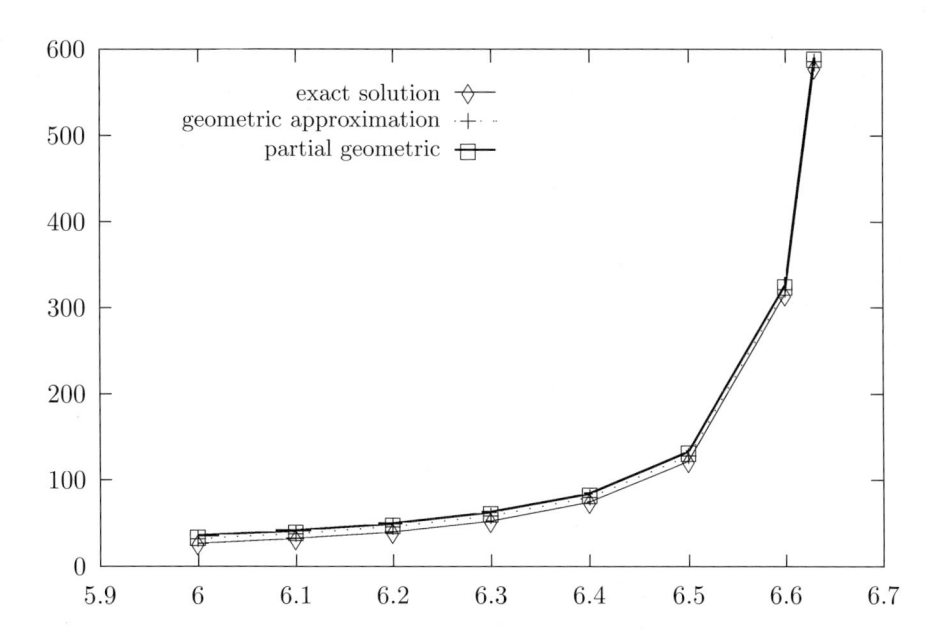

Fig. 4. Breakdowns and repairs: Average queue size against arrival rate
$N = 10$, $\mu = 1$, $\xi = 0.05$, $\eta = 0.1$

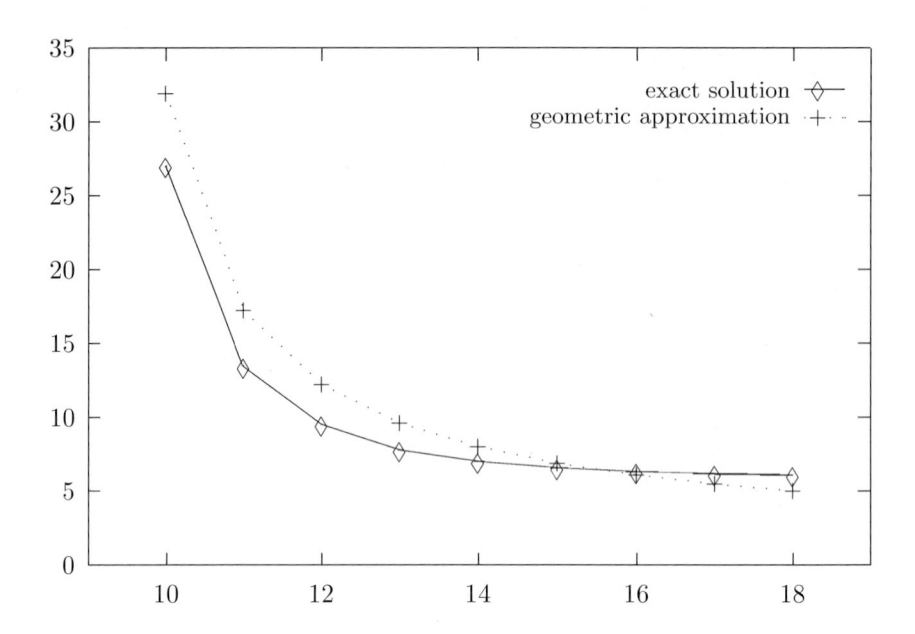

Fig. 5. Breakdowns and repairs: Average queue size against number of servers
$\lambda = 6$, $\mu = 1$, $\xi = 0.05$, $\eta = 0.1$

Since the marginal distribution of the environmental variable I is known to be given by (4), there is not much point in trying to approximate it. However, if the geometric approximation is nevertheless applied, e.g. to compute the average number of operative servers, then it is observed that the approximation improves when λ increases, even though the exact value of the average does not depend on λ.

In figure 5, the average queue size is evaluated for increasing number of servers, and hence decreasing load. This experiment disproves the conjecture that the geometric approximation always overestimates the exact values. Here the approximation starts off as an overestimate, but as N increases, it becomes an underestimate.

As in the previous model, when N becomes large (greater than about 30), the exact solution begins to warn of possible numerical problems due to ill-conditioned matrices; the geometric approximation does not display such symptoms.

5 Literature and Conclusion

The presentation in this tutorial is based on material from [14,13,11]. It is perhaps worth mentioning that there are two other solution techniques that can be used in the context of Markov-modulated queues. These are the matrix-geometric method (Neuts, [15]) and the generating functions method (as applied, for example, in [12]). However, we have chosen to concentrate on the spectral expansion solution method because it is versatile, readily implementable and efficient. A strong case can be made for using it, whenever possible, in preference to the other methods [13]. An additional point in its favour is that it provides the basis for a simple approximate solution.

Other examples of applications of the spectral solution method can be found in [3,4,5]

The geometric approximation is valid for a large class of heavily loaded systems. The arguments presented here do not rely on any particular model structure. One could relax the QBD assumption and allow batch arrivals and departures. As long as there is a spectral expansion solution with finitely many eigenvalues, there would be a single dominant eigenvalue and therefore the geometric approximation would be asymptotically exact in heavy traffic. Moreover, it *may* also be reasonable for moderate and light loads, as the examples in figures 3 and 5 illustrate.

References

1. Buzacott, J.A., Shanthikumar, J.G.: Stochastic Models of Manufacturing Systems. Prentice-Hall, Englewood Cliffs (1993)
2. Daigle, J.N., Lucantoni, D.M.: Queueing systems having phase-dependent arrival and service rates. In: Stewart, W.J. (ed.) Numerical Solutions of Markov Chains. Marcel Dekker, New York (1991)
3. Chakka, R., Do, T.V.: The MM $\sum_{k=1}^{K} CPP_k/GE/c/LG$-queue with Heterogeneous Servers: Steady state solution and an application to performance evaluation. Performance Evaluation 64, 191–209 (2007)
4. Chakka, R., Harrison, P.G.: A Markov modulated multi-server queue with negative customers – The MM CPP/GE/c/L G-queue. Acta Informatica 37, 881–919 (2001)
5. Do, T.V., Chakka, R., Harrison, P.G.: An Integrated Analytical Model for Computation and Comparison of the Throughputs of the UMTS/HSDPA User Equipment Categories. In: Procs, MSWiM 2007 (Modeling, Analysis, and Simulation of Wireless and Mobile Systems), Crete (2007)
6. Gail, H.R., Hantler, S.L., Taylor, B.A.: Spectral analysis of M/G/1 and G/M/1 type Markov chains. Adv. in Appl. Prob. 28, 114–165 (1996)
7. Gohberg, I., Lancaster, P., Rodman, L.: Matrix Polynomials. Academic Press, London (1982)
8. Jennings, A.: Matrix Computations for Engineers and Scientists. Wiley, Chichester (1977)
9. Konheim, A.G., Reiser, M.: A queueing model with finite waiting room and blocking. JACM 23(2), 328–341 (1976)
10. Latouche, G., Jacobs, P.A., Gaver, D.P.: Finite Markov chain models skip-free in one direction. Naval Res. Log. Quart. 31, 571–588 (1984)
11. Mitrani, I.: Approximate Solutions for Heavily Loaded Markov Modulated Queues. Performance Evaluation 62, 117–131 (2005)
12. Mitrani, I., Avi-Itzhak, B.: A many-server queue with service interruptions. Operations Research 16(3), 628–638 (1968)
13. Mitrani, I., Chakka, R.: Spectral expansion solution for a class of Markov models: Application and comparison with the matrix-geometric method. In: Performance Evaluation (1995)

14. Mitrani, I., Mitra, D.: A spectral expansion method for random walks on semi-infinite strips. In: IMACS Symposium on Iterative Methods in Linear Algebra, Brussels (1991)
15. Neuts, M.F.: Matrix Geometric Solutions in Stochastic Models. John Hopkins Press, Baltimore (1981)
16. Neuts, M.F., Lucantoni, D.M.: A Markovian queue with N servers subject to breakdowns and repairs. Management Science 25, 849–861 (1979)

Diffusion Approximation as a Modelling Tool

Tadeusz Czachórski[1] and Ferhan Pekergin[2]

[1] Institute of Theoretical and Applied Informatics
Polish Academy of Sciences
Baltycka 5, 44–100 Gliwice, Poland
tadek@iitis.gliwice.pl
[2] LIPN, Université Paris-Nord, 93430 Villetaneuse, France
pekergin@lipn.univ-paris13.fr

Abstract. Diffusion theory is already a vast domain of knowledge. This tutorial lecture does not cover all results; it presents in a coherent way an approach we have adopted and used in analysis of a series of models concerning evoluation of some traffic control mechanisms in computer, especially ATM, networks. Diffusion approximation is presented from engineer's point of view, stressing its utility and commenting numerical problems of its implementation. Diffusion approximation is a method to model the behavior of a single queueing station or a network of stations. It allows one to include in the model general sevice times, general (also correlated) input streams and to investigate transient states, which, in presence of bursty streams (e.g. of multimedia transfers) in modern networks, are of interest.

Keywords: diffusion approximation, queueing models, performance evaluation, transient analysis.

1 Single G/G/1 Station

1.1 Preliminaries

Let $A(x)$, $B(x)$ denote the interarrival and service time distributions at a service station. The distributions are general but not specified, the method requires only their two first moments. The means are: $E[A] = 1/\lambda$, $E[B] = 1/\mu$ and variances are $\text{Var}[A] = \sigma_A^2$, $\text{Var}[B] = \sigma_B^2$. Denote also squared coefficients of variation $C_A^2 = \sigma_A^2 \lambda^2$, $C_B^2 = \sigma_B^2 \mu^2$. $N(t)$ represents the number of customers present in the system at time t.
Define
$$\tau_k = \sum_{i=1}^{K} a_i,$$
where a_i are time intervals between arrivals. We assume that they are independent and indentically distributed random variables, hence, according to the central limit theorem, distribution of a variable
$$\frac{T_k - k\lambda}{\sigma_A \sqrt{k}}$$

tends to standard normal distribution as $k \to \infty$:

$$\lim_{k \to \infty} P[\frac{T_k - k\lambda}{\sigma_A \sqrt{k}} \leq x] = \Phi(x), \quad \text{where} \quad \Phi(x) = \frac{1}{\sqrt{2\Pi}} \int_{-\infty}^{x} e^{-\frac{\xi^2}{2}} d\xi.$$

hence for a large k: $P[T_k \leq x\sigma_A\sqrt{k} + k/\lambda] \approx \Phi(x)$. Denote $t = x\sigma_A\sqrt{k} + k/\lambda$, or $k = t\lambda - x\sigma_A\sqrt{k}\lambda$ and for large values of k, $k \approx t\lambda$ or $\sqrt{k} \approx \sqrt{t\lambda}$. Denote by $H(t)$ the number of customers arriving to the station during a time t; note that $P[H(t) \geq k] = P[T_k \leq t]$, hence

$$\Phi(x) \approx P[T_k \leq x\sigma_A\sqrt{k} + k/\lambda] = P[H(t) \geq k] =$$
$$= P[H(t) \geq t\lambda - x\sigma_A\sqrt{t\lambda}\,\lambda] = P\left[\frac{H(t) - t\lambda}{\sigma_A\sqrt{t\lambda}\,\lambda} \geq -x\right]$$

As for the normal distribution $\Phi(x) = 1 - \Phi(-x)$, then $P[\xi \geq -x] = P[\xi \leq x]$, and

$$P\left[\frac{H(t) - t\lambda}{\sigma_A\sqrt{t\lambda}\,\lambda} \leq x\right] \approx \Phi(x),$$

that means that the number of customers arriving at the interval of length t (sufficiently long to assure large k) may be approximated by the normal distribution with mean λt and variance $\sigma_A^2 \lambda^3 t$. Similarly, the number of customers served in this time is approximately normally distributed with mean μt and variance $\sigma_B^2 \mu^3 t$, provided that the server is busy all the time. Consequently, the changes of $N(t)$ within interval $[0, t]$, $N(t) - N(0)$, have approximately normal distribution with mean $(\lambda - \mu)t$ and variance $(\sigma_A^2 \lambda^3 + \sigma_B^2 \mu^3)t$.

Diffusion approximation [54,55] replaces the process $N(t)$ by a continuous diffusion process $X(t)$ the incremental changes of wich $dX(t) = X(t + dt) - X(t)$ are normally distributed with the mean βdt and variance αdt, where β, α are coefficients of the diffusion equation

$$\frac{\partial f(x,t;x_0)}{\partial t} = \frac{\alpha}{2} \frac{\partial^2 f(x,t;x_0)}{\partial x^2} - \beta \frac{\partial f(x,t;x_0)}{\partial x} \qquad (1)$$

which defines the conditional pdf $f(x,t;x_0) = P[x \leq X(t) < x + dx \mid X(0) = x_0]$ of $X(t)$. The both processes $X(t)$ and $N(t)$ have normally distributed changes; the choice $\beta = \lambda - \mu$, $\alpha = \sigma_A^2 \lambda^3 + \sigma_B^2 \mu^3 = C_A^2 \lambda + C_B^2 \mu$ ensures the same ratio of time-growth of mean and variance of these distributions.

More formal justification of diffusion approximation is in limit theorems for $G/G/1$ system given by Iglehart and Whitte [42,43]. If $\hat{N}_n$ is a series of random variables deriven from $N(t)$:

$$\hat{N}_n = \frac{N(nt) - (\lambda - \mu)nt}{(\sigma_A^2 \lambda^3 + \sigma_B^2 \mu^3)\sqrt{n}},$$

then the series is weakly convergent (in the sense of distribution) to ξ where $\xi(t)$ is a standard Wiener process (i.e. diffusion process with $\beta = 0$ i $\alpha = 1$) provided that $\varrho > 1$, that means if the system is overloaded and has no equilibrium state. In the case

of $\varrho = 1$ the series $\hat{N}_n$ is convergent to ξ_R. The $\xi_R(t)$ pocess is $\xi(t)$ process limited to half-axis $x > 0$:
$$\xi_R(t) = \xi(t) - \inf\left[\xi(u),\ 0 \le u \le t\right].$$
Service station with $\varrho \ge 1$ does not attein steady-state the number of customers is linearly growing, with fluctuatins around this deterministic trend. For service stations in equilibrium, with $\varrho < 1$, there is no similar theorems and we should rely on heuristic confirmation of the utility of this approximation.

The process $N(t)$ is never negative, hence $X(t)$ should be also restrained to $x \ge 0$. A simple solution is to put a *reflecting barrier* at $x = 0$, [48,49]. In this case

$$\int_0^\infty f(x,t;x_0)dx = 1, \quad \text{and} \quad \frac{\partial}{\partial t}\int_0^\infty f(x,t;x_0)dx = \int_0^\infty \frac{\partial f(x,t;x_0)}{\partial t}dx = 0\ ;$$

replacing the integrated function with the right side of the diffusion equation we get the boundary condition corresponding to the reflecting barrier at zero:

$$\lim_{x \to 0}\left[\frac{\alpha}{2}\frac{\partial f(x,t;x_0)}{\partial x} - \beta f(x,t;x_0)\right] = 0\ . \tag{2}$$

The solution of Eq. (1) with conditions (2) is [48]

$$f(x,t;x_0) = \frac{\partial}{\partial x}\left[\Phi\left(\frac{x - x_0 - \beta t}{\alpha t}\right) - e^{\frac{2\beta x}{\alpha}}\Phi\left(\frac{x + x_0 + \beta t}{\alpha t}\right)\right] \tag{3}$$

If the station is not overloaded, $\varrho < 1$ ($\beta < 0$), then steady-state exists. The density function does not depend on time: $\lim_{t \to \infty} f(x,t;x_0) = f(x)$, and partial differential equation (1) becoms an ordinary one:

$$0 = \frac{\alpha}{2}\frac{d^2 f(x)}{dx^2} - \beta\frac{d f(x)}{dx} \quad \text{with solution} \quad f(x) = -\frac{2\beta}{\alpha}e^{\frac{2\beta x}{\alpha}}\ . \tag{4}$$

This solution approximates the queue at $G/G/1$ system:

$$p(n,t;n_0) \approx f(n,t;n_0),$$

and at steady-state $p(n) \approx f(n)$; one can also choose e.g. $p(0) \approx \int_0^{0.5} f(x)dx$, $p(n) \approx \int_{n-0.5}^{n+0.5} f(x)dx$, $n = 1,2,\ldots$, [48].

The reflecting barrier excludes the stay at zero: the process is immediately reflected; Eqs. (3),(4) hold for $x > 0$ and $f(0) = 0$. Therefore this version of diffusion with reflecting barrier is a heavy-load approximation: it gives reasonable results if the utilisation ϱ of the investigated station is close to 1, i.e. probability $p(0)$ of the empty system is negligable. The errors are larger for small values of x as the mechanism of reflecting barrier does not correspond to the behaviour of a service station; some improvement may be achieved by renormalisation or [48] by shifting the barrier to $x = -C_B^2$ (for $C_B^2 \le 1$), [29].

This inconvenience may be removed by introduction of another limit condition at $x = 0$: *a barrier with instantaneous (elementary) jumps* [32]. When the diffusion process comes to $x = 0$, it remains there for a time exponentially distributed with a parameter λ_0 and then it returns to $x = 1$. The time when the process is at $x = 0$ corresponds to the idle time of the system. The choice of $\lambda_0 = \lambda$ is exact for Poisson input stream and approximate otherwise. Diffusion equation becomes

$$\frac{\partial f(x,t;x_0)}{\partial t} = \frac{\alpha}{2}\frac{\partial^2 f(x,t;x_0)}{\partial x^2} - \beta\frac{\partial f(x,t;x_0)}{\partial x} + \lambda p_0(t)\delta(x-1),$$

$$\frac{dp_0(t)}{dt} = \lim_{x \to 0}[\frac{\alpha}{2}\frac{\partial f(x,t;x_0)}{\partial x} - \beta f(x,t;x_0)] - \lambda p_0(t), \tag{5}$$

where $p_0(t) = P[X(t) = 0]$. The term $\lambda p_0(t)\delta(x-1)$ gives the probability density that the process is started at point $x = 1$ at the moment t because of the jump from the barrier. The second equation makes balance of the $p_0(t)$: the term $\lim_{x \to 0}[\frac{\alpha}{2}\frac{\partial f(x,t;x_0)}{\partial x} - \beta f(x,t;x_0)]$ gives the probability flow *into* the barrier and the term $\lambda p_0(t)$ represents the probability flow *out* of the barrier.

1.2 Steady State Solution

In stationary state, when $\lim_{t \to \infty} p_0(t) = p_0$, $\lim_{t \to \infty} f(x,t;x_0) = f(x)$, Eq.(5) becomes ordinary differential and its solution, if $\varrho = \lambda/\mu \neq 1$, may be expressed as:

$$f(x) = \begin{cases} \frac{\lambda p_0}{-\beta}(1 - e^{zx}) & \text{for} \quad 0 < x \leq 1, \\ \frac{\lambda p_0}{-\beta}(e^{-z} - 1)e^{zx} & \text{for} \quad x \geq 1, \end{cases} \quad z = \frac{2\beta}{\alpha}. \tag{6}$$

We get p_0 from normalisation: $p_0 = 1 - \varrho$, i.e. the exact result. The mean queue length

$$E[N] \approx \int_0^\infty x f(x) dx = \frac{\lambda p_0}{-\beta}\left(\int_0^1 x(1-e^{zx})dx + \int_1^\infty x(e^{-z}-1)e^{zx}dx\right) =$$

$$= \frac{\lambda p_0}{-\beta}\left(0.5 - \frac{1}{z}\right) = \left[0.5 + \frac{C_A^2\varrho + C_B^2}{2(1-\varrho)}\right]\varrho. \tag{7}$$

if we take $p(n) = \int_{n-1}^n f(x)dx$, $n = 1, 2, 3, \ldots$. then

$$E[N] = \left[1 + \frac{C_A^2\varrho + C_B^2}{2(1-\varrho)}\right]\varrho. \tag{8}$$

The solution (8) gives better results then (7) for small values of C_A^2, C_B^2 and small ϱ. The first discussion of errors, which are growing with C_A^2, C_B^2, was presented in [57].

1.3 Transient Solution

Consider a diffusion process with an absorbing barrier (absorbing barrier means that the process is finished when it attains the barrier) at $x = 0$, started at $t = 0$ from $x = x_0$. Its probability density function $\phi(x,t;x_0)$ has the following form, see e.g. [3]

$$\phi(x,t;x_0) = \frac{e^{\frac{\beta}{\alpha}(x-x_0) - \frac{\beta^2}{2\alpha}t}}{\sqrt{2\Pi\alpha t}}\left[e^{-\frac{(x-x_0)^2}{2\alpha t}} - e^{-\frac{(x+x_0)^2}{2\alpha t}}\right]. \tag{9}$$

The density function of first passage time from $x = x_0$ to $x = 0$ is

$$\gamma_{x_0,0}(t) = \lim_{x \to 0}\left[\frac{\alpha}{2}\frac{\partial}{\partial x}\phi(x,t;x_0) - \beta\phi(x,t;x_0)\right] = \frac{x_0}{\sqrt{2\Pi\alpha t^3}}e^{-\frac{(\beta t+1)^2}{2\alpha t}}. \quad (10)$$

Suppose that the process starts at $t = 0$ at a point x with density $\psi(x)$ and evry time when it comes to the barrier it stays there for a time given by a density function $l_0(x)$ and then reappears at $x = 1$. The total stream $\gamma_0(t)$ of mass probability that enters the barrier is

$$\gamma_0(t) = p_0(0)\delta(t) + [1 - p_0(0)]\gamma_{\psi,0}(t) + \int_0^t g_1(\tau)\gamma_{1,0}(t-\tau)d\tau \quad (11)$$

where

$$\gamma_{\psi,0}(t) = \int_0^\infty \gamma_{\xi,0}(t)\psi(\xi)d\xi, \qquad g_1(\tau) = \int_0^\tau \gamma_0(t)l_0(\tau - t)dt. \quad (12)$$

The density function of the diffusion process with instantaneous returns is

$$f(x,t;x_0) = \phi(x,t;\psi) + \int_0^t g_1(\tau)\phi(x,t-\tau;1)d\tau. \quad (13)$$

When Laplace transforms of these equations are considered, we have

$$\bar{\gamma}_0(s) = p_0(0) + [1 - p_0(0)]\bar{\gamma}_{\psi,0}(s) + \bar{g}_1(s)\bar{\gamma}_{1,0}(s),$$
$$\bar{g}_1(s) = \bar{\gamma}_0(s)\bar{l}_0(s) \quad (14)$$

where

$$\bar{\gamma}_{x_0,0}(s) = e^{-x_0\frac{\beta+A(s)}{\alpha}}, \qquad \bar{\gamma}_{\psi,0}(s) = \int_0^\infty \bar{\gamma}_{\xi,0}(s)\psi(\xi)d\xi,$$

and then

$$\bar{g}_1(s) = \left[p_0(0) + [1-p_0(0)]\bar{\gamma}_{\psi,0}(s)\right]\frac{\bar{l}_0(s)}{1 - \bar{l}_0(s)\bar{\gamma}_{1,0}(s)}. \quad (15)$$

Equation (13) in terms of Laplace transform becomes

$$\bar{f}(x,s;x_0) = \bar{\phi}(x,s;\psi) + \bar{g}_1(s)\bar{\phi}(x,s;1),$$

where

$$\bar{\phi}(x,s;x_0) = \frac{e^{\frac{\beta(x-x_0)}{\alpha}}}{A(s)}\left[e^{-|x-x_0|\frac{A(s)}{\alpha}} - e^{-|x+x_0|\frac{A(s)}{\alpha}}\right], \quad (16)$$

$$\bar{\phi}(x,s;\psi) = \int_0^\infty \bar{\phi}(x,s;\xi)\psi(\xi)d\xi, \qquad A(s) = \sqrt{\beta^2 + 2\alpha s}. \quad (17)$$

This approach was proposed in [7]. The inverse transforms of these functions could only be numerical and they may be obtained with the use of an inversion algorithm; we have used for this purpose the Stehfest's algorithm [59]. In this algorithm a function $f(t)$ is obtained from its transform $\bar{f}(s)$ for any fixed argument t as

$$f(t) = \frac{\ln 2}{2} \sum_{i=1}^{N} V_i \, \bar{f}\left(\frac{\ln 2}{t} i\right), \qquad (18)$$

where

$$V_i = (-1)^{N/2+i} \sum_{k=\lfloor \frac{i+1}{2} \rfloor}^{\min(i, N/2)} \frac{k^{N/2+1}(2k)!}{(N/2-k)!k!(k-1)!(i-k)!(2k-i)!}. \qquad (19)$$

N is an even integer end depends on a computer precision; we used $N = 12 - 40$.

Fig. 1a presents, for a certain t, a comparison of diffusion and exact results, known in case of $M/M/1$ station and expressed by a series of Bessel functions, e.g. [47]. If the diffusion results are below a certain level, the values of the diffusion density are automatically set to zero because of numerical errors of the applied Laplace inversion. The above transient solution of $G/G/1$ model assumes that parameters of the model are constant. If they are changing we should define the time periods where they are constant and solve diffusion equation within this intervals separately, transient solution obtained at the end of one serves as the initial condition for the next interval – see Fig. 1b. The curves "diffusion1", "diffusion2" correspond to mean queues computed with the use of $p(n,t) = f(n,t)$ and formulas (8).

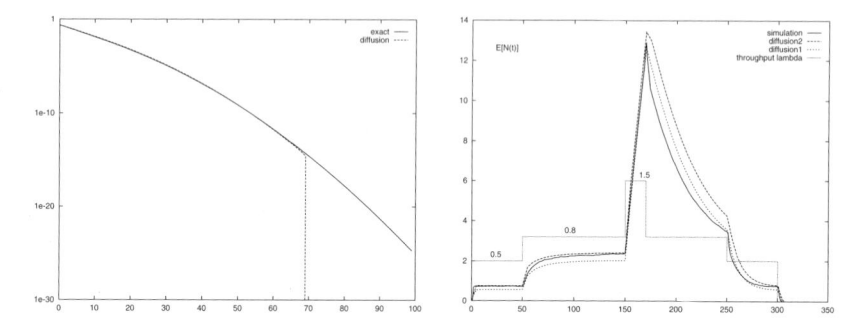

Fig. 1. (a) Exact distribution of M/M/1 queue, expressed by Bessel functions, and its diffusion approximation; $t = 70$, $\lambda = 0.75$, $\mu = 1$, queue is empty at the beginning; (b) The mean queue length following the changes of traffic intensity; diffusion approximation and simulation results; service time constant and equal 1 time unit; Poisson input traffic: M/D/1

1.4 Drawbacks of Stehfest Algorithm

The main drawback of Stehfest algorithm is a very large range of the values taken by the coefficients V_n in Eqs. (18) and (19) for even relatively small N. For instance, they are approximately in the range 10^0–10^{15} for $N = 20$. Fig. 3 presents the error functions shown in log–log–scale, for the shifted Heaviside (unit step) function $f(t) = \mathbf{1}(t-1)$.

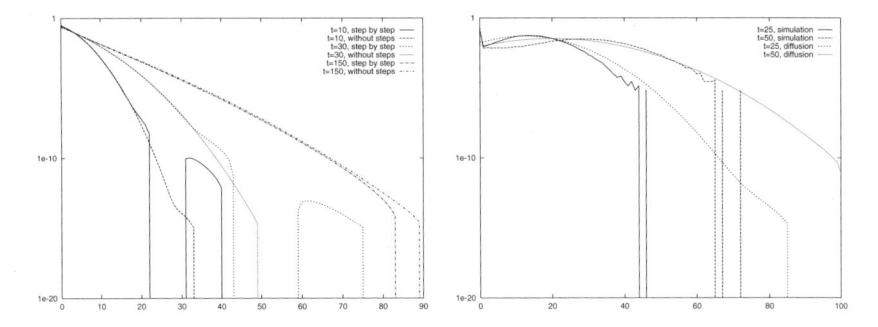

Fig. 2. Left: Comparison of diffusion approximations of the queue distribution at M/M/1 station having parameters $\lambda = 0.75$ and $\mu = 1$; the queue is empty at the beginning; results are obtained at $t = 10, 30, 150$ (a) with the initial condition fixed at $t = 0$ for all t and (b) computed separately for each interval of length $\Delta = 1$ with initial conditions determined at the end of previous interval. Right: Comparison of diffusion and simulation estimations of the queue distribution for $t = 25$, $t = 50$; the source has two phases with intensities $\lambda_1 = 1.6$, $\lambda_2 = 0.4$; the queue is empty at the beginning, computations are done within intervals $\Delta = 1$, simulation has 150 000 repetitions.

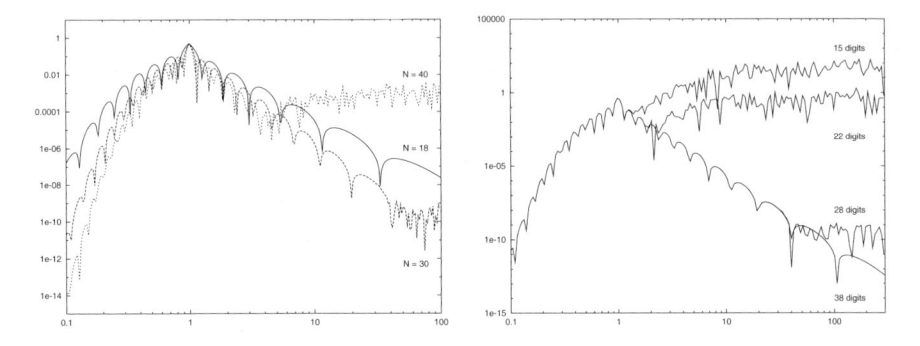

Fig. 3. Dependence of error function on value of N in Stehfest formula and on precision of numbers used for calculations

Its Laplace transform e^{-s}/s is inverted with Stehfest algorithm and compared with the original function. Because of a discontinuity in $t = 1$, the error in this point must be equal to 0.5. Fig. 3a shows the absolute error with respect to the summation limit N of Eq. (18). It is visible that for small values of N the error is relatively small and it achieves the theoretical value of 0.5 for $t = 1$. We can also notice that for $t < 1$ the absolute error is decreasing with the increase of N. The most important fact is that big values of both N and t cause totally erroneous output of the algorithm. Fig. 3b shows the influence of the used precision on the error function for the same example. A relatively high value of $N = 30$ is chosen to show the advantage of using larger mantissa. The increase of N makes narrower the interval of accurate calculations for t. The use of greater precision practically doesn't influence error function in usable range of t.

The errors of inversion algorithm are especially visible when the time axis is composed of small intervals and at each interval the density function is obtained from the results of previous interval (hence the errors in all former intervals influence the current results), see Fig. 2a. The transient queue distributions of the same service station but with the input stream of *on-off* type, given by diffusion and simulation are compared in Fig. 2b. The *on* and *off* phases have exponential distribution and their mean values are equal 100 time units. Although the simulation results were obtained by averaging 150 000 realizations of the random process, the tail of distribution where the probabilities have small values, was not accessible to simulation.

2 G/G/1/N Station

In the case of a queueue limited to N positions, the second barrier of the same type as at $x = 0$ is placed at $x = N$. The model equations become [32]

$$\frac{\partial f(x,t;x_0)}{\partial t} = \frac{\alpha}{2}\frac{\partial^2 f(x,t;x_0)}{\partial x^2} - \beta\frac{\partial f(x,t;x_0)}{\partial x} +$$
$$+ \lambda_0 p_0(t)\delta(x-1) + \lambda_N p_N(t)\delta(x-N+1),$$
$$\frac{dp_0(t)}{dt} = \lim_{x \to 0}\left[\frac{\alpha}{2}\frac{\partial f(x,t;x_0)}{\partial x} - \beta f(x,t;x_0)\right] - \lambda_0 p_0(t),$$
$$\frac{dp_N(t)}{dt} = \lim_{x \to N}\left[-\frac{\alpha}{2}\frac{\partial f(x,t;x_0)}{\partial x} + \beta f(x,t;x_0)\right] - \lambda_N p_N(t), \quad (20)$$

where $\delta(x)$ is Dirac delta function.

2.1 Steady State

In stationary state, when $\lim_{t\to\infty} p_0(t) = p_0$, $\lim_{t\to\infty} p_N(t) = p_N$, $\lim_{t\to\infty} f(x,t;x_0) = f(x)$, Eqs.(20) become ordinary differential ones and their solution, if $\varrho = \lambda/\mu \neq 1$, may be expressed as:

$$f(x) = \begin{cases} \frac{\lambda p_0}{-\beta}(1 - e^{zx}) & \text{for } 0 < x \leq 1, \\ \frac{\lambda p_0}{-\beta}(e^{-z} - 1)e^{zx} & \text{for } 1 \leq x \leq N-1, \\ \frac{\mu p_N}{-\beta}(e^{z(x-N)} - 1) & \text{for } N-1 \leq x < N, \end{cases} \quad (21)$$

where $z = \frac{2\beta}{\alpha}$ and p_0, p_N are determined through normalization

$$p_0 = \lim_{t\to\infty} p_0(t) = \{1 + \varrho e^{z(N-1)} + \frac{\varrho}{1-\varrho}[1 - e^{z(N-1)}]\}^{-1}, \quad (22)$$

$$p_N = \lim_{t\to\infty} p_N(t) = \varrho p_0 e^{z(N-1)}. \quad (23)$$

The steady-state solution does not depend on the distributions of the sojourn times in boundaries but only on their first moments.

Classes. We follow [33]. When the input stream λ is composed of K classes of customers and $\lambda = \sum_{k=1}^{K} \lambda^{(k)}$ (all parameters concerning class k have an upper index with brackets) then the joint service time pdf is defined as

$$b(x) = \sum_{k=1}^{K} \frac{\lambda^{(k)}}{\lambda} b^{(k)}(x),$$

hence

$$\frac{1}{\mu} = \sum_{k=1}^{K} \frac{\lambda^{(k)}}{\lambda} \frac{1}{\mu^{(k)}}, \quad \text{and} \quad C_B^2 = \mu^2 \sum_{k=1}^{K} \frac{\lambda^{(k)}}{\lambda} \frac{1}{\mu_{(k)}^2} (C_B^{(k)\,2} + 1) - 1. \quad (24)$$

We assume that the input streams of different class customers are mutually independent, the number of class k customers that arrived within sufficiently long time period is normally distributed with variance $\lambda^{(k)} C_A^{(k)\,2}$; the sum of independent randomly distributed variables has also normal distribution with variance which is the sum of composant variances, hence

$$C_A^2 = \sum_{k=1}^{K} \frac{\lambda^{(k)}}{\lambda} C_A^{(k)\,2}. \quad (25)$$

The above parameters yield α, β of the diffusion equation; function $f(x)$ approximates the distribution $p(n)$ of customers of all classes present in the queue: $p(n) \approx f(n)$ and the probability that there are $n^{(k)}$ customers of class k is

$$p_k(n^{(k)}) = \sum_{n=n^{(k)}}^{N} \left[p(n) \binom{n}{n^{(k)}} \left(\frac{\lambda^{(k)}}{\lambda}\right)^{n^{(k)}} \left(1 - \frac{\lambda^{(k)}}{\lambda}\right)^{n-n^{(k)}} \right], \quad k = 1, \ldots, K. \quad (26)$$

2.2 G/G/1/N, Transient Solution

The approach presented for $G/G/1$ station may be also used in case of two barriers with instantaneous returns, [7]. Consider a diffusion process with two absorbing barriers at $x = 0$ and $x = N$, started at $t = 0$ from $x = x_0$. Its probability density function $\phi(x, t; x_0)$ has the following form cf. [3]

$$\phi(x, t; x_0) = \begin{cases} \delta(x - x_0) & \text{for } t = 0 \\ \frac{1}{\sqrt{2\Pi\alpha t}} \sum_{n=-\infty}^{\infty} \left\{ \exp\left[\frac{\beta x'_n}{\alpha} - \frac{(x - x_0 - x'_n - \beta t)^2}{2\alpha t}\right] \right. \\ \left. - \exp\left[\frac{\beta x''_n}{\alpha} - \frac{(x - x_0 - x''_n - \beta t)^2}{2\alpha t}\right] \right\} & \text{for } t > 0, \end{cases} \quad (27)$$

where $x'_n = 2nN$, $x''_n = -2x_0 - x'_n$.

If the initial condition is defined by a function $\psi(x)$, $x \in (0, N)$, $\lim_{x \to 0} \psi(x) = \lim_{x \to N} \psi(x) = 0$, then the pdf of the process has the form $\phi(x, t; \psi) = \int_0^N \phi(x, t; \xi)\psi(\xi)d\xi$. The Laplace transform of $\phi(x, t; x_0)$ can be expressed as

$$\bar{\phi}(x, s; x_0) = \frac{\exp[\frac{\beta(x-x_0)}{\alpha}]}{A(s)} \sum_{n=-\infty}^{\infty} \left\{ \exp\left[-\frac{|x - x_0 - x'_n|}{\alpha} A(s)\right] \right.$$
$$\left. - \exp\left[-\frac{|x - x_0 - x''_n|}{\alpha} A(s)\right] \right\}, \quad (28)$$

where $A(s) = \sqrt{\beta^2 + 2\alpha s}$. For computational efficiency we rearranged the Eq. (28) to the form

$$\bar{\phi}(x, s; x_0) = \frac{\exp[\frac{\beta(x-x_0)}{\alpha}]}{A(s)} \left\{ 1_{(x \geq x_0)} \left[\exp\left(-\frac{xA(s)}{\alpha}\right) 2\sinh\left(\frac{x_0 A(s)}{\alpha}\right) \right] \right.$$
$$+ 1_{(x_0 < x)} \left[\exp\left(-\frac{x_0 A(s)}{\alpha}\right) 2\sinh\left(\frac{xA(s)}{\alpha}\right) \right]$$
$$\left. - 2\sinh\left(\frac{xA(s)}{\alpha}\right) 2\sinh\left(\frac{x_0 A(s)}{\alpha}\right) \times \sum_{n=1}^{\infty} \exp\left(-2nN\frac{A(s)}{\alpha}\right) \right\} (29)$$

Similarily, $\bar{\phi}(x, s; \psi) = \int_0^N \bar{\phi}(x, s; \xi)\psi(\xi)d\xi$.

The probability density function $f(x, t; \psi)$ of the diffusion process with elementary returns is composed of the function $\phi(x, t; \psi)$ which represents the influence of the initial conditions and of a spectrum of functions $\phi(x, t-\tau; 1)$, $\phi(x, t-\tau; N-1)$ which are pd functions of diffusion processes with absorbing barriers at $x = 0$ and $x = N$, started at time $\tau < t$ at points $x = 1$ and $x = N-1$ with densities $g_1(\tau)$ and $g_{N-1}(\tau)$:

$$f(x, t; \psi) = \phi(x, t; \psi) + \int_0^t g_1(\tau)\phi(x, t-\tau; 1)d\tau + \int_0^t g_{N-1}(\tau)\phi(x, t-\tau; N-1)d\tau. \quad (30)$$

Densities $\gamma_0(t)$, $\gamma_N(t)$ of probability that at time t the process enters to $x = 0$ or $x = N$ are

$$\gamma_0(t) = p_0(0)\delta(t) + [1 - p_0(0) - p_N(0)]\gamma_{\psi,0}(t) + \int_0^t g_1(\tau)\gamma_{1,0}(t-\tau)d\tau$$
$$+ \int_0^t g_{N-1}(\tau)\gamma_{N-1,0}(t-\tau)d\tau,$$

$$\gamma_N(t) = p_N(0)\delta(t) + [1 - p_0(0) - p_N(0)]\gamma_{\psi,N}(t) + \int_0^t g_1(\tau)\gamma_{1,N}(t-\tau)d\tau$$
$$+ \int_0^t g_{N-1}(\tau)\gamma_{N-1,N}(t-\tau)d\tau, \quad (31)$$

where $\gamma_{1,0}(t)$, $\gamma_{1,N}(t)$, $\gamma_{N-1,0}(t)$, $\gamma_{N-1,N}(t)$ are densities of the first passage time between corresponding points, e.g.

$$\gamma_{1,0}(t) = \lim_{x \to 0} \left[\frac{\alpha}{2} \frac{\partial \phi(x, t; 1)}{\partial x} - \beta\phi(x, t; 1)\right]. \quad (32)$$

For absorbing barriers

$$\lim_{x \to 0} \phi(x,t;x_0) = \lim_{x \to N} \phi(x,t;x_0) = 0,$$

hence $\gamma_{1,0}(t) = \lim_{x \to 0} \frac{\alpha}{2} \frac{\partial \phi(x,t;1)}{\partial x}$. The functions $\gamma_{\psi,0}(t)$, $\gamma_{\psi,N}(t)$ denote densities of probabilities that the initial process, started at $t = 0$ at the point ξ with density $\psi(\xi)$ will end at time t by entering respectively $x = 0$ or $x = N$.

Finally, we may express $g_1(t)$ and $g_N(t)$ with the use of functions $\gamma_0(t)$ and $\gamma_N(t)$:

$$g_1(\tau) = \int_0^\tau \gamma_0(t) l_0(\tau - t) dt, \qquad g_{N-1}(\tau) = \int_0^\tau \gamma_N(t) l_N(\tau - t) dt, \qquad (33)$$

where $l_0(x)$, $l_N(x)$ are the densities of sojourn times in $x = 0$ and $x = N$; the distributions of these times are not restricted to exponantial ones as it is in Eq. (20). Laplace transforms of Eqs. (31,33) give us $\bar{g}_1(s)$ and $\bar{g}_{N-1}(s)$:

$$\bar{g}_1(s) = \left\{ p_0(0) + \bar{\gamma}_{\psi,0}(s) + [p_N(0) + \bar{\gamma}_{\psi,N}(s)] \frac{\bar{l}_N(s) \bar{\gamma}_{N-1,0}(s)}{1 - \bar{l}_N(s) \bar{\gamma}_{N-1,N}(s)} \right\}.$$

$$\cdot \frac{\bar{l}_0(s)}{1 - \bar{l}_0(s) \bar{\gamma}_{1,0}(s)} \left[1 - \frac{\bar{l}_0(s) \bar{\gamma}_{1,N}(s)}{1 - \bar{l}_0(s) \bar{\gamma}_{1,0}(s)} \frac{\bar{l}_N(s) \bar{\gamma}_{N-1,0}(s)}{1 - \bar{l}_N(s) \bar{\gamma}_{N-1,N}(s)} \right]^{-1},$$

$$\bar{g}_{N-1}(s) = \frac{\bar{l}_N(s)}{1 - \bar{l}_N(s) \bar{\gamma}_{N-1,N}(s)} [p_N(0) + \bar{\gamma}_{\psi,N}(s) + \bar{g}_1(s) \bar{\gamma}_{1,N}(s)]$$

and the Laplace transform of the density function $f(x,t;\psi)$ is obtained as

$$\bar{f}(x,s;\psi) = \bar{\phi}(x,s;\psi) + \bar{g}_1(s) \bar{\phi}(x,s;1) + \bar{g}_{N-1}(s) \bar{\phi}(x,s;N-1). \qquad (34)$$

Probabilities that at the moment t the process has the value $x = 0$ or $x = N$ are

$$\bar{p}_0(s) = \frac{1}{s} [\bar{\gamma}_0(s) - \bar{g}_1(s)], \qquad \bar{p}_N(s) = \frac{1}{s} [\bar{\gamma}_N(s) - \bar{g}_{N-1}(s)]. \qquad (35)$$

3 State-Dependent Diffusion Parameters, G/G/N/N Transient Model

In the case of $G/G/N/N$ model, the value of the diffusion process corresponds to the number of active channels. The output stream depends on the number of occupied channels, hence the diffusion parameters depend also on the value of the diffusion process: $\alpha = \alpha(x,t)$, $\beta = \beta(x,t)$.

The diffusion interval $x \in [0, N]$ is divided into subintervals of unitary length and the parameters are kept constant within these subintervals. For each time- and space-subinterval with constant parameters, transient diffusion is obtained. The equations for space-intervals are solved together with balance equations for probability flows between neighbouring intervals. The results are obtained in the form of Laplace transforms of density functions of the investigated diffusion process and then inverted numerically.

If $n-1 < x < n$, it is supposed that n channels are busy, hence we choose

$$\alpha(x,t) = \lambda(t)C_A^2(t) + n\mu C_B^2, \qquad \beta(x,t) = \lambda(t) - n\mu \quad \text{for} \quad n-1 < x < n. \quad (36)$$

Jumps from $x = N$ to $x = N-1$ are performed with density μ.

In transient state we should balance the probability flows between neighbouring intervals with different diffusion parameters. We put imaginary barriers at the borders of these intervals and suppose that the diffusion process entering the barrier at n, $n = 1, 2, \ldots N-1$, from its left side (the process is growing) is absorbed and immediately reappears at $x = n + \varepsilon$. Similarly, the process which is diminishing and enters the barrier from its right side reappears at its other side at $x = n - \varepsilon$.

The density functions for the intervals are as follows:

$$f_1(x,t;\psi_1) = \phi_1(x,t;\psi_1) + \int_0^t g_1(\tau)\phi_1(x, t-\tau; 1)d\tau +$$

$$+ \int_0^t g_{1-\varepsilon}(\tau)\phi_1(x, t-\tau; 1-\varepsilon)d\tau,$$

$$f_n(x,t;\psi_n) = \phi_n(x,t;\psi_n) + \int_0^t g_{n-1+\varepsilon}(\tau)\phi_n(x, t-\tau; n-1+\varepsilon)d\tau +$$

$$+ \int_0^t g_{n-\varepsilon}(\tau)\phi_n(x, t-\tau; n-\varepsilon)d\tau, \qquad n = 2, \ldots N-1,$$

$$f_N(x,t;\psi_N) = \phi_N(x,t;\psi_N) + \int_0^t g_{N-1+\varepsilon}(\tau)\phi_N(x, t-\tau; N-1+\varepsilon)d\tau +$$

$$+ \int_0^t g_{N-1}(\tau)\phi_N(x, t-\tau; N-1)d\tau \qquad (37)$$

and the probability mass flows entering the barriers are:

$$\gamma_n^R(t) = g_{n-\varepsilon}(t), \qquad \gamma_n^L(t) = g_{n+\varepsilon}(t), \qquad n = 1, \ldots, N-1 \quad (38)$$

and $g_1(t), g_{N-1}(t)$ are the same as in $G/G/1/N$ model, Eq. (33). The densities $\gamma_{N_i}^R(t)$, $\gamma_{N_i}^L(t)$ are obtained in the similar way as in $G/G/1/N$, see Eq. (31):

$$\gamma_0(t) = p_0(0)\delta(t) + \gamma_{\psi_1,0}(t) + \int_0^t g_1(\tau)\gamma_{1,0}(t-\tau)d\tau +$$

$$+ \int_0^t g_{1-\varepsilon}(\tau)\gamma_{1-\varepsilon,0}(t-\tau)d\tau,$$

$$\gamma_1^L(t) = \gamma_{\psi_1,1}(t) + \int_0^t g_1(\tau)\gamma_{1+\varepsilon,1}(t-\tau)d\tau +$$

$$+ \int_0^t g_{2-\varepsilon}(\tau)\gamma_{2-\varepsilon,1}(t-\tau)d\tau,$$

$$\gamma_1^R(t) = \gamma_{\psi_2,1}(t) + \int_0^t g_{1+\varepsilon}(\tau)\gamma_{1+\varepsilon,1}(t-\tau)d\tau +$$

$$+ \int_0^t g_{2-\varepsilon}(\tau)\gamma_{2-\varepsilon,1}(t-\tau)d\tau ,$$

$$\gamma_n^L(t) = \gamma_{\psi_n,n}(t) + \int_0^t g_{n-1+\varepsilon}(\tau)\gamma_{n-1+\varepsilon,n}(t-\tau)d\tau +$$

$$+ \int_0^t g_{n-\varepsilon}(\tau)\gamma_{n-\varepsilon,n}(t-\tau)d\tau ,$$

$$\gamma_n^R(t) = \gamma_{\psi_{n+1},n}(t) + \int_0^t g_{n+\varepsilon}(\tau)\gamma_{n+\varepsilon,n}(t-\tau)d\tau +$$

$$+ \int_0^t g_{n+1-\varepsilon}(\tau)\gamma_{n+1-\varepsilon,n}(t-\tau)d\tau , \qquad n = 2, \ldots N-1$$

$$\gamma_N(t) = p_N(0)\delta(t) + \gamma_{\psi_N,N}(t) + \int_0^t g_{N-1+\varepsilon}(\tau)\gamma_{N-1+\varepsilon,N}(t-\tau)d\tau +$$

$$+ \int_0^t g_{N-1}(\tau)\gamma_{N-1,N}(t-\tau)d\tau , \tag{39}$$

where $\gamma_{i,j}(t)$ are the densities of the first passage time between points i,j and are obtained as in $G/G/1/N$ model, Eq.(32). This system of equations is subject to Laplace transformation and once again the Laplace transforms $\bar{f}_n(x,s;\psi_n)$ are obtained numerically, for a series of s values needed by the inversion algorithm for a specified t.

4 Open Network of $G/G/1$, G/G/1/N Queues, Steady State

The open networks of $G/G/1$ queues were studied in [33]. Let M be the number of stations and suppose at the beginning that there is one class of customers. The throughput of station i is, as usual, obtained from traffic equations

$$\lambda_i = \lambda_{0i} + \sum_{j=1}^M \lambda_j r_{ji}, \qquad i = 1, \ldots, M, \tag{40}$$

where r_{ji} is routing probability between station j and station i; λ_{0i} is external flow of customers coming from outside of network.

Second moment of interarrival time distribution is obtained from two systems of equations; the first defines C_{Di}^2 as a function of C_{Ai}^2 and C_{Bi}^2; the second defines C_{Aj}^2 as another function of $C_{D1}^2, \ldots, C_{DM}^2$:

1) The formula (41) is exact for $M/G/1$, $M/G/1/N$ stations and is approximate in the case of non-Poisson input [2]

$$d_i(t) = \varrho_i b_i(t) + (1-\varrho_i) a_i(t) * b_i(t), \qquad i = 1, \ldots, M, \tag{41}$$

where * denotes the convolution operation. From (41) we get

$$C_{Di}^2 = \varrho_i^2 C_{Bi}^2 + C_{Ai}^2(1-\varrho_i) + \varrho_i(1-\varrho_i). \tag{42}$$

2) Customers leaving station i according to the distribution $D_i(x)$ choose station j with probability r_{ij}: intervals between customers passing this way has pdf $d_{ij}(x)$

$$d_{ij}(x) = d_i(x)r_{ij} + d_i(x)*d_i(x)(1-r_{ij})r_{ij} + d_i(x)*d_i(x)*d_i(x)(1-r_{ij})^2 r_{ij} + \cdots \tag{43}$$

or, after Laplace transform,

$$\bar{d}_{ij}(s) = \bar{d}_i(s)r_{ij} + \bar{d}_i(s)^2(1-r_{ij})r_{ij} + \bar{d}_i(s)^3(1-r_{ij})^2 r_{ij} + \cdots = \frac{r_{ij}\bar{d}_i(s)}{1-(1-r_{ij})\bar{d}_i(s)},$$

hence

$$E[D_{ij}] = \frac{1}{\lambda_i r_{ij}}, \qquad C^2_{Dij} = r_{ij}(C^2_{Di} - 1) + 1. \tag{44}$$

$E[D_{ij}]$, C^2_{Dij} refer to interdeparture times; the number of customers passing from station i to j in a time interval t has approximately normal distribution with mean $\lambda_i r_{ij} t$ and variation $C^2_{Dij}\lambda_i r_{ij} t$. The sum of streams entering station j has normal distribution with mean

$$\lambda_j t = [\sum_{i=1}^{M} \lambda_i r_{ij} + \lambda_{0j}]t \qquad \text{and variance} \qquad \sigma^2_{Aj} t = \{\sum_{i=1}^{M} C^2_{Dij}\lambda_i r_{ij} + C^2_{0j}\lambda_{0j}\}t,$$

hence

$$C^2_{Aj} = \frac{1}{\lambda_j}\sum_{i=1}^{M} r_{ij}\lambda_i[(C^2_{Di} - 1)r_{ij} + 1] + \frac{C^2_{0j}\lambda_{0j}}{\lambda_j}. \tag{45}$$

Parameters λ_{0j}, C^2_{0j} represent the external stream of customers. For K classes od customers with routing probabilities $r^{(k)}_{ij}$ (let us assume for simplicity that the customers do not change their classes) we have

$$\lambda^{(k)}_i = \lambda^{(k)}_{0i} + \sum_{j=1}^{M} \lambda^{(k)}_j r^{(k)}_{ji}, \qquad i = 1,\ldots,M;\ k = 1,\ldots,K, \tag{46}$$

and

$$C^2_{Di} = \lambda_i \sum_{k=1}^{K} \frac{\lambda^{(k)}_i}{\mu^{(k)2}_i}[C^{(k)2}_{Bi} + 1] + 2\varrho_i(1-\varrho_i) + (C^2_{Ai} + 1)(1-\varrho_i) - 1. \tag{47}$$

A customer in the stream leaving station i belongs to class k with probability $\lambda^{(k)}_i/\lambda_i$ and we can determine $C^{(k)2}_{Di}$ in the similar way as it has been done in Eqs. (43-44), replacing r_{ij} by $\lambda^{(k)}_i/\lambda_i$:

$$C^{(k)2}_{Di} = \frac{\lambda^{(k)}_i}{\lambda_i}(C^2_{Di} - 1) + 1; \tag{48}$$

then

$$C^2_{Aj} = \frac{1}{\lambda_j}\sum_{l=1}^{K}\sum_{k=1}^{K} r^{(k)}_{ij}\lambda_i\left[\left(\frac{\lambda^{(k)}_i}{\lambda_i}(C^2_{Di} - 1)\right)r^{(k)}_{ij} + 1\right] + \sum_{k=1}^{K} \frac{C^{(k)2}_{0j}\lambda^{(k)}_{0j}}{\lambda_j}. \tag{49}$$

Eqs. (42), (45) or (47), (49) form a linear system of equations and allow us to determine C_{Ai}^2 and, in consequence, parameters β_i, α_i for each station.

5 Open Network of G/G/1, G/G/1/N Queues, One Class, Transient Solution

In the case of one class of customers, the time axis is divided into small intervals (equal e.g to the smallest mean service time) and at the beginning of each interval the equations (40),(42),(45) are used to determine the input parameters of each station based on the values of $\varrho_i(t)$ obtained at the end of the precedent interval, [26]. A software tool was prepared [15,16] and the examples below, see Fig. 4, are computed with its use.

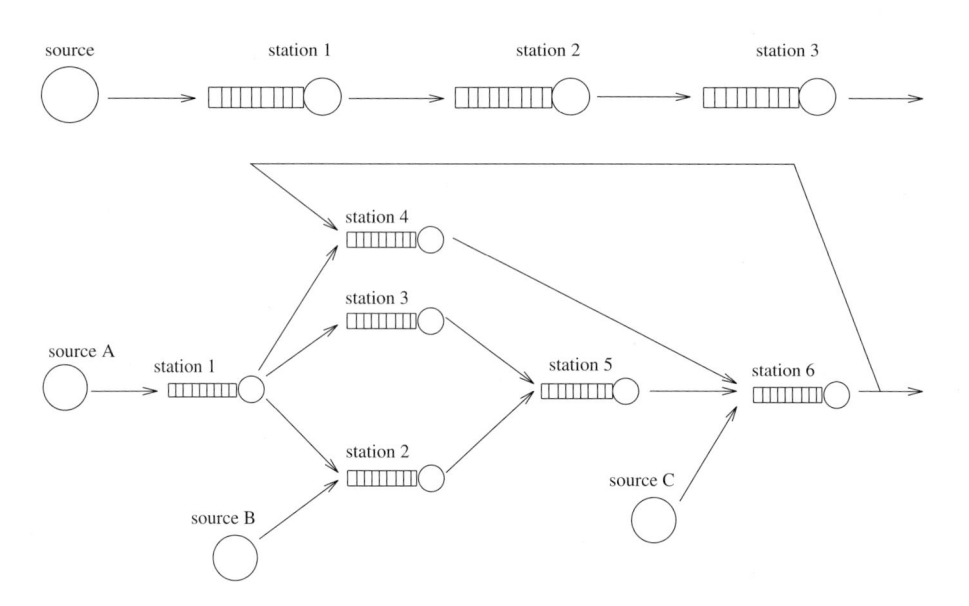

Fig. 4. Models 1 and 2

Model 1. The network is composed of the source and three stations in tandem. The source parameters are: $\lambda = 0.1\, t \in [0, 10]$, $\lambda = 4.0\, t \in [10, 20]$.
Parameters of all stations are the same: $N_i = 10$, $\mu_i = 2$, $C_{Bi}^2 = 1$, $i = 1, 2, 3$.

Fig. 5a presents mean queue lengths of stations in Model 1 as a function of time. Diffusion approximation is compared with simulation.

Model 2, its topologu is in Fig. 4. The characteristics of three sources and of one station are changing with time in the following pattern:
source A: $\lambda_A = 0.1$ for $t \in [0, 10]$, $\lambda_A = 4.0$ for $t \in [10, 21]$, $\lambda_A = 0.1$ for $t \in [21, 40]$,
source B: $\lambda_B = 0.1$ for $t \in [0, 11]$, $\lambda B = 4.0$ for $t \in [11, 20]$, $\lambda_B = 0.1$ for $t \in [20, 40]$,

source C: $\lambda_C = 0.1$ for $t \in [0, 15]$, $\lambda_C = 2.0$ for $t \in [15, 22]$, $\lambda_C = 4.0$ for $t \in [22, 30]$, $\lambda_C = 2.0$ for $t \in [30, 31]$, $\lambda_C = 0.1$ for $t \in [31, 40]$.
Station 6: $\mu_6 = 2$ for $t \in [10, 15]$ and $t \in [31, 40]$; $\mu_6 = 4$ for $t \in [15, 31]$.

Other parameters are constant: $N_1 = N_4 = 10$, $N_3 = 5$, $N_2 = N_6 = 20$, $\mu_1 = \cdots = \mu_5 = 2$. Routing probabilities are: $r_{12} = r_{13} = r_{14} = 1/3$, $r_{64} = 0.8$. Initial state: $N_1(0) = 5$, $N_1(0) = 5$, $N_2(0) = 10$, $N_3(0) = 10$, $N_4(0) = 5$, $N_5(0) = 5$, $N_6(0) = 10$. The results in the form of mean queue lengths are presented and compared with simulation in Figs. 5, 6.

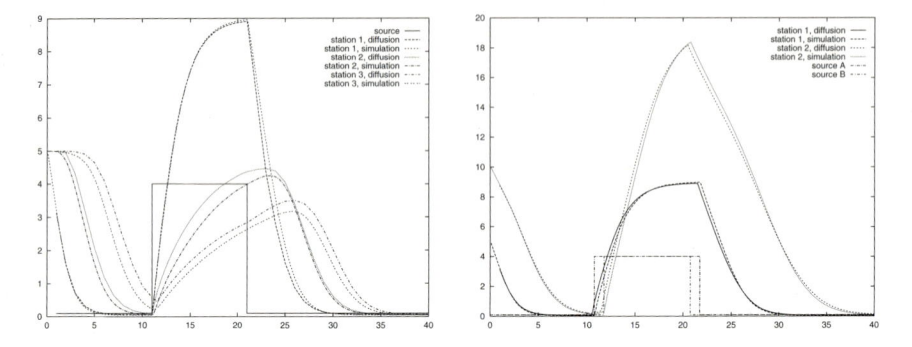

Fig. 5. (a) Model 1: mean queue lengths of station1, station2 and station3 as a function of time — diffusion and simulation (100 000 repetitions) results; the source intensity $\lambda(t)$ is indicated. (b) Model 2: Mean queue lengths of station1 and station2 as a function of time — diffusion and simulation (100 000 repetitions) results; the source intensities $\lambda_A(t)$, $\lambda_B(t)$ are indicated.

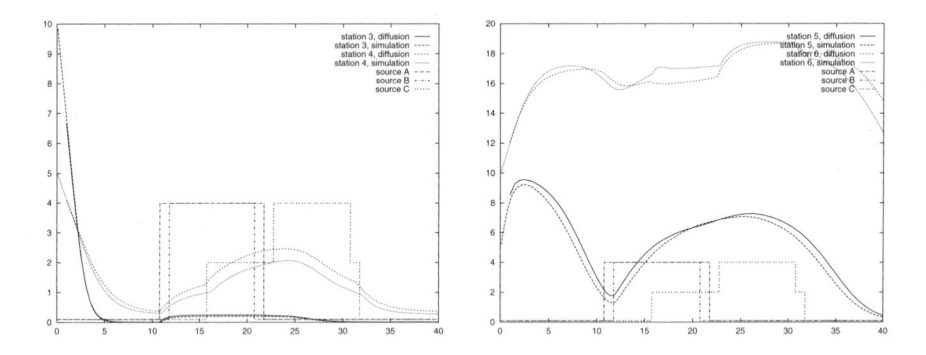

Fig. 6. Model 2: mean queue lengths of station3 and station4 (a) and of station5 and station6 (b)

6 Leaky Bucket Model

In the *leaky bucket* scheme, the cells, before entering the network, must obtain a token. Tokens are generated at constant rate and stocked in a buffer of finite capacity B. If there is a token available, an arriving cell consumes it and leaves the bucket. If not, it

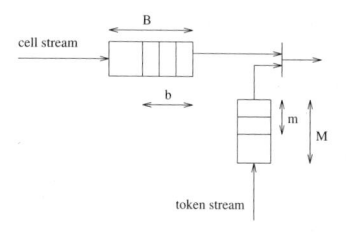

Fig. 7. Leaky bucket scheme

waits for the token in the cell buffer. The capacity of this buffer is also limited to M, Fig. 7. Tokens and cells arriving to full buffers are lost. The diffusion process $X(t)$ is defined on the interval $x \in [0, N = B + M]$ where B is the capacity of cell buffer and M is the capacity of token buffer, [12]. The current value of the process is defined as $x = b - m + M$, b and m being the current contents of the buffers.

Let us suppose that the cell interarrival time distribution has the mean $1/\lambda_c$ and squared coefficient of variation C_{Ac}^2. The tokens are generated with constant rate λ_t, hence $C_{At}^2 = 0$. Arrival of a cell increases the value of the process and arrival of a token decreases it, therefore we choose the parameters of the diffusion process as:

$$\beta = \lambda_c - \lambda_t, \qquad \alpha = \lambda_c C_{Ac}^2.$$

The process evolves between two barriers placed at $x = 0$ and at $x = M + B$; $x = 0$ represents the state where the whole token buffer is occupied and arriving tokens are lost; $x = M + B$ represents the state where the token buffer is empty and the cell buffer is full: arriving cells are lost.

The sojourn time at $x = M + B$ corresponds to the residual token interarrival time, i.e. the time between the moment when the cell buffer becomes full and the moment of the next token arrival. In [37] the density of holding time at the upper barrier of $M/D/1/N$ diffusion model was obtained; we follow this approach and assume that the density function $l_{M+B}(x)$ is

$$l_{M+B}(x) = \frac{\lambda_c e^{\lambda_c x}}{e^{\lambda_c/\lambda_t} - 1}. \tag{50}$$

The cell loss ratio $L(t)$ may be bounded by expression [37]

$$L(t) \le p_N(t) Pr[Arr(t, t + 1/\lambda_t) \ge 1]$$

where $Arr(t, t + 1/\lambda_t)$ is the number of cell arrivals during interval $[t, t + 1/\lambda_t]$.

If the cell stream is Poisson, the pdf $l_0(x)$ of the sojourn time at $x = 0$ is defined by the density of cell interarrival time; otherwise we take this density as an approximation of $l_0(x)$. Note that the sojourn times in boundaries are defined here by the densities $l_0(t)$, $l_N(t)$ and *are not restricted to exponential distributions.*

The values $x > M$ of the process correspond to states where cells are waiting for tokens, the value $x - M$ determines in this case the number of cells in the buffer; $x < M$ means that there are tokens waiting for cells and the value $M - x$ corresponds to the number of tokens in the buffer. Probability of b cells in the buffer at time

t is defined by $f(M+b,t)$; probability of the empty cell buffer is given by $p_t(t) = p_0(t) + \int_0^M f(x,t)dx$. Probability of m tokens in the buffer is given by $f(M-m,t)$ and probability of empty token buffer is determined by $\int_M^{M+B} f(x,t)dx + p_N(t)$ where $p_0(t) = Pr[X(t) = 0]$, $p_N(t) = Pr[X(t) = N]$.

The service time is constant, hence the density function of the cell waiting time for tokens (response time of leaky bucket) may be estimated as $r(x,t) = \lambda_t f(\lambda_t x + M, t)$.

Hence, using G/G/1/N model we obtain transient $f(x,t;\psi)$ and steady-state $f(x)$ distributions of the diffusion process for $0 \leq x \leq M+B$. This gives us the distribution of the number of tokens and cells in the leaky bucket, the response time distribution, the loss probabilities, the properties of the output stream, etc. The capacities of cell and token buffers may be null, so we are able to consider a number of leaky bucket variants.

The output process of the leaky bucket is the same as the cell input process provided, with probability $p_t(t)$, that there are tokens available and it is the same as token input process with probability $[1 - p_T 9t)]$ that tokens are not available; at the time moment t the pdf $d(x)$ of interdeparture times in the output stream is

$$d(x,t) = p_t(t)a(x,t) + [1 - p_t(x,t)]\delta(x - \frac{1}{\lambda_t}), \qquad (51)$$

where $a(x,t)$ is the time-dependent pdf of cell interarrival times distribution. Eq. (51) gives us the mean value and squared coefficient of interdeparture times distribution, i.e. whole information needed to incorporate one or multiple leaky-bucket stations (for example a cascade of leaky-buckets with different parameters) in the diffusion queueing network model of $G/G/1$ or $G/G/1/N$ stations. The principles of the latter model were given in [33].

Numerical example

At $t = 0$ the cell buffer is empty and the token buffer contains $M(0)$ tokens. The tokens are generated regularly each time-unit. The cell arrival stream is Poisson; the mean interarrival time is 0.5 time-unit for $0 \leq t < 100$ and 1.5 time units for $t \geq 100$, i.e. there is a traffic wave exceeding the accorded level during the first 100 units and then the traffic goes down below this level.

The buffer capacities are $B = M = 100$. Figure 8a displays the diffusion and simulation results concerning the output stream of leaky bucket for the initial number of tokens $M(0) = 0, 50$ and 100. The output dynamics given by simulation and by difusion model are very similar. Simulation results are obtained as a mean of 100 000 independent runs. If there is no tokens at the beginning, the cell stream is immediately cut to the level of token intensity (one cell per time unit), the excess of cells is stocked in the cell buffer and transmitted later, when $t > 100$ and input rate becomes smaller. If there are tokens in the token buffer, a part (for $M(0) = 50$) or almost totality (for $M(0) = 100$) of the traffic wave may enter into the network.

Figure 8b presents the comparison of mean number of cells in the cell buffer as a function of time, for different initial content of the token buffer $M(0) = 0, 50$ and 100, obtained by diffusion and simulation models. In Figure 9 the distributions of cell buffer contents obtained by simulation and by diffusion are presented for several moments, including $t = 100$, i.e. the end of high traffic period, when the congestion is the biggest. We see that although the the buffer is full is important (≈ 0.4). Note that we could mean

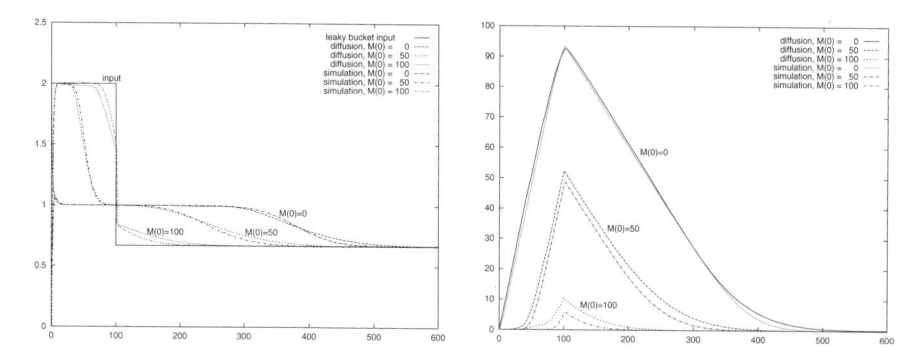

Fig. 8. (a) The input and output of leaky bucket as a function of time — the stream intensities for the initial number of tokens $M(0) = 0$, 50 and 100; diffusion and simulation results (b) Mean number of cells in the cell buffer as a function of time, $M(0) = 0$, 50 and 100; diffusion and simulation results

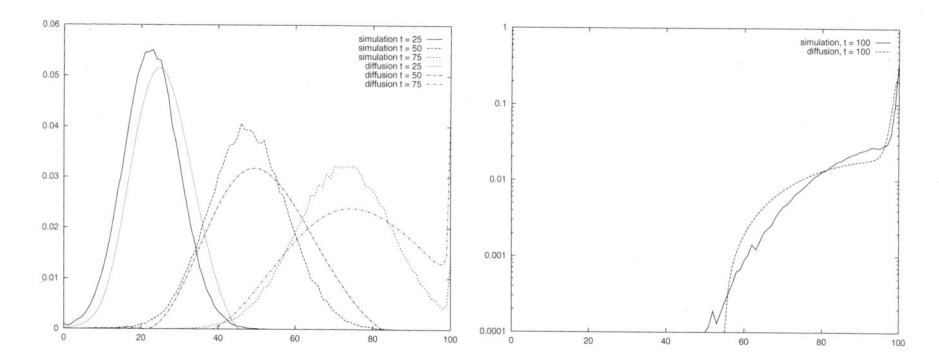

Fig. 9. Density of the number of cells during high source activity period, $t = 25, 50, 75$ (a), $t = 100$ (b); $M(0) = 0$; diffusion and simulation results

queue length is below the buffer capacity, the probability that not obtain this result with the use of fluid approximation even if the mean number of cells in the buffer predicted by diffusion and fluid approximations were similar.

7 Multiclass FIFO Queues, Output Streams Dynamics

Consider a queueing network model representing a computer network. Customers (packets of fixed size) are grouped into classes. Each class represents a connection between two points of the network and its description includes the features of the source and the itinerary across the network. Customer queues at servers correspond to the queues of cells waiting at a node to be sent further. At a queue exit the class of a customer should be known to determine its routing. In steady state queuing models this probability, for a certain class k, is given by the ratio of the throughput $\lambda^{(k)}$ of this class customers passing the node to the total througput λ of this node : $\lambda^{(k)}/\lambda$, where $\lambda = \sum_{k=1}^{K} \lambda^{(k)}$ and

K is the number of classes passing across the node. In transient state, the throughputs are a function of time and the probability $\lambda^{(k)}(t)/\lambda(t)$ as well. Morover, the flows at the entrance and the exit are not the same: $\lambda^{(k)}_{i,\,in} \neq \lambda^{(k)}_{i,\,out}$. The composition of output flow reflects the previous compositions at the entrance delayed by the response time of the queue.

To solve this problem, we choose the constant service time as the time unit, divide the time axis into intervals of unitary length and assume that the flow parameters are constant during that interval, e.g. $\lambda^{(k)}(\theta)$ denotes the class k flow at station i during an interval θ, $\theta = 1, 2, \ldots$

The input stream $\sum_k \lambda^{(k)}_{in}(\theta)$ reaches the output of the queue with a delay corresponding to the queue length in the buffer at the arrival time. The part $p(n, \theta) \sum_k \lambda^{(k)}_{i,\,in}(\theta)$ of this submitted load cannot be served and the corresponding flow cannot appear at the output before the time $t = \theta + n + 1$. So, taking into account these delays, the unfinished work ready to be processed at the time t in the station which is initially empty can be expressed as:

$$U^{(k)}(t) = \sum_{n=1}^{N} \lambda^{(k)}_{in}(t-n) \cdot p(n-1, t-n) \quad \text{for} \quad k = 1, \ldots, K.$$

A similar formulation may be easily derived for initially nonempty queue knowing its composition at $t = 0$. Remark that some accumulation periods (of high or quickly increasing load) may yield that ready work at t exceeds the server capacity: $\sum_k U^{(k)}(t) > 1$. Such a phenomenon introduces some additional delay in the transfer of the input stream to the output.

To compute the output throughput $\lambda^{(k)}_{out}(t)$, we find first the *ready time* ϑ_1 of the cells at the head of the queue. This is equal to the smallest value of θ such that:

$$\sum_{\tau=0}^{\theta} \sum_k U^{(k)}(\tau) \geq \sum_{\tau=0}^{t-1} \sum_k \lambda^{(k)}_{out}(\tau). \tag{52}$$

Then, we determine the smallest ϑ_2 for which

$$\sum_{\tau=0}^{\vartheta_2} \sum_k U^{(k)}(\tau) - \sum_{\tau=0}^{t-1} \sum_k \lambda^{(k)}_{out}(\tau) \geq 1 - p(0, t).$$

The output throughput $\lambda^{(k)}_{out}(t)$ is obtained as

$$\lambda^{(k)}_{out}(t) = \sum_{\theta=\vartheta_1}^{\vartheta_2} w_\theta \cdot U^{(k)}(\theta) \tag{53}$$

where w_{ϑ_1} represents the percentage of $U^{(k)}(\vartheta_1)$ that has not been sent yet, $w_\theta = 1$ for $\theta \neq \vartheta_1$ and ϑ_2, and w_{ϑ_2} is chosen such that $\sum_{\theta=\vartheta_1}^{\vartheta_2} w_\theta \cdot U(\theta) = 1 - p(0, t)$.

Numerical example. Consider a switch having 3 input streams end one output being the sum of the three input streams. The output queue has the capacity of 100 packets and is analysed with the use of $G/D/1/100$ multiclass diffusion model and $M/D/1$ multiclass fluid approximation model. The service time is constant, $\mu = 1$. The input streams are Poisson with parameters $\lambda^{(k)}(t)$ chosen as:

time t	0 – 10	10 – 20	20 – 30	30 – 40	40 – 80	> 80
$\lambda_{in}^{(1)}(t)$	0.1	0.8	0.8	0.1	0.1	0
$\lambda_{in}^{(2)}(t)$	0.2	0.2	0.2	0.2	0.2	0
$\lambda_{in}^{(3)}(t)$	0	0	0.5	0.5	0	0

Figures 10 – 12 present the resulting output stream (queue is empty at the beginning). These results, obtained with the use of diffusion approximation and fluid flow approximation, are compared with simulation results which represent the mean of 400 000 independent runs and practically may be considered as exact. The differences between input and output streams are clearly visible for the whole stream as well as for each class separately. They justify the need of the presented approach in the transient analysis of

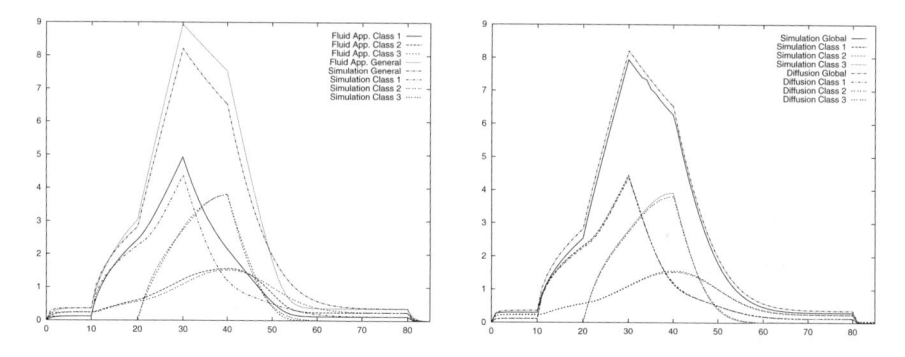

Fig. 10. The mean queue length: global and per classes as a function of time – simulation and fluid approximation results (a), simulation and diffusion approximation results (b)

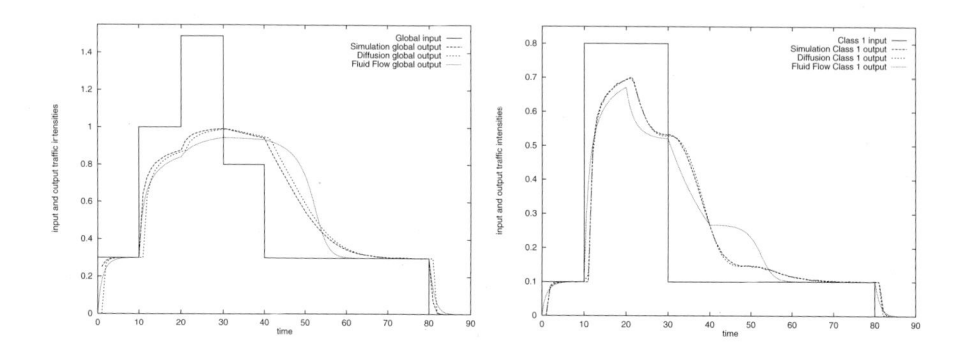

Fig. 11. The intensities $\lambda_{in}(t)$, $\lambda_{out}(t)$ of global input and output flows (a) and for class 1 traffic (b); the output flow is obtained by simulation, diffusion and fluid flow approximations

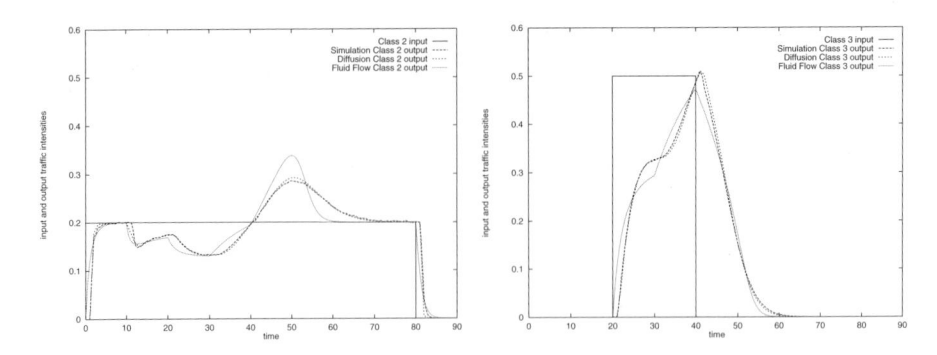

Fig. 12. The intensities of $\lambda_{in}^{(2)}(t)$, $\lambda_{out}^{(2)}(t)$ input and output flows for class 2 traffic (a) and for class 3 traffic (b); the output flow is obtained by simulation, diffusion and fluid flow approximations

networks: the impact the switch queue on the flow dynamics is important and cannot be neglected. In all considered cases the results of diffusion approximation are very close to simulation and clearly better that those obtained with the fluid flow approximation.

8 Switch with Threshold (Partial Buffer Sharing) Algorithm

In a node with *partial buffer sharing policy* the diffusion process represents the content of the cell buffer. The process is determined on the interval $x \in [0, N]$ where N is the buffer capacity, [12]. When the number of cells is equal or greater than the threshold N_1 ($N_1 < N$), only priority cells are admitted and ordinary ones are lost. Diffusion process represents the number cells of both classes, hence its parameters depend on their input and service parameters which are different for $x \leq N_1$ and $x > N_1$:

$$\beta(x) = \begin{cases} \beta_1 = \lambda^{(1)} + \lambda^{(2)} - \mu & \text{for } 0 < x \leq N_1, \\ \beta_2 = \lambda^{(1)} - \mu & \text{for } N_1 < x < N \end{cases} \quad (54)$$

and

$$\alpha(x) = \begin{cases} \alpha_1 = \lambda^{(1)} C_A^{(1)^2} + \lambda^{(2)} C_A^{(2)^2} + \mu C_B^2 & \text{for } 0 < x \leq N_1, \\ \alpha_2 = \lambda^{(1)} C_A^{(1)^2} + \mu C_B^2 & \text{for } N_1 < x < N. \end{cases} \quad (55)$$

We assume constant service time, hence $C_B^2 = 0$. Once again we use the Eq. 50 to represent the sojourn time in the barrier at $x = N$ and to determine μ_N as the inverse of the mean sojourn time.

Steady state solution. Let $f_1(x)$ and $f_2(x)$ denote the pdf function of the diffusion process in intervals $x \in (0, N_1]$ and $x \in [N_1, N)$. We suppose that
- $\lim_{x \to 0} f_1(x, t; x_0) = \lim_{x \to N} f_2(x, t; x_0) = 0$,
- $f_1(x)$ and $f_2(x)$ functions have the same value at the point N_1: $f_1(N_1) = f_2(N_1)$,
- there is no probability mass flow within the interval $x \in (1, N-1)$:

$$\frac{\alpha_n}{2} \frac{d f_n(x)}{d x} - \beta_n f_n(x) = 0, \quad x \in (1, N_1), n = 1 \quad \text{and} \quad x \in (N_1, N-1), n = 2,$$

and we obtain the solution of diffusion equations:

$$f_1(x) = \begin{cases} \dfrac{[\lambda^{(1)} + \lambda^{(2)}]p_0}{-\beta_1}(1 - e^{z_1 x}) & \text{for } 0 < x \leq 1, \\ \dfrac{[\lambda^{(1)} + \lambda^{(2)}]p_0}{-\beta_1}(1 - e^{z_1})e^{z_1(x-1)} & \text{for } 1 \leq x \leq N_1, \end{cases}$$

$$f_2(x) = \begin{cases} f_1(N_1)e^{z_2(x-N_1)} & \text{for } N_1 \leq x \leq N-1, \\ \dfrac{\mu p_N}{\beta_2}\left[1 - e^{z_2(x-N)}\right] & \text{for } N-1 \leq x < N, \end{cases} \quad (56)$$

where $z_n = \dfrac{2\beta_n}{\alpha_n}$, $n = 1, 2$. Probabilities p_0, p_N are obtained with the use of normalization condition. The loss ratio $L^{(1)}$ is expressed by the probability p_N, the loss ratio $L^{(2)}$ is determined by the probability $P[x > N_1] = \int_{N_1}^{N} f_2(x)dx + p_N$.

Numerical example. Fig. 13 presents in linear and logarithmic scale the steady-sate distribution given by Eqs. (56) of the number of cells present in a station. The buffer length is $N = 100$, the threshold value is $N_1 = 50$. Some of the values are compared with simulation histograms which we were able to obtain only for relatively large values of probabilities.

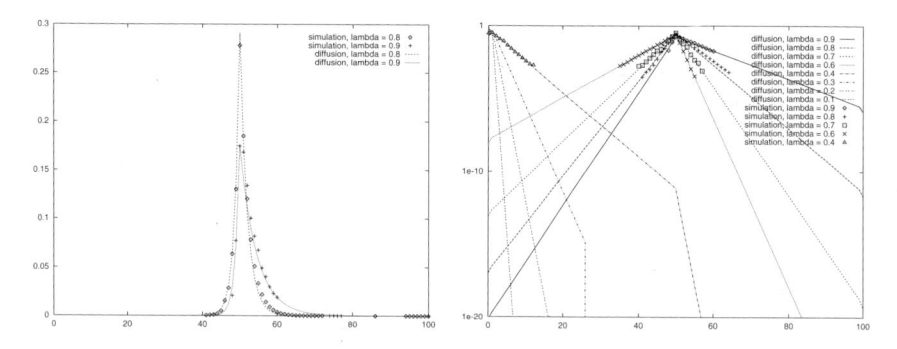

Fig. 13. Steady state distribution of the number of cells for traffic densities $\lambda^{(1)} = \lambda^{(2)} = \lambda = 0.9, 0.9$; diffusion and simulation results, normal (a) and logarithmic (b) scale

Transient solution. The transient solution is obtained with technique presented earlier for $G/G/N/N$ model. It makes use of the balance equations for probability flows crossing the barrier situated at the boundary between the intervals with different diffusion coefficients, i.e. at $x = N_1$. Let us consider two separate diffusion processes $X_1(t), X_2(t)$:

$X_1(t)$ is defined on the interval $x \in (0, N_1)$. At $x = 0$ there is a barrier with sojourn times defined by a pdf $l_0(t)$ and instantaneous returns to the point $x = 1$. At $x = N_1$ an absorbing barrier is placed. Denote by $\gamma_{N_1}^L(t)$ the pdf that the process enters the absorbing barrier at $x = N_1$. The process is reinitiated at $x = N_1 - \varepsilon$ with a density $g_{N_1-\varepsilon}(t)$.

$X_2(t)$ is defined on the interval $x \in (N_1, N)$. It is limited by an absorbing barrier at $x = N_1$ and by a barrier with instantaneous returns at $x = N$. The sojourn time at this barrier is defined by a pdf $l_N(t)$ and the returns are performed to $x = N - 1$. The process is reinitiated at $x = N_1 + \varepsilon$ with a density $g_{N_1+\varepsilon}(t)$. Denote by $\gamma_{N_1}^R(t)$ the pdf that the process $X_2(t)$ enters the absorbing barrier at $x = N_1$.

The interaction between two processes is given by equations

$$g_{N_1+\varepsilon}(t) = \gamma_{N_1}^L(t) \quad \text{and} \quad g_{N_1-\varepsilon}(t) = \gamma_{N_1}^R(t),$$

i.e. the probability density that one process enters to its absorbing barrier is equal to the density of reinitialization of the other process in the vicinity of the barrier.

Equations (31) and (33) form a set of eight equations with eight unknown functions. When we transform these equations with the use of Laplace transform, the convolutions of density functions become products of transforms and we have a set of linear equations where the unknown variables are: $\bar{g}_1(s)$, $\bar{g}_{N_1-\varepsilon}(s)$, $\bar{g}_{N_1+\varepsilon}(s)$, $\bar{g}_{N-1}(s)$, $\bar{\gamma}_0(s)$, $\bar{\gamma}_N(s)$, $\bar{\gamma}_{N_1-\varepsilon}(s)$, $\bar{\gamma}_{N_1+\varepsilon}(s)$. They may be expressed by all other functions, that means $\bar{\gamma}_{\psi_1,0}(s)$, $\bar{\gamma}_{\psi_1,N_1}(s)$, $\bar{\gamma}_{1,0}(s)$, $\bar{\gamma}_{1,N_1}(s)$, $\bar{\gamma}_{N_1-\varepsilon,0}(s)$, $\bar{\gamma}_{N_1-\varepsilon,N_1}(s)$, $\bar{\gamma}_{\psi_2,N_1}(s)$, $\bar{\gamma}_{\psi_2,N}(s)$, $\bar{\gamma}_{N_1+\varepsilon,N_1}(s)$, $\bar{\gamma}_{N_1+\varepsilon,N}(s)$, $\bar{\gamma}_{N-1,N_1}(s)$, $\bar{\gamma}_{N-1,N}(s)$ which are already determined by equations of type (32). This way we obtain the functions $\bar{g}_1(s)$, $\bar{g}_{N_1-\varepsilon}(s)$, $\bar{g}_{N_1+\varepsilon}(s)$, $\bar{g}_{N-1}(s)$ and use them in the pdfs (37). The time-domain originals $f_1(x,t;\psi_1)$, $f_2(x,t;\psi_2)$ are obtained numerically [59] from their transforms. The density of the whole process is

$$f(x,t;\psi) = \begin{cases} f_1(x,t;\psi_1) & \text{for } 0 < x < N_1, \\ f_2(x,t;\psi_2) & \text{for } N_1 < x < N. \end{cases}$$

To see the evolution of the number of cells belonging to a class, we have to consider the composition of input and output streams.

Numerical example. Let us suppose that at the beginning the buffer is empty and during the interval $t \in [0, 100]$ the input stream of priority cells has ratio $\lambda^{(1)} = 2$ cells per time unit and the one of low priority cells $\lambda^{(2)} = 1$; for $t > 100$ the ratio of high priority cells is $\lambda^{(1)} = 0.6667$, the ratio of low priority cells does not change. The service time is constant and equal one time unit. The buffer length is $N = 100$, the value of threshold varies between $N_1 = 50$ and $N_1 = 90$. The value of ε in Eqs. (37)–(39) was chosen $\varepsilon = 0.1$. In Fig. 14 the distributions of the number of cells at the buffer obtained by simulation and by diffusion model for chosen time moments are compared. Diffusion and simulation results are placed in separate figures to preserve their legibility. The shape of curves given by two models is very similar. At the end of second period ($t = 400, 500, 600$) the steady state distribution is attained.

Fig. 15 displays the mean number of cells in the buffer as a function of time. During the first 100 time units the congestion is clearly visible, the buffer quickly becomes saturated; during the second period the queue is also overcrowded, probability that the threshold is exceeded is near 0.7 but owing to the buffer sharing policy the probability that the buffer is inaccessible for priority cells remains negligible – Fig. 16.

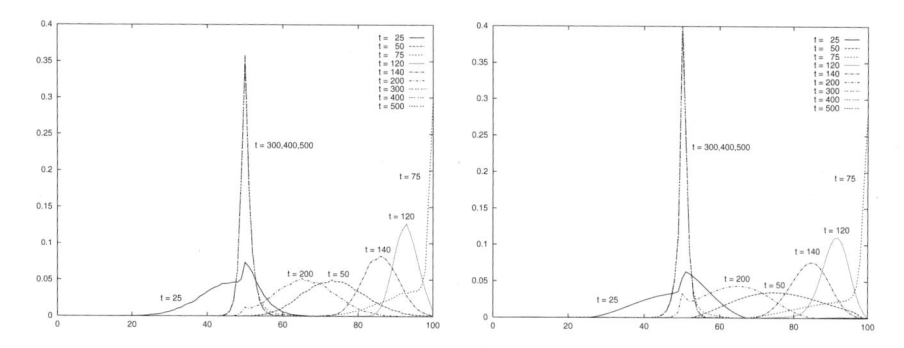

Fig. 14. Distribution of the number of all cells in the buffer for several time moments $t = 25 - 500$; buffer size $N = 100$, threshold $N_1 = 50$; simulation (a) and diffusion (b) results

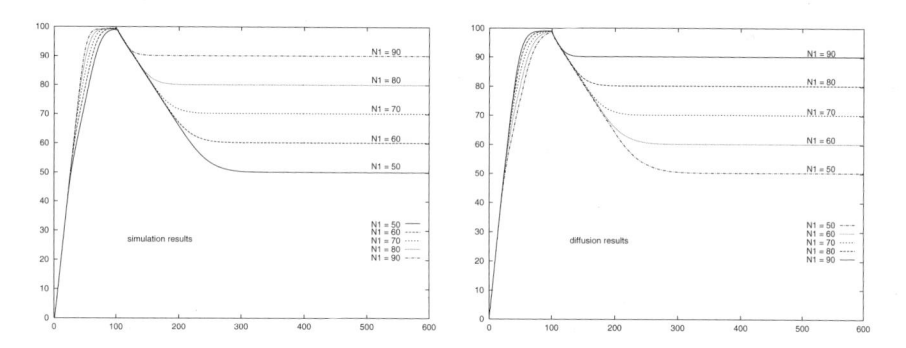

Fig. 15. Mean number of cells as a function of time, parametrized by the value $N_1 = 50, 60, 70, 80, 90$ of the threshold, simulation (a) and diffusion (b) results

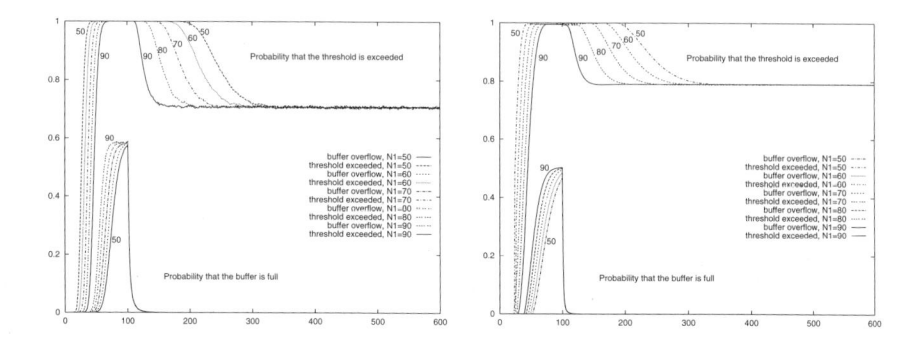

Fig. 16. Probability that the buffer of length $N = 100$ is full (priority cells are lost) and that the threshold is exceeded (ordinary cells are lost) as a function of time, parametrized by the threshold value $N_1 = 50, 60, 70, 80, 90$; simulation (a) and diffusion (b) results

The threshold value N_1 is a parameter of displayed curves. If N_1 increases the mean values of low priority cells increases (they have more space in the buffer, hence less of them is rejected) and the number of priority cells increases too (as there is more class 2 cells in the queue, class 1 cells wait longer).

Fig. 17 displays the mean number of high and low priority cells given by the approach we have described above and compared with simulation results. We see that the steady state mean value of class 2 cells is underestimated (because of overestimation of class 2 losses by diffusion approximation seen in Fig. 16) but the dynamics of class 2 cells vanishing from the queue during heavy saturation periods is well captured.

Some numerical problems were encountered when computing expressions of $\phi(x, t, \psi)$ and $\gamma_0(t)$ for very small values of $\lambda_{\text{eff}}^{(1)}(t)$, $\mu_{\text{eff}}^{(2)}(t)$ and forced us to very careful programming.

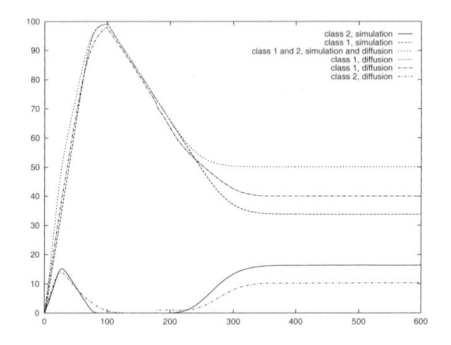

Fig. 17. Mean value of class 1 and class 2 cells as a function of time; $N = 100$, $N_1 = 50$

9 Conclusions

We describe how the diffusion approximation formalism is applied to study transient and behavior of G/G/1 and G/G/1/N network of queueus and we present some other models related to congestion control. The way we switch from one model to another demonstrates the flexibility of the method. Also the introduction of self-similar traffic is possible: as we change the diffusion parameters each small time-interval, we can modulate them to reflect self-similarity and long-term correlation of the traffic. Some other applications may be considered: recently we have used the diffusion approximation to estimate transfer times inside a sensor network [22], to study priority queues [24], the work of call centers [23], and to investigate the stability of TCP connections with IP routers having AQM queues [21]. Also the application of diffusion approximation to model wireless networks based on IEEE 802.11 standard gives promising results, [25].

Therefore we consider the diffusion approximation as a very convenient tool in the analysis of transient states queueing models in performance evaluation of computer and communication networks.

Acknowledgements

This research was partially financed by the Polish Ministry of Science and Education grant N517 025 31/2997.

References

1. Atmaca, T., Czachórski, T., Pekergin, F.: A Diffusion Model of the Dynamic Effects of Closed-Loop Feedback Control Mechanisms in ATM Networks. In: 3rd IFIP Workshop on Performance Modelling and Evaluation of ATM Networks, Ilkley, UK, July 4-7 (1995); rozszerzona wersja w. Archiwum Informatyki Teoretycznej i Stosowanej (1), 41–56 (1999)
2. Burke, P.J.: The Output of a Queueing System. Operations Research 4(6), 699–704
3. Cox, R.P., Miller, H.D.: The Theory of Stochastic Processes. Chapman and Hall, London (1965)
4. Czachórski, T.: A Multiqueue Approximate Computer Performance Model with Priority Scheduling and System Overhead. Podstawy Sterowania 10(3), 223–240 (1980)
5. Czachórski, T.: A diffusion process with instantaneous jumps back and some its applications. Archiwum Informatyki Teoretycznej i Stosowanej 20(1-2), 27–46
6. Czachórski, T., Fourneau, J.M., Pekergin, F.: Diffusion Model of the Push-Out Buffer Management Policy. In: IEEE INFOCOM 1992, The Conference on Computer Communications, Florence (1992)
7. Czachórski, T.: A method to solve diffusion equation with instantaneous return processes acting as boundary conditions. Bulletin of Polish Academy of Sciences, Technical Sciences 41(4) (1993)
8. Czachórski, T., Fourneau, J.M., Pekergin, F.: Diffusion model of an ATM Network Node. Bulletin of Polish Academy of Sciences, Technical Sciences 41(4) (1993)
9. Czachórski, T., Fourneau, J.M., Pekergin, F.: Diffusion Models to Study Nonstationary Traffic and Cell Loss in ATM Networks. In: ACM 2nd Workshop on ATM Networks, Bradford (July 1994)
10. Czachórski, T., Fourneau, J.M., Kloul, L.: Diffusion Approximation to Study the Flow Synchronization in ATM Networks. In: ACM 3rd Workshop on ATM Networks, Bradford (July 1995)
11. Czachórski, T., Fourneau, J.M., Pekergin, F.: The dynamics of cell flow and cell losses in ATM networks. Bulletin of Polish Academy of Sciences, Technical Sciences 43(4) (1995)
12. Czachórski, T., Pekergin, F.: Diffusion Models of Leaky Bucket and Partial Buffer Sharing Policy: A Transient Analysis. In: 4th IFIP Workshop on Performance Modelling and Evaluation of ATM Networks, Ilkley (1996); also in: Kouvatsos, D.: ATM Networks, Performance Modelling and Analysis. Chapman and Hall, London (1997)
13. Czachórski, T., Atmaca, T., Fourneau, J.-M., Kloul, L., Pekergin, F.: Switch queues – diffusion models (in polisch). Zeszyty Naukowe Politechniki /Sl/askiej, seria Informatyka (32) (1997)
14. Czachórski, T., Pekergin, F.: Transient diffusion analysis of cell losses and ATM multiplexer behaviour under correlated traffic. In: 5th IFIP Workshop on Performance Modelling and Evaluation of ATM Networks, Ilkley, UK, July 21-23 (1997)
15. Czachórski, T., Pastuszka, M., Pekergin, F.: A tool to model network transient states with the use of diffusion approximation. In: Performance Tools 1998, Palma de Mallorca, Hiszpania, wrzesie/n (1998)
16. Czachórski, T., Jedrus, S., Pastuszka, M., Pekergin, F.: Diffusion approximation and its numerical problems in implementation of computer network models. Archiwum Informatyki Teoretycznej i Stosowanej (1), 41–56 (1999)

17. Czachórski, T., Fourneau, J.-M., Kloul, L.: Diffusion Method Applied to a Handoff Queueing Scheme. Archiwum Informatyki Teoretycznej i Stosowanej (1), 41–56 (1999)
18. Czachórski, T., Pekergin, F.: Probabilistic Routing for Time-dependent Traffic: Analysis with the Diffusion and Fluid Approximations. In: IFIP ATM Workshop, Antwerp (1999)
19. Czachórski, T., Pekergin, F.: Modelling the time-dependent flows of virtual connections in ATM networks. Bulletin of Polish Academy of Sciences, Techical Sciences 48(4), 619–628 (2000)
20. Czachórski, T., Fourneau, J.-M., Jędruś, S., Pekergin, F.: Transient State Analysis in Cellular Networks: the use of Diffusion Approximation. In: QNETS 2000, Ilkley (2000)
21. Czachórski, T., Grochla, K., Pekergin, F.: Stability and dynamics of TCP-NCR(DCR) protocol in presence of UDP flows. In: García-Vidal, J., Cerdà-Alabern, L. (eds.) Euro-NGI 2007. LNCS, vol. 4396, pp. 241–254. Springer, Heidelberg (2007)
22. Czachórski, T., Grochla, K., Pekergin, F.: Un modèle d'approximation de diffusion pour la distribution du temps d'acheminement des paquets dans les réseaux de senseurs. In: Proc. of CFIP 2008, Les Arcs, Mars 25-28 (2008), Proceedings, edition electronique, http://hal.archives-ouvertes.fr/CFIP2008
23. Czachórski, T., Fourneau, J.-M., Nycz, T., Pekergin, F.: Diffusion approximation model of multiserver stations with losses. In: Proc. of Third International Workshop on Practical Applications of Stochastic Modelling PASM 2008, Palma de Mallorca, September 23 (2008); To appear also as an issue of Elsevier's ENTCS (Electronic Notes in Theoretical Computer Science)
24. Czachórski, T., Nycz, T., Pekergin, F.: Transient states of priority queues – a diffusion approximation study. In: Proc. of The Fifth Advanced International Conference on Telecommunications AICT 2009, Venice, Mestre, Italy, May 24-28 (2009)
25. Czachórski, T., Grochla, K., Nycz, T., Pekergin, F.: A diffusion approximation model for wireless networks based on IEEE 802.11 standard submitted to COMCOM Special Journal Issue on Heterogeneous Networks: Traffic Engineering and Performance Evaluation of Computer Communications
26. Duda, A.: Diffusion Approximations for Time-Dependent Queueing Systems. IEEE J. on Selected Areas in Communications SAC-4(6) (September 1986)
27. Feller, W.: The parabolic differential equations and the associated semigroups of transformations. Annales Mathematicae 55, 468–519 (1952)
28. Feller, W.: Diffusion processes in one dimension. Transactions of American Mathematical Society 77, 1–31 (1954)
29. Filipiak, J., Pach, A.R.: Selection of Coefficients for a Diffusion-Equation Model of Multi-Server Queue. In: PERFORMANCE 1984, Proc. of The 10th International Symposium on Computer Performance. North Holland, Amsterdam (1984)
30. Gaver, D.P.: Observing stochastic processes, and approximate transform inversion. Operations Research 14(3), 444–459 (1966)
31. Gaver, D.P.: Diffusion Approximations and Models for Certain Congestion Problems. Journal of Applied Probability 5, 607–623 (1968)
32. Gelenbe, E.: On Approximate Computer Systems Models. J. ACM 22(2) (1975)
33. Gelenbe, E., Pujolle, G.: The Behaviour of a Single Queue in a General Queueing Network. Acta Informatica 7(fasc. 2), 123–136 (1976)
34. Gelenbe, E.: A non-Markovian diffusion model and its application to the approximation of queueing system behaviour, IRIA Rapport de Recherche no. 158 (1976)
35. Gelenbe, E.: Probabilistic models of computer systems. Part II. Acta Informatica 12, 285–303 (1979)
36. Gelenbe, E., Labetoulle, J., Marie, R., Metivier, M., Pujolle, G., Stewart, W.: Réseaux de files d'attente – modélisation et traitement numérique, Editions Hommes et Techniques, Paris (1980)

37. Gelenbe, E., Mang, X., Feng, Y.: A diffusion cell loss estimate for ATM with multiclass bursty traffic. In: Kouvatsos, D.D. (ed.) Performance Modelling and Evaluation of ATM Networks, vol. 2. Chapman and Hall, London (1996)
38. Gelenbe, E., Mang, X., Önvural, R.: Diffusion based statistical call admission control in ATM. Performance Evaluation 27-28, 411–436 (1996)
39. Halachmi, B., Franta, W.R.: A Diffusion Approximation to the Multi-Server Queue. Management Sci. 24(5), 522–529 (1978)
40. Heffes, H., Lucantoni, D.M.: A Markov modulated characterization of packetized voice and data traffic and related statistical multiplexer parformance. IEEE J. SAC SAC-4(6), 856–867 (1986)
41. Heyman, D.P.: An Approximation for the Busy Period of the $M/G/1$ Queue Using a Diffusion Model. J. of Applied Probility 11, 159–169 (1974)
42. Iglehart, D., Whitt, W.: Multiple Channel Queues in Heavy Traffic, Part I-III. Advances in Applied Probability 2, 150–177, 355–369 (1970)
43. Iglehart, D.: Weak Convergence in Queueing Theory. Advances in Applied Probability 5, 570–594 (1973)
44. Jouaber, B., Atmaca, T., Pastuszka, M., Czachórski, T.: Modelling the Sliding window Mechanism. In: The IEEE International Conference on Communications, ICC 1998, Atlanta, Georgia, USA, czerwiec 7-11, pp. 1749–1753 (1998)
45. Jouaber, B., Atmaca, T., Pastuszka, M., Czachórski, T.: A multi-barrier diffusion model to study performances of a packet-to-cell interface, art. S48.5, Session: Special applications in ATM Network Management. In: International Conference on Telecommunications ICT 1998, Porto Carras, Greece, czerwiec 22-25 (1998)
46. Kimura, T.: Diffusion Approximation for an M/G/m Queue. Operations Research 31(2), 304–321 (1983)
47. Kleinrock, L.: Queueing Systems. Theory, vol. I. Computer Applications, vol. II. Wiley, New York (1975, 1976)
48. Kobayashi, H.: Application of the diffusion approximation to queueing networks, Part 1: Equilibrium queue distributions. J. ACM 21(2), 316–328 (1974); Part 2: Nonequilibrium distributions and applications to queueing modeling. J. ACM 21(3), 459–469 (1974)
49. Kobayashi, H.: Modeling and Analysis: An Introduction to System Performance Evaluation Methodology. Addison Wesley, Reading (1978)
50. Kobayashi, H., Ren, Q.: A Diffusion Approximation Analysis of an ATM Statistical Multiplexer with Multiple Types of Traffic, Part I: Equilibrium State Solutions. In: Proc. of IEEE International Conf. on Communications, ICC 1993, Geneva, Switzerland, May 23-26, pp. 1047–1053 (1993)
51. Kulkarni, L.A.: Transient behaviour of queueing systems with correlated traffic. Performance Evaluation 27-28, 117–146 (1996)
52. Lee, D.-S., Li, S.-Q.: Transient analysis of multi-sever queues with Markov-modulated Poisson arrivals and overload control. Performance Evaluation 16, 49–66 (1992)
53. Maglaris, B., Anastassiou, D., Sen, P., Karlsson, G., Rubins, J.: Performance models of statistical multiplexing in packet video communications. IEEE Trans. on Communications 36(7), 834–844 (1988)
54. Newell, G.F.: Queues with time-dependent rates, Part I: The transition through saturation. J. Appl. Prob. 5, 436-451 (1968); Part II: The maximum queue and return to equilibrium, 579–590 (1968); Part III: A mild rush hour, 591–606 (1968)
55. Newell, G.F.: Applications of Queueing Theory. Chapman and Hall, London (1971)
56. Pastuszka, M.: Modelling transient states in computer networks with the use of diffusion approximation (in polish), Ph.D. Thesis, Silesian Technical University (Politechnika Śląska), Gliwice (1999)

57. Reiser, M., Kobayashi, H.: Accuracy of the Diffusion Approximation for Some Queueing Systems. IBM J. of Res. Develop. 18, 110–124 (1974)
58. Sharma, S., Tipper, D.: Approximate models for the Study of Nonstationary Queues and Their Applications to Communication Networks. In: Proc. of IEEE International Conf. on Communications, ICC 1993, Geneva, Switzerland, May 23-26, pp. 352–358 (1993)
59. Stehfest, H.: Algorithm 368: Numeric inversion of Laplace transform. Comm. of ACM 13(1), 47–49 (1970)
60. Veillon, F.: Algorithm 486: Numerical Inversion of Laplace Transform. Comm. of ACM 17(10), 587–589 (1974); also: Veillon, F.: Quelques méthodes nouvelles pour le calcul numérique de la transformé inverse de Laplace, Th. Univ. de Grenoble (1972)
61. Zwingler, D.: Handbook of Differential Equations, pp. 623–627. Academic Press, Boston (1989)

Cross Layer Simulation: Application to Performance Modelling of Networks Composed of MANETs and Satellites

Riadh Dhaou[1], Vincent Gauthier[2], M. Issoufou Tiado[1], Monique Becker[2], and André-Luc Beylot[1]

[1] INPT-ENSEEIHT/IRIT
[2] GET-INT/SAMOVAR
{dhaou,tiado,beylot}@enseeiht.fr, {gauthier,becker}@int-evry.fr

Abstract. The Cross layer concept is a new way to see the quality of service in the network. It consists in adapting the current mechanisms at one level to the underlying levels and the definition of information to share between, not necessarily adjacent, levels and the global optimisation instead of multiple optimisations. Performance optimisation of the whole system is a crucial step in the process of design and validation of new systems. Models of large, complex and dynamic systems can often be reduced to smaller sub-models, for easier analysis, by processes known as decomposition or aggregation methods. Those techniques have to be implemented in a dynamic simulation tool. It has to be dynamic because simulations are driven by dynamic data and entail the ability to incorporate additional data (either archived or collected online) and reversely the simulator will be able to dynamically steer the measurement process.

Keywords: Cross layer, simulation, heterogeneous networks, network modelling.

1 General Introduction

The performance of wireless systems is extremely sensitive to the mobility, to the dynamicity and to the environment. Those systems may be extended on a large scale (long delay satellites, delay tolerant networks, IEEE802.20...) and are more and more complex.

The concept of layered network stack is the main idea which allowed the heterogeneous development of networks and standard protocols. But the specificity of the wireless networks, their interconnection with fixed wired networks and the use of adapted QoS mechanisms to wireless conditions have an impact on the performance.

The Cross layer concept is a new way to see the quality of service in the network. It does not consist in the addition of reservation mechanisms of bandwidth or any other Ad Hoc mechanism, but in adapting the current mechanisms at one level to the underlying levels; the definition of pertinent information, to share between not necessarily adjacent levels; and the global optimisation, instead of multiple optimisations.

Performance optimisation of the whole system is a crucial step in the process of design and validation of new systems. Models of large, complex and dynamic systems can often be reduced to smaller sub-models, for easier analysis, by processes known as decomposition or aggregation methods. Certain criteria for successful decomposition can be established. Models have to be designed on several levels and all the levels cannot be studied simultaneously. Decomposition has long been recognised as a powerful tool for the analysis of large and complex systems.

Those techniques have to be implemented in a dynamic simulation tool. It has to be dynamic because simulations are driven by dynamic data and entail the ability to incorporate additional data (either archived or collected online) and reversely the simulator will be able to dynamically steer the measurement process.

The first part of this paper provides an overview of the time, space and level dependence of highly complex networks. Then a state of the art of cross-layer design for today networks is presented.

The paper then designs cross-layer simulations which are aggregation methods, where an aggregate is a layer model. Those simulations are dynamic, autonomic and multi-layer. Examples of applications are presented: they concern the performance evaluation of MANETS. For the evaluation of routing algorithms MAC and network layers have to be simultaneously simulated. For transport layer MAC layer has to be modelled. Those cross-layer simulations happen to be difficult to operate, but they are promising methods.

In the second part we are interested in performances issues of routing in Ad Hoc networks and IP over GEO satellite links. Ad Hoc networks are collections of high rate randomly moving wireless devices within a particular area. These devices vary dynamically in their resource-richness. Such highly dynamic systems rise specific problems such as routing, media access control (MAC) and QoS mechanisms.

Many routing algorithms are proposed and standardisation is in process. Route is made difficult by the continuous moving of the devices as nodes in the networks. Access to the medium is made difficult by the so called hidden node problem. QoS mechanisms need to be designed by taking into account several techniques such as load balancing, MAC-level and IP-level differentiating mechanisms. Battery energy may be also saved by optimising the emission power. The distribution of the bandwidth in these mobile Ad Hoc Networks (MANETs) and GEO satellite networks can be very unfair. The behaviour of TCP and the introduction of QoS mechanisms at different levels and their interdependence remain to be studied more in detail in those systems.

Part 1 - Time, Space and Level Cyclic Dependence of Highly Complex and Dynamic Networks: Application to Dynamic Cross Layer Design of Mobile Systems

2 Introduction - Part 1

Performance evaluation is a crucial step in the process of designing and validating new complex telecommunications systems. Complexity comes from mobility, dynamicity, extension on a large scale and interaction between layers and elements. But performance evaluation of such systems is also a technical challenge. Out of the three classical methods for evaluating performances (analytical study, simulation study and experimental study) [26], only one may be applied in this context: the simulation. Indeed, analytical studies can only cope with systems of a reasonable complexity. This is definitely not the case for the new generations of telecommunications systems (satellite, mobile and sensor networks). And, as long as the whole system is not operational (this is obviously the case at the early stage of the design), experimental studies are either impossible, or restricted to small parts of the system.

Studies based on simulation techniques require both a computational model of the system and a simulation environment. Computational models are particularly difficult to elaborate because the system is large, complex and highly dynamic. At the physical level, for example, propagation effects are hard to model and often need to be accurately simulated in order to get realistic results. Furthermore, radio interfaces are now offering sophisticated mechanisms like power control or adaptive coding. Those particularities are added to the traditional complexity of mobile networks based on IP networks and protocols.

The motivation for improving the performance of networks has increased the interest in systems which rely on interactions between different levels. Cross layer systems appeared recently specially for wireless networks. The research, specific to cross layer systems, aims to optimise the overall performances of the protocol stack of a network rather than a local optimisation of each layer. The technique consists in taking into account, at the same time, information available from different levels [10], not necessarily adjacent, in order to create a system much more sensitive to wireless specificity.

This approach is advantageous with reserve to respect a minimal layered generic model [7], because the separation between layers allows a better interoperability between network elements (IP over everything) and allows to introduce relatively easily new local protocols without changing the other layers of the stack. There is a need to set up a generic element that interfaces with existing elements and using the current network interfaces [9].

Out of the challenges in designing cross layer systems, a major interest of performance evaluation community is:

- To design and develop a simulation environment introducing innovative features to ease model reuse and hierarchical modelling activities, as well as implementing powerful computing techniques;
- To define generic cross layer system network models, and implement their simulation within the framework; all mechanisms shall be addressed in order to provide a global model of the system;

- To identify and evaluate appropriate simulation techniques in cross layer network context.
- To implement mechanisms in one of the levels of the protocol architecture, while modelling in a rather accurate way the other levels (even if all the protocols are not yet standardised). For a study on network level, physical layer, data link and network level models must be developed; these models having to coexist (since the parameters of some models may influence the other hierarchical levels).

For these reasons, new simulation techniques called "Cross-layer Simulations" are set up. Those techniques have to be implemented in a dynamic simulation tool. It has to be dynamic because simulations are driven by dynamic data and entail the ability to incorporate additional data (either archived or collected online), and reversely the simulator will be able to dynamically steer the measurement process.

The generic cross layer network model, together with cross layer simulation model facilities offered by the tool, shall allow studies on cross layer systems.

This paper presents a state of the art on cross layer systems. Section 2 is emphasising on modelling problems that are specific to wireless cross layer networks simulation. Section 3 is presenting some original approaches that will be of use within a cross layer simulation environment.

3 Wireless Networks Complexity

Wireless networks have particularities that distinguish them from wired networks.

The wireless channels vary over time and space and has short term correlation due to multi-path. These variations are caused by either motion of the wireless device or changes in the surrounding physical environment, and lead to detector errors. This causes bursts of errors to occur during which packets cannot be successfully transmitted on the link.

Small scale channel variations due to fading are such that states of different channels can asynchronously switch from "good" to "bad" within a few milliseconds and vice versa. Furthermore, very strong forward error correction codes (i.e., very low rates) cannot be used to eliminate errors because this technique leads to reduce spectral efficiency [5].

In addition to small-scale channel variations, there are also spatiotemporal variations on a much greater time scale. Large scale channel variation means that the average channel rate conditions depend on user locations and interference levels. Thus, due to small scale and large scale changes in the channel access time than others based on their location or mobile velocity, even if their data rate requirement is the same as or less than other users.

4 Cross Layer Design

The concept of layered network stack is the main idea which allowed the heterogeneous development of networks and standard protocols. On these principles is that Internet networks could be born and with it the birth of all applications on IP and the true heterogeneous development of the networks. This model of this development gets many advantages such as:

- Each layer is independent.
- One generic model would be able to solve multiple problems
- The layers are quite interchangeable (different MAC layers exist)

But the growth of technologies in the field of the networks and the multiplication of the needs made the basic model more complex and these interactions (interfaces) which was formerly well structured, are now more complex and are not clearly defined.

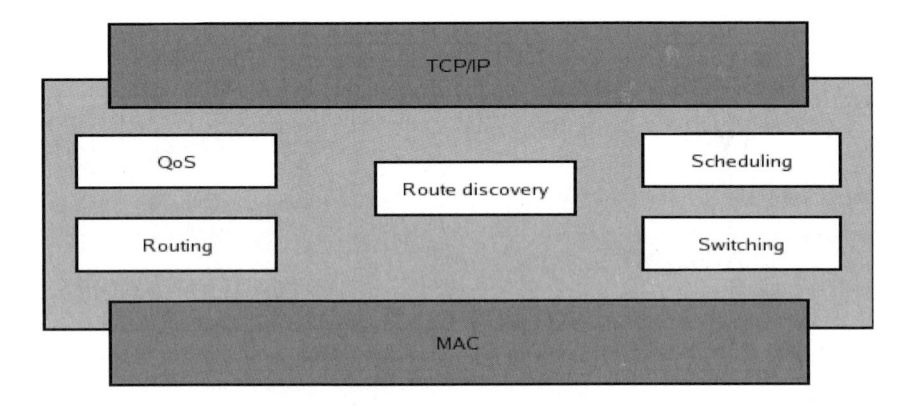

Fig. 1. Layers functionality

To avoid these problems in the GSM networks packets, multiple encapsulations is used to interface correctly GSM network and IP network. In the case of the ADSL access network, the same solution is applied to this problem. But, this technique increases enormously the overhead and the complexity the network topology. In the case of the MPLS networks, the interface between the networks layer and MAC layer was complex and then the limit between MAC layer and IP layer became very trouble (the IP addresses are used both by MAC layer and IP layer). Current researches in wireless networks, ad hoc networks and sensor networks find out new challenges which must be solve. In this type of networks, the problems of radio resources which are rare and time varying make problems which are very difficult to adapt with a layer structure where information between the different layers are independent and completely separate.

The evolution towards one Internet network which is able to interface with ad hoc networks and sensors networks in the same addressing map take the following problem: how to preserve the compatibility of the existing model taking in account the currently defined interfacing problems.

4.1 Inter-Layer Interactions

The diversity of the media and the different problems, which are likely to appear, cannot be solved in a generic and total way. Then it is necessary to create algorithms closer to the needs of each resource. The layers between themselves have to be able to exchange information about each problem. The concept of Cross-Layer makes possible to exchange information between the various networks layers. It allows improving the algorithms available at each network layer, thanks to information available in the other ones. One can for example use information about the link quality in the MAC Layer to create more powerful algorithms of routing. That's why the structure of networks stack must be modified so that information exchanges between the layers are possible. The layer structure itself does not have therefore been modified, that's why the creation of elements external to the networks stack, is used for this aim. Network Status is a component of the stack that interfaces the different layers between themselves. It acts as a database where each network layer can put or get information.

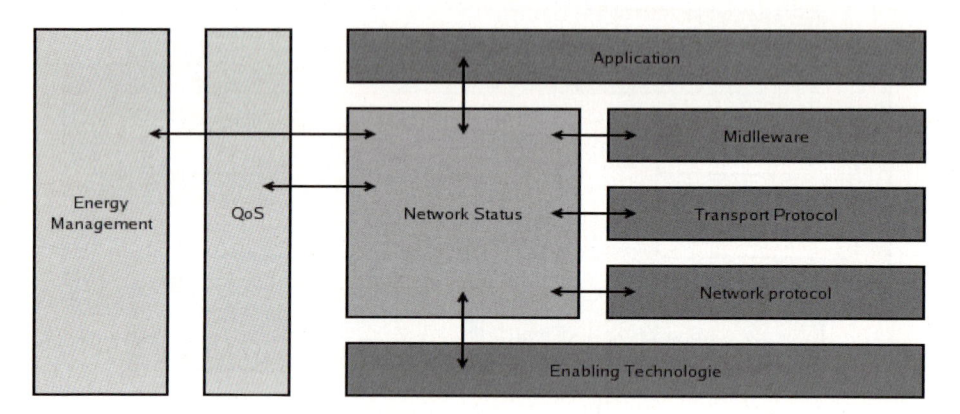

Fig. 2. Inter-layer interactions

The main think to be taken into account is to choose which information will be shared and which one will impact on the various algorithms (of routing for example). In wireless networks the element that we have to take into account is the radio channel quality and the energy consumption. But these elements cannot be managed simply at a local level but must be managed in a distributed way in the network. That's why a new network's metric must be created, instead of looking only to the number of hops between the transmitter and the receiver; it will also look at the quality of the radio channel along the entire path between the transmitter and the receiver. The assumption of channel quality in the metric can be considered as a new way of doing quality of service. Because it is now seen as an internal component of the networks, as for example the ad hoc layer in the grid systems. Its main advantage is the overall improving of the network capacity, and also the decrease of the QoS management complexity.

4.2 MANETS or Sensor Networks

The sensor networks are collaborative networks and very directed towards a precise application field, which implies that their structure is more easily controllable than the ad hoc network one. Because in the ad hoc networks the mobility and the very changeable nature of flows make very difficult any forecast on the nature of flows to transport. The structure of the network by its nature (mobile) makes the connectivity very uncontrollable. In the ad hoc networks the network stack is well defined and it is quite difficult to make changes to improve it. Whereas in the sensor networks, we can easily avoid this lack of flexibility by considering other models of protocol stack, more adaptable and arranged un-hierarchically. That's why the sensor networks are excellent candidates to implement a Cross-layer structure in their network stack for the exchange of information between the different layers.

It is obvious that cross layer design creates interactions, some intended, and others unintended. There is a need to examine the dependency relations, and to enforce time scale separation. Authors, in [7], argue that proposes of cross layer design must therefore consider the totality of the design, including the interactions with other layers, and also what other potential suggestions might be barred because they would interact with the particular proposal being made. They must also consider the architectural long term value of the suggestion. Cross layer design proposal must therefore be holistic rather than fragmenting.

Generally, properties of wireless networks are different from those of wired networks. In fact the interference created by the nodes emitting packets slow down the global throughput. The versatility of the wireless medium and the variability of the capacity of transmission of nodes have also an impact on the network throughput.

Another important factor in wireless networks is related to the problem of hidden station, where to avoid collisions a station that is emitting reduce the other neighbouring stations to silence [9].

Furthermore, routing algorithms used in wired networks such as shortest path are not adapted to wireless Ad Hoc networks [7].

Finally, the performances of the transport protocol TCP is not really adapted to wireless networks, or to Ad Hoc systems prone to high BER and mobility. Those factors induce connection interruptions, interpreted by TCP as a congestion indication, and decrease connections throughput. Congestion control mechanisms, including the congestion avoidance phase in all variants of TCP, have recently been shown to be distributed algorithms implicitly solving network utility to be distributed algorithms implicitly solving network utility maximisation problems, which are linearly constrained by link capacities that are assumed to be fixed quantities. However, network resources can sometimes be allocated to change link capacities, therefore change TCP dynamics and the optimal solution to network utility maximisation. For example, in CDMA wireless networks, transmit powers can be controlled to give different Signal to Interference Ratios (SIR) on the links, changing the attainable throughput on each link [1].

4.3 Examples of Cross Layers

The cross layers should allow to consult in real time available information on each network layer, nowadays this is not possible. Authors of [6] observe that some network functions, such as energy management, security and co-operation are cross layer by nature. The authors propose cross layer architecture, presenting a network status plan, which divides every shared network's functionality in different layers. Table 1 presents the levels implicated in different cross layer studies.

Table 1. Panel of cross layer studies at different levels

Physical layer	Channel state is useful to adapt the throughput to [1],[15],[20],[21].
MAC layer	The retransmission number at the MAC level may indicate the quality of the link, timer [2],[3],[4],[8],[11],[15],[20],[21],[23].
Network layer	Using quality of transmission information in routing algorithms [2],[3],[4],[8],[23].
Transport layer	The MAC layer could adapt the error control scheme among TCP retransmission timers and update the information on the delay for better QoS. [1], [2], [3], [11].

4.3.1 Physical-MAC Layers

A set of novel PHY-MAC mechanisms based on a cross layer dialogue was proposed. System efficiency improvement is achieved by means of automatic transmission rate adaptation, trading off generic packet switched CDMA access network [15]. The rate adaptation mechanism improves spectrum efficiency while keeping packet delay minimised. On the other hand, power dependent strategies reduce power consumption and inter-cell interference.

The joint effect of MAC and PHY layers on power efficiency was investigated. Specifically, Authors in [20] present a study of the link adaptation for a power efficient transmission by selecting a proper transmission mode and power level with the aid of a derived power efficiency model for IEEE 802.11a WLAN.

Another study, presented in [21] considers a reservation based medium access control MAC scheme where users reserve data channels through a slotted ALOHA procedure. The base station grants access to users in a Rayleigh fading environment using measurements at the physical layer and system information at the MAC layer.

4.3.2 MAC Layer

In [19] authors concentrate on the problem of medium access for wireless multi-hop networks. They first study CSMA/CA, and find that its performance strongly depends on the choice of the accompanying routing protocol. They then introduce two protocols that outperform CSMA/CA, both in terms of energy efficiency and achievable throughput. The Progressive Back-Off Algorithm (PBOA) performs medium access jointly with the power control. The Progressive Ramp-Up Algorithm (PRUA) sacrifices energy efficiency in favour of a tighter packing of transmissions and higher throughput.

The authors observe that: (i) power control can be very helpful in terms of energy efficiency, but its gains are limited in multi-hop environments. (ii) The First In First Out (FIFO) queuing discipline is sub-optimal for use in wireless ad hoc networks. A more relaxed rule can lead to a tighter packing of transmissions. (iii) The choice of routing protocol can significantly reduce the capacity of the network. Specifically, any routing strategy that assigns a single route to a node pair typically suffers a penalty in its performance. (iv) CSMA/CA is not well-suited for communication over weak links. Therefore, it should be coupled with routing protocols that avoid, if possible, such links. (v) Medium Access protocols that are distributed inherently carry a penalty in their performance since, to achieve capacity, it is necessary that nodes that are separated by an arbitrary large distance coordinate their transmissions. By definition, this is impossible in a distributed algorithm.

Authors in [5] observe that multi user diversity gains can substantially improve wireless network throughput. However when giving different types of QoS constraints, most of the studies so far are confined to single-cell scenario.

4.3.3 MAC-Network Layers

Channel reservation control packets employed at the MAC layer can be utilised at the physical layer in exchanging timely channel estimation information to enable an adaptive selection of a spectrally efficient transmission rate. In particular, the size of a digital constellation can be varied dynamically based on the channel condition estimated at the receiver which can be relayed to the transmitter via the control packets [2]. In addition this channel adaptive information gathered at the MAC layer can be communicated to the routing layer via different routing metrics for optimal route selection.

Many authors propose energy efficient schemes for wireless ad hoc and sensor networks. [4] proposes a scheme that utilises cross layer interactions between the network layer and MAC sub-layer to achieve energy conservation. In order to improve cross layer optimisation, [8] propose a multiple access collision avoidance protocol that combines RTS/CTS with scheduling algorithms to support the multicast routing protocol. The protocol avoids collision by including additional information in the RTS. Proposed scheme, together with extra benefits, such as power saving, reliable data transmission and higher channel utilisation compared with CSMA or multiple unicast, enables the support of multicast services.

Currently approaches solving problems associated with multicast for ad hoc networks solve them at the network layer. Authors in [8] propose a procedure for which, branch nodes are generated. In a simple multicast MAC protocol for branch node's transmitting multicast data is proposed, but there is no consideration of the hidden node problem. In [8], authors propose a multiple access collision avoidance protocol for multicast services that resolves the hidden node problem in ad hoc network.

The awareness of higher layer requirements is addressed in [16]. In fact, fundamentally applications look for services, not for IP addresses. This is especially so in ad hoc zero configuration application scenario. The "demand" for on-demand operation comes in the form of a service request from the user. The way the lower layers function should be determined by what the application requires. This suggests that having some information exchange across the layers would be of use. Specifically, the link-layer can use information about the "demand" to decide whether or not to form a particular link, or to decide which of a set of links to form.

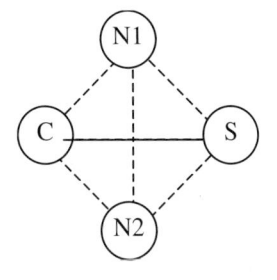

Fig. 3. Link layer topology

Consider the scenario depicted in figure 3. Node C is a client looking for a service S, N1 and N2 are two other nodes. All nodes are in physical proximity of each other. The link formation layer has to decide which links to form. Without any information on what the layer above requires, the algorithm to form links would be direction less. It could result in an inefficient route between C and S. Worse, it could unnecessarily activate extra links (the dotted ones in the figure), effectively voiding all the effort that has gone into the careful definition of low power link modes.

Path coupling involves MAC layer interactions that impact the performance of network layer paths that are otherwise disjoint. These interactions are shown to have significant impact on energy efficiency, throughput and delay [23].

4.3.4 MAC-Transport Layers

In [1] authors present a distributed power control algorithm that couple with the TCP protocol to increase the end-to-end throughput and energy efficiency of multi-hop transmissions in wireless ad hoc networks. The authors prove that the nonlinearly couples system converges to the global optimum of the joint congestion control and power control problem.

[3] Incorporates user feedback into the protocol stack by which the TCP throughput of desired set of applications running on the mobile host can be dynamically controlled. In addition the authors use the lower layer connection and disconnection information, to improve TCP performance.

4.3.5 MAC-Transport-Application Layers

In wireless networks, FEC protection is required at the application layer regardless of the underlying transport layer protocol in order to deliver high bit rate multimedia. [11] show that the amount of FEC overhead required, by transport protocol such as UDP Lite is considerably less than traditional UDP. Hence UDP Lite provides improvement in bandwidth utilisation in order to deliver loss less multimedia. [11] illustrates the suitability of cross layer protocol strategies supporting application-specific multimedia.

4.3.6 Application Layer

Cross layer information help also application level to understand network status. Many authors propose cross layer techniques for adaptive video streaming over wireless networks [12] and adaptive filter based on image content and dynamically changed threshold based on cross layer information [13].

It is shown that application adaptation can provide significant energy reductions over both fixed and adaptive hardware. Furthermore, [14] achieve joint hardware application adaptation while preserving the logical separation between layers. In this paper, authors propose a set of end to end application layer techniques for adaptive video streaming scheme over wireless networks. The adaptation is done both with respect to channel and data. Authors demonstrated the effectiveness of the application layer adaptivity combined with the RLP layer granularity.

4.3.7 Cross Layer Design

The paper [6] presents the IST-2001-38113 Mobile Man Project. The paper present the Mobile Man architecture (depicted on Figure 4). Some network functions, such as

```
┌─────────────┐    ┌──────────────────┐ ▲▲
│             │◄──►│  Applications    │ │ │
│             │    ├──────────────────┤ │ │
│             │◄──►│  Middleware      │ │ │
│  Network    │    ├──────────────────┤ │ │
│  Status     │◄──►│  Transport       │ │ │
│             │    ├──────────────────┤ │ │
│             │◄──►│  Network         │ │ │
│             │    ├──────────────────┤ │ │
│             │◄──►│  MAC and physical│ │ │
└─────────────┘    └──────────────────┘ ▼▼
```

Cross layer interaction   ◄──►   Stackwide features
Strict layer interaction   ↕
Security and cooperation   ◄═►
Energy management   ◄─►

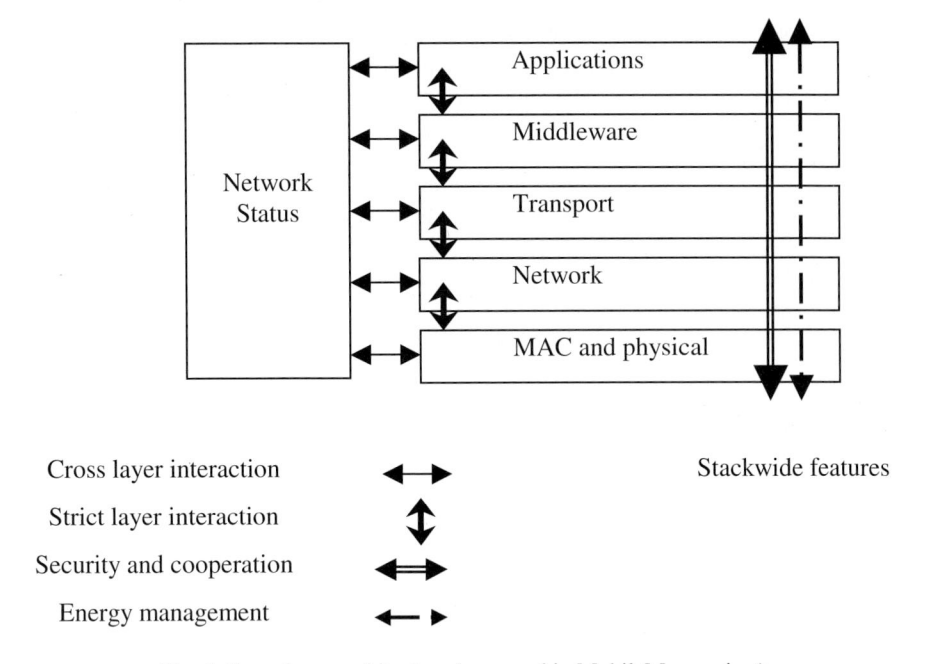

Fig. 4. Cross layer architecture (proposed in MobileMan project)

energy, management and security and cooperation are cross-layer by nature. Mobile Man seeks to extend cross-layering to all network functions through data sharing.

4.3.8 Cross Layer Information

Authors in [17] note that ad hoc networking is a multi-layer problem. The physical layer must adapt to rapid changes in link characteristics. Such as, the multiple access control layer needs to minimise collisions, allow fair access, and semi-reliably transport data over the shared wireless links in the presence of rapid changes and hidden or exposed terminals. The network layer needs to determine and distribute information used to calculate paths in a way that maintains efficiently when links change often and bandwidth is at a premium. Analysis above imply that each of protocol layer in ad hoc network is dependent each other, whether they directly connect or not. So a cross layer protocol design that supports adaptability and optimisation is needed.

However, authors note that, there remain many open questions in the understanding and implementation of this design philosophy. Such as, when building such cross

Table 2. Extend sense topology information

Layer	Information
Application layer	Topology control algorithm, Server location, Network Map
Transport layer	Congestion window, Timeout clock, Packet losses rate
Network layer	Routing affinity, Routing lifetime, Multiple routing
MAC/Link layer	Link bandwidth, Link quality, MAC packet delay
Physical layer	Node's location, Movement pattern, Radio transmission range, SNR information

layer architecture, complexity issues are introduced that, if not managed properly, can be insuperable. How to balance the power control and other system constraint such as delay, mobility and longevity in ad hoc network is still another problem.

4.3.9 Cross Layer Signalling

[22] propose an efficient, flexible and comprehensive scheme defined as Cross Layer Signalling Shortcuts (CLASS) and observe that layer by layer propagation approach of signals follows the data propagation mode. Consequently, the intermediate layers have to be involved even if only source layer and the destination layer are actually targeted. This causes unnecessary processing overhead and propagation latency. Direct signalling between layers adopted in CLASS is presented in the table 3.

Table 3. QoS adaptation and information exchange (for CLASS)

	APPLICATION	TRANSPORT	NETWORK	LINK	PHYSICAL
APPLICATION Real/non real time services	O	↔ Packet loss ratio, jitter, etc.	↔ delay constraints, etc.	↔ Desired value of QoS parameters	←
TRANSPORT TCP/UDP/RTP		O		↔ Joint error control using BER, Handoff notification	
NETWORK IP/ IntServ/ Diffserv	↔		O	← Joint delay control	
LINK Link quality; FEC/ARQ	↔ Environment measurements reports: SNR, RSS, etc.	↔	→	O	←
PHYSICAL Channel conditions	→ Environment measurements reports: SNR, RSS, etc.			→	O

5 Wireless and Mobile Cross Layer Systems Modelling

5.1 Modelling Goals

There is a need for a dynamic and global simulator. It has to be dynamic not only because nodes are moving fast and handoffs are often but also because simulation granularity has to be modified according to the level of the context. It has to be global because the topology is permanently changing and it is necessary to take into account every node. Also, models at different levels are required. This leads to the need for an extension of classical simulators, which were not designed for dynamic and global complex system.

The goal is to evaluate the capacity, the availability, the quality of service, and the spectral effectiveness of such complex systems. The metrics used to estimate these parameters are the following:

Capacity is estimated by the maximum instantaneous traffic, the traffic mean, the total user count, as well as the average number of connections established during a given period.

Availability is evaluated from a probability of communication establishment, or a probability of communication maintenance.

Quality of service is evaluated at several levels, through criterion of delay, jitter, BER, FER, ATM QoS parameters (CTD, CDV, CMR, CER, CLR, SECBR...), communication establishment overhead, connection outage time...

Evaluating the **spectral effectiveness** derived from an estimation of the supported traffic as a function of the number of subscribers, the geographical zone and service classes.

The studies concern the availability, effectiveness, the routing, the handoff, and the quality of service... We want to be able to simulate and compare several of these mechanisms and to integrate them in other studies. For example, a study of end-to-end QoS will integrate a handoff model but also a resource allocation model. Let us detail later these levels.

5.2 Main Modelling Problems

Models of large and complex systems can often be reduced to smaller sub-models, for easier analysis, by a process known as decomposition. Certain criteria for successful decomposition can be established. Models have to be designed on several levels and all the levels cannot be studied simultaneously. So an aggregation technique may be used [42]. The complex system is shared into subsystems, which are studied independently. Then the global system is studied taking into account the subsystem dependencies. For example, to study a handoff mechanism, the failure problems are neglected and to study propagation problems, we do not have to simulate the whole system. Nevertheless, most of the mechanisms and phenomena are correlated and in several ways.

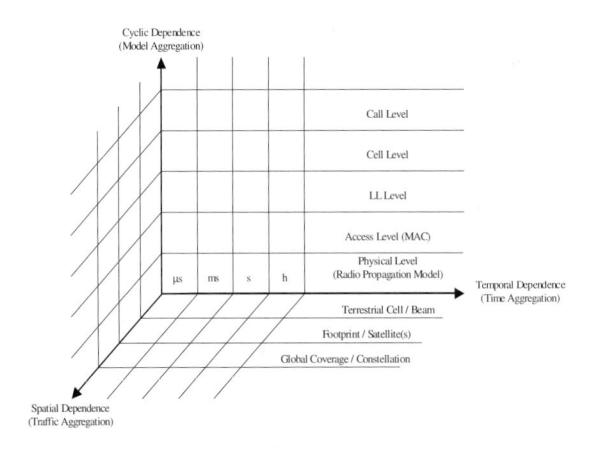

Fig. 5. Aggregation dimensions

5.2.1 Cyclic Dependence

A cyclic dependence occurs when several parts of the model, detailed on Figure 5, are correlated. For example, the performance of the radio channel depends on the traffic while the traffic itself depends on the performance of the radio channel, as shown on Figure 6. On another side, the interference calculation utilises the concurrent traffic.

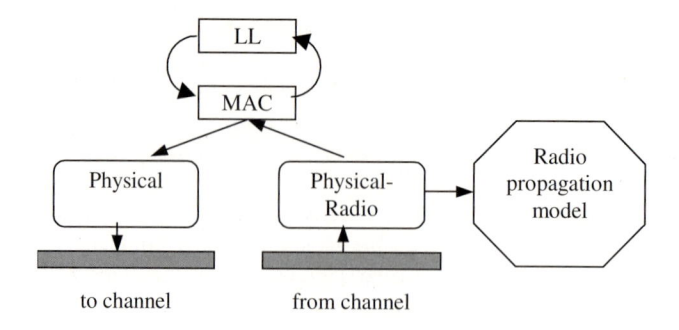

Fig. 6. Cyclic dependence

That leads to problems, for the representation of this concurrent traffic, for the relevance of its representation and the importance to take it into account.

In a limited space, many connections are handled simultaneously. Each of these connections generates a large amount of data, which results in a huge number of packets and a huge quantity of packets. But for some studies, the communication needs to be simulated both at the network and radio levels, because they are closely dependent. However, it is nearly impossible to carry out simulations, simultaneously, on two different levels.

In this case, the problem we have to solve is the choice of the model smoothness. The performance of a simulation has a significant impact on its feasibility, and its complexity has a significant impact on the confidence intervals. Studies focusing on a given level in general require a relatively coarse description of the other levels. For example, for a study on the MAC layer level, an evaluation of the BER is required from the radio layer. But it is not necessarily useful to include detailed models of orbitography in these BER computations: approximating the movement of nodes by simple equations makes it possible not to have to integrate movement equations and to replace bulky files handling by some simple calls to mathematical functions.

5.2.2 Temporal Dependence

The performance of the radio link at time t+1 depends on its performance at time t. This type of problem appears particularly when running step by step simulations with a very fine step and when running discrete event simulation with significantly correlated processes. That occurs when using detailed models of orbitography requiring, for their integration, a small step or when using long range dependence traffic models. In most cases, a trade-off has to be found between the smoothness of the model determining the step of the simulation and the desired result precision. Simulating process with long-term correlation is dangerous, because of the various time scales correlation. It is as dangerous as simulating non-stationary processes.

5.2.3 Space Dependence

The performance on site S at time t is related to the performance on a close site S' at t or near time t. The co-localised space zones are dependent, for instance, because they share a handoff mechanism or another mechanism or phenomenon. The complexity of

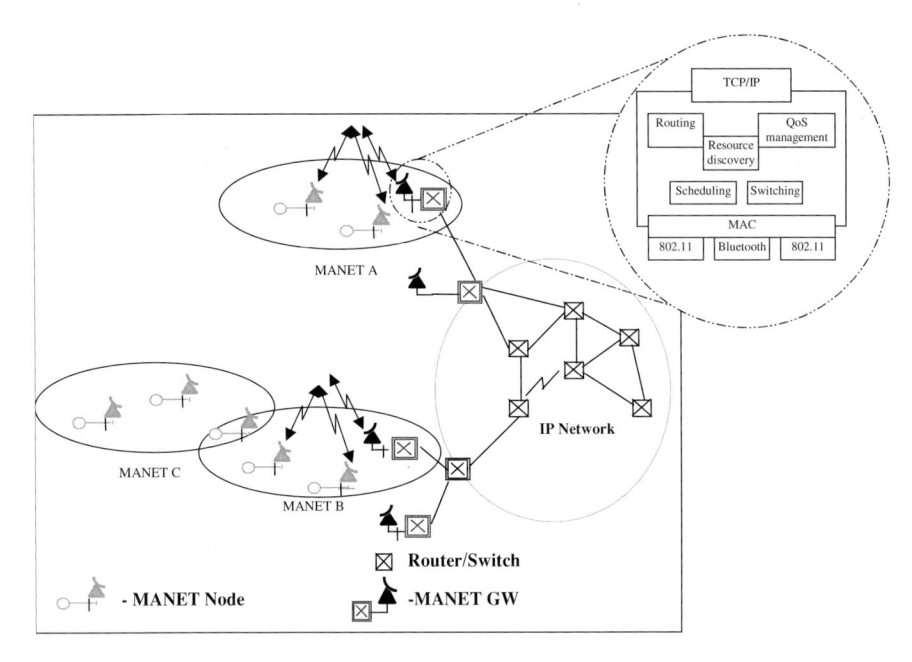

Fig. 7. Space dependence

global studies taking into account all the system can be reduced by undertaking the study at an elementary space granularity and by using the results of this study to estimate the parameters on higher levels: Iterative steps of traffic aggregation may be applied [24].

5.2.4 Multi-Scale Relations
There are also phenomena for which events at many scales of length or time do make contributions of comparable importance, or at least have a non negligible influence on each other. Any theory that describes these phenomena must take the entire spectrum of ranges into account in one way or another [25].

A typical example of multiple-time-level-space-scale dependence behaviour is provided by the highly complex, large and dynamic Ad Hoc and sensor systems. The high mobility of such systems make an experimental approach, based on measurements and testing, too much time consuming. Analytic or simulation models must be developed instead [26].

5.2.5 Choice of the Granularity
The choice depends on the studied topic. According to the type of study, various modelling levels of a same function or a same set of functions may be necessary. The modelling levels are represented on Figure 5. The multiple correlations are located; it is necessary here to identify these dependencies and treat them individually. We have to find and validate some efficient mechanisms in order to combine different time, space or model scales.

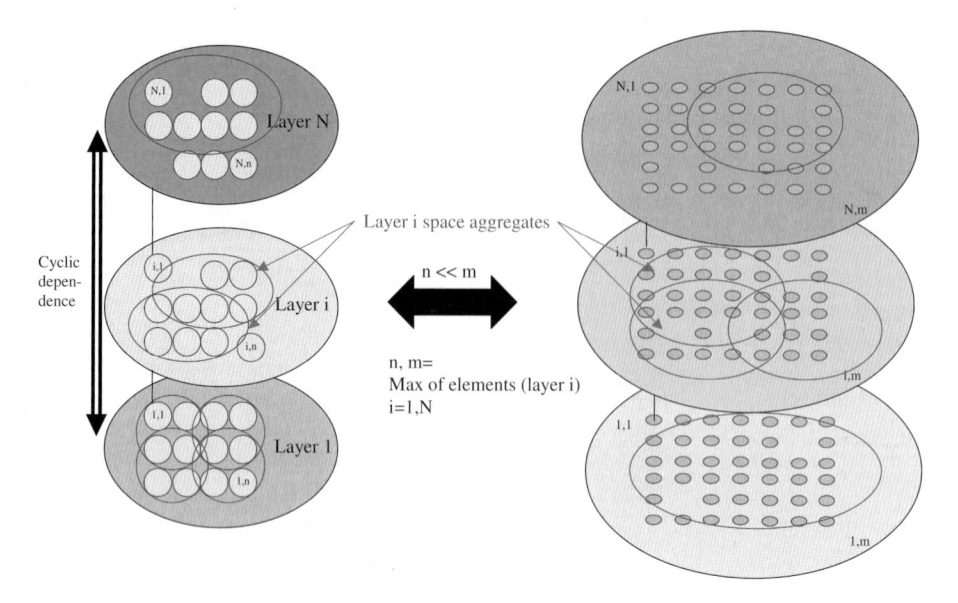

Fig. 8. Size and diversity of problems

Most of the problems we have mentioned are not treated in the traditional simulators (such as NS or OPNET), which leave the user to use some elementary block modules and to pile them up. But, the piled-up models cannot be simulated in a reasonable time to obtain realistic performance criteria. These problems are perhaps less crucial when local area networks or networks with a small number of nodes are simulated. Here the problem comes from the size (in terms of number of nodes or users) and from the diversity of the problems. It is thus advisable to consider them a priori to avoid disappointments.

Our point of view is that the only way to solve these problems is to provide for each object and modulus several visions that will depend on the level of necessary detail (time scale and space scale).

5.3 Cross Layer Simulation Approach

Setting up novel cross layer system results in the design of a completely novel in simulation environment. Usual approaches, for studying cross layer systems, are laborious and inefficient. In paragraph 0, we summarise and discuss these usual approaches. Then, in paragraph 0, we describe an approach that we propose to study cross layer systems.

5.3.1 Usual Approaches

When the need for simulating a cross layer system arises, the two usual approaches are either to build a new model of the targeted system on top of an existing simulation environment, or to build new and therefore proprietary simulation software, specifically designed for a given system. In the following, we give a short analysis of the advantages and drawbacks of these two approaches.

5.3.2 Reusing Existing Simulation Environment

Reusing an existing simulation environment, such as OPNET, STK, NS, Visualyze, COSSAP or SPW, has several advantages. First, existing environments usually include a set of common models or patterns out of which several may be reused. Second, the larger the user community of an environment is, the easier it is to find support and contributed models for this environment. Third, existing environments often come with a set of integrated or contributed tools, such as models animators and debuggers, plotters, data analysers, and so on, which improves their overall ergonomics and efficiency.

Unfortunately, cross layer systems are combining several aspects that used to be studied separately so far. Therefore, these separate kinds of studies lead to specific environments. Some are more specifically designed for network and protocol modelling, others for propagation and radio interference modelling, and others for space mechanics modelling. Of course, out of these specific environments, some have the ability to be extended to new areas. But, since a specific environment usually mean a specific and optimised design; such an extension is seldom easy: integrating new kinds of models often conflicts with the initial design and philosophy of the selected environment. The result is an added modelling complexity, as well as an added computing complexity, which are both critical points given the high complexity and large scale of the systems being studied.

5.3.3 Building Specific Simulation Software

Compared to the previous approach, building specific simulation software has the opposite advantages and drawbacks: the modelling complexity may be lowered to a minimal level and the computing complexity may be sharply optimised. But since the developments are specific and often proprietary, there is nothing or little to share, no community support, and a lot of additional effort is required to develop specific complementary tools or integrate existing ones.

5.3.4 Cross Layer Simulation Environment

Cross layer simulation environments have to provide, through an integrated user interface, all the functions required to achieve an experimental study based on simulation: model programming and assembly, experiment planning, simulation runtime support, and data analysis. This environment has to present several properties:

5.3.5 Elements of Design

Cross layer systems exhibit several levels of complexity: a complexity at a physical level with the radio transmissions, a complexity at an architectural level, with several kinds of nodes and links, a complexity at a scale level, with networks operating simultaneously up to thousands of terminals, and a complexity at a functional level, with several kinds of services (e.g.: real time or non real time services), several kinds of protocols (e.g.: IP stacks) and several kinds of procedures (e.g.: logon, power-control, CAC, hand-off). In order to cope with all these levels of complexity, the cross layer simulator has to implement and provide several innovative techniques.

5.3.6 Component Based Modelling Approach

The component based modelling approach is a key point in the cross layer design. This common pattern of object oriented programming [27], applied to the modelling area, allows for many interesting features, such as model genericity, variable granularity, and hierarchical modelling.

Genericity is a powerful property of cross layer models that allows the transparent reuse or exchange of any part of a given model in order to build new models. Variable granularity is the ability to switch the detail level of a given part of a model, from the most detailed level to the most approximate level. Hierarchical modelling is the ability to decompose complex parts of a model in simpler parts, until a reasonable complexity level is reached. An aggregation technique may then be used.

5.3.7 Open and Versatile Simulation Kernel

Cross layer simulator architecture is based on a framework approach. This will allow for a maximal flexibility and adaptability at all levels.

Typical complex simulators have shown to spend most of the computing time in event handling. Thus, this part of the simulator has to be carefully designed and optimised. But in the case of cross layer systems, the kind of designs and optimisations required depends on the part of the model being simulated. In order to allow several kinds of optimisation to be combined, cross layer simulator kernel will allow for transparent mixed-mode multi-level kernels.

6 Conclusions - Part 1

Cross layer is a promising research area, not yet completely exploited. Models based on cross layer could improve considerably wireless networks short term performance. Definition of pertinent information to share and global optimisation instead of multiple local optimisations are the key words for the future of those models. Nowadays research in this area concentrate on the model (cyclic) scale and have to take care of the cross layer design.

The Cross-Layer concept is a new way to see the quality of service in the network. It does not consist in the addition of reservation mechanism of band-width or any other ad hoc mechanism, but in adapting the current mechanisms of routing to the link. For that the sensor networks can take advantage from these exchange techniques of information, because their collaborative nature allows the use of information of other elements to improve the overall capacities of the networks.

Part 2 - Performance Issues of Routing in Ad Hoc Networks and IP over GEO Satellite Links

7 Introduction - Part 2

The ultimate aim of the Ad Hoc networks is to propose a fast deployment of new networks without presupposing the existence of an infrastructure. The aim is to have some machines successively assuming the function of host (transmitting and receiving packets) and of router to relay the packets towards other nodes of the network. The very active working IETF group MANET is interested in the standardisation of these networks and in particular in the routing algorithms. The ad hoc mobile networks are operational in local area networks, even if many studies on these algorithms are in progress...

Performance evaluation of ad hoc and satellite networks is a crucial issue. The specificity of wireless Ad hoc and long delay GEO networks has to be considered carefully. In this state of the art, we overview some routing techniques aiming to the minimisation and the optimisation the bandwidth use. Then, we present simulation results of some routing techniques considering several metrics such as fading, reception errors, overhead and delays. Finally we concentrate on the performance of IP on GEO satellite links. Section 7 is presenting a survey of ad hoc networks. Section 8 is presenting a performance comparison of the AODV and DSR routing protocols. Sections 9 and 10 are presenting some IP performance issues over GEO satellite links.

8 Ad Hoc Networks

Ad hoc network is a network spontaneously generated by a set of computers called mobile nodes to bring out their several mobility speed characteristic (pedestrian, automotive, plane, ship...). Mobile nodes use radio frequency broadcast/reception mechanism. To guaranty connectivity, those spontaneously network nodes need to work together: in the network, every node agrees to convey traffic information even if it is not the destination.

The possible mobility in space and in time of network nodes creates a dynamic topology that cause permanent route changing between any 2 nodes. This property generates the new "route installation" technique to exchange information between 2 mobile nodes. This technique allows activating the target location mechanism by the source. This "route installation" technique may be done at the network level through signalling. Over the traditional routing technique, it makes protocols to take in account environment parameters such as location, mobility, quality of service (QoS).

- Ad hoc networks are characterised by dynamic topology, limited bandwidth, energy consumption constraints, limited physical security with no centralised management. They are subject to many problems such as :
- Routing information technique choice;
- Configuration;

- Management;
- Security: signals detectors and passive receipts can spy radio communications if they are not protected;
- Low output: as the bandwidth is limited, the network management slice must be underplayed to let a maximum bandwidth to communications;
- Transmission errors : more frequent than in wired networks;
- Interference: wireless link are not isolated, two simultaneous broadcasting over the same frequency or two near frequencies came interfered. Some other equipment, like micro waves furnaces, may cause also interference;
- Changeable links: transmissions are subject to propagation condition. Link quality mandatory to operate properly radio communications;
- Signal strength: it decreases with the distance, it is subject to strong regulation by countries authorities;
- Limited radio capability: wave propagation depends on environment transmission (obstacles…);
- Hidden nodes: in some cases, for example in presence of obstacle between 2 nodes, they can broadcast simultaneously and thus can cause collision;
- No possibility to detect collision within a transmission : because nodes must broadcast and listen simultaneously;
- Energy : battery autonomy is limited;
- Mobility and dynamic topology;

The routing information technique choice problem and the configuration problem must ensure connectivity in the network. Routing information technique in ad hoc networks must be solved to ensure communication between mobile nodes.

8.1 Main Characteristics of Ad Hoc Network Routing Information Protocols

Ad hoc network protocols algorithms have to present some characteristics:

- be robust against topology changing
- be implemented over limited computing and memory equipments
- avoid routing loop
- work on limited bandwidth and try to use it in best
- be simple to be implemented

8.2 Ad Hoc Network Information Routing Mechanism Classification

Wired networks use regularly routing protocols to establish routes reaching one point of the network to another in advance, based on periodic routing information exchange between nodes. Those are pro-active protocols because they create routing tables in advance. They can be classified into two types: distant vector algorithm based protocols and link state algorithm based protocols, all of them are part of most short path and distributed routing classes.

A distance vector protocol is based on information exchange between adjacent nodes. A network node communicates with its immediate neighbours to exchange the routing information it detains. Those information incorporate nodes to be reached and each link cost. In normal operation, that scheme converges and produces the shortest path from one node to another of the network. In a link state protocol a node broadcasts its routing information to all network nodes. In normal operation, every node will get network global topology and must perform an additional algorithm to deduce all network paths with the best cost (example of Djikstra most short path algorithm).

Routing information in ad hoc network are similar to those used in wired networks. But because wireless networks are less efficient than wired networks in bandwidth, new signalling routing techniques have been designed to improve the bandwidth. Those techniques are bonded to dynamic network topology characteristic and are used on demand, route is installed when a node need to send an information to another node. In most of the cases, this can solve the problem of expiration of a route install in advance when the network topology has changed. The routing signalling is transported into new kind of protocols such as reactive protocols that establish routes when applications need them.

When a node has a packet to send to another node of the network, it sends a route request through the network to obtain a route that reaches the search destination. This protocol family contains: AODV (Ad hoc On Distant Vector) protocol [31] [32] [33], DSR (Dynamic Source Routing) protocol [40] [41] [42] [43], TORA (Temporally Ordered Routing Algorithm) protocol [34] [35], ABR (Associatively Based Routing) protocol [36] which have being studied by the IETF MANET group.

Contrary to reactive protocols, ad hoc network proactive protocols convey periodic control packets through the network to update their network topology knowledge. Actual works on this family of protocols regroup: DSDV (Destination Sequenced Distant Vector) protocol [28], OLSR (Optimised Link State Routing) protocol [29] [30], FSR (Fisheye State Routing) protocols [37] are submitted to the IETF MANET group [38] [39].

8.3 Ad Hoc Network Routing Protocols Comparative Classification

Table 4 summarise the main characteristics of main ah hoc network routing protocols. All protocols avoid loops either by using sequence numbers, or by source routing paradigm such as envisaged in DSR protocol. AODV is too much similar to DSR except the utilisation of the source routing notion in DSR. However, DSR offers several possible paths to a given destination.

None of the presented protocols considers security management issues, very important in wireless networks.

None of presented protocols assume QoS. Nevertheless, OLSR tries to avoid bad unreliable links. Finally none of the presented protocols takes into account the management of nodes power, this leads to an unbalanced network if some routes are much more used than others: battery exhaustion and links overload.

8.4 Two Examples of Routing Protocols

DSR: is a dynamic routing information protocol based on route discovery and route maintenance. Route discovery takes place when a node has a packet to send to another but does not know the path to use. The original node sends a route request, intermediate node that does not have the search path add their address in the address list field, intermediate that have the path or the destination node return the complete path to the sender.

Table 4. Characteristics of ad hoc routing protocols

Criterion	AODV	DSR	OLSR	FSR	CBRP	LANMAR	TBRPF	ZRP
Without loop	Yes	Yes	Yes	Yes	Yes	Yes	Yes	Yes
Many routes possible	No	Yes	No	No	Yes	No	No	Yes
Distributed	Yes	Yes	Yes	Yes	Yes	Yes	Yes	Yes
Kind	reactive	reactive	proactive	proactive	hybrid	hybrid	Proactive	hybrid
Security	No	No	No	No	No	No	No	No
Periodic messages control	No	No	Yes	Yes	Yes	Yes	Yes	Yes
Unidirectional links	No	Yes	Yes	Yes	Yes	No	No	Yes

CBRP: Cluster Based Routing Protocol.
LANMAR: Landmark Routing Protocol.
ZRP: Zone Routing Protocol.
TBRPF: Topology Broadcast Based on Reverse – Path Forwarding.

Route maintenance mechanism uses acknowledgement, route error messages, packet recovering by modifying source route of the packet and sending route error to the original sender, fragmentation to adjust packets to path size.

DSDV: is based on Bellman Ford method as RIP (Routing Information Protocol). Each node has a table containing destination nodes, number of hops to reach each destination, next hop to send packet, sequence number and last update date. Tables are regularly updated so that DSDV generates an important overhead. Another problem of DSDV is that the mobility must be estimated to find an adapted update time, but there is very less work over mobility. Table size is another problem for large networks.

8.5 MAC Routing vs. Network Routing

Mobile ad hoc routing can be located at different levels. In fact, HIPERLAN recommend a MAC routing whereas MANET recommends network routing. ANANAS is in favour of an intermediate level.

HIPERLAN (High Performance Radio Local Area Network) type 1 is an ETSI standard for local area networks without neither base station nor fixed infrastructure. HIPERLAN defines its own physical layer and routing algorithm, situated at the MAC layer. This European standard has reserved the frequency band 5.2-5.35 GHz and offer throughput attaining 23 Mbps. HIPERLAN defines its own MAC layer called EY-NPMA (Elimination Yield Non-pre-emptive Priority Multiple Access). This layer offers a QoS with five priority levels. The physical layer uses GMSK (Gaussian Minimum Shift Keying) as a frequency modulation technique. The routing, done a level 2, uses the physical address of the wireless cards. The routing table is constructed on the base of periodically disseminated neighbouring and topological information. The control information are diffused by the technique multihop relaying with higher priority with regard to data. The multihop relaying technique consists in diffusing a packet in the network using only a sub set of nodes in the network in order to save the bandwidth.

Advantages: from higher layers the network is seen as a local diffusing network. IP applications run normally without any change, and the HIPERLAN is transparent.

Drawbacks dedicated HIPERLAN cards have to be used. Otherwise, the implementation of the MAC ad hoc routing requires the modification of the drivers of the wireless cards.

MANET (Mobile Ad hoc NETwork) is an IETF working group. This group aims to specify and standardise the mobile ad hoc networks protocols, at the IP level. Protocols have to support the physical and MAC heterogeneity. In other words, the protocols have to be independent from lower layers. This is rather different from HIPERLAN that specify and consolidate the lower layers of the system.

The MANET network is defined as a network of mobile and autonomous plateforms. Thos plate-forms may have several shared hosts and interfaces. Thos plateforms may move freely without any constraint and should work as an autonomous network and support links to fixed networks and gateways. The MANET networks have dynamic multihop topologies and a variable size, from some ten hundreds nodes.

The first objective of this group is to choose one or several unicast routing protocols and the interaction between high and low layers. Then to study the problems of QoS and of multicast in mobile MANET environment.

ANANAS (A New Ad hoc Network Architectural Scheme) is an architecture that locate the management and the calculation of the routing at an intermediate level between the physical routing level and the IP routing level. The architecture of ANANAS is divided in three levels:

The physical level: it is the set of the network physical interfaces. Those interfaces may belong to different physical networks, using different technologies.

The ad hoc level: this level constructs a unique logical ad hoc network composed of the different technologies used by the physical layer cards. Each node has a unique identifier (ad hoc address). The ad hoc routing protocol is implemented at this level and is based on the unique identifiers of each node. The routing is considered as a commutation between physical interfaces and networks.

The IP level: at this level the network is seen as a classical Ethernet local network. IP see only the virtual interface that masks physical architecture. Thus, all applications work normally without any modification. For example, the packets intended to the address 255.255.255.255 reach all the destinations present in the network without any IP routing (packets intended to this address are not relayed by the IP routers).

This architecture do not specify any particular routing protocol, theoretically, it is possible to use one the proposed MANET protocols with light modifications in the packets format and in translation schemes between IP addresses and ad hoc addresses.

9 Performance Comparison of Routing Protocols

Both studies presented in this part are relatively old (1998 and 2000) but are referenced in several publications [43, 48, 53]:

The first study, realised by J.Broch, D.Maltz, D.Johnson and Y.Hu, was supported by the Monarch project [50]. It is really the first performance evaluation study on different ad hoc routing protocols. This study presents the four routing protocols behaviour: DSDV (proactive), TORA, DSR et AODV (reactive). The algorithms were implemented in ns-2 and the used mobility model is Random Waypoint. The choice of parameters is interesting because several ulterior studies were based on those parameters. The simulations are based on CBR traffic sources in order to control more easily le network total load. The number of nodes was fixed to 50, the number of sources is variable (10, 20 and 30), and the simulation duration is 900 seconds. All links are bi-directional.

The studied metrics are:

- **The loss rate:** important because the retransmission of the data is managed at transport level and consequently can influence the maximum throughput that the network supports
- **The routing overhead:** must be minimised to optimise the band-width of the network. It is measured as a number of packets
- **The relevance of the path:** the difference between the path taken by the data and the existing shortest path between the source and the destination. This metric shows the capacity of the protocol to find most efficient paths in terms of a number of intermediate nodes.

The degree of mobility selected to compare the four protocols is the time of break (parameter of the model of mobility). In terms of rate of losses, AODV and DSR are most effective whatever the number of sources. Concerning the control traffic, DSR is far better than the three other protocols. However, it should be underlined that the overhead is measured in number of packets. However, the size of the DSR header is considerably important because it contains all the paths that the packet have to follow. Considering the load, AODV becomes better except for a high mobility (very short time of break). The most powerful protocols as for the relevance of the path are DSDV and DSR. Moreover, contrary to TORA and AODV, this metric is relatively independent of the degree of mobility. To summarise, the results are favourable to DSR protocol, and to a lesser extent to AODV protocol (among the four, they are the only ones for which standardisation is in progress).

This study was considered, later on, in many simulations. For this reason, it was discussed on several points. The first one, inevitably, the selected degree of mobility is not representative of the network dynamics. The control traffic would have been compared to data traffic transferred by the protocol. Indeed, a protocol can generate more overhead but transport more data. Furthermore, the ratio (control volume/data volume) would have been more judicious as in [51]. Finally, to define the relevance of the path, the criterion of distance is rather reducing (no result relating to the delay is presented).

The second study, carried out by S.Das, C.Perkins and E.Royer, is more recent and compares more in detail the performance of AODV and DSR protocols [51]. As in the simulations of the first study, the selected environment is NS-2. The sources are CBR sources and the mobility model is Random Waypoint. Here, two surfaces of different simulations (1500*300 and 2200*600) are used respectively accommodating 50 and 100 mobile units with a varying number of sources from 10 to 40. The studied performance criteria are the rate of losses (as in the Monarch study), the end to end delay and the ratio overhead on information (in a number of packets). The estimator of mobility is always the duration of breaks. The links are bi-directional as in the preceding study. The two protocols are first tested on the configuration of 50 nodes:

- **The loss rate:** With 10 and 20 sources, both protocols are equivalent. AODV becomes more powerful with a more significant number of sources (30 and 40): DSR looses between 30 and 50 % of packets more than AODV for high mobility (short break time).
- **End to end delay:** The DSR is faster with 10 sources (a factor 4 compared to AODV) and 20 sources (but with a tighter difference). Beyond 30 sources, the tendency is reversed: AODV is better, in particular with short break times (AODV is twice faster).
- **Overhead/data:** DSR appears broad better in all the cases (at least 4 times better). When the number of sources increases, DSR preserves a stable ratio relatively to AODV even if its other criteria worsen. It as should be noticed that the delays of the two protocols are very long with 40 sources, even with low mobility. This is explained by the absence of load balancing mechanisms and by the policy of selection of the paths (for example, in the DSR, the shortest path in a number of nodes is always the optimal way).

With 100 mobile units, the results concerning the losses rate and the end to end delay are similar with those of the configuration of 50 stations. On the other hand, the difference between the ratio Overhead/data is weaker. Moreover, the ratio DSR ratio is no more stable.

In summary, let us notice that AODV protocol is better than DSR, as soon as the number of sources increases, in term of delay and losses rate. Even if the DSR remains better in term of overhead, these values were taken in a number of packets without taking into account their size. This behaviour can be explained because, in ad hoc networks, the access to the medium is much more expensive than to add some bytes with an already existing packet.

The behaviour of the two protocols is then observed while varying the load (from 0 to 800 Kbits/s) and by considering only 10 sources. The examined metrics are the end to end delay, the reception throughput and the overhead (in Kbits/s). It is noticed that the DSR throughput saturates from 325 Kbits/s because of a high losses rate while the AODV threshold is at 700Kbit/s. The time of the DSR is much better with weak load (at least a factor 2) but longer with high load. The overhead of both protocols are equivalent with a small advantage to the DSR. By taking 40 sources, the behaviours are identical but the thresholds of saturation are weaker (150 for the DSR and 300 for AODV) and the overhead of both protocols becomes higher than their reception throughput!

This study shows the problem of the two protocols, in particular the DSR, when the number of sources and the load of the network increase.

10 IP over GEO Satellite Links

This section presents in a few words performance issues of IP in a specific context; geostationary satellites. But lets first settle the context of IP communication and GEO satellite, before focusing on performance questions: why IP over satellite?

Even if service integration is not a brand new idea, it is still an area of research, since ADSL and cable operators fight to deliver over a same support access to the web, voice and television (Free, Tiscali...) at lower prices. All these services are based over high bandwidth medium and integrated via IP, which is unavoidable today.

As GEO satellites experience a new success with digital television, numerous projects study it as a medium for IP (IBIS [52], GEOCAST [53], DIPCAST [54] ...). Indeed its wide geographical coverage, its broadcasting nature, its unique capacity to feed remote areas and its high bandwidth are combined advantages contributing to its relevance for service integration. Therefore, satellite seems naturally an interesting support for IP, and more precisely for multicast or broadcast IP services.

However, high costs of satellite deployment and the standardisation issue have highly impaired investments. Our participation in a French project, DIPCAST (DVB for IP multicast, RNRT project), has leaded us to propose several solutions: architectures based on bent-pipe system, architectures with on-board processing, architectures based on a hybrid satellite (relevant to conventional bent-pipe systems as to the next-generation of GEO satellites).

In order to compare and evaluate IP over DVB architectures, performance studies are unavoidable. Nevertheless these studies are not quite simple, since satellite systems are really different from terrestrial networks. Indeed GEO satellite natural delays are so high that all over delay could be overlapped. Moreover access methods are difficult to evaluate since they are often private and specific to the system. A global standardisation issue in satellite domain has highly impaired the development of a real environment of performance evaluation.

In this short state of the art, we propose to present what are the standard IP over GEO satellite performance issues and how they are approached.

11 Satellite Performance Issues

11.1 Services over Satellite

One issue of communication over satellite is the service relevance and adaptation to such a system. Applications are designed, modified, adapted to be supported by the satellite media. In fact satellite specificity, advantages as drawbacks, imply a real study of services, more precisely real-time services and multimedia applications, and thus performance studies and QoS management [55].

11.2 TCP Studies and Transport Protocols

Since it has been a long time that satellites are used as backbone links, the question of TCP performance is a topic as old as relevant. Indeed TCP performance over satellite links known as bad [56] [57] since TCP is not relevant to:

Satellite high delay (long round trip time) impaired throughput usable as illustrated in figure 9. With GEO satellites, the propagation delay is thus on the order of 0.5 seconds. A total RTT value of 0.55 s would be appropriate [58]. The TCP sender must wait this length of time to receive ACKs, which is going to slow the throughput, hurting also interactive applications as well as TCP congestion control algorithms.

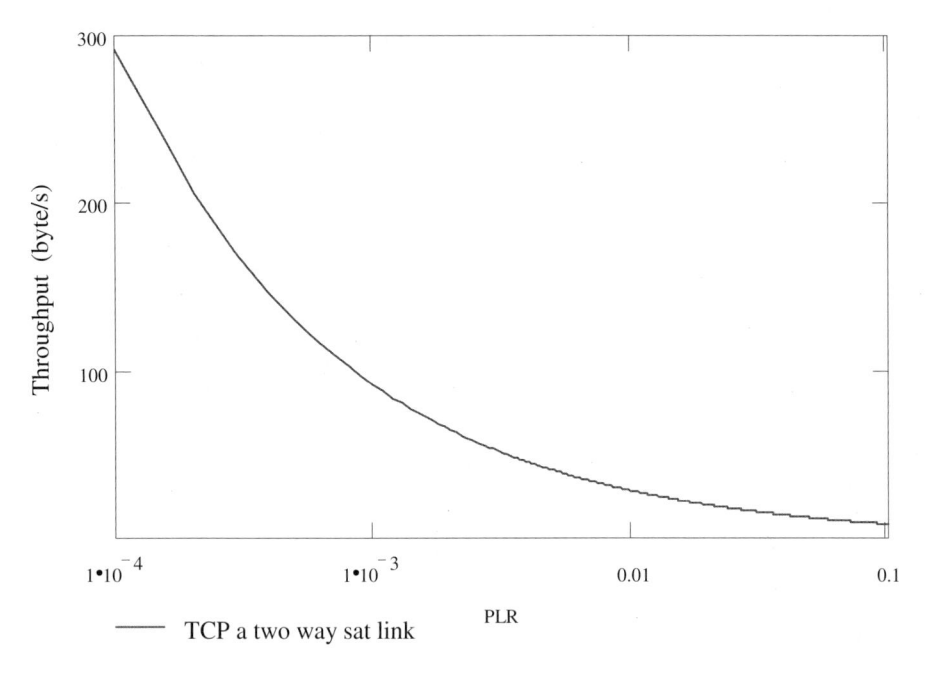

Fig. 9. TCP throughput over a symmetric satellite link

Transmission errors are more important in satellite channels than typical terrestrial network. Because TCP uses all packet drops as a congestion signal, its performance on satellite systems are decreased.

Asymmetric return link which are commonly used since end-to-end users cannot generally have their gateway. That may have an impact on TCP performance.

In this context many propositions have been made to fill TCP limits over satellite. Solutions and their studies are then a large work underway in IP over satellite domain. So many studies carry on focussing on performance evaluation of new solutions compared to old solutions. Here we present a panel of these solutions divided into two categories, external mechanisms and TCP improvements.

TCP improvements:

- CANIT algorithm in a satellite context in order to improve TCP congestion avoidance issue in satellite system [59].
- TCP-Proxy which enables full link throughput per connection [60].
- Fast retransmit.
- Fast recovery.
- SACK option.

External mechanisms:

- Path Maximum Transfer Unit Discovery [58] allows TCP to use the largest packet size without any risk of fragmentation. However its performances are criticised since it implies a longer delay before sending data.

- Forward Error Correction enables to lower PLR and thus to increase TCP maximum throughput. It seems a real opportunity nowadays [61] [62] but it is also used in other transport protocols [63] [64].
- Explicit Congestion Notification [65].

11.3 Access Mechanisms

Another general performance issue in IP over satellite is the access mechanisms and their studies. Indeed if the transmission delay in a satellite system is generally short compared to RTT, the connection to the media could induce an important and variable delay. Therefore many studies compare different access mechanisms and try to optimise them.

Such studies may integrate transport protocol issues [66].

11.4 Other Issues

There are many performance studies which seem interesting in the satellite domain. In a global architecture design, performance issues are numerous and must be considered step by step. For example in IP architecture over a next generation GEO satellite, On-board-processing performance has to be evaluated (switching, table updating...), then access mechanisms [67] and overheads. Indeed, overhead in IP over satellite is a common topic since in DVB solutions the common solution, MultiProtocol Encapsulation, has several limits [68] and satellite bandwidth remains precious. Moreover, IP encapsulation has to be rightly chosen as it often implies signalling protocols, which may impair routing schemes. Eventually, a global comparison between the new architecture and standard bent-pipes architecture have to be done, so as to validate routing schemes, on-board-processing interests...

12 Conclusions - Part 2

The distribution of the band-width in these Mobile Ad Hoc NETworks (MANETs) and GEO satellite can be very unfair. While the behaviour of TCP remains to be studied more in details in MANETs, it is necessary in addition to think about the introduction of quality of service mechanisms according to the routing between various users and according to the constraints of each one. A very dependent subject is the wireless Ad Hoc routing, in particular protocols such as DSR, AODV, OLSR..., which are conceived for the local area networks, but it is very interesting to study the impact for even larger networks (in term of space and of number of nodes). Many routing algorithms are proposed and standardisation is in process. However, QoS mechanisms at different levels and their interdependence is an open issue. Many interesting subjects, such as: (i) Scheduling and consensus algorithms for QoS in Ad Hoc networks. (ii) New architectures of Ad Hoc networks locate the management and the calculation of the routing in an intermediate level between the routing at the physical level and the routing on IP level.

The QoS mechanisms are very dependent on the underlying protocol layers. In particular, if we look at the evolutions of the standards of the IEEE 802.11 family, it appears that some mechanisms are now based on traffic differentiation. These mechanisms are not yet completely standardised. In addition, the physical layer in these types of networks consists in using a radio support which by nature will have a very variable quality in space and in time.

References

1. Chiang, M.: To layer or Not To Layer: Balancing Transport and Physical Layers in Wireless Multihop Networks. In: IEEE INFOCOM 2004 (2004) (to appear)
2. Yuen, W.H., Lee, H.-n., Andersen, T.D.: A Simple and Effective Cross Layer Networking System for Mobile Ad Hoc Networks. In: IEEE PIMRC 2002, pp. 1952–1956 (2002)
3. Raisinghani, V.T., Singh, A.K., Iyer, S.: Improving TCP performance over Mobile Wireless Environments using Cross Layer Feedback. In: IEEE ICPWC 2002, pp. 81–85 (2002)
4. Safwat, A., Hassanein, H., Mouftah, H.: Optimal Cross Layer Designs for Energy Efficient Wireless Ad-hoc and Sensor Networks. In: IEEE IPCCC 2003, pp. 123–128 (2003)
5. Shakkotai, S., Rappoaport, T.S., Karlson, P.C.: Cross Layer Design for Wireless Networks. IEEE Communiction Magazine, 74–80 (October 2003)
6. Conti, M., Maselli, G., Turi, G., Giordano, S.: Cross layering in mobile Ad Hoc Network Design, pp. 48–51. IEEE Computer Society, Los Alamitos (2004)
7. Kawadia, V., Kumar, P.R.: A Cautionary Perspective on Cross Layer Design. IEEE Wireless Communication Marazine (2004)
8. Lee, K.-H.: A Multiple Access Collision Avoidance Protocol for Multicast Services in Mobile Ad Hoc Networks. IEEE Communications Letters 7(10) (October 2003)
9. Gauthier, V.: Cross layer et QoS dans les réseaux ad hoc sans fil. GET-INT-RR (2004)
10. Kaixin Xu, M.G., Bae, S.: Effectiveness or RTS/CTS handshake in IEEE 802.11 based ad hoc networks. In: Globecom 2002 (2002)
11. Khayam, S.A., Karande, S., Krappel, M., Radha, H.: Cross layer protocol design for realtime multimedia applications over 802.11b networks. In: IEEE ICME 2003, pp. II-425–II-428 (2003)
12. Shan, Y., Zakhor, A.: Cross layer techniques for adaptive video streaming over wireless networks. In: IEEE ICME 2002 (2002)
13. Chen, J., Hsia: Joint cross layer Design for Wireless QoS Video Delivery. In: IEEE ICME 2003, pp. I-197–I-200 (2003)
14. Sachs, D.G., Adve, S.V., Jones, D.L.: Cross layer adaptive video coding to reduce energy on general purpose processors. In: IEEE 2003, pp. III-109–III-112 (2003)
15. Alonso, L., Agusti, R.: Automatic Rate Adaptation and Energy-Saving Mechanisms Based on Cross-Layer Information for Packet-Switched Data Networks. IEEE Radio Communications, S15–S19 (March 2004)
16. Raman, B., Bhagwat, P., Seshan, S.: Arguments for Cross Layer Optimisations in Bluetooth Scatternets. In: SAINT 2001, pp. 176–184. IEEE Computer Society, Los Alamitos (2001)
17. Li, X., Bao-yu, Z.: Study on Cross-layer Design and Power Conservation in Ad Hoc Network. In: IEEE PDCAT 2003, pp. 324–328 (2003)
18. Shan, Y., Zakhor, A.: Cross layer techniques for adaptive video streaming over wireless networks. In: IEEE ICME 2002, pp. 277–280 (2002)
19. Toumpis, S., Goldsmith, A.J.: Performance, optimisation, and Cross-Layer Design of Media Access Protocols for Wireless Ad Hoc Networks. In: IEEE ICC 2003, pp. 2234–2240 (2003)

20. Zhao, J., Guo, Z., Zhu, W.: Power Efficiency in IEEE 802.11a WLAN with Cross-Layer Adaptation. In: IEEE ICC 2003, p. 2030 (2003)
21. Maharshi, A., Tong, L.: Corss layer Designs of Multichannel Reservation MAC under Rayleigh Fading. IEEE Transactions on Signal Processing 51(8) (2003)
22. Wang, Q., Abu-Rgheff, M.A.: Cross layer signalling for next-generation wireless systems. In: IEEE WCNC 2003, pp. 1084–1089 (2003)
23. Fang, Y., Bruce McDonald, A.: Cross layer performance effects of path coupling in wireless ad-hoc networks: power and throughput implications of IEEE 802.11 MAC, pp. 281–290 (2002)
24. Becker, M., Beylot, A.-L., Dhaou, R.: Aggregation Methods for Performance Evaluation of Computer and Communication Networks. In: International Symposium: Performance Evaluation - Stories and Perspectives, Vienna, Austria (December 2003)
25. Courtois, P.-J.: On time and space decomposition of complex structures. Communication of the ACM 28(6) (June 1985)
26. Jain, R.: The art of computer performance analysis: Techniques for Experimental Design, Measurement, Simulation and Modelling. Wiley, Chichester (1991)
27. Gamma, E., Helm, R., Johnson, R., Vlissides, J.: Design Patterns: Elements of Reusable Object Oriented Software. Addison Wesley, Massachusetts (1994)
28. Perkins, C., Bhagwat, P.: Destination Sequenced Distance Vector (DSDV). Internet – Draft, IETF MANET Working Group (November 2001)
29. Jacquet, P., Clausen, T.: Optimized Link State Routing Protocol (OLSR). Internet – Draft, IETF MANET Working Group (October 2001)
30. Laouti, A., Muhlethaler, P., Najid, A., Plakoo, E.: Simulation Results of the OLSR Routing Protocol for Wireless Network. INRIA Rocquencourt (April 1998), http://menetou.inria.fr/~muhletha/medhoc.pdf
31. Royer, E.M., Perkins, C.E.: Multicast Operation of the Ad hoc On Demand Distance Vector Routing Protocol (August 2000)
32. Das, S., Perkins, C., Royer, E.: Ad hoc On Demand Distance Vector Routing (AODV). Internet – Draft, IETF MANET Working Group (November 2001)
33. Royer, E.M., Perkins, C.E.: Ad hoc On Demand Distance Vector Routing (AODV). In: Second Annual IEEE Workshop on Mobile Computing Systems and Applications, February 1999, pp. 90–100 (1999)
34. Corson, S., Vincent, P.: Temporally Ordered Routing Algorithm (TORA). Internet – Draft, IETF MANET Working Group (August 2001)
35. A performance comparison of TORA and ideal Link State Routing (February 1997), http://tonnant.itd.nrl.navy.mil/tora/tora_sim.html
36. Zinin, A.: Alternative OSPF ABR Implementations. Internet – Draft, IETF MANET Working Group (February 2001)
37. Gerla, M., Hong, X.: Alternative OSPF ABR Implementations. Internet – Draft, IETF MANET Working Group (December 2001)
38. Corson, S., Macker, J.: Mobile Ad hoc Networking (MANET): Routing Protocol Performance Issues and Evaluation Considerations (Janvier 1999)
39. Mobile Ad hoc Network (MANET), http://www.ietf.org/
40. Jetcheva, J.G., Hu, Y., Johnson, D., Maltz, D.: The Dynamic Source Routing Protocol for Mobile Ad hoc Networks (DSR). Internet Draft, IETF MANET working group (November 2001)
41. Johnson, D.B., Maltz, D.A.: Dynamic Source Routing in Ad hoc Wireless Networks (Mars 1998), http://www.monarch.cs.cmu.edu/monarch-papers/

42. Demir, T.: Simulation of Ad hoc Networks with DSR Protocol (May 2001), http://netlab.boun.edu.tr/papers/Iscis2001-DSR-TamerDEMIR+.pdf
43. The Dynamic Source Routing Protocol for Mobile Ad hoc Networks (DSR) (February 21, 2002), http://www.ietf.org/proceedings/02mar/I-D/draft-ietf-manet-dsr-07.txt
44. Farid, T.: Les modèles de mobilité dans les réseaux ad hoc. présentation interne
45. Ndoye, R.: Qos dans les réseaux mobiles ad hoc. présentation interne
46. Réseaux locaux et Internet, des protocoles à l'interconnexion, Laurent Toutain, Editions HERMES, série réseaux et télécommunications (1996)
47. RFC 2328, OSPF version 2 (April 1998), http://www.ietf.org/rfc/rfc2328.txt?number=2328
48. The Network Simulator – ns2: links to help getting started, http://www.isi.edu/nsnam/ns
49. The Network Simulator: Building ns, http://www.isi.edu/nsnam/ns/ns-build.htm
50. Broch, J., Maltz, D., Johnson, D.B., Hu, Y.-C., Jetcheva, J.: A Performance Comparison of Multi-Hop Wireless Ad Hoc Network Routing Protocols. In: Proc. ACM/IEEE International Conference on Mobile Computing and Networking (Mobicom 1998), pp. 85–97 (1998)
51. Das, S.R., Perkins, C.E., Royer, E.M.: Performance Comparison of Two On-demand Routing Protocols for Ad Hoc Networks. IEEE Personal Communications 8(1), 16–28 (2001)
52. Chacón, S., Casas, J.L., Cal, A., Rey, R., Prat, J., de la Plaza, J., Monzat, G., Carrere, P., Miguel, C., Ruiz, F.J.: Networking over the IBIS System. IST Mobile & Wireless Telecommunications Summit (June 2002)
53. sun, Z., Howarth, M.P., Cruickshank, H., Iyengar, S., Claverotte, L.: Networking Issues in IP Multicast over Satellite. International Journal of Satellite Communications and Networking (2003)
54. DIPCAST, "le Dvb comme support de l'IP multiCAst par SaTellite", RNRT project, http://www.dipcast-satellite.com
55. Mobasseri, M., Leung, V.C.M.: A new buffer management scheme for multimedia terminals in broadband satellite networks. In: Proc. 35th Hawaii International Conference on System Sciences, Hawaii (January 2002)
56. Akyildiz, I.F., Jeong, S.H.: Satellite ATM networks: A survey. IEEE Communications Magazine (1997)
57. Roddy, D.: Satellite communications, 3rd edn. (2001), ISBN 0-07-137176-1
58. Allman, M., Glover, D., Sanchez, L.: Enhancing TCP over Satellite Channels using Standard Mechanisms. RFC 2488 (January 1999)
59. Benadoud, H., Berqia, A., Mikou, N.: Enhancing TCP over satellite links using CANIT algorithm. International Journal of SIMULATION 3(3-4), 81–91 (2002)
60. Wu, L., Peng, F., Leung, V.C.M.: Dynamic Congestion Control for Satellite Networks Employing TCP Performance Enhancement Proxies. To be presented at IEEE PIMRC 2004, Barcelona, Spain (September 2004)
61. IETF Reliable Multicast Transport Group, http://www.ietf.org/rmt-charter.html, motlabs.com
62. Celandroni, N., Ferro, E., Potorti, F.: Goodput optimisation of long-lived TCP connections in a rain faded satellite channel. VTC (Spring 2004)
63. Nonnenmacher, J., Biersack, E.W., Towsley, D.: Parity-based loss recovery for reliable multicast transmission Networking. In: Proceedings of ACM SIGCOMM 1997 (1997)

64. Byers, H., Luby, M., Mitzenmacher, M., Rege, A.: A Digital Fountain Approach to Reliable Distribution of Bulk Data. In: Proceedings of ACM SIGCOMM 1998, Vancouver, BC (1998)
65. Ramakrishnan, K.K., Floyd, S.: A proposal to add explicit congestion notification (ecn) to IP. RFC 2481 (1999)
66. Li, Y., Jiang, Z., Leung, V.C.M.: Performance evaluations of PRR-CFDAMA for TCP traffic over geosynchronous satellite links. In: Proc. IEEE WCNC 2002, Orlando, FL (March 2002)
67. Acar, G., Rosenberg, C.: Performance Study of End-to-End Resource Management in ATM Geostationary Satellite Networks with On-Board Processing. Space Communications Journal 17(1-3), 89–106 (2001)
68. IETF working-group IP/DVB, http://www.erg.abdn.ac.uk/ip-dvb/

The Rare Event Simulation Method RESTART: Efficiency Analysis and Guidelines for Its Application

Manuel Villén-Altamirano[1] and José Villén-Altamirano[2,*]

[1] Technical University of Madrid
Dep. Ingeniería Sistemas Telemáticos, ETSIT, Ciudad Universitaria, 28040 Madrid, Spain
manolo.villen@dit.upm.es
[2] Technical University of Madrid
Dep. Matemática Aplicada, EUI, c/ Arboleda s/n, 28031 Madrid, Spain
jvillen@eui.upm.es

Abstract. This paper is a tutorial on RESTART, a widely applicable accelerated simulation technique for estimating rare event probabilities. The method is based on performing a number of simulation retrials when the process enters regions of the state space where the chance of occurrence of the rare event is higher. The paper analyzes its efficiency, showing formulas for the variance of the estimator and for the gain obtained with respect to crude simulation, as well as for the parameter values that maximize this gain. It also provides guidelines for achieving a high efficiency when it is applied. Emphasis is placed on the choice of the importance function, i.e., the function of the system state used for determining when retrials are made. Several examples on queuing networks and ultra reliable systems are exposed to illustrate the application of the guidelines and the efficiency achieved.

Keywords: Rare Event, Splitting, RESTART, Simulation, Performance, Reliability.

1 Introduction

Performance requirements of broadband communication networks and ultra reliable systems are often expressed in terms of events with very low probability. Probabilities of the order of 10^{-10} are often used to specify packet losses due to traffic congestion or system failures. Analytical or numerical evaluation of these probabilities is only possible for a very restricted class of systems. Simulation is an effective alternative, but acceleration methods are necessary because crude simulation requires prohibitive execution time for accurate estimation of very low probabilities.

One such method is importance sampling; see [1] for an overview. The basic idea behind this approach is to alter the probability measure governing events so that the formerly rare event occurs more often. A drawback of this technique is the difficulty of selecting an appropriate change of measure since it depends on the system being simulated. Researchers have, therefore, focused on finding good heuristics for particular types of models.

---

[*] His participation in this work was partially supported by the Comunidad of Madrid, Grant Riesgos CM (P2009/ESP-1685).

Another method is RESTART (REpetitive Simulation Trials After Reaching Thresholds). Let us roughly define the 'importance' of a state as the chance of the process entering the rare set after it has been in this state (a more precise definition of importance will be provided later). RESTART introduces a nested sequence of sets of states C_i ($C_1 \supset C_2 \supset ... \supset C_M$), which determines a partition of the state space Ω into regions $C_i - C_{i+1}$; the higher the value of i, the higher the importance of the states of regions $C_i - C_{i+1}$. A more frequent occurrence of the formerly rare event is achieved by performing a number of simulation retrials each time the process enters a set C_i. The retrials finish when they exit set C_i. Note that while in crude simulation the process spends most of its time in low importance regions, in RESTART simulation an oversampling is made in high importance regions to balance the time spent by the process in all the regions.

The sets C_i are defined by comparing the value taken by a function of the system state, the importance function, with certain thresholds. The application of this method for particular models requires the choice of a suitable importance function. The suitable importance function for a model is not as dependent on particular features of the model as the suitable change of measure required when importance sampling is applied. The paper shows formulas of the importance function for estimating overflow probabilities in Jackson and non-Jackson networks, and also for the study of highly dependable systems.

RESTART has a precedent in the splitting method described in [2]. Splitting also defines importance regions and performs retrials, but these are not made in the same way. They are only made the first time the process enters each set C_i, and they do not finish when they exit set C_i, but continue until the end of the simulation. Consequently, as indicated in [3], oversampling is performed not only in high importance regions, but also in low importance regions that are visited after the higher importance ones, leading to a loss of efficiency. This feature has limited its use to the simulation of processes in which a negligible amount of time is spent in low importance regions visited after the higher importance ones. This amount of time is only negligible in simulations made by means of short replicas, e.g., regenerative simulations of very simple systems, or short transient simulations.

RESTART was introduced by Bayes, A. J. in 1970 [4]. Villén-Altamirano, M. and Villén-Altamirano, J. coined in 1991 the name RESTART [5] and made a theoretical analysis that yields the variance of the estimator and the gain obtained with one threshold. The analysis was extended for multiple thresholds in 1994 [6]. The papers also derive optimal values of the parameters (thresholds and the number of retrials). By using these results, guidelines can be derived for optimizing the importance function and the parameter values. This analysis led to efficient applications of RESTART. While few applications with poor gains [4] or even failures [7] were reported before 1991, a large number of applications with dramatic gains have subsequently been reported. Examples of these applications are [8], [9], [10], [11], [12], [13], [14], [15], [16], [17], [18], [19], [20], [21], [22], [23], [24], [25], [26], [27], [28], [29], [30], [31], [32], [33] and [34].

The rest of the paper is organized as follows. Section 2 describes the method and Section 3 proves the unbiasedness of the estimator. Section 4 is devoted to show the efficiency of RESTART. Exact formulas for the variance of the estimator and the gain obtained are presented, as well as values for thresholds and the number of retrials that maximize the gain. Section 5 provides guidelines for an effective application of RESTART. It shows how the formula of the gain can be expressed as an ideal gain divided by four factors, which can be considered inefficiency factors. Guidelines are given to reduce each of the factors. Special emphasis is placed on the most critical factor, the one related to the chosen importance function. Section 6 exposes several examples on queuing networks and ultra reliable systems to illustrate the application of the guidelines and the efficiency achieved. Finally, conclusions are stated in Section 7.

2 Description of RESTART

Consider the simulation of a stochastic process $Z = (Z(t), t \geq 0)$, with discrete state space and either discrete or continuous parameter. The process may be Markovian or non-Markovian. As in any simulation, regardless of the use of RESTART, $Z(t)$ is simulated by means of a Markovian process $X(t)$ which includes, in addition to the state variables of $Z(t)$, those needed to determine $Z(t_1)$ for $t_1 > t$. These additional state variables include:

- The time of occurrence of any future event[1] that has already been scheduled at or before time t;
- The part of the history of the process that has to be incorporated into the system state at t to make $X(t)$ Markovian.

For a given process $Z(t)$, different ways of implementing the simulation model may lead to different processes $X(t)$. Although RESTART may be applied for any process $X(t)$, the application can be more efficient if $X(t)$ is defined following the guidelines given in Section 5.6. In the rest of the paper it is assumed that the process $X(t)$ is given.

Let Ω denote the state space of $X(t)$. A nested sequence of sets of states C_i, $(C_1 \supset C_2 \supset ... C_M)$ is defined, which determines a partition of the state space Ω into regions $C_i - C_{i+1}$; the higher the value of i, the higher the importance of the region $C_i - C_{i+1}$. These sets are defined by means of a function $\Phi : \Omega \to \Re$, called the importance function. Thresholds T_i ($1 \leq i \leq M$) of Φ are defined such that each set C_i is associated with $\Phi \geq T_i$.

The probability $\Pr\{A\}$ of the rare set A can be defined in many ways. For example, in a transient simulation, it can be defined as the probability that the system enters the rare set at least once in a given time interval. It is also often defined, both in transient and steady-state simulations, either as the probability of the system being in a state of the set A at a random instant or at the instant of occurrence of certain events denoted reference events. An example of a reference event is a packet arrival. If the rare set is a buffer being full, we are not usually interested in the probability of the buffer being full at a random instant but at a packet arrival. RESTART can be applied in all these

---

[1] In this paper the term event refers to a simulation event, i.e., an instantaneous occurrence that may change the state Z. The system state resulting from the change will be called system state at the event.

cases. However, for simplicity, the notation will only refer to the last definition. Analogously, the probability Pr{C_i} of set C_i is defined as the probability of the system being in a state of the set C_i at a reference event.

A reference event at which the system is in a state of the set A or set C_i is referred to as an event A or event C_i, respectively. Two additional events, B_i and D_i, are defined as follows:

B_i : event at which $\Phi \geq T_i$ having been $\Phi < T_i$ at the previous event;

D_i : event at which $\Phi < T_i$ having been $\Phi \geq T_i$ at the previous event.

RESTART works as follows:

- A simulation path, called main trial, is performed in the same way as if it were a crude simulation. It lasts until it reaches a predefined "end of simulation" condition.
- Each time an event B_1 occurs in the main trial, the system state is saved, the main trial is interrupted, and $R_1 - 1$ retrials of level 1 are performed. Each retrial of level 1 is a simulation path that starts with the state saved at B_1 and finishes when an event D_1 occurs.
- After the $R_1 - 1$ retrials of level 1 have been performed, the main trial continues from the state saved at B_1. Note that the total number of simulated paths $[B_1, D_1)$, including the portion $[B_1, D_1)$ of the main trial, is R_1. Each of these R_1 paths is called a trial $[B_1, D_1)$. The main trial, which continues after D_1, leads to new sets of retrials of level 1 if new events B_1 occur.
- Events B_2 may occur during any trial $[B_1, D_1)$. Each time an event B_2 occurs, an analogous process is set in motion: $R_2 - 1$ retrials of level 2, starting in B_2 and finishing in D_2, are performed, leading to a total number of R_2 trials $[B_2, D_2)$. The trial $[B_1, D_1)$, which continues after D_2, may lead to new sets of retrials of level 2 if new events B_2 occur.
- In general, R_i trials $[B_i, D_i)$ ($1 \leq i \leq M$) are performed each time an event B_i occurs in a trial $[B_{i-1}, D_{i-1})$. The number R_i is constant for each value of i.
- A retrial of level i also finishes if it reaches the "end of simulation" condition before the occurrence of event D_i. The term trial $[B_i, D_i)$, often used in the rest of the paper, indistinctively refers to a complete or to a prematurely finished trial $[B_i, D_i)$.
- In case that the process up crosses more than one threshold in a time step, it must be taken into account that several events B_i (with different values of i) simultaneously occur. If, for instance, an event at which $\Phi \geq T_{i+1}$ occurs having been $\Phi < T_i$ at the previous event, this event is both an event B_i and an event B_{i+1}. As it is an event B_i, R_i-1 retrials of level i have to be performed starting in this event B_i/B_{i+1} and finishing when an event D_i occurs. As the referred event is also an event B_{i+1} we have to consider that an event B_{i+1} has occurred in each of

the R_i trials $[B_i, D_i)$, thus $R_i(R_{i+1}-1)$ retrials of level $i+1$ have also to be performed, all of them starting in the referred event B_i/B_{i+1} and finishing when an event D_{i+1} occurs.

Figure 1 illustrates a RESTART simulation with $M = 3$, $R_1 = R_2 = 4$, $R_3 = 3$, in which the chosen importance function Φ also defines set A as $\Phi \geq L$. Bold, thin, dashed and dotted lines are used to distinguish the main trial and the retrials of level 1, 2 and 3, respectively.

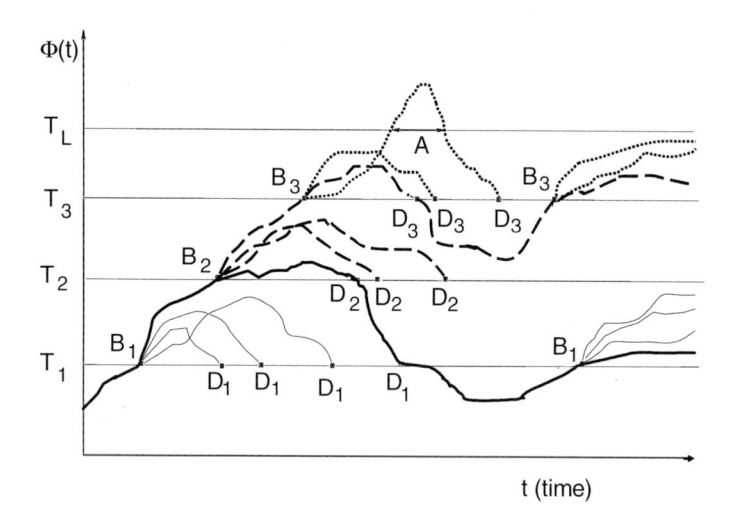

Fig. 1. Simulation with RESTART

Note that the oversampling made by RESTART in the region $C_i - C_{i+1}$ (C_M if $i = M$) is given by the accumulative number of trials:

$$r_i = \prod_{j=1}^{i} R_j \quad (1 \leq i \leq M).$$

Thus, for statistics taken on all the trials, the weight assigned to the occurrence of an event when it occurs in the region $C_i - C_{i+1}$ (C_M if $i = M$) must be $1/r_i$.

Although sets C_i must be usually chosen satisfying $A \subset C_M$, there are applications where a higher efficiency is achieved if $A \not\subset C_M$ (see [13], [16], [38]). For simplicity the formulas shown in this paper for the variance of the estimator and for the gain obtained only apply to the case in which $A \subset C_M$. Formulas for the variance of the estimator for the general case in which either $A \subset C_M$ or $A \not\subset C_M$ are provided in [36].

The "end of simulation" condition or the condition for the start or the end of a simulation portion (as e.g., the initial transient phase or a batch of a batch means simulation or a replica of a transient simulation) may be defined in the same way as in crude simulation. For example, the condition may be that a predefined value of the simulated time or of the number of simulated reference events is reached. These conditions hold for a trial when the sum of the time (or of the number of reference events) simulated in the trial and in all its predecessors reaches the predefined value.

Some more notations:

- $R_0 = 1$, $r_0 = 1$, $C_0 = \Omega$, $C_{M+1} = A$;
- $P_{h/i}$ ($0 \leq i \leq h \leq M+1$) : probability of the set C_h at a reference event, knowing that the system is in a state of the set C_i at that reference event. As $C_h \subset C_i$, $P_{h/i} = \Pr\{C_h\}/\Pr\{C_i\}$;
- $P_{A/i} = P_{M+1/i}$;
- $P = P_{M+1/0} = P_{A/0} = \Pr\{A\}$;
- N_A: total number of events A that occur in the simulation (in the main trial or in any retrial);
- N_A^0 : number of events A that occur in the main trial;
- N_i^0 ($1 \leq i \leq M$) : number of events B_i that occur in the main trial;
- N: number of reference events simulated in the main trial;
- a_i ($1 \leq i \leq M$): expected number of reference events in a trial $[B_i, D_i)$;
- X_i ($1 \leq i \leq M$): random variable indicating the state of the system at an event B_i;
- Ω_i ($1 \leq i \leq M$) : set of possible system states at an event B_i;
- P^*_{A/X_i} ($1 \leq i \leq M$) : importance of state X_i, defined as the expected number of events A in a trial $[B_i, D_i)$ when the system state at B_i is X_i. Note that P^*_{A/X_i} is also a random variable which takes the value P^*_{A/x_i} when $X_i = x_i$;
- $P^*_{A/i}$ ($1 \leq i \leq M$): expected importance of an event B_i :

$$P^*_{A/i} = E\left[P^*_{A/X_i}\right] = \int_{\Omega_i} P^*_{A/x_i} dF(x_i),$$

where $F(x_i)$ is the distribution function of X_i. Note that $P^*_{A/i} = E[N_A^0]/E[N_i^0]$ and that $P^*_{A/i} = a_i P_{A/i}$;

- $V(P^*_{A/X_i})$ ($1 \leq i \leq M$): variance of the importance of an event B_i :

$$V\left(P^*_{A/X_i}\right) = E\left[\left(P^*_{A/X_i}\right)^2\right] - \left(P^*_{A/i}\right)^2.$$

3 Unbiasedness of the Estimator

The estimator of the probability of the rare set A in a RESTART simulation depends on how this probability has been defined. For the definition adopted in this paper, the estimator for P is in the general case, in which either $A \subset C_M$ or $A \not\subset C_M$:

The Rare Event Simulation Method RESTART: Efficiency Analysis and Guidelines 515

$$\hat{P} = \frac{1}{N}\left(\sum_{i=0}^{M} \frac{N_{Ai}}{r_i}\right), \tag{1}$$

where N_{Ai} is the number of events A occurred in the set $C_i - C_{i+1}$ (C_M if $i = M$) in any trial and N takes a fixed value, which controls the "end of simulation" condition. Note that the weight assigned to N_{Ai} is $1/r_i$ given that N_{Ai} includes events occurred in all the trials, while the weight given to N is 1 since N only includes the reference events occurred in the main trial. In the case that $A \subset C_M$ formula (1) becomes:

$$\hat{P} = \frac{N_A}{N\ r_M}. \tag{2}$$

The unbiasedness of the estimator is proved in (35) for the case in which $A \subset C_M$ and in (36) for the general case. Let us see here the proof made in (35) for $A \subset C_M$. It is made by induction: the estimator of P in a crude simulation is $\hat{P} = N_A^0/N$, which is an unbiased estimator. As the crude simulation is equivalent to a RESTART simulation with $M=0$, and formula (2) becomes $\hat{P} = N_A^0/N$ for $M=0$, the estimator of P in a RESTART simulation is unbiased for 0 thresholds. Thus, it is enough to prove that if it is unbiased for M-1 thresholds, it is also unbiased for M thresholds.

Consider a simulation with M thresholds (T_1 to T_M). If the retrials of level 1 (and their corresponding upper-level retrials) are not taken into account, we have a simulation with M-1 thresholds (T_2 to T_M). Let N_A and $\hat{P}$ denote the number of events A and the estimator of P respectively in the simulation with M thresholds, and N_A^{M-1} and $\hat{P}^{M-1}$ the number of events A and the estimator of P in the simulation with M-1 thresholds.

Define α_m as the random variable which indicates the sum of the number of events A occurring in the mth trial $[B_1, D_1)$ performed from each event B_1 of the simulation counting all the events A occurring in the corresponding upper-level retrials. Note that, among the R_1 trials $[B_1, D_1)$ performed from each event B_1 in the M threshold simulation, only the one being a portion of the main trial belongs to the $M - 1$ threshold simulation. Assigning $m = 1$ to this trial:

$$N_A^{M-1} = \alpha_1; \quad N_A = \sum_{m=1}^{R_1} \alpha_m.$$

As the R_1 trials $[B_1, D_1)$ made from each event B_1 start with identical system state, $E[\alpha_1] = E[\alpha_2] = ... = E[\alpha_{R_1}]$. Thus, as R_1 is constant, $E[N_A] = R_1 E[N_A^{M-1}]$ and consequently:

$$E[\hat{P}] = \frac{E[N_A]}{N\prod_{i=1}^{M} R_i} = \frac{E[N_A^{M-1}]}{N\prod_{i=2}^{M} R_i} = E[\hat{P}^{M-1}] = P.$$

This proves that $\hat{P}$ is also unbiased in a RESTART simulation with M thresholds.

It is important to apply the same "end of simulation" condition to the main trial and to all the trials, as explained in Section 2. Otherwise the formula $E[\alpha_1] = E[\alpha_2] = ... = E[\alpha_{R_1}]$ would not be satisfied and thus the estimator would not be unbiased.

4 Efficiency of RESTART

The efficiency of an acceleration method is determined by the computational time required for estimating a certain rare event probability P with a given width of the confidence interval. As the width of the confidence interval depends on the variance of the estimator, formulas of this variance, $V(\hat{P})$, are shown in Section 4.1 and of the cost (in computational time) of the simulation in Section 4.2. In Section 4.3 the costs incurred by a RESTART simulation and by a crude simulation for estimating a same rare event probability with the same width of the confidence interval are compared to derive the efficiency gain obtained with the application of RESTART. As the efficiency gain depends on the number of thresholds and on the number of retrials used in the RESTART simulation, the values of these parameters that optimize the gain are shown in Section 4.4.

4.1 Variance of the Estimator

The variance of the estimator for the case in which $A \subset C_M$ was derived in [35] and for the general case in which the condition $A \subset C_M$ is not necessarily satisfied in [36]. Let us present here the formula of the variance when $A \subset C_M$ as well as the main steps followed in [35] to derive it. The variance of the estimator is also derived by induction: first a formula is derived for 0 thresholds (crude simulation) and generalized for M thresholds; then it is proved that if the generalized formula holds for M-1 thresholds, it also holds for M thresholds.

Variance for 0 Thresholds (Crude Simulation). In a crude simulation the variance of the estimator is given by:

$$V(\hat{P}) = V\left(\frac{N_A^0}{N}\right) = \frac{V(N_A^0)}{N^2} = \frac{P}{N} \frac{V(N_A^0)}{E[N_A^0]} = \frac{K_A P}{N},$$

where $K_A = V(N_A^0)/E[N_A^0]$. In simulations defined with a constant time duration t, K_A is the index of dispersion on counts, IDC(t), of the process of occurrence of events A for the time t simulated. In any case, K_A is a measure of the autocorrelation of the process of occurrence of events A. If the process is uncorrelated, K_A is close to 1 (exactly, $K_A = 1 - P$).

The definition of K_A also applies to a RESTART simulation, where N_A^0 is the number of events A in the main trial.

Variance for M Thresholds. The variance of $\hat{P}$ in a RESTART simulation with M thresholds is given by:

$$V(\hat{P}) = \frac{K_A P}{N}\left[\frac{1}{r_M} + \sum_{i=1}^{M} \frac{s_i P_{A/i}(R_i - 1)}{r_i}\right], \quad (3)$$

with:

$$s_i = \frac{1}{K_A P_{A/i}} \frac{V(E[N_A^0 | \chi_i])}{E[N_A^0]} \quad (1 \leq i \leq M), \quad (4)$$

where $\chi_i = \left(N_i^0, \left(X_i^1, X_i^2, \ldots, X_i^{N_i^0}\right)\right)$, N_i^0 being the random variable indicating the number of events B_i occurred in the main trial of a simulation randomly taken and $(X_i^1, X_i^2, \ldots, X_i^{N_i^0})$ being the vector of random variables describing the system states at those events B_i. A further development of formula (4) is shown later to gain insight on s_i.

The formula provided for 0 thresholds is an application of formula (3) to the case of $M = 0$ (where $r_M = r_0 = 1$). Let us now see that if formula (3) holds for $M-1$ thresholds it also holds for M thresholds.

Consider the two related $M-1$ and M threshold simulations described in Section 3. The variance of the estimator in the $M-1$ threshold simulation $V(\hat{P}^{M-1})$ and in the M threshold simulation $V(\hat{P})$ can be written as:

$$V(\hat{P}^{M-1}) = V\left(E[\hat{P}^{M-1}|\chi_1]\right) + E\left[V(\hat{P}^{M-1}|\chi_1)\right]. \quad (5)$$

$$V(\hat{P}) = V\left(E[\hat{P}|\chi_1]\right) + E[V(\hat{P}|\chi_1)]. \quad (6)$$

An intuitive explanation of formulas (5) and (6) is that the variance $V(\hat{P})$ for both $M-1$ and M threshold simulations can be considered to be the result of two contributions, reflected by the two terms of each of these formulas:

- The first term reflects the variance associated with the set of events B_1 occurred in the simulation. As the number of retrials made in B_1 does not affect this variance, the first term of the two formulas is equal for both crude and RESTART simulation.
- The second term reflects the variance of the number of events A occurred when the set of events B_1 is given. As this variance is reduced by performing retrials in B_1, the second term is R_1 times smaller for the M threshold simulation than for the $M-1$ threshold simulation.

This intuitive reasoning, which is confirmed in a rigorous way in [35], leads to:

$$V\left(E[\hat{P}|\chi_1]\right) = V\left(E[\hat{P}^{M-1}|\chi_1]\right). \tag{7}$$

$$E[V(\hat{P}|\chi_1)] = \frac{1}{R_1} E[V(\hat{P}^{M-1}|\chi_1)]. \tag{8}$$

Note that, as $V(\hat{P}^{M-1})$ does not depend on χ_1, an increment of the first term of formula (5) due to a different χ_1 leads to a decrement of the same value of the second one. Consequently an increment of the first term of formula (6) leads to a decrement R_1 times smaller of the second one. It means that a greater variance of the importance at events B_1 makes less efficient the application of RESTART. An explanation of this fact is that a greater variance of the importance at events B_1 leads to a higher correlation between trials made from a given B_1 and, as the retrials made from a given B_1 are less effective if they are correlated, the application of RESTART is less efficient. The same applies to a great variance of the importance at events B_i for any other given value of i.

Starting with the formula of $V(\hat{P}^{M-1})$, obtained by adapting formula (3) to the case of M-1 thresholds numbered from 2 to M, and using formulas (5), (6), (7) and (8), formula (3) for the variance of the estimator $V(\hat{P})$ is derived in [35].

Analysis of Factors. s_i. In order to gain insight on factor s_i, formula (4) of this factor has been further developed in [35], leading to:

$$s_i = \frac{a_i}{K_A}\left[K'_i + \frac{V(P^*_{A/X_i})}{(P^*_{A/i})^2}\gamma_i\right] \quad (1 \le i \le M), \tag{9}$$

where $K'_i = V(N_i^0)/E[N_i^0]$ $(1 \le i \le M)$ and:

$$\gamma_i = 1 + 2\sum_{m=1}^{\infty} \frac{E[Max(0, N_i^0 - m)]}{E[N_i^0]} \frac{ACV_m(P^*_{A/X_i})}{V(P^*_{A/X_i})} \quad (1 \le i \le M),$$

$ACV_m(P^*_{A/X_i})$ being the autocovariance of P^*_{A/X_i} at lag m.

Let us analyze formula (9):

- Factor K'_i: This factor is a measure of the autocorrelation of the process of the occurrence of events B_i in the main trial. If the process is uncorrelated, K'_i is close to 1 (exactly, $K'_i = 1 - P_{i/0}/a_i$). In most applications, the process has a weak positive autocorrelation and K'_i is slightly greater than 1.
- Factor γ_i: If the random variables X_i were independent all the covariances $ACV_m(P^*_{A/X_i})$ would be zero and thus $\gamma_i = 1$. In general, γ_i is a measure of the dependence of the importance of the system states X_i of events B_i occurring in

the main trial. In most practical applications, there may be some dependence between system states of close events B_i but this dependence is negligible for distant events B_j. Thus γ_i is usually close to 1 or at least of the same order of magnitude as 1.

- Ratio $V\left(P^*_{A/X_i}\right)/\left(P^*_{A/i}\right)^2$: It greatly depends on the chosen importance function and may have an important impact on the efficiency of RESTART. An ideal choice of the importance function and of the process $X(t)$ would lead to $V\left(P^*_{A/X_1}\right) = 0$ and thus $s_1 = K_1/K_A$, which is around 1 in many applications. Thus values of $s_1 \gg 1$ could indicate inefficiency in the application of RESTART due to an improper choice of the importance function.

4.2 Simulation Cost

Let us define the cost C of a simulation as the computational time required for the simulation, taking as time unit the average computational time per reference event in a crude simulation of the system. With this definition of time unit, the cost of a crude simulation with N reference events is $C = N$.

In a RESTART simulation, the average cost of a reference event is always greater as overheads are involved in the implementation of RESTART: (1) for each event, an overhead mainly due to the need to evaluate the importance function and to compare it with the threshold values, and (2) for each retrial, an overhead mainly due to the restoration of event B_i (which includes to restore the system state at B_i and, as explained in Section 5.6, to re-schedule the scheduled events). To account for these overheads, the average cost of a reference event in a RESTART simulation is inflated (1) by a factor $y_e > 1$ in any case and (2) by an additional factor $y_{ri} > 1$ if the reference event occurs in a retrial of level i.

Using the above definition of time unit, the average cost per reference event is $y_0 = y_e$ in the main trial and $y_i = y_e\, y_{ri}$ ($1 \le i \le M$) in a retrial of level i. As the expected number of reference events in the retrials of level i of a RESTART simulation (with N reference events in the main trial) is $N P_{i/0} r_{i-1}(R_i - 1)$, the expected cost of the simulation is:

$$C = N\left[y_0 + \sum_{i=1}^{M} y_i P_{i/0} r_{i-1}(R_i - 1) \right]. \tag{10}$$

Remark: Factors y_i affect the simulation cost when it is measured in terms of required computational time, but not when it is measured in terms of number of events to be simulated. In this case $y_i = 1$ ($0 \le i \le M$).

4.3 Simulation Gain with RESTART

A measure of the efficiency for computing $\hat{P}$ is given by the relative confidence-normalized cost, *RCNC*, which is defined as $C V(\hat{P})/P^2$. To compare the *RCNC* of several estimators is equivalent to comparing the computational costs for a fixed relative width of the confidence interval. RCNC is equal to K_A/P in crude simulation, given that $V(\hat{P}) = K_A P/N$ and $C = N$, and can be obtained from formulas (3) and (10) in RESTART simulation.

The gain G obtained with RESTART can be defined as the ratio of the RCNC with crude simulation to the RCNC with RESTART. Defining $s_0 = 0$, $s_{M+1} = 1$ and $y_{M+1} = 0$ the following formula of the gain is obtained:

$$G = \frac{1}{P\left(\sum_{i=0}^{M} \frac{s_{i+1} u_i (1 - P_{i+1/i})}{P_{i+1/0} r_i}\right)\left(\sum_{i=0}^{M} y_i v_{i+1} P_{i/0} r_i (1 - P_{i+1/i})\right)}, \quad (11)$$

where:

$$u_i = \frac{1 - \frac{s_i}{s_{i+1}} P_{i+1/i}}{1 - P_{i+1/i}} \; ; \; v_{i+1} = \frac{1 - \frac{y_{i+1}}{y_i} P_{i+1/i}}{1 - P_{i+1/i}} \quad (0 \leq i \leq M).$$

4.4 Quasi-Optimal Parameters

To maximize the gain G in formula (11), factors s_i and y_i must be minimized and optimal values for $P_{i/0}$ (or equivalently $P_{i+1/i}$) and r_i need to be derived. Let us focus in this section on the optimal values of $P_{i+1/i}$ and r_i. These optimal values, that are function of s_i and y_i, have been derived in [35]. However, in a practical application, the values of s_i and y_i are difficult to evaluate. Therefore, approximations of the optimal values of $P_{i+1/i}$ and r_i that are independent of s_i and y_i and given by simple expressions are recommended. As these approximations of the optimal parameters provide a gain close to that obtained with the optimal ones they are called quasi-optimal parameters. These parameters have also been derived in [35] assuming that the product $s_{i+1} u_i$ takes the same value for every i ($0 \leq i \leq M$) and that the same occurs for the product $y_i v_{i+1}$. With these assumptions, quasi-optimal parameters maximizing the gain have been derived from (11) in these three steps:

1. For fixed values of $P_{i+1/i}$, quasi-optimal values of r_i are derived. For deriving them the derivative of the gain in formula (11) with respect to r_i is made equal to zero for $1 \leq i \leq M$ and the resulting system of equations is solved. The solution obtained is:

$$r_i = \sqrt{\frac{1}{P_{i/0} P_{i+1/1}}} \quad (1 \leq i \leq M). \quad (12)$$

In practice, as the number of retrials R_i must be integer, a value close to that given by (12) that satisfies this restriction must be chosen for r_i

2. For these values of r_i quasi-optimal values of $P_{i+1/i}$ for a fixed number of thresholds are derived. For this purpose, r_i is substituted in (11) by the second term of

(12) and $P_{1/0}$ by $P \Big/ \prod_{i=1}^{M} P_{i+1/i}$. The derivative of the resulting expression of the gain with respect to $P_{i+1/i}$ is made equal to zero for $1 \leq i \leq M$, obtaining $P_{i+1/i} = P_{1/0}$. It means that for a fixed number of thresholds "quasi-optimal" gain is obtained when all the probabilities $P_{i+1/i}$ have the same value, which is:

$$P_{i+1/i} = P^{\frac{1}{M+1}} \qquad (0 \leq i \leq M). \tag{13}$$

3. For these values of r_i and $P_{i+1/i}$ quasi-optimal value of M is derived. Substituting also (13) in (11), we can observe that the larger the value of M, the greater the gain. Thus $P_{i+1/i}$ must be as close as possible to 1, i.e., the thresholds must be set as close as possible. In practice, there are two limitations on how close the thresholds can be set: one is due to the values that Φ can take when it is a discrete function; the other is due to the restrictions on the value of R_i derived from the chosen thresholds. This value must be an integer number greater than one, given that $R_i = 1$ means that T_i is not really a threshold.

The quasi-optimal gain, obtained when r_i and $P_{i+1/i}$ are given by (12) and (13) respectively and M tends to infinite, is given by:

$$G = \frac{1}{P(-AVG(s)\ln P + 1)(-AVG(y)\ln P + y_0)}, \tag{14}$$

where $AVG(s)$ and $AVG(y)$ are the arithmetical means of s_i and y_i $(1 \leq i \leq M)$ respectively.

5 Guidelines for an Effective Application of RESTART

The quasi-optimal gain given by formula (14) assumes that quasi-optimal parameters are used. In practice, quasi-optimal parameters are not possible: the importance function may be discrete and it prevents from setting thresholds with $P_{i+1/i}$ very close to 1; even when the importance function is continuous it is not possible to set infinite thresholds, as mentioned above; moreover, the evaluation of r_i is based on an estimation of $P_{i/0}$. Although this estimation can be made by means of pilot runs, there will be always some error in the estimation and thus r_i will not be exactly the quasi-optimal one. In addition the resulting R_i has to be rounded to an integer number. This section studies how the gain is affected by the errors and limitations in the setting of the optimal parameters as well as by the computer overhead produced by the implementation of RESTART and by the chosen importance function. Section 5.1 defines four factors reflecting the influence of these features in the gain and Sections 5.2 to 5.6 analyze each of the factors and provide guidelines for reducing them.

5.1 Factors Affecting the Efficiency of RESTART

As indicated in [35], the general formula of the gain (formula (11)) can be re-written as follows:

$$G = \frac{1}{f_V f_O f_R f_T} \frac{1}{P(-\ln P+1)^2},\qquad(15)$$

where:

$$f_V = \frac{\sum_{i=1}^{M}\frac{s_i(R_i-1)}{P_{i/0}r_i}+\frac{s_{M+1}}{Pr_M}}{\sum_{i=1}^{M}\frac{R_i-1}{P_{i/0}r_i}+\frac{1}{Pr_M}},\qquad f_O = \frac{\sum_{i=1}^{M}y_i P_{i/0}r_{i-1}(R_i-1)+y_0}{\sum_{i=1}^{M}P_{i/0}r_{i-1}(R_i-1)+1}.\qquad(16)$$

$$f_R = \frac{\left(\sum_{i=1}^{M}\frac{R_i-1}{P_{i/0}r_i}+\frac{1}{Pr_M}\right)\left(\sum_{i=1}^{M}P_{i/0}r_{i-1}(R_i-1)+1\right)}{\left(\sum_{i=0}^{M}\frac{1-P_{i+1/i}}{\sqrt{P_{i+1/i}}}+1\right)^2},\qquad f_T = \frac{\left(\sum_{i=0}^{M}\frac{1-P_{i+1/i}}{\sqrt{P_{i+1/i}}}+1\right)^2}{(-\ln P+1)^2}.\qquad(17)$$

The term $1/(P(-\ln P+1)^2)$ can be considered the ideal gain, which matches with the quasi-optimal gain (formula (14)) when $s_i = 1$ $(1 \le i \le M)$ and $y_i = 1$ $(0 \le i \le M)$. Factors f_V, f_O, f_R and f_T, all of them equal to or greater than 1 (with the exception of f_V which could be smaller than 1 in some cases), can be considered inefficiency factors that reduce the actual gain with respect to the ideal one. Each factor reflects:

- f_V: inefficiency due to the variance of the importance of the systems states at each B_i which in its turn is due to the non-optimal choice of the Markovian process $X(t)$ used for simulating the original process $Z(t)$ (see Section 2) and/or the non-optimal choice of the importance function;
- f_O: inefficiency due to the computer overhead produced by the implementation of RESTART;
- f_R: inefficiency due to the non-optimal choice of the number of retrials;
- f_T: inefficiency due to the non-optimal choice of the thresholds.

Note that the ideal gain $1/(P(-\ln P+1)^2)$ takes very high values, e.g., $4.6 \cdot 10^3$ for $P = 10^{-6}$, $1.7 \cdot 10^7$ for $P = 10^{-10}$ and $9.1 \cdot 10^{10}$ for $P = 10^{-14}$. Assuming a computational time of 0.1 msec. per reference event, to estimate these probabilities with crude simulation would require a computational time of 11 hours, 13 years and 127 millennia respectively. Applying RESTART these times are reduced, assuming that the ideal gain is achieved, to 9, 23 and 44 secs. respectively. In practice these times will be greater due to the inefficiency factors but, if these factors take moderate values, the resulting computational time may be low even though their values are not close to 1.

5.2 Analysis and Guidelines to Reduce Factor f_R

Let η_i denote the ratio of the actual value of r_i to its quasi-optimal value r_{iqo} given by (12), and $\eta_{\max}$ and $\eta_{\min}$ the maximum and minimum values of η_i respectively:

$$\eta_i = \frac{r_i}{r_{iqo}} \quad (1 \leq i \leq M); \quad \eta_0 = 1.$$

$$\eta_{max} = \max_{0 \leq i \leq M}(\eta_i); \quad \eta_{min} = \min_{0 \leq i \leq M}(\eta_i).$$

Based on left-hand formula (17) and on this notation, the following bound of f_R is derived in [35]:

$$f_R \leq \frac{\eta_{max}}{\eta_{min}}. \tag{18}$$

This bound allows providing guidelines for assigning values to r_i taking into account that R_i must be integer. For given thresholds, (assuming that a value has already been assigned to r_{i-1}) the value that must be assigned for r_i is: $r_i = r_{i-1} R_i$ where $R_i = r_{iqo}/r_{i-1}$ (rounded). R_i must be rounded to its integer part $\lfloor R_i \rfloor$ or to $\lfloor R_i \rfloor + 1$ depending on which alternative leads to the minimum value of $Max(\eta_i, 1/\eta_i)$.

Formula (18) also indicates that the impact on the gain of a non-optimal choice of r_i due to errors in the estimation of $P_{i/0}$ is moderate if the errors are not very large; thus a rough estimation of $P_{i/0}$ may be sufficient for this purpose.

5.3 Analysis and Guidelines to Reduce Factor f_T

Based on right-hand formula (17), the following bound of f_T is derived in [35]:

$$f_T \leq \frac{(1-P_{min})^2/P_{min}}{(\ln P_{min})^2}, \tag{19}$$

where:

$$P_{min} = \min_{0 \leq i \leq M}(P_{i+1/i}).$$

The value of f_T is moderate even for values of P_{min} far from 1. For example, $f_T \leq 1.04$ for $P_{min} = 0.5$, $f_T \leq 1.53$ for $P_{min} = 0.1$ and $f_T \leq 4.62$ for $P_{min} = 0.01$. It means that the impact on the gain of a discrete importance function is moderate except in the case that the thresholds have to been set very far each other.

Consequently the following guidelines may be provided for setting thresholds: if the importance function is continuous, thresholds should be set at a distance given by $P_{i+1/i} = 0.5$, given that it leads to $R_i = 2$ without need of rounding while f_T is only 1.04. If the importance function is discrete and the probability ratio of consecutive values of Φ is greater than 0.5, thresholds that lead to $R_i = 2$ with minimum rounding should be set. If the probability ratio of consecutive values of Φ is smaller than 0.5, a threshold should be set for each value of Φ.

5.4 Analysis and Guidelines to Reduce Factor f_O

According to right-hand formula (16), factors f_O is a weighted mean of y_i $(0 \leq i \leq M)$, all the weights being positive. Thus, the following bound of f_O can be defined:

$$f_O \leq \underset{0 \leq i \leq M}{\text{Max}} (y_i).$$

Factor f_O reflects the inefficiency due to the overhead produced by the implementation of RESTART, as explained in Section 4.2. The value taken depends on the system characteristics. For example, it is higher when the system state is described by many variables because it increases the overhead needed for restoring the system state at B_i.

Factor f_O can be reduced by the use of hysteresis, which reduces the number of events B_i in the simulation and by following some programming guidelines for reducing the overhead per event B_i. The use of hysteresis consists in defining for each threshold T_i an additional threshold $T_i' < T_i$ and extending the retrials of level i until $\Phi < T_i'$ (see, e.g., [37]). Guidelines to reduce the overhead per event B_i are explained in [37]. They are:

- To perform memory dump for saving or restoring the state at B_i instead of copying the system variables one by one;
- To perform a joint scheduling of all the pending events with negative exponentially distributed time of occurrence. When several of these events are pending to occur in the simulation, there are two programming options: to schedule all of them or to schedule only the one which will first occur. This second option is recommended when RESTART is used because it reduces the number of events simultaneous scheduled in the simulation and thus the number of them that, according to Section 5.6, must be re-scheduled at the beginning of each retrial.

Note that f_O is equal to 1 when the efficiency is measured in terms of the number of simulated events.

5.5 Analysis of Factor f_V

According to left-hand formula (16), factor f_V is a weighted mean of s_i $(1 \leq i \leq M+1)$, all the weights being positive. Thus, the following bound of f_V can be defined:

$$f_V \leq \underset{1 \leq i \leq M+1}{\text{Max}} (s_i).$$

As explained in Section 4.1, the term that may motivate a high value of s_i and thus of f_V is the variance of the importance of the system states at events B_i, $V(P^*_{A/X_i})$ or more precisely, the ratio $V(P^*_{A/X_i})/(P^*_{A/i})^2$. This ratio does not only depends on the chosen importance function but also on the characteristics of the process $X(t)$. The optimal importance function Φ is that for which each threshold T_i of Φ defines an importance

I_i such that, for any system state x, $\Phi(x) \geq T_i \Leftrightarrow P^*_{A/x} \geq I_i$. This importance function usually leads to very low values of $V(P^*_{A/X_i})/(P^*_{A/i})^2$ and thus of s_i and f_V. Nevertheless, if the process $X(t)$ may skip from a state to another of much higher importance, the importance of the events B_i may be much higher than I_i. For this type of processes s_i may take a high value even when the optimal importance function is chosen.

In order have a more meaningful bound of f_V the following bound of s_i has been derived in [3]:

$$s_i \leq \frac{a_i \, Max(K'_i, \gamma_i)}{K_A} \frac{Q^*_{A/i}}{P^*_{A/i}} \quad (1 \leq i \leq M), \tag{20}$$

where $Q^*_{A/i}$ is the supreme (or, in general, an upper bound) of P^*_{A/X_i}.

It is not necessary that P^*_{A/X_i} be bounded for finding a bound of s_i. If $Q^*_{A/i}$ is not a bound of P^*_{A/X_i} but $\Pr\{P^*_{A/X_i} > P^*_{A/x_i}\}$ for $P^*_{A/x_i} > Q^*_{A/i}$ decreases faster than $1/(P^*_{A/x_i})^{\beta_i}$ for some $\beta_i > 2$, that is, if

$$\Pr\{P^*_{A/X_i} > P^*_{A/x_i}\} \leq \frac{\Pr\{P^*_{A/X_i} > Q^*_{A/i}\}}{(P^*_{A/x_i}/Q^*_{A/i})^{\beta_i}} \quad (\beta_i > 2, \; \forall P^*_{A/x_i} > Q^*_{A/i}),$$

the following bound for s_i is derived in [3]:

$$s_i \leq \frac{a_i \, Max(K'_i, \gamma_i)}{K_A} \frac{Q^*_{A/i}}{P^*_{A/i}} \frac{\beta_i}{\beta_i - 2} \quad (1 \leq i \leq M). \tag{21}$$

Note that bound (20) is a particular case of bound (21) for β_i tending to infinity. Bound (21) is less restrictive because it does not require β_i tending to infinity (which implies a bounded P^*_{A/X_i}) but it only requires $\beta_i > 2$.

5.6 Guidelines to Reduce Factor f_V

As explained in Section 5.5, the value of factor f_V depends on the variance of the importance at events B_i. To reduce this variance, all the states x_i at events B_i must have similar importance; this is achieved by:

- Using a good importance function Φ, i.e., a function for which all the states on the threshold boundary $\Phi = T_i$ have similar importance.
- Reducing importance skipping, that is, avoiding that the process $X(t)$ may skip from a given state to another of much higher importance. Importance skipping may cause some events B_i to be far from the threshold boundary, thus having importance much higher than other events B_i on (or close to) the boundary.

We have treated in previous sections the way of optimizing RESTART for a given $X(t)$. However, for a given $Z(t)$, the process $X(t)$ may be different depending on how the simulation model is implemented. The extent of importance skipping is determined by the process $X(t)$, as explained below. Thus, a proper choice of the process $X(t)$ and the importance function Φ can reduce the factor f_V and hence increase the efficiency of RESTART. Let us see how to choose $X(t)$ and how to choose the importance function.

Guidelines for the Choice of $X(t)$. As explained in Section 4.1, to reduce the variance of the importance at events B_i for a given threshold i leads to reduce the correlation among trials made from a given B_i and vice versa. Although any one of these two reductions may be used as a criterion for selecting a proper process $X(t)$, both of them are used in this section to reinforce the reasoning.

As indicated in [3], when a simulation event occurs some random decisions may have to be taken, i.e., the values of some system variables (e.g., the number of packets in an arriving burst or the time scheduled for the occurrence of a future event) may have to be randomly determined. The definition of the process $X(t)$ depends on the way these random decisions are taken during the simulation, which determines the extent of importance skipping and correlations among trials $[B_i, D_i)$. Therefore, $X(t)$ also impacts the efficiency of RESTART, as shown in [38]. Let us illustrate this by using some examples.

Let us consider the two following options for determining the random number of packets in a burst that arrives at a queue: (a) to determine the entire length of the burst at the arrival of the first packet, and (b) to determine at the arrival of each packet whether it is the last packet or there are more packets in the burst. In option (b) $X(t)$ includes the number of packets in the burst arrived so far while in option (a) also includes the number of remaining (yet to arrive) packets in the burst.

Note that in option (a) only one random decision is made at the beginning of the burst, while in option (b) a number of sequential random decisions are made (conditioned on the number of packets in the burst arrived so far), one at the arrival of each packet in the burst. Clearly, the process $X(t)$ evolves at large increments in option (a), which may cause large importance skipping. On the other hand, in option (b) the process $X(t)$ evolves at small increments, which reduces importance skipping. Therefore, option (b) is recommended in a simulation in which RESTART is applied.

Also, note that $X(t)$ is Markovian in both options, since it contains sufficient information to execute the simulation of the system. However, in option (a) some future events are scheduled before they actually happen, while in option (b) no future events are scheduled unless necessary to continue the simulation. In the application of RESTART, option (a) will cause more sharing of future events (and hence more correlation) among trials made from a given B_i. This reinforces the reason given above for justifying why option (b) is favored over option (a) in the application of RESTART.

An alternative way of implementing option (b) is to determine the burst length at the arrival of the first packet (as in option (a)) and to determine it again at the arrival of each new packet by randomly generating the remaining burst length (conditioned on the number of packets arrived so far). This implementation of option (b) is equivalent to the previous one because it yields the same process $X(t)$.

Let us now consider the scheduling of future events in the simulation of a G/G/1 queue, where the rare set is defined as the queue length $q(t)$ greater than a certain threshold. If the times of occurrence of next arrival, t_{NA}, and of next service completion, t_{NC}, are scheduled only once at the previous arrival and at the start of the current service, respectively, then $X(t) = (q(t), t_{NA} - t, t_{NC} - t)$. This is similar to option (a) of the previous example: $X(t)$ includes the times of already scheduled future (arrival and service completion) events, which could cause importance skipping (e.g., when a high value is randomly assigned to $t_{NC} - t$). It also increases correlation among trials [B_i, D_i) due to sharing of future events. This can be avoided by using the process $X(t) = (q(t), t - t_{PA}, t - t_{CS})$ to simulate the system, where t_{PA} and t_{CS} are the times of the previous arrival and the start of the current service, respectively. This is similar to option (b) in the previous example. Here, arrival and service completion events can be rescheduled (conditioned on the elapsed times, $t - t_{PA}$ and $t - t_{CS}$, respectively) at the occurrence of every event. However, to avoid unnecessary overhead, it is sufficient to reschedule only at events B_i, at the beginning of each retrial. This rescheduling minimizes the sharing of future events by different trials from the same event B_i and hence reduces the correlation among them, which improves the efficiency of RESTART.

Guidelines for the Choice of the Importance Function. Once the variables required to describe $X(t)$ have been determined, an importance function, which is a function of these variables, must be chosen. With a proper importance function all the states on each of the threshold boundaries $\Phi = T_i$ have similar importance, and thus, if importance skipping is small, all the states $x_i \in \Omega_i$ also have similar importance for any i. It leads to small values of $V(P^*_{A/X_i})$ and thus also s_i for any i, and consequently to a small value of f_V.

In [20] it was pointed out that "the most challenging work for future research is to find and implement an efficient algorithm to determine good importance functions for defining thresholds".

In the case of one-dimensional systems, the choice of the importance function is straightforward, because the threshold boundary has only one state. Without importance skipping, this state is also the only state of Ω_i and thus $V(P^*_{A/X_i}) = 0$.

Also, for multidimensional systems, small values of s_i and thus of f_V are achieved if all the states $x_i \in \Omega_i$ have similar importance, but this condition is not strictly necessary. States $x_i \in \Omega_i$ with moderate probability of occurrence may have much lower importance than the most frequent ones without leading to high values of $V(P^*_{A/X_i})$ and s_i : consider that P^*_{A/x_i} is bounded and that the ratio of the probability of Ω_i^H, the set of states $x_i \in \Omega_i$ with importance P^*_{A/x_i} close to its supreme $Q^*_{A/i}$, to the probability of the whole set Ω_i is appreciable (greater than, say, 0.2 or 0.3). Then $P^*_{A/i}$, the mean importance of states $x_i \in \Omega_i$, is close to the mean importance of states $x_i \in \Omega_i^H$ (and thus also close to the supreme $Q^*_{A/i}$). Consequently, the ratio $Q^*_{A/i}/P^*_{A/i}$ is small and, according to formula (30), s_i is small.

In [43] it was stated that "in the case of multidimensional state spaces, good choices of the importance function for splitting are crucial, and are definitely non-trivial to obtain in general". It is non-trivial because an exact analytical evaluation of the importance of the states is not possible in most cases. Thus a combination of approximate analytical formulas, heuristic reasoning and feedback from simulation results must be used to choose an appropriate importance function.

An approach that can be used in queuing networks is to assume that the importance function is a linear combination of the queue lengths of the network nodes: $\Phi = \sum_{\forall i} a_i Q_i$. Several procedures can be used to assign values to the coefficients a_i:

- First of all, one of the coefficients can be made equal to 1 without loss of generality.
- If the network has few nodes, e.g., only two nodes or three nodes and, thus, only one or two coefficients, the simplest solution is to perform pilot runs to test several values of the coefficients and to choose the values for which the application is more efficient. For saving computational time of the pilot runs, they can be made for system parameter values for which the rare event is not so rare, given that the results obtained usually apply to the parameter values of interest.
- By means of heuristic reasoning the number of coefficients that have to be adjusted may be reduced. E.g., the coefficients can be made equal to zero for the nodes for which its queue length has not impact on the occurrence of the rare event or it is guessed that the impact is small. Another example is to assign the same value to coefficients corresponding to queue lengths with the same or similar impact on the occurrence of the rare event.
- If possible, approximate analytical formulas may be derived to roughly estimate the importance of some states. From these formulas the coefficient values may be evaluated by equating the importance function of states with the same importance. An alternative to this approach is to estimate the importance of some states by means of pilot runs.
- If values have been assigned to some of the coefficients, interpolation or extrapolation based on heuristic reasoning may be used to assign values to the remaining ones.
- All the assumptions or approximations made for assigning values can be checked by means of pilot runs (that usually can be made for system parameter values for which the rare event is not so rare). These pilot runs can also be used to introduce correction factors to a set of coefficients, previously obtained, to improve them. This approach may also be used when the set of parameters obtained for a model are going to be applied to another similar model.

Observe that the approximations are allowed for deriving the importance function because they could affect the efficiency of the method, but they do not affect the correctness of the estimates.

In the networks with few nodes the number of coefficients to be assigned is smaller but the values assigned to them could be more critical due to the strong dependency that usually exists among the nodes. However in more complex networks, though the number of coefficients is larger the accuracy of the values assigned to them is not so critical, as it is shown in the examples of Section 6. This fact may compensate the difficulty that complex networks could have due to the need of assigning many coefficient values. As we will see in Section 6, the two-queue Jackson tandem network was simulated in [41] defining the rare set as $q_2 \geq L$ and choosing the importance function $\Phi = q_2$. This importance function Φ led to a large value of f_V and, as reported in [41], to a very low efficiency. However in the multistage ATM switch studied in [16] an equivalent importance function led to a high efficiency. This switch has three stages of 8×8 switching elements (SE), eight of them in each stage. Each SE is an output buffered switch with eight separated buffers of size K. The rare set is defined as the overflow of a buffer of the third stage. The importance function Φ is defined as the queue length q of the buffer under study. An importance function equivalent to that used in the previous example was successful, despite the greater complexity of the system. This is because the cells in the buffers in the second stage do not need to go to the buffer under study in the third stage, but can go to any of the 64 buffers of this stage instead. As a result, the queue lengths of the buffers of the first or second stage have a small impact on the future queue length of a buffer of the third stage. Although the importance function used in [16] could be improved taking into account the queue length of the other queues (as will be seen in Section 6) the dependencies in this complex system are weak enough to be ignored without a significant impact on the efficiency achieved. In the two-queue tandem network, its simplicity notwithstanding, the dependence is strong and cannot be ignored.

In reliability problems, an importance function defined as a linear combination of variables representing the state of each component (1= failure, 0= operational) is not appropriate. The effect of the failure of a component is different depending which other components have also failed and this type of dependencies cannot be taken into account with a linear function of the states of the components. It is better in this case to obtain, based on some heuristic reasoning, a formula of the importance function that take into account these dependencies. In Section 6.3, a function of the states of the components obtained heuristically is proposed as importance function. Although the proposed importance function works well in all the cases studied, there are some cases in which it could be improved because, as indicated in that section, the proposed function does not account for all the features of the system state that may impact on the occurrence of the rare event. A possibility to improve it could be to obtain heuristically another function of the system state variables that accounts for those features of the system state that are not taken into account by the previous function. The final importance function could be a linear combination of the two functions. Thus the approach proposed for queuing networks consisting in the choice of an importance function built as a linear combination of variables of the system state could be generalized to the choice of a linear combination of functions of the system state.

6 Application Examples

Several examples on Jackson and non-Jackson queuing networks and on ultra reliable systems are shown in this section to illustrate the application of the guidelines given in Section 5 and the efficiency obtained. For evaluating the goodness of an application and its possibility of improvement, it is not only interesting to observe the required computational time but also the gain obtained and the values of the inefficiency factors. Section 6.1 explains how we can estimate the gain and the factors f_V and f_O. Factors f_R and f_T may be estimated by its bounds given by formulas (18) and (19) respectively.

In all the runs, the simulation length was adjusted to have a relative half width of the 95% confidence interval (relative error) equal to 10%. The interval width was evaluated using the batch means method. The experiments of the two-queue Jackson tamdem network were run on a Sun Ultra 5 workstation and the remainig ones on a Pentium(R) D CPU 3.01 GHz.

6.1 Jackson Networks

First we will see how to obtain the importance function for two-queue tandem networks by assigning the coefficient values of the linear combination of queue lengths by means of tests made with pilot runs. Then general Jackson networks are studied. As the method of assigning by means of pilot runs is not practicable when the number of nodes is large, the importance function is derived by means of approximate analytical formulas.

Two-Queue Jackson Tandem Network. In this network customers with Poisson arrival enter the first queue and, after being served, enter the second one. The mean arrival rate is λ and the service time is exponentially distributed in each queue with mean service rates μ_1 and μ_2, respectively. The load at each queue is $\rho_i = \lambda/\mu_i \ (i=1,2)$. The buffer space at each queue is assumed to be infinite. The system state $Z(t)$ is given by (q_1, q_2), where q_i is the number of customers at queue i. If rescheduling is made, the system state $X(t)$ is also given by (q_1, q_2). This model has received considerable attention in the rare event literature, e.g., [14], [17], [19], [22], [39], [40], [41] and [42].

The difficulty of applying accelerated simulation techniques arises when the first queue is the bottleneck and the rare set definition is related to the value of q_2. In order to cope with a difficult case the loads tested were $\rho_1 = 0.5$ and $\rho_2 = 0.33$.

The network was studied in [3] for the following three definitions of the rare set A: $Q_1 + Q_2 \geq L$; $Q_2 \geq L$ and $\min(Q_1, Q_2) \geq L$.

In these examples thresholds and number of retrials were determined in a similar manner to that explained later for general Jackson networks.

Rare Set Defined as $Q_1 + Q_2 \geq L$. For this definition of the rare set, the most "natural" importance function is $\Phi = Q_1 + Q_2$. Let us analyze if this function is appropriate. Assume that $L = 60$ and $T_i = 30$. The possible states at an event B_i are (0,30), (1,29), (2,28), ..., (29,1), (30,0). The importance of each of these states is different. The higher the value of Q_1 (for $Q_1 + Q_2 = 30$), the higher the importance of the state, given that a customer at Q_1 has to be served by both servers before leaving the system, while a customer at Q_2 has to be served only at the second one. Thus the supreme $Q^*_{A/i}$ is the importance of state (30,0). But, given that the first queue is the bottlenecck, the states with high value of Q_1 and low value of Q_2 have the highest probability. As these states have an importance close to the supreme, $\Pr\{\Omega_i^H|\Omega_i\}$ is high. Thus $\Phi = Q_1 + Q_2$ seems to lead to moderate values of s_i and therefore, also f_V. Simulation results confirm that this qualitative reasoning is valid: probabilities up to 10^{-66} were accurately estimated with less than 40 minutes of computational time. The very low values of f_V (smaller than 1.02) show that the choice of $\Phi = Q_1 + Q_2$ is appropriate and that the application is very close to the optimal one.

Let us see how to estimate the gain obtained and the values of factors f_V and f_O. The gain in events or the gain in time with respect to a crude simulation is estimated as follows: in a crude simulation with $L = 14$, thus $P = 1.22 \times 10^{-4}$, and the same remaining conditions, the number of reference events (arrivals in this case) and the computational time are measured. As $V(\hat{P}) = K_A P/N$ the measured values are extrapolated for the value of L for which we want to estimate the factors, e.g., 220, under the assumption of K_A taking the same value. The gain is the ratio between these extrapolated values and those measured in the simulation with RESTART. A gain in events equal to $3.4 \bullet 10^{61}$ and a gain in events equal to $5.0 \bullet 10^{60}$ are obtained. Then we compare the measured gain with the theoretical one derived from formula (15). If we assume $f_V = 1$ and $f_O = 1$ in (15), we obtain for $L = 220$ a gain equal to $P = 3.46 \bullet 10^{61}$ (taking and r_i given by formulas (13) and (12) respectively and thus taking $f_R = 1$ and f_T equal to its bound (19) evaluated for $P_{min} = P^{1/(L-1)}$). We see that the theoretical gain (for $f_V = 1$, $f_O = 1$) is 1.02 times the actual gain in events. Given that the gain in events is not affected by the factor f_O, the value 1.02 can be taken as an estimate of f_V for $L = 220$. Finally, the factor f_O can be estimated as the ratio between the gain in events and the gain in time. It leads to f sub $O = 6.8$.

Rare Set Defined as $Q_2 \geq L$. The simplicity of the system allows for simulating it by means of regenerative simulations and splitting, as in [40] and [41]. In [41] the chosen importance function is $\Phi = Q_2$. Let us consider that L and an intermediate threshold T_i take the values $L = 30$ and $T_i = 15$. At an event B_i, $q_2 = 15$ but Q_1 can take any value. It is clear that the probability of reaching A ($Q_2 \geq 30$) from an event B_i at which $Q_1 = 0, Q_2 = 15$ is very different from that of reaching A from an event B_i with, say, $Q_1 = 60, Q_2 = 15$. The supreme $Q^*_{A/i}$ is given by the limit of the importance of state $(Q_1, 15)$ when Q_1 tends to infinity. The probability of states $(Q_1, 15)$ with high value of Q_1 is much smaller than that of states with small value of Q_1, and thus $\Pr\{\Omega_i^H|\Omega_i\}$ is very small. Therefore this chosen function Φ leads to a large value of s_i and, as reported in [41], to a low efficiency.

It is clear that in the definition of Φ, Q_1 must be accounted for, since its value can affect the future evolution of Q_2. Some weight, albeit smaller than the weight given to Q_2, must be given to Q_1. Along this line of reasoning, we tested $\Phi = aQ_1 + Q_2$, with $0 \leq a \leq 1$. Pilot simulations for $L = 20$ with several values of a were run to determine the appropriate value of the coefficient a. The required computational times for $a = 0.8, 0.7, 0.6, 0.5$ and 0.4 were 48, 23, 15, 27 and 70 seconds, respectively. The value $a = 0.6$ is chosen, as it provides the best results.

Probabilities of the order of 10^{-29} and of 10^{-67} were accurately estimated with 7 and 200 minutes of computational time, respectively. Although the values of f_V (3.1 and 11.2 respectively) are not so small as in the previous case, they are moderate enough to accurately estimate very low probabilities at a reasonable computational time. These low values of f_V indicate that the application is close to the optimal one.

Rare Set Defined as $Min(Q_1, Q_2) \geq L$. For this case, the function Φ was defined in [40] as $\Phi = Min(Q_1, Q_2)$. This definition of Φ leads to high values of f_V and, as reported in that paper, low efficiency. For $T_i = 20$ (with $L > 20$), possible states at B_i are (100, 20), (20, 20) and (20, 100). The importance of, say, states (100, 20) or (20, 100) is much higher than that of state (20, 20). The supreme $Q^*_{A/i}$ is given by the limit of the importance of either state (20+j, 20) or state (20, 20+j) when j tends to infinity. The states with highest probability are (20+j, 20) or (20, 20+j) for low values of j. Consequently $Pr\{\Omega_i^H | \Omega_i\}$ is very low. Thus this definition of Φ does not seem to be appropriate.

For $Q_1 \leq L$ and $Q_2 \leq L$, we proposed to define Φ as a linear function of Q_1 and Q_2: $\Phi = aQ_1 + Q_2$. As the relative importance of the states depends on the system parameters, the appropriate value of the coefficient a may be greater, equal or smaller than 1 depending on the load values. Extending this definition for $Q_1 > L$ or $Q_2 > L$ does not appear to be appropriate: as the rare set is defined as $Q_1 > L$ and $Q_2 > L$, when Q_1 or Q_2 exceeds L we must give a lower weight to this excess. This effect may be taken into account by introducing a coefficient $b < 1$ in the definition of the importance function:

$$\Phi = a\Phi_1 + \Phi_2, \quad \text{where} \quad \Phi_i = \begin{cases} Q_i & \text{if } q_i \leq L \\ L + b(Q_i - L) & \text{if } q_i > L \end{cases}. \quad (22)$$

We tested the importance function (22) using pilot runs with $a = 1$ and different values of b. The best results were obtained for $b = 0.6$. Probabilities of the order of 10^{-32} and of 10^{-63} were accurately estimated with 7 and 72 minutes of computational time, respectively.

The low values of f_V (3.5 and 6.7, respectively) show that the application is very efficient and close to the optimal one. As the results were good enough, we did not investigated the improvement that could be obtained with other values of a.

This case was also studied in [19] by means of RESTART. They used the importance function $\Phi = Q_1 + 1.5Q_2$. It led to values of f_V (estimated by us based on their reported results) between 45 and 62 times the values reported in [3]. These results may be due to the fact that, in contrast to our approach, a lower weight was not given to Q_i-L when Q_i exceeds L.

In [40] the values of f_V are much higher. From their reported results, we have estimated a value of $f_V = 1600$ for $L = 10$. It is difficult to estimate f_V for larger values of L

because, as they claimed, the failure of their approach resulted in the underestimation of P and its relative error. Anyway this value of f_V is very huge and the tendency observed indicates that they must be much higher for higher values of L.

General Jackson Networks. Formulas for obtaining effective importance functions were provided for two-stage Markovian networks with any number of nodes in each stage in [25] and extended to general Jackson networks with any number of nodes in [32]. Jobs arrivals and departures are allowed in all the nodes. After being served in node l, jobs can go to any node m with probability p_{lm} or they can leave the network with probability p_{l0}. The steady-state probability of the number of jobs exceeding a level at a target node, $Q_{tg} \geq L$, was estimated.

Some approximations and assumptions were needed to derive the formulas of the importance function. First, it was assumed that the importance function is a linear function of the queue length of the nodes placed at a distance from the target node smaller than 3. Thus the queue length of a node was considered in the importance function only if the jobs leaving the node go directly to the target node (distance 1) or through only one intermediate node (distance 2). Then it was evaluated the importance of the extreme (also called boundary) states when the process enters sets C_i $\forall i$, that is, the system states at which only one queue is not empty. Finally, for calculating the coefficients of Q_i for each i in the importance function it was equated the importance of the extreme state corresponding to the target queue with the importance of each of the other extreme states. The goodness of the importance functions derived in the paper with such approximations was supported by the efficiency achieved in the simulations

The formula obtained for the two-queue Jackson tandem network with $\rho_1 > \rho_2$ and the rare set defined as $q_2 \geq L$ was: $\Phi = \dfrac{\ln \rho_1}{\ln \rho_2} Q_1 + Q_2$. For the loads above considered for this example the importance function given by the formula is: $\Phi = 0.63 Q_1 + Q_2$. Slightly better results were obtained with this formula than with the importance function $\Phi = 0.6 Q_1 + Q_2$ obtained heuristically.

Let us denote K the number of nodes with distance 1 and H the number of nodes with distance 2, for any value of K and H. Jobs with independent Poisson arrivals enter each node from the outside with arrival rates $\gamma_{1i}, i = 1,\ldots, H$ to the nodes with distance 2, $\gamma_{2j}, j = 1,\ldots, K$ to the nodes with distance 1 and γ_{tg} to the target node. The total arrival rates to each node (arrivals from the outside + arrivals from the other nodes) are denoted by: $\lambda_{1i}, i = 1,\ldots, H$, $\lambda_{2j}, j = 1,\ldots, K$ and λ_{tg}, respectively. The service times of all the nodes are assumed to be exponentially distributed with service rates $\mu_{1i}, i = 1,\ldots, H$, $\mu_{2j}, j = 1,\ldots, K$ and μ_{tg}, respectively. The buffer space in each queue is assumed to be infinite. Let us observe that When there are not nodes at a distance greater then 2, $\lambda_{1i} = \gamma_{1i} + \sum_{l=1}^{H} \lambda_{1l} p_{li} + \sum_{j=1}^{K} \lambda_{2j} p_{ji} + \lambda_{tg} p_{tgi}$, $i = 1,\ldots, H$. Analogous equations are obtained for $\lambda_{2j}, j=1,\ldots,K$ and λ_{tg}. The loads of the nodes are $\rho_{1i} = \lambda_{1i}/\mu_{1i}, i=1,\ldots,H$, $\rho_{2j} = \lambda_{2j}/\mu_{2j}, j=1,\ldots,K$ and $\rho_{tg} = \lambda_{tg}/\mu_{tg}$, respectively.

A general formula of the importance function valid for any Jackson network was derived in [32]. A simplified version of this formula (also given in that paper) that matches with the general one in almost all cases is the following:

$$\Phi = \sum_{i=1}^{H} Min\left\{1, \alpha_{1i}\frac{\ln\left(\rho_{tg}/\rho_{tgi}^{*}\right)}{\ln \rho_{tg}}\right\} Q_{1i} + \sum_{j=1}^{K} Min\left\{1, \alpha_{2j}\frac{\ln\left(\rho_{tg}/\rho_{tgj}^{\perp}\right)}{\ln \rho_{tg}}\right\} Q_{2j} + Q_{tg}, \quad (23)$$

where:

$$\rho_{tg} = \frac{\gamma_{tg} + \sum_{j=1}^{K}\lambda_{2j}P_{jtg} + \lambda_{tg}P_{tgtg}}{\mu_{tg}} = \frac{\lambda_{tg}}{\mu_{tg}},$$

$$\rho_{tgi}^{*} = \frac{\gamma_{tg} + \sum_{j=1}^{K}Min\{\lambda_{2j}+(\mu_{1i}-\lambda_{1i})p_{ij},\mu_{2j}\}P_{jtg} + \lambda_{tg}P_{tgtg}}{\mu_{tg}},$$

$$\rho_{tgi}^{\perp} = \frac{\gamma_{tg} + \mu_{2j}P_{jtg} + \sum_{l\neq j}\lambda_{2l}P_{ltg} + \lambda_{tg}P_{tgtg}}{\mu_{tg}},$$

$$\alpha_{1i} = 1 + \frac{\sum_{l\neq i}\gamma_{1l}\sum_{j=1}^{K}p_{lj}P_{jtg} + \sum_{j=1}^{K}\gamma_{2j}P_{jtg} + \gamma_{tg}}{\mu_{1i}\sum_{j=1}^{K}p_{ij}P_{jtg}},$$

$$\alpha_{2j} = 1 + \frac{\sum_{i=1}^{H}\gamma_{1i}\sum_{l\neq j}p_{il}P_{ltg} + \sum_{l\neq j}\gamma_{2l}P_{ltg} + \gamma_{tg}}{\mu_{2j}P_{jtg}}.$$

ρ_{tgi}^{*} and $\rho_{tgi}^{\perp}$ are, approximately, the loads of the target queue when a node i at distance 2 from the target node or a node j at distance 1, respectively, are not empty. It is more difficult to get insight of the meaning of α_{1i} and α_{2j} without following the derivation of formula (23). Nevertheless, the formulas are easy to apply because all their terms are parameters of the system.

Test Cases. Several simulation experiments on Jackson networks with different topologies and loads were conducted in [32]. The rare set A was defined in most cases as $Q_{tg} \geq 70$, where Q_{tg} is the number of customers at the target node. The steady state probability of A was of the order of 10^{-34} in those examples. The reason for simulating such small probabilities is to show the goodness of the importance functions obtained in the paper, given that if it is possible to estimate accurately such small probabilities with short or moderate computational time, it will take much less time to estimate more realistic probabilities.

Thresholds T_i were set for every integer value of Φ between 2 (in some cases 3) and a number varying between 71 and 75 depending on the case being simulated. Observe that, as $L=70$, the rare set A is not included in C_M given that $A \cap (C_i - C_{i+1}) \neq \phi$ if $T_i \geq 70$. Pilot runs (one or two for each case) were made to set the number of retrials. We proceeded as following: we set (for example) the thresholds 2, 3, 4, ... , 74 and we made a pilot simulation. This simulation derived the optimal number of retrials according to formula (12) following the guidelines given in Section 5.2 for rounding to integer values. If the derived value of the number threshold (in the pilot simulation) of retrials from a threshold was 1 was 1, such threshold was eliminated. If the number of retrials from the last threshold was greater than 5, an additional threshold was set. The number of retrials R_i finally was 2 or 3 in all cases.

Although it is not possible to simulate all the Jackson networks to prove that the importance function given by formula (23) is always effective, test cases that a priori could have some difficulties were selected in [32]. If the importance function is effective for these cases, it is supposed that it will be also effective for most Jackson networks. The systems simulated were the following:

- a two-queue Jackson tandem network;
- a three-queue Jackson tandem network;
- a three-stage network with 4 nodes in the first and second stage and 1 or 2 nodes in the third stage;
- a Jackson network with 7 nodes with 2 nodes at distance 1 from the target node, and 4 nodes at distance 2 from the target node.
- a Jackson network with 7 nodes with 2 nodes at distance 1 but with 2 nodes at distance 3 from the target node, 2 nodes at the target node. distance 2 and 2 nodes at distance 3.
- a large Jackson network with 15 nodes: 4 of them at distance 3 from the target node, 5 at distance 2 and 5 at distance 1.
- a 2-node Jackson network with strong feedback: jobs departing any of the two nodes join the other node with a probability of 0.8.
- a six-queue Jackson tandem network, in which the first 5 nodes have the same load (2/3) and the last (target) node has a lower load (1/3). This case and the two first ones are networks for which the dependency of the target queues on the queue length of the other queues is very high because all the customers of the other queues have to go to the target queue.

In all the networks, except the last one, probabilities of the order of 10^{-34} were estimated with short or moderate computational time with the importance function given by formula (23). For the six-queue Jackson tandem network thirty minutes of computational time was needed to estimate a probability of the order of 10^{-15} with that importance function (that does not take into account the queue length of the first 3 nodes). The importance function was improved heuristically giving the weights provided by the formula for the last 3 nodes and lower extrapolated weights to the nodes that are farther from the target node. With this importance function, which accounts for the dependence of the target node on all the nodes, an accurate estimation of the same probability was obtained with 10 minutes of computational time.

The results obtained in [32] show that the worst cases are networks with very high dependencies, in which the target queue has a much lower load than the other queues, and that the best cases are usually the most complex networks with a high number of nodes because there are usually weak dependencies in these cases. Although the importance function given by formula (23) can be improved for some specific networks, it seems to be good enough for estimating very low probabilities with short or moderate computational times for most (perhaps all) Jackson networks.

In order to illustrate the results of the simulations made in [32], those corresponding to a Jackson network with 7 nodes (with 2 nodes at distance 3 from the target node) will be reproduced here.

Jobs from the outside arrive at any node of the network at a rate $\gamma_i = 1$, $i = 1,\ldots,7$. After being served in each node, a job leaves the network with probability 0.2. Otherwise the job goes to another node in accordance to the following transition matrix:

	1	2	3	4	5	6	tg
1	0.2	0.2	0.2	0.2	0	0	0
2	0.2	0.2	0.2	0.2	0	0	0
3	0.1	0.1	0.1	0.1	0.2	0.2	0
4	0.1	0.1	0.1	0.1	0.2	0.2	0
5	0.1	0.1	0.1	0.1	0	0.1	0.3
6	0.1	0.1	0.1	0.1	0.1	0	0.3
tg	0.1	0.1	0.1	0.1	0.1	0.1	0.2

Let observe that jobs that leave nodes 1 or 2 have to visit nodes 3 or 4 and nodes 5 or 6 before entering the target node. We wished to check whether to ignore the impact of the queue lengths of the nodes at a distance greater than two on the queue length of the target node is a reasonable approximation. The results are summarized in Table 1.

For each group of loads of the nodes, three importance functions were used. The first one is given by formula (23), that is, without considering the queue lengths of nodes 1 and 2 placed at a distance 3 ($a = 0$ in the table). The second one also takes into account nodes 1 and 2. The values of their coefficients (called a in the table) were derived with the same methodology used to derive formula (23). The third importance function only takes into account the target node and the nodes that are at distance 1 from it ($a = b = 0$).

The best results were obtained with the second importance function, the function that takes into account the number of jobs in the seven nodes. However, the importance function given by formula (23) leads to very effective results: probabilities of the order of 10^{-34} were obtained with less than 10 minutes of computational time and the values of factor f_V, though slightly greater than those obtained with the other importance function, are very small. Consequently, it does not seem that it worth further complicating the importance function by taking into account the nodes that are at distance greater than 2 from the target node. Nevertheless, though it is not necessary in almost all the cases, if we want to improve the importance function taking into account more queue lengths in the function, it can be done deriving the coefficients of those nodes with the same methodology used to derive formula (23) or heuristically giving the weights provided by the formula to the nodes at distance lower than 3 and lower weights to the nodes that are farther from the target node. The results are much worse when using the third importance function, which only accounts for the target node and the nodes that are at distance 1 from the target node.

Table 1. Results for Jackson networks. Relative error = 0.1. Rare set probability:
$P(Q_{tg} \geq 70) = 3.13 \cdot 10^{-34}$. $\rho_{tg} = 0.3322$. $\Phi = a\sum_{i=1}^{2} Q_i + b\sum_{j=3}^{4} Q_j + c\sum_{k=5}^{6} Q_k + Q_{tg}$.

$\hat{P}$	ρ_i	ρ_j	ρ_k	a	b	c	Time (minutes)	f_V
$3.1 \cdot 10^{-34}$	0.62	0.51	0.41	0	0.31	0.47	9.6	3.6
$3.2 \cdot 10^{-34}$	0.62	0.51	0.41	0.07	0.31	0.47	7.5	2.7
$3.3 \cdot 10^{-34}$	0.62	0.51	0.41	0	0	0.47	518	204
$3.1 \cdot 10^{-34}$	0.47	0.51	0.41	0	0.31	0.47	5.2	1.7
$3.0 \cdot 10^{-34}$	0.47	0.51	0.41	0.09	0.31	0.47	4.2	1.4
$2.9 \cdot 10^{-34}$	0.47	0.51	0.41	0	0	0.47	207	55
$3.3 \cdot 10^{-34}$	0.31	0.31	0.31	0	0.51	0.61	3.3	1.1
$3.1 \cdot 10^{-34}$	0.31	0.31	0.31	0.15	0.51	0.61	3.0	1.0
$3.0 \cdot 10^{-34}$	0.31	0.31	0.31	0	0	0.61	6.1	2.4

6.2 Non-Jackson Networks

The importance function given by formula (23) has been derived equating the importance of one extreme state with the importance of each of the other extreme states. It is interesting to see whether the importance of the extreme states are affected in a similar manner when the interarrivals and/or services times are not exponentially distributed and, as a consequence, whether the importance function derived for Jackson networks fits for other networks.

In [44] networks with Poisson arrivals and Erlang service times were studied. Two of the networks above mentioned were analyzed: the three-queue tandem network and the first one of the two networks with seven nodes. In the first model, the service time at each node follows an Erlang distribution with shape parameter equal to α (2 or 3). Initially, the chosen importance function was $\Phi = aQ_1 + bQ_2 + Q_3$ evaluated according to formula (23). Then, the coefficients a and b of Φ were multiplied by a correction factor k, the same for both coefficients. The tested values of k were 0.6, 0.7, 0.8 and 0.9.

We observed that the best value of k was between 0.6 and 0.9, depending on the case. We also observed that for $\alpha = 2$, the value of k is closer to 1 than for $\alpha = 3$.

As the coefficient of variation of Pearson of the Erlang distribution is $1/\sqrt{\alpha}$, it seems that the more similar is this coefficient to that of the exponential distribution, the closer is the importance function to that given by the formula. It is also observed that lower values of the loads of the two first nodes lead to importance functions closer to those derived for the exponential distribution. Probabilities of the order of 10^{-15} were estimated with computational times between 0.4 and 21.8 minutes.

The values of factor f_O are much greater than those obtained in [32] for the same network but with exponential service times. The reason is that rescheduling is straightforward only for the exponential distribution due to the memoryless property of this model. Rescheduling service times of any other distribution is more time consuming. We can proceed as follows: a random value of the whole service time of a job is obtained. If that value is greater than the service time at the current time, the remaining service time of the job is obtained as the difference between the two amounts. Otherwise a new random value is obtained and so on. This procedure has the problem that the number of iterations is very huge when the actual service time is much larger than the mean service time, and it greatly increases factor f_O. As rescheduling is made to improve factor f_v but it is not strictly necessary, if after a fix number of trials, e.g., 50, the random value of the whole service time is always lower than the service time at the current time, the service time is not rescheduled. In this way factor f_O is significantly reduced and, as only 1 or 2% of the scheduled times are not rescheduled, the impact on f_V is negligible. Nevertheless, even with this improvement of the procedure, the value of factor f_O with Erlang service times is around four times greater than with exponential times for estimating a probability of the order of 10^{-15}.

Low or at least moderate values of factor f_V were obtained in all the cases. It shows that the application is not far from the optimal, at least for the tested cases. We observe that the worst results (greatest computational times and greatest values of factor f_V) were obtained when $\rho_3 < \rho_2 < \rho_1$, but even in these cases the computational times are moderate (21.8 minutes). The importance functions given by formula (23) lead to greater computational times, although these times are also moderate in all the cases, except in the case $\rho_3 < \rho_2 < \rho_1$, in which the computational time was greater than one day.

The second network studied in [44] is a network with 7 nodes with 2 nodes at distance 1 from the target node, and 4 nodes at distance 2, with Poison arrivals and Erlang service times.

The computational times needed for estimating probabilities of the same order of magnitude (10^{-15}) is much lower than in the previous network. The results are better in this network due to the weaker dependence between the queue lengths.

In the three cases of $\alpha = 2$ the best results were obtained with the importance function given by formula (23), while for $\alpha = 3$ the best results were obtained with coefficients of nodes at distance 1 and 2 around 10% lower than those given by formula (23), that is, with a correction factor $k = 0.9$. Nevertheless, very good results were also obtained without any correction factor. The very low values of f_V achieved (less than 1.7 in the six cases studied) show that the application is very close to the optimal one, at least for the tested cases.

In [34] the simulation study made in [44] was extended in a twofold direction. On the one hand it was also simulated two additional above mentioned networks: the large network with 15 nodes and the network with 2 nodes and very strong feedback, On the other hand we used hyperexponential and Erlang distribution for modelling the interarrival and/or service times.

In this paper a better method for improving formula (23) (formula derived for Jackson networks) is applied for non-Markovian networks: instead of multiplying some coefficients by a correction factor obtained heuristically, we also use formula (23) but the actual loads used in the formulas are substituted by "effective loads", defined as the loads ρ^e such that $\Pr\{Q \geq n\} = (\rho^e)^n$. For Jackson networks for a certain value of n. the "effective load" matches the actual load.

For the three-queue tandem network and for the network with 2 nodes and strong feedback, the efficiency obtained is much higher using effective loads than using actual loads. However similar efficiency is obtained using "effective loads" and actual loads (that is, using formula (23) without any correction) with the more complex networks of 7 and 15 nodes because the effective loads are similar to the actual loads in these networks. Overflow probabilities lower than those needed in practical problems (around 10^{-15}) were accurately estimated within short computational work. The worst results were obtained when the dependence of the target queue on the length of the other queues is very high (as occurs in a tandem network) and the load of the target queue is much lower than the others. For the worst case, less than 17 minutes of computational times were needed for estimating a probability of the order of 10^{-15}. In some of the cases the probability was estimated in less than one minute.

6.3 Ultra Reliable Systems

This section provides a simple importance function that can be useful for RESTART simulation of models of many highly dependable systems. Some examples from the literature illustrate the application of this importance function.

We consider generalized Machine Repairman Models. These models consist of multiple types of components with any number of components of each type, where each component can be in one of the following states: operational, failed, spare or dormant. An operational component becomes dormant if its operation depends upon the operation of other components and those components fail. General lifetime distributions and different failure rates can be specified for the operational, spare and dormant states. Dependencies among components and failure propagation (e.g., the failure of a component causes some other components to fail with given probabilities) are allowed. There is a set of repair services which repair failed components according to a general distribution and to some service discipline. The system is operational if certain combinations of components are operational. The concern is estimation of transient measures, such as system unreliability or unavailability at a given instant, and steady-state measures, such as steady-state unavailability and mean time between failures.

In a general system there are minimal cut sets with different cardinality. In a balanced system, where all the components have the same probability to fail, it is more probable that a system failure is due to the failure of all the components of a minimal cut set with the lowest cardinality. The "distance" to the system failure is related with the number of components that remain operational in the cut set with lowest the number of operational components. For this reason the importance function (at an instant t) is defined as:

$$\Phi(t) = cl - oc(t) ,\qquad(24)$$

where cl is the cardinality of the minimal cut set with the lowest cardinality and $oc(t)$ is the number of components that are operational at time t in the cut set with the lowest number of operational components. Thresholds T_i of Φ can be defined at 1, 2, ... , $cl-1$. For example, consider the network in Fig.2 that contains 8 links and 7 nodes. The system operates as long as there exists a path along operating links between node A and node B.

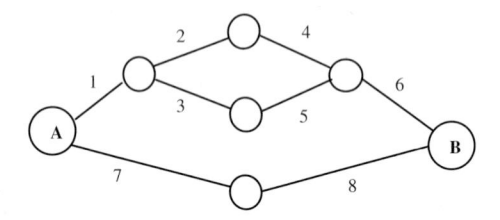

Fig. 2. Network with low redundancies

There are 4 minimal cut sets with 2 links: $(1,7),(1,8),(6,7),(6,8)$, and 8 minimal cut sets with 3 links: $(2,3,7),....,(3,4,8)$. In this network $cl = 2$ and we can define one threshold. The process is in the region C_1 (that is, $\Phi(t) \geq T_1$) if at least one component of any of the 4 cut sets with cardinality 2 or at least 2 components of any of the 8 cut sets with cardinality 3 are failed. As only one threshold can be defined, factor f_T would be high if the 8 links would be highly reliable.

The same importance function can be used for many unbalanced systems. The larger the difference among failure rates of the components, the greater the value of factor f_V. If the system is so unbalanced that factor f_V takes a very great value and, as a consequence, it is unfeasible to estimate the probability of interest within a reasonable computational effort, the importance function must be improved.

A limitation of the RESTART methodology for simulating highly-reliable systems is the difficulty to define thresholds close enough so that the probability of reaching the next threshold is reasonably large and, thus, close to the optimal. For this reason, L'ecuyer et al. [43] pointed out that this methodology is not appropriate for this type of systems and Xiao et al. [45] suggested that "importance splitting is hard to be adopted for dependability estimation of non-Markov systems, because thresholds function is hard to be presented under this situation". However, as it will be shown in the examples, probabilities up to the order of 10^{-16} can be accurately estimated within a reasonable computational effort.

We will describe three examples, two of them taken from [27] and the other one from [33]. Example 1, taken from [27] is the network of Fig.3 originally presented in [46], where it was simulated using importance sampling. The network contains 56 links, classified in 3 types, and a total of 107 components. Each type A link contains three identical components and fails when two components fail. Type B links contain one component. Each type C link contains two identical components and fails when one component fails. The mean lifetime is different for each type of component. The system operates as long as there exists a path along operating links between node 1 and node 20. There are 5 repair-persons, and repairs make components as good as new. Upon completing a repair, a repair-person selects the next component to repair randomly over the failed components in the network.

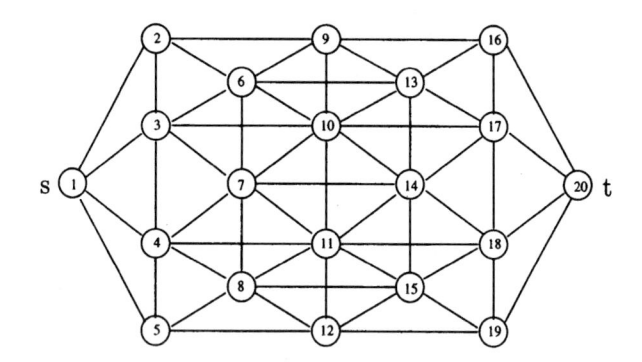

[Figure: Network diagram with nodes 1-20, where s=1 and t=20]

Type A links: (1,2), (1,3), (1,4), (1,5), (16,20), (17,20), (18,20), and (19,20)
Type B links: (2,3), (3,4), (4,5), (6,7), (7,8), (9,10), (10,11), (11,12), (13,14), (14,15), (16,17), (17,18), and (18,19).
All other links are type C.

Fig. 3. Network with redundancies

The system unreliability was estimated for different small values of intervals $(0, t_e)$. Simulations were made assuming first a Markovian model, that is assuming that component lifetimes and repair times are exponentially distributed, and second assuming that component lifetime distributions are Raleigh (Weibull distribution with shape parameter equal to 2) and that repair time distributions are Erlang with shape parameter equal to 3. The minimal cut sets are defined on the links (not on the components). The importance function was given by formula (24). As $cl = 4$ three intermediate thresholds could be defined.

Probabilities up to the order of 10^{-11} could be accurately estimated within short or moderate computational time. Nevertheless a high value of factor f_V was observed because, for a given value of i, the states at events B_i with more importance have smaller probability to occur, see Section 5.6. The system states at events B_i with more importance are those in which operational links have greater probability to fail. It is more unlikely that the operational links of a minimal cut set are those with greater probability to fail because it requires a previous failure of the other links of the same minimal cut set, which have lower probability to fail. As commented above, for unbalanced systems the factor f_V can take high values.

The models with exponential lifetimes and service times were exactly the same models simulated in [46] with importance sampling, and the estimates of the steady-state unavailability are very close in both cases. For simulating the Weibull-Erlang models with RESTART the same procedure as for the Markovian models could be used given that the same importance function is valid in both cases. With importance sampling only results for the Markovian case have been obtained, because the analytical study required for applying importance sampling to non-Markovian systems with significant redundancies is very complicated.

Example 2 of [27] is a computing system originally presented in [47], where it was studied using importance sampling, and also studied in many papers thereafter, e.g., [48]. A block diagram of the balanced version of the computing system considered is shown in Fig. 4.

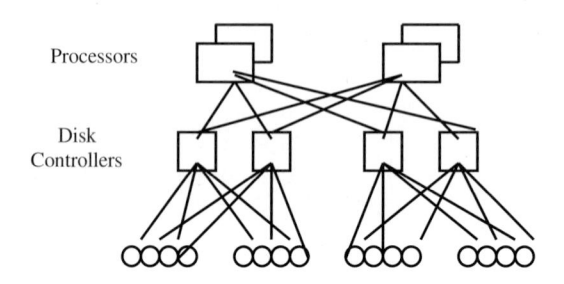

Fig. 4. Block diagram of a computing system

The system is composed of two types of processors each having passive redundancy 2; two types of disk controllers, each having active redundancy 2; and six sets of disk clusters, each having four disks. When a processor of one type fails, it causes a processor of another type to fail also with probability 0.01. The lifetime of all the components is assumed to be exponentially distributed with failure rates of processors, controllers, and disks of 1/2000, 1/2000, and 1/6000 per hour, respectively. It is assumed that each type of component can fail in one of two modes which occur with equal probability. The repair rates for all mode 1 and all mode 2 failures are 1 per hour and 0.5 per hour, respectively. There is a single repairman who fixes failed components in a random-order service. The system is defined to be operational if all data are accessible to both processor types, which means that at least one processor of each type, one controller of each type, and 3 out of 4 disk units in each of the 6 disk clusters are operational. This system was also studied in [27] with redundancy 3, i.e. the system has 3 processors and 3 controllers of each type and 5 disks in each cluster and it is necessary that 3 components of a type fail to have system breakdown, and with redundancy 4.

Probabilities of the order of 10^{-11} were estimated within reasonable computational times because the system is close to be balanced. It corroborates that the importance function $\Phi(t) = cl - oc(t)$ works quite well with balanced systems. Unlike with importance sampling, the simulation with RESTART of the same system but with higher redundancy does not require additional analytical effort.

There are two main ways in which a system may be made highly dependable in a cost-effective manner. The first is to use components that are "highly reliable" and have "low" built-in redundancies in the system. The second is to build "significant" redundancies in the system and use components that are just "reliable" instead of "highly reliable." Unlike with importance sampling, RESTART usually works better with higher redundancies (for estimating probabilities of the same order of magnitude) because it is possible to set more effective thresholds and thus have a lower value of factor f_T. In this sense, they could be considered complementary methods.

Example 3, taken from [33], studies dependability estimation for a consecutive-k-out-of-n: F repairable system with $(k-1)$-step Markov dependence. The system fails if and only if k or more consecutive components have failed. Exponential or Weibull distributions were considered for the lifetime of components and lognormal distribution for the repair time of a failed component. If there are h ($h < k$) consecutive failed components that precede the component i, the residual lifetime of component i will have failure rates that are greater as h increases. There is one repairman who gives priority to the most critical components. This model is an extension of that introduced in [45] to the case of non-exponential component lifetimes.

The importance function given by formula (24) was also used for simulating this system. In this model there are $(n-k+1)$ minimal cut sets. As all of them have the same cardinality (k), the definition of the importance function can be expressed as: "the number of components that are down at in the cut set with greatest number of failed components". The main differences between the importance, P^*_{A/X_i}, of the system states x_i at events B_i states are: i) whether the failed components of the cut set are consecutive or not, given that the importance is greater if the failed components of the cut set (with greatest number of failed components) are consecutive. And ii) the total number of components in the systems that are down when the process enters each set C_i. The greater is that number, the greater is the importance of the system state. It seems that the difference between the importance of these states could be relatively small. Thus, the variance $V\left(P^*_{A/X_i}\right)$ could be small. Simulation results corroborated this conjecture: the estimated values of factor f_V were very low in all the cases and unreliabilities up to the order of 10^{-16} and steady-state unavailabilities up to the order of 10^{-14} were accurately estimated with short computational effort (13 and 12 minutes, respectively).

In contrast with importance sampling, RESTART is not so dependent on particular features of the system and allows general component lifetime distributions and other generalizations of the model. Although the importance function depends on the system being simulated, the same importance function can be applied to different models without additional analytical effort regardless of the level of redundancy, the number of repairmen and of whether the model is Markovian in nature or not. This feature could extend the use of RESTART for dependability estimation to many other systems. The importance function given by formula (24) seems to lead to good simulation results, at least for balanced systems.

For very unbalanced systems the factor f_V, related with the chosen importance function, can take high values. Further investigation is needed for improving the importance function if it is unfeasible to estimate the probability of interest within a reasonable computational effort.

7 Conclusions

The method RESTART for accelerating rare event simulations has been presented. The paper, mainly based on the research activity of the authors, has described the method, has proved the unbiasedness of the estimator and has shown the formula of its variance. Then the formula of the gain has been obtained and quasi-optimal values for thresholds and the number of retrials that are easy to use in practical applications and lead to a gain close to the optimal one have been derived.

The paper has analyzed the factors that can affect the efficiency of RESTART and has focused on the most critical factor, the one related to the variance of the importance of the states that the system can have when each threshold is hit. As this variance depends on the chosen importance function, guidelines have been provided for the choice of a suitable importance function. The applications of these guidelines has been illustrated with several examples on queuing networks and ultra reliable systems.

In the queuing network examples, simulations of different types of Jackson and non-Jackson networks with different loads of the nodes were shown. The formula of the importance function, initially derived for Jackson networks by combining heuristic arguments with analytical results, could be easily adapted to non-Jackson networks. Buffer overflow probabilities much lower than those needed in practical problems have been accurately estimated within reasonable computational work. It has been shown that the efficiency of RESTART often improves with the complexity of the systems because the dependence of the target queue on the queue length of the other queues is weaker.

In the examples on ultra reliable systems, an importance function obtained heuristically has been applied to the estimation of transient and steady-state reliability measures of different systems. The same importance function has resulted to be valid in all these models regardless of the type of system, the level of redundancy, the number of repairmen and of whether the model is Markovian or non-Markovian.

The examples have shown that efficient applications of RESTART can be achieved though the importance function has been obtained heuristically or using analytical formulas derived with rough approximations. It has also been shown that an importance function derived for a system may be used, sometimes with small modifications, to other systems of the same family or to other time distributions. These two features make easier the use of RESTART and lead to a wide applicability of the method.

References

1. Rubino, G., Tuffin, B. (eds.): Rare event simulation using Monte Carlo methods. Wiley, Chichester (2009)
2. Kahn, H., Harris, T.E.: Estimation of Particle Transmission by Random Sampling. National Bureau of Standards Applied Mathematics Series, vol. 12, pp. 27–30 (1951)
3. Villén-Altamirano, M., Villén-Altamirano, J.: On the Efficiency of RESTART for Multidimensional Systems. ACM T. on Model. and Comput. Simul. 16(3), 251–279 (2006)
4. Bayes, A.J.: Statistical Techniques for Simulation Models. Australian Computer J. 2, 180–184 (1970)

5. Villén-Altamirano, M., Villén-Altamirano, J.: RESTART: A Method for Accelerating Rare Event Simulations. In: Cohen, J.W, Pack, C.D. (eds.) 13th International Teletraffic Congress. North Holland Studies in Telecommunication, vol. 15, pp. 71–76 (1991)
6. Villén-Altamirano, M., Martínez-Marrón, A., Gamo, J.L., Fernández-Cuesta, F.: Enhancement of the Accelerated Simulation Method RESTART by Considering Multiple Thresholds. In: Labetoulle, J., Roberts, J.W. (eds.) 14th International Teletraffic Congress. Teletraffic Science and Engineering, vol. 1a, pp. 797–810. Elsevier, Amsterdam (1994)
7. Hopmans, A.C.M., Kleijnen, J.P.C.: Importance Sampling in System Simulation: A Practical Failure? In: Mathematics and Computing in Simulation XXI, pp. 209–220 (1979)
8. Villén-Altamirano, M., Villén-Altamirano, J.: A Straightforward Method for Fast Simulation of Rare Event. In: 1994 Winter Simulation Conference, pp. 282–289. IEEE Press, Los Alamitos (1994)
9. Görg, C., Schreiber, F.: The RESTART/LRE method for Rare Event Simulation. In: 1996 Winter Simulation Conference, pp. 390–397. IEEE Press, Los Alamitos (1996)
10. Kelling, C.: A Framework for Rare Event Simulation of Stochastic Petri Net using RESTART. In: 1996 Winter Simulation Conference, pp. 317–324. IEEE Press, Los Alamitos (1996)
11. Kuhlmann, T., Kelling, C.: Case Studies on Multi-dimensional RESTART Simulations. Int. J. Electron. Commun. 52(3), 190–196 (1998)
12. Naldi, M., Calonico, F.: A Comparison of the GEVT and RESTART Techniques for the Simulation of Rare Events in ATM Networks. Int. J. of the Federation of Eur. Simul. Societies. 6, 181–186 (1998)
13. Villén-Altamirano, J.: RESTART Method for the Case where Rare Events Can Occur in Retrials from any Threshold. Int. J. Electron. Commun. 52(3), 183–190 (1998)
14. Garvels, M.J.J., Kroese, D.P.: A Comparison of RESTART Implementations. In: 1998 Winter Simulation Conference, pp. 601–609. IEEE Press, Los Alamitos (1998)
15. Görg, C., Fuß, O.: Comparison and Optimization of RESTART Run Time Strategies. Int. J. Electron. Commun. 52(3), 197–204 (1998)
16. Haraszti, Z., Townsend, J.K.: The Theory of Direct Probability Redistribution and its Application to Rare Event Simulation. ACM T. on Model. and Comput. Simul. 9(2), 105–140 (1999)
17. Garvels, M.J.J., Kroese, D.P.: On the Entrance Distribution in RESTART Simulation. In: RESIM 1999 Workshop, pp. 65–88. University of Twente, Enschede (1999)
18. Görg, C., Fuß, O.: Simulating Rare Event Details of ATM Delay-Time Distribution with RESTART-LRE. In: Key, P., Smith, D. (eds.) 16th International Teletraffic Engineering. Teletraffic Science and Engineering, vol. 3b, pp. 777–786. Elsevier, Amsterdam (1999)
19. Akyamac, A.A., Haraszti, Z., Townsend, J.K.: Efficient Rare Event Simulation using DPR for Multi-dimensional Parameter Spaces. In: 16th International Teletraffic Engineering. Teletraffic Science and Engineering, vol. 3b, pp. 767–776. Elsevier, Amsterdam (1999)
20. Tuffin, B., Trivedi, K.S.: Implementation of Importance Splitting Techniques in Stochastic Petri Net Package. In: Haverkort, B.R., Bohnenkamp, H.C., Smith, C.U. (eds.) TOOLS 2000. LNCS, vol. 1786, pp. 216–229. Springer, Heidelberg (2000)
21. Akin, O., Townsend, J.K.: Efficient Simulation of Delay in TCP/IP Networks using DPR-based Splitting. In: IEEE International Conference on Communication 2002, pp. 2619–2624 (2002)
22. Garvels, M.J.J., Kroese, D.P., Ommeren, J.K.C.W.: On the Importance Function in Splitting Simulation. Eur. T. on Telecom. 13(4), 363–371 (2002)

23. Radev, D., Iliev, M., Arabadjieva, I.: RESTART Simulation in ATM networks with tandem queue. In: International Conference on Automatics and Informatics, Sofia, pp. 37–40 (2004)
24. Elayoubi, S.E., Fourestie, B.: On Trajectory Splitting for Accelerating Dynamic Simulations in Mobile Wireless Networks. In: 19th International Teletraffic Congress, pp 1717–1726, Beijing (2005)
25. Villén-Altamirano, J.: Rare Event RESTART Simulation of Two-Stage Networks. Eur. J. of Oper. Res. 179(1), 148–159 (2007)
26. Cerou, F., Guyader, A.: Adaptive Multilevel Splitting for Rare Event Analysis. Stoch. Analysis and Applic. 25(2), 417–443 (2007)
27. Villén-Altamirano, J.: Importance Functions for RESTART Simulation of Highly-Dependable Systems. Simulation 83, 821–828 (2007)
28. Zimmermann, A.: Stochastic discrete event system. Springer, Berlín (2008)
29. Lagnoux, A.: Effective Branching Method Splitting under Cost Constraint. Stoch. Processes and their Application 18(10), 1820–1851 (2008)
30. Mykkeltveit, A., Helvik, B.E.: Application of the RESTART/Splitting Technique to Network Resilience Studies NS2. In: 19th IASTED International Conference (2008)
31. Dean, T., Dupuis, P.: The design and analysis of a generalized DPR/RESTART algorithm for rare event simulation. Annals of Oper. Res. (2010) (in Press)
32. Villén-Altamirano, J.: Importance Functions for RESTART Simulation of General Jackson Networks. Eur. J. of Oper. Res. 203(1), 156–165 (2010)
33. Villén-Altamirano, J.: Dependability Estimation for Non-Markov Consecutive-K-out-of-N: F Repairable Systems by RESTART Simulation. Reliab. Eng. Syst. Saf. 95(3), 247–254 (2010)
34. Villén-Altamirano, M., Villén-Altamirano, J., Vázquez-Gallo, E.: RESTART Simulation of non-Markovian Queuing Networks. In: RESIM 2010, Isaac Newton Institute, Cambridge (2010)
35. Villén-Altamirano, M., Villén-Altamirano, J.: Analysis of RESTART Simulation: Theoretical Basis and Sensitivity Study. Eur. T. on Telecom. 13(4), 373–385 (2002)
36. Villén-Altamirano, J., Villén-Altamirano, M.: Recent Advances in RESTART Simulation. In: RESIM 2008. IRISA - INRIA, Rennes (2008)
37. Villén-Altamirano, M., Villén-Altamirano, J.: Accelerated Simulation of Rare Event using RESTART Method with Hysteresis. In: ITC Specialists' Seminar on Telecommunication Services for Developing Economies, pp. 240–251. University of Mining and Metallurgy, Krakow (1991)
38. Villén-Altamirano, M., Villén-Altamirano, J., González-Rodríguez, J., Río-Martínez, L.: del: Use of Re-Scheduling in Rare Event RESTART Simulation. In: 5th St. Petersburg Workshop on Simulation, pp. 721–728 (2005)
39. Parekh, S., Walrand, A.: A Quick Simulation Method for Excessive Backlogs in Networks of Queues. IEEE T. on Aut. Control 34, 54–66 (1989)
40. Glasserman, P., Heidelberger, P., Shahabuddin, P., Zajic, T.: A Large Deviation Perspective on the Efficiency of Multilevel Splitting. IEEE T. on Aut. Control 43(12), 1666–1679 (1998)
41. Glasserman, P., Heidelberger, P., Shahabuddin, P., Zajic, T.: Multilevel Splitting for Estimating Rare Event Probabilities. Oper. Res. 47, 585–600 (1999)
42. Kroese, D.P., Nicola, V.F.: Efficient Simulation of a Tandem Jackson Network. ACM T. on Model. and Comput. Simul. 12(2), 119–141 (2002)
43. L'ecuyer, P., Demers, V., Tuffin, B.: Rare Events, Splitting and Quasi-Monte Carlo. ACM T. on Model. and Comput. Simul. 17 (2), Article 9 (2007)

44. Villén-Altamirano, J.: RESTART Simulation of Networks of Queues with Erlang Service Times. In: 2009 Winter Simulation Conference, pp. 251–279. IEEE Press, Austin (2009)
45. Xiao, G., Li, Z., Li, T.: Dependability Estimation for non-Markov Consecutive-k-out- of-n: F Repairable Systems by Fast Simulation. Reliab. Eng. Syst. Saf. 92(3), 293–299 (2007)
46. Alexopoulos, C., Shultes, B.C.: Estimating Reliability Measures for Highly-Dependable Markov Systems using Balanced Likelihood Ratios. IEEE T. on Reliab. 50(3), 265–280 (2001)
47. Goyal, A., Shahabuddin, P., Heidelberger, P.: A Unified Framework for Simulating Markovian Models of Highly Dependable Systems. IEEE T. on Comput. 41(1), 36–51 (1992)
48. Nicola, V., Shahabuddin, P., Nakayama, M.K.: Techniques for Fast Simulation of Models of Highly Dependable Systems. IEEE T. on Reliab. 50(3), 246–264 (2001)

An Algebraic Multigrid Solution of Large Hierarchical Markovian Models Arising in Web Information Retrieval

Udo R. Krieger[*]

Faculty Information Systems and Applied Computer Science
Otto-Friedrich University
Feldkirchenstr. 21, D-96052 Bamberg, Germany
udo.krieger@ieee.org

Abstract. In Web information retrieval stochastic link analysis provides important supplementary means to generate a ranking of searched objects. Considering a hierarchical algebraic description of a Web graph with host-oriented clustering of pages or a role-oriented perspective, we propose an efficient computation of the stationary distribution of the underlying homogeneous Markov chain of a random surfer by iterative aggregation/disaggregation procedures and algebraic multigrid methods. In particular, we discuss the application of an efficient multigrid variant of the multiplicative Schwarz iteration which can be performed on a single machine with limited storage requirements.

Keywords: Web information retrieval, page rank, Markov chain, algebraic multigrid.

1 Introduction

Nowadays, the Internet provides an efficient transport platform that integrates a rich and rapidly growing family of conventional text-based and new multi-media services built on top of an advanced Web technology. The provision of quick access to the required information and service sites constitutes an important item of the corresponding service delivery processes and requires advanced search and efficient retrieval services. For these purposes Web information retrieval techniques are applied and implemented in Web-based client-server architectures and their underlying tools like the Google or Yahoo search engines.

The performance of the service delivery chain crucially depends on the proper and efficient operation of these Web search, application and database servers. It is determined by their software architecture as well as the applied processing models and characterized by the underlying mathematical theory of ranking requested objects that are available somewhere in the Internet.

---

[*] The research was partly supported by the EU FP6-NoE project "EuroFGI" under contract 028022 and by the German Ministry of Education and Research (BMBF) under contract MDA 08/015.

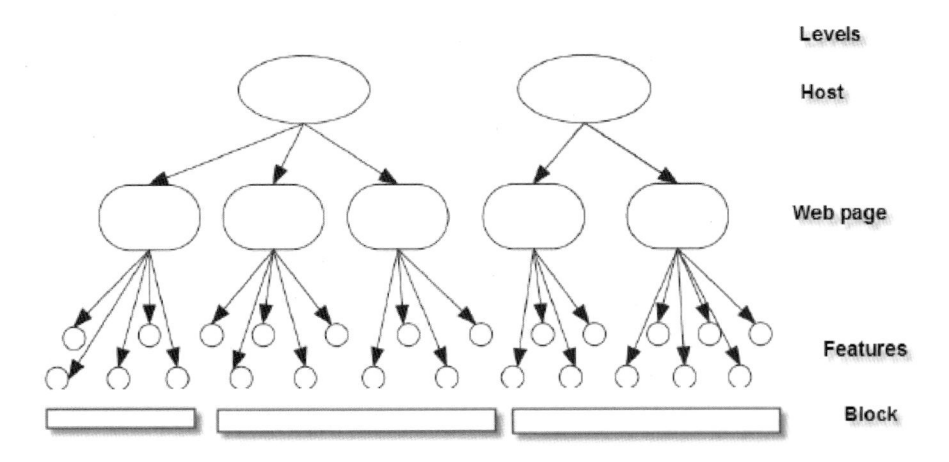

Fig. 1. Hierarchical Web structure

Considering Web information retrieval, basic operations are determined by effectively crawling the Web servers and the analysis of the observed pages and their hyperlink structures. The content, its semantic, size or location can be further used to group the structures into appropriate blocks, see Fig. 1 (cf. [8], [24]). Then a a related stochastic link analysis is performed on the derived Web graph which provides important supplementary means to generate a significant ranking of searched objects (cf. [1], [17], [18], [20]). The hierarchical structure of the Internet comprising different communities with their distinct interests, distributed Web servers at different domains with their Web pages stored along different paths and their rich content with different features should be comprehensively modelled as foundation of any efficient processing. For that purpose we propose as first step a hierarchical algebraic description of the Web graph with host-oriented clustering of pages in accordance with previous studies (cf. [8], [24]). Extending the classical approach of Brin, Page, Kleinberg and others (cf. [17]), it is then suggested to compute the stationary distribution of the underlying homogeneous Markov chain that describes the behavior of a random surfer on the Web by iterative aggregation/disaggregation procedures and related algebraic multigrid methods.

Nowadays, such iterative aggregation/disaggregation (IAD) procedures are a convenient and very efficient numerical solution technique which is well suited for hierarchically structured Markovian systems with a large but finite state space (cf. [11], [14], [15], [22]). Moreover, they are variants of algebraic multigrid (AMG) and domain decomposition methods. The latter have been successfully applied in numerical linear algebra to solve huge regular linear systems in various fields of engineering (cf. [6], [9], [16]).

In this paper we contribute to the computational framework of stochastic link analysis and sketch the application of multigrid techniques as efficient numerical solution method for finite Markov chains in Web information retrieval. First we formulate the problem of stochastic link analysis in a hierarchical manner.

Further, we apply this block level approach to derive a unified description of the randomized variant SALSA of Kleinberg's hubs/authority method HITS (cf. [1], [18], [20]). Then we show how iterative solution methods accelerated by an aggregation/disaggregation (IAD) technique can be applied and how the more general V-cycle procedure of a two-level algebraic multigrid (AMG) method can be used. Thereby, we point out an interesting way to exploit the great potential of very efficient multigrid schemes and their domain decomposition techniques such as the multiplicative Schwarz iteration.

This paper is organized as follows. In section 2 we present a unified hierarchical description of stochastic link analysis. Section 3 provides a simplified description of SALSA derived from the block level approach. Section 4 is devoted to the IAD-method and a two-level multigrid method given by the multiplicative Schwarz iteration. Finally the findings of the paper are summarized in the conclusion.

2 A Hierarchical Algebraic Description of the Link Structure of a Web Graph and Its Analysis

In this paper we adopt the notation of Berman and Plemmons [2, Chap. 2] with respect to vector and matrix orderings: let $x \in \mathbb{R}^n$, then $x \gg 0 \Leftrightarrow x_i > 0$ for each $i \in \{1,\ldots,n\}$; $x > 0 \Leftrightarrow x_i \geq 0$ for each $i \in \{1,\ldots,n\}$ and $x_j > 0$ for some $j \in \{1,\ldots,n\}$; $x \geq 0 \Leftrightarrow x_i \geq 0$ for each $i \in \{1,\ldots,n\}$. Let $\text{Diag}(x)$ denote the diagonal matrix generated by a vector $x \in \mathbb{R}^n$, $e \in \mathbb{R}^n$ be the vector with all ones and A^t denote the transpose of a matrix $A \in \mathbb{R}^{n \times n}$.

We consider a Web graph of pages and their hyperlink structure observed by a crawler on different hosts in the Internet. It is our objective to perform a stochastic link analysis of this structure taking into account the relationship among the machines hosting the collected Web pages.

Following the Google PageRank, HITS and SALSA approaches (cf. [1], [17], [18]), we represent the collected Web pages $p \in V$ on the identified hosts $h \in \mathcal{H} = \{h_1,\ldots,h_m\}$ and their hyperlink structure by a directed graph $G = (V, E)$ with $|V| = n$ nodes and $l = |E| \leq (n+1)n/2$ edges including potential loops. We can enumerate the pages and set $V = \{1, 2, \ldots, n\} \subset \mathbb{N}$. An edge $(i, j) \in E$ between nodes $i \in V$ and $j \in V$ is generated if page i refers by a hyperlink to page $j \in V$.

Following the Google and HITS concept (cf. [17], [20]), we define the associated *adjacency matrix* $A = A(G) \in \mathbb{R}^{n \times n}$ for each pair $(i, j) \in V \times V$ by

$$A_{i,j} = \mathbb{1}_E((i,j)) = \begin{cases} 1, & \text{if } (i,j) \in E \\ 0, & \text{else} \end{cases}$$

where $\mathbb{1}_E$ denotes the indicator function of the edges $E \subseteq V \times V$.

For any page $i \in V$ the *out-degree* k_i^{out} of i is defined by

$$k_i^{out} = \sum_{j=1}^n A_{i,j} = (A \cdot e)_i \qquad (1)$$

which represents the ith row sum of the adjacency matrix. For undirected graphs this term represents the degree of node i.

In a similar way, the *in-degree* k_i^{in} of i is defined by

$$k_i^{in} = \sum_{j=1}^{n} A_{j,i} = (e^t \cdot A)_i = (A^t \cdot e)_i \qquad (2)$$

which represents the ith column sum of the adjacency matrix A.

We provide a unique identification of the hosts $h \in \mathcal{H} = \{h_1, \ldots, h_m\}, m < n$ in terms of the pair (hostname, IP-address). This information can be extracted from the crawling results without extra cost. It may be stored during the crawling process on the fly in an appropriate data structure of the page relationship, e.g. adjacency lists, linked lists with reference encoding or hash tables. Then we can define a surjective function

$$H : V \longrightarrow \mathcal{H}$$
$$p \longmapsto H(p) = h$$

which assigns to each page $p \in V$ the corresponding host $H(p) \in \mathcal{H}$ storing its related content. If the same content is stored at different hosts, we can provide a unique identification in terms of the tuples (hostname, IP-address, page-version) describing the relation $H \subseteq \mathcal{H} \times V$.

Then we arrange the pages $p \in V$ according to their host relation H and provide a unique partition $\Gamma = \{J_1, \ldots, J_m\}$ of V by the inverse mapping $J_i = H^{-1}(\{h_i\}), 0 < |J_i| = n_i < n, i = 1, \ldots, m, \sum_{i=1}^{m} n_i = n$, i.e. $J_i \cap J_j = \emptyset, i \neq j$ and $V = \biguplus_{i=1}^{m} J_i$. By these means we can cluster the pages $p_i \in J_i$ in a natural way according to their storing hosts $H(p_i) = h_i$. We assume that the elements of these sets are enumerated in a consecutive order and that the elements of all vectors and matrices are arranged and numbered accordingly.

If we enumerate the pages according to the lexicographical ordering of the relation

$$(h_i, p_i) \in \mathcal{R} \subseteq \mathcal{H} \times V \Leftrightarrow h_i = H(p_i),$$

and embed $\Gamma \subseteq \mathcal{H} \times V$ by $\mathcal{R}$ we can partition the adjacency matrix $A = A(G)$ in the following manner

$$A = \begin{pmatrix} A_{11} & A_{12} & \cdots & A_{1m} \\ A_{21} & A_{22} & \ddots & A_{2m} \\ \vdots & \ddots & \ddots & \vdots \\ A_{m1} & \cdots & \cdots & A_{mm} \end{pmatrix} \in \mathbb{R}^{n \times n} \qquad (3)$$

into blocks $A_{i,j}, i, j \in \{1, \ldots, m\}$. Given arbitrary $p \in J_i, q \in J_j, (p, q) \in E$, the latter is arising from the corresponding block $(A_{i,j})_{p,q}$ related to the adjacency matrix of the subgraph $G_{i,j} = (J_i \uplus J_j, E \cap (J_i \uplus J_j))$. A reordering of nodes and a related decomposition of the Web graph according to the domain, host name, directory or any other local behavior of the search path yields a similar hierarchical block structure of the underlying scaled adjacency matrix (cf. [8]).

For each host $h_i \in \mathcal{H}$, $1 \leq i \leq m$, we define the J_ith block matrix row of the adjacency matrix by

$$A_{J_i,V} = [A_{i1}, A_{i2}, \ldots, A_{im}].$$

Following the Google method of Brin and Page (cf. [17]), we may assume the simplest case of uniform conditional distributions on the outgoing hyperlink structure of all pages $p_i \in J_i \subseteq V$ due to a lack of more detailed knowledge on the average clicks streams of the visitors. If we gain more insight into the click streams of the customers (see [12]), this knowledge can be incorporated into the structure by a general weight matrix $A \in \mathbb{R}^{n \times n}$. Since we follow the behavior of a random surfer who reaches a node $p_j \in J_j \subseteq V$ by a hyperlink from page $p_i \in J_i$ and changes instantaneously the site, the conditional probability of this movement is represented by the probability weights

$$P_{p_i,p_j} = \frac{A_{p_i,p_j}}{\sum_{k=1}^n A_{p_i,k}} = \frac{A_{p_i,p_j}}{(A \cdot e)_{p_i}} \in [0,1] \quad (4)$$

in the graph G if $k_{p_i}^{out} > 0$. Otherwise p_i is a dangling node.

If we scale the adjacency matrix A by the inverse of the diagonal matrix $\Delta = \text{Diag}(Ae)$, we get P. To guarantee the existence of Δ^{-1}, we can add a rank-one update provided by the dense substochastic matrix

$$S = b \cdot \left(\frac{1}{n} e^t\right) = b \cdot y^t > 0$$

with the uniform probability vector $0 \ll y = \frac{1}{n} e \in \mathbb{R}^n$, $y^t e = 1$, of a restart from the dangling nodes, called personalization vector. Here the binary vector $b \in \mathbb{R}^n$ has the elements:

$$b_i = \mathbb{1}_{\{0\}}((Ae)_i) = \begin{cases} 1, & \text{if } (Ae)_i = 0 \\ 0, & \text{if } (Ae)_i > 0 \end{cases}$$

We set $\Delta(b) = \text{Diag}(Ae + b)$, $\Delta = \Delta(0)$. Note that $0 \leq Se = b \leq e$.

Then we get the block partitioned stochastic matrix

$$\begin{aligned} P(S) &= (\text{Diag}((A+S)e))^{-1}(A+S) \\ &= (\text{Diag}(Ae+b))^{-1} A + (\text{Diag}(Ae+b))^{-1} S \\ &= P + \Sigma \\ &= \begin{pmatrix} P(S)_{11} & P(S)_{12} & \cdots & P(S)_{1m} \\ P(S)_{21} & P(S)_{22} & \ddots & P(S)_{2m} \\ \vdots & \ddots & \ddots & \vdots \\ P(S)_{m1} & \cdots & \cdots & P(S)_{mm} \end{pmatrix} \in \mathbb{R}^{n \times n} \end{aligned}$$

Its aggregation to the block level indicates the movement of the random surfer on the host graph $G_H = (V_H, E_H)$, $V_H = \{1, \ldots, m\}$, $e = (k, l) \in E_H \Leftrightarrow \exists p_k \in$

$J_k, p_l \in J_l$ such that $(p_k, p_l) \in E_\Sigma$ associated with the extended adjacency matrix $A + S$ and its matrix graph $G_\Sigma = (\Gamma, E_\Sigma)$ with the embedding $\Gamma \subseteq \mathcal{H} \times V \equiv \{1, \ldots, m\} \times \{1, \ldots, n\}$ of the node set V according to relation $\mathcal{R}$. Here we denote again the corrected scaled version $(\text{Diag}(Ae + b))^{-1} A$ of the adjacency matrix by P and set $\Sigma = (\text{Diag}(Ae + b))^{-1} b y^t$. Due to the construction the resulting matrix $P(S) \geq 0$ becomes stochastic, but it may still be reducible.

For each page $m \in J_i \subset V$ on host h_i the dangling node condition $k_m^{out} = (Ae)_m = 0$ determines a further block partition of the corresponding block row $A_{J_i,V}$ of the adjacency matrix. We can determine a permutation of the index set $\{1, \ldots, n\}$ such that the dangling nodes m are last numerated within each host set J_i. We assume that there are $0 \leq d_i \leq n_i$ dangling nodes and $0 < p_i = n_i - d_i \leq n_i$ nodes with positive out-degree $k_i^{out} > 0$. Using the binary vector b, we can then partition each host set $J_i = P_i \uplus D_i$ into the p_i positive and d_i dangling nodes and, accordingly, each block row $A_{J_i,V}$ in the following manner:

$$A_{J_i,V} = \begin{pmatrix} A_{i1p} & A_{i2p} & \ldots & A_{imp} \\ A_{i1d} & A_{i2d} & \ldots & A_{imd} \end{pmatrix}$$

$$= \begin{pmatrix} A_{i1p} & A_{i2p} & \ldots & A_{imp} \\ 0 & 0 & \ldots & 0 \end{pmatrix}$$

Consequently, block rows of the row stochastic matrix $P(S)$ inherit the following form

$$P(S)_{J_i,V} = \begin{pmatrix} P(S)_{i1p} & P(S)_{i2p} & \ldots & P(S)_{imp} \\ e \cdot (1/n) e^t \end{pmatrix} \in \mathbb{R}^{n_i \times n}$$

with $P(S)_{ijp} = [\text{Diag}(Ae)]^{-1}|_{P_i,P_i} A_{ijp} = [\text{Diag}(A_{P_i,V}e)]^{-1}|_{P_i,P_i} A_{P_i,J_j}$, $1 \leq j \leq m$ since the diagonal block $\Delta(b)_{J_i,J_i}$ of the scaling matrix coincides with the identity matrix on the set D_i of dangling nodes, i.e.

$$\Delta(b)_{J_i,J_i} = \begin{pmatrix} \text{Diag}(A_{P_i,V}e) & 0 \\ 0 & I_{d_i} \end{pmatrix}.$$

We realize that $P(S)$ arises from a perturbation of the substochastic matrix $P = [\text{Diag}((A+S)e)]^{-1}A$ by the substochastic matrix $\Sigma = \Sigma(n) = \frac{1}{n}\Delta(b)^{-1}be^t$, $\Sigma e = e$, which have zero block rows $P_{D_i,V}, \Sigma_{P_i,V}, 1 \leq i \leq m$. If we fuse all dangling and positive states $\mathcal{P} = \uplus_{i=1}^m P_i, \mathcal{D} = \uplus_{i=1}^m D_i$ of order $|\mathcal{P}| = \sum_{i=1}^m p_i = p$ and $|\mathcal{D}| = \sum_{i=1}^m d_i = d$, respectively, and perform a similarity transformation by a corresponding permutation matrix Π we get the representation

$$\Pi^{-1} P(S) \Pi = \begin{pmatrix} P_\mathcal{P} \\ 0 \end{pmatrix} + \begin{pmatrix} 0 \\ \Sigma_\mathcal{D} \end{pmatrix}$$

with the stochastic matrix $P_\mathcal{P} \in \mathbb{R}^{p \times n}$ and the stochastic rank-one matrix $\Sigma_\mathcal{D} = e \cdot y^t = \frac{1}{n} e \cdot e^t \in \mathbb{R}^{d \times n}$. Let $\|P\|_1 = \max_{1 \leq j \leq n} \sum_{i=1}^n |P_{i,j}| = \max_{1 \leq j \leq n} (e^t|P|)_j$

be the 1-norm of a matrix P. Then we get:

$$\|P_{\mathcal{P}}\|_1 \leq \|P(S)\|_1 = \|\begin{pmatrix} P_{\mathcal{P}} \\ 0 \end{pmatrix} + \begin{pmatrix} 0 \\ \Sigma_{\mathcal{D}} \end{pmatrix}\|_1$$
$$\leq \|P_{\mathcal{P}}\|_1 + \|\Sigma_{\mathcal{D}}\|_1$$

and

$$\|\Sigma_{\mathcal{D}}\|_1 = \max_{1 \leq j \leq n} e^t \cdot \left(\frac{1}{n} e \cdot e^t\right)_j$$
$$= \frac{1}{n} e^t \cdot e = \frac{\sum_{i=1}^m d_i}{n} = \frac{d}{n} = \chi(n).$$

Then the percentage of the dangling nodes $\chi(n) = \frac{d}{n}$ governs the behavior. It holds

$$\phi(n) = \frac{\max_{i \in V} k_i^{in}}{\max_{i \in \mathcal{P}} k_i^{out}} \leq \|P_{\mathcal{P}}\|_1$$
$$= \max_i \left(e^t \left([\text{Diag}(Ae)]^{-1}\right)_{\mathcal{P},\mathcal{P}} A_{\mathcal{P},V}\right)_i$$
$$\leq \max_{j \in V} \sum_{i \in \mathcal{P}} \frac{A_{i,j}}{k_i^{out}} \leq \frac{\max_{j \in V} k_j^{in}}{\max\{\min_{i \in \mathcal{P}} k_i^{out}, 1\}}$$
$$\leq \max_{j \in V} k_j^{in}.$$

Studies have shown that $d = O(0.75n)$, $\chi(n) = O(0.75)$, $\mathbb{P}\{k_i^{in} = k\} \propto C_i \cdot k^{-\gamma_i}$, $\mathbb{P}\{k_i^{out} = k\} \propto C_o \cdot k^{-\gamma_o}$, $\gamma_i \in \{1.94, 2.05, 2.1, 2.2, 2.3, 2.5, 2.63, 2.66\}$, $\gamma_o \in \{2.2, 2.5, 2.7, 2.82\}$, $\bar{k} = \mathbb{E}(k_i^{in}) = \mathbb{E}(k_i^{in}) \in \{7.22, 7.85\}$. Moreover, due to the small world effect $\bar{k} \sim n^{1/\bar{l}}$ holds for the mean shortest path length $\bar{l} \in [3, 20]$ of $A(G)$ and $\mathbb{E}(\max_{i \in V} k_i^{in}) \propto n \cdot \bar{k}$ (cf. [1], [5], [8]). Hence, we assume that $\phi(n) \in \Omega(n)$ and that this quantity is bounded from below, $0 < \psi \leq \phi(n)$. If we assume that $\chi(n) \in O(1)$ and that, in this manner, the behavior of $P_{\mathcal{P}} = P_{\mathcal{P}}(n)$ is not heavily affected by $|V| = n$ along an appropriate scale $n^{\phi(\bar{l})}$, $\phi(\bar{l}) = 1 + 1/\bar{l}$, then we conclude:

$$0 < \psi \leq \|P_{\mathcal{P}}\|_1 \leq \|P(S)\|_1 \leq \inf_{n \in \mathbb{N}} (\|P_{\mathcal{P}}\|_1 + \chi(n))$$
$$1 \sim \frac{\|P_{\mathcal{P}}\|_1}{n^{\phi(\bar{l})}} \leq \frac{\|P(S)\|_1}{n^{\phi(\bar{l})}} \leq \frac{\|P_{\mathcal{P}}\|_1}{n^{\phi(\bar{l})}} + \frac{\chi(n)}{n^{\phi(\bar{l})}} \simeq \frac{\|P_{\mathcal{P}}\|_1}{n^{\phi(\bar{l})}}$$

This property constitutes a weak coupling along the scale $n^{\phi(\bar{l})}$ among the strongly connected components that are only reached by the restart from dangling nodes $p \in \mathcal{D}$ via S (see also [8]).

According to the Google concept we define the stochastic teleportation matrix $E \gg 0$ in terms of the probability vector $v \in \mathbb{R}^n, v \gg 0, v^t \cdot e = 1$, as rank-one modification:

$$E = e \cdot v^t > 0, \quad e, v \in \mathbb{R}^n$$

Then we get the irreducible stochastic Google matrix

$$\widetilde{P} = (1-\omega)P(S) + \omega E = (1-\omega)P + \omega(E-\Sigma) + \Sigma$$

for any real $\omega \in (0,1)$. Hence, in general the block partitioned column-stochastic matrix

$$T_\omega = \widetilde{P}^t = (1-\omega)P(S)^t + \omega E^t \tag{5}$$

is primitive for any real $\omega \in (0,1)$ and has a simple Perron-Frobenius eigenvalue $\rho(T_\omega) = 1$ (cf. [2]). Thus, it determines a semiconvergent iteration scheme

$$x^{(k+1)} = T_\omega x^{(k)}, \quad k > 0 \tag{6}$$

for any start vector $x^{(0)} > 0, e^t \cdot x^{(0)} = 1$ which converges to the unique steady-state vector $\pi \in \mathbb{R}^n$:

$$T_\omega \pi = \pi \qquad \pi \gg 0, \quad e^t \cdot \pi = 1$$

We define the associated singular M-matrix $A(T)$ of the operator (5) by (cf. [2], [15]):

$$\begin{aligned} A(T) &= I - T_\omega \\ &= I - \omega E^t - (1-\omega)P(S)^t \\ &= I - \omega v e^t - (1-\omega)[\Sigma^t + P^t] \\ &= I - [\omega v e^t + (1-\omega)\Delta(b)^{-1} y b^t] - (1-\omega)P^t \end{aligned}$$

Let us choose the uniform distribution $v = y = 1/n \cdot e$ as personalization and teleportation vectors as before. Then $P(S)$ and $A(T)$ are determined by sub-stochastic perturbations $\Sigma = \Sigma(n) = \frac{1}{n}\Delta(b)^{-1} b e^t$, $E = E(n) = \frac{1}{n} e e^t$ of P such that after a permutation of states we get:

$$\begin{aligned} \Pi^{-1}\widetilde{P}\Pi &= (1-\omega)\left[\begin{pmatrix} P_\mathcal{P} \\ 0 \end{pmatrix} + \begin{pmatrix} 0 \\ \Sigma_\mathcal{D} \end{pmatrix}\right] + \omega \frac{1}{n} e e^t \\ &= \begin{pmatrix} (1-\omega)P_\mathcal{P} + \omega\frac{1}{n}ee^t \\ (1-\omega)\frac{1}{n}ee^t + \omega\frac{1}{n}ee^t \end{pmatrix} \\ &= \begin{pmatrix} (1-\omega)P_\mathcal{P} + \omega\frac{1}{n}ee^t \\ \frac{1}{n}ee^t \end{pmatrix} \\ &= \begin{pmatrix} \widetilde{P}_{\mathcal{P},V} \\ \widetilde{P}_{\mathcal{D},V} \end{pmatrix} \end{aligned}$$

After this permutation of the state space and block decomposition into sub-stochastic matrices we get the following column-stochastic iteration matrix:

$$\begin{aligned} \Pi^t T_\omega \Pi^{-t} &= (1-\omega)\Pi^t P(S)^t \Pi^{-t} + \omega E^t \\ &= \left((1-\omega)P_\mathcal{P}^t + \omega\frac{1}{n}ee^t, 0\right) + \left(0, \frac{1}{n}ee^t\right) \\ &= T_\omega^\mathcal{P} + T^\mathcal{D} =: T \end{aligned}$$

This decomposition can be exploited in the calculation steps.

3 Simplified Description of the Stochastic Link Structure and Its Markovian Analysis

Let us now consider the HITS approach of the analysis of a stochastic link structure and its ramification SALSA in more detail (cf. [17], [18]). We apply the sketched block level perspective to simplify its analysis. Both methods use the directed graph $G = (V, E)$, $V = \{1, 2, \ldots, n\} \subset \mathbb{N}$, of the crawled Web pages with their hyperlink structure $E \subseteq V \times V$ and the associated *adjacency matrix* $A = A(G) \in \mathbb{R}^{n \times n}$ with $A_{i,j} = \mathbb{1}_E((i, j))$, for each pair $(i, j) \in V \times V$.

Regarding HITS and SALSA the out-degree (1) and in-degree (2) of a page i are used to distinguish the role of the creator of a page: either he/she refers to other pages setting appropriate links, then he executes the *hub* role of a page; or he/she collects many links, then the page is provided by an *authority* on the corresponding subjects exhibited by the contents of the related page.

Inspired by Brin and Page's famous page rank model (cf. [17]), according to the SALSA concept a random surfer initiates a http session and travels through the Web sphere. She/he follows the links of the visited pages and will alternate between these two roles while she/he traverses a hyperlink of a page. Therefore, the directed random transitions across the related hyperlinks connecting a starting page i with its successors $j \in \{j_1, j_2, j_3, \ldots\}$ and so on can be modelled as random walk on a tensor space

$$V_S = \{H, A\} \times V = \{H, A\} \times \{1, 2, \ldots, n\}.$$

It represents the hub (H) and authority (A) roles of the creation and assessment processes and the actual position $i \in V \subset \mathbb{N}$ of the surfer. Accordingly, we split each node of the web graph and distinguish its H and A role, respectively. Therefore, we generate a weighted (bipartite) directed graph $G_S = (V_S, E_S, P)$ with $2n$ nodes $V_S = \{H, A\} \times V$. We consider a node $v = (H, i) \in V_S$ and adopt the output-oriented perspective of hubs when the surfer leaves a node i. Then the edges $e \in E$ are represented by the information encoded by the adjacency matrix A.

Due to a lack of more detailed knowledge on the average clicks streams of the visitors we may assume for all pages $i \in V$ the simplest case of uniform conditional distributions on the outgoing hyperlink structure. Since we follow the behavior of a random surfer who reaches a node j by a hyperlink from page i and changes instantaneously the role towards an authority A, the conditional probability of this movement is represented by the weights

$$P_{(H,i),(A,j)} = \frac{A_{i,j}}{\sum_{j=1}^{n} A_{i,j}} = \frac{A_{i,j}}{(A \cdot e)_i} \tag{7}$$

in the graph G_S.

If we examine a page j from the reverse perspective of an authority A, the input behavior of all supporting pages i is relevant. This inverse perspective is reflected by the transposed adjacency matrix A^t and the elements $(A^t)_{i,j} = A_{j,i}$.

An immediate jump will be governed by the in-degree of node j transfering the surfer back to i. The weights of this movement are reflected by

$$P_{(A,j),(H,i)} = \frac{A_{j,i}}{\sum_{j=1}^{n} A_{j,i}} = \frac{(A^t)_{i,j}}{(A^t \cdot e)_i} \tag{8}$$

in the weighted directed graph G_S.

In the following, we will use a general weighted (bipartite) directed graph model $G_S = (V_S, E_S, P(A))$ with an *arbitrary* nonnegative weight matrix $0 < A \in \mathbb{R}^{n \times n}$ and its probability weight matrix $P = P(A)$ in (7, 8). The latter, for instance, can reflect the values of counters $A_{i,j} \in \mathbb{N}$ of click streams traversing the hyperlinks from page $i \in V$ towards $j \in V$ during some measurement interval or any other preference relation, e.g. a histogram, describing the behavior of the customers visiting some host. When we partition the state space V_S by the roles and enumerate all states according to the anti-lexicographical ordering on strings $s \in \{A, H\}$, i.e. $H \prec A$, we get $V_S = (\{H\} \times V) \uplus (\{A\} \times V)$.

$\mathrm{Diag}(v) \in \mathbb{R}^{n \times n}$ denotes the diagonal matrix generated by the elements of a column vector $v \in \mathbb{R}^n$. The same construction $\mathrm{Diag}(v^t)$ is applied to the row vector v^t. If $v \gg 0$ is positive in each component, then $(\mathrm{Diag}(v))^{-1}$ exists and $(\mathrm{Diag}(v))^{-1} \cdot v = e$, $v^t \cdot (\mathrm{Diag}(v))^{-1} = e^t$, $(\mathrm{Diag}(v^t))^{-1} \cdot v = e$, $v^t \cdot (\mathrm{Diag}(v^t))^{-1} = e^t$, hold.

We assume that all pages have positive in- and out-degrees, i.e. $v = Ae \gg 0$, $e^t A \gg 0$. Then the diagonal matrices $(\mathrm{Diag}(Ae))^{-1}$, $(\mathrm{Diag}(A^t e))^{-1}$ with their positive diagonal elements exist. Otherwise, we restrict our analysis to the intersection of the corresponding subsets $\{H, A\} \times (\{i \in V | (Ae)_i > 0\} \bigcap \{j \in V | (A^t e)_j > 0\})$ of the state space V_S.

Then the row-stochastic transition probability matrix (tpm) $P \geq 0$ of the underlying alternating Markovian random walk of an arbitrary surfer can be described by a nonnegative $2n \times 2n$ matrix with a block level structure:

$$P = \begin{pmatrix} 0 & (\mathrm{Diag}(Ae))^{-1} \cdot A \\ (\mathrm{Diag}(A^t e))^{-1} \cdot A^t & 0 \end{pmatrix}$$

$Pe = e$ follows immediately since $(\mathrm{Diag}(Ae))^{-1} \cdot Ae = e$ and $(\mathrm{Diag}(A^t e))^{-1} \cdot A^t e = e$ hold by construction.

Let

$$C = C^t = [\mathrm{Diag}(e^t A)]^{-1} = [\mathrm{Diag}(A^t e)]^{-1} > 0$$
$$C^{1/2} = [\mathrm{Diag}(\sqrt{(e^t A)})]^{-1}$$
$$R = R^t = [\mathrm{Diag}(Ae)]^{-1} > 0$$
$$R^{1/2} = [\mathrm{Diag}(\sqrt{(Ae)})]^{-1}$$

denote the nonnegative inverse of the diagonal matrices of the node degrees or input/output weights and their modified variants with an element-wise square root of the degrees.

Let $W_r = R \cdot A = [\mathrm{Diag}(Ae)]^{-1} \cdot A > 0$ and $W_c = A \cdot C = A \cdot C^t = A \cdot [\mathrm{Diag}(A^t e)]^{-1} > 0$. Then W_r and W_c^t are row-stochastic since

$W_r e = [\text{Diag}(Ae)]^{-1} \cdot Ae = e, W_c^t e = [\text{Diag}(A^t e)]^{-1} \cdot A^t e = e$. In this manner the tpm is transformed into:

$$P = \begin{pmatrix} 0 & R \cdot A \\ C \cdot A^t & 0 \end{pmatrix} = \begin{pmatrix} 0 & W_r \\ W_c^t & 0 \end{pmatrix}$$

If we consider the random surfer only during the visits adapting the hub role, the underlying Markov chain is considered on the subspace $\{H\} \times V \subset V_S$. Then the tpm is the outcome of a block Gaussian elimination step and described by the Schur complement P_H of P on the latter subset:

$$\begin{aligned} P_H &= R \cdot A \cdot I^{-1} \cdot C \cdot A^t \\ &= [\text{Diag}(Ae)]^{-1} \cdot A \cdot [\text{Diag}(A^t e)]^{-1} \cdot A^t \\ &= W_r \cdot W_c^t \geq 0 \end{aligned}$$

Let us denote the spectrum of the nonnegative matrix $P \geq 0$ by $\sigma(P)$ and the spectral radius by $\rho(P) = \max\{|\lambda|, \lambda \in \sigma(P)\}$ (cf. [2]). Obviously, P_H is stochastic, hence $\rho(P_H) = 1$, $|\lambda| \leq 1, \lambda \in \sigma(P_H)$.

If we change the role and consider the surfer only during its visits adapting the authority role, we get the Schur complement P_A on $\{A\} \times V \subset V_S$ as appropriate description:

$$\begin{aligned} P_A &= C \cdot A^t \cdot I^{-1} \cdot R \cdot A \\ &= [\text{Diag}(A^t e)]^{-1} A^t \cdot [\text{Diag}(Ae)]^{-1} \cdot A \\ &= W_c^t \cdot W_r \geq 0 \end{aligned}$$

P_A is stochastic, too, and $\rho(P_A) = 1$.

Let $W = R^{1/2} \cdot A \cdot C^{1/2} \geq 0$. We can perform a similarity transformation of P by the following diagonal matrix $[\text{Diag}(\sqrt{(e^t A^t)}, \sqrt{(e^t A)})]^{-1}$ with positive diagonal elements:

$$\begin{aligned} \begin{pmatrix} R^{-1/2} & 0 \\ 0 & C^{-1/2} \end{pmatrix} \cdot P \cdot \begin{pmatrix} R^{1/2} & 0 \\ 0 & C^{1/2} \end{pmatrix} \\ = \begin{pmatrix} 0 & R^{1/2} A C^{1/2} \\ C^{1/2} A^t R^{1/2} & 0 \end{pmatrix} \\ = \begin{pmatrix} 0 & W \\ W^t & 0 \end{pmatrix} \end{aligned}$$

By this appropriate similarity transformation we obtain the following Schur complements:

$$\begin{aligned} R^{-1/2} \cdot P_H \cdot R^{1/2} &= R^{1/2} \cdot A \cdot C^{1/2} \cdot [R^{1/2} \cdot A \cdot C^{1/2}]^t \\ &= W \cdot W^t \\ C^{-1/2} \cdot P_A \cdot C^{1/2} &= [R^{1/2} \cdot A \cdot C^{1/2}]^t \cdot [R^{1/2} \cdot A \cdot C^{1/2}] \\ &= W^t \cdot W \end{aligned}$$

We note that $\sigma(WW^t) \setminus \{0\} = \sigma(W^tW) \setminus \{0\}$. Then both tpm's P_H and P_A are governed by the same nonzero eigenvalues $0 \neq \lambda \in \sigma(WW^t)$. We know that the nonnegative, symmetric, real square mutual enforcement matrix WW^t has only real eigenvalues $\sigma(WW^t) = \{\lambda_1, \ldots, \lambda_n\} = \sigma(P_H)$, $|\lambda_i| \leq 1, i = 1, \ldots, n$, and full eigenspaces (see [7]). Due to the Perron-Frobenius theorem for nonnegative operators there exists a maximal real eigenvalue $\lambda_1 = \rho(WW^t) > 0$ which is equal to the spectral radius $\rho(WW^t)$ of this nonnegative matrix (cf. [2]). Further, there exist nonnegative left and right eigenvectors $x > 0, y > 0$ with $y^t x = 1$ related to λ_1. The same holds for $\sigma(W^tW) = \sigma(P_A)$.

If the random surfer cannot leave a state $v = (s,i) \in V_S$ and stays there with probability $0 < \varepsilon < 1$, we perform a Bernoulli trial before each jump and implicitly modify the tpm P by a convex combination with the identity matrix (cf. [20]):

$$P(\varepsilon) = \varepsilon I + (1-\varepsilon) P \qquad (9)$$
$$= \begin{pmatrix} \varepsilon I & (1-\varepsilon) R \cdot A \\ (1-\varepsilon) C \cdot A^t & \varepsilon I \end{pmatrix}$$

Then the following corresponding Schur complements can be derived on the hubs and authority subspaces:

$$P_H(\varepsilon) = \varepsilon I + (1-\varepsilon) W_r W_c^t$$
$$P_A(\varepsilon) = \varepsilon I + (1-\varepsilon) W_c^t W_r$$

In the limit $\varepsilon \to 0$ they converge to $P_H = W_r \cdot W_c^t$ and $P_A = W_c^t \cdot W_r$, respectively.

If we assume that the adjacency matrix A and, hence, the diagonally scaled row-stochastic matrices $W_r \cdot W_c^t$ and $W_c^t \cdot W_r$ are irreducible, then for any $\varepsilon > 0$ the spectral radius $\rho(P_H(\varepsilon)) = \lambda_1(\varepsilon) = 1$ is a simple eigenvalue. The reason is that the corresponding nonnegative matrix is irreducible and aperiodic since it has a positive diagonal. In consequence, a unique, normalized pair of positive left and right eigenvectors $x(\varepsilon) > 0, y(\varepsilon) = e > 0$ exists such that

$$x^t(\varepsilon) \cdot P_H = \lambda_1(\varepsilon) \cdot x^t(\varepsilon) = x^t(\varepsilon) > 0$$
$$P_H \cdot y(\varepsilon) = y(\varepsilon)$$
$$y^t(\varepsilon) \cdot x(\varepsilon) = 1$$

In a similar manner, $\rho(P_A(\varepsilon)) = \lambda_1(\varepsilon) = 1$ is simple and has a normalized pair of positive left and right eigenvectors $u(\varepsilon) > 0, v(\varepsilon) = e > 0$ exists such that

$$u^t(\varepsilon) \cdot P_A = u^t(\varepsilon) > 0$$
$$P_A \cdot v(\varepsilon) = v(\varepsilon)$$
$$v^t(\varepsilon) \cdot u(\varepsilon) = 1$$

For $\varepsilon \to 0$ the corresponding vectors $x(\varepsilon), u(\varepsilon)$ on the subsets $\{H\} \times V, \{A\} \times V$ converge to the hubs and authority values $x(0), u(0)$ of the SALSA model.

In a way similar to the original SALSA model (cf. [18]), the steady-state distribution $\pi(\varepsilon)$ associated with the tpm (9) of the perturbed model can be computed in a straightforward manner. Let

$$\pi = \frac{1}{K}\begin{pmatrix} Ae \\ A^t e \end{pmatrix} \tag{10}$$

with the normalization constant

$$K = e^t Ae + e^t A^t e = 2e^t Ae$$

then we get:

$$\pi^t \cdot P(\varepsilon)$$
$$= \frac{1}{K}\left((Ae)^t, (A^t e)^t\right) \cdot \begin{pmatrix} \varepsilon I & (1-\varepsilon)R \cdot A \\ (1-\varepsilon)C \cdot A^t & \varepsilon I \end{pmatrix}$$
$$= \frac{1}{K}\left(\varepsilon(Ae)^t + (1-\varepsilon)(A^t e)^t[\mathrm{Diag}(A^t e)]^{-1}A^t,\right.$$
$$\left.(1-\varepsilon)(Ae)^t[\mathrm{Diag}(Ae)]^{-1}A + \varepsilon(A^t e)^t\right)$$
$$= \frac{1}{K}\left(\varepsilon(e^t A^t) + (1-\varepsilon)e^t A^t, (1-\varepsilon)e^t A + \varepsilon(e^t A)\right)$$
$$= \frac{1}{K}\begin{pmatrix} Ae \\ A^t e \end{pmatrix}^t = \pi^t$$

The vector π in (10) is the steady-sate solution of the perturbed model and its limit for $\varepsilon \to 0$ provided that a unique solution exists. The irreducibility of A is sufficient to guarantee this condition. In conclusion, the general hubs and authority rankings of the restricted models are parallel to the subvectors π_H, π_A and, thus, for $\varepsilon \to 0$ given by the normalized vectors:

$$x(\varepsilon) \to x(0) = \frac{Ae}{e^t Ae}, \qquad u(\varepsilon) \to u(0) = \frac{A^t e}{e^t Ae}$$

Now we consider a distributed system where the general output information of the hubs and the input information of the authorities is split and does not coincide any more. We describe the output relation of the hubs by an irreducible forward weight matrix A of the hubs and the backward perspective of the input relation of the authorities by an irreducible backward weight matrix B^t. Then the random walk on the Web graph of this *model with splitted information horizon* is described by a Markov chain with the tpm:

$$P^\sigma = \begin{pmatrix} 0 & (\mathrm{Diag}(Ae))^{-1}A \\ (\mathrm{Diag}(B^t e))^{-1}B^t & 0 \end{pmatrix}$$

We consider again the randomized variant of this model:

$$P^\sigma(\varepsilon) = \varepsilon I + (1-\varepsilon)P^\sigma$$
$$= \begin{pmatrix} \varepsilon I & (1-\varepsilon)(\mathrm{Diag}(Ae))^{-1}A \\ (1-\varepsilon)(\mathrm{Diag}(B^t e))^{-1}B^t & \varepsilon I \end{pmatrix}$$

The Schur complements generate the following tpm's on the subsets of hub and authority states:

$$P_H(\varepsilon) = \varepsilon I + (1-\varepsilon) W_r^\sigma (W_c^\sigma)^t = \varepsilon I + (1-\varepsilon) P_H^\sigma$$
$$P_A(\varepsilon) = \varepsilon I + (1-\varepsilon)(W_c^\sigma)^t W_r^\sigma = \varepsilon I + (1-\varepsilon) P_A^\sigma$$
$$P_H^\sigma = [\text{Diag}(Ae)]^{-1} \cdot A \cdot [\text{Diag}(B^t e)]^{-1} B^t$$
$$= W_r^\sigma \cdot (W_c^\sigma)^t \geq 0$$
$$P_A^\sigma = [\text{Diag}(B^t e)]^{-1} B^t \cdot [\text{Diag}(Ae)]^{-1} \cdot A$$
$$= (W_c^\sigma)^t \cdot W_r^\sigma \geq 0$$

This model does not exhibit a simple closed form steady-state vector π as before. Therefore, block structured numerical solution methods like those discussed in the next sections are required to compute the hub and authority rankings.

4 Stochastic Analysis of the Link Structure by an Algebraic Multigrid Approach

In this section we study the numerical solution of large hierarchically structured Markovian models arising from Web graph models. It is our objective to compute the steady-state vector π of the underlying discrete-time Markov chain (DTMC) Y_n of a random surfer on the Web graph by simple but more efficient algorithms than the well-known power iteration (6). For this purpose, we use the close relationship between the iterative aggregation-disaggregation (IAD) method and the algebraic multigrid (AMG) approach.

In the following, let $T = \widetilde{P}^t \in \mathbb{R}^{n \times n}$ denote the semiconvergent irreducible column-stochastic matrix associated with the transition probability matrix (tpm) $P(S)$ of this underlying finite homogeneous discrete-time Markov chain Y_n on the state-space Γ. It determines the edges $E_\omega = \{(i,j) \mid \exists \widetilde{P}_{i,j} > 0\}$ of a corresponding Web graph model $G_\omega = \{\Gamma, E_\omega, \widetilde{P}\}$. Let $A \equiv A(T) = I - T \in \mathbb{R}^{n \times n}$ be the irreducible Q-matrix with the simple eigenvalue zero associated with T, i.e. an irreducible singular M-matrix with $A \cdot e = 0$ (cf. [2], [21]).

If the original stochastic matrix $P(S)$ of the Web graph is not semiconvergent, i.e., there are further eigenvalues on the unit circle apart from the Perron-Frobenius eigenvalue $\rho(P(S)) = 1$ of index one, we may either use its modification by the standard construction of the teleportation matrix (5) or apply its semiconvergent extrapolated variant $T_\omega = \omega I + (1-\omega) P(S)^t$ for some $0 < \omega < 1$ if $P(S)$ is irreducible. The latter does not destroy the sparsity structure of the original adjacency matrix.

We denote the compact subset of all probability vectors in $\mathbb{R}^n$ by $\mathcal{P} = \{x \in \mathbb{R}^n \mid x > 0, e^t x = 1\}$. Let $\pi \in \mathcal{P}$ be the right eigenvector corresponding to the spectral radius $\rho(T) = 1$, i.e., the Perron-Frobenius vector $\pi > 0$ of the simple Perron-Frobenius eigenvalue of T (cf. [2]).

4.1 Description of the Iterative Aggregation-Disaggregation Method

As prominent variant of an algebraic multigrid approach the IAD-method has been employed as basic iterative solution method not only for nearly completely decomposable Markov chains, but also for all irreducible Markov chains (cf. [3], [15], [19], [22]). On the other hand, multigrid schemes have been applied successfully to solve large linear systems arising from various fields (cf. [6], [9], [11], [16]). First we sketch the IAD-method and interpret it in terms of the features of a structured Web graph G_ω. Then we show its relation to the V-cycle of a two-level algebraic multigrid method and illustrate its application in our context.

First we choose the host partition $\Gamma = \{J_1, \ldots, J_m\}$ of the state space $V = \{1, \ldots, n\}$ into $m \geq 2$ disjoint host sets $J_i = H^{-1}(\{h_i\}) \subset V$ with $n_i \geq 1$ elements each. The method also works for any other block partition derived from a more general feature extraction of the Web pages and their relationship (cf. [8]). Without loss of generality, we assume that the elements of these sets are enumerated in a consecutive order such that $i < j$ holds for $i \in J_l, j \in J_k, l < k$. Furthermore, we arrange and number the elements of T and $\pi^t = (\pi_1{}^t, \ldots, \pi_m{}^t)$ according to this state space partition and the lexicographical (host,page)-ordering.

Following the approach of Chatelin and Miranker [4], we can determine an aggregation $R \in \mathbb{R}^{m \times n}$ as membership matrix of the host partition Γ by

$$R_{ij} = \mathbb{1}_{J_i}(j) = \begin{cases} 1, & \text{if } j \in J_i \\ 0, & \text{otherwise} \end{cases} \quad 1 \leq i \leq m, 1 \leq j \leq n. \tag{11}$$

and a prolongation matrix $P_{(x)} \in \mathbb{R}^{n \times m}$ which depends on a probability vector $x \in \mathcal{P}$. Here $\mathbb{1}_{J_i}$ denotes the indicator function of set J_i.

To describe the standard iterative A/D-method by algebraic means (cf. [4], [10]), we determine the following entities. For an arbitrary, but fixed probability vector $x \in \mathcal{P}$, i.e.,

$$x = \begin{pmatrix} x_1 \\ \vdots \\ x_m \end{pmatrix} > 0 \quad \text{with} \quad e^t \cdot x = 1, \tag{12}$$

let the vector $y \equiv y_{(x)} = \begin{pmatrix} y_{(x)1} \\ \vdots \\ y_{(x)m} \end{pmatrix} > 0$ be given by

$$y_{(x)j} = \begin{cases} x_j/\alpha_{(x)j}, & \text{if } x_j > 0 \\ 1/n_j \cdot e, & \text{if } x_j = 0 \end{cases} \in \mathbb{R}^{n_j}$$

$$\alpha_{(x)j} = e^t \cdot x_j \in \mathbb{R}$$

for $j \in \{1, \ldots, m\}$. y is called the vector of intra-aggregate probabilities on the partition Γ since its components $y_{(x)j} > 0$ may be interpreted as conditional probabilities that the random surfer stays in the states of the aggregate J_j on host h_j.

According to the construction, the relations $e^t \cdot y_{(x)j} = 1$ and $x_j = \alpha_{(x)j} \cdot y_{(x)j}$ hold for each $j \in \{1, \ldots, m\}$. Defining $\alpha \equiv \alpha_{(x)} = (\alpha_{(x)1}, \ldots, \alpha_{(x)m})^t > 0$, relation (12) implies $e^t \cdot \alpha = 1$. Therefore, α is called the vector of inter-aggregate probabilities of the host set $\mathcal{H} \equiv \{1, \ldots, m\}$.

Then we define nonnegative diagonal matrices $\Delta_{(x)} = \mathrm{Diag}(x) \in \mathbb{R}^{n \times n}$ and $\Phi_{(x)} \in \mathbb{R}^{m \times m}$ by $\Delta_{(x)ii} = (x_l)_i$ for $i \in J_l, 1 \le l \le m$, use $\alpha_i = e^t x_i = (Rx)_i \ge 0$ and

$$\Phi_{(x)ii} = \mathbb{1}_{\{0\}}((Rx)_i) = \begin{cases} 1, & \text{for } (Rx)_i = \alpha_i = 0 \\ 0, & \text{for } \alpha_i > 0 \end{cases}$$

$1 \le i \le m$, respectively. Obviously, $\Phi_{(x)ii} = \mathbb{1}_{\{0\}}(\alpha_i)$, $1 \le i \le m$, holds. Furthermore, we define the nonnegative matrix $V_{(x)} = \Delta_{(x)} R^t + R^t \Phi_{(x)} \in \mathbb{R}^{n \times m}$.

Then the nonnegative prolongation matrix $P_{(x)}$ is defined by

$$P_{(x)} = V_{(x)} \cdot (R \cdot V_{(x)})^{-1} \in \mathbb{R}^{n \times m} \tag{13}$$

$$P_{(x)ij} = \begin{cases} (y_j)_i, & \text{if } i \in J_j \\ 0, & \text{otherwise} \end{cases} \quad 1 \le i \le n, 1 \le j \le m.$$

Hence, the matrices

$$R = \begin{pmatrix} e^t & 0 & \cdots & 0 \\ 0 & e^t & \ddots & \vdots \\ \vdots & \ddots & \ddots & 0 \\ 0 & \cdots & 0 & e^t \end{pmatrix}, P_{(x)} = \begin{pmatrix} y_1 & 0 & \cdots & 0 \\ 0 & y_2 & \ddots & \vdots \\ \vdots & \ddots & \ddots & 0 \\ 0 & \cdots & 0 & y_m \end{pmatrix}$$

have rank m and satisfy $e^t \cdot P_{(x)} = e^t$, $e^t \cdot R = e^t$ and $R \cdot P_{(x)} = I$.

By these means we define the nonnegative projection matrix:

$$\Pi \equiv \Pi_{(x)} = V_{(x)} \cdot (R \cdot V_{(x)})^{-1} \cdot R$$

$$= P_{(x)} \cdot R = \begin{pmatrix} y_1 e^t & 0 & \cdots & 0 \\ 0 & y_2 e^t & \ddots & \vdots \\ \vdots & \ddots & \ddots & 0 \\ 0 & \cdots & 0 & y_m e^t \end{pmatrix} \tag{14}$$

Π is a column-stochastic matrix of rank m and $\Pi_{(x)} \cdot P_{(x)} = P_{(x)}$, $R \cdot \Pi_{(x)} = R$ hold. Furthermore, $R \cdot x = \alpha_{(x)}$ and $P_{(x)} \cdot \alpha_{(x)} = x$ induce $\Pi_{(x)} \cdot x = x$. The column-stochastic matrix $\Pi_{(x)}$ is a projection onto the m-dimensional range $\mathrm{Im}(\Pi_{(x)}) = \mathrm{Im}(P_{(x)})$ of $\Pi_{(x)}$ or $P_{(x)}$, respectively, along the null space $\mathrm{Ker}(\Pi_{(x)}) = \mathrm{Ker}(R)$ (cf. [4, p. 20], [15]).

Now we define the projection of the iteration matrix T by $G = \Pi_{(x)} \cdot T \ge 0$. Hence, G is a column-stochastic matrix satisfying $\rho(G) = 1$ and $\mathrm{index}_1(G) = 1$. We define the nonnegative aggregated iteration matrix modelling the host relationship of the random surfer by

$$B \equiv B_{(x)} = R \cdot T \cdot P_{(x)} \in \mathbb{R}^{m \times m}. \tag{15}$$

Then B is column-stochastic and there exists a steady-state vector $0 < \alpha^*_{(x)} \in \mathbb{R}^m$ satisfying
$$B_{(x)} \cdot \alpha^*_{(x)} = \alpha^*_{(x)}, \qquad e^t \cdot \alpha^*_{(x)} = 1. \tag{16}$$
Supposing the steady-state condition $x = \pi$, these probabilities $(\alpha^*_{(\pi)})_i = 1/r_{ii}$ represent the inverse of the mean number of steps r_{ii} that are required by the random surfer to return to host h_i after starting there. Thus, after an appropriate permutation of the host index set $\{1, \ldots, m\}$ the probabilities α^*_i typically follow a Zipf-like law, i.e. $\alpha^*_i \sim c \cdot i^{-\gamma}$, $c \in \mathbb{R}$, $0.7 \leq \gamma \leq 1.4$, or, more generally, a Zipf-Mandelbrot law $\alpha^*_i \sim c \cdot (k+i)^{-\gamma}$ (cf. [1], [5], [13]). This ranking can be exploited to traverse the state space in an appropriate way.

The disaggregated vector $\tilde{x}^*_{(x)} \in \mathbb{R}^n$ defined by
$$\tilde{x}^*_{(x)} = P_{(x)} \cdot \alpha^*_{(x)} \geq 0 \tag{17}$$
fulfills $e^t \cdot \tilde{x}^*_{(x)} = 1$ and $\Pi_{(x)} \cdot \tilde{x}^*_{(x)} = P_{(x)} \cdot \alpha^*_{(x)} = \tilde{x}^*_{(x)}$. The solution $\tilde{x}^*_{(x)}$ fulfills $(I - \Pi_{(x)} \cdot T) \cdot \tilde{x}^*_{(x)} = 0$. As $\pi \in \mathcal{P}$ is the Perron-Frobenius eigenvector of T, we conclude that
$$(I - \Pi_{(x)} \cdot T) \cdot (\pi - \tilde{x}^*_{(x)}) = (I - \Pi_{(x)}) \cdot (\pi - x) \tag{18}$$
holds (cf. [4, (4.3), p. 31], [10, p. 954], [15]).

An A/D-step consists of an aggregation step (16) followed by a disaggregation step (17). Moreover, equation (18) provides a basic relationship between the errors before an A/D-step, $\epsilon = \pi - x$, and after an A/D-step, $\tilde{\epsilon} = \pi - \tilde{x}^*_{(x)}$ used in an error analysis.

In summary, the IAD algorithm to compute the steady-state ranking vector $\pi \in \mathbb{R}^n$ of the Markov chain Y_n reads as follows (cf. [15]):

IAD-Algorithm
Assumption: Select a host partition $\Gamma = \{J_1, \ldots, J_m\}$ of $V = \{1, \ldots, n\}$ into $2 \leq m < n$ disjoint sets. Let $T \in \mathbb{R}^{n \times n}$ be an irreducible semiconvergent column-stochastic matrix of a weighted Web graph model $G_\omega = (\Gamma, E_\omega, \widetilde{P})$.

Let $r(x) = \|(I - T) \cdot x\|_1$, $x \in \mathbb{R}^n$ be the continuous seminorm in $\mathbb{R}^n$. Select $L \in \mathbb{N}$.

(1) Initialization:
   Select an initial vector $x^{(0)} \gg 0$, $e^t x^{(0)} = 1$, an integer $\xi \in \mathbb{N}$, and three real numbers $\varepsilon, c_1, c_2 \in (0,1)$. Construct the matrices $\widetilde{P}$, $P_{(x^{(0)})} \in \mathbb{R}^{n \times m}$, $R \in \mathbb{R}^{m \times n}$, and $B_{(x^{(0)})} \in \mathbb{R}^{m \times m}$ according to (5), (13), (11), (15).
   Set $k = 0$.
(2) A/D step with reordering:
   For $j = 1$ to m do
   $$[y_{(x^{(k)})}]_j = \begin{cases} x_j^{(k)}/e^t \cdot x_j^{(k)}, & \text{if } x_j^{(k)} > 0 \\ 1/n_j \cdot e, & \text{if } x_j^{(k)} = 0 \end{cases}$$

For $i = 1$ to m do
$$\left(B_{(x^{(k)})}\right)_{ij} = e^t \cdot T_{J_i J_j} \cdot [y_{(x^{(k)})}]_j$$
$$= [y^t_{(x^{(k)})}]_{P_j} \widetilde{P}_{P_j, J_i} e + \frac{1}{n}[y^t_{(x^{(k)})}]_{D_j} e$$
endfor
endfor

Solve
$$B_{(x^{(k)})} \cdot \alpha^*_{(x^{(k)})} = \alpha^*_{(x^{(k)})}$$
subject to $\quad e^t \cdot \alpha^*_{(x^{(k)})} = 1, \; \alpha^*_{(x^{(k)})} > 0$
and compute
$$\tilde{x}^{(k)} = P_{(x^{(k)})} \cdot \alpha^*_{(x^{(k)})}$$
$$= \left([\alpha^*_{(x^{(k)})}]_j \cdot [y_{(x^{(k)})}]_j\right)_{j=1,\ldots,m}$$

If $\quad j \bmod L = 0$
then permute the order of the aggregates and states s.t.
$$(\alpha^*_{(x^{(k)})})_1 \geq (\alpha^*_{(x^{(k)})})_2 \geq \ldots \geq (\alpha^*_{(x^{(k)})})_m$$
endif

(3) Iteration step:

Compute $\quad x^{(k+1)} = T^\xi \cdot \tilde{x}^{(k)}$

(4) Test of the A/D-gain (optional step - enforcing convergence):

If $\quad r(x^{(k+1)}) \leq c_1 \cdot r(x^{(k)}) \quad$ then
$\qquad$ goto step (5)
else $\quad$ compute $x^{(k+1)} = T^m \cdot x^{(k)}$
$\qquad$ with $m = m(x^{(k)}) \in \mathbb{N}$ such that
$$r(x^{(k+1)}) \leq c_2 \cdot r(x^{(k)})$$
endif

(5) Termination test:

If $\quad \|x^{(k+1)} - x^{(k)}\|_\infty / \|x^{(k)}\|_\infty < \varepsilon \quad$ then
$\qquad$ goto step (6)
else $\quad k = k + 1$
$\qquad$ goto step (2)
endif

(6) Normalization:
$$\pi = \frac{x^{(k+1)}}{e^t \cdot x^{(k+1)}}$$

In step (2) the solution of the aggregated linear system on the host set can be performed by standard methods, e.g. an ILU approach or a Gauss-Seidel procedure.

It is possible to show that this IAD-algorithm converges globally to the steady-state probability vector $\pi \in \mathcal{P}$ of the Markov chain Y_n for any initial vector $x^{(0)} \gg 0$ with $e^t x^{(0)} = 1$ (cf. [15]). Moreover, the reduction factor is determined by $r(x^{(k+1)}) \leq \max(c_1, c_2) \cdot r(x^{(k)})$, $k \geq 0$.

4.2 The Multiplicative Schwarz AMG Method

The equivalence of the iterative A/D-method and the two-level AMG approach opens the way to a large class of successfully applied numerical solution methods for large sparse linear systems such as the weighted Web graph model $G_\omega = (\Gamma, E_\omega, \tilde{P})$ (cf. [6], [9], [16]). In particular, domain decomposition techniques such as the additive and multiplicative Schwarz iteration methods can be used to cope with the complexity of large state spaces (cf. [9, §11]).

In this section, we discuss a multiplicative Schwarz AMG method that can be employed to calculate efficiently the stationary distribution vector π of the ergodic finite Markov chain Y_n with the tpm $\tilde{P} = T^t \in \mathbb{R}^{n \times n}$.

As we used the hierarchical description by the (host, page) levels, we only state a basic V-cycle of a *two-level* AMG algorithm based on the principles of the multiplicative Schwarz iteration. It can be easily extended to an arbitrary number of levels if a more detailed hierarchical description of the Web pages is used (cf. [6], [8], [11]).

We consider again the singular linear system

$$0 = A(T) \cdot x = (I - T) \cdot x$$

with the semiconvergent irreducible column-stochastic matrix $T = \tilde{P}^t \in \mathbb{R}^{n \times n}$ and the associated singular M-matrix $A \equiv A(T)$ and first choose the host partition $\Gamma = \{J_1, \ldots, J_m\}$, $m \geq 2$, of the state space $\Gamma \subseteq \mathcal{H} \times V \equiv \{1, \ldots, m\} \times \{1, \ldots, n\}$ as domain decomposition.

Then we can define a family of aggregation matrices $R_l \in \mathbb{R}^{n_l \times n}$ and prolongation matrices $P_l \in \mathbb{R}^{n \times n_l}$, $1 \leq l \leq L = m$, on Γ satisfying $P_l \geq 0, e^t \cdot P_l = e^t, R_l \geq 0, e^t \cdot R_l = e^t, R_l \cdot P_l = I_{n_l}, n = \sum_{l=1}^{L} n_l$. Here I_{n_l} is the identity matrix of order n_l.

Moreover, we expect the condition $\mathbb{R}^n = \bigcup_{l=1}^{L} \text{Im}(P_l)$ to be fulfilled. As noticed before, both $P_l(\cdot)$ and $R_l(\cdot)$ may additionally depend on a common probability vector $x \equiv x^{(l-1)} \in \mathcal{P} \subset \mathbb{R}^n$ selected a priori. By these means we define a family of stochastic projection matrices $\Pi_l \equiv \Pi_l(x) = P_l(x) \cdot R_l(x) \in \mathbb{R}^{n \times n}$.

Let $\alpha_l^*(x) > 0$, $e^t \cdot \alpha_l^*(x) = 1$ be the unique normalized solution of $R_l(x) A P_l(x) \cdot \alpha_l = 0$, i.e., $\text{rank}(R_l(x) A P_l(x)) = n_l - 1$ and $\text{Ker}(R_l(x) A P_l(x)) = \text{span}\{\alpha_l^*(x)\}$ holds.

For each quadruple $(x^{(l-1)}, R_l, P_l, \Pi_l)$ we specify a corresponding V-cycle of a two-level AMG scheme by the error-correction matrix (cf. [9], [15], [16]):

$$\Phi_l(x) = I - P_l(x)(R_l(x) A P_l(x) + \alpha_l^*(x) \cdot e^t)^{-1} R_l(x) A$$

To define these corresponding matrices and to solve the related linear systems in an effective manner, we rearrange the elements of the state space Γ conceptually by a similarity transformation with an appropriate permutation matrix S_l and partition the resulting matrix

$$A_l = S_l^t A S_l = \begin{pmatrix} A_{11}^{(l)} & A_{12}^{(l)} \\ A_{21}^{(l)} & A_{22}^{(l)} \end{pmatrix} \begin{matrix} \Gamma \setminus \{J_l\} \\ J_l \end{matrix}$$

$$\begin{matrix} \Gamma \setminus \{J_l\} & J_l \end{matrix}$$

into blocks according to the rearranged host partition $\widetilde{\Gamma}_l = \{\Gamma \setminus \{J_l\}, J_l\}$.

To illustrate the basic idea of this scheme and its refined aggregation concept, we will only present the computational steps for $l = m$. In the considered case, $S_m = I$ is the identity, of course. In a way similar to (16) and (17), a single A/D-cycle on this last set J_m of $\widetilde{\Gamma}_m$ consists of three basic steps. First, a partial aggregation of the original partition Γ of V to an aggregated version $\widehat{\Gamma}_m = \{\widehat{J}_m, J_m\}$, $\widehat{J}_m = \{[1], \ldots, [m-1]\}$, of the reordered state space $\widetilde{\Gamma}_m = \widehat{J}_m \cup J_m$ is performed by lumping all states of each host set $J_i, i = 1, \ldots, m-1$, into a single state $[i]$. In the second step, the Schur complement of the corresponding singular M-matrix $A_m = R_m S_m^t A S_m P_m(x^{(m-1)})$ of this iterative operation on $\widetilde{\Gamma}_m$ is constructed on the host set J_m (see *solve* of AMG-step (2)) and used to determine the solution of a corresponding aggregated system in a way similar to (15) and (16). Then the solution is lifted to the original state space similar to (17). This sketched basic A/D-cycle is consecutively performed on all aggregates $J_1, \ldots, J_m$. The elements of the corresponding families of prolongation and aggregation matrices that are defined for the partitions $\widetilde{\Gamma}_l$ with their ramifications $\widehat{\Gamma}_l$ in a way similar to (13) and (11) determine the required matrices $\{P_l \in \mathbb{R}^{n \times n_l} \mid l = 1, \ldots m\}$, $\{R_l \in \mathbb{R}^{n_l \times n} \mid l = 1, \ldots m\}$.

Then the resulting error-correction matrix

$$\Phi(x^{(0)}) = \Phi_m(x^{(m-1)}) \cdot \Phi_{m-1}(x^{(m-2)}) \cdots \Phi_1(x^{(0)})$$

is recursively defined by the error corrections

$$\Phi_l(x^{(l-1)}) = I - S_l P_l (R_l S_l^t A S_l P_l + \alpha_l^*(x^{(l-1)}) e^t)^{-1} \cdot R_l S_l^t A$$

arising from the related two-level AMG schemes on the aggregates J_l with the outcomes $x^{(l)} = \Phi_l(x^{(l-1)}) \cdot x^{(l-1)}$ and the initial vectors $x^{(l-1)}$ for $l = 1, \ldots, m$.

Finally, we can define a multiplicative Schwarz-iteration scheme by this overall error-correction matrix

$$\Phi(x^{(0)}, \ldots, x^{(m-1)}) = \Phi_m(x^{(m-1)}) \cdot \Phi_{m-1}(x^{(m-2)}) \cdots \Phi_1(x^{(0)})$$

and the V-cylce algorithm of a basic two-level AMG scheme (cf. [9, §11.2.3]). The resulting scheme is stationary if all matrices $P_l(\cdot)$ and $R_l(\cdot)$ are constant, i.e. selecting $P_l(\cdot) = R^t$, otherwise it is nonstationary. In the following, we omit these functional dependencies and permutations to simplify the further exposition.

Following the approach of Douglas and Miranker [6, p. 380], we will now specify this basic V-cycle of a two-level algebraic multigrid scheme. It is applied to the irreducible semiconvergent column-stochastic matrix $T = T_\omega \in \mathbb{R}^{n \times n}$ of a Web graph and its associated singular M-matrix $A \equiv A(T) = I - T$ on the host partition $\Gamma = \{J_1, \ldots, J_m\}$ and its m reorderings $\widetilde{\Gamma}_l$ with the modifications $\widehat{\Gamma}_l = \{\widehat{J}_l, J_l\}$, $\widehat{J}_l = \{[i] \mid 1 \leq i \leq m, i \neq l\}$, $l = 1, \ldots, m$, respectively (cf. [16]).

AMG-Algorithm of the multiplicative Schwarz iteration

(1) Initialization:
Select an initial vector $x^{(0)} \gg 0, e^t x^{(0)} = 1$, an integer $\xi \in \mathbb{N}$, and three real numbers $\varepsilon, c_1, c_2 \in (0, 1)$.
Set $k = 0$.

(2) AMG-step:

Set $\qquad x_0^{(k)} = x^{(k)}$

For $\qquad l = 1$ to $m \qquad$ do
$$d_l^{(k)} = A x_{l-1}^{(k)}$$
Solve $\qquad R_l A P_l \cdot y_l^{(k)} = R_l \cdot d_l^{(k)}$
subject to $\qquad e^t \cdot y_l^{(k)} = 0$
Set $\qquad x_l^{(k)} = x_{l-1}^{(k)} - P_l \cdot y_l^{(k)}$

endfor

If $\qquad x_m^{(k)} > 0 \qquad$ then
set $\qquad \tilde{x}^{(k+1)} = x_m^{(k)}$
else set $\qquad \tilde{x}^{(k+1)} = x^{(k)}$
endif

(3) Iteration step:

Compute $\qquad x^{(k+1)} = T^\xi \cdot \tilde{x}^{(k+1)}$

(4) Test of the A/D-gain:

If $\qquad r(x^{(k+1)}) \leq c_1 \cdot r(x^{(k)}) \qquad$ then
goto step (5)
else compute $x^{(k+1)} = T^\eta \cdot x^{(k)}$
with $\eta = \eta(x^{(k)}) \in \mathbb{N}$ such that
$$r(x^{(k+1)}) \leq c_2 \cdot r(x^{(k)})$$
endif

(5) Termination test:

If $\qquad \|x^{(k+1)} - x^{(k)}\|_\infty / \|x^{(k)}\|_\infty < \varepsilon \qquad$ then
goto step (6)
else $\qquad k = k + 1$
goto step (2)
endif

(6) Normalization:

 If $x^{(k+1)} > 0$ then
    set $\pi = x^{(k+1)}/(e^t \cdot x^{(k+1)})$
 else print an error message and stop
 endif

As the fallback scheme $x^{(k+1)} = T \cdot x^{(k)}$ is convergent, the proposed multigrid algorithm converges globally for any initial vector $x^{(0)} \gg 0$ with $e^t x^{(0)} = 1$ to the steady-state probability vector $\pi \in \mathcal{P}$ of the Markov chain Y_n, provided that it stops successfully (cf. [16], [23, Theorem 4]).

5 Conclusion

In this paper we have studied stochastic link analysis techniques applied in Web information retrieval. First we have presented a hierarchical algebraic description of a Web graph with host-oriented clustering of pages and discussed a simplified analysis of SALSA derived from a block level description. Then we have discussed advanced numerical methods derived from an algebraic multigrid approach to compute the steady-state vector of the underlying large finite Markov chain of a random surfer on the Web graph. We have focussed on an iterative aggregation/disaggregation procedure and a multiplicative Schwarz variant of the multigrid domain decomposition technique.

It was our objective to contribute to the computational framework of Web information retrieval illustrating the application of efficient multigrid solution techniques. Based on the principles of domain decomposition and algebraic multigrid further effective distributed iterative solution procedure can be derived for the considered large Markovian Web graph models. They can be efficiently implemented on vector machines and distributed-memory multiprocessor architectures and illustrate the road towards the integration of numerical multi-linear algebra into Web information retrieval and its page ranking by the eigenvector approach.

In conclusion, we have pointed out a promising way to use the great potential of efficient multigrid schemes and provided a basic understanding of their effects.

References

1. Baldi, P., et al.: Modeling the Internet and the Web - Probabilistic Methods and Algorithms. John Wiley & Sons, Chichester (2003)
2. Berman, A., Plemmons, R.J.: Nonnegative Matrices in the Mathematical Sciences. Academic Press, New York (1979)
3. Cao, W., Stewart, W.J.: Iterative aggregation/disaggregation techniques for nearly uncoupled Markov chains. Journal of the ACM 32, 702–719 (1985)
4. Chatelin, F., Miranker, W.L.: Acceleration by aggregation of successive approximation methods. Linear Algebra and its Applications 43, 17–47 (1982)
5. Dorogovtsev, S.N., Mendes, J.F.F.: Evolution of Networks. Oxford University Press, New York (2003)

6. Douglas, C.C., Miranker, W.L.: Constructive interference in parallel algorithms. SIAM J. Numer. Analysis 25(2), 376–398 (1988)
7. Golub, G.H., van Loan, C.F.: Matrix Computations, 3rd edn. Johns Hopkins, Baltimore (1996)
8. Kumar, S. D., et al.: Exploiting the block structure of the Web for computing PageRank. Stanford University Technical Report (2003), http://citeseer.ist.psu.edu/kamvar03exploiting.html
9. Hackbusch, W.: Iterative Lösung großer schwachbesetzter Gleichungssysteme. Teubner, Stuttgart (1991)
10. Haviv, M.: Aggregation/disaggregation methods for computing the stationary distribution of a Markov chain. SIAM J. Numer. Analysis 24(4), 952–966 (1987)
11. Horton, G., Leutenegger, S.: A multilevel solution algorithm for steady-state Markov chains. In: Proc. SIGMETRICS 1994, Nashville, May 16-20 (1994)
12. Kammenhuber, N., Luxenburger, J., Feldmann, A., Weikum, G.: Web search clickstreams. In: Internet Measurement Conference 2006, pp. 245–250 (2006)
13. Krashakov, S.A., Teslyuk, A.B., Shchur, L.N.: On the universality of rank distributions of website popularity. Computer Networks 50(11), 1769–1780 (2006)
14. Krieger, U.R., Müller-Clostermann, B., Sczittnick, M.: Modeling and analysis of communication systems based on computational methods for Markov chains. IEEE Journal on Selected Areas in Communications 8(9), 1630–1648 (1990)
15. Krieger, U.R.: On a two-level multigrid solution method for finite Markov chains. Linear Algebra and its Applications 223/224, 415–438 (1995)
16. Krieger, U.R.: Numerical solution of large finite Markov chains by algebraic multigrid techniques. In: Stewart, W.J. (ed.) Computations with Markov Chains, pp. 403–424. Kluwer, Boston (1995)
17. Langville, A.N., Meyer, C.D.: Google's Pagerank and Beyond: The Science of Search Engine Rankings. Princeton University Press, Princeton (2006)
18. Lempel, R., Moran, S.: SALSA: the stochastic approach for link-structure analysis. ACM Trans. Information Systems 19, 131–160 (2001)
19. Mandel, J., Sekerka, B.: A local convergence proof for the iterative aggregation method. Linear Algebra and its Applications 51, 163–172 (1983)
20. Ng, A.Y., et al.: Stable algorithms for link analysis. In: Proc. 24th Ann. Int. ACM SIGIR Conf., pp. 258–266 (2001)
21. Schneider, H.: Theorems on M-splittings of a singular M-matrix which depend on graph structure. Linear Algebra and its Applications 58, 407–424 (1984)
22. Schweitzer, P.J.: A survey of aggregation-disaggregation in large Markov chains. In: Stewart, W.J. (ed.) Numerical Solution of Markov Chains, pp. 63–88. Marcel Dekker, New York (1991)
23. Schweitzer, P.J., Kindle, K.W.: An iterative aggregation-disaggregation algorithm for solving linear equations. Applied Mathematics and Computation 18, 313–353 (1986)
24. Wang, Q., et al.: Improving link analysis through considering hosts and blocks. In: Proc. IEEE/WIC/ACM Int. Conference on Web Intelligence 2006 (2006)

An Introduction to Modelling and Performance Evaluation for TCP Networks

Michele Pagano[1] and Raffaello Secchi[2]

[1] Dipartimento di Ingegneria dell'Informazione
Università di Pisa, Via Caruso, I-56122 Pisa, Italy
michele.pagano@iet.unipi.it
[2] Electronic Research Group
University of Aberdeen, Aberdeen AB24 3UE, United Kingdom
raffaello.secchi@erg.abdn.ac.uk

Abstract. The widespread diffusion of TCP over the Internet has motivated a significant number of analytical studies in TCP modelling, the most important of which are presented in this tutorial. The simplest approaches describe the dynamics of an individual source over a simplified network model (e.g. expressing the network behaviour in terms of average loss rate and latency). These models allow us to derive accurate estimations for the long-term TCP throughput under different network settings. More detailed techniques model the behaviour of a set of TCP connections over an arbitrary complex network. The latter are able to capture the network dynamics and effectively predict the closed-loop interaction between TCP and traffic management techniques. As an example the derivation of sufficient stability conditions for a network of RED queues is provided.

Keywords: TCP modelling, fluid models, stability conditions.

1 Introduction

The transmission control protocol (TCP) is used by the vast majority of Internet applications for reliable and ordered data transmission [1]. Since TCP is in charge of network congestion control and bandwidth fair sharing for these flows, it is in fact responsible for the stability of the entire Internet. Hence in the last decade, accurate modelling of its behaviour has attracted the interest of many researchers and a large amount of literature has been produced on this topic. This tutorial is a primer to the vast world of TCP modelling, which provides a first look to the most important contributions in this field. It is not intended to be comprehensive, but the interested reader may find here enough material to approach more advanced resources.

Accurate TCP models have been developed in the past with various purposes: lightweight simulation techniques, network dimensioning, systems requirement definition, design of new TCP-friendly protocols, etc. Simple mathematical models are today sometimes preferred to detailed network simulations to have a quick evaluation of capacity requirements and performance of TCP based services. As noted in [2], detailed computer simulations may be not able to scale to large network topologies or large number of concurrent flows. This rises concerns on what could be the consequences of the deployment of such a system over the Internet. By describing only the essential characteristics of congestion control, analytical models permit to produce accurate results for realistic scenarios in a reasonable short time.

TCP models proved useful also to design new standards for congestion control compatible with existing TCP implementations. As an example, the *throughput equation* of TCP has been used as the basis for the design of a congestion control algorithm for slowly variable flows [3], such as multimedia streaming. The analysis of other congestion control mechanisms at network layer, such as the active queue management (AQM) [4,5,6,7], has been successfully carried out by means of TCP models. As explained further in the tutorial, a closed control loop is formed between AQM and TCP, whose behaviour is well explained by analytical models.

TCP models find an application also in the solution of some network optimisation problems [8]. These techniques aim to assign flow rates to a set of flow sharing a common network in order to satisfy some optimisation criteria. These techniques are based on the definition of an *utility function* that describes the degree of satisfaction associated to a given flow rate. An optimisation algorithm (possibly distributed) is used to determine the solution that maximises the overall network utility. This idea, first proposed by Kelly in [9], has been elaborated upon by many researchers [8,10,11,12] and has been very useful for understanding the bandwidth sharing properties of TCP. For instance, Low identified [10] the utility function that well describe TCP Reno, TCP Reno with RED and TCP Vegas, while Massoulié et al. in [11] catalogued various utility functions that allow to achieve fairness, minimise delay and maximise throughput.

As far as the classification of TCP models is concerned, reference [13] distinguishes between TCP models for bulk file transfer and for short lived connections. In the first case, when the TCP steady state behaviour is analysed, accurate throughput formulas are obtained by simplifying some aspects of protocol dynamics, such as the connection start-up and the loss recovery phase, that have minor impact on the long term performance. On the other hand, models for short-lived connections, such as [14,15,16], focus on the start-up phase taking into account the connection establishment and the *Slow-Start* phase.

Here, we focus on TCP models for long lived connections that we divide into two families: *Single connection models* and *fluid models*. Although the TCP algorithms have evolved significantly over the years, the original mechanism based on additive increase multiplicative decrease (AIMD) paradigm has been preserved. Thus, capturing the essence of AIMD leads to a good approximation of TCP behaviour. Single connection models assume a network path with given characteristics, such as the mean round trip time (RTT) and the loss probability, and try to evaluate throughput as a function of these parameters [17,18,19,20,21,22,23,24,25]. On the other hand, fluid models [26,27,28,29] simplify the high complexity of the TCP algorithms using a fluid flow analogy. The discrete nature of packet transmission is neglected in favour of a representation of TCP dynamics using compact differential equations. In some case, fluid models are built introducing a sub-model of TCP protocol and a sub-model of TCP flow interaction over the network, and then iterating over the two sub-models by means of a fixed point procedure [26,27,30,31,32].

As a final consideration, we can say that TCP models have contributed significantly to the understanding of the nature of the Internet traffic. This approach is substantially different from the classical one, where traffic characterisation was achieved by fitting the corresponding time series to well-known statistical distributions [33,34]. Embedding a model of the source (TCP) into the mathematical framework allows us to explain the causes of the observed behaviour and potentially uncover better mechanisms.

The rest of the tutorial is organised as follows. In Section 2, we present models of individual connections and we derive throughput formulas for various operating conditions. In Section 3, the focus is on fluid models. The interaction between TCP and network layer for various congestion control settings is analysed. Finally, Section 4 ends the tutorial with the conclusions.

2 Single Connection Models

In this section we formulate a model for a single TCP connection with the intent of deriving a formula of TCP throughput. Later in the section, we will consider a model of wide-area traffic and analyse its implications. We start from a simple yet not accurate model and we proceed by a successive step of refinement.

Let us consider a TCP connection established over a lossy path [20] with large enough capacity and competing traffic so small that the queueing delay component of round trip time (RTT) can be assumed negligible. Let us also assume that the link introduces one drop every $1/p$ successful packet deliveries. Under these fairly strong assumptions, it is easy to obtain a closed-form formula for the TCP throughput. Indeed, after an initial transient, the TCP congestion window (*cwnd*), which corresponds to the number of in flight packets over the link, takes the periodic evolution illustrated in Figure 1. According to the AIMD principle, when the *cwnd* reaches its maximum (W) and a packet loss is met, the *cwnd* is backed off to $W/2$.

The speed of growth of the *cwnd* (the slope of the curve) depends on the way the receiver is acknowledging packets. Standard implementations require TCP receivers to

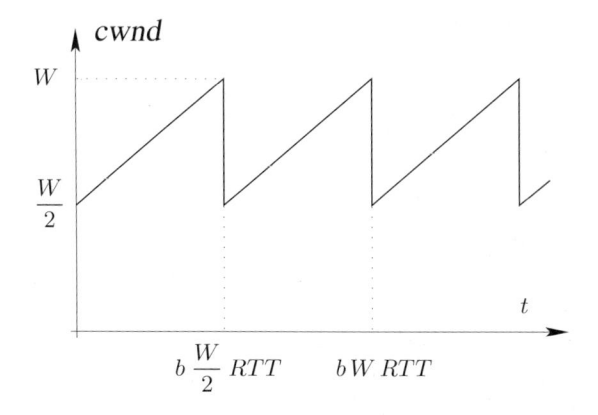

Fig. 1. Periodic evolution of TCP window

acknowledge the reception of every packet. However, some implementations adopt the delayed-ACK option [35], which consists of transmitting one cumulative ACK every two packets received. This roughly halves the number of ACKs transmitted, but it also slows down the growth of the *cwnd* during the Congestion Avoidance phase. Then, the duration of a cycle is $bW/2$ rounds, where b takes the value 2 or 1 depending on the delayed-ACK being enabled or disabled respectively. The duration of a cycle results

$$T_{\text{cycle}} = RTT \cdot b\frac{W}{2} \tag{1}$$

The total number of segments delivered within each cycle (A_{cycle}) is equivalent to the area under a period of the sawtooth, which is

$$A_{\text{cycle}} = b\frac{W}{4} \cdot \left(\frac{W}{2} + W\right) = b\frac{3W^2}{8}$$

packets per cycle. Since, by hypothesis, the number of packets in a cycle is $1/p$, we can solve for W obtaining

$$W = \sqrt{\frac{8}{3pb}} \tag{2}$$

The throughput achieved by the connection is the ratio between the total amount of data delivered in a cycle A_{cycle} and the duration of the cycle T_{cycle}. Substituting in the equation below, we get the mean throughput $\mathcal{B}$ in packet per second as:

$$\mathcal{B} = \frac{A_{\text{cycle}}}{T_{\text{cycle}}} = \frac{b\frac{3}{8}W^2}{RTT \cdot \frac{b}{2}W} = \sqrt{\frac{3}{2b}} \cdot \frac{1}{RTT\sqrt{p}} \tag{3}$$

This expression shows that the throughput is inversely proportional to RTT and the square root of loss probability. The constant of proportionality, which in this formula is $\sqrt{3/2b}$, is in general a function of TCP implementation and loss model. Slightly different values of this constant were derived in [19,21] or extrapolated from empirical data.

2.1 Stochastic Models of TCP Throughput

The TCP throughput formula presented above tends to overestimate the throughput for large loss rates, which is mainly due to not taking into account TCP retransmission timeouts. Indeed, timeouts occur when the number of packet drops in a window of data is such that the Fast Recovery algorithm [36] cannot detect the missing packet from duplicated ACKs. A refinement of the previous formulation which accounts for timeouts was given in [18], which is in part presented in the following.

Congestion Avoidance is described here in terms of rounds, each starting with the back-to-back transmission of a burst of packets equal to the congestion window and ending with the reception of the first ACK for the burst of delivered packets.

Since at each ACK reception the *cwnd* is incremented by $1/\lfloor cwnd \rfloor$ packets and an ACK acknowledges exactly b packets, at the beginning of the transmission of the next packet batch the value of *cwnd* is *cwnd* + 1/*b*. That is, during Congestion Avoidance and in absence of losses, the window size increases linearly with slope 1/*b*. To simplify the analysis, the duration of each round is supposed independent of the congestion window.

The TCP algorithm stops increasing *cwnd* and slows-down transmission when it detects a packet loss, because TCP associates packet drops with congestion indications. In order to emulate the loss behaviour induced by the drop-tail queueing discipline, which is quite common for Internet routers, we assume that the losses of packets within the same round are correlated. On the other hand, we can assume that packet losses in different rounds are independent. This is justified by the fact that packets in different rounds are separated by one RTT or more, and thus these are likely to encounter independent buffer states. Simulative studies [37] confirm this claim highlighting that packet drops tend to be grouped in bursts when the (drop tail) buffer is congested. Hence, we suppose that each packet may be dropped with probability p if the previous packet is not lost, and that all the packets following the first loss in a round are also lost.

Loss detection at the sender side can occur through either the reception of three duplicate ACKs or when a timeout expires. In the first case, after lost segments are retransmitted, the sender resumes the transmission with the *cwnd* set to one half of its previous value. On the contrary, when the loss is detected via a timeout T_0 expiration, the *cwnd* is reset to one and a packet is sent in the first round following the timeout. In case another timeout expires, the length of timeout is doubled. Doubling the timeout is repeated for each unsuccessful retransmission up to $64T_0$, after which the timeout remains constant. If L_k indicate the duration of a sequence of k consecutive timeouts, that is

$$L_k = \begin{cases} (2^k - 1)T_0 & \text{for } k \leq 6 \\ (63 + 64(k - 6))T_0 & \text{for } k \geq 7 \end{cases}$$

the mean length of timeouts can be easily calculated recalling that L_k has a geometric distribution due to the packet loss independence between consecutive rounds

$$T_{\text{timeout}} = \sum_{k=1}^{\infty} L_k \cdot p^{k-1}(1-p)$$
$$= T_0 \frac{1 + p + 2p^2 + 4p^3 + 16p^5 + 32p^6}{1 - p}$$

which means that the timeout period can be expressed in the form $T_{\text{timeout}} = T_0 \cdot f(p)$.

Figure 2 illustrates an example of the evolution of *cwnd*. The i-th TCP cycle consists of a sequence of n_i periods in which the *cwnd* has an AIMD behaviour (*cwnd-cycles*) and a sequence of one of more timeout periods. The first $n_i - 1$ *cwnd-cycles* end with a packet loss detection through three duplicate ACKs, while a timeout occurs in the last one. The throughput of a connection can be calculated as the ratio between the mean number of packets delivered within the cycle and the mean length of the cycle. If we assume that $\{n_i\}_i$ is a sequence of independent and identically distributed random variables with mean n, which are independent of the amount of data delivered in each *cwnd-cycle* and of the duration of the period, we can write the throughput as

$$\mathcal{B}' = \frac{n \cdot A_{\text{cycle}}}{n \cdot T_{\text{cycle}} + T_{\text{timeout}}} = \frac{A_{\text{cycle}}}{T_{\text{cycle}} + Q \cdot T_{\text{timeout}}} \qquad (4)$$

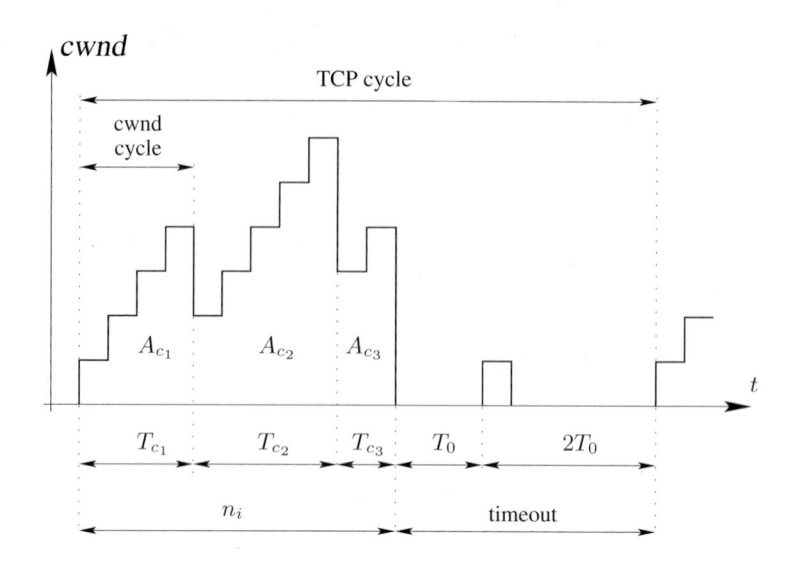

Fig. 2. TCP windowsize evolution

where $Q = 1/n$ denotes the probability that a loss indication is detected via a timeout. In order to complete the formula, we have to derive some closed-form expressions for A_{cycle}, T_{cycle} and Q. The T_{cycle} is given by (1), where W is substituted by the mean value of the steady-state window size, which is formally identical to equation (2) for small values of p:

$$E[W] = \sqrt{\frac{8}{3pb}} + o(1/\sqrt{p}) \qquad (5)$$

A formal proof of this can be found in [18].

We concentrate now in deriving an expression for Q. Let us consider a *cwnd-cycle* where a loss indication occurs and let w and k represent respectively the *cwnd* and the number of packets successfully delivered in a round. According to our loss model, if a packet is lost, so are all the packets that follow up to the end of the burst. Then, the probability $A(w, k)$ to have $k < w$ successful transmissions in this round is given by

$$A(w, k) = \frac{(1-p)^k p}{1 - (1-p)^w}.$$

Also, since the first k packets in the round are acknowledged, other k packets are delivered in the next round, which is the last round of this *cwnd-cycle*. Again, this round of transmission may have losses and the lost packets are the last transmitted in the burst. Indicating with m the number of packets successfully transmitted in the last round, the distribution of m can be expressed by

$$C(w, m) = \begin{cases} p(1-p)^m & \text{for } m \leq w - 1 \\ (1-p)^w & \text{for } m = w \end{cases}$$

For the duplicated ACKs in the last round of transmission, the TCP receiver disables the Delayed-ACK mechanism [38] and every packet following the lost packets is acknowledged. If the sender receives less than three duplicate ACKs, the *cwnd-cycle* ends with a timeout and the probability of this event is given by

$$\hat{Q}(w) = \begin{cases} 1 & \text{if } w \leq 3 \\ \sum_{k=0}^{2} A(w,k) + \sum_{k=3}^{w} \sum_{m=0}^{2} A(w,k) C(w,k) & \text{otherwise} \end{cases} \quad (6)$$

since, when a timeout occurs, either the number of packets transmitted in the second last round or the number of packets successfully delivered in the last round is less than three. Simple algebraic manipulations allow us to rewrite this expression as

$$\hat{Q}(w) = \min\left(1, \frac{(1-(1-p)^3)(1+(1-p)^3(1-(1-p)^{w-3}))}{1-(1-p)^w}\right).$$

Observing that $\hat{Q}(w) \approx 3/w$ when p tends to zero, we can get a good numerical approximation for $\hat{Q}$

$$\hat{Q}(w) \approx \min\left(1, \frac{3}{w}\right).$$

Finally, we can write the probability that a loss indication is given by a timeout and approximate it by

$$Q = \sum_{w=1}^{\infty} \hat{Q}(w) \cdot Pr\{W = w\} = E\{\hat{Q}\} \approx \hat{Q}(E[W]) \quad (7)$$

where $E[W]$ can be evaluated using expression (5). Now, let us concentrate on evaluating A_{cycle} by considering the i-th *cwnd-cycle*. We indicate with α_i the first packet lost during the i-th *cwnd-cycle*. As previously observed, after packet α_i, $W_i - 1$ more packets are sent in an additional round before the loss can be detected by the sender. Therefore, a total of $A_i = \alpha_i + W_i - 1$ packets are sent during the whole i-th cycle. It follows that

$$A_{\text{cycle}} = E[\alpha] + E[W] - 1$$

Based on our assumption on packet losses, the random process $\{\alpha_i\}_i$ is a sequence of independent and identically distributed random variables with distribution

$$Pr[\alpha = k] = p(1-p)^{k-1} \qquad k = 1, 2, \ldots$$

since the probability that $\alpha_i = k$ is equal to the probability that exactly $k-1$ packets are successfully acknowledged before a loss occurs. Thus, replacing the mean of α with $1/p$ we have

$$A_{\text{cycle}} = \frac{1-p}{p} + E[W] \quad (8)$$

where $E[W]$ is given by (5). Then, substituting (8), (1) and (7) into (4), we achieve an accurate estimate of the throughput of a TCP connection; after further simplifications, in [18] the following well known expression for TCP bandwidth (9) is given:

$$B' = \frac{1}{RTT\sqrt{\frac{2bp}{3}} + T_0 \min\left(1, 3\sqrt{\frac{3bp}{8}}\right) p(1+32p^2)} \quad (9)$$

where the approximation $f(p) \simeq 1 + 32p^2$ in the expression of T_{timeout} was applied.

It is worth noting that (9) holds as long as the throughput is less than $W_{\max}/RTT$, where $W_{\max}$ is the maximum buffer advertised by receiver, since more than $W_{\max}$ packets cannot be transferred every round trip. We observe that this model does not capture all the aspects of Fast Retransmit algorithm and disregard the Slow Start phase. However, the impact of these omissions is quite low and the measurements collected to validate the model indirectly validate the assumptions as well. Live experiments also suggest that this model is able to predict the throughput for a wider range of loss rates than (3). Indeed, real experiments carried out over narrow-band links show that, when the congestion window size is small, many TCP timeouts occur. This indicates that in reality Fast Retransmit does not detect all the loss events and an accurate formula for TCP throughput has necessarily to take into account TCP timeouts.

3 Fluid Models

Fluid modelling is a widely used technique in the analysis of network systems where packet flows are approximated by continuous (i.e. fluid) streams of data. The main advantage is the formulation of a system of equations to describe the behaviour of congestion-controlled sources, network elements, and flows interaction over the network. Experimental results [26] show that relatively simple systems of ordinary differential equations (ODEs) provide satisfactory predictions of the real system dynamics. Since performance evaluation results can be achieved with relatively low cost, both in terms of computational time (ODEs can be fast solved using numerical methods) and development effort, these techniques are often used during the first stage of protocol design. In fact, fluid modelling techniques are rugged enough to establish the effectiveness of a protocol and resolve dimensioning issues. In this section, we introduce various techniques for fluid modelling of TCP networks that recur in many research studies.

3.1 Throughput Formulas for WANs

The analysis that we are about to present was carried out in [21] to express the TCP throughput as a function of the bandwidth-delay product (BDP). The BDP is a very important and widely used metric influencing TCP performance. It corresponds, under ideal conditions, to the amount of data that a bulk TCP transfer should hold in flight to achieve high utilisation of link bandwidth. Thus, studying the TCP throughput as a function of the BDP is a natural way of expressing TCP performance. As several authors pointed out [20,21], the TCP throughput is so correlated to the BDP, that a direct relationship between the two metrics can be found. In the following we show an analytical way to achieve this goal.

It has been observed that the BDP is only relevant for wide area connections where the RTT may range from several tens to hundreds of milliseconds. In local area networks (LANs) the number of packets in flight is very small and the TCP throughput is limited by other factors, such as the loss rate induced by the medium access control scheme or the transmitter buffer size. Moreover, for long-haul connections the *cwnd* tends to open considerably and TCP performance are extremely sensitive to occasional packet losses due to transient congestion or packet corruptions. As it will be clarified in the following, TCP could not be able to *keep the pipe full*, i.e to achieve high capacity utilisation, in these cases.

Let us indicate with c the capacity of the link in packets per second, τ the round-trip propagation delay, and $T = \tau + 1/c$ the minimum observed RTT. For a wide area network (WAN) connection, the BDP, $c \cdot T$, is comparable in magnitude with the amount of packets queued at the bottleneck route.

Let us also assume that at a given time the bottleneck buffer is not empty. The packets are forwarded at rate c by the link server, ACKs are generated by the destination at rate c and, therefore, new packets can be released by the source every $1/c$ seconds[1]. Thus, the maximum possible number of unacknowledged packets is the sum of the packets in transit across the path, which is equal to cT, and of the packets in the buffer of size B. Therefore, if the size of the window exceeds $W_{\max} = cT + B$, a buffer overflow occurs. Actually, when the packet loss occurs, it is difficult to evaluate the exact *cwnd*, since it depends on the link capacity and on the RTT. However, as an approximation, we assume that TCP Reno, after retransmitting a lost packet, resumes Congestion Avoidance with the *cwnd* set to $(cT + B)/2$.

Using the fluid approximation, the *cwnd* dynamics can be easily written as a differential equation. Denoting with $a(t)$ the number of ACKs received by the source after t seconds spent in the Congestion Avoidance phase, the derivative of *cwnd* can be written as

$$\frac{dW}{dt} = \frac{dW}{da} \cdot \frac{da}{dt}.$$

If the *cwnd* is large enough, the bottleneck buffer is continuously backlogged, and the ACK rate is c. Otherwise, the ACK rate equals the sending rate W/T. In other words we have

$$\frac{da}{dt} = \min\{\frac{W}{T}, c\},$$

and recalling that during the Congestion Avoidance phase the *cwnd* is increased by $1/W$ for each ACK received, we have

$$\frac{dW}{da} = \frac{1}{W}$$

---

[1] For sake of simplicity we assume here and in the following that delayed-ACK is not implemented.

that is

$$\frac{dW}{dt} = \begin{cases} \dfrac{1}{T} & \text{if } W \leq cT \\ \dfrac{c}{W} & \text{if } W > cT \end{cases} \quad (10)$$

which means that the Congestion Avoidance phase consists of two sub-phases corresponding to $W \leq cT$ and to $W > cT$ respectively. We are now able to evaluate the duration of the first phase as

$$T_1 = T \cdot (cT - \tfrac{1}{2} W_{\max}) \quad (11)$$

and the number of packets successfully transmitted in this phase

$$\begin{aligned} N_1 &= \int_0^{T_1} \frac{W(t)}{T} \, dt = \frac{1}{T} \int_0^{T_1} \left(\frac{1}{2} W_{\max} + \frac{t}{T} \right) dt \\ &= \frac{1}{2T} \left(W_{\max} T_1 + \frac{T_1^2}{T} \right) \end{aligned} \quad (12)$$

When $W > cT$, the queue is increasing, the RTT as well increases and the *cwnd* opens more and more slowly. By integrating (10), in the second phase we get

$$W(t)^2 = 2c(t - T_1) + (cT)^2$$

and evaluating this expression for $t = T_1 + T_2$, we obtain the duration of the phase

$$T_2 = \frac{W_{\max}^2 - (cT)^2}{2c} \quad (13)$$

and the number of packets delivered

$$N_2 = c \cdot T_2, \quad (14)$$

since the link is fully utilised during this phase. Now, we can evaluate the throughput as a function of the ratio between the bottleneck buffer and the BDP

$$B'' = \frac{N_1 + N_2}{T_1 + T_2} = \frac{3c}{4} \cdot \frac{(1 + \frac{B}{cT})^2}{1 + \frac{B}{cT} + (\frac{B}{cT})^2}. \quad (15)$$

This expression, which holds for $\frac{B}{cT} \leq 1$, suggests that the performance of a bulk transfer over a WAN might be negatively affected by small (with respect to the BDP) bottleneck buffers. In order to fully exploit the link capacity, the buffer size should be as close as possible to the BDP, which is a rule often used to configure router buffers [39].

3.2 Throughput Formulas for WANs and Random Losses

Let us now consider the case where, in addition to buffer overflows, packets can be randomly lost over the bottleneck link with probability q. In such a case, it is possible to exactly compute the throughput through Markov chain analysis. In the following we sketch this method.

In absence of random losses, the evolution of *cwnd* of TCP Reno is entirely determined by the window size at the beginning of the cycle w, which is half of *cwnd* at the end of the previous cycle. Then, the evolution of *cwnd* can be described introducing the following functions:

- $W(n, w)$ the window size after n successful packet transmissions.
- $T(n, w)$ the time required to complete the transmission of n packets.

If the cycle terminates with losses due to buffer overflow, then $N_{\max}(w)$ represents the number of packets transmitted in this period; otherwise, the cycle terminate with a random loss after successfully transmitting $N \leq N_{\max}(w)$ packets. According to the random loss model, the distribution of N is given by

$$Pr\{N = n\} = \begin{cases} q(1-q)^n, & \text{for } n < N_{\max} \\ (1-q)^{N_{\max}}, & \text{for } n = N_{\max} \end{cases} \qquad (16)$$

For the i-th cycle, let w_i, N_i and T_i denote respectively the *cwnd* at the beginning of the Congestion Avoidance phase, the number of successful transmissions in the cycle, and the duration of the cycle. The TCP evolution can be expressed through the following recursive equations

$$\begin{cases} w_{i+1} = \frac{1}{2} W(N_i, w_i) \\ T_{i+1} = T_i(N_i, w_i) \end{cases} \qquad (17)$$

These equations, together with (16), define the transition probabilities for the continuous-time Markov chain $\{w_i\}$, whose solution gives the stationary distribution of w_i. The steady-state throughput is then given by

$$B''' = \frac{E[N_i]}{E[T_i]}$$

Since the exact solution of this Markov chain can be computationally expensive, an approximation of the previous method is also provided in [21].

An important result of the cited study is that the presence of random losses leads to significant throughput deterioration for large BDPs. In particular, it has been shown that the TCP throughput depends on the product of the loss probability and of the *square* of the BDP ($q \cdot (cT)^2$) and degrades rapidly when this parameter grows larger than one. This is due to the relatively earlier drops in the *cwnd-cycle* that lead to small initial values of the *cwnd*, thus requiring several transmission rounds to fill the pipe again. In other words, TCP over WAN networks suffers performance degradation in presence of non-congestion losses, such as the ones induced by competing real-time traffic or wireless links. Countermeasures, such as traffic differentiation or performance enhancing proxies (PEPs), are usually employed to eliminate this problem.

3.3 Interaction between TCP and AQM Mechanisms

So far, we presented various ways to relate TCP performance to network parameters. This was possible assuming a sort of independence between individual TCP flow state and network state. However, this assumption does not hold in many circumstances where the number of interacting flows is low or the network topology is not trivial.

We introduce now a different approach, still based on fluid modelling, where stochastic differential equations are uses to model TCP behaviour as well as network/flow interactions. This allows us to assess the benefits of the active queue management (AQM), and to select network parameters that enhance stability and performance.

As an example, we consider the random early detection (RED) technique [40], which is a highly popular AQM scheme within the Internet community. RED improves network performance reacting *earlier* (otherwise said pro-actively) to congestion. Similarly to other AQM schemes, RED randomly discards incoming packets with a probability that depends on the averaged queue size before reaching the full-buffer condition. This avoids the overstaying of congested situations in router buffers, which leads to a lower amount of congestion losses. Randomly selecting the packets to discard, RED also enables a flow management fairer than the drop-tail discipline. Indeed, the larger is the flow rate, the higher is the packet dropping probability, and the more frequent is the delivery of congestion signals. Eventually, this leads to a faster convergence to an equilibrium condition where fairness among flows is achieved.

Most of the IP routers today support the explicit congestion notification (ECN) option. This technique consists of marking a flag into the IP header instead of dropping the packet. The status of the flag is then copied into the returning ACK by the receiver to explicitly notify the sender the presence of congestion on the direct path. From a modelling point of view, the ECN marking scheme can be assimilated to the RED analysis, as the amount of packet losses is negligible with respect to the number of transmitted packets. Several enhancements of RED [5,41,42,43] have been proposed. These techniques consider more complicated packet dropping strategies to improve congestion control. Nevertheless, once the control law is known, fluid modelling techniques can be easily implemented to analyse the effects of these modifications. Figure 3 shows the classic dropping profile, which is used in the following analysis. It is a linear function between a lower $t_{\min}$ and an upper $t_{\max}$ threshold with a discontinuity at $t_{\max}$, which is analytically expressed by

$$p(x) = \begin{cases} 0 & 0 \leq x < t_{\min} \\ \dfrac{x - t_{\min}}{t_{\max} - t_{\min}} p_{\max} & t_{\min} \leq x \leq t_{\max} \\ 1 & t_{\max} < x \end{cases} \qquad (18)$$

3.4 The Model of the Network

The network is modelled as a set of L links with capacities c_l, $l \in \{1, 2, \ldots, L\}$. The links are shared by a set of S sources indexed by $s \in \{1, 2, \ldots, S\}$, each using a subset L_s of links. The sets L_s define a $L \times S$ routing matrix

$$A_{ls} = \begin{cases} 1 \text{ if } l \in L_s \\ 0 \text{ otherwise} \end{cases} \qquad (19)$$

which is a binary matrix where the 1s on a l-th row indicate the sources that share the link l and the 1s on s-th column represent the link crossed by source s. Each source is associated with its congestion window $W_s(t)$ at time t and each link l is associated with both its packet loss probability $p_l(t)$ (a scalar congestion measure) and with the instantaneous queue size $q_l(t)$.

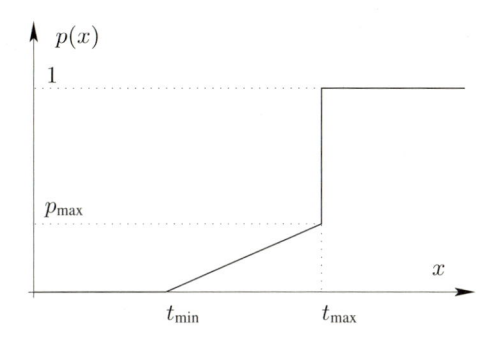

Fig. 3. RED drop function

The average RTT of the s-th TCP source at time t is approximated by

$$RTT_s(t) = \tau_s + \sum_{l \in L_s} \frac{q_l(t)}{c_l} \qquad s \in \{1, 2, \ldots, S\}, \qquad (20)$$

which is the sum of the round trip delay τ_s associated with the connection and the total queueing delay of the path. Considering the losses at the different queues as independent of each other (a reasonable assumption when modelling RED queues), we can express the packet loss probability $\hat{p}_s(t)$ of source s as

$$\hat{p}_s(t) = 1 - \prod_{l=1}^{L}(1 - A_{ls}p_l(t)) \simeq \sum_{l \in L} A_{ls}p_l(t) \qquad s = 1, 2 \ldots S, \qquad (21)$$

which corresponds to the end-to-end congestion measure for the s-th source.

The following is the differential version of the Lindley equation, describing the dynamic of l-th queue

$$\frac{dq_l(t)}{dt} = -1_{q_l(t)}c_l + \sum_{s=1}^{S} A_{ls}\frac{W_s(t)}{RTT_s(t)}.$$

Here, the derivative of the instantaneous queue length is the sum of two terms. The first one models the decrease of queue length, as long as it is greater than zero, due to service of packets at a constant rate. The second term corresponds to the increase in queue length due to the arrival of packets from the TCP flows that share the l-th queue. Since we are interested in a mean value analysis, we take the expectation[2] of both sides of equation (3.4):

$$\frac{d\bar{q}_l(t)}{dt} = E[-1_{q_l(t)}]\, c_l + \sum_{s=1}^{S} A_{ls} E\left[\frac{W_s(t)}{RTT_s(t)}\right]$$

$$\approx -E[1_{q_l(t)>0}]\, c_l + \sum_{s=1}^{S} A_{ls} \frac{\overline{W_s(t)}}{\overline{RTT_s(t)}}$$

In the derivation, we made the approximation $E[f(x)] \approx f(E[x])$, which is not strictly correct. However, simulation results suggest that at the steady-state the system reaches a quasi-periodic evolution where the random component of state variables, such as the instantaneous queue size, makes up a smaller and smaller contribution with respect to the deterministic one as the number of flows increases. Since the system is dominated by the deterministic evolution and the extent of fluctuations is small, the previous approximation is justified.

In order to approximate the term $E[1_q]$, we should consider that the bottleneck queues have $q_l(t) > 0$ with probability close to one, while the non-bottleneck queues are typically unloaded, which means $q_l(t) > 0$ with probability close to zero. On the basis of this observation, we approximate $E[1_{q(t)}] \simeq 1_{\bar{q}(t)}$, thus having

$$\frac{d\bar{q}_l(t)}{dt} = -1_{\bar{q}_l(t)} c_l + \sum_{s=1}^{S} A_{ls} \frac{\overline{W_s(t)}}{\overline{RTT_s(t)}} \tag{22}$$

The mean queue length is mapped by RED into a drop probability using the drop profile (18). We assume that RED estimates the average queue length using an exponential weighted moving average based on samples taken every T seconds. The smoothing filter (with α, $0 < \alpha < 1$, as weight) is described by

$$x_l[(k+1)T] = (1-\alpha)\, x_l[kT] + \alpha q_l[kT]$$

It is convenient to approximate the above equation with differential equation. Since this equation is a first order difference equation, the natural candidate is

$$\frac{dx_l(t)}{dt} = ax_l(t) + bq_l(t) \tag{23}$$

Recalling that in a sampled data system with $q(t) \equiv q[kT]$ in the interval $[kT,(k+1)T]$, $x_l[(k+1)T]$ is given by

$$x_l[(k+1)T] = e^{aT} x_l[kT] + b \int_{kT}^{(k+1)T} e^{a(kT-\mu)} d\mu \cdot q_l[kT], \tag{24}$$

and comparing the coefficients of (23) and (3.4), we obtain

---

[2] Throughout this section we will indicate, when possible, the mean value with a bar sign to simplify the notation.

$$a = -b = \frac{\ln(1-\alpha)}{T}.$$

Then, we rewrite the expression (23), describing the behaviour of $x(t)$, by taking the expected value of both sides:

$$\frac{d\bar{x}_l(t)}{dt} = \frac{\ln(1-\alpha)}{T}(\bar{x}_l(t) - \bar{q}_l(t)) \qquad (25)$$

3.5 The Model of the Source

The next step is to build a model of a TCP source. The model is based on the assumption that packet losses of a flow can be described by a Poisson counting process $\{N_s(t)\}$ with time varying rate $\lambda_s(t)$. The Poisson process is indeed suitable to represent the independent marking scheme used by AQM/RED. This process could be visualised imagining a flow of losses moving from network nodes towards TCP sources, whose rate is a function of the TCP flow rate.

If we denote with $N_s(t)$ the number of losses detected by the source s at time t, we can write the evolution of the congestion window as

$$dW_s(t) = \frac{dt}{RTT_s(t)} - \frac{W_s(t)}{2}dN_s$$

This equation only considers the AIMD behaviour of TCP. More specifically, the first term corresponds to the AI part, which increases the window size by one packet every RTT, and the second term corresponds to the MD part, which halves the congestion window immediately after the drop is detected by the sender ($dN_s(t) = 1$ in this case). Again, taking expectation, we obtain

$$\begin{aligned} dE[W_s(t)] &= E\left[\frac{dt}{RTT_s(t)}\right] - \frac{E[W_s(t)\,dN_s(t)]}{2} \\ d\overline{W}_s(t) &\simeq E\left[\frac{dt}{RTT_s(t)}\right] - \frac{\overline{W}_s(t)}{2}\lambda_s(t)dt \end{aligned} \qquad (26)$$

where $\lambda_s(t)$ is the rate of loss indication at the sender. Note that in (26) in order to split the term $E[W_s(t)\,dN_s(t)]$ in a product of two factors $E[W_s(t)]E[dN_s(t)]$, we have assumed that the terms $W_s(t)$ and $dN_s(t)$ are independent. This is not exact, but it is still able to capture the dynamics of AIMD.

In proportional marking schemes (such as RED) the rate of marking/dropping indications is proportional to the share of bandwidth of the connection. That is, if the bandwidth achieved by source s is $W_s(t)/RTT_s(t)$, the expected value for drop rate at link $l \in L_s$ is

$$p_l(t) \cdot \frac{W_s(t)}{RTT_s(t)} \qquad (27)$$

and equivalently the packet drop rate $\hat{p}_s(t)$ for a connection is calculated using (21).

However, we note that drops occur at the node about a RTT before they can be detected by the sender. This means that, in order to take into account the latency of feedbacks, we must shift the rate of congestion signals (27) forward in time of RTT_s seconds. Thus, from (26) the evolution of cwnd is given by

$$\frac{d\overline{W}_s(t)}{dt} = \frac{1}{\overline{RTT}_s(t)} - \frac{\overline{W}_s(t)}{2} \cdot \frac{\overline{W}_s(t - \overline{RTT}_s(t))}{\overline{RTT}_s(t - \overline{RTT}_s(t))} \hat{p}_s(t - \overline{RTT}_s(t)) \quad (28)$$

In conclusion we have $2L + S$ coupled equations (22), (25) and (28) in the unknowns $(\bar{x}, \bar{q}, \overline{W})$, that can be solved numerically. The solution yields an estimate of the average transient behaviour of the system, providing directly the window size of each connection and the queue size at each node and, from them, the loss rate and average RTT. The time needed to get accurate results through the use of the model is several order of magnitude less than that needed by simulations and the gain increases as the topology becomes more complex. To have an idea of the advantage, let consider that the solution of a system consisting of a thousand connections takes a few seconds, while the corresponding detailed simulation can take several hours to complete.

3.6 Linear Analysis: The Single Link Case

In the following, we accomplish the task of linearising the previous set of equations in the case of a single link topology. The linearised system is suitable to be studied through the classic tools of linear control theory and gives us many suggestions on the way to modify the algorithm to fulfil requirements of stability and robustness [44,45].

Let us consider then N identical TCP Reno connections (i.e. with the same RTT) sharing a common link with capacity C. If we can assume all the connections synchronised (i.e. $W_s(t) = W(t)$, $\tau_s = \tau$ and $RTT_s(t) = R(t)$), we can rewrite, for a generic TCP flow, equation (22), defining the evolution of the mean value of cwnd, and equation (28), concerning the dynamic of the queue[3]

$$\begin{cases} \dot{W}(t) = \frac{1}{R(t)} - \frac{W(t)W(t-R(t))}{2R(t-R(t))} p(t - R(t)) \\ \dot{q}(t) = \frac{W(t)}{R(t)} N - C \end{cases} \quad (29)$$

where the term $R(t) = \tau + \frac{q(t)}{C}$ represents the RTT for all the connections. When writing the equation for $\dot{q}(t)$, we assumed that the server is always transmitting packets, which is a reasonable assumption since we are studying the dynamics of the bottleneck. To complete the system of equations (29), we need also to specify the relationship between the queue size and the dropping probability, which depends on the employed AQM strategy. By borrowing the terminology from control-system language, we will refer to the AQM block as the *controller* and the rest of the system as the *plant*. A linear representation for the *plant* is derived in the following from the system (29), where the

---

[3] For ease of notation we will omit the sign of expectation and denote the temporal derivative of f as $\dot{f}$.

loss probability is considered as the input and the queue size as the output. Then, we will focus on the design of AQM controller using the tools of the linear control theory.

The main goal of AQM is to provide a stable closed-loop system. Beside the stability, there are other issues concerning control design. A feasible controller must possess an acceptable transient response and it must not be sensitive to the variations of model parameters and to disturbance factors. For instance, the presence of short lived connections in the queue has a noisy effect on long-lived TCP connections.

Small-signal analysis. The first step to linearise the system is to find the operating point (W_0, q_0, p_0), which is defined by $\dot{W} = 0$ and $\dot{q} = 0$. From (29) we have

$$\begin{cases} \frac{1}{R_0} - \frac{W_0^2}{2R_0} p_0 = 0 & \Rightarrow W_0 = \sqrt{\frac{2}{p_0}} \\ \frac{W_0}{R_0} N - C = 0 & \Rightarrow W_0 = \frac{R_0 C}{N} \end{cases} \quad (30)$$

where $R_0 = \frac{q_0}{C} + \tau$. The operating point is the state-space point, to which the system would converge if it would be globally stable. For the case here considered, this is simply found accounting $1/N$ of the available bandwidth to each flow. Since equation (28) omits modelling of retransmission timeouts, the long term throughput W_0/R_0 comes out as the *one-on-square-root-p* law (3), which has been said inaccurate for high values of p. To refine the model, a term accounting for timeouts could be added to the right hand side of equation (28), as it was done in [26] introducing additional assumptions on the packet drop model. However, the small-signal analysis for this case would be quite complicated and would escape the introductory purposes of this paper.

Introducing difference variables $(\delta W, \delta p, \delta q)$, we can linearise (29) around the operating point

$$\begin{cases} \delta \dot{W}(t) = -\frac{N}{R_0^2 C}(\delta W(t) + \delta W(t - R_0)) \\ \qquad - \frac{1}{R_0^2 C}(\delta q(t) - \delta q(t - R_0)) - \frac{R_0 C^2}{2N^2} \delta p(t - R_0) \\ \delta \dot{q}(t) = \frac{N}{R_0} \delta W(t) - \frac{1}{R_0} \delta q(t) \end{cases} \quad (31)$$

where equations (30) have been used to evaluate the derivatives. A simple expression for the eigenvalues of the linearised system can be estimated provided that

$$W_0 \gg 1 \quad \Rightarrow \quad \frac{N}{R_0^2 C} = \frac{1}{W_0 R_0} \ll \frac{1}{R_0}. \quad (32)$$

In this case indeed, the response-time of the aggregate of TCP flows is dominated by the time constant $R_0^2 C/N$, which is much larger than the round trip time R_0. This implies that in an interval R_0 the mean window size does not vary significantly with respect to its absolute value. Hence, we can approximate the system of equations by merging the terms $W(t)$ and $W(t - R_0)$, and neglecting the term $(\delta q(t) - \delta q(t - R_0))$

$$\begin{cases} \delta \dot{W}(t) = -\frac{2N}{R_0^2 C} \delta W(t) - \frac{R_0 C^2}{2N^2} \delta p(t - R_0) \\ \delta \dot{q}(t) = \frac{N}{R_0} \delta W(t) - \frac{1}{R_0} \delta q(t) \end{cases} \quad (33)$$

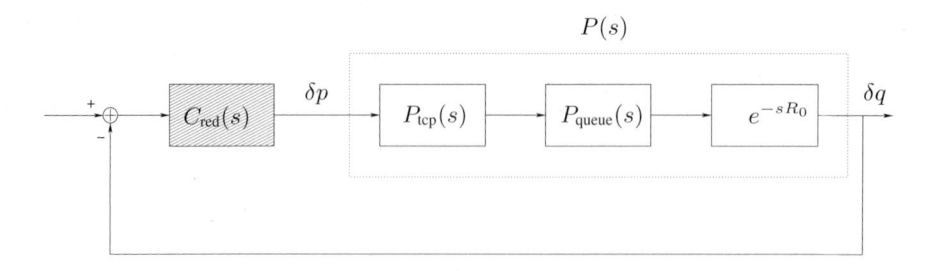

Fig. 4. Linearised block diagram

Figure 4 shows a block diagram for the system of equations (33), where $P_{\text{tcp}}(s)$ and $P_{\text{queue}}(s)$ are defined as

$$P_{\text{tcp}}(s) = \frac{\frac{R_0 C^2}{2N^2}}{s + \frac{2N}{R_0^2 C}}$$

$$P_{\text{queue}}(s) = \frac{\frac{N}{R_0}}{s + \frac{1}{R_0}}$$

The transfer function of the plant, relating the packet-marking probability to the queue length, is then given by:

$$P(s) = P_{\text{tcp}}(s)\, P_{\text{queue}}(s)\, e^{-sR_0} = \frac{\frac{R_0^3 C^3}{4N^2}}{(1 + s\frac{R_0^2 C}{2N})(1 + sR_0)}\, e^{-sR_0} \qquad (34)$$

Some remarks on the $P(s)$ expression can be done. The static gain $R_0^3 C^3/4N^2$ is proportional to the RTT and the capacity of the link and inversely proportional to the number of active flows. The static gain is in turn inversely proportional to the *gain margin*, which is the maximum static gain of the control system that would make the system unstable. The gain margin is indeed defined as the amplitude response at the point where the phase response is $-\pi$.

If the number of flows is small, the static gain increases reducing stability and leading to a more oscillatory response. As we could expect, when we have few TCP flows, the extent of rate variation due to multiplicative decrease is larger than in the case of many flows and this impacts the stability of the system. Moreover, an increase of R_0, which corresponds to a larger delay in the control loop, reduces the controllability of the system, as confirmed by its negative influence on the *gain margin*.

From (34), it could be also easily deduced the phase margin $\phi_m = \omega_{pm} - \pi$, where ω_{pm} is the phase response when the amplitude response is 0 dB. The phase margin in this context could be interpreted as the amount of additional delay in the RTT that the system would tolerate without becoming unstable.

Setting RED parameters. The majority of Internet routers uses drop tail buffers. This could be interpreted, from a control-theory point of view, as an ON/OFF control strategy, also known as relay-controller. Relay-controllers may cause instabilities in the system, such as an oscillating behaviour, and unpredictability due to non-linear effects. For instance, the queue size of a router can alternate between full and empty, which causes buffer overflows in one case and link underutilisation in the other. Mechanisms, such as RED or other AQM schemes, allow us to mitigate this condition.

Another objective of AQM is to reduce the mean queueing delays and delay variations. The smaller the queue-size, the less the time a packet spends in the queue and the less the end-to-end delay of a connection. However, forcing small queues may again lead to link underutilisation. A tradeoff between acceptable queueing delay and utilisation must be operated.

The queue control should be robust against variation of network parameters. This includes supporting a variable number of TCP sessions, variable RTTs and the presence of non congestion controlled traffic (that could be regarded as a sort of noise source on the TCP/AQM system). Here, we provide a condition that guarantees stability for a wide range of working conditions.

In order to find a Laplace representation of RED controller, we describe RED as the cascade of a smoothing filter (25) and a dropping function (18). Since the operating point falls between t_{min} and t_{max}, it is easy to find a linear relation between small variations of dropping probability δp and small variations of the averaged queue size δx. This is then substituted into the transfer-function of AQM/RED yielding to

$$C_{red}(s) = K \cdot \frac{1}{1 + \frac{s}{\beta}} \tag{35}$$

where

$$K = -\frac{\ln(1-\alpha)}{T}, \qquad \beta = \frac{p_{max}}{t_{max} - t_{min}} \tag{36}$$

$C_{red}(s)$ acts as a proportional controller with static gain K corresponding to the slope of the RED dropping profile. When we choose the $C_{red}(s)$, we need to take into account variations in the number of TCP sessions and RTT. In this case, the variations of the term R_0 are due to the propagation variable R_0.

We assume that the number of TCP sessions N is larger than a threshold N_{min} and the RTT R_0 is less than R_{max}. Our goal is to select the RED parameters K and β that stabilise the system for all the values of N and R_0 included in their definition intervals. A closed-loop control system is stable if the response to any bounded input is a bounded output. In this case we have no inputs, so the system is stable if the response to whatever initial condition converges exponentially to zero.

Let us consider the frequency response of the open-loop system

$$L_{red}(j\omega) = C_{red}(j\omega)P(j\omega) = \frac{K \frac{(R_0 C)^3}{(2N)^2} e^{-j\omega R_0}}{(\frac{j\omega}{\beta} + 1)(\frac{j\omega}{\frac{2N}{R_0^2 C}} + 1)(\frac{j\omega}{\frac{1}{R_0}} + 1)}$$

For the range of frequencies at least a decade lower than the minimum displacement of plant poles

$$\omega \leq \omega_g = \frac{1}{10} \min \left\{ \frac{2N_{min}}{(R_{max})^2 C}, \frac{1}{R_{max}} \right\} \tag{37}$$

the system frequency response can be approximated by

$$L_{\text{red}}(j\omega) \approx \frac{K \frac{(R_0 C)^3}{(2N)^2} e^{-j\omega R_0}}{\frac{j\omega}{\beta} + 1}$$

as the contribution of plant poles at the denominator is small. Then, we evaluate the amplitude of the system frequency response for $\omega = \omega_g$ and, under the hypothesis that $N \geq N_{\min}$, $R_0 \leq R_{\max}$, we find the following upper bound

$$|L_{\text{red}}(j\omega_g)| = \frac{K \frac{(R_0 C)^3}{(2N)^2}}{\sqrt{\frac{\omega_g^2}{\beta^2} + 1}} \leq \frac{K \frac{(R_{\max} C)^3}{(2N_{\min})^2}}{\sqrt{\frac{\omega_g^2}{\beta^2} + 1}}$$

Now, enforcing the condition

$$\frac{K(R_{\max} C)^3}{(2N_{\min})^2} \leq \sqrt{\left(\frac{\omega_g}{\beta}\right)^2 + 1}, \tag{38}$$

we have $|L_{\text{red}}(j\omega_g)| \leq 1$. This means that the unit gain cross-over frequency, which is unique as the frequency response amplitude is a decreasing function of ω, is upper bounded by ω_g.

In order to establish the closed loop stability, we invoke the Nyquist criterion: if the open loop system has not unstable roots, the closed loop system is stable if the curve $L_{\text{red}}(j\omega)$, $-\infty < \omega < \infty$ on the complex plane has not clockwise encirclement around $(-1 + 0j)$. Thus, recalling that $\omega_g R_0 \geq 0.1$ for (37), we have

$$\arg\{L_{\text{red}}(j\omega_g)\} \geq \arg\left\{\frac{K \frac{(R_{\max} C)^3}{(2N_{\min})^2}}{\frac{j\omega_g}{\beta} + 1}\right\} - \omega_g R_0 \geq -\frac{\pi}{2} - 0.1 > -\pi$$

Hence, being $|L_{\text{red}}(j\omega)| \leq 1$ for $\omega \geq \omega_g$, the curve does not encircle the point $(-1 + j0)$ and the system is stable. In conclusion, we found that, if K and β satisfy the condition (38), the linear system (34) with $C_{\text{red}}(s)$ as controller is stable for any $N \geq N_{\min}$ and $R_0 \leq R_{\max}$.

This example shows how the linear model can be used to build a robust design of RED, which accounts for the variation of the number of flows N and of the round trip time R. The rationale of this design is to force the controller to dominate the closed-loop behaviour. This is done by choosing a closed loop time-constant (close to ω_g) at least a decade higher than TCP time-constant or queue time-constant. The expression (37) leaves a degree of freedom in choosing the parameters (K, β) on the boundary of the set defined by (38). To determine the parameters, other constraints can be placed.

We could have chosen a multiplicative factor in (37) larger than 0.1. This would lead to a faster response time of the system, but would produce a controller with lower stability margins. In order to increase the bandwidth of the system, other strategies could be introduced as well, such as, for instance, the classical proportional-integral (PI) controller discussed in [6].

4 Conclusions

In this paper we introduced the most relevant works in the field of TCP analytical modelling. We considered models that describe the TCP behaviour both as an individual source (single connection models) or as a part of a network system described in terms of differential equations (fluid models).

The analysis of individual TCP connections in isolation led to expressions for the TCP throughput under given networking conditions. Since the analysis required the independence of the network state and the source state, these models are adequate when the network is loaded by a large number of flows. In fact, this condition is often verified. Asymptotic results [46,47] show indeed that the correlation between the state of two connections rapidly decreases as the number of connection increases.

In the second part, we faced the problem of analytically describing the interaction between TCP and AQM. In particular, we focused on networks with RED policy. We derived analytical results that allow a qualitative understanding of the transient behaviour of TCP over RED networks. Furthermore, we presented results on stability and robustness of the network system itself. As a final example of application of these techniques, we described a method proposed in literature to select the AQM parameters that lead to stable operations of the linear feedback control system.

This introduction did not address other important models available in literature. Some authors addressed for instance the modelling of short-lived connections [14,15,16] where other aspects of TCP, such as the Slow Start phase, dominate the TCP behaviour. These studies are well justified by the fact that many Internet flows are short and are completed before reaching a steady-state. As pointed out by Jacobson [48], the Slow Start phase should be seen as a mechanism for fast approaching of the equilibrium point, while the Congestion Avoidance as a mechanism for asymptotic stability. Thus, accurate modelling of this phase somehow complements the results provided by the analysis of long lived connections.

References

1. Postel, J.: Transmission Control Protocol. RFC 793 (Standard), Updated by RFCs 1122, 3168 (September 1981)
2. Paxson, V., Floyd, S.: Why We Don't Know How to Simulate the Internet. In: Proc. of the 29th Conference on Winter Simulation, Atlanta (US), pp. 1037–1044 (December 1997)
3. Floyd, S., Handley, M., Padhye, J., Widmer, J.: TCP Friendly Rate Control (TFRC): Protocol Specification. RFC 5348 (Proposed Standard) (September 2008)
4. Floyd, S., Fall, K.: Promoting the Use of End-to-End Congestion Control in the Internet. IEEE/ACM Transactions on Networking 7(4), 458–472 (1999)
5. Ott, T.J., Lakshman, T.V., Wong, L.H.: SRED: Stabilized RED. In: Proc. of IEEE INFOCOM, vol. 3, pp. 1346–1355 (1999)
6. Hollot, C.V., Misra, V., Towsley, D., Gong, W.: Analysis and Design of Controllers for AQM Routers Supporting TCP Flows. IEEE Transactions on Automatic Control 47(6), 945–959 (2002)
7. Chait, Y., Hollot, C.V., Misra, V., Towsley, D.F., Zhang, H., Lui, J.: Providing Throughput Differentiation for TCP Flows Using Adaptive Two Color Marking Multi-Level AQM. In: Proc. of IEEE INFOCOM, New York (US), pp. 23–27 (June 2002)

8. Kunniyur, S., Srikant, R.: End-to-End Congestion Control Schemes: Utility Functions, Random Losses and ECN Marks. In: Proc. of IEEE INFOCOM, Tel Aviv (Israel), vol. 3, pp. 1323–1332 (March 2000)
9. Kelly, F.P.: Charging and Rate Control for Elastic Traffic. European Transactions on Telecommunications 8, 33–37 (1997)
10. Low, S.H.: A Duality Model of TCP and Queue Management Algorithms. IEEE/ACM Transactions on Networking 11(4), 525–536 (2003)
11. Massoulié, L., Roberts, J.: Bandwidth Sharing: Objectives and Algorithms. IEEE/ACM Transaction on Networking 10(3), 320–328 (2002)
12. Mo, J., Walrand, J.: Fair End-to-End Window-based Congestion Control. IEEE/ACM Transactions on Networking 8(5), 556–567 (2000)
13. Khalifa, I., Trajkovic, L.: An Overview and Comparison of Analytical TCP models. In: Proc. of International Symposium on Circuits and Systems (ISCAS), Vancouver (Canada), vol. 5, pp. 469–472 (2004)
14. Heidemann, J., Obraczka, K., Touch, J.: Modeling the Performance of HTTP over Several Transport Protocols. IEEE/ACM Transactions on Networking 5(5), 616–630 (1997)
15. Cardwell, N., Savage, S., Anderson, T.: Modeling TCP Latency. In: Proc. of IEEE INFOCOM, vol. 3, pp. 1742–1751 (2000)
16. Mellia, M., Stoica, I., Zhang, H.: TCP Model for Short Lived Flows. IEEE Communications Letters 6(2), 85–87 (2002)
17. Paxson, V.: Empirically Derived Analytic Models of Wide-Area TCP Connections. IEEE/ACM Transactions on Networking 2(4), 316–336 (1994)
18. Padhye, J., Firoiu, V., Towsley, D.F., Kurose, J.F.: Modeling TCP Reno Performance: A Simple Model and its Empirical Validation. IEEE/ACM Transaction on Networking 8(2), 133–145 (2000)
19. Floyd, S.: Connections with Multiple Congested Gateways in Packet-Switched Networks Part 1: One-way Traffic. ACM SIGCOMM Computer Communication Review 21(5), 30–47 (1991)
20. Mathis, M., Semke, J., Mahdavi, J., Ott, T.: The Macroscopic Behaviour of the TCP Congestion Avoidance Algorithm. ACM SIGCOMM Computer Communication Review 27(3), 67–82 (1997)
21. Lakshman, T., Madhow, U.: The Performance of TCP/IP for Networks with High Bandwidth-Delay Products and Random Loss. IEEE/ACM Transactions on Networking 5(3), 336–350 (1997)
22. Kumar, A.: Comparative Performance Analysis of Versions of TCP in a Local Network with a Lossy Link. IEEE/ACM Transactions on Networking 6(4), 485–498 (1998)
23. Misra, A., Ott, T.J.: The Window Distribution of Idealized TCP Congestion Avoidance with Variable Packet Loss. In: Proc. of IEEE INFOCOM, New York (US), vol. 3, pp. 1564–1572 (1999)
24. Casetti, C., Meo, M.: A New Approach to Model the Stationary Behavior of TCP Connections. In: Proc. of IEEE INFOCOM, Tel Aviv (Israel), vol. 1, pp. 367–375 (2000)
25. Altman, E., Barakat, C., Ramos, V.M.R.: Analysis of AIMD Protocols over Paths with Variable Delay. Computer Networks 48(6), 960–971 (2005)
26. Misra, V., Gong, W.B., Towsley, D.F.: Fluid-based Analysis of a Network of AQM Routers Supporting TCP Flows with an Application to RED. ACM SIGCOMM Computer Communication Review 30(4), 151–160 (2000)
27. Hollot, C.V., Misra, V., Towsley, D.F., Gong, V.: A Control Theoretic Analysis of RED. In: Proc. of IEEE INFOCOM, Anchorage (Alaska), pp. 1510–1519 (April 2001)
28. Liu, Y., Presti, F.L., Misra, V., Towsley, D.F., Gu, Y.: Fluid Models and Solutions for Large-Scale IP Networks. ACM SIGMETRICS Performance Evaluation Review 31(1), 91–101 (2003)

29. Gu, Y., Liu, Y., Towsley, D.F.: On Integrating Fluid Models with Packet Simulation. In: Proc. of IEEE INFOCOM, Hong Kong (China), vol. 4, pp. 2856–2866 (March 2004)
30. Altman, E., Avrachenkov, K., Barakat, C.: A Stochastic Model of TCP/IP with Stationary Random Losses. In: ACM SIGCOMM, Stockholm (Sweden), pp. 231–242 (2000)
31. Bu, T., Towsley, D.: Fixed Point Approximations for TCP behavior in an AQM Network. ACM SIGMETRICS Performance Evaluation Review 29(1), 216–225 (2001)
32. Garetto, M., Lo Cigno, R., Meo, M., Marsan, M.A.: Closed Queueing Network Models of Interacting Long-lived TCP Flows. IEEE/ACM Transactions on Networking 12(2), 300–311 (2004)
33. Paxson, V., Floyd, S.: Wide-Area Traffic: The Failure of Poisson Modeling. IEEE/ACM Transactions on Networking 3(3), 226–244 (1995)
34. Park, K., Willinger, W.: Self-Similar Network Traffic and Performance Evaluation. Wiley-Interscience, Inc., Hoboken (2000)
35. Allman, M., Paxson, V., Stevens, W.: TCP Congestion Control. RFC 2581 (Proposed Standard) (April 1999)
36. Floyd, S., Henderson, T., Gurtov, A.: The NewReno Modification to TCP's Fast Recovery Algorithm. RFC 3782 (Proposed Standard) (April 2004)
37. Fall, K., Floyd, S.: Simulation-based Comparisons of Tahoe, Reno and SACK TCP. ACM SIGCOMM Computer Communication Review 26(3), 5–21 (1996)
38. Stevens, W.: TCP/IP Illustrated. The Protocols, vol. 1. Addison-Wesley, Reading (1994)
39. Villamizar, C., Song, C.: High Performance TCP in ANSNET. ACM SIGCOMM Computer Communications Review 24(5), 45–60 (1994)
40. Floyd, S., Jacobson, V.: Random Early Detection Gateways for Congestion Avoidance. IEEE/ACM Transactions on Networking 1(4), 397–413 (1993)
41. Lin, D., Morris, R.: Dynamics of Random Early Detection. ACM SIGCOMM Computer Communications Review 9(4), 127–137 (1997)
42. Feng, W., Kandlur, D.D., Saha, D., Shin, K.G.: A Self-Configuring RED Gateway. In: Proc. of IEEE INFOCOM, New York (USA), vol. 3, pp. 1320–1328 (1999)
43. Liu, S., Basar, T., Srikant, R.: Exponential-RED: A Stabilizing AQM Scheme for Low- and High-speed TCP Protocols. IEEE/ACM Transactions on Networking 5(5), 1068–1081 (2005)
44. Christiansen, M., Jaffay, K., Ott, D., Smith, F.D.: Tuning RED for Web Traffic. IEEE/ACM Transactions on Networking 9(3), 249–264 (2001)
45. Bonald, T., May, M., Bolot, J.C.: Analytic Evaluation of RED Performance. In: Proc. of IEEE INFOCOM, Tel Aviv (Israel), vol. 3, pp. 1415–1424 (2000)
46. Tinnakornsrisuphap, P., La, R.J.: Asymptotic Behavior of Heterogeneous TCP Flow and RED Gateways. IEEE/ACM Transactions on Networking 14(1), 108–120 (2006)
47. Tinnakornsrisuphap, P., Makowski, A.M.: Limit Behavior of ECN/RED Gateways Under a Large Number of TCP Flows. In: Proc. IEEE INFOCOM, San Francisco (US), vol. 2, pp. 873–883 (2003)
48. Jacobson, V.: Congestion Avoidance and Control. ACM SIGCOMM Computer Communication Review 18(4), 314–329 (1988)

Performance Modelling and Evaluation of a Mobility Management Mechanism in IMS-Based Networks[*]

Is-Haka M. Mkwawa[1], Demetres D. Kouvatsos[1], Wolfgang Brandstätter[2], Gerhard Horak[3], Alfons Geier[4], and Christoforos Kavadias[5]

[1] NetPEn - Networks and Performance Engineering Research Unit,
University of Bradford, UK
{I.M.Mkwawa1,D.Kouvatsos}@bradford.ac.uk
[2] Telekom Austria TA AG, Vienna, Austria
wolfgang.brandstaetter@telekom.at
[3] Alcatel-Lucent Austria AG
Gerhard.Horak@alcatel-lucent.com
[4] Nokia Siemens Networks, Germany
alfons.geier@nsn.com
[5] Teletel AS, Greece
C.Kavadias@teletel.eu

Abstract. The 3rd Generation Partnership Project (3GPP) for an IP multimedia subsystem (IMS) architecture defined a number of functional units, which exchange session initiation protocol (SIP) messages with register users and set up or terminate multimedia sessions. The processing of SIP messages, however, requires a significant amount of processing, queueing and transmission times with adverse implications on the overall performance of the IMS core architecture. This tutorial introduces a mobility management mechanism for an IMS testbed implemented by Nokia-Siemens as part of the EU IST VITAL project. The mechanism employs an open queueing network model (QNM) with priorities to represent the functional units and application servers of an IMS architecture during a handover process of SIP messages between different access networks. The QNM is analysed via the principle of maximum entropy and numerical experiments are employed to assess the performance impact of bursty traffic flows of SIP messages on the IMS architecture.

Keywords: IP multimedia subsystem (IMS), 3rd Generation Partnership Project (3GPP), Session Initiation Protocol (SIP), quality-of-service (QoS), performance modelling, performance evaluation, mobility, GSM, WLAN, Wi-Fi, serving call session control function (S-CSCF), handover, maximum entropy (ME), generalized exponential (GE), queueing network model (QNM).

---

[*] This work was supported in part by the EC NoE Euro-FGI (NoE 028022) and in part by the EC IST project VITAL (IST-034284 STREP).

1 Introduction

The co-existence of several radio access technologies in today's mobile devices has provided end users with a wider choice, which is mainly driven by low cost and quality of service (QoS)[1] guarantees. Wireless Local Area Networks (WLANs) can provide high speed connections to access networks at a very cheap rate as compared to those of cellular access networks. These are the two major factors that motivate during handover an ongoing voice call session of a cellular access network to discover a WLAN access point. With WLAN coverage restricted to a relatively small area compared to that of a cellular network, the handover from WLAN to cellular network for an ongoing session is crucial for seamless service continuity whenever there is a weakening of WLAN signal strength.

The interoperability between WLAN and cellular network is made possible with the use of the mobility management mechanism associated with the IP Multimedia Subsystem (IMS) architecture, based on the recommendations of the 3rd Generation Partnership Project [1]. The IMS is an overlay system that serves the convergence of mobile, wireless and fixed broadband data networks into a common network architecture, where all types of data communications will be hosted in all IP environments using the infrastructure of the Session Initiation Protocol (SIP) to exchange messages to register users and set up or terminate multimedia sessions. As the processing of SIP messages requires the creation of states, starting timers and execution of filtering criteria, these procedures consume a significant amount of processing, queueing and transmission times with adverse implications on the performance of the overall IMS core architecture.

Mobility management is one of the main challenges facing IMS which is designated to deploy its services over a mixture of network access technologies such as wireless, mobile and fixed accesses. IMS should be able to deal sufficiently with the important issue of terminals mobility management, that nowadays is tackled at physical interface level resulting into the weakening of services robustness and increased probability of disrupted communication.

IMS is logically divided into two main communication domains, namely i) The data traffic domain (i.e., real time protocol packets consisting of audio, video and data) and ii) The SIP signalling traffic domain. IMS has entry and exit functional units or proxies, such as proxy P-CSCF, serving S-CSCF and interrogating I-CSCF call session control functions. These proxies exchange SIP messages to register users and setup/terminate multimedia sessions.

The VITAL IMS testbed [2] has a mobility management application server (MMAS) for voice continuity call with dual mode handset (DMH) roaming between global system for mobile communications (GSM) and WLAN-IMS. The active call is handed over from GSM to WLAN and vice versa. Handover in IMS specifically refers to voice call continuity from a SIP and a Real Time Protocol (RTP) based call moving from WLAN-IMS to cellular, or vice-versa. GSM-WLAN-IMS handover denotes the ability to change the access network from GSM to WLAN-IMS and vice versa, while a voice call is ongoing. This

---

[1] A list of all the acronyms used in this tutorial paper can be seen in Appendix I.

voice continuity call (VCC) feature applies to users with a GSM-IMS subscription and DMH. It is performed by the MMAS and the client on the DMH.

In this VITAL IMS architecture, the MMAS acts as back to back user agent (B2BUA). Whenever a handover is initiated for an existing call, the DMH first performs a new call that is routed to the MMAS. Then the MMAS operating as B2BUA connects the new call leg with the existing call by using a SIP Re-Invite [3]. Finally, the MMAS deletes the old call leg that is no longer used. The details of the handover concept are described below based on four specific handover scenarios:

- An IMS-IMS call is ongoing. Then the A-Party performs an IMS originating handover: IMS → GSM. The result is a GSM-IMS call.
- An IMS-IMS call is ongoing. Then the B-Party performs an IMS terminating handover: IMS→GSM. The result is an IMS-GSM call.
- A GSM-IMS call is ongoing. Then the A-Party performs an IMS originating handover: GSM→IMS. The result is an IMS-IMS call.
- An IMS-GSM call is ongoing. Then the B-Party performs an IMS terminating handover: GSM'IMS. The result is an IMS-IMS call.

In all handover scenarios, the following basic rules apply:

- On the IMS terminating side (for UE-B), the MMAS is the last IMS application server in the path.
- On the IMS originating side (for UE-A), the MMAS is the first IMS application server in the path.
- The voice application server (VAS) is invoked between originating MMAS and terminating MMAS.
- In the case of UE-A handover, the MMAS on the originating side performs the handover control (e.g. breaks the call).
- In the case of UE-B handover, the MMAS on the terminating side performs the handover control (e.g., breaks the call).
- The MMAS and the VAS need to be included into the SIP signalling path at the call setup such that a handover can be performed during the call. The MMAS cannot be included into the SIP signalling path after the call setup.

The handover switch time is significantly felt when a cellular/WLAN signal either is substantially weakened or abruptly disappears. In this context, the handover switching time is mainly due to the control messages traversing through a control path in order to setup a new call leg without disconnecting an ongoing call session.

The VITAL mobility specification [2] shows that the control functions in the path process several SIP messages per single handover request. With SIP messages often exceeding more than 1000 bytes, they are likely to have an adverse impact on the overall performance of the IMS architecture. Therefore, performance modelling and analysis may guarantee efficient and smooth IMS architecture operations whilst providing QoS guarantees. Within this framework, the design and implementation of IMS architecture will be well planned if performance metrics, such as server utilization, throughput and response time, are predicted quickly and efficiently.

Most of the ongoing research in the IMS field focuses on improving the development, engineering and performance impact of SIP messages. Cortes et al [4] have shown that SIP messages face significant challenges in SIP servers and, hence, special attention should be devoted on the design handling and parsing of SIP messages functionalities. Batterman et al [5] proposed SIP messages prioritization method in SIP servers. It was shown that the prioritized SIP messages can be processed without a major delay. A SIP offload engine was introduced in [6], whereby the scheme transforms SIP messages in binary format at the front end in order to parse the binary in the SIP stack. Rajagopal and Devetsikiotis [7] have proposed a modelling methodology that uses real life workload characterization, queueing analysis and optimization. The proposed methodology gives a systematic way for the selection of system design parameters in order to guarantee network performance whilst maximizing the overall system utility. Moreover, studies relating to the registration and session setup for accessing application servers in IMS [8] and the performance modelling and evaluation of handover mechanisms in IMS [9], respectively, were shown to specifically comply with the operational characteristics of an open central server queueing network model (QNM) [10]. Finally, Forte and Schulzrinne [11] introduced a new compression mechanism, based on the concept of templates that may be used in wireless networks by cellular operators. Such mechanism makes it possible to achieve the delay requirements of most time-critical applications, such as push to talk over cellular (PoC) in IMS.

This tutorial paper has its roots in the IMS performance evaluation studies reported in [8,9] relating to the VITAL network architecture [2] and the analytic works on the entropy maximisation and open GE-type QNMs with arbitrary configuration (c.f., [12,13]. More specifically, it introduces a mobility management mechanism for an IMS testbed, which was implemented by Nokia-Siemens as part of the deliverables of the VITAL project [2]. It also studies and evaluates the bursty traffic flows of SIP messages during a handover process involving WLAN and GSM access networks. Moreover, it models the functional units and application servers of an IMS architecture with SIP messages as an open (QNM) with finite capacity, generalised exponential (GE) interarrival times, head-of-line priorities and complete buffer sharing management scheme under a repetitive service blocking with random destination (RS-RD) (c.f., [13]). Finally, typical numerical experiments are included to validate the credibility of the analytic ME results against those devised by discrete event simulation and assess the performance impact of traffic flows of SIP messages on the functional IMS units during handover between different IMS access networks.

Note that this work assumes that all user equipments (UEs) are subscribers of IMS and associated cellular networks whilst the terms user, client, DMH and UE are used interchangeably. Moreover, it deals with the first scenario only, i.e., an IMS-IMS call is ongoing. Then the A-Party performs an IMS originating handover IMS→GSM. The result is a GSM-IMS call.

The rest of the paper is organised as follows: Section 2 introduces GE-type distribution whilst Section 3 reviews the maximum entropy (ME) methodology as applied to the analysis of arbitrary open queueing network models (QNMs) with RS-RD blocking. Section 4 describes the message flows during the handover process. Section 5 devises a custom made open QNM of the functional units and application servers of an IMS network architecture during a handover session. Sections 6 explains the simulation setup used for the validation of the QNM and presents some typical numerical experiments, which validate the credibility of the ME performance metrics against simulation and also assess the performance impact of bursty GE-type traffic flows of SIP messages on the IMS architecture. Conclusions follow Section 7 and a list of acronyms used throughout the manuscript is placed in Appendix I.

2 The GE-Type Distribution

The GE-type distribution is a mixed interevent-time distribution of the form (c.f., Fig. 1)

$$F(t) = P(A \leq t) = 1 - \tau e^{-\sigma t}, \quad t \geq 0, \tag{1}$$

$$\tau = \frac{2}{C^2 + 1} \tag{2}$$

$$\sigma = \tau \nu \tag{3}$$

where A is a mixed-time random variable (rv) of the interevent-time and $(1/\nu, C^2)$ are the mean and squared coefficient of variation (SCV) of rv A (c.f., [12]- [13]). The GE-type distribution is versatile, possessing pseudo-memoryless properties which make the solution of many GE-type queueing systems and networks analytically tractable.

For $C^2 > 1$, the GE is an extremal case of the family of Hyperexponential-2 (H_2) distributions with the same (ν, C^2) having a corresponding counting process equivalent to a compound Poisson process (CPP) with parameter $2\nu/(C^2 + 1)$

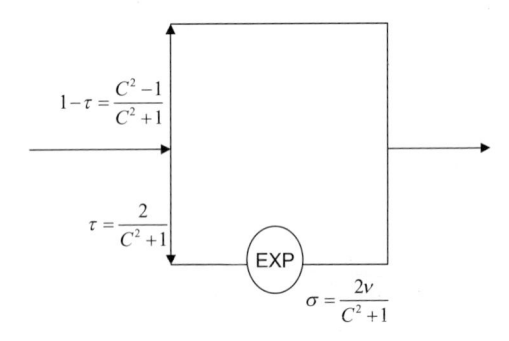

Fig. 1. The GE distribution with parameters τ and $\sigma (0 \leq \tau \leq 1)$

and geometrically distributed bulk sizes with mean, $(1+C^2)/2$ and SCV, $(C^2-1)/(C^2+1)$. The CPP is expressed by

$$P(N_{cp}=n) = \begin{cases} \sum_{i=1}^{n} \frac{\sigma^i}{i!} e^{-\sigma} \binom{n-1}{i-1} \tau^i (1-\tau)^{n-i}, & n \geq 1 \\ e^{-\sigma}, & n = 0 \end{cases} \quad (4)$$

where N_{cp} is a CPP rv of the number of events per unit time corresponding to a stationary GE-type interevent rv.

The choice of the GE distribution is further motivated by the fact that measurements of actual interarrival or service times may be generally limited and so only few parameters can be computed reliably. Typically, only the mean and variance may be relied upon, and thus, a choice of a distribution which implies least bias (i.e., introduction of arbitrary and therefore, false assumptions) is that of GE-type distribution.

3 Entropy Maximization and QNMs RS-RD Blocking: A Review

This section presents a review of an extended product-form approximation and a related queue-by-queue decomposition algorithm, based on the principle of ME, for open QNMs with arbitrary topology and RS-RD blocking [13]).

3.1 A Product Form Approximation

Consider an arbitrary open QNM at equilibrium with M single GE-type server queues, R $(R > 1)$ distinct HOL priority classes (indexed from 1 to R in descending order of priority), GE-type external inter-arrival times, random routing with class switching, CBS buffer management scheme and RS-RD blocking. Each queueing station k $(k = 1, 2, \ldots, M)$ is assumed to be modelled by a building block GE/GE/1/N_k/HOL/CBS queue k with finite capacity N_k ($N_k \geq 1$).

Notation

For each queue k $(k = 1, 2, \ldots, M)$ and job class i $(i = 1, 2, \ldots, R)$, let

λ_{ki}, C^2_{aki} be the mean rate and SCV of the overall actual inter-arrival process of class i jobs at queue k, respectively,

μ_{ki}, C^2_{ski} be the mean rate and SCV of the actual service process of class i jobs at queue k, respectively,

$n_{ki}(0 \leq n_{ki} \leq N_k)$ be the number of jobs of class i at queue k waiting and receiving service,

$\mathbf{n}_k = (n_{k1}, n_{k2}, \ldots n_{km})$ be the aggregate joint state the network,

π_{ki} be the blocking probability that a completer from any queue m, $m \neq k$ of class i is blocked by queue k,

π_{cki} be the blocking probability that a completer of class i will be blocked by a downstream queue,

$\{a_{kio}, a_{kimj}\}$ be the transition probabilities (first order Markov chain) that a class i job transmitted from queue k leaves the network or attempts to join queue m as a class j job, respectively,

$\{\lambda_{0ki}, C^2_{a\,0ki}\}$ be the mean arrival rate and SCV of the actual external inter-arrival process of class i jobs at queue k, respectively,

$\{\lambda_{kimj}, \hat{\lambda}_{kimj}\}$ and $\{\hat{C}^2_{a\,kimj}, C^2_{a\,kimj}\}$ be the mean arrival rates and SCVs of the actual and effective, respectively, inter-arrival processes of class i jobs transmitted from queue k to queue m as class j jobs,

$\{\pi_{kimj}\}$ be the blocking probabilities that a job of class i upon its service completion from queue k will be blocked by queue m, as class j,

$\{\pi_{0ki}\}$ be the blocking probabilities that an external arrival of class i is blocked by queue k.

A credible universal product-form ME approximation of the joint state probability $\{P(\mathbf{n}), \mathbf{n} = (\mathbf{n}_1, \mathbf{n}_2, \ldots \mathbf{n}_M)\}$, subject to normalization and marginal (per queue and class) constraints of server utilization, busy state probability, mean queue length and full buffer state probability (when a class i job is in service at queue k) can be established (c.f., [13]), namely

$$P(\mathbf{n}) = \prod_{k=1}^{M} P_k(\mathbf{n}_k), \qquad (5)$$

where $\{P_k(\mathbf{n}_k), k = 1, 2, \ldots, M\}$ are the marginal (class) ME state probabilities at queue k.

3.2 ME Queue-by-Queue Decomposition Algorithm

A ME queue-by-queue decomposition algorithm for the approximate analysis of the aforementioned arbitrary open QNMs is summarized in Algorithm 1.

A pictorial presentation of the flow streams and queue-by-queue decomposition can be seen in Fig. 2.

Remarks

– The GE-type distribution is used to approximate the effective inter-arrival and inter-departure time processes for each class $i (i = 1, 2, \ldots, R)$ at each queue $k (k = 1, 2, \ldots, M)$ of the network. The algorithm incorporates a feedback correction of the original service parameters $\{\mu_{ki}, C^2_{ski}, \forall\, k, i\}$ in order to mitigate the strong underlying assumption that arrival streams per class within the network can be modelled via renewal CPPs. Note that, under the RS-RD blocking mechanism, the ME algorithm utilizes stochastic closed-form expressions for the calculation of the effective service time parameters $\{\hat{\mu}_{ki}, \hat{C}^2_{ski}\}$ (c.f., [13]). Since RS-RD blocking imposes a dependence

Algorithm 1. The ME Decomposition Algorithm

Input Data

- $M, R,$
- $\{N_k, \lambda_{0ki}, C^2_{a\,0ki}, \mu_{ki}, C^2_{s\,ki}, a_{kimj}\}$ $k = 1, 2, \ldots, M, m = 0, 1, \ldots, M,$
  $i, j = 1, 2, \ldots, R.$

Begin

Step 1 Feedback correction

Step 2 Initialize π_{0ki} & π_{kimj} to any value in $(0,1)$, $\forall k, m = 1, 2, \ldots, M$ and $\forall i, j = 1, 2, \ldots, R$; Set $C^2_{dki} = 1, \forall k, i$;

Step 3 Solve the system of the non-linear equations of blocking probabilities $\{\pi_{0ki}, \pi_{kimj}, \forall k, m, i, j\}$;

  Step 3.1 For each censored GE/GE/1/N/HOL/CBS queueing station $k, k = 1, \ldots, M$ under RS blocking, calculate the effective flow transition probabilities $\{\hat{a}_{kimj}, \forall k, m, i, j\}$;

  Step 3.2 Calculate effective inter-arrival time message flow balance equations for $\{\hat{\lambda}_{0ki}, \hat{\lambda}_{ki}, \forall k, i\}$;

  Step 3.3 Calculate the effective service-time parameters, $\{\hat{\mu}_{ki}, \hat{C}^2_{s\,ki}, \forall k, i\}$ under RS-RD blocking mechanism;

  Step 3.4 Calculate the overall GE-type inter-arrival-time parameters, $\{\lambda_{ki}, C^2_{a\,ki}, \forall k, i\}$;

  Step 3.5 Obtain new values for $\{\pi_{0ki}\,\pi_{mjki}, \forall k, i\}$, by applying Newton Raphson method;

Step 4 Calculate GE-Type inter-departure parameters $\{\lambda_{dki}, C^2_{d\,ki}, \forall k, i\}$;

Step 5 Obtain a new value for the overall inter-arrival-time SCVs, $\{C^2_{a\,ki} \forall, k, i\}$;

Step 6 Return to Step 3 until convergence of $\{C^2_{a\,ki}, \forall\,k, i\}$;

Step 7 Obtain GE-type performance metrics of interest.

End

relationship on the actual routing of jobs from one queueing station to another, it is necessary to create and adopt an effective transition probability matrix $\hat{A} = (\hat{\alpha}_{ki0}, \hat{\alpha}_{kimj})$ in the solution process.

- The ME algorithm describes the computational process of solving the non-linear equations for job loss, $\{\pi_{0ki}\}$ and blocking, $\{\pi_{kimj}\}$, probabilities under GE-type flow formulae [12, 13] for the determination of the first two moments of merging, splitting and departing streams. The main computational cost of the algorithm is of $O\{kR^2M^2\}$, where k is the number of iterations in step 3 and R^2M^2 is the number of operations for inverting the associated Jacobian matrix of the system of non-linear equations $\{\pi_{0ki}, \pi_{kimj}, \forall\,k, m, i, j\}$ via a quasi-Newton numerical method.

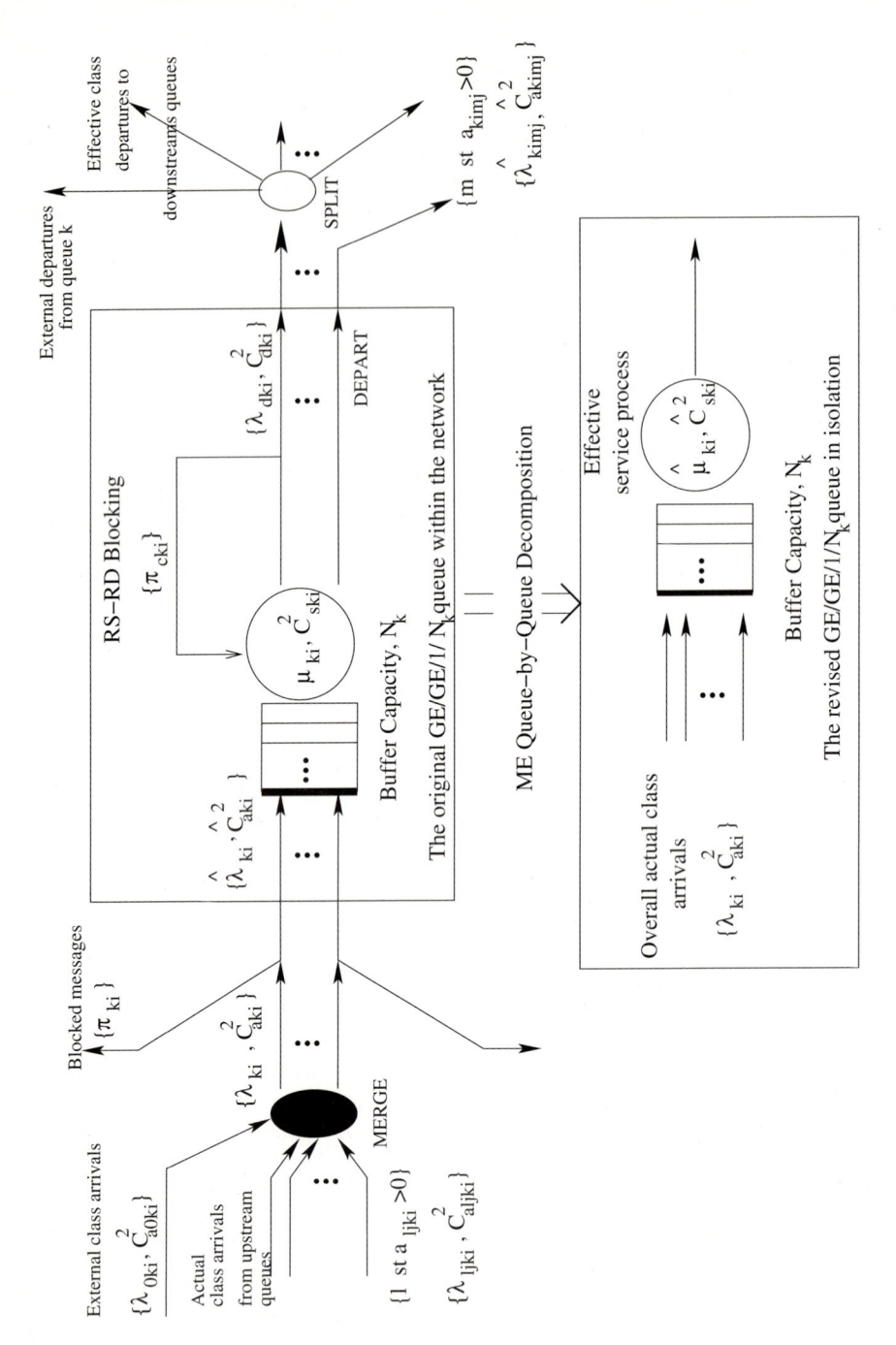

Fig. 2. Flow streams and ME queue-by-queue decomposition

4 Message Flows

The message flows considered in this paper consist of voice calls with the following profile: Whilst an IMS-IMS voice call is ongoing between GSM/IMS subscribers with DMH, the UE-A performs an IMS originating handover from WLAN/IMS to GSM/IMS (c.f., [2]). The outcome of these operations is a GSM-IMS call.

Before the handover occurs, a call is established from UE-A to UE-B. This is illustrated in numbers 1 to 6 of Fig. 3 The handover is initiated by the DMH roaming (c.f., number 7 in Fig. 3). This may be based on measurements of the WLAN signal strength in the DMH that triggers the handover procedure when the signal becomes weaker or abruptly disappears. The handover procedure may also be triggered manually (e.g., when pressing a button on the DMH).

The handover is performed by the following steps: The UE-A initiates a voice call via the GSM network. The call setup request is addressed to UE-A (the mobile basically makes a call to itself). The originating mobile switching centre (MSC) of the visited GSM network forwards the call Prefix-Routing based on the Calling-Party (A-Party) address to the media gateway control function (MGCF). When the MMAS receives the call setup request (i.e., the SIP Invite with Roaming Number Originating (RN-O) in the SIP Request-URI) in number 12 (c.f., Fig. 3), then a handover is required. This is because the MMAS already has a SIP call in IMS ongoing from UE-A and thus, it is aware that UE-A wants to roam from WLAN to GSM (while the call is ongoing). To actually perform the handover, the MMAS (acting as B2BUA) sends a SIP Re-Invite message to UE-B (n.b. this Re-Invite message is not shown in Fig. 3). The Re-Invite traverses along the SIP signalling path of the existing call and carries the session description protocol (SDP) with the IP address and port information used for the voice bearer of the call in the media gateway (MGW). The MMAS obtained this SDP from the MGCF with the SIP Invite message (c.f., numbers 10-12 of Fig. 3). When the SIP Re-Invite arrives at the UE-B, then the UE-B redirects the voice bearer traffic from UE-B to the MGW and answers the request with a 200-OK, which carries the SDP of UE-B. At the end of the procedure, the voice bearer is exchanged between the MGW and the UE-B. Finally, the MMAS closes the WLAN/IMS based call leg between the UE-A and the MMAS by sending a SIP Bye message via the S-CSCF to the UE-A. This completes the handover procedure. The new call now follows the signalling path indicated in Fig. 3.

The MMAS of the UE-A, the Voice AS and the MMAS of UE-B remain within the SIP signalling path after the handover. This is in order to i) control another handover of UE-A ii) control a handover of UE-B and iii) provide Supplementary Services (SS) based on VAS.

Note that the aforementioned handover mechanism is only applicable to a single session. Therefore, the DMH cannot have several voice calls active at the same time, which may then get successfully handed over from WLAN/IMS to

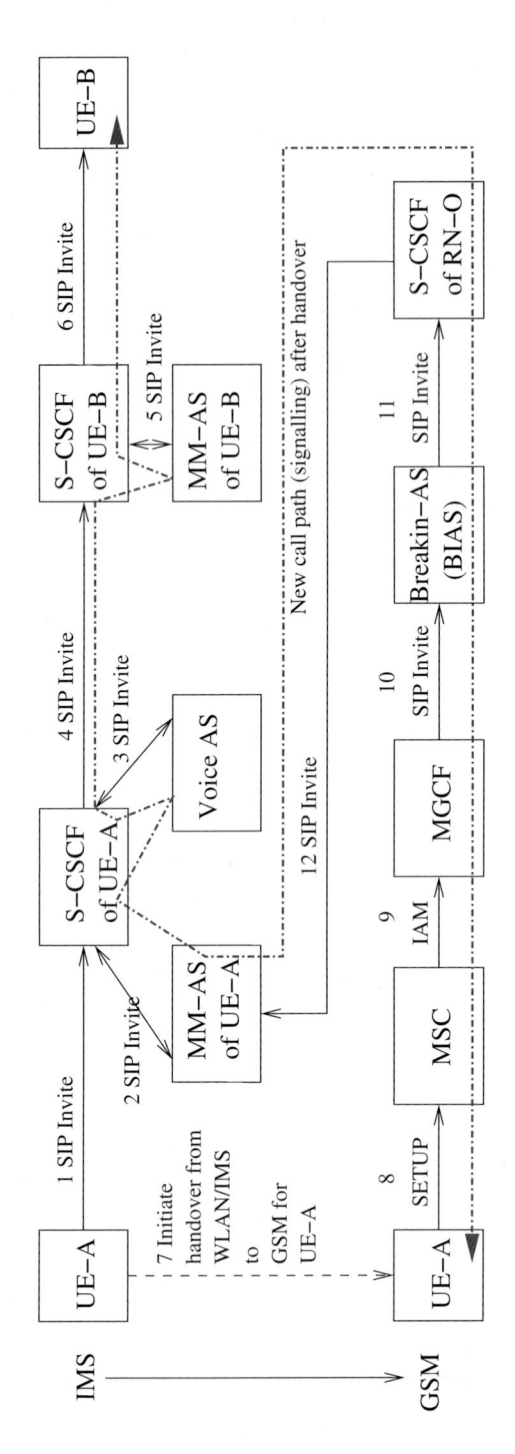

Fig. 3. An IMS originating handover from WLAN/IMS to GSM/IMS

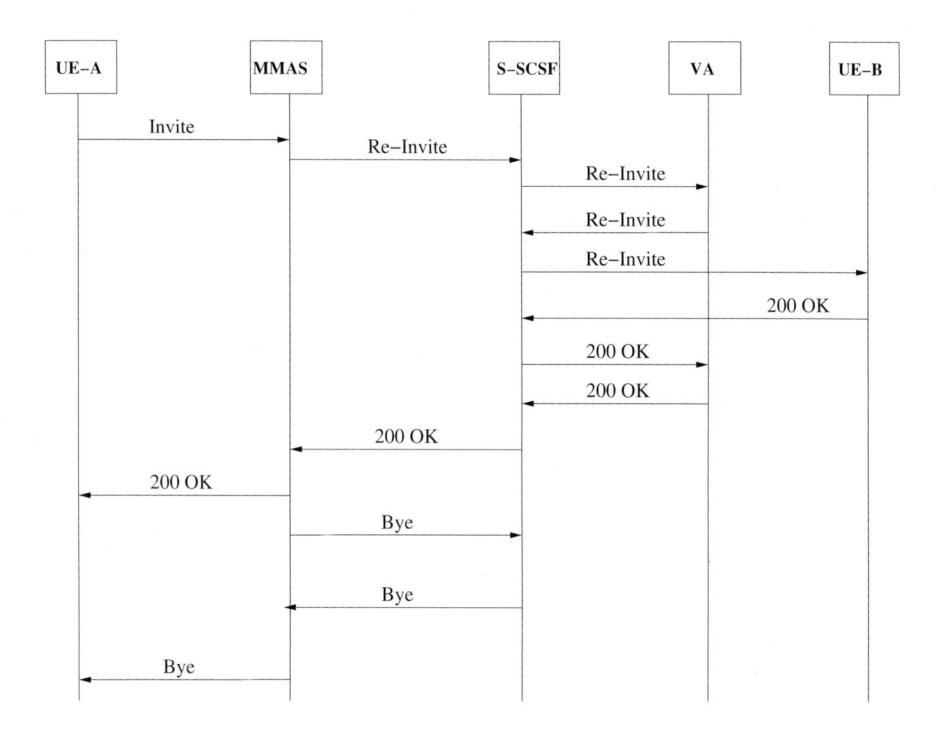

Fig. 4. Flows of SIP messages during WLAN/IMS to GSM/IMS handover mechanism

GSM/GSM and vice versa. This limitation was made to simplify the logic in the MMAS and DMH.

Fig. 4 depicts the flow chart of SIP messages during a handover mechanism from WLAN/IMS to GSM/IMS.

5 An Open QNM of the IMS Functional Units and Application Servers

A diagrammatic illustration of an open QNM of the IMS functional units (P-CSCF,S-CSCF and I-CSCF) and application servers (VAS and MMAS) during a handover process is displayed in Fig.5. There is an additional application server for future use that represents an IP Television (IPTV).

Note that $\hat{\mu}_{ki}$ and $\hat{C}^2_{ski}, \forall ki$ are the effective service time parameters, λ_{ki} and C^2_{aki} are overall actual class arrivals. In the context of this paper, $R = 2$, i.e., two distinct HOL priority classes are considered. The high priority is given to SIP Re-invite messages that trigger the handover process. The rest of the SIP messages and other traffic flows are given the low priority. Note that Real-Time Transport Protocol (RTP) traffic does not flow and pass through the IMS core network.

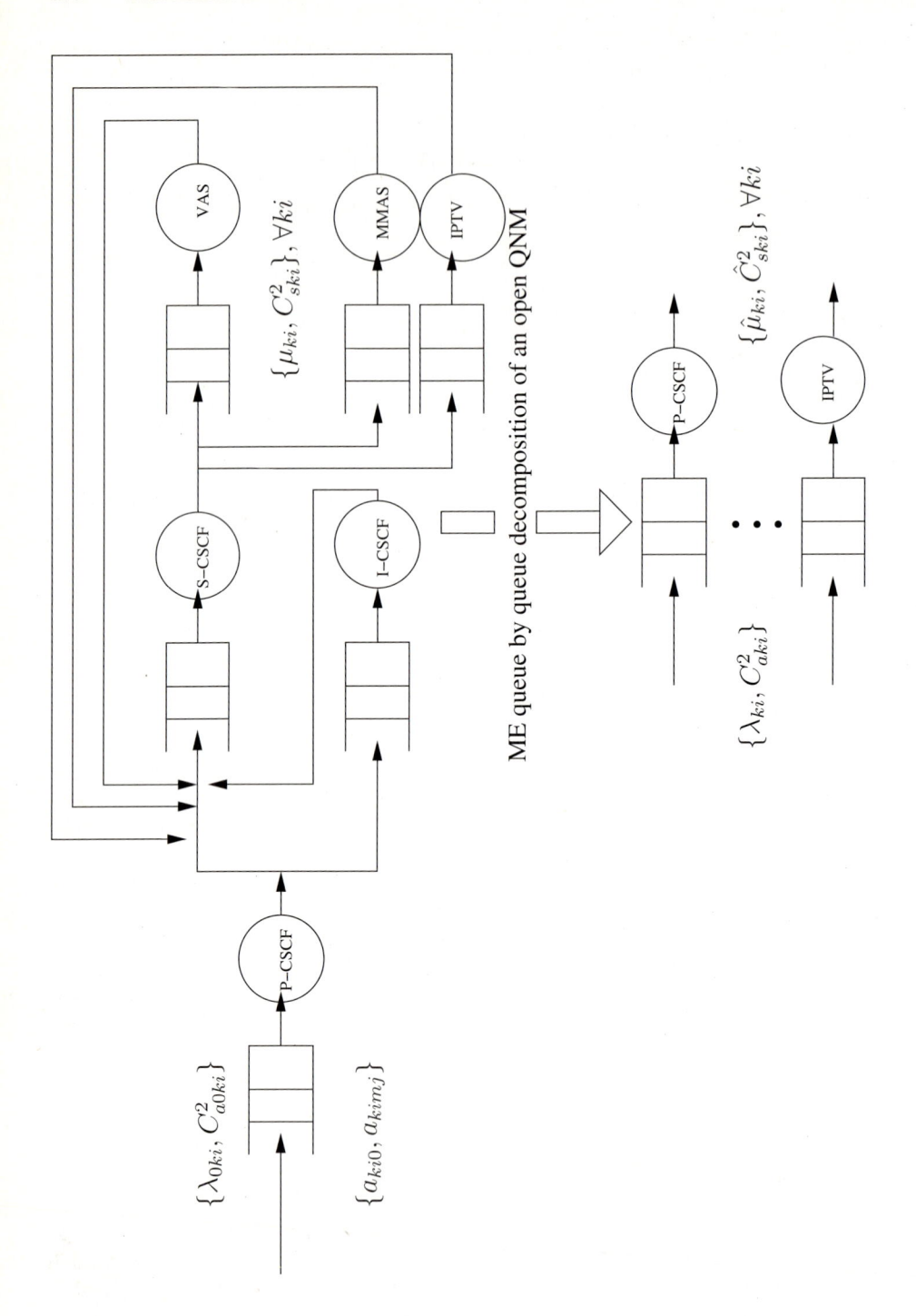

Fig. 5. An open QNM for IMS Functional Units and Application Servers

6 Simulation Setup and Numerical Results

This section presents numerical validation results based on the ME analysis and simulation of the proposed open QNM in Fig. 5. The input data used in the simulation setup in order to validate the open QNM for the IMS handover management mechanism against simulation are presented in Table 1 at 95% confidence intervals.

Table 1. Numerical Inputs and Results for the Open QNM of Fig. 5

Number of Classes=2					
$\lambda_{011}=0.7$	$\lambda_{012}=0.5$	$C^2_{a011}=5.0$	$C^2_{a012}=5.0$		
Buffer capacity at each link=5					
$\mu_{11}=2$	$C^2_{s11}=3$	$\mu_{12}=3$	$C^2_{s12}=5$	$\mu_{21}=1$	$C^2_{s21}=2$
$\mu_{22}=3$	$C^2_{s22}=4$	$\mu_{31}=2$	$C^2_{s31}=5$	$\mu_{32}=3$	$C^2_{s32}=4$
$\mu_{51}=1$	$C^2_{s51}=4$	$\mu_{52}=2$	$C^2_{s52}=5$	$\mu_{41}=2$	$C^2_{s41}=4$
$\mu_{42}=1$	$C^2_{s42}=2$	$\mu_{61}=3$	$C^2_{s61}=4$	$\mu_{62}=1$	$C^2_{s62}=5$
Transition Matrix					
$a_{2151}=0.5$	$a_{2252}=0.5$	$a_{2141}=0.5$	$a_{2242}=0.5$	$a_{1121}=0.5$	$a_{1222}=0.5$
$a_{3222}=0.5$	$a_{3121}=0.5$	$a_{4222}=0.5$	$a_{4121}=0.5$	$a_{1131}=0.5$	$a_{1232}=0.5$
$a_{2151}=0.5$	$a_{2252}=0.5$	$a_{5222}=0.5$	$a_{5121}=0.5$	$a_{6212}=0.5$	$a_{6111}=0.5$

The simulation was implemented using for communication purposes the Java Remote Method Invocation (RMI) package. The main program was developed using Java SDK 6 whilst the S-CSCF, P-CSCF, I-CSCF, MMAS, VAS and an extra application server for IPTV were simulated using Linux 2.6.9-42.0.2.ELsmpi686 Athlon i386 machine with 4GB of RAM. Other background traffic flows were introduced in the network consisting of file transfer protocols (FTPs) and hyper text transfer protocol (HTTP). It was assumed that 4 UE-Bs and 10 UE-As were available.

Without loss of generality, the comparative study is based on marginal performance metrics of mean response time and aggregate mean queue length of SIP Re-Invite handover messages by varying the SCV of the inter-arrival times. It can be seen from Fig. 6 that the mean response time of SIP messages during a handover process begins to deteriorate for increasing values of inter-arrival time SCV.

Fig. 7 illustrates the effect of varying inter-arrival SCV on the aggregate mean queue length of SIP messages. In Fig. 8, each functional unit of the IMS is compared to each other in terms of the corresponding mean response time of SIP Re-Invite messages against the the SCV of the inter-arrival traffic. From the analytical and simulation results, it can be seen that S-CSCF provides the most pessimistic mean response time.

It can be observed in Figs. 6-8 that the analytic ME results compare favourably with those of the simulation. Moreover, the variability of the SCV of the inter-arrival times has an adverse impact on the performance of the IMS functional units. Typically, the S-CSCF is the bottleneck in the IMS core network because it processes a large number of SIP messages.

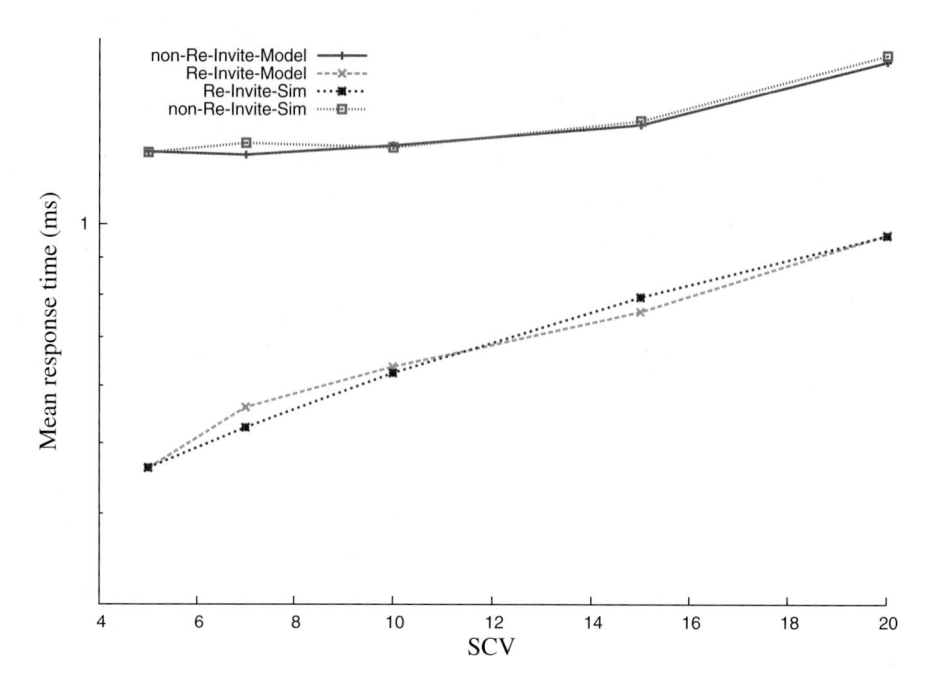

Fig. 6. The effect of traffic variability on the mean response time of SIP messages

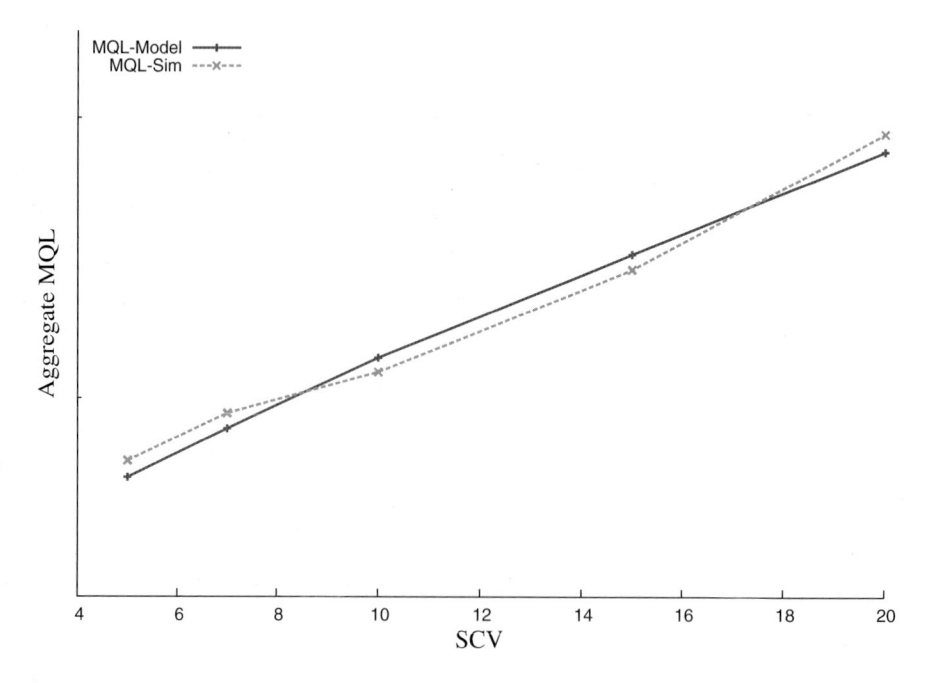

Fig. 7. The effect of traffic variability on the aggregate mean queue length

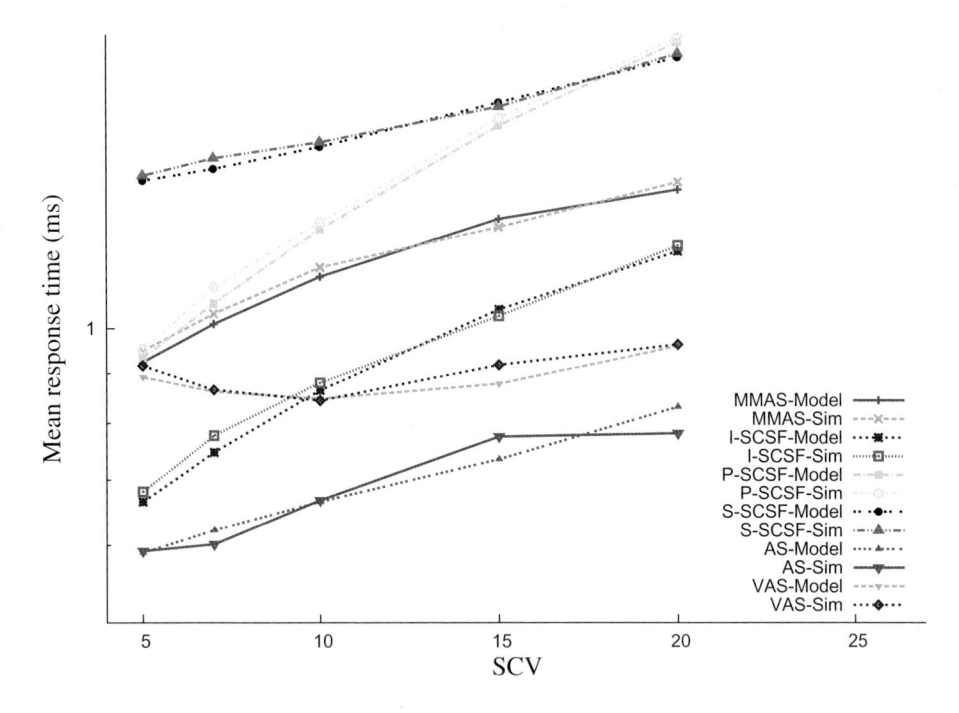

Fig. 8. The mean response time at each IMS functional unit

7 Conclusions

A mobility management mechanism was introduced for an IP multimedia subsystem (IMS) testbed implemented by Nokia-Siemens as part of the EU IST VITAL project [2]. The GE-type distribution [12] was used to characterise the bursty traffic flows of SIP messages during the handover process between the WLAN and GSM access networks. The functional units and application servers of the IMS architecture were presented in the context of an open QNM with HoL priority classes for the SIP messages and complete buffer sharing under RS-RD [13].

The quantitative analysis of the open QNM was based on the universal ME product-form approximation and related queue-by-queue decomposition algorithm as well as discrete event simulation. Typical numerical experiments showed that the analytic ME solutions were very comparable to the corresponding simulation results. Moreover, the performance prediction study revealed that the variability of bursty traffic described by the interarrival time SCV had an adverse impact on the performance of the IMS functional units. In particular, the S-CSCF proxy was identified as the bottleneck device of the IMS core network architecture processing a large number of SIP messages during handover.

References

1. 3GPP: Technical Specification Group Services and System Aspects, IP Multimedia Subsystem (IMS), Stage 2, TS 23.228, 3rd Generation Partnership Project (2006), Website, http://www.3gpp.org/specs/specs.htm
2. VITAL: Enabling Convergence of IP Multimedia Services Over Next Generation Networks Technology (2006), Website, http://www.ist-vital.eu
3. Rosenberg, J., Schulzrinne, H., Camarillo, G., Johnston, A., Peterson, J., Sparks, R., Handley, M., Schooler, E.: SIP: Session Initiation Protocol. RFC 3261 (2002)
4. Cortes, M., Ensor, R., Esteban, J.: On SIP Performance. Bell Labs Technical Journal 9(3), 155–172 (2004)
5. Batterman, H., Meeuwissen, E., Bemmel, J.: SIP Message Prioritization and its Application. Bell Labs Technical Journal 11(1), 21–36 (2006)
6. Zou, J., Xue, W., Liang, Z., Zhao, Y., Yang, B., Shao, L.: SIP Parsing Offload. In: Global Telecomminications Conference (Globecom), pp. 2774–2779 (2007)
7. Rajagopal, N., Devetsikiotis, M.: Modeling and Optimization for the Design of IMS Networks. In: 39th Annual Simulation Symposium, pp. 34–41 (2006)
8. Mkwawa, I.M., Kouvatsos, D.D.: Performance Modelling and Evaluation of IP Mmultimedia Subsystems. In: HET-NETs 2008 International Working Conference on Performance Modelling and Evaluation of Heterogeneous Networks, pp. B07.1–B07.7 (2008)
9. Mkwawa, I.M., Kouvatsos, D.D.: Performance Modelling and Evaluation of Handover Mechanism in IP Multimedia Subsystems. In: ICSNC 2008: Proceedings of the 2008 Third International Conference on Systems and Networks Communications, pp. 223–228. IEEE Computer Society Press, Washington, DC, USA (2008)
10. Baskett, F.: Mathematical Models of Multiprogrammed Computer Systems. TSN-17, Computer Centre, The University of Texas, Austin, Texas (1971)
11. Forte, G., Schulzrinne, H.: Template-Based Signaling Compression for Push-To-Talk over Cellular (PoC). In: Schulzrinne, H., State, R., Niccolini, S. (eds.) IPTComm 2008. LNCS, vol. 5310, pp. 296–321. Springer, Heidelberg (2008)
12. Kouvatsos, D.D.: Entropy Maximization and Queueing Network Models. Annals of Operation Research 48, 63–126 (1994)
13. Kouvatsos, D.D., Awan, I.: Entropy Maximisation and Open Queueing Networks with Priorities and Blocking. Perform. Eval. 51(2-4), 191–227 (2003)

Appendix I   Table of Acronyms

3GPP	3rd Generation Partnership Project (3GPP)
AS	Application Server
B2BUA	Back to Back User Agent
DMH	Dual Mode Handset
GE	Generalised Exponential
GSM	Global System for Mobile communications
HoL	Head of Line
I-CSCF	Interrogating Call Session Control Function
IMS	IP Multimedia Subsystem
IP	Internet Protocol
IPTV	IP Television
ME	Maximum Entropy
MGW	Media Gateway
MMAS	Mobility Management Application Server
MGCF	Media Gateway Control Function
MSC	Mobile Switching Center
P-CSCF	Proxy Call Session Control Function
PoC	Push to talk over Cellular
QNM	Queueing Network Model
RN-O	Roaming Number-Originating
RN-T	Roaming Number-Terminating
RS-RD	Repetitive Service blocking with Random Destination
RTP	Real-time Transport Protocol
S-CSCF	Serving Call Session Control Function
SCV	Squared Coefficient of Variation
SDP	Session Description Protocol
SIP	Session Initiation Protocol
SS	Supplementary Services
UE	User Equipment
URI	Uniform Resource Identifier
VAS	Voice Application Server
VCC	Voice Continuity Call
WLAN	Wireless Local Area Network

Generalized QBD Processes, Spectral Expansion and Performance Modeling Applications

Tien Van Do[1] and Ram Chakka[2]

[1] Department of Telecommunications
Budapest University of Technology and Economics
H-1117, Magyar tudósok körútja 2., Budapest, Hungary
do@hit.bme.hu

[2] Meerut Institute of Engineering and Technology (MIET),
Meerut, India 250005
ramchakka@yahoo.com

Abstract. This paper suggests new queuing models, in the Markovian framework, which can tackle the presence of burstiness in the traffic and autocorrelations among the inter-arrival times of packets in the performance evaluation of next generation networks. These models are essentially based on certain generalizations of the Quasi Birth-Death (QBD) processes. Efficient steady state solution of these new queuing models, along with some illustrative applications, is presented. The proposed models and their further evolutions have the potential to be useful tools for the performance evaluation of modern telecommunication networks.

1 Introduction

The concept of Quasi Birth-Death (QBD) processes, as a generalization of the classical birth and death processes (e. g. the $M/M/1$ queue) was first introduced in the late sixties by [55] and [26]. A QBD process is a Markov process on a two-dimensional lattice, finite in one dimension (finite or infinite in the other). A state is described by two integer-valued random variables: the one in the finite dimension is the phase and the other is the level [37, 39, 41]. Transitions in a QBD process are possible within the same level or between adjacent levels. It is observed that QBD processes create a useful framework for the performability analysis pertaining to many problems occurring in telecommunications and computer networks [4, 8, 18, 20, 23, 36, 40, 42, 45, 56, 57].

In a QBD process, if the nonzero jumps in levels are not accompanied with changes in a phase, then these processes can be known as Markov modulated Birth and Death processes. The large or infinite number of states involved makes the solution of these models nontrivial. There are several methods of solving these models, either the whole class of models or some of the subclasses.

Seelen has analyzed a Ph/Ph/c queue in the QBD frame work [46]. Seelen's method is an approximate one where the Markov chain is first truncated to a finite state Markov chain. By exploiting the structure an efficient iterative solution algorithm can be applied. The second method is to reduce the infinite-states problem to a linear equation involving vector generating function and some unknown probabilities. The latter are then determined with the aid of the singularities of the coefficient matrix. A comprehensive treatment of that approach, in the context of a discretetime process with a general M/G/1 type structure, is presented in [30]. The third way of solving these models is the well known matrix geometric method, first proposed by Evans [26, 41]. In this method a nonlinear matrix equation is first formed from the system parameters and the minimal nonnegative solution R of this equation is computed by an iterative method. The invariant vector is then expressed in terms of the powers of R. Neuts claims this method has a probabilistic interpretation for the steps in computation. However, this method suffers from the fact that there is no way of knowing how many iterations are needed to compute R to a given accuracy. It can also be shown that for certain parameter values the computation requirements are uncertain and formidably large. The fourth method is known as spectral expansion method [5, 6, 39]. It is based on expressing the invariant vector of the process in terms of eigenvalues and left eigenvectors of a certain matrix polynomial. The generating function approach and the spectral expansion method are closely related. However, the latter computes steady state probabilities directly using an algebraic expansion while the former provides them through a transform.

It is confirmed by a number of works that the spectral expansion method is better than the matrix geometric method in a number of aspects [5, 32, 33, 39]. It is observed that the spectral expansion method is proved to be a mature technique for the performance analysis of various problems [5, 6, 7, 8, 9, 10, 11, 12, 13, 14, 15, 16, 17, 20, 21, 22, 23, 24, 25, 27, 28, 31, 32, 38, 39, 49, 50, 51, 53, 54, 52, 58].

Due to heterogeneous requirements concerning network technology and services that next generation networks (NGN) [34] are required to support, the issue of modeling the packet traffic and nodes in modern communication networks has become complicated because of the existence of burstiness (time varying arrival or service rates, arrivals or services of packets in *batches*) and important correlations among inter-arrival times [44]. In addition the traffic arriving at a node is often the superposition of traffic from a number of sources (homogeneous or heterogeneous), which further complicates the analysis of the system. Self-similar traffic models such as the FBM [43] can represent both burstiness and auto-correlations, but they are not analytically tractable in a queuing context.

The CPP[1], defined in [19] and employed in [14, 29] and the $\sum_{k=1}^{K} CPP_k$ (superposition of K independent CPPs) traffic models often give a good representation of the burstiness (batch size distribution) of the traffic from one or more sources (along with mathematical tractability), but not the auto-correlations of the inter-arrival times (of batches) observed in real traffic. The usefulness and applicability of these models has been validated by measurements for example in [8, 23].

Recently, we have proposed two new queuing models, the MM $\sum_{k=1}^{K} CPP_k$/GE/c/L G-queue [10] with homogenous servers (the *Sigma* queue) and

---

[1] Throughout this paper when we refer to CPP we mean the compound Poisson process with independent and geometric batch-sizes, this is for convenience of referring.

with heterogeneous servers (the *HetSigma* queue [8]). We have also developed some transformations which, when applied to the steady state balance equations, result in QBD-M type computable form. These models do provide a large flexibility to accommodate geometric as well as non-geometric batch sizes in both arrivals and services, and hence are capable of emerging as generalized Markovian node models. In these queues, the GE service time distribution is widely used, which is motivated by the fact that only the mean and variance may be computed reliably from the measurements. Therefore a choice of the distributions which implies least bias is that of the GE distribution [35,47]. The parameters of the GE distribution are estimated from the real traffic trace in our numerical study. Moreover, the accommodation of *large* or *unbounded* batch-sizes, with efficient steady state queuing solutions, is a definitive advantage of our models besides the ability to accommodate negative customers.

In this paper, we present the spectral expansion methodology for the QBM-M queue. Then we give the short overview of the *HetSigma* queuing model, efficient computation of its steady state performance, possible extensions, along with some non-trivial applications to the performance evaluation of some problems in telecommunications networks.

The rest of the paper is organized as follows. Section 2 gives a brief overview of the important stochastic processes and distributions used in the new queuing model. It also presents the spectral expansion for QBD-M processes. The *HetSigma* queue is described in Section 3. The required steady state solution and an application are given in Sections 4 and 5, respectively. Some extensions are discussed in Section 6. Future directions to this research and conclusions are dealt in Section 7.

2 A Brief Overview of the Stochastic Processes and Distributions Involved

2.1 The QBD-M Process

The QBD-M (Quasi Simultaneous-Bounded-Multiple Births and Simultaneous-Bounded-Multiple Deaths) process is a two-dimensional Markov process on a finite or semi-infinite lattice strip [5,6,39]. The state at any time t is denoted by two integer valued random variables, $I(t)$ and $J(t)$. $I(t)$ takes a finite set of values (*phases*) $\{1, 2, \ldots, N\}$, and $J(t)$ takes a set of values (*levels*) $\{0, 1, \ldots, L\}$, where L can be finite or infinite. We assume that the Markov process, Y if L is infinite, and $\overline{Y}$ if L is finite, is denoted by $\{[I(t), J(t)]; t \leq 0\}$ and is irreducible.

The possible transitions underlying this Markov process are given by the following transition rate matrices, each of size $N \times N$:

A_j : purely lateral (phase) transitions – $A_j(i, k)$ is the transition rate from state (i, j) to state (k, j) $(i \neq k; 0 \leq i, k \leq N; j = 0, 1, \ldots, L)$.

$B_{j,j+s}$: bounded s–step upward transitions – $B_{j,j+s}(i, k)$ is the transition rate from the state (i, j) to state $(k, j + s)$ $(0 \leq i, k \leq N; 1 \leq s \leq y_1; j = 0, 1, \ldots, L - 1)$. $B_{j,j+s} = 0$ if $j + s > L$.

$C_{j,j-s}$: bounded s−step downward transitions − $C_{j,j-s}(i,k)$ is the transition rate from state (i,j) to state $(k, j-s)$ $(0 \leq i, k \leq N; 1 \leq s \leq y_2; j = 1, 2, \ldots, L)$. $C_{j,j-s} = 0$ if $j - s < 0$.

There is a threshold T such that, $A_j = A$ $(j \geq T), B_{j,j+s} = B_s$ $(j \geq T - y_1), C_{j,j-s} = C_s$ $(j \geq T)$, thus these matrices are independent of j.

The spectral expansion solution of the QBD-M process is based on the observation that the steady state balance equations can be written in the form

$$\sum_{k=0}^{y} \mathbf{v}_{j+k} Q_k = 0 \qquad (T - y_1 \leq j \leq L - y - 1; \; y = y_1 + y_2), \qquad (1)$$

where the coefficient matrices Q_k can be obtained from system parameters, following the methodology in [6, 8, 14].

Therefore, when L is finite, the probability invariant vector $\mathbf{v}_j$ is given by [5, 6, 39]

$$\mathbf{v}_j = \sum_{l=1}^{y_1 N} a_l \psi_l \lambda_l^{j-T+y_1} + \sum_{l=1}^{y_2 N} b_l \gamma_l \xi_l^{L-1-j} \qquad (T - y_1 \leq j \leq L - 1), \qquad (2)$$

where (λ_k, ψ_k) are the left-eigenvalue and eigenvector pairs of the characteristic, quadratic matrix-polynomial $Q(l) = \sum_{k=0}^{y} Q_k \lambda^k$ pertaining to the Markov process Y, and (ξ_k, γ_k) are the left-eigenvalue and eigenvector pairs of the characteristic, quadratic matrix polynomial $\overline{Q}(l) = \sum_{k=0}^{y} Q_{y-k} \lambda^k$. a_l and b_l are the constants, which can be determined with the aid of the state-dependent balance equations [5].

When L is infinite (unbounded) and the ergodicity condition is satisfied, then the above solution reduces to

$$\mathbf{v}_j = \sum_{l=1}^{N y_1} a_l \psi_l \lambda_l^{j-T+y_1} \quad (j = T - y_1, T - y_1 + 1, \ldots). \qquad (3)$$

2.2 The QBD-U Process

In a QBD-M process, if y_1 or y_2 is *unbounded*, it becomes a QBD-U process (Quasi Simultaneous-Unbounded-Multiple Births and Simultaneous-Unbounded-Multiple Deaths). QBD-U processes are very useful in performance modeling of NGN as we shall see in the rest of this paper. Only in certain special cases of the QBD-U processes, there have been efficient, exact steady state solution methods [14, 15].

2.3 The Generalized Exponential (GE) Distribution

Excellent treatment of the GE distribution, its usefulness and applications are available in [35, 47]. The GE distribution is versatile, possessing pseudo-memoryless properties. This makes the solution of many queuing systems and networks employing GE distribution analytically tractable [35]. The GE distribution is given in the following form:

$$F(t) = P(W \leq t) = 1 - (1-\phi)e^{-\mu t} \quad (t \geq 0), \tag{4}$$

where W is the GE random variable with parameters μ, ϕ. Thus, the GE parameter estimation can be obtained by $1/\nu$, the mean, and C_{coeff}^2, the squared coefficient of variation of the inter-event time of the sample as

$$1 - \phi = 2/(C_{coeff}^2 + 1) \;;\; \mu = \nu(1-\phi). \tag{5}$$

For $C_{coeff}^2 > 1$, the GE model is a mixed-type probability distribution, e.g. Hyperexponential-2 having the same mean and coefficient of variation, and with one of the two phases having zero service time, or, a bulk type distribution with an underlying counting process equivalent to a Batch (or Bulk) Poisson Process (BPP) with batch-arrival rate μ and geometrically distributed batch size with mean $1/(1-\phi)$ and SCV $(C_{coeff}^2 - 1)/(1 + C_{coeff}^2)$ (c.f. [47]). It can be observed that there is an infinite family of BPPs with the same GE-type inter-event time distribution. It is shown that, among them, the BPP with geometrically distributed bulk sizes (referred as the CPP through this paper) is the only one that constitutes a renewal process (the zero inter-event times within a bulk/batch are *independent* if the bulk size distribution is geometric [35]).

The choice of the GE distribution is often motivated by the fact that measurements of actual inter-arrival or service times may be generally limited in accuracy, and so only a few parameters (for example the mean and variance) can be computed reliably. Typically, when only the mean and variance can be relied upon, a choice of a distribution which implies least bias (bias means, introduction of arbitrary and false assumptions) is that of GE-type distribution [35, 47].

2.4 The CPP, MMCPP and MM$\sum_{k=1}^{K} CPP_k$ Processes

Though BPP and CPP are synonymous, when we refer to a CPP in this paper, we actually refer to a CPP with independent and geometric batch-sizes, for convenience of referring.

When the parameters of a CPP are modulated by an external Markov chain, we obtain the MMCPP (Markov modulated CPP) process. Let the generator matrix of the modulating CTMC (continuous time Markov chain) be given by,

$$Q = \begin{bmatrix} -q_1 & q_{1,2} & \cdots & q_{1,N} \\ q_{2,1} & -q_2 & \cdots & q_{2,N} \\ \vdots & \vdots & \ddots & \vdots \\ q_{N,1} & q_{N,2} & \cdots & -q_N \end{bmatrix},$$

where $q_{i,k}(i \neq k)$ is the instantaneous transition rate from phase i to phase k, $q_{i,i} = 0 \; \forall i$, and $q_i = \sum_{j=1}^{N} q_{i,j}$, $(i = 1, \ldots, N)$. Let $\boldsymbol{r} = (r_1, r_2, \ldots, r_N)$ be the vector of equilibrium probabilities of the modulating phases. Then, $\boldsymbol{r}$ is uniquely determined by the equations, $\boldsymbol{r}Q = 0$; $\boldsymbol{r}\boldsymbol{e}_N = 1$, where $\boldsymbol{e}_N$ stands for the column vector with N elements, each of which is unity. In the MMCPP arrival process, the inter-arrival time distribution of customers, in phase i, is GE with parameters (σ_i, θ_i).

The MM$\sum_{k=1}^{K} CPP_k$ is obtained by Markov modulation of the parameters of the superposition of K independent CPP streams. That is, all the K independent CPPs are jointly Markov modulated.

3 The *HetSigma* Queuing Model

We introduce the terminology *HetSigma* to denote the MM$\sum_{k=1}^{K} CPP_k$/GE/c/L G-queue with *heterogenous* servers. In the queue, the effective customer arrival process is MM $\sum_{k=1}^{K} CPP_k$ in which the superposed K CPPs are independent and their parameters are jointly Markov modulated. The same modulating process also modulates the parameters of the service time and those of the CPP of the negative customers, as we shall see below in detail.

3.1 The Arrival Process

The arrival and service processes are modulated by the same continuous time, irreducible Markov phase process with N states. Let Q be the generator matrix of this process. The arrival process, in any given modulating phase, is the superposition of K independent CPP arrival streams of customers (or packets, in packet-switched networks) and an independent CPP of *negative* customers. Customers of different arrival streams are not distinguishable. The parameters of the GE inter-arrival time distribution of the kth ($1 \leq k \leq K$) customer arrival stream, in modulating phase i, are $(\sigma_{i,k}, \theta_{i,k})$, and (ρ_i, δ_i) are those of the negative customers. That is, the inter-arrival time probability distribution function is $1 - (1 - \theta_{i,k})e^{-\sigma_{i,k}t}$, in phase i, for the kth stream of customers, and $1 - (1 - \delta_i)e^{-\rho_i t}$ for the negative customers. Thus, in a given phase, all the $K + 1$ arrival *point*-processes are Compound Poisson, with batches arriving at each point having geometric size distribution. Specifically, in phase i, the probability that a batch is of size s is $(1 - \theta_{i,k})\theta_{i,k}^{s-1}$ for the kth stream of customers, and $(1 - \delta_i)\delta_i^{s-1}$ for the negative customers. Strictly during a given phase i, the effective arrival process is $\sum_{k=1}^{K} CPP_{i,k}$, where $CPP_{i,k}$ is the kth CPP arrival process in the modulating phase i.

Let $\sigma_{i,.}, \overline{\sigma_{i,.}}$ be the average arrival rate of customer batches and customers in phase i respectively. Let $\sigma, \overline{\sigma}$ be the overall average arrival rate of batches and customers respectively. Then, it can be written

$$\sigma_{i,.} = \sum_{k=1}^{K} \sigma_{i,k}, \quad \overline{\sigma_{i,.}} = \sum_{k=1}^{K} \frac{\sigma_{i,k}}{(1-\theta_{i,k})}, \quad \sigma = \sum_{i=1}^{N} \sigma_{i,.} r_i, \quad \overline{\sigma} = \sum_{i=1}^{N} \overline{\sigma_{i,.}} r_i \quad (6)$$

3.2 The GE Multi-server

The *HetSigma* queue is the extension of the *Sigma* queue in [10], where the service facility has c *heterogeneous* servers in parallel. The servers are numbered just as their service priorities, *i.e.* $1, 2, \ldots, c$, without loss of generality. The GE-distributed service time parameters of server n, in phase i, are $\mu_{i,n}, \phi_{i,n}$ ($n = 1, 2, \ldots, c$). A number of scheduling policies can be thought of. Though, in principle, a number of scheduling policies can indeed be modeled by following our methodology, the one that we have adopted in this paper, for illustration and detailed study, is as follows. A set of service priorities is chosen by giving each server a unique service priority, 1 is the highest and c is the lowest. This set can be chosen arbitrarily from the $c!$ different possible ways. However, the impact of choosing service priorities can be very high on the performance measures, whose study is not in the scope of this paper. The optimal allocation of service priorities can be an interesting research item for investigation.

The service discipline is FCFS (First Come First Scheduled, for service) and each server serves at most one positive customer at any given time. Customers, on their completion of service, leave the system. When the number of customers in the system, j, (including those in service if any) is $\geq c$, then only c customers are served with the rest $(j - c)$ waiting for service. When $j < c$, only the first j servers, (*i.e.*, servers numbered $1, 2, \ldots, j$), are occupied and the rest are idle. This is made possible by what is known as customer switching. Thus, when server n becomes idle, an awaiting customer would be taken up for service. If there is no awaiting customer, then a customer that is being served by the lowest possible priority server (*i.e.*, among servers $(c, c-1, \ldots, n+1)$) switches to server n. In such a switching, the (batch) service time is governed by either *resume or repeat with resampling*, thus preserving the Markovian property. The switching is instantaneous or the switching time is treated negligible. Negative customers neither wait in the queue, nor are served.

The operation of the GE server is similar to that described above in the *Sigma* case [10]. The batch size associated with a service completion is bounded by one more than the number of customers waiting to commence service at the departure instant. When $c \leq j < L + 1$, the maximum batch size at a departure instant obviously is $j - c + 1$, only one server being able to complete a service period at any one instant under the assumption of exponentially distributed batch-service times. Thus, in phase i, the probability that a departing batch is of size s can be shown as, $\sum_{n=1}^{c} \frac{\mu_{i,n}(1-\phi_{i,n})\phi_{i,n}^{s-1}}{\mu_{i.}}$ for $1 \leq s \leq j - c$ and $\sum_{n=1}^{c} \frac{\mu_{i,n}\phi_{i,n}^{j-c}}{\mu_{i.}}$ for $s = j - c + 1$, where $\mu_{i.} = \sum_{n=1}^{c} \mu_{i,n}$. However, when $1 \leq j \leq c$, the departing batch has size 1 since each customer is already engaged by a server and there are no customers waiting to commence service.

3.3 Negative Customer Semantics

A negative customer removes a positive customer in the queue, according to a specified *killing discipline*. A number of different killing disciplines are indeed possible, suitable in different contexts.

The RCE killing discipline. We consider here a variant of the RCE killing discipline (removal of customers from the end of the queue), where the most recent positive arrival is removed, but which does *not* allow a customer actually in service to be removed: a negative customer that arrives when there are no positive customers waiting to start service has no effect. We may say that customers in service are immune to negative customers or that the service itself is *immune servicing*. Such a killing discipline is suitable for modeling e.g. load balancing, where work is transferred from overloaded queues but never work, that is, actually in progress.

When a batch of negative customers of size l ($1 \leq l < j - c$) arrives, l positive customers are removed from the end of the queue leaving the remaining $j - l$ positive customers in the system. If $l \geq j - c \geq 1$, then $j - c$ positive customers are removed, leaving none waiting to commence service (queue length becomes c). If $j \leq c$, the negative arrivals have no effect since all customers are in service.

$\overline{\rho_i}$, the average arrival rate of negative customers in phase i and $\overline{\rho}$, the overall average arrival rate of negative customers are given by

$$\overline{\rho_i} = \frac{\rho_i}{1 - \delta_i} \; ; \quad \overline{\rho} = \sum_{i=1}^{N} r_i \overline{\rho_i}. \tag{7}$$

Other killing disciplines. Apart from the RCE with *immune servicing*, there are two other popular killing disciplines, the RCE-*inimmune servicing* and the RCH killing disciplines. The applicability of the killing disciplines rather depends on the situation and the purpose, and hence depending on these, many more killing disciplines are theoretically possible. Our methodology can easily be extended to many other killing disciplines also, this is explained briefly in this section.

The RCE-inimmune servicing - In this, the negative customer removes the most recent positive arrival regardless of whether it is in service or waiting. This is the traditional killing discipline suited to the modeling of killing signals in speculative parallelism. It can also be used to model cell losses caused by the arrival of a corrupted cell or one encountering a full buffer, when the preceding cells of a packet would be discarded.

The RCH discipline - Another popular killing discipline is the RCH (Removal of customers from the head of the queue) killing discipline. This is appropriate for modeling server breakdowns, where a customer in service will be lost for sure and may be also a portion of queue of waiting customers. The RCH killing discipline is already applied to the case of the MM CPP/GE/c/1 G-queue in [14].

3.4 Condition for Stability

When L is finite, the system is ergodic since the representing CTMC is irreducible. Otherwise, i.e. when the queuing capacity is unbounded, the overall average departure rate increases with the queue length, and its maximum (the overall average departure rate when the queue length tends to ∞) can be determined as,

$$\overline{\mu} = \sum_{n=1}^{c} \sum_{i=1}^{N} \frac{r_i \mu_{i,n}}{1 - \phi_{i,n}}. \tag{8}$$

Hence, the necessary and sufficient condition for the existence of steady state probabilities is, $\overline{\sigma} < \overline{\rho} + \overline{\mu}$.

3.5 The Markov Model

The state of the system at any time t can be specified completely by two integer-valued random variables, $I(t)$ and $J(t)$. $I(t)$ varies from 1 to N, representing the phase of the modulating Markov chain, and $0 \leq J(t) < L+1$ represents the number of positive customers in the system at time t, including any in service. The system is now modeled by a CTMC $\overline{Y}$ (Y if L is infinite), on a rectangular lattice strip. Let $I(t)$, the phase, vary in the horizontal direction and $J(t)$, the queue length or *level*, in the vertical direction. We denote the steady state probabilities by $\{p_{i,j}\}$, where $p_{i,j} = \lim_{t \to \infty} Prob(I(t) = i, J(t) = j)$, and let $v_j = (p_{1,j}, \ldots, p_{N,j})$. The process $\overline{Y}$ evolves due to the following instantaneous transition rates:

(a) $q_{i,k}$ – purely lateral transition rate – from state (i,j) to state (k,j), for all $j \geq 0$ and $1 \leq i, k \leq N$ ($i \neq k$), caused by a phase transition in the modulating Markov process ($q_{i,i} = 0$);

(b) $B_{i,j,j+s}$ – s-step upward transition rate – from state (i,j) to state $(i, j+s)$, caused by a new batch arrival of size s of positive customers in phase i. For a given j, s can be seen as bounded when L is finite and unbounded when L is infinite;

(c) $C_{i,j,j-s}$ – s-step downward transition rate – from state (i,j) to state $(i, j-s)$, $(j-s \geq c+1)$, caused by either a batch service completion of size s or a batch arrival of negative customers of size s, in phase i;

(d) $C_{i,c+s,c}$ – s-step downward transition rate – from state $(i, c+s)$ to state (i, c), caused by a batch arrival of negative customers of size $\geq s$ or a batch service completion of size s ($1 \leq s \leq L - c$), in phase i;

(e) $C_{i,c-1+s,c-1}$ – s-step downward transition rate, from state $(i, c-1+s)$ to state $(i, c-1)$, caused by a batch departure of size s ($1 \leq s \leq L-c+1$), in phase i;

(f) $C_{i,j+1,j}$ – 1-step downward transition rate, from state $(i, j+1)$ to state (i, j), ($c \geq 2$; $0 \leq j \leq c-2$), caused by a single departure, in phase i.

Notice that $\overline{Y}$ and Y (i.e., when $L = \infty$) are essentially QBD-U processes.

3.6 The Transition Rate Matrices

The transition rate matrices and parameters can be obtained as [8].

$$B_{i,j-s,j} = \sum_{k=1}^{K} (1-\theta_{i,k})\theta_{i,k}^{s-1}\sigma_{i,k} \quad (\forall i \, ; \, 0 \leq j-s \leq L-2 \, ; \, j-s < j < L) \, ;$$

$$B_{i,j,L} = \sum_{k=1}^{K} \sum_{s=L-j}^{\infty} (1-\theta_{i,k})\theta_{i,k}^{s-1}\sigma_{i,k} = \sum_{k=1}^{K} \theta_{i,k}^{L-j-1}\sigma_{i,k} \quad (\forall i \, ; \, j \leq L-1) \, ;$$

$$C_{i,j+s,j} = \sum_{n=1}^{c} \mu_{i,n}(1-\phi_{i,n})\phi_{i,n}^{s-1} + (1-\delta_i)\delta_i^{s-1}\rho_i$$

$$(\forall i \, ; \, c+1 \leq j \leq L-1 \, ; \, 1 \leq s \leq L-j)$$

$$= \sum_{n=1}^{c} \mu_{i,n}(1-\phi_{i,n})\phi_{i,n}^{s-1} + \delta_i^{s-1}\rho_i \quad (\forall i \, ; \, j=c \, ; \, 1 \leq s \leq L-c)$$

$$= \sum_{n=1}^{c} \phi_{i,n}^{s-1}\mu_{i,n} \quad (\forall i \, ; \, j=c-1 \, ; \, 1 \leq s \leq L-c+1)$$

$$= 0 \quad (\forall i \, ; \, c \geq 2 \, ; \, 0 \leq j \leq c-2 \, ; \, s \geq 2)$$

$$= \sum_{n=1}^{j+1} \mu_{i,n} \quad (\forall i \, ; \, c \geq 2 \, ; \, 0 \leq j \leq c-2 \, ; \, s=1) \, ;$$

Define,

$$B_{j-s,j} = \text{Diag}\,[B_{0,j-s,j}, B_{1,j-s,j}, \ldots, B_{N,j-s,j}] \quad (j-s < j \leq L) \, ;$$

$$B_s = B_{j-s,j} \quad (j < L)$$

$$= \text{Diag}\left[\sum_{k=1}^{K} \sigma_{0,k}(1-\theta_{0,k})\theta_{0,k}^{s-1}, \ldots, \sum_{k=1}^{K} \sigma_{N,k}(1-\theta_{N,k})\theta_{N,k}^{s-1}\right] \, ;$$

$$\Sigma_k = \text{Diag}\,[\sigma_{0,k}, \sigma_{1,k}, \ldots, \sigma_{N,k}] \quad (k=1,2,\ldots,K) \, ;$$
$$\Theta_k = \text{Diag}\,[\theta_{0,k}, \theta_{1,k}, \ldots, \theta_{N,k}] \quad (k=1,2,\ldots,K) \, ;$$
$$\Sigma = \sum_{k=1}^{K} \Sigma_k \, ;$$
$$R = \text{Diag}\,[\rho_0, \rho_1, \ldots, \rho_N] \, ; \quad \Delta = \text{Diag}\,[\delta_0, \delta_1, \ldots, \delta_N] \, ;$$

$$M_n = \text{Diag}\,[\mu_{0,n}, \mu_{1,n}, \ldots, \mu_{N,n}] \quad (n=1,2,\ldots,c) \, ;$$

$$\Phi_n = \text{Diag}\,[\phi_{0,n}, \phi_{1,n}, \ldots, \phi_{N,n}] \quad (n=1,2,\ldots,c) \, ;$$

$$C_j = \sum_{n=1}^{j} M_n \quad (1 \leq j \leq c) ;$$

$$= \sum_{n=1}^{c} M_n = C \quad (j \geq c) ;$$

$$C_{j+s,j} = \text{Diag}\,[C_{0,j+s,j}, C_{1,j+s,j}, \ldots, C_{N,j+s,j}] ;$$

$$E = \text{Diag}(\mathbf{e}'_N) .$$

Then, we get,

$$B_s = \sum_{k=1}^{K} \Theta_k^{s-1}(E - \Theta_k)\Sigma_k ; \quad B_1 = B = \sum_{k=1}^{K}(E - \Theta_k)\Sigma_k ;$$

$$B_{L-s,L} = \sum_{k=1}^{K} \Theta_k^{s-1}\Sigma_k ;$$

$$C_{j+s,j} = \sum_{n=1}^{c} M_n(E - \Phi_n)\Phi_n^{s-1} + R(E - \Delta)\Delta^{s-1}$$

$$(c+1 \leq j \leq L-1 ; \; s = 1, 2, \ldots, L-j) ;$$

$$= \sum_{n=1}^{c} M_n(E - \Phi_n)\Phi_n^{s-1} + R\Delta^{s-1} \quad (j = c ; \; s = 1, 2, \ldots, L-c) ;$$

$$= \sum_{n=1}^{c} M_n \Phi_n^{s-1} \quad (j = c-1 ; \; s = 1, 2, \ldots, L-c+1) ;$$

$$= 0 \quad (c \geq 2 ; \; 0 \leq j \leq c-2 ; \; s \geq 2) ;$$

$$= C_{j+1} \quad (c \geq 2 ; \; 0 \leq j \leq c-2 ; \; s = 1) .$$

4   Steady State Solution

The method presented in this Section is done for sufficiently large L such that $L \geq 2c + K + 3$. When $L < 2c + K + 3$, then the Markov process $\overline{Y}$ can be solved by traditional methods [48].

First, the steady state balance equations are obtained [8]. Let the term $<j>$ denote the vector balance equation for level j. A novel methodology is developed to solve these equations *exactly and efficiently*. First these complicated equations are *transformed* to a computable form by using mathematically oriented transformations. The resulting transformed equations are of the QBD-M type and hence can be solved.

Define the functions, $F_{K,l}$ ($l = 1, 2, \ldots, K$) and $H_{c,n}$ ($n = 1, 2, \ldots, c$) using their properties and recursions as given below.

$$F_{k,0} = E \ , \ F_{k,k} = \prod_{i=1}^{k} \Theta_i \ (k = 1, 2, \ldots, K);$$

$$F_{k,l} = 0 \ (k = 1, 2, \ldots, K; \ l < 0) \ ; \ F_{k,l} = 0 \ (k = 1, 2, \ldots, K; \ l > k) \ ;$$

$$F_{1,0} = E \ ; \ F_{1,1} = \Theta_1 \ ;$$

$$F_{k,l} = F_{k-1,l} + \Theta_k F_{k-1,l-1} \ (2 \leq k \leq K \ , \ 1 \leq l \leq k-1) \ ; \tag{9}$$

$$H_{m,0} = E \ , \ H_{m,m} = \prod_{i=1}^{m} \Phi_i \ (m = 1, 2, \ldots, c);$$

$$H_{m,n} = 0 \ (m = 1, 2, \ldots, c; \ n < 0) \ ; \ H_{m,n} = 0 \ (m = 1, 2, \ldots, c; \ n > m) \ ;$$

$$H_{1,0} = E \ ; \ H_{1,1} = \Phi_1 \ ;$$

$$H_{m,n} = H_{m-1,n} + \Phi_m H_{m-1,n-1} \ (2 \leq m \leq c \ , \ 1 \leq n \leq m-1) \ . \tag{10}$$

Please note E is the Identity matrix of size $N \times N$. The parameters Θ_i are the same as in [8].

Transformation 1. *Modify simultaneously the balance equations for levels j ($L - 2 - c \geq j \geq c + K + 1$), by the transformation:*

$$<\mathbf{j}>^{(1)} \longleftarrow <\mathbf{j}> + \sum_{l=1}^{K} (-1)^l <\mathbf{j-1}> F_{K,l} (c + K + 1 \leq j \leq L - 2 - c);$$

$$<\mathbf{j}>^{(1)} \longleftarrow <\mathbf{j}> \qquad\qquad (j > L - 2 - c \ or \ j < c + K + 1).$$

Apply the second transformation to the resulting equations.

Transformation 2. *Modify simultaneously the balance equations for levels j ($L - 2 - c \geq j \geq c + K + 1$), by the transformation:*

$$<\mathbf{j}>^{(2)} \longleftarrow <\mathbf{j}>^{(1)} + \sum_{n=1}^{c} (-1)^n <\mathbf{j+n}>^{(1)} H_{c,n}$$

$$(c + K + 1 \leq j \leq L - 2 - c);$$

$$<\mathbf{j}>^{(2)} \longleftarrow <\mathbf{j}>^{(1)} \qquad (j > L - 2 - c \ or \ j < c + K + 1).$$

Apply the third and final transformation to the resulting equations.

Transformation 3. *Modify simultaneously the balance equations for levels j ($L - 2 - c \geq j \geq c + K + 1$), by the transformation:*

$$<\mathbf{j}>^{(3)} \longleftarrow <\mathbf{j}>^{(2)} - <\mathbf{j+1}>^{(2)} \Delta \qquad (c + K + 1 \leq j \leq L - 2 - c);$$

$$<\mathbf{j}>^{(3)} \longleftarrow <\mathbf{j}>^{(2)} \qquad (j > L - 2 - c \ or \ j < c + K + 1).$$

Theorem 1. *With these above three transformations, the transformed balance equation, $<\mathbf{j}>^{(3)}$'s, for the rows ($c + K + 1 \leq j \leq L - 2 - c$), will be of the form:*

$$\mathbf{v}_{j-K} Q_0 + \mathbf{v}_{j-K+1} Q_1 + \ldots + \mathbf{v}_{j+c+1} Q_{K+c+1} = 0$$

$$(j = L - 2 - c, L - 1 - c, \ldots, c + K + 1), \tag{11}$$

where $Q_0, Q_1, \ldots, Q_{K+c+1}$ are $K+c+2$ number of j-independent matrices which can be derived algebraically from the system parameters

Proof. With Transformation 1, we get

$$<\mathbf{j}>^{(1)} \longleftarrow <\mathbf{j}> + \sum_{l=1}^{K}(-1)^l<\mathbf{j}-\mathbf{l}>F_{K,l}. \tag{12}$$

Applying Transformation 2 to the jth row, from the above (12), we get

$$<\mathbf{j}>^{(2)} \longleftarrow <\mathbf{j}>^{(1)} + \sum_{n=1}^{c}(-1)^n<\mathbf{j}+\mathbf{n}>^{(1)} H_{c,n}. \tag{13}$$

Expanding the terms, equation (13) can be written as

$$<\mathbf{j}>^{(2)} \longleftarrow <\mathbf{j}> + \sum_{l=1}^{K}(-1)^l<\mathbf{j}-\mathbf{l}>F_{K,l}$$
$$+ \sum_{n=1}^{c}(-1)^n\left[<\mathbf{j}+\mathbf{n}> + \sum_{l=1}^{K}(-1)^l<\mathbf{j}-\mathbf{l}+\mathbf{n}>F_{K,l}\right]H_{c,n}. \tag{14}$$

Applying Transformation 3 to the jth row, and substituting from the above (14), for $<\mathbf{j}+\mathbf{1}>^{(2)}$

$$<\mathbf{j}>^{(3)} \longleftarrow <\mathbf{j}> + \sum_{l=1}^{K}(-1)^l<\mathbf{j}-\mathbf{l}>F_{K,l}$$
$$+ \sum_{n=1}^{c}(-1)^n\left[<\mathbf{j}+\mathbf{n}> + \sum_{l=1}^{K}(-1)^l<\mathbf{j}-\mathbf{l}+\mathbf{n}>F_{K,l}\right]H_{c,n}$$
$$- \left[<\mathbf{j}+\mathbf{1}> + \sum_{l=1}^{K}(-1)^l<\mathbf{j}+\mathbf{1}-\mathbf{l}>F_{K,l}\right]\Delta$$
$$- \sum_{n=1}^{c}(-1)^n\left[<\mathbf{j}+\mathbf{1}+\mathbf{n}> + \sum_{l=1}^{K}(-1)^l<\mathbf{j}+\mathbf{1}-\mathbf{l}+\mathbf{n}>F_{K,l}\right]H_{c,n}\Delta. \tag{15}$$

Expanding and grouping the terms together, equation (15) can be written as

$$<\mathbf{j}>^{(3)} \longleftarrow \sum_{m=-c-1}^{K}<\mathbf{j}-\mathbf{m}>G_{K,c,m}, \tag{16}$$

where

$$G_{K,c,m} = \sum_{\substack{l-n=m \\ l=-1,\ldots,K \\ n=0,\ldots,c}} (-1)^{l+n}[F_{K,l}H_{c,n} + F_{K,l+1}H_{c,n}\Delta]$$

$$= \sum_{n=0}^{c} (-1)^{m+2n}[F_{K,m+n} + F_{K,m+n+1}\Delta]H_{c,n}$$

$$= (-1)^m \sum_{n=0}^{c} [F_{K,m+n} + F_{K,m+n+1}\Delta]H_{c,n} \quad (m=-1-c,\ldots,K).$$

(17)

The balance equations $<\mathbf{j+c+1}>,\ldots,<\mathbf{j}>,\ldots,<\mathbf{j-1}>,\ldots,<\mathbf{j-K}>$, respectively are given by,

$$\sum_{s=1}^{j+c+1}\sum_{k=1}^{K}\mathbf{v}_{j+c+1-s}\Theta_k^{s-1}(E-\Theta_k)\Sigma_k + \mathbf{v}_{j+c+1}[Q-\Sigma-C_{j+c}-R]$$
$$+ \sum_{s=1}^{L-j-c-1}\mathbf{v}_{j+c+1+s}C_{j+c+1+s,j+c+1} = 0;$$

$$\vdots$$

$$\sum_{s=1}^{j}\sum_{k=1}^{K}\mathbf{v}_{j-s}\Theta_k^{s-1}(E-\Theta_k)\Sigma_k + \mathbf{v}_j[Q-\Sigma-C_j-R]$$
$$+ \sum_{s=1}^{L-j}\mathbf{v}_{j+s}C_{j+s,j} = 0;$$

$$\vdots$$

$$\sum_{s=1}^{j-l}\sum_{k=1}^{K}\mathbf{v}_{j-l-s}\Theta_k^{s-1}(E-\Theta_k)\Sigma_k + \mathbf{v}_{j-l}[Q-\Sigma-C_{j-l}-R]$$
$$+ \sum_{s=1}^{L-j+l}\mathbf{v}_{j-l+s}C_{j-l+s,j-l} = 0;$$

$$\vdots$$

$$\sum_{s=1}^{j-K}\sum_{k=1}^{K}\mathbf{v}_{j-K-s}\Theta_k^{s-1}(E-\Theta_k)\Sigma_k + \mathbf{v}_{j-K}[Q-\Sigma-C_{j-K}-R]$$
$$+ \sum_{s=1}^{L-j+K}\mathbf{v}_{j-K+s}C_{j-K+s,j-K} = 0.$$

Substituting or applying the above to (16), for the coefficients (Q_{K-m}) of $\mathbf{v}_{j-m}$ in $<\mathbf{j}>^{(3)}$, we get

$$Q_{K-m} = \sum_{l=-1-c}^{m-1}\left[\sum_{n=1}^{K}\Theta_n^{m-l-1}(E-\Theta_n)\Sigma_n\right]G_{K,c,l} + [Q-\Sigma-C_{j-m}-R]G_{K,c,m}$$

$$+ \sum_{l=m+1}^{K}[C_{j-m,j-l}]G_{K,c,l}$$

$$(m = j-L,\ldots,-2,-1,0,\ldots,K,\ldots,j). \tag{18}$$

Also, for $m = -1-c, 0, \ldots, K$, substituting $C_{j-m} = C$ and $C_{j-m,j-l} = $

$$C_{j-l+l-m,j-l} = \sum_{n=1}^{c} M_n(E-\Phi_n)\Phi_n^{l-m-1} + R(E-\Delta)\Delta^{l-m-1} \text{ in (18), we get}$$

$$Q_{K-m} = \sum_{l=-1-c}^{m-1}\left[\sum_{n=1}^{K}\Theta_n^{m-l-1}(E-\Theta_n)\Sigma_n\right]G_{K,c,l} + [Q-\Sigma-C-R]G_{K,c,m}$$

$$+ \sum_{l=m+1}^{K}\left[\sum_{n=1}^{c}M_n(E-\Phi_n)\Phi_n^{l-m-1} + R(E-\Delta)\Delta^{l-m-1}\right]G_{K,c,l}$$

$$(m = -1-c,\ldots,0,\ldots,K). \tag{19}$$

Using the above, the required Q_l's can be computed easily. Notice the above Q_l's in equation (19) are j- independent. The other coefficients, *i.e.* those of $\mathbf{v}_{j-K-1}, \mathbf{v}_{j-K-2}, \ldots, \mathbf{v}_0$ and of $\mathbf{v}_{j+c+2}, \mathbf{v}_{j+c+3}, \ldots$, can be shown to be zero, case wise, by using computer programs in *Mathematica* or other symbolic manipulation languages. A rigorous proof is indeed possible, but it is beyond the scope of the present paper □

It is observed that the solution of the HetSigma queue can be performed within the framework of the QBD-M processes with the following threshold parameters $y_1 = K; y = K+c+1; T_1 = c+K+1$.

After obtaining $F_{K,l}$'s and $H_{c,n}$'s thus, $G_{K,c,k}, (k = -1-c, \ldots, K)$ can be computed from (17). Then, using them directly in (19), the required Q_l ($l = 0, 1, \ldots, K+c+1$) can be computed. An alternate way of computing the $G_{K,c,l}$'s is by the following properties and recursion which are obtained from (9),(10) and (17) as

$$G_{k,n,l} = G_{k,n-1,l} - \Phi_n G_{k,n-1,l+1}$$

$$(2 \leq k \leq K,\ -1 \leq l+c \leq k+n \leq k+c),$$

$$G_{k,c,l} = G_{k-1,c,l} - \Theta_k G_{k-1,c,l-1}\ (2 \leq k \leq K,\ -1 \leq l \leq k+c). \tag{20}$$

Thus, the resulting equations (11) corresponding to the rows from $c+K+1$ to $L-2-c$, are of the same form as those of the QBD-M processes and hence have an efficient solution by several alternative methods such as the spectral expansion method, Bini-Meini's method [3] and the matrix-geometric methods with folding and block-size enlargement. In our implementation, we have used the spectral expansion method. The required unknowns in the spectral expansion solution and the steady state probabilities can be computed using the remaining balance equations and the normalisation equation.

5 Applications

The *HetSigma* queue has been applied to evaluate performance problems in telecommunication systems [8,23]. In this section, we present the application of the *HetSigma* queue to the performance analysis of MPLS networks which are being used in the backbone of IP networks.

5.1 A Model for a Multipath Routing in MPLS Networks

An example of an MPLS domain with routers and links is illustrated in Figure 1. Traffic demands traversing the MPLS domain are conveyed along pipes called Label Switched Paths (LSPs). When a packet arrives at the ingress router – a Label Edge Router (LER) of the MPLS domain, the LER classifies incoming IP (or other packets for example Ethernet or even MPLS) packets to the appropriate FEC (Forward Equivalence Class) and encapsulates the packets within MPLS packets. Labels are automatically assigned with the use of appropriate protocols, though labeling can also be allocated manually by the network administrator. A routing table (as the result of assignments of labels) in the LER is used to switch packets in MPLS networks.

In what follows, we describe the proposed model for an ingress-egress router pair illustrated in Figure 2. Several paths can be defined and determined between a given ingress-egress (IE) node pair in the MPLS network according to some predefined criteria of the MPLS traffic engineering that is applied (e.g.: paths with disjoint edges) for a single service class. Assume that there are c paths with different bandwidths to be established between the IE node pair for a specific service class. We model the c distinct paths (LSPs) in the system as the c distinct heterogeneous servers in parallel in the corresponding queuing model that we

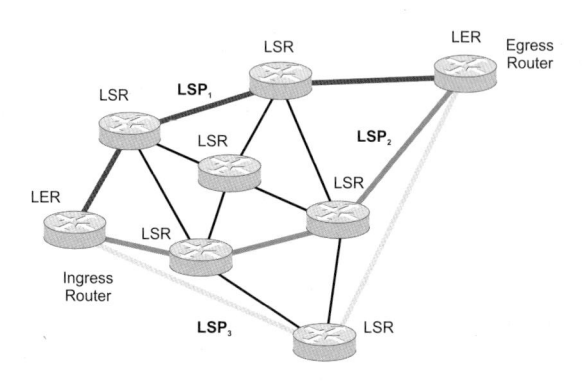

Fig. 1. MPLS domain

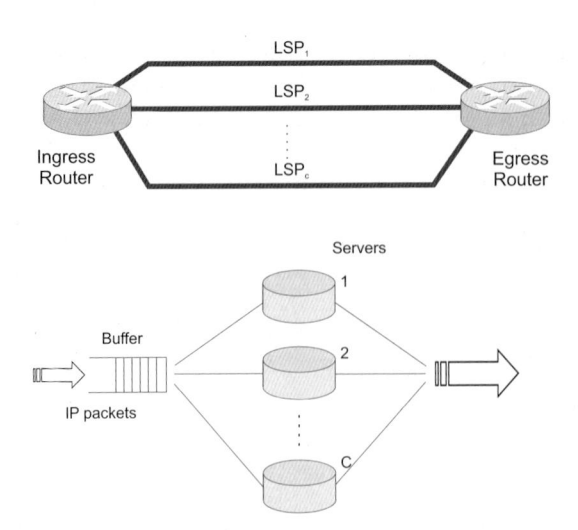

Fig. 2. Model of an ingress node performing load balancing

are introducing here for modeling the communication between the IE pair. The LSPs in the network and the corresponding servers in the model are numbered in the same order. The GE-distributed service time parameters of the nth server ($n = 1, 2 \ldots, c$) are denoted by (μ_n, ϕ_n) when the server is functional. L is the queuing capacity (finite or infinite), in all phases, including the customers in service, if any.

The packet arrival stream is represented by the $MM \sum_{k=1}^{K} CPP_k$ arrival process, this can accommodate traffic-burstiness, correlations among inter-arrival times and correlations among batch sizes. It has been shown in the recent work [21, 23] that the CPP is accurate enough to model real traffic (when CPP parameters are estimated from the captured traffic) and can be used for the performance evaluation of real systems. The arrival process(the $MM \sum_{k=1}^{K} CPP_k$) is inherently modulated by a CTMC X with N_1 states, with generator matrix Q_X.

LSPs are prone to failures because of various reasons (e.g.: unreliable equipment, hardware failures, software bugs or cable cut). Such faults and failures may affect the operation of LSPs and cause packet losses and delays. In case of failures, the load balancing mechanism can move packets that are queued for the affected LSPs to the unaffected LSPs.

When the failed link is repaired, the repaired LSP can be used again. Repair strategy defines the order of repairing failed links, when there are more failed links than repairmen (it is assumed simultaneous repairs can happen). Repair strategies can be preemptive or non-preemptive. However, it is not very realistic to apply a preemptive-priority repair strategy for the operation of networks in the case of link cuts because of the travel cost of a maintenance team. In this paper, we assume that one maintenance team is available to repair failures in the network. However, the analysis can be easily extended to the case of multiple maintenance facilities as well.

We consider four repair strategies, FCFS (First-Come-First-Served), LCFS (Last- Come-First-Served non-preemptive) and two based on link-priorities. In the strategies based on link-priorities, links with higher priorities should be set into repair sooner, even if these failures have occurred later. The priority list of links may then be constructed in a greedy way, with a view to repair earlier the link which would fetch larger gain in performance.

We assume the LSP states (or LSP configurations) that arise due to failures and repairs of the LSPs can be described by a CTMC called Z. These LSP configurations would indeed correspond to all possible multi-server configurations (also termed, operative states of the multi-server) with functional as well as failed servers, in the corresponding queuing model. This can well be so when exponential or phase-type failures and repairs are assumed. Let N_2, Q_Z denote the number of server-configurations and the generator matrix of the arising CTMC Z. Indeed N_2 and Q_Z would depend on the parameters of the parallel servers, number of repairmen and the repair strategy.

In the presence of one repair team, we would need $N_2 = \sum_{l=0}^{c} Pe(c, l) = 2^{c+1}$ operative states for FCFS and LCFS repair strategies, where $Pe(c, l)$ is the number of permutations of c distinct elements taken l at a time. $Pe(c, l) = \frac{c!}{(c-l)!}$. Let ξ_k and η_k be the failure and repair rates respectively, of the kth LSP (that is, k^{th} server in the multi-server model). Then Q_Z can be determined (illustration for $c = 3$ is presented below). Many other repair strategies can also be modeled, however, that is not in the scope of the present paper.

Both the arrival and the service processes can be thought of being jointly modulated by the same continuous time, irreducible Markov process, with N states where $N = N_1 \cdot N_2$. That generator matrix of this joint modulating process is denoted by Q where Q is determined as

$$Q = Q_Z \bigoplus Q_X. \tag{21}$$

Let the random variable $I_1(t)$ ($1 \leq I_1(t) \leq N_1$) represent the phase of the modulating process X at any time t. We introduce $I_2(t)$ – an integer-valued random variable to describe the server configuration of the model (which corresponds to the network state) at time t. We define the following function,

$$\gamma(I_2(t), n) = \begin{cases} 1 & \text{the } n^{th} \text{ server is functional} \\ 0 & \text{otherwise} \end{cases}. \tag{22}$$

The state $I(t)$ ($1 \leq I(t) \leq N$; $N = N_1 \cdot N_2$) of the joint modulating process is constructed by lexicographically sorting the two variables $(I_1(t), I_2(t))$ as illustrated in Table 1.

Table 1. Order of the phase variable

I	1	2	...	N_1	N_1+1	...	$2N_1$	...	$N_1 N_2 - N_1 + 1$	...	$N_1 N_2$
(I_1, I_2)	(1,1)	(2,1)	...	$(N_1, 1)$	(1,2)	...	$(N_1, 2)$	...	$(1, N_2)$	...	(N_1, N_2)

From table 1, we can write the following equations,

$$I(t) = I_1(t) + (I_2(t) - 1)N_1, \; I_2(t) = f_2(I(t)) = \left\lfloor \frac{I(t)-1}{N_1} \right\rfloor + 1,$$
$$I_1(t) = f_1(I(t)) = [(I(t)-1) \; mod \; N_1] + 1. \quad (23)$$

Consequently, the parameters of the arrival process are mapped as follows. The parameters of the GE inter-arrival time distribution of the kth ($1 \le k \le K$) customer arrival stream in phase i ($i = 1, ..., N$) are $(\sigma_{f_1(i),k}, \theta_{f_1(i),k})$. The service time parameters of the nth server ($n = 1, 2, ..., c$) in phase i ($1 \le i \le N$), denoted by $(\mu_{i,n}, \phi_{i,n})$ can be determined as,

$$\mu_{i,n} = \gamma(f_2(i), n)\mu_n \; , \; \phi_{i,n} = \gamma(f_2(i), n)\phi_n. \quad (24)$$

5.2 Numerical Results

An elaborate case study is carried out to determine the performance of a specific ingress-egress node pair in an European Optical Network topology [1]. The network in Figure 3 contains 19 nodes and 78 optical links.

Traffic to the ingress node to be carried to the egress node, shown in the figure, is assumed follow the ON- OFF process with two states ($ON - OFF$). The ON and OFF periods are exponentially distributed with mean 0.4 and 0.5 s respectively. The distribution of inter-arrival times in state ON is GE with parameters (σ = 150.6698990 (1/s), θ = 0.526471). In state OFF no packets arrive. These parameters of the arrival stream are indeed derived from the inter-arrival times of the recorded samples of the real Bellcore traffic trace BC-pAug89 [2], by matching the mean and variance.

Assume that there are three LSPs established between node 11 (ingress) and node 4 (egress) of the IP/MPLS network as seen in Figure 3. The first LSP is routed through nodes 10 and 5. The second LSP goes through nodes 17 and 16, while the third one is through node 9. Thus, we have, in the model, $c = 3$ servers with GE service time parameters (μ_i, ϕ_i ($i = 1, 2, 3$)) given by, $\mu_1 = 96$ (1/s), $\mu_2 = 128$ (1/s), $\mu_3 = 160$ (1/s); $\phi_1 = \phi_2 = \phi_3 = 0.109929$. Note that these parameters are obtained, based on the service capacity of the LSPs and the packet lengths from the trace.

A specific LSP fails when a link through which the LSP is routed becomes inoperative (e.g.: cable cut). Normally, the failure rate of each link can be assumed to depend on its length. Using simple calculations, the failure rates are obtained as $\xi_1 = 0.001$, $\xi_2 = 0.001$ and $\xi_3 = 0.0006$. The repair rate values (η_1, η_2 and η_3) are chosen in such a way to have the availability of the connectivity ensured by the LSPs between the ingress and egress node to be 99.9%.

Four repair strategies are compared in this section as follows. Note that after the identifications of LSP configurations in each strategy, the corresponding Q_Z matrices can be easily determined.

FCFS repairs. It can be seen that there would be 16 LSP configurations or operative states represented as: $(0, 0, 0)_{1,2,3}$, $(0, 0, 0)_{1,3,2}$, $(0, 0, 0)_{2,1,3}$, $(0, 0, 0)_{2,3,1}$,

$(0,0,0)_{3,1,2}$, $(0,0,0)_{3,2,1}$, $(0,0,1)_{1,2}$, $(0,0,1)_{2,1}$, $(0,1,0)_{1,3}$, $(0,1,0)_{3,1}$, $(1,0,0)_{2,3}$, $(1,0,0)_{3,2}$, $(0,1,1)$, $(1,0,1)$, $(1,1,0)$ and $(1,1,1)$. Here, the kth bit from left within the brackets, is 0 when LSP k is broken, 1 when operative. Also, the suffix indicates the order in which the failed LSPs are to be repaired, if they are greater than one. If the above order is numbered from 0 to 15, then the nonzero and off- diagonal elements of matrix Q_Z can be given by, $Q_Z(0,10) = Q_Z(1,11) = Q_Z(6,13) = \eta_1$, $Q_Z(8,14) = Q_Z(12,15) = \eta_1$, $Q_Z(2,8) = Q_Z(3,9) = Q_Z(7,12) = \eta_2$, $Q_Z(10,14) = Q_Z(13,15) = \eta_2$, $Q_Z(4,6) = Q_Z(5,7) = Q_Z(9,12) = \eta_3$, $Q_Z(11,13) = Q_Z(14,15) = \eta_3$, $Q_Z(10,3) = Q_Z(11,5) = Q_Z(13,7) = \xi_1$, $Q_Z(14,9) = Q_Z(15,12) = \xi_1$, $Q_Z(8,1) = Q_Z(9,4) = Q_Z(12,6) = \xi_2$, $Q_Z(14,11) = Q_Z(15,13) = \xi_2$, $Q_Z(6,0) = Q_Z(7,2) = Q_Z(12,8) = \xi_3$, and $Q_Z(13,10) = Q_Z(15,14) = \xi_3$.

In the above, for example, the state $(1,0,0)_{2,3}$ means that LSP 1 is functional, LSPs 2 and 3 are broken, the single repairman is working on LSP 2. From this state, the possible next states are $(0,0,0)_{2,3,1}$ which happens by the failure of LSP 1, or, $(1,1,0)$ which happens by the repair of LSP 2. These are shown in Q_Z as, $Q_Z(10,3) = \xi_1, Q_Z(10,14) = \eta_2$. All other elements of Q_Z can be explained in a similar way.

LCFS. In this LCFS non-preemptive repair strategy, the repairman chooses the failed server according to the LCFS discipline, and a repair is never preempted by a new breakdown. Following the same representation and order as in FCFS strategy, the non zero and off-diagonal elements of matrix Q_Z, in this case, can be obtained as, $Q_Z(0,10) = Q_Z(1,11) = Q_Z(6,13) = \eta_1$, $Q_Z(8,14) = Q_Z(12,15) = \eta_1$, $Q_Z(2,8) = Q_Z(3,9) = Q_Z(7,12) = \eta_2$, $Q_Z(10,14) = Q_Z(13,15) = \eta_2$, $Q_Z(4,6) = Q_Z(5,7) = Q_Z(9,12) = \eta_3$, $Q_Z(11,13) = Q_Z(14,15) = \eta_3$, $Q_Z(10,2) = Q_Z(11,4) = Q_Z(13,7) = \xi_1$, $Q_Z(14,9) = Q_Z(15,12) = \xi_1$,

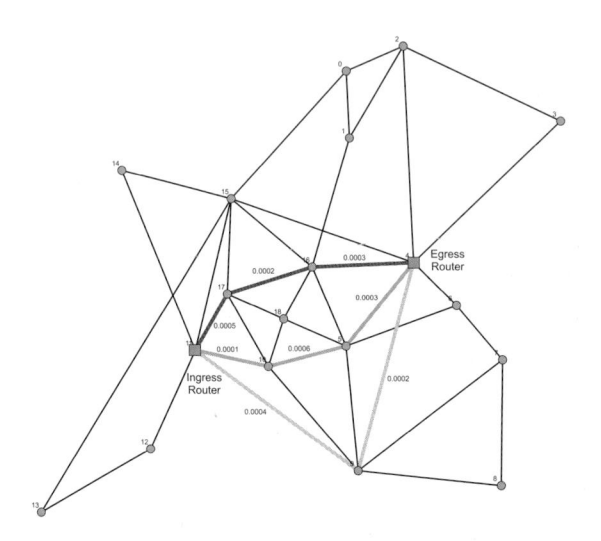

Fig. 3. Network topology

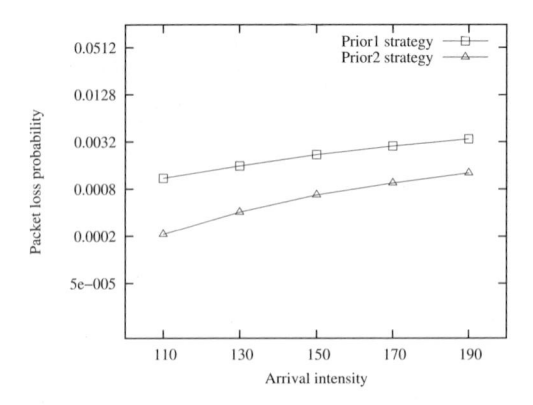

Fig. 4. Packet loss probability versus arrival rate

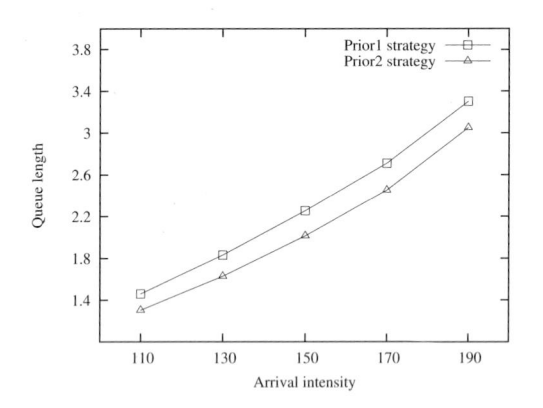

Fig. 5. Mean queue length versus arrival intensity

$Q_Z(8,0) = Q_Z(9,5) = Q_Z(12,6) = \xi_2$, $Q_Z(14,11) = Q_Z(15,13) = \xi_2$, $Q_Z(6,1) = Q_Z(7,3) = Q_Z(12,8) = \xi_3$, and $Q_Z(13,10) = Q_Z(15,14) = \xi_3$.

Priority-based Strategy. In a LSP-priority based repair strategy, it can be crucial to determine the priority of links according to which failed links are to be repaired. In this section, we try to derive a decision function based on system parameters to decide upon the priorities of the failed links.

Let f_i and r_i be the failure and repair rates of link i respectively and b_i, its bandwidth. The bandwidth of a link is computed from the summation of the LSPs bandwidths passing through this particular link. This approach ensures that links used by several LSPs may get higher priority, over the links used only by one LSP, when the former has larger bandwidth contribution. The w_i, the weight of link i is set to

$$wi = \frac{r_i}{f_i} * b_i. \tag{25}$$

A repair strategy called *Prior*1 is based entirely on the failure rate of the links. That is, when there are several failed links, the link with a smallest failure rate

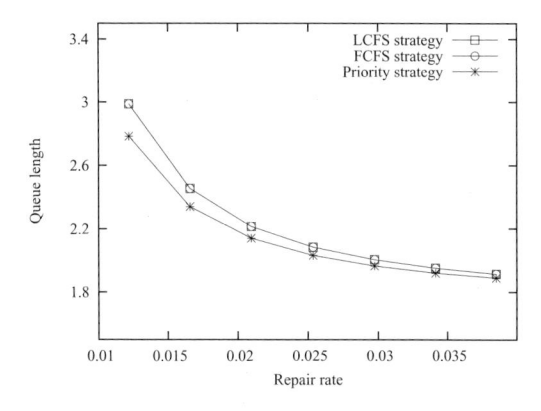

Fig. 6. Mean queue length versus the repair rate

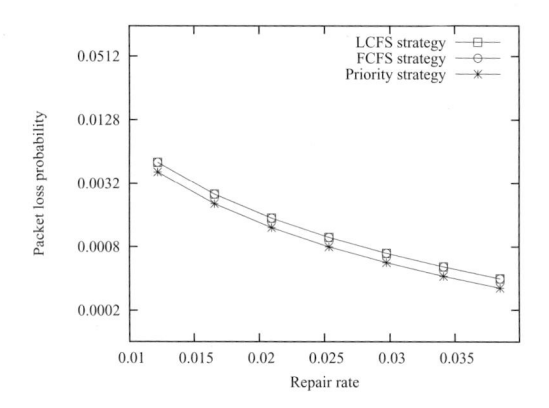

Fig. 7. Packet loss probability versus the repair rate

is set into repair by the single repair facility available. Another strategy called $Prior2$ is based on the decision function (w_i). If $w_i > w_k$, then link i is given higher repair priority than link k in $Prior2$.

To compare $Prior1$ and $Prior2$, we plot the curves concerning packet loss and average number of customers waiting in the queue are versus the arrival rate of packets in Figure 4 and 5, respectively. We can observe that applying $Prior2$ repair strategy results in a better performance characteristics than $Prior1$. The $Prior2$ strategy takes more information into account about the network links towards the decision of repair order which can explain the better results. It balances not only failure rates, but also the repair rates and the bandwidth of links used by the passing-through LSPs. Therefore, we will only investigate strategies FCFS, LCFS and $Prior2$ (will be referred as priority strategy) in what follows.

We now examine the effect of the repair rate of optical links on the performance of the system. In order to ensure that the availability of service between the ingress and egress node to lie between 99.9% and 99.999%, it can be shown

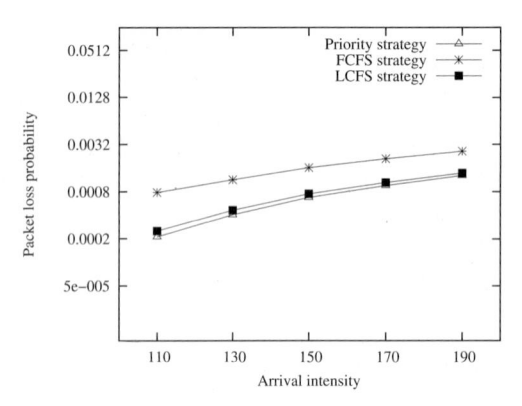

Fig. 8. Packet loss probability versus arrival rate

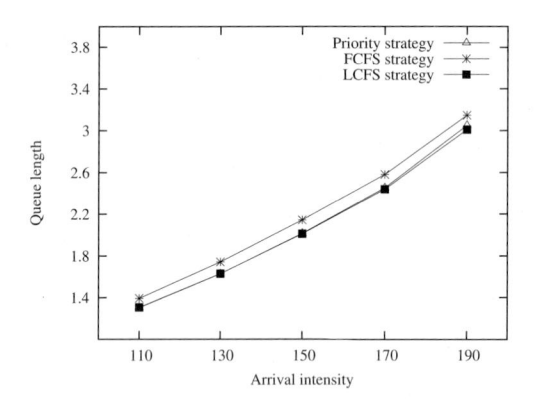

Fig. 9. Mean queue length versus arrival intensity

that the repair rate should be in the interval [0.0397703, 0.405355]. We investigate the effect of the repair rate of link within that range, on the performance. The expected queue length was reduced only by a small extent by the increase of repair rate (see Figure 6). The packet-loss probability varied nearly in inverse-proportion to the repair rate (see Figure 7). Of course, the results would actually depend on many other factors too including the load in the network. Accurate estimation of the returns on investment that is used to increase repair rate can be made, only if the cost-patterns involved in increasing the repair rate are known. It may also be observed, in the considered range of experimentation, the various repair strategies considered do not have significant impact on the performance and the availability of the service in the network (in the range of repair rates that are considered).

Figures 8 and 9, where the strategies are compared versus the arrival rate also confirm an observation that the priority based repair outperforms the LCFS and FCFS ones.

6 Extensions

6.1 Other Killing Disciplines

Apart from the killing discipline that was used above, there are two other popular killing disciplines, the RCE-*inimmune servicing* and the RCH killing disciplines. The applicability of the killing disciplines rather depends on the situation and the purpose, and hence depending on these, many more killing disciplines are theoretically possible. Our methodology can easily be extended to many other killing disciplines also, this is explained briefly in this section.

The RCE-*inimmune* discipline. In the RCE-*inimmune* servicing, the negative customer removes the most recent positive arrival regardless of whether it is in service or waiting; thus a negative arrival has no effect only when it encounters an empty queue and all servers idle. This is the traditional killing discipline, suited to the modeling of killing signals in speculative parallelism, for example. It can also be used to model cell losses caused by the arrival of a corrupted cell or one encountering a full buffer, when the preceding cells of a packet would be discarded. In this case, the later and the upward transitions remain as before, but the downward transition rates become [13]

$$\begin{aligned}
C_{j+s,j} &= C(E-\Phi)\Phi^{s-1} + R(E-\Delta)\Delta^{s-1} \\
&\quad (c \leq j \leq L-1 \,;\, s = 1, 2, \ldots, L-j) \\
&= C\Phi^{s-1} + R(E-\Delta)\Delta^{s-1} \quad (c > 1 \,;\, j = c-1 \,;\, 1 \leq s \leq L-c+1) \\
&= C\Phi^{s-1} + R\Delta^{s-1} \quad (c = 1 \,;\, j = c-1 = 0 \,;\, 1 \leq s \leq L) \\
&= C_{j+1} + R(E-\Delta) \quad (1 \leq j \leq c-2 \,;\, s = 1) \\
&= R(E-\Delta)\Delta^{s-1} \quad (1 \leq j \leq c-2 \,;\, 2 \leq s \leq L-j) \\
&= R\Delta^{s-1} \quad (c > 1 \,;\, j = 0 \,;\, 2 \leq s \leq L) \\
&= C_1 + R \quad (j = 0 \,;\, s = 1).
\end{aligned}$$

Thus, after obtaining the required transition rates, e.g. lateral, upward and downward, and the balance equations in this case, the same procedure can be used, perhaps with slight appropriate modifications, to transform the balance equations to a suitable form (QBD-M type equations) for solving by the existing methods.

The RCH discipline. Another popular killing discipline is the RCH (Removal of customers from the head of the queue) killing discipline. This is appropriate for modeling server breakdowns, where a customer in service will be lost for sure and may be also a portion of queue of waiting customers. Here too, the A matrices and the B matrices remain unchanged, however the C matrices would be different and can be determined. For example, to obtain $C_{j+s,j}$, consider the system (say, a single server system with $c = 1$) in state $(i, j+s)$, where $j + s \leq L$. The rate at which a batch of l negative customers arrives is $(1 - \delta_i)\delta_i^{l-1}\rho_i$ ($l = 1, 2, \ldots$). If $l \geq j + s$, then all the jobs will be removed by the negative customers. If $l < j + s$, then the job *in service* plus $l - 1$ jobs waiting for service would be removed, leaving only $j+s-l$ jobs. However, due to the nature of the generalized exponential service times, a certain number of customers may be 'leaked out', i.e. serviced *instantaneously* just after the killing takes place. Alternatively, if we redefine the operation of the system such that this leakage does not occur, i.e. immediately after a negative arrival, the next customer in the queue (if any) *cannot* skip service, then the equilibrium state probabilities would be same as in the case of the RCE-inimmune servicing.

Assuming leakage, the rate matrices $C_{j+s,j}$ can be derived as [13],

$$C_{j+s,j} = (E - \Phi)\Phi^{s-1}M + (E - \Delta)R(E - \Phi)\{\Phi\Delta\}_{s-1}$$
$$(1 \leq j \leq L-1 \,;\, s \geq 1) \,;$$
$$= \Phi^{s-1}M + (E - \Delta)R\{\Phi\Delta\}_{s-1} + R\Delta^s \quad (j = 0 \,;\, s \geq 2) \,;$$
$$= M + R \quad (j = 0 \,;\, s = 1) \,;$$

where,

$$\{\Phi\Delta\}_{s-1} = \sum_{k=0}^{s-1} \Phi^{s-1-k}\Delta^k \quad (s \geq 2) \,;$$
$$= E \quad (s = 1) \,.$$

This case also can be modeled following the previous procedures, perhaps with minor modifications.

6.2 Truncated-CPP(t-CPP)

By incorporating several CPP's and superposing them, non-geometric batch size arrivals could be modeled, to certain extent and with certain limitations. That flexibility can be increased, in fact greatly, by replacing the CPP's by truncated-CPP's (or t-CPP's). The method of transforming the balance equation for efficient solution may be applicable in principle, but with some modifications, to this case as well and hence can be extended with some effort.

Let the parameters of the kth t-CPP in the ith phase be $(\sigma_{i,k}, \theta_{i,k}, b_{i,k}, d_{i,k})$, which means the batch size is geometrically distributed and *bounded* by $b_{i,k}$ as the minimum batch size and $d_{i,k}$ as the maximum. Also, the inter-arrival time between successive batches, during that modulating phase, is exponential with parameter $\sigma_{i,k}$. The probability of batch size being s of this t-CPP is then given by $\frac{(1-\theta_{i,k})\theta_{i,k}^{s-1}}{\theta_{i,k}^{b_{i,k}-1} - \theta_{i,k}^{d_{i,k}}}$ if $b_{i,k} \leq s \leq d_{i,k}$, 0 otherwise. Hence, the overall batch size distribution in phase i arrivals is then modified as

$$\pi_{l/i} = \sum_{k=1}^{K} \frac{\sigma_{i,k}}{\sigma_{i,.}} \frac{(1-\theta_{i,k})\theta_{i,k}^{l-1}}{\theta_{i,k}^{b_{i,k}-1} - \theta_{i,k}^{d_{i,k}}} f_{i,k,l}, \qquad (26)$$

where $f_{i,k,l} = 1$ if $b_{i,k} \leq l \leq d_{i,k}$, else 0. And, the overall batch size distribution of arrivals is

$$\pi_{l/.} = \sum_{i=1}^{N} r_i \pi_{i,l}. \qquad (27)$$

Clearly, the superposition of t-CPP's (26, 27), because of larger number of parameters and batch size bounds, offers much more flexibility than the superposition of CPP's in order to generate/model certain given non-geometric batch size distributions, by parameter tuning.

6.3 GE with Batch Size Truncation (GE-t)

The service time distribution, GE-t (General Exponential with batch size truncation), is essentially a batch-exponential service, with geometric and bounded batch size distribution. This is quite analogous to the t-CPP. In the case of homogeneous servers the parameters of the GE-t, in phase i, are of the form $(\mu_i, \phi_i, g_i, h_i)$ where μ_i is the batch-service rate, g_i, h_i respectively are lower and upper bounds of the batch size which is geometrically distributed with parameter ϕ_i. However, as far as the service is concerned, the actual batch size served also depends on c, j, where j is the number of jobs in the system. If $j < g_i$, then there will not be any services. If $j \geq h_i$, then the probability that the size of the batch being served is s would be $\frac{(1-\phi_i)\phi_i^{s-1}}{\phi_i^{g_i-1}-\phi_i^{h_i}}$ where $g_i \leq s \leq h_i$. If $g_i < j < h_i$, then the batch size distribution of the served batch can be estimated if the service is clearly defined in such a case. Using these expressions and the steady state probabilities in this case, it is possible to estimate the effective batch size distribution of the served batches. This extension can be incorporated into the main model with a viable solution following same procedures as before, perhaps with minor modifications. This modification can result in, by appropriate parameter tuning, incorporating certain given non-geometric batch size services.

That flexibility can be enhanced even further if we use heterogeneous servers. In such a case, let the parameters of server n ($n = 1, 2, \ldots, c$) in phase i be, $(\mu_{i,n}, \phi_{i,n}, g_{i,n}, h_{i,n})$. Extrapolating the earlier analyis to this case, there are no services in phase i if $j < Min(g_{i,1}, g_{i,2}, \ldots, g_{i,c})$. When $j \geq Max(h_{i,1}, h_{i,2}, \ldots, h_{i,c})$ in phase i, then the probability that the batch size is s for the next service can be derived as $\sum_{n=1}^{c} \frac{\mu_{i,n}(1-\phi_{i,n})\phi_{i,n}^{s-1}}{\mu_{i.}(\phi_{i,n}^{g_{i,n}-1}-\phi_{i,n}^{h_{i,n}})}$ where $Min(g_{i,1}, g_{i,2}, \ldots, g_{i,c}) \leq s \leq Max(h_{i,1}, h_{i,2}, \ldots, h_{i,c})$, and $\mu_{i.} = \sum_{n=1}^{c} \mu_{i,n}$. For all other ranges or values of s, the expressions for the probability distribution can be obtained.

6.4 Towards Abritrary Batch Size Distributions

The consideration of abritrary batch size distributions is most desirable in the performance evaluation of modern telecommunication networks. To achieve this now, is not a very far off thing, owing to the above suggested viable extensions. And thus, we arrive at a much more general and useful model, that is, *the MM $\sum_{k=1}^{K} t - CPP_k/GE$-$t/c/L$ G-queue with heterogeneous servers*. Further work is being carried out on this model.

7 Conclusions

The *HetSigma* queuing model is developed as a generalization of the QBD processes. In this queue, it is possible to accommodate inter-arrival time correlations, service time correlations, batch-size correlations, large and unbounded batch sizes. All these aspects are useful in order to model emerging communication systems. Certain non-trivial transformations are conceived on the balance equations. The transformed balance equations are of the QBD-M type, hence there is a fast solution using the spectral expansion method. The queue is applied for the performance evaluation of MPLS networks with unreliable nodes and can be used to model various problems in telecommunication network as well. It is possible to extend or further generalize these models to develop highly generalized Markovian node models for the emerging Next Generation Networks. That work is underway.

Acknowledgement. Ram Chakka thanks the Chancellor of Sri Sathya Sai University, Prasanthi Nilayam, India, for constant encouragement, guidance and inspiration throughout this work.

References

1. Network Research Topologies. The European optical network (EON), http://www.optical-network.com/topology.php
2. The Internet traffic archive, http://ita.ee.lbl.gov/index.html
3. Bini, D., Meini, B.: On the solution of a nonlinear matrix equation arising in queueing problems. SIAM Journal on Matrix Analysis and Applications 17(4), 906–926 (1996)
4. Blondia, C., Casals, O.: Statical Multiplexing of VBR Source: A Matrix-Analytic Approach. Performance Evaluation 16, 5–20 (1992)
5. Chakka, R.: Performance and Reliability Modelling of Computing Systems Using Spectral Expansion. PhD thesis, University of Newcastle upon Tyne (Newcastle upon Tyne) (1995)
6. Chakka, R.: Spectral Expansion Solution for some Finite Capacity Queues. Annals of Operations Research 79, 27–44 (1998)
7. Chakka, R., Do, V.T.: The $MM\sum_{k=1}^{K} CPP_k/GE/c/L$ G-Queue and Its Application to the Analysis of the Load Balancing in MPLS Networks. In: Proceedings of 27th Annual IEEE Conference on Local Computer Networks (LCN 2002), Tampa, FL, USA, November 6-8, pp. 735–736 (2002)
8. Chakka, R., Do, V.T.: The MM $\sum_{k=1}^{K} CPP_k/GE/c/L$ G-Queue with Heterogeneous Servers: Steady state solution and an application to performance evaluation. Performance Evaluation 64, 191–209 (2007)
9. Chakka, R., Do, V.T., Pandi, Z.: Generalized Markovian queues and applications in performance analysis in telecommunication networks. In: Kouvatsos, D.D. (ed.) The First International Working Conference on Performance Modelling and Evaluation of Heterogeneous Networks (HET-NETs 2003), pp. 60/1–10 (July 2003)
10. Chakka, R., Do, V.T., Pandi, Z.: A Generalized Markovian Queue and Its Applications to Performance Analysis in Telecommunications Networks. In: Kouvatsos, D. (ed.) Performance Modelling and Analysis of Heterogeneous Networks, pp. 371–387. River Publisher (2009)

11. Chakka, R., Ever, E., Gemikonakli, O.: Joint-state modeling for open queuing networks with breakdowns, repairs and finite buffers. In: 15th International Symposium on Modeling, Analysis, and Simulation of Computer and Telecommunication Systems (MASCOTS), pp. 260–266. IEEE Computer Society, Los Alamitos (2007)
12. Chakka, R., Harrison, P.G.: Analysis of MMPP/M/c/L queues. In: Proceedings of the Twelfth UK Computer and Telecommunications Performance Engineering Workshop, Edinburgh, pp. 117–128 (1996)
13. Chakka, R., Harrison, P.G.: The Markov modulated CPP/GE/c/L queue with positive and negative customers. In: Proceedings of the 7th IFIP ATM Workshop, Antwerp, Belgium (1999)
14. Chakka, R., Harrison, P.G.: A Markov modulated multi-server queue with negative customers - the MM CPP/GE/c/L G-queue. Acta Informatica 37, 881–919 (2001)
15. Chakka, R., Harrison, P.G.: The MMCPP/GE/c queue. Queueing Systems: Theory and Applications 38, 307–326 (2001)
16. Chakka, R., Mitrani, I.: Multiprocessor systems with general breakdowns and repairs. In: SIGMETRICS, pp. 245–246 (1992)
17. Chakka, R., Mitrani, I.: Heterogeneous multiprocessor systems with breakdowns: Performance and optimal repair strategies. Theor. Comput. Sci. 125(1), 91–109 (1994)
18. Chien, M., Oruc, Y.: High performance concentrators and superconcentrators using multiplexing schemes. IEEE Transactions on Communications 42, 3045–3050 (1994)
19. Kouvatsos, D.: Entropy Maximisation and Queueing Network Models. Annals of Operations Research 48, 63–126 (1994)
20. Do, V.T.: Comments on multi-server system with single working vacation. Applied Mathematical Modelling 33(12), 4435–4437 (2009)
21. Do, V.T., Chakka, R., Harrison, P.G.: An integrated analytical model for computation and comparison of the throughputs of the UMTS/HSDPA user equipment categories. In: MSWiM 2007: Proceedings of the 10th ACM Symposium on Modeling, Analysis, and Simulation of Wireless and Mobile Systems, pp. 45–51. ACM, New York (2007)
22. Do, V.T., Do, N.H., Chakka, R.: Performance Evaluation of the High Speed Downlink Packet Access in Communications Networks Based on High Altitude Platforms. In: Al-Begain, K., Heindl, A., Telek, M. (eds.) ASMTA 2008. LNCS, vol. 5055, pp. 310–322. Springer, Heidelberg (2008)
23. Do, V.T., Krieger, U.R., Chakka, R.: Performance modeling of an apache web server with a dynamic pool of service processes. Telecommunication Systems 39(2), 117–129 (2008)
24. Do, V.T., Papp, D., Chakka, R., Truong, M.T.: A Performance Model of MPLS Multipath Routing with Failures and Repairs of the LSPs. In: Kouvatsos, D. (ed.) Performance Modelling and Analysis of Heterogeneous Networks, pp. 27–43. River Publisher (2009)
25. Drekic, S., Grassmann, W.K.: An eigenvalue approach to analyzing a finite source priority queueing model. Annals OR 112(1-4), 139–152 (2002)
26. Evans, R.V.: Geometric Distribution in some Two-dimensional Queueing Systems. Operations Research 15, 830–846 (1967)
27. Ever, E., Gemikonakli, O., Chakka, R.: A mathematical model for performability of beowulf clusters. In: Annual Simulation Symposium, pp. 118–126. IEEE Computer Society, Los Alamitos (2006)

28. Ever, E., Gemikonakli, O., Chakka, R.: Analytical modelling and simulation of small scale, typical and highly available beowulf clusters with breakdowns and repairs. Simulation Modelling Practice and Theory 17(2), 327–347 (2009)
29. Fretwell, R.J., Kouvatsos, D.D.: ATM traffic burst lengths are geometrically bounded. In: Proceedings of the 7th IFIP Workshop on Performance Modelling and Evaluation of ATM & IP Networks, Antwerp, Belgium. Chapman and Hall, Boca Raton (1999)
30. Gail, H.R., Hantler, S.L., Taylor, B.A.: Spectral analysis of M/G/1 type Markov chains. Technical Report RC17765, IBM Research Division (1992)
31. Grassmann, W.K.: The use of eigenvalues for finding equilibrium probabilities of certain markovian two-dimensional queueing problems. INFORMS Journal on Computing 15(4), 412–421 (2003)
32. Grassmann, W.K., Drekic, S.: An analytical solution for a tandem queue with blocking. Queueing System (1-3), 221–235 (2000)
33. Haverkort, B., Ost, A.: Steady State Analysis of Infinite Stochastic Petri Nets: A Comparing between the Spectral Expansion and the Matrix Geometric Method. In: Proceedings of the 7th International Workshop on Petri Nets and Performance Models, pp. 335–346 (1997)
34. ITU-T Recommendation Y. 2001. General overview of NGN, Geneve, Switzerland (2005)
35. Kouvatsos, D.D.: A maximum entropy analysis of the G/G/1 Queue at Equilibrium. Journal of Operations Research Society 39, 183–200 (1998)
36. Krieger, U.R., Naoumov, V., Wagner, D.: Analysis of a Finite FIFO Buffer in an Advanced Packet-Switched Network. IEICE Trans. Commun. E81-B, 937–947 (1998)
37. Latouche, G., Ramaswami, V.: Introduction to Matrix Analytic Methods in Stochastic Modeling. ASA-SIAM Series on Statistics and Applied Probability (1999)
38. Mitrani, I.: Approximate solutions for heavily loaded markov-modulated queues. Perform. Eval. 62(1-4), 117–131 (2005)
39. Mitrani, I., Chakka, R.: Spectral expansion solution for a class of Markov models: Application and comparison with the matrix-geometric method. Performance Evaluation 23, 241–260 (1995)
40. Naoumov, V., Krieger, U.R., Warner, D.: Analysis of a Multi-Server Delay-Loss System With a General Markovian Arrival Process. In: Chakravarthy, S.R., Alfa, A.S. (eds.) Matrix-analytic methods in stochastic models. Lecture Notes in Pure and Applied Mathematics, vol. 183. Marcel Dekker, New York (1996)
41. Neuts, M.F.: Matrix Geometric Soluctions in Stochastic Model. Johns Hopkins University Press, Baltimore (1981)
42. Niu, Z., Takahashi, Y.: An Extended Queueing Model for SVC-based IP-over-ATM Networks and its Analysis. In: Proceedings of IEEE GLOBECOM, pp. 490–495 (1998)
43. Norros, I.: A Storage Model with Self-similar Input. Queueing Systems and their Applications 16, 387–396 (1994)
44. Paxman, V., Floyd, S.: Wide-area traffic: The failure of Poisson modelling. IEEE/ACM Transactions on Networking 3(3), 226–244 (1995)
45. Rosti, E., Smirni, E., Sevcik, K.C.: On processor saving scheduling policies for multiprocessor systems. IEEE Trans. Comp. 47, 47–52 (1998)
46. Seelen, L.P.: An Algorithm for Ph/Ph/c queues. European Journal of Operational Research 23, 118–127 (1986)

47. Skianis, C., Kouvatsos, D.: An Information Theoretic Approach for the Performance Evaluation of Multihop Wireless Ad Hoc Networks. In: Kouvatsos, D.D. (ed.) Proceedings of the Second International Working Conference on Performance Modelling and Evaluation of Heterogeneous Networks (HET-NETs 2004), Ilkley, UK, vol. P81/1–13 (July 2004)
48. Stewart, W.J.: Introduction to Numerical Solution of Markov Chains. Princeton University Press, Princeton (1994)
49. Tran, H.T., Do, T.V.: An iterative method for queueing systems with batch arrivals and batch departures. In: Proceedings of the 8th IFIP Workshop on Performance Modelling and Evaluation of ATM&IP Networks, Ilkley, UK, vol. 80/1–13 (2000)
50. Tran, H.T., Do, T.V.: A new iterative method for systems with batch arrivals and batch departures. In: Proceedings of CNDS 2000, San Diego, USA, pp. 131–137 (2000)
51. Tran, H.T., Do, T.V., Ziegler, T.: Analysis of MPLS-compliant Nodes Deploying Multiple LSPs Routing. In: Proceedings of the First International Working Conference on Performance Modelling and Evaluation of Heterogeneous Networks, HET-NETs, pp. 26/1–26/10 (July 2003)
52. Tran, H.T., Do, T.V.: Computational Aspects for Steady State Analysis of QBD Processes. In: Periodica Polytechnica, Ser. El. Eng., pp. 179–200 (2000)
53. Tran, H.T., Do, T.V.: Generalised invariant Subspace based Method for Steady State Analysis of QBD-M Proceses. Periodica Polytechnica, Ser. El. Eng. 44, 159–178 (2000)
54. Tran, H.T., Do, T.V.: Comparison of some Numerical Methods for QBD-M Processes via Analysis of an ATM Concentrator. In: Proceedings of 20th IEEE International Performance, Computing and Communications Conference, IPCCC 2001, Pheonix, USA (2001)
55. Wallace, V.L.: The Solution of Quasi Birth and Death Processes Arising from multiple Access Computer Systems. PhD thesis, University of Michigan (1969)
56. Wierman, A., Osogami, T., Harchol-Balter, M., Scheller-Wolf, A.: How many servers are best in a dual-priority M/PH/k system? Perform. Eval. 63(12), 1253–1272 (2006)
57. Wüchner, P., Sztrik, J., de Meer, H.: Finite-source M/M/S retrial queue with search for balking and impatient customers from the orbit. Computer Networks 53(8), 1264–1273 (2009)
58. Zhao, Y., Grassmann, W.K.: A numerically stable algorithm for two server queue models. Queueing Syst. 8(1), 59–79 (1991)

Some New Markovian Models for Traffic and Performance Evaluation of Telecommunication Networks

Ram Chakka[1] and Tien Van Do[2]

[1] Meerut Institute of Engineering and Technology (MIET),
Meerut, India 250005
ramchakka@yahoo.com

[2] Department of Telecommunications
Budapest University of Technology and Economics
H-1117, Magyar tudósok körútja 2., Budapest, Hungary
do@hit.bme.hu

Abstract. A new queueing model is proposed, namely the $MM\sum_{k=1}^{K}CPP_k/GE/c/L$ G-Queue, also known as the *Sigma* queue. A steady state solution for this queue in a computable form is presented, based on an analytic methodology applying certain non-trivial transformations on the balance equations, which turn out to be of the QBD-M (Quasi Simultaneous-Multiple-Bounded Births and Deaths) type. As a consequence, existing methods such as the spectral expansion method or the matrix-geometric method can be used to obtain the steady state probabilities. The utility of this queue is demonstrated through the performance analysis of an optical burst switching (OBS) node. Numerical results show the impact of the burstiness of arrival process on the performance of the OBS multiplexer.

1 Introduction

It is well known that the traffic in today's telecommunication systems often exhibits burstiness, that is batches of transmission units (e.g. packets) arriving together, with correlations among the inter-arrival times. Several models have been proposed to model such arrival, service processes and the queues and networks with these processes. These include the compound Poisson process (CPP) in which the inter-arrival times have generalized exponential (GE) probability distribution [4], the Markov modulated Poisson process (MMPP) and the self-similar traffic models such as the Fractional Brownian Motion (FBM) [15]. The CPP and the $\sum_{k=1}^{K}CPP_k$ traffic models often give a good representation of the burstiness (batch size distribution) of the traffic from one or more sources [9], but not the auto-correlations of the inter-arrival times observed in real traffic. Conversely, the MMPP models can capture the auto-correlations but not the burstiness [11,16]. The self-similar traffic models such as the FBM can represent both burstiness and auto-correlations, but they are analytically intractable in a queuing context. The arriving traffic to a node is often the superposition of traffic from a number of sources, which complicates the system analysis further.

In order to make the modeling capability vastly flexible and also to accommodate the superposition of multiple arrival streams, a new traffic and queuing model, the *Markov modulated* $\sum_{k=1}^{K} CPP_k/GE/c/L\,G$-queue is introduced in this paper. This is a homogeneous multi-server queue with c servers, GE service times and with the superposition of K independent positive and an independent negative[1] customer arrival streams each of which is a CPP, i.e. a Poisson point process with batch arrivals of geometrically distributed batch size. In other words, inter-arrival times of each of these $K+1$ arrival streams are also independent GE random variables. Also, all the $K+2$ GE distributions (K positive and 1 negative customer inter-arrival times, and the service time) are jointly modulated by a continuous time Markov chain (CTMC), also termed as the modulating process. The notational representation of this new queue is the $MM \sum_{k=1}^{K} CPP_k/GE/c/L\,G$-queue. It is also termed as the *Sigma* queue, for easier reference. Thus, the *Sigma* queue and its extensions can capture a large class of traffic and queuing models applicable to todays Internet in the Markovian framework.

We propose the $MM \sum_{k=1}^{K} CPP_k/GE/c/L\,G$-Queue and present a methodology for the solution of the new queue. The new methodology applies certain transformations to the balance equations to produce a computable form (of the QBD-M type). As a consequence, the steady state solution is possible by the spectral expansion method [3,12] or an appropriate extension of Naoumov's method [13] for QBD-M processes, based on the matrix-geometric solution [14].

The queuing model and its variants were successfully used to model [5,6] High-speed Downlink Packet Access (HSDPA [1]) terminal categories. In this paper, we demonstrate the utility of this new queue through the performance analysis of an optical burst switching (OBS) node. We revisit the problem of performance analysis of an OBS multiplexer that was attempted by Turner [18]. Briefly, the multiplexer assigns arriving bursts to channels in a link with c available data channels (known as wavelengths, in OBS terminology) and storage locations for $L-c$ bursts. An arriving burst is diverted to a storage location if all c data channels are in use when it arrives. A burst will be discarded when it arrives if all c channels are busy and all $L-c$ burst storage locations are being used by bursts that have not yet been assigned a channel. However, once a stored burst has been assigned to an output channel, its storage location becomes available for use by an arriving burst, since the stored burst will vacate space. Turner has proposed a birth-death process to analyze this problem [18]. However, Turner's model has some limitations like the assumption of exponential inter-burst arrival process and constant burst size. The use of the proposed new queue overcomes those limitations. Extensive numerical study is carried out with the use of the

---

[1] Negative customers remove (positive) customers from the queue and have been used to model random neural networks, task termination in speculative parallelism, faulty components in manufacturing systems and server breakdowns [8].

new model. The results quantitatively show the impact of the burstiness of the arrival process on the performance of an optical burst switching multiplexer. It is worth emphasizing that the performance analysis of an optical burst switching node is only one of many applications for the new generalized queuing model.

The rest of the paper is organized as follows. We propose the $MM \sum_{k=1}^{K} CPP_k/GE/c/L$ G-queue in section 2. Next, we present a solution technique for the steady state joint probability distribution in section 3. We derive the departure size distribution in section 4. We show some numerical results in 5. The paper then concludes in section 6.

2 The $MM \sum_{k=1}^{K} CPP_k/GE/c/L$ G-Queue

2.1 The Arrival Process

The arrival and service processes are modulated by the same continuous time, irreducible Markov phase process with N states. Let Q be the generator matrix of this process, given by

$$Q = \begin{bmatrix} -q_1 & q_{1,2} & \cdots & q_{1,N} \\ q_{2,1} & -q_2 & \cdots & q_{2,N} \\ \vdots & \vdots & \ddots & \vdots \\ q_{N,1} & q_{N,2} & \cdots & -q_N \end{bmatrix},$$

where $q_{i,k}$ ($i \neq k$) is the instantaneous transition rate from phase i to phase k, and

$$q_i = \sum_{j=1}^{N} q_{i,j} \qquad (i = 1, \ldots, N).$$

In the above, $q_{i,i} = 0$. Let $\mathbf{r} = (r_1, r_2, \ldots, r_N)$ be the vector of equilibrium probabilities of the modulating phases. Then, $\mathbf{r}$ is uniquely determined by the equations:

$$\mathbf{r}Q = 0 \quad ; \quad \mathbf{r}\mathbf{e}_N = 1 ,$$

where $\mathbf{e}_N$ stands for the column vector with N elements, each of which is unity.

The arrival process of the customers (positive customers) ($MM \sum_{k=1}^{K} CPP_k$) is the superposition of K independent CPP arrival streams of packets (customers) in any given modulating phase. Also, in each modulating phase, there is a CPP arrivals of negative customers. The packets of different arrival streams are not distinguishable. The parameters of the GE inter-arrival time distribution of the kth ($1 \leq k \leq K$) positive customer arrival stream in phase i are ($\sigma_{i,k}, \theta_{i,k}$), and ($\rho_i, \delta_i$) are those of the negative customers. That is, the inter-arrival time probability distribution function is $1 - (1 - \theta_{i,k})e^{-\sigma_{i,k}t}$, in phase i, for the kth stream of positive customers and $1 - (1 - \delta_i)e^{-\rho_i t}$ for the negative customers. Thus, all the $K + 1$ arrival point-processes are Poisson, with batches arriving at each point having geometric size distribution. Specifically, the probability that a batch is of size s is $(1-\theta_{i,k})\theta_{i,k}^{s-1}$, in phase i, for the kth stream of positive customers, and $(1-\delta_i)\delta_i^{s-1}$ for the negative customers.

Let $\sigma_{i,.}, \overline{\sigma_{i,.}}$ be the average arrival rate of customer batches and customers in phase i respectively. Let $\sigma, \overline{\sigma}$ be the overall average arrival rate of batches and customers respectively. Then, we get

$$\sigma_{i,.} = \sum_{k=1}^{K} \sigma_{i,k}, \quad \overline{\sigma_{i,.}} = \sum_{k=1}^{K} \frac{\sigma_{i,k}}{(1-\theta_{i,k})}, \quad \sigma = \sum_{i=1}^{N} \sigma_{i,.} r_i, \quad \overline{\sigma} = \sum_{i=1}^{N} \overline{\sigma_{i,.}} r_i. \tag{1}$$

Because of the superposition of many CPP's, the overall arrivals in phase i can be considered as bulk-Poisson ($M^{[x]}$) with arrival rate $\sigma_{i,.}$ and with a batch size distribution $\{\pi_{i,l}\}$, that is, more general than mere geometric. The probability that the batch size is l ($\pi_{i,l}$) and the overall batch size distribution ($\pi_{.,l}$), can be given by

$$\pi_{i,l} = \sum_{k=1}^{K} \frac{\sigma_{i,k}}{\sigma_{i,.}} (1-\theta_{i,k})\theta_{i,k}^{l-1}, \quad \pi_{.,l} = \sum_{i=1}^{N} r_i \pi_{i,l}. \tag{2}$$

Clearly, by choosing K and other parameters appropriately, it may then be possible to approximate $\{\pi_{i,l}\}$ or $\{\pi_{.,l}\}$ to suit certain given classes of batch size distribution, however this is a matter for further research only.

2.2 The GE Multi-server

The service facility has c homogeneous servers in parallel, each with GE-distributed service times with parameters (μ_i, ϕ_i) in phase i. The service discipline is FCFS and each server serves at most one positive customer at any given time. Negative customers neither wait in the queue, nor are served. The operation of the GE server is similar to that described for the CPP arrival processes above. L denotes the queuing capacity, in all phases, including the customers in service, if any. L can be finite or infinite. We assume, when the number of customers is j and the arriving batch size of positive customers is greater than $L - j$ (assuming finite L), that only $L - j$ customers are taken in and the rest are rejected.

However, the batch size associated with a service completion is bounded by one more than the number of customers waiting to commence service at the departure instant. For queues of length $c \leq j < L+1$ (including any customers in service), the maximum batch size at a departure instant is $j - c + 1$, only one server being able to complete a service period at any one instant under the assumption of exponentially distributed batch-service times. Thus, the probability that a departing batch has size s is $(1-\phi_i)\phi_i^{s-1}$ for $1 \leq s \leq j-c$ and ϕ_i^{j-c} for $s = j - c + 1$. In particular, when $j = c$, the departing batch has size 1 with probability one, and this is also the case for all $1 \leq j \leq c$ since each customer is already engaged by a server and there are then no customers waiting to commence service.

It is assumed that the first positive customer in a batch arriving at an instant when the queue length is less than c (so that at least one server is free) *never* skips service, i.e. always has an exponentially distributed service time. However, even without this assumption the methodology described in this paper is still applicable.

2.3 Negative Customer Semantics

A negative customer removes a positive customer in the queue, according to a specified *killing discipline*. We consider here a variant of the RCE killing discipline (removal of customers from the end of the queue), where the most recent positive arrival is removed, but which does *not* allow a customer actually in service to be removed: a negative customer that arrives when there are no positive customers waiting to start service has no effect. We may say that customers in service are immune to negative customers or that the service itself is *immune servicing*. Such a killing discipline is suitable for modeling of load balancing where work is transferred from overloaded queues but never work, that is, actually in progress.

When a batch of negative customers of size l ($1 \leq l < j - c$) arrives, l positive customers are removed from the end of the queue leaving the remaining $j - l$ positive customers in the system. If $l \geq j - c \geq 1$, then $j - c$ positive customers are removed, leaving none waiting to commence service (queue length equal to c). If $j \leq c$, the negative arrivals have no effect.

$\overline{\rho_i}$, the average arrival rate of negative customers in phase i and $\overline{\rho}$, the overall average arrival rate of negative customers are given by

$$\overline{\rho_i} = \frac{\rho_i}{1 - \delta_i} \quad ; \quad \overline{\rho} = \sum_{i=1}^{N} r_i \overline{\rho_i}. \tag{3}$$

2.4 Condition for Stability

When L is finite, the system is ergodic since the representing Markov process is irreducible. Otherwise, i.e. when $L = \infty$, the overall average departure rate increases with the queue length, and its maximum (the overall average departure rate when the queue length tends to ∞) can be determined as

$$\overline{\mu} = c \sum_{i=1}^{N} \frac{r_i \mu_i}{1 - \phi_i}. \tag{4}$$

Hence, the necessary and sufficient condition for the existence of steady state probabilities is

$$\overline{\sigma} < \overline{\rho} + \overline{\mu}. \tag{5}$$

2.5 The Steady State Balance Equations

The state of the system at any time t can be specified completely by two integer-valued random variables, $I(t)$ and $J(t)$. $I(t)$ varies from 1 to N, representing the phase of the modulating Markov chain, and $0 \leq J(t) < L+1$ represents the number of positive customers in the system at time t, including any in service. The system is now modeled by a continuous time discrete state Markov process, $\overline{Y}$ (Y if L is infinite), on a rectangular lattice strip. Let $I(t)$, the phase, vary in the horizontal direction and $J(t)$, the queue length or *level*, in the vertical direction. We denote the steady state probabilities by $\{p_{i,j}\}$, where $p_{i,j} = \lim_{t \to \infty} Prob(I(t) = i, J(t) = j)$, and let $\mathbf{v}_j = (p_{1,j}, \ldots, p_{N,j})$.

The process $\overline{Y}$ evolves due to the following instantaneous transition rates:

(a) $q_{i,k}$ – purely lateral transition rate – from state (i,j) to state (k,j), for all $j \geq 0$ and $1 \leq i, k \leq N$ ($i \neq k$), caused by a phase transition in the Markov chain governing the arrival phase process ($q_{i,i} = 0$);

(b) $B_{i,j,j+s}$ – s-step upward transition rate – from state (i,j) to state $(i, j+s)$, for all phases i, caused by a new batch arrival of size s positive customers. For a given j, s can be seen as bounded when L is finite and unbounded when L is infinite;

(c) $C_{i,j,j-s}$ – s-step downward transition rate – from state (i,j) to state $(i, j-s)$, $(j - s \geq c+1)$ for all phases i, caused by either a batch service completion of size s or a batch arrival of negative customers of size s;

(d) $C_{i,c+s,c}$ – s-step downward transition rate – from state $(i, c+s)$ to state (i, c), for all phases i, caused by a batch arrival of negative customers of size $\geq s$ or a batch service completion of size s ($1 \leq s \leq L - c$);

(e) $C_{i,c-1+s,c-1}$ – s-step downward transition rate, from state $(i, c-1+s)$ to state $(i, c-1)$, for all phases i, caused by a batch departure of size s ($1 \leq s \leq L - c + 1$);

(f) $C_{i,j+1,j}$ – 1-step downward transition rate, from state $(i, j+1)$ to state (i,j), ($c \geq 2$; $0 \leq j \leq c-2$), for all phases i, caused by a single departure;

where

$$B_{i,j-s,j} = \sum_{k=1}^{K} (1 - \theta_{i,k}) \theta_{i,k}^{s-1} \sigma_{i,k}$$

$$(\forall i \,;\, 0 \leq j - s \leq L - 2 \,;\, j - s < j < L) \,;$$

$$B_{i,j,L} = \sum_{k=1}^{K} \sum_{s=L-j}^{\infty} (1 - \theta_{i,k}) \theta_{i,k}^{s-1} \sigma_{i,k} = \sum_{k=1}^{K} \theta_{i,k}^{L-j-1} \sigma_{i,k}$$

$$(\forall i \,;\, j \leq L - 1) \,;$$

$$C_{i,j+s,j} = (1-\phi_i)\phi_i^{s-1}c\mu_i + (1-\delta_i)\delta_i^{s-1}\rho_i$$

$$(\forall i \; ; \; c+1 \le j \le L-1 \; ; \; 1 \le s \le L-j) \; ;$$
$$= (1-\phi_i)\phi_i^{s-1}c\mu_i + \delta_i^{s-1}\rho_i$$

$$(\forall i \; ; \; j = c \; ; \; 1 \le s \le L-c) \; ;$$
$$= \phi_i^{s-1}c\mu_i$$
$$(\forall i \; ; \; j = c-1 \; ; \; 1 \le s \le L-c+1) \; ;$$
$$= 0 \quad (\forall i \; ; \; c \ge 2 \; ; \; 0 \le j \le c-2 \; ; \; s \ge 2) \; ;$$
$$= (j+1)\mu_i$$
$$(\forall i \; ; \; c \ge 2 \; ; \; 0 \le j \le c-2 \; ; \; s = 1) \; .$$

Define,

$$B_{j-s,j} = \text{Diag}\,[B_{1,j-s,j}, B_{2,j-s,j}, \ldots, B_{N,j-s,j}]$$
$$(j-s < j \le L) \; ;$$
$$B_s = B_{j-s,j} \quad (j < L)$$
$$= \text{Diag}\left[\ldots, \sum_{k=1}^{K}\sigma_{i,k}(1-\theta_{i,k})\theta_{i,k}^{s-1}, \ldots\right] \; ;$$
$$\Sigma_k = \text{Diag}\,[\sigma_{1,k}, \sigma_{2,k}, \ldots, \sigma_{N,k}] \; (k = 1, 2, \ldots, K) \; ;$$
$$\Theta_k = \text{Diag}\,[\theta_{1,k}, \theta_{2,k}, \ldots, \theta_{N,k}] \; (k = 1, 2, \ldots, K) \; ;$$
$$\Sigma = \sum_{k=1}^{K}\Sigma_k \; ;$$
$$R = \text{Diag}\,[\rho_1, \rho_2, \ldots, \rho_N] \; ;$$
$$\Delta = \text{Diag}\,[\delta_1, \delta_2, \ldots, \delta_N] \; ;$$
$$M = \text{Diag}\,[\mu_1, \mu_2, \ldots, \mu_N] \; ;$$
$$\Phi = \text{Diag}\,[\phi_1, \phi_2, \ldots, \phi_N] \; ;$$
$$C_j = jM \quad (0 \le j \le c) \; ;$$
$$= cM = C \quad (j \ge c) \; ;$$
$$C_{j+s,j} = \text{Diag}\,[C_{1,j+s,j}, C_{2,j+s,j}, \ldots, C_{N,j+s,j}] \; ;$$
$$E = \text{Diag}(\mathbf{e}'_N) \; .$$

Then, we get,

$$B_s = \sum_{k=1}^{K}\Theta_k^{s-1}(E-\Theta_k)\Sigma_k \; ;$$

$$B_1 = B = \sum_{k=1}^{K}(E-\Theta_k)\Sigma_k \; ;$$

$$B_{L-s,L} = \sum_{k=1}^{K} \Theta_k^{s-1} \Sigma_k \; ;$$

$$C_{j+s,j} = C(E-\Phi)\Phi^{s-1} + R(E-\Delta)\Delta^{s-1}$$

$$(c+1 \leq j \leq L-1 \; ; \; s = 1,2,\ldots,L-j) \; ;$$
$$= C(E-\Phi)\Phi^{s-1} + R\Delta^{s-1}$$
$$(j = c \; ; \; s = 1,2,\ldots,L-c) \; ;$$
$$= C\Phi^{s-1}$$
$$(j = c-1 \; ; \; s = 1,2,\ldots,L-c+1) \; ;$$
$$= 0 \; (c \geq 2 \; ; \; 0 \leq j \leq c-2 \; ; \; s \geq 2) \; ;$$
$$= C_{j+1} \; (c \geq 2 \; ; \; 0 \leq j \leq c-2 \; ; \; s = 1) \; .$$

The steady state balance equations are
- for the Lth row or level:

$$\sum_{s=1}^{L} \mathbf{v}_{L-s} B_{L-s,L} + \mathbf{v}_L \left[Q - C - R\right] = 0 \; ; \qquad (6)$$

- for the jth row or level:

$$\sum_{s=1}^{j} \mathbf{v}_{j-s} B_s + \mathbf{v}_j \left[Q - \Sigma - C_j - R I_{j>c}\right] +$$
$$\sum_{s=1}^{L-j} \mathbf{v}_{j+s} C_{j+s,j} = 0 \quad (0 \leq j \leq L-1) \; ; \qquad (7)$$

- normalization

$$\sum_{j=0}^{L} \mathbf{v}_j \mathbf{e}_N = 1 \; . \qquad (8)$$

Note that $I_{j>c} = 1$ if $j > c$ else 0, and $\mathbf{e}_N$ is a column vector of size N with all ones.

2.6 Performance Measures

Some performance measures can be derived as follows:
- Average number of bursts in the system

$$E(j) = \sum_{j=0}^{L} j \mathbf{v}_j \mathbf{e}. \qquad (9)$$

- Burst loss probability

$$\sum_{i=1}^{N} \sum_{j=0}^{L} \sum_{l=L-j+1}^{\infty} p_{i,j} \pi_{i,l} \frac{(l - (L-j))\sigma_{i,\cdot}}{\sigma}. \qquad (10)$$

3 Solution Methodology and Technique

3.1 Transforming the Balance Equations

When $L < c + K + 4$, then the number of the states of the Markov process Y is not large and can be solved by traditional methods [17]. However, when L is large, computationally more efficient other methods are available.

In this Section the balance equations are essentially transformed into a computable form (of QBD-M type). The necessary mathematical transformations are based on profound theoretical analysis and proofs. Moreover, they are very convenient for (software or program) implementation.

Let the balance equations for level j be denoted by $<\mathbf{j}>$. Hence, $<\mathbf{0}>$, $<\mathbf{1}>$, ..., $<\mathbf{j}>$, ..., $<\mathbf{L}>$ are the balance equations for the levels $0, 1, \ldots, j, \ldots, L$ respectively. Substituting $B_{L-s,L} = \sum_{k=1}^{K} \Theta_k^{s-1} \Sigma_k$ and $B_s = \sum_{k=1}^{K} \Theta_{.k}^{s-1}(E - \Theta_k) \Sigma_k$ in (6, 7), we get the balance equations for level L and for all the other levels as:

$$<\mathbf{L}> : \qquad \sum_{s=1}^{L} \sum_{k=1}^{K} \mathbf{v}_{L-s} \Theta_k^{s-1} \Sigma_k + \mathbf{v}_L [Q - C - R] = 0;$$

$$<\mathbf{L-1}> : \qquad \sum_{s=1}^{L-1} \sum_{k=1}^{K} \mathbf{v}_{L-1-s} \Theta_k^{s-1}(E - \Theta_k) \Sigma_k + \mathbf{v}_{L-1}[Q - \Sigma - C_{L-1} - R] + \mathbf{v}_L C_{L,L-1} = 0;$$

$$<\mathbf{L-2}> : \qquad \sum_{s=1}^{L-2} \sum_{k=1}^{K} \mathbf{v}_{L-2-s} \Theta_k^{s-1}(E - \Theta_k) \Sigma_k + \mathbf{v}_{L-2}[Q - \Sigma - C_{L-2} - R]$$
$$+ \sum_{s=1}^{2} \mathbf{v}_{L-2+s} C_{L-2+s,L-2} = 0;$$

$$<\mathbf{L-3}> : \sum_{s=1}^{L-3} \sum_{k=1}^{K} \mathbf{v}_{L-3-s} \Theta_k^{s-1}(E - \Theta_k) \Sigma_k + \mathbf{v}_{L-3}[Q - \Sigma - C_{L-3} - R]$$
$$+ \sum_{s=1}^{3} \mathbf{v}_{L-3+s} C_{L-3+s,L-3} = 0;$$

$$\vdots$$

$$<\mathbf{j}> : \qquad \sum_{s=1}^{j} \sum_{k=1}^{K} \mathbf{v}_{j-s} \Theta_k^{s-1}(E - \Theta_k) \Sigma_k \mathbf{v}_j [Q - \Sigma - C_j - R]$$
$$+ \sum_{s=1}^{L-j} \mathbf{v}_{j+s} C_{j+s,j} = 0 \quad (j = L-4, L-5, \ldots, c+K+2);$$

$$<\mathbf{c+K+1}> \ : \ \sum_{s=1}^{c+K+1}\sum_{k=1}^{K} \mathbf{v}_{c+K+1-s}\Theta_k^{s-1}(E-\Theta_k)\Sigma_k +$$

$$\mathbf{v}_{c+K+1}\left[Q - \Sigma - C_{c+K+1} - R\right] +$$

$$\sum_{s=1}^{L-c-K-1} \mathbf{v}_{c+K+1+s}C_{c+K+1+s,c+K+1} = 0;$$

$$\vdots$$

$$<\mathbf{j}> \ : \ \sum_{s=1}^{j}\sum_{k=1}^{K} \mathbf{v}_{j-s}\Theta_k^{s-1}(E-\Theta_k)\Sigma_k + \mathbf{v}_j\left[Q - \Sigma - C_j - RI_{j>c}\right]$$

$$+ \sum_{s=1}^{L-j} \mathbf{v}_{j+s}C_{j+s,j} = 0 \quad (j = c+K, c+K-1, \ldots, 0).$$

Define the functions, $F_{K,l}$, $(l = 1, 2, \ldots, K)$ as,

$$F_{K,l} = \sum_{\substack{1 \leq k_1, k_2, \ldots, k_l \leq K \\ k_1 \neq k_2 \neq \ldots \neq k_l}} \Theta_{k_1}\Theta_{k_2}\ldots\Theta_{k_l} \quad (l = 1, 2, \ldots, K)$$

$$= E \quad \text{if } l = 0$$

$$= 0 \quad \text{if } l \leq -1 \text{ or } l > K \tag{11}$$

From the above, for example,

$$K = 1: \quad F_{1,0} = E; F_{1,1} = \Theta_1$$
$$K = 2: \quad F_{2,0} = E; F_{2,1} = \Theta_1 + \Theta_2; F_{2,2} = \Theta_1\Theta_2$$
$$K = 3: \quad F_{3,0} = E; F_{3,1} = \Theta_1 + \Theta_2 + \Theta_3;$$
$$F_{3,2} = \Theta_1\Theta_2 + \Theta_2\Theta_3 + \Theta_3\Theta_1; F_{3,3} = \Theta_1\Theta_2\Theta_3.$$

The required $F_{K,l}$'s can be computed easily using the following properties and the recursion which also can be considered as an alternate definition for $F_{K,l}$'s. Let the arrival streams be numbered as $1, 2, \ldots, K$ in any order. Define $F_{k,l}$ be the function in equation (11) when only the first k streams of customer arrivals are present and the rest are absent. Then, we get

$$F_{k,0} = E \ , \ F_{k,k} = \prod_{n=1}^{k}\Theta_n \quad (k = 1, 2, \ldots, K);$$

$$F_{k,l} = 0 \ (k = 1, 2, \ldots, K; l < 0)$$

$$F_{k,l} = 0 \ (k = 1, 2, \ldots, K; l > k) \tag{12}$$

The recursion, then, is

$$F_{1,0} = E \ ; \ F_{1,1} = \Theta_1 \ ;$$
$$F_{k,l} = F_{k-1,l} + \Theta_k F_{k-1,l-1}$$
$$(2 \leq k \leq K, \ 1 \leq l \leq k-1) \ . \tag{13}$$

Transformation 1. *Modify simultaneously the balance equations for levels j ($L - 1 \geq j \geq K$), by the transformation:*

$$<\mathbf{j}>^{(1)} \longleftarrow <\mathbf{j}> + \sum_{l=1}^{K} (-1)^l <\mathbf{j}-\mathbf{l}> F_{K,l}$$
$$(K \leq j \leq L - 1)$$
$$<\mathbf{j}>^{(1)} \longleftarrow <\mathbf{j}> \qquad (j = L \text{ or } j < K).$$

Apply the following transformation to the resulting equations.

Transformation 2. *Modify simultaneously the balance equations for levels j ($L - 3 \geq j \geq c + K + 1$), by the transformation:*

$$<\mathbf{j}>^{(2)} \longleftarrow <\mathbf{j}>^{(1)} -$$
$$<\mathbf{j}+\mathbf{1}>^{(1)}(\Phi + \Delta) + <\mathbf{j}+\mathbf{2}>^{(1)}\Phi\Delta$$
$$(c + K + 1 \leq j \leq L - 3)$$
$$<\mathbf{j}>^{(2)} \longleftarrow <\mathbf{j}>^{(1)}$$
$$(j > L - 3 \text{ or } j < c + K + 1).$$

With these above two transformations, the transformed balance equation, $<\mathbf{j}>^{(2)}$'s, for the rows ($c + K + 1 \leq j \leq L - 3$), will be of the form:

$$\mathbf{v}_{j-K} Q_0 + \mathbf{v}_{j-K+1} Q_1 + \ldots + \mathbf{v}_{j+2} Q_{K+2} = 0$$
$$(j = L - 3, L - 4, \ldots, c + K + 1), \tag{14}$$

where $Q_0, Q_1, \ldots, Q_{K+2}$ are $K+3$ number of j-independent matrices and can be easily derived algebraically from the system parameters by following the above mentioned transformation procedures.

Thus, the resulting equations (14) corresponding to the rows from $c + K + 1$ to $L - 3$, are of the same form as those of the QBD-M processes and hence have an efficient solution by several methods such as the spectral expansion method, Bini-Meini's method or the matrix-geometric method with folding or block size enlargement [10].

3.2 Spectral Expansion Solution of the Balance Equations

The set of equations (14) concerning the levels $c + K + 1$ to $L - 3$ have the coefficient matrices $Q_0, Q_1, \ldots, Q_{K+2}$ that are independent of j and hence have an efficient solution by the spectral expansion method [12]. These Q_l's ($l = 0, 1, \ldots, K + 2$) can be obtained quite easily following the computational procedure in Appendix A, A.1.

Define the matrix polynomials $Z(\lambda)$ and $\overline{Z}(\xi)$ as,

$$Z(\lambda) = Q_0 + Q_1 \lambda + Q_2 \lambda^2 + \ldots + Q_{K+2} \lambda^{K+2}, \tag{15}$$
$$\overline{Z}(\xi) = Q_{K+2} + Q_{K+1} \xi + Q_K \xi^2 + \ldots + Q_0 \xi^{K+2}. \tag{16}$$

Then, the spectral expansion solution for $\mathbf{v}_j$ $(c+1 \leq j \leq L-1)$ is given by

$$\mathbf{v}_j = \sum_{l=1}^{KN} a_l \psi_l \lambda_l^{j-c-1} + \sum_{l=1}^{2N} b_l \gamma_l \xi_l^{L-1-j} \quad (c+1 \leq j \leq L-1), \tag{17}$$

where λ_l $(l = 1, 2, \ldots, KN)$ are the KN eigenvalues of least absolute value of the matrix polynomial $Z(\lambda)$ and ξ_l $(l = 1, 2, \ldots, 2N)$ are the $2N$ eigenvalues of least absolute value of the matrix polynomial $\overline{Z}(\xi)$. ψ_l and γ_l are the left-eigenvectors of $Z(\lambda)$ and $\overline{Z}(\xi)$ respectively, corresponding to the eigenvalues λ_l and ξ_l respectively. a_l's and b_l's are arbitrary constants to be determined later.

It can be proved that the matrix $\sum_{l=0}^{K+2} Q_l$ is singular, so that $\lambda = 1$ is an eigenvalue on the unit-circle for both $Z(\lambda)$ and $\overline{Z}(\xi)$. If (5) is satisfied, the number of eigenvalues of $Z(\lambda)$ that are strictly within the unit circle is KN. If (5) is not satisfied, that number is $KN - 1$.

These and also certain other properties of these eigenvalues, Eigenvectors, also the relevant spectral analysis are dealt with (some of them are proved, others explained in detail) in [12,2]. Some of them are summarized below. Let the rank of Q_0 be $N - n_0$ and that of Q_{K+2} be $N - n_{K+2}$. Then,

(a) $Z(\lambda)$ would have $d = (K+2)N - n_{K+2}$ eigenvalues of which n_0 are zero eigenvalues (also referred to as null eigenvalues), whereas $\overline{Z}(\xi)$ would have n_{K+2} zero eigenvalues and $(K+2)N - n_0 - n_{K+2}$ non-zero eigenvalues.
(b) If $(\lambda \neq 0, \psi)$ is a non-zero eigenvalue-eigenvector pair of $Z(\lambda)$, then there exists a corresponding non-zero eigenvalue-eigenvector pair, $(\xi = \frac{1}{\lambda}, \gamma = \psi)$ for $\overline{Z}(\xi)$. Thus, the non-zero eigenvalues of these two matrix polynomials are mutually reciprocal.
(c) The KN eigenvalues of least absolute value of $Z(\lambda)$ and the $2N$ eigenvalues of least absolute value of $\overline{Z}(\xi)$ lie either strictly inside, or on, their respective unit-circles, but not outside.
(d) There is no problem posed by multiple eigenvalues, i.e. independent eigenvectors having coincident eigenvalues, since each pair (λ, ψ) is *distinct*.

If the unknowns a_l's and b_l's are determined in such a way that all the balance equations are satisfied, then the vectors $\mathbf{v}_j$ $(c+1 \leq j \leq L-1)$ can be computed from the steady state solution (17). Hence, the unknowns are the scalars, $a_1, a_2, \ldots, a_{K \cdot N}, b_1, b_2, \ldots, b_{2N}$, and the vectors $\mathbf{v}_0, \mathbf{v}_1, \ldots, \mathbf{v}_c, \mathbf{v}_L$. These are totally, $(c+2)N + K \cdot N + 2N = (c+K+4)N$ scalar unknowns. In order to solve for them, we still have the transformed balance equations concerning the levels $0, 1, \ldots, c+K, L-2, L-1, L$ and also the equation (8). These are $(c+K+4)N+1$ linear simultaneous equations in the above $(c+K+4)N$ scalar unknowns. Of these equations only $(c+K+4)N$ equations (including equation (8)) are independent. Hence, all these unknowns can be solved for.

As far as the computational complexity is regarded the solution of $(c+K+4)N$ equations requires $O(c^3N^3 + K^3N^3)$ work and a generalized nonsymmetric eigenvalue-problem of dimension $(K+2)N$ is solved at the expense of $O(K^3N^3)$ amount of work. Therefore, the computational complexity of the proposed method is $O(c^3N^3 + K^3N^3)$.

3.3 System with Infinite Queuing Capacity

So far the analysis has been for the case of finite L. A corresponding analysis for the case of infinite queuing capacity yields the spectral expansion solution as follows:

$$\mathbf{v}_j = \sum_{l=1}^{KN} a_l \psi_l \lambda_l^{j-c} \quad (j = c+1, c+2, \ldots). \tag{18}$$

Here, we need only the KN relevant eigenvalues-eigenvectors of $\Sigma(\lambda)$, and the KN a_k's (which exist as real or complex conjugate pairs) are to be computed. Notice that the equation (18) is the same as (17) when the limit $L \to \infty$ is taken. Notice also that the computation time for this case would be much less than that for finite L.

4 Notes on the Departure Process

4.1 Departure Batch Size Distribution

From Sect. 3.2, we have the solution for the steady state probabilities, $p_{i,j}$. The marginal probabilities, $p_{i.}$ and $p_{.j}$ are then defined as:

$$p_{i.} = \sum_{j=0}^{L} p_{i,j} \;; p_{.j} = \sum_{i=0}^{L} p_{i,j}. \tag{19}$$

Now consider the system in the state (i,j), where $j > c$. Here, all the c servers are busy, with $j-c$ unattended positive customers in the queue. In this state, the departure rate associated with a batch size of s is $(1-\phi_i)\phi_i^{s-1}c\mu_i$ for $1 \le s \le j-c$ and for $s = j - c + 1$. Hence, the average rate at which batches of size n, for $2 \le n \le L - c + 1$, depart from the queue is

$$\nu_n = \sum_{i=1}^{N} \sum_{j=c+n}^{L} p_{i,j}(1-\phi_i)\phi_i^{n-1} c\mu_i + \sum_{i=1}^{N} p_{i,c+n-1} \phi_i^{n-1} c\mu_i \quad (n = 2, \ldots, L-c+1). \tag{20}$$

The average rate of single departures is

$$\nu_1 = \sum_{i=1}^{N} \sum_{j=1}^{c} p_{i,j} j \mu_i + \sum_{i=1}^{N} \sum_{j=c+1}^{L} p_{i,j}(1-\phi_i) c\mu_i. \tag{21}$$

Thus, by the Law of Large Numbers for Markov chains, $\frac{\nu_n}{\sum_{s=1}^{L-c+1} \nu_s}$ is the equilibrium probability that the batch size is n $(n = 1, 2, \ldots, L - c + 1)$.

The number of batch departures per unit time ν is

$$\nu = \sum_{s=1}^{L-c+1} \nu_s, \tag{22}$$

and the average departure rate of positive customers $\overline{\nu}$

$$\overline{\nu} = \sum_{s=1}^{L-c+1} s\nu_s. \tag{23}$$

Hence the loss rate of positive customers due to either overflow or being killed by negative customers is $\overline{\sigma} - \overline{\nu}$.

4.2 System at Departure Epochs

Let $\beta_{i,j}$ be the probability that the state of the system is (i,j) immediately after a (batch) departure epoch. Then, $\beta_{i,j}$ is proportional to the probability flux into state (i,j) due to a departure, i.e. $\beta_{i,j} \propto f_{i,j}$ where, for $1 \leq i \leq N$,

$$f_{i,j} = \sum_{n=1}^{L-j} p_{i,j+n} c\mu_i (1-\phi_i)\phi_i^{n-1} \qquad (j \geq c)$$

$$f_{i,c-1} = \sum_{n=1}^{L-c+1} p_{i,c-1+n} c\mu_i \phi_i^{n-1}$$

$$f_{i,j} = p_{i,j+1}(j+1)\mu_i \qquad (0 \leq j \leq c-2).$$

The normalisation constant is the reciprocal of the sum of all the $f_{i,j}$, i.e. the reciprocal of

$$\sum_{i=1}^{N} \left[\sum_{j=c}^{\infty} \sum_{n=1}^{L-j} p_{i,j+n} c\mu_i (1-\phi_i)\phi_i^{n-1} + \sum_{n=1}^{L-c+1} p_{i,c-1+n} c\mu_i \phi_i^{n-1} + \sum_{j=0}^{c-2} p_{i,j+1}(j+1)\mu_i \right].$$

5 Numerical Results

5.1 Case Study

In this section we present some numerical results to demonstrate the capabilities of the proposed model. We consider an optical burst switching[2] multiplexer that assigns arriving bursts to channels in a link with c available data channels (wavelengths) and storage locations for $L-c$ bursts. An arriving burst is diverted to a storage location if all c data channels are in use when it arrives. A burst will be discarded if it arrives when all c channels are busy and all $L-c$ burst storage locations are being used by bursts that have not yet been assigned a channel. However, once a stored burst has been assigned to an output channel, its storage location becomes available for use by an arriving burst, since the stored burst will vacate space at the same rate that the arriving burst occupies it. It is apparent

---

[2] Packet and burst switching have been proposed for optical networks because they are better to accommodate bursty traffic generated by IP applications. Optical burst switched (OBS) networks switch packets of constant or variable length while the payload data stays in the optical domain [18,19,21,7,18,20].

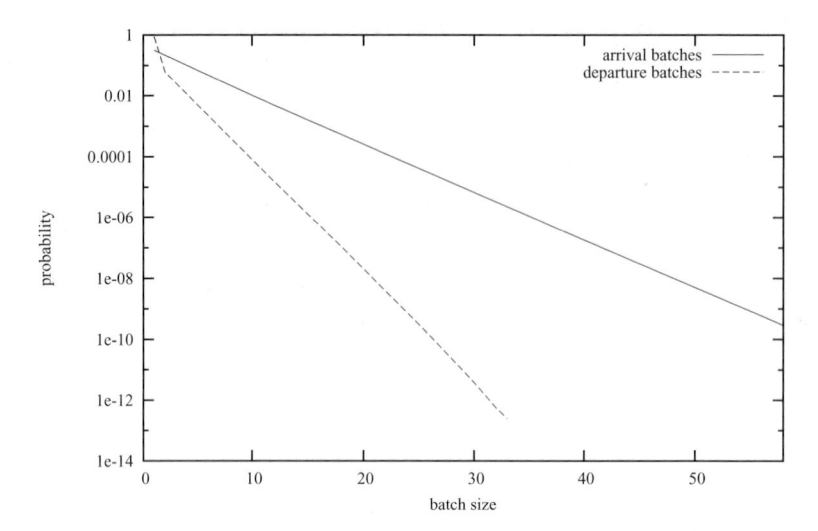

Fig. 1. Arrival and departure batch size distribution at load=0.5, $c = 32$, $L = 64$

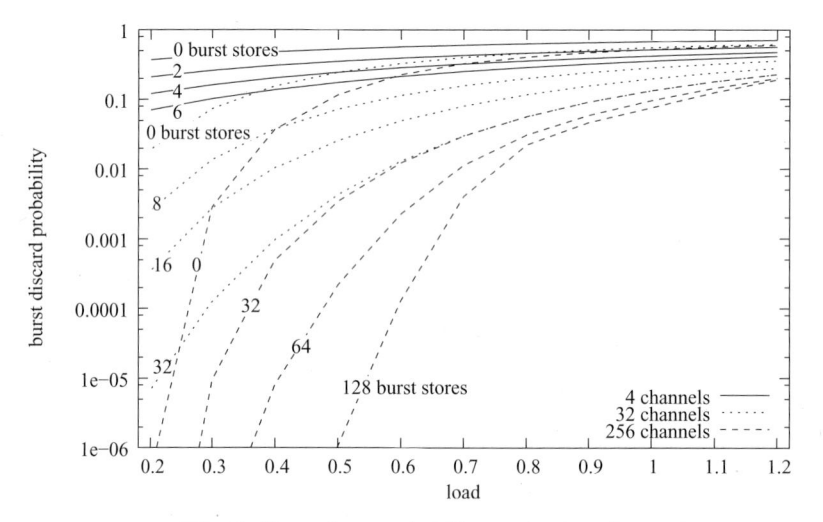

Fig. 2. Burst loss probability vs load and c

that an optical burst switching multiplexer can be modeled as a queue of c servers and a buffer of size $L - c$. In the literature, Turner has proposed one dimensional birth-death process to analyze this problem [18]. However, Turner's model has some limitations like the assumption of exponential burst arrival process and constant burst size. Our model overcomes these limitations.

The aim is to show the impact of the burstiness of the offered traffic on the performance of the multiplexer. To compare with Turner's model, we produced the similar figure as presented in [18]. However, the numerical results are obtained with different arrival and serving processes, and the assumption of no negative customers. The following system parameters were used:

$$[q_{i,j}] = \begin{bmatrix} -0.2 & 0.2 \\ 0.9 & -0.9 \end{bmatrix};$$

$$[\sigma_{i,k}] = \begin{bmatrix} 1 & 2 \\ 2 & 2.5 \end{bmatrix}; \quad [\theta_{i,k}] = \begin{bmatrix} 0.65 & 0.7 \\ 0.65 & 0.7 \end{bmatrix};$$

$$[\phi_i] = \begin{bmatrix} 5 & 5 \end{bmatrix}; \quad [\mu_i] = \begin{bmatrix} 0.5 & 0.5 \end{bmatrix},$$

where all $\sigma_{i,k}$ was scaled as appropriate to set the system load to the examined values and the number of positive sources (K) was chosen to be 2. The batch size distribution was chosen to be geometric (see Fig. 1).

The departure batch size, however, is a function of steady-state system behaviour, and thus all system parameters. Its distribution has a finite support set consisting of the states $\{1, 2, ..., L - c + 1\}$ as illustrated in Figure 1.

Figure 2 plots the burst discard probabilities for different channel numbers with different space available for burst storage. Space allocated for burst storage is printed on the lines in the figure. It can be observed that batch arrivals can be better handled by increasing the storage space (at the expense of some queuing delay) than by increasing the number of channels. The performance of 256 channels with no burst storage space is worse than that of 32 channels with a storage space for 8 bursts in our example for relative load values above 0.4.

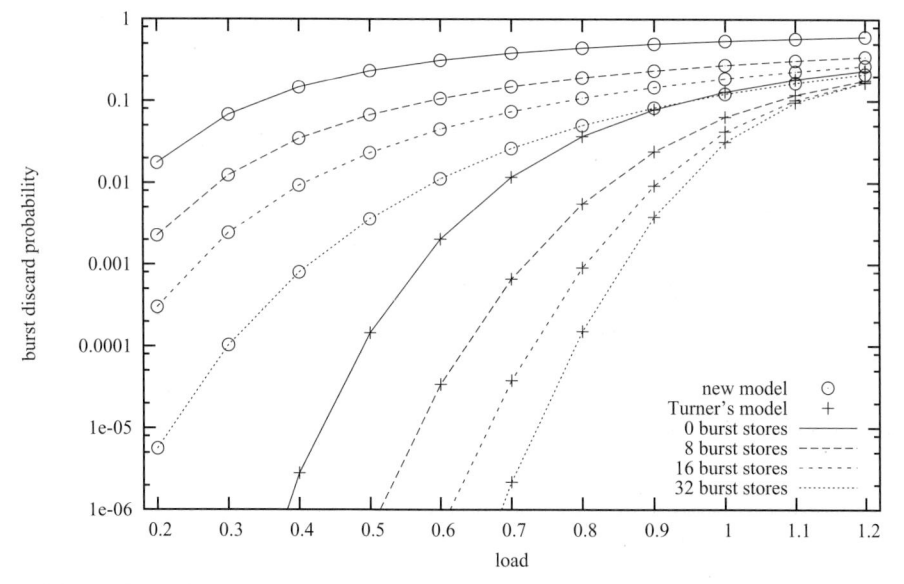

Fig. 3. Burst loss probability vs load: comparison with Turner's model ($c = 32$)

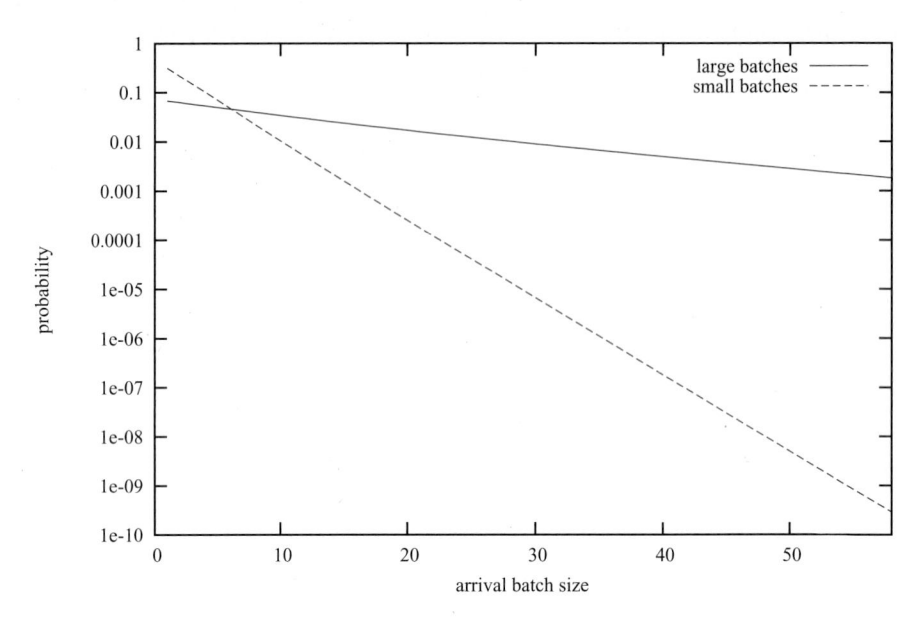

Fig. 4. Arrival batch distributions

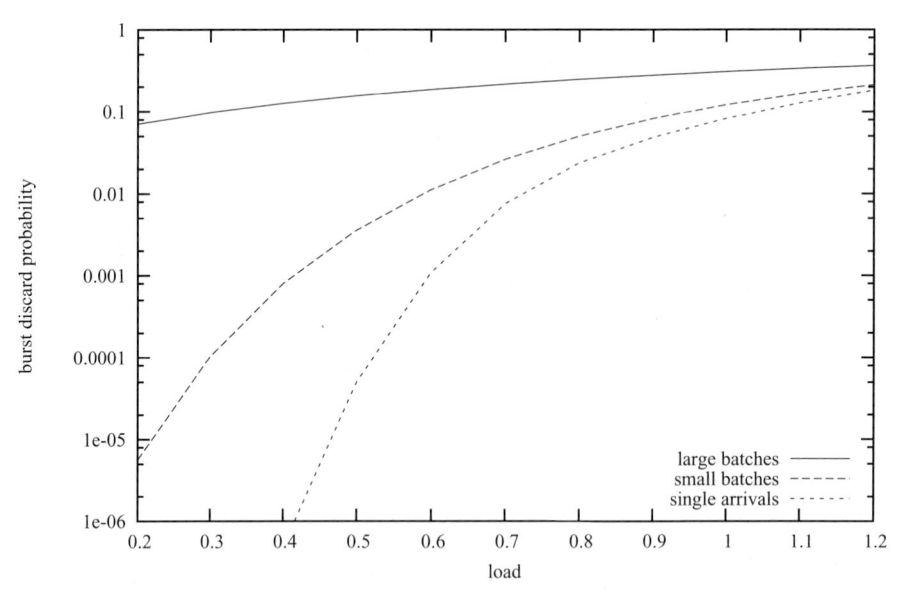

Fig. 5. Burst loss probability vs load: impact of arrival burstiness ($c = 32, L = 64$) with the arrival batch distributions in Fig. 4

Figure 3 depicts the results obtained with Turner's model and our model in the case of 32 channels. Turner's model shows that burst discard probability may be

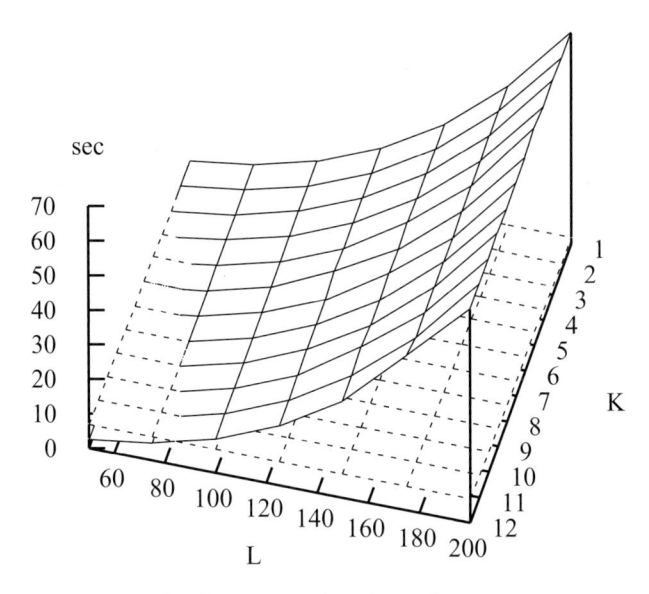

Fig. 6. Runtimes of different methods for different input parameters

kept under 10^{-6} for load values almost as high as 0.7 with a buffer of adequate size. However, batch arrivals make the situation worse (high utilization is very hard to be realized). The impact of burst arrival is further demonstrated in Figures 4 and 5, where different burst arrival distributions are used. The figures show that when the higher the probability of long batches is, the more significant performance loss is encountered at the same system for the same load values.

Figure 6 demonstrates the trade-off between the two solution methods. The implementations were run on a SUN Ultra 60 Workstation for different values of L and K, while c was always 3. The displayed times were averaged over five runs. It can be observed that the performance of the spectral expansion method overcomes the traditional method in a large ranges of parameters, moreover and it does not depends on the buffer size. Moreover, it takes even less time in the case of infinite buffer.

Further reduction in the computation complexity can be obtained for the new method based on the spectral expansion if we express the vectors $\mathbf{v}_0, \mathbf{v}_1, \ldots, \mathbf{v}_c, \mathbf{v}_L$ as the functions of the unknowns a_l's and b_l's. At the present the implementation of this idea is in progress.

6 Conclusions

One of the research aims in the performance evaluation of telecommunications networks is to find analytically tractable queuing models with the capability of capturing the burstiness and correlation of the traffic. In this context, the first contribution of this paper is the derivation of the *exact* result for the equilibrium state probabilities of the $MM \sum_{k=1}^{K} CPP_k/GE/c/LG$-queue, which can capture the burstiness and correlation of the traffic and take into account environmentally sensitive service times. We also provide ideas to extend the modeling capability of the new queue.

The queuing model and its variants were successfully used to model [5,6] High-speed Downlink Packet Access HSDPA terminal categories. In this paper, we have applied the new queue for the performance analysis of multiplexers in optical burst switching networks.

Acknowledgement

Ram Chakka thanks the Chancellor of Sri Sathya Sai Institute of Higher Learning, Prasanthi Nilayam, India, for constant encouragement, guidance and inspiration throughout this work.

References

1. 3GPP Technical Report 25.214, version 7.0.0. Physical layer procedures (FDD). The 3GPP project (March 2006)
2. Chakka, R.: Performance and Reliability Modelling of Computing Systems Using Spectral Expansion. PhD thesis, University of Newcastle upon Tyne (Newcastle upon Tyne) (1995)
3. Chakka, R.: Spectral Expansion Solution for some Finite Capacity Queues. Annals of Operations Research 79, 27–44 (1998)
4. Kouvatsos, D.: Entropy Maximisation and Queueing Network Models. Annals of Operations Research 48, 63–126 (1994)
5. Do, T.V., Chakka, R., Harrison, P.G.: An integrated analytical model for computation and comparison of the throughputs of the UMTS/HSDPA user equipment categories. In: MSWiM 2007: Proceedings of the 10th ACM Symposium on Modeling, Analysis, and Simulation of Wireless and Mobile Systems, pp. 45–51. ACM, New York (2007)
6. Do, T.V., Do, N.H., Chakka, R.: Performance Evaluation of the High Speed Downlink Packet Access in Communications Networks Based on High Altitude Platforms. In: Al-Begain, K., Heindl, A., Telek, M. (eds.) ASMTA 2008. LNCS, vol. 5055, pp. 310–322. Springer, Heidelberg (2008)
7. El-Bawab, T.S., Shin, J.-D.: Optical Packet Switching in Core Networks: Between Vision and Reality. IEEE Communications Magazine, 60–65 (September 2002)
8. Fourneau, J.M., Gelenbe, E., Suros, R.: G-networks with Multiple Classes of Positive and Negative Customers. Theoretical Computer Science 155, 141–156 (1996)
9. Fretwell, R.J., Kouvatsos, D.D.: ATM traffic burst lengths are geometrically bounded. In: Proceedings of the 7th IFIP Workshop on Performance Modelling and Evaluation of ATM & IP Networks, Antwerp, Belgium. Chapman and Hall, Boca Raton (1999)

10. Haverkort, B., Ost, A.: Steady state analyses of infinite stochastic Petri nets: A comparison between the spectral expansion and the matrix geometric methods. In: Proceedings of the 7th International Workshop on Petri Nets and Performance Models, Saint Malo, France, pp. 335–346 (1997)
11. Meier-Hellstern, K.S.: The analysis of a queue arising in overflow models. IEEE Transactions on Communications 37, 367–372 (1989)
12. Mitrani, I., Chakka, R.: Spectral expansion solution for a class of Markov models: Application and comparison with the matrix-geometric method. Performance Evaluation 23, 241–260 (1995)
13. Naoumov, V., Krieger, U., Wagner, D.: Analysis of a Multi-server Delay-loss System with a General Markovian Arrival Process. In: Chakravarthy, S.R., Alfa, A.S. (eds.) Matrix-Analytical Methods in Stochastic Models, pp. 43–66. Marcel Dekker, New York (1997)
14. Neuts, M.F.: Matrix Geometric Soluctions in Stochastic Model. Johns Hopkins University Press, Baltimore (1981)
15. Norros, I.: A Storage Model with Self-similar Input. Queueing Systems and their Applications 16, 387–396 (1994)
16. Shah-Heydari, S., Le-Ngoc, T.: MMPP models for multimedia traffic. Telecommunication Systems 15(3-4), 273–293 (2000)
17. Stewart, W.J.: Introduction to Numerical Solution of Markov Chains. Princeton University Press, Princeton (1994)
18. Turner, J.S.: Terabit Burst Switching. Journal of High Speed Networks 8, 3–16 (1999)
19. Venkatesh, T., Sankar, A., Jayaraj, A., Siva Ram Murthy, C.: A Complete Framework to Support Controlled Burst Retransmission in Optical Burst Switching Networks. IEEE Journal on Selected Areas in Communications 26(S-3), 65–73 (2008)
20. Yao, S., Xue, F., Mukherjee, B., Ben Yoo, S.J., Dixit, S.: Electrical Ingress Buffering and Traffic Aggregation for Optical Packet Switching and Their Effect on TCP-Level Performance in Optical Mesh Networks. IEEE Communications Magazine, 66–72 (2002)
21. Zhang, Z., Liu, L., Yang, Y.: Slotted Optical Burst Switching (SOBS) networks. Computer Communications 30(18), 3471–3479 (2007)

Appendix

A Obtaining the Matrices Q_l's

Consider any row j where $c + K + 1 \leq j \leq L - 3$. With Transformation 1, we get

$$<\mathbf{j}>^{(1)} \longleftarrow <\mathbf{j}> + \sum_{l=1}^{K} (-1)^l <\mathbf{j-1}> F_{K,l}$$

$$<\mathbf{j+1}>^{(1)} \longleftarrow <\mathbf{j+1}> + \sum_{l=1}^{K} (-1)^l <\mathbf{j+1-l}> F_{K,l}$$

$$<\mathbf{j+2}>^{(1)} \longleftarrow <\mathbf{j+2}> + \sum_{l=1}^{K} (-1)^l <\mathbf{j+2-l}> F_{K,l}. \qquad (24)$$

Applying Transformation 2 to the jth row, and substituting from the above (24) for $<\mathbf{j+1}>, <\mathbf{j+2}>$, we get

$$<\mathbf{j}>^{(2)} \longleftarrow <\mathbf{j}> + \sum_{l=1}^{K} (-1)^l <\mathbf{j-1}> F_{K,l}$$

$$- \left[<\mathbf{j+1}> + \sum_{l=1}^{K} (-1)^l <\mathbf{j+1-l}> F_{K,l} \right] (\Phi + \Delta)$$

$$+ \left[<\mathbf{j+2}> + \sum_{l=1}^{K} (-1)^l <\mathbf{j+2-l}> F_{K,l} \right] \Phi \Delta. \qquad (25)$$

Expanding and grouping the terms together, equation (25) can be written as

$$<\mathbf{j}>^{(2)} \longleftarrow \sum_{l=-2}^{K} <\mathbf{j-1}> G_{K,l}, \qquad (26)$$

where

$$G_{K,l} = (-1)^l [F_{K,l} + (\Phi + \Delta) F_{K,l+1} + \Phi \Delta F_{K,l+2}] \quad (l = -2, -1, \ldots, K). \qquad (27)$$

The balance equations $<\mathbf{j+2}>, <\mathbf{j+1}>, <\mathbf{j}>, \ldots, <\mathbf{j-1}>, \ldots, <\mathbf{j-K}>$, respectively are given by,

$$\sum_{s=1}^{j+2} \sum_{k=1}^{K} \mathbf{v}_{j+2-s} \Theta_k^{s-1} (E - \Theta_k) \Sigma_k + \mathbf{v}_{j+2} [Q - \Sigma - C_{j+2} - R]$$

$$+ \sum_{s=1}^{L-j-2} \mathbf{v}_{j+2+s} C_{j+2+s,j+2} = 0 \,;$$

$$\sum_{s=1}^{j+1}\sum_{k=1}^{K}\mathbf{v}_{j+1-s}\Theta_k^{s-1}(E-\Theta_k)\Sigma_k + \mathbf{v}_{j+1}\left[Q-\Sigma-C_{j+1}-R\right]$$
$$+ \sum_{s=1}^{L-j-1}\mathbf{v}_{j+1+s}C_{j+1+s,j+1} = 0;$$

$$\sum_{s=1}^{j}\sum_{k=1}^{K}\mathbf{v}_{j-s}\Theta_k^{s-1}(E-\Theta_k)\Sigma_k + \mathbf{v}_{j}\left[Q-\Sigma-C_{j}-R\right]$$
$$+ \sum_{s=1}^{L-j}\mathbf{v}_{j+s}C_{j+s,j} = 0;$$

$$\vdots$$

$$\sum_{s=1}^{j-l}\sum_{k=1}^{K}\mathbf{v}_{j-l-s}\Theta_k^{s-1}(E-\Theta_k)\Sigma_k + \mathbf{v}_{j-l}\left[Q-\Sigma-C_{j-l}-R\right]$$
$$+ \sum_{s=1}^{L-j+l}\mathbf{v}_{j-l+s}C_{j-l+s,j-l} = 0;$$

$$\vdots$$

$$\sum_{s=1}^{j-K}\sum_{k=1}^{K}\mathbf{v}_{j-K-s}\Theta_k^{s-1}(E-\Theta_k)\Sigma_k + \mathbf{v}_{j-K}\left[Q-\Sigma-C_{j-K}-R\right]$$
$$+ \sum_{s=1}^{L-j+K}\mathbf{v}_{j-K+s}C_{j-K+s,j-K} = 0;$$

Substituting or applying the above to (26), for the coefficients (Q_{K-m}) of $\mathbf{v}_{j-m}$ in $<\mathbf{j}>^{(2)}$, we get

$$Q_{K-m} = \sum_{l=-2}^{m-1}\left[\sum_{n=1}^{K}\Theta_n^{m-l-1}(E-\Theta_n)\Sigma_n\right]G_{K,l} + \left[Q-\Sigma-C_{j-m}-R\right]G_{K,m}$$
$$+ \sum_{l=m+1}^{K}\left[C_{j-m,j-l}\right]G_{K,l}$$
$$(m = j-L,\ldots,-2,-1,0,\ldots,K,\ldots,j). \qquad (28)$$

Also, for $m = -2,-1,0,\ldots,K$, substituting $C_{j-m} = C$ and $C_{j-m,j-l} = C_{j-l+l-m,j-l} = C(E-\Phi)\Phi^{l-m-1} + R(E-\Delta)\Delta^{l-m-1}$ in (28), we get

$$Q_{K-m} = \sum_{l=-2}^{m-1} \left[\sum_{n=1}^{K} \Theta_n^{m-l-1}(E - \Theta_n) \Sigma_n \right] G_{K,l} + [Q - \Sigma - C - R] G_{K,m}$$

$$+ \sum_{l=m+1}^{K} \left[C(E - \Phi)\Phi^{l-m-1} + R(E - \Delta)\Delta^{l-m-1} \right] G_{K,l}$$

$$(m = -2, -1, 0, \ldots, K). \tag{29}$$

Using the above, the required Q_l's can be computed easily as described in the subsection below. Notice the above Q_l's in equation (29) are j- independent.

The other coefficients, i.e. those of $\mathbf{v}_{j-K-1}, \mathbf{v}_{j-K-2}, \ldots, \mathbf{v}_0$ and of $\mathbf{v}_{j+3}, \mathbf{v}_{j+4}, \ldots$, can be shown to be zero. This is elaborated in the next section for a numerical value of K, and a general proof is given later on.

A.1 Computation

After obtaining $F_{K,l}$'s thus, $G_{K,l}$ ($l = -2, -1, \ldots, K$) can be computed from (27). Then, using them directly in (29), the required Q_l ($l = 0, 1, \ldots, K+2$) can be computed.

An alternate way of computing the $G_{K,l}$'s is by the following properties and recursion which are obtained from (12,13,27)

$$G_{1,-2} = \Phi\Delta \; ; \; G_{1,-1} = -(\Phi + \Delta) - \Phi\Delta\Theta_1 \; ;$$
$$G_{1,0} = E + (\Phi + \Delta)\Theta_1 \; ; \; G_{1,1} = -\Theta_1 \; ;$$
$$G_{k,l} = 0 \; (l \leq -3) \; ; \; G_{k,l} = 0 \; (l \geq k+1) \; ;$$
$$G_{k,l} = G_{k-1,l} - \Theta_k G_{k-1,l-1} \; (2 \leq k \leq K, \; -2 \leq l \leq k). \tag{30}$$

Another interesting property of $G_{k,l}$'s is

$$\sum_{l=-2}^{k} G_{k,l} = \sum_{l=-2}^{k} G_{k-1,l} - \Theta_k \sum_{l=-2}^{k} G_{k-1,l-1}$$

$$= \sum_{l=-2}^{k-1} G_{k-1,l} - \Theta_k \sum_{l-1=-2}^{k-1} G_{k-1,l-1}$$

(since $G_{k-1,k} = G_{k-1,-3} = 0$)

$$= \left[\sum_{l=-2}^{k-1} G_{k-1,l} \right] (E - \Theta_k). \tag{31}$$

Also, from (30), we have $\sum_{l=-2}^{1} G_{1,l} = (E - \Phi)(E - \Delta)(E - \Theta_1)$. Hence, we get the following result combining the above results:

$$\sum_{l=-2}^{k} G_{k,l} = (E - \Phi)(E - \Delta) \prod_{n=1}^{k} (E - \Theta_n). \tag{32}$$

Modelling and Analysis of a Dynamic Guard Channel Handover Scheme with Heterogeneous Call Arrival Processes

Lan Wang, Geyong Min, Demetres D. Kouvatsos, and Xiangxiang Zuo

Department of Computing, School of Computing, Informatics and Media,
University of Bradford, Bradford, BD7 1DP, UK
lanwang100@googlemail.com,
{g.min,d.kouvatsos,x.x.m.zuo}@bradford.ac.uk

Abstract. This chapter presents a detailed review of existing handover schemes and focuses on an analytical model developed for a Dynamic Guard Channel Scheme (DGCS), which manages adaptively the channels reserved for handover calls depending on the current status of the handover queue. The Poisson process and Markov-Modulated-Poisson-Process (MMPP) are used to model the arrival processes of new and handover calls, respectively. The accuracy of this model is demonstrated through the extensive comparisons of the analytical results against those obtained from discrete-event simulation experiments. Moreover, the analytical model is used to assess the effects of the number of channels originally reserved for handover calls, the number of dynamic channels and the burstiness of handover calls on the performance of the DGCS scheme.

Keywords: Wireless Cellular Networks, Handover, Guard Channel Scheme, Performance Modelling, Markov-Modulate-Poisson-Process (MMPP).

1 Introduction

In communication industry, wireless networks have undergone the fastest growth in the past decades. Mobile cellular systems have evolved from the first generation analogue system, e.g. Advanced Mobile Phone System (AMPS) and European Total Access Cellular System (ETACS), to the second generation digital systems, e.g. Global System for Mobile (GSM) and IS-95 cdmaOne. The increasing demands for parallel, rapid internet-based services drives the development of third generation mobile systems (3G). With the deployment of increasing wireless applications and access constraints, a tremendous development to improve the efficient usage of the limited radio spectrum resources in academic and industry fields is required. The reason for the cellular network topology is to enable frequency reuse.

Cellular networks deploy smaller cells in order to achieve high system capacity due to the limited spectrum. The frequency band is divided into smaller bands which are reused in non interfering cells [1-3]. Smaller cells cause an active mobile station (MS) to cross several cells during an ongoing conversation. This active call should be transferred from one cell to another in order to achieve call continuation during boundary crossing.

Handover represents a process of changing the channel associated with the current connection while a call is in progress. Handover is an essential component of cellular communication systems since it enables two adjacent base stations (BS) to guarantee the continuity of a conversation. A well designed handover scheme should minimize the probability of dropped handover calls because forced termination of new calls is more favoured than termination of handover calls in the point view of mobile users. Therefore, the design of an efficient handover scheme requires to minimize the handoff call blocking probability as well as to maintain the new call blocking probability and the utilisation of precious wireless resources at acceptable levels.

Prioritizing handover calls is an effective strategy towards these goals. There are two popular schemes for handover prioritization: Guard Channel Scheme (GCS) and Handover Queueing Scheme (HQS) [1]. The basic idea behind GCS is to give high priority to handover calls by exclusively reserving a number of dedicated (or, guard) channels. In HQS, the forced termination probability of handover calls can be reduced by allowing handover requests to be queued temporarily. Both schemes can reduce the handover call dropping probability effectively in wireless mobile networks [4].

The existing GCS and HQS schemes often reduce the blocking probability of handover calls at the expense of increasing the blocking probability of new calls and the delay of handover calls. With the aim of reducing the deteriorating effects of GCS and HQS on the blocking probability of new calls and the delay of handover calls, this chapter firstly presents an efficient handover scheme, referred to as Dynamic GCS (DGCS) which can dynamically manage the channels reserved for handover calls depending on the current status of the handover queue. Secondly, a three-dimensional Markov model is developed for performance analysis of this dynamic handover scheme. The new and handover call arrivals are modelled by the non-bursty Poisson process and bursty Markov-Modulate-Poisson-Process (MMPP), respectively. Moreover, the effects of users' mobility are taken into consideration by modelling the dwell times of both calls spent in a cell by an exponential distribution. Thirdly, the analytical mode is used to investigate the effects of the number of channels originally reserved for handover calls, the number of dynamic channels, and burstiness of handover calls on the performance of the DGCS scheme.

The rest of the chapter is organized as follows. Section 2 provides a review of handover schemes. Section 3 presents the analytical model and derives the expressions of the performance measures including the mean number of calls in the system, aggregate response time, aggregate call blocking probability, handover call blocking probability, new call blocking probability and handover delay. Section 4 validates the model and conducts extensive performance evaluation. Finally, conclusions are drawn in Section 5.

2 Review of Handover Schemes

Handover presents a process of changing channels associated with an existing connection which is always initiated either by a worse quality of received signal on the current channel or by crossing a cell boundary [1-4]. A detailed and comprehensive overview of handover is presented in this section.

2.1 Handover Type

Handovers can be categorized as hard and soft handovers according to the number of connections [5]. In the process of hard handover, a radio link is established to the new base station and the radio link of the old base station is released. This means that a mobile node is only allowed to maintain connection to one base station at one time. The handover is initiated based on a hysteresis imposed on the current link. Based on the link measurements, the target BS is selected and executes the handover, and the active connection is transferred to the target BS instantly. The connection experiences a brief interruption during the actual transfer because MS can only connect to one BS at a time. Hard handover does not have advantage of diversity gain opportunity during the process of handover where the signals from two or more BSs arrive at comparable strengths. However, it is a simple and inexpensive way to implement handover.

On the contrary, a soft handover enables a mobile node to maintain radio connections with more than two base stations simultaneously while it does not release any radio links unless the signal strength drops below a specified threshold value [5]. Soft handover completes when the MS selects the best available candidate BS as the target cell. Soft handover is a type of mobile assisted handover. The disadvantage of soft handover is that it is complex and expensive to implement. Moreover, forward interference actually increases since several BSs are used in soft handover simultaneously to connect the MS. The increase in forward interference can become a problem if the handover region is large such that there are many MSs in soft handover mode.

2.2 Channel Assignment

Channel assignment is a key phase of handover process and consists of the allocation of resources to calls in the new base station. When a call is admitted to access the network, in order to, make a decision of acceptance or rejection by a call admission control (CAC) algorithm, several characters including QoS of the existing connection, and the amount of available resource versus QoS requirements [7] should be considered. The success of handover process is also affected by the radio technology of the channel assignment process. Various channel assignment strategies (e.g. non-prioritized scheme, the guard channel scheme, and queuing priority scheme) have been proposed to reduce the forced termination of calls at the expense of increasing the number of dropped or blocked calls.

Channel assignment strategy can either be fixed or dynamic. Fixed channel assignment strategy allows each cell to contain a specific and fixed number of channels. New arriving calls are served by unused channels in that cell. If all the channels are in use, any new arriving calls will be blocked. Several variations of the fixed assignment strategies are available. One approach, referred to as channel borrowing strategy, allows a cell to borrow channels from a neighbouring cell if all of its own channels are in use. The MSC controls the borrowing procedures and ensures that this procedure does not disrupt or interfere with services in the donor cell.

Dynamic channel assignment strategy does not assign channels to any cells permanently. Instead, when a call arrives, the serving BS requests a channel from the MS. The MS then allocates that channel to the cell and uses an algorithm to calculate the likelihood of future blocking within the cell, the reuse distance of the channel, and other cost functions. The MS only assigns a channel if that channel is not in use in the cell or any other cells which falls within the minimum restricted distance of channel reuse to avoid co-channel interference [8]. Dynamic channel assignment strategies decrease the likelihood of blocking and increase the trunking capacity of the system. Dynamic channel assignment strategy requires the MS to collect real time data on channel occupancy, traffic distribution, and radio signal strength indications (RSSI) of all channels on a continuous basis. This increases the storage and computational loads on the system but increases the channel utilization and decreases the blocking probability of calls.

2.3 Handover Schemes

Since ongoing communications are very sensitive to interruption, the handover dropping probability must be minimized. In what follow, two well-known prioritization schemes [9] are presented.

Guard Channel Scheme. GCS is the most popular way of assigning priority to handover calls by reserving a fixed number of channels out of the total number of available channels for handover calls [6, 10-15]. The high priority is assigned to handover calls over calls originated from the serving cell in GCS, e.g. h channels are assigned to handover calls only from a total of N channels in a cell. New calls and handover calls share the remaining $n=N-h$ channels together. If the number of available channels in the cell is less than or equal to h, any arrivals of new calls are rejected/blocked because the remaining idle channels are used by handover calls exclusively. A handover call is blocked if there is no channel available in the cell at all. The flowchart for GCS is show in Fig. 1.

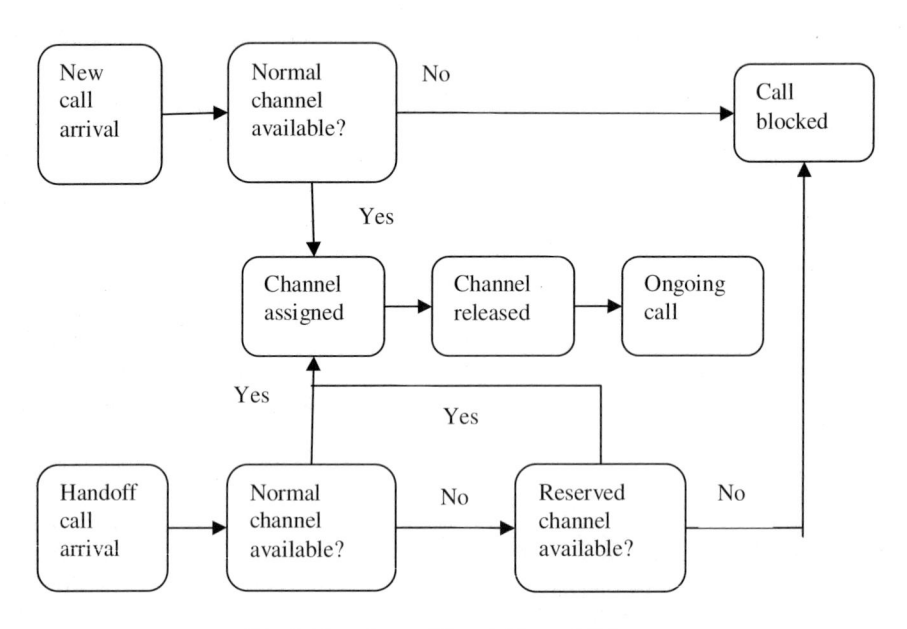

Fig. 1. Flowchart of Guard Channel Scheme

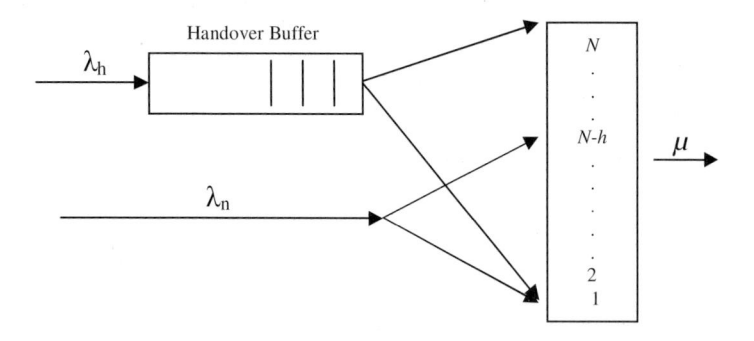

Fig. 2. System Model with Priority and Handover Queueing Scheme

Handover Queueing Scheme. In the queuing handover calls prioritization scheme, if all of the channels are occupied in a cell, arrival handover calls are stored in a queue [9, 16-20]. The handover calls stored in the queue can be served whenever a channel becomes idle. In order to decrease the blocking probability of new calls, queuing scheme can also be applied to new calls. Fig. 2 shows the queuing scheme that store newly arriving handover calls when all channels are not available.

2.4 Existing Analytical Models of Handover Schemes

One of the earliest analytical frameworks for GCS has been developed by Guerin in [21]. Zhang, Soong and Ma [22] have proposed a novel approximation approach using a simplified one-dimensional Markovian model to evaluate the performance of a guard channel scheme. Vazuqez-Avila, Cruz-Perez and Ortigoza-Guerrero [23] have developed a Markovian model for Fractional Guard Channel (FGC) which accepts an arriving new call with a probability calculated on the base of the number of busy channels but accepts an arriving handover call unless there are no channels available. Ogbonmwan and Li [24] have extended GCS to support voice traffic by using three dynamic thresholds to, respectively, provide priority to handover data calls, new voice calls and handover voice calls. They have developed a two-dimensional Markov chain to analyze this new bandwidth allocation scheme, namely, multi-threshold bandwidth reservation scheme.

In addition to the fixed channel allocation schemes, GCS and its variants have also been analyzed with the dynamic allocation schemes in [25]. Moreover, several recent studies [26-29] have investigated the performance of GCS with a First-In-First-Out (FIFO) queue of handover calls via simulation or analytical modelling. Louvros, Pylarinos and Kotsopoulos [30] have proposed a multiple queue model for handover in microcellular networks with a dedicated queue for each transceiver in the cell. Xhafa and Tonguz [14] have presented an analytical framework which employs two queues for the two priority classes of handover calls and incorporates a priority transition between handover calls in the queue. However, the arrivals of both new calls and handover calls were modelled by the traditional non-bursty Poisson process in the aforementioned references. Only a few studies [10, 31] have adopted the Markov-Modulated-Poisson-Process (MMPP) [32] to capture the bursty properties of handover call arrivals.

3 Analytical Model

3.1 System Description

The dynamic handover scheme, DGCS, is illustrated in Fig. 3. An FIFO finite capacity queue is adopted to accommodate handover calls waiting for a free channel. Let N be the total number of channels available in a cell and k be the buffer capacity. When the buffer is empty, as shown in Fig. 3(a), the maximum number of channels, which can be used to transmit new calls, is n ($n \leq N$). Consequently, all remaining channels, h ($h = N - n$), are reserved for handover calls. In the dynamic handover scheme, the number of channels allocated to handover calls, h, changes according to the queue status. Specifically, when the buffer of handover calls is not empty, t more channels are allocated to handover calls in order to reduce the delay and loss probability of handover calls. In this case, the maximum numbers of channels which can be used for new calls and handover calls are changed to $n' = (n - t)$ and $h' = (h + t)$, respectively, as shown in Fig. 3(b).

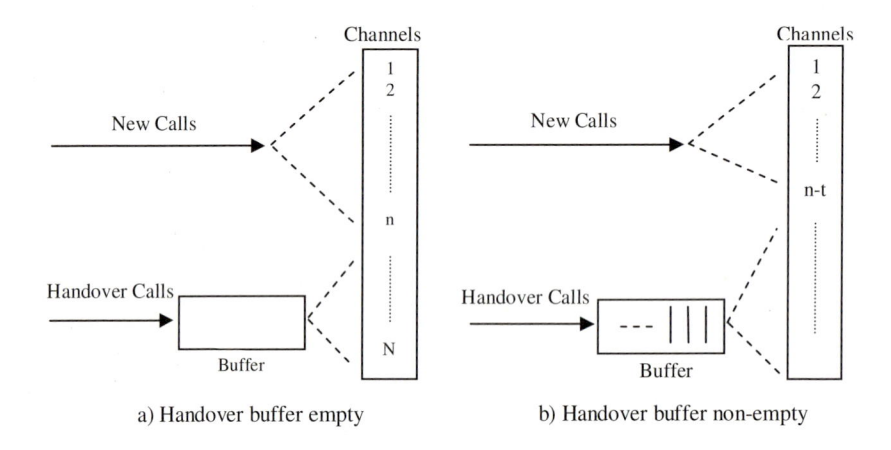

Fig. 3. The dynamic guard channel scheme

The arrivals of new calls follow a Poisson process with average arrival rate λ. The arrivals of handover calls are modelled by a two-state MMPP with the infinitesimal generator $Q = \begin{bmatrix} -\delta_1 & \delta_1 \\ \delta_2 & -\delta_2 \end{bmatrix}$ and rate matrix $\Lambda = \begin{bmatrix} \lambda_1 & 0 \\ 0 & \lambda_2 \end{bmatrix}$, respectively, where λ_m ($m = 1, 2$) is the arrival rate when the MMPP is at state m and δ_m is the intensity of transition from state m to the other. The call duration time, which is the amount of time that the call remains in progress without forced termination, is assumed to be exponentially distributed with mean μ_{du}^{-1}. Moreover, the cell dwell time, i.e., the time duration that a mobile user resides in the cell before crossing the cell boundary, follows an exponential distribution with mean μ_{dw}^{-1}. Therefore, the channel holding time defined as the time spent in a cell follows an exponential distribution with mean $\mu^{-1} = (\mu_{du} + \mu_{dw})^{-1}$. Furthermore, a handover call waiting in the queue is forced to terminate if the mobile station moves out of the radio coverage of the handover area. The corresponding handover dwell time is exponentially distributed with mean d^{-1}.

3.2 System State Transition Diagram

This section presents the state transition diagram of the system with the dynamic handover scheme. As shown in Figure 4, the three-dimensional Markov chain is constructed from two 2-dimensional Markov chains with one in the front layer and the other in the back (shaded) layer. The transition between the corresponding states from one layer to the other represents the change between the two states of the MMPP.

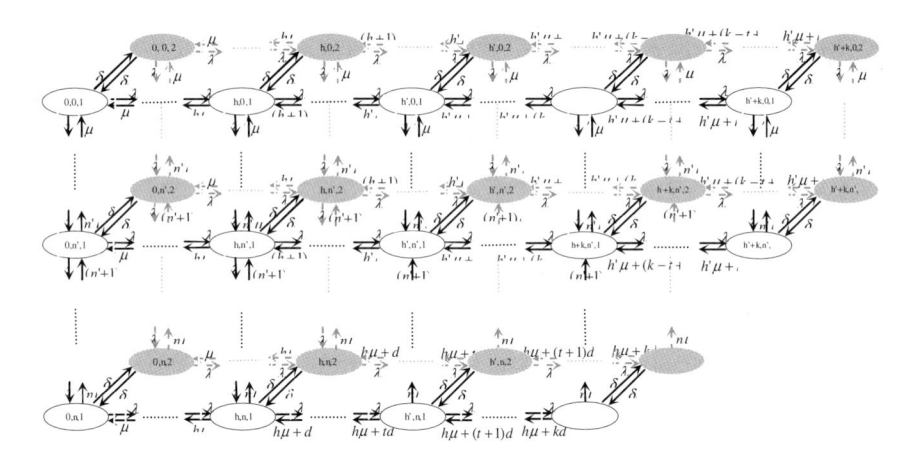

Fig. 4. State transition diagram of a three dimensional Markov chain for modelling the handover scheme

Each state (i, j, m) in the three-dimensional Markov chain corresponds to the case where there are i, $(0 \leq i \leq (h'+k))$, handover calls and j, $(0 \leq j \leq n)$, new calls in the system and the two-state MMPP is at state m, $(m = 1, 2)$. The transitions from state (i, j, m) to $(i+1, j, m)$, $(0 \leq i < h'+k, \ 0 \leq j \leq n, m = 1, 2)$, and from state (i, j, m) to $(i, j+1, m)$, $(0 \leq i \leq h'+k, \ 0 \leq j < n, m = 1, 2)$ imply that a handover call and a new call enters into the system, respectively. Therefore, the transition rates out of state (i, j, m) to $(i+1, j, m)$ and to $(i, j+1, m)$ are λ_m and λ, respectively. On the other hand, the transitions from state (i, j, m) to $(i-1, j, m)$, $(1 \leq i \leq h'+k, \ 0 \leq j \leq n, \ m = 1, 2)$, and from state (i, j, m) to $(i, j-1, m)$, $(0 \leq i \leq h'+k, \ 1 \leq j \leq n, m = 1, 2)$ represents that a handover call and a new call departs from the system, respectively. Therefore, the transition rate out of state (i, j, m) to $(i, j-1, m)$ is $j \times \mu$. If a state (i, j, m) indicates that the buffer is empty (i.e., $i \le h'$ and $i + j \le N$), the transition rate from (i, j, m) to $(i-1, j, m)$ is $i \times \mu$. Otherwise, the transition rate is $(N-j) \times \mu + (i-N+j) \times d$ and $h'\mu + (i-h')d$ when $n' \le j \le n$ and $j < n'$, respectively.

Let p_{ijm} denote the joint probability of state (i, j, m) in the three-dimensional Markov chain. Let **P** be the steady-state probability vector of this Markov chain, $\mathbf{P} = (p_{001}, \cdots, p_{0n1}, \cdots, p_{(h'+k)01}, \cdots, p_{(h'+k)n'1}, \cdots, p_{002}, \cdots, p_{0n2}, \cdots)$. The infinitesimal generator matrix, **Z**, of this Markov chain is of size $(2 \times ((h'+k+1) \times (n+1) - t(t+1)/2)) \times (2 \times ((h'+k+1) \times (n+1) - t(t+1)/2))$ and can be given according to the state transition rates of the Markov chain presented in the previous paragraph. The steady-state probability vector, **P**, satisfies the following equations.

$$\begin{cases} \mathbf{PZ} = 0 \\ \mathbf{Pe} = 1 \end{cases} \quad (1)$$

where $\mathbf{e} = (1, 1, \cdots, 1)^T$ is a unit column vector of length $(2 \times ((h'+k+1) \times (n+1) - t(t+1)/2))$. Solving Equation (1) using the approach presented in [32-34] yields the steady-state probability vector, **P**, as

$$\mathbf{P} = \boldsymbol{\alpha}(\mathbf{I} - \mathbf{X} + \mathbf{e}\boldsymbol{\alpha})^{-1}. \quad (2)$$

where matrix $\mathbf{X} = \mathbf{I} + \mathbf{Q}/\beta$, $\beta \le \min\{\mathbf{Q}_{ii}\}$ and $\boldsymbol{\alpha}$ is an arbitrary row vector of **X**.

The aggregate and marginal state probabilities, p_x and p_x^q, that there are x calls in the system and in the queue, respectively, can be calculated based on the joint state probability p_{ijm}. The probabilities p_x and p_x^q can be written as.

$$p_x = \begin{cases} \sum_{j=0}^{x} \sum_{m=1}^{2} p_{(x-j)jm} & 0 \le x \le n \\ \sum_{j=0}^{n} \sum_{m=1}^{2} p_{(x-j)jm} & n < x \le h'+k \\ \sum_{j=x-h'-k}^{n} \sum_{m=1}^{2} p_{(x-j)jm} & h'+k < x \le N+k \end{cases} \quad (3)$$

$$p_x^q = \begin{cases} \sum_{m=1}^{2} (\sum_{i=0}^{h} \sum_{j=0}^{n} p_{ijm} + \sum_{i=h+1}^{h'} \sum_{j=0}^{N-i} p_{ijm}) & x = 0 \\ \sum_{m=1}^{2} (\sum_{j=0}^{n'} p_{(h'+x)jm} + \sum_{j=n'+1}^{n} p_{(N-j+1)jm}) & 1 \le x \le k \end{cases} \quad (4)$$

The probabilities, p_x and p_x^q, are essential for the derivation of the aggregate and marginal performance metrics below.

3.3 Derivation of Performance Metrics

Analytical expressions are derived to evaluate the mean number of calls in the system $\overline{L}$, aggregate response time $\overline{R}$, aggregate call blocking probability $\overline{CLP}$, handover call blocking probability $\overline{CLP_H}$, handover delay $\overline{D}$ and new call blocking probability $\overline{CLP_N}$.

The mean number of calls in the system $\overline{L}$ can be calculated as follows

$$\overline{L} = \sum_{x=0}^{N+k}(x \times p_x). \tag{5}$$

The expressions for the mean aggregate response time can be derived using Little's Law [35].

$$\overline{R} = \frac{\overline{L}}{\overline{T}}. \tag{6}$$

where $\overline{T}$ is the aggregate throughput and is equal to the sum of the throughput of handover calls and new calls, $\overline{T_H}$ and $\overline{T_N}$. Due to the equilibrium of the rates of incoming flows and outgoing flows in the steady state, the mean marginal throughputs $\overline{T_H}$ and $\overline{T_N}$ are equal to the corresponding arrival rate multiplied by the probability that a handover or new call are not lost.

$$\overline{T} = \overline{T_N} + \overline{T_H}. \tag{7}$$

$$\overline{T_H} = \sum_{m=1}^{2} \lambda_m \left(\sum_{i=0}^{h'+k-1} \sum_{j=0}^{n'} p_{ijm} + \sum_{j=n'+1}^{n} \sum_{i=0}^{N-j+k-1} p_{ijm} \right). \tag{8}$$

$$\overline{T_N} = \lambda \sum_{m=1}^{2} \left(\sum_{i=0}^{h} \sum_{j=0}^{n-1} p_{ijm} + \sum_{i=h+1}^{h'} \sum_{j=0}^{N-i-1} p_{ijm} + \sum_{i=h'+1}^{h'+k} \sum_{j=0}^{n'-1} p_{ijm} \right). \tag{9}$$

As a blocked call does not contribute to the throughput, the call blocking probability is calculated by

$$\overline{CLP} = \frac{\overline{\lambda_H} + \lambda - \overline{T}}{\overline{\lambda_H} + \lambda}. \tag{10}$$

The average arrival rate, $\overline{\lambda_H}$, of handover calls, modelled by an MMPP-2 can be given by

$$\overline{\lambda_H} = \frac{\lambda_1 \delta_2 + \lambda_2 \delta_1}{\delta_1 + \delta_2}. \tag{11}$$

Similarly, the handover call blocking probability and new call blocking probability can be written as

$$\overline{CLP_H} = \frac{\overline{\lambda_H} - \overline{T_H}}{\overline{\lambda_H}}. \tag{12}$$

$$\overline{CLP_N} = \frac{\lambda - \overline{T_N}}{\lambda}. \tag{13}$$

Moreover, the expression for the handover delay $\overline{D}$ can also be derived using Little's Law and is given by [35].

$$\overline{D} = \frac{\overline{L_q}}{\overline{T_H}}. \tag{14}$$

Finally, the mean number of handover calls in the queue can be easily calculated using the probability p_x^q

$$\overline{L_q} = \sum_{x=0}^{k}(i \times p_x^q). \tag{15}$$

4 Model Validation and Performance Analysis

To investigate the accuracy of the above developed performance model, a discrete-event simulator has been developed for the dynamic handover scheme using JAVA programming language. Extensive numerical experiments have been performed for several combinations of different degrees of burstiness of handover calls, various numbers of channels (h) originally reserved for handover calls and numbers of dynamic channels (t). The figures presented below in this section reveal that the simulation results closely match those predicted by the analytical model. Moreover, we investigate the effects of the number of channels originally reserved to handover calls, the number of dynamic channels and the burstiness of handover calls on the system performance.

Figs. 5-10 illustrate, respectively, the mean number of calls in the system, aggregate response time, aggregate call blocking probability, handover call blocking probability, handover delay and new call blocking probability against the number of dynamic channels t with different degrees of burstiness of handover calls and the original number of channel reserved for handoff calls $h = 10, 15$. Performance results depicted in Figs. 5-10 are presented for the following cases: The total number of available channels in the cell is $N = 30$. The buffer capacity, k, is set to be 10. Additionally, the call duration time, cell dwell time and handover dwell time are exponentially distributed with mean $\mu_{du}^{-1} = 0.67$, $\mu_{dw}^{-1} = 10$, and $d^{-1} = 100$, respectively. The mean arrival rate of new calls λ is set to be 20. Furthermore, handover calls with high burstiness is generated using the infinitesimal generator matrix $\mathbf{Q} = \begin{bmatrix} -0.001 & 0.001 \\ 0.001 & -0.001 \end{bmatrix}$ and the rate matrix $\lambda = \begin{bmatrix} 1 & 0 \\ 0 & 30 \end{bmatrix}$. In addition, handover calls with low burstiness is generated using the infinitesimal generator matrix $\mathbf{Q} = \begin{bmatrix} -0.97 & 0.97 \\ 0.97 & -0.97 \end{bmatrix}$ and the rate matrix $\lambda = \begin{bmatrix} 1 & 0 \\ 0 & 30 \end{bmatrix}$. The burstiness of 2-state MMPP which is characterized by the infinitesimal generator $\mathbf{Q} = \begin{pmatrix} -\delta_1 & \delta_1 \\ \delta_2 & -\delta_2 \end{pmatrix}$ and rate matrix $\Lambda = \begin{pmatrix} \lambda_1 & 0 \\ 0 & \lambda_2 \end{pmatrix}$ is given by Equation (16).

$$c^2 = 1 + \frac{2\delta_1\delta_2(\lambda_1 - \lambda_2)^2}{(\delta_1 + \delta_2)^2(\lambda_1\lambda_2 + \lambda_2\delta_1 + \lambda_1\delta_2)}. \tag{16}$$

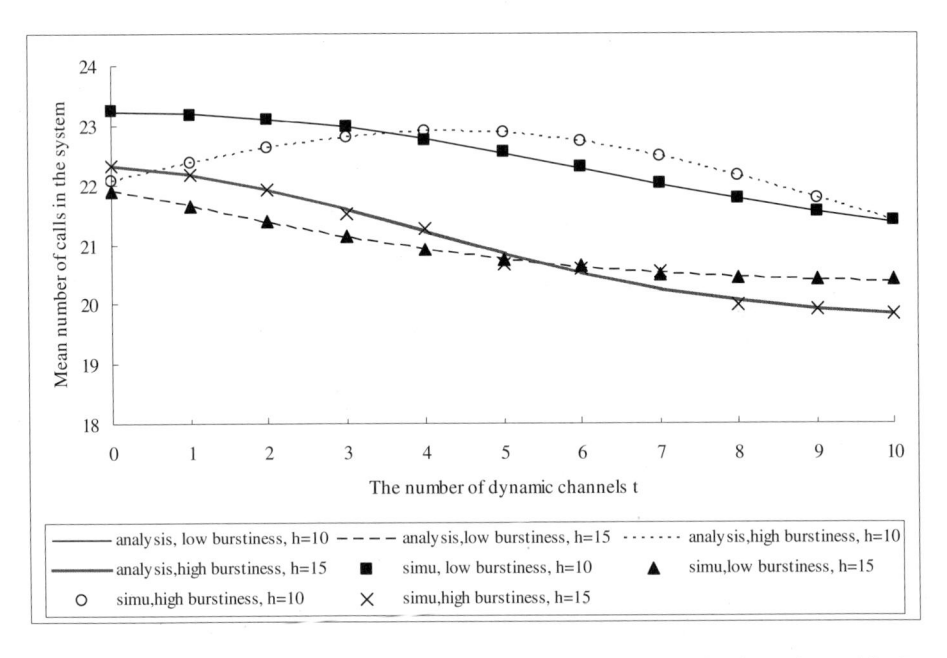

Fig. 5. Mean number of calls in the system vs. the number of dynamic channels t with the number of channels reserved for handover calls $h = 10, 15$ and different burstiness of handover calls

As the number of dynamic channels (t) increases, more and more handover calls are accepted whilst more and more new calls are denied. The decrease in the mean number of calls in the system, on condition that the burstiness of handoff calls is low

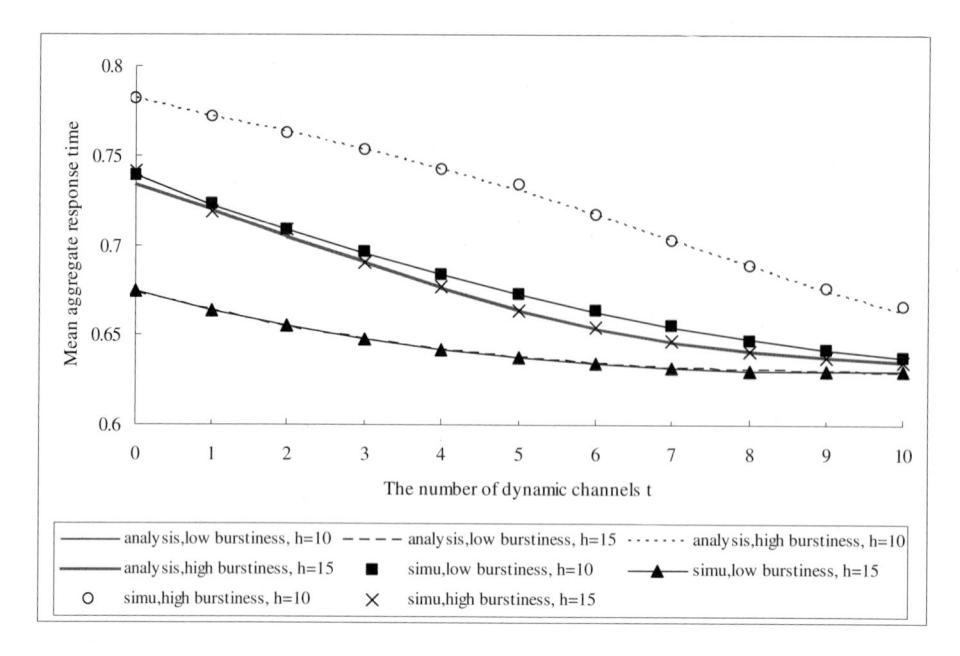

Fig. 6. Mean aggregate response time vs. the number of dynamic channels t with the number of channels reserved for handover calls $h = 10, 15$ and different burstiness of handover calls

or on condition that h is small (i.e., $h = 10$) and the burstiness of handover calls is high, indicates that the decrease of the number of new call arrivals holds the leading position in affecting the aggregate number of calls in the system. However, Fig. 5 depicts that when the burstiness of handover calls and h are high (i.e., $h = 10$), the mean number of calls that can be accommodated in the system tends to increase before it reaches a maximum value as t increases. This trend represents that the system accepts more handover calls although the number of accepted new calls decreases. When the number of calls in the system reaches its maximum value, the number of handover calls in the system is at the saturation point and the decrease of the number of new call arrivals consequently holds the leading position in affecting the aggregate number of calls in the system.

Furthermore, Fig. 5 shows the negative effect of high burstiness of handover calls, for instance, when h is fixed as 10, high burstiness results in an increased mean aggregate number of calls in the system if t is smaller than 4. Once the number of dynamic channels t exceeds 4 the decrease in the number of new calls plays more important role than burstiness in affecting the aggregate number of calls in the system. Moreover, the mean aggregate number of calls in the system when the h is small (i.e., $h = 10$) is larger than that when h is big (i.e., $h = 15$).

Figs. 6-9 reveal that, respectively, the increase of t remarkably reduces the mean aggregate response time, aggregate call blocking probability, handover call blocking probability and handover delay, in particular, with high burstiness and less number of channels originally reserved for handover calls (i.e., $h = 10$), in that more channels

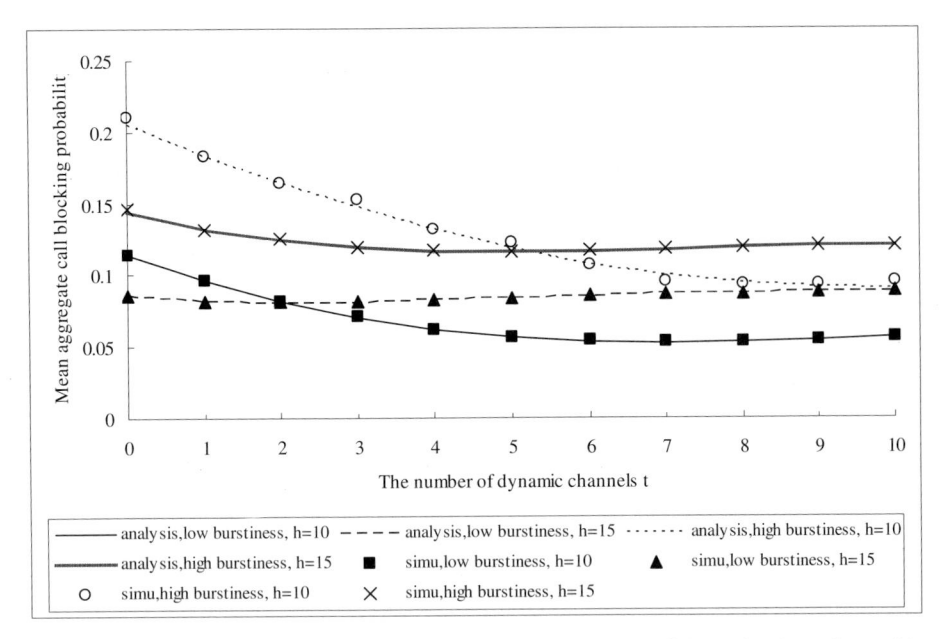

Fig. 7. Mean aggregate call blocking probability vs. the number of dynamic channels t with the number of channels reserved for handover calls $h = 10, 15$ and different burstiness of handover calls

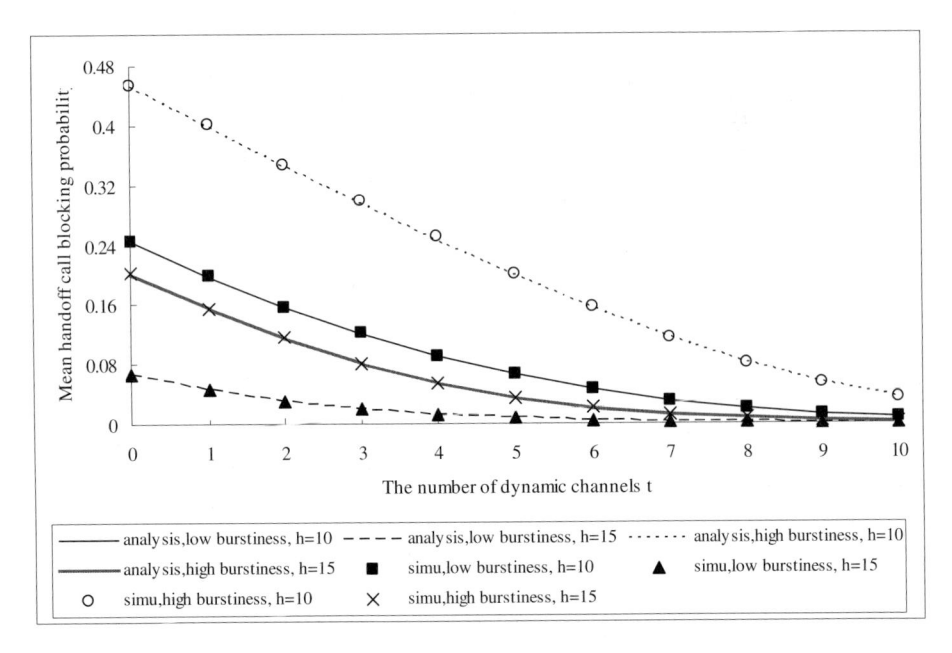

Fig. 8. Mean handover call blocking probability vs. the number of dynamic channels t with the number of channels reserved for handover calls $h = 10, 15$ and different burstiness of handover calls

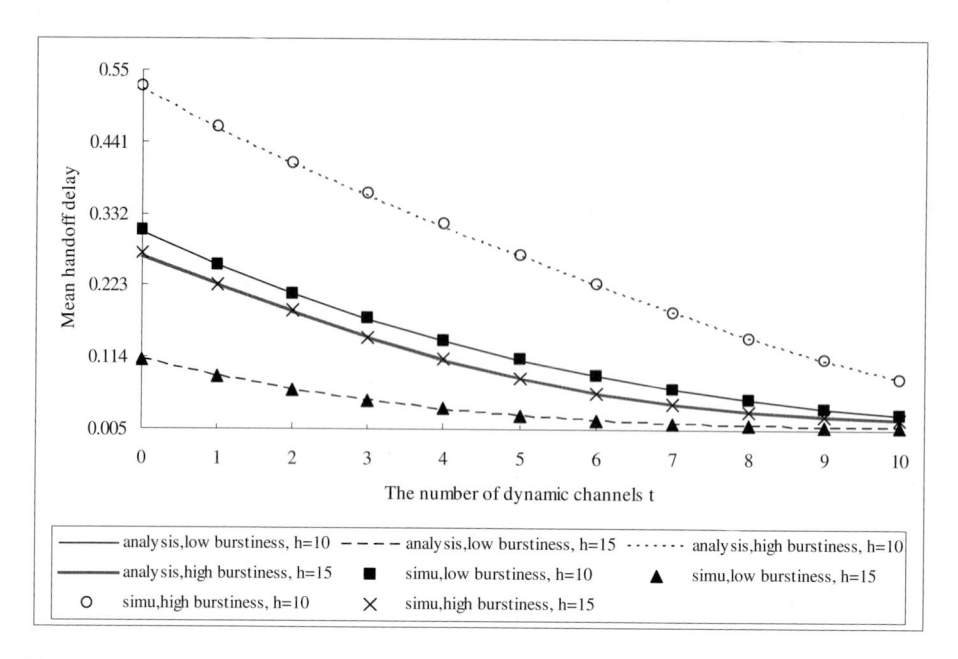

Fig. 9. Mean handover delay vs. the number of dynamic channels t with the number of channels reserved for handover calls $h = 10, 15$ and different burstiness of handover calls

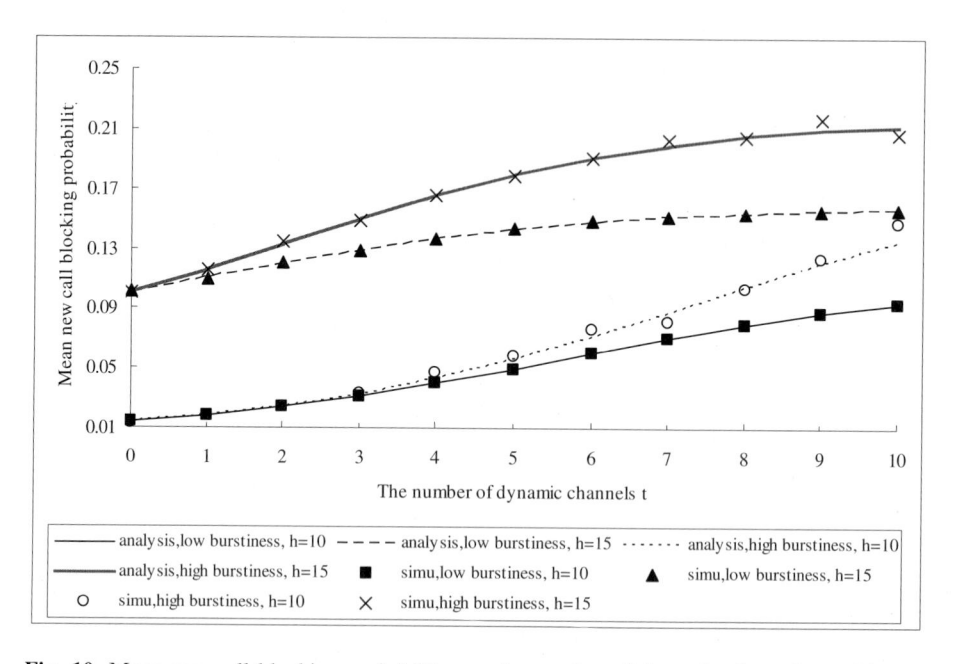

Fig. 10. Mean new call blocking probability vs. the number of dynamic channels t with the number of channels reserved for handover calls $h = 10, 15$ and different burstiness of handover calls

are allocated to handover calls. In addition, long aggregate response time, high aggregate call blocking probability, high handover call blocking probability and long handover delay are demonstrated in Figs. 6-9, respectively, as a result of increasing the burstiness of handover calls. On the other hand, it is easily understandable that the increase of h reduces the number of handover calls in the buffer, consequently decreases the aggregate response time, handover call blocking probability and handover delay. Meanwhile, Fig. 7 denotes that, when t is small (i.e., $t \leq 2$ under low burstiness, $t \leq 5$ under high burstiness), the aggregate call blocking probability increases as h grows. However, for a large t, the aggregate call blocking probability with a big h (i.e., $h = 15$) is higher than that with small one (i.e., $h = 10$).

As more and more handover calls are accepted, Fig. 10 illustrates that the new call blocking probability increases as t rises. High burstiness results in the growth of new call blocking probability. Such a trend becomes remarkable as t increases. Finally, as shown in Fig. 10, if more channels is originally reserved for handover calls (i.e., $h = 15$), new call blocking probability, the number of channels used for new call transmission decreases and consequently, more arriving new calls are to be blocked.

5 Conclusions

This chapter presented a detailed review of existing handover schemes for wireless communication networks. Particular emphasis was given to a proposed handover scheme DGCS, which can assign dynamically the reserved channels in order to reduce both the handover call blocking probability and mean handover delay. In this context, an analytical model was devised to assess the performance of this handover scheme and the MMPP was used to capture the impact of bursty and correlated arrival patterns of handover calls. The credibility of the model was illustrated by comparing favourably the analytic results against those obtained through extensive simulation experiments. Moreover, it was shown that the proposed handover scheme can effectively reduce the handover blocking probability and mean handover delay.

The analytical model can be employed to obtain a desirable number of dynamic channels in order to establish an optimal trade-off amongst the blocking probabilities of new calls and handover calls as well as the aggregate blocking probability of all calls.

References

1. Tekinay, S., Jabbari, B.: Handover and Channel Assignment in Mobile Cellular Networks. IEEE Comm. Mag. 29(11), 42–46 (1991)
2. Pollioni, G.P.: Trends in Handover Design. IEEE Comm. Mag. 34(3), 82–90 (1996)
3. Marichamy, P., Chakrabarti, S., Maskara, S.L.: Overview of Handoff Schemes in Cellular Mobile Networks and Their Comparative Performance Evaluation. In: 50th IEEE Vehicular Technology Conference Fall, pp. 1486–1490. IEEE Press, Amsterdam (1999)
4. Chaudhary, V., Tripathi, R., Shukla, N.K., Nasser, N.: A New Channel Allocation Scheme for Real-Time Traffic in Wireless Cellular Networks. In: IEEE Performance, Computing, and Communications Conference, pp. 551–555. IEEE International, New Orleans (2007)

5. Nasser, N., Hasswa, A., Hassanein, H.: Handoffs in Fourth Generation Heterogeneous Networks. IEEE Comm. Mag. 44(10), 96–103 (2006)
6. Yavuz, E.A., Leung, V.C.M.: Computationally Efficient Method to Evaluate the Performance of Guard-Channel-Based Call Admission Control in Cellular Networks. IEEE Trans. Veh. Tech. 55(4), 1412–1424 (2006)
7. Nasser, N., Hassanein, H.: Adaptive Call Admission Control for Multimedia Wireless Networks with QoS Provisioning. In: International Conference on Parallel Processing Workshops, Canada, pp. 30–37 (2004)
8. Rappaport, T.: Wireless Communications: Principles and Practice, 2nd edn. Prentice-Hall, Englewood Cliffs (2002)
9. Ekiz, N., Salih, T., Kucukoner, S., Fidanboylu, K.: An Overview of Handoff Techniques in Cellular Networks. International J. Information Technology 2(2) (2005)
10. Niyato, D., Hossain, E., Alfa, A.S.: Performance Analysis of Multi-service Wireless Cellular Networks with MMPP Call Arrival Patterns. In: IEEE GLOBECOM, pp. 3078–3082 (2004)
11. Louvros, S., Pylarinos, J., Kotsopoulos, S.: Mean Waiting Time Analysis in Finite Storage Queues for Wireless Cellular Networks. J. Wireless Personal Communications 40(2), 145–155 (2007)
12. Ye, Z., Law, L.K., Krishnamurthy, S.V., Xu, Z., Dhirakaosal, S., Trpathi, S.K., Molle, M.: Predictive Channel Reservation for Handoff Prioritization in Wireless Cellular Networks. J. Computer and Telecommunications Networking 51(3), 798–822 (2007)
13. Tung, D., Wong, C., Kong, P.: Wireless Broadband Networks. John Wiley & Sons, Inc., Chichester (2009)
14. Xhafa, A.E., Tonguz, O.K.: Dynamic Priority Queueing of Handover Calls in Wireless Networks: an Analytical Framework. IEEE J. on Selected Areas in Communications 22(5), 904–916 (2004)
15. Pati, H.K.: A Distributed Adaptive Guard Channel Reservation Scheme for Cellular Networks. Int. J. Commun. Syst. 20(9), 1037–1058 (2007)
16. Stojmenovi, I.: Handbook of Wireless Networks and Mobile Computing. John Wiley & Sons, Inc., Chichester (2002)
17. Akyildiz, I.F., McNair, J., Ho, J.S.M., Uzunalioglu, H., Wang, W.: Mobility Management in Next-Generation Wireless Systems. Proceedings of the IEEE 87(8), 1347–1384 (1999)
18. Znati, T.F., Kim, S.J.: Adaptive Channel Management Schemes for Wireless Communication Systems. International Journal of Communication Systems 13(6), 435–460 (2009)
19. Xhafa, A.E., Tonguz, O.K.: Dynamic Priority Queueing of Handoff Requests in PCS. In: IEEE International Conference on Communications, Finland, pp. 341–345 (2001)
20. Pandey, V., Ghosal, D., Mukherjee, B., Wu, X.: Call Admission and Handoff Control in Multi-Tier Cellular Networks: Algorithms and Analysis. Wireless Personal Communications 43(3), 857–878 (2007)
21. Guerin, R.: Queuing Blocking System with Two Arrival Streams and Guard Channels. IEEE Trans. Communications 36(2), 153–163 (1988)
22. Zhang, Y., Soong, B., Ma, M.: Approximation Approach on Performance Evaluation for Guard Channel Scheme. Electronics Letters 39(5), 465–467 (2003)
23. Vazuqez-Avila, J.L., Cruz-Perez, F.A., Ortigoza-Guerrero, L.: Performance Analysis of Fractional Guard Channel Policies in Mobile Cellular Networks. IEEE Trans. Wireless Communications 5(2), 301–305 (2006)
24. Ogbonmwan, S., Li, W.: Multi-Threshold Bandwidth Reservation Scheme of An Integrated Voice/Data Wireless Network. J. Computer Communications 29(9), 1504–1515 (2006)

25. Zheng, J., Regentova, E.: QoS-Based Dynamic Channel Allocation for GSM/GPRS Networks. In: Jin, H., Reed, D., Jiang, W. (eds.) NPC 2005. LNCS, vol. 3779, pp. 285–294. Springer, Heidelberg (2005)
26. Hong, D., Rapaport, S.: Traffic Model and Perfromance Analysis for Cellular Mobile Radio Telephone Systems with Prioritized and Nonprioritized Handoff Procedures. IEEE Trans. Vehicular Technology 35(3), 77–92 (1986)
27. Du, W., Lin, L., Jia, W., Wang, G.: Modeling and Performance Evaluation of Handover Service in Wireless Networks. In: Lu, X., Zhao, W. (eds.) ICCNMC 2005. LNCS, vol. 3619, pp. 229–238. Springer, Heidelberg (2005)
28. Wang, Z., Mathiopoulos, P.: On the Performance Analysis of Dynamic Channel Allocation with FIFO Handover Queuing in LEO-MSS. IEEE Trans. Communications 53(9), 1443–1446 (2005)
29. Ma, X., Cao, Y., Liu, Y., Trivedi, K.S.: Modeling and Performance Analysis for Soft Handoff Schemes in CDMA Cellular Systems. IEEE Trans. Vehicular Technology 55(2), 670–680 (2005)
30. Louvros, S., Pylarinos, J., Kotsopoulos, S.: Handoff Multiple Queue Model in Microcellular Networks. Computer Communications 30(2), 396–403 (2007)
31. Farahani, M.A., Guizani, M.: Markov Modulated Poisson Process Model for Hand-off Calls in Cellular Systems. In: Proc. IEEE Wireless Communications and Networking Conference, pp. 1113–1118 (2000)
32. Fischer, W., Meier-Hellstern, K.: The Markov-modulated Poisson Process (MMPP) Cookbook. Performance Evaluation 18(2), 149–171 (1993)
33. Paige, C.C., Styan, G.P.H., Wachter, P.G.: Computation of the Stationary Distribution of A Markov Chain. J. Statist. Comput. Simulation. 4, 173–186 (1975)
34. Min, G., Ould-Khaoua, M.: A Performance Model for Wormhole-switched Interconnection Networks under Self-similar Traffic. IEEE Transactions on Computers 53(5), 601–613 (2004)
35. Kleinrock, L.: Queueing Systems: Compute Applications, vol. 1. John Wiley & Sons, New York (1975)

On the Performance Modelling and Optimisation of DOCSIS HFC Networks

Neelkamal P. Shah[1], Demetres D. Kouvatsos[1], Jim Martin[2], and Scott Moser[2]

[1] School of Computing, Informatics and Media
University of Bradford, UK
{N.P.Shah,D.Kouvatsos}@bradford.ac.uk
[2] School of Computing
Clemson University, USA
{Jim.Martin,S.Moser}@cs.clemson.edu

Abstract. The DOCSIS protocol defines the MAC and physical layer operations governing two-way transmission of voice, video and multimedia data over HFC cable networks, thus constituting a complex system with many interdependent parameters. This tutorial employs simulation and analytic methodologies for the performance modelling and optimisation of DOCSIS 1.1/2.0 HFC networks with particular focus on the contention resolution algorithm, upstream bandwidth allocation strategies, flow-priority scheduling disciplines, QoS provisioning and TCP applications. In this context two performance evaluation case studies are reviewed in detail - based respectively on two open queueing network models of DOCSIS 1.1/2.0 HFC networks. The first study evaluates, via 'ns' simulation, the effect of carrying TCP/IP traffic on network performance whilst the second optimises analytically the upstream network bandwidth allocation. It is expected that many of the performance affecting operational behaviours exhibited by former releases of DOCSIS-based HFC networks will also exist under the latest DOCSIS 3.0 protocol and future extensions.

Keywords: Data-over-cable service interface specification (DOCSIS), hybrid fibre coax (HFC), Media Access Control (MAC), Quality of Service (QoS), broadband access, scheduling, contention resolution algorithms, bandwidth allocation, Transmission Control Protocol (TCP), performance modelling and evaluation, queueing network models (QNM's), network decomposition.

1 Introduction

Community antenna television (CATV) systems were introduced as a way to deliver television content to households located in hilly terrain that could not receive broadcast television. Over the years CATV companies began offering Internet access, data and telephony services to their customers in addition to television channels as a means of increasing revenue. Initially cable operators deployed proprietary systems. To stay competitive with other access technologies such as the digital subscriber line (DSL), it was decided to open the cable modem (CM) market by creating a single standard hoping to make CM's commodity items. The industry converged on the DOCSIS standard [1] which defines the MAC and physical layers that are used to provide high speed data communication over HFC cable networks. By pushing fibre further to the subscriber, fewer amplifiers are needed, noise is less of a problem and two-way data communication is possible.

In the early 1990's, the cable industry developed a large number of schemes for supporting two-way data over cable and several competing standards emerged:

1.1 IEEE 802.14

In 1994 the IEEE 802.14 working group was chartered to develop a MAC layer that would support both the asynchronous transfer mode (ATM) and IP over HFC networks. This working group has since disbanded. The upstream channel was time-slotted to enable multiple access with a slot size of 8 bytes. ATM's constant bit rate (CBR), variable bit rate (VBR), available bit rate (ABR) and unspecified bit rate (UBR) services were supported over the HFC network. Primarily due to time constraints, the standard did not obtain vendor support.

1.2 Multimedia Cable Network System's (MCNS's) DOCSIS

In response to competition from DSL, key multiple system operators (MSO's) in the early 1990s formed the MCNS to define a standard system for providing data and services over a CATV infrastructure. In 1997 MCNS released version 1 of DOCSIS (DOCSIS 1.0). The upstream channel was time-slotted with a configurable slot size (referred to as a minislot). This standard was quickly endorsed by the cable industry. The DOCSIS standard is now managed by CableLabs, a non-profit research and development group funded by cable industry vendors and providers.

1.3 DAVIC/DVB

The non-profit Swiss organisation Digital Audio Visual Council (DAVIC) was formed in 1994 to promote the success of digital audio-visual applications and services. The organisation produced the DAVIC 1.2 and the very similar Digital Video Broadcast Return Channel for Cable (DVB-RCC) radio frequency (RF) CM standards that defined the physical and MAC layers for bidirectional communications over CATV HFC networks. The DVB-RCC standard was popular in Europe for several years. However, to benefit from the economies of scale, the European cable industry moved towards the EuroDOCSIS standard.

Fig. 1 illustrates a simplified DOCSIS cable network. A Cable Modem Termination System (CMTS) interfaces with hundreds or possibly thousands of CM's. A Cable Operator allocates a portion of the RF spectrum for data usage and assigns a channel to a set of CM's[1]. A downstream RF channel of 6MHz (8MHz in Europe) is shared by all CM's in a one-to-many bus configuration (i.e. the CMTS is the only sender).

---

[1] A group of CM's that share an RF channel connect to an Optical/Electrical (O/E) node with a coaxial cable using a branch-and-tree topology.

The cable industry is undergoing a period of rapid change. Fuelled primarily by demand for voice over IP (VoIP) (requiring a more symmetric service) and IPTV services (requiring greater downstream bandwidth), the operations support systems of MSO's are being upgraded. In DOCSIS 1.0, only one QoS class was supported, that is, 'best effort' (BE), for data transmission in the upstream direction. Upstream data rates were limited to 5.12Mbps. DOCSIS 1.1 provides a set of ATM-like QoS guarantees. In addition, the physical layer supports an upstream data rate of up to 10.24 Mbps. DOCSIS 2.0 further increases upstream capacity to 30.72 Mbps through more advanced modulation techniques and by increasing the RF channel allocation to 6.4MHz. DOCSIS 3.0 supports hundreds of Mbps in both the upstream and downstream channels through channel bonding techniques.

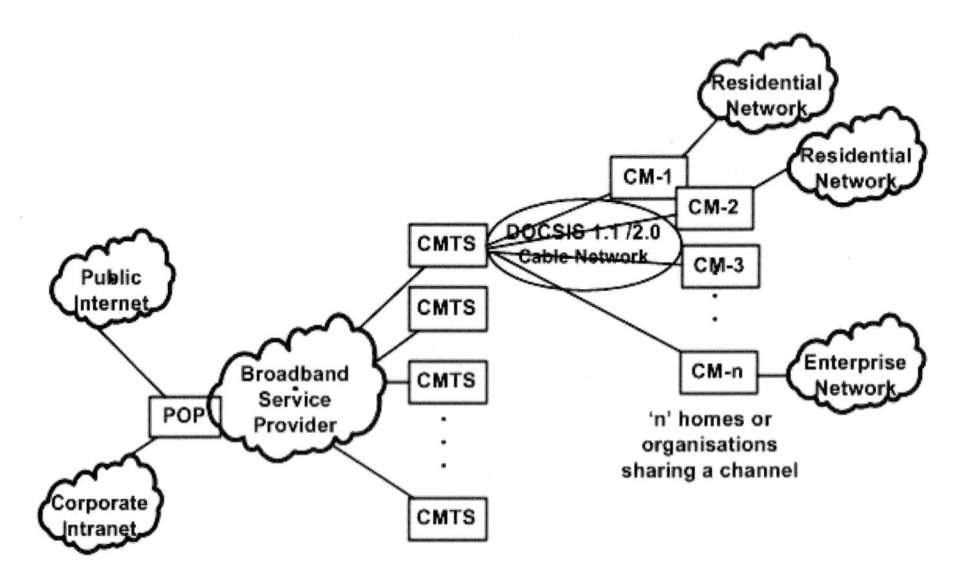

Fig. 1. Simplified DOCSIS cable network

In DOCSIS HFC networks, there are many configuration parameters and it is difficult to know a priori how particular combinations of parameters and different traffic mixes impact the network performance. This tutorial is a development of [2] and is motivated by the need to unveil the aforementioned mysteries by solving the numerous intriguing operational design, bandwidth allocation and performance evaluation problems and opportunities presented by HFC networks particularly under the complex DOCSIS protocol. The performance modelling reviewed herein relates to HFC cable networks operating under the DOCSIS 1.1 or 2.0 protocols (shortened to 'DOCSIS 1.1/2.0 HFC networks' throughout the manuscript) and it is expected that in many situations the models and their solutions will be applicable to the cases of DOCSIS 3.0 and future releases. In addition to performance modelling characterised by the underlying aims of response, throughput and/or utilisation improvement, DOCSIS has received attention in the context of monetary cost-reduction efforts. This has occurred via proposals to couple cable broadband networks to fixed wireless networks in order to eliminate last mile cabling or couple mobile communication systems such as 3G UMTS networks with these wired networks to minimise the need to build backhaul networks for the mobile systems [3, 4].

This tutorial is organised as follows: An overview of the DOCSIS 1.1/2.0 protocol is presented in Section 2. A review of selected DOCSIS HFC network performance models and their solutions is carried out in Section 3. Two quantitative DOCSIS HFC network performance evaluation studies (simulation and analytic) are detailed in Section 4. Conclusions and areas of future work follow in Section 5. Finally a list of the acronyms used throughout the manuscript is provided in Appendix I after the references.

2 The DOCSIS 1.1/2.0 Protocol

DOCSIS 1.1/2.0 is presented successively through descriptions of its general operation, QoS provision and security issues and finally an illustration of its MAC transmission layer through a QNM.

2.1 General Operation

Once powered on, the CM establishes a connection to the network and maintains this connection until the power to it is turned off. Registration of the CM onto the network involves acquiring upstream and downstream channels and encryption keys from the CMTS and an IP address from the ISP. The CM also determines propagation time from the CMTS in order to synchronise itself with the CMTS (and in effect the network) and finally logs in and provides its unique identifier over the secure channel. Due to the shared nature of these cable networks, transmissions are encrypted in both the upstream and downstream directions [5].

DOCSIS 1.1/2.0 specifies an asymmetric data path with downstream and upstream data flows on two separate frequencies. The upstream and downstream carriers provide two shared channels for all CM's. On the downstream link the CMTS is a single data source and all CM's receive every transmission. On the upstream link all CM's may transmit and the CMTS is the single sink.

Packets sent over the downstream channel are broken into 188 byte MPEG frames each with 4 bytes of header and a 184 byte payload. Although capable of receiving all frames, a CM is typically configured to receive only frames addressed to its MAC address or frames addressed to the broadcast address. In addition to downstream user data, the CMTS will periodically send management frames. These frames control operations such as ranging, channel assignment, operational parameter download, CM registration and so on. Additionally, the CMTS periodically sends MAP messages over the downstream channel that identify future upstream time division multiple access (TDMA) slot assignments over the next MAP time. The CMTS makes these upstream CM bandwidth allocations (bandwidth grants) based on CM requests and QoS policy requirements.

The upstream channel is divided into a stream of time division multiplexed 'minislots' which, depending on system configuration, normally contain from 8 to 32 bytes of data. The CMTS must generate the time reference to identify these minislots. Due to variations in propagation delays from the CMTS to the individual CM's, each CM must learn its distance from the CMTS and compensate accordingly such that all CM's will have a system wide time reference to allow them to accurately identify the proper location of the minislots. This is called ranging and is part of the CM initialisation process.

Ranging involves a process of multiple handshakes between the CMTS and each CM. The CMTS periodically sends sync messages containing a timestamp. The CMTS also sends periodic bandwidth allocation MAPs. From the bandwidth allocation MAP the CM learns the ranging area from the starting minislot number and the ranging area length given in the message. The CM will then send a ranging request to the CMTS. The CMTS, after evaluating timing offsets and other parameters in the ranging request, returns to the CM a ranging response containing adjustment parameters. This process allows each CM to identify accurately the timing locations of each individual minislot.

In addition to generating a timing reference so that the CM's can accurately identify the minislot locations, the CMTS must also control access to the minislots by the CM's to avoid collisions during data packet transmissions. Fig. 2 illustrates a hypothetical MAP allocation that includes allocated slots for contention requests, user data and management data. For BE traffic, CM's must request bandwidth for upstream transmissions. There are several mechanisms available: contention bandwidth requests, piggybacked bandwidth requests and bandwidth requests for concatenated packets.

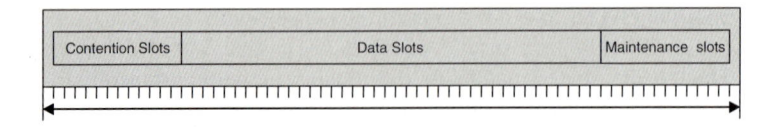

Fig. 2. Example upstream MAP allocation

Contention Bandwidth Requests. The CMTS must periodically provide transmission opportunities for CM's to send a request for bandwidth to the CMTS. As in slotted Aloha networks [6], random access bandwidth request mechanisms are inefficient as collisions will occur if two (or more) CM's attempt to transmit a request during the same contention minislot. Most implementations will have a minimum number of contention minislots to be allocated per MAP time, and in addition, any unallocated minislot will be designated as a contention minislot.

DOCSIS 1.1/2.0 specifies a truncated binary exponential backoff (tBEB) collision resolution algorithm (CRA) and it can be described as follows. When a packet arrives at the CM that requires upstream transmission, the CM prepares a contention-based bandwidth request by computing the number of minislots that are required to send the packet including all framing overhead. The contention algorithm requires the CM to randomly select a number of contention minislots to skip before sending this request (an initial backoff). This number is drawn from a range between 0 and a value that is provided by the CMTS in each MAP. The values sent are assumed to be a power of 2, so that a 5 would indicate a range of 0 – 31. After transmission, if the CM does not receive an indication that the request was received, the CM must randomly select another number of contention minislots to skip before re-sending the request. The CM is required to exponentially backoff the range following each collision with the maximum backoff specified by a maximum backoff range parameter contained in each MAP. The CM drops the packet after it has attempted to send the request 16 times.

As an example of tBEB, assume that the CMTS has sent an initial backoff value of 4, indicating a range of 0 – 15, and a maximum backoff value of 10, indicating a range of 0 – 1023. The CM, having data to send and looking for a contention minislot to use to request bandwidth, will generate a random number within the initial backoff range. Assume that an 11 is randomly selected. The CM will wait until eleven available contention minislots have passed. If the next MAP contains 6 contention minislots, the CM will wait. If the following MAP contains 2 contention minislots, a total of 8, the CM will still continue to wait. If the next MAP contains 8 contention minislots the CM will wait until 3 contention minislots have passed, 11 in total, and transmit its request in the fourth contention minislot in that MAP.

The CM then looks for either a Data Grant from the CMTS or a Data Acknowledge. If neither is received, the CM assumes a collision has occurred. The current backoff range is then doubled, i.e. the current value is increased from 4 to 5 making the new backoff range 0 – 31, and the process is repeated. The CM selects a random value within this new range, waits the required number of contention minislots, and resends its request. The backoff value continues to be incremented, doubling the range, until it reaches the maximum backoff value, in this example 10, or a range of 0 – 1023. The current backoff range will then remain at this value for any subsequent iterations of the loop. The process is repeated until either the CM receives a Data Grant or Data Acknowledge from the CMTS, or the maximum number of 16 attempts is reached.

Piggybacked Bandwidth Requests. To minimise the frequency of contention-based bandwidth requests, a CM can piggyback a request for bandwidth on an upstream data frame. For certain traffic dynamics, this can completely eliminate the need for contention-based bandwidth requests. This takes place via the Extended Header field in the MAC frames which can be used to request bandwidth for additional upstream transmissions during the current data transmission. Thus requests for bandwidth can be made outside of the contention process thereby reducing the frequency of collisions and consequently the access delay.

Bandwidth Requests for Concatenated Packets. DOCSIS 1.1/2.0 provides both Fragmentation MAC Headers, for splitting large packets into several smaller packets, and Concatenation MAC Headers, to allow multiple smaller packets to be combined and sent in a single MAC burst. One bandwidth request is presented to the CMTS for the group of packets undergoing concatenation. Concatenation can also be used to reduce the occurrence of collisions by reducing the number of individual transmission opportunities needed. Concatenation is the only method for transmitting more than one data packet in a single transmission opportunity. The CMTS, receiving the Concatenation MAC Header, must then 'unpack' the user data correctly. The Concatenation MAC Header precludes the use of the Extended Header field and therefore piggybacking of future requests cannot be done in a concatenated frame.

2.2 Quality of Service (QoS)

DOCSIS 1.1/2.0 manages bandwidth in terms of service flows that are identified by service flow IDs (SID's). Traffic arriving at either the CMTS or the CM for transmission over the DOCSIS 1.1/2.0 network is mapped to an existing SID and treated based on the profile. A CM will have at least 2 SID's allocated, one for downstream BE traffic and a second for upstream BE traffic. The upstream SID's at the CM are implemented as FIFO queues. Traffic-types extra to BE, such as VoIP, might be assigned to a different SID that supports a different scheduling service e.g. Unsolicited Grant Service (UGS) for toll quality telephony. The DOCSIS 1.1/2.0 protocol purposely does not specify the upstream bandwidth allocation algorithms so that vendors are able to develop their own solutions. DOSCIS requires CM's to support the following set of scheduling services: UGS, real-time polling service (rtPS), unsolicited grant service with activity detection (UGS-AD), non-real-time polling service (nrtPS) and BE service.

Unsolicited Grant Service (UGS). Designed to support real-time data flows generating fixed size packets on a periodic basis. For this service the CMTS provides fixed-size grants of bandwidth on a periodic basis. The CM is prohibited from using any contention requests. Piggybacking is prohibited. All CM upstream transmissions must use only the unsolicited data grants.

Real-Time Polling Service (rtPS). Designed to support real-time data flows generating variable size packets on a periodic basis. For this service the CMTS provides periodic unicast request opportunities regardless of network congestion. The CM is prohibited from using any contention requests. Piggybacking is prohibited. The CM is allowed to specify the size of the desired grant. These service flows effectively release their transmission opportunities to other service flows when inactive [1], demonstrating more efficient bandwidth utilisation than UGS flows at the expense of delay, which is worse.

Unsolicited Grant Service with Activity Detection (UGS-AD). Designed to support UGS flows that may become inactive for periods of time. This service combines UGS and rtPS with only one being active at a time. UGS-AD provides unsolicited grants when the flow is active and reverts to rtPS when the flow is inactive.

Non-Real-Time Polling Service (nrtPS). Designed to support non real-time data flows generating variable size packets on a regular basis. For this service the CMTS provides timely unicast request opportunities regardless of network congestion. The CM is allowed to use contention request opportunities.

BE Service. Designed to provide efficient service for the remaining flows. The CM is allowed to use contention or piggybacking to transmit bandwidth requests.

In the downstream direction, arriving packets are classified into a known SID and treated based on the configured service definition. For BE traffic, the service definition is limited to a configured service rate. For downstream traffic, the CMTS provides prioritisation based on SID profiles, where each SID has its own queue. Management frames, in particular MAP frames, are given highest priority. Telephony and other real-time traffic would be given next priority. BE traffic would share the remaining available bandwidth. There is also a single downstream transmission queue. Queuing occurs at the SID queues only if downstream rate control is enabled. All downstream queues are FIFO with the exception that MAP messages are inserted at the head of the transmission queue.

2.3 Security

Historically, cable systems have had an image as being insecure. The 'always-on' capability attracts attacks on subscriber networks. Subscriber networks with machines running Microsoft Windows with improper security settings have caused significant problems[2]. The security of cable networks has also been questioned since, as in a bus-based Ethernet LAN, data is received by all CM's. By default, a CM is placed in non-promiscuous mode; however it is possible for a subscriber to change the configuration and to have the CM receive all data sent over the RF channel. Further, it is possible to increase the provisioned service rates by modifying the configuration. To counter the theft of service, CableLabs extended the Baseline Privacy Interface (BPI) security service described in the DOCSIS 1.0 specification to Baseline Privacy Interface Plus (BPI+) in the DOCSIS 1.1/2.0 releases.

BPI+ addresses two areas of concern: securing the data as it travels across the network and preventing the theft of service. Both BPI and BPI+ require encryption of the frames essentially forming a virtual private network for all transmissions between the CMTS and the CM, in order to protect the customer's data as it traverses the coaxial cable. The Data Encryption Standard cipher algorithm is specified to be used in cipher block chaining mode for encryption of both the upstream and downstream MAC frame's packet data in both the BPI and BPI+ security services. Public key encryption is used by the CM to securely obtain the required keys from the CMTS. Each CM must contain a key pair for the purpose of obtaining these keys from the CMTS.

To prevent the theft of service BPI+ requires the use of secure modem authentication procedures to verify the legitimacy of a particular CM. CM's download their firmware from the service provider each time they boot. BPI+ requires the CM to successfully boot only if the downloaded code file has a valid digital signature. When a CM makes an authorisation request to the CMTS it must provide a unique X.509 digital certificate. After receiving a properly signed X.509 certificate and verifying the 1024 bit key pair the CMTS will encrypt an authorisation key using the corresponding public key and send it to the CM. A trust chain is developed by using a three level certificate hierarchy. At the top level is the root certification authority (CA) which belongs to CableLabs. The root CA uses its certificate to sign a manufacturer's CA certificate at the second level. The manufacturer CA certificates are then used to sign individual certificates for each CM produced by that particular manufacturer. This process ensures that a given CM is legitimate and that the keys for encrypting the user's data are only distributed to trusted CM's.

Although DOCSIS 1.1/2.0 specifies the use of these security procedures to protect both the service provider and the customer, like all security measures, the system's

---

[2] The security vulnerability occurs when a subscriber configures his/her network with file or print sharing. There are many reports of how dangerous this can be for example http://cable-dsl.navasgroup.com/netbios.htm#Scour.

defence is jeopardised if they are not used. Prior to 2005, polls and press reports indicated that the majority of cable network operators had not enabled the security methods required by DOCSIS 1.1/2.0.

2.4 A Queueing Network Model (QNM)

This section presents a QNM of a general DOCSIS 1.1/2.0 HFC network. The main contributors to the delay experienced by data that arrives to the CM's and requires onward routing from the DOCSIS network are the buffering delay at the CM, contention delay via the CRA, scheduling delay at the CMTS (bandwidth requests or periodic grants) and transmission delay (of the bandwidth requests (when applicable) and data transmissions).

The QNM draws on modelling aspects from a QNM of part of a DOCSIS 1.1/2.0 HFC network given in [7] and an open QNM of a DOCSIS 1.1/2.0 HFC network in [8]. It models transmission in both the upstream and downstream channels.

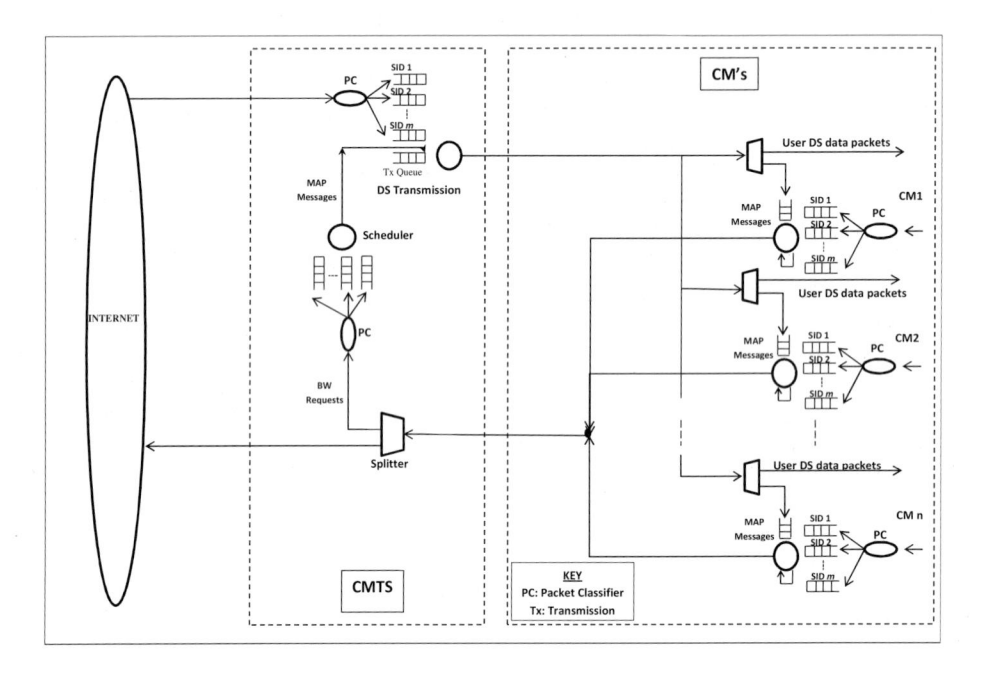

Fig. 3. QNM of a general DOCSIS 1.1/2.0 HFC network

The QNM illustrates that the bottleneck in a DOCSIS HFC network is upstream transmission due to the many-to-one access topology. In addition the upstream channels are restricted in their capacity to transport packets at high rates. This upstream packet rate limitation impacts both downstream and upstream throughput.

Most of this QNM (Fig. 3) is considered to be comprehensible in light of the overview of the protocol operation given in the previous sub-sections and the DOCSIS specification. The blocking at the CM schedulers represents contention and/or scheduling delay of data packets at the CM service station while they await an allocation of a transmission opportunity either in response to an (aperiodic) bandwidth request sent to the CMTS by the CM scheduler or via a pre-arranged periodic grant.

3 Performance Modelling of DOCSIS HFC Networks

This section summarises existing approaches and results of the performance modelling and evaluation of DOCSIS networks and centres on the following operational aspects among others: CRA's, upstream bandwidth/slot allocation algorithms, flow-priority scheduling disciplines and QoS provisioning. Finally an overview of research on the effect on DOCSIS network performance of running TCP applications is presented. Corresponding research surrounding HFC networks under the IEEE 802.14 protocol is included on some occasions due to its conceptual and operational similarity to DOCSIS in many respects and the implied opportunities for the cross-pollination of modelling approaches and their solutions between the two transmission technologies. Performance models of DAVIC/DVB HFC networks were not reviewed due to time limitations.

Similarities can be expressed as the existence of the following in both protocols: data encryption in both directions, CM ranging, time-slotted upstream channel, the use of random access methods for registering with and requesting bandwidth from the headend and employment of request-grant procedures for upstream bandwidth allocation. Piggybacking may be used by stations to make bandwidth requests. QoS support to differentiated flows by enabling the assignment of a subset of contention slots to a particular class of CM's and for reduced access delay both allow use of contention slots for short information transfers (immediate access). In addition both use the MPEG-2 format for downstream packet transmission. One of the major differences is in the variability of transmitted packet size and others include the implementation of the above algorithms for example the CRA, security and transport mechanisms and QoS support. The reader is directed to the following publications for detailed comparisons between these two protocols as well as the DAVIC/DVB standard and all their implementations: [9-11].

Overall, it was found that for DOCSIS-compliant HFC networks, performance was evaluated predominantly using simulation as opposed to analysis and little queue modelling has been carried out. This can be attributed to its inherently complex architecture and operation characterised by interdependence between several system-parameters. It was found that in all the cases of performance modelling i.e. simulation and analysis, simplifying assumptions were made to aid evaluation. Verification of models against real network data was starkly atypical.

3.1 Collision Resolution Algorithms (CRA's)

In order to control the performance experienced by packets arriving to the CM's due to the CRA, DOCSIS allows dynamic adjustment of the initial and maximum backoff range parameters. Therefore CRA performance models must essentially include these two parameters. If the initial backoff parameter is too low, then the frequency of collisions rises resulting in repeated attempts and hence greater delays whereas if either or both of the parameters are too large then minislots and thus bandwidth is wasted.

A great amount of analysis on the performance bounds experienced by contention requests under tBEB has been carried out in the literature and this has been helpfully summarised in [12]. Kwak et al carried out throughput and mean delay analysis of contention requests under a generalised exponential backoff CRA (where tBEB arose as a special case of the general version) [12]. The performance metrics were evaluated in terms of the above two backoff parameters for a fixed number of stations in a saturated network (i.e. one where each CM always has a packet to transmit) and where the collision resolution process was taken to be in equilibrium. Independence between successive collisions was assumed facilitating tractable mathematical analysis which was dominated by probabilistic arguments. The mean contention delay of requests in contention was taken as the mean total number of backoff time slots that a customer tarries while contending for transmission. Wang and Qiu [13] evaluated via simulation a proposed improvement to the tBEB algorithm that was claimed to offer improved delays to requests in contention.

The 802.14 CRA is based on a sophisticated n-ary tree splitting algorithm and analogous to tBEB it too has faced proposed extensions for example that by van den Broek et al [14]. Of the few queueing models encountered during the literature review on the performance modelling of HFC networks, one involved modelling two variants of a ternary tree-splitting contention algorithm using queueing models by Boxma et al [15]. The authors found that the first two moments of the contention delay in the free access contention tree algorithm are closely modelled by the sojourn time of a conventional finite-capacity machine repair network model with Random Order of Service (ROS)[3] whereas those of the blocked-access variant were found to match the sojourn time moments of the aforementioned network with a delay prior to the ROS queue. Noteworthy is the observation that unlike tBEB, the ternary-tree algorithm cannot be relied upon to preserve priority assignments when resolving collisions [16].

Lin et al [17] compared via simulation the request access delay (RAD) and throughput of HFC networks operating under early versions of the 802.14 (draft 2) and DOCSIS protocols. RAD was defined as the time from arrival of a data customer to the CM to the receipt of CMTS acknowledgement of the request for bandwidth and thus it is a measure of the CRA efficiency. Fair comparison was achieved by examining the performance measures of the network operating under the fine-tuned parameter settings and the same minislot allocation strategy of the respective protocols. Throughput was found to be very similar and RAD was at most about 22% longer in the DOCSIS network with notable differences occurring in the load range of about 40 – 75%. For the rest of the load range, the delay was very similar. This difference in network performance during moderate load was attributed to the more efficient first transmission policy in 802.14 which reduces the numbers of requests contending the same minislot cluster.

3.2 Bandwidth/Slot Allocation

As mentioned previously in Section 2.4, the performance bottleneck in DOCSIS 1.1/2.0 HFC networks is the uplink and the upstream throughput is highly dependent

---

[3] They assert that indeed any work-conserving service discipline produces the same results as the ROS scheduling discipline as shown by a prior result employing the PS service discipline.

on the ratio of contention capacity to total upstream channel capacity [18]. Too high a ratio and the transmission capacity is limited resulting in reduced upstream bandwidth utilisation. On the other hand, too low a ratio and the opportunities to gain transmission grants decrease resulting in longer delays at the CM's. Several mechanisms have been proposed to alleviate these limitations and to this end studies (both analytic and simulation) have also been carried out to determine the optimum ratio. It can be seen intuitively that allocating unused minislots for contention (corresponding to periods of light loading) and maintaining a minimum number of minislots for contention during periods of high load helps to reduced contention delay [2, 18].

Lambert et al [8, 19] modelled the upstream transmission (i.e. contention and reservation) in DOCSIS 1.1/2.0 HFC networks by an open QNM comprising two processor share (PS) queues in tandem and employed a decomposition technique to solve the network. It was found that a ratio of between 10%-15% yielded the least access delay for a wide range of inter-arrival time correlation levels in the arrival processes of data packets to the CM's. The work of Lambert et al is detailed in Section 4.2.

Cho et al [20] observed from simulation experimentation using the common simulation framework[4] version 13.0 (CSF v13) that the optimal ratio of contention to total upstream capacity, resulting in the highest throughput and least access delay was 0.15 for a MAP size of 2ms. It must be pointed out that though the assumed arrival process of customers to the CM's was not stated in [20], their result is accepted here because of support from Lambert et al's observed invariance of the ratio to correlation-levels in the arrival process [19] and thus it is thought invariance to the arrival process.

3.3 Flow-Priority Scheduling and Quality of Service (QoS) Provisioning

The scheduling of the DOCSIS-defined priority-services in networks is open to vendor-implementation. Several scheduling mechanisms have been designed and evaluated in the literature and feature priority queueing mechanisms such as the weighted fair queueing (WFQ) policy or its variants, Pre-emptive Resume (PR) priority scheduling and the earliest deadline first (EDF) policy among others. Existing works pertaining to the provision of QoS in DOCSIS HFC networks are summarised below.

It was found by Xiao and Bing [21] in a very sparsely-populated experimental DOCSIS network implemented using $Arris^{TM}$ CM's and CMTS that for CBR traffic flows to the CM's, the network performance varied with packet length in the following way: in both upstream and downstream directions the larger the packet, the greater the throughput and the less the delay and packet loss. This expected observation is attributable to the lower ratio of overhead to transmitted data for larger packets compared to that when smaller packets are being transmitted. Additionally, as expected, loss began to occur in both directions when the rate of sending of packets reached the corresponding link capacity and this coincided with levelling off of throughput and delay. It was found that the level of network performance experienced

---

[4] A DOCSIS simulation program created by OPNET Technologies Inc. in conjunction with CableLabs.

by customers to different CM's belonging to a particular priority-class was the same. For fixed packet lengths, all classes of CM's experienced the same level of network performance for increasing sending rates until they approached to within a close margin of a 'breakout' value, after which the throughputs levelled-off, loss rates ramped at different speeds and delay rose extremely quickly to different maximum values. Finally a simulation of the experimental DOCSIS network was carried out in OPNET with a standard WFQ CMTS scheduling mechanism where priorities were implemented by setting appropriate probabilities of being served. The simulation revealed that in the context of performance this scheduling discipline accurately modelled the behaviour shown by the $Arris^{TM}$ CMTS.

Hawa and Petr [22] proposed a preliminary new CMTS scheduling architecture to enable the five DOCSIS-defined QoS services satisfy the bandwidth and delay guarantees of CBR, VBR and BE traffic. The complex queueing station design had multiple buffers and grouped the five flows into three classes: Type 1 represented UGS data grants and unicast requests for rtPS and nrtPS flows, Type 2 represented flows with minimum bandwidth reservations and Type 3 related to flows with no bandwidth reservations. Type 1 flows were prioritised over Types 2 and 3 flows via a semi-pre-emptive mechanism (whereby customers were allowed to complete service when their deadlines preceded those of new arrivals to the queue). Flows within the latter two classes were differentiated through a priority WFQ system. Both random early detection (RED) and multi-priority RED were proposed for use in buffer management due to their support for TCP traffic. The authors stated that they were in the process of evaluating the scheduler's performance via simulation and analysis[5].

Zhenglin and Chongyang [23] modelled scheduling at the CMTS analytically under various simplifying assumptions including two traffic flows namely real-time CBR traffic and non-real-time data traffic under UGS and BE DOCSIS contention services respectively. The authors considered the real-time flow having higher priority over data and implemented this prioritisation using the PR scheduling discipline. The arrivals of the bandwidth requests of the two classes to the CMTS scheduler were assumed to be independent Poisson processes with different rates and the service times of these two classes were assumed to be independent and generally distributed with different means. Concatenation and immediate access were not modelled. The fixed contention slot allocation scheme was used with contention slots grouped at the end of each MAP. Mean performance metrics at the CMTS were derived via the celebrated P-K formula and stochastic and queueing theoretic arguments. Neither the analytic model nor formulae were verified against simulation or actual measurement. The experiments carried out showed, as expected, that the use of the PR scheduling discipline in this specific context enabled the CBR real-time traffic to meet its stringent time constraints obviously at the expense of non-real-time traffic whose requests could be timed-out. Finally it was shown that larger packets exhibited better bandwidth utilisation efficiency. This was attributed to the fact that larger packets use a relatively smaller physical overhead.

Lin and Lee [24] designed and evaluated an admission control policy at the CMTS to service time-sensitive flows. This QoS scheduling mechanism permitted flows

---

[5] At the time of publication of this tutorial, a later publication by Hawa and Petr presenting performance evaluation of their CMTS scheduling queueing architecture was not found after a brief search of the internet.

based on available bandwidth and delay guarantee provision using the tolerated jitter parameter. The EDF policy was employed at the CMTS scheduler and it was claimed via analysis and simulation that both delay and throughput were enhanced in this admission control policy compared to several existing scheduling schemes.

Unfortunately these performance gains come at the expense of rejection of flows whose QoS requirements cannot be fulfilled.

Droubi et al [25] proposed a CMTS architecture that provided bandwidth guarantees using the existing self-clocked weighted fair queueing (SCFQ) scheme. This scheme was chosen because it provides the least computation and implementation complexity among the WFQ algorithms. The SCFQ scheme can be implemented as a head of the line queue where customer priorities are their finish times calculated using the negotiated transmission rates thus incorporating bandwidth guarantees. In addition an implementation was proposed for providing the UGS service for delay-sensitive CBR traffic. The architectures were verified via simulation using the CSF package. Two sets of experiments were conducted over a sparsely populated network of 15 CM's characterised by Poisson then Markov-modulated Poisson process (MMPP) arrivals respectively.

Bushmitch et al [26] proposed a new upstream service flow scheduling service, UGS with piggybacking and showed via simulation, using real video traces, that it improved both the overall upstream bandwidth utilisation and delay experienced by real-time upstream VBR video packets when compared to the existing UGS (low delay, CBR allocation) and rtPS (good bandwidth utilisation for both CBR and VBR but higher delay) service flow provisioning. This came at the expense of more complex implementation and the degraded QoS experience of lower priority SID flows e.g. BE flows. It must be noted that in DOCSIS 1.1/2.0 piggybacking is not permitted with UGS nor are any other contention mechanisms and therefore the aim of this proposal was to highlight possible areas of improvement to the DOCSIS 1.1/2.0 specification. The application of the proposed scheduling service assumed that the real-time VBR traffic had been 'smoothed' to reduce burstiness. The authors referred to works which state that compressed digital video and other types of video streams are long range dependent exhibiting burstiness over multiple time scales. Several 'smoothing' techniques of video streams were described, most of which result in video streams comprising a significant CBR component and an additional bursty component which cannot be avoided. It is this CBR component that was supported by the UGS part of the scheduling discipline and the piggyback requests accommodated the extra bandwidth required for the bursts, while maintaining better delay constraints than when using rtPS.

Golmie et al [16] on the other hand proposed facilitating differentiated service by a three-pronged approach. Firstly priorities were assigned to contention slots such that a flow with a new request waited for a group of contention slots with its own priority in order to transmit and did so with probability one within this range of slots. The second scheme involved customising the tBEB backoff range offering such that for high priority requests the maximum backoff parameter was set to the number of contention slots reserved for that priority. Thus high priority customers retransmitted in the assigned contention slots in the next available MAP with probability one and this way delays were minimised. Finally the ratio of data to contention slots was adjusted dynamically according to an algorithm that was a slight modification of an existing one. The authors evaluated via simulation a DOCSIS network comprising up to 200 CM's servicing Poisson arrivals and it was clearly observed that indeed differentiated service was provided in terms of access delays to the different classes of customers. This occurred however at the expense of the delays experienced by the lower priority classes and extra processing at the CMTS and CM's.

Sdralia et al [27] evaluated via simulation using CSF v12, a priority-FCFS scheduling mechanism at the CMTS in the interest of providing reference statistics against which the performance of the CMTS under other scheduling mechanisms could be compared. Here requests for bandwidth that arrived at the CMTS were queued in their respective priority buffers. The authors simulated a network comprising 200 CM's and eight priorities while ignoring concatenation. It was found that the maximum upstream throughput efficiency was about 77% with larger packet sizes of 1.5 kB and only 61% for smaller packet sizes of 100 bytes. These conservative maxima are due to MAC and physical layer overheads, unused capacity and the MAP structure. The authors also found that small packet sizes exhibited high access delay which could be reduced with concatenation. They asserted that large packet sizes make more efficient use of bandwidth but under saturation even large packets suffer and thus justified the inclusion of prioritisation.

A proposal for network-wide QoS provisioning via a QoS management device connected to the CMTS on the one hand and to proposed QoS controllers in the CM's on the other was made by Adjih et al [28]. The QoS management device was designed to fulfil QoS levels to requesting subscribers by logging the network usage statistics and when bandwidth is limited, negotiating with the QoS controllers more suitable traffic demands. In this case the authors proposed a network of adaptable CM's sensitive to bandwidth availability.

This QoS support however, comes at the expense of implementation complexity and increased network traffic due to the additional management packets traversing the network.

Lin et al [17] compared via simulation the performance of a DOCSIS HFC network under three upstream scheduling disciplines: shortest job first (SJF), longest job first (LJF) and modified-SJF. Here the size of job (i.e. short or long) refers to the amount of bandwidth requested by the CM. The SJF discipline showed poorer RAD but lower data transfer delay (DTD), a measure of the efficiency of the upstream transmission scheduling algorithm and defined as the time between receipt of bandwidth request at the CMTS and subsequent receipt of full data packets there. The larger RAD was attributed to the shorter DTD that results in larger proportions of time that the CM is empty and consequently a larger proportion of arrivals to the CM having to contend for channel transmission via the CRA. The modified-SJF was sought to help to alleviate this limitation by splitting data grants allocated to a single CM into smaller sizes and distributing these across several minislots. The network running under the modified-SJF scheduling discipline exhibited the most balanced performance of the three disciplines.

3.4 TCP Applications over DOCSIS HFC Networks

The intended use of DOCSIS cable networks for IP transmission necessitates the study of the behaviour of TCP traffic over the DOCSIS network as this transmission service forms a major proportion of Internet traffic. Further, the behaviour of TCP over asymmetric paths in other infrastructures such as wireless systems [29-32] has implications on its effect on the performance of DOCSIS networks.

A network exhibits bandwidth asymmetry when running TCP applications if achieved throughput is not solely a function of the link and traffic characteristics of the forward direction but in fact depends on the impact of transmission in the reverse direction too. Most of the prior work focused on highly asymmetric paths with respect to bandwidth where the normalised asymmetry level (the ratio of raw bandwidths to the ratio of packet sizes in both directions) typically would be of the order of 2-4 [29]. In DOCSIS HFC networks the upstream channel exhibits packet rate asymmetry due to low upstream packet rates with respect to downstream capacity. However the problem symptoms are similar. Various methods have been proposed to alleviate the TCP over asymmetric path problems including header compression and modified upstream queue policies (drop-from-front, TCP acknowledgement prioritisation, TCP acknowledgement filtering). Some of these ideas can be applied to DOCSIS networks. For example, a CM that supports TCP acknowledgement filtering could drop 'redundant' TCP acknowledgements that are queued. While this would increase the TCP acknowledgement rate, it would also increase the level of TCP acknowledgement compression. TCP acknowledgement reconstruction could be implemented in the CMTS to prevent the increased level of TCP acknowledgement compression from affecting network performance.

Liao and Ju [33] designed and evaluated two novel mechanisms to improve TCP transmission that is downstream-heavy via an ns-2 simulation of a small DOCSIS network comprising 30 CM's. In the first, bandwidth requests were sent faster in order to reduce the asymmetry ratio thereby helping to reduce upstream access delay while maintaining TCP downstream data transmission rates. In the mechanism 'piggybacked' bandwidth requests were sent in reserved unicast minislots at the front of the MAP if the data grant was at the backend of its MAP and the new transmission cycle had not yet started. If the new packet arrived before the start of the data grant but after the start of the reserved minislot, the CM would send the request via piggybacking in the normal way.

Naturally, this can be seen to occur at the expense of additional contention delay for new stations attempting to begin transmitting and additional implementation complexity.

In the second mechanism, upstream TCP acknowledgements were prioritised by reducing the sending rates of the larger data packets compared to those of the (smaller) TCP acknowledgements. This had an adverse effect on upstream TCP transfer latency though not significantly, it was claimed, and additional implementation complexity at the CMTS (only).

Elloumi et al [34] found that TCP throughput over an 802.14 network was low primarily due to TCP acknowledgement compression. The authors proposed two solutions: one involving piggybacking and a second involving TCP rate smoothing by controlling the TCP acknowledgement spacing. It was found that piggybacking helped reduce the burstiness associated with the TCP acknowledgement stream in certain situations. However it was limited in its ability to effectively match offered load over a range of operating conditions. The authors' second solution was to control the TCP sending rate by measuring the available bandwidth and calculating an appropriate TCP acknowledgement rate and allowing the CM to request a periodic grant that would provide sufficient upstream bandwidth to meet the required TCP acknowledgement rate.

Cohen and Ramanathan observed that an HFC network presents difficulties for TCP due to the asymmetry between upstream and downstream bandwidth's and due to high loss rates (the authors assumed channel loss rates as high as 10-50%) [35]. Because of the known problems associated with TCP/Reno in these environments [36-38], the authors proposed a 'faster than fast' retransmit operation where a TCP sender assumes that a packet is dropped when the first duplicate TCP acknowledgement is received (rather than the usual triple duplicate TCP acknowledgement indication).

4 Case Studies

In this section, two performance modelling studies of DOCSIS 1.1/2.0 HFC networks, based on open QNM's and evaluated via simulation and analysis respectively are detailed.

The first study evaluates the performance of a DOCSIS 1.1/2.0 HFC network via and *'ns'* simulation [2]. In light of the previously observed impact of DOCSIS network configurations on performance [39, 40], two sets of experiments were conducted to investigate the impact of different network configurations and those of different upstream bandwidth allocation strategies on the network performance when carrying TCP/IP traffic. Parameter values were discovered that showed a marked improvement in the network performance characterised by almost perfect downstream utilisation, significantly reduced access delay and lower collision rates and web response times (WRT's).

The second performance model is a high level abstraction of a DOCSIS 1.1/2.0 HFC network represented by an open QNM with blocking [8]. This was solved approximately by decomposing the QNM into two dependent sub-models: a closed QNM and a group of single-server queues. An optimum range of ratios of contention channel capacity to entire uplink channel capacity that minimised the mean time for packets to exit the cable network (equal to the mean time to access the wide area network/Internet) was derived.

4.1 Simulation

The simulation modelled the behaviour of the DOCSIS 1.1/2.0 MAC and physical layers as defined in [1] over a cable network which is illustrated in Fig. 4. A detailed discussion of the validation of the model is presented in [41]. The implementation of the simulation network model and associated web traffic models were based on the "flexbell" model with user-session variables characterised by heavy-tailed distributions with infinite variance as defined in [42]. These user-session variables include inter-arrival times of web pages, number of objects per webpage and the size of objects and so on. Withstanding the challenges of simulating real networks with web traffic characterised by not only self-similar (mono-fractal) but also multi-fractal properties at small time-scales, the flexbell model with session variables satisfying different Pareto distributions has been found to provide reasonable estimates to real

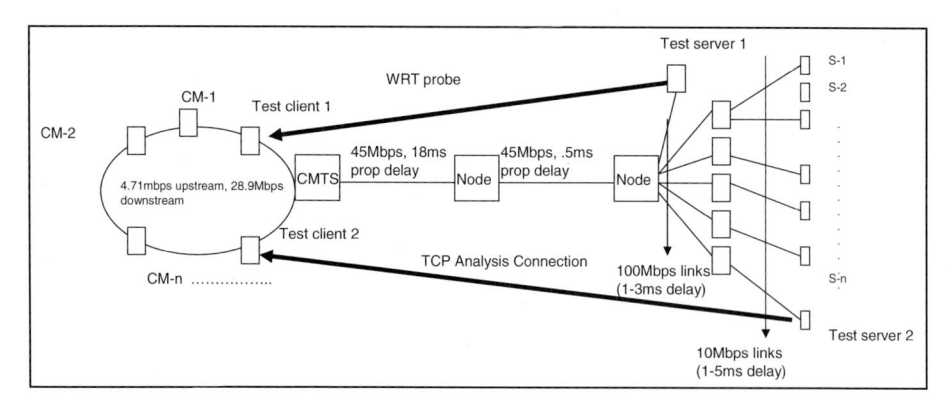

Fig. 4. Simulation network model

network behaviour, exhibiting self-similar properties despite not convincingly bearing multi-fractal scaling [42]. This makes such a model an attractive basis for use in studying network transmission technologies with current multimedia traffic profiles.

The "flexbell" topology represents numerous clients (which in the case of DOCSIS networks are the CM's) at one end of the network connected via a single bottleneck link to numerous sets of servers at the other. Transmission to a particular set of servers is through a single node as illustrated at the right-hand side of the simulation network model in Fig. 4 [42].

The simulation modelled the CM contention process, TDMA upstream bandwidth allocation and upstream and downstream packet transmission with all the nodes in the simulation network model (Fig. 4) acting as delay-stations, modelled as finite-capacity queues. The maximum size of each queue was a simulation parameter.

All experiments involved a variable number of CM's (i.e. CM-1 through CM-n in the simulation network model) that interacted with a set of servers (S-1 through S-n). The RTT from the CM's to the servers was randomly selected in the range between 42 – 54 ms.

Downstream web traffic was simulated via a four-dimensional traffic model where each constituent component (i.e. each dimension) modelled a different user-session variable satisfying a heavy-tailed infinite variance distribution. These variables were simulated using different Pareto distributions whose parameters values are given below. In addition to downstream web traffic, 5% of the CM's were configured to generate downstream low speed UDP streaming traffic (i.e. a 56Kbps audio stream), 2% of the CM's downstream high speed UDP streaming traffic (i.e. a 300Kbps video stream) and 5% of the CM's to generate downstream P2P traffic. The P2P model (based on [43]) incorporated an exponential on/off TCP traffic generator that periodically downloaded on average 4MB of data with an average idle time of 5s between each download.

The limitations of the simulation were as follows: i) CM's were confined to a single default BE service flow and a single UGS or rtPS flow; ii) the model was limited to one upstream channel for each downstream channel; iii) the model did not support dynamic service provisioning; iv) physical layer impairments were not modelled; v) the model assumed that the CMTS and the CM clocks were synchronised.

The model accounted for MAC and physical layer overhead including forward error correcting (FEC) data in both the upstream and downstream directions. For the simulations an FEC overhead of 4.7% (8% in the upstream direction) was assumed and this was modelled by reducing channel capacity accordingly[6]. The downstream and upstream channels supported an optional service rate. Service rates were implemented using token buckets where the rate and maximum token bucket size were simulation parameters.

Traffic arriving at either the CMTS or the CM for transmission over the DOCSIS 1.1/2.0 HFC network was mapped to an existing SID and treated based on the profile. In this DOCSIS HFC network model, when a CM session began, it registered itself with the CMTS which established the default upstream and downstream SID. A CM had an upstream FIFO queue for each SID. In the downstream direction there were per SID queues as well as a single transmission queue. Queuing occurred at the SID queue only if downstream rate control was enabled. All downstream queues were FIFO with the exception that MAP messages were inserted at the head of the transmission queue.

The scheduler had a configured MAP time (through a MAP_TIME parameter) which was the amount of time covered in a MAP message. The MAP_FREQUENCY parameter specified how often the CMTS sent a MAP message. Usually these two parameters were set between 1 – 10 ms. The scheduling algorithm supported dynamic MAP times through the use of a MAP_LOOKAHEAD parameter which specified the maximum MAP time the scheduler could 'look ahead'. If this parameter was 0, MAP messages were limited to MAP_TIME amount of time in the future. If set to 255 the scheduler could allocate up to 255 slots in the future. This was only used on BE traffic and only if there were no conflicting periodic UGS or rtPS allocations.

The grant allocation algorithm (i.e. the scheduling algorithm) modelled requests as jobs of a non-pre-emptive soft real-time system [44]. The system could hold two types of jobs: periodic and aperiodic. Periodic jobs resulted in UGS periodic data grants and rtPS periodic unicast request grants. Aperiodic jobs were in response to rtPS and BE requests for upstream bandwidth. Every job had a release time, a deadline and a period. The release-time denoted the time after which the job could be processed. The deadline denoted the time before which the job had to have been processed. For periodic jobs, the period was used to determine the next release time of the job.

The scheduler maintained four queues of jobs where a lower number queue had priority over a higher number queue. The first and second queues contained UGS and rtPS periodic jobs respectively both operating under the EDF policy. UGS jobs were unsolicited grants and rtPS jobs were unsolicited polls to CM's for bandwidth requests. The third queue contained all the bandwidth requests that were in response to previous unicast request grants. Similarly, the fourth queue contained the bandwidth requests that arrived successfully from the contention request process with the latter two queues serviced according to the FIFO discipline. The CMTS processed jobs from the four queues in strict priority order with no pre-emption.

---

[6] To account for FEC overhead the upstream channel capacity was reduced by 8%. This approximation was suggested by CISCO Systems Inc. (www.cisco.com). The DOCSIS 1.1/2.0 framing overhead adds an additional 30 bytes to an IP packet. A system tick of 6.25 μs and an effective channel capacity of 4.71Mbps lead to 18 bytes of data per slot for a total of 85 slots required for a 1500 byte IP packet.

The parameters associated with a UGS service flow included the grant size, the grant interval and the maximum tolerated jitter. When a CM registered a UGS flow with the CMTS, the CMTS released a periodic job in the system with release time set to the current time and the deadline set to the release time plus maximum tolerated jitter. Finally, the period was set to the grant interval. After every period, a new instance of the job was released.

The same algorithm was used for rtPS except that the maximum poll jitter was used to determine the deadline. Requests for bandwidth allocations from BE contention or from rtPS polling were treated as aperiodic jobs. Periodic jobs with the earliest deadline were serviced first. Remaining bandwidth was then allocated to aperiodic jobs. The scheduler had an additional parameter PROPORTION that was used to establish a relative priority between rtPS allocations and BE allocations.

The two sets of simulation experiments (I and II) were based on the network depicted in Fig. 4 and the respective simulation delay model in Fig. 5 below. The second set differed from the first set in several significant ways: i) the scheduler allocated unused slots for contention requests; ii) the number of IP packets allowed in a concatenated frame was no longer limited to two; iii) the buffer size at the CMTS downstream queue was increased from 50 to 300 packets; iv) the number of system ticks per slot was increased from 4 to 5 which decreased the number of slots per map from 80 to 64.

The underlying simulation delay model is illustrated below in Fig. 5.

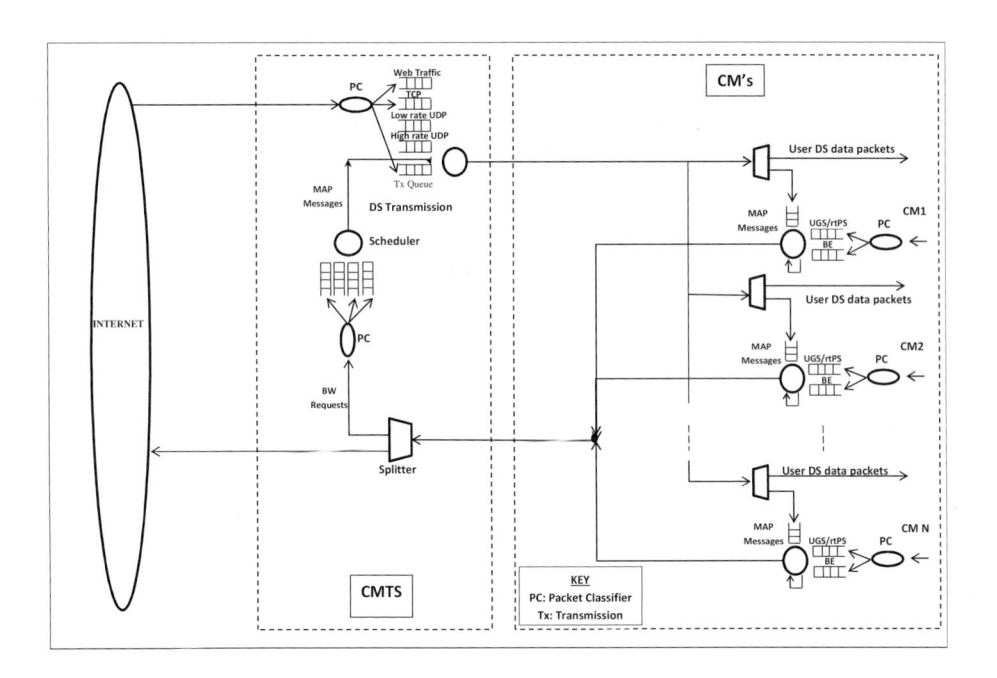

Fig. 5. Simulation delay model

The simulation model and downstream traffic parameter-values are given, respectively, in Table 1 and Table 2 below. Both sets of experiments were conducted over a range of settings of MAP_TIME and for a given MAP_TIME setting the number of CM's was varied from 100 to 500[7]. This was carried out over six MAP_TIME settings ranging from .001 to .01 s. The default MAP time setting was 2 ms (80 minislots per MAP).

Table 1. Simulation model parameter settings

Parameter	Value
upstream bandwidth	5.12 Mbps
downstream bandwidth	30.34 Mbps
Preamble	80 bits
Ticks per minislot	4
Fragmentation	OFF
Concatenation	ON
MAP_LOOKAHEAD	255 slots
Backoff Start	8 slots
Backoff stop	128 slots
Contention slots	12
Management slots	3
Simulation time	1000 s

Table 2. Downstream web traffic model parameter values

Traffic component	Pareto Mean	Pareto Shape Parameter
Inter-page interval	10	2
Objects per page	3	1.5
Inter-object interval	0.5	1.5
Object size	12 (segments)	1.2

For each experiment the following statistics were obtained:

Collision rate. Each time a CM detected a collision it incremented a counter. The collision rate was the ratio of the number of collisions to the total number of upstream packet transmissions attempted.

Downstream and upstream channel utilisation. At the end of a run, the CMTS computed the ratio of the total bandwidth consumed to the configured raw channel bandwidth. The utilisation value reflects the MAC and physical layer overhead including FEC bits.

Average upstream access delay. All CM's kept track of the delay from when an IP packet arrived at the CM in the upstream direction until it got transmitted. This statistic is the mean of all of the samples.

---

[7] Many providers provision a downstream RF channel by assigning 2000 households per channel which made this range of active CM's reasonable.

Web response time (WRT). A simple TCP client server application was run between Test Client 1 and the Test Server 1. Test Server 1 periodically sent 20KB of data to Test Client 1. With each iteration the client obtained a response time sample. The iteration delay was set to 2 s. At the end of the test, the mean of the response times was computed. The mean WRT was linked to end user perceived quality by using a very coarse rule of thumb which proposes that end users are bothered by lengthy download times characterised by WRT > 1s. This value was not advocated to be an accurate measure of end user quality of experience but rather it was used to simply provide a convenient network performance reference.

Experiment Set I. When the dominant application is web browsing (which uses the TCP service of TCP/IP's Transport Layer) the majority of data travels in the downstream direction. However, for certain configurations, the system can become packet rate bound in the upstream direction which can limit downstream throughput due to a reduced TCP acknowledgement rate. For the first set of experiments, piggybacking and concatenation were enabled however the maximum number of packets that could be concatenated into a single upstream transmission was limited to two.

Fig. 6 shows that the collision rates got extremely high as the number of active CM's increased. When only 100 users were active, the collision rate was about 50%. As the load increased, the collision rate approached 90-100% depending on the MAP_TIME setting. The behaviour of the system was influenced by several MAC protocol parameters. First, the number of contention slots assigned per map (i.e. the CONTENTION_SLOTS) directly impacted the collision rates at high loads. This set of experiments used a fixed number of contention slots, 12 per MAP which, as illustrated in Fig. 6, was insufficient at high loads. The set of curves in Fig. 6 illustrate the collision rate at different MAP_TIME settings. The collision rate was roughly 10 percent higher for the largest MAP_TIME than for the smallest MAP_TIME. This was a direct result of the MAP allocation algorithm which allocated a fixed number of contention slots each map time. If the scheduler's behaviour was altered so as to assign all unused data slots for contention, the collision rate would have been significantly lower. As the MAP_TIME was increased the bandwidth allocated for contention requests was effectively reduced.

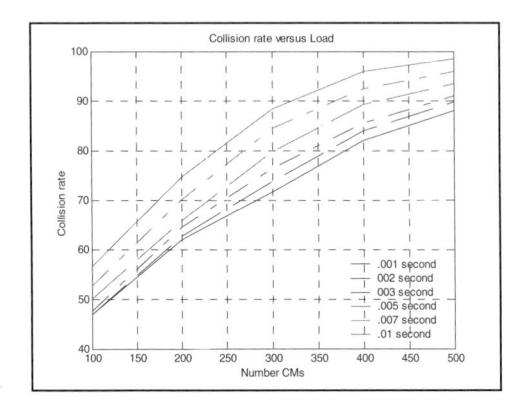

Fig. 6. Upstream collision rates as the number of CM's increase. (Experiment set I).

Fig. 7 a and b plot the channel utilisation as the load (i.e. number of active CM's) was increased. The downstream utilisation reached a maximum of about 64% with a MAP_TIME setting of .001 s. In this case, 12 contention slots per MAP were sufficient. For smaller MAP_TIME values, the downstream utilisation ramped up to its maximum value and then decreased at varying rates as the load was increased. As the collision rate increased, downstream TCP connection throughput decreased. Larger MAP_TIME values resulted in fewer contention-slot allocations leading to higher collision rates and reduced downstream utilization.

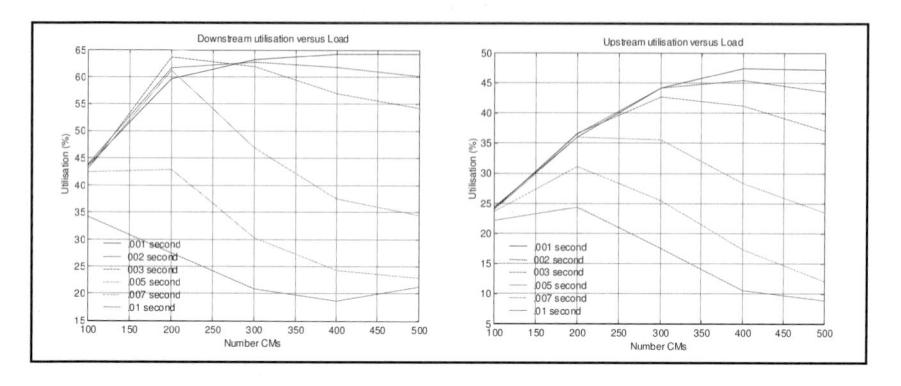

Fig. 7a. Downstream channel utilisation. (Experiment set I).

Fig. 7b. Upstream channel utilisation

Further illustrating this behaviour, Fig. 8 a shows that the average upstream access delay became very large at high loads when configured with large MAP_TIME settings. Even for lower MAP_TIME values, the access delay was significant. For a MAP_TIME of .002 s, the access delay exceeded .5s at the highest load level. To assess the end-to-end cable network performance WRT's were monitored. Using the rule of thumb described earlier, Fig. 8 b indicates that for MAP_TIME settings less than .005, up to 300 active users were accommodated before performance became bothersome. This result is clearly not generally applicable as it depends on the specific choice of simulation parameters.

Rather than making the full channel capacity available to subscribers, MSO's typically offer different service plans where each plan is defined by a service rate. For example, Charter communications offers a 3Mbps downstream rate and 512Kbps upstream rate [45]. While reduced service rates prevent customers from consuming more than their fair share of bandwidth at the expense of other customers, they offer little benefit when the network becomes congested. Fig. 9 a and b illustrate the results of an experiment that was identical to the web congestion scenario of Fig. 8 a and b except that CM's were restricted to a 2Mbps downstream service rate. Fig. 9 a shows the average upstream access delay was almost identical to that observed in the scenario without rate control. The WRT results shown in Fig. 9 b further suggest that a 2Mbps downstream service rate limit was of little use.

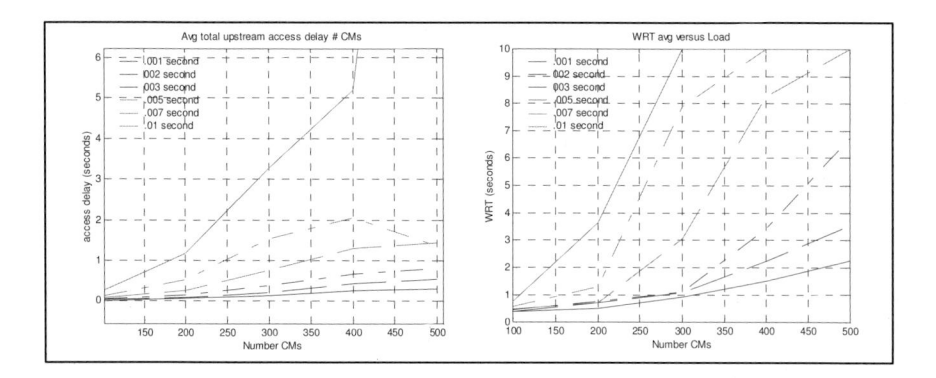

Fig. 8a. Upstream access delay (no rate control) (Experiment set I).

Fig. 8b. WRT

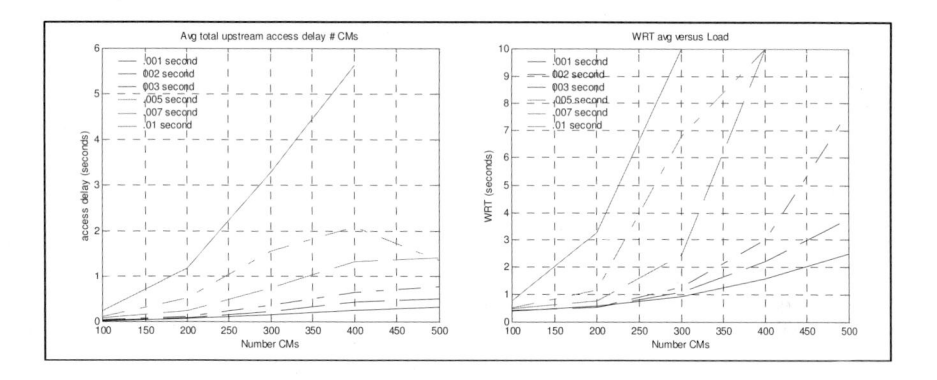

Fig. 9a. Upstream access delay (with rate control) (Experiment set I).

Fig. 9b. WRT

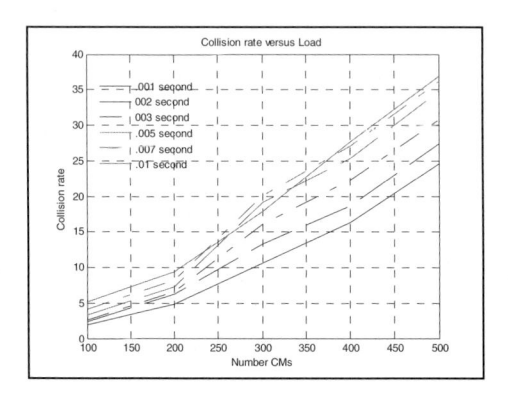

Fig. 10. Upstream collision rates. (Experiment set II).

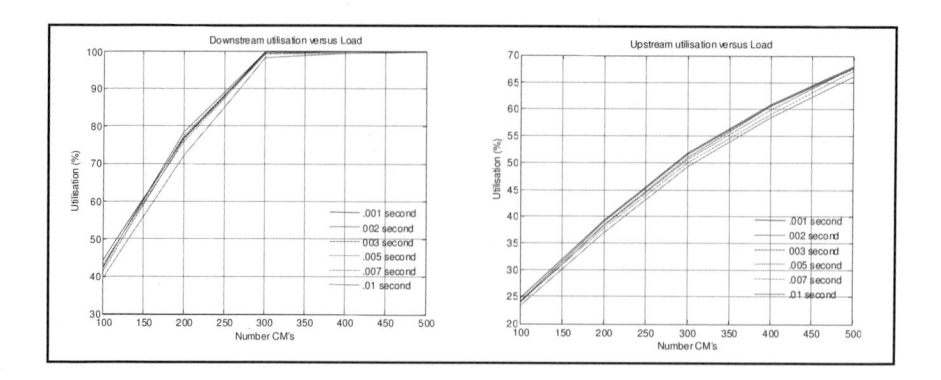

Fig. 11a. Downstream channel utilisation (Experiment set II)

Fig. 11b. Upstream channel utilisation

Experiment Set II. In the second set of experiments, the change that had the most impact was the increased bandwidth allocated for upstream contention requests. Fig. 10 shows that the collision rate ranged from 2% - 37% compared to 50% - 100% for set I. Collision rates were lowest for the runs with smaller MAP times. As the offered load to the system increased, the number of unused slots became smaller consequently reducing the number of contention slots. Hence the proportion of bandwidth allocated for contention slots was greater for small MAP times.

Fig. 11 a and b show that the utilisations in both directions were higher with a marked increase in the downstream utilisation and both were not affected by MAP times, unlike the first set of experiments. The invariance of upstream utilisation to MAP size was attributed to the profitable use of piggybacking by the runs with larger MAP times thus countering the adverse impact of their larger collision rates. The upstream rates of TCP acknowledgement packets in turn correspondingly affect the downstream rates, explaining the invariance to MAP times in the downstream direction.

The increased upstream utilisation was attributed to the increase in number of packets permitted to be concatenated and the greater number of transmission grants as a consequence of more slots available for bandwidth requests. Higher upstream TCP acknowledgement rate in turn has a positive effect on the downstream utilisation as does the larger downstream buffer. The increased downstream efficiency in turn leads to greater upstream utilisation.

Fig. 12 illustrates that from 40% - 90% of all packets sent upstream used a piggyback bandwidth request depending on the MAP size used. The runs with large MAP times were able to take advantage of piggybacking more than the runs with small MAP times because there was more time for packets to accumulate while waiting for a data grant.

The experiments were repeated with the concatenation limit relaxed and similar results were obtained with the exception that extreme levels of TCP acknowledgement-packet compression occurred. It has been shown that TCP acknowledgement compression leads to higher loss rates and that it makes it difficult for protocols that estimate bottleneck bandwidth's or that monitor packet delays to operate correctly [46-48]. Also, concatenation significantly increases access delay experienced by packets at other CM's therefore it is avoided by MSO's. However since all nodes in the network model were configured with adequate buffers in the simulation, network performance was not adversely impacted by the bursty traffic profiles caused by the TCP acknowledgement compression.

Piggybacking is less effective than concatenation for primarily downstream TCP traffic. It tends to be more advantageous for scenarios involving constant upstream traffic or backlogged upstream flows as it ensures that one packet is transmitted every cycle.

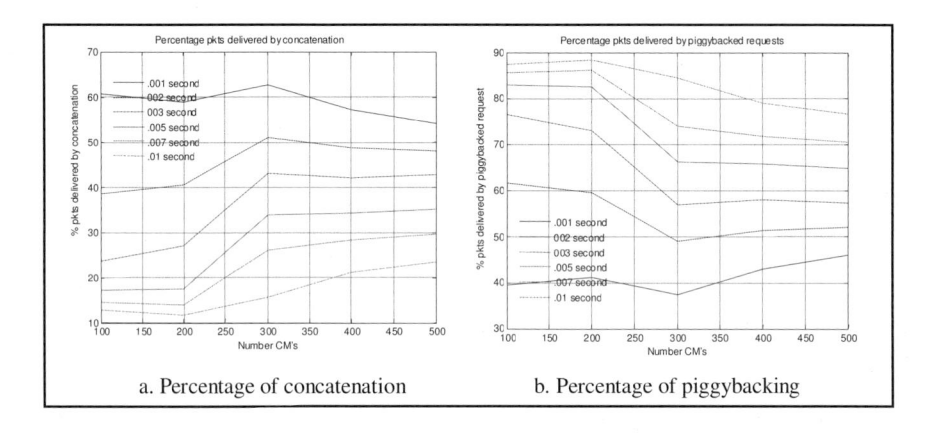

Fig. 12. Proportions of packets delivered by concatenation or due to piggybacked requests. (Experiment set II).

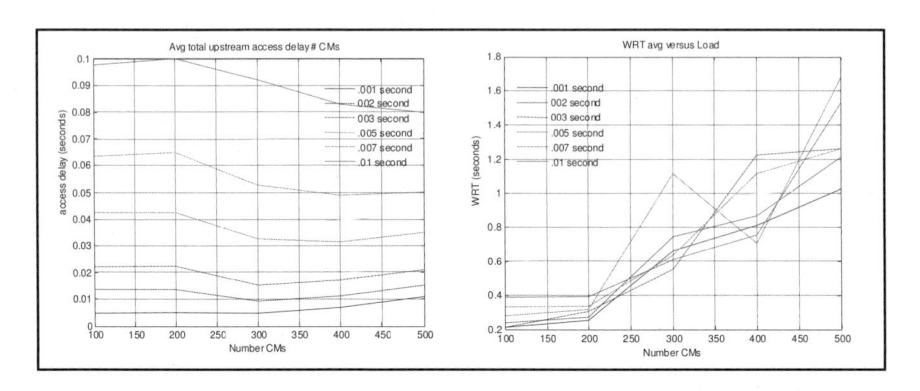

Fig. 13a. Upstream access delay (Experiment set II)

Fig. 13b. WRT

The upstream access delay was an order of magnitude lower than the first set of experiments, attributable to the lower collision rate which was in turn due to more bandwidth available for contention. The WRT too, was around one order of magnitude lower because of less frequent TCP retransmissions due to reduced packet-loss, which in turn was a direct consequence of the larger downstream buffer (Fig. 13).

The network performance improvements gained by transitioning from the parameter-settings of experiment set I to II were dramatic and they can be summarised as follows:

- As expected, the collision rate decreased dramatically due primarily to higher levels of bandwidth allocated for contention.
- The utilisation in both directions was higher with a marked increase in the downstream utilisation and both were not affected by MAP times, unlike the first set of experiments due to concatenation restriction relaxation and the dynamic contention bandwidth allocation.
- The access delay is more than an order of magnitude lower because of the reduced collision rate.
- The WRT metric was also around one order of magnitude lower because of fewer TCP retransmissions due to lower packet loss.

These performance gains can be seen to be achieved at the expense of increased implementation complexity and at the cost of providing greater downstream buffer capacity.

4.2 Analysis

In this section an overview is given of the optimisation of the upstream bandwidth allocation in DOCSIS 1.1/2.0 HFC networks via an open QNM and its approximate analytic solution [8]. Lambert et al [8] abstracted the upstream contention and transmission processes as an open QNM and derived the optimum ratio of contention to total uplink channel capacity that minimised the sum of the mean access and mean transmission times. The authors also assessed the effect of varying several system parameters on this ratio, for example the number of CM's, initial backoff parameter and so on.

The open QNM operates with blocking and comprises a group of single server finite-capacity queues (modelling the CM's) connected to the first of two single-server finite-capacity PS queues in tandem. The first PS queue (referred to as the contention queue) models the contention delay and the second (referred to as the reservation queue) models the transmission time of the data packets. The latter queue models the complement of the contention channel capacity, called the reservation channel and in this case the PS service discipline is chosen for its ability to appropriately capture the DOCSIS MAC design principle of distributing the transmission capacity fairly among active stations. The service rate of the contention queue was taken to be the saturation throughput of tBEB at equilibrium according to established modelling practice within the 802.11 environment and this was derived analytically.

The following network diagram (Fig. 14) has been inferred from the open QNM's textual description in [8]:

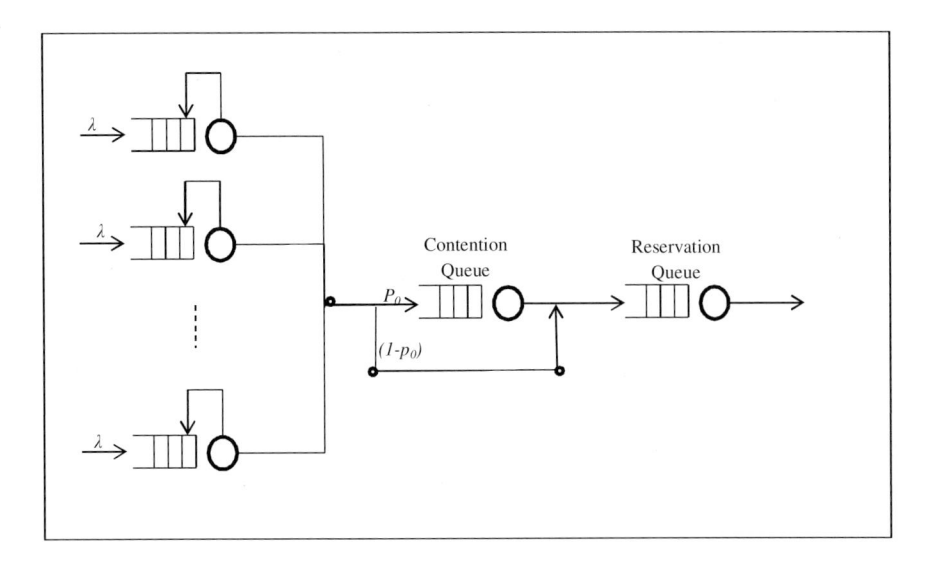

Fig. 14. DOCSIS HFC network abstracted open QNM

The QNM is constrained by the condition that there is only one customer per CM in either the contention or reservation queues at any one time therefore a customer at the head of the waiting line in a CM queue is blocked until its previous counterpart leaves the open network.

The arrival process to the network i.e. to the CM's was modelled as Poisson with rate λ packets/ms. It was claimed in [8] that Poisson arrivals to the CM queues result in a pessimistic prediction for the amount of contention channel needed because such arrivals operating with piggybacking forfeit a significant amount of performance gain experienced by bursty arrivals sending bandwidth requests via piggybacking. All arriving packets to nonempty CM's rely on piggybacking to send their bandwidth request and thus they bypass the contention process.

The solution involved decomposing this open QNM into two interdependent sub-models (a closed QNM and a collection of independent single-server queues) and then evaluating these sub-models in repeated succession of each other whereby certain input values required for the solution of one of the sub-models was taken from the most recent solution of the other. Thus the intermediate solutions converge and the iterative solution process stopped when the desired level of accuracy is achieved.

Sub-Model I. The closed QNM was obtained by removing the buffers (waiting rooms) of the CM queues and forming an unbroken loop as shown in Fig. 15 below. The packet length distribution is irrelevant as the performance measures of such a network are insensitive to the service-time distribution of its constituent PS service centres. The mean residence times at the contention and reservation queues $E[R_{Cont}]$ and $E[R_{Res}]$ respectively are required for solving the second subsystem and they can be calculated using any algorithm used to solve closed QNM's with a fixed level of multiprogramming. Here p_0 is the probability of having an empty CM queue.

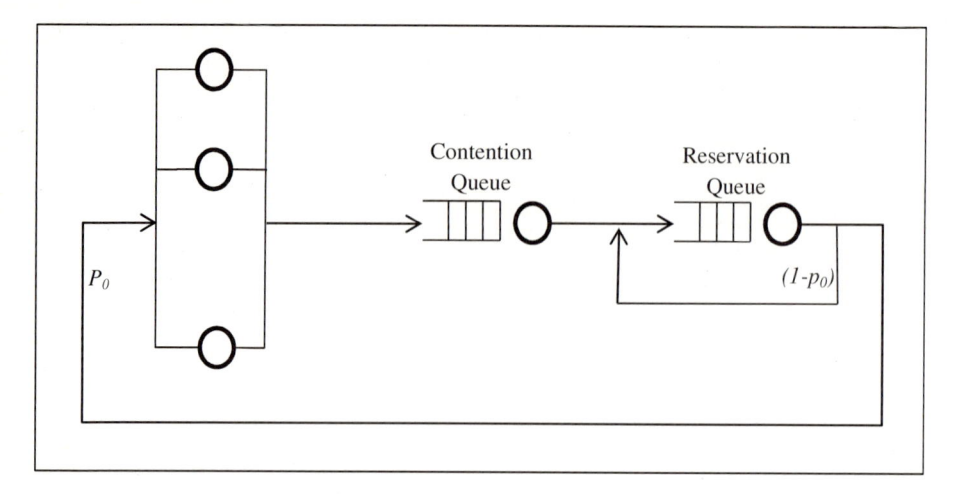

Fig. 15. Sub-model I: closed QNM

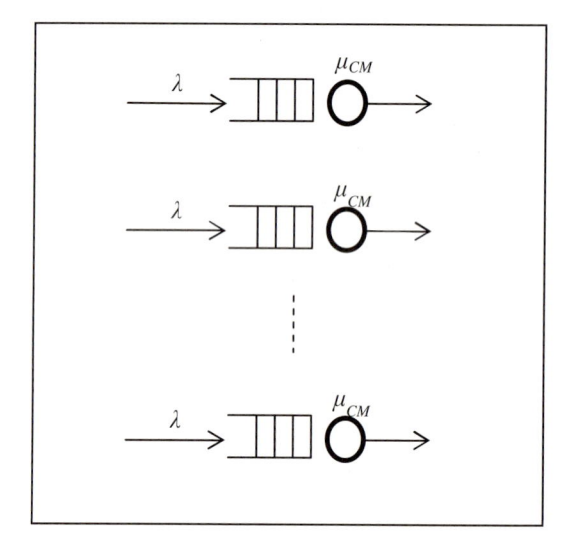

Fig. 16. Sub-model II: Collection of single-server queues

Sub-Model II. The independent single-server queues comprising this sub-model are standard M/M/1/N queues operating under the FCFS scheduling discipline and they represent the CM's (Fig. 16.). In a later publication in order to assess the impact of a correlated arrival process on the optimal fraction, the authors employed a multiple class Markovian arrival process with marked transitions (the MMAP process). The MMAP arrival process is a non-renewal point-process capable of modelling correlation with analytic tractability and it generalises a large group of inter-arrival time distributions for example those characterising the MMPP, the phase-type renewal process and their respective superpositions [19].

The mean service time of each queue, $1/\mu_{CM}$ is the mean time spent by the customer contending for access and transmitting its data and it is calculated as follows:

$$\frac{1}{\mu_{CM}} = E[R_{Cont}]p_0 + E[R_{Res}].$$

Subsequently p_0 is updated via the steady-state probability distribution for an M/M/1/N queue and this new value of p_0 is fed into the evaluation of the first subsystem. This iterative process continues till the desired accuracy of p_0 is achieved.

It was found that the analytic results of the saturation throughput and mean access and transmission times of the network compared reasonably with the simulation of the original open QNM with increasing numbers of CM's and arrival rates as well as increasing ratio of contention to total upstream channel capacity. This implies that the decomposition technique used provides a good means for solving complex open QNM's of this kind.

It was deduced that assigning 10 – 15% of the upstream minislots for contention yields near optimum results. Interestingly an identical range was discovered as optimal in the advanced system with a wide range of levels of correlation in the arrival process. In other words the fraction of contention to total upstream channel capacity was found to be invariant to the level of correlation in the arrival process at the CM's [19]. The exact value (within this range) depends on the specific system-parameter values such as data load level, minimum contention window and number of CM's among others.

It would be interesting to discern the level of accuracy of the open QNM by say comparing its simulation against actual network measurements or another independently constructed simulation.

5 Conclusions and Future Work

The DOCSIS 1.1/2.0 protocol over HFC cable networks constitute a complex system with many interdependent parameters, the intricacies of which are further heightened by the presence of bursty and/or self-similar input traffic flows characteristic of current internet traffic. This has often necessitated the performance evaluation of DOCSIS 1.1/2.0 HFC networks via simulation rather than analytic modelling especially when the nature and extent of the interdependence among several network characteristics is being studied. On the other hand analytic methodologies provide a cost-effective means to derive optimal (or optimal narrow ranges of) parameter-estimates for dimensioning a limited number of operational aspects of DOCSIS HFC networks. This tutorial has shown how both simulation and analytic approaches can successfully be used to optimise the performance of DOCSIS 1.1/2.0 HFC network configurations. Moreover the respective tradeoffs encountered as a consequence of performance improvements to DOCSIS networks were identified.

The DOCSIS protocol continues to evolve and cable network equipment implementing the latest DOCSIS 3.0 standard is now being deployed. This latest version supports Internet Protocol version 6 and achieves much higher service rates in both the upstream and downstream directions as multiple channels can be 'bonded' together and thus deliver more packets simultaneously [49]. While channel bonding can greatly increase raw capacity, limitations may still be experienced say in downstream DOCSIS network throughput of TCP traffic, which is directly affected by the rate at which TCP acknowledgements can be transported upstream by the cable modems. DOCSIS 3.0 does take steps to reduce this bottleneck by for example allowing individual cable modems to have multiple requests outstanding at any given time [49].

It is expected that many of the performance impacting behaviours observed in the former releases of DOCSIS over HFC networks will also broadly exist under DOCSIS 3.0 [49] and future extensions. In this context the performance models and their quantitative analyses reviewed in this tutorial could also be used, with appropriate enhancements to evaluate and predict the performance of new and future releases of DOCSIS protocols over HFC networks.

References

1. Cable Television Laboratories Inc.: Radio Frequency Interface Specification, Data Over Cable Service Interface Specifications DOCSIS 2.0, http://www.cablelabs.com/cablemodem/specifications/specifications20.html
2. Shah, N., et al.: A tutorial on DOCSIS: protocol and performance models. In: Third International Working Conference on the Performance Modelling and Evaluation of Heterogeneous Networks, Ilkley, UK (2005)
3. Shirali, C., Shahar, M., Doucet, K.: High-bandwidth interface for multimedia communications over fixed wireless systems. IEEE Multimedia 8(3), 87–95 (2001)
4. Elfeitori, A.A., Alnuweiri, H.: Network architecture and medium access control for deploying third generation (3G) wireless systems over CATV networks: Research Articles. Wirel. Commun. Mob. Comput. 5(2), 139–152 (2005)
5. Tanenbaum, A.S.: Computer Networks, 4th edn. Prentice Hall PTR, Englewood Cliffs (2003)
6. Abramson, N.: THE ALOHA SYSTEM: another alternative for computer communications. In: Proceedings of the Fall Joint Computer Conference, November 17-19, pp. 281–285. ACM, Houston (1970)
7. Laubach, M.E., Farber, D.J., Dukes, S.D.: Delivering Internet Connections over Cable: Breaking the Access Barrier. John Wiley, New York (2001)
8. Lambert, J., Van Houdt, B., Blondia, C.: Dimensioning the Contention Channel of DOCSIS Cable Modem Networks. In: Boutaba, R., Almeroth, K.C., Puigjaner, R., Shen, S., Black, J.P. (eds.) NETWORKING 2005. LNCS, vol. 3462, pp. 342–357. Springer, Heidelberg (2005)
9. Lin, Y.-D., Yin, W.-M., Huang, C.-Y.: An investigation into HFC MAC protocols: Mechanisms, implementation, and research issues. IEEE Communications Surveys & Tutorials 3(3), 2–13 (2000)
10. Ali, M.T., et al.: Performance evaluation of candidate MAC protocols for LMCS/LMDS networks. IEEE Journal on Selected Areas in Communications 18(7), 1261–1270 (2000)
11. Perkins, S., Gatherer, A.: Two-way broadband CATV-HFC networks: state-of-the-art and future trends. Computer Networks 31(4), 313–326 (1999)
12. Kwak, B.-J., Song, N.-O., Miller, L.E.: Performance analysis of exponential backoff. IEEE/ACM Trans. Netw. 13(2), 343–355 (2005)
13. Wang, B., Qiu, K.: Study of Collision Resolution Algorithms of DOCSIS. Journal of University of Electronic Science and Technology of China 32(33), 293–295 (2003)

14. van den Broek, M.X., et al.: A novel mechanism for contention resolution in HFC networks. In: Twenty-Second Annual Joint Conference of the IEEE Computer and Communications, INFOCOM 2003, IEEE Societies, Los Alamitos (2003)
15. Boxma, O., Denteneer, D., Resing, J.: Delay models for contention trees in closed populations. Performance Evaluation 53(3-4), 169–185 (2003)
16. Golmie, N., Mouveaux, F., Su, D.H.: Differentiated services over cable networks. In: Global Telecommunications Conference, GLOBECOM 1999 (1999)
17. Lin, Y.-D., Huang, C.-Y., Yin, W.-M.: Allocation and scheduling algorithms for IEEE 802.14 and MCNS in hybrid fiber coaxial networks. IEEE Transactions on Broadcasting 44(4), 427–435 (1998)
18. Chu, K.-C., et al.: A novel mechanism for providing service differentiation over CATV network. Computer Communications 25(13), 1214–1229 (2002)
19. Lambert, J., Van Houdt, B., Blondia, C.: Queues in DOCSIS cable modem networks. Computers & Operations Research 35(8), 2482–2496 (2008)
20. Cho, S.-H., Kim, J.-H., Park, S.-H.: Performance evaluation of the DOCSIS 1.1 MAC protocol according to the structure of a MAP message. In: IEEE International Conference on Communications, ICC 2001 (2001)
21. Xiao, C., Bing, B.: Measured QoS performance of the DOCSIS hybrid-fiber coax cable network. In: The 13th IEEE Workshop on Local and Metropolitan Area Networks, LANMAN 2004 (2004)
22. Hawa, M., Petr, D.W.: Quality of service scheduling in cable and broadband wireless access systems. In: Tenth IEEE International Workshop on Quality of Service (2002)
23. Zhenglin, L., Chongyang, X.: An Analytical Model for the Performance of the DOCSIS CATV Network. The Computer Journal 45(3), 278–284 (2002)
24. Lin, J.-T., Lee, W.-T.: Bandwidth admission control mechanism for supporting QoS over DOCSIS 1.1 HFC networks. In: 10th IEEE International Conference on Networks, ICON 2002 (2002)
25. Droubi, M., Idirene, N., Chen, C.: Dynamic bandwidth allocation for the HFC DOCSIS MAC protocol. In: Proceedings of the Ninth International Conference on Computer Communications and Networks (2000)
26. Bushmitch, D., et al.: Supporting MPEG video transport on DOCSIS-compliant cable networks. IEEE Journal on Selected Areas in Communications 18(9), 1581–1596 (2000)
27. Sdralia, V., et al.: Performance characterisation of the MCNS DOCSIS 1.0 CATV protocol with prioritised first come first served scheduling. IEEE Transactions on Broadcasting 45(2), 196–205 (1999)
28. Adjih, C., et al.: An Architecture for IP Quality of Service Provisioning Over CATV Networks. In: Roger, J.-Y., Stanford-Smith, B., Kidd, P.T. (eds.) Business and Work in the Information Society: New Technologies and Applications, pp. 368–374. IOS Press, Ohmsha (1999)
29. Balakrishnan, H., Padmanabhan, V.N., Katz, R.H.: The effects of asymmetry on TCP performance. Mobile Networks and Applications 4(3), 219–241 (1999)
30. Lakshman, T.V., Madhow, U., Suter, B.: Window-based error recovery and flow control with a slow acknowledgement channel: a study of TCP/IP performance. In: Proceedings of Sixteenth Annual Joint Conference of the IEEE Computer and Communications Societies, IEEE INFOCOM 1997 (1997)
31. Jacobson, V.: Compressing TCP/IP headers for low-speed serial links (1990)
32. Kalampoukas, L., Varma, A., Ramakrishnan, K.K.: Improving TCP throughput over two-way asymmetric links: analysis and solutions. In: Proceedings of the 1998 ACM SIGMETRICS Joint International Conference on Measurement and Modeling of Computer Systems, pp. 78–89. ACM, Madison (1998)

33. Liao, W., Ju, H.-J.: Adaptive slot allocation in DOCSIS-based CATV networks. IEEE Transactions on Multimedia 6(3), 479–488 (2004)
34. Elloumi, O., et al.: A simulation-based study of TCP dynamics over HFC networks. Computer Networks 32(3), 307–323 (2000)
35. Cohen, R., Ramanathan, S.: TCP for high performance in hybrid fiber coaxial broad-band access networks. IEEE/ACM Trans. Netw. 6(1), 15–29 (1998)
36. Elloumi, O., Afifi, H., Hamdi, M.: Improving congestion avoidance algorithms for asymmetric networks. In: 1997 IEEE International Conference on Communications, ICC 1997, Towards the Knowledge Millennium, Montreal (1997)
37. Fall, K., Floyd, S.: Simulation-based comparisons of Tahoe, Reno and SACK TCP. SIGCOMM Comput. Commun. Rev. 26(3), 5–21 (1996)
38. Hoe, J.C.: Improving the start-up behavior of a congestion control scheme for TCP. In: Conference Proceedings on Applications, Technologies, Architectures, and Protocols for Computer Communications, pp. 270–280. ACM, Palo Alto (1996)
39. Martin, J., Shrivastav, N.: Modeling the DOCSIS 1.1/2.0 MAC protocol. In: Proceedingsof the 12th International Conference on Computer Communications and Networks, ICCCN 2003 (2003)
40. Martin, J.: The Interaction Between the DOCSIS 1.1/2.0 MAC Protocol and TCP Application Performance. In: Second International Working Conference on the Performance modelling and Evaluation of Heterogeneous Networks, Ilkley, UK (2004)
41. Martin, J., Westall, J.: A Simulation Model of the DOCSIS Protocol. SIMULATION 83(2), 139–155 (2007)
42. Feldmann, A., et al.: Dynamics of IP traffic: a study of the role of variability and the impact of control. SIGCOMM Comput. Commun. Rev. 29(4), 301–313 (1999)
43. Saroiu, S., Gummadi, P., Gribble, S.: A Measurement Study of Peer-to-Peer File Sharing Systems. In: Multimedia Computing and Networking. San Jose, CA, USA (2002)
44. Stankovic, J., et al.: Deadline Scheduling for Real-time Systems - EDF and Related Algorithms. Kluwer Academic Publishers, Dordrecht (1998)
45. Charter Communications, http://www.charter.com/
46. Balakrishnan, H., et al.: TCP behavior of a busy Internet server: analysis and improvements. In: Proceedings of the Seventeenth Annual Joint Conference of the IEEE Computer and Communications Societies, INFOCOM 1998, IEEE, Los Alamitos (1998)
47. Paxson, V.: Measurements and Analysis of End-to-End Internet Dynamics. In: Computer Science Division. University of California, Berkeley (1997)
48. Martin, J., Nilsson, A., Injong, R.: Delay-based congestion avoidance for TCP. IEEE/ACM Transactions on Networking 11(3), 356–369 (2003)
49. Cable Television Laboratories Inc.: MAC and Upper Layer Protocols Interface Specification, Data-Over-Cable Service Interface Specifications DOCSIS 3.0, http://www.cablelabs.com/cablemodem/specifications/specifications30.html

Appendix I: List of Acronyms

ABR	Available bit rate
ATM	Asynchronous transfer mode
BE	Best effort
BPI	Baseline privacy interface
BPI+	Baseline privacy interface plus
CA	Certification authority
CATV	Community antenna television
CBR	Constant bit rate
CM	Cable modem
CMTS	Cable modem termination system
CRA	Collision resolution algorithm
CSF vX.Y	Common simulation framework version X.Y
DAVIC	Digital Audio Visual Council
DOCSIS	Data-over-cable service interface specification
DSL	Digital subscriber line
DTD	Data transfer delay
DVB-RCC	Digital video broadcast return channel for cable
EDF	Earliest deadline first
FEC	Forward error correcting
HFC	Hybrid fibre coax
LJF	Longest job first
MAC	Media Access Control
MCNS	Multimedia cable network system
MMAP	Markovian arrival process with marked transitions
MMPP	Markov-modulated Poisson process
MSO	Multiple system operator
nrtPS	Non-real-time polling service
PR	Pre-emptive resume
PS	Processor share
QNM	Queueing network model
QoS	Quality of Service
RAD	Request access delay
RED	Random early detection
RF	Radio frequency
ROS	Random order of service
rtPS	Real-time polling service
SCFQ	Self-clocked weighted fair queueing
SID	Service flow ID
SJF	Shortest job first
tBEB	Truncated binary exponential backoff
TCP	Transmission Control Protocol
TDMA	Time division multiple access
UBR	Unspecified bit rate
UGS	Unsolicited grant service
UGS-AD	Unsolicited grant service with activity detection
VBR	Variable bit rate
VoIP	Voice over IP
WFQ	Weighted fair queueing
WRT	Web response time

Mobility Models for Mobility Management

Vicente Casares-Giner*, Vicent Pla, and Pablo Escalle-García

ITACA-Universidad Politécnica de Valencia
Camino de Vera s/n, 46022 Valencia Spain
vcasares@dcom.upv.es, vpla@dcom.upv.es, pescalle@upvnet.upv.es

Abstract. The main goals of today's wireless mobile telecommunication systems are to provide both, mobility and ubiquity to mobile terminals (MTs) with a required quality of service. By ubiquity we understand the ability of a MT to be connected to the network anytime, anywhere, regardless of the access channel's characteristics. In this chapter we deal with mobility aspects. We provide some basic background on mobility models that are being used in performance evaluation of relevant mobility management procedures, such as handover and location update. For handover, and consequently for channel holding time, we revisit the characterization of the cell residence time. Then, based on those previous results, models for the location area residence time are built. Cell residence time can be seen as a micro-mobility parameter while the latter can be considered as a macro-mobility parameter; and both have a significant impact on the handover and location update algorithms.

1 Introduction

Mobility models play an important role in the performance evaluation of wireless communication systems. They are needed in the planning and design of wireless networks, including a plethora of technologies such as Local Area Networks (WLAN), Mobile Ad-hoc Networks (MANET), Vehicular Networks (VANET), Cellular, Cordless and Satellites. The main purpose of mobility models is the characterization of terminal movements that could help in the prediction of its position within the coverage area of the wireless network. Good mobility models are required to cope many fundamental issues such as handover, location update, registration, paging, management, and the like.

Attached mobile terminals (MTs) that are roaming in a wireless service area can be in-session or out-of-session [56]. In the first case a call is in progress and handover procedures are needed in order to maintain the continuity of the service with a certain QoS. In the second case no call is in progress that is, the MT is in idle period but it must be aware of possible incoming sessions by permanently listening to the control channel of the cell at which the MT is camped. For this second case, location management procedures are necessary.

---

* Correspondence to: Vicente Casares-Giner, Departamento de Comunicaciones, Universidad Politécnica de Valencia. Edificio ETSIT-UPV. 46022 Valencia, Spain.

Handover is the procedure by which a call in progress is switched from its current bearer radio resource, a frequency (FDMA), a time slot (TDMA) or an spreading code (CDMA) to a new bearer radio resource that offers better quality of connection than the old radio resource (the terms radio resource and radio channel are used without distinction). Handover procedure is initiated when the quality of the call in progress deteriorates below certain threshold [21]. Handovers can be classified as intra-handover and inter-handover. By intra-handover it is understood as the change of a radio resource within the same cell that is handling the call in progress. By inter-handover it is understood as the action to switch from an old radio resource to a new one that is associated to a new base station. Then, handover management procedures allows to maintain the quality of an ongoing connection as the MT moves along the wireless service area. Several handovers in succession may be necessary while a call is in progress.

By location management it is understood as the set of procedures by which the system keeps track the geographical position of the MT. These procedures are called location update (LU) and call delivery (CD). For LU the MT must report to the system database its current position in the cell coverage area and from time to time it must be maintained up to date while the MT is attached. Then, location management procedure allows to locate the MT at anytime, anywhere, to deliver incoming calls.

Both, handoff management and location management are mobility management procedures. The aim of this chapter is to provide an overview of some mobility models that can be of great help in the design of suitable inter-handoff and location management procedures. Of course, it is not the author's intention to overview the huge amount of excellent contributions in this field. For sure this is rather prohibitive, therefore many valuable publications are not cited unintentionally; we apologize for that. Being aware of that, the author's intention is to provide some basic backgrounds on mobility models that could helps as an initiating or starting point in the subject.

The structure of the chapter is as follows. In section 2 we present a classification of scenarios and mobility modes with major use in the literature. It is followed by the analysis of the cell residence time in arbitrary geographical areas, section 3. The terms residence, sojourn and dwell times are used indistinctly. In section 4 analytical results about classical studies on cell residence time are revisited. In many cases, closed form expressions from those analytical models are not possible. Then, in section 5 some numerical methods are sketched. Analytical expressions of probability density functions (pdf) or numerical methods to obtain these pdfs must be approximated by manageable fitting distribution. This task is discussed in section 6. In section 7 we review basic concepts about synthetic mobility models. The chapter concludes in section 8 with an overview of some key elements of transportation theory that are close related to the mobility management for wireless communication networks.

2 Scenarios and Mobility Models

2.1 Type of Cells

A wireless cellular networks is composed by a group of cells covering a certain geographical area. Each cell is covered by a base station with omnidirectional or sectorial antennas. In the first case we assume cells with circular shape although the approach with hexagonal or square cells is also been used. In the second case a single base station serves to more than one cell, since antennas can radiate over circular sectors of, for instance, 60° or 120°. Then, we assume cells with circular sector shape as a portion of a circle enclosed by two radii and an arc, although triangular shape can also been considered. Other coverage areas are segments of streets, main road and highways. In these geographical scenarios, cells can adopt the form of rectangular cells, some times quite narrow and hence the name of cigar cells because of the beam along the coverage area.

According to the size, cells are classified as macro, micro and pico cells. Macro-cells are circular in shape, between 1 and 40 Km; typically they are covering big rural areas and are used as umbrellas in urban areas. Micro-cells, between 100 m and 1 Km, serves streets, main roads, avenues, adopting the so called cigar shape. Pico-cells, between 10 and 100 meters are less regular in shape. In general they are for indoors areas, installed in airports, railway stations, business and residential areas. In practice, cell have irregular shapes due to main features such as propagation, shadowing, etc. For modeling and performance analysis regular shapes are considered quite often: triangle, rectangle, hexagon and circle cells. In this chapter the circular shape for cells is mostly considered but also some comment about rectangular and other special shapes found in the handover area are reported.

2.2 Mobility Models

In addition to geographical environment and cell geometry, cell residence time depends on other movement related parameters such as speed and trajectory of the mobile user. These parameters, among others are related to mobility models. Mobility models are widely used in the simulation -based performance analysis of mobile networks [52]. Currently, they are classified in two types: traces and synthetic; see the excellent work by Camp *et al.* in [40]. Traces are mobility patterns that are captured from the observation of realistic human trip movements in the daily life behaviors. Traces are quite complex to parameterize and their results are rather difficult to compare. Synthetic models attempt to realistically represent the behaviors of mobile terminals. They are based on simple driven-parameters, then easier to manage compared with trace models [37]. Random Walk (RW), Random Waypoint (RWP), Random Direction (RD) and Gauss-Markov (GM) mobility models are of most common use.

Also, mobility models can be classified according to the cell size and geographical scenario under study. Then we talk about both, microscopic or macroscopic levels. In the first case the studies are carried out at cell level and they become very useful in the design of handoff algorithms. Microscopic models are mostly related with the characterization of individual movement behavior. Without exclusions, RW, RWP, RD and GM are good representative tools used in those scenarios. On the other hand, macroscopic models mainly concern with the flow of vehicles and person that moves in between medium and large geographical areas. In other words, macroscopic models attempt to characterize the aggregated traffic of people moving on their own car or by public transportation. In the analysis of macroscopic scenarios the gravity model as a representative component of transportation theory [26], [28] is a very useful tool in the planning and design of location and paging areas.

At this point it is worth mentioning that there is no exclusive use of certain mobility model for specific scenarios. For instance, fluid flow can be used at cell level in the calculus of handover rates by using the model described in section 3.2. Also, Markovian tools can be used at macroscopic level, for instance in the evaluation of location update rates and paging load of cellular systems under certain mobility patterns assumptions [58].

3 Residence Time in Arbitrary Geographical Areas

3.1 Memoryless Model

Exponential residence time is the most simple model used in one-dimensional (1-D) and two-dimensional (2-D) scenarios. Its pdf, $f_T(t)$, is

$$f_T(t) = \mu e^{-\mu t}. \tag{1}$$

Due to its memoryless property, the exponential pdf has been used in many performance studies on wireless networks. To name a few, in [3] Hong and Rappaport proposed prioritized and non prioritized handoff procedures and consider exponential residence time for both sojourn times, in the cell and in the handoff area. In [10] McMillan considers exponential cell residence time in the open queuing network model that takes into account the mobility of users and hence handover requirements. Also, in [19] Bar-Noy et al. use the exponential distribution as cell residence time, in the performance evaluation of dynamic location update algorithms.

3.2 Fluid Flow Model

In 1987 [6], Morales-Andrés and Villen-Altamirano derived a fluid flow model of subscriber mobility for cellular radio networks. Considering that $v(x,y)$ is the velocity magnitude in a point (x,y) of the plane, then the normal velocity $v_n(x,y)$ when point (x,y) lies in a specific closed contour C is $v_n(x,y) = v(x,y)\cos\omega$, being ω the angle between the velocity and the normal vector to the tangent to the closed contour at point (x,y). Then, the total number of subscribers crossing this point that are leaving the area contained in the closed contour C, $(\omega \in (-\pi/2, \pi/2))$, is

$$v_n(x,y) = \int_{-\pi/2}^{\pi/2} v(x,y)\cos\omega \frac{1}{2\pi} d\omega = \frac{v(x,y)}{\pi}$$

where it is assumed that direction or angle of mobile's trajectory is uniformly distributed in the interval $\omega \in (-\pi, \pi)$.

Then, the rate of users leaving the given area is

$$v_n(x,y) = \int_C \sigma(x,y) v_n(x,y) \mathrm{d}l = \int_C \sigma(x,y) \frac{v(x,y)}{\pi} \mathrm{d}l$$

where C is the closed contour that limits the area under consideration. If the density and the velocity are constants in the closed contour C, $\sigma(x,y) = \sigma$, $v(x,y) = v$ and being L the length of the perimeter, the above expression is reduced to

$$h = \frac{\sigma v L}{\pi} \qquad (2)$$

Later in 1988 [7], Thomas et al. took into account the distribution of the velocity of the subscribers and obtain a slight generalized result. They consider an infinitesimal element of the border of the cell, $\mathrm{d}l$ and the traffic flow moves from inside to outside of the cell. Let ω be the angle formed by the velocity vector $v(x,y)$ and the perpendicular to the border element $\mathrm{d}l$. Then, assuming a general pdf $f_V(v(x,y))$ for $v(x,y)$, the number of mobile users crossing this border element is

$$\mathrm{d}M((x,y), \mathrm{d}t, \mathrm{d}l, \mathrm{d}v, \mathrm{d}\omega) = f_V(v(x,y))\sigma(x,y)v(x,y)\cos\omega \mathrm{d}t\mathrm{d}l\mathrm{d}v \frac{\mathrm{d}\omega}{2\pi} \qquad (3)$$

$$\mathrm{d}M((x,y), \mathrm{d}t, \mathrm{d}l) = \int_0^\infty \int_{-\pi/2}^{\pi/2} \mathrm{d}M((x,y), \mathrm{d}t, \mathrm{d}l, \mathrm{d}v, \mathrm{d}\omega) \mathrm{d}v\mathrm{d}\omega$$
$$= \frac{\sigma(x,y)E[V(x,y)]}{\pi} \mathrm{d}t\mathrm{d}l. \qquad (4)$$

If $\sigma(x,y)$ and $E[V(x,y)]$ are independent of the coordinates (x,y), $\sigma(x,y) = \sigma$ and $E[V(x,y)] = E[V]$, we have

$$h = \frac{\mathrm{d}M}{\mathrm{d}t} = \int_C \frac{\sigma(x,y)E[V(x,y)]}{\pi} \mathrm{d}l = \frac{\sigma E[V]L}{\pi}. \qquad (5)$$

Biased speed at boundaries In [14] Xie and Kuek show that the pdf of the cell-crossing terminal velocity is different from the pdf of the terminal speed in cells. Xie and Kuek's result can be derived as follows. From (3) we can write

$$\mathrm{d}M((x,y), \mathrm{d}t, \mathrm{d}l, \mathrm{d}v) = \int_{-\pi/2}^{\pi/2} \mathrm{d}M((x,y), \mathrm{d}t, \mathrm{d}l, \mathrm{d}v, \mathrm{d}\omega) \mathrm{d}\omega$$
$$= \frac{\sigma(x,y)v(x,y)f_V(v(x,y))}{\pi} \mathrm{d}t\mathrm{d}l\mathrm{d}v. \qquad (6)$$

Then, accounting for all velocities, we have

$$\mathrm{d}M((x,y),\mathrm{d}t,\mathrm{d}l) = \int_0^\infty \mathrm{d}M((x,y),\mathrm{d}t,\mathrm{d}l,\mathrm{d}v)\mathrm{d}v = \frac{\sigma(x,y)E[V(x,y)]}{\pi}\mathrm{d}t\mathrm{d}l. \quad (7)$$

The ratio between (6) and (7) represents the portion of outgoing terminals with speed $v(x,y)$. As a result we have

$$\frac{\mathrm{d}M((x,y),\mathrm{d}t,\mathrm{d}l,\mathrm{d}v)}{\mathrm{d}M((x,y),\mathrm{d}t,\mathrm{d}l)} = \frac{v(x,y)f_V(v(x,y))}{E[V(x,y)]}\mathrm{d}v = f_V^*(v(x,y))\mathrm{d}v \quad (8)$$

and simplifying notation in (8)

$$f_V^*(v) = \frac{v f_V(v)}{E[V]}; \quad \text{for } v \in [0,\infty]. \quad (9)$$

The result of (9) is quite intuitive. When terminals that cross the contour C are pooled in a random way, the density associated with the velocity of the selected terminals is given in terms of the original density, $f_V(v)$, as Kleinrock explains in [1]-formula (5.8). Notice that sampling at the boundary always favors high speeds.

Biased direction at boundaries. In [15] Xie and Goodman show that the pdf of the cell-crossing terminal direction has a direction bias towards the normal. Xie and Goodman's result can be derived as follows. Let $f_\Omega(\omega)$ be the pdf of the direction of all terminals in a given point (x,y) of the contour C. The pdf of the direction of all boundary crossing (entering) terminals is given by

$$f_\Omega^*(\omega) = \frac{f_\Omega(\omega)\cos\omega}{A}; \quad \text{with } A = \int_{-\pi/2}^{\pi/2} f_\Omega(\omega)\cos\omega \, \mathrm{d}\omega. \quad (10)$$

When the direction of all terminals are uniformly distributed in $(0,2\pi)$, i.e. $f_\Omega(\omega) = 1/2\pi$, equation (10) becomes

$$f_\Omega^*(\omega) = \frac{\cos\omega}{2}; \quad \text{for } \omega \in [-\pi/2, \pi/2] \quad (11)$$

3.3 Markovian versus Fluid Flow Models

Markovian and fluid flow mobility model are commonly used in the performance evaluation of mobility management. They are mathematically expressed by, respectively equations (1) and (5). Both models have been related, among other authors, by Thomas [7] et al. and Bauman and Niemegeers [17] as follows. We divide the global rate equation, (5) by the total number of users in the area A. Then we get the individual rate

$$\mu = \frac{\sigma E[V]L}{\pi \sigma A} = \frac{E[V]L}{\pi A}. \quad (12)$$

Expression (13) gives a map between physical parameters (the velocity of the MT and the cell size) and the rate μ of an exponential distribution that models the cell sojourn time. It can be used to identify the individual rate in a triangular, rectangular, hexagonal or circular areas; that is

$$\mu = \frac{E[V]L}{\pi A} = \begin{cases} 2\,\dfrac{2E[V]}{\pi R}; \text{ for triangle cell} \\ \sqrt{2}\,\dfrac{2E[V]}{\pi R}; \text{ for square cell} \\ \dfrac{2}{\sqrt{3}}\,\dfrac{2E[V]}{\pi R}; \text{ for hexagonal cell} \\ \dfrac{2E[V]}{\pi R}; \text{ for circular cell} \end{cases} \quad (13)$$

3.4 Location Area Residence Time

For location management, location areas (LAs) are defined in radio cellular systems. A LA is a cluster of cells that intercommunicate each other. An attached MT in the "out-of-session" state will send a LU message as soon as it leaves it actual LA.

LAs can be designed in many different ways. The first studies in cellular systems start with the assumption of regular scenarios. The triangle, the square and the hexagons in their regular shapes are the three polygons used as cell coverage areas. They have the nice property that perfectly fills the plane with no overlaps and no gaps. Then, we talk about a tessellation of the plane with triangles, with squares or with hexagons.

In regular hexagonal scenarios, LAs are quite often identified with Mosaic Graph T or M, [12], [13]. Mosaic graphs are constructed by arranging cells in concentric cycles around a starting point or cell. When the center is a point or vertex, we get the so called Mosaic Graphs M_n, where n denotes the ring number or number of cycles around the center. For M_n, $n = 0$ corresponds to a cluster of 3 hexagons. When, instead of a point, the center is a cell we get dual mosaic graphs, T_n. For T_n, $n = 0$ corresponds to a single hexagon, and for $n = 1$ corresponds to a cluster of 7 hexagons. See figures 1 and 2.

The LA residence time of a MT gives the rate of LU messages triggered by that MT. Clearly, it depends on the mobility patterns of the MT. In many studies on location management, a Semi-Markov random walk mobility model on mosaic graphs M_n or T_n is commonly assumed. It is based on the following principles. The cell residence time is characterized by a pdf $f_{cell}(t)$ with Laplace Transform (LT) denoted as $f^*_{cell}(s)$. The shape of $f_{cell}(t)$ is treated in section 4. When the MT leaves a cell, it visits one of its 6 neighbor cells with probability 1/6, a random walk model. The random walk mobility model is discussed and extended in section 7.

Then, the pdf of the dwell time of a MT in ring i, $i = 0, 1, 2, ..., D-1$, in the LT domain, denoted by $f^*_{ring\,i,X}(s)$, can be expressed as [48]

Fig. 1. Mosaic graph M_2, [13]

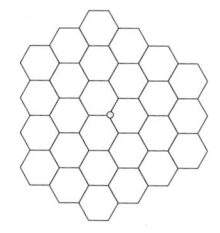

Table 1. Number of cells in ring n (K_n) and accumulated value (L_n)

Mosaic		$n=0$	Ring n $n=1$ ...	$n=m$
Mosaic M_m	K_n	3	9 ...	$6m+3$
	L_n	3	12 ...	$3m^2+6m+3$

Fig. 2. Mosaic graph T_2, [13]

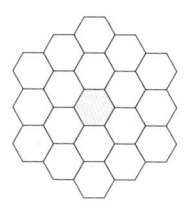

Table 2. Number of cells in ring n (K_n) and accumulated values (L_n)

Mosaic		$n=0$	Ring n $n=1$ ...	$n=m$
Mosaic T_m	K_n	1	6 ...	$6m$
	L_n	1	7 ...	$3m^2+3m+1$

$$f^*_{ring\,0,X}(s) = \begin{cases} f^*_{cell}(s) & \text{, for } X = T \\ \dfrac{2f^*_{cell}(s)}{3 - f^*_{cell}(s)} & \text{, for } X = M \end{cases}$$

$$f^*_{ring\,i,X}(s) = \dfrac{2f^*_{cell}(s)}{3 - f^*_{cell}(s)} \text{, for } X = T \text{ or } M,$$
$$\text{and } i = 1, 2, ..., D-1 \qquad (14)$$

In (14) taking the first derivative w.r.t. s at the origin, we get the mean values

$$\overline{T}_{ring\,0,X} = -f^{*'}_{ring\,0,X}(s)_{s=0} = \begin{cases} \dfrac{1}{\mu_{cell}} & \text{, for } X = T \\ \dfrac{3}{2\mu_{cell}} & \text{, for } X = M \end{cases}$$

$$\overline{T}_{ring\,i,X} = -f^{*'}_{ring\,i,X}(s)_{s=0} = \dfrac{3}{2\mu_{cell}} \text{, for } X = T \text{ or } M,$$
$$\text{and } i = 1, 2, ..., D-1 \qquad (15)$$

Our interest is to characterize the LA residence time configured by mosaic graphs M_n or T_n. To that purpose, first, due to the symmetry of the random walk model and for the sake of tractability, the 2-D space has been usually mapped into a 1-D space, by mapping each ring of the mosaic graph, M_n or T_n, into a single state [22]. Fig. 3 shows the referred mapping for LA T_n. Table 3 gives the transition probabilities for LA M_n and T_n. They are obtained by geometric considerations, [22].

Table 3. Transition probabilities $p_{j,j-1}$ and $p_{j,j+1}$ from a cell in ring j to a cell in rings $j-1$ and $j+1$ respectively, [22]

Mosaic graph M								
$p_{0,1}$	$p_{1,0}$	$p_{1,2}$	$p_{2,1}$	$p_{2,3}$	$\cdots$	$p_{j,j-1}$	$p_{j,j+1}$	$\cdots$
$\frac{2}{3}$	$\frac{2}{9}$	$\frac{4}{9}$	$\frac{4}{15}$	$\frac{6}{15}$	$\cdots$	$\frac{2j}{6j+3}$	$\frac{2(j+1)}{6j+3}$	$\cdots$
Mosaic graph T								
$p_{0,1}$	$p_{1,0}$	$p_{1,2}$	$p_{2,1}$	$p_{2,3}$	$\cdots$	$p_{j,j-1}$	$p_{j,j+1}$	$\cdots$
1	$\frac{1}{6}$	$\frac{3}{6}$	$\frac{3}{12}$	$\frac{5}{12}$	$\cdots$	$\frac{2j-1}{6j}$	$\frac{2j+1}{6j}$	$\cdots$

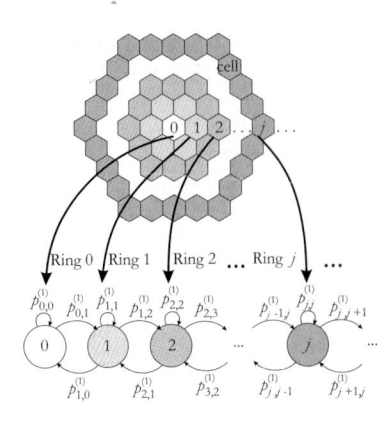

Fig. 3. Mapping mosaic graph T_j into 1D Markov Chain, [22]

Then, assuming that our target MT starts visiting the LA at ring n, the residence time in that LA, in the LT domain can be expressed by the following recursion [39]

$$f^*_{LA\,n,n,X}(s) = \frac{\alpha_n f^*_{ring\,n,X}(s)}{1 - \omega_n f^*_{ring\,n,X}(s) f^*_{LA\,n-1,n-1,X}(s)}; \text{ for } X = T \text{ or } M. \quad (16)$$

The notation on $f^*_{LA\,k,n,X}(s)$ means that we refers to a LA with $n+1$ consecutive rings, from ring 0 to ring n. The index k means that our target MT start roaming at ring k, with $k \leq n$. Also $\alpha_k + \omega_k = 1$. Therefore in (16) we assume $k = n$. Taking the first derivative at the origin in (16) we have the mean values

$$\overline{T}_{LA\,n,n,X} = \frac{\overline{T}_{ring\,n,X} + \omega_n \overline{T}_{LA\,n-1,n-1,X}}{\alpha_n}; \text{ for } X = T \text{ or } M \quad (17)$$

with the initial condition

$$\overline{T}_{LA\,0,0,X} = \overline{T}_{ring\,0,X}; \text{ for } X = T \text{ or } M. \quad (18)$$

Then, by induction we finally get

$$\overline{T}_{LA\,n,n,X} = -f^{*'}_{LA\,n,n,X}(s)_{s=0} = \begin{cases} \dfrac{3n^2+3n+1}{2n+1}\dfrac{1}{\mu_{cell}}, & \text{for } X = T \\ \dfrac{3n^2+6n+3}{2n+2}\dfrac{1}{\mu_{cell}}, & \text{for } X = M \end{cases} \quad (19)$$

On the other hand we can see that (19) is in full agreement with the fluid flow model approach. In fact we can use equation (12) to a LA of $n-ring$ size, that is:

$$\mu_{LA\,n,n,X} = \frac{E[V]L_{LA}}{\pi A_{LA}} = \begin{cases} \dfrac{E[V]6(2n+1)R}{\pi(3n^2+3n+1)A_H} = \dfrac{2n+1}{3n^2+3n+1}\mu_H, & \text{for } X = T \\ \dfrac{E[V]12(n+1)R}{\pi(3n^2+6n+3)A_H} = \dfrac{2n+2}{3n^2+6n+3}\mu_H, & \text{for } X = M \end{cases} \quad (20)$$

where μ_H is the rate of the single hexagon cell, i.e. the last equality in (20) is obtained when (13) is inserted into (20). Hence we realize that (20) and (19) are indeed the same equations.

4 Cell Residence Time: Analytical Models

In this part we overview basic backgrounds already derived for the cell dwelling time in cellular systems. In general, for the sake of simplicity in the performance analysis task, many researchers have assumed the cell dwelling time to be an exponentially distributed random variable. However this simplicity should be avoided if more precise quantitative results are needed. This is the case of handovers, where the channel holding time has a strong impact. The channel holding time is defined as the time occupancy of a given resource since the instant the resource was hunted by the call (or session) until the call ends or the mobile terminal abandon the actual roaming cell, whichever occur firs. Then, distinction is made between "new calls" and "handover calls". A "new call" is considered when a call stars inside a cell. A "handover call" always starts at the perimeter of the cell, see Fig. 4.

Several papers have contributed to those characterization. For instance, in a seminal paper by Hong and Rappaport [3] analytical expressions of the sojourn time in circular cells are derived. In [23] Yeung and Nanda address the cell sojourn time analysis for micro and macrocells where a more precise distribution for both, the speed and direction distribution of the in-cell mobile terminals are obtained when the "biased sampling problem" is considered. This biased sampling problem is also taken into consideration by Schweigel in [46] where analytical expressions for the cell sojourn time are derived for a rectangular cell geometry.

In this section, we draft the analysis reported in [23], since we consider it provides the basic background of the geometric dwelling time analysis. We start with the cell dwell time for "new calls" and it is followed by the cell dwell time for "handover calls" in circular cell of radius R.

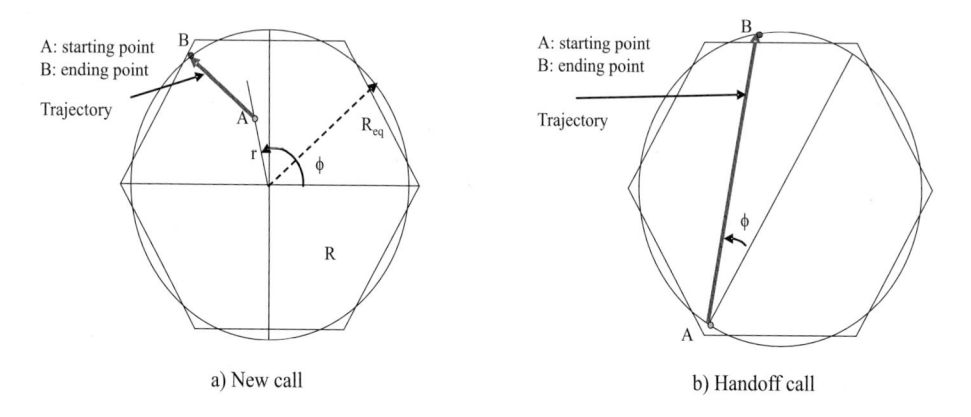

Fig. 4. a) New call, b) Handoff call

Time to leave a cell. Call originated in a cell. The probability density function (pdf) and the cumulative distribution function (CDF) are given, respectively by, [3], [23], see Fig. 4-a)

$$f_X(x) = \begin{cases} \dfrac{2}{\pi R^2}\sqrt{R^2 - \left(\dfrac{x}{2}\right)^2} & ; \text{for } 0 \le x \le 2R \\ 0 & ; \text{elsewhere} \end{cases} \quad (21)$$

$$F_X(x) = \begin{cases} \dfrac{2}{\pi R^2}\left[\dfrac{x}{2}\sqrt{R^2 - \left(\dfrac{x}{2}\right)^2} + R^2 \arcsin\left(\dfrac{x}{2R}\right)\right] & ; \text{for } 0 \le x \le 2R \\ 1 & : \text{elsewhere} \end{cases} \quad (22)$$

Let V denote the speed of the mobile terminal within a cell, selected from a random variable with pdf $f_V(v)$. Let T_X denote the MT sojourn time in the call initiated cell. Clearly T_X given by $T_X = X/V$ and, as a function of two independent random variables, T_X is a random variable with the following pdf, see [8]-section 4.6

$$f_{T_X}(t) = \int_{-\infty}^{\infty} f_{T_X}(t/V = v) f_V(v) dv = \int_{-\infty}^{\infty} |v| f_X(tv/V = v) f_V(v) dv =$$
$$= \int_{-\infty}^{\infty} |v| f_{X,V}(tv, v) dv = \int_{-\infty}^{\infty} |v| f_X(tv) f_V(v) dv \quad (23)$$
$$= \int_{0}^{2R/t} \dfrac{2v}{\pi R^2} \sqrt{R^2 - \left(\dfrac{tv}{2}\right)^2} f_V(v) dv; \quad \text{for } t \ge 0$$

and its CDF is

$$F_{T_X}(t) = \int_0^t f_{T_X}(s) ds$$
$$= \int_0^t \int_0^{2R/s} \dfrac{2v}{\pi R^2} \sqrt{R^2 - \left(\dfrac{sv}{2}\right)^2} f_V(v) dv ds; \quad \text{for } t \ge 0. \quad (24)$$

Then, the expected mean sojourn time is expressed as, after some algebra

$$E[T_X] = \int_0^{\infty} t f_{T_X}(t) dt = \int_0^{\infty} \int_0^{2R/t} \dfrac{2vt}{\pi R^2} \sqrt{R^2 - \left(\dfrac{tv}{2}\right)^2} f_V(v) dv dt$$
$$= \dfrac{1}{\pi R^2} \int_0^{\infty} \dfrac{8R^3}{3v} f_V(v) dv = \dfrac{8R}{3\pi} E\left[\dfrac{1}{V}\right]. \quad (25)$$

Time to cross a cell. Handover case. Here we are interested in the pdf of the cell dwell time of a mobile terminal that cross a circular cell. As in the previous cases it is derived in two steps. We have to keep in mind that direction and velocity on the circular contour C of the cell are biased random variables, respectively given by (9) and (11).

First we are interested in the pdf of the distance y from point A to point B, [3], [23], see Fig. 4-b). Since $y = 2R\cos\omega$ and $2f_\Omega(\omega)|d\omega| = f_Y(y)|dy|$ then

$$f_Y(y) = \begin{cases} \dfrac{y}{2R\sqrt{4R^2 - y^2}} & ; \text{for } 0 \leq y \leq 2R \\ 0 & ; \text{elsewhere} \end{cases} \qquad (26)$$

Therefore its CDF is expressed as

$$F_Y(y) = \begin{cases} 1 - \dfrac{\sqrt{4R^2 - y^2}}{2R} & ; \text{for } 0 \leq y \leq 2R \\ 1; & \text{elsewhere} \end{cases} \qquad (27)$$

As before, let T_Y denote the MT sojourn time in the call initiated cell. Obviously T_Y given by $T_Y = Y/V^*$ where this time V^* is the biased the random variable of V, see expression (9). T_Y is a function of two independent random variables and, from (9) and (26), its pdf is given by, see [8]-section 4.6

$$\begin{aligned} f_{T_Y}(t) &= \int_{-\infty}^{\infty} f_{T_Y}(t/V = v) f_V^*(v) dv = \int_{-\infty}^{\infty} |v| f_Y(tv/V = v) f_V^*(v) dv = \\ &= \int_{-\infty}^{\infty} |v| f_{Y,V}(tv, v) dv = \int_{-\infty}^{\infty} |v| f_Y(tv) f_V^*(v) dv = \\ &= \frac{1}{E[V]} \int_0^{2R/t} \frac{v^3 t}{2R\sqrt{4R^2 - (tv)^2}} f_V(v) dv; \quad \text{for } t \geq 0. \end{aligned} \qquad (28)$$

and its CDF is

$$\begin{aligned} F_{T_Y}(t) &= \int_0^t f_{T_Y}(s) ds \\ &= \int_0^t \frac{1}{E[V]} \int_0^{2R/s} \frac{v^3 s}{2R\sqrt{4R^2 - (sv)^2}} f_V(v) dv ds; \quad \text{for } t \geq 0. \end{aligned} \qquad (29)$$

Then, the expected mean sojourn time is expressed as, after some algebra

$$E[T_Y] = \int_0^\infty t f_{T_Y}(t) dt = \frac{\pi R}{2E[V]}. \qquad (30)$$

In [25] Zonoozi and Dassanayake have formulated a systematic tracking of the random movement of a mobile station in a cellular environment together with a simulation tool that obtains the behavior of different mobility related parameters. They have studied the cell residence time with some detail and conclude that the generalized gamma distribution is adequate to describe the cell residence time distribution of both "new calls" and "handover calls". The generalized gamma distribution is given by

$$f_T(t; a, b, c) = \frac{c}{b^{ac}\Gamma(a)} t^{ac-1} e^{-(t/b)^c}. \tag{31}$$

where $\Gamma(a)$ is the gamma function, defined as $\Gamma(a) = \int_0^\infty x^{a-1} e^{-x} dx$ for any real and positive number a.

In [5] Guerin describes and analyze a mobility model to characterize the cell residence time in a cell. This model has been considered in [45] as the foundation for a number of mobility models. In the original Guerin's model the mobile terminal starts its travel from an arbitrary point of a circular cell with an initial direction from a uniform distribution over $[0, 2\pi]$. A non zero velocity V is chosen from a distribution $f_V(v)$. The mobile terminal travels along a straight line at a constant speed V for a period of time selected from an exponential distribution $f_T(t) = \mu_t e^{-\mu_t t}$. If the mobile terminal did not exit the circular cell after this first trip, a new travel direction and a new period of time are obtained from their respective pdfs. The constant velocity remains the same as in the first trip. Then, the process is repeated until the mobile terminal abandon the cell. This mobility models is discussed with some detail in Guerin's paper in the context of channel holding time. Combined with the call holding time, assumed to be of exponential duration, Guerin's model shows a very good agreement between the simulated distributions and the "fitted" exponential model for channel time occupancy, tested with the use of the Kolmogorv-Smirnov goodness-of-fit test.

5 Cell Residence Time: Numerical Methods

In practical applications, very few models on cell residence time have a known closed-form solution. As we have seen in a previous section the distribution of the sojourn time of a mobile in a cell can be computed by a double integral of the moving distance and the velocity of the mobile. But, some times the mentioned double integral cannot be solved analytically when the geometric distance is not simple. In those situations, simulations or approximating methods have to be considered to solve these problems. Some work has been done in this direction. For instance, in [29], Cho et al. propose a simple an accurate method to derive the dwelling time distribution of a mobile in a cell by a numerical integration approach using for instance, a midpoint formula. The midpoint formula is easy to use, it shows stable for roundoff error accumulation, and provides more accurate that the popular trapezoidal rule. Similar approaches to [29] have been derived in [41]. In the next lines we provide the basic details.

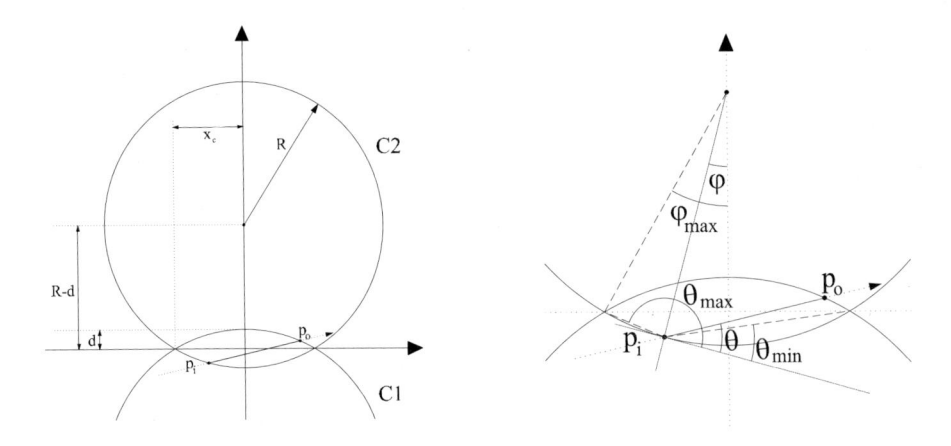

Fig. 5. a) General diagram. b) Diagram and domain for angles φ and θ.

5.1 Handoff Area Residence Time: Numerical Methods

In [41] Pla and Casares-Giner model and study the distribution of the sojourn time inside the handover area. The handover area is defined as the common overlap area between two cells of circular shape, see figure 5. The two cells, source and destination, are considered to determine handoff initiation and termination. The distance between the centers of two adjacent cells (Fig. 5) is $2(R-d)$. In the area where the two circles overlap (HA), the MT can be served by any (or both) BSs. It is assumed that the MT moves from the current cell (C1) toward the target cell (C2). Also, as in previous models, it is assumed that no changes in the the movement occur, ether in speed or direction, while the MT is inside the HA.

Although the traveled distance through the HA is completely specified by p_i and p_o we introduce variables φ and θ (see Fig. 5) for convenience. The quantities x_c, φ_{max}, θ_{min} and θ_{max}, which are geometrically defined in Fig. 5, have the following expressions

$$x_c = \sqrt{d(2R-d)} \qquad \varphi_{max} = \arctan\left(\frac{x_c}{R-d}\right)$$

$$\theta_{min}(\varphi) = \varphi + \arctan\left(\frac{d - R(1-\cos\varphi)}{x_c + R\sin\varphi}\right)$$

$$\theta_{max}(\varphi) = \varphi + \pi - \arctan\left(\frac{d - R(1-\cos\varphi)}{x_c - R\sin\varphi}\right).$$

Then, for a given cell layout, i.e., a fixed pair of values for R and d, the point p_i is defined by φ and while p_o is defined by θ (given φ). In practice p_o is numerically computed by means of the bisection method. Therefore, the distance Z can be expressed as a function of both angles, $Z = dist(p_i, p_o) = f(\varphi, \theta)$.

Once the traveled distance (Z) is known, the sojourn time is given by $T_d = Z/V$, where V is the speed of the MT, which is assumed to be constant while in the HA.

In order to obtain the distribution of T_d, the distribution of Z is first derived as described below. Then, the residence time distribution is computed by means of a slightly modified version of the method in [29].

Moving distance distribution. Let us assume that the angle distributions are known and let their CDF's be $F_\varphi(\varphi)$ and $F_\theta^\varphi(\theta)$. Note that the distribution of θ depends on the value of φ. Let $I_\varphi = \{\varphi_1, \ldots, \varphi_n\}$ be a set of n values for r.v. φ distributed appropriately. Analogously, let us define $I_\theta(\varphi_i) = \{\theta_1^i, \ldots, \theta_m^i\}$. These sets can be obtained as $I_\varphi = F_\varphi^{-1}(U)$ and $I_\theta(\varphi) = (F_\theta^\varphi)^{-1}(U)$, where U is a set (of the appropriate size) the elements of which are evenly spaced in the interval $[0, 1]$, e.g., if N is the size of U, then: $U = \{0, 1/N, \ldots, (N-1)/N\}$. Now for every pair of values (φ_i, θ_j^i) the corresponding distance $z_{ij} = f(\varphi_i, \theta_j^i)$ can be computed and, $I_Z = \{z_{ij} \setminus i = 1, \ldots, n; \ j = 1, \ldots, m\}$ is thus obtained. If n and m are sufficiently large then I_Z will be a representative sample of r.v. Z and, therefore, a discrete sampling of functions $F_Z(z)$ and $f_Z(z)$ can be approximated using the data in I_Z.

6 Fitting of Residence Time Distributions

For tractability reasons, analytical expressions or numerical data (obtained as a result of a simulation or a numerical method as the one previously described) that are candidates to characterize the residence time of a MT in a certain area, have to be approximated by a probability distribution that is manageable.

The exponential distribution is the first and most convenient candidate as the fitting distribution. It has been used in [3] and in [5]. Considering exponential duration for the message holding time, Hong and Rappaport [3] compared the survivor function of the channel holding $F_{T_H}^c(t)$ with the survivor function of the exponential distribution of rate $\mu_H = 1/\overline{T}_H$. The "goodness of fit" for this approximation was measured by

$$G = \frac{\int_0^\infty |F_{T_H}^c(t) - \hat{F}^c(t)| dt}{2 \int_0^\infty F_{T_H}^c(t) dt} \tag{32}$$

with $\hat{F}^c(t) = e^{-\mu_H t}$. A value of $G = 0$ indicates an exact fit and a value of $G = 1$ specifies no correlation.

In models such as those reported in [3] and in [5] the velocity and travel direction of the MTs may be hard to characterize. Moreover, in practical systems, the shape of cell and handover areas are irregular [34]. Those are aspects that could make the exponential distribution unable to match with synthetic models or field data, in other word, to be far from the observed reality. Hence, it seems more appropriated to directly model the cell and handover residence time as a suitable random variable with its pdf able to capture the combined effect of the shape of area and MTs mobility patterns [34]. In this direction lies the work by Zonoozi and Dassanayake [25]. In the same direction, Orlik and Rappaport [32] use a SOHYP (Sum Of HYper-exPonentials) to model the cell residence time. The SOHYP is quite simple, although acceptable for macrocells, but not for micro and picocells.

Its pdf has the following form

$$f_T(t) = \sum_{i=1}^{M_1} \alpha_{1i} m_{1i} \eta_{1i} e^{-m_{1i}\eta_{1i}t} * \sum_{i=1}^{M_2} \alpha_{2i} m_{2i} \eta_{2i} e^{-m_{2i}\eta_{2i}t} * \cdots$$

where '*' denotes the convolution operator. In [30] Ruggieri et al. propose a model of the handover area dwell time in circular cells with good agreement between modeling and simulation results. The asymptotic behavior of the resulting distribution is fitted with a truncated Gaussian function. The goodness of fit is assessed through the indicator G (see Eq. (32)) in various test cases showing a satisfactory fit.

The use of either a generalized Gamma pdf [25] or truncated Gaussian pdf [30] leads to the loss of the Markovian property, which makes both quite intractable for analytical modeling purposes. Having in mind the tractability of the memoryless and a proper fit of field data, Fang and Chlamtac [34] suggest the Hyper-Erlang distribution. According to the results in [34] the Hyper-Erlang distribution is quite simple and versatile enough to meet the aforementioned desirable properties. Its pdf is

$$f_{exp}(t) = \sum_{i=1}^{M} \alpha_i \frac{(m_i \eta_i)^{m_i} t^{m_i-1}}{(m_i - 1)!} e^{-m_i \eta_i t}$$

According to [34], the Cox distribution [1,2] could be used as a good fitting distribution. However, the Cox models contain too many parameters to be identified, hence the statistical fitting using a Cox model becomes excessively complex in the general case.

In [41], several known distributions were adjusted to match the numerical results obtained for the dwell time in the handover area. The candidate distributions considered there were: *exponential, double exponential, Erlang-k, Erlang-jk, hiper Erlang -jk* and *generalized gamma*; and the goodness of fit was assessed through the indicator G (see Eq. (32)).

7 Advanced Synthetic Mobility Models

Hong and Rappaport's model, [3], does not allow that the direction of the MT changes in the walk area. The same occurs in the Pla and Casares-Giner's model, [41]. On the other hand, in Guerin's and Zonoozi and Dassayanake's models, [5] and [25] respectively, the direction changes can be performed randomly, that is, anywhere anytime in the walk area. From that point of view, [5] and [25] can be seen as partial generalizations of previous models [3] and [41].

The models in [3], [5], [25] and [41] can be seen as particular cases of Random Way Point (RWP) and Random Direction (RD) models. In the next lines we sketch the fundamental ideas behind those models.

7.1 Random Way Point and Random Direction Mobility Models

Random Way Point model (RWP) was originally proposed to study the performance of ad hoc routing protocols, [50]. A RWP can be described as follows [49], [54]. Initially, the node or the MT is placed at the point P_0 chosen at random from a uniform distribution over a finite area A. Then a destination point (also called waypoint) P_1 is chosen from a uniform distribution over A and the node moves along a straight line from P_0 to P_1 with constant velocity V_1 drawn independently of the location from a velocity distribution with pdf $f_V(v)$. Once the node reaches P_1, a new destination point, P_2, is drawn independently from a uniform distribution over A and velocity V_2 is drawn from $f_V(v)$ independently of the location and V_1 The node again moves at constant velocity V_2 to the point P_2, and the process repeats. Formally, the RWP process is defined as an infinite sequence of triples (P_0, P_1, V_1), (P_1, P_2, V_2), (P_2, P_3, V_3), ... Alternatively, when considering an open area $A \in R^2$ each destination points P_i can be identified from the previous reached point P_{i-1} using two independent random variables, angle Θ and distance D with pdfs $f_\Theta(\theta)$ over $[0, 2\pi]$ and $f_D(d)$ over $[0, \infty]$, respectively. Then, the RWP is defined as an infinite sequence of 4-plas

$$(P_0, \Theta_1, D_1, V_1), \quad (P_1, \Theta_2, D_2, V_2), \quad (P_2, \Theta_3, D_3, V_3), \quad \ldots \tag{33}$$

As a synthetic model, the RWP has been studied in many specific scenarios. For instance in [54] it is analyzed the stationary spatial distribution of a node moving according to the RWP model in a given convex area. For this, an explicit expression is given, which is in the form of a 1-D integral giving the density of the nodes up to a normalization constant. This result is also generalized to the case where the waypoints have a non-uniform distribution.

A Random Direction model (RD) can be described as follows [55]. Initially, the node or MT is placed at the point P_0 chosen at random from a uniform distribution over an area A. Then a direction of the trip is chosen, Θ_1, from a uniform distribution with pdf $f_\Theta(\theta) = 1/2\pi$ over $[0, 2\pi]$ and the node moves along a straight line for a period of time T_1 (the duration of the trip) that is chosen from a distribution with pdf $f_T(t)$ with constant velocity V_1 drawn independently of the location from a velocity distribution with pdf $f_V(v)$. After this first trip, a new direction Θ_2, a new time T_2 and a new velocity V_2 for the second trip is chosen. Θ_2, T_2 and V_2 are drawn independently each other. Also they are independent of the actual location P_1 and previous locations. Then, the node again moves along the direction Θ_2, at velocity V_2 for a period of time T_2, and the process repeats. In a parallel way as before, the RD process is defined as an infinite sequence of 4-plas

$$(P_0, \Theta_1, T_1, V_1), \quad (P_1, \Theta_2, T_2, V_2), \quad (P_2, \Theta_3, T_3, V_3), \quad \ldots \tag{34}$$

We remark that in the RWP model the sequence Θ_i (D_i, V_i) are independent and identically distributed random variables with pdf $f_\Theta(\theta)$ $(f_D(d), f_V(v))$. Also, in the RD model the sequence Θ_i (T_i, V_i) are independent and identically distributed random variables with pdf $f_\Theta(\theta)$ $(f_T(t), f_V(v))$. It is also worth remarking that in RWP and RD velocity and direction remain constant for the whole duration of the trip.

As a synthetic model, the RD has been widely used in many scientific areas. For instance, in [4] Stadje obtained the exact probability distribution of a simple model for the locomotion of microorganisms on surfaces. In this model, a particle moves in straight-line paths at constant speed, and changes its direction after exponentially distributed time intervals, where the lengths of the straight-line paths and the turn angles are independent, the angles being uniformly distributed. For mobile communications research, in [52] Gloss et al. propose a random direction model with location dependent parameterization (RD-LDP) which extends the RD model.

RWP and RD mobility models have been widely used over finite, mostly rectangular area, to simulate Ad-Hoc and MANET network scenarios. RWP and the RD models can be extended by considering a *pause* time interval between two consecutive trips. That is, the MT or node alternates periods of movements (*move* phase) to period during which it pauses (*pause* phase). The durations of *move* and *pause* phases are, in general, distributed according to independent random variables. The pdf of the *move* phase follows from the previous RWP and RD model descriptions. The duration of the *pause* phase is drawn from the pdf $f_P(p)$. For instance in [59] a RD model over an infinite bidimensional domain is considered; in which the duration of the trip (*move* phase), T, and the duration of the pause (*pause* phase), P, follow exponential distributions, respectively with rates μ and λ. Clearly, when $\lambda \to \infty$ the duration of the *pause* phase tends to zero.

7.2 Brownian Motion

Here we describe the Brownian motion (described mathematically by Einstein in 1926, [37]). It belongs to several families of well understood stochastic processes: (i) Markov processes, (ii) Martingales, (iii) Gaussian processes and (iv) Levy processes, [33]. Let us consider that a MT moves in the x, y plane according to the following rule. A trip is a concatenated set of M movements. Each movement is independent of the others, and is defined by a pair $(\theta_k, a_k,)$, $k = 1, 2, ..., M$ in which θ_k is the angular direction, or phase, in movement k and a_k is the travel distance in movement k. The sequence of angles $\{\theta_k\}$ are *i.i.d.* with probability density function $f_\Theta(\theta)$ in the interval $(-\pi, \pi)$ or $(0, 2\pi)$ and the sequence of distances $\{a_k\}$ are *i.i.d.* with pdf $f_A(a)$ in the interval $(0, \infty)$. Then, after a given number of movements, M, the MT location in complex coordinates is

$$z = x + jy = \sum_{k=1}^{M} a_k e^{j\theta_k} \tag{35}$$

This model can be viewed as particular case of the RWP model, section 7.1, equation (33), where $f_A(a) \equiv f_D(d)$, the velocity distribution of the MT is ignored and the number of trips is limited to M. The model has been widely studied in the literature. For instance, in [16] Barber obtains a formula for the joint pdf of the angle and length of the resultant of an M-step non-isotropic two dimensional random walk for arbitrary step angle and travel distance probability density and for any fixed number of steps. Also, in [36]-appendix B, it is proved that when the phases of the individual components, θ_k, are uniformly distributed and if the number of terms M is allowed to increase without bound, $M \to \infty$, the joint characteristic function of the real and imaginary parts asymptotically approach a circularly symmetrical Gaussian function. That is, the real and imaginary parts of the resulting random walk are *joint* Gaussian random variables

$$p(x,y) \to \frac{1}{4\pi A^2} e^{-\frac{(x^2+y^2)}{4A^2}}$$

It is also worth to mention the one-dimensional version of Brownian motion used by Rose and Yates in [27] when discussing the location uncertainty in mobile networks. They consider that the infinity horizontal axis is divided in regular intervals of size Δx. The mobile starts at space interval $i = 0$. After a short period of time, Δt, with probability p the mobile will visits interval $i = 1$, with probability q the mobile will visit interval $i = -1$ and with probability $1 - p - q$ the mobile will remain in the same interval $i = 0$. From this description, is straightforward to obtain expressions for $f_\Theta(\theta)$ and $f_A(a)$. Then, if the drift velocity and the diffusion constant are defined as, respectively

$$v = (p-q)\Delta x/\Delta t; \quad D = 2[(1-p)p + (1-q)q + 2pq]\frac{(\Delta x)^2}{\Delta t}$$

the pdf of the MT location for Δx and Δt very small is a Gaussian pdf

$$p(x,t) = \frac{1}{\sqrt{\pi D t}} e^{-\frac{(x-vt)^2}{Dt}}$$

Considering the regular cell layout scenarios described in section 3.4 mobile terminal moves, after a certain sojourn time in the cell, from an initial cell to one of its three (triangle) four (square) or six (hexagon) cells in the vicinity. Clearly, it make sense to assume that the length distance of each movement or trip is constant, i.e. $a_k = A$.

This type of random walk has been studies along the years, since 1656 when Pascal and Fermat discussed the problem that today is known as the Gambler's Ruin problem [47]. Since then, a huge amount of contributions to solve it have been reported. See for instance in [42] Kmet and Petkovsek where the gambler's ruin problem was studied in several dimensions.

In dynamic location management schemes, when the movement-based location update algorithm is implemented, it is useful to predict the cell position of a target MT after M movements (M visits to cells) since the last interaction or contact with the network [19], [58]. The MT's position is described by the complex random variable z defined in (35). Obviously, the higher the value of M is the higher the uncertainly of the MT's position will be. For large M the distribution of z asymptotically approaches Gaussian function. In [57], Ollila has studied the circularity property (or properness) or lack of it (noncircularity, nonproperness) of a complex random variable based on second-order moments, called circularity quotient, ϱ_z, and gave an intuitive geometrical interpretation: the modulus and phase of its principal square-root are equal to the eccentricity and angle of orientation of the ellipse defined by the covariance matrix of the real and imaginary part of z.

Let $E[Z] = m_z$ denote the mean value of the random variable $z = x + jy$ of (35). We have

$$m_z = E[\sum_{k=1}^{M} Ae^{j\theta_k}] = \sum_{k=1}^{M} AE[e^{j\theta_k}] = AME[e^{j\theta}] \tag{36}$$

with modulus equal to

$$|m_z|^2 = \sum_{k=1}^{M}\sum_{m=1}^{M} A^2 E[e^{j\theta_k}]E[e^{-j\theta_m}] = A^2 M^2 |E[e^{j\theta}]|^2. \tag{37}$$

The *variance*, σ_z^2, can be written as

$$\sigma_z^2 = E[(\sum_{k=1}^{M} Ae^{j\theta_k} - \sum_{k=1}^{M} AE[e^{j\theta_k}])(\sum_{m=1}^{M} Ae^{-j\theta_m} - \sum_{m=1}^{M} AE[e^{-j\theta_m}])] =$$
$$= A^2 M \left(1 - |E[e^{j\theta}]|^2\right). \tag{38}$$

and the *pseudo-variance*, τ_z, as

$$\tau_z = E[(\sum_{k=1}^{M} Ae^{j\theta_k} - \sum_{k=1}^{M} AE[e^{j\theta_k}])^2] = E[(\sum_{k=1}^{M} Ae^{j\theta_k})^2] - (\sum_{k=1}^{M} AE[e^{j\theta_k}])^2 =$$
$$= A^2 M \left(E[e^{j2\theta}] - (E[e^{j\theta}])^2\right). \tag{39}$$

Variance (38), together with pseudo-variance (39), carry all the second order information, [57]

$$\sigma_x^2 = \frac{\sigma_z^2 + \mathrm{Re}[\tau_z]}{2} = \frac{A^2 M}{2}\left(1 - |E[e^{j\theta}]|^2 + \mathrm{Re}\big[E[e^{j2\theta}] - (E[e^{j\theta}])^2\big]\right)$$

$$\sigma_y^2 = \frac{\sigma_z^2 - \mathrm{Re}[\tau_z]}{2} = \frac{A^2 M}{2}\left(1 - |E[e^{j\theta}]|^2 - \mathrm{Re}\big[E[e^{j2\theta}] - (E[e^{j\theta}])^2\big]\right)$$

$$\sigma_{xy} = \frac{\mathrm{Im}[\tau_z]}{2} = \frac{A^2 M}{2}\left(\mathrm{Im}\big[E[e^{j2\theta}] - (E[e^{j\theta}])^2\big]\right). \tag{40}$$

Table 4. Table for some continue distributions

Distribution	$m_z(\text{x}AM)$	$\sigma_z^2(\text{x}A^2M)$	$\tau_z(\text{x}A^2M)$	$\sigma_x^2(\text{x}A^2M)$	$\sigma_y^2(\text{x}A^2M)$	$\sigma_{xy}(\text{x}A^2M)$
Uniform	0	1	0	$\frac{1}{2}$	$\frac{1}{2}$	0
Triangular	$\frac{4}{\pi^2}$	$1 - \frac{16}{\pi^2}$	$-\frac{16}{\pi^2}$	$\frac{1}{2}\left(1 - \frac{32}{\pi^4}\right)$	$\frac{1}{2}$	0
Raised cosine	$\frac{\beta}{2}$	$1 - \frac{\beta^2}{4}$	$-\frac{\beta^2}{4}$	$\frac{1}{2}\left(1 - \frac{\beta^2}{2}\right)$	$\frac{1}{2}$	0

Then the circularity quotient of a r.v. (with finite variance) is defined as the quotient between the pseudo-variance and the variance. In our case

$$\varrho_z = \frac{\text{cov}(Z, Z^*)}{\sqrt{\text{var}(z)\text{var}(z^*)}} = \frac{\tau_z}{\sigma_z^2} = \frac{E[e^{j2\theta}] - \left(E[e^{j\theta}]\right)^2}{1 - \left|E[e^{j\theta}]\right|^2}. \tag{41}$$

The polar representation of (41) induces two quantities. Its modulus is called the *circularity coefficient* and its argument the *circularity angle*. Hence, eccentricity ϵ and orientation α of the ellipse can be calculated as, respectively

$$\epsilon = \sqrt{|\varrho_z|}; \quad \text{and} \quad \alpha = \arg[\varrho_z]/2. \tag{42}$$

Since $|\varrho_z| = \lambda_1 - \lambda_2$ and $\sigma_z^2 = \lambda_1 + \lambda_2$, the major axis (minor axis) of the ellipse are proportional to $\sqrt{\lambda_1} e_1$ ($\sqrt{\lambda_2} e_2$)

Of course, we can consider continuous pdf for the angle θ. For instance

$$f_\Theta(\theta) = \begin{cases} \frac{1}{2\pi}; & \theta \in (-\pi, \pi) & \text{Uniform} \\ \frac{\pi - |\theta|}{\pi^2}; & \theta \in (-\pi, \pi); & \text{Triangular distribution} \\ \frac{1 + \beta\cos\theta}{2\pi}; & \theta \in (-\pi, \pi), \; |\beta| \le 1 & \text{Raised cosine} \end{cases} \tag{43}$$

Table 4 shows the expression of the first and second order statistic parameters.

Table 5. Table for some discrete distributions, $d(\alpha) = \alpha^2 + \alpha + 1$ and $\sigma_{xy} = 0$ in all cases

Distribution	$m_z \text{x}AM$	$\sigma_z^2 \text{x}A^2M$	$\tau_z \text{x}A^2M$	$\sigma_x^2 \text{x}A^2M$	$\sigma_y^2 \text{x}A^2M$
1-D scenario	$\frac{(\alpha-1)^2}{(\alpha+1)^2}$	$\frac{4\alpha}{(\alpha+1)^2}$	0	$\frac{2\alpha}{(\alpha+1)^2}$	$\frac{2\alpha}{(\alpha+1)^2}$
Square cells	$\frac{(\alpha-1)^2}{(\alpha+1)^2}$	$\frac{4\alpha}{(\alpha+1)^2}$	0	$\frac{2\alpha}{(\alpha+1)^2}$	$\frac{2\alpha}{(\alpha+1)^2}$
Hexagonal cells	$\frac{\alpha^2-1}{d(\alpha)}$	$\frac{\alpha(2\alpha^2+5\alpha+2)}{d(\alpha)^2}$	$-\frac{\alpha(\alpha-1)^2}{d(\alpha)^2}$	$\frac{\alpha(\alpha^2+7\alpha+1)}{d(\alpha)^2}$	$\frac{3\alpha}{2d(\alpha)}$

But our interest reside in the cell positioning of our target MT, i.e. when the phase follows a discrete pdf. We have considered three particular cases. Notice that 1-D scenario is a particular case. Then, using the notation of Dirac's delta,

$$f_\Theta(\theta) = \begin{cases} \sum_{i=0}^{1} p_i \delta(\theta - i\pi); & \theta \in (0, 2\pi) \quad \text{1-D scenario} \\ \sum_{i=0}^{3} p_i \delta(\theta - i\pi/2); & \theta \in (0, 2\pi) \quad \text{2-D square scenario} \\ \sum_{i=0}^{5} p_i \delta(\theta - i\pi/3); & \theta \in (0, 2\pi) \quad \text{2-D hexagonal scenario} \end{cases} \quad (44)$$

Table 5 shows these three interesting cases: 1-D, 2-D square and 2-D hexagonal scenarios, respectively. For $\alpha \in [0, \infty)$

$$\text{1-D}: \quad p_0 = \frac{\alpha}{\alpha+1} = p; \quad p_1 = \frac{1}{\alpha+1} = 1 - p = q;$$

2-D Square

$$p_0 = \frac{\alpha^2}{(\alpha+1)^2} = p^2; \quad p_1 = p_3 = \frac{\alpha}{(\alpha+1)^2} = pq; \quad p_2 = \frac{1}{(\alpha+1)^2} = q^2;$$

2-D Hexagone

$$p_0 = \frac{\alpha^3}{D(\alpha)}; \quad p_1 = p_5 = \frac{\alpha^2}{D(\alpha)}; \quad p_2 = p_4 = \frac{\alpha}{D(\alpha)}; \quad p_3 = \frac{1}{D(\alpha)}$$

with $D(\alpha) = (\alpha^3 + 2\alpha^2 + 2\alpha + 1) = (\alpha+1)(\alpha^2 + \alpha + 1) = (\alpha+1)d(\alpha)$

also $\quad p_0 = \frac{p^3}{1-pq}; \quad p_1 = p_5 = \frac{p^2 q}{1-pq}; \quad p_2 = p_4 = \frac{pq^2}{1-pq}; \quad p_3 = \frac{q^3}{1-pq}.$

Clearly, the parameter α provides a wide range on the directionality of the MT's movement. For instance, if $\alpha = 1$ we obtain the standard isotropic random walk model and no drift movement exists, since $m_z = 0$. When $\alpha \to \infty$ or $\alpha \to 0$ the MT will always travel in the same direction. Intermediate levels of directionality can be obtained by varying the tuning parameter α between 0 and 1 or between 1 and ∞.

Fig. 6 shows a random sample of length $n = 100$ for $\alpha = 5$, M=8, case a), and $\alpha = 10$, M=8 case b).

7.3 Gauss-Markov

In reallity, mobile nodes or MTs movements are conditioned by the geographical areas (urban, rural areas) they are roaming. Directions and velocities are restricted by physical laws and by human regulations that impose time correlations. On the other hand, Brownian motion models are basically memoryless

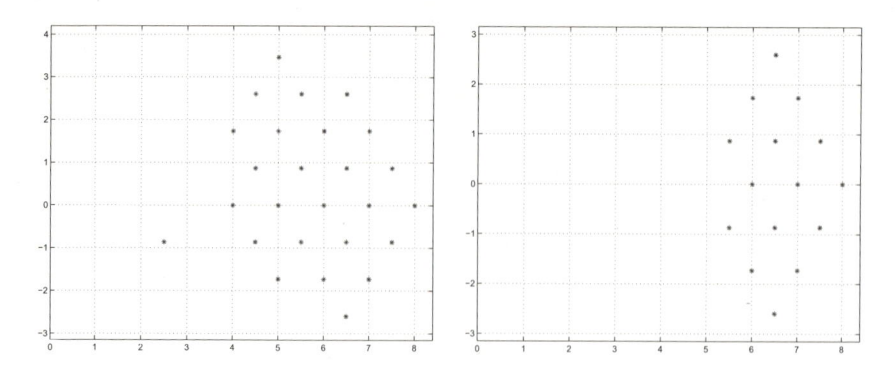

Fig. 6. Random sample of length $n = 100$, a) $\alpha = 5$ and M=8, b) $\alpha = 10$ and M=8

processes. They cannot represent the time correlation in future direction and MT's velocity. In [43], Liang and Haas introduced a mobility model based on the Gauss-Markov process. They postulate that it can be used to model the mobile movement in a Personal Communication Service (PCS) network, since it capture the essence of the correlation of a MT's velocity in time. Gaussian-Markov mobility model was designed to adapt to different levels of randomness via one tuning parameter α, [40]. The Gaussian-Markov model is a temporally dependent mobility model in which the degree of dependency is tuned by the memory level parameter α. The driving equations are

$$v_t = \alpha v_{t-1} + (1-\alpha)\overline{v} + \sqrt{1-\alpha^2} w_{s,t-1};$$
$$\theta_t = \alpha \theta_{t-1} + (1-\alpha)\overline{\theta} + \sqrt{1-\alpha^2} w_{\theta,t-1};$$
(45)

where v_t and θ_t are the velocity and the angle of direction of the MT trajectory at time instant t, $\overline{v}$ and $\overline{\theta}$ are constants representing the mean value of velocity v and angle of direction θ, $w_{v,t-1}$ and $w_{\theta,t-1}$ are random variables that follows Gaussian pdf with standard deviations σ_v and σ_θ and α is the tuning parameter to reflect the randomness of the Gauss-Markov process, $0 \leq \alpha \leq 1$. Notice that in (45) it is assumed uncorrelation between velocity and angle. There is no difficult to rewrite (45) to represent the more general case of correlated dimensions and to consider different tuning parameters for velocity and angle of direction.

Liang and Haas [43] observed that the parameter set $(\sigma_v, \sigma_\theta, \overline{v}, \overline{\theta}, \alpha)$ can be tuned to obtain other well known mobility models such as the Brownian motion (or Random Walk (RW)), the Fluid Flow (FF) and the Random Way Point (RWP) models. Hence, the RW model with drift, can be obtained from (45) with $\alpha = 0$

$$v_t = \overline{v} + w_{s,t-1};$$
$$\theta_t = \overline{\theta} + w_{\theta,t-1};$$
(46)

while the FF model with constant velocity, linear motion, is achieved from (45) with $\alpha = 1$:

$$v_t = v_{t-1};$$
$$\theta_t = \theta_{t-1};$$
(47)

8 Transportation Theory

Transportation theory deals with approaches for modeling human movement behaviors when using public transportation (trains, buses, aircraft, ferries, ..) or private vehicles (cars, vans,..) or as pedestrian. Models describing the fluid of traffic can be classified into two categories: microscopic and macroscopic [11], [28]. Microscopic models mainly concern with individual behaviors at small geographical scale, therefore of great importance in the design of handover algorithms. Macroscopic models can be seen as aggregates of the behavior observed in microscopic models, hence at large geographical scale, therefore with impact in the location update procedures. In the next lines we will indistinctly use the terms vehicle or mobile terminal (MT).

For both, microscopic and macroscopic models, the main characteristics are *Flow*, *Speed* and *Density* of the MTs. The typical measurement parameter for flow is the time headway between two consecutive vehicles in the microscopic model and the flow rates for the macroscopic models. Speed is measured in an individual way or as an average of all vehicles, for micro and macroscopic models respectively. Finally, density can be measured as the distance headway between two consecutive vehicles or as an average global density for micro and macroscopic models, respectively.

8.1 Microscopic Models

The time headway is the elapsed time between the arrival of two consecutive vehicles (or pedestrian, rail, water, air transportation, ...) at a given observation point. This microscopic parameter is of paramount importance in the traffic flow characteristic that affects the safety, quality of service, driver behavior and capacity of a transportation system. Time headway is a positive random variable that depends of the observation scenario.

According to flow level conditions several distributions have been considered in the literature. Hence negative exponential pdf for very low flow level, distribution with fitting parameters for intermediate flow level or deterministic pdf for high flow level conditions. These conditions are also named as random or free flow, intermediate and constant headway, respectively. The free flow speed corresponds to very low vehicle density conditions, where vehicle speed is not limited by safety distance but by the road characteristics (street width, driver visibility, etc) [26].

Free flow condition was considered by El-Dolil et al. [9] where the authors use negative exponential pdf for the time headway to address the teletraffic analysis of highway microcells with overlay macrocell. For intermediate flow level conditions, the Pearson type III distribution has been postulated as a good pdf candidate [11]. Its pdf is

$$f_F(t) = \frac{\lambda}{\Gamma(a)}[\lambda(t-\alpha)]^{a-1}e^{-\lambda(t-\alpha)} \qquad (48)$$

where $\Gamma(a)$ is the gamma function defined in (31).

But, at a given observation point along the urban areas, one realizes that the vehicles moves in platoon [20]. That is, due to the traffic lights(semaphores) "green-reed" sequence, the passing of vehicles can be seen as a "on-off" like process. During the "on" period a platoon of vehicles is observed, being the platoon size a discrete random variable. The duration of the "on" period is a continuous random variable. The "off" period duration, measured as the time interval between the arrival of the last vehicle in a platoon and the arrival of the first vehicle in the next platoon, is also continuous random variable. In [20] Alfa and Neuts have studied and modeled this scenario by means of a discrete time analysis, with easy generalization to its continuous time analogue. In particular in [20] it assumed a discrete phase type distribution for i) the intraplatoon time headway (time distance between vehicles that are in the same platoon) ii) the interplatoon intervals and iii) the numbers of vehicles in a platoon. For a nice description on phase type distributions see, for instance [35].

Also we cite the work by Leung et al. [18]. In [18] a deterministic fluid model and two stochastic traffic models were introduced; being the scenario a highway with multiple entrances and exits. The deterministic model ignores the behavior of individual vehicles and treats them as a continuous fluid. On the other hand the stochastic traffic models consider the random behavior of each vehicle. The common denominator of all three models is that they use the same two coupled partial differential equations or ordinary differential equations that describe the system evolution.

8.2 Macroscopic Models

Gravitation theory is being used as a macroscopic fluid of traffic model. It reflects the Newton's gravitation law. Gravity models have been applied in many transportation research scenarios, from regions of varying size to city models and to national and international models [24]. In this model, two areas are attracted following certain Newton's law like. Then, the rate of transitions between two attraction points i and j, $T_{i,j}$ is expressed as

$$T_{i,j} = K_{i,j} \frac{P_i P_j}{d^\gamma}. \tag{49}$$

In (49) P_i is the population size in area i and γ takes into account the inverse dependence with the distance between areas i and j. Hence, for $\gamma = 2$ (49) is analogous to Newton's gravitational Law. According to [24] it is also possible to interpret P_i as the "attractivity" of area i and $T_{i,j}$ as the probability of movements between i and j for a MT, therefore describing the individual movement behavior.

Macroscopic traffic flow theory relates traffic flow, density and running speed. Flow rate (or volume) is the important macroscopic flow characteristic and is defined as the number of vehicles or MT passing a point in a given period of time. Traffic density is a key macroscopic characteristic defined as the number of vehicles occupying a length of a roadway. Macroscopic speed characteristics

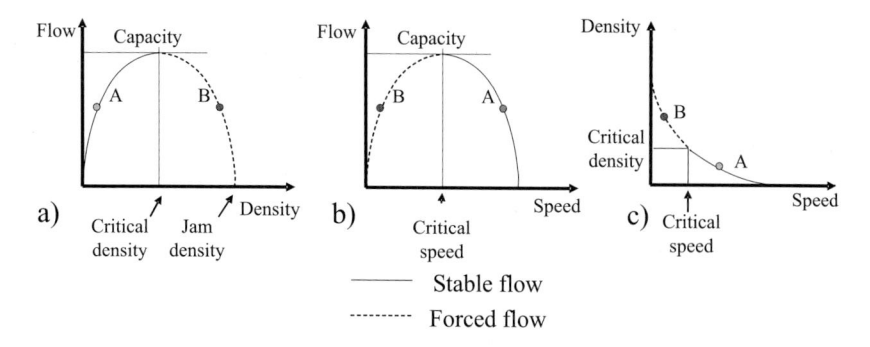

Fig. 7. Relationship among Flow, Density and Speed (from reference [51])

are those of MTs group passing a point or short segment during a specific period of time. The fundamental relationship between Flow F, Density D and speed or velocity V is

$$F = D \cdot V. \qquad (50)$$

Equation (50) says that for a certain average speed V the flow rate increases with the density. And vice versa, for a certain density D of vehicles the flow rate increases with the speed. Notice that the fluid flow model presented in section 3.2, except for a multiplicative constant which is scenario dependent, follows equation (50). However, expression (50) has some limitations in practical scenarios. Otherwise it would make the mathematical interpretation of traffic flow unintelligible [51]. We recognize that when the density increases the distance headway decreases and due to human behavior it turns to a velocity reduction. As a consequence, the resulting net effect is a reduction of the flow rates, as in figure 7-a) is depicted. In this figure, while the slop of the curve is positive we are in the *stable* flow, represented by the solid line portion of the curve. When the slop is zero the maximum capacity is reached. The dashed portion of the curve shows *unstable* or *forced* flow. The same arguments can be said about figure 7-b) and figure 7-c).

9 Conclusions

Mobility models are very essential in the analysis and simulation of mobile terminal's trajectories that are under the coverage area of wireless mobile networks. They have strong influence in the mobility management procedures of cellular systems, such as handover, location update, registration, paging, an the like. The chapter intends to provide a basic background about models on the residence time at several geographical areas, with granularity differences, that is, at microscopic and at macroscopic levels. On one hand, we have reviewed the residence time models at the cell level and at the handover area level. On the other hand, we have study the residence time at the location area level, a coarser granularity. For microscopic scenarios we have pointed out several theoretical and numerical methods, mainly based on memoryless and probabilistic models. Both, analytical expressions or numerical data are quite often required to be replaced by manageable probability distribution functions. A glance on "fitting" techniques have been reported to that purpose. In the macroscopic scenarios we have derived some analytical models that characterize the sojourn time in location areas an hence allowing the evaluation of location update rates. We also have paid attention to the usefulness of other advanced mobility models such as the Random Way Point, the Random Direction and the Gauss-Markov models. Finally the main characteristics about transportation theory has been indicated, as a basic tool to model human movement behaviors at large geographical scale. We believe that our attempt to conduct this survey on mobility models, up to a certain extend, has been achieved and that its reflects the state-of-the art in this field. We also believe that this material can be of great help to researchers in the area of mobility management.

Acknowledgments

This study has been realized under the national project numbers *TSI2001-66869-C02-02* and *TIN2010-21378-C02-02* and under the Euro-NF Network of Excellence *(FP7,IST 216366)*. Thanks are given to *Ministerio de Ciencia e Innovación* (www.micinn.es) for the financial support.

References

1. Kleinrock, L.: Queuing theory. John Wiley Sons, Chichester (1975)
2. Augustin, R., Buscher, K.J.: Characteristics of the COX-distribution. ACM Sigmetrics Performance Evaluation Review 12(1), 22–32 (Winter 1982-1983)
3. Hong, D., Rappaport, S.S.: Traffic model and performance analysis for cellular mobile radiotelephone system with prioritized and non prioritized handoff procedures- version 2a. Technical report No. 773 State University of New Your at Stony Brook; Previously published at IEEE Trans. on Vehicular Technology 35(3), 877–922 (August 1986)
4. Stadje, W.: The exact probability distribution of a two-dimensional random walk. Journal of Statistical Physics 46(1/2), 207–216 (1987)
5. Guerin, R.A.: Channel occupancy time distribution in a cellular radio system. IEEE Trans. on Vehicular Technology 36(3), 89–99 (1987)
6. Morales-Andres, G., Villen-Altamirano, M.: An approach to modelling subscriber mobility in cellular radio networks. In: Proceedings of the Forum Telecom 1987, Geneva, Switzerland, December 1987, pp. 185–189 (1987)
7. Thomas, R., Gilbert, H., Maziotto, G.: Influence of the moving of the mobile stations on the performance of a radio mobile cellular network. In: Proceedings of the 3rd Nordic Seminar on Digital Land Mobile Radio Communications, Paper 9.4, Copenhagen Denmark (September 1988)
8. Leon-Garcia, A.: Probability and random processes for electrical engineering. Addisson-Wesley, Reading (1989)
9. El-Dolil, S.A., Wong, W.C., Steelet, R.: Teletraffic performance of highway microcells with overlay macrocell. IEEE Journal Selected Areas Communications 7(1), 71–78 (1989)

10. McMillan, D.: Traffic modelling and analysis for cellular mobile networks. In: Proceedings of the ITC 13th, Copenhaguen, June 19-26, pp. 627–632 (1991)
11. May, A.: Traffic flow fundamentals. Prentice-Hall, Englewood Cliffs (1990)
12. Harborth, H.: Concentric cycles in mosaic graphs. In: Berum, G.E., et al. (eds.) Applications of Fibonnacci Numbers, pp. 123–128. Kluwer Academic Publishers, Dordrecht (1990)
13. Alonso, E., Meier-Hellstern, K.S., Polini, G.P.: Influence of cell geometry on handover and registration rates in cellular and universal personal telecommunication networks. In: Proceedings of the 8th ITC Specialist Seminar on Universal Personal Telecommunications, Santa Margherita Ligure, Genova, Itay, October 12-14, pp. 261–260 (1992)
14. Xie, H., Kuek, S.: Priority handoff analysis. In: Proceedings of the 43th IEEE VTC, pp. 855–888 (1993)
15. Xie, H., Goodman, D.J.: Mobility models and the biased sampling problem. In: Proceedings of the 2nd ICUP, Ottawa, Canada, pp. 855–888 (1993)
16. Barber, B.C.: The non-isotropic two-dimensional random walk. Waves in Random Media 3, 243–256 (1993)
17. Bauman, F.V., Niemegeers, I.G.: An evaluation of location management procedures. In: Proceedings of the 3rd ICUPC, San Diego, California, September 27-October 1, pp. 359–364 (1994)
18. Leung, K.K., Massey, W.A., Whitt, W.: Traffic models for wireless communication networks. IEEE Journal Selected Areas in Communications 12(8), 1353–1634 (1994)
19. Bar-Noy, A., Kessler, I., Sidi, M.: Mobile users: To update or not to update? Wireless Networks 1(2), 175–185 (1995)
20. Alfa, A.S., Neuts, M.F.: Modelling vehicular traffic using the discrete time Markovian arrival process. Transportation Science 29(2), 109–117 (1995)
21. Polini, G.P.: Trends in handover design. IEEE Communications Magazine 34(3), 82–90 (1996)
22. Akyildiz, I.F., Ho, J.S.M., Lin, Y.-B.: Movement based location update and selective paging for PCS networks. IEEE/ACM Trans. on Networking 4(4), 629–638 (1996)
23. Yeung, K.L., Nanda, S.: Channel management in microcell - macrocell cellular radio systems. IEEE Trans.on Vehicular Technology 45(4), 601–612 (1996)
24. Lam, D., Cox, D.C.: Teletraffic modeling for personal communications service. IEEE Communications Magazine 35(2), 79–87 (1997)
25. Zonoozi, M.M., Dassanayake, P.: User mobility modeling and characterization of mobility patterns. IEEE Journal Selected Areas in Communications 15(7), 1239–1252 (1997)
26. Markoulidakis, J.G., Lyberopoulos, G.L., Tsirkas, D.F., Sykas, E.D.: Mobility modeling in third-generation mobile telecommunications systems. IEEE Personal Communications 4(4), 41–56 (1997)
27. Rose, C., Yates, R.: Location uncertainty in mobile networks. A theoretical framework. IEEE Communications Magazien 35(2), 94–101 (1997)
28. Lieu, H.: Revised monograph on traffic flow theory. U.S. Department of Transportation Federal Highway Administration Research, Development, and Technology. Transportation Research Board. Turner-Fairbank Highway Research Center. 6300 Georgetown Pike McLean, Virginia (November 1997)
29. Cho, M., Kim, K., Szidarovszky, F., You, Y., Cho, K.: Numerical analysis of the dwell time distribution in mobile cellular communication systems. IEICE Trans. Commun. E81-B(4), 715–721 (1998)

30. Ruggieri, M., Graziosi, F., Santucci, F.: Modeling of the handover dwell time in cellular mobile communications systems. IEEE Trans. Vehiclar Technology 47(2), 489–498 (1998)
31. Norris, J.R.: Markov Chain. University of Cambridge (1998)
32. Orlik, P.V., Rappaport, S.S.: A model for teletraffic performance and channel holding time characterization in wireless cellular communication with general session and dwell time distributions. IEEE Journal on Selected Areas in Communications 16(5), 788–803 (1998)
33. Burdzy, K.: Brownian motion. A tutorial. PPT presentation. University of Washington
34. Fang, Y., Chlamtac, I.: Teletraffic analysis and mobility modeling of PCS networks. IEEE Trans. on Communications 47(7), 1062–1072 (1999)
35. Latouche, G., Ramaswami, V.: Introduction to matrix analytic methods in stochastic modeling. ASA-SIAM serie (1999)
36. Goodman, J.W.: Statistical Optics. John Wiley, Chichester (2000)
37. Sánchez, M., Manzoni, P.: ANEJOS: a Java based simulator for ad hoc networks. Best of Websim 1999 in Future Generation Computer Systems 17(5), 573–583 (2001)
38. Bettstetter, C.: Mobility modeling in wireless networks: Categorization, Smooth Movement, and Border Effects. SIGMOBILE Mob. Comput. Commun. Rev. 5(3), 55–66 (2001)
39. Casares-Giner, V.: Variable bit rate voice using hysteresis thresholds. Telecommunication Systems 17(1-2), 31–62 (2001)
40. Camp, T., Boleng, J., Davies, V.: A survey of mobility models for ad hoc network research. Wireless Communications and Mobile Computing 2(5), 483–502 (2002)
41. Pla, V., Casares-Giner, V.: Analytical-Numerical study of the handoff area sojourn time. In: Proceedings Globecom 2002, pp. 886–890 (2002)
42. Kmet, A., Petkovsek, M.: Gambler's ruin problem in several dimensions. Advances in Applied Mathematics, vol. 28, pp. 107–117 (2002)
43. Liang, B., Haas, Z.J.: Predictive distance-based mobility management for multidimensional PCS networks. IEEE/ACM Trans. on Networking 11(5), 718–732 (2003)
44. Ashbrook, D., Starner, T.: Using GPS to learn significant locations and predict movement across multiple users research. Personal Ubiquitous Comput. 7(5), 275–286 (2003)
45. Jardosh, A., BeldingRoyer, E.M., Almeroth, K.C., Suri, S.: Towards realistic mobility models for mobile Ad hoc networks. In: Proceedings of the MobiCom Conference, San Diego, California, USA, September 14-19 (2003)
46. Schweigel, M.: The Cell residence time in rectangular cells and its exponential approximation. In: ITC 18th, Berlin, August 31- September 5 (2003)
47. Hald, A.: History of probability and statistics and their applications before 1750. John Wiley, Chichester (2003)
48. Casares-Giner, V., García-Escalle, P.: An hybrid movement–distance based location update strategy for mobility tracking. In: Proceedings of the European Wireless 2004, Barcelona, Spain, February 24-27, pp. 121–127 (2004)
49. Bettstetter, C., Hartenstein, H., Pérez-Costa, X.: Stochastic properties of the random waypoint mobility model. ACM/Kluwer Wireless Networks: Special Issue on Modeling and Analysis of Mobile Networks 10(5) (September 2004)
50. Hyytia, E., Lassila, P., Nieminen, L., Virtamo, J.: Spatial node distribution in the random waypoint mobility model. COST279TD(04)029 (September 2004)
51. Roess, R.P., Prassas, E.S., McShare, W.R.: Traffic engineering. Pearson, Prentice Hall (2004)

52. Gloss, B., Scharf, M., Neubauer, D.: A more realistic random direction mobility model. In: TD(05)052, 4th Management Committee Meeting, Würzburg, Germany, October 13-14 (2005)
53. Yoon, J., Noble, B.D., Liu, M., Kim, M.: Building realistic mobility models from coarse-grained traces. In: Proceedings of the MobiSys 2006, Uppsala, Sweden, June 19-22, pp. 177–189 (2006)
54. Hyytia, E., Lassila, P., Virtamo, J.: Spatial node distribution of the random waypoint mobility model with applications. IEEE Trans. on Mobile Computing 5(6), 680–694 (2006)
55. Garetto, M., Leonardi, E.: Analysis of random mobility models with partial differential equations. IEEE Trans. on Mobile Computing 6(11), 1204–1217 (2007)
56. Sricharan, M.S., Vaidehi, V.: Mobility patterns in macrocellular wireless. In: IEEE - ICSCN 2007, February 22-24, pp. 128–132. MIT Campus, Anna University, Chennai, India (2007)
57. Ollila, E.: On the circularity of a complex random variable. IEEE Signal Processing Letters 15, 841–844 (2008)
58. Casares-Giner, V., Garcia-Escalle, P.: On Movement-Based Location Update. A Lookahead Strategy. In: Proceedings of the NGI 2009, Aveiro, Portugal (July 2009)
59. Carofiglio, G., Chiasserini, C.F., Garetto, M., Leonardi, E.: Route stability in MANETs under the random direction mobility model. IEEE Trans. on Mobile Computing 8(6), 1167–1179 (2009)

Wireless Ad Hoc Networks: An Overview

David Remondo*

Dept. of Telematics Eng., EPSC, Tech. Univ. of Catalonia,
Avda. del Canal Olimpic s/n, 08860 Castelldefels, Barcelona, Spain
remondo@entel.upc.es

Abstract. This tutorial provides a general view on the research field of ad hoc networks. After a definition of the concept, the discussion concentrates on enabling technologies, including physical and medium access control layers, networking and transport issues. We find discussions on the adequacy of enabling technologies for wireless multihop communication, specifically in the case of the pervasive Bluetooth and IEEE 802.11. Then, a variety of dynamic routing protocols are presented and specific issues that are relevant in this context are highlighted. After a short discussion on TCP issues in this context, we look at power awareness, which is a very important issue in this scenario. Finally, we discuss proposals that aim at maintaining Service Level Agreements in isolated ad hoc networks and ad hoc networks connected to fixed networks.

Keywords: Ad hoc networks, multihop wireless networks, wireless networks, mesh networks.

1 Introduction

Wireless ad hoc networks are formed by devices that are able to communicate with each other using a wireless physical medium without having to resort to a pre-existing network infrastructure. These networks, also known as mobile ad hoc networks (MANETs), can form stand-alone groups of wireless terminals, but (some of) these terminals could also be connected to a cellular system or to a fixed network. A fundamental characteristic of ad hoc networks is that they are able to configure themselves on-the-fly without the intervention of a centralized administration.

Terminals in ad hoc networks can function not only as end systems (executing applications, sending information as source nodes and receiving data as destination nodes), but also as intermediate systems (forwarding packets from other nodes). Therefore, it is possible that two nodes communicate even when they are outside of each others transmission ranges because intermediate nodes can function as routers. This is why wireless ad hoc networks are also known as multi-hop wireless networks.

---

* This work was partially supported by the Euro-NGI Network of Excellence of the European Commission and the Spanish CICYT Project TSI2007-66637-C02-01, which is partially funded by FEDER.

Compared to cellular networks, ad hoc networks are more adaptable to changing traffic demands and physical conditions. Also, since the attenuation characteristics of wireless media are nonlinear, energy efficiency will be potentially superior and the increased spatial reuse will yield superior capacity and thus spectral efficiency. These characteristics make ad hoc networks attractive for pervasive communications, a concept that is tightly linked to heterogeneous networks and 4G architectures.

The need for self-configurability and flexibility at various levels (for example, dynamic routing or distributed medium access control) poses many new challenges in wireless ad hoc networks. Cross-layer optimization can significantly improve system performance and thus we will discuss some cross-layer issues here.

This tutorial concentrates on enabling technologies, including physical and medium access control layers, networking and transport. Research in middleware and security in this context is not considered.

Depending on their communication range, wireless ad hoc networks can be classified into Body (BAN), Personal (PAN) and Wireless Local (WLAN) Area Networks. A BAN is a set of wearable devices that have a communication range of about 2 m. The second type, PANs, refers to the communication between different BANs and between a BAN and its immediate surroundings (within approximately 10 m). WLANs have communication ranges of the order of hundreds of metres. The main existing technology for implementing BANs and PANs is Bluetooth, while for WLANs the main option is the family of standards IEEE 802.11. Although ad hoc networks are not restricted to these technologies, most of the current research assumes Bluetooth or IEEE 802.11 to be the underlying technologies.

After a general introduction, this tutorial discusses the main characteristics of Bluetooth, also considering open issues such as scatternet formation and real-time traffic support. The IEEE 802.11 technology is also considered: first we look at the basic functioning of the system in ad hoc mode and then we elaborate on its shortcomings for multi-hop communication, namely the lack of efficiency of the RTS/CTS mechanism and the impact of the difference between transmission and carrier sense ranges. Some proposed solutions are discussed briefly.

Routing is the most active research field in ad hoc networking. In this context, it is closely related with different communication layers. Minimizing the number of hops is no longer the objective of a routing algorithm, but rather the optimization of multiple parameters, such as packet error rate over the route, energy consumption, network survivability, routing overhead, route setup and repair speed, possibility of establishing parallel routes, etc. We compare different types of proposed routing algorithms and as a means of example we illustrate the functioning of a non-location based on-demand unicast routing protocol: DSR. Thereafter, we describe other algorithms of the same type (AODV), some proactive protocols (e.g. OLSR) and some location based schemes with their associated forwarding mechanisms (e.g. DREAM, LAR, Greedy Forwarding). During the discussion, we also point out specific issues that have to be considered in wireless ad hoc networks: for instance, two disjoint routes may have mutual influence if a node in one route is within the transmission range of a node in the other route, which has an impact on the construction of parallel routes.

The use of TCP over wireless links is known to present many problems. Communication over wireless multi-hop networks inherits these problems but also introduces some additional issues: the nodes mobility introduces unfairness between TCP flows, route failures lead to unnecessary congestion control and MAC contention reduces throughput in long routes. We also look at the proposed solutions briefly.

Since most wireless terminals can be expected to have limited energy storage, power awareness is very important. This subject spans across several communication layers. We pay attention to different power saving approaches. Objectives are not only the reduction of transmission power, but also the management of sleep states or the extension of network survivability through energy aware routing.

It may seem incoherent to deal with Quality of Service (QoS) support in such dynamic systems with unreliable wireless links. However, some authors have presented proposals to support QoS in isolated ad hoc networks, including QoS oriented MAC protocols suitable for distributed systems, QoS aware dynamic routing protocols, DiffServ in wireless multi-hop networks and resource reservation protocols such as INSIGNIA and SWAN. We study and compare these schemes and discuss new proposals for end-to-end QoS support in ad hoc networks attached to fixed networks through inter-network cooperation.

2 Enabling Technologies

In this section we find a discussion on some of the main enabling technologies for ad hoc networks, i.e. Bluetooth, IEEE 802.11 and Ultra-Wide Band radio.

2.1 Bluetooth

Bluetooth [33] is a single-chip, low-cost, radio-based wireless network technology suited for ad hoc networks, with communication ranges in the order of 10 m. The single-chip design makes this technology specially useful for small terminals with low energy storage capacity. It operates in the unlicensed industrial, scientific and medical (ISM) band at 2.40 to 2.48 GHz.

The physical channels are separated by fast Frequency Hopping Spread Spectrum (FHSS), using 79 carriers in most countries. Hopping slots have a duration of 625 us. In order to save power, this technology incorporates a powerful energy management architecture that comprises four different power consumption states.

Terminals arrange themselves in piconets, sets of terminals with one device functioning as master and up to other seven terminals functioning as slaves. In principle, any device can become a master or a slave. The master determines a hopping sequence and all the slaves use this hopping sequence to communicate with the master. Different piconets have different hopping sequences. Direct communication between slaves is not possible.

Terminals that are within the coverage areas of two or more masters may work as connections between different piconets. Such a terminal, called gateway, can belong to different piconets simultaneously on a time-division basis. A set of connected piconets forms a scatternet.

Links use Time-Division Duplex (TDD). The communication links are of two types depending on the arrangement of the time slots:

- Synchronous Connection-Oriented Link (SCO). The master reserves two consecutive time-slots for the forward and return directions respectively. Each SCO link supports 64 Kb/s on each direction with optional forward error correction (FEC) and no retransmission.
- Asynchronous Connectionless Link (ACL). The master uses a polling scheme, where one, three or five consecutive slots can be allocated to a link. FEC is optional. Data rates are up to 433.9 Kb/s per direction in symmetric links and up to 723.2 Kb/s / 57.6 Kb/s in asymmetric links. Headers are used to enable fast retransmission.

The scheduling of polling intervals for ACL links in view of service differentiation and provisioning of delay bounds is not specified in the system and is thus open for research (see, for example, [5]).

The topology of multihop networks largely depends on which terminals function as gateways between picocells. Scatternet formation is thus a relevant and difficult research issue because it affects topology significantly and has a large impact on the system performance. We can find a recent overview of scatternet formation and optimization protocols in [4].

Another research issue with a growing interest, in view of the future 4G vision of heterogeneous networking, is the coexistence between this technology and IEEE 802.11 (see e.g. [32]).

2.2 IEEE 802.11

The family of standards IEEE 802.11 [1] comprises the standards IEEE 802.11, IEEE 802.11b, IEEE 802.11g and IEEE 802.11a, amongst other standards that deal with specific issues such as security, service differentiation, etc. IEEE 802.11 provides wireless and infrared connectivity, but all implemented products use the unlicensed radio bands of 5 GHz (for IEEE 802.11a) and 2.4 GHz (for the rest).

The physical layer offers a number of channels. A set of terminals that use the same channel and are within the communication range of (some of) the other terminals of the set is called Basic Service Set (BSS). The number of physical channels depends on whether BSSs are multiplexed by using FHSS, Direct Sequence Spread Spectrum (DSSS) or Orthogonal Frequency Division Multiplexing (OFDM) techniques. Also, the communication rates per channel depend on the modulation, multiplexing, and forward error correction coding rates, ranging from 1 Mb/s to 54 Mb/s.

In the context of ad hoc networks, users belonging to the same BSS share the medium by means of a distributed random access mechanism called Distributed

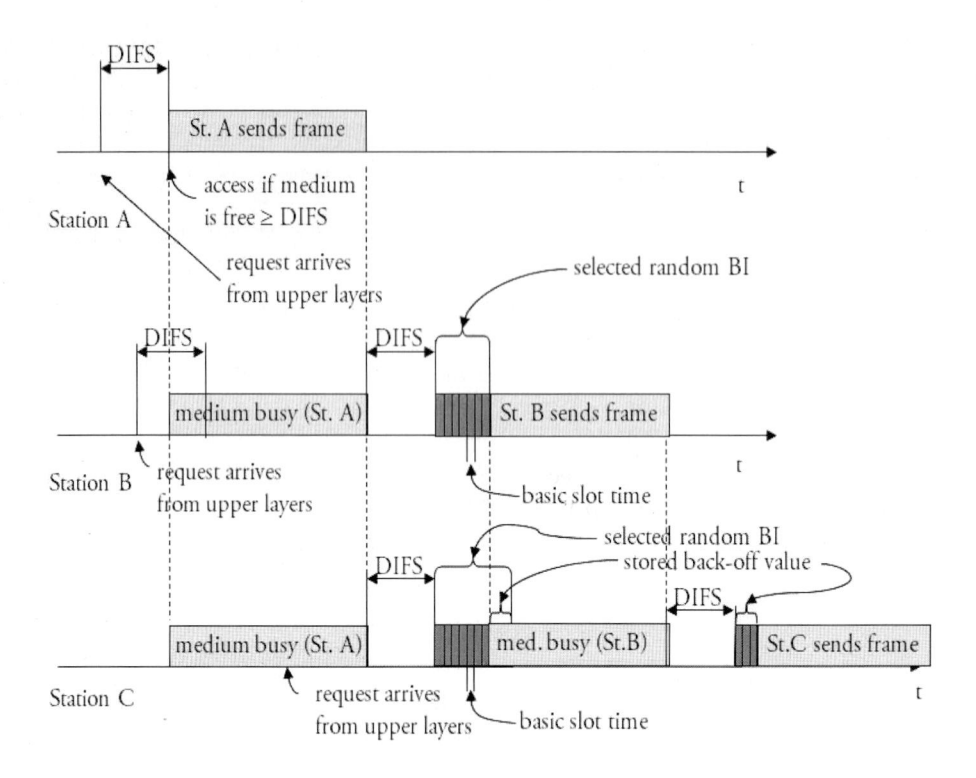

Fig. 1. Basic functioning of DCF

Coordination Function (DCF), basically a Carrier-Sense Medium Access with Collision Avoidance (CSMA/CA) technique.

Fig. 1 illustrates the basic functioning of DCF. There, we see how three stations behave as a function of time if they are all within reach of each other. When a Mobile Station (MS) gets a frame from upper layers to transmit, it first senses the channel to determine whether another MS is transmitting. If the MS has sensed the channel to be idle for a period of time equal to the DCF Inter Frame Space (DIFS), which is a quantity equal for all stations, then it starts transmitting the frame. Otherwise, as soon at it senses the channel to be busy, it will defer the transmission. When deferring, the station will continue sensing the channel.

At the point in time when the medium becomes idle again, the station will continue sensing and it will wait for the period DIFS to elapse again. If the medium becomes busy during this period, the station will go back to the deferring state again. However, if the medium remains idle for this DIFS period, the station will go to the back-off state.

When entering the back-off state, the MS selects a Back-off Interval (BI) randomly between zero and a Contention Window period (CW). The quantity CW is an integer number of basic time slots. If the medium remains idle for the duration of BI, then the station transmits the frame. However, if the medium

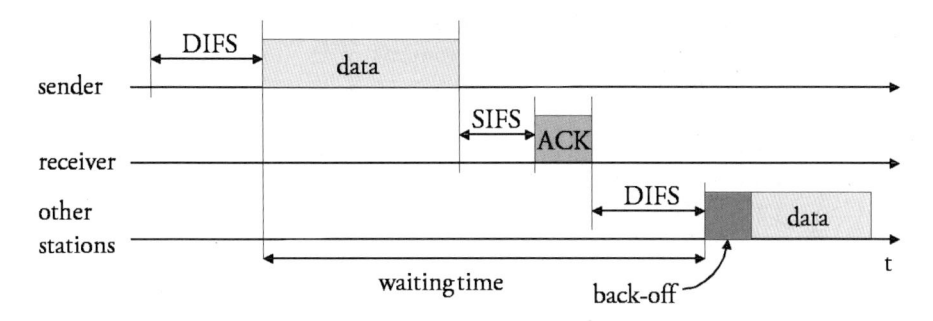

Fig. 2. Acknowledgements for DCF

becomes busy before the BI elapses, then the MS stores the remaining BI time (that is, the value of the chosen BI minus the elapsed time since entering the backoff state).

A collision will occur if two or more MSs select the same BI (provided the condition stated above, that the frames coexist spatially at one or more of the receiving stations). When a collision occurs, the stations that have caused the collision sense the medium again for DIFS and go again to the back-off state, selecting a new BI randomly with the value of CW doubled. The other stations, which stored their remaining BI times, also wait for DIFS and then go to the back-off state with BI equal to the stored value.

The value of CW is doubled every time that a station tries to transmit a given frame and a collision occurs, until a maximum CW (denoted CWmax) is reached. When this maximum value is reached, the BI will be randomly selected out of the interval [0, CWmax].

There is a maximum number of retransmission attempts: if the station has tried to transmit the frame this number of times and a collision has always occurred, then the station gives up trying to transmit the frame.

The radio transmission channel is relatively unreliable and the probability of transmitting a frame successfully is highly variable, even if there is no competition from other stations. Therefore, in IEEE 802.11 DCF, frames that have a single destination (which will be the case we will concentrate on) have a corresponding reception acknowledgement. This is illustrated on Fig. 2, where we show the behaviour of one transmitting station, the corresponding station (receiver) and a third station that has received a request to transmit from upper layers after the first station. After a station has transmitted the data frame, it will wait for an acknowledge frame (ACK) to arrive after a time period SIFS (Short IFS). The size of SIFS is unique for all stations and is smaller than DIFS. In this way, we guarantee that the first frame that is transmitted after the data frame is the acknowledgement of that frame and not any other data frame (unless the transmission medium fails temporarily).

An optional feature of DCF is the Request-To-Send and Clear-To-Send extension (RTS/CTS), which is common in commercial implementations. This option prevents many collisions induced by the *hidden-terminal problem* and the *exposed-terminal problem.* These problems are of major relevance in multihop wireless networks. The former problem consists in the following. Assume that Station A intends to send a frame to Station B. Station A will be able to hear only some but not necessarily all transmissions from other stations affecting Station B, thereby assuming that the medium is free when it is busy at its intended destination node. The exposed-terminal problem is complementary: Station A could refrain from sending data when hearing transmissions that do not arrive to Station B. To overcome these problems, the MS first transmits an RTS message and waits for a CTS message from the recipient before beginning data transmission. RTS and CTS frames also include the size of the data to be sent, so that a station hearing an RTS/CTS frame granting access to a different station refrains from sending an RTS to the medium for the duration of the indicated transmission time.

Synchronization and management of power saving modes are distributed but they suffer from scalability problems. This is because they are based on the periodical transmission of beacons that have to compete for medium access.

DCF provides best-effort service to higher layers, there is no differentiation between types of traffic. There have been many proposals for introducing service differentiation in DCF, some of them incorporated to the standard IEEE 802.11e [2].

DCF is known to be rather inefficient for wireless multihop communication. In particular, the RTS/CTS mechanism does not fully counteract the hidden-terminal problem because the power needed for interrupting a packet reception is much lower than that of delivering a packet successfully [3]. Some proposed solutions involve using busy tones, adjustable power or directional antennas, but this may not always be practical. For example, a recently proposed solution is to send the CTS only if the received power of the corresponding RTS is larger than a certain threshold [3], but this reduces the effective transmission range. For the design of new MAC mechanisms, more research is needed to analyze the effect of the presence of multiple stations within the transmission, reception or interference ranges of a given station.

2.3 Ultra-Wide Band Radio

The development of Ultra-Wide Band radio (UWB) is progressing quickly. The idea behind UWB, a technology that has also received the names of *baseband, carrier-free* or *impulse* radio, is to use electromagnetic signals with a very wide spectrum (with -10dB bandwidth in excess of 25% of the central frequency, typically in the order of several GHz) so that the power spectral density is so low that the system can coexist with existing licensed spectral bands. Typically, low power, low range communications are a natural context for UWB. This is especially true in environments where multipath propagation is important, such as indoor channels. This technology also has the property of supplying accurate ranging information between UWB devices, which can be used to improve communication at higher layers (e.g. routing). UWB technology has many options in IEEE 802.15.3 group's work towards low-power BANs with 10 m communications range and data rates above 100 Mbit/s, and on IEEE 802.11.4 group's activities aimed at low data rate support with ranging functionality.

The Federal Communications Commission (FCC) of the U.S.A. allowed for commercial operation of products using UWB-RT with at most -41.3 dBm/MHz between 3.1 and 10.6 GHz, differentiating between indoor and outdoor operation in neighbouring bands. The European Telecommunications Standards Institute (ETSI) is also regulating an identical power level between the mentioned frequency values, but with more restrictive values in neighbouring bands.

Several projects have investigated aspects of UWB with the support of the European Commission, such as Ultra-wideband Concepts for Ad hoc Networks (UCAN), Ultra Wideband Audio Video Entertainment System (ULTRAWAVES) or Pervasive Ultra-wideband Low Spectral Energy Radio Systems (PULSERS), with the participation of some companies such as Philips, Acorde, Telefónica, VTT or IMST. In the U.S.A., several companies have been developing UWB radio chips, including Freescale, Time-Domain, Alereon, Multispectral Solutions, Intel and Texas Instruments. Major Japanese companies, such as Matsushita, Sony, Sharp, JVC, Pioneer, NEC and Mitsubishi, considered UWB as an ultra-fast wireless interface.

Given the fact that UWB systems are supposed not to interfere with other communication systems, we can expect that UWB will be complementing systems such as IEEE 802.11 in future. IEEE 802.11 systems have larger communication ranges but lack precise ranging features; also, the impact of obstacles and multipath propagation is different in both physical layers. Therefore, the symbiosis between both systems could significantly boost the performance perceived by the user of wireless multihop networks.

3 Routing Issues

Routing protocols for wireless multihop networks are dynamic due to the potential node and link mobility.

Unicast routing protocols can be classified in the following way:

- Proactive vs. reactive. Proactive protocols periodically maintain the routing information so that any node can use a existing route to reach a destination at any time. This is the rule in fixed networks, but in mobile ad hoc networks this would require a very frequent update of routing information for all nodes, which implies a lot of overhead. Reactive protocols, on the contrary, obtain the necessary routing information only when a route is needed between two nodes; the route is maintained only when the route is active. This is why reactive protocols are also called *on-demand* protocols. Reactive protocols imply lower overhead than proactive ones, but they suffer from route setup delays when a communication flow is to start. There also exist hybrid protocols, which combine proactive mechanisms within a local scope and reactive mechanisms within global scope (e.g. the Zone Routing Protocol [19]).
- Location-based vs. non location-based. Location-based are protocols where some means exist by which nodes can obtain some knowledge about their

relative physical (or geographic) position with respect to other nodes, such as distance or angle. Non location-based protocols do not rely on this information: nodes only know which links are active. Similarly to routing protocols in fixed networks, non location-based protocols spread topology information about which pairs of nodes are immediate (one hop) neighbours. In contrast to this, networks using location-based protocols can make use of the geographical information to significantly improve the efficiency of the route setup process in terms of speed and overhead, as we will see. In practice, the major drawback of location-based protocols is that nodes are required to incorporate a system that provides information about their physical position, such as the Global Positioning System (GPS). We find a good overview of location based protocols in [20].

- Hierarchical and flat. Especially in large ad hoc networks, arranging nodes in clusters for routing purposes can increase the efficiency of the routing protocol. Also, introducing hierarchies of routing protocols can be applied to distinguish routes pertaining to the ad hoc network only from routes linking the ad hoc networks with a gateway to a fixed network. Clustering has a long research history starting from the times of packet radio, but it is still an active research field: the formation of clusters can be made according to many different criteria, such as nodes' mobility patterns, traffic patterns, nodes' capabilities (e.g. energy storage, processing power, etc.). An example of a hierarchical on-demand routing protocol is the Cluster-Based Routing Protocol (CBRP) [15].

To illustrate the functioning of dynamic routing protocols, we consider the Dynamic Source Routing (DSR) protocol [6], [7]. DSR is, together with the Ad Hoc Distance Vector protocol (AODV) [8], the most widely studied routing protocol for mobile ad hoc networks. DSR does not require location information and it is relatively simple because it is a flat, on-demand protocol. We consider networks where all nodes have identical capabilities and responsibilities.

In DSR, when a source needs a route to a destination, it initiates a route discovery process to locate the destination node. The source node floods a Route Request packet (RREQ) requesting a route for the destination. A Route Reply (RREP) packet is sent back to the source either by the destination node or by any node that knows how to reach the destination. The addresses of intermediate nodes are accumulated on the RREQ and RREP packets. Every node in the network uses the information in the RREQ and RREP packets to learn about routes to other nodes in the network. This information is stored in route caches.

Once a source node receives an RREP, it knows the entire route to the destination. If a link contained in the route breaks during the transmission of data packets, the transmitting-side node uses a different path if it has an alternate route cached; otherwise, it reports an error back to the source and leaves it to the source to establish a new route.

The already mentioned AODV protocol is similar to DSR, but it is table based rather than source based. AODV is restricted to networks of symmetric links because the RREP packets are sent via the reverse route. However, a one-hop RREP acknowledgement packet can be used to counteract this problem [9], [10]. Expiry timers are used to keep the route entries fresh.

DSR and AODV maintain the needed routes dynamically, have a relatively low overhead and avoid the formation of routing loops. A drawback is their relatively low scalability with the number of nodes. DSR and AODV are not shortest path algorithms in the sense of least number of hops: since nodes reply to the first arriving RREQ, these protocols select the route to the destination that is fastest at the moment that the route is set up.

DSR and AODV require a relatively low processing power, since they do not resort to cost functions for optimal route search. This relieves the nodes from calculating such a function every time a routing packet is forwarded. However, this is an obstacle for using multiple parameters for route optimization. In practice, in ad hoc networks we could be interested in routing for maximum route stability, minimum energy consumption, minimum number of hops, maximum link reliability, QoS support, etc.

Many improvements have been proposed for DSR and AODV [9]. For example, a route can be repaired faster if the node at the transmitting side of the broken link starts a route discovery process by itself. In this case of repairing routes, Query Localization limits flooding to nodes close to the original route. Also, control overhead can be reduced by limiting the area that is flooded during a route discovery process (Expanding Ring Search). The size of this area can be calculated by using information gathered from previous source-destination data flows, a central idea of the Relative Distance Micro-Diversity Ad Hoc Routing protocol (RDMAR) [11]. In RDMAR there is no need for location information systems: the source-destination distance is estimated from the number of hops used in the previous data flow, the time elapsed since the previous data flow finished, the velocity of the source and the destination nodes and the transmission range.

Flooding for route setup can cause many collisions, which is known as the *Broadcast Storm Problem*. Many heuristics have been proposed to counteract this problem, such as staggering the route search packets at intermediate nodes or re-broadcasting with probability $p<1$ (see, e.g. [28]).

Many other non location-based, flat, reactive routing protocols exist. The Associativity-Based Routing protocol (ABR) [12] and the Signal Stability Adaptive (SSA) [13] protocol are source-initiated protocols that tend to select the routes with the most stable links. Although routes are built on demand, a periodic beaconing mechanism is needed for establishing the stability of the links. Protocols based on ant colony algorithms have been proposed, but they have poor scalability properties with the number of nodes and data flows.

Some other algorithms do not rely on flooding control messages. The Temporary Ordered Routing Algorithm (TORA) is based on the construction of a directed acyclic graph (DAG) for each destination [14]. In this way, topology changes induce a very limited amount of control messages. TORA is specially indicated for highly mobile networks.

Most proactive, flat, non location-based routing protocols are derived from existing routing protocols for fixed networks. This is the case of the Destination-Sequenced Distance Vector (DSDV) protocol [16], which uses a distance vector shortest-path (in terms of hops) algorithm where incremental changes are exchanged more frequently than full routing information. Another protocol of this kind is the Optimized Link State Routing (OLSR) protocol [17], a link state protocol where the amount of control messages is reduced by restricting the rebroadcast of control messages to a subset of nodes (the *Multipoint Relays*). OLSR is possibly the most scalable proactive, flat, non-location based protocol [18].

Location based protocols can have the location information stored in some of the networks nodes or in all of them. Also, the stored information can comprise the location of all the nodes or only of a subset. Depending on which option is taken, a location service is denominated *some-for-all, all-for-all, all-for-some* or *some-for-some*.

The Distance Routing Effect Algorithm for Mobility (DREAM) [21] is an all-for-all location service where all the nodes spread its location information periodically. The frequency and range of the information dissemination depends on the mobility of a node and the relative distance to the receivers of the information ('distance effect'); DREAM also includes forwarding. The Grid Location Service (GLS) is a some-for-all scheme where location information is stored according to a hierarchical cartesian grid. Other some-for-all location services are based in the concept of *quorum:* the network is divided subsets of nodes with non-empty intersections. The route setup process can be made with Greedy Packet Forwarding (GPF), which has several variations [20]. DREAM forwards route discovery packets by flooding within a zone defined by the transmitting node's position and the expected destinations area. Location Aided Routing (LAR) [22] is a forwarding strategy with two versions: the first is similar to DREAM, where flooding is restricted to an area that depends on the nodes' relative distances and velocities; the second allows route discovery packets to be forwarded only if the receiving node is closer to the destination node than the transmitting node. We can mention two hierarchical location based forwarding strategies: Terminodes [23] routing and Grid [24] routing. Terminodes combines proactive distance vector routing for local scope and reactive GPF.

Some protocols exist that are based on the construction of clusters. Usually, a cluster has a leader node. Different protocols differ in the way that clusters are built, how the cluster is chosen and the responsibilities assigned to the leader node. The Core-Extraction Distributed Ad Hoc Routing (CEDAR) [25] is a hierarchical routing protocol where a subset of nodes, called the core, is selected such that all nodes are at most one hop from the nodes of this subset. Core nodes execute a link state protocol where each core node knows the state of local links and stable, high-bandwidth links far away. A route is found on-demand by the core nodes, but this does not mean that the route itself has to traverse core nodes.

In unicast routing, it can be interesting to have multiple routes between two nodes. Reasons for this can be to speed up the route repair process, to increase the reliability of data delivery by sending duplicates of data packets along different routes or to distribute traffic according to QoS requirements. There are diverse proposals for multipath routing. For example, AOMDV [26] is a modified AODV protocol for multipath routing that seeks link-disjoint routes to the destination, that is, routes that have no common links but may share common intermediate nodes. Basically, RREQ packets include a field that indicates the first node traversed after the source node (that is, the immediate neighbour of the source they have passed). Upon reception of an RREQ, a node only re-broadcasts the packet if the mentioned field indicates a different first-hop node than other RREQ packets that have already arrived. AOMDV yields better end-to-end and lower routing overhead, especially with high traffic loads.

The role of wireless ad hoc networks as access networks is gaining interest. A relevant research issue therein is routing when the source or destination is a gateway to a fixed network. We can find relatively new contributions within this area, such as Load Balancing AODV (LB-AODV) [27], where nodes are arranged into groups in order to reduce the routing overhead. Nodes belonging to different groups may not forward packets originated in nodes of other groups. Another multipath routing protocol is Gossip [28].

Multicast routing protocols for wireless ad hoc networks can be classified into tree-based and mesh-based in general. Mesh based schemes are more robust because they yield multiple redundant routes, but resources are wasted as a result of unnecessary forwarding of duplicate data. In tree based schemes resource usage is optimized, but network mobility induces major reconstruction overhead and latency.

An example of a tree-based multicast protocol is MAODV [9]. In MAODV, a node joins a multicast group through RREQ packet flooding. When an RREQ packet arrives at a member of the multicast tree, it responds with an RREP. Since more than one node of the multicast tree may be reached by the RREQ, the source sends a Multicast Activation (MACT) packet along the selected route so that the involved nodes know that they have become part of the multicast tree.

A widely studied mesh-based multicast protocol is the On-Demand Multicast Routing Protocol (ODMRP) [29]. A node wishing to send multicast packets floods a Join Data packet throughout the network periodically. On receiving a Join Data packet, each multicast group member broadcasts a Join Table packet to all its neighbours. Multiple routes from a sender to a multicast receiver may exist due to the mesh structure created by the forwarding group members. There is no explicit join or leave procedure.

4 Transport Issues

The issues associated with the use of TCP on wireless channels have been widely studied. Random errors may cause Fast Retransmit, which implies halving the congestion window size. When errors are not frequent this is not necessary and it reduces the throughput. Errors may even cause transmission timeouts and thus a severe reduction of the congestion window. But in ad hoc networks, TCP is affected not only by wireless transmission errors, but also medium access contention in neighbouring hops and route failures due to mobility.

In general, the throughput of a flow is reduced when we increase the number of hops on one route from 1 to 3, due to contention at MAC level. Beyond 3 hops there should be no further throughput degradation, but in TCP flows there is a reduction in throughput beyond 3 hops due to contention between TCP data and acknowledgements. A measure to counteract this is to reduce the number of transmitted acknowledgements.

Experiments have also shown that increasing the mobility also has a negative impact on the throughput of TCP flows. Mobility induces route failures. While the route is repaired, packets and acknowledgements are en route are lost. At a given moment, no more packets are transmitted. If the TCP sender has not timed out before the route has been repaired, the first retransmission will not occur until the time out. If the route repair process is slower, it can happen that the TCP sender times out before there is a route available and thus the timer will be doubled. In conclusion, large route repair delays have a severe impact on TCP performance.

An idea for improving the performance of TCP is to use network feedback. This consists in letting TCP know that there has been a route failure and informing TCP that a route has been repaired [30]. The use of route caching in on-demand routing protocols may have a negative impact on TCP performance: although caching reduces route repair times, it may cause a flow to use stale routes [30].

Another issue that arises when a route is broken is how to choose the TCP window size and retransmission timeout value after a route has been repaired.

5 Energy Awareness

We should realize that issues such as QoS support, TCP performance, speed of routing repair processes, etc. are secondary if nodes have a high probability of running out of energy resources. As mentioned above, energy awareness in wireless ad hoc networks spans across several communication layers.

Battery technology has advanced very slowly if we compare it with the results achieved in integrated circuit technology and it certainly cannot be compared to the rate of growth in communication speeds. Therefore, saving transmission power will represent one of the most significant factors in the performance of wireless systems in the long term.

Several proposals in literature relate routing to energy awareness. In [35] we find two routing protocols designed for scenarios where the nodes can adjust their transmission power dynamically according to the effect of link layer error rates and consequent packet retransmissions. Such considerations motivate a routing protocol [36] based on a cost function that comprises the link error rate and the energy required for a single transmission attempt across the link.

The work in [37] and [38] compares routing schemes that aim at minimizing the transmission power when selecting a route to routing schemes that try to maximize the lifetime of the nodes in the network as a whole.

In [39] it is proposed to introduce the battery characteristics directly into a routing protocol using the remaining battery capacity as metric of the lifetime of each host.

The two objectives of minimizing the total transmission energy for a route and for all the network can lead to contradiction, for example in the case that several minimum energy routes have a common host, then the battery power of this host will be exhausted quickly. In [37] and [38] a new routing scheme is presented that aims at satisfying the two constraints simultaneously: the Minimum Battery Cost Routing algorithm (MBCR). This protocol aims at finding a route with the maximum total remaining capacity. Let us define $f_i(c_i^t)$ as a battery cost function of host n_i, where c_i^t represents the battery capacity of the host at time t. We can choose the cost function to be for example

$$f_i(c_i^t) = 1/c_i^t \qquad (1)$$

The battery cost R_J for a selected route J will be then:

$$R_J = \sum_{i=0}^{D_J-1} f_i(c_i^t), \qquad (2)$$

where D_J is the number of nodes belonging to the route J. To select a route with the maximum total remaining capacity, one should choose the route m that has the minimum battery cost:

$$R_m = min\{R_J \mid J \in A\}, \qquad (3)$$

where A is the set containing all possible routes.

The Simple Energy Aware Dynamic Source Routing (SEADSR) [40] is a protocol that improves the network survivability while maintaining the simplicity of DSR. The basic idea behind this algorithm is as follows. When an intermediate node in an ad hoc network decides to forward a RREQ message (in the DSR fashion) that it has received, it introduces an additional delay τ before re-transmitting this message:

$$\tau = (C_{max} - C)\tau_{max}/C_{max}, \qquad (4)$$

where C_{max} is the battery capacity, C is the current battery level and τ_{max} is a design parameter that represents the maximum delay introduced. We can appreciate that τ takes a value between 0 and τ_{max} and is directly proportional to the energy consumed by the node. As in DSR, the route selection will depend on the previously mentioned factors, but this additional delay establishes interdependency between the route selection and the battery levels of the nodes. The parameter τ_{max} plays an important role in the route selection. The larger the parameter τ_{max}, the larger the influence of the battery level will be against the other factors.

The use of directional antennas has the potential of reducing transmission power and also may increase the communication capacity of the network due to the higher spatial reuse. However, it introduces many new challenges for MAC and routing protocols. Research on directional antennas for ad hoc networking is relatively incipient. A good overview can be found in [31].

Besides reducing transmission power, there is also research in the direction of reducing energy consumption for reception. This is already incorporated to Bluetooth and IEEE 802.11 technologies, as well as contemplated in other alternative technologies such as ETSIs HIPERLAN family of standards. The most widely used strategies are schemes that allow terminals to switch off their transceivers temporarily. The power saving scheme in IEEE 802.11 is one of this kind, but it scales very poorly in ad hoc mode. A recent proposal is the Power-Aware Multi-Access protocol (PAMAS) [34]. This protocol conserves battery power by powering off a node when a neighbour is transmitting packets to another node. PAMAS uses a separate control channel for a node to probe whether the data channel is busy.

Energy consumption can be also reduced by adjusting the transmission power so that only just the necessary power to reach the receiver is employed. In IEEE 802.11, however, it is not straightforward to have nodes transmitting with different power because it would produce many collisions. A simple proposal is the Power Controlled Multiple Access (PCMA) protocol [41]. In this protocol, nodes use a busy tone to let their neighbours know what level of interference they can tolerate. If a node R can tolerate an interference level N, it will transmit a busy tone with power C/N, where C is a constant. This tone will be received at a neighbouring node X with power $g \cdot C/N$, where g is the gain of the link R-X. Node X will be allowed to transmit with a power not larger than $C/(g \cdot C/N)$. This implies that the power received at node R from node X will be smaller than N, assuming that the link is symmetrical in terms of gain. Despite the drawback of requiring the transmission of busy tones, PCMA improves aggregate throughput and reduces power consumption.

6 QoS Support

Providing QoS is a challenging area of future research in wireless ad hoc networks. The networks ability to provide QoS depends on the characteristics of all the network components, from transmission links to the MAC and network layers. In these networks, links have a relatively low, highly variable capacity and high loss rates. Besides, mobility provokes frequent link breakages. Finally, link layers typically use unlicensed spectral bands, making it more diffcult to provide strong QoS guarantees. If the nodes are highly mobile, even statistical QoS guarantees may be impossible to attain, due to the lack of suffciently accurate knowledge of the network states. Furthermore, since the available network resources (e.g., MAC congestion levels or battery state) varies with time, present QoS architectures for wired networks are unsuitable.

Important QoS components include: QoS aware medium access control, QoS oriented routing and resource-reservation signalling. QoS aware MAC protocols solve the problems of medium contention, support reliable unicast communications and provide resource reservation for real-time traffic in a distributed wireless environment. Among numerous MAC protocols and improvements that have been proposed, a protocol that can provide QoS guarantees to real time traffic in a distributed wireless environment is Black-Burst (BB) [43]. This protocol is built upon IEEE 802.11 DCF and has good QoS characteristics as far as the traffic flows have constant bit rates. An overview of proposed modifications to IEEE 802.11 for QoS support at MAC level, specifically providing traffic differentiation, can be found in [2].

QoS routing refers to the discovery and maintenance of routes that can satisfy QoS objectives under given resource constraints, while QoS signalling is responsible for flow admission control as well as resource reservation along the established route. INSIGNIA is the first QoS signalling protocol specifically designed for resource reservation in ad hoc environments [44]. It supports in-band signalling by adding a new option field in the IP header to carry the signalling control information. Like RSVP, the service granularity supported by INSIGNIA is per-flow. If the required resource is unavailable, the flow will be degraded to best-effort service. QoS reports are sent to the source node periodically to report network topology changes, as well as QoS statistics (loss rate, delay and throughput).

SWAN is an alternative to INSIGNIA with improved scalability properties. SWAN is a stateless network scheme specifically designed for wireless ad hoc networks employing a best-effort distributed wireless MAC [45]. Intermediate nodes do not keep any per-flow information and thus avoid complex signalling and state control mechanisms and make the system more simple and scalable. It distinguishes between two traffic classes: real-time UDP traffic and best-effort UDP and TCP traffic. A classifier differentiates between real-time and best-effort traffic. Then, a leaky-bucket traffic shaper handles best-effort packets at a previously calculated rate, applying an AIMD (Additive Increase Multiplicative Decrease) rate control algorithm. Every node measures the per-hop MAC delays locally and this information is used as feedback to the rate controller. Every T seconds, each device increases its transmission rate gradually (additive increase with increment rate of c bit/s) until the packet delays at the MAC layer become excessive. As soon as the rate controller detects excessive delays, it reduces the rate of the shaper with a decrement rate (multiplicative decrease of $r\%$).

Rate control restricts the bandwidth for best-effort traffic so that real-time applications can use the required bandwidth. On the other hand, the bandwidth not used by real-time applications can be efficiently used by best-effort traffic. The total best-effort and real-time traffic transported over a local shared channel is limited below a certain 'threshold rate' to avoid excessive delays.

SWAN also uses sender-based admission control for real-time UDP traffic. The rate measurements from aggregated real-time traffic at each node are employed as feedback. This mechanism sends an end-to-end request/response probe to estimate the local bandwidth availability and then determine whether a new real-time session should be admitted or not. The source node is responsible for sending a probing request packet toward the destination node. This request is a UDP packet containing a "bottleneck bandwidth" field. All intermediate nodes between the source and destination must process this packet, check their bandwidth availability and update the bottleneck bandwidth field in the case that their own bandwidth is less than the current value in the field. The available bandwidth can be calculated as the difference between an admission threshold and the current rate of real-time traffic. The admission threshold is set below the maximum available resources to enable that real-time and best-effort traffic are able to share the channel efficiently. Finally, the destination node receives the packet and returns a probing response packet with a copy of the bottleneck bandwidth found along the path back to the source. When the source receives the probing response it compares the end-to-end bandwidth availability and the bandwidth requirement and decides whether to start a real-time flow accordingly. If the flow is admitted, the real-time packets are marked as RT (Real-Time packets) and they bypass the shaper mechanism at the intermediate nodes and are thus not regulated.

The traffic load conditions and network topology change dynamically so that real-time sessions might not be able to maintain the bandwidth and delay bound requirements and they must be rejected or readmitted. For this reason it is said that SWAN offers soft QoS. The Explicit Congestion Notification mechanism (ECN) regulates real-time sessions as follows. When a mobile node detects congestion or overload conditions, it starts marking the ECN bits in the IP header of the real-time packets. The destination monitors the packets with the marked ECN bits and informs the source sending a regulate message. Then the source node tries to re-establish the real-time session with its bandwidth needs accordingly.

QoS routing is in charge of setting up the route for successful resource reservation by QoS signaling. This is a difficult task because optimal QoS routing requires frequent updates on link state information such as delay, bandwidth, cost, loss rate or error rate. This can result in a large amount of control overhead, which can be prohibitive for bandwidth constrained ad hoc environments. In addition, the dynamic nature of wireless ad hoc networks makes the maintenance of the precise link state information extremely difficult. Even after resource reservation, the QoS levels still cannot be guaranteed due to the frequent link failures and topology changes. Several QoS routing algorithms were published recently with a variety of QoS requirements and resource constraints [46].

There has been little research on the support of QoS when a wireless ad hoc network is attached to a fixed IP network. In this context, co-operation between the ad hoc network and the fixed network can facilitate the end-to-end QoS support. In [47], a new protocol is proposed that is based on the co-operation between a resource reservation protocol within the ad hoc network and a DiffServ domain in the fixed network. The resource reservation protocol is similar to SWAN, but it uses adaptive parameters according to feedback signals sent from the closest edge router in the DiffServ domain. In this way, the end-to-end delay of variable bit-rate, real-time traffic can be controlled efficiently.

Access networks in the Internet of the future can be expected to be heterogeneous. These networks will comprise a multiplicity of wireless and optical technologies, for example Passive Optical Networks (PONs), IEEE 802.11 or IEEE 802.16 [49]. Ad hoc network functionality can be also considered as a part of these heterogeneous access networks. In fact, some of the wireless nodes can be static and form what is called *mesh networks,* which will have higher efficiency thanks to the increased network stability.

A good overview on QoS support in wireless ad hoc networks can be found in [48].

7 Cross-Layer Issues

An example of the benefits we can obtain from a cross-layer approach is the work in [42]. The idea behind this recently developed protocol is that using power control such as in the PCMA medium access scheme [41] (discussed above) has an obvious impact on routing. This protocol contains a series of mechanisms in order to perform efficient power control and, at the same time, use an appropriate transmission power for establishing the route (i.e. the power used for transmission of RREQ packets in DSR or AODV).

Another example of the implications that protocol design has on other layers is the way that route maintenance is done in reactive unicast flat routing protocols (such as DSR or AODV) and whether the traffic is carried with UDP or TCP [9]. For UDP flows, re-constructing the route from the source may result in excessive packet loss and a route repair strategy closer to the broken link would be more effective in this case.

Due to the broadcast nature of radio, if omnidirectional antennas are used, two disjoint routes may have mutual influence if a node in one route is within the transmission range of a node in the other route. This has an impact on the construction of parallel routes with the purpose of distributing traffic load evenly or according to QoS requirements. Also, in the DS-SWAN scheme [47], this fact has to be considered when selecting the nodes that have to adapt their traffic shaping parameters.

In the ABR routing protocol [12], the periodic exchange of packets for determining the degree of associativity between nodes may constitute an obstacle for scheduling sleep modes in the terminals, thereby increasing power consumption.

Obviously, the interaction between TCP and the link layer is a very significant cross-layer issue and several proposals have appeared. Also, the effect of transmission power control on TCP is non-negligible, since the former yields routes with a larger number of hops and the latter behaves better with shorter routes in general.

8 Conclusions

This tutorial aims at giving some light on basic concepts and research challenges in wireless ad hoc networking. There are many aspects to point out in this relatively new research field.

Research in wireless ad hoc networks is receiving growing interest. This is a multidisciplinary subject, where the interaction between protocols at different layers is of paramount importance. In many cases, legacy protocols from fixed networks are not adequate for this type of networks and in most cases, new protocols are needed. Despite the relatively large amount of contributions in some areas, such as routing, many new challenges continue appearing.

References

1. IEEE Standard 802.11. Wireless LAN Medium Access Control (MAC) and Physical Layer (PHY) Specifications (1999)
2. Lindgren, A., Almquist, A., Schelén, O.: Quality of Service Schemes for IEEE 802.11 Wireless LANs An Evaluation. Mobile Networks and Applications 8, 223–235 (2003)

3. Xu, K., Gerla, M., Bae, S.: Effectiveness of RTS/CTS handshake in IEEE 802.11 based ad hoc networks. Ad Hoc Networks 1, 107–123 (2003)
4. Melodia, T., Cuomo, F.: Ad hoc networking with Bluetooth: key metrics and distributed protocols for scatternet formation. Ad Hoc Networks 2, 185–202 (2004)
5. Yaiz, R.A., Heijenk, G.: Providing delay guarantees in Bluetooth. In: 23rd International Conference on Distributed Computing Systems, pp. 722–728 (2003)
6. Johnson, D.B., Maltz, D.A.: Dynamic Source Routing in Ad Hoc Wireless Networks. In: Imielinski, T., Korth, H. (eds.) Mobile Computing, pp. 153–181. Kluwer Academis Publishers (1996)
7. Johnson, D.B., Maltz, D.A.: DSR: The Dynamic Source Routing Protocol for Multihop Wireless Ad Hoc Networks. In: Perkins, C.E. (ed.) Ad Hoc Networking. Addison Wesley, Boston (2001)
8. Perkins, C.E., Belding-Royer, E., Das., S.: Ad hoc On-demand Distance Vector (AODV) Routing. IETF Internet-Draft (November 2001)
9. Belding-Royer, E.M., Perkins, C.: Evolution and future directions of the ad hoc on-demand distance-vector routing protocol. Ad Hoc Networks 1, 125–150 (2003)
10. Marina, M.K., Das, S.R.: Routing performance in the presence of unidirectional links in multihop wireless networks. In: 3rd Symposium of Mobile Ad Hoc Networking and Computing (MobiHoc), Lausanne, Switzerland (2002)
11. Aggelou, G., Tafazolli, R.: RDMAR: a bandwidth-efficient routing protocol for mobile ad hoc networks. In: ACM International Workshop on Wireless Mobile Multimedia (WoWMoM), pp. 26–33 (1999)
12. Toh, C.: A novel distributed routing protocol to support ad-hoc mobile computing. In: IEEE 15th Annual International Phoenix Conference, pp. 480–486 (1996)
13. Dube, R., Rais, C., Wang, K., Tripathi, S.: Signal stability based adaptive routing (SSA) for ad hoc mobile networks. IEEE Personal Communications 4(1), 36–45 (1997)
14. Park, V.D., Corson, M.S.: A Highly Adaptive Distributed Routing Algorithm for Mobile Wireless Networks. In: INFOCOM 1997 (1997)
15. Jiang, M., Ji, J., Tay, Y.C.: Cluster based routing protocol. Internet Draft, draft-ietf-manet-cbrp-spec-01.txt, work in progress (1999)
16. Perkins, C.E., Watson, T.J.: Highly dynamic destination sequenced distance vector routing (DSDV) for mobile computers. In: ACM SIGCOMM 1994 Conference on Communications Architectures, London, U.K (1994)
17. Jacquet, P., Muhlethaler, P., Clausen, T., Laouiti, A., Qayyum, A., Viennot, L.: Optimized link state routing protocol for ad hoc networks. In: IEEE INMIC, Pakistan (2001)
18. Abolhasan, M., Wysocki, T., Dutkiewicz, E.: A review of routing protocols for mobile ad hoc networks. Ad Hoc Networks 2, 1–22 (2004)
19. Haas, Z.J., Pearlman, R.: Zone routing protocol for ad-hoc networks. Internet Draft, draft-ietf-manet-zrp-02.txt, work in progress (1999)
20. Mauve, M., Widmer, A., Hartenstein, H.: A survey on position-based routing in mobile ad hoc networks. IEEE Network 15(6), 30–39 (2001)
21. Basagni, S., Chlamtac, I., Syrotivk, V.R., Woodward, B.A.: A distance effect algorithm for mobility (DREAM). In: 4th Annual ACM/IEEE International Conference on Mobile Computing and Networking (Mobicom 1998), Dallas, TX, U.S.A (1998)
22. Ko, Y.-B., Vaidya, N.H.: Location-aided routing (LAR) in mobile ad hoc networks. In: 4th Annual ACM/IEEE International Conference on Mobile Computing and Networking (Mobicom 1998), Dallas, TX, U.S.A (1998)

23. Blazevic, L., Buttyan, L., Capkun, S., Iordano, S., Hubaux, J., Le Boudec, J.: Self-organization in mobile ad-hoc networks: the approach of Terminodes. IEEE Communications Magazine (2001)
24. Li, J., Jannotti, J., De Couto, D.S.J., Karger, D.R., Morris, R.: A scalable location service for geographic ad hoc routing. In: 6th Annual ACM/IEEE International Conference on Mobile Computing and Networking (MOBICOM 2000), Boston, MA, U.S.A, pp. 120–130 (2000)
25. Sivakumar, R., Sinha, P., Bharghavan, V.: CEDAR: a core-extraction distributed ad hoc routing algorithm. IEEE Journal on Selected Areas in Communications 17(8), 1454–1465 (1999)
26. Marina, M.K., Das, S.R.: On-demand multipath distance vector routing in ad hoc networks. In: 9th International Conference on Network Protocols, pp. 14–23 (2001)
27. Song, J.-H., Wong, V.W.S., Leung, V.C.M.: Efficient on-demand routing for mobile ad-hoc wireless access networks. In: Global Communications Conference (GLOBECOM 2003), San Francisco, CA, U.S.A (2003)
28. Haas, Z.J., Halpern, J.Y., Li, L.: Gossip-based ad-hoc routing. In: IEEE INFOCOM 2002, New York City, NY, U.S.A (2002)
29. Lee, S.-J., Su, W., Gerla, M.: Wireless ad hoc routing with mobility prediction. ACM/ Kluwer Mobile Networks and Applications 6(4), 351–360 (2001)
30. Holland, G., Vaidya, N.: Impact of routing and link layers on TCP performance in mobile ad hoc networks. In: IEEE Wireless Communications and Networking Conference (WCNC), pp. 1323–1327 (1999)
31. Ramanathan, R.: On the performance of ad hoc networks with beamforming antennas. In: ACM MobiHoc 2001, Long Beach, CA, U.S.A (2001)
32. de Morais Cordeiro, C., Agrawal, D.P.: Employing dynamic segmentation for effective co-located coexistence between Bluetooth and IEEE 802.11 WLANs. In: IEEE Global Telecommunications Conference (GLOBECOM 2002), vol. 1, pp. 195–200 (2002)
33. Specification of the Bluetooth System, Version 1.1, Bluetooth SIG (2001)
34. Singh, S., Raghavendra, C.S.: PAMAS-Power Aware Multi-Access protocol with Signalling for Ad hoc Networks. ACM Communication Review (1998)
35. Misra, A., Banerjee, S.: MRPC: Maximizing Network Lifetime for Reliable Routing in Wireless Environments. In: IEEE Wireless Communications and Networking Conference (WCNC), Orlando, Florida, U.S.A (2002)
36. Banerjee, S., Misra, A.: Minimum Energy Paths for Reliable Communication in Multi-hop Wireless Networks. In: Mobihoc 2002, Lausanne, Switzerland (2002)
37. Toh, C.-K., Cobb, H., Scott, D.: Performance Evaluation of Battery-Life-Aware Routing Schemes for Wireless Ad Hoc Networks. In: IEEE International Conference on Communications (IEEE ICC), Helsinki, Finland (2001)
38. Toh, C.-K.: Maximum Battery Life Routing to Support Ubiquitous Mobile Computing in Wireless Ad Hoc Networks. IEEE Communications Magazine 39(6), 138–147 (2001)
39. Singh, S., Raghavendra, C.S.: Power-Aware Routing in Mobile Ad hoc networks. In: MOBICOM 1998, Dallas, TX, U.S.A (1998)
40. Domingo, M.C., Remondo, D., León, O.: A Simple Routing Scheme for Improving Ad Hoc Network Survivability. In: Global Communications Conference (GLOBECOM), San Francisco, CA, U.S.A (2003)
41. Monks, J.P., Bharghavan, V., Hwu, W.-M.W.: A power controlled multiple access protocol for wireless packet networks. In: 20th Annual Joint Conference of the IEEE Computer and Communications Societies (INFOCOM 2001). Proceedings IEEE, vol. 1, pp. 219–228 (2001)

42. Krunz, M., Muqattash, A.: A power control scheme for MANETs with improved throughput and energy consumption. In: 5th International Symposium on Wireless Personal Multimedia Communications 2002, pp. 771–775 (2002)
43. Sobrinho, J.L., Krishnakumar, A.S.: Quality-of-service in ad hoc carrier sense multiple access wireless networks. IEEE Journal on Selected Areas in Communications 17(8) (1999)
44. Lee, S.B., Campbell, A.: INSIGNIA. Internet Draft (May 1999)
45. Ahn, G.-S., Campbell, A.T., Veres, A., Sun, L.-H.: SWAN. Internet Draft, draft-ahn-swan-manet-00.txt (February 2003)
46. Perkins, D.D., Hughes, H.D.: A survey on quality of service support in wireless ad hoc networks. Journal of Wireless Communication & Mobile Computing (WCMC), Special Issue on Mobile Ad Hoc Networking: Research, Trends and Application 2(5), 503–513 (2002)
47. Domingo, M.C., Remondo, D.: Analysis of VBR VoIP traffic for ad hoc connectivity with a fixed IP network. In: IEEE Vehicular Technology Conference (VTC) Fall, Los Angeles, CA, U.S.A (2004)
48. Chakrabarti, S., Mishra, A.: Quality of service challenges for wireless mobile ad hoc networks. Wireless Communications and Mobile Computing 4, 129–153 (2004)
49. Remondo, D., Sargento, S., Cesana, M., Nunes, M., Filipini, I., Triay, J., Agustí, A., De Andrade, M., Gutiérrez, L.I., Sallent, S., Cervelló-Pastor, C.: Integration of Optical and Wireless Technologies in the Metro-Access: QoS Support and Mobility Aspects. In: 5th Euro-NGI Conference on Next Generation Internet Networks (NGI 2009), Aveiro, Portugal (2009)

Broadcasting Methods in MANETS: An Overview*

Is-Haka M. Mkwawa and Demetres D. Kouvatsos

NetPEn - Networks and Performance Engineering Research Unit,
Informatics Research Institute (IRI),
University of Bradford, Bradford, BD7 1DP, UK
{I.M.Mkwawa1,D.Kouvatsos}@bradford.ac.uk

Abstract. Broadcasting in mobile ad hoc networks (MANETs) is an information dissemination process of sending a message from a source node to all other nodes of the network. Even though it has been studied extensively for wired networks, broadcasting in MANETs poses more challenging problems because of the variable and unpredictable characteristics of its medium as well as the fluctuation of the signal strength and propagation with respect to time and environment. Furthermore, node mobility creates a continuously changing communication topology in which routing paths break and new ones form dynamically. In this context, efficient broadcasting in Mobile Ad hoc networks is crucial for providing control and routing information for multicast and point to point communication protocols. This paper presents an overview on the state of the art of broadcasting methods in MANETs and makes recommendations on how to improve the efficiency and performance of tree and cluster based broadcasting methods.

Keywords: MANET, broadcasting, wireless network, binomial tree, performance evaluation, system decomposition.

1 Introduction

A mobile Ad Hoc Network (MANET) is a special type of temporary wireless mobile network of nodes (routers) without the aid of a fixed network infrastructure or centralized administration. It is made up of wireless nodes (routers) that can move around freely and cooperate in relaying packets on the behalf of one another. If a source node is unable to send a message directly to its destination node due to limited transmission range, the source node uses intermediate nodes to forward the message towards the destination node. MANET's applications range from civilian use to emergency rescue sites, such as military battlefields.

The main design and operational challenges of a MANET are reliability, bandwidth and battery power. The network has unpredictable characteristics such as dynamically changing topology, fluctuation of signal strengths with time and

---

* This work was supported in part by the EC NoE Euro-FGI (NoE 028022) and in part by the EC IST project VITAL (IST-034284 STREP).

environment, failures of existing communication routes and on the fly formation of new ones. Thus, communication algorithms and protocols should be very light in computational and storage needs in order to conserve energy and bandwidth (c.f., [1,2,3,4,5]).

Broadcasting in a MANET is an information dissemination process of sending a message from a source node to all other nodes of the network. This process is important to routing information discovery protocols, which use broadcasting to establish routes such as dynamic source routing (DSR) [6], ad hoc on demand distance vector (AODV) [7], zone routing protocol (ZRP) [2,8,9] and location aided routing (LAR) [10].

Broadcasting in MANETs poses more challenges than those in wired networks due to node mobility and scarce system resources. Because of the mobility aspect, there is no single optimal scheme for all scenarios. An important feature characterizing the quality of a MANET is the ability to effectively disseminate the information amongst the nodes of the network. Thus, efficient message dissemination is a key component in achieving high performance in MANETs.

In this context, two main broadcast problems are described below, where a graph $G(U, L)$ is used to represent a MANET with U and L as the set of nodes and links, respectively.

- **One-to-all broadcast problem (broadcasting):** Let us assume that a node u, $u \in U$, knows a piece of information $I(u)$, which is unknown to all other nodes in U. The problem is to find a communication scheme such that all nodes in G learn the piece of information $I(u)$ (c.f., Fig. 1).
- **All-to-all broadcast problem (gossiping):** Given the graph $G(U, L)$ and for all $u \in U$, let $I(u)$ be a piece of information residing in each u. The problem is to find a communication scheme such that each node from U learns the whole cumulative message.

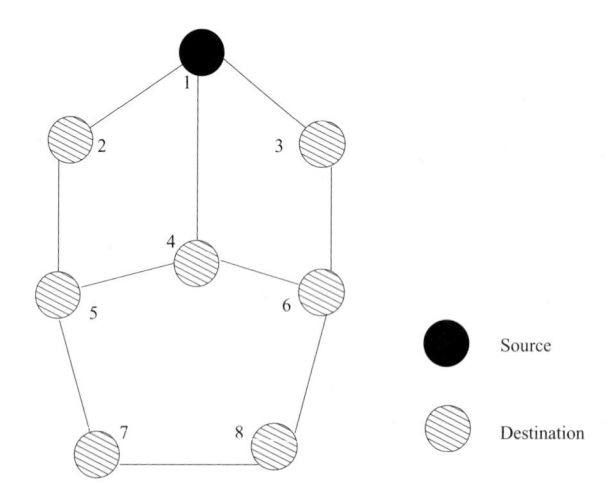

Fig. 1. One-to-All broadcasting

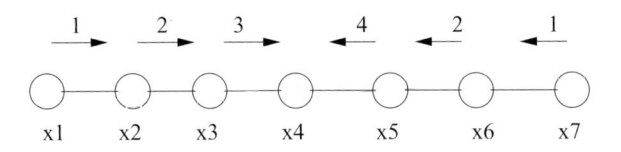

Fig. 2. Half duplex mode

The broadcast problem is to spread the knowledge of one node to all other nodes in the MANET and the gossip problem is to accumulate the knowledge of all nodes in each node of the MANET. Each communication scheme involves a sequence of steps. The term 'communication mode' describes the way how links/nodes may be used or may not be used in one communication step. There are several communication modes investigated in the literature but half and full duplex, single and all port modes have received a lot of interest and are described below.

- **Half duplex:** In this mode and for each single step each node may be active only via one of its adjacent links either as a sender or a receiver. This means that the information flow is one way. Fig. 2 shows an accumulation for the path of 7 nodes and the node $x4$ is depicted. More specifically, in the first step the node x1 sends a message to $x2$ and $x7$ sends a message to $x6$. In the second step, $x2$ sends a message to $x3$ and $x6$ sends a message to $x5$. In the third step $x3$ sends a message to $x4$, and in the fourth step $x5$ sends to $x4$. In these steps one can see that the properties of the half duplex have been satisfied at the last two steps.
- **Full duplex:** In this mode, in a single step, each node may be active only via one of its adjacent links and, if it is active, then it simultaneously sends a message and receives a message through a given, active communication link, i.e., if one link is used for communication, the information flow is bidirectional.
- **Single port:** Each node of the network can access at most one link that is incident to it i.e., only one port can be used at a time. Note that Fig. 2 also satisfies the properties of single port mode.
- **All ports:** Each node of the network can simultaneously access all links incident to it i.e., all ports can be used at a time.

A fast and efficient broadcasting scheme is a fundamental prerequisite underpinning the computational algorithms of MANETs. This is done by a sequence of calls over the links of the MANET, subject to the following two constraints:

- Each communication from one node to another requires a unit time step.
- A node can only communicate with an adjacent node.

For a given graph $G(U, L)$ with m nodes and a node u, $u \in U$, let $b(u, G)$ be the necessary and sufficient number of required steps to solve from node u the

```
                          broadcast
                              |
              ┌───────────────┴───────────────┐
          probalistic                    deterministic
         ┌────┼────┐              ┌────┬────┬────┬────┐
```
distance-based | location-based | counter-based | self-pruning | scalable broadcasting | ad hoc broadcasting | cluster-based | simple flooding

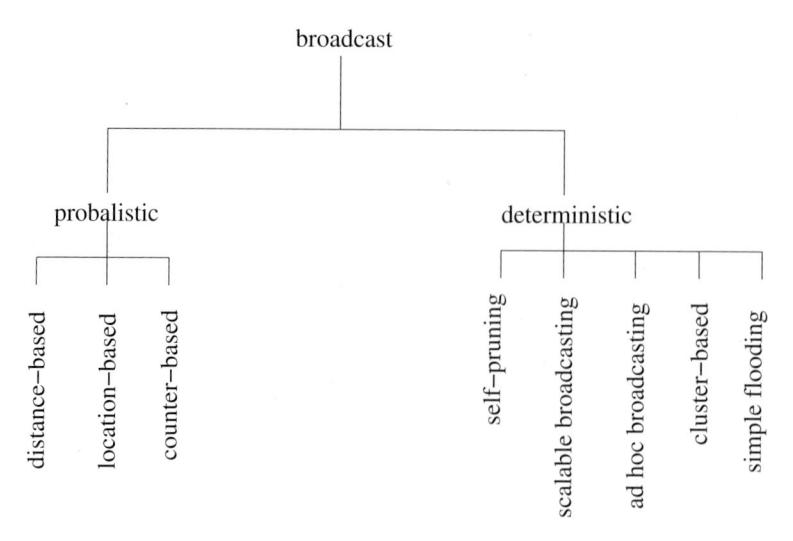

Fig. 3. Broadcasting methods

broadcast problem for graph G in the half duplex communication mode. The broadcast complexity of G in the half duplex mode of communication is defined by $b_{max}(G) = \max\{b(u,G)|u \in U\}$ whilst the lower bound of the broadcast complexity is defined by $b_{min}(G) = \min\{b(u,G)|u \in U\}$ and clearly, the following relation holds: $b(G) \geq \min b(G) \geq \lceil log_2 m \rceil$, where m is the number of nodes in the set U of graph $G(U,L)$. Note that the graphs $G(U,L)$ satisfying the property $b_{min}(G) \geq \lceil log_2 m \rceil$ are called minimal broadcast graphs.

This paper has its roots in a tutorial by Kouvatsos and Mkwawa [11] and provides an overview of some of the main broadcasting techniques in MANETs (c.f., Fig. 3). Moreover, it makes recommendations to improve the efficiency and performance of the current and future broadcasting techniques by using the concept of the binomial tree (c.f., Lo et al [12]) and a decomposition criterion for the design of complex systems (c.f., Kouvatsos [13]).

The paper is organised as follows: Section 2 reviews the probabilistic and deterministic broadcasting methods. Section 3 and 4 review the cluster and tree based broadcasting methods, respectively. The proposed enhancements of cluster and binomial tree-based broadcasting methods are presented in Section 5. Conclusions follow in Section 6.

2 Broadcasting Methods

The broadcasting methods highlighted in this section are based on statistical and geometrical models, which estimate the additional coverage of re-broadcasting. These methods have been categorised into four families utilising the IEEE802.11 MAC specifications [14].

1. The Simple Flooding Method [15, 16] requires each node in a MANET to rebroadcast all packets.

2. The Probability-based Methods [17] assign probabilities to each node to rebroadcast depending on the topology of the network.
3. The Area-based Methods [17] assume a common transmission distance and rebroadcasts a node if there is sufficient coverage area.
4. The Neighbourhood-based Methods [18, 19, 20, 21, 22] maintain the state on the neighbourhood and the information obtained from the neighbouring nodes is used for rebroadcast.

Comparisons amongst these broadcasting methods can be seen in [23, 24]. Apart from Simple Flooding Method, all broadcasting methods aim to optimise energy and bandwidth by minimising message retransmission. Some of these broadcasting methods are described in the following subsections.

2.1 Simple Flooding (SF) Method

According to the SF method, a source node of a MANET disseminates a message to all its neighbouring nodes, each of which will check whether it received this message at an earlier transmission. If yes, the message will be dropped, if no the message will be re-disseminated at once to all their neighbours. The process goes on until all nodes have received the message.

Although this method is very reliable for a MANET with low density nodes and high mobility, nevertheless it is very harmful and unproductive as it causes severe network congestion and a quickly exhaustion of battery power. In a MANET of size m, the number of messages is of the magnitude $\bigcirc(m^2)$ and is depicted in Fig. 4.

2.2 Probability Based Methods

Probability-based (P-B) Approach. The P-B approach tries to solve some of the drawbacks associated with the simple flooding method. Each node $i \in N$ is given a predetermined probability p_i for re-broadcasting. Consequently, as there will be a number of nodes that do not rebroadcast, there will be a related decrease in the degree of the network's congestion and collisions . However, there is a danger that some nodes will not receive the broadcast message. Note that if $\forall i$ such that $p_i = 1$, the probability based approach is reduced to a simple flooding approach. More efficient broadcasting reduces p_i as the number of neighbour density increases and vise versa.

Counter-based (C-B) Scheme. Under the C-B scheme, a random assessment delay (RAD) is set, a threshold K is determined and a counter $k \geq 1$ is formed on the number of times the broadcast message is received. During RAD, the counter k is incremented by one for each redundant message received and if $k > K$ when RAD expires, the message is dropped. Otherwise, it is rebroadcasted. In this approach, some nodes will not rebroadcast in denser MANETs, whilst in a less dense MANETs all nodes will rebroadcast.

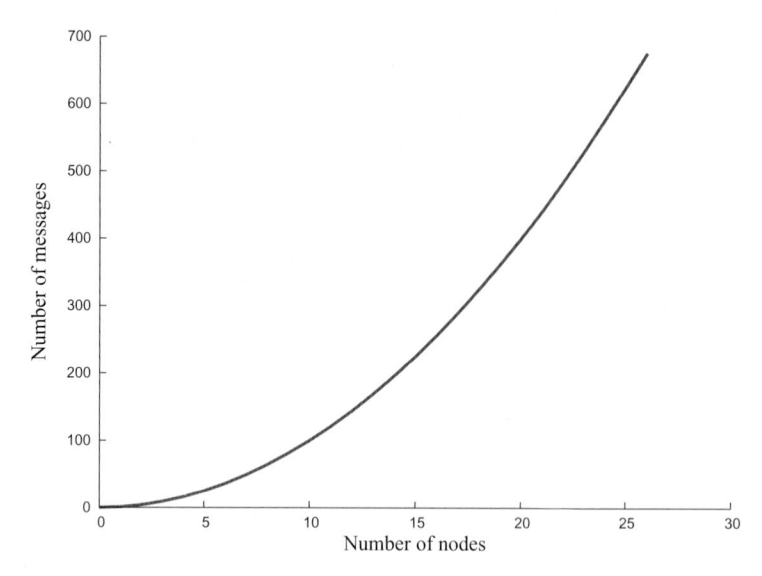

Fig. 4. $O(n^2)$ number of messages in a simple flooding method

2.3 Area-Based Methods

Distance-Based (D-B) Approach. The D-B Approach utilises the distance between a receiving node and it's neighbours to decide either to drop a message or to rebroadcast. Let d be the distance between the receiving node and the source node. If d is small, then the rebroadcast coverage of the receiving node is also small. If d is large, then the rebroadcast coverage is large. If $d = 0$, then the rebroadcast coverage is also 0.

A receiving node will normally determine the threshold distance D and set the RAD. Redundant messages will be stored until the RAD expires, in which case all distances from source nodes will be checked and if $d < D$ then the received messages will be dropped, otherwise the messages will be rebroadcast. Ni et al [17] suggested that the signal strengths can be used to calculate the distance from the source node. The role of distance can even be directly replaced by the signal strength by setting the signal strength threshold.

Location-Based (L-B) Approach. According to the L-B approach, each node must have the means to establish it's own location in order to estimate the additional coverage more precisely. This approach can be supported by the global positioning system (GPS) [25]. More specifically, each node in a MANET will add its own location to the header of each message it sends or rebroadcasts. When a node received a message, it will note the location of the sender and compute the additional coverage area to rebroadcast. If this is less than the given threshold, the message is dropped when the RAD expires, otherwise the message will be rebroadcasted.

The problem of the location based approach is the cost of calculating additional coverage areas, which is calculating many intersections among many circles. This will drain the scarcely available energy.

2.4 Neighbour Knowledge Methods

Self Pruning (SP). Each node under SP is required to have knowledge of it's neighbours, which can be achieved by periodic "Hello" messages. The receiving node will first compare its neighbour's lists to the list of the sender node and then it will rebroadcast, if the additional nodes could be reached, otherwise it will drop the message. This is the simplest approach of the neighbour knowledge methods.

As it can be seen in Fig. 5, after receiving a message from node 2, node 1 will rebroadcast the message to node 4 and node 3 as they are its only additional nodes. Note that node 5 also will rebroadcast the same message to node 4 as it is the only additional node. Thus, message redundancy still takes place under SP.

Scalable Broadcasting (SB) Approach. The SB approach improves the SP approach by reducing the number of the chances of message retransmission. Under SB approach, all nodes in a MANET have knowledge of their neighbouring nodes up to a two hop distance. This knowledge is established by "Hello" messages. In this approach, each node has a two hop topology information. In Fig. 5, node 1 receives a message from node 2, since node 2 is a neighbour, node 1 has knowledge of all its own neighbours and also node's 2 neighbours, which have received the broadcast message. The additional nodes of node 2 will receive the message rebroadcast by node 2 with the aid of RAD. Note that still node 4 will receive the redundant message.

Pen and Lu [20] dynamically adjusted RAD according to a given MANET conditions. Each node will look for a neighbour with maximum degree in it's

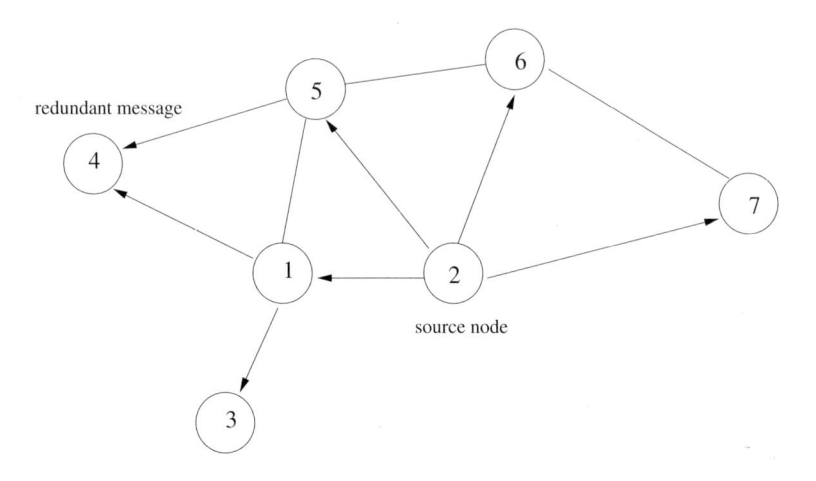

Fig. 5. Self Pruning approach

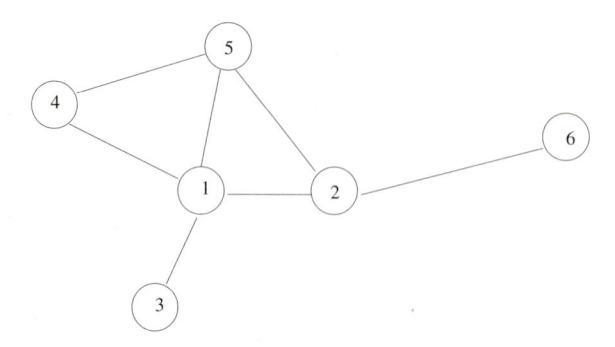

Fig. 6. Ad Hoc broadcasting (AHB) Approach

knowledge base. When a neighbour with the maximum number degree δ_i is found, the RAD is computed based on the ratio $\frac{\delta_i}{\delta}$, where δ is the degree of the current node. Nodes with large RAD will always rebroadcast first.

Ad Hoc Broadcasting (AHB) Approach. According to the AHB approach, only nodes selected as gateway and header nodes are allowed to rebroadcast a message. The AHB approach can be described as follows:-

1. Locate all two hop neighbours that can only be reached by a one hop neighbour and select these nodes as gateways.
2. Calculate the cover set that will receive the message from the current gateway set.
3. Out of the neighbours not yet in the gateway set, select randomly one node that would cover at most two hop neighbours not in the cover set. Set this node as a gateway.
4. Repeat processes 2 and 3 until all two hop neighbours are covered.
5. When a node receives a message and is a gateway, this node determines which of its neighbours already received the message in the same transmission. These neighbours are considered already covered and are dropped from the neighbour used to select the next hop gateways.

In Fig. 6, node 2 has 1, 5 and 6 nodes as one hop neighbours, 3 and 4 nodes has two hop neighbours. Node 3 can be reached through node 1 as a one hop neighbour of node 2. Node 4 can be reached through node 1 or node 5 as one hop neighbours of node 2. Node 2 selects node 1 as a gateway to rebroadcast the message to nodes 3 and 4. Upon receiving the message node 5 will not rebroadcast the message as it is not a gateway.

2.5 Shortcomings of Existing Broadcasting Methods

The shortcomings of the aforementioned broadcasting methods for MANETs are deduced below from a detailed comparative study carried out by Williams and Camp [23].

1. All methods apart from the Neighbour Knowledge Methods, require more rebroadcasts, with respect to the number of retransmitting nodes [17].
2. Methods making use of RAD, such as the Probability-based C-B scheme and the Area-based D-B and L-B methods, underperform in high density MANETs unless a mechanism to dynamically adjust RAD to its network conditions is developed.
3. The AHB approach runs into difficulties in a very high mobile MANET due to not making any use of local information to decide whether to rebroadcast or not.

Based on the comparative studies in [23], none of the existing broadcasting protocols are satisfactory for wide ranging MANET environments. Because of its adaptive nature, the SB approach has significant improvements over the non adaptive approaches.

Due to these shortcomings there is a need to develop new efficient broadcasting approaches with the common goal of conserving the available scarce resource in MANETs.

3 Cluster-Based Methods

In this section clustering based methods, which are founded on graphic theoretic concepts are reviewed. Note that the clustering approach has been used to address traffic coordination schemes [26], routing problems [26] and fault tolerance issues [27]. Note that the cluster approach proposed in Ni et al [17] was adopted by Jiang et al [28] in order to reduce the complexity of the storm broadcasting problem.

Each node in a MANET periodically sends "Hello" messages to advertise its presence. Each node has a unique identification (ID). A cluster is a set of nodes formed as follows: A node with a local minimal ID will elect itself as a cluster head. All surrounding nodes of a head are members of the cluster identified by the heads ID. Within a cluster, a member that can communicate with a node in another cluster is a gateway. To take mobility into account, when two heads meet, the one with a larger ID gives up its head role. This cluster formation is depicted in Fig. 7.

More specifically, Ni et al [17] assumed that the cluster formed in a MANET will be maintained regularly by the underlying cluster formation algorithm. In a cluster, the heads' rebroadcast can cover all other nodes in it's cluster. To rebroadcast message to nodes in other clusters, gateway nodes are used, hence there is no need for a non-gateway nodes to rebroadcast the message. As different clusters may still have many gateway nodes, these gateways will still use any of the broadcasting approaches described in Section 2 to determine whether to rebroadcast or not. In particular Ni et al [17] showed that the performance of the L-B cluster-based method compared favourably to the original location based scheme. The method saved a lot more rebroadcasts and lead to shorter average broadcast latencies. For low density MANETs , however, its reachability was poor.

X Gateway
H Head

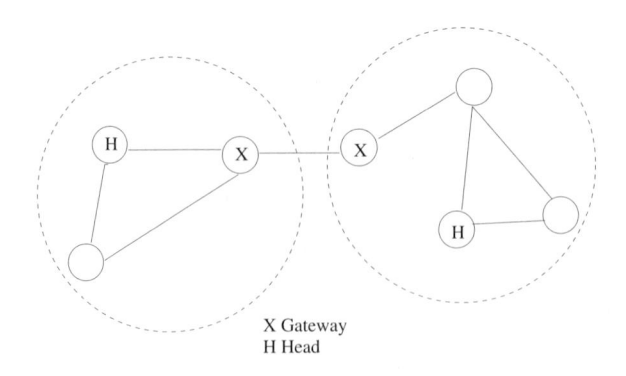

Fig. 7. Clustered MANET

4 Tree-Based Methods

Although broadcasting using tree methods in wired networks is a well known and widely used technique, it is typically claimed to be inappropriate for MANETs because of their dynamic change in network topologies. To the contrary, Juttner and Magi [29] have shown that a tree based method is an efficient, reliable and stable even in case of the ever changing network structure of the MANETs.

The tree constructed in [29] was a spanning tree (i.e., a spanning tree of $G(U,L)$ is a selection of links of $G(U,L)$ that form a tree spanning every node of $G(U,L)$). The broadcasting using this tree was achieved by forwarding a broadcast message not to all neighbours but only to those who are neighbours of this tree. Since a tree is acyclic, each message is received only once by each node, giving advantages over the existing methods.

Several proposed algorithms in the literature may be used for constructing and maintaining trees such as the spanning tree algorithm of bridged Ethernet networks [30]. However, most of these algorithms are developed to work for stable networks and not in the constantly changing topology of a MANET. The authors in [31,32,33,34,35] focused their work into the study of generic multicast trees, whilst Perkins et al [36] dealt the analysis of multicast trees in MANETs. These algorithms, however, are not suitable for handling the ever changing topologies of MANETs. Other algorithms although involve the construction of a spanning tree, they are not appropriate for the maintenance of the tree in a dynamically changing topology [29].

A feature that differentiates the tree based method reported in Juttner and Magi [29] from the other methods is that it specifically optimises the concept of one-to-one transmissions. In this way the many drawbacks of local broadcasts do not affect the algorithm and thus, the method is very well suited to perform reliable broadcasts. Moreover, whilst all other methods require extra messaging to keep up network states, the proposed method in [29] has been designed to minimise this extra signalling traffic. More details on how the algorithm constructs and maintains the spanning tree can be seen in [29].

5 Enhancements of the Cluster and Tree-Based Methods

5.1 A Cluster-Decomposition-Based Method

Some of the shortcomings in the existing cluster-based broadcasting methods in MANETs (c.f., Section 3) can be addressed by adopting an information theoretic decomposition criterion proposed by Kouvatsos (c.f., [13]) for the hierarchical design of a complex system, S. In this context, system S was represented by a graph $G(U, L)$, where U is the set of m nodes and L is the set of links.

The decomposition criterion aims to decompose system S (or, graph $G(U, L)$) into most independent subsystems $\{S_1, S_2, \ldots, S_\mu\}$ (or subgraphs $G_1, G_2, \ldots, G_\mu$) and thus, minimise the information transfer amongst them (c.f., [13]). It is based on i) a multivariate binary probability measure, constructed on a succession of multiple linear regression models ii) the information theoretic concept of entropy function (c.f., Shannon [37]) and iii) the Gibbs Theorem (c.f., Watanabe [38]), leading into the minimization of the information transfer amongst $\mu (1 < \mu \leq m)$ subgraphs $G_1, G_2, \ldots, G_\mu$ of a graph $G(U, L)$ representing subsystems $\{S_1, S_2, \ldots, S_\mu\}$, respectively. The theoretical and practical results in [13] revealed that the logical structure of a complex system, S, represented as a graph $G(U, L)$, may be expressed in the form of a hierarchical tree-like structure (c.f., [39]).

Each partition π belongs in a specific partition-type $\Pi = \{\pi_1, \pi_2, \ldots, \pi_\mu\}$, which is the set of all partitions $\{\pi_i, i = 1, 2, \ldots, \mu, 1 < \mu \leq m\}$, each of which imposes, respectively, the same cardinality in each subsystem in a subset $\{S_1, S_2, \ldots, S_\mu\}$. The optimum partition can be determined by employing the aforementioned decomposition criterion, which is expressed by $\min_{\pi} \left\{ \sum_{\pi \in \Pi} \rho_{ij}^2 \right\}$, where ρ_{ij}, ($|\rho_{ij}| \leq 1$) is the correlation coefficient of binary random variables $\{z_i, z_j, i \neq j, i, j = 1, 2 \ldots, m\}$ associated with each pair of nodes (i, j) of graph $G(U, L)$. Note that each $z_i, i = 1, 2, \ldots, m$ cuts the domain of all different design solutions (i.e, forms) for the system S into two sets such that z_i takes the value 0 with probability p_i, for the set of forms that the node i fits and the value 1 with probability $q_i = 1 - p_i$ for the set of forms that it doesn't fit (c.f., Alexander [39]). Moreover, $\sum_{\pi \in \Pi} \rho_{ij}^2$ is the sum of ρ_{ij}^2 of the links $\{u_i \to u_j\}$ cut by the partition π, $\pi \in \Pi$, whilst $\sum_{\pi \in \Pi} \rho_{ij}^2$ expresses the degree of interconnection amongst the subsystems.

In the context of MANETs, the decomposition criterion can be interpreted by analogy as the $\min_{\pi} \left\{ \sum_{\pi \in \Pi} \omega_{ij}^2 \right\}$, where $\sum_{\pi \in \Pi} \omega_{ij}^2$ is the sum of the squares of average aggregate signal strengths on the links $\{u_i \to u_j\}$ of set L cut by the partition. Each partition type Π offers a number of possible decompositions of graph $G(U, L)$. Within each partition type Π, that partition π with the smallest $\sum_{\pi \in \Pi} \omega_{ij}^2$ is chosen. However, it is not feasible to identify the optimum $\sum_{\pi \in \Pi} \omega_{ij}^2$ amongst the available partition types which cut different number of links according to the cardinality of $G_1, G_2, \ldots, G_\mu$ and the topology of graph $G(U, L)$. Hence, the optimum partition π over all partition types can be determined by

(a)   (b)

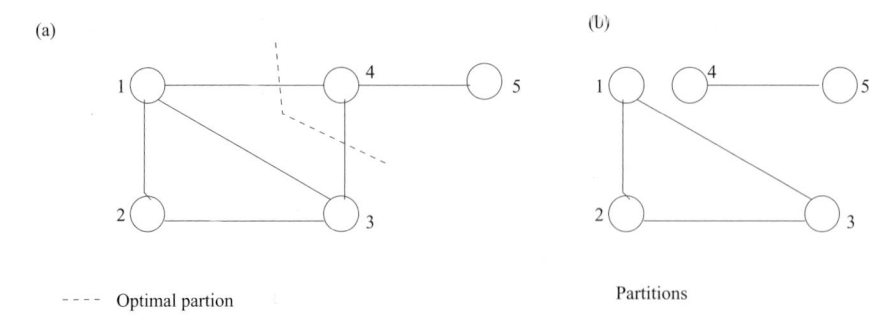

---- Optimal partion   Partitions

Fig. 8. Decomposition of graph G into most independent subgraphs

applying a normalisation on $\sum_{\pi \in \Pi} w_{ij}^2$ namely $N(\pi) = \dfrac{\min_{\pi} \sum_{\pi \in \Pi} w_{ij}^2 - E(\sum_{\pi \in \Pi} w_{ij}^2)}{(var(\sum_{\pi \in \Pi} w_{ij}^2))^{\frac{1}{2}}}$, where $E(\sum_{\pi \in \Pi} w_{ij}^2)$ and $var(\sum_{\pi \in \Pi} w_{ij}^2)$ are the mean and variance of $\sum_{\pi \in \Pi} w_{ij}^2$ respectively (c.f., [39, 40]). An illustration of the decomposition of a graph with five vertices is displayed in Fig. 8.

The normalisation procedure leads to the optimal decomposition of a MANET viewed as a system S into subsystems $\{S_1, S_2, \ldots, S_\mu\}$ whilst repeated applications of the decomposition process for each subsystem $\{S_i, i = 1, 2, \ldots, \mu\}$ will identify a tree structure with one root and m leaves. The root component represents the MANET as an entire design problem, the leaves are the individual nodes and the intermediate components (i.e., subsystems) are the subsets of nodes representing the most independent design subproblems. Thus, using the normalisation process to implement the decomposition criterion, the logical structure of the MANET has a tree-like representation, which can be constructed in a top-down manner. In this sense, the strongest interactions between the nodes of the MANET (i.e., system's design requirements) are satisfied at an early stage in the procedure before any irreversible decisions have been taken, and moreover, any system failure can be rectified within a subsystem (or, module) causing the least possible disturbance to the rest of the system.

The decomposition of the MANET's nodes into clusters of strongly (in terms of signal strengths) connected nodes will solve the problem of reachability of nodes within clusters and between gateway and head nodes with their respective cluster nodes. The proposed graph decomposition-based method can be employed to enhance the existing cluster-based schemes described in Section 3.

5.2   A Binomial Tree-Based Method

In the context of one-to-one transmission mechanism (c.f., [29]), a spanning tree will not give an optimal time. Therefore, it seems appropriate to design and develop efficient broadcasting schemes that aim to enhance the tree based methods (c.f., Section 4) and improve the broadcasting time in MANETs. In this section, the concept of a binomial tree is suggested for this purpose.

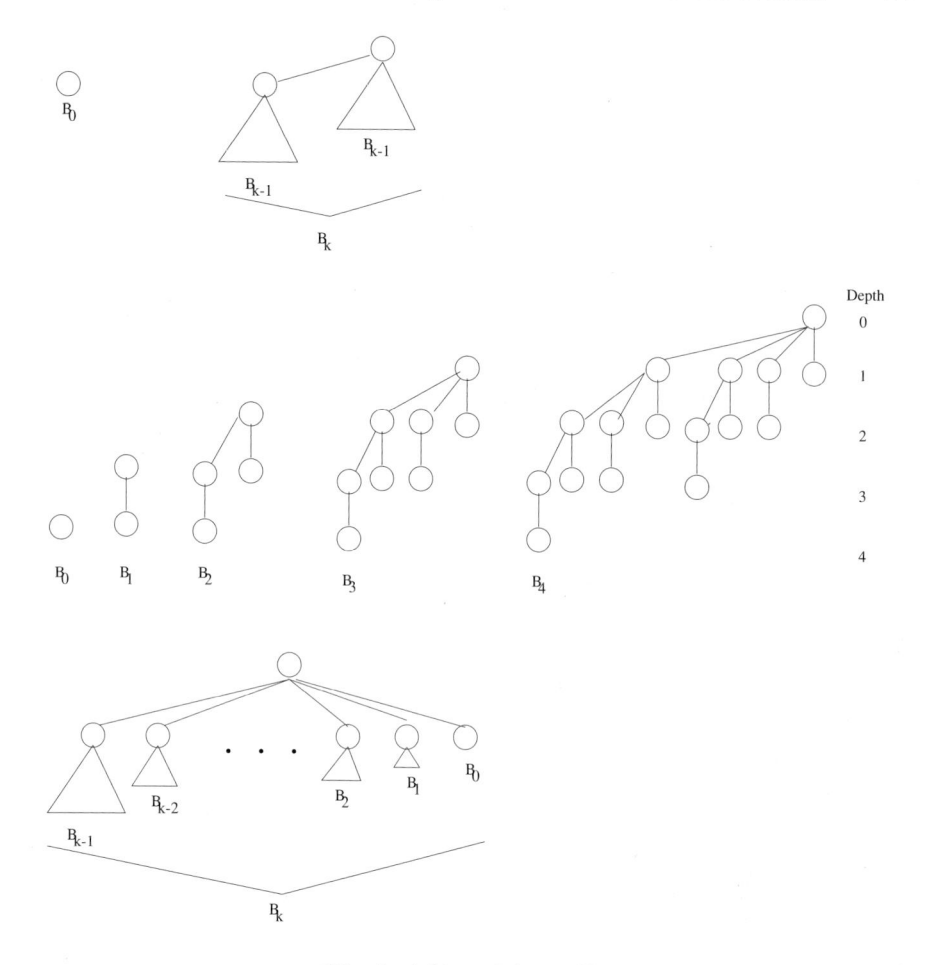

Fig. 9. A binomial tree B_k

A binomial tree structure is one of the most frequently used tree structures for parallel applications in various systems. Lo et al [12] has identified the binomial tree as an ideal computation structure for parallel divide and conquer algorithms. Binomial trees can easily be embedded into a fully connected graph with constant dilation 1 (c.f., [41]). In an one-to-one transmission, the binomial tree at each step at most doubles the informed number of nodes and thus, it gives an optimal broadcasting time.

A binomial tree B_k is an ordered tree defined recursively (c.f., Fig. 9). For the binomial tree B_k, there are $n = 2^k$ nodes, the height of the tree is k and the root has degree k. The maximum degree of any node in a binomial tree is $log_2 n$ and all nodes are labelled as binary string of length k.

For a binomial tree, there are exactly $\binom{n}{i}$ nodes at depth i for $i = 0, 1, \ldots, k$, and the root has degree k, which is greater than that of any other node. Moreover, two nodes are connected if their corresponding strings differ precisely in one position. The root node (i.e., a node at depth $i = 0$) of the tree is labelled as

000...0 where k binomial trees are attached. Nodes at depth $i = 1$ will have a single 1 in their labels in ascending order from left to right (i.e., $u_1 < u_2 < u_3 < \ldots u_{\binom{n}{i}}$). Nodes at depth $i = 2$ will have a two 1s in their labels in ascending order from left to right. This goes on to the last node of the tree at depth k with $111\ldots 1$ as a label.

In the context of a one hop ad hoc network, the binomial tree will reduce the number of forwarding nodes into at most $\frac{N_1(u)}{2} - 1$ as there $\frac{N_1(u)}{2}$ leaf nodes.

The binomial tree can easily reconfigure as nodes leave and join the transmission range. A node joining the transmission range will be connected to depth 1 of the tree provided that $|i| < \binom{n}{i}$, where $i = 1$ and $|i|$ represents the number of nodes at depth i. If $|i| = \binom{n}{i}$ for $i = 1$, then the joining node will start a new level 0.

If any node at any depth i, $0 \leq i \leq k$ leaves the transmission range, the far right node of depth 1 (i.e., u_j, $1 \leq j \leq \binom{n}{i}$) will take the position of the leaving node. No action will be taken if the far right node of depth 1 leaves the transmission range.

For a multi hop MANET, the tree structure will be a heap of binomial trees. The MANET will look like a cluster of binomial trees. A heap of binomial trees is a grouping of binomial trees that has the following binomial heap properties:-

1. No two binomial trees in the group have the same size.
2. Each node of each tree has a unique key.
3. Each binomial tree in the group is heap ordered, i.e., each node except a root node to have a key less than the key of its parent.

Each node keeps in a heap of binomial trees the info fields:-

1. Its own field key.
2. Its own field degree.
3. Child pointer field, this points to its left most child.
4. Sibling pointer field, this points to the right most sibling.
5. Parent pointer field, this points to the parent node.

For broadcasting in a MANET, the following operations with their corresponding worst case time complexities (represented by big O notation which establishes an upper bound on time complexity and big Θ notation which describes the case where the upper and lower bounds are on the same order of magnitude) are expected in a heap of binomial trees:-

1. Creating of a new heap, $\{\Theta(1)\}$.
2. Searching for the minimum key, $\{\Theta(log_2 n)\}$.
3. Joining two binomial heaps, $\{O(log_2 n)\}$.
4. Inserting of a node, $\{O(log_2 n)\}$.
5. Decreasing a key, $\{\Theta(log_2 n)\}$.
6. Removal of a node, $\{\Theta(log_2 n)\}$.

The proofs of the above time complexities can be found in Ralf [42];

6 Conclusions

Broadcasting is one of the most essential operations of MANETs and, thus, it is imperative to utilize the most efficient broadcast methods possible as well as design and develop novel ones in order to ensure high network reliability and performance.

This tutorial paper was based on the overview of some major broadcasting methods for MANETs suggested in the literature focusing on their functionalities and shortcomings. Moreover, the adoption of a graph theoretic decomposition criterion and the concept of the binomial tree were suggested in order to enhance, respectively, the efficiency and robustness of the cluster and tree based methods. These broadcasting methods may be adopted in the context of the decentralised optimal control of Robotic mobile wireless Ad hoc NETworks (RANETs), leading into performance enhancement (c.f., [43]), robustness to local failures, scalability and a wide range of applications (e.g., [44]).

Due to the dynamic changes of MANET's topology and its scarce resource availability, however, there is no at present single optimal algorithm that could be applicable to all relevant scenarios. Towards this goal, further analytic investigations are recommended into the proposed enhancements of the cluster and tree based broadcasting methods for MANETs in conjunction with related experimental performance evaluation studies.

References

1. Park, V., Corson, S.: A Highly Adaptive Distributed Routing Algorithm for Mobile Wireless Networks. In: INFOCOM 1997, pp. 1407–1415 (1997)
2. Haas, Z.: A New Routing Protocol for Reconfigurable Wireless Networks. In: ICUPC 1997, pp. 562–566 (1997)
3. Perkins, C., Royer, E.: Ad Hoc on Demand Distance Vector Routing. In: 2nd IEEE Workshop on Mobile Computing Systems and Applications, pp. 3–12 (1999)
4. Sivakumar, R., Sinha, P., Bharghavan, V.: Cedar: A Core Extraction Distributed Ad Hoc Routing Algorithm. In: INFOCOM 1999, pp. 202–209 (1999)
5. Johnson, D., Maltz, D.: Dynamic Source Routing in Ad Hoc Wireless Networks. In: Mobile Computing, pp. 153–181. Academic Publishers, New York (1996)
6. Johnson, D., Maltz, D., Hu, Y.: The Dynamic Source Routing Protocol for Mobile Ad Hoc Networks. Internet Draft: draft-ietf-manet-dsr-09.txt (2003)
7. Perkins, C., Beldig-Royer, E., Das, S.: Ad Hoc on Demand Distance Vector (AODV) Routing. Request for Comments 3561 (2003)
8. Haas, Z., Liang, B.: Ad Hoc Mobility Management With Randomized Database Groups. In: Proceedings of the IEEE International Conference on Communications, pp. 1756–1762 (1999)
9. Haas, Z., Pearlman, M.: The Performance of Query Control Schemes for the Zone Routing protocol. IEEE/ACM Transactions on Networking 9, 427–438 (2001)
10. Ko, Y., Vaidya, N.: Location-Aided Routing (LAR) in Mobile Ad Hoc Networks. In: Proceedings of the ACM/IEEE International Conference on Mobile Computing and Networking (MOBICOM), pp. 66–75 (1998)

11. Kouvatsos, D.D., Mkwawa, I.M.: Broadcasting Methods in Mobile Ad Hoc Networks: An Overview. In: Proceedings of the 3rd International Working Conference on the 'Performance Modelling and Evaluation of Heterogeneous Networks' (HET-NETs 2005), pp. T09/1 – T09/14. Networks UK Publishers (2005)
12. Lo, V.M., Rajopadhye, S., Gupta, S., Keldesn, D., Mohamed, M.A., Telle, J.: Mapping Divide and Conquer Algorithms to Parallel Architectures. In: Proc. of the 1990 International Conference on Parallel Processing, pp. 128–135 (1990)
13. Kouvatsos, D.D.: Decomposition Criteria for the Design of Complex Systems. International Journal of Systems Science 7, 1081–1088 (1976)
14. I. S. Committee. Wireless LAN Medium Access Control (MAC) and Physical Layer specifications. IEEE 802.11 Standard. IEEE, New York (1997)
15. Ho, C., Obraczka, K., Tsudik, G., Viswanath, K.: Flooding for Reliable Multicast in Multi-Hop Ad Hoc Networks. In: International Workshop in Discrete Algorithms and Methods for Mobile Computing and Communication, pp. 64–71 (1999)
16. Jetcheva, J., Hu, Y., Maltz, D., Johnson, D.: A Simple Protocol for Multicast and Broadcast in Mobile Ad Hoc Networks. Internet Draft, draft-ietf-manet-simple-mbcast-01.txt (2001)
17. Tseng, Y., Ni, S., Chen, Y., Sheu, J.: The Broadcast Storm Problem in a Mobile Ad Hoc Network. In: International Workshop on Mibile Computing and Networks, pp. 151–162 (1999)
18. Lim, H., Kim, C.: Multicast Tree Construction and Flooding in Wireless Ad Hoc Networks. In:Proceedings of the ACM International Workshop on Modeling, Analysis and Simulation of Wireless and Mobile Systems, MSWIM (2000)
19. Peng, W., Lu, X.: Efficient Broadcast in Mobile Ad Hoc Networks Using Connected Dominating Sets. Journal of Software (1999)
20. Peng, W., Lu, X.: On the Reduction of Broadcast Redundancy in Mobile Ad Hoc Networks. In: Proceedings of MOBIHOC (2000)
21. Peng, W., Lu, X.: Ahbp: An Efficient Broadcast Protocol for Mobile Ad Hoc Networks. Journal of Science and Technology (2002)
22. Sucec, J., Marsic, I.: An Efficient Distributed Network-Wide Broadcast Algorithm for Mobile Ad Hoc Networks. CAIP Technical Report 248 - Rutgers University (2000)
23. Williams, B., Camp, T.: Comparison of Broadcasting Techniques for Mobile Ad Hoc Networks. In: Proceedings of the ACM Symposium on Mobile Ad Hoc Networking and Computing (MOBIHOC), pp. 194–205 (2002)
24. Karthikeyan, N., Palanisamy, V., Duraiswamy, K.: Performance Comparison of Broadcasting Methods in Mobile Ad Hoc Network. International Journal of Future Generation Communication and Networking 2, 47–58 (2009)
25. Kaplan, E.D.: Understanding GPS: Principles and Applications. Artech House, Boston (1996)
26. Gerla, M., Tsai, J.T.: Multicluster, Mobile, Multimedia Radio Network. ACM-Baltzer Journal of Wireless Networks 1, 255–265 (1995)
27. Alagar, S., Venkatesan, S., Cleveland, J.: Reliable Broadcast in Mobile Wireless Networks. In: Military Communications Conference, MILCOM Conference Record, vol. 1, pp. 236–240 (1995)
28. Jiang, M., Li, J., Tay, Y.C.: Cluster Based Routing Protocol Functional Specification. Internet Draft (1998)
29. Juttner, A., Magi, A.: Tree Based Broadcast in Ad Hoc Networks. MONET Special Issue on WLAN Optimization at the MAC and Network Levels (2004)
30. IEEE. IEEE Specification 820.1d MAC Bridges (1989)

31. Ballardie, A., Francis, P., Crowcroft, J.: Core Based Trees (CBT) an Architecture for Scalable Interdomain Multicast Routing. In: SIGCOMM 1993, San Francisco, pp. 85–95 (1993)
32. Estrin, D.: Protocol Independent Multicast Sparse Mode (PIM-SM). Protocol Specification RFC 2362 (1998)
33. Adams, A., Nicholas, J., Siadak, W.: Protocol Independent Multicast-Dense Mode (PIM-DM). Protocol Specification (Revised), IETF Internet-Draft, draft-ietf-pim-dm-new-v2-02.txt (2002)
34. Waitzman, D., Partridge, C., Deering, S.: Distance Vector Multicast Routing Protocol. RFC 1075 (1988)
35. Moy, J.: Multicast Routing Extensions for OSPF. CACM 37, 61–66 (1994)
36. Perkins, C., Royer, E., Das, S.: Ad Hoc On-Demand Distance Vector (AODV) Routing. IETF Internet-Draft, draft-ietf-manet-aodv-11.txt (2002)
37. Shannon, C.E.: A Mathematical Theory of Communication. The Bell System Technical Journal 27(1), 379–623 (1948)
38. Watanabe, S.: Information Theoretical Analysis of Multivariate. The Bell System Technical Journal 27(1), 379–623 (1948)
39. Alexander, C.: Notes on the Synthesis of the Form. Harvard University Press, Cambridge (1967)
40. Kouvatsos, D.D.: Mathematical Methods for Modular Design of Complex Systems. Computers and Operations Research 4(1), 55–63 (1977)
41. Wu, J., Eduardo, B., Luo, Y.: Embedding of Binomial Trees in Hypercubes with Link faults. Journal of Parallel and Distributed Computing 54, 59–74 (1998)
42. Ralf, H.: Explaining Binomial Heaps. J. Funct. Program 9, 93–104 (1999)
43. Kouvatsos, D.D.: Performance Modelling and Evaluation of RANETs. Invited PP Presentation, First European Telecommunications Standards Institute (ETSI) Workshop on 'Networked Mobile Wireless Robotics', Municon Conference Centre, Munich Airport, Germany, October 8 (2010)
44. Wang, Z., Zhou, M., Ansari, N.: Ad-Hoc Robot Wireless Communication. In: IEEE International Conference on System, Man and Cybernetics, vol. 4, pp. 4045–4050 (2003)

ROMA: A Middleware Framework for Seamless Handover

Adrian Popescu[1], David Erman[1], Karel de Vogeleer[1], Alexandru Popescu[1,2], and Markus Fiedler[1]

[1] Blekinge Institute of Technology
371 79 Karlskrona, Sweden
adrian.popescu@bth.se
[2] University of Bradford
Bradford, West Yorkshire BD7 1DP, United Kingdom

Abstract. The chapter reports on a new middleware architecture suggested for seamless handover. The middleware is part of an architectural solution suggested by Blekinge Institute of Technology (BTH) for seamless handover, which is implemented at the application layer. This architecture is subject for the PERIMETER STREP and MOBICOME projects, granted by the EU FP7 and EUREKA, respectively. The suggested middleware, called ROMA, represents a software system with two sets of Application Programmer Interface (API), one for application writers and another one for interfacing various overlay and underlay systems. ROMA thus provides a transport-agnostic platform for future Internet applications. The paper provides a short description of the ROMA middleware, with particular focus on API design and address translation.

1 Introduction

Future mobile networks are expected to be all-IP-based heterogeneous networks that allow users to use any system anytime and anywhere. They consist of a layered combination of different access technologies, e.g., UMTS, WLAN, WiMAX, WPAN, connected via an IP-based core network to provide interworking. These networks are expected to provide high usability (anytime, anywhere, any technology), support for multimedia services, and personalization. Technologies such as seamless handover, middleware, multicarrier modulation, smart antenna techniques, OFDM-MIMO techniques, cooperative communication services and local/triangular retransmissions, software-defined radio and cognitive radio are expected to be used.

There are three possibilities to handle movement: at the link layer (L2), network layer (L3) and application layer (L5) in the TCP/IP protocol stack. The complexity of handover is large and demands for solving problems of different nature. Accordingly, a number of standard bodies have been working on handover, e.g., IEEE, 3GPP, 3GPP2, WiMAX, IETF.

The main requirements for handover are in terms of service continuity, context-awareness, security, flexibility, cost, quality, user transparency and a system architecture that is independent of the access technology. Furthermore, future mobile applications demand for robust fault-tolerance algorithms, able to adapt to imperfect network environments and to provide QoS. This means that elements like middleware are increasingly demanded to provide, among others, support for robust algorithms to guarantee fault-tolerance for mobile applications, to react very adaptively to environment changes and to provide the requested QoS guarantees. The mobile middleware is also requested to integrate different services and resources in a distributed execution environment and supply the users with open and consistent APIs.

Today, there is a quite large number of standardization forums for mobile middleware. Because of this, the situation is quite diffuse, with the consequence of less focus in defining particular middleware solutions for particular situations.

In this context, a new architecture has recently been advanced by Blekinge Institute of Technology (BTH), which is able to provide soft QoS guarantees on top of Peer-to-Peer (P2P) networks. This is an ongoing research, where the main research challenges are on middleware, overlay routing, mobility modeling and prediction, and handover implemented above the transport layer [1].

The rest of the paper is as follows. Section II is about the existent seamless handover solutions. Section III is about middleware requirements. Section IV describes a new architecture suggested for seamless mobility, which is implemented at the application layer. Section V presents the main concepts and results of the suggested middleware as well as some important open issues. Finally, section VI concludes the paper.

2 Seamless Handover

Most of the existent solutions attempt to solve the handover problem at L2 (access and switching) and L3 (IP) with particular consideration given to L4 (transport) [1]. Some of the most important requirements are on seamless handover, efficient network selection, security, flexibility, transparency with reference to access technologies and provision of QoS.

Typically, the handover process involves the following phases: handover initiation; network and resource discovery; network selection; network attachment; configuration (identifier configuration; registration; authentication and authorization; security association; encryption); and media redirection (binding update; media rerouting).

The basic idea of L2/L3 handover is using Link Event Triggers (LET) fired at Media Access Control (MAC) layer, and sent to a handover management functional module such as L3 Mobile IP (MIP) or L3 Fast MIP (FMIP) or IEEE 802.21 Information Server (IS). LET is used to report on changes with regard to L2 or L1 conditions, and to provide indications regarding the status of the radio channel. The purpose of these triggers is to assist IP in handover preparation and execution.

The type of handover (horizontal or vertical) as well as the time needed to perform it can be determined with the help of neighbor information provided by the Base Station (BS) or Access Point (AP) or the IEEE 802.21 Media Independent Handover Function (MIHF) Information Server (IS).

Given the extreme diversity of the access networks, the initial model was focused on developing common standards across IEEE 802 media and defining L2 triggers to make Fast Mobile IP (FMIP) work well. Connected with this, media independent information needs to be defined to enable mobile nodes to effectively detect and select networks. Furthermore, appropriate ways need to be defined to transport the media independent information and the triggers over all 802 media.

In reality, however, the situation is much more challenging. This is because of the extreme diversity existent today with reference to access networks, standard bodies and standards as well as architectural solutions. Other problems are because of the lack of standards for handover interfaces, lack of interoperability between different types of vendor equipment, lack of techniques to measure and assess the performance (including security), incorrect network selection, increasing number of interfaces on devices and the presence of different fast handover mechanisms in IETF, e.g., MIPv4, Fast MIPv6 (FMIPv6), Hierarchical MIPv6 (HMIPv6), Fast Hierarchical MIPv6 (FHMIPv6). Furthermore, implementing make-before-break handovers is very difficult at layers below the application layer.

IETF anticipated L2 solutions in standardized form (in the form of triggers, events, etc), but today the situation is that we have no standards and no media independent form. Other problems are related to the use of L2 predictive trigger mechanisms, which are dependent on L1 and L2 parameters. Altogether, the consequence is in form of complexity and dependence on the limitations of L1, L2 and L3. The existent solutions are generally not yet working properly, which may result in service disruptions.

Today, user mobility across different wireless networks is mainly user centric, thus not allowing operators a reasonable control and management of inherently dynamic users. Furthermore, the traditional TCP/IP protocol stack was not designed for mobility but for fixed computer networks. The responsibility of individual layers is ill-defined with reference to mobility. The main consequence is that problems in lower layers related to mobility may create bigger problems in higher layers. Higher layer mobility schemes are therefore expected to better suit Internet mobility. This kind of solutions opens up for research and development of new architectural solutions for handover based on movement, possibly implemented at L5 in the TCP/IP protocol stack.

3 Middleware Requirements

The main elements of a software architecture for mobile protocol stack are the Operating System (OS), TCP/IP Protocol Stack (TCP/IP), Mobile Middleware (MM) and User Interaction Support (UIS). Different applications obtain services from these entities through Application Programming Interfaces (API). Mobile applications however are distributed and they demand for particular standard protocols that the applications can use in their interactions. A mobile middleware therefore represents an execution environment where a set of generic service elements like configuration management, service discovery and event notification are defined. Figure 1 shows a typical example of mobile middleware and the generic service elements [2].

Some of the most important requirements for middleware are:

- Provide support for multiple types of mobile platforms, e.g., mobile phones, PDAs, laptops.
- Address the diversity of mobile devices and provide a consistent programming environment across them with high level modeling approaches.
- Provide different types of profile with reference to, e.g., network interface, access network, flow description.
- Provide implementation style (i.e., local, client-server and P2P) invisible to applications.
- Provide support for context-awareness, which means that the mobile applications should be able to adapt to diverse user context changes regarding, e.g., user preferences, terminal and network facilities, user environment and user mobility.
- Provide support for fault-tolerance, which means that the mobile applications should be able to adapt to the particular churn situation existent in the network by using adaptive overlay routing.
- Provide support for lightweight protocols, able to adapt, with minimum of resources, the mobile applications to different domain and environment needs.
- Reduce the gap between the performance of external communications among hosts and internal communication among processes residing on the same machine or within local clusters under common administrative domains.
- Provide security facilities like, e.g., application security, device security, firewall facilities and hosted server policies in the case of using hosted services.
- Provide diverse management facilities like, e.g., backup and restore, mobile system policy control, Over-the-Air (OTA) provisioning, software and hardware inventory management.
- Allow the developers to create applications through an interactive process of selecting the elements of the user interface and the objects they manipulate. Further, local emulation of mobile devices should be offered to developers to test the particular application without installing the particular software on the device.

4 ROMA Architecture

We suggest a new architectural solution for seamless handover, which is implemented at the application layer [1]. Compared to the existent handover solutions, implemented at the link layer and network layer, this solution offers the advantage of less dependence on physical parameters and more flexibility in the design of architectural solutions. By this, the convergence of different technologies is simplified. Furthermore, by using an architecture based on middleware and overlays, we have the possibility to combine the services offered by different

[Figure: Mobile middleware architecture diagram showing Open APIs at top, with UI Support on the left side, and generic service elements (Environment Monitoring, Event Notifications, Service Discovery, ..., Configuration Management, Mobile Data Management, Trust and Privacy Support, Context Modeling Tools) grouped as "Generic service elements". Below these are Distributed Execution Environment, Internet Protocol Suite, Operating System, and Computing and Communication Hardware. The middle block is labeled "Mobile Middleware".]

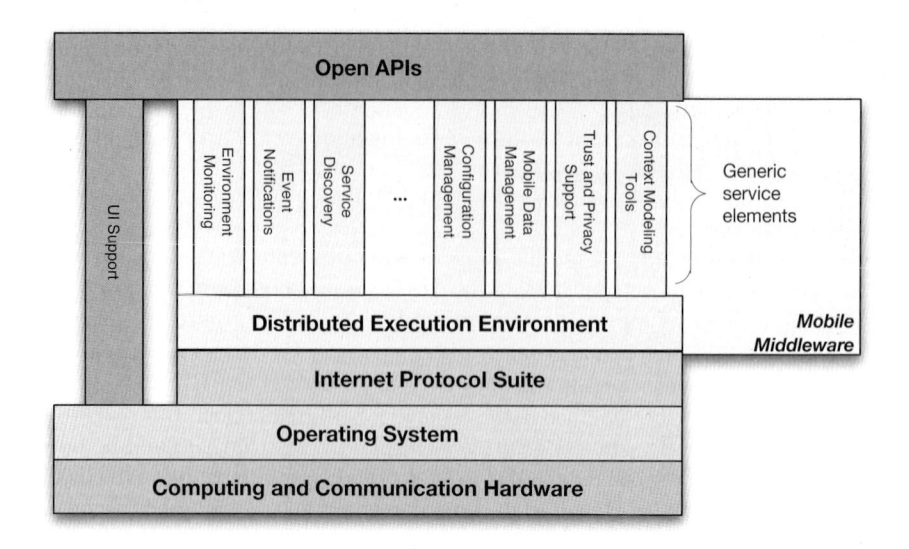

Fig. 1. Mobile middleware and the generic service elements [1]

overlays. This offers the advantage of flexibility in the development of new services and applications. The suggested architecture resembles the Android mobile development platform developed by Google [3], opening thus up for similar architectural solutions developed in the terminal and in the network. By this, new applications and services can be easily designed and developed.

The suggested architectural solution is shown in figure 2. It is based on using a middleware (with a common set of Application Programming Interfaces (APIs)), a number of overlays and a number of underlays. By middleware, we refer to software that bridges and abstracts underlying components of similar functionality and exposes the functionality through a common API. On the other hand, by overlay we refer to any network that implements its own routing or other control mechanisms over another already existing substrate, e.g., TCP/IP, Gnutella. Finally, by underlays we refer to substrates, which are abstracted networks. Bu substrate, we refer to any network (e.g., IP overlay or "normal" IP network) that can perform the packet transportation function between two nodes. Thus, ROMA implements a transport-agnostic enabler for any kind of application.

The underlays can be either structured or unstructured. Structured overlays are networks with specific type of routing geometry decided by the Distributed Hash Table (DHT) algorithm they use. Structured underlays use keys for addressing like, e.g., Chord [4]. In unstructured overlays the topology can be viewed as emergent instead of being decided before hand. Unstructured overlays can use IP addresses or other forms of addressing, e.g., Gnutella, which uses Universal Unique IDs (UUIDs) for addressing.

An important goal of the middleware is to abstract structured and unstructured underlays as well as overlays. This API architecture is used in several projects, relating to overlay routing with soft QoS, and seamless roaming [1].

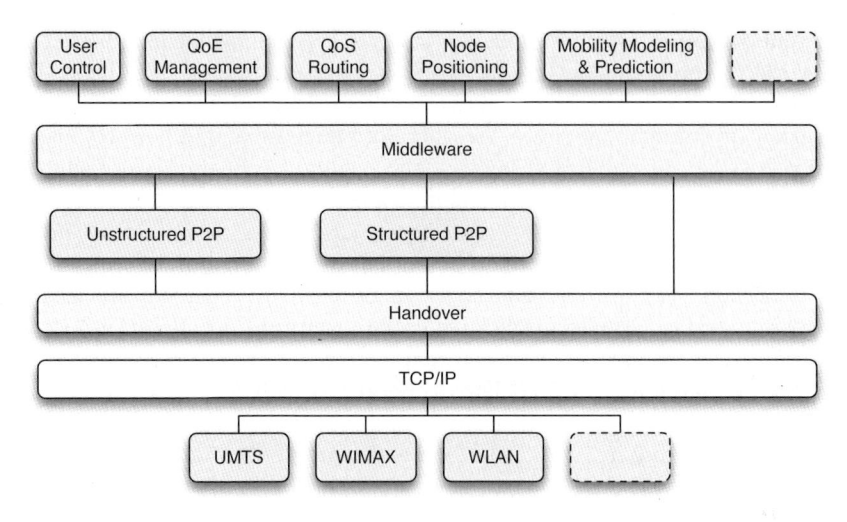

Fig. 2. ROMA architecture

We use the application layer protocol Session Initiation Protocol (SIP) [5] for mobility management. This solution has the advantage of eliminating the need for a mobility stack in mobile nodes and also does not demand for any other mobility elements in the network, beyond SIP servers. Simple IP is used in this case together with a SIP protocol stack. The drawback however is because the existing client frameworks do not accommodate IETF SIP [5] and 3GPP SIP [6] within the same framework. The consequence is that today one needs two different sets of client frameworks on the mobile, one for the mobile domain (e.g., UMTS) and the other one for the fixed domain (e.g., fixed broadband access in combination with WLAN).

BTH has also co-developed an interesting solution for vertical handover, called the Network Selection Box (NSB) [7,12]. NSB encapsulates the raw packet in a UDP datagram and sends it over a real network. A tunneling concept is used to send the packets over the interfaces encapsulated in UDP. The NSB can today be used for the transport over WLAN, UMTS and GPRS. While the NSB does solve the issue of inter-technology handovers, ROMA provides a comprehensive support structure for interoperation of both applications and various transport substrates.

5 ROMA Middleware

The main goal of the project is to develop a testbed to facilitate the development, testing, evaluation and performance analysis of different solutions for user-centric mobility, while requiring minimal changes to the applications using the platform. In other words, we implement a software system with two sets of APIs, one for application writers and another one for interfacing various overlay and underlay systems.

Current overlay implementations are built with incompatible language specific frameworks on top of the low level networking abstractions, e.g., YOID, i3, JXTA [4,8]. This complicates the design of overlays and their comparison as well as the integration of different overlays. We therefore suggest a middleware based on the Key-Based Routing (KBR) layer of the common API framework suggested in [8]. By doing so, independent development of overlay protocols, services and applications is facilitated.

The middleware is intended to work on top of both structured and unstructured underlays. Structured underlays can be used to construct services such as Distributed Hash Tables (DHT), scalable group multicast/anycast and decentralized object location. The advantage is that they support highly scalable, resilient, distributed applications like cooperative content distribution and messaging. Unstructured overlays do not have such facilities, but they tend to have less overhead in handling churn and keyword searches [9].

By using a common API, we can develop applications by using combinations of arbitrary overlays and underlays. This facility allows us to design a testbed where we can investigate interoperability issues and performance of different combinations of protocols. This also allows us to have overlays that export APIs that other overlays can use. For instance, we can have the "Quality of Experience (QoE) Management" export an API that can be used by the "Quality of Service (QoS) Routing" overlay and "Handover" underlay.

5.1 API Design

The ROMA API is based on the KBR layer of the common API framework suggested in [8]. Our approach differs in one major aspect. In [8] it is assumed that the KBR API runs on top of a structured underlay only. In our case, the ROMA API operates on top of both structured and unstructured underlays. This means that both structured and unstructured overlays are able to use the ROMA API.

This feature allows for the development of applications using combinations of arbitrary, ROMA-based, overlays and underlays. This allows us to design testbeds that investigate the interoperability and performance of various protocol combinations. For instance, given a new congestion control algorithm, one can test how it performs when running on top of Chord or Content Addressable Network (CAN) or BitTorrent or IP, respectively.

Furthermore, since the ROMA API is an enabler for seamless handover, the API is designed to provide the ability to run several overlays and several underlays simultaneously. The advantage is that one can have some overlays export APIs that other overlays can use, and this is valid for underlays as well. For example, one can have a measurement overlay exporting an API that can be used by a resource reservation overlay to reserve a path satisfying specific constraints. The resource reservation overlay can in turn export an API, which is used by a video conference application that requires some QoS guarantees between endpoints. This means that basic services can be chained or combined into one or more complex services, which is an important part of the dynamic composability expected of future Internet services.

The API core is the KBR abstract base class. The KBR class specifies the API functions in terms of abstract class methods. The API functions are the same as those presented in [8], and we refer to this for the semantics of each function. It is important to mention that the KBR class does not implement these functions, but it expects derived classes to implement them.

Given the software implementation of an API underlay, say CAN, one may need to adjust it so it can easily provide the functionality required be the ROMA API. This implies wrapping the CAN implementation into a class that translates between function calls supported by the ROMA API and functions supported by the CAN API. The wrapper class inherits (in the OOP-sense) the KBR class and it is forced to provide all API functions. We call these wrapper classes shim layers and expect to have one shim layer for each underlay we wish to plug into the API.

An overlay can either extend (OOP inheritance) one shim layer or act as a container for several shim layers, depending on the intended application. By exporting the API of an overlay we mean merely providing header files that other overlay can use to instantiate objects of the exporting overlay and obtaining services from it. Although one can discover such APIs at runtime, through introspection for instance, this is not currently in the scope of our project.

5.2 Architecture

Asynchronous system calls are used within the middleware. This is achieved by means of the *boost* framework [10]. In particular the *Asio* package is utilized, which provides us with portable networking solutions, including sockets, timers, hostname resolution and socket iostreams. The middleware does not block upon I/O system calls, e.g., sockets writing and reading, as a consequence of the asynchronous properties of the boost's *Asio* package. Hence the middleware does not show any CPU hogging behaviour and can operate in a conventional fashion with slow I/O streams.

Table 1. Overview of the Middleware's API

Function	Description
init()	Initializes the application.
run()	Starts the execution of the application.
route()	Routes a message towards a destination.
deliver()	Upcall from the middleware to deliver a message.
forward()	Upcall from the middleware to forward a message.
addUnderlay()	Adds an middleware to the application.
addOverlay()	Adds an application to the middleware.
lookup()	Checks whether a specific service is bound to the middleware (service discovery).

The middleware consists of a shim layer, `kbr`, that is inherited by the classes wanting to utilize the functionalities of the middleware. The shim layer, an Application Programming Interface (API), provides basic functions by which the middleware functionalities can be exploited. Table 1 elaborates the API accessible for applications.

Messages originating from an application are exchanged between another middleware entity and then forwarded to the destination application. Applications running on top of the middleware are identified by *keys*. The *Keys* can be used as identifiers for applications but also to refer to other entities in the middleware stack such as the middleware core itself or to identify files. These *keys* are formated as standardized UUIDs by the ITU-T [11], 128 bit or 16 bytes long.

5.3 Address Translation

The most important tasks of the ROMA middleware are to provide an addressing and routing substrate, pseudo-asynchronous event management and non-blocking TCP connection management.

The KBR API defines a 160 bit addressing scheme and a small set of functionality for routing messages. It fits the purpose of the ROMA architecture well. If, during the development of the ROMA architecture, the KBR API is found to be lacking, then this will be extended or modified to suit the purposes of ROMA.

ROMA layers address nodes by using keys. However, different overlays (e.g., Chord, CAN, Kademlia) use different keys or key formats. This may demand that a layer knows how to convert its key format to the key format of any other layer it wants to talk to. This means that whenever a new ROMA layer is designed one must extend all other layers to understand the new key format. This is clearly undesirable. To avoid this, we introduce a common key format, defined in [8]. Whenever a layer needs to communicate an address to another layer (above or below), it converts its own key to the common key format. If layers follow this rule, then whenever a new layer is added, the particular layer needs only to know how to convert from its own key format to the common key format and viceversa. By this, none of the other layers need to be changed.

A common key is defined as a 160-bit string [8]. Specific keys can be strings that are longer or shorter. Roughly speaking, when the specific key is shorter than the common key, the unused high bits of the common key are masked. In the case the specific key is longer than the common key, then the specific key is truncated to fit into the common key. When a specific key is mapped by truncation to an existing (used) common key, then the next common key is chosen. If the common key is used as well, then the process is repeated until an unused common key is found. We assume that it is unlikely that as many as 2^{160} keys can be active simultaneously. However, the truncation may mean that no guarantees can be provided that the common keys are unique throughout the network. Therefore, we require that keys are unique only within the scope of a ROMA stack on one host. This is not necessarily a problem since either IP addresses or a DHT with specific keys can provide uniqueness.

The general rule is that all common keys are created by the bottom layer (closest to the physical layer). Layers in between learn about keys when their *getKey()*, *route()*, *forward()* and *deliver()* functions are being called. The last three functions have the usual semantic as described in the KBR API [8]. The *underlay.getKey(Socket)* function is called when a layer needs to obtain a common key for the IP address in the *Socket* variable. The function is called by each underlay until it reaches the bottom layer. The bottom layer checks if the socket is in use, in which case it returns the corresponding common key. If the socket is not in use, a new common key is mapped to the IP address and returned to the layer above. Each layer can then create a specific key corresponding to the common key, before returning from the function call. This procedure guarantees the uniqueness of the common key across the stack.

5.4 Situation Today

The basic functions of the middleware have today been implemented. The middleware is able to transmit and receive message in an asynchronous manner. The middleware maintains a set of connections used by the applications that are linked to the middleware. Data streams are automatically relayed by the middleware with a smart switch between application and connection and vice versa.

The current ROMA implementation is in the C++ language, with in-house developed wrapper classes around the glib library and the *epoll()* linux system call for event handling. This makes the middleware tied to the linux kernel. We have therefore changed this to use the Boost asynchronous IO library instead, to make the middleware more platform-agnostic [10].

5.5 Open Issues

An important design assumption for the ROMA middleware is to have access to more than one network interface with IP functionality. In addition to this, each interface may correspond to more than one underlay, by using compositions of underlay shims. This means that, for an overlay making use of the ROMA API, data can be received and transmitted on several underlays. The consequence is that the middleware may be required to perform multiplexing/demultiplexing of the dataflows through the middleware.

In the current incarnation of the middleware, each application linking to the middleware gets its own ROMA instance. This means that flow multiplexing/demultiplexing does not need to be performed between applications. However, this also means that the functionality is duplicated for each instance as well as that addresses are not guaranteed to be unique between different ROMA applications.

A network interface switch running in the kernel will therefore be used to switch between technologies and it is currently under development. The middleware is able to communicate with this in-kernel switch and can retrieve vital information for decision making and handovers.

6 Conclusions

The paper has reported on a new middleware architecture suggested for seamless handover. The middleware is part of an architectural solution suggested by Blekinge Institute of Technology for seamless handover, which is implemented at the application layer. The paper has particularly focused on the main design elements, namely API design and address translation.

In the future, we will further implement the suggested middleware as well as test it in real mobile environments. The expected results will be used in the EU FP7 projects PERIMETER and MOBICOME, in which BTH is actively participating.

References

1. Popescu, A., Ilie, D., Erman, D., Fiedler, M.: An Application Layer Architecture for Seamless Roaming. In: 6th International Conference on Wireless On-demand Network Systems and Services, Snowbird, Utah, USA (February 2009)
2. Raatikainen, K.: Recent Developments in Middleware Standardization for Mobile Computing, Nokia Research Center, Finland
3. Android, An Open Headset Alliance Project, http://code.google.com/android/
4. Stoica, I., Morris, R., Karger, D., Kaashoek, F., Balakrishnan, H.: Chord: A Scalable Peer-to-Peer Lookup Service for Internet Applications. In: ACM SIGCOMM 2001, San Diego, USA (August 2001)
5. Rosenberg, G., Schulzrinne, H., Camarillo, G., Johnston, A., Peterson, J., Sparks, R., Handley, M., Schooler, E.: SIP: Session Initiation Protocol. IETF RFC 3261, http://www.ietf.org
6. ETSI/3GPP, Universal Mobile Telecommunication System (UMTS): Signaling Interworking Between the 3GPP Profile of the Session Initiation Protocol (SIP) and non-3GPP SIP Usage, 3GPP TR 29.962 version 6.1.0 Release 6, http://www.3gpp.org/ftp/specs/html-info/29962.htm
7. Isaksson, L.: Seamless Communications Seamless Handover Between Wireless and Cellular Networks with Focus on Always Best Connected. PhD thesis, BTH, Karlskrona, Sweden (March 2007)
8. Dabek, F., Zhao, B., Druschel, P., Kubiatowicz, J., Stoica, I.: Towards a Common API for Structured Peer-to-Peer Overlays. In: Kaashoek, M.F., Stoica, I. (eds.) IPTPS 2003. LNCS, vol. 2735. Springer, Heidelberg (2003)
9. Chawathe, Y., Ratnasamy, S., Breslau, L., Lanham, N., Shenker, S.: Making Gnutella-Like P2P Systems Scalable. In: ACM Conference on Applications, Technologies, Architectures and Protocols for Computer Communications, Karlsruhe, Germany (2003)
10. Boost C++ Libraries (2009), http://www.boost.org
11. Generation and registration of Universally Unique Identifiers (UUIDs) and their use as ASN.1 Object Identifier components, ITU-T Rec. X.667 | ISO/IEC 9834-8
12. Chevul, S., Isaksson, L., Fiedler, M., Lindberg, P.: Network selection box: An implementation of seamless communication. In: García-Vidal, J., Cerdà-Alabern, L. (eds.) Euro-NGI 2007. LNCS, vol. 4396, pp. 171–185. Springer, Heidelberg (2007)

Seamless Roaming: Developments and Challenges

Adrian Popescu[1], David Erman[1], Dragos Ilie[1], Markus Fiedler[1],
Alexandru Popescu[1,2], and Karel de Vogeleer[1]

[1] Blekinge Institute of Technology
371 79 Karlskrona, Sweden
adrian.popescu@bth.se
[2] University of Bradford
Bradford, West Yorkshire BD7 1DP, United Kingdom

Abstract. The chapter reports on recent developments and challenges focused on seamless handover. These are subject for the research projects MOBICOME and PERIMETER, recently granted by the EU EUREKA and EU STREP FP7, respectively. The research projects are considering the recently advanced IP Multimedia Subsystem (IMS), which is a set of technology standards put forth by the Internet Engineering Task Force (IETF) and two Third Generation Partnership Project groups, namely 3GPP and 3GPP2. The foundation of seamless handover is provided by several components, the most important ones being the handover, mobility management, connectivity management and Internet mobility. The paper provides an intensive analysis of these components.

1 Introduction

There are many types of handover systems existing today, which can be partitioned in different ways. Several dimensions can be used in partitioning the handover systems. These are, e.g., regarding the domain, the system, the overlay and the technology [1].

For instance, handover systems can be partitioned with reference to technology, which can be similar or different. In the first case we have homogeneous handovers and in the second case we have heterogeneous handovers. Handover systems can be also partitioned with reference to the place of the access points, which can be within the same network or in different ones. The first case refers to horizontal handover and the second case to vertical handover. The vertical handover can in turn be of two classes, which are the upward handover and the downward handover.

Another dimension is the domain. Handover systems can in this case be of two classes, namely intra-domain handover and inter-domain handover. Intra-domain handover means that the mobile node can roam within the same network domain. Inter-domain handover means that the mobile node can cross from one domain to another one.

Finally, the last dimension is the system. An inter-system handover refers to the case that a mobile node hands off between two independent systems controlled by different operators. An intra-system handover refers to the situation where the both domains are deserved by the same system.

The IETF document RFC 3753 on "Mobility Related Terminology" is perhaps one of the best documents that defines terms for mobility related terminology [2]. The document covers specific terminology used in handover as well as in mobile ad-hoc networking. All types of handover are considered to facilitate seamless roaming in a heterogeneous environment formed by highly-coupled and heterogeneous networks.

There are three possibilities to handle movement: at the link layer (L2), network layer (L3) and application layer (L5) in the TCP/IP protocol stack. The complexity of handover is large and demands for solving problems of different nature. Accordingly, a number of standard bodies have been working on handover, e.g., IEEE, 3GPP, 3GPP2, WiMAX, IETF.

L2 mobility across different access technologies is covered by 3GPP, 3GPP2 and WiMAX in a number of documents. L3 mobility is addressed by IETF. Therefore, the IP Multimedia Subsystem (IMS), which is acting as a service layer, does not need to cover mobility issues related to access but other mobility issues.

The rest of the paper is as follows. Section II briefly overviews the main characteristics and most important technologies of the fourth generation mobile communication systems. Section III is about seamless handover and the solutions existent today with a particular focus on their limitations. Section IV describes the main elements involved in mobility management. Section V shortly describes the algorithms that can be used for connectivity management in connection with mobility. Section VI is about Internet mobility and the most important solutions used to solve this. Sections VII and VIII present short overviews of the network layer mobility and application layer mobility, respectively. Finally, section IX concludes the paper.

2 Vision

Future mobile networks are expected to be all-IP-based heterogeneous networks that allow users to use any system anytime and anywhere. They consist of a layered combination of different access technologies, e.g., UMTS, WLAN, WiMAX, WPAN, which are connected via a common IP-based core network to provide interworking. These networks are expected to provide high usability (anytime, anywhere, any technology), support for multimedia services, and personalization. Key features are user friendliness and personalization as well as terminal and network heterogeneity [3].

User friendliness refers to the way the user interacts with the terminal, which must be simple and friendly. User personalization refers to the way users configure the operational mode of the terminal based on personal preferences. Given the large spectrum of existent users with different preferences, experiences and background, the consequence is that user friendliness and personalization should be able to offer a high degree of granularity such as the huge amount of information is selected in an appropriate way.

Terminal heterogeneity refers to the different types of terminals existent today and expected to appear in the future. This heterogeneity refers to, e.g., energy consumption, bandwidth, display size, weight, portability, complexity. On the other hand, network heterogeneity refers to the increasing heterogeneity of networks, e.g., UMTS, WLAN, WiMAX and Bluetooth. This heterogeneity mainly refers to technology, coverage area, data rate, latency, and loss rate. One of the biggest challenges is therefore to provide communication services with the best QoS and the best price irrespective of the type of terminal and network involved in the communication process.

The most important technologies of the future mobile networks are multicarrier modulation, use of smart antenna techniques, use of OFDM-MIMO techniques, use of adaptive modulation and coding with time-slot scheduler, use of cooperative communication services and local/triangular retransmissions, software-defined radio and cognitive radio [4].

With reference to handover, the main requirements are in terms of service continuity, provision of horizontal and vertical handover, provision of security, policy-based handover, flexibility, making the heterogeneous network transparent to user and design of system architecture such as it is independent of the (wireless) access technology. Connected to this, particular focus must be given to mobility management aspects (e.g., access network location, paging and registration) as well as provision of QoS, user and network security [5].

3 Seamless Handover - Situation Today

There are three possibilities to handle movement, namely at the link layer (L2), network layer (L3) and application layer (L5). Most of the existent solutions attempt to solve the handover at L2 (access and switching) and L3 (IP) with particular consideration given to L4 (transport). Some of the most important requirements are on seamless handover, efficient network selection, security, flexibility, transparency with reference to access technologies and provision of QoS.

Typically, the handover process involves the following phases:

- Handover initiation
- Network and resource discovery
- Network selection
- Network attachment
- Configuration (identifier configuration; registration; authentication and authorization; security association; encryption)
- Media redirection (binding update; media rerouting)

The basic idea of L2/L3 handover is using Link Event Triggers (LET) fired at Media Access Control (MAC) layer, and sent to a handover management functional module such as L3 Mobile IP (MIP) or L3 Fast MIP (FMIP) or IEEE

802.21 Information Server (IS). LET is used to report on changes with regard to L2 or L1 conditions, and to provide indications regarding the status of the radio channel. The purpose of these triggers is to assist IP in handover preparation and execution.

The type of handover (horizontal or vertical) as well as the time needed to perform it can be determined with the help of neighbor information provided by the Base Station (BS) or Access Point (AP) or the IEEE 802.21 Media Independent Handover Function (MIHF) Information Server (IS).

Based on the type of handover, one or more layers may be involved in the handover procedure, as shown in table 1. This table shows an example on how the basic handover functions are handled at the layers L2, L3 and L5 in an IP-based handover environment [6].

Table 1. Handover operations at L2, L3 and L5 [6]

Handover operation	L2	L3	L5
Discovery	Scanning	Router advertisement	Domain advertisement
Authentication	EAPoL	IKE, PANA	S/MIME
Security association	802.11i	IPSEC	TLS SRTP
Configuration	ESSID	DHCP stateless	URI
Address uniqueness	MAC address	ARP DAD	SIP registration
Binding update	Cache update	Update CN, HA	SIP re-invite
Media routing	IAPP	Encapsulation tunneling	Direct media routing

Given the extreme diversity of the access networks, the initial model was focused on developing common standards across IEEE 802 media and defining L2 triggers to make Fast Mobile IP (FMIP) work well. Connected with this, media independent information needs to be defined to enable mobile nodes to effectively detect and select networks. Furthermore, appropriate ways need to be defined to transport the media independent information and the triggers over all 802 media.

In reality, however, the situation is much more challenging. This is because of the extreme diversity existent today with reference to access networks, standard bodies and standards as well as architectural solutions [7]. Other problems are because of the lack of standards for handover interfaces, lack of interoperability between different types of vendor equipment, lack of techniques to measure and assess the performance (including security), incorrect network selection, increasing number of interfaces on devices and the presence of different fast handover mechanisms in IETF, e.g., MIPv4, Fast MIPv6 (FMIPv6), Hierarchical MIPv6 (HMIPv6), Fast Hierarchical MIPv6 (FHMIPv6).

IETF anticipated L2 solutions in standardized form (in the form of triggers, events, etc), but today the situation is that we have no standards and no media independent form [7]. Other problems are related to the use of L2 predictive trigger mechanisms, which are dependent of L1 and L2 parameters. Altogether, the consequence is in form of complexity of the existent solutions and dependence on the limitations of L1, L2 and L3. The existent solutions are simply not yet working properly, which may result in service disruptions. Because of this, it is important to develop cross-layer architectural solutions where cooperation is established between L2 and L3 to assist the IP handover process and to improve the performance. Even better would be to develop architectural solutions where IP has control over specific L2 handover-related actions.

Today, user mobility across different wireless networks is mainly user centric, thus not allowing operators a reasonable control and management of inherently dynamic users. This is the reason for why the IEEE 802.21 Working Group is doing an effort to ratify the Media Independent Handover (MIH) standard, to enhance the user centric mobility handovers and enable network controlled handovers across heterogeneous networks [8]. In parallel to this, IETF addresses the IP level support for mobile heterogeneous access like, e.g., the Working Group (WG) on "The Mobility for IP: Performance, Signaling and Handoff Optimization (MISHOP)". This WG regards the delivery of information for MIH services at L3 or above. The L3 discovery component is also defined. The target is to enable MIH services even in the absence of the corresponding L2 support. The security issue is addressed as well.

IEEE 802.21 defines a framework to support information exchange regarding mobility decisions, which is irrespective of media. The goal is to facilitate handovers among heterogeneous access networks. Handover decisions are taken based on information collected from both mobile nodes and network, e.g., link type, link identifier, link availability, link quality.

The core of the IEEE 802.21 framework is the Media Independent Handover Function (MIHF), which provides abstracted services to higher layers by means of a unified interface. This unified interface provides service primitives that are independent of the access technology. This interface is called Service Access Point (SAP).

IEEE 802.21 MIH is targeted at optimizing L3 and above handovers. It acts across 802 networks and extends to cellular networks like 802.3, 802.11, 802.16. 802.21 MIHF Information Server (IS) has information about location of PoA, list of available networks, cost, L2 information (neighbor maps), higher layer services (e.g., ISP, MMS) and others. Key benefits are optimum network selection, seamless roaming and low power operation for multi-radio devices.

Furthermore, it is important to point out that the traditional TCP/IP protocol stack was not designed for mobility but for fixed computer networks. This is particularly shown by the fact that the responsibility of individual layers is ill-defined with reference to mobility. The main consequence is that problems in lower layers related to mobility may create bigger problems in higher layers. Higher layer mobility schemes are therefore expected to better suit Internet mobility.

Better prediction mechanisms and especially some form of movement prediction would definitely improve the handover performance in the sense that this could compensate for errors connected with delay in the handover process and the associated service disruptions. One should also keep in mind that this kind of solutions opens up for research and development of new architectural solutions for handover based on movement, possibly implemented at L5 in the protocol stack like, e.g., the application layer architecture developed by the Blekinge Institute of Technology (BTH) research group [9].

4 Mobility Management

Mobility management refers to the problem of managing the mobility of users in the context of diverse computing and networking environments. Considerations must be given in this case to elements like location-aware services, system capacity and application demands.

There are two major elements involved in mobility management, i.e., handover management and location management [5]. Handover management refers to the way the network acts to keep mobile users connected when they move and change their position and access points in the network. For instance, in the case of UMTS, there are two types of handover: intra-cell handover and inter-cell handover. Intra-cell handover refers to the situation when the mobile user changes the communication channel to one with a better signal strength at the same Base Station (BS). Inter-cell handover occurs when a user moves from one cell to another. In this case, another BS takes over the control of the user connection.

The procedure for intra-cell and inter-cell handovers is as follows:

- The user initiates a handover procedure
- The network or the mobile unit provides necessary information
- The routing operation associated with the handover is performed
- All subsequent calls to the user are transferred from the former connection to the later one

Location management refers to the process used by a network to find out the current attachment point of a mobile user and provide call delivery. There are two phases involved in location management, namely location registration or update and paging. Location registration means that the mobile user periodically notifies the network about the new access point and the network uses this information to authenticate users and to update the location profile. Paging means that the network is queried for the user location profile so that the current position is found.

The standard solution existent today for Location Area (LA) based location update does not allow adaptation to the mobility characteristics of the mobile node. Many research efforts have therefore been done over the last years to improve the performance by designing dynamic location update mechanisms and paging algorithms. The basic idea is that these mechanisms take into consideration user mobility and accordingly optimize the signaling cost associated with location update and paging. The goal is to reduce the costs associated with these mechanisms to a minimum. Examples of such algorithms are [5]:

- Distance-based location update approach
- Time-based location update approach
- Movement-based location update approach
- Movement threshold scheme
- Information theoretic approach

A very important research issue is therefore regarding location modeling and mobility modeling and prediction. Location modeling refers to how to describe the position of a mobile user, whether it is a one-dimension or two-dimension or three-dimension system. Different methods can be used for location modeling, which depend upon the specific network infrastructure. Usually, the position of a mobile user can be specified at three levels: location area, cell ID and the position inside the cell. Furthermore, one should also mention that a more precise location modeling (i.e., within a cell or a WLAN rather than finding the residing cell) may demand for solving a so-called geo-location problem.

Mobility modeling and prediction strongly influences the choice and performance of other resource management elements like call admission control, routing and handover. A precise model for mobility offers the possibility of improving the performance of mobility prediction, with positive effects on performance. Diverse criteria can be used for mobility modeling like, e.g., dimension, scale, randomness, geographical constraints and change of parameters. The most popular models are [5]:

- Fluid-flow models
- Random-walk models and derivatives
- Random-waypoint models and derivatives
- Smooth random-mobility models
- Gaussian-Markov models
- Geographic-based models
- Group-mobility models
- Kinematic mobility models

These models have specific advantages and drawbacks, and each of them is usually used in specific cases only.

5 Connectivity Management

The extreme heterogeneity existing today with reference to access networks and network technologies has had as a consequence that the problem of mobility management has now become more complex. The fact that a handover procedure is not directly related to physical parameters like coverage and movement speed has had as a consequence that the mobility has now become a logical concept rather than a physical one. This means that today mobility refers not only to the user geographic position but also to the change of a logical location with respect to network access points. The consequence is that mobility management becomes more of a connectivity management procedure.

There are two aspects that must be considered in vertical handover. These are regarding handover at device level and handover at flow level [10]. Device level handover refers to the situation when data transfers are switched over from one network interface to another within the same mobile node. On the other hand, flow level handover refers to the situation when the network interface is selected based on the specific traffic flow and every individual traffic flow takes own handover decisions. Multi-homing handover is possible in this case when multiple network connections are simultaneously used.

There are two general classes of algorithms used in the vertical handover, which are based on [10]:

- Traditional algorithms, and
- Context based algorithms

Traditional algorithms are typically used in horizontal handover and focus mainly on L1 and L2 parameters like link quality conditions, e.g., Received Signal Strength Indicator (RSSI), Signal to Noise Ratio (SNR), frame error rate and base station workload. These parameters can be used in vertical handover as well. The target in this case is to minimize the number of unnecessary handovers while maintaining throughput and latency constraints.

Context based algorithms target at always providing best possible QoS and user-perceived Quality of Experience (QoE). High level information like user preferences, cost, application features, device capacity, bandwidth, security are considered in this case. The target is to provide the so-called "Always Best Connected (ABC)" paradigm in the handover procedure.

There are three categories of context based algorithms [10]:

- Traffic flow based algorithms
- Simple Additive Weighting (SAW) algorithms, and
- Advanced Multiple Criteria Decision Making (MCDM) algorithms

Traffic flow algorithms classify the packets based on their traffic class field, IP address, port number and protocol. Different network interfaces are assigned to different traffic flows based on the characteristics of applications, e.g., real-time and non-real-time services.

SAW-based algorithms use weights assigned to parameters considered relevant for a specific handover mechanism. Weighted sums are computed based on all normalized factor values for the specific parameters. Based on this, individual scores are computed and the network interfaces are ranked based on the scores resulted from the evaluation [11].

MCDM-based algorithms are quite sophisticated. The handover decision is treated in this case as a MCDM problem, which is solved using classical MCDM methods and including techniques like Analytic Hierarchy Process (ARP), Technique for Order Preference by Similarity to Ideal Solution (TOPSIS) and Grey Relation Analysis (GRA) [10].

It is important to mention that, in the handover decision making algorithm, the evaluation and decision are processes that can be local or distributed. Especially the case of distributed algorithms is very challenging, given that it is not only the decision making algorithm itself that must be solved but also other control mechanisms that are typical for distributed algorithms, e.g., synchronization, causality. The target in this case is to come to an optimal global decision with reference to a set of local and distributed requirements.

6 Internet Mobility

Internet mobility refers to providing support for communication continuity when an IP-based mobile node moves to different networks and it changes the point of attachment. There are in this case several basic requirements on the TCP/IP protocol stack and networks. These requirements refer to handover and location management, support for multihoming, support for current services and applications as well as security. Other important requirements related to mobility refer to minimum changes to applications, avoidance of using third-party for routing and security purposes as well as easy integration in the existent infrastructure.

The traditional TCP/IP protocol stack and networks have been designed and developed for fixed computer networks. This means that a number of limitations must be solved when further developing the system to provide support for mobility. The main limitations are particularly because of physical and link layer, IP layer, lack of cross-layer awareness and cooperation, transport layer and applications [12].

Today, wireless access techniques are typically providing mobility of homogeneous networks at link layer only. On the other hand, Internet mobility across heterogeneous networks demands for mobility support provided in higher layers as well. Furthermore, radio channels typically show limitations when compared to fixed networks. They are characterized by lower bandwidth, higher bit error rates, faded and interfered signal. These limitations degrade the performance of transport protocols.

The main limitation related to IP layer is because IP addresses play the roles of both locator and identifier. In a mobile environment the IP address of a mobile node must be changed when moving to another network to reflect the change of the point of attachment. This feature is in conflict with the situation at fixed networks, where the IP addresses never change.

Other important limitations are because of the lack of cross-layer awareness and cooperation. For instance, the congestion control mechanism of TCP is not able to distinguish packet losses due to link properties from those due to handover. Because of this, TCP does not perform well for seamless roaming. In a similar way, the lack of L2/L3 cross-layer interaction further deteriorates the performance. Another fundamental limitation of transport protocols is because they can not deal with mobility on their own.

Limitations due to improper design of applications for mobile environments are important as well. For instance, applications like Domain Name System (DNS) and Session Initiation Protocol (SIP) have characteristics that are not favorable for mobility. The best example is given by DNS, where the Fully Qualified Domain Name (FQDN) is usually statically bound to an IP address of a node. This is not favorable in the case of mobility, where mobile nodes change IP addresses. Further, the main limitation of SIP is because of the relatively large delays associated with SIP transactions.

A number of solutions have been suggested and developed to solve the problem of Internet mobility. They can be partitioned into four classes:

– Mobility support at L3, e.g., MIPv4, MIPv6, Location Independent Network Architecture for IPv6 (LIN6)
– Mobility support at L4 of type improving TCP performance for mobility (e.g., Mobile TCP - MTCP) or mobility extension to TCP (e.g., MSOCK, Mobile UDP - MUDP, Mobile SCTP - MSCTP)
– New layer between L3 and L4, where the Internet mobility is deployed, e.g., Host Identity Protocol (HIP), Multiple Address Service for Transport (MAST)
– Mobility support at L5, e.g., Dynamic Updates to DNS (DDNS), Session Initiation Protocol (SIP), MOBIKE

Detailed description of these protocols, together with their limitations, is provided in [12].

Table 2 presents an example of functions provided by different solutions existent for Internet mobility at L3, L4, new layer between L3 and L4, and L5. The following functions (needed for mobility) have been considered:

– Handover management (HO)
– Location management (LO)
– Multihoming (MH)
– Support for current services and applications (APP)
– Security protection (SEC)

It is observed that none of the available solutions fulfills all requirements for mobility. For instance, the network layer solutions do not support multihoming,

Table 2. Internet mobility and limitations [12]

	L3		L4			New layer		L5
	MIP	LIN6	TCP	UDP	SCTP	HIP	MAST	SIP
HO	★	★	★	★	★	★	★	
LO	★	★				★	★	★
MH					★	★	★	★
APP	★	★	★	★	★			
SEC	★	★	★			★	★	★

the transport layer solutions do not support location management, application layer solutions are only appropriate for specific applications and so on.

7 Network Layer Mobility

L3 mobility means that the network layer handles mobility and it can be either mobile controlled or network controlled. In the first case, the mobile node is equipped with a mobility stack and interacts with remote entities like Home Agent (HA). Network controlled mobility means that there are networking units in the network that interact with HA and perform handover related functions. It is important to mention that, even in the case of network controlled mobility, the mobile node still assists the mobility function by providing information about, e.g., signal-to-noise ratio and other specific measurement related information.

In the case of mobile controlled mobility done with, e.g., the Client MIPv6 (CMIPv6), the mobility stack in the mobile node sets up a tunnel between the Mobile Node (MN) and HA. The mobile node sends a binding update to the HA and the Correspondent Node (CN), which maps the new Care-of-Address (CoA) for the mobile node with its own home address. With the help of some route optimization procedure, the CN updates its own cache and sends traffic directly to MN instead of via HA.

Network controlled mobility avoids the overhead associated with tunneling. The price is in different forms, e.g., limited mobility domain (like in the case of cellular IP, HAWAII), use of proxies in the network like the so-called Proxy Mobile Agents (PMA) [13]. The solution with limited mobility domain still does require a mobility stack in MN. On the other hand, PMA does not demand for mobility stack in MN but rather uses the proxies on the edge routers to help performing mobility functions like binding updates to HA.

The Proxy MIPv6 (PMIPv6) based mobility is preferred when mobility is confined within a domain and also when avoiding overload of mobile nodes by setting up tunnels between MN and HA. Mobile overload means that extra processing is added and bandwidth constraints are set to the wireless hop.

8 Application Layer Mobility

Application layer mobility refers to using the application protocol Session Initiation Protocol (SIP) [13,14]. This solution offers the advantage of eliminating the need for mobility stack in mobile nodes and also does not demands for any other mobility elements in the network. Simple IP is used in this case together with a SIP protocol stack. No additional elements are needed to support application layer mobility. This solution is very suitable for applications like VoIP.

SIP-based handover has several drawbacks. These are mainly because SIP is an application protocol and therefore involves large delays in handover, due to application layer processing. There are several solutions to reduce the handover delays in this case, and one of the most efficient is to develop a tight-coupled interworking architecture like, e.g., in the case where the WLAN Access Points are integrated into the UMTS network architecture [4].

Another drawback is that this solution is not suitable for non SIP-based applications like FTP and Telnet based applications. SIP can be used to support RTP and TCP based applications. Furthermore, another drawback is because the TCP connection must be kept alive even when the underlying IP address is changed. This means that better solutions must be used in this case for TCP, like TCP Migrate [5,12]. Furthermore, it is very important that prediction is used in this case to reduce the negative effects of changing the IP address.

It is important to mention that things may become quite complicated when the mobile node and the network have different mobility protocols. The mobile node may for instance support simple IP without any mobility stack or it can be equipped with SIP or, alternatively, it can be equipped with a MIPv6 protocol stack. The network in this case needs to complement the mobile node protocol. In the case of IP protocol in the mobile node, the network does not need any other protocol. In the case of MIPv6 in the mobile node, then the network must have this protocol stack as well.

9 Conclusions

The chapter reported on several important developments and challenges related to seamless handover. These are regarding L2/L3 handover, mobility management, connectivity management, Internet mobility, network layer mobility and application layer mobility.

References

1. Vidales, P.: Seamless Mobility in 4G Systems, technical report UCAM-CL-TR-656, University of Cambridge, UK (November 2005), ISSN 1476-2986
2. Manner, J., Kojo, M.: Mobility Related Terminology, IETF RFC 3753, http://www.ietf.org
3. Frattasi, S., Fathi, H., Fitzek, F.H.P., Prasad, R., Katz, M.D.: Defining 4G Technology from the User's Perspective, IEEE Network (January/February 2006)
4. Garg, V.K.: Wireless Communications and Networking. Morgan Kaufmann, San Francisco (2007)
5. Katsaros, D., Nanopoulos, A., Manolopoulos, Y.: Wireless Information Highway. IRM Press (2005)
6. Dutta, A., Lyles, B., Schulzrinne, H., Chiba, T., Yokota, H., Idoue, A.: Generalized Modeling Framework for Handoff Analysis. In: Annual IEEE International Symposium on Personal, Indoor and Mobile Radio Communications (PIMRC 2007), Athens, Greece (September 2007)
7. Gupta, V., Williams, M.G., Johnston, D.G., McCann, S., Barber, P., Ohba, Y.: 802.21 - Overview of Standard for Media Independent Handover Services. IEEE 802 tutorial, http://ieee802.org/802_tutorials/index.html
8. IEEE, Draft IEEE Standard for Local and Metropolitan Area Networks: Media Independent Handover Services, IEEE P802.21/D04.00. IEEE (February 2007)

9. Popescu, A., Ilie, D., Erman, D., Fiedler, M., Popescu, A., De Vogeleer, K.: An Application Layer Architecture for Seamless Roaming. submitted to the Sixth International Conference on Wireless On-Demand Network Systems amd Services (WONS 2009), Snowbird, Utah, USA (February 2009)
10. Sun, J.-Z.: A Review of Vertical Handoff Algorithms for Cross-Domain Mobility. In: International Conference on Wireless Communications, Networking and Mobile Computing (WiCOM), Shanghai, China (September 2007)
11. Isaksson, L.: Seamless Communications Seamless Handover Between Wireless and Cellular Networks with Focus on Always Best Connected. PhD thesis, BTH, Karlskrona, Sweden (March 2006)
12. Le, D., Fu, X., Hogrefe, D.: A Review of Mobility Support Paradigms for the Internet. IEEE Communications Surveys and Tutorials 8(1), 1st Quarter (2006)
13. Chiba, T., Yokota, H., Idoue, A., Dutta, A., Das, S., Lin, F.J., Schulzrinne, H.: Mobility Management Schemes for Heterogeneity Support in Next Generation Wireless Networks. In: 3rd Euro-NGI Conference, Trondheim, Norway (May 2007)
14. Rosenberg, G., Schulzrinne, H., Camarillo, G., Johnston, A., Peterson, J., Sparks, R., Handley, M., Schooler, E.: SIP: Session Initiation Protocol. IETF RFC 3261, http://www.ietf.org

Optical Metropolitan Networks: Packet Format, MAC Protocols and Quality of Service

Tülin Atmaca[1] and Viet-Hung Nguyen[2]

[1] Telecom SudParis, 9 rue Charles Fourier, 91011 Evry, France
`tulin.atmaca@it-sudparis.eu`
[2] Itron, 52 rue Camille Desmoulins, 92130 Issy-les-Moulineaux, France
`viet-hung.nguyen@itron.com`

Abstract. Optical Packet Switching (OPS) is among the most promising solutions for next generation metropolitan networks. The increase of packet-based services (video on demand, etc.) is pushing metropolitan networks providers to renew their infrastructures. Today, metropolitan networks are based on SONET/SDH circuit-switched networks, which are becoming inefficient and costly to support new requirements of quality of service and bandwidth of sporadic packet-based traffic. To solve this problem, many new network solutions are proposed recently, including Next Generation SONET/SDH, Resilient Packet Ring, etc. Among others, the optical networking technology appears a good choice thanks to its following benefits: huge transmission capacity, high reliability, and high availability. This paper is devoted to provide an overview of the metropolitan network infrastructure and particularly to its evolution towards OPS networks. It also highlights performance issues in terms of optical packet format, medium access control protocol and quality of service, as well as traffic engineering issues.

Keywords: Metropolitan Area Network, Optical Packet Switching, Optical Packet Format, Medium Access Control, Quality of Service, Dynamic Intelligent MAC, Circuit Emulation Service, Time Division Multiplexing, Fairness.

1 Introduction

Today's metropolitan area networks (MANs) are faced with a significant challenge; to maintain traditional circuit services (e.g. voice) while, at the same time, enabling new, value-added packet-based services (i.e. video and data) to be carried over the same packet-based network infrastructure. This challenge is the result of the unprecedented proliferation of packet-based services, which in turn has led to a rapid growth in demand in terms of bandwidth and sophisticated quality of service (QoS) requirements in metropolitan areas. MAN service providers must therefore renew their network infrastructures to adapt to these service requirements as well as deliver the bandwidth demanded.

Transformation of metro networks infrastructure can be accomplished by increasing capacity, but more importantly by introducing new technologies with high performance gains relative to infrastructure cost. Several solutions have been proposed to enable the deployment of next-generation of metro optical networks, which allow a gradual increasing of capacity in a scalable and cost effective manner. Some of them focus on enhancing and adapting existing Synchronous Digital Hierarchy / Synchronous Optical NETwork (SDH/SONET) technologies, while others are designed specifically to serve as a substitute for SDH/SONET. While the standardization in the area of SDH/SONET is not specific to metro networks, the importance of developing new network solutions for metropolitan area is reflected by the large number of recently initiated standardization activities and industry forums as Internet Engineering Task Force (IETF) working group for IP over RPR (IPoRPR), IEEE 802.17 Resilient Packet Ring working group (RPRWG), Metro Ethernet Forum (MEF) and Resilient Packet Ring alliance. That is, an important trend in networking in metropolitan area is the migration of packet-based technology from Local Area Networks (LANs) to MANs. Therefore, in the next generation of metro networks, transport functions will migrate from SDH/SONET architecture to packet-switching transport networks, and will complement service layer features to satisfy the full range of infrastructure and service-specific requirements.

The design of a substitute for actual SDH/SONET architecture is complex because the solutions are diverse and thus the election of the best approach remains a controversial issue. Over the past few years, many studies have been dedicated to the design of architecture for the next generation of metropolitan area networks. Unfortunately, their diverging approaches and viewpoints highlight the complexity of the problem. However, the major standardization works, as supported by Institute of Electrical and Electronics Engineers (IEEE), IETF and MEF have in common the following elements. Firstly, the bandwidth access should be dynamic, flexible and provide a large spectrum of granularity. The metro networks need to handle fine granularity traffic with highly variable characteristics. Moreover, a metro network must directly interoperate with a large range of protocol types (e.g., IP, ATM, Gigabit Ethernet, Frame Relay, etc.) and thus the aggregation is a more significant function than the transport. Secondly, the network should be capable to transport all types of traffic with specific QoS. Here, the TDM support (i.e. backward compatibility) is needed despite the huge growth in data traffic. Finally, the network resiliency will play an important role in electing the best approach. There are efforts to develop solutions capable to offer a sub-50 ms network resiliency as in SDH/SONET.

In this context, the optical networking technology appears a technology of choice for the next generation MAN. The main benefit of optical technology can be summarized in the following terms: *huge transmission capacity*, *high reliability*, and *high availability*. This tutorial is devoted to provide an overview of the MAN infrastructure and particularly to its evolution towards optical packet networks during the last decades. It also highlights performance issues in optical networking in metro area in terms of optical packet format, medium access control (MAC) protocol and quality of service (QoS), as well as traffic engineering issues. We begin with a brief state-of-the-art and perspective on optical metro network. Next, we provide a number of arguments for an answer to the problem of the choice of packet format to be adopted in future metropolitan optical packet switching networks. Here, we provide performance comparison between fixed length packet and variable length packet approaches. Then, we explore the performance issues at MAC layer and present some improvements for MAC protocol. This is followed by the discussion about how to guarantee QoS in multiservice optical packet switching (OPS) metro network, illustrated by some possible mechanisms allowing the transport of circuit-based TDM traffic on packet-based networks. Finally, we finish with some conclusions.

2 Overview and Perspective on Optical Metro Network

2.1 Evolution of Optical Metro Network

In an end-to-end connectivity perspective, a MAN is a network that interconnects many LANs and provides connections with other MANs through the backbone WAN (Fig. 1). A MAN typically provides access and services in a metro region (i.e., a large campus or a city).

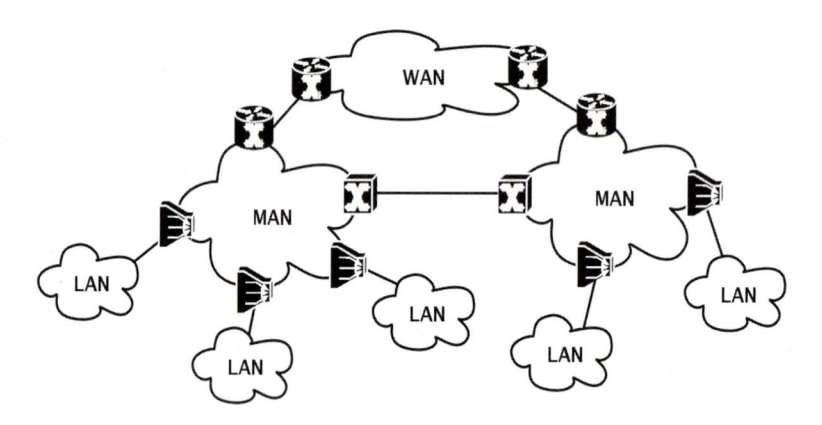

Fig. 1. Global view of metropolitan area network

Designed to be a network that services different access networks and high-bandwidth end-users (e.g., financial organizations, banking, large Internet Service Providers (ISP), etc.), a MAN needs to provide much higher bandwidth capacity and connectivity than a LAN. It should provide reliable and available services as well. The reliability and availability requirements for MAN can be expressed through the very fast restoration of service in the event of failures, today in around 50 ms. Added to that, a MAN should support services with end-to-end security such as virtual private networks (VPN) or virtual LAN (VLAN). With the ever-growing of data and video traffic today, a MAN should also offer high scalability to accommodate rapid change in terms of bandwidth demand in the metro area.

The advent of optical technology is changing the face of the MAN. There can be no doubt that optical networks offer promises to build excellent MANs that meet many of the requirements stated above. The main interest of optical technology is that it provides huge capacity for the network. In addition, optical fibers offer a much more reliable medium of transmission than copper cables. Therefore a MAN based on optical technology will provide a common and solid infrastructure over which varied services could be delivered. The evolution of optical networks is summarized in Fig. 2.

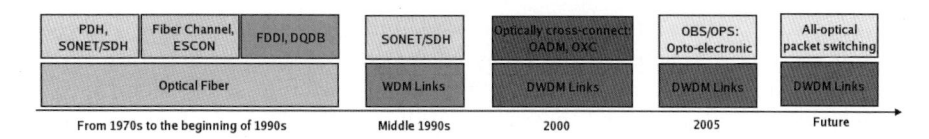

Fig. 2. Evolution of optical networks

In the 1980s, to transmit data at higher bit rate, most carriers (or service providers) deployed unidirectional optical fibers in their backbone infrastructures with point-to-point mesh topology. The processing of switching and multiplexing all remained in electronic domain. During the first half of the same decade, the Plesiochronous Digital Hierarchy (PDH) network was widely deployed. This technology has primarily focused on multiplexing digital voice circuits. The Synchronous Optical Network (SONET) systems were introduced later in North America. A closely related standard, Synchronous Digital Hierarchy (SDH), has been adopted in Europe, Japan and for most submarine links. These systems were developed to overcome some drawbacks of the PDH systems such as lack of multiplexing flexibility, point-to-point limited supervision, etc. Besides, at the end 1980s and the beginning of the 1990s, a number of standards for MAN has been proposed, namely Fiber Distributed Data Interface (FDDI) token ring, an ANSI standard [1], and Distributed Queue Dual Bus (DQDB) IEEE standard [2]. Those networks are packet-based and use ring topology. Other enterprise networks such as Fiber Channel and ESCON were developed to offer storage services to enterprises. They constitute another class of networks, existing in parallel with SONET/SDH systems.

As the demand for wider bandwidth has been increasing continuously, the Wavelength Division Multiplexing (WDM) technology was introduced in optical networks at the middle of the 1990s. WDM increases transmission bit rate by several orders of magnitude. Moreover, in addition to the considerable increase of transmission rate, the optical Add/Drop Multiplexer (OADM), providing wavelength-based add/drop function, and the optical cross-connect (OXC), offering wavelength-based cross-connect function, are the key elements of this generation of optical networks. OADM and OXC were introduced at the very beginning of the year 2000.

Recently, many researchers have studied the feasibility of performing multiplexing and switching optically. In reality, it is advantageous to adopt optical switching and routing means over electronic counterparts, as they are more amenable at higher bit rates and do not require optical-to-electrical (O/E) and electrical-to-optical (E/O) conversions. The feasibility of optical packet switching and optical burst switching networks has been studied (e.g., [3], [4], [5], and [6]).Thus the latest generation of optical networks that could appear at the end of this decade should be all-optical packet switching networks.

2.2 Challenges in Optical Metro Network

The strong increase of demand in terms of bandwidth and value-added services in both access and metropolitan networks today is pushing the service providers to innovate their existing network infrastructures. Although WDM provides huge transmission capacity, there is always a need of providing more sophisticated services to customers such as interactive games and video-on-demand and telephony over IP, which generate much more revenues than raw-bandwidth service offered by WDM. As new applications are constantly being generated, notably in the metro area in which several organizations and enterprises are directly involved, a new generation of MAN is strongly required to replace the obsolete and inappropriate MAN.

The main requirements which are challenging the metro network limits are: a sporadic incoming traffic from LANs with different QoS constraints, different client formats, cost sensitivity and finally the necessity for a scalable architecture to support increasing traffic. The rapid increasing volume of data traffic in metro network is challenging the flexibility of existing transport infrastructure based on circuit-oriented technologies like SDH/SONET. In spite of its high reliability, high QoS and standardization, SDH/SONET lacks the efficiency of transporting data traffic. On one hand, the solutions developed to adapt packet-based traffic to circuit-switching environment (i.e. Packet over SDH/SONET (PoS) and IP over ATM (IPoATM)) use inefficiently the transport infrastructure bandwidth. On the other hand, although ATM over SDH/SONET (ATMoS) offers a predictable end-to-end transport service, it leads to complex interfaces (i.e. SAR algorithms), 10 percent of capacity consumed with transport overhead, and some bandwidth loss when ATM cells are placed in fixed size SDH/SONET frames. Furthermore, the complex algorithms used to transport packet-based traffic over such architecture become difficult to implement at high-speed processing.

Over the last decade, the introduction of optical technology in metropolitan area was studied intensively. The fibre technology is considered the response to the problems of bandwidth requirements and QoS because of its huge capacity. Optical networking technology, through WDM, may represent the solution for meeting increasing bandwidth demands. Unfortunately, the introduction of all-optical WDM equipments and architectures in the metro environment for replacement of the traditional architectures, as initially envisioned, has been slowed down significantly by the current economic situation. As a result, the cost of components and the design of the network as a whole on the other hand has become extremely important and cost versus performance trade-offs now play the central role in the design and engineering of metro networks. The metropolitan networking environment with its constant and rapidly changing bandwidth requirements, cost sensitivity and different customer traffic need present a challenge for the deployment of the latest optical technologies.

2.2.1 Traffic Evolution

The traffic is changing in communication networks, notably in the metro regions. The evolution of the global traffic is led by many elements. First of all, the voice traffic obviously plays an important role in the global traffic today. Some recent observations have pointed out that although the volume of mobile voice and voice over IP continuously increases, the total volume of voice traffic tends to decrease due to the strong decrease of fixed voice traffic.

Beside this traditional traffic, the volume of multimedia video traffic becomes more and more important. The emergence of new multimedia applications such as high quality video conferencing, high-definition television on demand, etc, leads to a considerable growth of video traffic volume (e.g., video traffic is about 60% of the total traffic volume). As a result, there is a need of upgrading the access networks to very high-speed in order to offer enough bandwidth for multimedia end-users. For instance, in Asia, notably in Japan, one plans to provide up to 1 Gbs per user using FTTH (Fiber-To-The-Home) technology, in which 80% bandwidth will be reserved for high quality multimedia applications. The MAN, which connects those access networks, must be upgraded in terms of capacity and quality of service consequently.

Another element that considerably contributes to the evolution of traffic today is the evolution of Pear-to-Pear (P2P) traffic. Recent measures shown that P2P volume continues to grow rapidly (e.g., around 20% of global internet traffic in the year 2004 [7]). This growth may break the asymmetric bandwidth assumption, and also shifts traffic exchange to domestic networks due to P2P nature. This means that most of traffic exchanges in the future may be local exchange (i.e., the major traffic will circulate inside the same MAN), and may does not reach backbone network.

Last, but not least, we cannot ignore the growth of Internet traffic over the world, which mainly consists of data traffic such as WEB browsing, file transfer, hosting, etc. Globally, Internet traffic volume doubles each year. But the Internet growth rate is different in each continent. For example, in the year 2004, the Internet growth rate is about 400% in Asia, 200% in America but only 80% in Europe [8]. The proliferation of this traffic, in addition to the P2P traffic, causes an unprecedented growth of data traffic in the global traffic.

All this goes to prove that the traffic pattern is excessively changing, the voice traffic volume becomes minor compared to multimedia and data traffic volume. The MAN is facing with a real challenge: mainly transporting sporadic video and data traffic, while still being able to offer high level of quality of service for the voice traffic as it represents most revenues of service providers.

2.2.2 Limits of Traditional Circuit Switching Networks

Traditionally, metro networks are based on SONET/SDH rings employing circuit switching technology. Professional users are usually connected directly to SONET/SDH rings through their ADMs, while residential users are connected to those same rings indirectly via the central offices and the access networks. It is a well-known fact that SONET/SDH rings combined with TDM circuit switching and WDM wavelength switching (i.e., SONET over WDM) technologies can provide a huge network capacity with reliable services for metropolitan users. All the same, owing to the ever-growing of video and data traffic that has surpassed voice traffic today, SONET/SDH is facing with great challenges because of its disadvantages in supporting sporadic traffic.

The most important limitation of SONET/SDH is the lack of flexibility and scalability to deal with new demands of data service today. Essentially designed for transporting voice traffic, SONET/SDH networks are unsuitable for the transport of data traffic. Thanks to the constant rate and predictable behaviour of voice traffic, the establishment and provision of circuits were easy. Unfortunately, the volume of data traffic today exceeds that of voice traffic. This tendency is likely to continue for at least the next several years. In contrast to voice traffic, the data traffic is usually sporadic, changing and unpredictable. As a consequence, SONET/SDH must be able to provision and upgrade circuits *efficiently* and *rapidly* to support data traffic. However, this capability did not exist in the traditional design of SONET/SDH.

More specifically, SONET/SDH connections are usually established and dimensioned for a long term contract (months or years). This means that the traffic transported by these connections must be predictable and supposed to be unchanged. Such conditions are not applicable for data traffic due to its sporadic nature. Besides, these connections in most cases use dedicated circuits or wavelengths, with the granularity of PDH hierarchy (about several tens of Mbs) or wavelength bit rate (about several Gbs) in case of wavelength switching. These granularities are too coarse compared to the large variety of data rates that may be required by end-users. Furthermore, even if those circuits are reserved for voice traffic only, they will not be used during periods of inactivity of related users, whereas new users are asking for bandwidth. The implication is that the bandwidth reserved for each circuit/wavelength is often wasted or is used inefficiently.

New techniques are currently being developed to address many of these limitations. Generic Framing Procedure (GFP) [9] has been developed as a new framing for data accommodation into SDH/SONET and optical transport network (OTN). Virtual Concatenation (VC) [10] has been standardized for flexible bandwidth assignment of SDH/SONET paths. Link Capacity Adjustment Scheme (LCAS) [10] has been discussed for dynamic bandwidth allocation in support of virtual concatenation. One of the most important objectives of these new technologies is to enable flexible and reliable data transport over SDH/SONET, which is referred to as data over SDH/SONET (DoS).

The solution listed above may be sufficient to achieve a high utilization in backbone networks where the traffic flows are aggregates of many individual flows and are relatively smooth. In metro networks, however, the traffic is bursty/variable and it is desired to efficiently share the available capacity between the nodes at the time scale of individual packets (packet switching) or bursts of packets (burst switching). While the standardization efforts in the area of SONET/SDH are not specific to metro networks, the importance of metro gap is reflected by the large number of recently initiated standardization activities and industry forums such as IETF working group for IP over RPR (IPoRPR), the IEEE 802.17 resilient packet ring working group (RPRWG), the Metro Ethernet Forum (MEF), and the resilient packet ring alliance.

Last, but not least, SONET/SDH needs a very long time to provision a new circuit. As a matter of fact, it takes weeks to months to fulfill a new bandwidth request, and requires long-term contractual agreements as well. This way of deploying services now becomes obsolete and inefficient. The data traffic today is continuously increasing and varying, constantly generating new service requests. Consequently, the modern MAN should be able to deploy new services rapidly without long-term contracts. Of course, this entails that the modern MAN should provide finer granularity in terms of switching and upgrading relative to SONET/SDH. In this context, sub-wavelength switching technologies such as optical burst switching and optical packet switching, which provide fine switching granularity at a burst and packet levels, are widely studied by many researchers and service providers recently, in order to improve, not to say replace, their existing infrastructures.

2.3 Towards Optical Packet Switching Networks

The need of fine switching granularity for next generation MAN leads researchers to investigate the optical packet switching (OPS) technology that provides the finest granularity. Many recent projects regrouping both academic and industrial researchers have focused on the analysis of OPS, such as KEOPS in the year 1998 [3], WASPNET in the year 1999 [11], DAVID in the year 2003 [4] and ROM-EO [6] in the year 2005.

In OPS networks, packets are switched optically without being converted to electrical signal. The main advantage of this method is to bypass the electronic switching bottleneck and provide enabled switching solution that matches with WDM transmission capability. Fig. 3 shows the generic architecture of an OPS switch. Readers interested in the detail of OPS switch are warmly invited to refer to [3], [11], [12], [13], [4] and [5].

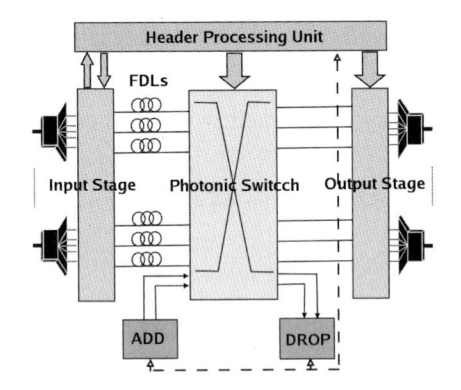

Fig. 3. A generic all-optical switch in OPS network

The main benefit of OPS relative to circuit switching approaches is that OPS results in a better network utilization since it can easily achieve a high degree of statistical multiplexing and effectively be amenable for traffic engineering. Moreover it enables packet-switching capability at high rates that cannot be contemplated using electronic devices. Nevertheless, statistical multiplexing property, which makes OPS attractive over circuit switching, introduces some important effects. Among others, the need of buffering optical packets in case of contention is typical. Contention occurs whenever two or more optical packets are trying to leave a switch from the same output port on the same wavelength. This cannot happen to circuit switching networks as an output port is always reserved for a specific circuit. How the contention is resolved has significant impact on the overall network performance.

Contention is easy to resolve in the electronic world, where electronic random access memory (RAM) is available. OPS, however, is still lacking mature technologies addressing contention issues. Owing to the lack of optical RAM, OPS typically uses fiber delay line (FDL) to delay the packets in contention and send them out at a later time. As fiber delay lines are strict fist-come-first-serve (FIFO) queues with fixed delays, they are far less efficient than electronic RAM. Switched fiber delay lines are recently developed to more efficiently resolve contention, but they require large number of fiber delay lines. Handling a large number of fibers in a switch appears to be a difficult task today. Therefore, optical RAM still remains in an early stage until now.

Other approaches are also used to address contention issue in OPS, such as deflection routing and spectral contention resolution using wavelength conversion. Deflection routing means that a packet in contention is send to a free output port other than the desired port. The disadvantage of deflection routing is that it introduces extra propagation delays and causes packets to arrive out of order. Spectral contention resolution consists in the use of wavelength converters to change the wavelength of the packets in contention. By consequent, it multiples packets can be sent simultaneously to the same output port as multiple wavelengths are generally available at each output port.

It is worthwhile to note that there currently exists another switching technology that is also gaining attention of service providers: the optical burst switching (OBS). OBS is a solution that lies between optical circuit switching and optical packet switching. It offers the switching granularity between a circuit and a packet. A burst is composed of several packets having the same destination. It provides high node throughput, hence high network utilization, and at the same time, it can be implemented with currently available technologies. It promises to be a short-term practical choice as an alternative to traditional circuit switching, before OPS becomes mature.

3 OPS Ring Network for Future MAN

Through the previous subsections of this tutorial, a clear trend one can recognize is that service providers are investigating much effort to construct common packet-based network infrastructures for the future MAN. The key elements of these infrastructures, in the long term, will be undoubtedly all-optical technologies, namely OPS and DWDM. This section is devoted to specifically introduce a good candidate for the future MAN: the optical packet switching networks using ring topology (OPSR). We begin with a discussion about the reason why the ring topology is widely adopted in MAN. Then we describe an example of OPSR network that has been studied by a number of industrials recently.

3.1 Motivation for Ring Topology

Packet switching networks using ring topology are the most common architectures in metropolitan areas. For instance, at the end of 1980s and the beginning of 1990s, there were a number of standards for MAN using ring topology such as FDDI token ring and DQDB IEEE standard. More recently, a ring architecture known as Hybrid Optoelectronic Ring Network (HORNET) has been proposed in the year 2002 [14] and, in the year 2003, a new standard for optical MAN has been approved by IEEE: the Resilient Packet Ring (RPR) [15].

One may wander why ring topology is widely used in MAN. There are actually several factors that have made ring the topology of choice. First of all, bidirectional rings inherently provide fast restoration time after events of outage such as link cut or node failure. For example, in case of a single link cut, packets can be simply redirected to transmit on the other direction of the ring to reach destinations. The highest record of restoration time in today's MAN is around 50 ms, which is typically guaranteed by SONET/SDH rings. RPR also provides this resilience feature. The fast restoration time plays an important role in providing good quality of services (QoS) to customers. It actually contributes to the degree of network availability, a typical requirement that MAN's providers must take into account when designing their networks.

There is a second argument that cannot be ignored, namely the *statistical multiplexing* of data traffic flowing from different nodes over the shared medium of transmission (also known as *collecting function* of a ring). As a MAN typically interconnects a variety of enterprises and organizations of small, medium or big size, its overall traffic is rather sporadic and heterogeneous. If the MAN uses point-to-point or meshed topology to connect their nodes, there will be cases where some links are underused, while the others are overloaded. Thus it would be better if the underused links could support part of traffic from the overloaded links. The ring topology with shared medium of transmission can effectively deal with this problem without complicated routing algorithms. As a matter of fact, since the total bandwidth of the network is shared by ring nodes, the effect of statistical multiplexing will clearly improve the resource utilization. In addition to this feature, a similar utility that a ring could provide is spatial reuse. The idea is roughly described as follows: bandwidth occupied by traffic of upstream nodes will be released at destination nodes, hence will be reused by other nodes after those destination nodes.

As far as the cost of network infrastructure is concerned, ring topology in the metro areas provides very efficient utilization of optical fibers, hence effectively reduces the infrastructure cost by using small number of optical fibers. In point-to-point or mesh topologies, the number of optical fibers is usually proportional to the number of links between nodes, whereas only one fiber may be sufficient to connect all nodes in a ring topology.

Finally, it is worthwhile to note that ring topology simplifies the routing functionality of ring nodes. Since there are either one (for unidirectional rings) or two (for bidirectional rings) paths for routing traffic, the complexity of routing function is obviously reduced relative to mesh topologies.

3.2 Example of an OPS Ring

We now introduce the architecture of an OPSR that uses opto-electronic components. The concept of this architecture mainly comes from on a recent experimental architecture, the Dual Bus Optical Ring Network (DBORN) [16]. The idea is to design an optical network that satisfies requirements for the next generation MAN, while

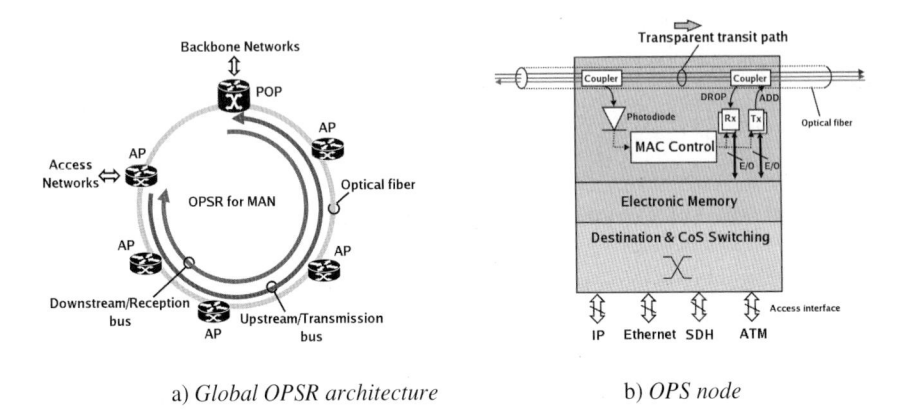

a) *Global OPSR architecture*   b) *OPS node*

Fig. 4. OPSR architecture

tremendously reducing the network cost by employing advanced technologies in optical networking.

As shown in Fig. 4, the OPSR logically consists of two unidirectional buses: a transmission (upstream) bus that provides a shared transmission medium for carrying traffic from several access point (AP) nodes to a point of presence (POP) node; and a reception (downstream) bus carrying traffic from the POP node to all AP nodes. Thus, an AP node always "writes" to the POP node employing the transmission bus and "listens" for the POP node using the reception bus. The traffic emitted on the transmission bus by an AP node is first received by the POP node, then is either switched to the reception bus to reach its destination node, or is routed to other MAN or backbone networks. The easiest way to implement these separated transmission/reception buses is to use two optical fibers: one for the transmission bus, the other for the reception bus. However, this approach will require two more fibers for the purpose of protection in case of fiber cut. Another way is to employ different wavelength bands. In this case, only two optical fibers are needed, one main fiber for normal usage and another for protection.

In this architecture, each AP node consists of two parts (Fig. 4). The electronic part has interfaces with client access networks such as IP, ATM or Ethernet. It is equipped with electrical memory where electronic packets are buffered before being converted to optical signals. The optical part mainly consists of Packet-OADM performing packet add/drop function optically. On the transmission bus, transit optical packets flowing from upstream AP nodes pass through downstream AP nodes optically (i.e., without O/E/O conversion), thanks to the use of optical couplers (Fig. 4.b). Hence transit packets at an AP node are not received or processed by the node. This is the well-known concept of passive optical network (PON), which highly reduces the number of transmitters and receivers at AP nodes. An AP node uses burst mode transmitters (Tx), which works in an asynchronous mode, to insert its optical packet on the transmission bus. Similarly, it employs burst mode receivers (Rx) to drop packets on the reception bus. Therefore, AP nodes in this architecture only need equipment (i.e., optical couplers and burst mode transponders - BMT) to process the packets to and from its local customers. On the contrary, conventional architectures, notably the ones using O/E/O conversion, such as RPR, require significantly more equipment in AP nodes because each node must receive, process and retransmit all optical packets that pass through. In addition to the advantage of offering low number of equipments, this architecture offers an important feature: Tx and Rx in an AP node are out of the transit line, making the upgrade of the node easier, and lessening the service interruption's probability.

4 Optical Packet Format: Fixed versus Variable

In optical packet switching networks, the client layer information is adapted from its original format to optical payloads and then transported transparently through the optical area. At the outgoing border, the client information are extracted from optical payloads, reconverted to their initial format, and forwarded to per-hop or final destination. Therefore, the influence of packet format is to be discussed at two levels: at the border edge and in the core of the optical networks.

4.1 Packet Format in the Core of Optical Network

The impact of packet format in the core of optical networks was intensively studied in the last few years. Numerous works have analyzed the interaction between the packet format and the performance of optical switch fabrics situated at the heart of optical networks. When the access protocol is based on variable length optical packets, the optical payload corresponds to an Ethernet Layer 2 frame. When using fixed format, the optical payload corresponds to one or several Ethernet Layer 2 frames. According to the optical packet creation policies, several upper layer services may be multiplexed onto single optical transport unit. This implies additional electronic information for packet delineation at outgoing edge. The Ethernet PDU trailers can be used for packet delineation. We note that in case of fixed format optical packet, the optical payload should be large enough to accommodate the maximum PDU (e.g., Ethernet MTU).

Here, the format of optical packet raises two questions: how to manage "optical" memories with the presence of fixed and variable packet formats and the hardware limits of optical technology in handling different formats of optical packet.

Regarding the resources contention, the packet format influences the design of Contention Resolution Mechanisms (CRMs) and sizes optical memories. In case of fixed format, the CRMs latency is limited to exploiting the WDM dimension and to exploiting multi-terabit switching planes. The variable format adds a supplementary temporal dimension to contention resolution algorithms. When all output resources are occupied, the optical payloads are buffered by mean of optical memory organized as feed-back or recirculation lines (i.e., FDL). When the optical packets have a fixed format, the FDL is equivalent to a normal queue of which traffic issues are well managed. Variable format approach leads to more complex issue [17], where the packet loss probability has shown to be a convex function of the delay granularity of FDL. Therefore, the optimum value of elementary delay line depends on distribution of optical packet sizes.

The optical processing requires a strict delineation of optical packets. When the network is slotted there is a necessity for synchronization in core nodes. The precision of synchronization is very important and can be achieved only in order of nanosecond. The utilization of variable format brings no constraints for synchronization during the optical processing. In case of switching optical payloads, the packet format with fixed size seems to be more efficient since a synchronous approach offers the possibility to adopt large switching plane organized in "pages" of constant size (at least it is the case in fast-switching ATM fabrics).

4.2 Packet Format at Edge of Optical Network

In case of optical payloads with fixed format, the complexity is placed at the edge. Optical payloads are filled with the client packets usually structured as packets with variable format. If the length of the optical packet payload is too short, the client layer packet is segmented and sent in several optical packets. At the outgoing edge, the packets are reassembled. Bandwidth loss comes from the additional electronic header carrying the additional information needed to reassemble the packet, and also from the last optical packet presumably partial filled. On the contrary, if the optical packet is large enough, several blocks are aggregated in it. The details of the optical payload content should be stored in the optical header which leads to bandwidth loss and some bandwidth from optical payload may not be used. An additional delay is incurred, corresponding to the time to fill the optical packet.

A large optical packet seems to be a good solution since 50 percent of Internet packets are small packets, so the aggregation is an advantage as it does not constrain the capacity of core nodes. This solution improves the effective bandwidth used. However this solution may prevent the clients from efficiently using the network resources in case of low and unbalanced network load conditions, when a timer mechanism is used, this avoids the increase in delay introduced by optical packet creation process. There are many timer-based mechanisms that were proposed in the literature ([18], [19] and [20]).

In case of variable length optical packet, segmentation may still be needed when the upper layer packet size is larger than the network Maximum Transmission Unit (MTU). But the use of variable format avoids the complex processes of aggregation (at egress node) and packet extraction from optical payload (at ingress node). The gain in throughput is presumably related to the optical header size. The optical packet creation delay is given by the electronic to optical conversion and depends on the wavelength bit rate.

As far as the E/O interface is concerned, an optical packet with fixed format guarantees better performance than the variable format. First of all, this solution provides a good efficiency of adapting client information to optical layer only when large optical payloads are used. Therefore, a SAR mechanism was proposed in order to assure good utilization of network resources regardless the size of optical payload. Here, the segmentation is done "on demand" implying lower overhead and less complexity. Next, a fixed size of optical payloads may avoid scalability problems when the pattern of electronic packet size changes. Furthermore, an eventual system upgrade (i.e. passing from a wavelength modulated at 10 Gbps to 40 Gbps) leads to an increase in utilization of network resources.

Unfortunately, the optical systems employing fixed size of transport unit contain several parameters to be configured and this is not an easy task in presence of traffic with different QoS constraints. One important point is defining the size of optical packet which has an impact on the performance. Works done in [21], [22], [23] highlighted that the delay exhibited by clients leads to a complex behavior, where the queuing delay is a convex function of system utilization. At low loads, the optical packet creation may lead to poor delay characteristics. Furthermore, queuing delay grows up as optical payload increases. For example, a voice flow of 64 Kbps fills up a payload of 5 μs (i.e. 50 Kbit at 10 Gbps) in 781 ms! To avoid such situation, timer mechanisms should be deployed in order to efficiently prevent electronic information from starvation. Yet, a timer mechanism introduces a high quantity of overload [23] and hence underutilizes system resources. In congestion region, optical payloads partially filled introduce an overload that influences negatively the system performance. Thus, it is essential to use large optical payloads.

A variable format of optical packet leads to simple E/O interface. There are two parameters involved in the performance of optical packet creation. The first one is represented by the distribution of packet sizes in client networks. We have shown [23] that the performance of the interface based on transport unit with variable format is very sensitive to electronic packet size distribution. An eventual change in this distribution may lead to serious problems of scalability. The second one is the wavelength capacity and hence depends on transport infrastructure. Here, an increase in optical channel capacity reduces the efficiency of transporting upper layer clients and thus requires an upgrade of system resources.

The last step in electing the format of packet to be deployed in next generation of metropolitan optical packet switching networks, questions about the interaction between the packet format and the end-to-end network performance. Therefore, it is expressed by the interaction between packet format and the performance of MAC protocol. In this context, the discussion is reduced to an analysis of slotted and unslotted schemes of access protocol [24].

5 Performance Issues on MAC Protocol

5.1 Optical MAC Protocol

5.1.1 Traditional MAC

The MAC protocol is specially required to control the access of multiple nodes to the shared resources. Its main function is to detect or avoid collisions in packet insertion process to efficiently exploit the available bandwidth of the medium. Globally, there are two major categories of MAC protocols that are extensively deployed in the networks, namely static channel allocation using Time Division Multiplexing (TDM), and dynamic channel allocation using Carrier Sense Multiple Access (CSMA). In TDM, bandwidth is divided into fixed time slots which are allocated to different users. Every user can transmit on his allocated time slots only. This leads to a waste of bandwidth when reserved time slots are not used because the user has no data to transmit. TDM also requires global clock synchronization on the network. What is more, this protocol may do not work well with data traffic due to its sporadic nature.

The CSMA protocol is used when a node can listen for the occupation state of the carrier (transmission medium) and act accordingly. This protocol can work either in the synchronous mode (i.e., *slotted* CSMA) or in the asynchronous mode (i.e., *unslotted* CSMA). There are several versions of this protocol, including the CSMA with Collision Detection (CSMA/CD) and CSMA with Collision Avoidance (CSMA/CA). The main advantage of those protocols is their simplicity that translates into low cost of implementation, management and maintenance. Moreover, they are appropriate to support sporadic data traffic. In particular, the unslotted CSMA offers the capability of supporting variable length packets without complex and costly segmentation/assembly processes.

5.1.2 Optical Unslotted (OU-)CSMA/CA Protocol

The OU-CSMA/CA protocol is a derivation of the CSMA/CA, which has been widely used in wireless networks. In addition to the simplicity property inherited from CSMA, the CSMA/CA protocol provides more efficient resource utilization thanks to avoidance of collision. Recently, there is a number of works that extends this protocol to the optical domain, such as in HORNET's first version [14] and DBORN [16].

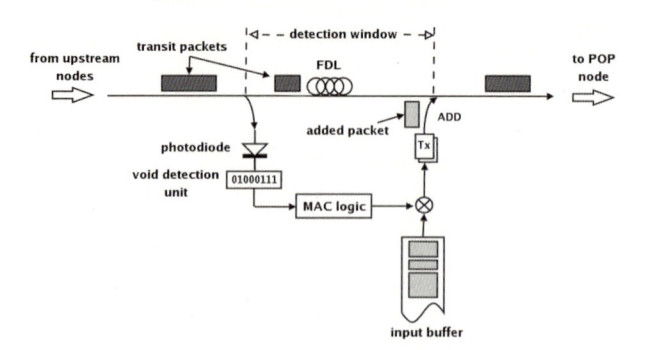

Fig. 5. OU-CSMA/CA schema

The network that uses OU-CSMA/CA protocol obviously supports variable length packets thanks to its asynchronous nature.

Fig. 5 schematizes a possible implementation of the OU-CSMA/CA protocol. The OU-CSMA/CA protocol functions based on the detection of idle periods (we also call it *voids*) on the transmission wavelength shared by several nodes. To detect activity on one wavelength, a node uses low bit rate photodiode, typically 155 MHz. Thanks to an optical coupler; the incoming signal is separated into two identical signals: the main transit signal that goes straightforwardly, and its copy that is received by the photodiode. The photodiode sends control information to the MAC logic which will inform the transmitters about whether it can transmit. The avoidance of collision is performed by employing a fiber delay line (FDL), which creates on the transit line a fixed delay between the control and the add/drop functions. This FDL should be long enough to provide the MAC logic with sufficient time to listen and to measure the medium occupancy. The FDL storage capacity should be at least larger than the maximum transmission unit (MTU) of the transport protocol used. For instance, if the network is used to transport Ethernet packets, then the storage capacity of the FDL should be greater than 1500 bytes, which is the Ethernet MTU.

The OU-CSMA/CA is appropriate to be used in a ring network such as the OPSR network above described, where there is problem of bandwidth sharing among ring nodes (e.g., on the transmission bus). Due to interdependence among ring nodes, an exact performance analysis of OPSR networks using OU-CSMA/CA protocol is difficult. However approximate methods may be used to assess the performance of such system. Among others, authors of [25], [26] have approximately analyzed the performance of this system using priority queuing theory. Globally, the authors have identified two main performance characteristics of OU-CSMA/CA protocol in OPSR networks. The first one is unfairness among ring nodes due to positional priority: upstream nodes (i.e., the nodes closest to the beginning of the shared transmission medium) might monopolize all the bandwidth, and prevent downstream nodes from transmitting. For instance, Fig. 6 (extracted from [25]) shows that the mean response time is likely to increase rapidly as we move downstream on the shared bus.

The second one is the bandwidth fragmentation due to the asynchronous nature of the OU-CSMA/CA protocol, resulting in low resource utilization. Indeed, the asynchronous transmission of packets at upstream nodes may fragment the global

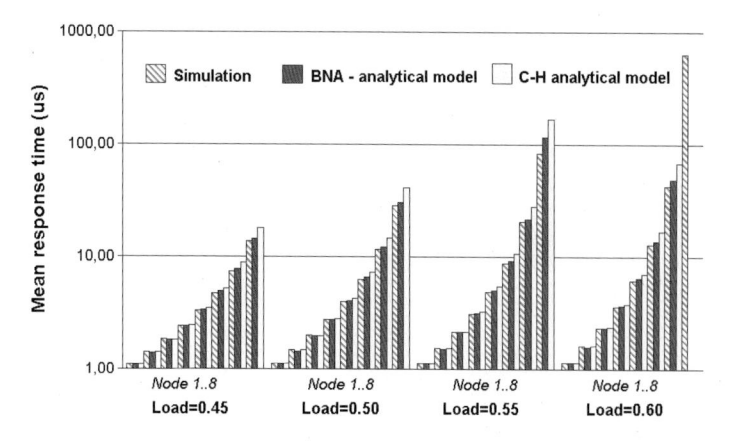

Optical Metropolitan Networks: Packet Format, MAC Protocols and Quality of Service 823

Fig. 6. Mean response time at ring nodes as function of ring load: "real-life" Internet traffic assumption

bandwidth into small voids unusable for the transmission of downstream nodes. The implication is that the network resource is used inefficiently, and the acceptable ring load of such system is limited to some 60% of the total network bandwidth as shown in Fig. 6.

5.2 Enhanced MAC Protocol

Since the combination of OU-CSMA/CA and OPSR network provides limiting performance as shown above, many researches have been carried out to find new access schemes enhancing the performance of such system. This subsection describes some new enhanced access schemes that provide remarkably higher network performance.

5.2.1 Packet Concatenation Mechanism

The transmission efficiency obviously appears one of great importance in exploiting network transmission resources. The transmission efficiency can be roughly defined as the ratio of useful bandwidth (i.e., the bandwidth occupied by client payload) to the effective bandwidth that the network must use to transport client payload. Of course, the effective bandwidth includes the client payload as well as possible overheads (headers and guard bands) needed for routing, signaling and header processing purposes. Thus, in order to increase the transmission efficiency of a given network, it is essential to reduce the total volume of overheads used to transport the volume of client data.

In the OPSR network considered, ring nodes use burst mode transceivers (BMT) to only communicate with the POP node on the transmission bus. Thus the optical header is reduced to a simple bit pattern for synchronization purpose. Therefore, in [27], a new mechanism called Modified Packet Bursting (MPB) has been proposed to increase the transmission efficiency of such network by suppressing unnecessary optical overheads. The concept of MPB mechanism relies on the basic concept of packet bursting in Gigabit Ethernet (GPB [28]), but with further improvements.

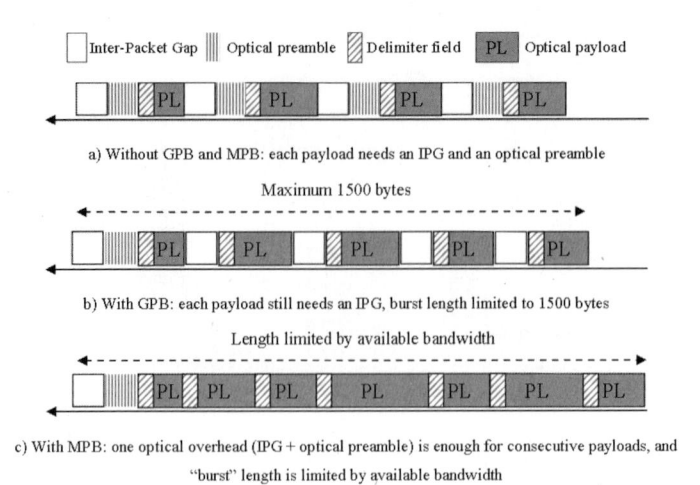

Fig. 7. Sequence of payloads transmitted consecutively with and without MPB

With MPB, in order to maximize the transmission efficiency, payloads (having the same destination) transmitted by a given node are concatenated to form a big "optical burst" consisting of one optical overhead followed by several payloads.

The principle of MPB is explained in Fig. 7. With MPB, if a node has a number of payloads to transmit, the first payload is transmitted with an optical overhead to initialize the "burst"; other payloads are then transmitted consecutively without any gap or optical overhead. Only one delimiter field is required at the beginning of each payload to separate two consecutive payloads. At the receiver side, the optical overhead is processed and consecutive payloads are extracted thanks to that delimiter field. An important point in this concept is that the "burst" length can be extended as long as the node has client payloads to transmit (including those arrive after the transmission of the first payloads) and the bandwidth is still available. The "burst" ends only when the node buffer becomes empty or there is no more available bandwidth.

Additionally, MPB also introduces the *bursting timer* principle to improve network performance. Actually, MPB uses a *bursting timer* (with BT_size parameter) defining the period of gathering client payloads for MPB. The longer the duration of this *gathering period*, the higher the probability of having more than one client payload cumulated in the buffer of the transmission node, hence the higher the transmission efficiency thanks to the concatenation of these payloads. Of course, a trade-off between transmission efficiency and access delay introduced by the bursting timer should be considered in MPB.

Regarding the performance of MPB versus OU-CSMA/CA scheme, it was pointed out that MPB effectively reduce the wasted bandwidth due to optical overhead, while providing satisfying performance for all ring nodes. For example, the performance results shown in Fig. 8 indicate that with a big enough *bursting timer* value, MPB may provide almost the same small mean response time at all nodes, while successfully transferring all client offered traffic with a negligible volume of optical overheads (e.g. around 0.4% against 5% in case with OU-CSMA/CA).

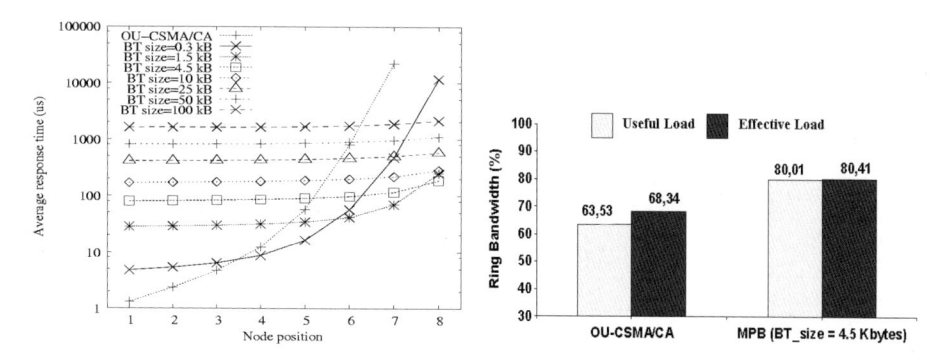

a) Avg. resp. time at each node vs. *BT_size*   b) Useful and eff. loads: CSMA/CA vs. MPB

Fig. 8. MPB performance under offered ring load of 0.80 and real-life Internet traffic assumption

5.2.2 Fair Access Protocols

In previous sections we have identified that OU-CSMA/CA scheme suffers from unfair bandwidth sharing due to positional priority, and bandwidth fragmentation problem due to asynchronous packet insertion. For unslotted ring networks using CSMA-type protocols, there are few works in the literature addressing their fairness issue, due to the complexity in analyzing such systems by either simulation or mathematical methods. There are generally two ways of performing fairness control in a network: *centralized* and *distributed fairness schemes*. The former usually requires ring nodes to exchange control information among them. This may need dedicated control channel for transmitting feedback messages (e.g. [29]) or ring nodes must be able to handle transit traffic (e.g. [15] and [30]). This may also require a master node that controls the overall network state and parameterizes the operation of ring nodes (e.g. [26]). The main drawback of the centralized scheme for controlling fairness is that it may take a long time (due to control information exchange) for reacting to the change of network state.

Regarding the distributed fairness scheme, there are few works employing this approach, mostly owing to the difficulty of guaranteeing fairness among network nodes without knowing the global network state. Indeed, a distributed fairness scheme requires each node in the network to be able to improve the global network fairness while operating independently based on local knowledge only. The advantage of such approach is that a distributed fairness algorithm may immediately and adaptively act according to the change of the node state. This property makes distributed fairness scheme attractive to be used in modern MAN network where the traffic and the bandwidth demand are excessively changing/growing at an unprecedented rate.

An example of a distributed fairness scheme is the *Dynamic Intelligent MAC (DI-MAC)* protocol described in [31]. This protocol addresses both unfairness and bandwidth fragmentation issues in the OPSR network in question. The main idea of DI-MAC is to avoid inefficient bandwidth fragmentation by intelligently spacing out the transmission of local packets at upstream nodes, hence preserving usable voids for the transmission of downstream nodes.

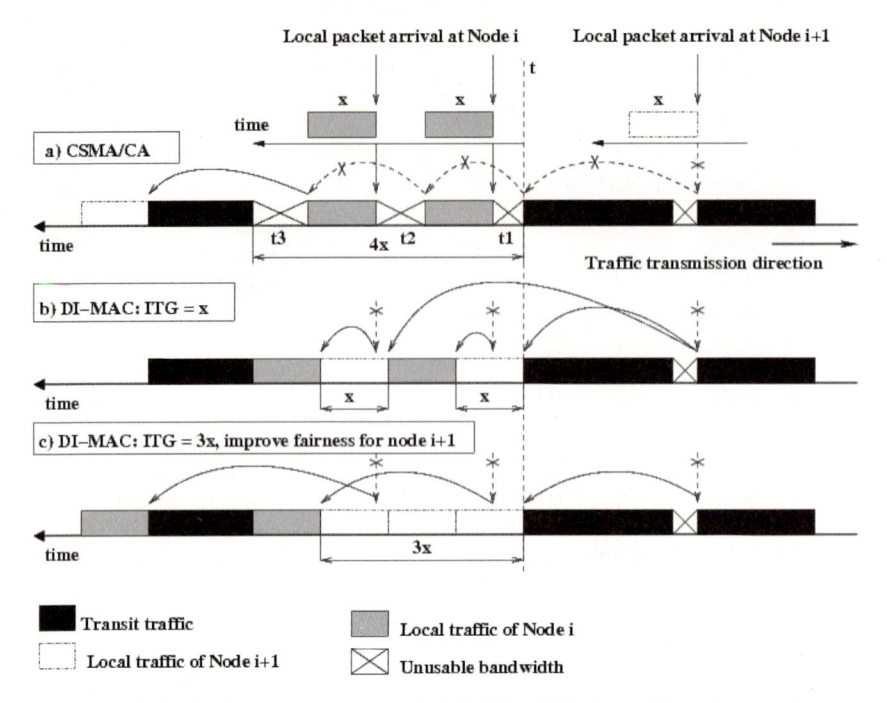

Fig. 9. Packet insertion process with OU-CSMA/CA (a) and DI-MAC (b, c)

The packet insertion process using OU-CSMA/CA at upstream nodes fragments the shared bandwidth into voids of variable lengths. Fig. 9 shows a state of the shared bandwidth at a given moment t when the node i observes a void of size $4x$, where x is the MTU of the transmission protocol used. Note that x is a metric that can be interpreted in time or bits. We suppose that in this example node i and node $i+1$ transmit packets of size x only, and at instant t their local buffers are empty. In OU-CSMA/CA case (Fig. 9-a), the asynchronous insertion of two packets at node i fragments the big void $4x$ into three small voids (whose size is t_1, t_2 and $t_3 < x$ respectively) that are not usable for the node $i+1$ to insert its packets of size x.

In Fig. 9-b, DI-MAC at node i intelligently fragments the big void into usable voids for the packet insertion of downstream nodes. Indeed, the transmission of the first local packet (the packet arrives at $t + t_1$) of node i is delayed, and it is transmitted only when a large enough void (e.g., equal to x) has been reserved for the downstream node $i+1$. The terms "large enough" mean that the void reserved by node i should allow the insertion of a maximum size packet of node $i+1$, ... The transmission of the second packet, in turn, is also delayed and carried out only when node i reserves a void equal to x after the successful transmission of the first packet. In such a way, the node $i+1$ becomes able to transmit its packets. The void reserved by

DI-MAC, through the control of the inter-transmission of local packets, is called Inter-Transmission Gap (ITG). ITG is a local parameter of DI-MAC implemented at each ring node. In Fig. 9-b, ITG is set to x. It is clear that ITG should at least equal to the MTU of the transmission protocol used, since this helps a node to release a void that is always usable for downstream nodes.

DI-MAC also provides a simple way to address fairness issue in unslotted OPSR network, without any additional control wavelength or synchronization operation. Indeed, the fairness enhancement can be done by simply increasing the ITG value at upstream nodes. Fig. 9-c shows an example with ITG equal to $3x$, where downstream nodes have larger bandwidth for transmitting their traffic.

It is worth mentioning that to effectively provide more bandwidth for downstream nodes, the first upstream node plays an important role in the operation of DI-MAC. Indeed, the first upstream node begins the fragmentation of the shared bandwidth. Thus the packet insertion operation of all other nodes strongly depends on the length of voids left between the inter-departures of local packets at the most upstream node. For instance, it is easy to see that the bandwidth state viewed be node 2 in case with DI-MAC will be a series of occupied and free bandwidth (voids) with void length greater or equal to ITG value of node 1.

Clearly, the choice of ITG value at each node strongly influences the performance of DI-MAC. Authors of [31] have proposed a dynamic distributed algorithm for controlling the ITG value at each node. Its principle is summarized as follows. Each node must always try to increase its ITG value whenever possible to release more bandwidth for downstream nodes (if any); at the same time, a node must try to guarantee that the waiting time of its client packets does not exceed a predefined threshold W_{max}, which may be freely chosen or deduced from the QoS parameters offered by service providers (e.g., from Service Level Agreement - SLA). This means that if a node observes that its client packets have been waited for a time considered too long, it must decrease the value of its own ITG instead of increasing it. The implication is

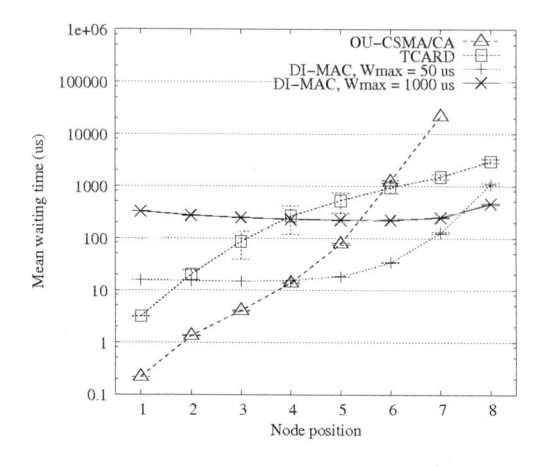

Fig. 10. Mean waiting time at each node with different access schemes: offered ring load = 0.80, uniform traffic pattern on the ring

that each ring node will be able to adjust its own ITG value that will have neither too big value nor too small value due to the ITG increasing process alternating with the ITG decreasing process. With this algorithm, DI-MAC remains fully-distributed since it requires only knowledge about the standard QoS parameters offered by service providers. And, it possesses the dynamic property since its ITG value changes with the variation of local packets waiting time, which clearly reflects the variation of bandwidth occupation state and of local packer arrival process.

Performance analysis of DI-MAC has shown that DI-MAC considerably improves the network performance (much higher than that obtained with OU-CSMA/CA) both in terms of resource utilization and performance parameters such as loss and delay. For instance, Fig. 10 plots the mean waiting time at each node obtained with DI-MAC, with OU-CSMA/CA and with a centralized mechanism TCARD [26]. With a big value of W_{max} (e.g., 1 ms), DI-MAC provides almost the same small mean waiting time for all nodes, hence offering the highest fairness level (and performance as well) to the network. Furthermore, this mechanism renders the network performance more stable and almost insensitive to network configuration and traffic change [31].

6 Guarantee of QoS in Multiservice Optical Metro Ring

Delivering services with guaranteed quality to users is an essential requirement for a metropolitan network. A metropolitan network typically interconnects enterprises, organizations and academic campus generating a variety of applications that might require quality-guaranteed services such as delay-sensitive (real-time) applications (voice, video conferencing, ...), loss-sensitive applications (banking transaction, critical data transfer, ...), or quality-non-guaranteed services such as best-effort Internet applications. The volume of data traffic today largely exceeds that of voice traffic. However, most revenues of service providers are still generated by voice service. Hence it points to a burning topic that MAN service providers should seek to construct novel common packet-based network infrastructure, which possesses the capability of supporting data traffic efficiently while delivering the same quality for voice service as in traditional TDM-based networks (e.g. SONET/SDH) to metropolitan users. This section is devoted to discuss some issues on the guarantee of QoS in a multiservice OPSR network.

6.1 QoS Architecture and Service Mapping

The global metropolitan traffic can be divided into two main categories: real-time traffic (e.g. TDM) and non-real-time traffic (e.g. data). The non-real-time traffic can be divided, in turn, into QoS-guarantee traffic (e.g. loss-sensitive data applications) and QoS-non-guarantee traffic (e.g. Internet best-effort traffic). These three traffic categories (or classes of service - CoS) usually cover all types of traffic that a MAN might be supposed to transport.

A metropolitan network like OPSR might interconnect many types of client networks, such as IP, SONET/SDH, ATM, Frame Relay..., each provides its own quality of service. Thus it is important to define a rule for the mapping of the services

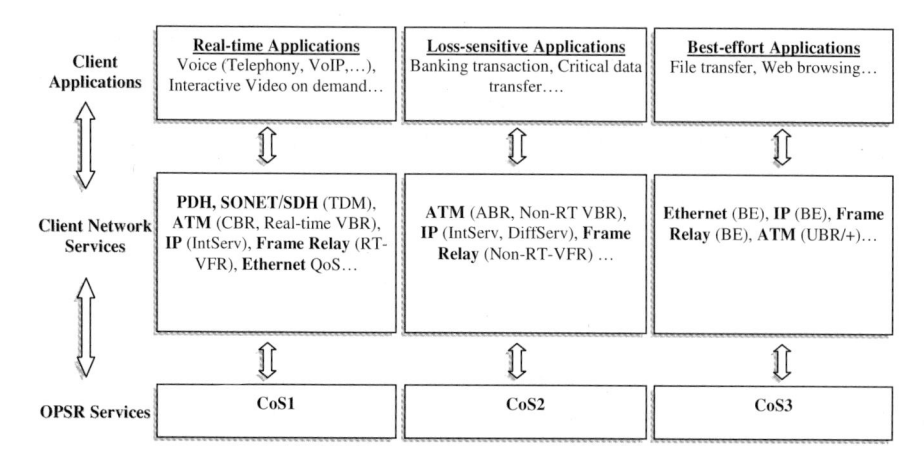

Fig. 11. Service mapping in OPSR network

	Real-time Applications	Loss-sensitive Applications	Best-effort Applications
Client Applications	Voice (Telephony, VoIP,…), Interactive Video on demand…	Banking transaction, Critical data transfer….	File transfer, Web browsing…
Client Network Services	PDH, SONET/SDH (TDM), ATM (CBR, Real-time VBR), IP (IntServ), Frame Relay (RT-VFR), Ethernet QoS…	ATM (ABR, Non-RT VBR), IP (IntServ, DiffServ), Frame Relay (Non-RT-VFR) …	Ethernet (BE), IP (BE), Frame Relay (BE), ATM (UBR/+)…
OPSR Services	CoS1	CoS2	CoS3

provided by those client networks to the services offered by the OPSR network itself. Based on the assumptions on QoS architecture stated above, a possible service mapping solution is described in Fig. 11.

6.2 Transport of TDM Circuit across Packet-Based Networks

Currently, two main efforts exist in telecommunication research community, trying to guarantee the QoS to some types of traffic that a metro network may transport, mainly the real-time TDM traffic. The design of metropolitan architecture employing ring topology and OU-CSMA/CA protocol faces the fairness issue where the traffic transmitted at upstream nodes influences the performance of downstream nodes. To protect TDM-based flows performance from the influence of data traffic circulating on the bus, new protocols may be needed. Thus, hybrid protocols, which combines feature of CSMA/CA and token-passing-type, may represent a possible solution [32].

Another approach being supported by many standard organisations is to consider that the convergence between circuit- and packet-switched services should be done at the "packet" layer and therefore they concentrate on defining the functionalities and requirements at the interface between these two worlds. The technology allowing the convergence of circuit- and packet-switched services over a packet-based network is called Circuit Emulation Service (CES). This section focuses on this technology.

6.2.1 Circuit Emulation Service

Globally, Circuit Emulation Service is a technology allowing the transport of TDM service such as PDH (E1/T1/E3/T3) as well as SONET/SDH circuit over packet switching networks. The main intention of CES is to make the packet switching network behave as a standard TDM-based SONET/SDH/PDH network as seen from the customer's point of view. Thus CES should allow customers to be able to use the same existing TDM equipment, regardless of whether their traffic is carried by standard SONET/SDH/PDH network or a packet switching network using CES.

CES is supported by a number of organisms, including the Internet Engineering Task Force (IETF) [33], the Metro Ethernet Forum (MEF) [34], the International Telecommunication Union (ITU) and the Multi-Protocol Label Switching, Frame Relay and ATM alliance (MFA forum) [35], [36]. There are no important differences between the CES standards being defined by these organisms. They actually address different layers within the network (e.g. IP, MPLS, Ethernet...), and emphasize different aspects of the CES depending on the specific services they are concerned with. It is worth mentioning that using CES to transport voice across an IP network is not like voice over IP (VoIP) technology. Indeed, VoIP is used for transporting voice only, and it requires complex signaling protocols such as H.323 or SIP (Session Initial Protocol). In contrast CES can be used for transporting voice, video, and data over many types of packet switching network (not necessary IP network). Moreover it generally does not need gateway signaling as in VoIP, and requires low processing latency. Many believe that CES will displace VoIP in the future thanks to its efficiency and flexibility.

The reference model for CES on the metro OPSR network is based on the global model for circuit emulation described in IETF RFC3985 [37] and MEF3 specification [34]. Fig. 12 presents the reference model of CES for OPSR network. Generally speaking, we have two TDM customers' edges (CE) communicating via OPSR. One CE is connected to an AP node (ingress CE), the other CE is connected to the POP node (egress CE). TDM service generated by ingress CE is emulated across the OPSR network to egress CE. The emulated TDM service between two CEs is managed by two inter-working functions (IWF) implemented at appropriate nodes.

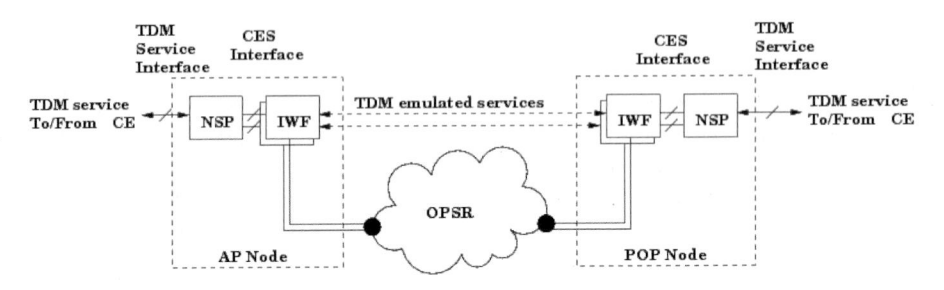

Fig. 12. Reference model for CES on metropolitan OPSR network

Operation Modes. CES has two principal modes of operation. In the first one, called "unstructured" or "structure-agnostic" emulation mode, the entire TDM service bandwidth is emulated transparently, including framing and overhead present. The frame structure of TDM service is ignored. The ingress bit stream is encapsulated into an emulated TDM flow (also called CES flow) and is identically reproduced at the egress side. The second mode, called "structured" or "structure-aware" emulation, requires the knowledge of TDM frame structure being emulated. In this mode, individual TDM frames are visible and are byte aligned in order to preserve the frame structure. "Structured" mode allows frame-by-frame treatment, permitting overhead stripping, and flow multiplexing/demultiplexing. This means that a single "structured" TDM service may be decomposed into two or more CES flows, or two or more "structured" TDM services may be combined to create a single CES flow as well.

Functional Blocks. In the reference model of CES, the Native Service Processing (NSP) block performs some necessary operations (in TDM domain) on native TDM service such as overhead treatment or flow multiplexing /demultiplexing, terminating the native TDM service coming/going from/to CE. For instance, as the "unstructured" TDM service does not need framing treatment, it might not be handled by the NSP and can pass directly to the IWF block for emulation. However, the "structured" TDM service should be treated by the NSP block before going to the IWF block. Actually, the NSP could be the standard SONET/SDH framer or map per, or other propriety products.

The Inter-Working Function (IWF) block could be considered as an adaptation function that interfaces the CES application to the OPSR layer. This means that the CES technology could be considered as an application service that uses the OPSR network as a virtual wire between two TDM networks. Thus, the IWF block is responsible for ensuring a good operation of the emulated service. The main functions of IWF are to encapsulate TDM service in transport packets, to perform TDM service synchronization, sequencing, signalling, and to monitor performance parameters of emulated TDM service. Each TDM emulated service requires a pair of IWF installed respectively at ingress and egress sides.

TDM Frame Segmentation. A TDM frame would ideally be relayed across the emulated TDM service as a single unit. However, when the combined length of TDM frame and its associated header exceeds the MTU length of the transmission protocol used, a segmentation and re-assembly process should be performed in order to deliver TDM service over the OPSR network. By this consideration therefore, two segmentation mechanisms have been proposed in [38]. The first one, called *dynamic segmentation*, fragments a TDM frame into smaller segments according to void length detected on the transmission wavelength. This approach promises a good use of wavelength bandwidth, but is technically complex to implement. The second segmentation method is *static segmentation*, which segments the TDM frame according to a predefined threshold. This technique is simple to implement, and it provides resulting TDM segments with predictable size. Thus current TDM monitoring methods could be reused, simplifying the management of CES. Additionally, in [39] authors have recommended some rules to determine the segmentation threshold for TDM frames. First, the segmentation threshold should be either an integer multiple or an integer divisor of the TDM payload size. Second, for unstructured E1 and DS1 services, the segmentation threshold for E1 could be 256 bytes (i.e., multiplexing of 8 native E1 frames), and for DS1 could be 193 bytes (i.e., multiplexing of 8 native DS1 frames).

6.2.2 Performances of Circuit Emulation

Fig. 13 illustrates performance parameters of the OPSR network measured at the offered network load of 0.80 using different access schemes ([25]). Globally, OU-CSMA/CA and TCARD schemes may provide satisfying QoS for TDM service, but at the expense of degrading or even losing low priority traffic. However, DI-MAC scheme seems to support better all types of service. It effectively guarantees expected QoS for TDM service, and, at the same time provides low loss and low packet delay for lower priority services.

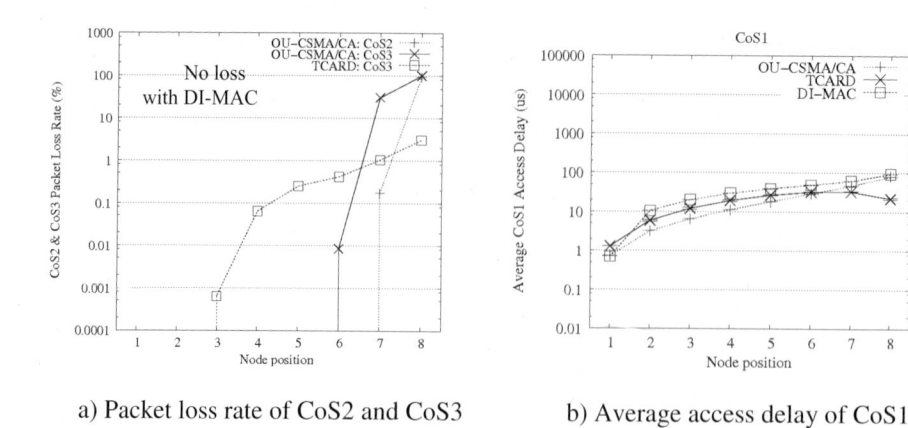

a) Packet loss rate of CoS2 and CoS3   b) Average access delay of CoS1

Fig. 13. Performance results on CES: offered ring load = 0.80, TDM segmentation threshold = 810 bytes

7 Summary

We have provided in this paper an overview of the optical metropolitan networks. The introduction of new optical networking into the future MAN to improve, not to say replace, the current circuit-based networks is becoming more and more urgent due to the rapid change of traffic and of user demand in today's MAN. Among many possible solutions, the OPS technology was identified as a good candidate for the future MAN thanks to its flexibility, scalability and cost-efficiency.

Many technological aspects of OPS networks have been investigated in this paper, namely the optical packet format issues, the performance improvements of MAC protocol and the guarantee of QoS in a multi-service environment. Globally, the choice of optical format has important impact on the network performance. Numerous studies have shown that there is no a unique optimal choice for optical packet format, but this choice depends on a number of factors: the traffic profile, the different QoS requirements, the network segment (edge or core), etc. After the packet format, comes the MAC protocol that is an essential issue in the design of a MAN. The aim of a MAC protocol is to control the multiple accesses to shared resources in such a way that the network resources are exploited best. A simple and cost-efficient MAC protocol that was proposed for metro OPS ring networks is the optical unslotted CSMA/CA protocol. Due to its limit in terms of performance, this protocol needs further improvements to overcome some drawbacks, namely the inefficient bandwidth utilization due to bandwidth segmentation and the unfairness among ring nodes. These issues can be resolved using packet concatenation mechanisms (e.g. Modified Packet Bursting) or / and fairness control schemes (e.g. Dynamic Intelligent MAC), which were proved efficient and robust in increasing network performance. Finally, these schemes combined with the Circuit Emulation Service technology are able to provide service distinction for multi-service optical metro networks. More specifically, with CES technology, a metro OPS network is able to support sporadic packet-based services while guaranteeing TDM-like QoS for circuit-based services.

References

1. FDDI Token Ring – Media Access Control. ANSI X3. 139-1987 American National Standards Institute (1986)
2. IEEE standard for Local and Metropolitan Area Networks - Distributed Queue Dual Bus (DQDB) Subnetwork of a Metropolitan Area Network (MAN). IEEE STD 802.6-1990 (1991)
3. Gambini, P., Renaud, M., Guillemot, C., Callegati, F., Andonovic, I., Bostica, B., Chiaroni, D., Corazza, G., Danielsen, S.L., Gravey, P., Hansen, P.B., Henry, M., Janz, C., Kloch, A., Krähenbühl, R., Raffaelli, C., Schilling, M., Talneau, A., Zucchelli, L.: Transparent Optical Packet Switching: Network Architecture and Demonstrators in the KEOPS Project. IEEE Journal on Selected Areas in Communications 16(7) (1998)
4. Stavdas, A., Sygletos, S., O'Mahoney, M., Lee, H.L., Matrakidis, C., Dupas, A.: IST-DAVID: Concept Presentation and Physical Layer Modeling of the Metropolitan Area Network. Journal of Lightwave Technology 21(2) (2003)
5. Dixit, S.: IP over WDM: Building the Next Generation Optical Internet. John Wiley & Sons, New Jersey (2003)
6. ROM-EO – Réseau Optique Multi-Service Exploitant des Techniques Electro-optiques et Opto-electroniques,
   http://www.telecom.gouv.fr/rnrt/rnrt/projets/res_02_52.htm
7. Karagiannis, T., Broido, A., Faloutsos, M., Claffy, K.C.: Transport Layer Identification of P2P Traffic. In: Internet Measurement Conference 2004 (2004)
8. TeleGeography Report of Global Internet Geography,
   http://www.telegeography.net
9. Generic Frame Protocol (GFP). G. 7041 ITU-T Standard (2001)
10. Link Capacity Adjustment Scheme (LCAS). G.7042 ITU-T Standard (2001)
11. Hunter, D.K., Nizam, M.H., Chia, M.C., Andonovicand, I., Guild, K.M., Tzanakaki, A., O'Mahony, M.J., Bainbridge, J.D., Stephens, M.F.C., Penty, R.V., White, I.H.: WASPNET A Wavelength Switched Packet Network. IEEE Communications Magazine 37(3), 120–129 (1999)
12. Yao, S., Ben Yoo, S.J., Mukherjee, B., Dixit, S.: All-Optical Packet Switching for Metropolitan Area Networks: Opportunities and Challenges. IEEE Communications Magazine (2001)
13. Jourdan, A., Chiaroni, D., Dotaro, E., Eilenberger, G.J., Masetti, F., Renaud, M.: The Perspective of Optical Packet Switching in IP-Dominant Backbone and Metropolitan Networks. IEEE Communications Magazine (2001)
14. White, I.M.: A New Architecture and Technologies for High-Capacity Next Generation Metropolitan Networks. Ph.D. Dissertation, Department of Electrical Engineering of Stanford University, CA (2002)
15. Resilient Packet Ring. IEEE 802.17-2004 IEEE Standard (2004)
16. Le Sauze, N., Dotaro, E., Dupas, A., et al.: DBORN: A Shared WDM Ethernet Bus Architecture for Optical Packet Metropolitan Network. In: Photonic in Switching Conference (2002)
17. Callegati, F.: Optical Buffers for Variable Length Packets. IEEE Communication Letters 4(9), 292–294 (2000)
18. Nguyen, T.D., Eido, T., Atmaca, T.: Impact of fixed-size packet creation timer and packet format on the performance of slotted and unslotted bus-based optical MAN. In: ICDT 2008, Bucharest, Romania (2008)
19. Nguyen, T.D., Eido, T., Atmaca, T.: DCUM: Dynamic creation of Fixed-Size Containers in Multiservice Synchronous OPS Ring Networks. In: QoSim 2009, Roma, Italy (2009)

20. Nguyen, D.T., Eido, T., Atmaca, T.: Performance of a Virtual Synchronization Mechanism in an Asynchronous Optical Networks. In: AICT 2009, Venise, Italy (2009)
21. Popa, D., Atmaca, T.: On Optical Packet Format and Traffic Characteristics. In: IEEE/Euro-NGI SAINT 2005 Workshop on Modelling and Performance Evaluation for QoS in Next Generation Internet, Trento, Italy, pp. 292–296 (2005)
22. Popa, D., Atmaca, T.: Unexpected Queueing Delay Behaviours in Optical Packet Switching Systems. In: Euro-NGI HET-NET 2005 Conference, Ilkley, UK (2005)
23. Popa, D.: Performance Issues in Metropolitan Optical Networks: Packet Format, MAC Protocol and QoS. PhD Dissertation, Telecom Sud-Paris, France (2005)
24. Atmaca, T., Kotuliak, V., Popa, D.: Performance Issues in All-Optical Networks: Fixed versus Variable Packet Format. In: Photonics in Switching 2003 (PS 2003), Versaille, France, pp. 292–296 (2003)
25. Nguyen, V.H.: Performance Study on Multi-service Optical Metropolitan Area Network: MAC Protocols and Quality of Service. PhD Dissertation, Telecommunication Networks and Services Department, Telecom Sud-Paris, France (2006)
26. Bouabdallah, N., Beylot, A.L., Dotaro, E., Pujolle, G.: Resolving the Fairness Issue in Bus-Based Optical Access Networks. IEEE Journal on Selected Areas in Communications 23(8) (2005)
27. Nguyen, V.H., Atmaca, T.: Performance analysis of the Modified Packet Bursting mechanism applied to a metropolitan WDM ring architecture. In: IFIP Open Conference on Metropolitan Area Networks: Architecture, protocols, control and management, pp. 199–215. HCMC, Viet Nam (2005)
28. IEEE 802.3z and 802.3ab - Gigabit Ethernet. IEEE Standard (1998)
29. Bhargava, A., Kurose, J.F., Towsley, D.: A hybrid media access protocol for high-speed ring networks. IEEE Journal on Selected Areas in Communications 6(6), 924–933 (1988)
30. Tang, H., Lambadaris, I., Mehrvar, H., et al.: A new access control scheme for metropolitan packet ring networks. In: IEEE Global Telecommunications Conference Workshops (GlobeCom Workshops 2004), pp. 281–287 (2004)
31. Nguyen, V.H., Atmaca, T.: Dynamic Intelligent MAC Protocol for Metropolitan Optical Packet Switching Ring Networks. In: IEEE International Conference on Communications – ICC 2006, Istanbul, Turkey (2006)
32. Popa, D., Atmaca, T.: Performance Evaluation of a Hybrid Access Protocol for an All-Optical Double-Bus Metro Ring Network. In: ICOCN Conference, Hong Kong (2004)
33. Structure-agnostic TDM over Packet, SAToP (2005), http://www.ietf.org/internet-drafts/draft-ietf-pwe3-satop-03.txt
34. Circuit Emulation Service definitions, framework and requirements in metro Ethernet networks (2004), http://www.metroethernetforum.org/PDFs/Standards/MEF3.pdf
35. TDM Transport over MPLS using AAL1 Implementation Agreement (2003), http://www.mfaforum.org/tech/MFA4.0.pdf
36. Emulation of TDM Circuits over MPLS Using Raw Encapsulation Implementation Agreement (2004), http://www.mfaforum.org/tech/MFA8.0.0.pdf
37. Pseudo Wire Emulation Edge-to-Edge (PWE3) Architecture, RFC1985 (2005), http://www.ietf.org/rfc/rfc3985.txt?number=3985
38. Nguyen, V.H., Ben Mamoun, M., Atmaca, T., et al.: Performance evaluation of Circuit Emulation Service in a metropolitan optical ring architecture. In: de Souza, J.N., Dini, P., Lorenz, P. (eds.) ICT 2004. LNCS, vol. 3124, pp. 1173–1182. Springer, Heidelberg (2004)
39. Structure-aware TDM Circuit Emulation over Packet Switched Network, CESoPSN (2005), http://www.ietf.org/internet-drafts/draft-ietf-pwe3-cesopsn-06.txt

Performance of Multicast Packet Aggregation in All Optical Slotted Networks

Hind Castel-Taleb, Mohamed Chaitou, and Gerard Hébuterne

INSTITUT TELECOM, TELECOM SudParis
9, rue Charles Fourier 91011 Evry Cedex, France
hind.castel@it-sudparis.eu

Abstract. The study focuses on packet aggregation mechanisms on the edge router of an optical network. The device works as an interface between the electronic and optical domains : it takes IP packets coming from client layers and converts them into optical packets to be sent into the optical network. An efficient aggregation mechanism supporting QoS (Quality of Service) requirements of IP flows is presented. A timer is implemented to limit the aggregation delay. Analytical models based on Markov chains are presented in order to study the packetisation efficiency (filling ratio) and the mean time of packetisation for each class. Numerical results show the effect of IP packet lengths distribution on aggregation delay and efficiency. Also, we show the importance of the timer in bounding the delay of transmitted packets without much altering the aggregation efficiency.

Keywords: Optical networking, Packet aggregation, Bandwidth efficiency, Quality of service (QoS), Traffic engineering.

1 Introduction

In recent years, considerable research has been devoted to design IP full optical backbone networks, based on Wavelength Division Multiplexing (WDM) technology [10], in order to relieve the capacity bottleneck of classical electronic-switched networks. In a long-term scenario, the optical packet switching (OPS), based on fixed-length packets and synchronous node operation, can provide a simple transport platform based on a direct IP over WDM structure which can offer high bandwidth efficiency, flexibility, and fine granularity [26]. In [24], both client/server and Peer to Peer traffics are studied in order to study the impact of traffic profiles on the performance of the system. Two major challenges face the application of packet switching in an optical domain. First, the adaptation of IP traffic, which mainly consists of asynchronous and variable length packets, with the considered synchronous OPS network. Second, the handling of QoS (Quality of Service) requirements in the context of a multi-service packet network. To cope with the first problem, IP packet aggregation at the interface of the optical network ([8,15,6]) presents an efficient solution among few other proposals in literature ([22,2]).

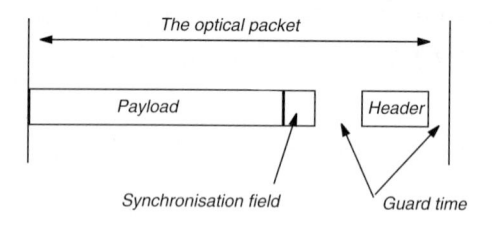

Fig. 1. Optical packet format

This is because in the current OPS technology, a typical guard time of 50 ns must be inserted between optical packets [12]. Also, a synchronisation field is to be provided, so that the optical packet is built as in Fig.1. In any packet switching network, information is carried and processed in blocks, incorporating the useful *payload* and the *header* needed to process and forward the packet through the network. Different packet formats are defined, which impact the mean delays in the optical MAN [7]. As the header has no usefulness from the end-to-end viewpoint, it corresponds to a portion of the total bandwidth which may be seen as wasted and thus its relative importance has to be minimized, which calls for rather long packets.

A possible issue is the aggregation of several IP packets into a single macro-packet with fixed size, wich represents an aggregate optical packet. Furthermore, it is necessary to perform the aggregation process regardless of the destinations of IP packets. This is due to the permanent increase in the number of IP networks, and consequently, in the number of destinations, which leads to a poor filling ratio of the optical packet if the aggregation process is performed by destination (i.e., IP packets with same destinations are aggregated together). The QoS problem is treated by adopting a class-based scheme in the edge nodes, which simplifies the core of the optical network by pushing the complexity towards the edge nodes.

According to the above discussion, this paper proposes aggregation mechanisms in order to aggregate variable-length IP packets into the payload of an optical packet. In the first mechanism proposed, the aggregation cycle ends if the aggregated packet cannot accommodate more IP packets. We present the analytical based on Markov chains in order to evaluate performance measures as packetisation delay and efficiency. Analytical results show clearly the influence of IP packet lengths on performance measures. Secondly, we introduce in the aggregation mechanism the QoS requirements of IP flows, and timer. In this second mechanism, variable-length and multi-CoS (Class of Service) IP packets are aggregated regardless of their final destinations.

The idea consists of the separation of IP traffic into $\{J \geq 2\}$ prioritized queues according to the desired CoS. At each fixed interval of time (τ), an aggregation cycle begins and an aggregate packet is constructed from several IP packets belonging to the different queues. In addition, we apply an aggregation priority mechanism by collecting IP packets, at the beginning of each aggregation cycle, from higher priority queues before those of lower priority. However, in order to relieve the drawbacks of such strict priority discipline, an hybrid version of the probabilistic priority algorithm presented in [16,17] may be used. Since the size of an IP packet at the head of a queue i, $\{1 \leq i \leq J\}$ may be greater than the gap remaining into the aggregate packet, the IP packet can be segmented in this case.

The second aggregation technique exhibits two particular characteristics. First, an aggregate packet is generated at regular time intervals, which may vary dynamically in order to sustain a prefixed amount of the filling ratio [19]. Second, a multicast optical packet is constructed. We present possible applications of such aggregation method in MANs and WANs respectively. As candidate applications in MANs, we mention the family of slotted ring networks deploying destination stripping such as the network studied in [21]. In addition, a broadcast network, which is a slotted version of the multi-channel packet-switched network called DBORN (Dual Bus Optical Ring Network) [20], represents an important application. This is because DBORN matches very well the multicast nature of the generated optical packets without the addition of any complexity in the node architecture. Furthermore, DBORN, coupled with the aggregation technique, can be adapted to use an access scheme based on TDMA, but avoids the lack of efficiency exhibited by the latter in the case of unbalanced traffic.

The present paper is organized as follows. The first aggregation mechanism is presented in section 2. Using a mathematical analysis, we give the packetisation efficiency and delay. We present analytical results in order to see the impact of packet length distribution on the packetisation mechanism. In section 3, we introduce the second aggregation mechanism with different QoS levels. Moreover, a timer is implemented in order to limit the aggregation delay. We present the analytical model in order to compute the aggregation delay for each class. Some numerical examples are presented and commented. Sections 4 and 5 explain how to apply the aggregation technique to MANs and WANs, respectively. Finally, Section 6 concludes the paper.

2 The First Packet Aggregation Mechanism

The study considers the aggregation of blocks of data in a single packet. The typical application is the building of optical packets in the edge router of optical backbone: IP packets, of variable length, are put together in fixed-size packets before being sent in the network. Most often an Ethernet link is used so that IP packets are segmented according to the maximum size of Ethernet frames. For convenience, the entering packets from the client layer are named "blocks", while the term of "packet" is reserved to the constant-size optical packet. The operating mode is as follows:

- Individual blocks, with variable length, arrive in the aggregation unit. The block is tentatively inserted in the packet under building.
- If there is enough room for the block, the packet remains in the unit until next block. If the block cannot be inserted because of its size, then the

current packet is considered as being completed and is sent in the network. The arriving block is denoted as a *trigger block*, as it provokes the packet sending . The part of the packet which is still not filled is wasted. At the same time a new packet is created, and the trigger block is inserted in it.
- When a new packet is created, a time-out is initialized. If the time-out fires before the packet is completed, the packet is immediately sent, whichever its current filling.

There are two problems related with this process, namely the choice of the packet size and the choice of the time-out value. The point is that long packets yield an optimal use of the available bandwidth, at the price of an increasing end-to-end delay, while too short packets increase the burden of headers and wasted bandwidth resource.

2.1 Mathematical Analysis

Let the individual blocks have their lengths distributed according to a common probability distribution function F, with density f. The successive block sizes are independent and identically distributed (i.i.d.). Let m be the average block length. The constant size of the optical packet is denoted as K. The units for measuring K or the block lengths may be either bits, or bytes, or larger units. For the analysis of delay performance, one has to specify the arrival process of the client blocks. When needed, one assumes they arrive according to a Poisson process, with rate λ.

In a time interval T, λT blocks arrive, which represent a total amount of $\lambda T m$ bits. They will be sent in $N(T)$ packets, carrying both the useful data, the header information and the padding bits. Let E denote the average pad length. The average number of blocks per packet, p, gives the average useful payload pm (using Wald's relation proves this result), which is related with the average waste, by the relation : $E + pm = K$.

The packetisation efficiency ϵ is defined as the ratio of the average filled payload to the total payload length :

$$\epsilon = \frac{pm}{K} \qquad (1)$$

$N(T)$ is given by the simple relation :

$$N(T) = \frac{\lambda T}{p} \qquad (2)$$

The average packetization delay θ is related with p: the packet begins with block 1 and is sent when block $p+1$ arrives, representing thus p interarrival periods. So the average delay θ is equal to:

$$\theta = \frac{p}{\lambda} \qquad (3)$$

If D represents the bandwith (in bit/second), then $D = \lambda m$, and we obtain that:

$$\theta = \frac{pm}{D} \qquad (4)$$

A model for packet length. The study considers the process (X_n) of the cumulated packet size after the arrival of the n-th block (whether the optical packet is sent or not). Let $H_n(x)$ denote the probability distribution function of the (X_n). The process (X_n) is a discrete-time Markov process, which obeys the following recurrence relation (f_n stands for the size of the n-th arriving block):

$$X_n = \begin{cases} X_{n-1} + f_n & \text{if } X_{n-1} + f_n \leq K \\ f_n & otherwise \end{cases} \quad (5)$$

The above recurrence yields the following relation between pdf's :

$$P\{X_n \leq x\} = \int_{y=0}^{x} f_n(y) P\{X_{n-1} > K - y\}\, dy \\ + \int_{y=0}^{x} f_n(y) P\{X_{n-1} \leq x - y\}\, dy \quad (6)$$

Now, we assume that a stationary limit exists. This gives the fundamental relation:

$$H(x) = \int_{y=0}^{x} f(y)[1 + H(x-y) - H(K-y)]\, dy \quad (7)$$

From the solution of the precedent equation, the major performance figures are derived. Especially, the distribution of the packet size is of prime importance. Let $Q(x)$ be the probability that a packet is sent with size lower or equal to x. First, the probability Π that a packet is sent at epoch n is equal to the probability that the n-th arriving block is a "trigger block". So:

$$\Pi = \int_{y=0}^{K} f(y)[1 - H(K-y)]\, dy \quad (8)$$

Now, Q is derived from H: this is the probability that the X_n is lower than x, given that a packet is actually sent:

$$Q(x) = P(X_{n-1} \leq x \text{ and } f_n \text{ trigger block} | \text{a packet is sent at } n-1) \quad (9)$$

So:

$$Q(x) = \frac{1}{\Pi} \int_{u=0}^{x} \int_{y=K-u}^{K} dy\, H(u) f(y)\, dy = \frac{1}{\Pi} \int_{y=K-x}^{K} f(y)\left[1 - H(K-y)\right] dy \quad (10)$$

The case where block sizes have a discrete distribution. When the f_n take only discrete values, a simpler set of equations may be used. First, one can assume that the distribution of the arriving block is an integer multiple of a basic block (a byte, or more – e.g. 64 bytes). In this case, all the problem is reformulated using the basic block as unit. The packet size K is then expressed as an integer number of basic blocs. The recurrence relation (5) is still valid, but it allows a more tractable one-dimension Markov chain formulation. Especially, $H(x)$ is given by:

$$H(x) = \sum_{y=0}^{x} P\{f = y\}[1 + H(x-y) - H(K-y)] \quad (11)$$

As (X_n) can take only discrete values, the following relation holds: $P\{X_n = x\} = H(x) - H(x-1)$, which is equal to:

$$\sum_{y=0}^{x} P\{f_n = y\} \, P\{X_n = x - y\} + P\{f_n = x\} P\{X_n > K - x\} \qquad (12)$$

This relation corresponds to the local balance equation on state $\{X_n = x\}$. In other words, in the case of discrete values of ingoing blocks, the relations given for $H(x)$ generate the resolution of the discrete Markov chain which governs the evolution of (X_n). Let us explain now how to compute performance measures, such as the mean size of the payload of optical packets, and the average packetisation delay. We must give $Q(x)$ in the case of discrete size of packets : it will represent the probability that a packet of size $x \in N$ will be sent. The relations given in the continuous case still hold. First let us write Π, the packetization probability, which becomes in the discrete case:

$$\Pi = \sum_{y=0}^{K} P\{f = y\} [1 - H(K - y)] \qquad (13)$$

So $Q(x)$ is given by the following relation :

$$Q(x) = \frac{\sum_{y > K-x} P\{f = y\} P\{X = x\}}{\Pi} \qquad (14)$$

Let N be the mean size of the effective payload in optical packets :

$$N = \sum_{x=0}^{K} x \, Q(x) \qquad (15)$$

The packetisation efficiency ϵ is :

$$\epsilon = \frac{N}{K} \qquad (16)$$

The average packetization delay θ is :

$$\theta = \frac{N}{m\lambda} \qquad (17)$$

which is equivalent to :

$$\theta = \frac{N}{D} \qquad (18)$$

2.2 Numerical Results

In this section, we give some numerical results in order to see the performance of packet aggregation. We suppose that the block sizes have a discret distribution, and we compute the packetisation efficiency and the mean time of packetisation, using equations given in section 2.1.

Table 1. Block length distributions

C_i	64 bytes	576 bytes	1600 bytes
C_1	0.25	0.75	0
C_2	0.25	0.5	0.25
C_3	0.25	0.25	0.5
C_4	0.25	0	0.75

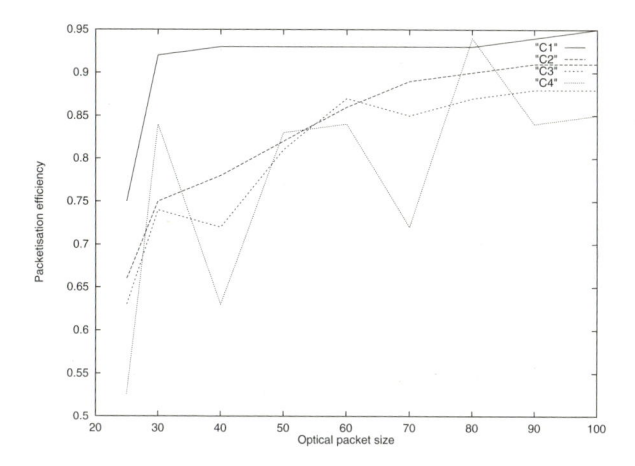

Fig. 2. Packetisation efficiency versus optical packet size

In Fig.2, we have plotted several curves representing the packetisation efficiency versus the optical packet size. For all the experiments, we have supposed that the arrival rate λ of clients IP blocks is $10/\mu s$, and the unit for measuring the block lengths and optical packet size K is 64 bytes. Performance measures are computed for optical packet sizes varying from 1600 bytes to 6400 bytes (25 packet units to 100 packet units in figures). We have supposed different block length probability distributions for input IP packet flows.

The curve "C_i" for $i \in [1\ldots 4]$ corresponds to the C_i configuration for the block length distribution. We have chosen block lengths equal to : 64 bytes, 576 bytes, and 1600 bytes. In table 1, we give block length distributions of IP packet flows for each C_i.

We notice that when the ratio of blocks whith a high length increases, then the packet efficiency decreases. For the curve "C_4", we can remark an oscillation phenomenon : as we have often arrivals of large length block then we have two cases : the packet is either well filled, or not (he has an important free places but not enough to be filled with the block). Except for this case, we can see that the block length distribution has not an important impact on the packetisation efficiency. As it seems that current IP flows exhibit a rather high proportion of short IP packets (acknowledgements typically), configurations "C_1" and "C_2" are the most likely to be observed. However, the partial filling effect is to be mixed with the bandwidth waste related to guard bands and headers. Assume

Table 2. Global performance figures

Payload length	Packet duration	Global efficiency
1600 bytes	1.48 μs	0.57
3200 bytes	2.76 μs	0.77
6400 bytes	5.32 μs	0.87

for instance that they result in a 200 ns (100 ns for guard times and 100 ns for the header), and that the link speed is 10 Gbits/s. In table.2 we give the following global performance figures in the case of the (probably most typical) configuration "C_2".

In Fig.3, we show the mean packetisation delay (in μ s) versus the optical size. As can be expected, the longer the packet, the longer the aggregation delay. Also, as the aggregation efficiency increases, so does the delay-since more data is needed to fill the packet. Therefore, efficiency is obtained at the price of additional delay. We can see in this figure, that for an optical packet size of 1600 bytes, the packetisation delay varies from 0.26 μs (curve C1) to 0.069 μs (curve C4). And for 6400 bytes it varies from 1.358 μs (curve C1) to 0.45 μs (curve C4). So we deduce that the delay is quite negligible for λ=10 blocks/μs, but it increases if λ decreases (for λ=10 blocks/s, the delay will have the same values but given in secondes). Suppose now that the flow to carried through the aggregation process is 1Gbits/s (so D=1Gbits/s), we obtain the following values of the mean packetisation delay given in table 3.

We can see that for D=1 Gbit/s, the delay is quite negligible, but as for λ, if D decreases, the delay increases.

We have proposed in this section a first mechanism for packet aggregations, and we have evaluate the performance using mathematical models. This system is not very complex, We have seen the impact of packet length distributions on

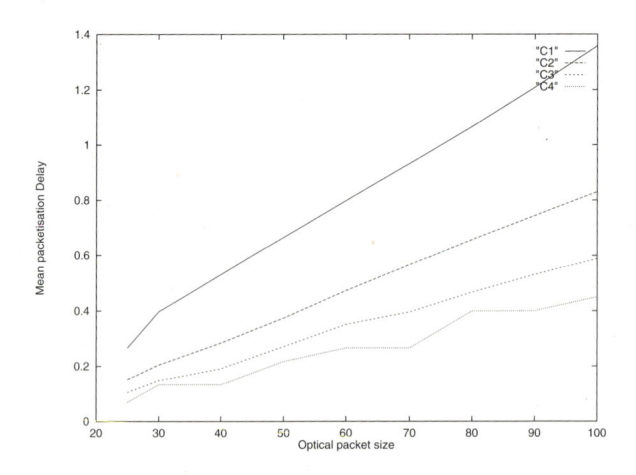

Fig. 3. Average packetisation delay in μs versus optical packet size

Table 3. Packetisation delay in μs for 1 Gbit/s flow

Payload length	C1	C2	C3	C4
1600 bytes	1.19	1.06	1.01	0.84
3200 bytes	2.98	2.64	2.60	2.64
6400 bytes	6.09	5.84	5.67	5.48

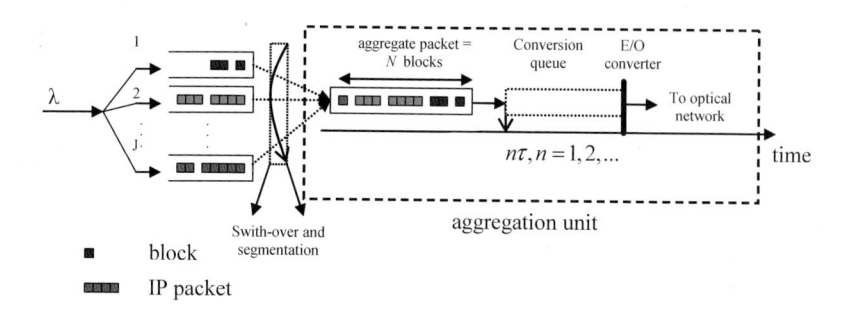

Fig. 4. The aggregation mechanism at the optical interface

packetisation delay and efficiency. Furthermore, we need to complexifiy the aggregation mechanism by introducing a timer parameter, and taking into account QoS of different packet flows.

3 The Second Packet Aggregation Mechanism

Let there be J classes of packets (throughout the paper the term "packet" stands for "IP packet"), where packets with a smaller class number have a higher priority than packets with a larger class number. Each class of packets has its own queue and the buffer of the queue is infinite. Packets in the same queue are served in FCFS fashion.

3.1 Mathematical Model

Each packet is modeled by a batch of blocks having a fixed size of b bytes. Let X be the batch size random variable with probability generating function (PGF) $X(z)$, and probability mass function (pmf) $\{x_n = P(X = n), n \geq 1\}$. The size of the aggregate packet is fixed to N blocks ($N > \max(X)$), and a timer with a time-out value τ is implemented as shown in Fig. 4. At each timer expiration (i.e at instants $\{n\tau, n = 0, 1, 2, \ldots\}$) the aggregation unit takes $\min(N,$ the whole queue 1 length) blocks to attempt filling the aggregate packet. If a gap still remains, the aggregation unit attempts to fill it from queue 2, then from queue 3,..., until the aggregate packet becomes full or until queue J is reached. If the whole of the IP packet cannot be inserted (e.g., the packet at the head of queue J in Fig. 4), only a part of it, needed to fill the aggregate packet, is transferred to the aggregation unit (a segmentation interface performs packets segmentation just before the aggregation unit). The aggregate packet is then sent to a queue (called conversion queue) preceding the stage of the electronic to optical conversion (E/O), and a new aggregation cycle is performed by polling queue 1 again. Note that in the discipline described above, it may take several times for the aggregation unit to switch from one queue to the other to actually perform packets aggregation in each aggregation cycle. Throughout the rest of this paper, we assume that the switch-over time can be much small as compared to the sojourn time and will not be taken into account in the analysis. It is worthwhile mentioning that the discrimination between successive IP packets, contained in an aggregate packet, is performed by using a delineation protocol such as the protocol proposed in [11].

In the analytical study, we first begin by considering only two classes of packets ($J = 2$). However, the analysis can be extended easily to a multi-class system, as it will be shown later. The following assumptions and notations are used throughout this paper. We assume that packets arrive at the corresponding queues according to independent Poisson processes with rates λ_1 and λ_2, i.e. the total arrival rate is $\lambda_0 = \lambda_1 + \lambda_2$. We define $\{A_t^c, c = 0, 1, 2\}$ as the number of blocks arriving at queue c, (for $c = 0$ the queue corresponds to the combination of queues 1 and 2), during an interval of time t, and we design by $\{A_t^c(z) = e^{\lambda_c t(X(z)-1)}\}$ its PGF and $\{a_t^c(n), n \geq 0\}$ its pmf.

3.2 Blocks Number Pre-departure Probabilities and Filling Ratio

We define $\{Y^c(t), c = 0, 1, 2\}$ by the number of blocks in queue c at time t, and we suppose that $Y_n^c = Y^c(t_n^-)$. We choose a set of embedded Markov points as those points in time which are just before timer expirations. Let $t_0, t_1, \ldots, t_n, \ldots$, be the epochs of timer expirations. Since the whole system and queue 1 behave in a similar way, the steady state distribution for $\{Y_n^c, n = 0, 1, 2, \ldots\}$ is obtained by the same manner for $\{c = 0, 1\}$:

$$y_k^c = \lim_{n \to \infty} P(Y_n^c = k), \qquad k \geq 0 \tag{19}$$

The following state equation holds:

$$Y_{n+1}^c = \mid Y_n^c - N \mid^+ + A_\tau^c \tag{20}$$

where $\mid c \mid^+$ denotes $max(0, c)$. The equilibrium queue length distribution (in blocks number) at an arbitrary time epoch is then described by the probability generating function $Y^c(z)$, which can be derived in (21) by a straightforward and well-known fashion [9]. It is given by :

$$Y^c(z) = \frac{A_\tau^c(z)(z-1)(N - E[A_\tau^c])}{z^N - A_\tau^c(z)} \prod_{k=1}^{N-1} \frac{z - z_k}{1 - z_k} \tag{21}$$

where, $z_1, z_2, \ldots, z_{N-1}$ are the $N-1$ zeros of $z^N - A_\tau^c(z)$ inside the unit circle of the complex plane, and $E[\ldots]$ is the expectation value of the expression between square brackets. Using the inverse fast fourier transform (ifft) of MATLAB, y_n^c can be derived from (21) in few seconds. Equation (21) allows us to obtain the pmf of the aggregate packet filling value (i.e the number of blocks in the aggregate packet). Let F be the filling value random variable, and define the filling ratio random variable by $F_r = F/N$ (which is equivalent to the bandwidth efficiency). If we denote by $\{f_n = P(F = n), \quad 0 \leq n \leq N\}$ the pmf of F, we obtain:

$$f_n = \begin{cases} y_n^0 & \text{if } 0 \leq n < N-1 \\ 1 - \sum_{i=0}^{N-1} y_i^0 & \text{if } n = N \end{cases} \qquad (22)$$

To obtain the steady state distribution for $\{Y_n^2, n = 0, 1, 2, \ldots\}$, the state equation can be written as:

$$Y_{n+1}^2 = \mid Y_n^2 - G \mid^+ + A_\tau^2 \qquad (23)$$

where G represents the gap random variable. It is given by $G = N - F$, and hence, its pmf defined by $\{g_n, n = 0, 1, 2, \ldots, N\}$ can be obtained easily from (22). The PGF of Y^2 is then given by (see [9]):

$$Y^2(z) = \frac{A_\tau^2(z)(z-1)(N - E[U])}{z^N - U(z)} \prod_{k=1}^{N-1} \frac{z - z_k}{1 - z_k} \qquad (24)$$

where U is the random variable defined by: $U = N + A_\tau^2 - G$, and $z_1, z_2, \ldots, z_{N-1}$ are the $N-1$ zeros of $z^N - U(z)$ inside the unit circle.

Blocks number random instant probabilities. Let $\{K^c, c = 1, 2\}$ denote the number of blocks, at a random instant t, in queues 1 and 2 respectively, and let $\{q_k^c = P(K^c = k), k \geq 0\}$ be its pmf.

Lemma 1. K^c is related to Y^c by:

$$K^c(z) = Y^c(z) \frac{1 - e^{-\lambda \tau (X(z) - 1)}}{\lambda \tau (X(z) - 1)} \qquad (25)$$

Proof: The proof is given for $c = 2$, the case of $c = 1$ is a particular case obtained by replacing the random variable G (the gap) with the constant parameter N (the length of the aggregate packet). Let T_e be the elapsed time since the last timer expiration. By conditioning on T_e, the following state equation can be obtained:

$$K^2 \mid (T_e = t) = \mid Y^2 - G \mid^+ + A_t^2 \qquad (26)$$

Then (superscript 2 is omitted for simplicity), if we denote by $K^* = K(z \mid T_e = t)$ is given by:

$$\begin{aligned} K^* &= \sum_{k=0}^{\infty} P[\mid Y - G \mid^+ + A_t = k] z^k \\ &= \sum_{k=0}^{\infty} \sum_{i=0}^{N} g_i P[\mid Y - i \mid^+ + A_t = k] z^k \\ &= \sum_{k=0}^{\infty} \sum_{i=0}^{N} g_i \left\{ \sum_{j=0}^{i} y_j P[A_t = k] + \sum_{j=i+1}^{\infty} y_j P[A_t = k - j + i] \right\} z^k \\ &= A_t(z) \left\{ \sum_{i=0}^{N} g_i \left(\sum_{j=0}^{i} y_j + z^{-i} \left(Y(z) - \sum_{j=0}^{i} y_j z^j \right) \right) \right\} \end{aligned} \qquad (27)$$

With (see [9])

$$\left(\sum_{i=0}^{N} g_i \sum_{j=0}^{i} y_j - \sum_{i=0}^{N} g_i z^{-i} \sum_{j=0}^{i} y_j z^j\right) = \frac{Y(z)\left(1 - A_\tau(z)\sum_{i=0}^{N} g_i z^{-i}\right)}{A_\tau(z)} \Rightarrow$$
$$K(z \mid T_e = t) = \frac{A_t(z)}{A_\tau(z)} Y(z) \tag{28}$$

To obtain $K(z)$, it is sufficient to remove the condition on T_e, thus,

$$K(z) = \int_{t=0}^{\tau} \frac{A_t(z)}{A_\tau(z)} Y(z) \left(\frac{dt}{\tau}\right) \tag{29}$$

and the proof is completed.

3.3 Mean Delay Analysis

In this section we present a method to obtain the mean aggregation delay of an IP packet belonging to class 1 or to class 2. The aggregation delay random variable of a class c packet, $\{c = 1, 2\}$, is represented by $\{D^c, c = 1, 2\}$, and it is defined as the time period elapsed between the arrival instant of the packet to its corresponding queue, and the instant when the last block of the packet leaves the queue. The delay can be decomposed in two parts: the waiting time of the packet first block until it becomes at the head of the queue, and the delay due to the packet segmentation when the packet cannot be inserted directly into the remaining gap of the aggregate packet. The decomposition is written, in term of mean delays, as:

$$E[D^c] = E[D_b^c] + D_s^c \tag{30}$$

where D_b^c denotes the packet first block waiting time in the queue and D_s^c stands for the packet segmentation delay. By using the Little theorem we get:

$$E[D_b^c] = \frac{E[K^c]}{\lambda_c \times E[X]} \tag{31}$$

where $E[K^c]$ can be obtained by putting $z = 1$ in the first derivative of $K^c(z)$, and its given by:

$$E[K^c] = E[Y^c] - \frac{\lambda_c \tau E[X]}{2}$$
$$= \left\{ E[A_\tau^c] + \left[\frac{VAR[C(U, A_\tau^1)]}{2(N - E[C(U, A_\tau^1)])} - \frac{E[C(U, A_\tau^1)]}{2} + \frac{1}{2} \sum_{k=1}^{N-1} \frac{1+z_k^c}{1-z_k^c} \right] \right\} - \frac{\lambda_c \tau E[X]}{2} \tag{32}$$

with,

$$C(x, y) = \begin{cases} x & \text{if } c = 1 \\ y & \text{if } c = 2 \end{cases} \tag{33}$$

$VAR[X]$ is the variance of X, $\{z_k^1, 1 \leq k \leq N - 1\}$ and $\{z_k^2, 1 \leq k \leq N - 1\}$ are the roots inside the unit circle of the two respective following equations: $\{z^N - A_\tau^1(z)\}$ and $\{z^N - U(z)\}$, with $U(z)$ defined in section 3.2. Note that the roots are found using a numerical method with a precision of 10^{-10}.

Computation of D_s^1. We consider the case where $N > \max(X)$, where X denotes the batch size random variable (i.e the size of an IP packet in term of blocks), which implies that a class 1 packet cannot be segmented more than once. If we denote by $p_s^{1,n}$ the probability that a packet is segmented n times, then the segmentation delay of a class 1 packet will be given by:

$$D_s^1 = \sum_{n=1}^{\infty} p_s^{1,n} \times (n\tau) = p_s^{1,1} \times \tau \qquad (34)$$

The following is a method to obtain $p_s^{1,1}$: let N_s be the random variable representing the number of blocks that enters the aggregation unit before the first block of a random packet, given that the latter (the first block of the packet) has entered the aggregation unit. The pmf of N_s, $\{P(N_s = n, n = 0, \ldots, N-1\}$, can be obtained as follows:

$$\begin{aligned} P[N_s = n] &= \sum_{k=0}^{\infty} P\left[K^{1,a} = kN + n\right] \\ &= \sum_{k=0}^{\infty} \sum_{i=0}^{\infty} k_i^{1,a} \delta(i - kN - n) \\ &= \sum_{i=0}^{\infty} k_i^{1,a} \sum_{k=-\infty}^{\infty} \delta(i - kN - n) \end{aligned} \qquad (35)$$

where $K^{1,a}$ is the number of blocks presented in queue 1 seen at the arrival of a random packet. PASTA property [25] implies that $K^{1,a} = K^1$. $\delta(n)$ is the Kronecker delta function, which equals 1 for $n = 0$ and 0 for all other n, and $\{k_i^1 = 0, \text{ for } i < 0\}$. Now we make use of the following identity:

$$\sum_{k=-\infty}^{\infty} \delta(i - kN - n) = \frac{1}{N} \sum_{s=0}^{N-1} a^{s(i-n)} \qquad (36)$$

with: $a = exp\left(j\frac{2\pi}{N}\right)$. In words: the right-hand side of (36) equals zero unless the integer $i - n$ is a multiple of N, when it equals unity. Thus,

$$\begin{aligned} P[N_s = n, 0 \le n \le N-1] &= \sum_{i=0}^{\infty} k_i^1 \frac{1}{N} \sum_{s=0}^{N-1} a^{s(i-n)} \\ &= \frac{1}{N} \sum_{s=0}^{N-1} a^{-sn} K^1(a^s) \end{aligned} \qquad (37)$$

where $K^1(a^s)$ is $K^1(z)$ evaluated at $z = a^s$.

Obtaining the pmf of N_s by using (37) for each value of n leads to a very long computation time, especially when N becomes large. However, we give an equivalent matrix equation for this relation. This approach reduces the computation time considerably since it gives the pmf of N_s by using only one matrix equation. If P_{N_s} denotes the row vector representing the pmf of N_s, i.e $P_{N_s} = (P(N_s = 0) P(N_s = 1) P(N_s = N-1))$, and if we define R_{K^1} by: $\left(K^1(a^0) K^1(a^1) \ldots K^1(a^{N-1})\right)$, we will have:

$$P_{N_s} = \frac{R_{K^1}}{N} \times \begin{pmatrix} a^0 & a^0 & a^0 & \ldots & a^0 \\ a^0 & a^{-1} & a^{-2} & \ldots & a^{-(N-1)} \\ a^0 & a^{-2} & a^{-4} & \ldots & a^{-2(N-1)} \\ \ldots & & & & \\ a^0 & a^{-(N-1)} & a^{-2(N-1)} & \ldots & a^{-(N-1)^2} \end{pmatrix} \qquad (38)$$

where the last matrix in 38 is an $N \times N$ matrix. Now, $p_s^{1,1}$ can be obtained easily by: $p_s^{1,1} = P(X > N - N_s)$.

Computation of D_s^2. Unlike the class 1 case, where a packet cannot be segmented more than once, a class 2 packet may encounter several segmentations before its complete transmission. This is because when a class 2 packet is segmented for the first time, its remaining blocks cannot enter the aggregation unit unless queue 1 is polled again.

If we denote by $p_s^{2,n}$ the probability that a packet is segmented n times, we obtain:

$$D_s^2 = \sum_{n=1}^{\infty} p_s^{2,n} \times (n\tau) \qquad (39)$$

However, considering only the first two terms of (39) is sufficient as we will see in the numerical examples. To obtain the pmf of N_s we proceed as in the previous section with the difference that here the number of blocks that enter the aggregation unit before a class 2 packet is equal to the number of blocks presented in queue 1 and queue 2 when the packet arrives (i.e, $K^{1,a} + K^{2,a}$), plus the number of class 1 blocks that arrive during the waiting time of the packet first block (represented by the random variable $N_{b,2}^1$). Now we use the PASTA property $\{K^{c,a} = K^c, c = 1,2\}$ and we approximate $N_{b,2}^1$ by its mean ($E[N_{b,2}^1] = \lambda_1 E[X]E[D_b^2]$) thanks to Little theorem). Then, by replacing the random variable K^1 with $\{K^1 + K^2 + E[N_{b,2}^1]\}$ in the method presented in the previous section, and by supposing that queue 1 and queue 2 are independent (i.e, K^1 and K^2 are two independent random variables), we can express the pmf of N_s by the following:

$$P\left[N_s = n, 0 \leq n \leq N-1\right] = \frac{1}{N} \sum_{s=0}^{N-1} a^{-sn} \left((a^s)^{E[N_{b,2}^1]} K^1(a^s) K^2(a^s)\right) \qquad (40)$$

The last term between parentheses in (40) is the z-transform of $\{K^{1,a} + K^{2,a} + E[N_{b,2}^1]\}$, ($= z^{E[N_{b,2}^1]} K^1(z) K^2(z)$), evaluated at $z = a^s$. Now the first two terms of 39 can be obtained as follows:

$$p_s^{2,1} = P(X > N - N_s) \quad \text{and} \quad p_s^{2,2} = P\left(X - (N - N_s) > G\right) \qquad (41)$$

where G stands for the gap random variable.

3.4 Extension to $J, \{J > 2\}$, Classes

We explain how to extend the analysis when more than two packet classes are desired.

The filling ratio can be obtained by the same manner used in the case of two classes. However, for the delay analysis of packets belonging to queue i, $\{i = 2, \ldots, J\}$, we combine queues $\{1, 2, \ldots, i-1\}$ in a single queue and the analysis is reduced to two queues with respective arrival rates: $\lambda_1 = \sum_{k=1}^{i-1} \lambda_k$, and $\lambda_2 = \lambda_i$.

3.5 The Probabilistic Priority Discipline

In order to overcome the drawbacks of the strict priority discipline that has been used in the analytical model, one may apply a probabilistic algorithm such as the one presented in [16]. For this purpose, a parameter p_i, $1 \leq i \leq J$, $0 \leq p_i \leq 1$, is assigned to each arrival queue i, and a relative weight $r_i = p_i \prod_{j=1}^{i}(1 - p_{j-1})$, where $p_0 = 0$, is computed. Then, the following steps are applied.

1. At each aggregation cycle, the set of non-empty queues, NQ is determined, and a normalized relative weight, $\tilde{r}_i = \frac{r_i}{\sum_{j \in NQ} r_j}$ is calculated. The latter is regarded as the probability with which queue i is served among all non-empty queues in an aggregation cycle.
2. Fill the aggregate packet from the polled queue in Step 1. If the aggregate packet becomes full, send it to the conversion queue and apply Step 1 to the next aggregation cycle, elsewhere exclude the polled queue from NQ and apply Step 1 to the same aggregation cycle if NQ is not empty, or apply Step 1 to the next aggregation cycle if NQ is empty.

3.6 Numerical Examples

We give some numerical examples showing the usefulness of the model. All the computations (probabilities and means) have been done in double precision. The mean delay and the mean filling ratio are obtained from their corresponding pmfs. We consider the following assumptions.

1. From experimental measurements [23], the size distribution of IP packets has three predominant values: 40 bytes, 552 bytes and 1500 bytes, with the corresponding probabilities 0.6, 0.25 and 0.15 respectively. To discretize the size distribution, we suppose that the size of a random packet is a batch of 40 bytes blocks, and hence the packet size PGF is given by: $\{X(z) = 0.6z + 0.15z^{14} + 0.25z^{38}\}$, where each power of z represents the first integer greater or equal to the division quotient of the corresponding packet size by the block size (e.g., $\lceil \frac{552}{40} \rceil = 14$, $\lceil . \rceil$ is the first integer greater or equal to .).
2. Each node has two classes (real-time applications and non-real-time applications) with proportions 0.6 and 0.4, respectively. The arrival processes of the two classes are Poisson processes with rates λ_1 (packet/s) and λ_2 (packet/s) respectively; the total arrival rate is $\lambda_0 = \lambda_1 + \lambda_2$. In the examples, we suppose that the arrival rates are represented, in Mb/s, by: $\{\theta_c, c = 0, 1, 2\}$, where $c = 0$ represents the case of the combination of queues 1 and 2. It is easy to verify that λ_c is related to θ_c by: $\lambda_c = \frac{\theta_c \times 10^6}{8bE[X]}$. In the rest of the paper, we use θ and θ_0 interchangeably. Note that the choice of two classes in the numerical examples is adopted for comparison purposes only since the model is scalable regardless of the number of classes. Furthermore, Poisson traffic is considered because we aim at proving that the multicast aggregation technique increases the filling ratio of an optical packet. Clearly, this conclusion remains true regardless of the traffic profile.

3. The size of the aggregate packet (N blocks) is supposed to be fixed by the operator, and the maximal arrival rate $\theta_{0,max}$ (or simply θ_{max}) is supposed to be known a priori (by effectuating measurements over several time-scales).

Nodes stability condition and stability region. The stability condition of the node is respected if (see 21): $N > E[A_\tau^0]$, with $E[A_\tau^0] = \lambda_0 \tau E[X]$.

stability region. Let the parameter a be the following: $a = E[A_\tau^0]$ (this is the mean of blocks number that arrive during τ). For a given value of N and θ_0 (or simply θ), the parameter a must be strictly less than N to maintain the stability and then we must choose τ according to:

$$\tau_\theta < \frac{N}{\lambda_0 E[X]} \qquad (42)$$

where τ_θ is the value of the time-out when operating at θ. 42 defines what we call the stability region at the arrival rate θ. To obtain a desired value of a inside the stability region we choose τ_θ according to: $\tau_\theta = \frac{a}{\lambda_0 E[X]}$. For instance, for $N = 76$ blocks, and $\theta = 900$ Mb/s, (42) implies that the limit of the stability region is $\tau_{max} = 27$ μs.

Impact of the time-out. We present in Fig. 5 the impact of τ on the filling ratio. The pmf of the filling ratio is shown for two values of τ ($\tau = 12.5\mu s$ and $\tau = 25\mu s$), with $\theta = 900$ Mb/s and $N = 76$. We observe that when τ increases, the probability that the aggregate packets are sent with better filling ratio increases. This is because when τ increases, while remaining inside the stability region, the number of packets presented in queues 1 and 2 at a random epoch increases. Note that for $\tau = 25\mu s$ the aggregate packets are sent full with probability 0.82952. This is because τ is very close from the stability limit ($27\mu s$).

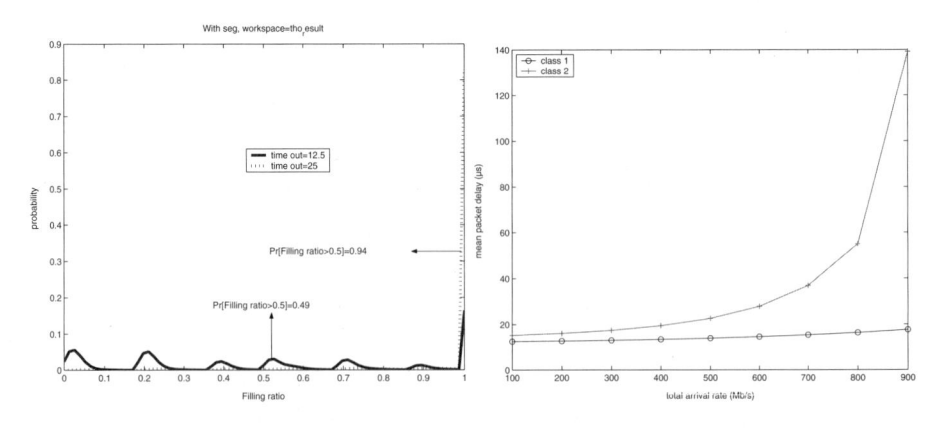

Fig. 5. pmf of the filling ratio

Fig. 6. The mean packet delay as function of the arrival rate θ

Fig. 7. The mean filling ratio as function of the arrival rate θ

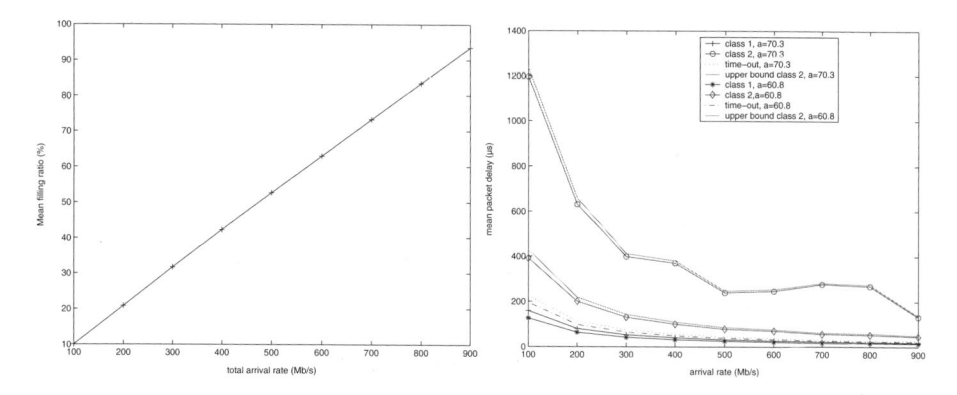

Fig. 8. The mean packet delay in the case of the adaptive τ scenario

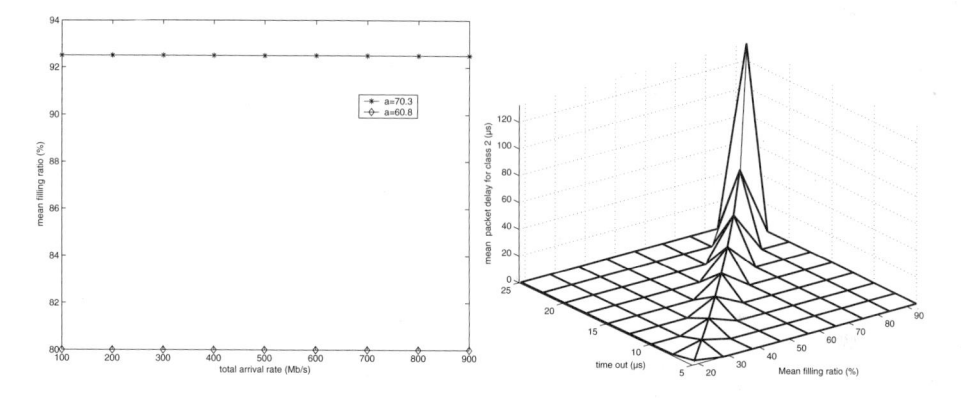

Fig. 9. The mean filling ratio in the case of the adaptive τ scenario

Fig. 10. Class 2 mean packet delays

Effect of the arrival rate. The impact of the arrival rate on both mean packet delay and mean filling ratio is given in Fig. 6 and Fig. 7 respectively. N is fixed to 76 and τ to $25\mu s$. It can be observed that the mean delay of class 1 packets remains approximately unchanged when θ varies, while the mean delay of class 2 packets degrades when θ increases. The mean filling ratio increases also as θ increases. At $\theta = 900$ Mb/s, the mean filling ratio attains 93.506% since at this arrival rate, $\tau = 25$ μs becomes very close to the limit of the stability region. Note that when θ decreases, the enhancement in the delay is compensated by a loss in the filling ratio. This is because τ remains constant. To avoid this limitation, we can adapt the value of τ with respect to the variation of θ in order to conserve a fixed value of the parameter a (obtained from nodes stability), which represents the mean size (in blocks) of the packets that arrive between two consecutive timer expirations. In Fig. 8 we show the effect of a on the mean packet delay (we take $a = 70.3$ blocks and $a = 60.8$ blocks). It is clear that when a decreases the delay decreases (decreasing a means that for a given θ the time-out τ has a less value which implies less delays). The advantage of the adaptive τ scenario is to maintain the mean filling constant as shown in Fig. 9. We can also deduce that reducing a leads to a decrease in the mean filling ratio as expected (because of the decrease in τ). Finally, Fig. 10 represents how an industrial operator can make use of the aggregation model. If a certain value of the mean filling ratio is desired, then by using the curve we can obtain the value of class 2 mean packet delay (z-axis) and the value of the time-out to setup (y-axis). We can adapt the value of the filling ratio to obtain a desired value of class 2 mean packet delay, while class 1 mean packet delay remains below τ (as shown in Fig. 8).

4 Application to MANS

4.1 Hub Stripping Network: DBORN

The term hub stripping refers to the case where a single node in the ring network, called the hub, drops optical packets. This is the case of DBORN, where the hub can be regarded as the single destination of ring nodes in the transmission phase. DBORN has been initially proposed for asynchronous systems [20,3,4]. However, we consider in this work an hybrid slotted version of the original system. DBORN is an optical metro ring architecture connecting several edge nodes, e.g., metro clients like enterprise, campus or local area networks (LAN), to a regional or core network. The ring consists of a unidirectional fiber split into downstream and upstream channels spectrally disjointed (i.e., on different wavelengths) [3]. The upstream wavelength channels are used for writing (transmitting), while downstream wavelength channels are used for reading (receiving). In a typical scenario, the metro ring has a bit-rate 2.5 Gb/s, 10 Gb/s or 40 Gb/s per wavelength. In order to keep the edge node interface cards as simple as possible, all traffic (external and intra-ring) has to pass the hub. Specifically, no edge node receives or even removes traffic on upstream channels or inserts traffic on downstream channels. Thus, both upstream and downstream channels can be modeled as shared unidirectional buses. Packets circulate around the ring without any electro-optic conversion at intermediate nodes. The hub is responsible of terminating upstream wavelengths and hence, the first node in the upstream bus receives always free slots from the hub. Moreover, the hub electronically processes the packets, which may leave to the backbone or go through the downstream bus to reach their destinations. In the latter case, the ring node must pick up a copy of the signal originating from the hub by means of a splitter in order to recover its corresponding packets by processing them electronically. In the following, we consider that each node is equipped with one fixed transmitter and as many fixed receivers as reception channels, i.e., for each node we assign only one transmission wavelength at the upstream bus.

Two MAC protocols may be used. The empty slot procedure and a slot reservation mechanism.

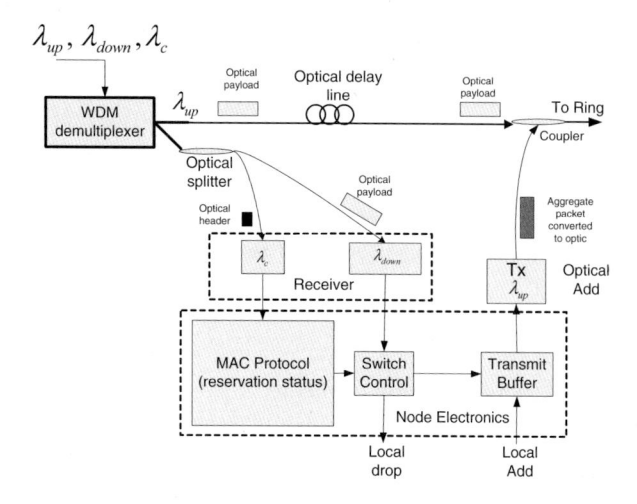

Fig. 11. The ring node architecture of slotted DBORN

In the case of empty slot procedure, the slot header is detected to determine the status of the slot (i.e., empty/full) and a node may transmit on every empty slot.

In the case of the reservation approach, we implement a slot reservation mechanism at the upstream bus. Indeed, a fixed number of slots are reserved for each node. We suppose that the slot assignment is performed in the hub which writes the address of the node, for which the slot is reserved, in the slot's header. Hence, when a node receives an incoming slot, it detects the header to determine the status of the slot (empty/full) and the reservation information to decide whether to transmit or not. In addition to its reserved slots, the node can use empty slots reserved to any of its upstreams. This is because all slots are emptied by the hub before being sent on the upstream bus. This interesting feature of DBORN makes it possible to avoid the inefficiency behavior of a TDMA scheme. For instance, if a node i does not fill a reserved slot because its local queue is empty, then any downstream node which transmits on channel λ_{e_i} can use this slot for transmission because the hub will empty the slot before it reaches node i again. This enables efficient use of the available bandwidth in the case of non uniform traffic. Furthermore, at overloaded conditions, i.e., when the local queues of all nodes are always non empty, the MAC protocol reduces to a TDMA scheme, since each node will consume all its reserved slots for transmission.

Note that in order to process the control information, only the control channel is converted to the electrical domain at each ring node, while the bulk of user information remains in the optical domain until it reaches the hub which is viewed as the destination of upstream data. This is in conformity with the notion of all-optical (or transparent) networks in literature (e.g., [14]). The corresponding network node architecture is given in Fig. 11, where λ_{up}, λ_{down} and λ_c represent the upstream bus, the downstream bus and the control channel respectively. The slots in the control channel have a locked timing relationship to the data slots and arrive earlier at each node by a fixed amount of time (this is achieved via an optical delay line, see Fig. 11) allowing the process of the control slot content.

At the downstream reading bus, ring nodes preserve the same behavior initially proposed in DBORN and hence, the optical signal is split and IP packets are recovered at each node. The latter drops packets which are not destined to it. This means that DBORN supports multicast without addition of any complexity in its MAC protocol and hence, it represents an interesting application of the multicast packet aggregation method presented in this work.

5 Application to WANs

The multicast aggregation technique analyzed throughout this paper can be easily adapted to the studied MANs. One one hand, the support of multicast traffic is facilitated due to the simplicity of ring architecture. On the other hand, aggregation of IP packets belonging to different QoS classes in one aggregate packet is justified since there is no packet loss inside such ring networks. This is because in-transit traffic has always higher priority than local traffic at intermediate nodes.

The problem of multicast handling and QoS support in WAN may introduce some adjustments to the proposed aggregation techniques because wide area networks have a mesh topology in general. The scenario proposed in [18]

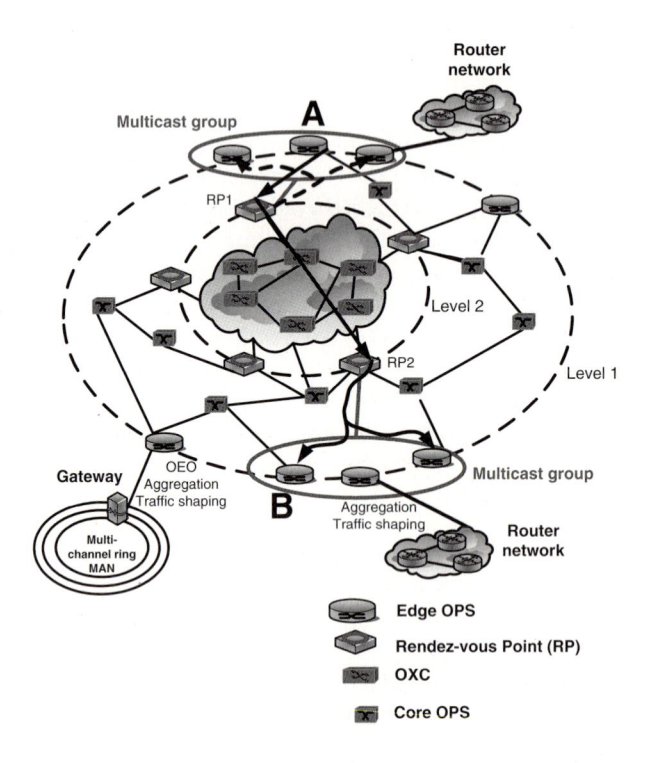

Fig. 12. Possible WAN architecture

is adopted. Indeed, optical packet switching provided through optical packet switches (OPSs) and circuit switching ensured by OXCs (Optical Cross-Connects), coexist within the network. The former is deployed in areas where granularity is below the wavelength level, while the latter interconnects high-capacity points that will fully utilize the channel capacity in the core of the network. To do this, some optical channels (wavelength paths) may interconnect OXCs. Other channels might be reserved to OPSs to support optical packet transmission. Once the technology matures and the need for a more flexible fully IP-centric network is dominant, optical packet switches may replace the OXCs, or alternatively reduce their size and cost significantly, as the wavelength channels are more efficiently used and hence, equipment requirements are reduced. Furthermore, we propose a network with two hierarchy levels. The first level is constituted from edge and core OPSs. Edge OPSs differ from core OPSs in that they are connected to client layer networks, such as IP networks and to metropolitan area networks such as the networks studied in this work. Edge OPSs are responsible of performing IP packet aggregation in order to improve bandwidth efficiency. Indeed, optical packets received through MANs must be converted to electronic in order to join a new aggregation process. The latter must take into consideration the two fundamental questions (multicast and QoS) in the context of the WAN architecture. This is may be given as follows. In WANs the number of edge nodes is much greater than that in MANs and hence, aggregating IP packets regardless of their destinations leads to generating a big number of broadcast optical packets which may waste the network resources. Instead, we suppose that edge nodes are separated in multicast groups. Each multicast group has a designed router, called Rendez-vous point (RP).

The OPS is responsible of communicating with the level-1 hierarchy nodes, while the interconnection between RPs through lightpaths between OXCs makes the level-2 hierarchy. An IP-centric control plane is responsible of four major tasks. 1) Constructing the multicast groups by using the IGMP protocol [5]. 2) Electing an RP for each group by using an approach similar to that of the PIM-SM multicast protocol [13]. 3) Constructing lightpaths between RPs. 4) Constructing a shared multicast tree between nodes of each group and fixed lightpaths between RPs. GMPLS [1] is a candidate to perform these tasks after doing the necessary extensions. Now each edge node performs a separate aggregation process per RP. That is, IP packets destined to edge nodes belonging to the same multicast group are aggregated together. If the multicast optical packet is destined to the designated RP of the multicast group, the RP multicasts it on the shared multicast tree, elsewhere the RP sends the packet to the corresponding RP through a lightpath on the level-2 hierarchy and the latter multicasts it on the shared tree. For instance, in Fig. 12, suppose that node A wishes to send two optical packets: one for its group and another one for node B and other nodes in a different multicast group. In the former case, the packet is forwarded to RP_1 which multicasts it through its multicast tree (dashed arrows). In the latter case, RP_1 sends the packet to RP_2 which multicasts it on its shared tree to get the destination nodes including node B.

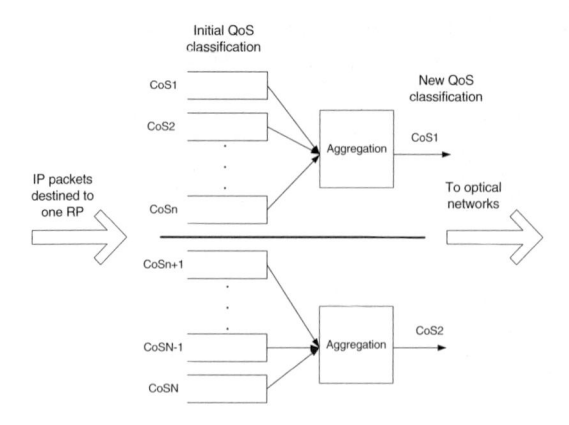

Fig. 13. A possible QoS classification

The QoS support must consider that OPSs have limited optical buffering capacity, in terms of fiber delay lines[1] and hence it may be important to generate aggregate packets with several classes of service, contrary to the aggregation in MANs, in order to reduce packet loss of higher priority traffic. Typically, four QoS classes may be sufficient. However, as the number of classes increases as the complexity of the RPs increases where separate control information, such as restoration of lightpaths in case of failure, must be guaranteed for each class. In order to reduce this complexity, two QoS classes may be adopted as shown in Fig. 13.

6 Conclusion

We propose and analyze a novel approach for efficiently supporting IP packets in a slotted WDM optical layer with several QoS requirements. A simple analytical model, allowing the evaluation of IP packets aggregation delay and the bandwidth efficiency, has been presented. The results showed an increase in the bandwidth efficiency when using the aggregation compared to the standard approach. Concerning IP packets aggregation delay, a trade-off with the bandwidth efficiency results by modifying the value of the parameter a. The QoS support mechanism based on assigning higher aggregation priorities for higher QoS classes, has been evaluated by means of an analytical study. The application of the aggregation technique has been shown in the context of MANs and WANs, respectively.

References

1. Banerjee, A.: Generalized multiprotocol label switching: An overview of signaling enhancements and recovery techniques. IEEE Comm. Mag. 39(7), 144–151 (2001)
2. Bengi, K.: Access protocols for an efficient and fair packet-switched IP-over-WDM metro network. Computer Networks 44(2), 247–265 (2004)

---

[1] Fiber delay lines are used due to the absence of random access memory in optic.

3. Bouabdallah, N., et al.: Resolving the fairness issue in bus-based optical access networks. IEEE Commun. Mag. 42(11) (November 12-18, 2004)
4. Bouabdallah, N., Beylot, A.-L., Dotaro, E., Pujolle, G.: Resolving the Fairness Issues in Bus-Based Optical Access Networks. IEEE J. Select. Areas Commun. 23(8) (August 2005)
5. Cain, B., et al.: Internet Group Management Protocol, Version 3, RFC 3376, http://www.ietf.org/rfc/rfc3376.txt?number=3376
6. Careglio, D., Pareta, J.S., Spadaro, S.: Optical slot size dimensioning in IP/MPLS over OPS networks. In: 7th International Conference on Telecommunications. ConTEL 2003, Zagreb, Croatia, pp. 759–764 (2003)
7. Castel, H., Chaitou, M.: Performance Analysis of an Optical Man ring. In: 7th Informs Telecommunication Conference, Boca Raton Florida (2004)
8. Chaitou, M., Hébuterne, G., Castel, H.: On Aggregation in Almost All Optical Networks. In: Second IFIP International Conference on Wireless and Optical Communications Networks, WOCN 2005, Dubai, United Arab Emirates UAE, pp. 210–216 (2005)
9. Chaitou, M.: Performance of multicast packet aggregation with quality of service support in all-optical packet-switched ring networks. PHD thesis, Evry, France (2006)
10. DeCusatis, C.: Optical Wavelength Division Multiplexing for data communication networks. Fiber Optic Data Communication, 134–215 (2002)
11. Detti, A., Eramo, V., Listanti, M.: Performance Evaluation of a New Technique for IP Support in a WDM Optical Network: Optical Composite Burst Switching (OCBS). IEEE J. Lightwave Technol. 20(2), 154–165 (2002)
12. Dittmann, L., et al.: The European IST project DAVID: a viable approach towards optical packet switching. IEEE J. Select. Areas Commun. 21(7) (September 2003)
13. Estrin, D., et al.: Protocol Independent Multicast-Sparse Mode (PIM-SM): Protocol Specification. RFC 2362, http://www.ietf.org/rfc/rfc2362.txt?number=2362
14. Gambini, P., et al.: Transparent optical packet switching: network architecture and demonstrators in the KEOPS project. IEEE J. Select. Areas Commun. 17(7) (September 1998)
15. Hébuterne, G., Castel, H.: Packet aggregation in all-optical networks. In: Proc. First Int. Conf. Optical Commun. and Networks, Singapore, pp. 114–121 (2002)
16. Jiang, Y., Tham, C., Ko, C.: A probabilistic priority scheduling discipline for high speed networks. In: IEEE Workshop on High Performance Switching and Routing, May 2001, pp. 1–5 (2001)
17. Jiang, Y., Tham, C., Ko, C.: A probabilistic priority scheduling discipline for multiservice networks. Computer Communications 25(13) (August 2002)
18. O'Mahony, M.J., Politi, C., Klonidis, D., Nejabati, R., Simeonidou, D.: Future optical networks. IEEE J. Lightwave Technol. 24, 4684–4696 (2006)
19. Mortensen, B.B.: Packetisation in Optical packet switch fabrics using adaptative timeout values. In: Workshop on High Performance Switching and Routing, Poznan, Poland (June 2006)
20. Sauze, N.L., et al.: A novel, low cost optical packet metropolitan ring architecture. In: European Conference on Optical Communication (ECOC 2001), Amesterdam, October 2001, vol. 3, pp. 66–67 (2001)
21. Scheutzow, M., Seeling, P., Maier, M., Reisslein, M.: Multicast capacity of packet-switched ring WDM networks. In: Proc. IEEE INFOCOM, vol. 1, pp. 706–717 (2005)

22. Srivatsa, A., et al.: CSMA/CA MAC protocols for IP-HORNET: an IP over WDM metropolitan area ring network. In: Proceedings of GLOBECOM 2000, San Francisco, CA (2000)
23. Thompson, K., Miller, G.J., Wilder, R.: Wide-area Internet Traffic Patterns and Characteristics. IEEE Network 11(6), 10–23 (1997)
24. Uscumlic, B., Gravey, A., Morvan, M., Gravey, P.: Impact of Peer-to-Peer traffic on the Efficiency of optical packet rings. In: IEEE Broadnets 2008, UK, London, September 8-11, pp. 146–155 (2008)
25. Wolff, R.W.: Stochastic modeling and the theory of queues. Prentice-Hall, London (1989)
26. Yao, S., Mukherjee, B., Dixit, S.: Advances in photonic packet switching:an overview. IEEE Commun. Mag. 38(2), 84–94 (2000)

Performance Modelling and Traffic Characterisation of Optical Networks

Harry Mouchos, Athanasios Tsokanos, and Demetres D. Kouvatsos

NetPen - Networks and Performance Engineering Research Unit
Informatics Research Institute (IRI),
University of Bradford, Bradford
West Yorkshire, BD7 1DP, UK
harry.mouchos@networks-simulation.com, atsokanos@teilam.gr,
d.kouvatsos@bradford.ac.uk

Abstract. A review is carried out on the traffic characteristics of an optical carrier's OC-192 link, based on the IP packet size distribution, traffic burstiness and self-similarity. The generalised exponential (GE) distribution is employed to model the interarrival times of bursty traffic flows of IP packets whilst self-similar traffic is generated for each wavelength of each source node in the optical network. In the context of networks with optical burst switching (OBS), the dynamic offset control (DOC) allocation protocol is presented, based on the offset values of adapting source-destination pairs, using preferred wavelengths specific to each destination node. Simulation evaluation results are devised and relative comparisons are carried out between the DOC and Just-Enough-Time (JET) protocols. Moreover parallel generators of optical bursts are implemented and simulated using the Graphics Processing Unit (GPU) and the Compute Unified Device Architecture (CUDA) and favourable comparisons are made against simulations run on general-purpose CPUs.

Keywords: Wavelength division multiplexing (WDM), Synchronous Optical Networking (SONET), optical burst switching (OBS) protocol, Just Enough Time (JET) protocol, Generalised Exponential Distribution (GE), bursty traffic, self-similar traffic, Compute Unified Device Architecture (CUDA), Parallel Processing, Wavelength Division Multiplexing (WDM), Dense-wavelength Division Multiplexing (DWDM), Optical Packet Switching (OPS), Optical Burst Switching (OBS), Self-Similarity, Long-Range Dependence (LRD), Generalised Exponential (GE) Distribution, Graphics Processing Unit (GPU).

1 Introduction

Optical networks with wavelength division multiplexing (WDM) have recently received considerable attention by the research community, due to the increasing bandwidth demand, mostly driven by Internet applications such as peer-to-peer networking and voice over IP traffic. In this context, several routing and wavelength reservation schemes, applicable to present and future optical networks, have been proposed [1], [2], [3], [4].

More specifically, optical burst switching (OBS) [1] and [2] was proposed as an alternative to current schemes like SONET. The main feature of OBS is the separation of data burst transmission and the corresponding control information entitled Burst Head Packet (BHP). Each burst is preceded by its own BHP, which travels slightly ahead, configuring the switches and reserving a wavelength path for the upcoming burst. Several OBS protocols define the transmission time delay, called the offset, of a data burst following the BHP. A protocol such as the JET (Just Enough Time) [3] appears to outperform the TAG (Tell and Go)-based and JIT (Just in Time) protocols [3], [4]. However, there is still scope for further exploration in the OBS realm such as burst loss reduction and quality of service (QoS) provisioning. There is a requirement to develop networking protocols to efficiently use the raw bandwidth provided by the WDM optical networks [5].

In this tutorial a novel allocation protocol entitled the dynamic offset control (DOC) protocol is proposed in an attempt to tackle efficiently the aforementioned performance and QoS issues arising in the context of OBS networks. Similar proposals that provide feedback of the blocking probabilities to adapt the offset time have been proposed in the past, like [6]. However, even though [6] mentions Long-Range Dependence, it is never actually taken into consideration during the experiments. The paper assumes, exponentially distributed Optical Bursts arriving into the core optical network, with exponential Optical Burst sizes. The interarrival time of Optical Bursts is assumed to be exponentially distributed, and it depends on the number of available wavelengths (each of which would have OC-192 traffic Load - i.e. an aggregation of OC-192 traffic load per wavelength for each ingress edge node), the aggregation strategy employed at each edge node (based on the running protocol), the IP packet interarrival time distribution as well as Self-Similar Long-Range dependent traffic load, as well as the IP packet size distribution. Having multiple OC-192 traffic streams would definitely cause certain optical bursts to arrive into the core optical network simultaneously. This makes the Poissonian assumption incorrect. Even though it would be more logical to use GE [7] (albeit with different parameters) to model optical burst interarrival in the short time scales than the exponential distribution, this paper makes no assumptions on the bursts' interarrival times and aggregates and assembles optical bursts based on the measured IP packet sizes, the measured IP Packet interarrival times and the aggregation strategy chosen.

What [6] accomplishes (with the assumptions it makes), is to achieve fairness in the blocking probability for all source-destination pairs, regardless of the amount of hops in the light-path. This however, is accomplished at the expense of the lightpaths that don't have many hops (i.e. it increases the blocking probability in the bursts with a small number of hops to achieve fairness). This is accomplished by simply increasing offset times for the bursts that need to travel more hops on average, thus achieving less blocking. This, however, would have significant impact at the edge nodes' buffer length, which is not taken into account into the paper's analysis.

Other authors have considered to proactively drop bursts at the edge node, also based on blocking probability estimates [8]. However, burst arrivals are assumed to follow a Poisson process as well, and the blocking probability estimates are based on the Erlang-B formula. Even though [8] suggests that the blocking probability estimates need to use a different method if the arrival process is not Poisson, the paper does not provide nor does it investigate realistic traffic.

A Poissonian model is frequently considered in the literature. Yu, X. et al in [9] make the first proper attempt to get the characteristics of assembled traffic. The paper investigates several scenarios however the Poissonian assumption is again followed. No IP packet arrives at exactly the same time, IP packet interarrival time distribution and a fixed IP packet size is assumed. Even in the case where a variable IP packet size is considered, it assumes to be exponentially distributed instead of actually using the measured IP packet size distribution. Although [9] actually considers LRD in certain scenarios it is still affected by incorrect IP packet sizes and interarrival times. Realistic IP packet sizes and interarrival times are of vital importance when investigating characteristics of optical assembly.

The proposed algorithm in this paper does not rely solely on modifying offset times, but introduces the concept of having a preferred wavelength per source-destination pair, which adapts according to the network performance. Three different network scenarios were simulated, an arbitrary complex network, UK's educational backbone network JANET and the US NSFNET network with significant results.

The proposed DOC algorithm is extremely hard to model analytically as it makes absolutely no assumptions on IP packet sizes, IP interarrival times (even though GE is used to model time slots of less than 1 second, traffic load is self-similar in the simulation experiment). The nature of self-similarity and the proposed feedback algorithm makes it extremely difficult to track analytically.

The paper is organized as follows: Section 2 carries out traffic characterization of network traces taken by an OC-192 backbone network. Section 3 describes the OBS Network architecture considered in this paper. Section 4 presents the DOC OBS protocol. In section 5 the simulation and associated numerical results are shown. Section 6 describes the advantages of running simulations using a GPU. Conclusions and remarks for further work follow in Section 7.

2   Backbone Core Network Traffic Characterisation

Since 1995 many papers revealed and worked on Internet Traffic's self-similar characteristics. Since then, the research community employing simulations required generators of synthetic traces to match those revealed characteristics. It is proven that the traffic load per unit of time (bin) is self similar [10], [11]. Nevertheless, little research has been made on how to convert this self-similar traffic load into interarrival times for simulation purposes.

A couple of methods on how to convert Self-Similar traffic load into interarrival times are presented in [12]. However these methods resort to exponentially distributed packet arrivals (albeit with various mean rates), underestimating burstiness in the short timescales.

Some light to this problem was also shed in [14] which introduces the following findings:

Packet arrivals appear Poisson at sub-second time scales: The packet's interarrival time follows an exponential distribution. In addition, packet sizes and interarrival times appear uncorrelated.

Internet traffic exhibits long-range dependence (LRD) at large time-scales: In agreement with previous findings, Internet traffic is proven to be LRD at scales of seconds and above.

The measurements of the above-mentioned paper were taken on CAIDA monitor located at a SONET OC-48 (2488.32 Mbps) link that belongs to MFN, a US Tier 1 Internet Service Provider (ISP).

In essence, the general consensus is that although the traffic load is self-similar, traffic at the sub-second level could easily be modelled by Poisson or Exponential (for interarrival times) distribution, but with different parameters per time slot (usually 1-second slots).

Conveniently, many researchers use the exponential or a fixed-rate arrival to model incoming Internet traffic to an optical edge node, taking advantage of the fact that IP packet arrivals could be Poisson distributed at sub-second time scales however some ignore LRD altogether. A good example is [15]. The Poissonian assumption however, remains simplistic, since it doesn't take into account packets that arrive simultaneously to the switch, i.e. burstiness in the small time domain (sub-second). (We consider simultaneous arrivals, packet arrivals that happen within one microsecond of each other).

An investigation into a high-bandwidth backbone core network was required to attempt to analyze the packet size distribution and interarrival times. Anonymized traces were downloaded from CAIDA of an OC-192 link (9953.28 Mbps) [16].

The equinix-chicago Internet data collection monitor is located at an Equinix datacenter in Chicago, IL, and is connected to an OC192 backbone link (9953.28 Mbps) of a Tier1 ISP between Chicago, IL and Seattle, WA.

2.1 Hardware

The infrastructure consists of 2 physical machines, numbered 1 and 2. Both machines have a single Endace 6.2 DAG network monitoring card. A single DAG card is connected to a single direction of the bi-directional backbone link. The directions have been labeled A (Seattle to Chicago) and B (Chicago to Seattle) [16]. Both machines have 2 Intel Dual-Core Xeon 3.00GHz CPUs, with 8 GB of memory and 1.3 TB of RAID5 data disk, running Linux 2.6.15 and DAG software version dag-2.5.7.1. In a test environment both machines dropped less than 1% of packets with snaplen 48 at 100% OC192 line utilization, using a Spirent X/4000 packet generator sending packets with a quadmodal distribution, with peaks at 40, 576, 1500 and 4283 bytes [16].

2.2 Time Synchronization

Both physical machines are configured to synchronize their hardware clock via NTP. The DAG measurement cards have their own internal high-precision clock, that allows it to timestamp packets with 15 nanosecond precision. Every time a traffic trace is taken, The DAG internal clock gets synchronized to the host hardware clocks right before measurement starts. NTP accuracy is typically in the millisecond range, so at initialization the clocks on the individual DAG measurement cards can be off by a couple of milliseconds relative to each other. The precision (i.e. timing within the packet trace) within a single direction of trace data is 15 nanosecond for the DAG files, and 1 microsecond for PCAP files.

2.3 CAIDA Traffic Analysis

IP Packet Size Distribution. In 2003, official packet size distribution measurements were monitored and recorded over standard IP internet traffic on specific dates at specific times [17]. In particular, three Cisco routers have been used for this purpose and the measurements have been taken at a daily peak on a five minute average on March 2003, whilst another, but more reliable, IP packet length distribution has been captured from 39 trace files between May 13th 1999 at 19:13:46 PDT and May 19th 1999 at 13:02:20 PDT. The latest distribution of IP packet sizes was seen at the NASA Ames Internet Exchange (AIX) by CAIDA and since they contain contributions from the different workloads carried by the network at different times of day, they should represent more of an average picture of the packet size distribution than any other individual trace. However, both of the above sources demonstrate very important aspects of internet traffic. Specifically, the majority of the packets seen are 40, 576 and 1500 bytes and all the internet traffic packets are traced at the range between 23 and 1500 bytes. More than 65% of the packets have been traced at a smaller than 576 bytes size and 50% of the total byte volume belongs to the 1500 bytes packet size.

The analysis, however, of the data collected by the equinix-chicago Internet data collection monitor of an OC192 backbone link (9953.28 Mbps) prove that the IP packet size distribution has changed significantly.

Although the IP Packet size distribution is way too granular to describe in detail, roughly, about 50% of the IP packets appear to have a size of around 40 bytes, about 30% of the IP packets appear to have a size of roughly 1500 bytes, with the remaining ranging from 40 to 1500 bytes. The actual Probability Density and the Cumulative Probability of the Packet Sizes are shown (c.f., Fig. 1 and Fig. 2).

Burstiness. As mentioned before there is a probability that a number of packets will arrive simultaneously (i.e. with a time difference of less than 1 micro second) forming batches of IP packets arriving. Distribution Fitting using MATLAB revealed that the batch sizes throughout the traces' duration can be described by the Geometric Distribution:

$$CDF = 1 - (1-p)^k \tag{1}$$

and

$$PDF = (1-p)^{k-1}p, k \in \{1,2,3,...\} \tag{2}$$

Interarrival Times. Fitting of the Exponential Distribution to the interarrival times has been conducted with MATLAB is shown in (c.f., Fig. 4). It was proven that the exponential distribution accurately fits the measurements: The fitting

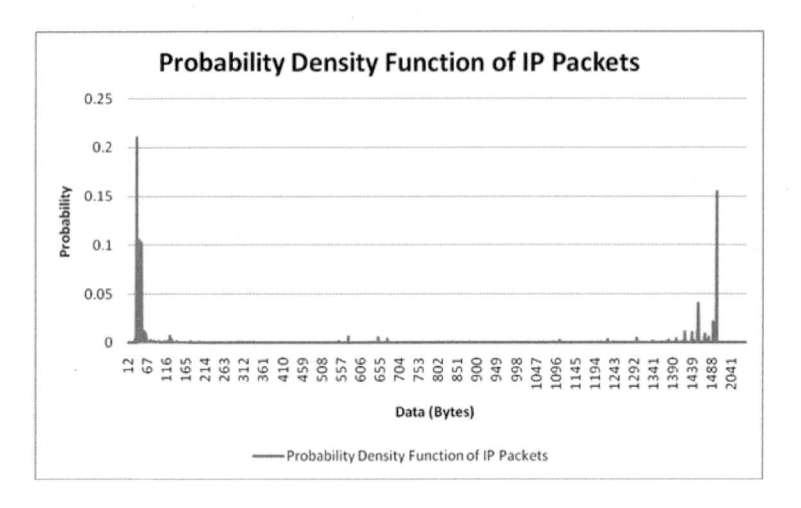

Fig. 1. Probability Density of IP Packet Sizes resulted from the analysis of the OC-192 traffic traces

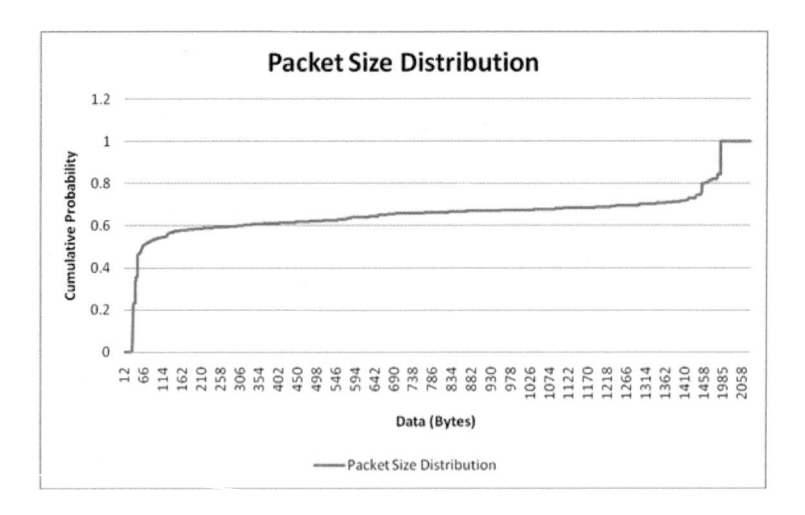

Fig. 2. Cumulative IP Packet Size Distribution resulted from the analysis of the OC-192 traffic traces

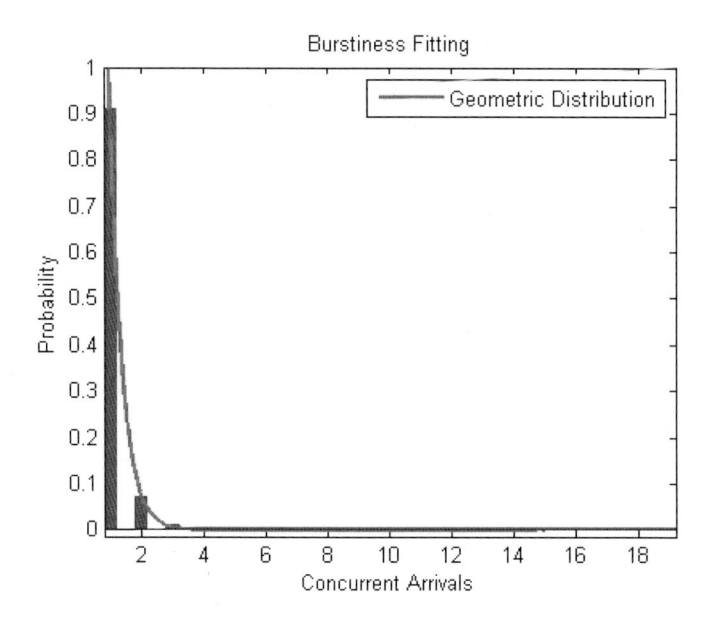

Fig. 3. Burstiness fitting with the Geometric distribution

shown in the image above, was calculated to have 99.98% accuracy (i.e. $R^2 = 0.999878489995854$)[1].

The average batch size estimated over the entire trace duration is 2.235122.

Distribution fitting was conducted for each 1-second duration slot in the trace duration. With an accuracy of above 98% the exponential distribution fits the interarrival time distribution albeit with different mean for each 1-second interval.

Therefore at sub-second scales we have exponentially distributed batch arrivals, with a geometrically distributed batch size. These findings suggest that the Generalized Exponential Distribution [7], [18] is ideal to model arrival of IP Internet Traffic on a high bandwidth backbone link, at sub-second intervals.

Load Traffic Self-Similarity. It has been well established in the literature that network traffic load shows self - similar traffic characteristics [19], [20], [21]. It is, therefore, required to analyse the data provided by CAIDA [16] for Self-Similarity.

The cumulative traffic (in Bytes) per second for over 3700 seconds of cumulative traffic was analysed. Using the wavelet method to estimate the self-similarity degree (Hurst Parameter), resulted in a highly self-similar LRD traffic with Hurst = 0.935 (95% Confidence Interval [0.876, 0.994]) (c.f., Fig. 5).

---

[1] The coefficient of determination, is a good measure of how well the chosen distribution fits the given data. It must lie between 0 and 1, and the closer it is to 1, the better the fit. It is symbolised as R^2.

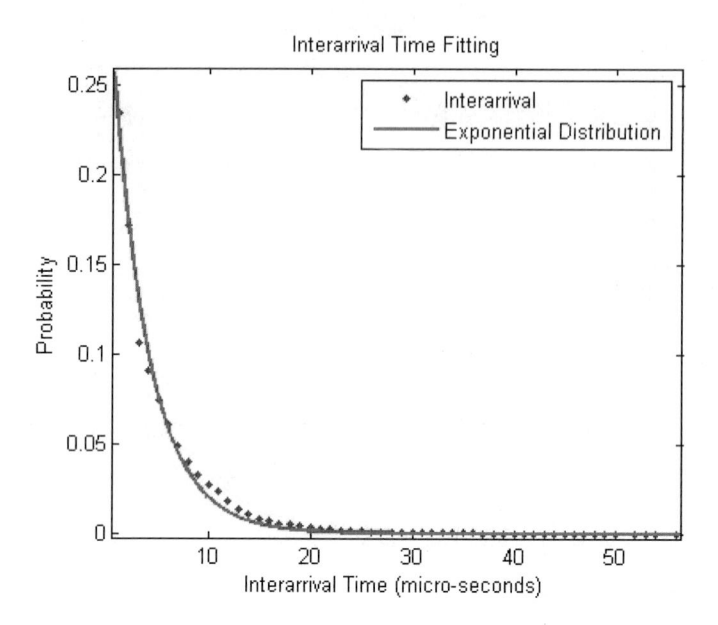

Fig. 4. Interarrival time fitting with the Exponential distribution

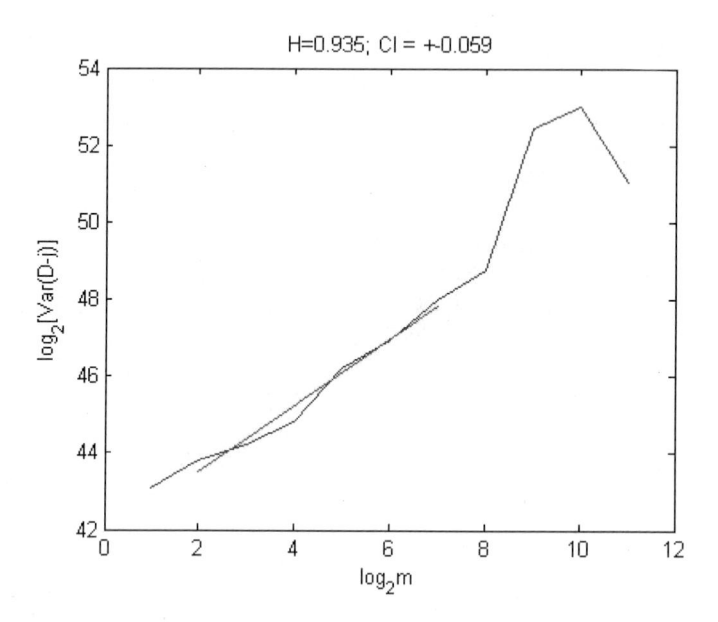

Fig. 5. Estimation of the Hurst parameter using the wavelet method

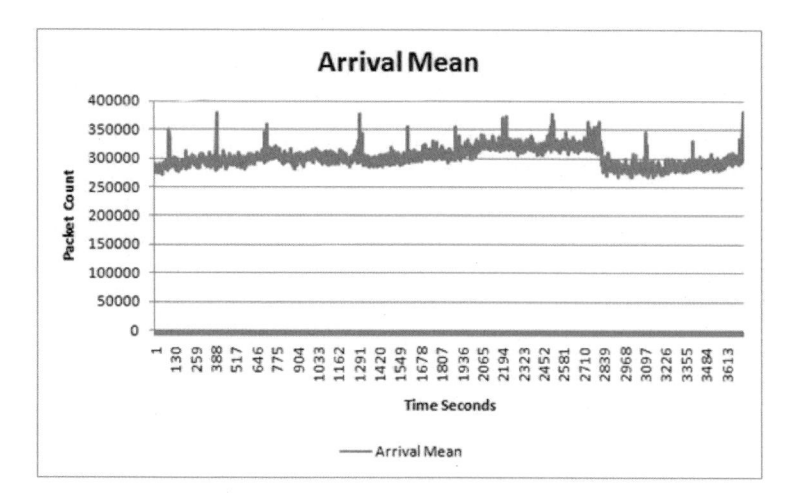

Fig. 6. Estimated MAR per time slot over the entire duration of the trace *packetcount/second*

This confirms previous findings of past research attempts by the community [22], [23], [24], [25].

General Observations. Fitting of the exponential distribution on the interarrival times was conducted for each 1-second duration slot of the trace duration, as well as fitting of the geometric distribution for the batch sizes for the same 1-second duration slots.

The mean arrival rate (MAR) and the Squared Coefficient of Variation (SCV) is calculated for each second of the entire duration of the backbone traces. The estimations are shown below:

Arrival Mean Estimation (per slot). In (c.f., Fig. 6) the estimations of the MAR lambda (measured in Bytes/second) that resulted from the exponential distribution fitting is shown for the entire duration of the trace. The total traffic load self-similar characteristics require a variable mean rate which is proven by the CAIDA traffic trace analysis. Researchers may use this variable rate to accurately model the interarrival times of packets for each time slot. However, the exponential distribution is not sufficient to model batches of packets arriving at the same time.

Each estimation of the SCV was calculated with accuracy of above 98%[2]. In (c.f., Fig. 6) the estimations are shown for the SCV C_α^2, that resulted from the Geometric Distribution fitting during the entire duration of the trace. The fitting was performed on the average batch size (i.e., simultaneous arrivals)[3] for each time slot.

---
[2] With $R^2 > 0.98$.
[3] Arrivals are considered simultaneous if the occur within 1 μsec.

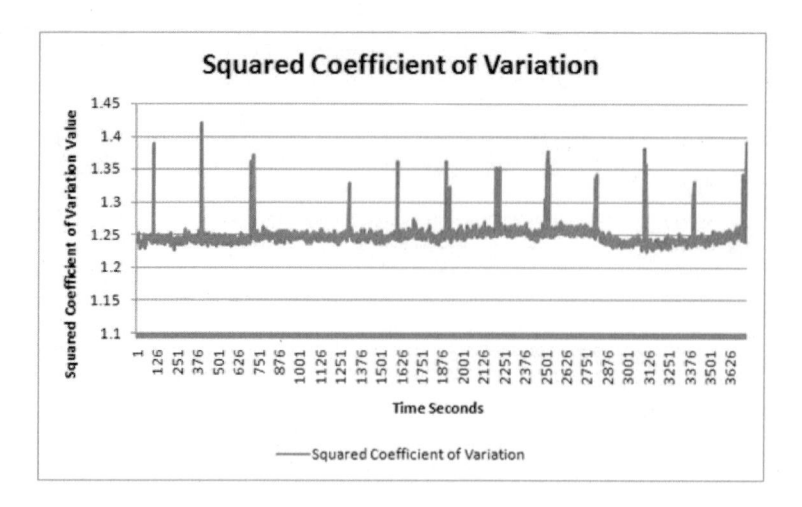

[Figure: Squared Coefficient of Variation plot, y-axis from 1.1 to 1.45, x-axis Time Seconds from 1 to 3626]

Fig. 7. Estimated SCV per time slot over the entire duration of the trace (value/second). Values of greater than 1, indicate burstiness in the short-time scales.

Summary. Traffic Characterisation was conducted on traces taken by CAIDA of an OC-192 Backbone Core Network. Results have shown that even in the short time scales (sub-second duration) some burstiness persists. This could be better modelled by the Generalised Exponential Distribution as it maintains the analytical tractability of the Exponential Distribution but does not underestimate burstiness. For each time slot, the estimated SCV is greater than 1, indicating batches of simultaneous arrivals in the node. On average over the entire period the value of the SCV was estimated at 1.534340.

To validate the estimation of λ and C_α^2, (c.f., Fig. 7) shows a comparison between the measured traffic from the CAIDA traces and the traffic resulted by the estimation. It is obvious that the traffic loads match accurately. Multiple aggregated streams arrive to an optical backbone node, converted, demultiplexed and transmitted through several wavelengths over the same fiber. Potentially this could increase the average batch size of simultaneous arrivals significantly. The GE distribution is proven to be ideal for modelling heavier traffic loads by simply using higher values for the C_α^2.

3 OBS Network Architecture

An OBS network uses one-way reservation protocols, sending a control packet to configure the switches along a path, followed by a data burst without waiting for an acknowledgement for a successful connection establishment. Source nodes have an offset time depending on the destination node. OBS networks employ several different protocols for bandwidth reservation and vary in architectures.

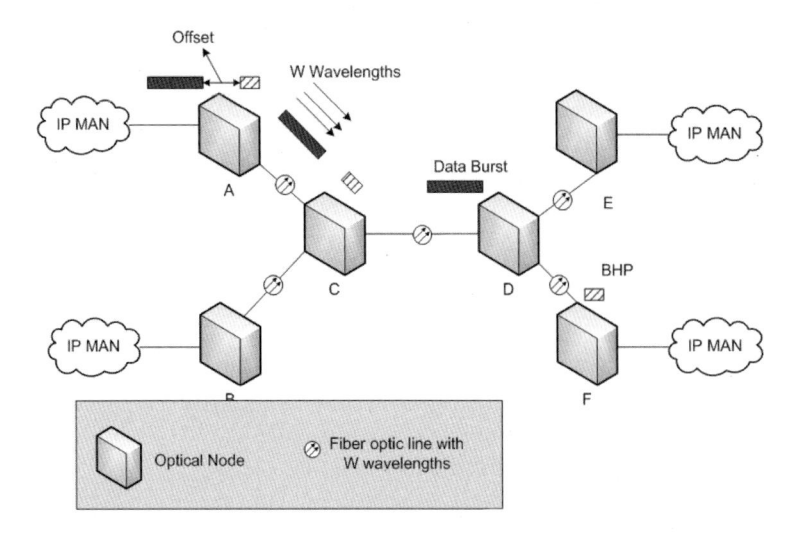

Fig. 8. A simple OBS network architecture

3.1 Just-Enough-Time (JET)

For illustration purposes, a simple OBS network model with two source and two destination nodes is displayed in (c.f., Fig. 8). Source nodes A and B transmit data bursts into the network using the OBS JET protocol with random destinations (E and F). Every source node transmits data bursts after an offset time $Offset_{JET}$ since the transmission of the control packet (also called Burst Header Packet - BHP).

In JET no buffering is required at the intermediate nodes, due to the fact that the bursts are stored in buffers at the source node, and they are transmitted after an offset time large enough to allow the control packet to be processed.

Another attractive feature of JET is the delayed reservation. Bandwidth is reserved only from the moment the burst arrives till the moment it departs the intermediate node. These times are specified in the control packet. In JET the control packet needs to include the burst length and the remaining value of the offset time $Offset_{JET}$. The control packet is timestamped with its arrival time at each intermediate node along with its expected transmission. In addition the control packet will carry an up-to-date value of $Offset_{JET}$ to the next node.

In case, Fiber Delay Lines exist in the intermediate nodes, they can be entirely utilized for the purpose of resolving conflicts instead of waiting for the control packet to be processed [26].

However, due to the high costs, it is more likely that no buffering will be available in the core optical network. Because of the lack of buffering burst dropping is quite probable. To tackle this, classifying bursts according to a priority system is suggested in [26]. This suggestion involves adding an additional offset time T' to the base $Offset_{JET}$ of the burst of a high-priority class. To completely isolate one priority class from another, the authors suggest that the high priority class'

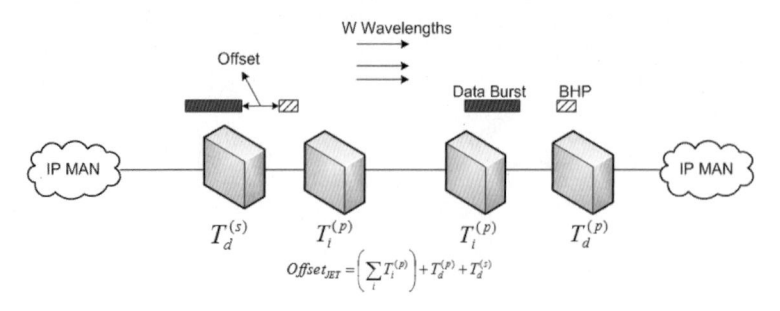

Fig. 9. Offset Time Calculation

offset time to have a value of a multiple of the lower priority class' average burst size. Blocking (or loss) of data bursts occurs in node C.

Under the JET protocol, a fixed offset time is used which is calculated in a manner that takes into account the processing delays of the BHP at the intermediate switches. Let T_i be the processing delay of the BHP at an intermediate switch i, $T_d^{(p)}$ be the processing delay of the BHP at the destination switch and $T_d^{(s)}$ denote the time to setup and configure the destination switch. Clearly, the $Offset_{JET}$ for JET is determined by

$$Offset_{JET} = \left(\sum T_i^{(p)}\right) + T_d^{(p)} + T_d^{(s)} \qquad (3)$$

Let $\{i\}$ be the set of intermediate switching nodes belonging to a path from a source node to a destination node. The offset time calculation for the JET protocol is illustrated in Fig. 9 for a path that includes two intermediate switching nodes between the source and the destination of the data burst. The offset time needs to be long enough to account for the processing time of the BHP at the two intermediate nodes and the destination, plus the setup time at the destination [27].

3.2 Routing and the Control Wavelength

Let's assume that every optical fiber link of the network has W wavelengths. One wavelength is used for the transmission of BHP and control packets while the (W-1) remaining wavelengths are used to transmit bursts.

The control wavelength uses Time Division Multiplexing (TDM) for every node in the network, including the intermediate and destination nodes, in order to avoid collisions by assigning a time slot to every node. A source node transmits a BHP containing information about the destination address, burst size and offset of the upcoming burst. To achieve the wavelength reservation for the upcoming burst, the BHP travels through the control wavelength to every intermediate node on its path until it reaches its destination.

3.3 Incoming Traffic Characteristics

For each Optical Network Topology considered, all edge nodes act as both ingress and egress nodes. This means that all nodes situated at the edge of the optical network, both transmit and receive optical bursts. Each node is considered to have 4 incoming and 4 output ports. Each port-pair is transmitting and receiving on a separate wavelength. Each fiber in the optical network is considered to have 4 wavelengths.

After the analysis of the CAIDA traffic, no assumptions are made during the simulation experiments regarding optical burst interarrival times. For each ingress edge node, OC-192 traffic is assumed for each wavelength. Self-similar traffic load for each wavelength is generated based on the traffic measurements, and converted into interarrival times. The IP packets arriving, are assembled into optical bursts according to destination, and are stored in electronic buffers (each buffer associated with a specific destination node) prior to their conversion into the optical domain. The optical burst assembly strategy chosen is using both a time limit as well as a burst length limit. After the optical bursts are assembled they are transmitted into the core optical network according to the specific OBS protocol.

4 The Dynamic Offset Control (DOC) Allocation Protocol

Like JET, the Dynamic Offset Control (DOC) Allocation Protocol, possesses the qualities that made JET so attractive. DOC requires no buffering at the intermediate nodes and it has the delayed reservation feature [38]. However, due to its control packet mechanism, no additional information needs to be embedded in the control packet at each node on the path.

Each source node transmits a BHP containing information about the destination address, burst size and offset of the upcoming burst. The initial offset at every source node is calculated in the same manner as in the JET protocol [26], [30]. When a burst arrives at its destination or when a collision is detected at some intermediate node, the destination or intermediate node will, respectively, transmit a control packet in their own time slot to the source node with information regarding the arriving or blocked burst. This information includes the burst's destination, offset time and size. In this fashion, the source node calculates progressively a blocking percentage for every source - destination pair.

4.1 Problem Statement

Protocol Formulation
Notation
Given:

- $LimPer_{sd}$: The Blocking Tolerance for the source - destination pair (s, d).
- $OffsetStep$: This value is used to increment or decrement the offset value for each source - destination Pair $Offset_{sd}$ up to its $MAX_{Offset_{sd}}$ value or down to $MIN_{Offset_{sd}}$ value.
- $Hurst$: Self-Similarity Degree.

- $TopologyInfo$: Topology information which includes: actual topology, number of wavelengths and routing.

Variables:

- $BPer_{sd}$: Blocking percentage for the source - destination pair (s, d).
- $Offset_{sd}$: The current offset value for the source - destination pair (s, d).
- $INIT_{Offset_{sd}}$: The default/initial offset value for the routing path of the source - destination pair (s, d). The initial offset value is a value between the minimum offset value and the maximum offset value for DOC. In the JET protocol, the initial offset value is also the minimum.
- $MIN_{Offset_{sd}}$: The minimum offset value for the routing path of the source - destination pair (s, d). The minimum offset value is the minimum time required for all intermediate nodes to process the BHP packets.
- $MAX_{Offset_{sd}}$: The maximum offset value for the routing path of the source - destination pair (s, d). The maximum is calculated based on the QoS requirements of the applications running on the network.
- $PWave_{sd}$: Describes the preferred wavelength to use to transmit bursts from the source node s to the destination Node d. This value is adapted according to each source - destination pair's traffic parameters by the DOC protocol.

Objectives:
Achieve Higher Throughput
Maintain Mean Queue Lengths
Higher Utilisation

4.2 Outline of the Algorithm

A summary relating to the operational aspects of the DOC protocol is presented in a stepwise fashion below.
Begin
Input Data: $TopologyInfo$, $LimPer_{sd}$, $OffsetStep$, $Hurst$

Step 1: Calculate initial offset $INIT_{Offset_{sd}}$, minimum offset $MIN_{Offset_{sd}}$ and maximum offset $MAX_{Offset_{sd}}$ at each source node for each destination node as in JET protocol; for each node i, i = 1, 2, ... , N. Use the first available wavelength to transmit bursts through.

Step 1.1: Generate Self-Similar Traffic Load and employ the GE distribution to model the interarrival times of data bursts per time slot;

Step 1.2: Define at each source node initial offset values for each destination node;

Step 1.3: Bursts are transmitted through the preferred wavelengths $PWave_{sd}$ unless they are busy transmitting, in which case the next available one is chosen.

Step 2: If a data burst arrives successfully at a destination node, then a control packet is sent to the source node to update $BPer_{sd}$;

Step 2.1: If $BPer_{sd} < LimPer_{sd}$
 Step 2.1.1: If $Offset_{sd} > MIN_{Offset_{sd}}$ then $Offset_{sd} = Offset_{sd} - OffsetStep$ otherwise $Offset_{sd} = MIN_{Offset_{sd}}$
Step 2.2: If $BPer_{sd} = LimPer_{sd}$
 Step 2.2.1: $Offset_{sd} = INIT_{Offset_{sd}}$
Step 2.3: If $BPer_{sd} > LimPer_{sd}$
 Step 2.3.1: If $Offset_{sd} < MAX_{Offset_{sd}}$ then $Offset_{sd} = Offset_{sd} + OffsetStep$ otherwise
 Step 2.3.2: $Offset_{sd} = INIT_{Offset_{sd}}$ and set a new preferred wavelength $PWave_{sd}$
Step 3: If a data burst is blocked at an intermediate node, then a control packet is sent to the source node to update $BPer_{sd}$;
 Step 3.1: If $BPer_{sd} < LimPer_{sd}$
 Step 3.1.1: If $Offset_{sd} > MIN_{Offset_{sd}}$ then $Offset_{sd} = Offset_{sd} - OffsetStep$ otherwise $Offset_{sd} = MIN_{Offset_{sd}}$
 Step 3.2: If $BPer_{sd} = LimPer_{sd}$
 Step 3.2.1: $Offset_{sd} = INIT_{Offset_{sd}}$
 Step 3.3: If $BPer_{sd} > LimPer_{sd}$
 Step 3.3.1: If $Offset_{sd} < MAX_{Offset_{sd}}$ then $Offset_{sd} = Offset_{sd} + OffsetStep$ otherwise
 Step 3.3.2: $Offset_{sd} = INIT_{Offset_{sd}}$ and set a new preferred wavelength $PWave_{sd}$
End.

4.3 DOC Algorithm Description

The key idea of the proposed DOC protocol is to keep the loss rate between the source-destination pair to the minimum [38]. Optical bursts are assembled and transmitted into the core optical network. Should a burst arrive at the destination node, or if a blocked data burst at an intermediate node occurs, a control packet will provide new data for the source node to calculate the new $BPer_{sd}$. If the blocking percentage drops under $LimPer_{sd}$ in case of a successfully transmitted data burst, it will start decreasing $Offset_{sd}$ by $OffsetStep$ unless the offset is equal to the $MIN_{Offset_{sd}}$ value for the source - destination pair. However, if the blocking percentage exceeds $LimPer_{sd}$ in case of a blocked data burst, then the node sets the pair's offset value to the initial offset value and increases it progressively by $OffsetStep$ until $BPer_{sd}$ drops again under $LimPer_{sd}$ or the offset reaches $MAX_{Offset_{sd}}$.

The "delay time to reaction" is a well-known problem which accompanies any feedback-based algorithm. To tackle this, source nodes in DOC adjust their offset values and preferred wavelengths based on percentage values of the performance metrics in question and thresholds that are specified as input to the algorithm. These percentage values, change according to feedback received and when these values cross certain thresholds the algorithm adjusts the offset values and preferred wavelengths. This "slow" adaptation prevents sudden reactions to delayed feedback where the situation has changed by the time the feedback is received.

Thus, $Offset_{sd}$ may be defined by

$$Offset_{sd} = \begin{cases} INIT_{Offset_{sd}} \leq Offset_{sd} \leq MAX_{Offset_{sd}}, \\ \quad if BPer_{sd} > LimPer_{sd} \\ MIN_{Offset_{sd}} \leq Offset_{sd} < INIT_{Offset_{sd}}, \\ \quad if BPer_{sd} \leq LimPer_{sd} \end{cases} \quad (4)$$

Although increasing the offset between the BHP packet and the burst helps towards lowering blocking, it may prove that it is not sufficient to decrease blocking between the two nodes. To this end, the source is required to choose an alternate route, which is implemented by introducing the concept of *preferred wavelength*.

4.4 Preferred Wavelength

The network transmits each burst on the first available wavelength through its corresponding output port, however, in DOC the concept of *preferred wavelength* is introduced. Each source node maintains a preferred wavelength for each destination node. The source node will first attempt to transmit the burst through the specified preferred wavelength. If the preferred wavelength is occupied (because of a transmission of another burst), the first available is chosen instead in a cyclic round-robin fashion.

Initially, the source nodes choose the first wavelength as the preferred for each destination node. If changing the offset is not enough to keep the Burst Loss percentage $LimPer_{sd}$ to a low value, then a new preferred wavelength $PWave_{sd}$ is chosen for this destination node. The whole purpose of choosing a preferred wavelength is for the various source-destination pairs to reach a state of balance that minimizes collisions.

5 The Simulation and Experimental Results

An event-driven simulator in CUDA (using an NVIDIA 8800 GTX), MATLAB and C# .NET was developed for the quantitative analysis of three types of optical network topologies, an arbitrary topology, JANET and NSFNET with source nodes transmitting towards all others having four wavelengths each and no wavelength conversion capabilities. The simulator uses a non approximation methodology which takes into account the actual self-similar traffic characteristics of an optical network analyzed in the beginning of this paper to demonstrate the effectiveness of the newly proposed DOC protocol and get unbiased results. The link lengths vary and are specified in each topology description. The IP packet traffic load is self-similar and generated for each wavelength of each source node. The IP packet interarrival times are distributed according to the GE distribution, with mean and SCV calculated on a time-slot basis. The Optical Burst Assembly Strategy employed is the Time - Length Constraint strategy with the size of the SONET OC-192 frame. Taking advantage of the GPU's processing power [29] facilitated simulation of the scenarios in high detail and speed.

The simulator generates bursts randomly for the entire duration of the simulation run, after aggregating incoming IP packets, with random destinations. During the simulation, the blocking percentage of every pair, the throughput and the buffer length at the source nodes are calculated.

5.1 Experimental Results

Network Topology Scenarios. Various topologies were used to perform a comparison between the JET and the DOC protocol. An arbitrary topology (tested in two modes, a simple mode with three source nodes and three destination nodes and a bidirectional mode with all edge nodes being source and destination nodes), JANET the UK Education Backbone network and the US' NSFNET network.

Arbitrary Topology. In the arbitrary topology three source nodes were initially considered (shown in c.f., Fig. 10 as A, B and C). Each source node had a choice of three destination nodes (shown in c.f., Fig. 10 as K, L and M).

Bidirectional Topology. The topology shown in c.f., Fig. 10 was used as the bidirectional topology. In this scenario, all edge nodes (A, B, C, K, L and M) were considered as both ingress and egress nodes. Each node could transmit bursts towards all other edge nodes.

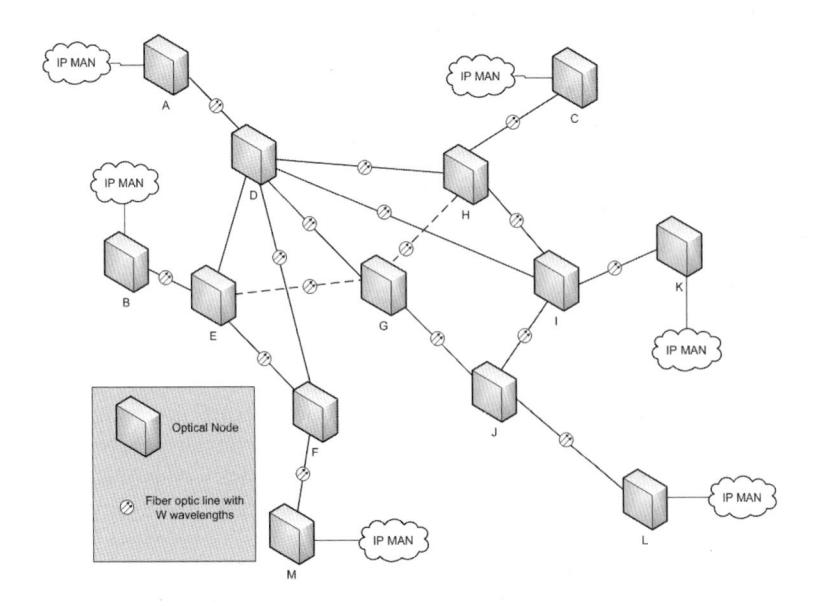

Fig. 10. Arbitrary Topology used for simulation purposes. This topology was also used in a bidirectional mode.

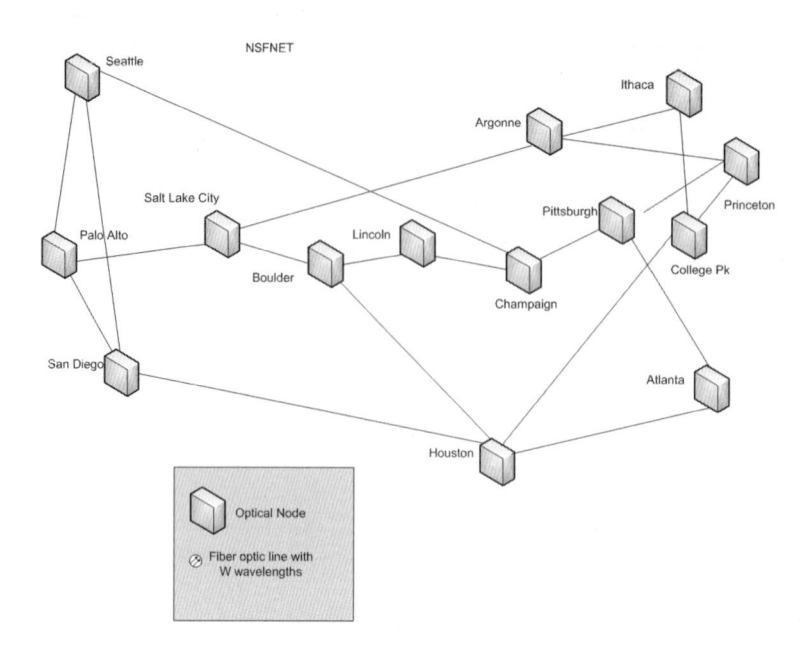

Fig. 11. The NSFNET T1 Network Topology

NSFNET Network Topology. The NSFNET Topology was also used as a network scenario (c.f. Fig. 11). All nodes in the network were considered as both ingress and egress nodes.

JANET Topology. Similarly to NSFNET, the JANET scenario (c.f., Fig. 12) used all nodes as both ingress and egress nodes.

Overall Blocking and Throughput. The mean Overall Blocking and Overall Throughput for all Topologies are shown in Fig. 13 and Fig. 14. It is obvious that for every topology scenario considered, the DOC protocol performed considerably better than the JET protocol. In Fig. 13 and Fig. 14 the average blocking and average throughput over the entire network are shown.

Port (Wavelength) Utilisation. Each source node transmits a burst towards its destination over a specific wavelength through an output port. The source ingress nodes each have 4 ports that transmit the bursts. Under the JET protocol, they transmit simply through the first available port. Certain wavelengths are over-utilised in comparison to the remaining ones and as a consequence there is a higher chance of collision, since a great number of bursts are traversing the same wavelengths.

The newly proposed DOC protocol, achieves fairness in terms of port utilisation by using various preferred wavelengths for each destination node. It is

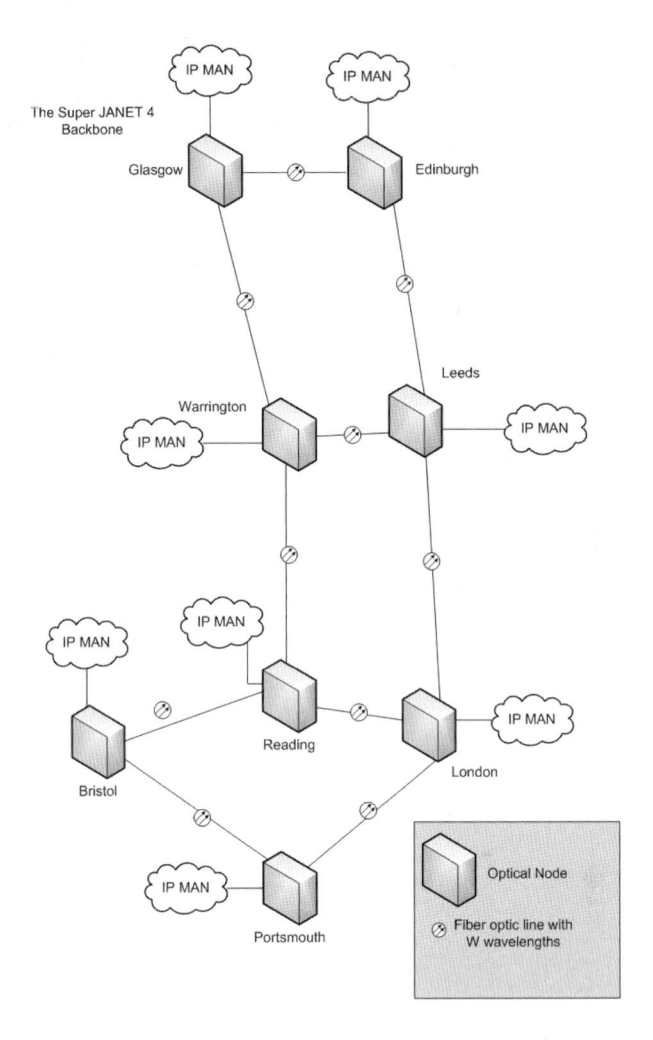

Fig. 12. The UKs Educational Network Super JANET 4

shown in Fig. 15 that on average most wavelengths will be better utilised under the DOC protocol in all the topologies investigated.

Mean Queue Length. Measurements were taken for each network topology scenario on the mean buffer length of each source node (measured in IP packet counts) under both protocols investigated, in order to investigate the effect of the variable offset values to the edge node buffers. It is obvious by the results (c.f., 16), that under DOC, the mean buffer lengths were not significantly affected by the offset value variability in comparison to JET. The blocking probability decrease achieved by DOC and the option of using preferred wavelengths for transmission, allows the offset values to remain at their minimum for the greatest part of the simulation duration.

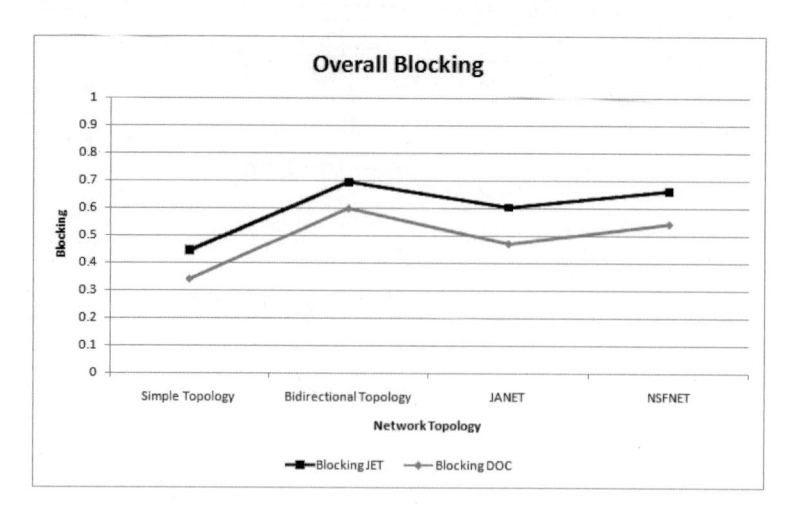

Fig. 13. Overall blocking probability comparison between JET and DOC for various network topologies

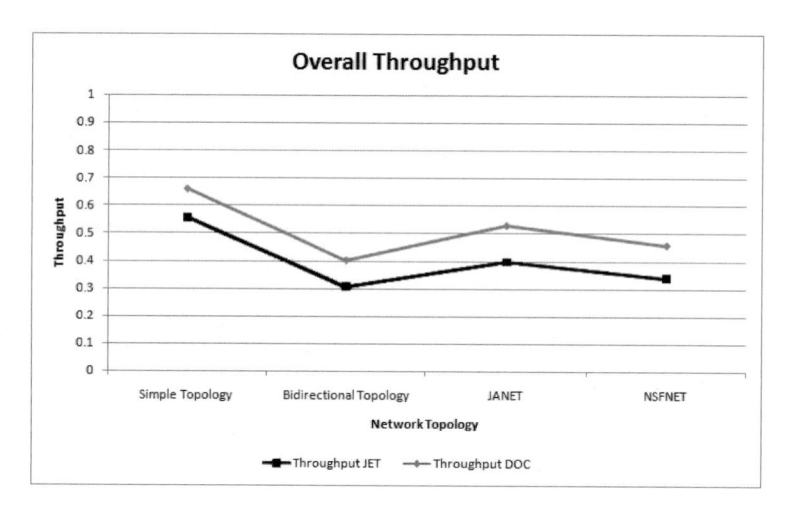

Fig. 14. Overall throughput probability comparison between JET and DOC for various network topologies

Blocking and Throughput Probability. Apart from the overall blocking, blocking measurements were taken on a source - destination pair basis. It is obvious (c.f., 17) that in most pairs, DOC decreased the blocking probability significantly since under DOC, source destination pairs adjust their offset values and preferred wavelengths based on performance metrics and feedback from the network.

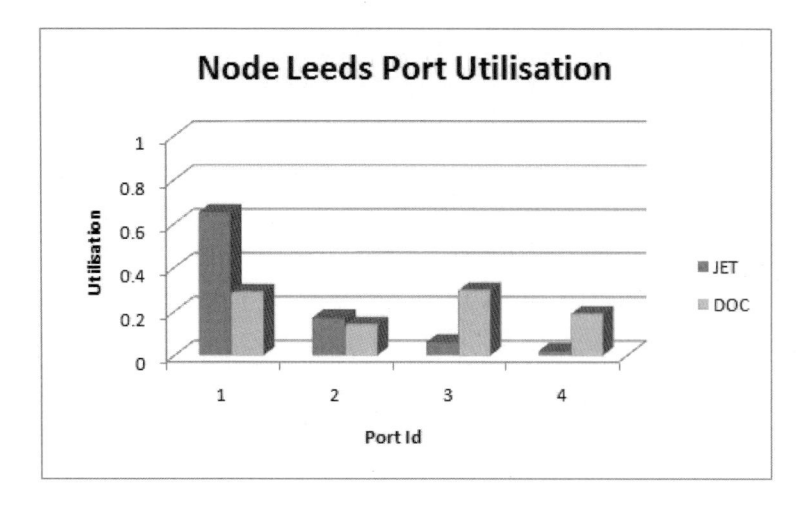

Fig. 15. JANET topology wavelength utilisation for the 'Leeds' source node

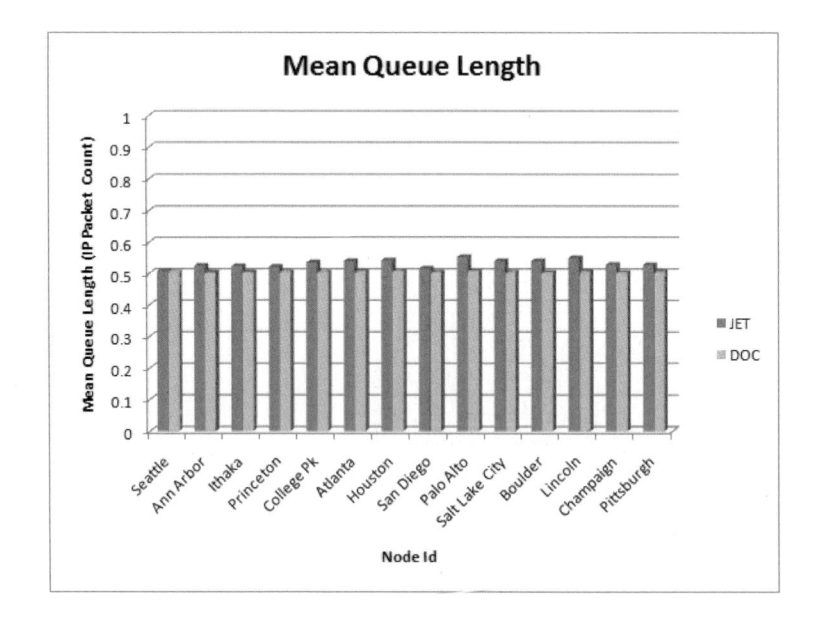

Fig. 16. Mean queue length for source nodes in NSFNET Topology network

Similarly, throughput measurements on a source-destination pair basis were calculated. All the network topologies achieved much higher throughput when running under the proposed DOC protocol (c.f., 18).

DOC vs. Multiclass OBS JET networks. Note that depending on QoS policies for the various networks, it might be justifiable for JET to incorporate

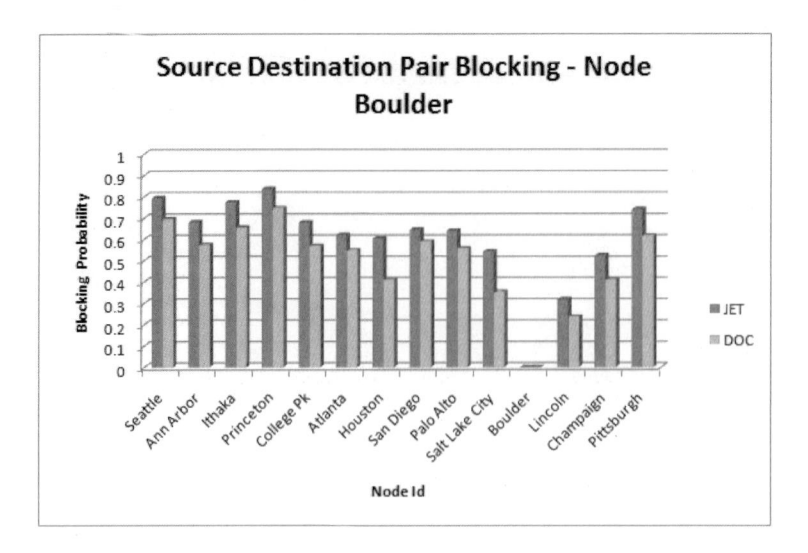

Fig. 17. Source - destination Pair Blocking Probability Comparison in NSFNET topology scenario for the source node at 'Boulder' between the JET and DOC protocol

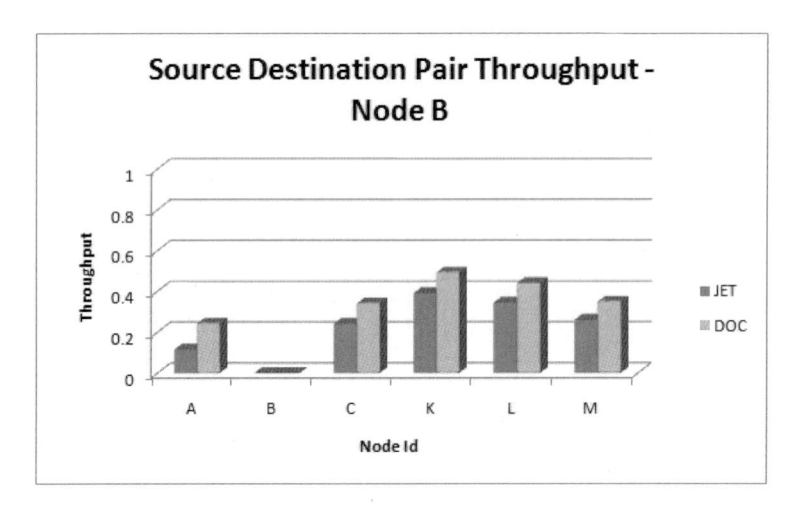

Fig. 18. Source - destination Pair Throughput Probability Comparison in NSFNET topology scenario for the source node at 'Boulder' between the JET and DOC protocol

higher offset values in order to allow control packets added time to be more effectively processed so that network resources are properly allocated, as for example in the case of high-priority bursts [31].

Below there are sample comparisons between JET and DOC, where JET's offset value is higher than the minimum, for demonstrative purposes. It is demonstrated that DOC outperforms JET under such a scenario as well. It's worth

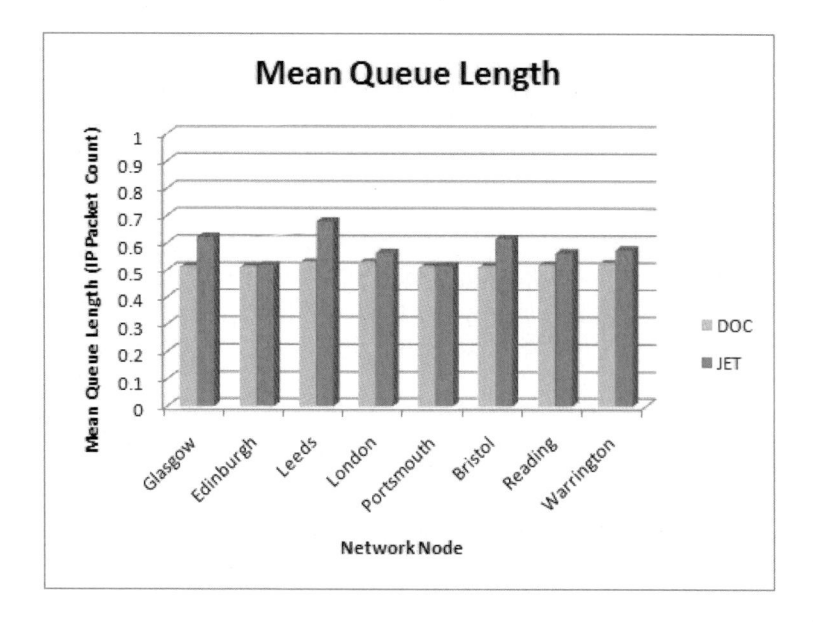

Fig. 19. Mean queue length for source nodes in JANET Topology network

noting that although the Mean Queue Length rises when JET uses a higher offset value, DOC manages to keep the Mean Queue Length lower while also achieving a decrease in blocking.

6 Simulation with the GPU

Due to self-similarity's high complexity and the associated analytical difficulty, simulation became the most favourable tool to evaluate the performance of networks [39]. Consequently, many self-similar trace generators were invented towards the creation of synthetic traces of network traffic such as On/Off Sources and $M/G/\infty$ [37].

The inherent complexity of optical networks is an additional obstacle to the speed and duration of the simulations. Due to WDM / DWDM, current optical fiber speeds in WANs exceed several Tbps, when each wavelength's speed is 10 - 40 Gbps. A simulation would practically need millions of simulated optical bursts and for each optical burst a large number of IP Packets. Therefore, a Uniform Random Number Generator with an extended period is necessary. One such huge-period generator was proposed in [32], called Mersenne Twister (MT), which has a massive prime period of $2^{19937} - 1$.

However, the computational requirements remain extremely taxing when running an optical network simulation that is fed by self-similar LRD traffic flows on an average home computer. Thus, supercomputers and other parallel systems are usually chosen for these types of simulations that potentially have very high costs and are not always readily available.

6.1 Simulating on a GPU

Modern CPUs have multiple cores, which means in essence that more than one processor exists within the same chipset. Today, CPUs with four cores have increased performance by allowing more processes to be handled simultaneously. However, even if the first steps towards multiple processors in one chipset have been made, it is hardly enough.

Alternatively, graphics cards created by NVidia [33] may be employed for the simulations to run on, instead of the actual CPU of the system. These Graphics Processing Units (GPUs) have proven to be a good cost-effective solution to the lack of processing power problem.

Differences between CPU and GPU. In the past five years, a lot of progress has been made in the field of graphics processing, giving birth to graphics cards that have a large amount of processors as well as memory sufficient to allow their use to other fields than simply processing and depicting graphics. The processing power (floating-point operations per second) superiority of modern graphics cards (GPUs) to that of the CPU with NVidia's GeForce 8800 GTX reaches almost 340 GFlops whilst newer models like the GeForce GTX 280, can reach almost 900 GFlops(c.f., [29]). Nevertheless, even if the GPU appears to have significantly more power than the conventional CPU, it is worth noting that the CPU is capable of handling different kinds of processes quickly, while the GPU is only capable of processing a specific task very fast. This latter task needs to be in the form of a problem composed of independent elements, due to the large parallelization of GPUs (c.f., [29]).

Note that the GeForce 8800 GTX is equipped with 128 scalar processors divided into 16 groups - called multiprocessors - of 8. The calculation power of such GPU was shown in [34] to be clearly superior to the CPU. However, the difference is not that large unless this power is efficiently utilized [39].

GPU Architecture. Via the CUDA framework, the GPU is exposed as a parallel data streaming processor, which consists of several processing units. CUDA applications have two segments. One segment is called "kernel" and is executed on the GPU. The other segment is executed on the host CPU and controls the execution of kernels and transfer of data between the CPU and GPU [29].

Several threads that run on the GPU run a kernel. These threads belong to a group called block. Threads within the same block may communicate with each other using shared memory and may not communicate with threads of another block. There is a hierarchical memory structure, where each memory level has different size, restrictions and speed [29].

A disadvantage of the GPU was that it adopted a 32-bit IEEE floating-point numbers, which is well lower than the one supported by a general-purpose CPU. However, the recently released NVIDIA 280 series supports double precision floating point numbers [39].

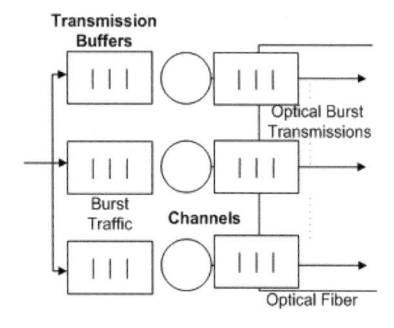

Fig. 20. Ingress node that transmits optical bursts into the core optical network, multiplexing several wavelengths (decomposed)

6.2 Numerical Results

Simulating a decomposed optical edge (c.f., Fig. 20) node would require separate self-similar generators for each transmission rate. This transforms optical burst generation into a problem that can be parallelized into multiple processor cores. However, running each of these self-similar generators on a single CPU, would introduce a significant overhead as the self-similar generators can be extremely time-consuming. Assuming the time required for these generators can be minimized, this would greatly increase the efficiency of optical packet/burst generation simulations.

The aims of the experiments, are to observe whether the use of a GPU for traffic generators can improve their efficiency and minimize their time requirements. Three types of generators were chosen, a generator based on the GE distribution for lightpath session arrivals, an LRD generator based on the $M/G/\infty$ delay system and an LRD generator based on On/Off sources. The length of the generated samples was limited specifically to reveal potential bottlenecks when running these generators on the GPU versus the CPU.

Simulations of a decomposed optical edge (c.f., Fig. 20) node were conducted on an NVidia 8800 GTX. For this investigation many different scenarios were considered, in order to provide a deeper insight on the GPU's capabilities on various test cases.

Lightpath session arrivals were generated based on the GE-type distribution, so that batches of sessions are taken into consideration. A total of 24,002,560 lightpath sessions were created, initially on one block on the GPU and on three types of general purpose CPUs, single, dual and quad core (c.f., Fig. 21). Initially, measurements were taken by increasing the number of threads that simultaneously ran on the same block.

The results showed that by using only one block (i.e., one multiprocessor) on the GPU, the overall simulation duration was significantly longer when the thread number was lower than 8. As the threads were increasing, the performance improvement was significant, surpassing even a Quad core CPU.

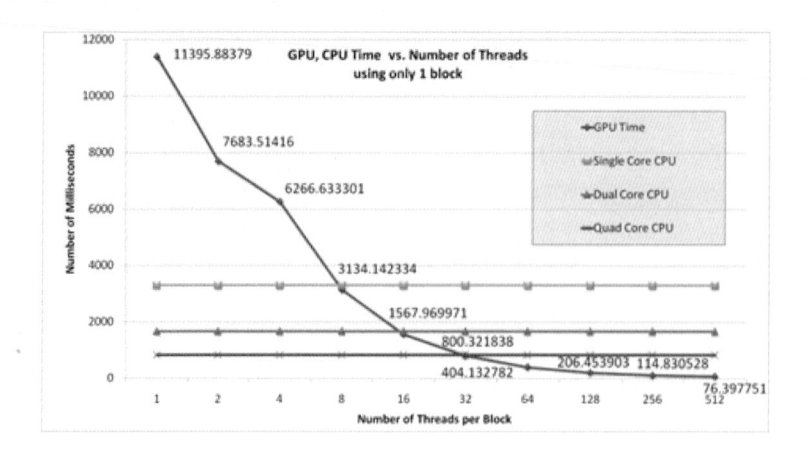

Fig. 21. Simulation duration in milliseconds, of GPU and CPU simulations vs. number of simultaneous threads per block

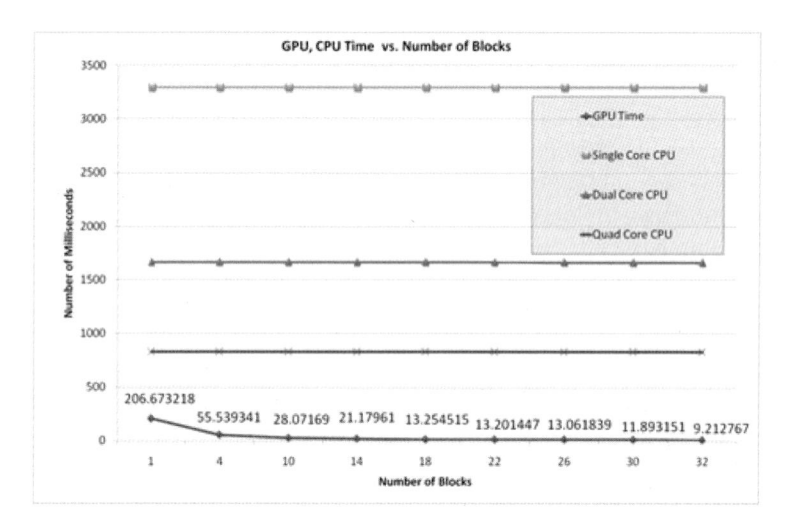

Fig. 22. Performance comparisons between the GPU and different types of CPUs. Graph shows duration in milliseconds versus number of blocks

The same experiment was conducted, while increasing the number of blocks processing the kernel at the same time. This essentially increased the parallelization of the generator, increasing dramatically the performance. To this end, a parallel version of Mersenne Twister was employed to allow for simultaneous random number generators on each of the 190 wavelengths [35].

As the number of blocks is increasing, significant performance improvement is observed, as more blocks use more multiprocessors (c.f., Fig. 22). Note that a total of 16 blocks are required for all multiprocessors of the NVidia 8800 GTX to be used and up to 32 to be properly utilized for maximum performance.

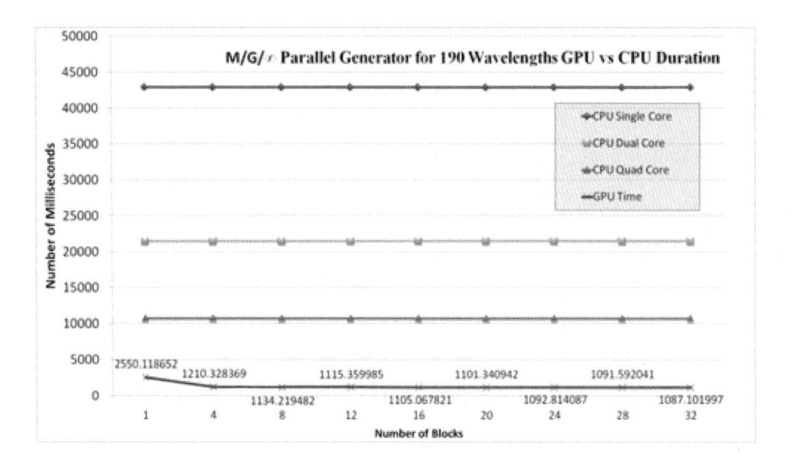

Fig. 23. Performance comparisons between the GPU and different types of CPUs. Graph shows duration in Milliseconds versus number of blocks. Number of blocks start with 1 block using 16 Threads and continues increasing the blocks employing 128 threads per block.

However, as mentioned in the beginning of this tutorial, the Poisson arrival process and the compound Poisson arrival process are not suitable to simulate network traffic for an OBS Network. Therefore, self-similarity generators need to be implemented since most are computational intensive. The $M/G/\infty$ generator was chosen specifically because it is not easily implemented on a parallel system, to allow increased complexity of calculations for each wavelength. To this end, an optical edge node is being simulated having 190 wavelengths. Each wavelength produces streams of optical bursts, with a traffic load that exhibits self-similar properties. For demonstration purposes the generated samples per wavelength were limited to 302084, for a total of 58000128 bursts for the entire node. For each wavelength a series of self-similar traces is generated based on simulating individual $M/G/\infty$ delay systems. The results shown in (c.f., Fig. 23) indicate that the GPU performs even better when the complexity of the problem is high in comparison to the CPU, provided it has been parallelized enough to efficiently utilize the GPU's resources. This observation was also validated in [34].

An experiment with On/Off Sources was also conducted. This method is inherently easy to implement on a parallel system. It involves generating traffic from each wavelength of the fiber independently on a parallel system and then aggregating them into a multiplexed stream. Aggregation, however, needs to run on a single core since it requires access to all generated traffic streams. Essentially this creates a bottleneck, which limits the overall performance of the GPU. To illustrate this, the amount of traffic for each wavelength was limited. As shown in Fig. 24 by increasing the number of threads per block and by using only one block, the GPU eventually surpasses the quad core CPU once the number of threads exceeds 128. Nevertheless, the performance difference is not that great. This is due to the aggregation overhead.

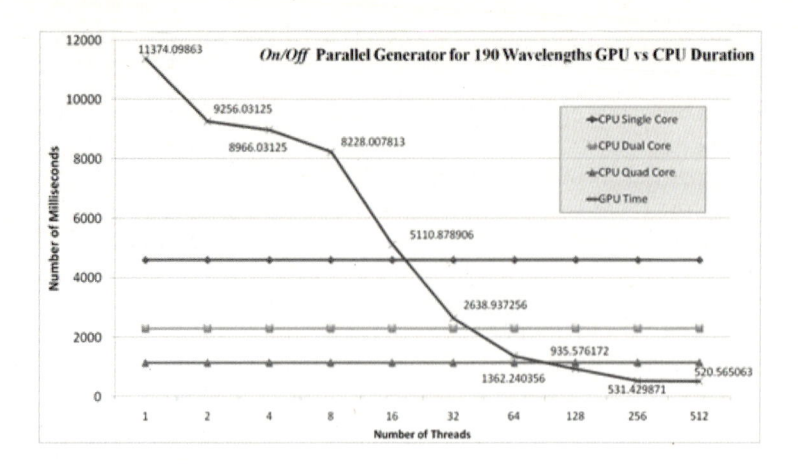

Fig. 24. Performance Comparisons between the GPU and different types of CPUs. Graph shows duration in Milliseconds versus number of threads per block.

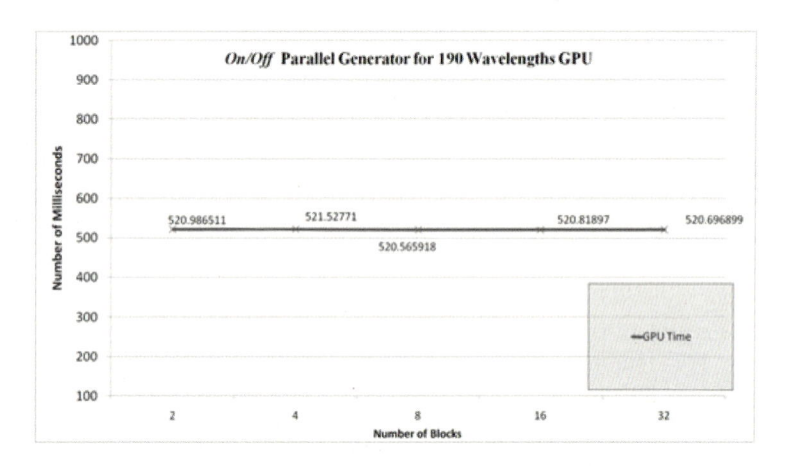

Fig. 25. Performance of GPU while increasing blocks, for a small amount of traffic traces

Furthermore, it is shown in Fig. 25 that by increasing the number of blocks, no real difference in performance, is observed since the number of samples is too small whereas the aggregation overhead too great.

Increasing the number of samples would actually increase the performance difference between the CPU and the GPU, while at the same time diminishing the aggregation's overhead effect. This leads to the conclusion that without proper parallelization, the GPU remains significantly underutilized, thus making the general-purpose CPU the main choice for processes that were programmed without considering a parallel system.

7   Conclusions and Further Work

The traffic characterization of an OC-192 backbone network link was conducted, from traces retrieved by CAIDA [16]. The suitability of the GE distribution, with variable parameters based on the self-similar traffic slot bin was demonstrated and differences in the IP packet size distribution were presented. These traffic characteristics were used in the OBS protocol simulations to prevent biased results. The exposition showed that although the traffic load is self-similar, nevertheless burstiness still occurs in the short-time scales.

The performance of the proposed DOC allocation protocol for high performance OBS networks without wavelength converters was also investigated. Different traffic demands amongst the nodes of the optical network were taken into account and a dynamic updating of the offset was adopted based on the occurrence of blocked bursts and successful transmissions. Preferred wavelengths were also chosen on a source - destination basis. Numerical evaluation results based on simulation were devised focusing on the performance metrics of throughput, mean queue length and blocking probability. Self-similar traffic was generated for each wavelength of every source node in all network topologies employed. Moreover, the MAR and SCV of GE-type interarrival times of IP packets were calculated for each time slot. It was observed in many measurements taken that both the overall and each source - destination pair blocking probabilities under the DOC protocol were greatly decreased in comparison to those associated with the JET protocol. In addition, the DOC protocol achieved fairness in terms of wavelength utilisation and moreover, it was experimentally shown that this protocol, based on a dynamic control of the offset and preferred wavelength, did not increase buffer length requirements at each source node in comparison to those of the JET protocol.

The DOC protocol could be extended to take into consideration the case of broken links of its topology. This will necessitate the creation of a new feature that could be added on top of the variable offset values and the preferred wavelength system. This will cause the OBS network under DOC to pick a new preferred route for each source - destination pair if neither the varying offset values nor the preferred wavelength managed to keep a low value for the associated blocking probability. Choosing an alternate route may allow an OBS network to bypass the broken links in the network achieving, therefore, higher throughput in contrast to other OBS networks. Moreover, statistical analysis of the offset values needs to be conducted whilst further investigation is required under other scenarios where the thresholds of the performance metrics of the network are also adjusted according to their attributes, as appropriate.

The simulations in this tutorial were executed on an NVidia 8800 GTX graphics card acting as a parallel system. Results showed a GPU can provide significantly faster results in comparison to the CPU, provided that the optical node simulation is decomposed properly and the simulation is designed for a parallel system. This allows independent systems based on different wavelengths to be simulated simultaneously. Note that simulations need to have high complexity and duration in order to justify the use of a GPU. However, it is required that simulation designers develop their simulations efficiently; otherwise worse performance may be experienced. In the future, more and more processing cores are expected to appear and the research community needs to create cost-effective algorithms that take advantage of their fast multiprocessing potential.

Moreover, it is essential to develop an optimal a hybrid simulation approach that utilises in full the available technology and allows multithreaded segments to be calculated on the GPU and single core bottlenecks on the CPU, respectively. In addition a new royalty-free standard, named Open Computing Language (OpenCL), needs further consideration as it is gaining popularity for cross-platform, parallel programming of modern processors found in personal computers, servers and handheld/embedded devices. OpenCL is nowadays adopted by more and more operating systems and platforms, whilst drivers were issued by NVIDIA for the OpenCL to work with CUDA. It can be used to greatly improve speed and responsiveness for a wide spectrum of applications in numerous market categories from gaming and entertainment to scientific and medical software [36].

References

1. Amstutz, S.R.: Burst Switching - an introduction. IEEE Communications Magazine, 36–42 (1983)
2. Listanti, M., Eramo, V., Sabella, R.: Architectural and Technological Issues for Future Optical Internet Networks. IEEE Communications Magazine 38(9), 82–92 (2000)
3. Yoo, M., Qiao, C.: Just-Enough-Time (JET): a high speed protocol for bursty traffic in optical networks. Vertical-Cavity Lasers, Technologies for a Global Information Infrastructure, WDM Components Technology, Advanced Semiconductor Lasers and Applications, Gallium Nitride Materials, Processing and Devices 11(15), 26–27 (1997)
4. Dolzer, K., Gauger, C., Spath, J., Bodamer, S.: Evaluation of reservation mechanisms in optical burst switching networks. AEU Int. J. of Electronics and Communications 55(1) (2001)
5. Verma, S., Chaskar, H., Ravikanth, R.: Optical Burst Switching: A Viable Solution for Terabit IP Backbone. IEEE Network, 48–53 (2000)
6. Tan, K., Mohan, G., Chua, K.C.: Link Scheduling State Information Based Offset Management for Fairness Improvement in WDM Optical Burst Switching Networks. Computer Networks 45(6), 819–834 (2004)
7. Kouvatsos, D.D.: Entropy Maximization and Queueing Network Models. Annals of Operation Research 48, 63–126 (1994)
8. Zhang, Q., Vokkarane, V., Jue, J., Chen, B.: Absolute QoS Differentiation in Optical Burst-Switched Networks. IEEE Journal on Selected Areas in Communications 22(9) (2004)
9. Yu, X., Chen, Y., Qiao, C.: Study of traffic statistics of assembled burst traffic in optical burst switched networks. In: Proceedings, Optical Networking and Communication Conference (OptiComm), Boston, MA (2002)
10. Park, K., Willinger, W.: Self-Similar Network Traffic and Performance Evaluation. John Wiley and Sons, Chichester (2000)

11. Willinger, W., Taqqu, M., Erramilli, A.: A Bibliographical Guide to Self-Similar Traffic and Performance Modeling for Modern High-Speed Networks. Stochastic Networks: Theory and applications. Oxford University Press, Oxford (1996)
12. Jeong, H.-D.J., Pawlikowski, K., McNickle, D.C.: Generation of Self-Similar Processes for Simulation Studies of Telecommunication Networks. Mathematical and Computer Modelling 38, 1249–1257 (2003)
13. Jeong, H.-D.J., Pawlikowski, K., McNickle, D.C.: Generation of Self-Similar Processes for Simulation Studies of Telecommunication Networks. Mathematical and Computer Modelling 38, 1249–1257 (2003)
14. Karagiannis, T., Molle, M., Faloutsos, M., Broido, A.: A Nonstationary Poisson View of Internet Traffic. In: INFOCOM 2004, 23rd Annual Joint Conference of the IEEE Computer and Communications Societies, vol. 3, pp. 1558–1569 (2004)
15. de Vega Rodrigo, M., Goetz, J.: An analytical study of optical burst switching aggregation strategies. In: Proceedings of the Third International Workshop on Optical Burst Switching, WOBS (2004)
16. Shannon, C., Aben, E., Claffy, K., Andersen, D.: The CAIDA Anonymized 2008 Internet Traces - 19/06/2008 12:59:08 - 19/06/2008 14:01:00 (2008), CAIDA, http://www.caida.org/passive/passive_2008_dataset.xml (retrieved December 10, 2008)
17. CAIDA. The Cooperative Association for Internet Data Analysis (2003), http://www.caida.org/research/traffic-analysis/AIX/ (Retrieved from CAIDA)
18. Kouvatsos, D.D.: Maximum Entropy and the G/G/1 Queue. Acta Informatica 23, 545–565 (1986)
19. Crovella, M., Bestavros, A.: Self-Similarity in World-Wide Web Traffic: Evidence and Possible Causes. In: Proc. ACM SIGMETRICS 1996 (1996)
20. Paxson, V., Floyd, S.: Wide Area Traffic: The Failure of Poisson Modeling. In: Proc. ACM SIGCOMM 1994 (1994)
21. Popescu, A.: Traffic Analysis and Control in Computer Communications Networks. Blekinge Institute of Technology, Stockholm (2007) (preprint)
22. Beran, J.: Statistics for Long-Memory Processes. Chapman and Hall, Boca Raton (1994)
23. Abry, P., Veitch, D.: Wavelet Analysis of Long Range Dependent Traffic. IEEE Transactions on Information Theory (1998)
24. Stallings, W.: High-Speed Networks and Internets: Performance and Quality of Service, 2nd edn. Prentice-Hall, Englewood Cliffs (2002)
25. Taqqu, M., Teverovsky, V.: On Estimating the Intensity of Long-Range Dependence. Finite and Infinite Variance Time Series. Boston University USA, Boston (1996)
26. Qiao, C., Yoo, M.: Choices, Features and Issues in Optical Burst Switching (OBS). Optical Networking Magazine 2 (1999)
27. Xu, L., Perros, H.G., Rouskas, G.N.: A Simulation Study of Access Protocols for Optical Burst-Switched Ring Networks. In: Proceedings of Networking (2002)
28. Aysegul, G., Biswanath, M.: Virtual-Topology Adaptation for WDM Mesh Networks Under Dynamic Traffic. IEEE/ACM Transactions on Networking 11, 236–247 (2003)
29. NVIDIA. NVIDIA CUDA Compute Unified Device Architecture, Programming Guide. NVIDIA (2007), http://www.nvidia.com/object/cuda_develop.html
30. Dolzer, K., Gauger, C.: On burst assembly in optical burst switching networks - a performance evaluation of Just-Enough-Time. In: Proceedings of the 17th International Teletraffic Congress (ITC 17), Salvador (2001)

31. Barakat, N., Sargent, E.H.: Analytical Modeling of Offset-Induced Priority in Multiclass OBS Networks. IEEE Transactions on Communications 53(8), 1343–1352 (2005)
32. Matsumoto, M., Nishimura, T.: Mersenne Twister: A 623-Dimensionally Equidistributed Uniform Pseudo-Random Number Generator. ACM Transactions on Modelling and Computer Simulation (TOMACS) 8(1), 3–30 (1998)
33. Hu, G., Dolzer, K., Gauger, C.: Does Burst Assembly Really Reduce the Self-Similarity. In: Optical Fiber Communications Conference, OFC 2003, vol. 1, pp. 124–126 (2003)
34. Triolet, D.: NVidia CUDA: Preview - BeHardware (2007), http://www.behardware.com/art/imprimer/659
35. Feldmann, A.: Characteristics of TCP Connection Arrivals. In: Self-Similar Network Traffic and Performance Evaluation, pp. 367–397. Wiley Interscience, Hoboken (2000)
36. Khronos Group.: OpenCL - The open standard for parallel programming of heterogeneous systems, http://www.khronos.org/opencl/
37. Park, K., Willinger, W.: Self-Similar Network Traffic: An Overview. In: Park, K., Willinger, W. (eds.) Self-Similar Network Traffic and Performance Evaluation, pp. 1–38 (2000)
38. Mouchos, C., Tsokanos, A., Kouvatsos, D.D.: Dynamic OBS Offset Allocation in WDM Networks. Computer Communications (COMCOM) - The International Journal for the Computer and Telecommunications Industry, Special Issue on 'Heterogeneous Networks: Traffic Engineering and Performance Evaluation 31(suppl. 1) (to appear mid, 2010), (in Press Corrected Proof), ISSN 0140-3664, doi: 10.1016/j.comcom.2010.04.009, http://www.sciencedirect.com/science/article/B6TYP-4YVY769-1/2/e3903ceb381e6d5f30adf33d1824281a (available online April 16, 2010)
39. Mouchos, C., Kouvatsos, D.D.: Parallel Traffic Generation of a Decomposed Optical Edge Node on a GPU, in Traffic and Performance Engineering for Heterogeneous Networks. In: Kouvatsos, D.D. (ed.) Performance Modelling and Analysis of Heterogeneous Networks, ch. 20, Aalborg, Denmark. Series of Information Science 7 Technology, vol. 2, pp. 417–439. River Publishers (2009), ISBN 978-87-92329-18-9

The Search for QoS in Data Networks: A Statistical Approach

Pablo Belzarena and María Simon

Universidad de la República
Montevideo, Uruguay
{belza,msimon}@fing.edu.uy

Abstract. New Internet services like video on-demand, high definition IPTV, high definition video conferences and some real time applications have strong QoS requirements regarding losses, delay, jitter, etc. This work addresses the challenge of guaranteeing quality of service (QoS) in the Internet from a statistical point of view. Three lines of work are proposed. The first one is about the estimation of the QoS parameters from traffic traces (in the context of large deviation theory and effective bandwidth). The second one, address the admission control problem from results of the many sources and small buffer asymptotic. Finally, the third line focuses on the estimation of QoS parameters seen by an application based on end-to-end active measurements and statistical learning tools.

Keywords: large deviations, statistical learning, admission control, active measurements, quality of service.

1 Introduction and Motivation

Internet services with high quality of service requirements like video on-demand, high quality video conferences, high definition IPTV, telematic services with real time requirements, etc. have grown at a smaller rate than the initially hoped. One possible cause is the difficulty that exists in order to guarantee end-to-end quality of service (QoS) in IP networks. Another possible cause is that the operators have not deploy the different proposals developed during the last 15 years in order to assure QoS (IntServ, DiffServ, etc.). These difficulties have been recently increased by the heterogeneity of the access networks (xdsl, cablemodem, wifi, wimax, 2G, 3G, mesh networks, etc.). End-to-end QoS leads to another issue; in the general case, the end-to-end performance parameters can not be estimated from the performance parameters of each individual router in the path. This problem becomes even worst when the service operator offers its service over multiple domains. In this case, the nodes of the path are under the administration of different network operators.

An important issue in this context is the network admission control based on end-to-end QoS. In a network of "premium" services this kind of admission control allows the operator to control the end-to-end QoS. This issue is one of the main motivations of this work.

The focus of this work is on the estimation of the admission control region. We look for a simple and efficient procedure for such estimation which can be applied on line. A control admission tool using this estimation can decide if it accepts or not a new service request.

The admission control mechanisms proposed in the literature are mainly based on one link analysis [1]. We start analyzing admission control mechanisms where the link analysis is based on Large Deviation Theory (LDT). In the analysis of networks performance using LDT [2] three main asymptotic regimes have been described. These are the large buffer regime, the many sources asymptotic and the many source and small buffer asymptotic. In the first case the convergence rate to zero of some QoS parameter (e.g. loss probability) when the buffer size goes to infinity is studied. In the second one it is also studied the convergence rate to zero of the loss probability but when there are many independent and identically distributed sources arriving at the link and the link capacity and the buffer size both increases at the same rate as the number of sources. In the third case, there are many independent sources, the link capacity grows with the number of sources but the buffer size grows slower than the number of sources. The large buffer asymptotic can be applied only in the access networks where there are few sources and the buffer per source can be considered big. However, the large buffer asymptotic can only be applied to one isolated link because the output of that link does not verify the assumptions needed to apply the asymptotic to the following link in the path.

In networks such as an internet backbone, the many sources asymptotic approach is more reasonable than the large buffer one. In fact, in this kind of backbone, large numbers of flows from different sources arrive, the capacities are high and the buffer sizes per source are in general small, because they are intended to serve many sources but not many bursts at the same time.

We start analyzing in Section 2 a link based admission control. This mechanism is based on the many sources asymptotic and in particular on the effective bandwidth notion [3]. In this work we address an important issue for an on-line admission control mechanism: the estimation of the QoS performance parameters (particulary the buffer overflow probability) based on traffic traces.

Although we analyze the estimation of buffer overflow probability in the many source asymptotic, the results of this work can also be applied to estimate the large deviation rate function in the many sources and small buffer asymptotic introduced by Ozturk et al.[4]. The small buffer asymptotic presents more interesting results in order to analyze the end-to-end QoS and not only the QoS of an isolated link. For a service provider the most interesting issue is about admission control mechanisms based on the end-to-end QoS. In Section 3 we analyze end-to-end QoS applying the so called "fictitious network model". This model is based on the many sources and small buffer asymptotic. We will show that this model allows simple and on-line estimations of end-to-end QoS parameters, which will be in turn used to decide which flows can access the network.

Ozturk et al. find a useful way to analyze the overflow probability in a network interior link and show that when the fictitious network model is applied, an overestimation is obtained. The fictitious network analysis gives then a simple and efficient yet conservative way to implement on-line admission control mechanisms. However, the overestimation can translate into wasted network resources. If a flow is admitted, its QoS is guaranteed but the link capacities can be under-used. In this work we analyze in detail the fictitious network model and we find conditions to assure that the fictitious network analysis in an interior link gives the same overflow probability than the real network analysis, being much simpler. We also find a method to bound the overestimation when these conditions are not fulfilled. In addition, since no model is assumed for the input traffic, we define an estimator of the end-to-end Loss ratio based on traffic measurements. We show that this estimator is a good one, i.e. it is consistent and verifies a Central Limit Theorem (CLT). These results allow us to define an admission control mechanism based on the expected end-to-end Loss Ratio that a flow traversing the network will obtain.

However, the many sources and small buffer asymptotic can only be applied to analyze an end-to-end path in a backbone network. If the end points are end users this asymptotic cannot be applied because the path goes through the backbone but also through the access network where the many sources asymptotic is not valid. The research community does not have yet a model in order to analyze an end-to-end path including the access and the backbone network. Therefore, a different approach must be applied if the access control mechanism must take a decision based on end-to-end QoS.

Some authors propose end-to-end admission control mechanism based on active measurements [5]. In the third part of this tutorial, in Section 4, we propose a different approach for an end-to-end admission control. This approach is based on active measurements and statistical learning tools. We analyze the application of an statistical learning approach in order to predict the quality of service seen by an application.

Although the end-to-end admission control problem is our main motivation, the different issues analyzed in this work can be applied to many other network operation and management problems like for example to share resources in a network, to continuous monitoring a Service Level Agreement (SLA), etc..

2 Effective Bandwidth and Link Operation Point Estimation

2.1 Introduction

One of the main issues in QoS admission control is the estimation of the resources needed for guaranteed VBR communications, which cannot be the peak rate nor the mean rate of the service. Indeed, the mean rate would be a too optimistic estimation, that would cause frequent losses. On the other side, the peak rate would be too pessimistic and would lead to resource waste.

Effective bandwidth (EB) defined by F. Kelly in [3] is an useful and realistic measure of channel occupancy. The EB is defined as follows:

$$\alpha(s,t) = \frac{1}{st} \log \mathbf{E}\left(e^{sX_t}\right) \qquad 0 < s, t < \infty. \tag{1}$$

where X_t is the total amount of work arriving from a source in the time interval $[0,t]$, which is supposed to be a stochastic process with stationary increments. $\alpha(s,t)$ lies between the mean rate (for $s \to 0$) and the peak rate (for $s \to \infty$) of the input process.

Parameters s and t are referred to as the space and time parameters respectively. When solving for a specific performance guarantee, these parameters depend not only on the source itself, but on the context on which this source is acting. More specifically, s and t depend on the capacity, buffer size and scheduling policy of the multiplexer, the QoS parameter to be achieved, and the actual traffic mix (i.e. characteristics and number of other sources). The EB concept can be applied to sources or to aggregated traffic, as we find in a network's core link.

Under the *many sources asymptotic regime* discussed in [6], where it is assumed that, as the number of sources feeding a switch grows, the switch capacity and buffer size increase proportionally, the EB is related with the stationary loss probability through buffer overflow by the so called *inf sup* formula:

$$\Gamma = \inf_{t \geq 0} \sup_{s \geq 0} ((B + Ct)s - Nst\alpha(s,t))$$

where C is the link capacity, B is its buffer size and N the number of incoming multiplexed sources of effective bandwidth $\alpha(s,t)$. If Q_N represents the stationary amount of work in the queue, the buffer overflow probability or loss probability is approximately given by:

$$\log \mathbf{P}(Q_N > B) \approx -\Gamma$$

We call s^* and t^* to the values of parameters s and t in which the *inf sup* is attained. These values s^* and t^* are called the link's *operating point*.

Therefore, a good estimation of s^* and t^* is useful for the network's design, for the Connection Admission Control (CAC) function, or for optimal operating.

We point out the need of a good estimation of the bandwidth in order to optimize resource sharing.

In section 2.2 we show how the operating point of a link can be estimated from its EB, the consistency of this estimation and its confidence interval. We observe that other well known estimators fit the necessary conditions for the validity of the theorem.

Analytical results are compared with numerical data in section 2.3. These numerical data were obtained independently from the analytical work from simulations models that are also explained in this section. In this framework, overflow probability estimation is a key topic, which makes necessary EB and link's operating point estimation.

2.2 Estimation

Estimating the operating point of a link, as defined in section 2.1 is closely related with its defining equation which we rewrite here on a *per source* basis:

$$\gamma = \inf_{t \geq 0} \sup_{s \geq 0} ((b + ct)s - st\alpha(s, t)) \quad (2)$$

where γ is the asymptotic decay rate of the overflow probability as the number of sources increases, c and b are the link's capacity and buffer size *per source* and $\alpha(s,t)$ the effective bandwidth function (equation 1). With the present notation, stationary overflow probability in a switch multiplexing N sources, having capacity $C = Nc$ and buffer size $B = Nb$ verifies:

$$\lim_{N \to \infty} \frac{1}{N} \log \mathbf{P}(Q_N > B) = -\gamma \quad (3)$$

In general, the effective bandwidth function $\alpha(s,t)$ is unknown, and shall be estimated from measured traffic traces. The problem is how to estimate the moment generating function $\Lambda(s,t) = \mathbf{E}\left(e^{sX_t}\right)$ of the incoming traffic process X_t for each s and t.

Different approaches have been presented to solve this problem. One of them, presented in [7] and [8] is to estimate the expectation $\mathbf{E}\left(e^{sX_t}\right)$ as the time average given by:

$$\Lambda_n(s,t) = \frac{1}{n} \sum_{k=1}^{n} e^{s(X_{kt} - X_{(k-1)t})} \quad (4)$$

which is valid if the process increments are stationary and satisfy any weak dependence hypothesis that guarantees ergodicity. To estimate $\Lambda(s,t)$ a traffic trace of length $T = nt$ is needed. We can construct an appropriate estimator of the EB as $\alpha_n(s,t) = \frac{1}{st} \log(\Lambda_n(s,t))$.

When a model is available for incoming traffic, a parametric approach can be taken. In the case of a Markov Fluid model, i.e. when the incoming process is modulated by a continuous time Markov chain which dictates the rate of incoming work, explicit computation can be made as shown by Kesidis et al. in [9]. In this case, an explicit formula is given for $\Lambda(s,t)$ and $\alpha(s,t)$ in terms of the infinitesimal generator or Q-matrix of the Markov chain. In a previous work of our group [10], and based on the maximum likelihood estimators of the Q-matrix parameters presented in [11], an EB estimator and confidence intervals are developed.

Having an estimator of the function $\alpha(s,t)$, it seems natural to estimate γ, and the operating point s^*, t^* substituting the function $\alpha(s,t)$ by $\alpha_n(s,t)$ in equation (2) and solving the remaining optimization problem. The output would be some values of γ_n, s_n^* and t_n^*, and the question is under what conditions these values are good estimators of the real γ, s^* and t^*.

Therefore, we may discuss two different problems concerning estimation. The first one is, given a "good" estimator $\alpha_n(s,t)$ of $\alpha(s,t)$, find sufficient conditions

under which the estimators s_n^*, t_n^* and γ_n^* obtained by solving the optimization problem:
$$\gamma_n = \inf_{t \geq 0} \sup_{s \geq 0} ((b+ct)s - st\alpha_n(s,t)) \tag{5}$$
are "good" estimators of the operating point s^*, t^* and the overflow probability decay rate γ of a link. This affirmation is not an obvious result because s^* and t^* are found from a non linear and implicit function. We remark that the reasoning applied to s^* and t^* can be also applied to other parameters that are deduced from the EB. Further in the article the parameters B and C are also studied.

The second problem is finding this good estimator of the EB and determining whether the conditions are met, so that the operating point can be estimated using equation (5).

The remaining part of the section addresses the first problem, where a complete answer concerning consistency and Central Limit Theorem (CLT) properties of estimators is given by theorem 1, based on regularity conditions of the EB function. At the end of the section we discuss the validity of the theorem for some known estimators and in section 2.3 we compare our analytical results with numerical ones.

Let us define:
$$g(s,t) = s(b+ct) - st\alpha(s,t)$$
which can be rewritten in terms of $\Lambda(s,t) = \mathbf{E}\left(e^{sX_t}\right)$. We have that $\frac{\partial}{\partial s}g(s,t) = 0$ if and only if:
$$\frac{\partial}{\partial s}g(s,t) = b + ct - \frac{\frac{\partial}{\partial s}\Lambda(s,t)}{\Lambda(s,t)} = 0 \tag{6}$$

Assuming that for each t there exists $s(t)$ such that, $\frac{\partial}{\partial s}g(s(t),t) = 0$, it is easy to show that $\sup_{s \geq 0} g(s,t) = g(s(t),t)$ because $g(s,t)$ is convex as a function of s. In that case, $\gamma = \inf_{t \geq 0} g(s(t),t)$, and:
$$\frac{\partial}{\partial t}g(s(t),t) = \frac{\partial}{\partial s}g(s(t),t)\dot{s}(t) + \frac{\partial}{\partial t}g(s,t)\bigg|_{s=s(t)}$$

If there exists t^* such that: $\frac{\partial}{\partial t}g(s(t^*),t^*) = 0$ and the infimum is attained, it follows that: $\gamma = g(s(t^*),t^*)$.

If we define $s^* = s(t^*)$, we have that $\gamma = g(s^*,t^*)$ where:
$$\frac{\partial}{\partial s}g(s^*,t^*)\dot{s}(t^*) + \frac{\partial}{\partial t}g(s^*,t^*) = 0 \quad \text{and} \quad \frac{\partial}{\partial s}g(s^*,t^*) = 0$$

and then we have the relations:
$$\frac{\partial}{\partial s}g(s^*,t^*) = \frac{\partial}{\partial t}g(s^*,t^*) = 0 \tag{7}$$

Since:
$$\frac{\partial}{\partial t}g(s,t) = cs - \frac{\frac{\partial}{\partial t}\Lambda(s,t)}{\Lambda(s,t)} \tag{8}$$

it follows from (6), (7) and (8) that the operating point must satisfy the equations:

$$b + ct^* - \frac{\frac{\partial}{\partial s}\Lambda(s^*,t^*)}{\Lambda(s^*,t^*)} = 0 \quad \text{and} \quad cs^* - \frac{\frac{\partial}{\partial t}\Lambda(s^*,t^*)}{\Lambda(s^*,t^*)} = 0 \qquad (9)$$

If we make the additional assumptions that interchanging the order of the differential and expectation operators is valid, and that $\dot{X}_t$ exists for almost every t we can write:

$$\frac{\partial}{\partial s}\Lambda(s,t) = \mathbf{E}\left(X_t e^{sX_t}\right) \qquad \frac{\partial}{\partial t}\Lambda(s,t) = \mathbf{E}\left(s\dot{X}_t e^{sX_t}\right) \qquad (10)$$

Replacing the expressions of (10) in equations (9) we deduce an alternative expression for the solutions s^* and t^*:

$$b + ct^* - \frac{\mathbf{E}\left(X_{t^*} e^{s^*X_{t^*}}\right)}{\mathbf{E}\left(e^{s^*X_{t^*}}\right)} = 0 \quad \text{and} \quad cs^* - \frac{\mathbf{E}\left(s^* \dot{X}_{t^*} e^{s^*X_{t^*}}\right)}{\mathbf{E}\left(e^{s^*X_{t^*}}\right)} = 0 \qquad (11)$$

Therefore, we can reformulate the optimization problem presented in (2). The operating point of the link can be calculated solving the system of equations (9), or (11) if the additional assumptions are valid. The first formulation, which is more general, is the one used in the main result of this work, which follows:

Theorem 1. *If $\Lambda_n(s,t)$ is an estimator of $\Lambda(s,t)$ such that both are C^1 functions and:*

$$\Lambda_n(s,t) \xrightarrow[n]{} \Lambda(s,t) \quad \frac{\partial}{\partial s}\Lambda_n(s,t) \xrightarrow[n]{} \frac{\partial}{\partial s}\Lambda(s,t) \quad \frac{\partial}{\partial t}\Lambda_n(s,t) \xrightarrow[n]{} \frac{\partial}{\partial t}\Lambda(s,t) \qquad (12)$$

almost surely and uniformly over bounded intervals, and if we denote s_n^* and t_n^* the solutions of:

$$b + ct_n^* - \frac{\frac{\partial}{\partial s}\Lambda_n(s_n^*,t_n^*)}{\Lambda_n(s_n^*,t_n^*)} = 0 \quad cs_n^* - \frac{\frac{\partial}{\partial t}\Lambda_n(s_n^*,t_n^*)}{\Lambda_n(s_n^*,t_n^*)} = 0 \qquad (13)$$

then (s_n^*, t_n^*) are consistent estimators of (s^*, t^*). Moreover, if a functional Central Limit Theorem (CLT) applies to $\Lambda_n - \Lambda$, i.e,

$$\sqrt{n}\left(\Lambda_n(s,t) - \Lambda(s,t)\right) \xrightarrow[n]{w} G(s,t),$$

where $G(s,t)$ is a continuous gaussian process, then:

$$\sqrt{n}\left((s_n^*, t_n^*) - (s,t)\right) \xrightarrow[n]{w} N(\mathbf{0}, \Sigma) \qquad (14)$$

where $N(\mathbf{0}, \Sigma)$ is a centered bivariate normal distribution with covariance matrix Σ.

Proof. See [12].

Remark 1. The computation of Σ is not trivial. However, if replication is possible (for instance by taking large traces of weak-dependent signals), the previous result allows the estimation of Σ in terms of empirical covariances. Arguments of this type are used in section 2.3.

Remark 2. Since the convergence assured by theorem 1 is uniform over bounded intervals, it is also assured that γ_n given by:

$$\gamma_n = s_n^*(b + ct_n^*) - \Lambda_n(s_n^*, t_n^*)$$

inherits the properties of the s_n^* and t_n^* estimators. That is, $\gamma = F(s^*, t^*, \Lambda)$ where F is a differentiable function. Also, $\gamma_n = F(s_n^*, t_n^*, \Lambda_n)$. Therefore, if the estimator Λ_n verifies a functional CLT we have for γ_n:

$$\sqrt{n}\,(\gamma_n - \gamma) \underset{n}{\overset{w}{\Longrightarrow}} N(0, \sigma^2)$$

Remark 3. In a many source environment, expressions for the buffer size b and the link capacity c obtained by Courcoubetis [7] are similar to the *inf sup* equation. Therefore, the reasoning used in the previous theorem extends consistency and CLT results to b^* and c^*. Also, confidence intervals for these design parameters can be constructed in this way.

We address now the second question posed at the beginning of the section. As we can see, for the validity of theorem 1 it is necessary that the estimator $\Lambda_n(s,t)$ converge uniformly to the moment generating function over bounded intervals, as well as its partial derivatives. These conditions are reasonably general, and it can be verified that they are met by the estimator (4) presented in [7] and [8], and by the estimator for Markov Fluid sources presented in [10]. In both cases a CLT can be obtained so the CLT conclusion of the theorem is also valid. It should be noticed that a consistent but non-smooth estimator can be used with this procedure, if it is previously regularized by convolution with a suitable kernel.

2.3 Simulation and Numerical Results

EB and operation point estimation. In order to validate the results obtained in the previous section, we simulated traffic using a two state (ON-OFF) Markov Fluid model. In that model, a continuous time Markov chain drives the process. When the chain is in the ON state, the workload is produced at constant rate h_0, and when it is in the OFF state no workload is produced ($h_1 = 0$). Denoting by Q the Markov chain infinitesimal generator, by $\boldsymbol{\pi}$, its invariant distribution, and by H, the diagonal matrix with the rates h_i in the diagonal. The effective bandwidth for a source of this type is [9][3]:

$$\alpha(s,t) = \frac{1}{st} \log \{\boldsymbol{\pi} \exp\left[(Q + Hs)t\right] \mathbf{1}\} \tag{15}$$

where $\mathbf{1}$ is a column vector of ones.

In our simulations we generated three hundred traffic traces of length T samples, with the following Q-matrix:

$$Q = \begin{pmatrix} -0.02 & 0.02 \\ 0.1 & -0.1 \end{pmatrix}$$

The effective bandwidth for this process calculated through equation (15).

For each traffic trace we estimated EB using the following procedure. We divided the trace in blocks of length t and constructed the following sequence:

$$\tilde{X}_k = \sum_{i=(k-1)t}^{kt} x(i) \qquad 1 \leq k \leq \lfloor T/t \rfloor$$

where $x(i)$ is the amount of traffic arrived between samples and $\lfloor c \rfloor$ denotes the largest integer less than or equal to c.

EB can then be estimated by the time average proposed in [7], [8] as

$$\alpha_n(s,t) = \frac{1}{st} \log \left[\frac{1}{\lfloor T/t \rfloor} \sum_{j=1}^{\lfloor T/t \rfloor} e^{s\tilde{X}_j} \right] \qquad (16)$$

where $n = \lfloor T/t \rfloor$. This is merely an implementation of the time average estimator in equation (4) based on a finite length traffic trace. When the values of t verify that $t \ll T$, the number of replications of the increment process within the trace is good enough to get a good estimation.

In order to find the operating point (s^*,t^*) of the theoretical Markov model, and its estimator (s_n^*,t_n^*) for each simulated trace, we solve the *inf sup* optimization problem of equation (2). In our case $\alpha(s,t)$ will be the previous theoretical equation (15) for the Markovian source or the $\alpha_n(s,t)$ estimated for each trace. The numerical solution has two parts. First, for a fixed t we find the $s^*(t)$ that maximize $g(s,t)$ as a function of s. It can be shown that $st\alpha(s,t)$ is a convex function of s. This convexity property is used to solve the previous optimization problem, that is reduced to find the maximum difference between a convex function and a linear function of s, and it can be done very efficiently. After the $s^*(t)$ is found for each t, it is necessary to minimize the function $g(s^*(t),t)$ and find t^*. For this second optimization problem, there are no general properties that let us make the search algorithm efficient and a linear searching strategy is used.

An important issue is to develop a confidence region for (s^*,t^*). We simulated 300 traces of length 100000(T) samples and constructed, for each simulated trace indexed by $i = 1,\ldots,K$ the corresponding estimator $(s_n^*(i), t_n^*(i))$. By theorem 1 the vector $\sqrt{n}((s_n^*,t_n^*) - (s^*,t^*))$ is asymptotically bivariate normal with $(0,0)$ mean and covariance matrix Σ. We estimated the matrix Σ using the empirical covariances of the observations $\{\sqrt{n}((s_n^*(i), t_n^*(i)) - (s^*(i), t^*(i)))\}_{i=1,\ldots,K}$ given by:

$$\Sigma_K = \frac{n}{K} \begin{pmatrix} \sum_{i=1}^{K} \left(s_n^*(i) - \overline{s_n^*}\right)^2 & \sum_{i=1}^{K} \left(s_n^*(i) - \overline{s_n^*}\right)\left(t_n^*(i) - \overline{t_n^*}\right) \\ \sum_{i=1}^{K} \left(s_n^*(i) - \overline{s_n^*}\right)\left(t_n^*(i) - \overline{t_n^*}\right) & \sum_{i=1}^{K} \left(t_n^*(i) - \overline{t_n^*}\right)^2 \end{pmatrix}$$

where $\overline{s_n^*} = \frac{1}{K}\sum_{i=1}^{K} s_n^*(i)$ and $\overline{t_n^*} = \frac{1}{K}\sum_{i=1}^{K} t_n^*(i)$.

Therefore, we can say that approximately: $(s_n^*, t_n^*) \approx N\left((s^*, t^*), \frac{1}{n}\Sigma_K\right)$, from where a level α confidence region: $R_\alpha = (s_n^*, t_n^*) + \frac{A_K^t B\left(\mathbf{0}, \sqrt{\chi_\alpha^2(2)}\right)}{\sqrt{n}}$, being A_K the matrix that verifies $A_K^t A_K = \Sigma_K$, while $B(x, r)$ is the ball of center x and radius r.

To verify our results, we calculated the theoretical operating point (s^*, t^*) and simulated another 300 traces independent of those that were used to estimate Σ_K. We constructed then the 95% confidence region. If the results are right, approximately 95% of the times, (s^*, t^*) must fall inside that region, or equivalently and easier to check, approximately 95% of the simulated (s_n^*, t_n^*) must fall inside the region $R = (s^*, t^*) + \frac{1}{\sqrt{n}} A_K^t B\left(\mathbf{0}, \sqrt{\chi_{0.05}^2(2)}\right)$. Numerical results, plotted in figure 1 (left), verify that the confidence level is attained, 95.33% of the estimated values fall inside the predicted region.

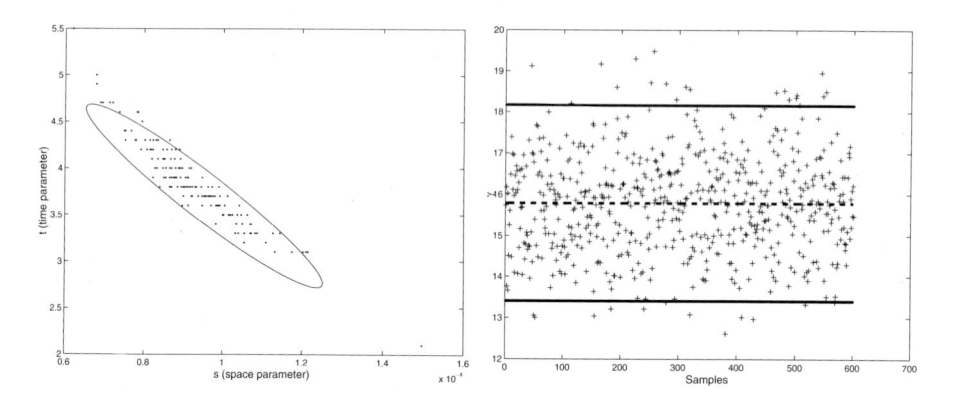

Fig. 1. Estimated operating points (left), γ_n and theoretical γ (right) and its confidence regions

QoS parameters estimation. We estimate the link operating point in order to estimate loss probability. As was said in section 2.2, if we have an EB estimator that verifies the hypotheses of theorem 1, then

$$\gamma_n = \inf_t \sup_s ((b + ct)s - st\alpha_n(s, t)) \tag{17}$$

is a consistent estimator and has CLT properties. From this estimator loss probability could be approximated by

$$q_n = P_n(Q_N > B) \approx \exp^{-N\gamma_n} \qquad (18)$$

where Q_N is the queue size and N is the number of sources. Figure 1 (right) shows the estimations of γ_n for 600 simulated traces, its theoretical value and its confidence interval. Numerical results show that in this case 94.8% of the values fall in the 95% confidence interval.

3 End-to-End QoS, the Fictitious Network Analysis

3.1 Introduction

As we have explained in the previous section, using Large Deviations Theory and in the many sources asymptotic Wischik [6] proves the following formula (called *inf sup* formula) for the overflow probability:

$$\log \mathbf{P}(Q_N > B) \approx -\inf_{t \geq 0} \sup_{s \geq 0}((B+Ct)s - Nst\alpha(s,t))$$

where Q_N represents the stationary amount of work in the queue, C is the link capacity, B is the buffer size and N is the number of incoming multiplexed sources of effective bandwidth $\alpha(s,t)$.

Wischik also shows in [13] that in the many sources asymptotic regime the aggregation of independent copies of a traffic source at the link output and the aggregation of similar characteristics at the link input, have the same effective bandwidth in the limit when the number of sources goes to infinity. This result allows to evaluate the end to end performance of some kind of networks like "in-tree" ones. Unfortunately this analysis can not be extended to networks with a general topology.

A slightly different asymptotic with many sources and small buffer characteristics was proposed by Ozturk, Mazumdar and Likhanov in [4]. They consider an asymptotic regime defined by N traffic sources, link capacity increasing proportionally with N but buffer size such that $\lim \frac{B(N)}{N} \to 0$. In their work they calculate the rate function for the buffer overflow probability and also for the end to end loss ratio. This last result can be used to evaluate the end to end QoS performance in a network backbone in contrast with the Wischick result explained before, where it is necessary to aggregate at each link N i.i.d. copies of the previous output link.

Ozturk et al. also introduce the "fictitious network" model. The fictitious network is a network with the same topology than the real one, but where each flow aggregate goes to a link on its path without being affected by the upstream links until that link. The fictitious network analysis is simpler and so, more adequate to on-line performance evaluation and traffic engineering. Ozturk et al. show that the fictitious network analysis overestimates the overflow probability. In this work we analyze when, for an interior network link, the overflow probability calculated using the fictitious network is equal to the overflow probability of the real network.

In the next section we summarize Ozturk et al. main results.

3.2 Many Sources and Small Buffer Asymptotic Performance Model

Ozturk, Mazumdar and Likhanov work. Consider a network of L links which is accessed by M types of independent traffic. Consider a discrete time fluid FIFO model where traffic arrives at time $t \in Z$ and is served immediately if buffer is empty and is buffered otherwise. Each link k has capacity NC_k and buffer size $B_k(N)$ where $B_k(N)/N \to 0$ with $N \to \infty$. Input traffic of type m=1,...,M, denoted $X^{m,N}$ is stationary and ergodic and has rate $X_t^{m,N}$ at time t (workload at time t of N sources of type m).

Let $\mu_m^N = \mathbf{E}(X_0^{m,N})/N$ and $X^{m,N}(t_1, t_2) = \sum_{t=t_1}^{t_2} X_t^{m,N}$. We assume that $\mu_m^N \xrightarrow[N \to \infty]{} \mu_m$ and $X^{m,N}(0,t)/N$ satisfies the following Large Deviation Principle (LDP) with *good rate function* $I_t^{X^m}(x)$:

$$- \inf_{x \in \Gamma^o} I_t^{X^m}(x) \leq \liminf_{N \to \infty} \frac{1}{N} \log \mathbb{P} \left(\frac{X^{m,N}(0,t)}{N} \in \Gamma \right) \tag{19}$$

$$\leq \limsup_{N \to \infty} \frac{1}{N} \log \mathbb{P} \left(\frac{X^{m,N}(0,t)}{N} \in \Gamma \right) \leq - \inf_{x \in \overline{\Gamma}} I_t^{X^m}(x) \tag{20}$$

where $\Gamma \subset \mathbb{R}$ is a Borel set with interior Γ^o and closure $\overline{\Gamma}$ and $I_t^{X^m}(x) : \mathbb{R} \to [0, \infty)$ is a continuous mapping with compact level sets. We also assume the following technical condition: $\forall \ m$ and $a > \mu_m$,

$$\liminf_{t \to \infty} \frac{I_t^{X^m}(at)}{\log t} > 0$$

Type m traffic has a fixed route without loops and its path is represented by the vector $\mathbf{k}^m = (k_1^m, ..., k_{l_m}^m)$, where $k_i^m \in (1,..,L)$. The set $\mathcal{M}_k = \{m : k_i^m = k, 1 \leq i \leq l_m\}$ denotes the types of traffic that goes through link k. To guarantee system stability it is assumed that

$$\sum_{m \in \mathcal{M}_k} \mu_m < C_k \tag{21}$$

The main result of Ozturk et al. work is the following theorem.

Theorem 2. *Let $X_{k,t}^{m,N}$ be the rate of type m traffic at link k at time t. There exist a continuous function $g_k^m : \mathbb{R}^M \to \mathbb{R}$ relating the instantaneous input rate at link k for traffic type m to all of the instantaneous external input traffic rates such that:*

$$\frac{X_{k,0}^{m,N}}{N} = g_k^m \left(\frac{X_0^{1,N}}{N}, ..., \frac{X_0^{M,N}}{N} \right) + o(1) \tag{22}$$

The buffer overflow probabilities are given by:

$$\lim_{N\to\infty} \frac{1}{N} \log P(\text{overflow in link } k) = -\mathbf{I}_k =$$

$$-\inf\left\{\sum_{m=1}^{M} I_1^{X^m}(x_m) : x = (x_m) \in \mathbb{R}^M, \sum_{m=1}^{M} g_k^m(x) \geq C_k\right\} \quad (23)$$

In (22), $o(1)$ verifies that $\lim_{N\to\infty} o(1) = 0$ since $\frac{B_k(N)}{N} \underset{N\to\infty}{\to} 0$. The function $g_k^m(x)$ is constructed in the proof of the theorem. Ozturk et al. prove that the continuous function relating the instantaneous input rate at link i for traffic m to all of the instantaneous external input traffic rates is the same function relating these variables in a no buffers network. The function relating the instantaneous output rate at link i for traffic m to all of the instantaneous input traffic rates at this link is:

$$f_i^m(x, C_i) = \frac{x_m C_i}{\max(\sum_{j\in\mathcal{M}_i} x_j, C_i)} \quad (24)$$

In a feed-forward network the function $g_k^m(x)$ can be written as composition of the functions of type (24) in a recursive way. Using equation (24) the buffer overflow probability can be calculated for any network link, by solving the optimization problem of equation (23). We need to know the network topology, the link capacities and, for each arrival traffic type m, the rate functions $I_1^{X^m}$.

Ozturk et al. define also the total (end to end) loss ratio as the ratio between the expected value of lost bits at all links along a route and the mean of input traffic in bits, for stream m identified by X^m. With the previous definition they find the following asymptotic for the loss ratio $\mathbf{L}^{m,N}$:

$$\lim_{N\to\infty} \frac{1}{N} \log \mathbf{L}^{m,N} = -\min_{k\in k^m} \mathbf{I}_k \quad (25)$$

The main problem of this approach is that the optimization problem of equation (23) could be very hard to solve for real-size networks. The calculation of the function $g_k^m(x)$ is recursive and so, when there are many links it becomes complex. In addition, the virtual paths can change during the network operation. Therefore, it is necessary to recalculate on-line the function $g_k^m(x)$. To solve equation (23), it is also necessary to optimize a nonlinear function under nonlinear constraints. In order to simplify this problem, Ozturk et al. introduce the "fictitious network" concept, that is simpler and gives conservative results. In the next section we find conditions to assure that there is no overestimation in the calculus of the link overflow probability in the fictitious network analysis. We also find a bound for the error (difference between the rate function calculated for the real network and the fictitious one) in those cases where the previous condition is not satisfied.

The aim of our work is to define an admission control mechanism. Such a mechanism is simple a set of rules to accept or reject a flow that intend to access the network. This can be done by defining an acceptance region, i.e. which is the set of flows that can access the network. In [4] an acceptance region based on end-to-end QoS guarantees, is defined. This acceptance region is the traffic mix that can flow through the network without QoS violation. Assume that $X^{m,N}$ is the sum of Nn_m i.i.d. process. More formally, the acceptance region noted by $\mathcal{D}$ is the mix or collection $\{n_m\}_{m=1}^M$ of sources which can be flowing through the network while the QoS (loss ratio) for each class is met, that is:

$$\mathcal{D} = \{(n_m), m = 1, ..., M : \lim_{N \to \infty} \frac{1}{N} \log \mathbf{L}^{m,N} < -\gamma_m\} \quad \text{with} \quad \gamma_m > 0 \quad (26)$$

We will concentrate then in the estimation of this acceptance region. We aim not only to do it in a efficient way but also in a simple one in order to apply it on-line.

3.3 Fictitious Network Analysis

We analyze an interior network link k under the same assumptions that in Ozturk et al. work. $\mathcal{M}$ is the set of traffic types that access the network and $\mathcal{M}_i$ is the set of traffic types that go through link i. We suppose that the network is feed-forward, this means that each traffic type has a fixed route without loops. In the real network, the link k overflow probability large deviation function (or rate function) is given by:

$$I_k^R = \inf \left\{ \sum_{i \in \mathcal{M}} I_1^{X^i}(x_i) : x = (x_i)_{i \in \mathcal{M}}, \sum_{i \in \mathcal{M}} g_k^i(x) \geq C_k \right\} \quad (27)$$

In the fictitious network this function is given by

$$I_k^F = \inf \left\{ \sum_{i \in \mathcal{M}_k} I_1^{X^i}(x_i) : x = (x_i)_{i \in \mathcal{M}_k}, \sum_{i \in \mathcal{M}_k} x_i \geq C_k \right\} \quad (28)$$

In the following it is assumed that each traffic type is an aggregate of N i.i.d sources. This implies that each rate function $I_1^{X^i}$ is convex and $I_1^{X^i}(\mu_i) = 0$ for all i. Then, (27) and (28) are convex optimization problems under constraints. The second one has the advantage that the constraints are linear and there are well known fast methods to solve it. The functions $I_1^{X^i}$ are continuous, so we solve the following problems corresponding to the real and fictitious network respectively.

$$P_R \begin{cases} \min \sum_{i \in \mathcal{M}} I_1^{X^i}(x_i) \\ \sum_{i \in \mathcal{M}} g_k^i(x) \geq C_k \end{cases} \qquad P_F \begin{cases} \min \sum_{i \in \mathcal{M}_k} I_1^{X^i}(x_i) \\ \sum_{i \in \mathcal{M}_k} x_i \geq C_k \end{cases}$$

Definition 1. *Consider two optimization problems*

$$P_1 \begin{cases} \min f_1(x) \\ x \in D_1 \end{cases} \quad \text{and} \quad P_2 \begin{cases} \min f_2(x) \\ x \in D_2 \end{cases}$$

P_2 *is called a relaxation of* P_1 *if* $D_1 \subseteq D_2$ *and* $f_2(x) \leq f_1(x)$, $\forall x \in D_1$.

Proposition 1. P_F *is a relaxation of* P_R.

Proof. Since the functions $I_1^{X^i}$ are non negatives, it is clear that $\sum_{i \in \mathcal{M}_k} I_1^{X^i}(x_i) \leq \sum_{i \in \mathcal{M}} I_1^{X^i}(x_i)$ $\forall x = (x_i)_{i \in \mathcal{M}}$. Then, we have to prove that

$$\left\{ x : \sum_{i \in \mathcal{M}} g_k^i(x) \geq C_k \right\} \subseteq \left\{ x : \sum_{i \in \mathcal{M}_k} x_i \geq C_k \right\}$$

By definition, $g_k^i(x) = 0 \ \forall \ i \notin \mathcal{M}_k$ and $g_k^i(x) \leq x_i \ \forall \ i \in \mathcal{M}_k$ (since g_k^i can be written as composition of functions of type (24)) then

$$\sum_{i \in \mathcal{M}} g_k^i(x) = \sum_{i \in \mathcal{M}_k} g_k^i(x) \leq \sum_{i \in \mathcal{M}_k} x_i$$

and therefore $\sum_{i \in \mathcal{M}_k} g_k^i(x) \geq C_k$, implies $\sum_{i \in \mathcal{M}_k} x_i \geq C_k$.

Remark 4. If an optimum of the fictitious problem P_F verifies the real problem P_R constraints and the objective functions take the same value at this point, then it is an optimum of the real problem too.

The following theorem gives conditions over the network to assure that link k overflow probability rate function for the real and for the fictitious network are equal ($E = I_k^R - I_k^F = 0$). Since the network is feed forward, it is possible to establish an order between the links. We say that link i is "previous to" or "less than" link j if for one path, link i is found before than link j in the flow direction.

Theorem 3 (Sufficient Condition). *If* $\tilde{x} = (\tilde{x}_i)_{i \in \mathcal{M}_k}$ *is optimum for* P_F, *and the following condition is verified for all links* i *less than* k:

$$C_k - \sum_{j \in \mathcal{M}_k \setminus \mathcal{M}_i} \mu_j \leq C_i - \sum_{j \in \mathcal{M}_i \setminus \mathcal{M}_k} \mu_j \quad \forall \ i < k \tag{29}$$

then x^* *defined by:*

$$(x^*)_i = \begin{cases} \tilde{x}_i & \text{if } i \in \mathcal{M}_k \\ \mu_i & \text{if } i \notin \mathcal{M}_k \end{cases}$$

is optimum for P_R.

Proof. See [14].

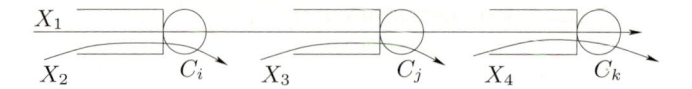

Fig. 2.

Example 1. Consider a network like in figure 2. We analyze the overflow probability at link k.
If condition (29) is attained for link k, then $E = I_k^R - I_k^F = 0$. This condition is:

$$\begin{cases} C_k - \mu_4 \leq C_i - \mu_2 \\ C_k - \mu_4 \leq C_j - \mu_3 \end{cases}$$

Sufficient condition in terms of available bandwidth

Definition 2. *For a traffic type m in a link j, it is defined the available bandwidth ABW_j^m as the difference between the link j capacity and the mean value of the transmission rate of the other traffic types in j.*

In terms of the previous definition, the theorem condition (29) assures that the overflow probability rate function at link k on real and fictitious network are the same if for all link $j < k$, and for all m traffic type in $\mathcal{M}_j \cap \mathcal{M}_k$, $ABW_j^m > ABW_k^m$. This condition is represented in figure 3 for a simple network with two links.

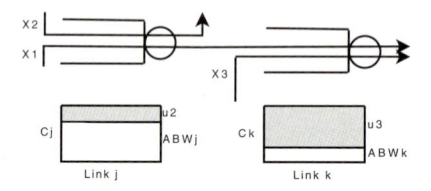

Fig. 3. Sufficient condition in terms of available bandwidth

Sufficient but not necessary condition. The theorem condition (29) is sufficient to assure that the overflow probability rate function at link k on real and fictitious networks are the same, but it is not a necessary condition. In fact, if $\tilde{x}$ is optimum for the fictitious problem, and if x^* defined as:

$$(x^*)_i = \begin{cases} \tilde{x}_i & \text{si } i \in \mathcal{M}_k \\ \mu_i & \text{si } i \notin \mathcal{M}_k \end{cases} \quad (30)$$

satisfies the real problem constraints, then x^* is optimum for the real problem. If x^* verifies the following condition

$$\sum_{j \in \mathcal{M}_i} (x^*)_j \leq C_i \quad \forall i < k \quad (31)$$

it also verifies the real problem constraints and therefore is optimum for the real problem.

Therefore, in the case that the theorem condition is not fulfilled, if we found $\tilde{x}$ optimum for the fictitious problem, then is easy to check if the rate functions are equal or no. It is enough to check (31), where x^* is defined in (30).

Error bound. Since the functions $I_1^{X^i}$ are non negatives, it is clear that the rate function for the real problem is always greater than the fictitious one. Then the error $E = I_k^R - I_k^F$ is always non negative. This implies that the fictitious network overestimates the overflow probability. We are interested in finding an error bound for the overestimation of the fictitious analysis when conditions (29) and (31) are not satisfied. A simple way to get this bound is to find a point x which verifies the real problem constraints. In this case, we have that:

$$E = I_k^R - I_k^F \leq \sum_{i \in \mathcal{M}} I_1^{X^i}(x_i) - \sum_{i \in \mathcal{M}_k} I_1^{X^i}(\tilde{x}_i)$$

To assure that x verifies the real problem constraints, we have already seen that it is enough to show that $\sum_{j \in \mathcal{M}_i} x_j \leq C_i \ \forall \ i < k$ and $\sum_{j \in \mathcal{M}_k} x_j \geq C_k$. Therefore, we have to solve this inequalities system. It can be seen that the optimum of the fictitious problem is in the boundary of the feasible region ($\sum_{i \in \mathcal{M}_k} \tilde{x}_i = C_k$). Since we are looking for a point near the optimum of the fictitious problem in the sense that the error bound be as small as possible, we solve the following system:

$$\begin{cases} \sum_{j \in \mathcal{M}_i} x_j \leq C_i \quad \forall \, i < k \\ \sum_{j \in \mathcal{M}_k} x_j = C_k \end{cases} \quad (32)$$

For the interesting cases, where there are losses at link k, this system always has a solution. In the following an algorithm to find a solution of this system is defined. We define the following point:

$$(x^*)_j = \begin{cases} \tilde{x}_j & \text{if } j \in \mathcal{M}_k \\ 0 & \text{if } j \notin \mathcal{M}_k \end{cases}$$

If x^* verifies the conditions (32), we find a point that verifies the real problem constraints. In some cases this is not useful because $I_1^{X^j}(0) = \infty$ and we have that the error bound is infinite. If $P(X_1^{j,N} \leq 0) \neq 0$, the function $I_1^{X^j}(0) < \infty$ and a finite error bound is obtained. If x^* is not solution for system (32), then we redefine (by some small value) the coordinates where $\sum_{j \in \mathcal{M}_i} x_j > C_i$ in such a way that $\sum_{j \in \mathcal{M}_i} x_j = C_i$. The second equation must be verified too and, since some coordinates were reduced, others coordinates have to increase to get the total sum equal to C_k. Since the system is compatible, following this method, a solution is always found. There is no guarantee that the solution given by this method minimizes the error bound. However, this method has a very simple implementation and gives reasonable error bounds as we can see in the numerical examples of the last section.

3.4 End-to-End Loss Ratio Evaluation

In the previous section we found sufficient conditions to assure that results on the fictitious and on the real network analysis coincide for an interior link. However, to define an admission control mechanism based on the end-to-end quality of service, we must find a condition that guarantees that the end-to-end loss ratio coincides for both networks. A natural answer is that the sufficient condition found in theorem 3 must be verified for all links in the considered path. However, as equation 25 suggest, we will show that this is not necessary since it is enough that the sufficient condition is verified for the link with minimum overflow probability rate function. This link must be then identified, and clearly we aim to do it within the fictitious network context. We must then be sure that the link with minimum rate function is the same for the real and the fictitious network. In the sequel we address this two issues.

Proposition 2. *Let k_f be the link with minimum overflow probability rate function in the fictitious network for traffic type m: $\overline{I}_{k_f} = \min_{k_i \in \mathbf{k}^m} \overline{I}_{k_i}$*

If the conditions of theorem 3 are verified for link k_f, the minimum overflow probability rate function for traffic type m in the real network is also attained at link k_f.

Proof. See [14].

Proposition 3. *Let k be the link where $I_k = -\min_{k \in \mathbf{k}^m} \mathbf{I}_k^m$ for the real network, i.e. the link where the minimum rate function of traffic type m is attained. Let $\overline{I}_k$ be the rate function of the same link k for the fictitious network. If the sufficient conditions of theorem 3 are verified for link k then $\mathbf{L}^m = \overline{\mathbf{L}}^m$, i.e. the end-to-end loss ratio for real and fictitious network coincide.*

Proof. See [14].

Remark 5. Previous propositions show that to evaluate the end-to-end loss ratio $\mathbf{L}^m$, it is enough that sufficient conditions of theorem 3 are verified by the link k where the minimum rate function of traffic type m path is attained. In this case, it results that $\mathbf{L}^m = \overline{\mathbf{L}}^m = \overline{I}_k$. If sufficient conditions are not verified, then the error bound obtained for the one link case can be applied.

3.5 Rate Function Estimation

In previous sections we show how to evaluate the end-to-end loos ratio in terms of the rate function for the fictitious network. In order to implement an on-line admission control based on this information, we must be able to accurately estimate the corresponding rate function. In this section we analyze how this estimation can be done using traffic traces of the input traffic.

Let $X_k^{m,N}(0,t)$ be the traffic type m workload at link k during the time interval $(0,t)$. We suppose that $X_k^{m,N}$ is the sum of $N\rho_m$ independent sources of type m:

$$X_k^{m,N}(0,t) = \sum_{i=1}^{N\rho_m} \widetilde{X}_k^{m,i}(0,t)$$

In this case, the instantaneous rate of traffic type m at time t is given by:

$$X_{k,t}^{m,N} = \sum_{i=1}^{N\rho_m} \widetilde{X}_{k,t}^{m,i}$$

Given the stationarity of the traffic, we can replace the t-index by 0 and for simplicity we omit the link index k. Then the instantaneous rate of total input traffic at link k is:

$$Z_0^N = \sum_{m \in \mathcal{M}_k} X_0^{m,N} = \sum_{m \in \mathcal{M}_k} \sum_{i=1}^{N\rho_m} \widetilde{X}_0^{m,i} = \sum_{j=1}^{N} \widetilde{Z}^j$$

where the variables $\widetilde{Z}^j$ are independent and identically distributed (*iid*) random variables. Each variable $\widetilde{Z}^j$ has the distribution of a mix of the variables $\widetilde{X}_0^{m,i}$ (given by the proportions ρ_m of each traffic type m present at link k). This means that instantaneous rate of input traffic at link k is the sum of N *iid* random variables and Cramer theorem (see for example [6]) can be applied. The variable $\frac{Z_0^N}{N}$ verifies then a large deviation principle with rate function:

$$I_t^Z(x) = \sup_{\theta \geq 0}\{\theta x - \Lambda(\theta)\} = \sup_{\theta \geq 0}\{\theta x - \log \mathbf{E}\left(e^{\theta \widetilde{Z}^1}\right)\} \qquad (33)$$

Given the rate function of the LDP, $I_t^Z(x)$, we can calculate I_k^F:

$$I_k^F = \inf\{I^Z(z) : z \geq C_k\} = \inf_{z \geq C_k} \sup_{\theta \geq 0}\{\theta z - \Lambda(\theta)\}$$

$$= \sup_{\theta \geq 0}\{\theta C_k - \Lambda(\theta)\} \qquad (34)$$

Before solving the optimization problem 34, we must calculate or estimate $\Lambda(\theta)$. If some model is assumed for the traffic, $\Lambda(\theta)$ can be calculated analytically. In case no model is assumed as in our case, it must be estimated from measurements i.e. from traffic traces. A possible and widely used approach [7,8] is to estimate the expectation as a temporal average of a given traffic trace $\{\widetilde{Z}^N(t)\}_{t=1:n}$:

$$\mathbf{E}\left(e^{\theta \widetilde{Z}^1}\right) = \mathbf{E}\left(e^{\theta \frac{Z_0^N}{N}}\right) \approx \frac{1}{n}\sum_{t=1}^{n} e^{\theta Z^N(t)/N}$$

Then $\Lambda(\theta)$ can be estimated by $\Lambda_n(\theta) = \log\left(\frac{1}{n}\sum_{t=1}^{n} e^{\theta Z^N(t)/N}\right)$

Now, the rate function I_k^F can be estimated as: $I_{k,n}^F = \sup_{\theta \geq 0}\{\theta C_k - \Lambda_n(\theta)\}$.

However it remains unclear how good is this estimation. We will show that if $\Lambda_n(\theta)$ is a *good* estimator of $\Lambda(\theta)$, then $I_{k,n}^F$ is also a *good* estimator for the rate function I_k^F.

Theorem 4. *If $\Lambda_n(\theta)$ is an estimator of $\Lambda(\theta)$ such that both are C^1 functions and:*

$$\Lambda_n(\theta) \xrightarrow[n]{} \Lambda(\theta) \qquad \frac{\partial}{\partial \theta}\Lambda_n(\theta) \xrightarrow[n]{} \frac{\partial}{\partial \theta}\Lambda(\theta)$$

where the convergence is almost surely and uniformly over bounded intervals, then $I_{k,n}^F$ is a consistent estimator of I_k^F. Moreover, if a functional Central Limit Theorem (CLT) applies to $\Lambda_n - \Lambda$, i.e, $\sqrt{n}\,(\Lambda_n(\theta) - \Lambda(\theta)) \xrightarrow[n]{w} G(\theta)$, where $G(\theta)$ is a continuous gaussian process, then: $\sqrt{n}\left(I_{k,n}^F - I_k^F\right) \xrightarrow[n]{w} N(0,\sigma)$, where $N(0,\sigma)$ is a centered normal distribution with variance σ.

Proof. See [14].

From the previous analysis we conclude that the rate function and then the admission control region can be accurately estimated from traffic traces in a simple way. As we claimed before, this can be used in the definition of an admission control mechanism based in the end-to-end quality of service expected by the traffic.

3.6 Numerical Example

In this section we present a numerical example to validate our results. Additional numerical examples can be found in [14].

There are many issues that could be evaluated to analyze the performance of an admission control mechanism. However, since the overall performance of our proposition depends on how accurate are the results obtained when the fictitious network model is considered, we will concentrate here only in this aspect.

Example 2. Consider a network like in figure 4. We analyze the overflow probability at link k, assuming that $C_i > C_k$.

If condition (29) is attained for link k, then $E = I_k^R - I_k^F = 0$.
This condition is: $C_k \leq C_i - \mu_2$.

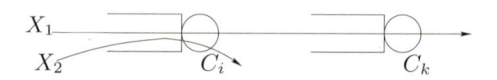

Fig. 4. Example 2-Network topology

If this condition is not satisfied, since $\tilde{x} = C_k$ is optimum for P_F, we first verify if $x^* = (C_k, \mu_2)$ is optimum for (P_R). It is sufficient to show that x^* verifies the real problem constraints, i.e:
$$\begin{cases} C_k + \mu_2 \leq C_i \\ C_k = C_k \end{cases}$$

If $C_k + \mu_2 > C_i$, we look for $x^* = (x_1^*, x_2^*)$ that verifies $\begin{cases} x_1^* + x_2^* \leq C_i \\ x_1^* = C_k \end{cases}$

It is possible to choose $x_1^* = C_k$ and $x_2^* = C_i - C_k > 0$ resulting in the following error bound:

$$E \leq I_1(C_k) + I_2(C_i - C_k) - I_1(\tilde{x}_1) = I_2(C_i - C_k) \tag{35}$$

In the following numerical example, we calculate the overflow probability rate function for the real and fictitious network. Let $C_i = 16kb/s$ per source and C_k growing from 4 to $15.5kb/s$ per source. All traffic sources are on-off Markov processes. For X_1, the bit rate in the on state is $16kb/s$, and average times are $0.5s$ in the on state and $1.5s$ in the off state. For X_2, the bit rate in the on state is $16kb/s$, and average times are $1s$ in the on state and $1s$ in the off state. Since $\mu_1 = 4kb/s$ the stability condition is $C_k > \mu_1 = 4kb/s$. Using these values, the sufficient condition (29) is, $C_k \leq 8kb/s$. Figure 5 shows that while this condition is satisfied both functions match, but after $C_k \geq 8kb/s$ they separate. Figure 5 also shows the overestimation error ($E = I_k^R - I_k^F$) and the error bound (35) described before. In this case, the error bound is exactly the error.

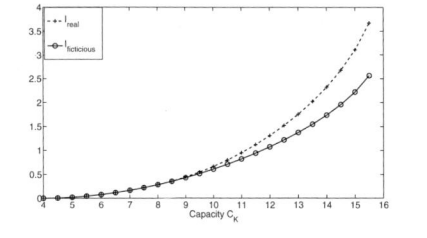

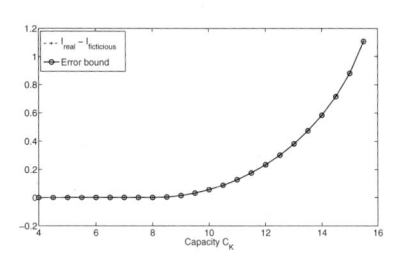

Fig. 5. Example 2-Rate functions and error bound

4 End-to-End QoS Prediction Based on Active Measurements and Statistical Learning

4.1 Introduction

The many sources and small buffer asymptotic analyzed in the previous section, can only be applied to analyze an end-to-end path in a backbone network. If the end points are end users this asymptotic cannot be applied because the path goes through the backbone but also through the access network where the many sources asymptotic is not valid. Therefore, a different approach must be applied if the access control mechanism must take a decision based on end-to-end QoS.

In this section we analyze another approach based on measurements and statistical learning in order to evaluate the end-to-end QoS parameters seen by applications like a video on demand service.

A possible measurement technique for such tool is to send the application traffic (a video for example) and to measure the video QoS parameters at the receiver. However, in many cases these application flows may have bandwidth requirements that are not negligible compared with links capacity. This technique could overload a congested link degrading the QoS perceived by clients using the system. This QoS degradation can be tolerated during short periods but the previous methodology cannot be used if the operator requires a permanent or frequent network monitoring.

Other measurement techniques estimate the QoS parameters seen by an application using light probe packets and without considering the particular characteristics of the application. These probe packets do not overload the network but this procedure assumes, for example, that the delay of a specific application can be approximated by the probe packets delay. This previous assumption is not always true because the QoS parameters depend on the statistical behavior of each traffic. Therefore, in many cases, this kind of estimation yields inaccurate results.

We propose a methodology that is an intermediate point between both approaches (to send a multimedia flow during long periods or to send light probe packets during short periods) and provides an accurate estimation of QoS parameters seen by an application without overloading the network during long periods.

Our goal is to learn the relation between the QoS parameters seen by an application and the probe packets interarrival times statistic. This statistic characterizes the network state. Once the model is learned, in order to predict the QoS parameters, it is necessary only to send light probe packets.

More formally, we consider the regression model

$$Y = \Phi(X) + \varepsilon \qquad (36)$$

where X, Y and ε are random variables. The random variable X is an estimation of the state of network path, the response Y is the QoS parameter seen by the application (delay, jitter, loss rate) and ε is a centered random variable which represents an error, where ε and X are independent.

The previous formulation evidences two problems to be addressed in this work. First, it is necessary to find an accurate estimation of the state of the network path (the variable X). Second, it is necessary to estimate the function Φ. We propose to estimate this function learning Φ from samples of the random variables Y and X.

In order to estimate the state of the network path, we analyze a functional random variable X that is the empirical distribution of the probe packets interarrival times.

In order to estimate the function Φ we propose a statistical learning approach based on the Nadaraya-Watson estimator. Nadaraya-Watson, first introduced for real data [15], is used in this work mainly for functional regression [16]. We propose also an extension of theoretical results about the Nadaraya-Watson functional estimator in a nonstationary context. This non-parametric approach is based on mapping data obtained from probe packets and any QoS parameter seen by an application.

4.2 Problem Formulation and Proposed Solution

We first consider the case of a path with a single link. The multilink case is discussed later. We assume that the cross traffic, the link capacity and the buffer size are unknown. The QoS parameter seen by the application is called Y and it is a function of the link and traffic characteristics: $Y = F(X_t, V_t, C, B)$.

where X_t is the cross traffic stochastic process, V_t is the video or other application traffic stochastic process, C is the link capacity and B is the buffer size. The link capacity C and the buffer size B are not known but it is assumed that both have constant values during the monitoring process. As the goal is to estimate a QoS parameter over the known process V_t (a video sequence for example), V_t can be considered as an input to our problem. Taking into account the previous considerations, we can say that $Y = F(X_t)$. At the end of this section we discuss these assumptions about C, B, and V_t.

The previous formulation pose two different problems that should be addressed. On one hand the estimation of the function that relates the cross traffic and the QoS parameter and on the other the estimation of the cross traffic process X_t. In order to take into account the multilink case, the last estimation is what we call the estimation of the state of the network path.

In order to estimate the cross traffic we send probe packets from the user equipment and measure the interarrival times. When two consecutive probe packets are queued in the same busy period at the link queue, as shown in figure 6 (left), the interarrival time is equal to $\frac{X_i}{C} + \frac{K}{C}$, where X_i is the amount of cross traffic that arrived at the queue between probe packets i and $i+1$, K is the probe packets size and C is the link capacity. Then, during the busy periods, the interarrival times are proportional to the cross traffic volume at least up to a constant.

In the case where the packet $i+1$ is queued after the packet i leaves the queue, as we infer the cross traffic volume from the values t_{out}^i, we can conclude that there is a cross traffic volume larger than the real one.

Baccelli et al. [17] present a rigorous probabilistic approach to active probing methods for cross traffic estimation. They analyze the system identifiability and show that different cross traffic types can give rise to the same sequence of observed probe delays. Therefore, it is not always possible to determine the

Fig. 6. Probe packets, probe video (video) and cross traffic (ct)

distribution of any desired aspect of the cross traffic using probes. However, we are not looking for an accurate estimator of the cross traffic. We are actually looking for an estimation of Y. Therefore, our interest is only in finding an estimator that allow us to distinguish between possible states of the network. This state is represented by a variable X that is a function of the probe packets interarrival times.

We will estimate the function Φ in the regression model of equation 36 from the observations of the pairs (X, Y).

We divide the experiment in two phases. The first phase is called the learning phase. In the learning phase we send a burst of probe packets. The probe packets interdeparture time is a fixed value t_{in} and the packets have a fixed size K. Immediately after the probe packets we send a video sequence training sample during a short period.

This procedure is repeated periodically sending the probe packets and the video sequence alternatively as shown in figure 6 (right).

We build the variable X_j by measuring for each experiment j the interarrival times of the probe packets burst. We also measure the QoS parameter Y_j of the corresponding video sequence and we have a pair (X_j, Y_j) for each experiment. The problem is how to estimate the function $\Phi : \mathcal{D} \to \mathbb{R}$ by $\widehat{\Phi}$ from these observations, where $X \in \mathcal{D}$ and $\mathbb{R}$ is the real line.

The second phase is called the monitoring phase. During the monitoring phase we send only the probe packets. We build the variable X in the same way as in the learning phase. The QoS parameter $\widehat{Y}$ of the video sequence is estimated using the function $\widehat{\Phi}$ built in the learning phase by $\widehat{Y} = \widehat{\Phi}(X)$. We remark that this procedure does not load the network because it avoids sending the video sequence during the monitoring phase.

Remark 6. The previous discussion is based on the single link case. We discuss now some considerations about the multilink case. First, we must highlight that the multilink case can be reduced to the single link one in many important escenarios. For example, when the application service is offered by a server located at the ISP backbone (for example a video on demand server) and the user access is a cellular link or an ADSL link. In these cases the bottleneck is normally located at the access since the backbone is overprovisioned and it behaves as a single link.

However, there are cases where the packets must wait in more than one queue. In these cases the different queues modify the variable X that we use to characterize the cross traffic. This means that we estimate a variable X where the influence of all queues are accumulated. Nevertheless even in this case our method will work fine if it is possible to distinguish with this variable between different cross traffic processes observed in the path.

Remark 7. Another assumption was that the network path, the link capacities and the buffer sizes are fixed. For link capacities and buffer sizes this assumption is reasonable. However, the route between two points on the network can change. This problem can be solved because it is possible to verify periodically the route between two points using for example an application like trace-route. If a new route is detected two circumstances can arise. If the system has learned information about the new route, this information can be used for the estimation. If the system has not learned information about the new route it is necessary to trigger a learning phase. Finally, we remark that in some cases a change in the route does not affect the measures, for example when the bottleneck is in the access link and the backbone is overprovisioned.

Remark 8. In this section we work with the assumption that the system is trained with a unique kind of video (we assume that V_t is a fixed sequence). This is not really an issue since the video QoS parameters depend on a set of characteristics like coding, bit-rate, frame-rate and motion level. Therefore, we can train the system with a set of video sequences that represent the different classes of videos. Later the system will use the corresponding training samples depending on the specific video that we want to monitor.

4.3 Statistical Learning, the Nadaraya-Watson Estimator

In this section we discuss the mathematical tool selected to estimate Φ. We present a brief review of current results about Nadaraya-Watson estimations. We consider the regression model of equation 36. It is not assumed an explicit form for the function Φ that relates the state of the network with the QoS parameters, and it is not assumed either any particular probability distribution for the random variables involved in the model. For this reason the model is nonparametric.

There are several results on nonparametric regression for real random variables and for random variables in $\mathbb{R}^d$ since the works of Nadaraya and Watson [18]. The Nadaraya-Watson estimator for the real case is

$$\widehat{\Phi}_n(x) = \frac{\sum_{i=1}^{n} Y_i K\left(\frac{||x-X_i||}{h_n}\right)}{\sum_{i=1}^{n} K\left(\frac{||x-X_i||}{h_n}\right)} = \frac{\sum_{i=1}^{n} Y_i K_n(X_i)}{\sum_{i=1}^{n} K_n(X_i)} \qquad (37)$$

K is a Kernel, which is a positive function that integrates one and $K_n(X_i) = K\left(\frac{||x-X_i||}{h_n}\right)$. h_n is a sequence that tends to zero and it is called the kernel bandwidth. This estimator is a weighted average of the samples $Y_1, \ldots, Y_n$. The weights are given by $K_n(X_i)$ taking into account the distance between x and each point of the sample $X_1, \ldots, X_n$.

4.4 The Empirical Distribution of the Probe Packets Interarrival Times

In this section we select for the variable X, the empirical distribution of the probe packets interarrival times. This lead us to a functional regression model. In last years some theoretical results on the functional Nadaraya-Watson estimator were developed.

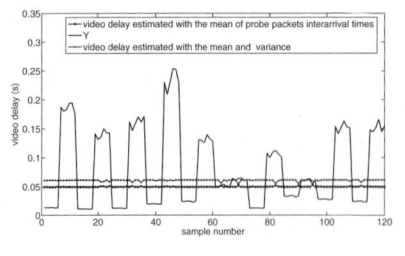

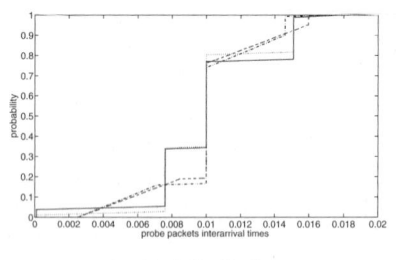

| Estimation of Y using mean and variance | Empirical distributions |

Fig. 7. Probe packets inter-arrival times

Why functional regression? We try to use as first option for the variable X the mean and/or the variance of the probe packets interarrival times.

In figure 7 (left) it can be observed the estimation of Y using these possible choices for X. We develop many experiments with simulated data and with data taken from operational networks and the estimations of Y are in all cases inaccurate. It is not possible to estimate Y from the mean and the variance.

In figure 7 (right) we show four empirical distribution functions for simulated data. Two of them were obtained in the presence of high cross traffic and the others with low cross traffic. These empirical distribution functions capture some network characteristics that allow us to distinguish between them. In the next section we analyze the QoS estimation using these empirical distribution functions.

Functional Nadaraya-Watson estimator. For functional random variables, i.e. when the regressor X is a random function Ferraty et al. [16] introduce a Nadaraya-Watson type estimator for Φ, defined by equation (37), where the difference with the real case is that $\|\cdot\|$ is a seminorm on a functional space $\mathcal{D}$. One of the main issues in the functional approach is the "curse of dimensionality". The estimation $\widehat{\Phi}_n(x)$ will be accurate if there are enough training samples near x. This issue becomes crucial when the observations come from an infinite dimensional vector space. This problem is addressed in the literature and we refer for example to [16,19] for different approaches. These works state the convergence and the asymptotic distribution of the estimator for stationary and weakly dependent (for example mixing) functional random variables.

Extensions to the nonstationary case. The cross traffic X_t on the Internet is a dependent and non-stationary process. This topic has been studied by many authors during last ten years. Zhang et al. [20] show that many processes on the Internet (losses for example) can be well modelled as independent and identical distributed (i.i.d.) random variables within a "change free region", where stationarity can be assumed. They describe the overall network behavior as a series of piecewise-stationary intervals.

The nonstationarity has different causes. In all cases it is very important to have estimators that can be used with nonstationary traffic.

As our data comes from Internet data traffic and it is typically nonstationary, we extend previous results about functional nonparametric regression to this case. Instead of considering random variables X equally distributed we consider a model introduced by Perera in [21] defined by $X_i = \varphi(\xi_i, Z_i)$ where ξ_i takes values in a seminormed vector space with a seminorm $||\cdot||$, and Z_i is a real random variable that takes values in a finite set $\{z_1, z_2, \ldots, z_m\}$. For each $k = 1, \ldots, m$ the sequence $(\varphi(\xi_i, z_k))_{i \geq 1}$ is weakly dependent and equally distributed, but the sequence Z_i may be nonstationary as in [21]. The model represents a mixture of weakly dependent stationary process, but the mixture is nonstationary and dependent. Here ξ represents the usual variations of the traffic, and the variable Z selects between different traffic regimes, and represents types of network traffic.

With this model two main theoretical issues appear: the convergence and the asymptotic distribution of the estimator. We prove in [23] the almost sure convergence of the estimator. The asymptotic distribution of the estimator for this model is discussed in [22].

4.5 First Application to Simulated Data

In this section we analyze the accuracy of estimations with functional Nadaraya-Watson applied to simulated data. We analyze the estimation procedure by simulations using the ns-2 simulator software [24]. We simulate a link fed with a video trace, a simulated cross traffic and probe packets. The cross traffic corresponds to a model $X = \varphi(\xi, Z)$. We have two Markovian ON-OFF sources and Z is a random variable that takes values in $\{0, 1\}$ selecting periodically between this two sources. Fixing the value of Z we obtain stationary processes $\varphi(\xi, 0)$ and $\varphi(\xi, 1)$.

The first source (source 0) generates Markovian ON-OFF traffic corresponding to $\varphi(\xi, 0)$ with average bit rate varying from 150 Mb/s to 450 Mb/s and average time Ton in the ON state and Toff in the OFF state varying from 100 to 300 ms. The second source (source 1) generates Markovian ON-OFF traffic corresponding to $\varphi(\xi, 1)$ with average bit rate varying from 600 Mb/s to 900 Mb/s and average time Ton in the ON state and Toff in the OFF state varying from 200 to 500 ms. For each period an independent random variable is sampled to select the average bit rate. The payload of probe packets is 20 bytes and for the video packets is 1400 bytes. The video sequence has an average bit rate of 480 kbps. The link capacity is 1.6 Mbps.

We send this cross traffic to a network link together with the probe packets and the simulated video sequence. Each test consists on a probe packet burst with fixed interdeparture time t_{in}^*. After this burst we send a simulated video traffic (a video traffic trace). For each test j we compute from the probe packets the empirical distribution function of interarrival times X_j and we measure the average delay Y_j of the video packets.

The kernel is $K(x) = \begin{cases} (x^2 - 1)^2 & \text{if } x \in [-1, 1] \\ 0 & \text{if } x \notin [-1, 1] \end{cases}$
and we use the L^1 norm for the distance between the empirical distribution functions.

Concerning the time scales in our experiment the probe traffic is sent with fixed time t between consecutive probe packets. The aim is to find some criterion for choosing the best time scale in order to have accurate estimates. We consider different sequences of observations for a finite set of time scales $\{t_1, t_2, \ldots, t_r\}$. In practice, as we send bursts of probe traffic with fixed time t between packets we have observations with time scales in the set $\{t, 2t, \ldots, rt\}$. Consider $n + m$ observations for each time scale $\left\{(X_i^{t_j}, Y_i^{t_j}) : 1 \leqslant i \leqslant n + m, \ 1 \leqslant j \leqslant r\right\}$.

By dividing the sequence for a fixed time scale in two we can estimate the function Φ^{t_j} (for the time scale t_j) by $\widehat{\Phi}_n^{t_j}$ with the first n samples.

We then compute the difference $\sigma^2_{t_j}(n, m) = \frac{1}{m} \sum_{i=1}^{m} \left(\widehat{\Phi}_n^{t_j}(X_{n+i}^{t_j}) - Y_{n+i}^{t_j} \right)^2$, that gives a measure of the estimator performance for the time scale t_j. We choose $t_{n,m}^*$ such that minimize $\sigma^2_{t_j}(n, m)$.

The kernel bandwidth is selected with a similar procedure.

In the simulations we have 360 values of (X, Y) and we divide the sample in two parts. The estimation of Φ is obtained from the last 300 samples and the accuracy of the estimation is evaluated over the first 60 samples by comparing $\Phi_n(X_j)$ with the measured average delay Y_j for $j = 1, \ldots, 60$. The relative error in each point j is computed by $\frac{|\widehat{\Phi}_n(X_j) - Y_j|}{Y_j}$. Figure 8 show the estimated and the measured value of the average delay, showing a good fitting.

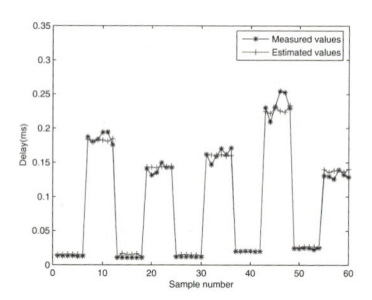

Fig. 8. Average delay estimation for simulated data

4.6 Experimental Results

In this section we show results of the procedures presented in this paper applied to different operational networks. The experiments were done with a measurement software tool specially developed for this purpose. In order to evaluate the practical limits of this methodology we analyze different scenarios that have different levels of complexity. In this paper we show only measurements using a cellular access network.

We analyze a cellular connection used with a PC and an cellular modem. The video sequences are downloaded from a server located at Facultad de Ingeniería, Universidad de la República. In this case the videos were codified at an average rate of 96 kbps. First of all, we take the first 30 samples in order to select the

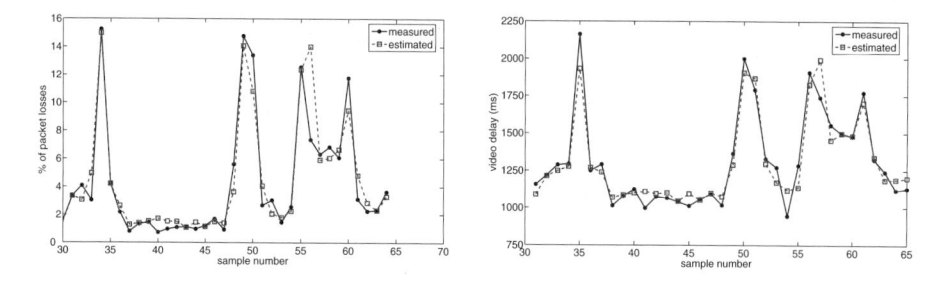

Fig. 9. Video packet losses (left) and mean delay (right) in the cellular case

model. Next, we take the other 35 points not used to select the model in order to validate the model.

Figure 9 left and right show the video losses and its mean delay for the 35 points of the validation sample. The accuracy of the estimation is reasonable taking into account the variability of the data.

5 Conclusions

This work addresses the challenge of guaranteeing quality of service (QoS) in the Internet from a statistical point of view. First, we have discussed the end-to-end QoS parameters estimations based models from the Large Deviation Theory. Later, we have analyzed the estimation of QoS parameters seen by an application based on end-to-end active measurements and statistical learning tools. We have discussed how these methodologies can be applied to different parts of the network in order to analyze its performance. We have obtained tight estimations applying both methodologies.

Acknowledgments. This work was partially supported by a grant of CSIC-UDELAR. The authors want to thank the following members of ARTES research group: L. Aspirot, B. Bazzano, P. Bermolen, P. Casas, A. Ferragut F. Larroca and G. Perera for their contributions to this work.

References

1. Breslau, L., Jamin, S., Shenker, S.: Comments on the performance of measurement-based admission control algorithm. In: IEEE INFOCOM 2000, Tel Aviv, Israel, pp. 1233–1242 (2000)
2. Dembo, A., Zeitouni, O.: Large Deviations Techniques and its Applications. Jones and Bartlett, New York (1993)
3. Kelly, F.: Notes on Effective Bandwidth. In: Kelly, Zachary, Ziedins (eds.) Stochastic Networks: Theory and Applications. Oxford University Press, Oxford (1996)
4. Ozturk, O., Mazumdar, R., Likhanov, N.: Many sources asymptotics in networks with small buffers. Queueing Systems (QUESTA) 46(1-2), 129–147 (2004)

5. Más, N., Karlsson, G.: Probe-based admission control for a differentiated-services internet. Computer Networks 51, 3902–3918 (2007)
6. Wischik, D.: Sample path large deviations for queues with many inputs. Annals of Applied Probability (11), 389–404 (2000)
7. Courcoubetis, C., Siris, V.A.: Procedures and tools for analysis of network traffic measurements. Elsevier Science, Amsterdam (2001)
8. Rabinovitch, P.: Statistical estimation of effective bandwidth. M.Sc.thesis, University of Cambridge (2000)
9. Kesidis, G., Walrand, J., Chang, C.S.: Effective Bandwidths for Multiclass Markov Fluids and Other ATM Sources. IEEE/ACM Trans. Networking (1), 424–428 (1993)
10. Pechiar, J., Perera, G., Simon, M.: Effective Bandwidth estimation and testing for Markov sources. In: Kouvatsos, D.D. (ed.) Performance Evaluation. Elsevier, New Holland (2002)
11. Lebedev, E.A., Lukashuk, L.I.: Maximum likelihood estimation of the infinitesimal matrix of a Markov chain with continuous time (Russian, English summary). Dokl. Akad. Nauk Ukr. SSR, Ser. A (1), 12–14 (1986)
12. Aspirot, L., Belzarena, P., Bermolen, P., Ferragut, A., Perera, G., Simon, M.: Quality of service parameters and link operating point estimation based on effective bandwidths. Performance Evaluation 59(2-3), 103–120 (2005)
13. Wischik, D.: The output of a switch or effective bandwidths for network. Queueing Systems 32(4), 383–396 (1999)
14. Belzarena, P., Bermolen, P., Simon, M., Casas, P.: End-to-End Quality of Service-based Admission Control Using the Fictitious Network Analysis. Computer Communications (COMCOM) - The International Journal for the Computer and Telecommunications Industry, Special issue on Heterogeneous Networks: Traffic Engineering and Performance Evaluation (2010) (to appear)
15. Nadaraya, E.A.: Nonparametric estimation of probability densities and regression curves. Mathematics and its Applications (Soviet Series), vol. 20. Kluwer Academic Publishers Group, Dordrecht (1989)
16. Ferraty, F., Vieu, P.: Nonparametric Functional data analysis: Theory and Practice. Springer Series in Statistics. Springer, Heidelberg (2006)
17. Machiraju, S., Veitch, D., Baccelli, F., Nucci, A., Bolot, J.: Theory and practice of cross-traffic estimation. In: SIGMETRICS, pp. 400–401 (2005)
18. Nadaraya, E.A.: On estimating regression. Theory of Probability and its Applications 9(1), 141–142 (1961)
19. Masry, E.: Nonparametric regression estimation for dependent functional data: asymptotic normality. Stochastic Process. Appl. 115, 155–177 (2005)
20. Zhang, Y., Duffield, N.: On the constancy of internet path properties. In: IMW 2001: Proceedings of the 1st ACM SIGCOMM Workshop on Internet Measurement, pp. 197–211 (2001)
21. Perera, G.: Irregular sets and central limit theorems. Bernoulli 8, 627–642 (2002)
22. Bertin, K., Aspirot, L.: Asymptotic normality of the Nadaraya-Watson estimator for non-stationary data. To appear in Journal of Nonparametric Statistics (2009)
23. Aspirot, L., Belzarena, P., Bazzano, B., Perera, G.: End-To-End Quality of Service Prediction Based On Functional Regression. In: Proc. Third International Working Conference on Performance Modelling and Evaluation of Heterogeneous Networks (HET-NETs 2005), Ilkley, UK (2005)
24. McCanne, S., Floyd, S.: ns network simulator, http://www.isi.edu/nsnam/ns/ (accessed March 2009)

Transform-Domain Analysis of Packet Delay in Network Nodes with QoS-Aware Scheduling

Stijn De Vuyst, Sabine Wittevrongel, and Herwig Bruneel

SMACS* Research Group, Department of Telecommunications and Information Processing, Ghent University, Sint-Pietersnieuwstraat 41, 9000 Gent, Belgium
{sdv,sw,hb}@telin.ugent.be

Abstract. In order to differentiate the perceived QoS between traffic classes in heterogeneous packet networks, equipment discriminates incoming packets based on their class, particularly in the way queued packets are scheduled for further transmission. We review a common stochastic modelling framework in which scheduling mechanisms can be evaluated, especially with regard to the resulting per-class delay distribution. For this, a discrete-time single-server queue is considered with two classes of packet arrivals, either delay-sensitive (1) or delay-tolerant (2). The steady-state analysis relies on the use of well-chosen supplementary variables and is mainly done in the transform domain. Secondly, we propose and analyse a *new* type of scheduling mechanism that allows precise control over the amount of delay differentiation between the classes. The idea is to introduce N reserved places in the queue, intended for future arrivals of class 1.

1 Introduction

In heterogeneous packet-based networks, the Quality of Service (QoS) perceived by a particular application highly depends on the presence of and interaction with *other* flows, due to statistical multiplexing and queueing in the intermediate network nodes. This is particularly the case in situations of high network load or congestion in intermediate network nodes. In a sense, managing a large heterogeneous network successfully, is a many-facetted problem, not unlike governing a society. Different groups of people have different characteristics, desires and aspirations, often conflicting with each other, but all of them are part of the same, highly interactive system. The actions of one group will always affect the others to some degree and it is the responsibility of the authorities to issue clear rules and laws as to how human interactions should take place, as well as to distribute the limited (monetary) resources among the different interest groups. Ideally, the primary aim of government's legislation and budget is to maximise prosperity and individual well-being. In this sense, traffic flows in a network are similar to people in society in many respects. Just like people, they share the same infrastructure, are unpredictable and may exhibit erratic or even malevolent behaviour. Therefore, enforcing strict rules and intelligent resource distribution are

---
* SMACS: Stochastic Modeling and Analysis of Communication Systems.

required to achieve maximal QoS satisfaction. Hence, there is a continuing trend of introducing QoS-awareness into the equipment of network nodes, with the aim of sharing available network resources (e.g. bandwidth) among the traffic flows in a more intelligent and deliberate way, tailored as closely as possible to the needs of each flow. For this, many different techniques are used, both at the application level (traffic shaping, congestion notification) and at the level of the network (congestion avoidance, active queue management, specific packet scheduling and packet discarding mechanisms), see e.g. [25] for an overview. In answer to the scalability problems that arise when trying to meet the QoS demands of each separate flow in the network, the Differentiated Services (DiffServ) architecture [4] has been proposed. DiffServ is packet-based instead of flow-based. This means the individual flows are aggregated into a limited number of service classes and nodes decide on the per-hop behaviour of each packet based on its class. A simulation study of DiffServ with both delay and loss QoS classes is found in [44]. A *controllable* DiffServ mechanism is Proportional Differentiated Services [18], which aims to quantitatively differentiate the QoS among the classes relative to some predefined parameters. Many other scheduling mechanisms that limit the delay of a selected flow keep track of a 'deadline' for each queued packet, such as earliest deadline first (EDF) [43,38] and alternative best-effort [26] or use virtual clocks [52]. An overview can be found in [25,40]. Another approach to accommodate delay-sensitive traffic, is to keep queues small by discarding packets before congestion can arise. Several packet-discarding strategies have been presented and analysed in literature, such as push-out buffer (POB) [11], partial buffer sharing (PBS) [10,27], random early detection (RED) [39,23] and their variants. Aside from policies deciding whether or not a packet is admitted to the buffer, there is also a host of *scheduling mechanisms* that are responsible for determining which packet from the buffer is the next to be transmitted once the output link becomes available. QoS-*aware* scheduling mechanisms aim at providing somehow a better service to selected flows with higher importance, at the expense of a worse service delivered to the other flows. A theoretically ideal and fair way to share the server capacity over different flows is Generalised Processor Sharing (GPS) [41,51], but this mechanism is difficult to apply in packet-based networks, so adaptations for packet scheduling are needed. For instance in the framework of ATM [3], weighted-round-robin (WRR) and weighted-fair-queueing (WFQ) [13] were proposed to achieve GPS-like weighted throughput and fairness [33]. For these mechanisms, there are separate queues for each type of traffic and the server 'visits' each queue in a cyclic, weighted and/or timed manner [42]. In the DiffServ framework, WFQ and WRR were recycled to 'class-based' WFQ and WRR, meaning that the flows (sometimes several thousands) are aggregated into a limited number of service classes between which a service differentiation is desirable.

We first review a common stochastic modelling framework by which the performance of some basic scheduling mechanisms can be evaluated, especially with regard to the resulting per-class delay distribution. A discrete-time single-server queueing model is considered with *two* classes (or 'types') of packet arrivals, either *delay-sensitive* (type 1) or *delay-tolerant* (type 2). The queue operates in slotted time and all packets are assumed to require a single slot of service. We make abstraction of any packet loss and assume the queue to have infinite capacity. The arrival process is time-independent, although the numbers of 1- and 2-packets that arrive within a slot may be correlated. The equilibrium delay distribution experienced by each class is obtained via some standard intermediary steps. The first ① is to identify a minimal but sufficiently large state space by which the system can be described as a Markov process. As we will see, the number of required supplementary variables [12,28] or 'dimensions' in the state space specifically depends on the considered scheduling mechanism. Secondly ② the equilibrium distribution of the system's state over this space is calculated as a joint probability generating function (pgf), which in turn is required for the final step, ③ obtaining the per-class delay distribution, also in the form of a generating function. This *generating-functions* approach, also outlined in e.g. [29,49,5], has proven to be successful in application to an impressively broad range of queueing models. We here specifically apply it to assess the performance of an alternative scheduling mechanism called the MR (*Multiple Reservation*) mechanism, which we introduce in the next section.

2 Class-Based Scheduling Mechanisms

The quintessential QoS-flat scheduling mechanism (or 'queueing discipline') is FIFO (First-In First-Out) where packets are served in their arrival order, without regard of their class. Occasionally useful disciplines are also LIFO (Last-In First-Out) [2] and ROS (Random Order of Service) [30,22]. The latter not only neglects the class but also the arrival order of the packets. Clearly, no significant differentiation in the per-class delay distribution is expected under any of these disciplines. On the other hand, the most *extreme* way of service differentiation is AP (Absolute Priority) or 'HOL priority' (Head of Line), either preemptive or non-preemptive, see [45,47,24,9,48] as well as the contribution of Walraevens et al. in this volume. Under AP, the next scheduled packet is the one that (1) belongs to the set of queued packets with highest priority and (2) has the longest waiting time of the packets in the set. If there are M traffic classes with lower classes having higher priority, then a class-i packet only has a transmission opportunity in the complete absence of packets from classes 1 to $i-1$. Hence, AP guarantees the *lowest* possible delay for class-1 traffic, at the cost of increasing the delay for the other classes. The drawback is that classes with lower priority may experience dramatically high delay, especially when the partial load of the higher-priority classes is high, an effect known as *packet starvation*. Therefore, AP may be a bridge too far in many situations. Several amendments to AP have been proposed that try to soften the severe strictness of the policy. For example, in [46] a Probabilistic Priority scheme is discussed that assigns a small probability p_i to each class by which the server may skip service to class $i+1$, even though class-i packets are available. In case of two priority classes, [35,36] analyse different ways of implementing priority jumps (PJ), as proposed in [32].

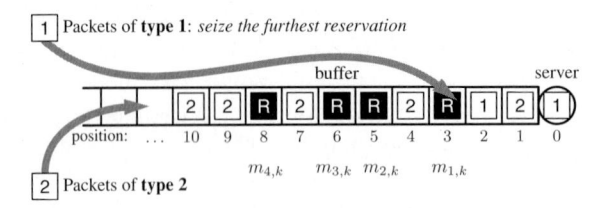

Fig. 1. A queue operating under MR with $N=4$. The queue spaces are numbered as indicated, and the positions of the reservations at the start of slot k are given by the variables $m_{j,k}$ ($j=1,\ldots,N$).

The idea is to allow class-2 packets (with lower priority) to promote to class 1 on some occasions and 'jump' to the tail of the high-priority queue. Considered jumping schemes are: HOL-PJ where class-2 packets jump after some delay threshold, HOL-MBP (Merge-by-probability) where the *entire* class-2 queue jumps with some probability β, HOL-JOS (Jump-or-serve) where jumps occur at the same rate as the output link, HOL-JIA (Jump-if-Arrival) where jumps are synchronised with class-2 arrivals, and so on. We here specifically focus on the performance analysis of a new queue scheduling mechanism that can realise a statistic reduction of the delay for type 1 traffic at the cost of increasing the delay for type 2 traffic. This delay differentiation is achieved by introducing a total of N ($N \geqslant 1$) *reserved spaces* in the queue that will be occupied by future arriving packets of type 1. We refer to this scheduling mechanism as the *Multiple Reservation* (MR) discipline. The basic idea was first coined by Burakowski and Tarasiuk in [8], where a rudimentary estimation of the mean delay for both traffic types is given in case of Poisson arrival flows. In our previous work [14,15], we have obtained exact expressions for the equilibrium distribution of the delay of both packet types, in case of a single reserved position (i.e. $N=1$), as well as for geometric service times in [19]. This was subsequently extended in [16,17] to the general case $N>1$ where efficient numerical procedures were presented to calculate the pgf, mean and tail distribution of the delay experienced by both types of packets. As a convention, let us number the positions of the spaces in the queue as in Fig. 1, i.e. the server has position 0, while the queue spaces have numbers that increase with their distance to the server. The policy for storing and scheduling packets in the queue is then the following:

▶ Initially, there are N reserved positions (R's) in the queue, on positions 1 to N. This is also the case every time the queue is empty or 'idle'.
▶ Of all the packets arriving in a slot, the 1-packets are stored in the queue first and then the 2-packets.
▶ If a 1-packet is stored, it seizes the most advanced R in the queue (i.e. the one with the lowest position number) and makes a *new* R at the end of the queue. A 2-packet on the other hand, is stored at the end of the queue in the usual **FIFO** manner.

▶ As long as it is not seized by a 1-packet, an R behaves as a normal packet in the sense that it shifts one place to the right every time a packet leaves the server. However, an R cannot enter the server at position 0, nor can it leave the queue.

It is clear that the number of R's always remains equal to N, due to the fact that each 1-packet seizes an R but at the same time also creates an R at the end of the queue. Note that MR is work-conserving and maintains the per-type packet ordering, just like FIFO and AP. The differentiation in delay arises from the fact that 1-packets may jump over already stored 2-packets. For example, in the situation of Fig. 1, the next arriving 1-packet will take position 3, thereby jumping over 4 packets of type 2. Clearly, the higher we choose N the more the delays of type 1 and type 2 packets will differ. More so, if $N \to \infty$ the queue operates in exactly the same way as AP, providing the maximum possible delay differentiation. In this way, the number of reservations N in the queue can be seen as a parameter by which the *amount* of delay differentiation can be carefully controlled.

3 Discrete-Time Queueing Model

In the following, we propose a discrete-time queueing model by which the per-class delay distribution under a specific scheduling mechanism can be assessed quantitatively. We assume all packets are stored in a single queue with infinite capacity and that each packet requires one slot service. In general, packets belong to one of M service classes with class i having more strict delay requirements (higher priority) than class $i+1$. Let $a_{i,k}$ denote the number of packets of type i ($1 \leqslant i \leqslant M$) that arrive in the queue during slot k. We assume that the numbers of arrivals in subsequent slots are independent and identically distributed (iid), with their joint distribution given by the pgf

$$\mathcal{A}(z_1, \ldots, z_M) \triangleq \mathrm{E}[z_1^{a_{1,k}} \cdot \ldots \cdot z_M^{a_{M,k}}]. \tag{1}$$

This allows for the numbers of arrivals of different classes during the same slot to be correlated. We assume *delayed access* for the packets, meaning that any arrivals during slot k are not stored in the queue until the end of that slot. Consequently, no packet can be served during the slot in which it arrived. Let the random variable $v_{i,k}$ denote the number of packets of type i present in the system at the *beginning* of slot k, summing up to $u_k = v_{1,k} + \ldots + v_{M,k}$, denoting the total system content at that time. In case of QoS-flat disciplines, it is often logical to single out a traffic class i and regard the other as an aggregated class $\bar{\imath}$. Let $a_{\bar{\imath},k} = \sum_{j \neq i} a_{j,k}$, then $A(x,z) \triangleq \mathrm{E}[x^{a_{k,\bar{\imath}}} z^{a_{k,i}}] = \mathcal{A}(\underset{1}{x}, \ldots, \underset{i-1}{x}, \underset{i}{z}, \underset{i+1}{x}, \ldots, \underset{M}{x})$ is the joint distribution of the arrivals in slot k of classes $\bar{\imath}$ and i. This is equivalent to considering $M=2$, with type-1 packets having delay priority over type-2 packets, and joint pgf $A(z_1, z_2) = \mathrm{E}[z_1^{a_{1,k}} z_2^{a_{2,k}}]$. We will also use the marginal distributions $A_i(z)$ of the number of arrivals per slot of type $i = 1, 2$, which are given by

$$A_1(z) \triangleq \mathrm{E}[z^{a_{1,k}}] = A(z,1), \qquad A_2(z) \triangleq \mathrm{E}[z^{a_{2,k}}] = A(1,z). \tag{2}$$

The *total* number of arrivals (both of type 1 and 2) during slot k is denoted by $a_{T,k}=a_{1,k}+a_{2,k}$ with pgf $A_T(z)=\mathrm{E}[z^{a_{T,k}}]=A(z,z)$. The mean number of arrivals per slot (arrival rate) of type i follows from the moment-generating property of pgfs as $\lambda_i = \mathrm{E}[a_{i,k}] = A_i'(1)$, while the total arrival rate is $\lambda_T = \lambda_1 + \lambda_2 = A_T'(1)$. Some higher-order derivatives evaluated for $z=1$ are $\lambda_i' = A_i''(1)$ and $\lambda_T' = A_T''(1)$, as well as the mixed moment $\lambda_{12} = \frac{\partial^2}{\partial z_1 \partial z_2}A(1,1) = \mathrm{E}[a_1 a_2]$. To keep the forthcoming expressions tractable, we use the following notation for the distribution of the type-1 arrivals in the probability domain. Let the mass function of the number of 1-arrivals per slot be

$$\beta_i \triangleq \mathrm{Prob}[a_{1,k}=i], \quad i \geq 0, \quad \text{and} \quad \alpha \triangleq \beta_0 = A_1(0) = A(0,1), \qquad (3)$$

such that $A_1(z)=\sum_{i=0}^{+\infty} \beta_i z^i$. Additionally, define

$$A_{i*}(z) \triangleq \mathrm{E}[z^{a_{2,k}}\{a_{1,k}=i\}], \quad i \geq 0, \qquad (4)$$

which is the pgf of the number of 2-arrivals in case there are exactly i arrivals of type 1 in the same slot, and where $\mathrm{E}[X\{Y\}]=\mathrm{E}[X|Y]\mathrm{Prob}[Y]$. Note also that $\beta_i = A_{i*}(1)$, $i \geq 0$.

4 Equilibrium State of Multi-class Queues

The 'system' we are considering is the queue as specified above, containing packets of both classes. Our final goal is to obtain the distribution of the per-class delay. Formally, let us consider an *arbitrary* class-i packet $(i=1,2)$ and tag it as packet $\mathcal{P}$. Denote the arrival slot of $\mathcal{P}$ as slot I. We define the delay d_i of $\mathcal{P}$ as the number of slots between the end of slot I and the end of the slot in which $\mathcal{P}$ departs from the queue. It is clear that for most scheduling disciplines, d_i heavily depends on the state of the system when $\mathcal{P}$ arrives. For example, if the system content u_I is large at the moment $\mathcal{P}$ arrives, its delay will typically be large as well, because all of the already present packets have to be served prior to $\mathcal{P}$'s transmission. This is the reason why intermediate step ② of obtaining the exact equilibrium system state is required.

As mentioned in section 1, the first step ① however is to identify a suitable description of the system's state during an arbitrary slot k which contains enough information so that the per-class delay distribution can be derived from it. The variables $\langle s_1, \ldots, s_n \rangle$ that are chosen to be part of the system state are called the *system state variables* and we characterise their distribution in slot k by their joint pgf P_k. The primary concern here, is to choose the collection of system state variables in such a way that it is Markovian, in the sense that P_{k+1} can be determined from P_k without relying on any information pertaining to slots other than k. We assume that the system reaches equilibrium for $k \to \infty$, i.e. that appropriate stability conditions for the system's parameters are met. Whenever we deal with random variables assumed in this equilibrium regime, we may drop the time index k where appropriate. For the equilibrium distribution of the system state we thus write $P = \lim_{k \to \infty} P_k$. For a general arrival process, note however that the distribution P_I of the system state in the arrival slot of $\mathcal{P}$ may be *different* from the overall equilibrium distribution P. Nevertheless, it is known [5] that due to the temporally uncorrelated (iid) nature of the packet arrival process we assumed in section 3, the system state as 'seen' by an *arbitrary arriving packet* of either type has the same distribution as the system state in an *arbitrary slot*, i.e. $P_I = P$. The results we obtain in this section can therefore immediately be used for the delay analysis in sections 5 and 6.

So let us apply steps ① and ② to the QoS-flat disciplines first. For 2-class FIFO, it suffices to maintain the total system content $\langle u_k \rangle$ as per-slot information, i.e. the only required system state variable is u_k. The evolution of the system content is governed by the well-known Lindley equation:

$$u_{k+1} = (u_k - 1)^+ + a_{T,k}, \tag{5}$$

where $(\cdot)^+$ is the operator $\max(\cdot, 0)$. The equilibrium pgf of the system state then follows as

$$U(z) \triangleq \lim_{k \to \infty} \mathrm{E}[z^{u_k}] = A_T(z)\mathrm{E}[z^{(u-1)^+}] = (1-\lambda_T)\frac{A_T(z)(z-1)}{z - A_T(z)}, \tag{6}$$

see [5,6,29]. Applying the moment-generating property of pgfs on (6), as well as applying de l'Hôpital's rule, we obtain the mean equilibrium system content as

$$\mathrm{E}[u] = U'(1) = \lambda_T + \frac{\lambda_T'}{2(1-\lambda_T)}. \tag{7}$$

The importance of (6) and (7) is the fact that they hold for most work-conserving disciplines. The *total* system content u, i.e. *all* packets irrespective of their class, is distributed as in (6) for all the scheduling mechanisms we consider in this paper. Whether the total system content alone suffices as system state description is another matter of course. For LIFO, no system state needs to be maintained at all (i.e. $\langle\rangle$) since the delay of $\mathcal{P}$ is independent of the number of packets u_I already stored in the system. For ROS on the other hand, $\langle u_k \rangle$ is again sufficient.

For QoS-aware scheduling mechanisms, the delay of $\mathcal{P}$ usually depends on how many of the present packets in the system are of type 1 and how many of type 2. In case of AP [47] however, it turns out that keeping track of the total system content u_k alone is sufficient for the delay analysis, i.e. we require $\langle u_k \rangle$, leading again to (5)–(7). Nevertheless, the joint pgf of v_1 and v_2 can be obtained using the system equations

$$v_{1,k+1} = (v_{1,k} - 1)^+ + a_{1,k}, \qquad v_{2,k+1} = \begin{cases} (v_{2,k} - 1)^+ + a_{2,k} & \text{if } v_{1,k}=0, \\ v_{2,k} + a_{2,k} & \text{if } v_{1,k}>0. \end{cases} \tag{8}$$

The joint pgf of v_1 and v_2 then found as [47]

$$U(z_1, z_2) = \mathrm{E}[z_1^{v_1} z_2^{v_2}] = (1-\lambda_T)\frac{A(z_1,z_2)(z_2-1)}{z_1 - A(z_1,z_2)}\frac{(z_1 - X(z_2))}{z_2 - X(z_2))}, \tag{9}$$

where the function $X(z)$ is implicitly defined by $X(z) = A(X(z), z)$ and $X(1)=1$. For the variants of mechanisms with priority jumps (PJ) [35,36] mentioned in the introduction however, the required state space is usually $\langle v_{1,k}, v_{2,k} \rangle$, leading to different variants of (9).

For the Multiple Reservation mechanism (MR), recall that the jth packet in an arriving batch of 1-packets seizes the jth reserved space if $j \leq N$. Hence, besides the queue content u_k we also need to keep track of the precise positions of all N reservations in the queue. We number the R's from 1 to N, with the first reservation being the one *closest* to the server, i.e. the one with the smallest position number. Conversely, we may also say that the jth reservation ($j = 1, \ldots, N$) has *order* j. At the start of slot k, we denote the position of the order j reservation by $m_{j,k}$, as is illustrated in Fig. 1. As a 1-packet always seizes the R with the lowest position number, it is seen that there can be no 1-packets in the queue on positions larger than $m_{1,k}$. In other words, all packets behind the first reservation must be of type 2. So what variables need to be added to the system state in order for the per-class delay distribution to be derived? If $\mathcal{P}$ is of type 1 and the jth 1-packet of an arriving batch, it will seize the jth order reservation ($j \leq N$). Each time, a new reservation is made at the end of the queue, at position $u_k + N$. Therefore, it is clear that the positions of all R's together with the queue content u_k need to be part of the system's state variables, i.e. $\langle m_{j,k}, j=1,\ldots, N; u_k \rangle$.

The following system equations establish the value of the system state variables in slot $k+1$, for all possible values of those variables in slot k. The working method is to start from a certain state at the start of slot k. Then consider every possible event *during* this slot in terms of arrivals, storage, scheduling and departures and finally, write down the new system state this results in at the start of slot $k+1$. In principle, this yields a function from one $(N+1)$-dimensional space to another, although this space can be somewhat reduced by ruling out states than can never be reached. For instance, because of their physical meaning, we know that the system state variables must satisfy the constraint

$$1 \leq m_{1,k} < m_{2,k} < \ldots < m_{N,k} \leq (u_k - 1)^+ + N, \tag{10}$$

which for the position of the jth reservation individually results in

$$j \leq m_{j,k} \leq (u_k - 1)^+ + j, \qquad j=1,\ldots, N. \tag{11}$$

As a convention, let j indicate any value from 1 to N, unless stated otherwise. In our analysis it turns out that, instead of the variables $m_{j,k}$, it is often more convenient to work with the variables

$$\hat{m}_{j,k} \triangleq m_{j,k} - j, \tag{12}$$

which all have 0 as their minimal value instead of j. Therefore, the constraints (10) and (11) now respectively become

$$0 \leq \hat{m}_{1,k} \leq \hat{m}_{2,k} \leq \ldots \leq \hat{m}_{N,k} \leq (u_k - 1)^+, \tag{13}$$

$$0 \leqslant \hat{m}_{j,k} \leqslant (u_k - 1)^+. \tag{14}$$

Obviously, knowledge of the value or distribution of $m_{j,k}$ implies that of $\hat{m}_{j,k}$ and vice versa, so we may interchangeably use both as system state variables.

For the system equations we can distinguish between four cases, in all of which the new system content u_{k+1} is determined by (5). Observe also that $a_{2,k}$ appears in (5) but not in any of the following equations for $\hat{m}_{j,k+1}$ where only the number of arrivals of type 1 is of importance. Assuming that the system is not empty to begin with, the events during slot k can generally be summarised as follows. First, there are $a_{1,k}$ 1-arrivals to be stored. One by one they seize the first R they see and make a new one at the end. As such, seen as a group, the first N of these 1-arrivals take R's that existed before slot k, while any remaining 1-arrivals seize an R that was created by a previous 1-arrival in slot k. At the end of the slot, after the 2-packets have been stored as well, the packet in the server terminates its service and leaves the queue. Then, at the start of slot $k+1$ a new packet will enter service, at least if there are any left in the system. It is the first packet (the one with lowest position number p) that will jump over any R's at positions 1 to $p-1$ into the server at position 0. Then, since position p is free now, all packets and reservations on positions larger than p shift one position towards the server. As we have said, these considerations lead us to distinguish four groups of system equations as follows.

▶ $u_k = 0$ (empty system)
First of all, in case of an empty system we know that the N reservations are grouped together on positions 1 to N, so

$$u_k = 0 \quad \Rightarrow \quad \hat{m}_{1,k} = \hat{m}_{2,k} = \ldots = \hat{m}_{N,k} = 0.$$

Therefore, we have in slot $k+1$:

$$\hat{m}_{j,k+1} = (a_{1,k} - 1)^+. \quad \textbf{(Empty)}$$

▶ $u_k > 0$, $a_{1,k} = 0$ (no 1-arrivals)
In this and the remaining cases, we know that the system is not empty in slot k. Since $a_{1,k} = 0$ there are no arrivals of type 1 here. None of the reservations will be seized so they all survive to the next slot. However, after the packet in service during slot k has left, they will be shifted by one position as far as the lower constraint in (14) is not violated. We have

$$\hat{m}_{j,k+1} = (\hat{m}_{j,k} - 1)^+. \quad \textbf{(Keep)}$$

▶ $u_k > 0$, $a_{1,k} = i$ with $1 \leqslant i < N$
In this case, the number of 1-arrivals i is smaller than the number of reservations N. These i arrivals seize the first i reservations, i.e. those at positions $m_{1,k}$ up to $m_{i,k}$ and make i new reservations at the end of the queue. So the last i reservations in the next slot will be newly created and positioned together at the end of the queue. Then, accounting for the fact that one packet will leave, it turns out that we have

$$\hat{m}_{j,k+1} = u_k + i - 2, \quad \text{if} \quad j = N - i + 1, \ldots, N. \quad \textbf{(AtEnd)}$$

On the other hand, the *first* $N-i$ R's in the new slot are reservations that were not seized and have survived. Their ordering number has simply decreased by i or in other words, they have 'i-shifted':

$$\hat{m}_{j,k+1} = \hat{m}_{j+i,k} + i - 1, \quad \text{if} \quad j = 1, \ldots, N-i. \quad \textbf{(i-shift)}$$

▶ $u_k > 0$, $a_{1,k} = i$ with $i \geqslant N$

Now, there are at least as many 1-arrivals as there are reservations. In this case all new reservations will be grouped at the end of the queue, since none of the old R's survive. Equation **(AtEnd)** now applies for *all* new reservation positions, i.e.

$$\hat{m}_{j,k+1} = u_k + i - 2, \quad j = 1, \ldots, N. \quad \textbf{(AtEnd)}$$

Now we use the above equations to obtain the equilibrium distribution of the system state $\langle m_{j,k}, j=1,\ldots,N; u_k \rangle$, of which we define the joint pgf as

$$P_k(y_1, y_2, \ldots, y_N; z) \triangleq \mathrm{E}[y_1^{\hat{m}_{1,k}} y_2^{\hat{m}_{2,k}} \cdots y_N^{\hat{m}_{N,k}} z^{u_k}]. \tag{15}$$

The equations **(Empty)**, **(Keep)**, **(i-shift)** and **(AtEnd)** allow to relate the joint pgf P_{k+1} of the system state in slot $k+1$ to the distribution P_k in slot k. We perform separate calculations for the four cases above, splitting up the joint pgf into four terms as

$$P_{k+1}(y_1, y_2, \ldots, y_N; z) = \mathrm{E}[y_1^{\hat{m}_{1,k+1}} y_2^{\hat{m}_{2,k+1}} \cdots y_N^{\hat{m}_{N,k+1}} z^{u_{k+1}}]$$
$$= \mathrm{E}[\ldots \{u_k = 0\}] + \mathrm{E}[\ldots \{u_k > 0, a_{1,k} = 0\}] \tag{16}$$
$$+ \sum_{i=1}^{N-1} \mathrm{E}[\ldots \{u_k > 0, a_{1,k} = i\}] + \sum_{i=N}^{+\infty} \mathrm{E}[\ldots \{u_k > 0, a_{1,k} = i\}].$$

For the first term, **(Empty)** applies, as well as (5) such that

$$\mathrm{E}[y_1^{\hat{m}_{1,k+1}} y_2^{\hat{m}_{2,k+1}} \cdots y_N^{\hat{m}_{N,k+1}} z^{u_{k+1}} \{u_k = 0\}]$$
$$= \mathrm{E}[(y_1 y_2 \cdots y_N)^{(a_{1,k}-1)^+} z^{a_{1,k}+a_{2,k}} \{u_k = 0\}]$$
$$= \ldots$$
$$= \frac{p_{0,k}}{y_1 y_2 \cdots y_N} \left[(y_1 y_2 \cdots y_N - 1) A(0, z) + A(y_1 y_2 \cdots y_N z, z) \right], \tag{17}$$

where, $p_{0,k} = U_k(0) = P_k(0, 0, \ldots, 0; 0)$ is the probability that the system is empty at the beginning of slot k. The second term of (16) can be further developed with **(Keep)**, which yields

$$\mathrm{E}[y_1^{\hat{m}_{1,k+1}} y_2^{\hat{m}_{2,k+1}} \cdots y_N^{\hat{m}_{N,k+1}} z^{u_{k+1}} \{u_k > 0, a_{1,k} = 0\}]$$
$$= \mathrm{E}[y_1^{(\hat{m}_{1,k}-1)^+} y_2^{(\hat{m}_{2,k}-1)^+} \cdots y_N^{(\hat{m}_{N,k}-1)^+} z^{u_k - 1 + a_{2,k}} \{u_k > 0, a_{1,k} = 0\}]$$
$$= A(0, z) \mathrm{E}[y_1^{(\hat{m}_{1,k}-1)^+} y_2^{(\hat{m}_{2,k}-1)^+} \cdots y_N^{(\hat{m}_{N,k}-1)^+} z^{u_k - 1} \{u_k > 0\}]$$

$$= A(0,z) \sum_{n=1}^{+\infty} \sum_{j_1=0}^{n-1} \sum_{j_2=j_1}^{n-1} \sum_{j_3=j_2}^{n-1} \cdots \sum_{j_N=j_{N-1}}^{n-1} y_1^{(j_1-1)^+} \cdots y_N^{(j_N-1)^+} z^{n-1} p_k(j_1, \ldots, j_N; n)$$

$$= A(0,z) \sum_{n=1}^{+\infty} z^{n-1} \Bigg[$$

$$\sum_{j_1=1}^{n-1} \sum_{j_2=j_1}^{n-1} \sum_{j_3=j_2}^{n-1} \sum_{j_4=j_3}^{n-1} \cdots \sum_{j_N=j_{N-1}}^{n-1} y_1^{j_1-1} y_2^{j_2-1} y_3^{j_3-1} \cdots y_N^{j_N-1} p_k(j_1, \ldots, j_N; n)$$

$$+ \sum_{j_2=1}^{n-1} \sum_{j_3=j_2}^{n-1} \sum_{j_4=j_3}^{n-1} \cdots \sum_{j_N=j_{N-1}}^{n-1} y_2^{j_2-1} y_3^{j_3-1} \cdots y_N^{j_N-1} p_k(0, j_2, \ldots, j_N; n)$$

$$+ \sum_{j_3=1}^{n-1} \sum_{j_4=j_3}^{n-1} \cdots \sum_{j_N=j_{N-1}}^{n-1} y_3^{j_3-1} \cdots y_N^{j_N-1} p_k(0, 0, j_3, \ldots, j_N; n)$$

$$+ \ldots$$

$$+ \sum_{j_N=1}^{n-1} y_N^{j_N-1} p_k(0, 0, \ldots, 0, j_N; n)$$

$$+ p_k(0, 0, \ldots, 0; n) \Bigg]$$

$$= \frac{A(0,z)}{z} \Bigg[\frac{1}{y_1 y_2 y_3 \cdots y_N} \Big(P_k(y_1, y_2, y_3, \ldots, y_N; z) - P_k(0, y_2, y_3, \ldots, y_N; z) \Big)$$

$$+ \frac{1}{y_2 y_3 \cdots y_N} \Big(P_k(0, y_2, y_3, \ldots, y_N; z) - P_k(0, 0, y_3, \ldots, y_N; z) \Big)$$

$$+ \ldots$$

$$+ \frac{1}{y_N} \Big(P_k(0, 0, \ldots, 0, y_N; z) - P_k(0, 0, \ldots, 0, 0; z) \Big)$$

$$+ P_k(0, 0, \ldots, 0, 0; z) - p_{0,k} \Bigg], \tag{18}$$

where we have used the following notation for the mass function of the system state distribution in slot k:

$$p_k(j_1, j_2, \ldots, j_N; n) \triangleq \text{Prob}[\hat{m}_{1,k} = j_1, \hat{m}_{2,k} = j_2, \ldots, \hat{m}_{N,k} = j_N, u_k = n]. \tag{19}$$

In the third term of (16), we must apply (*i*-**shift**) for the new reservations of order 1 to $N-i$ and (**AtEnd**) for the remaining reservation positions. We have for $1 \leq i < N$, using (4):

$$\mathrm{E}[\underbrace{y_1^{\hat{m}_{1,k+1}} \cdots y_{N-i}^{\hat{m}_{N-i,k+1}}}_{(i\text{-shift})} \cdot \underbrace{y_{N-i+1}^{\hat{m}_{N-i+1,k+1}} \cdots y_N^{\hat{m}_{N,k+1}}}_{(\text{AtEnd})} \cdot z^{u_{k+1}} \{u_k > 0, a_{1,k} = i\}]$$

$$= \mathrm{E}[y_1^{\hat{m}_{i+1,k}+i-1} \cdots y_{N-i}^{\hat{m}_{N,k}+i-1} y_{N-i+1}^{u_k+i-2} \cdots y_N^{u_k+i-2} z^{u_k-1+i+a_{2,k}} \{u_k > 0, a_{1,k} = i\}]$$

$$= (zy_1 y_2 \cdots y_{N-i})^{i-1} (y_{N-i+1} \cdots y_N)^{i-2} A_{i*}(z)$$

$$\mathrm{E}[y_1^{\hat{m}_{i+1,k}}\cdots y_{N-i}^{\hat{m}_{N,k}}(zy_{N-i+1}\cdots y_N)^{u_k}\{u_k>0\}]$$
$$=(zy_1y_2\cdots y_{N-i})^{i-1}(y_{N-i+1}\cdots y_N)^{i-2}A_{i*}(z)$$
$$\left[P_k(1,1,\ldots,\underset{i}{1},y_1,y_2,\ldots,y_{N-i};zy_{N-i+1}\cdots y_N)-p_{0,k}\right]. \qquad (20)$$

Finally, in the last term of (16), we find for $i \geqslant N$ using **(AtEnd)**,
$$\mathrm{E}[y_1^{\hat{m}_{1,k+1}}y_2^{\hat{m}_{2,k+1}}\cdots y_N^{\hat{m}_{N,k+1}}z^{u_{k+1}}\{u_k>0,a_{1,k}=i\}]$$
$$=z^{i-1}(y_1y_2\cdots y_N)^{i-2}A_{i*}(z)\left[U_k(zy_1y_2\cdots y_N)-p_{0,k}\right]. \qquad (21)$$

Adding up the terms (17), (18), (20) and (21), we get the right-hand side of (16). If equilibrium kicks in, we can single out the function $P(y_1,\ldots,y_N;z)$ and find our basic functional equation for the equilibrium distribution of the MR system state with N reservations:

$$(z\tilde{y}_1 - A(0,z))P(y_1, y_2, \ldots, y_N; z)$$
$$= zp_0\Big[(\tilde{y}_1-1)A(0,z) + A(z\tilde{y}_1,z)\Big]$$
$$+ A(0,z)\Big[(y_1-1)\,P(0,y_2,\ldots,y_N;z)$$
$$\qquad\qquad + y_1(y_2-1)\,P(0,0,y_3,\ldots,y_N;z)$$
$$\qquad\qquad + y_1y_2(y_3-1)\,P(0,0,0,y_4,\ldots,y_N;z)$$
$$\qquad\qquad + \cdots$$
$$\qquad\qquad + y_1y_2\cdots y_{N-1}(y_N-1)\,P(0,0,\ldots,0;z) - \tilde{y}_1 p_0\Big]$$
$$+ \sum_{i=1}^{N-1}\frac{(z\tilde{y}_1)^i}{\tilde{y}_{N-i+1}}A_{i*}(z)\Big[P(1,\ldots,\underset{i}{1},y_1,y_2,\ldots,y_{N-i};z\tilde{y}_{N-i+1})-p_0\Big]$$
$$+ \sum_{i=N}^{+\infty} z^i \tilde{y}_1^{i-1} A_{i*}(z)\Big[U(z\tilde{y}_1)-p_0\Big], \qquad (22)$$

where we define $\tilde{y}_j$ as the product $y_j y_{j+1}\cdots y_N$. This functional equation completely determines the equilibrium distribution P, although we see that a lot of unknown functions have yet to be determined. Nevertheless, all these unknowns can be resolved by using relation (22) only, as we will demonstrate. Observe that there are a total of $2N$ unknown functions on the right-hand side of (22). Let us designate a shorthand to each of these functions and order them in a list as follows

1. $\boxed{1} = P(0, y_2, y_3, y_4, \ldots, y_N; z)$
2. $\blacksquare 1 = P(1, y_2, y_3, y_4, \ldots, y_N; z)$
3. $\boxed{2} = P(0, 0, y_3, y_4, \ldots, y_N; z)$
4. $\blacksquare 2 = P(1, 1, y_3, y_4, \ldots, y_N; z)$
5. $\boxed{3} = P(0, 0, 0, y_4, \ldots, y_N; z)$
6. $\blacksquare 3 = P(1, 1, 1, y_4, \ldots, y_N; z)$

$$\vdots$$

$$2N-1. \quad \boxed{N} = P(0,0,0,0,\ldots,0;z)$$
$$2N. \quad \boxed{N} = P(1,1,1,1,\ldots,1;z) = U(z) \qquad (23)$$

Note that we could have included the probability p_0 in this list as well, although it easily follows as $p_0 = 1 - \lambda_T$ by imposing the normalisation condition on $U(z)$. The unknown functions can be determined in this order by performing the appropriate substitutions in (22). In fact, the functional equation is able to provide each of the unknowns in the list as a function of unknowns *further* in the list. To make clear how this is done, let us denote by ■ the function $P(y_1, y_2, \ldots, y_N; z)$ in an *explicit* form, i.e. equal to (22) but with all unknowns (those in list (23)) resolved. On the other hand, we represent by ■ ? $\boxed{1}\boxed{1}\boxed{2}\boxed{2}\ldots\boxed{N}\boxed{N}$ a relation determining $P(y_1, y_2, \ldots, y_N; z)$, but in which the functions after the question mark are still unresolved. Obviously, this is the functional equation in the form given by (22). Clearly, the final explicit expression for the equilibrium distribution of the system state we are looking for is ■. Even for small N though, obtaining ■ is an enormous task to do by hand, so we only explain *how* to do this, rather than actually doing it.

As we have said, (22) holds the key to determining all the unknown functions explicitly by evaluating it for the right arguments. In what follows, we describe a 'binary tree backtracking' scheme that shows us the way. There are *two types* of substitutions that yield relevant information. The first one is to let

$$y_1 \to 1, \ y_2 \to 1, \ \ldots, \ y_{n-1} \to 1, \ y_n \to 1, \qquad (24)$$

for some $n = 1, \ldots, N$. This directly gives the relation $\boxed{n}$? $\boxed{n+1}\boxed{n+1}\ldots\boxed{N}\boxed{N}$. The second type of substitution is to let

$$y_1 \to 1, \ y_2 \to 1, \ \ldots, \ y_{n-1} \to 1, \ y_n \to \frac{A(0,z)}{y_{n+1} y_{n+2} \cdots y_N z}, \qquad (25)$$

which is mostly the same as (24), except for the last step. Note that if (25) is performed on (22), the left-hand side vanishes. Based on the fact that probability generating functions are bounded for arguments lying in the unit disc and by using a similar argumentation as in [14,15], we know that the right-hand side has to vanish as well. This provides the relation $\boxed{n}$? $\boxed{n}\boxed{n+1}\boxed{n+1}\ldots\boxed{N}\boxed{N}$.

We can arrange the substitutions of type (24) and (25) in a binary tree as shown on the left side of Fig. 2. Branches going down correspond to substitutions $y_j \to 1$, while branches to the right indicate a substitution $y_j \to A(0,z)/\tilde{y}_{j+1}z$. Hence, every path in this tree corresponds to either (24) or (25), depending on the last branch. In other words, each path represents a sequence of substitutions which, if applied to the functional equation, yield the relation indicated on the node the path ends in. Starting from the top of the tree, we can progressively determine the relations on each node, until finally, we obtain $\boxed{N} = U(z)$ explicitly. Note that we have included the latter function in the list of unknowns notwithstanding the fact that it is known to be (6). On the right side of Fig. 2, we show

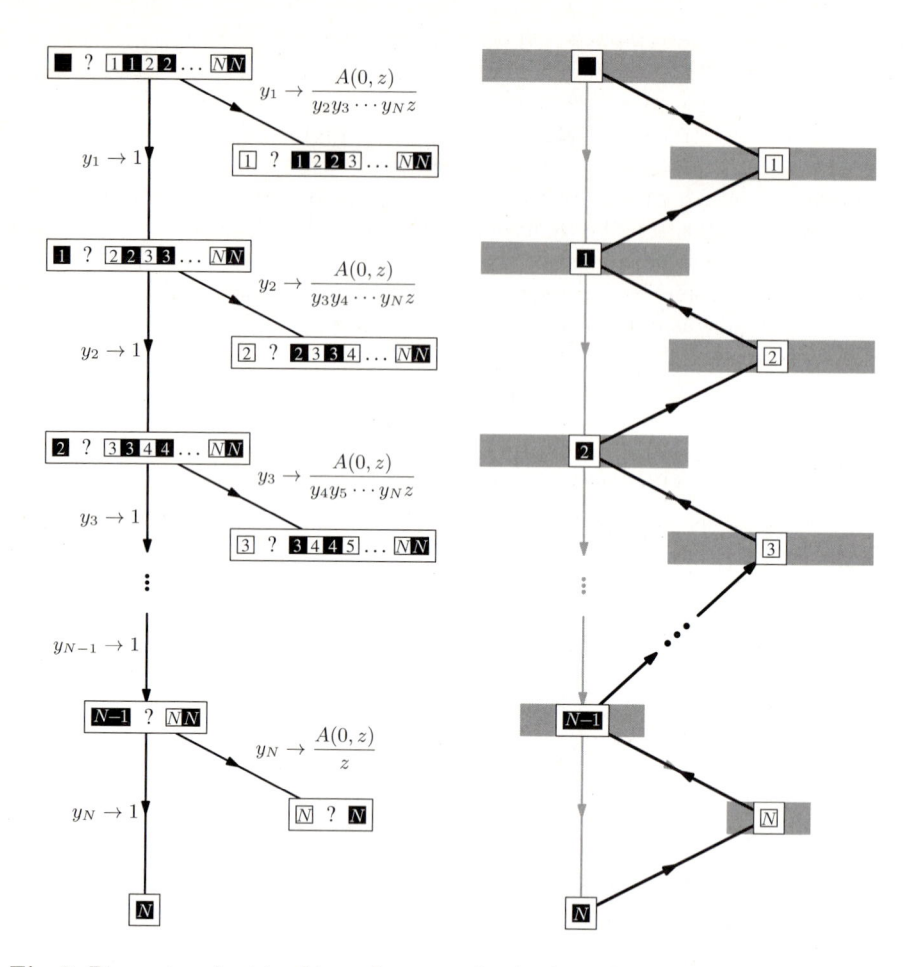

Fig. 2. Binary tree backtracking scheme to obtain the unknown functions for the MR system state distribution. First, we go from top to bottom executing the substitutions indicated on the branches. This progressively yields relations for each unknown in the list as a function of all unknowns *further* in the list. Then, it is possible to backtrack from the bottom to the top which allows to fully resolve the function on each node.

the second part of the calculation scheme. Starting from the bottom node, we work our way to the top by backtracking the previously obtained unresolved relations. Indeed, once we have ▨, we can use this in the node with relation ▨ ? ▨ to resolve ▨. In turn, the explicit expressions ▨ and ▨ allow to obtain ▨-1 in the node with relation ▨-1 ? ▨▨, and so forth until we reach the top. At this point, we have an explicit expression for ■, as well as for every other unknown in the list (23). Note that in case $N=1$, the scheme exists of only one stage containing substitutions $y_1 \to 1$ and $y_1 \to A(0,z)/z$, which we have used in [14,15] to obtain $U(z)$ and $P(0,z)$ respectively.

We now discuss an important property regarding the behaviour of the MR system which may not readily be apparent from the analysis so far. Let us first introduce the following notation: $m_{j,k}^{[N]}$ is the position of the jth reservation at the beginning of slot k in a system with N reservations. Corresponding to (12), let also $\hat{m}_{j,k}^{[N]} = m_{j,k}^{[N]} - j$. Of course, if it is clear from the context that we are considering a system with N reservations, the superscript $[N]$ may be dropped. In what follows, we use this notation for other quantities as well, to indicate the number of reservations in the system they are related to. The following theorem is crucial to our analysis of the packet delay distribution.

Theorem 1 (Reservation Theorem). *If a queue with N reservations and a queue with $N-1$ reservations are both empty in slot 0 and are both subjected to the same number of arriving 1- and 2-packets in each of the following slots 0 to $k-1$, then we have that*

$$m_{j,k}^{[N]} = m_{j-1,k}^{[N-1]} + 1, \quad \text{or equivalently,} \quad \hat{m}_{j,k}^{[N]} = \hat{m}_{j-1,k}^{[N-1]}, \qquad (26)$$

at the beginning of slot k, for $j=2,\ldots,N$.

Proof. As we assume that both systems are empty in slot 0, the variables $\hat{m}_{j,0}^{[N]}$ and $\hat{m}_{j,0}^{[N-1]}$ are all equal to 0 such that (26) holds. Now, suppose that (26) holds in slot k for all $j=2,\ldots,N$. If we can show that

$$\hat{m}_{j,k+1}^{[N]} = \hat{m}_{j-1,k+1}^{[N-1]}, \qquad j=2,\ldots,N, \qquad (27)$$

then by induction, this proves the theorem. For certain values of the reservation positions in slot k, the system equations **(Empty)**, **(Keep)**, **(AtEnd)** and (**i-shift**) provide the new reservation positions in slot $k+1$. Therefore, we must compare these equations to their equivalent in case of a system with only $N-1$ reservations. Doing so, assuming that (26) holds, it can be checked easily that (27) holds as well, for every possible value of u_k and $a_{1,k}$. This completes the proof.

Together with the fact that the MR systems with N and $N-1$ reservations also hold the same total number of packets, theorem (26) leads to the following corollary concerning the joint pgfs of the system state of both systems. Let $P_k^{[N]}$ be the system state distribution (15) for a system with N reserved spaces. It is now easily seen that

$$\begin{aligned} P_k^{[N]}(1, y_2, \ldots, y_N; z) &= \mathrm{E}[y_2^{\hat{m}_{2,k}^{[N]}} y_3^{\hat{m}_{3,k}^{[N]}} \cdots y_N^{\hat{m}_{N,k}^{[N]}} z] \\ &= \mathrm{E}[y_2^{\hat{m}_{1,k}^{[N-1]}} y_3^{\hat{m}_{2,k}^{[N-1]}} \cdots y_N^{\hat{m}_{N-1,k}^{[N-1]}} z] \\ &= P_k^{[N-1]}(y_2, \ldots, y_N; z). \end{aligned} \qquad (28)$$

If we let the arguments y_2 to y_n assume the value 1, then it directly follows from (28) for some $0 \leqslant n < N$ that

$$P_k^{[N]}(1, \ldots, 1, y_{n+1}, \ldots, y_N; z) = P_k^{[N-n]}(y_{n+1}, \ldots, y_N; z). \qquad (29)$$

This property says that the distribution of the last $N-n$ reservation positions in a system with N reservations is equal (up to a fixed shift) to that of reservation positions in a system with only $N-n$ reservations.

5 Delay Analysis in Multi-class Queues

Before continuing, let us remind the usefulness of Little's law [21,34], which states that $\mathrm{E}[u] = \lambda_T \mathrm{E}[d]$. The mean of the delay d of an arbitrary packet (regardless of type) can therefore immediately be obtained from (7). For a 2-class system, this can be written as

$$\mathrm{E}[u] = \mathrm{E}[v_1] + \mathrm{E}[v_2] = \lambda_T \mathrm{E}[d] = \lambda_1 \mathrm{E}[d_1] + \lambda_2 \mathrm{E}[d_2]. \tag{30}$$

Additionally, the law also applies for packets of a single type only, i.e. $\mathrm{E}[v_i] = \lambda_i \mathrm{E}[d_i]$, $i = 1, 2$. So for the work-conserving scheduling disciplines we consider here, once the mean delay of one type of packets is known, that of the other type directly follows from (7) and (30). That said, we now perform the per-class delay analysis ③ for the QoS-flat scheduling mechanisms. As we will see, contrary to intuition, the distributions of d_1 and d_2 are *not* necessarily the same, even though the scheduling mechanism does not distinguish between classes [50]. For FIFO, the delay of a packet of class i is $d_i = (u-1)^+ + f_i + 1$, where f_i is the number of packets that arrive in the same slot as $\mathcal{P}$ and are stored in the queue *before* $\mathcal{P}$. Note that we used u instead of u_I since they have the same distribution anyway. Let a_1^I and a_2^I be the total numbers of arrivals of type 1 and 2 respectively during the arrival slot I of $\mathcal{P}$. Note that these variables do *not* have the same joint distribution as a_1 and a_2. If $\mathcal{P}$ is of type i, the joint mass function of the numbers of arrivals of type 1 and 2 during the arrival slot of $\mathcal{P}$ is given by [5]

$$\mathrm{Prob}[a_1^I = j_1, a_2^I = j_2] = \frac{j_i}{\lambda_i} \mathrm{Prob}[a_1 = j_1, a_2 = j_2] \quad, i = 1, 2. \tag{31}$$

Taking into account that the arrivals in slot I are stored in totally *random* order and that whenever $a_1^I + a_2^I = h$, $\mathcal{P}$ is stored in any of the h possible positions with equal probability, one finds for the pgf $F_i(z)$ of f_i:

$$F_i(z) = \frac{1}{\lambda_i} \frac{1}{1-z} \int_z^1 \frac{\partial}{\partial z_i} A(x, x) \mathrm{d}x, \quad \text{with } \mathrm{E}[f_i] = \frac{\lambda_{12} + \lambda_i'}{2\lambda_i}. \tag{32}$$

Hence, the delay of a packet of type i has pgf $D_i(z) = \mathrm{E}[z^{(u-1)^+ + f_i + 1}]$ given by

$$D_i(z) = z F_i(z) \frac{(1-\lambda_T)(z-1)}{z - A_T(z)}, \quad \text{with } \mathrm{E}[d_i] = 1 + \frac{\lambda_T'}{2(1-\lambda_T)} + \frac{\lambda_{12} + \lambda_i'}{2\lambda_i}, \tag{33}$$

the latter complying to (30). In case of LIFO, the delay of $\mathcal{P}$ is determined by the number of packets f_i^* stored *after* $\mathcal{P}$ in slot I instead of before. Clearly, f_i and f_i^* have the same distribution (32). In the following slots, the queue content needs to drop by f_i^* levels before $\mathcal{P}$ can be served. We therefore have $d_i = 1 + \sum_{n=1}^{f_i^*} \eta_n$ where the η_n are iid random variables with common pgf $Y_T(z)$, known as *sub-busy periods* and defined as $\eta = \min\{h > 0 : u_{k+h} = u_k - 1\}$ for any slot k where $u_k > 0$. In other words, a sub-busy period η is the time it takes to end up with one less packet in the system as when the period started. By a recursive probabilistic argument [5,47], the pgf $Y_T(z)$ of η can be obtained as the solution of $Y_T(z) = zA_T(Y_T(z))$ and $Y_T(1) = 1$, from which $Y'_T(1) = \frac{1}{1-\lambda_T}$. The delay of an i-packet under LIFO is therefore

$$D_i(z) = z\mathrm{E}[(Y_T(z))^{f_i^*}] = zF_i(Y_T(z)), \quad \text{with } \mathrm{E}[d_i] = 1 + \frac{\lambda_{12}+\lambda'_i}{2(1-\lambda_T)\lambda_i}, \quad (34)$$

again with the latter complying to (30), as can be checked. Note that this derivation does not require knowledge of the distribution of u, as we already mentioned in section 4. For ROS, the order in which the arrivals of slot I are stored does not matter, since this mechanism does not keep any order in the queue. Both classes of packets have therefore exactly the same distribution of which the mean directly follows from (7) and (30). Higher-order moments can be derived as well, see [30].

For the per-class delay analysis of the QoS-aware mechanisms AP and MR, packets of different types arriving in slot I are not stored in the queue as a random mix so that (32) can not be used. Instead, first all 1-packets are stored and then all 2-packets. Therefore, supposing $\mathcal{P}$ is of type i, it is useful to define ℓ_i as the number of i-packets arriving in slot I that are stored *before* (and including) $\mathcal{P}$. Its pgf $L_i(z)$ is found as

$$L_i(z) = \frac{z(1-A_i(z))}{\lambda_i(1-z)}, \quad i=1,2, \quad (35)$$

again by considering (31) and the fact that $\mathcal{P}$ could be *any* of the a_i^I arrivals with equal probability. Likewise, if $\mathcal{P}$ is of type 2, the joint pgf of a_1^I and ℓ_2 is found to be

$$F(x,y) = \mathrm{E}[x^{a_1^I} y^{\ell_2}] = \frac{y}{1-y} \frac{A_1(x) - A(x,y)}{\lambda_2}. \quad (36)$$

Now, in case of the QoS-aware absolute priority mechanism AP, the delay of 1-packets can be retrieved fairly easy. Note that the 1-packets are entirely insensitive to the presence of 2-packets. Therefore, the delay d_1 is the same as in a FIFO system where only 1-arrivals occur, i.e. where $a_2 = 0$ and hence $A_T(z) = A_1(z)$. For such an arrival process, (32) reduces to $F_1(z) = \frac{1-A_1(z)}{\lambda_1(1-z)}$ and (33) to

$$D_1^{\mathrm{AP}}(z) = \frac{1-\lambda_1}{\lambda_1} z \frac{A_1(z)-1}{z-A_1(z)}. \quad (37)$$

Note that the system content distribution (6) for a FIFO-system with only 1-arrivals indeed equals the marginal distribution $U(z,1)$ of (9), see (49). For the delay of 2-packets however, we need to take into account that d_2 is affected by 1-arrivals occurring *after* slot I [47]. We have

$$d_2 = 1 + \sum_{n=1}^{(u-1)^+ + a_1^I + \ell_2 - 1} \eta_{1,n}, \quad (38)$$

where the $\eta_{1,n}$ are iid variables denoting a sub-busy period *pertaining to 1-packets only*, i.e. the time required for $u_{1,k}$ to be reduced by one unit. Seen differently however, $\eta_{1,n}$ is also the time it takes for $\mathcal{P}$ to advance one position in the queue. Unlike before with LIFO, now only 1-arrivals can impede the advancement of $\mathcal{P}$ and the pgf $Y(z)$ of $\eta_{1,n}$ is therefore implicitly given by $Y(z)=zA_1(Y(z))$ with $Y(1)=1$. Hence, from (38) and using (36),

$$D_2^{\text{AP}}(z) = \frac{1-\lambda_T}{\lambda_2} z \frac{A_T(Y(z)) - A_1(Y(z))}{Y(z) - A_T(Y(z))}. \tag{39}$$

The delay moments can directly be derived from (37) and (39), as well as expressions for the tail behaviour of these distributions. Concerning the latter, we note that $D_2^{\text{AP}}(z)$ may exhibit non-exponential decay in some situations, see [37,47] for a discussion. The delay analysis under MR of type-1 packets is given in the following section. Once $\text{E}[d_1]$ is calculated, the mean delay $\text{E}[d_2]$ can then directly be obtained from (7) and Little's law (30). However, for a *full* analysis of the delay distribution experienced by packets of type 2, we refer to our previous work [16,17].

6 Type-1 Delay of the Reservation Mechanism

Our purpose here is to obtain the delay d_1 experienced by a packet $\mathcal{P}$ of type 1 as it goes through the MR system with N reserved spaces. Let us introduce

$$w_j = \text{Prob}[\ell_1 = j] = \frac{1}{\lambda_1} \sum_{i=j}^{+\infty} \beta_i, \qquad j \geqslant 1, \tag{40}$$

for the mass function of ℓ_1, such that according to (35), $L_1(z) = \sum_{j=1}^{+\infty} w_j z^j$. What is of importance is the exact position in which $\mathcal{P}$ will be stored at the end of slot I. If $\mathcal{P}$ is the first of the batch ($\ell_1 = 1$) it will seize the first reservation at position m_1, if $\ell_1 = 2$ then it takes position m_2, and so forth. If ℓ_1 is larger than N however, $\mathcal{P}$ will seize a reservation created by one of the $\ell_1 - 1$ previous 1-arrivals in slot I, located somewhere at the end of the queue. We find that

$$d_1 = \begin{cases} m_j & \text{if } \ell_1 = j \leqslant N, \\ (u-1)^+ + \ell_1 & \text{if } \ell_1 > N. \end{cases} \tag{41}$$

Taking the z-transform and using (40), we get

$$\begin{aligned} D_1(z) &= \sum_{j=1}^{N} w_j \text{E}[z^{m_j}] + \text{E}[z^{(u-1)^+}] \sum_{j=N+1}^{+\infty} w_j z^j \\ &= \sum_{j=1}^{N} w_j \text{E}[z^{\hat{m}_j}] z^j + \text{E}[z^{(u-1)^+}] \left(L_1(z) - \sum_{j=1}^{N} w_j z^j \right), \end{aligned} \tag{42}$$

where $\mathrm{E}[z^{(u-1)^+}]$ is given in (6). The marginal distribution $\mathrm{E}[z^{\hat{m}_j}]$ appearing in (42) on the other hand, is more difficult to obtain. However, this is where Theorem 1 and its corollary (29) come into play. For $j=1,\ldots,N$ we have

$$\mathrm{E}[z^{\hat{m}_j^{[N]}}] = P^{[N]}(1,1,\ldots,1,\underset{j}{z},1,1,\ldots,1;1) = \mathrm{E}[z^{\hat{m}_1^{[N-j+1]}}]. \tag{43}$$

The conclusion is that instead of having to calculate the marginal distributions of all reservation positions $\hat{m}_j^{[N]}$ ($j=1,\ldots,N$), it is sufficient to obtain the marginal distributions of $\hat{m}_1^{[N-j+1]}$, i.e. only of *the first reservation positions* in the systems with 1 up to N reservations.

6.1 Distribution of the First Reservation Position m_1

The marginal distribution of $\hat{m}_1$ can be obtained from the *full* system state distribution in slot I determined by functional equation (22). We derive a recursive relation for $\mathrm{E}[z^{\hat{m}_1}]$ by using substitutions similar to those in the first stage of Fig. 2. Specifically, let all arguments in (22) be equal to 1 except for the first one, for which we take $y_1 \to z$. Using Theorem 1 again in the form of (43), we find

$$(z-\alpha)P^{[N]}(z,1,1,\ldots,1;1) = p_0(z-1)f_N(z) + \alpha(z-1)P^{[N]}(0,1,\ldots,1;1) \\ + \sum_{i=1}^{N-1}\beta_i z^i P^{[N-i]}(z,1,1,\ldots,1;1), \tag{44}$$

where

$$f_n(z) \triangleq \frac{1}{z-A_T(z)}\sum_{i=n}^{+\infty}\beta_i z^i, \quad n \geqslant 1. \tag{45}$$

Note that the factor $(z-A_T(z))^{-1}$ is entirely due to the last term of (22) where $U(z)$ appears under the mentioned substitution. Let us also define

$$\Phi_n(z) \triangleq f_n(z) - f_n(\alpha) = \frac{1}{z-A_T(z)}\sum_{i=n}^{+\infty}\beta_i z^i - \frac{1}{\alpha-A_T(\alpha)}\sum_{i=n}^{+\infty}\beta_i \alpha^i. \tag{46}$$

In (44), the probability $P(0,1,\ldots,1;1)$ that $\hat{m}_1 = 0$ can be obtained from the functional equation by evaluating it for the right arguments. First, let $z \to \alpha$ in (44) such that the left-hand side vanishes. As we have explained before, the other side must be equal to 0 then as well, which results in

$$P^{[N]}(0,1,\ldots;1) = -\frac{p_0}{\alpha}f_N(\alpha) + \frac{1}{\alpha(1-\alpha)}\sum_{i=1}^{N-1}\beta_i\alpha^i P^{[N-i]}(\alpha,1,\ldots;1). \tag{47}$$

Plugging this into (44) yields the desired recursive relation for the distribution of the first reservation position. If we first introduce a shorter notation for this distribution, $\Omega_n(z) = P^{[n]}(z,1,\ldots,1;1) = \mathrm{E}[z^{\hat{m}_1^{[n]}}]$, then we finally find

$$\Omega_N(z) = p_0(z-1)\frac{\Phi_N(z)}{z-\alpha} + \sum_{i=1}^{N-1}\beta_i\frac{z^i\Omega_{N-i}(z) - \frac{z-1}{\alpha-1}\alpha^i\Omega_{N-i}(\alpha)}{z-\alpha}. \tag{48}$$

In principle, our work is done now, since this relation determines all $\Omega_n(z)$, $n=1,\ldots,N$, being the distributions of the first reservation position in systems with 1 up to N reservations. Through (42) and (43), this directly provides the pgf of d_1. Indeed, (48) can be solved in an iterative way since $\Omega_N(z)$ appears only on the left-hand side while the other side only depends on $\Omega_1(z)$ to $\Omega_{N-1}(z)$. However, the problem is that we also have to determine the constant $\Omega_n(\alpha)$ in step n ($n=1,\ldots,N-1$). Since $\Omega_n(z)$ is a pgf and $\alpha<1$, we know there must be a solution for the quantities $\Omega_n(\alpha)$ lying between 0 and 1, but obtaining these values in a direct analytic way proves to be difficult. The main issue here is that if (48) is solved by iteration, the complexity doubles with each step, producing expressions of exponentially increasing length and requiring the n-fold use of de l'Hôpital's rule. Fortunately, the quantities $\Omega_n(\alpha)$ can be obtained numerically using an entirely different approach, discussed in the next paragraph.

6.2 Obtaining the Unknowns $\Omega_n(\alpha)$

It is seen that for $N=\infty$, the packets in the MR system behave the same as in the AP system. Indeed, the MR queue will decouple into two logical sub-queues in this case: one for the 1-packets at the front closest to the server and one containing a swarm of 2-packets at the far end. We call these sub-queues the 1-queue and the 2-queue respectively. In between these sub-queues there is an impenetrable barrier containing an infinite number of R's that causes a 'decoupled' operation of the system. Arrivals of type 1 will always be stored in the first R of the barrier and thus find connection to the logical 1-queue. On the other hand, the MR discipline dictates that the arriving 2-packets are stored at the end of the queue and therefore become part of the logical 2-queue. If the server becomes available, the next packet that is scheduled for service is the one positioned closest to the server. Since the 1-packets are grouped on the first positions, the server always schedules a 1-packet if one is available. Only if there are no 1-packets present, a 2-packet will jump over the barrier to be served next. We can use this resemblance of the logical 1-queue in an MR system with $N=\infty$ to the high-priority packets in the queue under AP, to obtain the limiting distribution $\Omega_\infty(z)$ of the position $\hat{m}_1^{[\infty]}$ of the first reservation. As before, note that since for AP the 1-packets are not in any way affected by the 2-packets, we know that v_1 is distributed as for FIFO given that there are no 2-arrivals. Hence, similar to (6), we have

$$U_1^{\text{AP}}(z) = (1-\lambda_1)\frac{A_1(z)(z-1)}{z - A_1(z)}. \tag{49}$$

Now, assuming that v_1 is also the number of 1-packets in the logical 1-queue, we have

$$\hat{m}_1^{[\infty]} = m_1^{[\infty]} - 1 = (v_1 - 1)^+. \tag{50}$$

With (49), it then easily follows that

$$\Omega_\infty(z) = (1-\lambda_1)\frac{z-1}{z-A_1(z)}, \quad \text{and} \quad \Omega_\infty(\alpha) = (1-\lambda_1)\frac{\alpha-1}{\alpha-A_1(\alpha)}. \tag{51}$$

Whereas calculating the functions $\Omega_n(z)$, $n = 1, 2, \ldots$ iteratively from (48) is difficult, taking the transform of this sequence is much easier. Specifically, let us define $\Omega(x, z) = \sum_{n=1}^{+\infty} \Omega_n(z) x^n$, which is the generating function of the generating functions $\Omega_n(z)$. From (48) we find a closed-form expression for $\Omega(x, z)$:

$$\Omega(x, z) = \frac{p_0(z-1)\,\Phi(x, z) - \frac{z-1}{\alpha-1}(A_1(\alpha x) - \alpha)\,\Omega(x, \alpha)}{z - A_1(zx)}, \qquad (52)$$

where we have used the x-transform of the functions $\Phi_n(z)$ as well, $\Phi(x, z) = \sum_{n=1}^{+\infty} \Phi_n(z) x^n$, which from (46) is calculated as

$$\Phi(x, z) = \frac{x}{1-x}\left[\frac{A_1(z) - A_1(zx)}{z - A_T(z)} - \frac{A_1(\alpha) - A_1(\alpha x)}{\alpha - A_T(\alpha)}\right]. \qquad (53)$$

The function $\Omega(x, z)$, being the transform of probability generating functions, is known to be analytic for x and z lying in the unit disc. If we could find a pair (x, z) in that region for which the denominator in (52) becomes zero, then we know the numerator should be zero as well. Fortunately, one can invoke Rouché's theorem to show that if $|x| < 1$, there always exists a unique $\hat{Y}(x)$ for which $|\hat{Y}(x)| < 1$ and that satisfies

$$\hat{Y}(x) = A_1\bigl(x\,\hat{Y}(x)\bigr), \qquad \text{with} \quad \hat{Y}(1) = 1. \qquad (54)$$

Note that $\hat{Y}(z) = Y(z)/z$ where $Y(z)$ was used in (39). Now, if we let $z \to \hat{Y}(x)$ in (52), the numerator must vanish, which yields

$$\Omega(x, \alpha) = p_0(\alpha - 1)\,\frac{\Phi(x, \hat{Y}(x))}{A_1(\alpha x) - \alpha}. \qquad (55)$$

This, together with (53), allows us to write (52) as

$$\Omega(x, z) = p_0 \frac{x}{1-x}\frac{z-1}{z - A_1(zx)}\left[\frac{A_1(z) - A_1(zx)}{z - A_T(z)} - \frac{A_1(\hat{Y}(x)) - \hat{Y}(x)}{\hat{Y}(x) - A_T(\hat{Y}(x))}\right], \qquad (56)$$

which determines the sequence of pgfs $\Omega_1(z), \Omega_2(z), \ldots$. Let us assume a fixed (complex) value of z, then it is possible to obtain $\Omega_n(z)$ by inverting the x-transform (56). There exist many numerical methods to obtain the coefficients $[x^n]\Omega(x, z)$ of a generating function and most of them involve the evaluation of $\Omega(x, z)$ on a number of discrete points on a contour C around the origin in the x-plane. For instance, the inversion method in [1] uses a circular contour C_r of radius $0 < r < 1$ around $x = 0$. However, the problem with the evaluation of $\Omega(x, z)$ now is that the function $\hat{Y}(x)$ appearing in (56) is *not known explicitly*. Indeed, we know that $\hat{Y}(x)$ exists and is unique, but we only have the implicit relation (54) to determine it. This complicates matters a bit, since every time we want to evaluate $\Omega(x, z)$ for a certain x on C_r (and a certain z, of course), we also have to determine $\hat{Y}(x)$ numerically from (54). To find this value, one can choose any complex root-finding algorithm to find the root $z^* = \hat{Y}(x)$ of $z^* - A_1(z^* x)$. We can apply this numerical inversion method particularly in case $z = \alpha$, i.e. to obtain the quantities $\Omega_n(\alpha)$, $n = 1, 2, \ldots$ which we will need to obtain the mean value of the type-1 packet delay.

6.3 Mean Value of the Type-1 Packet Delay

For the pgf $D_1(z)$ of the type-1 packet delay, we now have from (42) and (43) that

$$D_1(z) = \sum_{j=1}^{N} w_j z^j \Omega_{N-j+1}(z) + p_0 \frac{z-1}{z - A_T(z)} \left(L_1(z) - \sum_{j=1}^{N} w_j z^j \right), \quad (57)$$

where the functions $\Omega_n(z)$ follow from the discussion in the previous paragraph. As such, $D_1(z)$ can be evaluated numerically for any particular z. The mean packet delay follows from (57) as

$$\mathrm{E}[d_1] = D_1'(1) = \sum_{j=1}^{N} w_j \, \Omega'_{N-j+1}(1) + L_1'(1) + \frac{\lambda_T'}{2(1-\lambda_T)} \sum_{j=N+1}^{+\infty} w_j, \quad (58)$$

where $\Omega'_n(1) = \mathrm{E}[\hat{m}_1^{[n]}]$, $n = 1, \ldots, N$ and where we have used the fact that $p_0 = 1 - \lambda_T$. Clearly, the problem at hand is now to determine the mean value $\mathrm{E}[\hat{m}_1^{[n]}]$ of the first reservation position in a system with n reservations, $n = 1, \ldots, N$. In order to do so, we assume that the quantities $\Omega_n(\alpha)$ are available. As we discussed, they can either be obtained analytically by iterating (48) and taking the limit $z \to \alpha$ in each step, or they follow from the numerical inversion method discussed in the previous paragraph. Differentiating (48) to z and taking the limit $z \to 1$ on both sides, we find after some straightforward manipulations and using (46):

$$\mathrm{E}[\hat{m}_1^{[N]}] = \frac{1}{1-\alpha} \left[\lambda_1 - 1 + \frac{\lambda_T'}{2(1-\lambda_T)} \sum_{i=N}^{+\infty} \beta_i - \frac{p_0}{\alpha - A_T(\alpha)} \sum_{i=N}^{+\infty} \beta_i \alpha^i \right.$$
$$\left. + \frac{1}{1-\alpha} \sum_{i=1}^{N-1} \beta_i \alpha^i \Omega_{N-i}(\alpha) \right] + \sum_{i=1}^{N-1} \frac{\beta_i}{1-\alpha} \mathrm{E}[\hat{m}_1^{[N-i]}]. \quad (59)$$

As was the case with (48), this relation can be solved iteratively. However, still assuming that we know the sequence $\Omega_n(\alpha)$, it is possible to provide a *direct* solution of the expected values $\mathrm{E}[\hat{m}_1^{[n]}]$ from (59). For the sake of clarity, let us define the following shorthands

$$\mu_n \triangleq \mathrm{E}[\hat{m}_1^{[n]}] = \Omega'_n(1), \qquad n = 1, \ldots, N, \quad (60)$$

$$\delta_i \triangleq \frac{\beta_i}{1-\alpha}, \qquad i > 0, \quad (61)$$

$$\Gamma_n \triangleq \frac{1}{1-\alpha} \left[\lambda_1 - 1 + \frac{\lambda_T'}{2(1-\lambda_T)} \sum_{i=n}^{+\infty} \beta_i - \frac{p_0}{\alpha - A_T(\alpha)} \sum_{i=n}^{+\infty} \beta_i \alpha^i \right.$$
$$\left. + \frac{1}{1-\alpha} \sum_{i=1}^{n-1} \beta_i \alpha^i \Omega_{n-i}(\alpha) \right]. \quad (62)$$

This reduces (59) to

$$\mu_N = \Gamma_N + \sum_{i=1}^{N-1} \delta_i \mu_{N-i}. \qquad (63)$$

In this relation, the quantities Γ_n, $n = 1, \ldots, N$ and δ_i, $i \geq 1$ are fully known whereas the quantities μ_n are the mean values we seek. One can already deduce from (63) that each μ_n will be a linear combination of the quantities Γ_1 up to Γ_n with coefficients being a function of δ_1 up to δ_n. In order to find these coefficients we proceed as follows. Let us first arrange the values obtained from (62) in a $N \times 1$ matrix $\boldsymbol{\Gamma}$,

$$\boldsymbol{\Gamma}^T \triangleq \begin{bmatrix} \Gamma_1 & \Gamma_2 & \Gamma_3 & \ldots & \Gamma_N \end{bmatrix}. \qquad (64)$$

Secondly, we use the values (61) to define the following $N \times N$ matrix,

$$\mathbf{H} \triangleq \begin{bmatrix} \delta_1 & 1 & 0 & 0 & \cdots & 0 \\ \delta_2 & 0 & 1 & 0 & \cdots & 0 \\ \delta_3 & 0 & 0 & 1 & \cdots & 0 \\ \vdots & & & & \ddots & \\ \delta_{N-1} & 0 & 0 & 0 & \cdots & 1 \\ \delta_N & 0 & 0 & 0 & \cdots & 0 \end{bmatrix}. \qquad (65)$$

This matrix $\mathbf{H}$ is an instance of what is known as a *Leslie matrix* [31] due to P.H. Leslie who used this kind of matrices in 1945 for the study of population growth. In addition, let $\mathbf{e}$ be the row matrix $\mathbf{e} \triangleq \begin{bmatrix} 1 & 0 & 0 & \ldots \end{bmatrix}$ of appropriate size. Now, it can be verified that μ_n is obtained by calculating the $(n-1)$th power of $\mathbf{H}$, i.e. the solution of (63) is

$$\mu_n = \mathbf{e}\, \mathbf{H}^{n-1}\, \boldsymbol{\Gamma}, \qquad n = 1, \ldots, N. \qquad (66)$$

This provides the mean values $\mu_n = \Omega'_n(1)$ in the expression for the mean delay (58) which now becomes

$$E[d_1^{[N]}] = \mathbf{e}\Big(\sum_{j=1}^{N} \omega_j\, \mathbf{H}^{N-j} \Big) \boldsymbol{\Gamma} + \frac{\lambda'_T}{2(1-\lambda_T)}\Big(1 - \sum_{j=1}^{N} \omega_j\Big) + 1 + \frac{\lambda'_1}{2\lambda_1}. \qquad (67)$$

This expression allows us to calculate the mean delay of type 1 in the system with N reservations by means of $N-1$ matrix multiplications. However, in doing so, it is possible to arrange the calculations in such a way that the mean values $E[d_1^{[n]}]$ of the delay in the corresponding systems with *less* than N reservations are produced as well. In other words, calculating the delay in a system with one additional reservation requires only one additional matrix multiplication. The following algorithm shows how this can be achieved.

▶ For $n = 1, \ldots, N$, calculate the values $\Omega_n(\alpha)$, either analytically or numerically, as explained before. Note that for high n, one could consider approximating $\Omega_n(\alpha)$ by the limiting value $\Omega_\infty(\alpha)$ given in (51).

- For $n=1,\ldots,N$, calculate the entries Γ_n in the matrix $\boldsymbol{\Gamma}$ using (62).
- Now construct the matrix $\mathbf{H}$ as in (65) and define the starting values

$$\psi_0 = \frac{\lambda'_T}{2(1-\lambda_T)} + 1 + \frac{\lambda'_1}{2\lambda_1} \quad \text{and} \quad \mathbf{Q}_0 = \mathbf{0}. \qquad (68)$$

Then, for $n=1,\ldots,N$, calculate

$$\psi_n = \psi_{n-1} - \frac{\lambda'_T}{2(1-\lambda_T)} \omega_n, \quad \text{and} \quad \mathbf{Q}_n = \mathbf{Q}_{n-1}\mathbf{H} + \omega_n \mathbf{I}, \qquad (69)$$

where $\mathbf{I}$ is the $N \times N$ identity matrix. As the mean value of $d_1^{[n]}$ follows from (67) for $N=n$, we can now see that after the nth step in this iteration, the mean delay of type 1 in a system with n reservations is given by

$$E[d_1^{[n]}] = \psi_n + \mathbf{e}\,\mathbf{Q}_n\,\boldsymbol{\Gamma}. \qquad (70)$$

6.4 Tail Distribution of the Type-1 Packet Delay

Another important characteristic of the delay distribution besides the mean value, is its tail distribution. We use the *dominant pole approximation* which is known to yield very accurate results, see e.g. [5,7]. Specifically, from the inversion formula for z-transforms, it follows that the probability mass function $\mathrm{Prob}[d_1=n]$ can be expressed as a weighted sum of negative nth powers of the poles of $D_1(z)$. Since all these poles have a modulus larger than 1, $\mathrm{Prob}[d_1=n]$ is dominated by the contribution of the pole z_d with the smallest modulus. It was shown that this 'dominant' pole z_d must necessarily be real and positive in order to ensure a nonnegative probability mass function. As such, the probability for a 1-packet to experience a delay of n slots can be expressed by the following geometric form for sufficiently large values of n:

$$\mathrm{Prob}[d_1^{[N]}=n] \cong -\theta_1^{[N]} z_d^{-n-1}, \qquad (71)$$

where z_d is the pole of $D_1(z)$ with smallest modulus and $\theta_1^{[N]}$ is the residue in z_d:

$$\theta_1^{[N]} = \mathrm{Res}_{z_d} D_1(z) = \lim_{z \to z_d} (z-z_d) D_1(z). \qquad (72)$$

The first thing to do therefore, is to identify the dominant pole z_d of $D_1(z)$. After careful inspection of the expression (57), one can prove that its dominant pole can only originate from the factor $(z - A_T(z))^{-1}$ appearing in the second term, but also present in each $\Omega_n(z)$ through (46) and (48). Note that the multiplicity of this factor is equal to 1 in all of these terms. As such, z_d can be obtained numerically as the smallest real root larger than 1 of

$$z - A_T(z) = 0. \qquad (73)$$

This value is *independent* of N and identical to the dominant pole we have in case of FIFO, see (6). Secondly, we have to evaluate the limit in (72) to obtain the residue $\theta_1^{[N]}$. Fortunately, not *all* terms in $D_1(z)$ as given in (57) have z_d as a pole. Consequently, all the contributions to $D_1(z)$ that do not, will vanish when taking the limit $z \to z_d$, due to the factor $(z-z_d)$. Therefore, if we only consider in (48) the *contributions that have a pole in* z_d, we hope that the recursion becomes easier to solve. This approach is still exact, because we only neglect terms that would vanish under the limit (72) anyway. Let us define $\Omega_n^*(z)$, $n=1,\ldots,N$ as these 'modified' versions of the original functions $\Omega_n(z)$ determined by (48), such that

$$\mathrm{Res}_{z_d} \Omega_n^*(z) = \lim_{z\to z_d}(z-z_d)\Omega_n^*(z) = \lim_{z\to z_d}(z-z_d)\Omega_n(z) = \mathrm{Res}_{z_d}\Omega_n(z). \quad (74)$$

Note that these modified functions are no longer pgfs. Their only correct interpretation is having the same residue in z_d as the original pgfs. (48) yields

$$\Omega_N^*(z) = p_0(z-1)\frac{f_N(z)}{z-\alpha} + \sum_{i=1}^{N-1} \beta_i \frac{z^i}{z-\alpha} \Omega_{N-i}^*(z), \quad (75)$$

where we recall that $f_n(z)$ is defined in (45). The required residues (74) are therefore determined by the recursion

$$\mathrm{Res}_{z_d}\Omega_N(z) = \frac{p_0}{1-A_T'(z_d)}\frac{z_d-1}{z_d-\alpha}\sum_{i=N}^{+\infty}\beta_i z_d^i + \sum_{i=1}^{N-1}\beta_i \frac{z_d^i}{z_d-\alpha}\mathrm{Res}_{z_d}\Omega_{N-i}(z), \quad (76)$$

where we have used de l'Hôpital's rule and definition (45). This relation can be represented much simpler if we introduce

$$\mu_n^* \triangleq \mathrm{Res}_{z_d}\Omega_n(z), \qquad n=1,\ldots,N, \quad (77)$$

$$\delta_i^* \triangleq \frac{\beta_i}{z_d-\alpha} z_d^i, \qquad i>0, \quad (78)$$

$$\Gamma_n^* \triangleq \frac{z_d-1}{z_d-\alpha}\frac{p_0}{1-A_T'(z_d)}\sum_{i=n}^{+\infty}\beta_i z_d^i, \qquad n=1,\ldots,N, \quad (79)$$

similar to (60)–(62). Relation (76) then becomes

$$\mu_N^* = \Gamma_N^* + \sum_{i=1}^{N-1}\delta_i^* \mu_{N-i}^*, \quad (80)$$

which is symbolically *exactly* the same as (63) and therefore has the same kind of solution:

$$\mu_n^* = \mathrm{Res}_{z_d}\Omega_n(z) = \mathbf{e}\,(\mathbf{H}^*)^{n-1}\,\boldsymbol{\Gamma}^*, \qquad n=1,\ldots,N. \quad (81)$$

Here, the matrix $\mathbf{H}^*$ is the same as $\mathbf{H}$, but with every entry δ_i replaced by δ_i^*. Using this solution, we finally find from (57) for the residue $\theta_1^{[N]}$:

$$\theta_1^{[N]} = \mathbf{e}\Big(\sum_{j=1}^{N}\omega_j\, z_d^j\,(\mathbf{H}^*)^{N-j}\Big)\boldsymbol{\Gamma}^* + \frac{p_0 z_d}{\lambda_1}\frac{A_1(z_d)-1}{1-A_T'(z_d)} + \frac{p_0(1-z_d)}{1-A_T'(z_d)}\sum_{j=1}^{N}\omega_j\, z_d^j. \quad (82)$$

As with the mean value of the delay, the calculation of the residue $\theta_1^{[N]}$ can be performed in such a way that the equivalent residues $\theta_1^{[n]}$ for systems with fewer than N reservations are produced as well. The following algorithm implements this.

▸ For $n=1,\ldots,N$, calculate the entries Γ_n^* in the matrix $\boldsymbol{\Gamma}^*$ using (79).
▸ Now determine the values δ_i^* as in (78) and populate the matrix $\mathbf{H}^*$. The residues $\theta_1^{[1]}$ to $\theta_1^{[N]}$ can now progressively be obtained as follows. Define the starting values

$$\psi_0^* = p_0\,\frac{z_d}{\lambda_1}\frac{A_1(z_d)-1}{1-A_T'(z_d)}, \quad \text{and} \quad \mathbf{Q}_0^* = \mathbf{0}. \quad (83)$$

Then, for $n=1,\ldots,N$, calculate

$$\psi_n^* = \psi_{n-1}^* + p_0\,\frac{1-z_d}{1-A_T'(z_d)}\,\omega_n\, z_d^n, \quad \text{and} \quad \mathbf{Q}_n^* = \mathbf{Q}_{n-1}^*\,\mathbf{H}^* + \omega_n\, z_d^n\,\mathbf{I}. \quad (84)$$

After each step, the residue $\theta_1^{[n]}$ then follows from (82) as $\theta_1^{[n]} = \psi_n^* + \mathbf{e}\,\mathbf{Q}_n^*\,\boldsymbol{\Gamma}^*$.

7 A Comparative Example

Let us consider a specific example to demonstrate the delay differentiation realised between the two packet types by MR as compared to FIFO and AP. We choose the distribution of the arrivals as

$$A(z_1, z_2) = \frac{1}{1+\lambda_1 - z_1\lambda_1}\cdot e^{\lambda_2(z_2-1)}, \quad (85)$$

i.e. the numbers of arrivals per slot of type 1 and 2 are independent and have a geometric and Poisson distribution respectively, with partial loads λ_1 and λ_2. In Fig. 3 we plotted $\mathrm{E}[d_1^{[N]}]$ and $\mathrm{E}[d_2^{[N]}]$ as functions of the traffic mix λ_1/λ_T and for a fixed total load $\lambda_T = 0.9$. The mean delay under FIFO and the mean delay for 1- and 2-packets under AP are shown as well. Note that for FIFO we plotted only *one* curve, indicating the delay of an arbitrary packet regardless of its type. We see that the higher N, the more the mean delays of both types deviate from FIFO and the closer they get to their respective AP limits. On the far left side of the plot, there are but few 1-packets among a multitude of 2-packets. As a consequence, the queue contains mainly 2-packets and always has almost all of its reservations positioned directly in front of the server. Therefore, a rare arriving 1-packet can generally jump over the whole queue content and be served directly in the next slot. So even if there is only one reserved space, the behaviour under the MR mechanism is equal to that under AP for very low

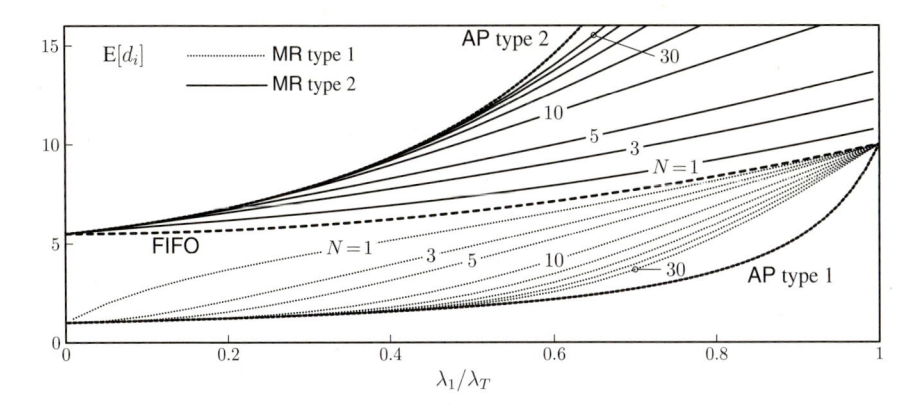

Fig. 3. Mean delay of both 1- and 2-packets versus the traffic mix λ_1/λ_T in case of $N = 1, 3, 5, 10, 15, 20, 25$ and 30 reservations in the queue. The arrivals of type 1 and type 2 are independent and have a geometric and Poisson distribution respectively with total load $\lambda_T = 0.9$.

λ_1/λ_T, resulting in a maximal delay differentiation. On the far right of the plot, most of the traffic is of type 1, while 2-packets arrive only very rarely. From the point of view of the 1-packets, there is no difference between FIFO, AP or the intermediate MR mechanism if λ_1/λ_T is very high. However, the delay of a rare 2-packet is influenced a great deal by the queueing discipline in this case. While such a packet is almost sure to stay in the queue forever under AP, its delay under MR increases from the FIFO value more or less linearly with N. In our opinion, this is where the main strength compared to AP emerges. If only a small part of the traffic consists of low-priority traffic, their delay can be chosen at an arbitrary level by changing N, whereas the delay is almost infinite under AP (packet starvation).

8 Conclusions

We have reviewed an approach for modelling and analysing multi-class queues in heterogeneous packet network architectures (e.g. DiffServ). In particular, a discrete-time single-server queue with infinite capacity is considered. In case there are two traffic classes, type 1 with delay-sensitive and type 2 with delay-tolerant packets, we show the effectiveness of a transform domain analysis with carefully chosen supplementary variables to assess the per-class packet delay distributions. The achieved QoS differentiation mainly depends on the scheduling mechanism in the queue. We considered, partially or in whole, First-In First-Out (FIFO), Last-In First-Out (LIFO), Random Order of Service (ROS), Absolute Priority (AP) and a novel approach based on Multiple Reservations (MR). For the latter, we discussed in detail the delay analysis of the type 1 packets in the queue.

References

1. Abate, J., Whitt, W.: Numerical Inversion of Probability Generating Functions. Operations Research Letters 12(4), 245–251 (1992)
2. Abate, J., Whitt, W.: Limits and Approximations for the M/G/1 LIFO Waiting-time Distribution. Operations Research Letters 20, 199–206 (1997)
3. Bae, J.J., Suda, T.: Survey of Traffic Control Schemes and Protocols in ATM Networks. Proceedings of the IEEE 79(2), 170–189 (1991)
4. Blake, S., et al.: An Architecture for Differentiated Services. Internet RFC 2475 (December 1998)
5. Bruneel, H., Kim, B.G.: Discrete-Time Models for Communication Systems Including ATM. Kluwer Academic Publishers, Boston (1993)
6. Bruneel, H.: Performance of Discrete-Time Queueing Systems. Computers & Operations Research 20(3), 303–320 (1993)
7. Bruneel, H., Steyaert, B., Desmet, E., Petit, G.: Analytic Derivation of Tail Probabilities for Queue Lengths and Waiting Times in ATM Multiserver Queues. European Journal of Operational Research 76, 563–572 (1994)
8. Burakowski, W., Tarasiuk, H.: On New Strategy for Prioritising the Selected Flow in Queuing System. In: Proceedings of the COST 257 11th Management Committee Meeting, Barcelona, Spain, January 20-21 (2000); COST-257 TD(00)03
9. Choi, B.D., Choi, D.I., Lee, Y., Sung, D.K.: Priority Queueing System with Fixed-length Packet-train Arrivals. IEE Proceedings-Communications 145(5), 331–336 (1998)
10. Chuang, L., Wanming, L., Baoping, Y., Chanson, S.: A Dynamic Partial Buffer Sharing Scheme for Packet Loss Control in Congested Networks. In: Proceedings of the International Conference on Communication Technology, ICCT 2000, Beijing, China, August 21-25, vol. 2, pp. 1286–1293 (2000)
11. Cidon, I., Georgiadis, L., Guerin, R., Khamisy, A.: Optimal Buffer Sharing. IEEE Journal on Selected Areas in Communications 13, 1229–1240 (1995)
12. Cox, D.: The Analysis of non-Markovian Stochastic Processes by the Inclusion of Supplementary Variables. Proceedings of the Cambridge Philosophical Society 51, 433–441 (1955)
13. Demers, A., Keshav, S., Shenker, S.: Analysis and Simulation of a Fair Queueing Algorithm. In: Proceedings of the ACM Symposium on Communications Architectures & Protocols, SIGCOMM 1989, Austin, TX, USA, September 19-22, pp. 1–12 (1989)
14. De Vuyst, S., Wittevrongel, S., Bruneel, H.: Delay Differentiation by Reserving Space in Queue. Electronics Letters 41(9), 69–70 (2005)
15. De Vuyst, S., Wittevrongel, S., Bruneel, H.: Place Reservation: Delay Analysis of a Novel Scheduling Mechanism. Computers and Operations Research, Special Issue on "Queues in Practice" 35(8), 2447–2462 (2008)
16. De Vuyst, S., Wittevrongel, S., Bruneel, H.: Parametric Delay Differentiation Between Packet Flows Using Multiple Reserved Spaces. In: Proceedings of VALUETOOLS 2006, the First International Conference on Performance Evaluation Methodologies and Tools, Pisa, Italy, October 11-13 (2006)
17. De Vuyst, S., Wittevrongel, S., Fiems, D., Bruneel, H.: Controlling the Delay Trade-off Between Packet Flows Using Multiple Reserved Places. Performance Evaluation 65(6-7), 484–511 (2008)
18. Dovrolis, C., Stiliadis, D., Ramanathan, P.: Proportional Differentiated Services: Delay Differentiation and Packet Scheduling. ACM Computer Communications Review 29(4), 109–120 (1999)

19. Feyaerts, B., De Vuyst, S., Wittevrongel, S., Bruneel, H.: Analysis of a Discrete-time Priority Queue with Place Reservations and Geometric Service Times. In: Proceedings of DASD 2008, the 6th Symposium on Design, Analysis and Simulation of Distributed Systems, Edinburgh, UK, June 16-19, pp. 140–147 (2008)
20. Feyaerts, B., Wittevrongel, S.: Performance analysis of a priority queue with place reservation and general transmission times. In: Thomas, N., Juiz, C. (eds.) EPEW 2008. LNCS, vol. 5261, pp. 197–211. Springer, Heidelberg (2008)
21. Fiems, D., Bruneel, H.: A Note on the Discretization of Little's Result. Operations Research Letters 30(1), 17–18 (2002)
22. Flatto, L.: The Waiting Time Distribution for the Random Order Service M/M/1 Queue. Annals of Applied Probability 7, 382–409 (1997)
23. Floyd, S., Jacobson, V.: Random Early Detection Gateways for Congestion Avoidance. IEEE-ACM Transactions on Networking 1(4), 397–413 (1993)
24. Gail, H.R., Hantler, S.L., Taylor, B.A.: On a Preemptive Markovian Queue with multiple servers and two priority classes. Mathematics of Operations Research 17(2), 365–391 (1992)
25. Gevros, P., Crowcroft, J., Kirstein, P., Bhatti, S.: Congestion Control Mechanisms and the Best Effort Service Model. IEEE Network, 16–26 (May-June 2001)
26. Hurley, P., Le Boudec, J.-Y., Thiran, P., Kara, M.: ABE: Providing a Low-Delay Service within Best Effort. IEEE Network, 60–69 (May-June 2001)
27. Kausha, S., Sharma, R.K.: Modeling and Analysis of Adaptive Buffer Sharing Scheme for Consecutive Packet Loss Reduction in Broadband Networks. International Journal of Computer Systems Science and Engineering 4(1), 8–15 (2007)
28. Kosten, L.: Stochastic Theory of a Multi-entry Buffer (1). Delft Progress Report 1, 10–18 (1974)
29. Kontovasilis, K., Wittevrongel, S., Bruneel, H., Van Houdt, B., Blondia, C.: Performance of Telecommunication Systems: Selected Topics. In: Proceedings of the 17th IFIP World Computer Congress, Montreal, August 25-30, pp. 61–93 (2002)
30. Laevens, K., Bruneel, H.: Delay Analysis for ATM Queues with Random Order of Service. Electronic Letters 31(5), 346–347 (1995)
31. Leslie, P.H.: On the Use of Matrices in Certain Population Mathematics. Biometrika 33, 183–212 (1945)
32. Lim, Y., Kobza, J.E.: Analysis of a Delay-dependent Priority Discipline in an Integrated Multiclass Traffic Fast Packet Switch. IEEE Transactions on Communications 38(5), 659–685 (1990)
33. Liu, K.Y., Petr, D.W., Frost, V.S., Zhu, H.B., Braun, C., Edwards, W.L.: Design and Analysis of a Bandwidth Management Framework for ATM-Based Broadband ISDN. IEEE Communications Magazine 35(5), 138–145 (1997)
34. Little, J.D.C.: A Proof of the Queuing Formula $L = \lambda W$. Operations Research 9(3), 383–387 (1961)
35. Maertens, T., Walraevens, J., Bruneel, H.: On Priority Queues with Priority Jumps. Performance Evaluation 63, 1235–1252 (2006)
36. Maertens, T., Walraevens, J., Bruneel, H.: Performance Comparison of Several Priority Schemes with Priority Jumps. Annals of Operations Research 162, 109–125 (2008)
37. Maertens, T., Walraevens, J., Bruneel, H.: Non-exponential Tail Probabilities in Queuing Systems. In: Proceedings of ITC-19, 19th International Teletraffic Congress, August 29-September 2, pp. 1155–1164 (2005)

38. Menth, M., Schmid, M., Heiss, H., Reim, T.: MEDF – A Simple Scheduling Algorithm for Two Real-time Transport Service Classes with Application in the UTRAN. In: Proceedings of INFOCOM 2003, San Francisco, USA, March 30-April 3 (2003)
39. Van Mieghem, P., Steyaert, B., Petit, G.H.: Performance of Cell Loss Priority Management Schemes in a Single Server Queue. International Journal of Communication Systems 10, 161–180 (1997)
40. Mowbray, M., Karlsson, G., Köhler, T.: Capacity Reservation for Multimedia Traffic. Distrib. Syst. Engng. 5, 12–18 (1998)
41. Parekh, A.K., Gallager, R.G.: A Generalized Processor Sharing Approach to Flow Control in Integrated Services Networks: The Single Node Case. IEEE/ACM Transactions on Networking 1(3), 344–357 (1993)
42. Semeria, C.: Supporting Differentiated Service Classes: Queue Scheduling Disciplines, Juniper Networks, White Paper (2001)
43. Sivaraman, V., Chiussi, F.: Providing End-to-End Statistical Delay Guarantees with Earliest Deadline First Scheduling and Per-Hop Traffic Shaping. In: Proceedings of INFOCOM 2000, Tel Aviv, Israel, March 26-30 (2000)
44. Striegel, A., Manimaran, G.: Packet Scheduling with Delay and Loss Differentiation. Computer Communications 25, 21–31 (2000)
45. Takine, T., Sengupta, B., Hasegawa, T.: An Analysis of a Discrete-Time Queue for Broadband ISDN with Priorities Among Traffic Classes. IEEE Transactions on Communications 42(2-4), 1837–1845 (1994)
46. Tham, C.-K., Yao, Q., Jian, Y.: A Multi-class Probabilistic Priority Scheduling Discipline for Differentiated Services Networks. Computer Communications 25, 1487–1496 (2002)
47. Walraevens, J., Steyaert, B., Bruneel, H.: Performance Analysis of a Single-Server ATM Queue with a Priority Scheduling. Computers & Operations Research 30(12), 1807–1829 (2003)
48. Walraevens, J., Wittevrongel, S., Bruneel, H.: A Discrete-Time Priority Queue with Train Arrivals. Stochastic Models 23(3), 489–512 (2007)
49. Wittevrongel, S., Bruneel, H.: Discrete-time ATM Queues with Independent and Correlated Arrival Streams. In: Performance Evaluation and Applications of ATM Networks, ch. 16, pp. 387–412. Kluwer Academic Publishers, Boston (2000)
50. Wittevrongel, S., Bruneel, H.: Per-source Mean Cell Delay and Mean Buffer Contents in ATM Queues. Electronics Letters 33(6), 461–462 (1997)
51. Yashkov, S.F.: Processor Sharing Queues: Some Progress in Analysis. Queueing Systems 2, 1–17 (1987)
52. Zhang, L.: Virtual Clocks: a New Traffic Control Algorithm for Packet Switching Networks. In: Proceedings of the ACM Symposium on Communications Architectures & Protocols, SIGCOMM 1990, Philadelpia, PA, USA, September 24-27, pp. 19–29 (1990)

IP Networking and Future Evolution

Zhili Sun

Centre for Communication Systems Research (CCSR), University of Surrey,
Guildford, Surrey, GU2 7XH, U.K.
Z.Sun@surrey.ac.uk

Abstract. Over the recent years, IP networking has been developing in a large range of telecommunication areas such as network technologies, protocols, new services and applications. This chapter presents an overview of the fundamental concepts of IP networking and architectures, future evolution of the Internet and its protocols, applications and services, current technologies and integration implications and convergence of IP and other multi-service networks. Particular attention is devoted towards the exposition of research issues of current and future Internets, including some important mechanisms to control and manage diversity and network resources to enhance network performance and Quality-of-Service (QoS).

Keywords: Internet Protocol (IP), IP networking, Quality-of-Service (QoS), network protocol, multicast, security, IP Version 6 (IPv6), future Internet.

1 Introduction

In the early years, the Internet was developed and used mainly by universities, research institute, industry, military and government. The main network technologies were campus networks with dial-up terminals and server networks interconnected by backbone networks. The main applications were email, file transfer, news groups and remote login. The explosion of interest in Internet started in mid-1990s, when the World Wide Web (WWW) provided a new application with interface to ordinary users without knowing anything about the Internet technology and thinking that the Web was the Internet. The impact was far beyond people's imagination and the use of the Internet started reach our daily lives, such as information access, communications, entertainment, e-commerce and e-government. New applications and services have been developed based on web technology every day.

In the meantime, the technologies and industries started to comprehend that computer, telephony; broadcast, mobile and fixed networks cannot longer be separated from each other. Due to the popularity and new applications and services, the original design of the Internet cannot meet the increasing new demands and requirements that the Internet Engineering Task Force (IETF) started to work on the next generation of networks called Internet Protocol (IP) version 6 (IPv6). The IPv6 is a result of the development of the next generation of Internet networks. The 3rd generation mobile networks, Universal Mobile Telecommunications Systems (UMTS), have also planned to have an all-IP protocol for the mobile communication networks. Of course, this paper will not be able to address all these issues, but will be able to provide an overview of next and future generation Internets including IPv6 from view points of protocol, performance, Traffic Engineering (TE) and Quality-of-Service (QoS) support for future Internet applications and services. Development and evolution of next generation Internet continue since the advent of IPv6, but the fundamental issues of the future Internet are still the same that these issues are covered herewith.

The tutorial paper is organised as follows: The basic concepts of IP networking are introduced in Section 1. The IP at the Network Layer is described in Section 3. The Transmission Control Protocol (TCP) and User Datagram Protocol (UDP) are presented at the Transport Layer in Section 4. The next generation Internet is explored in Section 5. Future directions and concluding remarks are included in Sections 6 and 7, respectively. Finally, a list of acronyms employed throughout this chapter can be seen in Appendix 1.

2 Basic Concepts of IP Networking

IP networking is an outcome of evolution of computer and data networks. Initially, there are many technologies available to support different data services and applications using different methods for different types of networks. The network technologies include Local Area Network (LAN), Metropolitan Area Network (MAN) and Wide Area Networks (WAN) using star, bus ring, tree and mesh topologies and different media access control mechanisms.

LAN is used to connect computers together in a room, building or campus. Sometimes bridges are used to extend the reach of the LAN. MAN is a high-speed network to connect computers together in a city. WAN, the same as Internet is used in a country, continent and worldwide. Different types of networks may use different protocols and interconnecting them together to form a larger network. There were many solutions to the challenge in the past, but IP networking is the solution most widely used today.

2.1 Protocol Hierarchies

Protocol hierarchy is an important concept to deal with the complexity in the network design. It consists of a stack of different layers or level. Each layer consists of entities or peers communicating with each other using a protocol define for that layer. Each layer also offers services to the layer above through its interface, and makes use of the services provided by the protocol below though the lower layer interfaces. The following is an explanation of the terms [1-3]:

- Layer or level: each offers certain services to the higher layers and shields details of how the services are actually implemented.
- Protocol: it provides the rules and conventions used for communications between the communicating parties.
- Peer: it is an entity of corresponding layer on different machine.
- Interface: It is an access point between the layers. A peer can make use of services provided from a layer below or provide services to the layer above through the interface.

- Architecture: it is a set of layers and protocols.
- Protocol stack: it consists of list of protocols, one protocol per layer.

2.2 Network Connection-Oriented and Connectionless Services

There are two types of connections in networks: connection-oriented and connectionless services. In connection-oriented service, a connection is set up before communications to reserve resources from source to destination; the connection will be granted by the network when there are enough resource is available to support the required QoS for both the new connection and the ongoing connections, then data can be exchanged between the communication parties; other wise the connection set-up will fail hence there is no data exchange; the connection is closed down after using. Telephone network is a typical example of connection-oriented service.

In connectionless service, there is no need to set up a connection. Each packet carries the destination address of the data. Each network node or router makes use of the routing protocols forwards the packet to the next router until the packet reaches its destination. A typical example of connectionless service is the postal service. IP at network layer also provides a connectionless service.

2.3 The OSI Reference Model

The Open System Interconnection (OSI) reference model is a proposed standard developed by the International Standard Organisation (ISO). It deals with connecting open systems, i.e., systems are open for communication with other systems. The OSI reference consists of seven layers [1]:

- Application layer
- Presentation layer
- Session layer
- Transport layer
- Network layer
- Data link layer
- Physical layer

The following list the principles that were applied to arrive at the seven layers model:

- A layer is a different level of abstraction;
- Each layer performs a well defined function;
- The function of each layer should be chosen to lead to internationally standardised protocols;
- The layer boundaries should be chosen to minimise information flow across the interface;
- The number of layers should be large enough but not too large;

It is a complete reference model has been discussed in almost every book on computer and data networks [1-3]. Every network protocol, such as IEEE 802.3 LAN, IEEE 802.11 LAN, Asynchronous Transfer Mode (ATM) or Frame Relay, tries to use it as reference, for comparison or for explanation the functions of each layer of the protocol. But the OSI model is not much used today, but IP is everywhere.

2.4 The IP Reference Model

Fig. 1 illustrates the early Internet reference model. It is can be seen that the main applications in the early development of the Internet are telecommunications network (telnet), File Transfer Protocol (FTP), email and Domain Name System (DNS) at application layer of the OSI model. The corner stones of Internet protocol are the TCP and UDP at the transport and IP at network layers, respectively [4].

Application	Telnet	FTP	Email	DNS
Transport	TCP		UDP	
Network	IP			
Link & Physical	ARPANET	SATNET	Packet Radio	LAN

Fig. 1. The early Internet reference model

The actual networks can be LAN, radio, satellite or any other type of networks. It can be seen that the Internet protocol reference model defines the protocol, and doesn't define the type network or its technology, but to support IP the network has to provide physical layer for real data transmission and data link layer to carry IP packets.

3 The IP at Network Layer

The Internet is also referred to as a network of networks [4]. It defines the format of the Internet packets and functions of the routers to deliver the packets from source to destination. For simplicity purpose, the IEEE data transmission standard 802.3 for LANs (known as Ethernet) is used as an example to explain the Internet concept.

In an Ethernet, each computer can be uniquely identified at physical and link layer by its Ethernet address. It can send data to another computer or other computers in the Ethernet by broadcasting using the other hosts' addresses or Ethernet broadcasting address. Each host can have a unique IP address of IP network identifier (net-id) and IP hot identifier (host-id) in the Internet.

All the computers in the Ethernet have the same IP net-id forming a sub-network (subnet). The subnet can be interconnected together by using routers to become a part of a larger subnet. Then eventually all the subnets together forms the Internet. The routers attached to the same subnet (autonomous system) can exchange information using routing protocols to figure out the topology of the Internet and calculate the best router which to forward the packet to reach its destination.

Clearly, the host can send a packet to the other host within the same subnet. If the other host is outside of the network, the host can send the packet to a router attached to the subnet. The router can forward the packet to the next one until the packets reach their destinations.

Therefore, the fundamental components of the Internet should include IP address, address resolution between the IP address and link/physical address such as Ethernet and routing protocol.

3.1 IP Packet Format

In the Internet reference model, there is only one network layer protocol that is associated with IP. It is a unique protocol making use of the transmission services provided by the different types of networks below, and providing end-to-end network layer service to the transport layer protocols above.

The IP packets may be carried across different type of networks, but their IP format stays the same. Therefore, any protocol above IP layer can only see the IP packet that the differences of the networks are screened out by the IP layer.

Fig. 2 shows the format of the IP packet and Fig. 3 the options. The following is a brief discussion of each field of the IP packet header [4].

Version	IHL	Type of services	Total length			
Identification			-	DF	MF	Fragment offset
Time to Live		Protocol	Header checksum			
Source address						
Destination address						
Options						
Payload						

Fig. 2. The IP packet header

- The version field keep track of which version of the protocol the datagram belong to. The current version is 4, also called IPv4 [4]. IPv5 is an experimental version. The next version to be introduced into the Internet is IPv6 [5].
- The IP header length (IHL) field is the length of the header in 32-bit words. The minimum value is 5 and maximum 15, which limits the header to 60 bytes.
- The type of service field allows the host to tell the network what kind of service it wants. Various combinations of Delay, Throughput and reliability are possible.
- The total length includes both header and data. The maximum value is 65535.
- The identification field is needed to allow destination host to determine which datagram a newly arrived fragment belong to. Every IP packet in the network is identified uniquely.
- DF: Don't Fragment. This is to tell the network not to fragment the packet, as a receiving party may not be able to reassemble the packet.
- MF: More Fragments. This is to indicate that more fragments to come as a part of the IP packet.
- The fragment offset tells where in the current datagram this fragment belongs.
- The time to live is a counter used to limit packet lifetime to prevent the packet staying in the network forever.

- The protocol field tells it which transport process to give it to. It can be TCP or UDP. It can also be possible to carry data of other transport layer protocols.
- The checksum field verifiers the IP heard only.
- The source and destination addresses indicate the network number and host number.
- Options are variable length. Five functions are defined: security, strict routing, loose source routing, record route and timestamp.

Option	Description
Security	Describe how secret the IP packet is
Strict source routing	Give a complete path to be followed
Loose source routing	Gives a list of routers to be visited by the packet
Record route	Make each router append its IP address
Timestamp	Make each router append its IP address & timestamp

Fig. 3. Option fields of the IPv4 packet header

3.2 IP Address

The IP address used in the source and destination address fields of the IP packet is 32 bits long. It can have up to three parts. The first part identifies the class of the network address from A to E, the second part is the net-id and the third part is the host-id.

In class A and B addresses, there are a large number of host-ids. The hosts can be grouped into subnets and each subnet is identified by using some of the high order host-id bits. A subnet mask is introduced to indicate the split between net-id + sub-id and host-id.

Similarly, there is a large number of net-id in the class C addresses. Some of the lower order bits of the net-id can be grouped together to form a super-net. This is also called Classless Inter-Domain Routing (CIDR) addressing. Routers do not need to know anything within the super-net or the domain.

Class A and B addresses identify the attachment point of the hosts. Class D addresses identify the multicast address (like radio channel) but not an attachment point in the network. Class E is reserved for future use.

3.3 Address Resolution Type Protocols

The Address Resolution protocol (ARP) is a protocol used to find the mapping between the IP address and network address such as Ethernet address. Within the network, a host can ask for the network address giving an IP address to get the mapping. If the IP address is outside the network, the host will forward the IP address to a router (it can be a default or proxy).

The reverse ARP (RARP) is the protocol solved the reverse problem, i.e., to find the IP address giving a network address such as Ethernet. This is normally resolved by introducing a RARP server. The server keeps a table of the address mapping. An example of using RARP is when a booting machine has not got an IP address and should contact a server to get an IP address to be attached to the Internet.

3.4 IP Routing Protocols

As the Internet becomes larger and larger, it is impractical or impossible to configure the routing table manually though in the early days and small network manual configuration of network was carried out for convenience but error prone. Protocols have to be developed to configure the Internet automatically and dynamically.

A part of the Internet owned and managed by a single organisation or by a common policy can form a domain or autonomous system (AS). The interior gateway routing protocol is used for IP routing within the domain. Between domains, the exterior gateway routing protocol has to be used as political, economical or security issues often need to be taken into account.

3.4.1 The Interior Gateway Routing Protocol

The original routing IP (RIP) was distance vector [6, 7]. Within the domain, each router has a routing table of the next router leading the destination network. The router periodically exchanges its routing information with its neighbour routers and updates its routing table based on the new information received.

Due to its slow convergence problem, a new routing protocol is introduced in 1979, called link state protocols. Instead of getting routing information from its neighbour, the link state protocol collects and exchanges information on their links. Every router in the network will have the same set of link state information and calculate independently the routing table. This solves the problems of the RIP.

In 1988, the IETF began work on a new interior gateway routing (IGR) protocol, called Open Shortest Path First (OSPF). The OSPF became a standard in 1990. It is based on the link state protocol. It is also based on published in Open literatures, and designed to supports a variety of distance metrics, adaptive to changes in topology automatically and quickly, support routeing based on type of services, and real time traffic, support load balancing, support for hierarchical systems and some levels of security, and also deals with routes connected to the internet via a tunnel.

The OSPF supports three kinds of connections and networks, namely Point-to-point lines between exactly two routers, multicast networks (such as LANs) and multi-access networks without broadcasting (such as WANs).

When booting, a router to contact its neighbour routers by sending contact message for information. Adjacent routers (designated routers in the each LANs) exchange information. Each router periodically floods link state information to each of its adjacent routers. Database description messages include the sequence numbers of all the link state entries, sent at IP packets. Using flooding, each router informs all the other neighbour routers. This allows each router to construct the graph for its domain and compute the shortest path to form a routing table.

3.4.2 The Exterior Gateway Routing Protocol

An IGR protocol moves packets as efficiently as possible but this is not the case with an exterior gateway protocol (EGP) protocol [8], as EGP needs to deal with external subnets, which may have different ownerships, policies and network technologies. The Border Gateway Protocol (BGP) [9] is created to replace EGP, and is fundamentally a distance vector protocol, but quite different from most others, such as RIP. Each BGP router keeps track of the exact path used. This also solves the problem of RIP. BGP works similarly to RIP but overcomes the weakness of RIP by tracking the path.

4 TCP and UDP Protocols at Transport Layer

The transport layer protocols appear only on the hosts. When an IP packet arrives in a host, it decides which application process to handle the data, as the IP packet can carry data for email, telnet, FTP or WWW. It may also add additional functions on this layer such as reliability, flow control and congestion control. There are two protocols at the transport layer, namely TCP [10] and UDC [11], defined in the Internet reference model.

4.1 The TCP at Transport Layer

The TCP is a connection-oriented, end-to-end reliable protocol [10]. It provides for reliable inter-process communication between pairs of processes in host computers. Very few assumptions are made as to the reliability of the network layer protocols below the TCP layer. TCP assumes it can obtain a simple, potentially unreliable datagram service from the lower level protocols (such as IP). In principle, the TCP should be able to operate above a wide spectrum of communication systems ranging from hard-wired connections to packet-switched or circuit-switched networks.

4.1.1 The TCP Segment Header Format
Fig. 4 illustrates the TCP segment header. It includes the following fields [10]:

- Source Port: 16 bits - The source port number.
- Destination Port: 16 bits - The destination port number.
- Sequence Number: 32 bits - The sequence number of the first data octet in this segment (except when the synchronised (SYN) control bit is present). If SYN is present the sequence number is the initial sequence number (ISN) and the first data octet is ISN+1.
- Acknowledgment Number: 32 bits - If the Acknowledgment (ACK) control bit is set this field contains the value of the next sequence number the sender of the segment is expecting to receive. Once a connection is established this is always sent.
- Data Offset: 4 bits - The number of 32 bit words in the TCP Header. This indicates where the data begins. The TCP header (even one including options) is an integral number of 32 bits long.
- Reserved: 6 bits - Reserved for future use (must be zero).
- Control Bits: 6 bits (from left to right):
    - URG: Urgent Pointer field significant
    - ACK: Acknowledgment field significant
    - PSH: Push Function
    - RST: Reset the connection
    - SYN: Synchronize sequence numbers
    - FIN: No more data from sender

- Window: 16 bits - The number of data octets beginning with the one indicated in the acknowledgment field which the sender of this segment is willing to accept.
- Checksum: 16 bits - The checksum field is the 16 bit one's complement of the one's complement sum of all 16 bit words in the header and text. If a segment contains an odd number of header and text octets to be check summed, the last octet is padded on the right with zeros to form a 16 bit word for checksum purposes. The pad is not transmitted as part of the segment. While computing the checksum, the checksum field itself is replaced with zeros.
- URG Pointer: 16 bits - This field communicates the current value of the urgent pointer as a positive offset from the sequence number in this segment.
- Options: variable.
- Padding: variable.

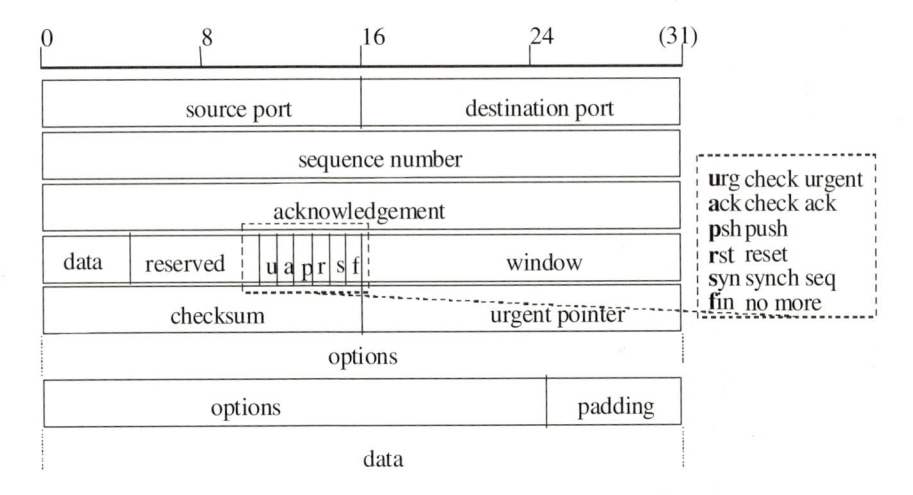

Fig. 4. The TCP segment header

To identify the separate data streams that a TCP may handle, the TCP provides a port identifier. Since port identifiers are selected independently by each TCP they might not be unique. To provide for unique addresses within each TCP, we concatenate an internet address identifying the TCP with a port identifier to create a socket which will be unique throughout all networks connected together.

A connection is fully specified by the pair of sockets at the ends. A local socket may participate in many connections to different foreign sockets. A connection can be used to carry data in both directions, that is, it is "full duplex".

TCPs are free to associate ports with processes however they choose. However, several basic concepts are necessary in any implementation.

Well-known sockets are a convenient mechanism for a priori associating a socket address with a standard service. For instance, the "Telnet-Server" process is permanently assigned to a socket number of 23, FTP-data 20 and FTP-control 21, Trivial

FTP (TFTP) 69, Simple Mail Transfer Protocol (SMTP) 25, Post Office Protocol 3 (POP3) 110 and www-http 80.

4.1.2 Connection Set-Up and Data Transmission

A connection is specified in the OPEN call by the local port and foreign socket arguments. In return, the TCP supplies a (short) local connection name by which the user refers to the connection in subsequent calls. There are several things that must be remembered about a connection. To store this information we imagine that there is a data structure called a Transmission Control Block (TCB). One implementation strategy would have the local connection name be a pointer to the TCB for this connection. The OPEN call also specifies whether the connection establishment is to be actively pursued, or to be passively waited for [10].

The procedures of establishing connections utilize the SYN control bit (flag) and involve an exchange of three messages. This exchange has been termed a three-way hand shake. The connection becomes "established" when sequence numbers have been synchronized in both directions. The clearing of a connection also involves the exchange of segments, in this case carrying the FIN control flag.

The data that flows on a connection may be thought of as a stream of octets. The sending user indicates in each SEND call whether the data in that call (and any preceding calls) should be immediately pushed through to the receiving user by the setting of the PSH flag.

A sending TCP is allowed to collect data from the sending user and to send that data in segments at its own convenience, until the push function is signalled, then it must send all unsent data. When a receiving TCP sees the PSH flag, it must not wait for more data from the sending TCP before passing the data to the receiving process.

There is no necessary relationship between push functions and segment boundaries. The data in any particular segment may be the result of a single SEND call, in whole or part, or of multiple SEND calls.

4.1.3 Congestion Control and Flow Control

One of TCP's functions is end-host based congestion control for the Internet. This is a critical part of the overall stability of the Internet [12]. In the congestion control algorithms, TCP assumes that, at the most abstract level, the network consists of links and queues. Queues provide output-buffering on links that are momentarily oversubscribed. They smooth instantaneous traffic bursts to fit the link bandwidth.

When demand exceeds link capacity long enough to fill the queue, packets must be dropped. The traditional action of dropping the most recent packet ("tail dropping") is no longer recommended, but it is still widely practiced.

TCP uses sequence numbering and ACKs on an end-to-end basis to provide reliable, sequenced, once-only delivery. TCP ACKs are cumulative, i.e., each one implicitly ACKs every segment received so far. If a packet is lost, the cumulative ACK will cease to advance.

Since the most common cause of packet loss is congestion, TCP treats packet loss as a network congestion indicator (but such assumption is not applicable in wireless or satellite networks where packet loss is more likely caused by transmission errors). This happens automatically, and the subnetwork need not know anything about IP or

TCP. It simply drops packets whenever it must, though some packet-dropping strategies are fairer than others.

TCP recovers from packet losses in two different ways. The most important is by a retransmission timeout. If an ACK fails to arrive after a certain period of time, TCP retransmits the oldest unacknowledged packet. Taking this as a hint that the network is congested, TCP waits for the retransmission to be acknowledged before it continues, and it gradually increases the number of packets in flight as long as a timeout does not occur again (slow start).

A retransmission timeout can impose a significant performance penalty, as the sender will be idle during the timeout interval and restarts with a congestion window of 1 following the timeout. To allow faster recovery from the occasional lost packet in a bulk transfer, an alternate scheme known as "fast recovery" was introduced.

Fast recovery relies on the fact that when a single packet is lost in a bulk transfer, the receiver continues to return ACKs to subsequent data packets, but they will not actually ACK any data. These are known as "duplicate acknowledgments" or "dupacks". The sending TCP can use dupacks as a hint that a packet has been lost, and it can retransmit it without waiting for a timeout. Dupacks effectively constitute a negative acknowledgement (NAK) for the packet whose sequence number is equal to the acknowledgement field in the incoming TCP packet. TCP currently waits until a certain number of dupacks (currently 3) are seen prior to assuming a loss has occurred; this helps avoid an unnecessary retransmission in the face of out-of-sequence delivery.

The TCP congestion control algorithm is the end-system congestion control algorithm used by TCP [12]. This algorithm maintains a congestion window (cwnd), which controls the amount of data which TCP may have in flight at any given point in time. Reducing cwnd reduces the overall bandwidth obtained by the connection; similarly, raising cwnd increases the performance, up to the limit of the available bandwidth.

TCP probes for available network bandwidth by setting cwnd at one packet and then increasing it by one packet for each ACK returned from the receiver. This is TCP's "slow start" mechanism. When a packet loss is detected (or congestion is signalled by other mechanisms), cwnd is set back to one and the slow start process is repeated until cwnd reaches one half of its previous setting before the loss. Cwnd continues to increase past this point, but at a much slower rate than before. If no further losses occur, cwnd will ultimately reach the window size advertised by the receiver.

4.2 The User Datagram Protocol (UDP)

This UDP is defined to make available a datagram mode of packet-switched computer communication in the environment of an interconnected set of computer networks [11]. This protocol assumes that the IP is used as the underlying protocol.

This protocol provides a procedure for application programs to send messages to other programs with a minimum of protocol mechanism. The protocol is connectionless and does not provide any guarantee on delivery and duplicate protection. Applications requiring ordered reliable delivery of streams of data should use the TCP.

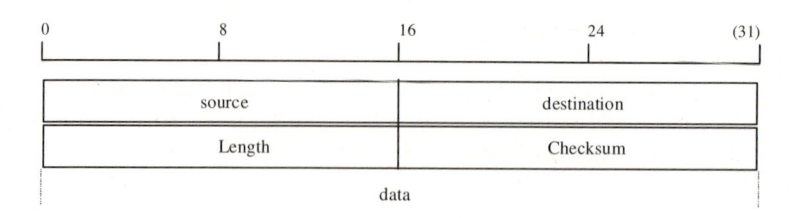

```
0           8              16              24           (31)
|_____|_____|_____|_____|
|        source             |        destination         |
|        Length             |        Checksum            |
|                        data                            |
```

Fig. 5. The UDP datagram header format

Fig. 5 illustrates the UDP datagram header format. The fields of the UDP datagram header include:

- Source Port is an optional field, when meaningful, it indicates the port of the sending process, and may be assumed to be the port to which a reply should be addressed in the absence of any other information. If not used, a value of zero is inserted.
- Destination Port has a meaning within the context of a particular internet destination address.
- Length is the length in octets of this user datagram including this header and the data. (This means the minimum value of the length is eight).
- Checksum is the 16-bit one's complement of the one's complement sum of a pseudo header of information from the IP header, the UDP header, and
- The data, padded with zero octets at the end (if necessary) to make a multiple of two octets.

The major uses of this protocol are the Internet Name Server and the Trivial File Transfer, and recently for real time applications such as Voice over IP (VoIP) where retransmission of lost data is undesirable. The well-know ports are defined in the same way as the TCP.

5 Next Generation Internet

It can be seen from the IP networking basics that the Internet was originally designed to transfer data between computers. The applications and services including email, ftp and telnet don't require real time QoS which is required by telephony and provided by telephony networks for a century. More and more people own portable computers and mobility becomes a new requirement to give people more freedom of movement taking their computer with them. More and more business transactions, commercial and public uses of the Internet make the security of a very important issue of the Internet. All these new requirements and the large scale of the Internet today were not taken into consideration of the original design of the Internet. In addition, there was also concern that the IPv4 may soon run out of IP address which trig the development of the next generation of Internet Protocol. IPv6 is a result of the new development. Though the IPv6 has started to address these issues, there is still a long way from a perfect solution to the problems. The following is a brief discussion of these issues before discussing the future directions in the next section.

5.1 QoS and Network Performance

Two aspects need to be taking into consideration: user application aspect describes what level of QoS is considered to be acceptable and network aspect provides mechanisms to meet the QoS requirement of the user and makes efficient use of network resources [13, 14].

5.1.1 Best Effort Service and Elastic Applications

In the future, networks will carry at least two types of applications. Some applications (which are already common in the Internet) are relatively insensitive to the performance they receive from the network. For example, a file transfer application would prefer to have an infinite bandwidth and zero end-to-end delay. On the other hand, it works correctly, through with degraded performance, as the available bandwidth decreases and the end-to-end delay increases. In other words, the performance requirements of such applications are elastic, they can adapt to the resources available. Such applications are called best-effort applications, because the network promises them only to attempt to deliver their packets, without guaranteeing them any particular performance bound. Note that best effort service, which is the service provided for best effort applications, does not require the network to reserve resources for a connection.

5.1.2 Guaranteed Service and Inelastic Applications

Besides best effort application, we expect future networks to carry traffic from applications that do require a bound on performance. For example an application that carries voice as a 64 Kbit/s stream become nearly unusable if the network provide less that 64Kbit/s on end-to-end path. More over, if the application is two-way and interactive, human ergonomic constraints require the round-trip delay to be smaller than around 150 ms. If the network want to support a perceptually "good" two-way voice application, it must guarantee, besides a bandwidth of 64 Kbit/s, a round trip delay of less than 150 ms. The performance requirements of such applications are inelastic. Thus the application, and other application of its kind, demands a guarantee of service quality from the network. We call these applications guaranteed-service applications. Guaranteed service applications require the network to reserve resources on their behalf.

5.1.3 QoS Provision and Network Performance

Best effort service is the default service a network would give to a datagram between the source and destination in today's Internet networks. Among other implications, this means that if a datagram is changed to a best effort datagram, all flow control that is normally applied to best effort datagrams is applied to that datagram too.

The *controlled load service* is intended to support a broad class of applications which have been developed for use in today's Internet, but are highly sensitive to overloaded conditions. Important members of this class are the "adaptive real-time applications" currently offered by a number of vendors and researchers. These applications have been shown to work well on unloaded nets, but to degrade quickly under overloaded conditions. A service which mimics unloaded nets serves these applications well.

Guaranteed service guarantees that datagrams will arrive within the guaranteed delivery time and will not be discarded due to queue overflows, provided the flow's traffic stays within its specified traffic parameters. This service is intended for applications which need a firm guarantee that a datagram will arrive no later than a certain time after it was transmitted by its source. For example, some audio and video "playback" applications are intolerant of any datagram arriving after their play-back time. Applications that have hard real-time requirements will also require guaranteed service.

In the playback applications, datagrams often arrive far earlier than the delivery deadline and will have to be buffered at the receiving system until it is time for the application to process them.

5.2 IP Mobility

Mobile IP [15] is proposed as a standard by a working group within the IETF designed to solve this problem by allowing the mobile node to use two IP addresses: a fixed home address and a care-of address that changes at each new point of attachment.

To maintain existing transport-layer connections as the mobile node moves from one place to another, but keeping its IP address the same. Most of the Internet applications today are based on the TCP. TCP connections are indexed by a quadruplet of the IP addresses and port numbers of sources and destination. Changing any of these four numbers will cause the connection to be disrupted and lost. On the other hand, correct delivery of packets to the mobile node's current point of attachment depends on the network number contained within the mobile node's IP address, which changes at new points of attachment. To change the routing requires a new IP address associated with the new point of attachment.

In Mobile IP, the home address is static and is used, for instance, to identify TCP connections. The care-of-address changes at each new point of attachment and can be thought of as the mobile node's topologically significant address; it indicates the network number and thus identifies the mobile node's point of attachment with respect to the network topology. The home address makes it appear that the mobile node is continually able to receive data on its home network, where Mobile IP requires the existence of a network node known as the home agent. Whenever the mobile node is not attached to its home network (and is therefore attached to what is termed a foreign network), the home agent gets all the packets destined for the mobile node and arranges to deliver them to the mobile node's current point of attachment.

When the mobile node moves to a new place, it registers its new care-of address with its home agent. To get a packet to a mobile node from its home network, the home agent delivers the packet from the home network to the care-of address. The further delivery requires that the packet be modified so that the care-of address appears as the destination IP address. This modification can be understood as a packet transformation or a redirection. When the packet arrives at the care-of address, the reverse transformation is applied so that the packet once again appears to have the mobile node's home address as the destination IP address.

When the packet arrives at the mobile node, addressed to the home address, it will be processed properly by TCP.

In Mobile IP the home agent redirects packets from the home network to the care-of address by constructing a new IP header that contains the mobile node's care-of address as the destination IP address. This new header then shields or encapsulates the original packet, causing the mobile node's home address to have no effect on the encapsulated packet's routing until it arrives at the care-of address. Such encapsulation is also called tunnelling bypassing the usual effects of IP routing.

Mobile IP, then, is best understood as the cooperation of three separable mechanisms:

- Discovering the care-of address: agent advertisement and agent solicitation.
- Registering the care-of address: the registration process begins when the mobile node, possibly with the assistance of a foreign agent, sends a registration request with the care-of address information. When the home agent receives this request, it (typically) adds the necessary information to its routing table, approves the request, and sends a registration reply back to the mobile node. The registration is authenticated by using Message Digest 5.
- Tunnelling to the care-of address: the default encapsulation mechanism that must be supported by all mobility agents using Mobile IP is IP-within-IP. Minimal encapsulation is slightly more complicated than that for IP-within-IP, because some of the information from the tunnel header is combined with the information in the inner minimal encapsulation header to reconstitute the original IP header. On the other hand, header overhead is reduced.

5.3 IP Security

Internet security is a very important problem but also a very difficult one, as the Internet covers the world across political and organisational boundaries [16, 17].

It also involves how and when communicating parties (such as users, computer, services and network) can trust each another, as well as understanding the network hardware and protocols.

The basic mechanics in the Internet at network layer used to provide security including authentication using public key systems, privacy using public and secret key systems and access control using firewalls, and at transport layer include Secure Socket Layer (SSL). A brief discussion of the topics is presented below.

The IETF has provided security standards for the Internet known as IP Security (IPSec). The IPSec protocol suite is used to provide interoperable cryptographically based security services (i.e. confidentiality, authentication and integrity) at the IP layer. It is composed of the Authentication Header (AH) protocol, the Encapsulated Security Payload (ESP) confidentiality protocol and it also includes an Internet Security Association Establishment and Key Management Protocol (ISAKMP).

IP AH and IP ESP may be applied alone or in combination with each other. Each protocol can operate in one of two modes: Transport mode or tunnel mode. In transport mode, the security mechanisms of the protocol are applied only to the upper layer data and the information pertaining to IP layer operation as contained in the IP header is left unprotected. In tunnel mode, both the upper layer protocol data and the IP header of the IP packet are protected or 'tunnelled' through encapsulation. The transport mode is intended for end-to-end protection that can be implemented only by the source and destination hosts of the original IP datagram. Tunnel mode can be used between firewalls.

Filters can also be set up in the firewalls to block some IP packets from entering the network based on the IP addresses and port numbers.

A firewall consists of two routers doing IP packet filtering and an application gateway for higher layer checking. The inside one checks outgoing packets. The outside one checks incoming packets. An application gateway, between the routers, carries out further examination of higher layer protocol data including TCP, UDP, Email, WWW and other application data. This configuration is to make sure that no packets get in or out without having to pass through the application gateway. Packet filters are table driven and check the raw packets. The application gateway checks contents, message sizes, and headers.

SSL builds a secure connection between two sockets including:

- Parameter negotiation between client and server.
- Mutual authentication of client and server.
- Secure communication.
- Data integrity protection.

5.4 IP Multicast

We review the IP multicast technology [18]. Multicast allows a communications network source to send data to multiple destinations simultaneously whilst transmitting only a single copy of the data on to the network. The network then replicates the data and fans it out to recipients as necessary. Multicast can be considered as part of a spectrum of three types of communications:

- Unicast: transmitting data from a single source to a single destination (for example, downloading a web page from a server to a user's browser; or copying a file from one server to another);
- Multicast: transmitting data from a single source to multiple destinations. The definition also encompasses communications where there may be more than one source (i.e. multipoint-to-multipoint). Videoconferences provide an example of this latter, where each participant can be regarded as a single source multicasting to the other participants in the videoconference.
- Broadcast: transmitting data from a single source to all receivers within a domain (for example within a LAN; or from a satellite to all receivers within a satellite spotbeam).

The advantages of multicast are as follows:

- Reduced network bandwidth usage: for example, if data packets are being multicast to 100 recipients the source only sends a single copy of each packet. The network forwards this to the destinations, only making multiple copies of the packet when it needs to send packets on different network links to reach all destinations. Thus only a single copy of each packet is transmitted over any link in the network, and the total network load is reduced compared to 100 separate unicast connections. This is particularly beneficial on satellite systems where resources are limited and expensive.
- Reduced source processing load: the source host does not need to maintain state information about the communications link to each individual recipient.

Multicast can be either best effort or reliable. "Best effort" means that there is no guarantee that the data sent by any multicast source is received by all or any receivers, and is usually implemented by a source transmitting UDP packets on a multicast address (the addressing mechanism is described in further detail below). "Reliable" means that mechanisms are implemented to ensure that all receivers of a multicast transmission receive all the data that is sent by a source: this requires a reliable multicast protocol.

5.4.1 IP Multicast Addressing

Each terminal or host in the Internet is uniquely identified by its IP address. In IP Version 4, this consists of a 32-bit address space, divided into a network number and a host number which respectively identify a network and the terminal attached to the network. A normal unicast IP datagram includes a source address and destination address in the IP packet header; routers use the destination address to route the packet from the source to the destination. Such a mechanism cannot be used for multicast purpose, since the source terminal may not know when, where and which terminals will try to receive the packet [18].

From the IP networking basics we know that a range of addresses is defined for multicast purposes only. The range of addresses, called Class D addresses, is from 224.0.0.0 to 239.255.255.255. Unlike Classes A, B and C, these addresses are not associated with any physical network number or host number, but instead are associated with a multicast group that is like a radio channel; members of the group receive multicast packets sent to this address, and the address is used by multicast routers to route IP multicast packets to users that register for a multicast group. The mechanism by which a terminal registers for a group, Internet Group Membership Protocol (IGMP) [19], is described below.

5.4.2 Multicast Group Management

In order to make efficient use of network resources, the network sends multicast packets only to those networks and subnets that have users belonging to the multicast group. The IGMP [19] allows hosts or terminals to declare an interest in receiving a multicast transmission and supports three main types of messages: Report, Query and Leave.

A terminal wishing to receive a multicast transmission issues an IGMP join Report, which is received by the nearest router. This Report specifies the IP multicast class D address of the group being joined. The router then uses a multicast routing protocol (described below) to determine a path to the source. To confirm the state of terminals receiving multicast, a router occasionally issues an IGMP Query to terminals on its network/sub-network. When a terminal receives such a query, it sets a separate timer for each of its (potentially many) group memberships. When each timer expires, the terminal issues an IGMP Report to confirm that it still wishes to receive the multicast transmission. However, in order to suppress duplicate reports for the same Class D group address, if a terminal has already heard a report for that group from another terminal it stops its timer and does not send a Report. This has the benefit of avoiding flooding the subnetwork with IGMP Reports.

When a terminal wishes to finish receiving the multicast transmission it issues an IGMP Leave request. If all the members of a group in a subnet have left, the router does not forward any more multicast packets to that subnet.

5.4.3 IP Multicast Routing

In a normal IP router used for unicast, the routing table contains information that specifies paths that lead to a given IP destination addresses. However, this routing table is not useful for IP multicast since multicast packets do not contain information about the location of the packet's destinations. Therefore different routing protocols and routing tables have to be used. Multicast routing protocols address the issue of identifying a route for data to be transmitted across a network from a source to all its destinations, while minimising the total network resources required for this.

In IP multicast, the routing table is effectively built from destinations to the sources rather than from sources to destinations, since only the source address in the IP datagram corresponds to a single physical location. Tunnelling techniques may also be used to support multicast over routers that do not have multicast capabilities.

A number of multicast routing protocols have been developed by the IETF. These include Multicast Extensions to OSPF (M-OSPF) [20], Distance Vector Multicast Routing Protocol (DVMRP) [21], Protocol-Independent Multicast - Sparse Mode (PIM-SM) [22] and PIM-Dense Mode (PIM-DM) [23] and Core-Based Tree (CBT) [24].

Here we briefly review the underlying principle of operation of two protocols. DVMRP and PIM-DM are "flood and prune" algorithms: in these protocols, when a source starts sending data, the protocols flood the network with the data. All routers that have no multicast recipients attached send a prune message back towards the source (they know they have no receivers because they have received no IGMP join Reports). These protocols have the disadvantage that a "prune" state is required in all routers (i.e. "I have pruned on this multicast address"), including those routers with no multicast recipients downstream.

Flood and prune protocols use Reverse Path Forwarding (RPF) to forward multicast packets from a source to the recipients: the RPF interface for any packet is the interface that the router would use to send unicast packets to the packet source. If a packet arrives on the RPF interface it is flooded to all other interfaces (unless they have been pruned), but if the packet arrives on any other interface it is silently discarded. This ensures efficient flooding and prevents packet looping.

DVMRP uses its own routing table to compute the best path to the source, whereas PIM-DM uses an underlying unicast routing protocol.

5.4.4 IP Multicast Scope

Scoping is the mechanism that controls the geographical scale of a multicast transmission, by making use of the time to live (IP networking) (TTL) field in the IP header. It tells the network how far (in terms of router hops) any IP packet is allowed to propagate, allowing IP multicast sources to specify whether packets should be sent only to the local sub-network, or to larger domains or the whole Internet [18]. This is achieved by each router reducing the TTL by 1 when forwarding the packet to the next hop, and discarding the packet if the TTL is 0. Each subnet may additionally have filters or firewall to discard some packets according its security policy that may be beyond the control of the multicast source.

It can be seen in a satellite network that even with a small TTL value, IP multicast packets can reach a very large number of members of a multicast group scattered in a very large geographical area.

5.4.5 Address Mapping and Configuration

Different network technologies may use different addressing scheme for assigning addresses, also called physical addresses, to devices. For example, an IEEE 802 LAN uses a 48-bit address for each attached device, and Integrated Services Digital Network (ISDN) uses the International Telecommunication Union-Telecommunication (ITU-T) E.164 address scheme. Similarly, in a satellite network each ground earth station or gateway station has a physical address to be used for circuit connections or packet transmissions. However, the routers that are interconnected by the satellite network know only the IP addresses of the other routers. Therefore, address mapping between each IP address and its related physical address is required, so that packet exchanges between the routers can be carried out through the satellite network using the physical addresses. The precise details of this mapping depend on the underlying data link layer protocols used over the satellite.

5.5 IPv6

The IPv6 [5] packet header format can be seen in Fig. 6. This IPv6 is designed to address the drawbacks of IPv4, namely to

- support more host address.
- reduce the size of the routing table.
- simplify the protocol to allow routers to process packets fast.
- have better security (authentication and privacy).
- provide different service to different Type of service including real time data.
- aid multicasting (allow scopes).
- allow mobility (roam without changing address).
- facilitate the protocol's evolution.
- permit co-exists of old and new protocols.

Version	Priority	Flow label		
Payload length			Next header	Hop limit
Source address (16 bytes)				
Destination address (16 bytes)				
Payload				

Fig. 6. IPv6 packet header format

Comparing to the IPv4, it can be seen that a significant change has been made on the IP packet format to achieve the objectives of the IP network layer functions.

- The *version* field has the same function as the IPv4. It is 6 for IPv6 and 4 for IPv4.
- The *traffic class* field is used to distinguish between packets with different real time delivery requirement.

- The *flow label* field is used to allow source and destination to set up a pseudo-connection with particular properties and requirements
- The *payload* field is the number of bytes following the 40-byte header, instead of total length in IPv4.
- The *next header* field tells which transport handler to pass the packet to, like the protocol field in the IPv4.
- The *hop limit* field is a counter used to limit packet lifetime to prevent the packet staying in the network forever, like the TTL field in IPv4.
- The *source* and *destination addresses* indicate the network number and host number, 4 times larger that IPv4
- There are also *extension headers* like the *options* in IPv4. Table 1 shows the IPv6 extension header.

Each extension header consists of *next header* field, and fields of *type*, *length* and *value*.

Table 1. IPv6 Extension Headers

Extension Header	Description
Hop by hop options	Miscellaneous information for routers
Destination options	Additional information for destination
Routing	Loose list of routers to visit
Fragmentation	Management of datagram fragments
Authentication	Verification of the sender's identity
Encrypted security payload	Information about encrypted contents

6 Future Directions

New research and developments have been carried out in every aspects of the IP networking including user terminal, access networks, core networks, applications and services, and network technologies.

In user terminals, convergence has been taking place that mobile phone can be used to surf the web, to send email message, video and picture in addition to telephony service. In access networks, we have seen the development of wireless LAN (WLAN), x-Digital Subscriber Line (xDSL) and cable, optical fibre and other broadband access in local loops. In core network, the UMTS has adapted the IP core networks as a part of its standards. IP multicast has been exploited for broadcasting services whilst digital video broadcasting (DVB) is also trying to support future Internet services.

It is clear that the IPv6 has addressed some of the important issues on network protocols in the network layers, but there are still many issues beyond the scope of IPv6. We also see the convergence different levels and different sections of the networks.

In the future, the terminals will be different from what we have today, the access networks will be different, the core networks, new application will be developed and new technologies deployed. We will not be able to know what exactly what will be the terminals, access network, core networks, new application and new network technologies and how much traffic will be generated onto the networks, but will be able to develop new tools, new algorithms and new platform and new methodology to characterise the traffic generated by the application, plan and dimension the network, engineering the traffic to utilise the network resources efficiently, meet the QoS of individual application, and also for future development of network to meet the increased demand of interconnecting different types of networks.

It is always difficult and sometimes impossible to make a prediction of the future. Many tried but failed. I do not intend to make a prediction, but to observe the directions we are going. Everyone should be able to get a sense of direction by observe the direction we going. But the direction may change when time goes by.

Following the principles of the networking tradition, the direction of future could develop vertically in the areas terminal services and applications, access networks, core networks, network technologies and horizontally in the areas of end-to-end issues such as TE, QoS, IP security, network performance modelling and simulations.

6.1 IP Services and Applications

The open problems include what the future applications and services are and how to characterise them quantitatively in term of traffic flows. The typical traffic will generated by the following applications and services:

- Web services,
- IP telephone and IP Multimedia applications,
- Distribution services: video on demand, broadcasting, news papers will be distributed over IP networks,
- Middleware and Global Resource Information Database (GRID), Distributed, Pervasive and Ubiquitous Computing,
- Disconnected and nomadic computing, tiny and ubiquitous computers,
- Distributed storage management, high performance servers and networking, special-purpose and embedded systems,
- User behaviours and charge policy.

6.2 IP Access Network

There are many types of access networks available including Ethernet LAN, dial-up connection, broadband access and WLAN, Mobile and satellite networks. All these converge towards IP based access.

The migration of all services over IP, and the integration of IP and different technology, make the IP-based paradigm the common platform for both traditional data and new real-time and multimedia services that will be developed in both wireless and wire-line environments. It is envisaged that future networks (called "Beyond 3rd

Generation" systems) will be composed of an IPv6 backbone and multiple access network using different technologies (e.g. UMTS, Local Multipoint Distribution System (LMDS) and WLAN. This architecture has to offer advanced traffic control capabilities and QoS guarantees to the end users. Providing quality of service to user that can access an IP backbone network via different access technologies is a challenging task, because the user has not to be aware of the underlying access technologies.

The access unawareness is the key concept of multi-access network and that has an impact on the requirements the network should have. Moreover new functionalities have to be defined:

- Enhanced QoS management: in a multi-access scenario the functions of IP QoS control (centralized or distributed) and Resource Management for each access have to be tightly related to one another, in order to perform a more accurate admission control for each user.
- QoS adaptivity: as different access networks may provide different QoS and bit rates to user applications, some new adaptation schemes that help services and applications to adapt to changing QoS conditions, are needed.
- Seamless mobility: the user has to be able to move among different accesses networks in a seamless manner in order to avoid service interruptions.

More convergence is expected relating to evolving High speed LAN, WLAN, UMTS, dial-up link, broadband wireless and satellite networks.

6.3 IP Core Networks

The Internet is modelled as a group of interconnected ASs or as a group of interconnected routers. The BGPv4 routing tables characterize at the AS level the topology of the Internet.

The Internet topology can be characterized using the information obtained from real BGP tables.

Expected developments in core networks relate to Inter-domain routing, inter-domain routing, QoS, Multi-Protocol Label Switching (MPLS), congestion control, TE and resource management [25].

Concerning Intra-domain routing, IP networks typically use distributed protocols for intra-domain routing which take routing decisions locally, according to simple algorithms of path computation. In these protocols the path computation is based on static metrics such as hop-count, delay, or bandwidth, independently from how loaded the links may be. This simple and decentralized approach allows an efficient deployment of IP networks, avoiding complex planning procedures that other routing technologies have to face.

However, this simplicity of path selection algorithms introduces problems related to convergence and control of traffic flows. In an IP network, administrators actually have little control over end-to-end paths, which favours traffic concentration in a few links and, therefore, bandwidth misuse. That is why some proposals are trying to solve some of the major problems in intra-domain IP routing, avoiding to lose its advantages. The objective is to deploy mechanisms to improve network performance without an excess of complexity.

For multipath Routing, a mono-path routing protocol is usually unable to do a proper allocation of resources in presence of heavy load, because only one path can be used to send the traffic. In this context, there is a risk that sections with good metric concentrate traffic on few links, which, consequently, are prone to saturate even with light global loads. Besides, if metrics are computed dynamically according to the network situation, mono-path routing can lead to serious instabilities, since it does not allow a gradual reaction in case of overload. Multipath routing addresses these problems by using simultaneously more than one next-hop choice for a given destination.

Dynamic Routing [26] allows flows to spread out in the network in the most efficient way, adapting to changes in demand and avoiding points of congestion. In this way, global network performance would be less vulnerable to sudden changes in demand, as the network would be able to smooth its effects. Most of the dynamic routing solutions that are currently being proposed for IP TE follow an MPLS-based model. A different approach is to modify present Interior Gateway Protocols (IGPs) such as OSPF or Intermediate System to Intermediate System (IS-IS) OSI-compliant routing protocols [27]. Different contributions have developed this idea. In these proposals, distributed processing is used and decisions are taken locally, thus leading to a better behaviour and allowing an incremental deployment strategy. Nevertheless solutions based on this approach have to be carefully designed to avoid problems of instability.

Concerning combining level 2 and level 3 routing, MPLS has been proposed as a ubiquitous solution for TE and routing problems. MPLS [25] as well as Generalised MPLS (GMPLS) [28] introduce complexity and present both scalability issues and difficulties for TE support in large networks [25]. Due to these facts, an "all IP" solution, based in the improvement of existing IP routing protocols and the introduction of simple QoS mechanisms, is a new direction to pursue given that it seems the best approach to improve the performance of the IP networks.

For QoS Routing [25], the upcoming gigabit-per-second high-speed networks are expected to support a wide range of communication-intensive real-time multimedia applications. The QoS requirements of multimedia applications raise new challenges for next-generation IP networks. In particular, one of the key-issues is QoS routing, a class of solutions that are aimed at identifying cost-efficient network paths that are sufficient resources to meet certain performance and reliability constraints. QoS routing allows selecting network routes with enough resources for the requested QoS parameters. The goal of routing solutions is twofold: first, it must satisfy the QoS requirements for every admitted connection and, second, achieving global efficiency in resource utilization. QoS based routing, as well called constraint based routing, is needed in all aspects of routing, e.g., routing algorithms, route optimization algorithms, routing management, etc. The main areas of QoS routing, which still require broad and deep investigations, are the following:

- Architecture for QoS routing (c.f., Differential Services (Diffserv) over MPLS [29, 30]).
- QoS routing protocols [31] (c.f., source routing, distributed routing, hierarchical routing).
- QoS routing algorithms [32] (c.f., Path computation algorithms, Path selection algorithms, Path recovery algorithms, Path optimization algorithms).

Concerning TE and QoS in core networks, there are different approaches to handle QoS. The easiest one is over-dimensioning where the capacities must be upgraded when the traffic exceeds a predetermined threshold. A more flexible methods results from providing over-provisioning only for some service classes in combination with DiffServ. In these cases not any QoS parameter can be guaranteed but due to weighted queuing and low load on the physical link, real time service is differentiated from best effort one. Another method consists by limiting the number of service categories, and provides virtual tubes. Current proposals establish a number of categories close to three types: elastic, semi-elastic and non-elastic flow [33]. A strong TE scheme results from MPLS in combination with forward equivalent classes, or by a so-called bandwidth broker [34].

The future work has to focus on the comparison of the advantages and disadvantages of the different traffic engineering schemes. For this purpose various methods under different service scenarios and in combination with different network types must be considered. This requires software tools for network design and dimensioning and performing analysis. The analytical models used in these tools have to be evaluated by corresponding simulation tools.

6.4 Network Technology

Wireless and Optical network technologies represent a unique opportunity because of their almost unlimited potential bandwidth. Providing better support for QoS in Internet environments, however, raises fundamental challenges including [35 - 37]:
- IP over wireless access networks.
- IP over Wavelength Division Multiple Access (WDMA) optical core networks.
- Complexity of multi-layer architectures.
- New switching techniques, integration of routing and wavelength assignment, dynamic provisioning of optical channels.
- New architectural alternatives to explored the full potential of optical networks.
- Flexible and cost effective use of network resources and designed to serve diverse user demands.

6.5 Traffic Engineering

Future network infrastructures will have to handle a huge amount of IP-based data traffic, which supports several types of services, including a significant portion of real-time services [38, 39]. The multi-service characteristics of such network infrastructure rise up a clear requisite: the ability to support different classes of services demanding for different QoS requirements. Moreover, Internet traffic is much variable with time, with respect to traditional voice traffic, and it is not easily predictable [40]. This means that networks have to be flexible enough to react adequately to traffic changes. Besides the requirements of flexibility and multi-service capabilities that lead to respect different levels of QoS requirements, there is another key aspect that needs to be taken into account: cost effectiveness. In fact, global IP traffic will increase by a factor of five from 2008 to 2013, approaching 56 exabytes per month in 2013, compared to approximately 9 exabytes per month in 2008 [41]. But the revenues will not expect to approach anywhere near to a factor of five. Actually, revenues coming from voice-based services are quite higher with respect to the ones derived by current Internet services. Therefore, to obtain cost effectiveness it is necessary to design networks that make an effective use of bandwidth or, in a broader sense, of network resources.

TE is a solution that enables the fulfilment of all those requirements, since it allows network resources to be used when necessary, where necessary and for the desired amount of time [42]. Practically speaking, TE can be regarded as the ability of the network to dynamically control traffic flows in order to prevent congestions, to optimize the availability of resources, to choose routes for traffic flows while taking into account traffic loads and network state, to move traffic flows towards less congested paths, to react to traffic changes or failures timely.

6.6 New Concepts of Network Architecture

It is necessary to search for balance between different types of services and applications, between different layers of the protocols, between quality and quantitative, between efficiency and complexity, between keep the protocols and break the protocols, and between keep the balance and break the balance.

The balance point is also instable point due to change of applications and technologies. Therefore, the future is about to search for a new balance as soon as lost the old or existing balance, hence progress will be made.

7 Concluding Remarks

An overview was presented covering the fundamentals of IP network architecture evolution, applications and services, current technologies, technology integration, convergence of IP and other multi-service networks and future directions. Particular attention is devoted to the description of some important mechanisms that control and manage diversity and network resources, towards the enhancement of network performance and QoS.

Acknowledgements. The author acknowledges gratefully the support from EU research programme and contributions from colleagues within the FP6 Euro-NGI NoE consortium.

References

1. Tanenbaum, A.S.: Computer Networks, 4th edn. Prentice Hall PTR, Englewood Cliffs (2003), ISBN 0-13-3499945-6
2. Comer, D.E.: Computer Networks and Internet, 3rd edn. Prentice-Hall, Englewood Cliffs (1999), ISBN 0-13-3499945-6
3. Stalling, W.: Data and Computer Communication, 6th edn. Prentice-Hall, Englewood Cliffs (2000), ISBN 013-048370-9
4. RFC 791, INTERNET PROTOCOL, IETF (1981)
5. RFC 2460, INTERNET PROTOCOL, Version 6 (IPv6) Specification, IETF (2003)
6. RFC 1058, Routing Information Protocol (RIP), IETF (1988)
7. RFC 2453, RIP Version 2, IETF (1998)

8. RFC 904, Exterior Gateway Protocol (EGP) Formal Specification, IETF (1984)
9. RFC 4217, A Border Gateway Protocol 4 (BGP-4), IETF (2006)
10. RFC 793, Transmission Control Protocol, IETF (1981)
11. RFC 768, User Datagram Protocol, IETF (1980)
12. RFC 5681, TCP Congestion Control, IETF (2009)
13. RFC 2475, An Architecture for Differentiated Services, IETF (1998)
14. RFC 2998, A Framework for Integrated Services Operation over Diffserv Networks, IETF (2000)
15. RFC 2002, IP Mobility Support, IETF (1996)
16. RFC 4301, Security Architecture for the Internet Protocol, IETF (2005)
17. RFC 4346, The Transport Layer Security (TLS) Protocol Version 1.1, IETF (2006)
18. RFC 3170, IP Multicast Applications: Challenges and Solutions, IETF (2001)
19. RFC 3376, Internet Group Management Protocol, Version 3, IETF (2002)
20. RFC 1584, Multicast Extensions to OSPF, IETF (1994)
21. RFC 1075, Distance Vector Multicast Routing Protocol, IETF (1988)
22. RFC 4602, Protocol Independent Multicast - Sparse Mode (PIM-SM), IETF (2006)
23. RFC 3973, Protocol Independent Multicast - Dense Mode (PIM-DM): Protocol Specification (Revised), IETF (2005)
24. RFC 2189, Core Based Trees (CBT version 2) Multicast Routing, IETF (1997)
25. RFC 3031, Multiprotocol Label Switching Architecture, IETF (2001)
26. RFC 2328, Open Shortest Path First (OSPF) Version 2, IETF (1998)
27. RFC 5340, OSPF for IPv6, IETF (2008)
28. RFC 5145, Framework for MPLS-TE to GMPLS Migration, IETF (2008)
29. RFC 2475, An Architecture for Differentiated Services, IETF (1998)
30. RFC 3270, Multi-Protocol Label Switching (MPLS)Support of Differentiated Services, IETF (2002)
31. RFC 1633, Integrated Services in the Internet Architecture: an Overview, IETF (1994)
32. RFC 4655, A Path Computation Element (PCE)-Based Architecture, IETF (2006)
33. RFC 4594, Configuration Guidelines for DiffServ Service Classes, IETF (2006)
34. RFC 2638, A Two-bit Differentiated Services Architecture for the Internet, IETF (1999)
35. IEEE Std 802.11-2007, Part 11: Wireless LAN Medium Access Control (MAC) and Physical Layer (PHY) Specifications, IEEE (2007)
36. IEEE Std 802.16-2009, Part 16: Air Interface for Broadband Wireless Access Systems, IEEE (2009)
37. RFC 3717, IP over Optical Networks: A Framework, IETF (2006)
38. RFC 5151, Resource Reservation Protocol-Traffic Engineering (RSVP-TE) Extensions, IETF (2008)
39. RFC 3272, Overview and Principles of Internet Traffic Engineering, IETF (2002)
40. Sun, Z., He, D., Liang, L., Cruickshank, H.: Internet QoS and Traffic Modelling. IEE Proceedings Software "Special Issue on Performance Engineering 151(5), 248–255 (2004), ISSN 1462-5970
41. http://en.wikipedia.org/wiki/Internet_traffic
42. http://en.wikipedia.org/wiki/Teletraffic_engineering

APPENDIX 1: List of Acronyms

ACK - Acknowledgment
AH - Authentication Header
ARP - Address Resolution Protocol
ARPANET - Advanced Research Projects Agency Network
AS - Autonomous System
ATM - Asynchronous Transfer Mode
CBT - Core-Based Tree
BGP - Border Gateway Protocol
BGPv4 - BGP version 4
CIDR - Classless Inter-Domain Routing
DF - Don't Fragment Flag
Diffserv - Differential Services
DNS - Domain Name System
DVB - Digital Video Broadcasting
DVMRP - Distance Vector Multicast Routing Protocol
EGP - Exterior Gateway Protocol
ESP - Encapsulated Security Payload
FIN - Finish
FTP - File Transfer Protocol
GRID - Global Resource Information Database
Host-id - Host Identifier
IETF - Internet Engineering Task Force
IGMP - Internet Group Membership Protocol
IGP - Interior Gateway Protocol
IHL - Internet Header Length
IP - Internet Protocol
ISAKMP - Internet Security Association Establishment and Key
    Management Protocol
ISDN - Integrated Services Digital Network
IS-IS - Intermediate System to Intermediate System
ISN - Initial Sequence Number
ISO - International Standard Organisation
IPSec - IP Security
IPvn - IP Version n, n = 4, 5 and 6
ITU-T - International Telecommunication Union-Telecommunication
LAN - Local Area Network
LMDS - Local Multipoint Distribution System
MAN - Metropolitan Area Networks
MF - May Fragment Flag
M-OSPF - Multicast Extensions to OSPF
MPLS - Multi-Protocol Label Switching
NAK - Negative Acknowledgement
Net-id - Network Identifier
OSI - Open System Interconnection
OSPF - Open Shortest Path First

PIM - Protocol-Independent Multicast
PIM-DM - PIM-Dense Mode
PIM-SM - PIM-Sparse Mode
PSH - Push
POP3 - Post Office Protocol 3
QoS - Quality-of-Service
RARP - Reverse ARP
RIP - Routing IP
RPF - Reverse Path Forwarding
RST - Reset
SATNET - Satellite Network
SMTP - Simple Mail Transfer Protocol
SSL - Secure Socket Layer
SYN - Synchronization
TCB - Transmission Control Block
TCP - Transmission Control Protocol
Telnet – a protocol over TCP/IP for accessing remote computers
TE - Traffic Engineering
TFTP - Trivial FTP
TTL - Time to Live (IP networking)
UDP - User Datagram Protocol
UMTS - Universal Mobile Telecommunications Systems
URG - Urgent
VoIP - Voice over IP
xDSL - x-Digital Subscriber Line
WAN - Wide Area Networks
WDMA - Wavelength Division Multiple Access
WLAN - Wireless LAN
WWW - World Wide Web

Content Distribution over IP: Developments and Challenges

Adrian Popescu[1], Demetres D. Kouvatsos[2],
David Remondo[3], and Stefano Giordano[4]

[1] Blekinge Institute of Technology, Sweden
adrian.popescu@bth.se
[2] University of Bradford, UK
[3] University of Catalunia, Spain
[4] University of Pisa, Italy

Abstract. This chapter focuses on the multimedia distribution over Internet IP under the auspices of the NoE Euro-NGI research project "Routing in Overlay Networks (ROVER)". The multimedia distribution is supported by several components such as services, content distribution chain, protocols and standards whilst Internet is used for content acquisition, management and delivery as well as an Internet Protocol Television (IPTV) infrastructure with QoS facilities. As the convergence between fixed and mobile services of wide and local area networks is also expected to take place in the home networking, this puts an extra burden on multimedia distribution, which requires the different types of wireless access solutions (e.g., WiMAX). In this context, the ROVER research project adopts the IP Multimedia Subsystem (IMS), which offers a wide range of multimedia services over a single IP infrastructure such as authentication and, for wireless services, roaming capabilities. The research project also considers overlay routing as an alternative solution for content distribution.

1 Introduction

The telecommunication industry is actually facing two serious challenges with implications on future architectural solutions. The first challenge is regarding the irreversible move towards IP-based networking. The second challenge is regarding the deployment of broadband access in the form of diverse Digital Subscriber Line (DSL) technologies based on optical fiber and high-capacity cable but also the WiMAX access (IEEE 802.16 Worldwide Interoperability for Microwave Access) [20] to provide high bandwidth access to home networks as well as to small and medium-sized businesses. Altogether, these developments offer the opportunity for more advanced and more bandwidth-demanding multimedia applications and services, e.g., Internet Protocol Television (IPTV), Voice over IP (VoIP), online gaming. A plethora of QoS requirements and facilities are associated with these applications, e.g., multicast facilities, high bandwidth, low delay/jitter, low packet loss. Furthermore, a very important issue is regarding the perceived

QoS and the standards associated, e.g., as defined in ITU-T BT.500.11, ITU-T P.862. Even more difficult is for the service provider to develop a networking concept and to deploy an infrastructure able to provide end-to-end (e2e) QoS for applications with completely different QoS needs. On top of that, the architectural solution must be a unified one, which is independent of the access network and content management (i.e, content acquisition, storage and delivery). Other facilities like billing and authentication must be provided as well.

The foundation of multimedia distribution is provided by several components, the most important ones are services, content distribution chain, protocols and standards. The fundamental idea is to use the Internet for content acquisition, creation, management and delivery. Furthermore, an important goal is to offer the end user the so-called Triple Play, which means grouping together Internet access, TV and telephone service into one subscription on a broadband connection. Other important issues are billing and content protection, e.g., copyright issues, encryption and authentication (Digital Rights Management).

The convergence between fixed and mobile services that is actually happening in the wide and local area networking is expected to happen in the home networking as well. This puts an extra burden on multimedia distribution, which means that wireless access solutions of different types (e.g., WiMAX) must be considered as well. The consequence of throwing Triple Play into wireless services is the upcome of Quadruple Play.

It is therefore important to consider mechanisms and protocols put forth by the Internet Engineering Task Force (IETF) to provide a robust and systematic design of the basic infrastructure, and protocols like Session Initiation Protocol (SIP), IP DiffServ (RFC 2474/2475), together with Multi Protocol Label Switching (MPLS) and traffic engineering (RFC 3031), should be taken into consideration as possible solutions for the QoS control in core networks. Another important IETF initiative is regarding content distribution issues, which are addressed in the IETF WG for Content Distribution Networks (CDN) and Content Distribution Internetworking (CDI). Furthermore, new developments within wireless communications like IP Multimedia Subsystem (IMS) [7,16] are highly relevant for such purposes. Similarly, the new paradigms recently developed for content delivery application-based routing (e.g., based on Peer-to-Peer (P2P) solutions) can be considered as alternative solutions for the provision of QoS on an e2e basis, without the need to replace the IPv4 routers with IP DiffServ routers. The main challenge therefore is to develop an open architectural solution that is technically feasible, open for future updating and services and cost-effective.

2   State of the Art

There are several important components involved in multimedia distribution over IP. The most important ones are Internet Protocol Television (IPTV), multimedia-related protocols (e.g., Session Initiation Protocol (SIP), Common Open Policy Service (COPS), Real Time Streaming Protocol (RTSP)), P2P networking, overlay routing, strategies for content management, billing, and authentication.

A general belief existing today is that the emergence of the IPTV system and the associated protocols represents a tremendous opportunity for carriers to push for new advantageous bussiness models and for customers to obtain new exciting multimedia services. IPTV is expected to offer new services like live programming over the network, VoD, two-ways interactive communication, personalization, digital video creation and recording as well as integration with computer platforms. Allthought IPTV is still evolving, the promises are huge. For instance, according to recent delivered figures, the pay-TV market provided more than 55 billion dollars in subscriber revenue in 2004 [9]. The same study shows that it is forecasted that the Asia/Pacific market alone could reach as much as 20 million IPTV subscribers in 2009.

IPTV is a method for distributing television content over IP. It describes a system where a digital television is delivered to subscribing consumers by using the Internet Protocol over a broadband connection [11]. IPTV is not a protocol but a service that covers both live TV (multicasting) and stored video, i.e., Video-on-Demand (VoD). IPTV uses a two-way broadcast signal sent through the provider's backbone network and servers, and allowing consumers to select content on demand, to timeshift and other interactive options, e.g., on-demand video gaming. The consumer must have either a Set-Top Box (STB) or a personal computer to send and receive different requests.

IPTV operates on a different premise than the traditional broadcast, cable or satellite television in the sense that only the selected content is delivered to the consumer. On the other hand, in the traditional TV system, all channels are permanently pushed to the consumer rather than on a per-selection basis. This feature offers important advantages for IPTV as the number of channels is unlimited in this case whereas the number of channels offered on a cable or satellite network is limited by the allocated spectrum. IPTV primarily uses the protocols Internet multicasting with Internet Group Management Protocol version 2 (IGMPv2) [10] for live television broadcasts and Real-Time Streaming Protocol (RTSP) [17] for on-demand programs. Alternative solutions use multicast overlay routing implemented at the application layer, but this is still under research [4,15,8].

One of the most important questions for telecommunication companies developing IPTV is regarding a successful digital video strategy and the associated challenges related to network architecture, content acquisition and management, storage and delivery. At the same time, it is important to consider the newly started developments towards the research and development of the IP Multimedia Subsystem (IMS) [7,19,16].

IMS represents a new framework, basically specified for mobile networking, to provide IP-based telecommunication services. Fundamental elements in IMS are the convergence of voice, data and multimedia services, integration of mobile and Internet domains as well as facilities created to allow consumers to access, create, consume and share digital content by using interoperable devices.

IMS represents in fact a culmination of technology standards put forth by the Internet Engineering Task Force (IETF) and two third Generation Partnership Project groups (3GPP and 3GPP2). Based on that, more and more telecommunication carriers and equipment vendors (e.g., Ericsson, Lucent Technologies, Motorola, Nokia, Alcatel, British Telecommunications) are releasing equipments and services according to IMS recommendations [7].

The IMS architecture provides basicaly a framework to integrate a range of protocols and media types. Some of the most important functionalities include IP connectivity-based development, access-independent processing, QoS guarantees for multimedia, policy control for efficient use of media resources, user and data security and authentication using SIP, charging capabilities, roaming support, service control across the network and service development with API support.

Another important aspect is regarding the models for content delivery existing today. The Internet was initially developed as a simple model for content delivery, in which the network does the routing and the end-system does the control. The ubiquitous client-server computing model together with the World Wide Web content delivery have created the fundamental infrastructure for content delivery that exists today. Tremendous effort has also been put in place in the development of systems to provide networks with Quality of Service guarantees (IP QoS). In spite of big research and development efforts, the limitations of such systems have now become clear, especially in terms of scalability, failure to emerge as an open end-to-end service, and difficulties in developing suitable models for charging. Furthermore, provisioning of end-to-end QoS for a traffic flow that traverses multiple Autonomous Systems (ASs) has been proven to be difficult due to difficulties in arranging cooperation among ASs.

At the same time, new paradigms for content delivery have emerged, where the main point is that widely-distributed applications are making their own forwarding decisions. New classes of applications include content distribution networks [13], robust routing overlays [1], Peer-to-Peer (P2P) file sharing [18], network-embedded storage [14], scalable object location [3], and scalable event propagation [5]. An important characteristic of these applications is that specific facilities are created for the convergence, with different degrees, of networking, distributed computing and applications.

Over the last years, such systems have evolved to be some of the major traffic contributors in the Internet [12]. P2P applications have now become immensely popular in the Internet community, due to characteristics like communication among equals (computers are acting as both clients and servers, so-called "servents") as well as pooling and sharing of exchangeable resources such as storage, bandwidth, data and CPU cycles. Although an exact definition of "P2P systems" is still debatable, such a system typically represents a distributed computing paradigm where a spontaneous, continuously changing group of collaborating computers act as equals in supporting applications such as resource redundancy, content distribution, and other collaborative actions.

In most cases the peers act from the network's edge instead of core, and they can dynamically join and leave the network, discover each other and form ad-hoc collaborative environments. Each of the participating peers is sharing and exploiting the resources brought collectively to the network pool. The resources needed for the execution of a specific (application) task are dynamically aggregated for the required time period, e.g., by swarming techniques [6]. Beyond that, the allocated resources return to the network pool. These features allow the P2P system to still provide services even when losing resources, in contrast to the classical client-server concept where failures in the system may completely disrupt the service.

3 Content Distribution Networks

Content Distribution Networks (CDNs) are networking solutions where high-layer network intelligence is used to improve the performance in delivering media content over the Internet, e.g., static or transaction-based Web content, streaming media, real-time video, radio. The fundamental concept is based on distributing content to cache servers located close to end users, resulting so in better performance, e.g., maximize bandwidth, minimize content latency and jitter, improve accessibility. CDNs are composed by multiple Points of Presence (PoP) with clusters (so-called surrogate servers) that maintain copies of (identical) content, resulting so in better balance between cost for content providers and QoS for customers. CDN nodes are deployed in multiple locations, in most cases placed in different backbones. They cooperate with each other, transparently moving content so as to optimize the delivery process and to provide users the most current content. The optimization process may result, e.g., in reducing the bandwidth cost, improving availability and improving QoS.

The client-server communication flow is replaced in CDN by two communication flows, namely between the origin server and the surrogate server and between the surrogate server and the client. On top of that, questions related to QoS, content multicasting and multipath routing heavily complicate the picture. Requests for content delivery are intelligently directed to nodes that are optimal with reference to some parameter of interest, e.g., minimum number of hops, or networks, away from the requester.

Organizations offering content to geographically distributed clients sign a contract with a CDN provider and distribute the content over the selected CDN by using a specific overlay model. Some of the most popular commercial CDN providers are Akamai, Nexus, Mirror Image Internet and LimeLight Network.

It is also important to mention that content distribution can be done by using IP multicast as well. In such a case, specific code is deployed in IP routers or switches such as they are able to recognize specific application types and make forward decision of their own that are based on predefined policies.

In practice, there are several challenges that must be solved in order to offer high-quality distribution at reasonable prices. Some of the most important questions are related to where to place the surrogate servers, which content to outsource, which practice to use for the selected content outsourcing, how to exploit data mining over CDN to improve the performance and what model to use for CDN pricing.

For instance, it is very important to choose the best network placement for surrogate servers since this is critical for the content outsourcing performance. A good placement solution may also have other positive effects, e.g., by reducing the number of surrogate servers needed to cover a specific CDN. Actually, several placement algorithms have been suggested, e.g., Greedy, Hot Spot and Tree-Based Replica, each of them with own advantages and drawbacks.

Another challenge is the selection of the content that should be outsourced in order to meet the customers needs. An adequate management strategy for content outsourcing should consider grouping the content based on correlation figures or access frequency and replicate objects in units of content clusters. Furthermore, given a specific CDN infrastructure with a given set of surrogate servers and selected content for delivery, it is important to select an adequate policy for content outsourcing, e.g., cooperative push-based, uncooperative pull-based, cooperative pull-based. These policies are associated with different advantages and drawbacks, today however most of the commercial CDN providers (Akamai, Mirror Image) use uncooperative pulling. This is done in spite of non-optimal solutions used to select the optimal server from which to serve the content. The challenge is to provide an optimal trade-off between cost and user satisfaction and new techniques like caching, content personalization and data mining can be used to improve the QoS and performance of CDN.

An important parameter is related to the CDN pricing. Today, some of the most significant factors affecting the pricing of CDN services are bandwidth cost, traffic variations, size of content replicated over surrogate servers, number of surrogate servers, and security cost associated with outsourcing content delivery. It is well known that cost reduction occurs when technology investments allow for delivering services with fewer resources. The situation is however more complex in the case of CDN since higher bandwidth and lower bandwidth cost also have as a side effect that customers develop more and more resource-demanding applications with harder and harder demands for QoS guarantees.

The fundamental entities of a CDN are network infrastructure, content management, content routing and performance measurement. Content management is about the entire content workflow, from media encoding and indexing to content delivery at edges including ways to secure and manage the content. On the other hand, content routing is about delivering the content from the most appropriate server to the client requesting for it. Finally, performance measurement is considered as part of network management and it is regarding measurement technologies used to measure the performance of the CDN as a whole.

4 Routing in Overlay Networks

Overlay networks recently emerged as a viable solution to the problem of content distribution with multicasting and QoS facilities. Overlay networks are networks operating on the inter-domain level, where the edge hosts learn of each other and, based on knowledge of underlying network performance, they form loosely coupled neighboring relationships. These relationships are used to induce a specific graph, where nodes are representing hosts and edges are representing neighboring relationships. Graph abstraction and the associated graph theory can be used to formulate routing algorithms on overlay networks. The main advantage of overlay networks is that they offer the possibility to augment the IP routing as well as the Quality of Service (QoS) functionality of the Internet.

One can state that, generally, every P2P network has an overlay network at the core, which is mostly based on TCP or HTTP connections. The consequence is that the overlay and the physical network can be completely separated from each other as the overlay connections do not reflect the physical connections. This is due to the abstraction offered by the TCP/IP protocol stack at the application layer. Furthermore, by means of cross-layer communication, the overlay network can be matched to the physical network if necessary. This offers important advantages in terms of reduction of the signaling traffic.

Overlay networks allow designers to develop own routing and packet management algorithms on top of the Internet. A similar situation happened in fact with the Internet itself. The Internet was developed as an overlay network on top of the existing telephone network, where long-distance telephone links were used to connect IP routers. Overlay networks operate in a similar way, by using the Internet paths between end-hosts as "links" upon which the overlay routes data, building so a virtual network on top of the network. The result is that overlay networks can be used to deploy new protocols and functionality atop of IP routers without the need to upgrade the routers. New services can be easily developed, with own routing algorithms and policies.

Generally, there are two classes of overlay networks, i.e., routing overlays and storage and lookup overlays. Routing overlays operate on inter-domain IP level and are used to enhance the Border Gateway Protocol (BGP) routing and to provide new functionality or improved service. However, the overlay nodes operate, with respect to each other, as if they were belonging to the same domain on the overlay level. QoS guarantees can be provided as well.

On the other hand, storage and lookup overlays focus on techniques to use the power of large, distributed collections of machines, like in the case of Chord and Akamai. These overlays are actually used as a support for a number of projects on large distributed systems. The distinction between the two classes of overlays has become more and more blurred over the last years.

Strategies for overlay routing describe the process of path computation to provide traffic forwarding with soft QoS guarantees at the application layer. There are three fundamental ways to do routing. These are source routing, flat (or distributed) routing and hierarchical routing. Source routing means that nodes are required to keep global state information and, based on that, a feasible path is computed at every source node. Distributed routing relies on a similar concept but with the difference that path computation is done in a distributed fashion. This may however create problems like distributed state snapshots, deadlock and loop occurrence. There are better versions that use flooding but at the price of large volumes of traffic generated. Finally, hierarchical routing is based on aggregated state maintained at each node. The routing is done in a hierarchical way, i.e., low level routing is done among nodes in the neighborhood of a logical node and high level routing is done among logical nodes. The main problem with hierarchical routing is related to imprecise states.

Notably, overlay routing exploits knowledge of underlying network performance and adapts the end-to-end performance to asymmetry of nodes in terms of, e.g., connectivity, network bandwidth and processing power as well as the lack of structure among them. Overlay routing has the possibility to offer soft QoS provisioning for specific applications while retaining the best-effort Internet model. It can for instance bypass the path selection of BGP to improve performance and fault tolerance.

A specific challenge with overlay routing is related to the presence of high churn rates in P2P networks. The consequence is that the topology is very dynamic, which makes it difficult to provide hard QoS guarantees. Similar situations do exist in wireless ad-hoc networks.

There are two main categories of routing protocols for overlay networks, namely proactive protocols and reactive protocols. Proactive protocols periodically update the routing information, i.e., independent of traffic arrivals. On the other hand, reactive protocols update the routing information on-demand, i.e., only when routes need to be created or adjusted due to changes in routing topology or other conditions (e.g., traffic must be delivered to an unknown destination). Proactive protocols are generally better at providing QoS guarantees for real-time traffic like multimedia. The drawback lies in the traffic volume overhead generated by the protocol itself. Reactive protocols scale better, but they experience higher latency when setting up a new route.

A number of research activities are being carried out worldwide focusing on overlay routing for services like streaming and on-demand. Important research questions are, e.g., on scalability, data search and retrieval, load balancing, churn handling, QoS provisioning with multicast or multipath facilities. It is our ambition to give our contribution to answering these questions.

5 Conclusions

The chapter reported on some of the recent developments in multimedia distribution over Internet IP. The multimedia distribution is supported by several components such as services, content distribution chain, protocols and standards whilst Internet is used for content acquisition, management and delivery. Given that the convergence between fixed and mobile services of wide and local area networks is expected to also take place in the home networking, this puts an extra burden on multimedia distribution. In this context, the ROVER research project adopted the IP Multimedia Subsystem (IMS), which offers a wide range of multimedia services over a single IP infrastructure such as authentication and, for wireless services, roaming capabilities. The research project also considers overlay routing as an alternative solution for content distribution.

References

1. Andersen, D., Balakrishnan, H., Kaashoek, F., Morris, R.: Resilient Overlay Networks. In: 18th ACM Symposium on Operating Systems Principles (SOSP), Banff, Alberta, Canada (October 2001)
2. Akamai Technologies, http://www.akamai.com
3. Balazinska, M., Balakrishnan, H., Karger, D.R.: INS/Twine: A scalable peer-to-peer architecture for intentional resource discovery. In: Mattern, F., Naghshineh, M. (eds.) PERVASIVE 2002. LNCS, vol. 2414, p. 195. Springer, Heidelberg (2002)
4. Biersack, E.W.: Where is Multicast Today? ACM SIGCOMM Computer Communication Review 35(5) (October 2005)
5. Castro, M., Druschel, P., Kermarrec, A.-M., Rowstron, A.: Scribe: A Large Scale and Descentralized Application-Level Multicast Infrastructure. IEEE Journal on Selected Areas in Communications 20(8) (October 2002)
6. Cohen, B.: BitTorrent, http://www.bittorent.com
7. Geer, D.: Building Converged Networks with IMS Technology. IEEE Computer (November 2005)
8. Hamra, A.A., Felber, P.A.: Design Choices for Content Distribution in P2P Networks. ACM SIGCOMM Computer Communication Review 35(5) (October 2005)
9. Harris, A., Ireland, G.: Enabling IPTV: What Carriers Need to Know to Succeed, White Paper, IDC, http://www.bitpipe.com/detail/RES/1123080212_477.html
10. Internet Group Management Protocol version 2 (IGMPv2), RFC 2236, IETF, http://www.ietf.org
11. IPTV, Wikipedia, http://en.wikipedia.org/wiki/IPTV
12. Karagiannis, T., Broido, A., Brownlee, N., Claffy, K., Faloustos, M.: File Sharing in the Internet: A Characterization of P2P Traffic in the Backbone. Technical Report, University of California at Riverside (2003)
13. Krishnamurthy, B., Wills, C., Zhang, Y.: On the Use and Performance of Content Distribution Networks. In: ACM SIGCOMM IMW 2001, San Francisco, CA, USA (November 2001)
14. Kubiatowicz, J., Bindel, D., Chen, Y., Czerwinski, S., Eaton, P., Geels, D., Gummadi, R., Rhea, S., Weatherspoon, H., Weimer, W., Wells, C., Zhao, B.: Ocean Store: An Architecture for Global-Scale Persistent Storage. In: 9th International Conference on Architectural Support for Programming Languages and Operating Systems (ASPLOS 2000), Cambridge, MA, USA (November 2000)
15. Neumann, C., Roca, V., Walsh, R.: Large Scale Content Distribution Protocols. ACM SIGCOMM Computer Communication Review 35(5) (October 2005)
16. IP multimedia - a new era in communications, White Paper, Nokia Networks, http://www.nokia.com
17. Real Time Streaming Protocol (RTSP)), RFC 2326, IETF, http://www.ietf.org
18. Rowstron, A., Druschel, P.: Storage Management and Caching in PST, A Large-Scale Persistent Peer-to-Peer Storage Utility. In: 18th ACM Symposium on Operating Systems Principles (SOSP), Banff, Alberta, Canada (October 2001)
19. Tadault, M., Soormally, S., Thiebaut, L.: Network Evolution Towards IP Multimedia Sybsystem, Strategy White Paper, Alcatel, 4th quarter 2003/1st quarter, Alcatel Telecommunications Review (2004)
20. WiMAX, Wikipedia, http://en.wikipedia.org/wiki/WiMAX

Implementation and Evaluation of Network Intrusion Detection Systems

Monis Akhlaq[1], Faeiz Alserhani[1], Irfan Awan[1,2], John Mellor[1], Andrea J. Cullen[1], and Abdullah Al-Dhelaan[2]

[1] Informatics Research Institute, University of Bradford,
Bradford, BD7 1DP, United Kingdom
{m.akhlaq2,f.m.f.alserhani,i.u.awan,
j.e.mellor,a.j.cullen}@bradford.ac.uk
[2] Computer Science, KSU Saudi Arabia
dhelaan@ksu.edu.sa

Abstract. Performance evaluation of Network Intrusion Detection Systems (NIDS) has been carried out to identify its limitations in high speed environment. This has been done by employing evasive and avoidance strategies simulating real-life normal and attack traffic flows on a sophisticated Test-Bench. Snort, an open source Intrusion Detection System, has been selected as an evaluation platform. In this paper, Snort has been evaluated on host and virtual configurations using different operating systems and hardware implementations. Evaluation methodology is based on the concept of stressing the system by injecting various traffic loads (packet sizes, bandwidth and attack signatures) and analyzing its packet handling and detection capacity. We have observed few performance issues with Snort which has resulted into packet drop and low detection rate. Finally, we have analyzed the factors responsible for it and have recommended techniques to improve systems packet handling and detection capability.

Keywords: Host platform, Network Intrusion Detection Systems (NIDS), operating systems (OS), packet drop, performance evaluation, Snort, Traffic, virtual platform.

1 Introduction

The recent era has witnessed a tremendous increase in the usage of computer network applications. The rapid inception of these systems in almost every circle of our life has made them attractive target to malicious activities. In response, various techniques have been designed to ensure security and stability of these systems among them Network Intrusion Detection Systems (NIDS) has gained substantial importance. NIDS are designed to prevent intrusion into a computer system or a network by providing information on a hostile activity that may have incurred to challenge[1] Security Attributes. Intrusion Detection Systems (IDS) of any type (host based or network

---

[1] Security Attributes: Confidentiality, Integrity, Privacy and Non Repudiation of data exchanged in a computer network.

based) operate via network or system users. They can also be controlled via sensors configured to report suspected activity back to the central system or administrator. IDS generally fall into two key categories: anomaly-based (behaviour-based) and signature-based (knowledge-based) [1]. Today, variety of open source and commercial NIDS are available, this can be said that the NIDS has established their importance, however; their performance remains a debatable issue.

Anomaly-based IDS apply various application logics to determine the network threats. They work by establishing a "normal" profile for a system or a network and later generating alerts by observing the deviations from this norm. Signature-based IDS use predefined attack signatures to detect threats and generate alerts. The signature definitions represent known system and network vulnerabilities, for example specific viruses/worms or known patterns of malicious activity like log file editing etc. These systems need continuous update and evaluation by IDS vendors or administrators [2].

A typical scenario of employing NIDS in a network is its implementation on the server with minimum active services. This set-up is quite susceptible to insider attacks especially in high speed environments. The current NIDS are also threatened by resource crunch attempts such as Distributed Denial of Service (DDoS) attacks which has increased from few megabits in the year 2000 to 40 Gbps in 2010 [3]. The performance criteria of NIDS demands that every single packet (header and payload) passing through the network need to be evaluated with the same link speed; however the massive increase in speed and throughput has generated performance issues. Sending a large amount of traffic or using computationally expensive techniques like fragmentation can also compromise NIDS and force it to start dropping the packets.

NIDS can be implemented as software-based or hardware-based. Software-based NIDS are more configurable, easy to update and require less maintenance; however their performance is quite slow. On the other hand, hardware-based NIDS can handle a large volume of traffic. They are expensive, require more maintenance and are hard to update. The choice between the two is a trade-off between cost and performance. We have felt a need to investigate the performance of software-based systems in current day high speed conditions using different implementations scenarios. Quite few efforts have also been made to measure the performance of NIDS. Most of the evaluation methodologies are based on testing the IDS in a moderate traffic conditions. Furthermore, some of these approaches have used previously saved data sets instead of real-time traffic which looks quite unrealistic. The actual performance of the systems was gauged under limited conditions and under a simulated non realistic network flow. The results obtained during these investigations could not portray actual performance output of NIDS.

We have endeavored to evaluate the system on realistic network conditions providing NIDS different levels of hardware support in order to analyze its performance more practically. The recent development of multi-core systems has also added few more opportunities to deploy software-based system; these shall also be investigated in this paper.

The aim of our work is to provide the answers to the following questions:

- Is it possible to deploy the current software-based NIDS such as Snort [4] at rate more than 500 Mbps using commodity hard-ware? Also to identify the limits of incoming traffic, a system can handle effectively in terms of packet loss.

- Is the use of different operating systems (OS) (in normal desktop and server machines), hardware resource (single and multi-core processors) and configurations (host and virtual) affect the NIDS performance?
- Do the software-based NIDS can catch up with the advancement in network speed once implemented on normal hardware or there is no space for software-based NIDS anymore?

Our research has focused on signature-based IDS with an emphasis on evaluating its performance in high-speed traffic conditions. We have selected Snort [4] as a test platform because of its popularity and status as a *de-facto* IDS standard. We are confident that the results obtained in this research would also be applicable to other IDS available in the market. The test environments selected for the research has a significant edge over [5] and our results develop a new understanding of IDS performance limitations.

Our evaluation methodology is based on the concept of analyzing the system capacity in terms of its packet handling capability by implementing it into different hardware configurations and testing platforms. We have achieved this by establishing three different Test-Benches where every setup has been assigned specific evaluation task. Test-Bench I implements the Snort on mid range commodity hardware (limited processing power and system memory). The results obtained on this platform describe the efficacy of NIDS implementation at this level. The Test-Bench II and III utilize high range commodity hardware built on Intel Xeon Dual Quad-Core processor using 4.0 GB RAM. These Test-Benches analyzed the system performance on host and virtual configurations, respectively. We have also analyzed the system capability by observing its response to known attacks in Test-Bench I; however this criterion has not been considered for other Test-Benches.

This paper is organized into sections. Section 2 gives an overview of Snort. Sections 3, 4 & 5 describe Test-Benches I, II and III respectively. Section 6 presents the analysis of results obtained and finally in Section 7 few solutions are recommended to improve the performance of the system in high speed conditions. Conclusions follow in Section 8. Finally, a list of acronyms can be found in Appendix I and string matching algorithms are revised in Appendix II.

2 Overview of Snort

Open source software has gained tremendous popularity and acceptance among academia and research community. Apart from being free of cost there are several other qualities which has made them popular. Few advantages of open source software are access to source code, detailed documentation and online forum to discuss implementation and operational issues. This research has focused on a widely accepted open source software tool, Snort which has received great acceptance in the IDS market.

Snort is capable of performing real-time traffic analysis and packet logging on the network. It performs protocol analysis and can detect variety of network threats by using content/ signature matching algorithms. Snort can be configured as a packet sniffer, packet logger and NIDS. As packet sniffer, it reads the packets off the network and in a packet logger mode, it logs packets to the storage device. NIDS mode enables the Snort to analyze the network traffic against set of defined rules in order to detect intrusion threats.

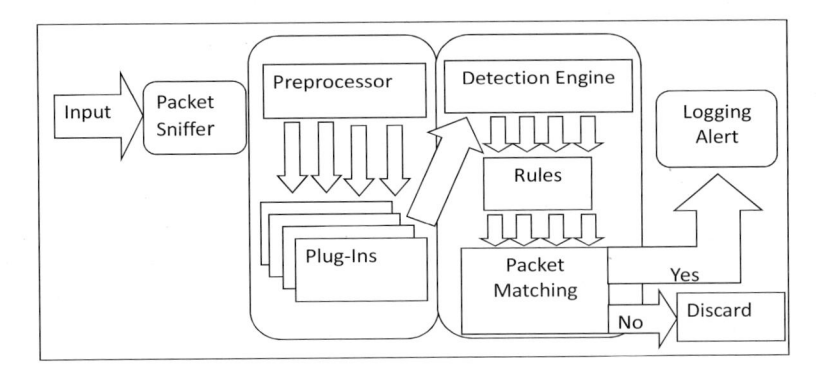

Fig. 1. Snort Architecture

Snort architecture consists of four basic components: sniffer, pre-processor, detection-engine and output as shown in Fig. 1. In its most basic form, Snort is a packet sniffer. However, it is designed to receive packets and process them through its pre-processors. A pre-processor takes the raw packets and evaluates them against certain plug-ins for example the RPC (Remote Procedure Call) plug-in, Hypertext Transfer Protocol (HTTP) plug-in, etc. The plug-ins check for certain types of behaviour by undertaking packet inspection. Once the packet is determined to have a particular "behaviour", it is then sent to the detection engine. Snort handles many kinds of pre-processors and their affiliated plug-ins, including IP fragmentation handling, port scanning and flow control. It also includes analysis of richly featured protocols such as the HTTPinspect pre-processor handles etc [6].

Snort being a signature based IDS uses rules to check for hostile packets in the network. Rules are sets of requirements used to generate an alert and have a particular syntax. For example, one rule that checks for peer-to-peer file sharing services looks for the string "GET" in connection to the service running on any port other than Transport Control Protocol (TCP) port 80. If a packet matches the rule, an alert is generated. Once an alert is triggered, the alert can be sent to multiple places such as a log file, a database or generates a Simple Network Management Protocol (SNMP) trap [7]. On successful detection of a hostile attempt, the detection engine sends an alert to a log file through a network connection into the required storage (output) [6]. Snort can also be used as Intrusion Prevention System (IPS) [6]. Snort 2.3.0 RC1 integrated this facility via Snort-inline into the official Snort project [4].

The main objective of Snort and other NIDS is to effectively analyze all packets passing through the network without any loss. The performance of majority of running applications depends upon memory and processing power. In the context of NIDS, this performance dependency includes Network Interface Cards (NIC), Input and Output (I//O) disk speed and OS. In recent years, technologies have advanced in both hardware and software domains, multi-core systems have been introduced to offer powerful processing functionality also software are progressing well in providing more functionalities.

Snort, which was developed in early 1990's, doesn't support multi-threading. Detection engine component of Snort represent the critical part, where pattern matching function is performed. Recent Virtual Resource Tree (VRT) rules library contains more than 8000 rules; this augments the need of an effective pattern matcher. Snort is using three different pattern matching algorithms: Aho-Corasick, modified Wu-Manber and low memory key-word trie (lowmem) [6,8]. Further details of these algorithms can be found in Appendix II. Modifications have been made in these algorithms to analyze their performances. We have conducted memory usage and performance comparison tests on different pattern matching algorithms and the results are shown in Table 1. This can be concluded that the pattern matching component of IDS is a major hurdle in the performance issues once related to system hardware. Quite significant amount of memory is utilized to match approximately 8000 rules in a noticeable time.

Table 1. Performance Measurements – Pattern Matching Algorithms

Algorithms (8,296 Rules)	Memory usage Mega Bytes (MB)	Packet processing time(seconds)
Aho-Corasick (full) [6]	640	620
Aho-Corasick (sparse) [6]	240	714
Aho-Corasick (std) [6]	1080	665
Wu-Manber [8]	130	635
Wu-Manber (low) [8]	75	655

3 Test-Bench I

The network is composed of six machines using Pro-Curve Series 2900 switch [9] as shown in Fig. 2 and it is similar to that adopted in [10]. The Test-Bench I comprises a number of high performance PCs running open source tools to generate background traffic, run attack signatures and monitor network performance.

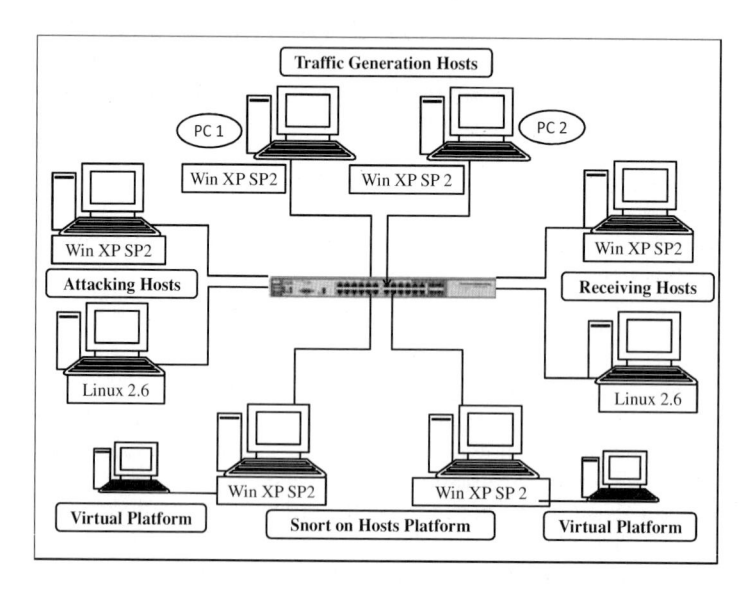

Fig. 2. Test-Bench I

The hardware description of the network is shown in Table 2. The network components are described below.

Table 2. Network Description- Test-Bench 1

Machine Type	Hardware Description	Tools Used
Network traffic/ back ground traffic generator (Win XP SP2)	Dell Precision T3400, Intel Quad-Core, Q6600 2.40 GHz. 2 GB RAM, PCIe, IGbps RJ45, Network Card (Broadcom NetXtremo Gigabit Ethernet).	Traffic Generators: NetCPS [10], Tfgen [11], Http Traffic Gen [12], LAN Trrafic Version 2 [13] and D-ITG Version 2.6 [14]
Network traffic/ back ground traffic generator (Linux 2.6)	Dell Precision T3400, Intel Quad-Core, Q6600 2.40 GHz. 2 GB RAM, PCIe, IGbpss RJ45, Network Card (Broadcom NetXtremo Gigabit Ethernet).	Traffic Generators: D-ITG Version 2.6 [14] and hping Version 2 [15]
Attack Machine • Win XP SP2 • Linux 2.6	Dell Precision T3400, Intel Quad-Core, Q6600 2.40 GHz. 2 GB RAM, PCIe, IGbps RJ45, Network Card (Broadcom NetXtremo Gigabit Ethernet).	Attack Traffic Generator: Metasploit framework [16]
IDS Machines • Snort – Win XP SP2 • Snort – Linux 2.6	Dell Precision T3400, Intel Quad-Core, Q6600 2.40 GHz. 2 GB RAM, PCIe, IGb/s RJ45, Network Card (Broadcom NetXtremo Gigabit Ethernet).	• IDS:Snort [4], Traffic Monitor: Bandwidth Monitor [17] on Win XP SP2 • IDS:Snort and Traffic Monitor: nload [18] on Linux 2.6.
Switch	ProCurve Series 2900 [9], 10Gbps switch with 24x1 Gbps ports and 2x10 Gbps 3CR17762-91-UK ports.	

3.1 Traffic Generators

Two machines are configured to generate network traffic on Windows XP SP 2 [20] and Linux 2.6 [21] respectively as shown in Fig. 2. Distribution of network traffic is TCP (70%), User Datagram Protocol (UDP) (20%) and Internet Control Message Protocol (ICMP) (10%).

3.2 Attacking Host

Two machines are configured to generate attacks/ exploits on Windows XP SP 2 and Linux 2.6 as shown in Fig. 2.

3.3 IDS Machine (Snort)

In the Test-Bench I, Snort is operated on both host and virtual machines for both Windows and Linux platforms. This has been done to analyze the performance of Snort using the limited resources of a virtual machine and also with the full processing capability of host computer. We have selected Snort version 2.8.3 [4] for evaluation.

Snort was also tested for its accuracy on the different OS platforms (Windows XP SP2 & Linux 2.6). The platforms were tested by injecting a mixture of heavy network traffic and scripted attacks through the Snort host. *Snort.conf* file in its default configuration was selected for evaluation. The performance of Snort was also been evaluated under the following variant conditions:

- Generating attacks from different operating system hosts.
- Varying traffic payload, protocol and attack traffic in different scenarios, named as Alpha, Bravo, Charlie, Delta and Echo (as shown in Table 3).
- Subjecting it to hardware constraints of virtual machine configurations.

Table 3. Test-Bench Scenarios

Scenario	Network Traffic Generator PC1	Network Traffic Generator PC2	Attack Traffic Generator – Metasploit	IDS: Snort
Alpha	Host Win XP SP2	Host Win XP SP2	Host Linux 2.6	Virtual Win XP SP2
Bravo	Host Win XP SP2	Host Win XP SP2	Host Linux 2.6	Virtual Linux 2.6
Charlie	Host Win XP SP2	Host Win XP SP2	Host Linux 2.6	Host Win XP SP2
Delta	Host Win XP SP2	Host Win XP SP2	Host Linux 2.6	Host Linux 2.6
Echo	Host Win XP SP2	Host Win XP SP2	Host Win XP SP2	Host Linux 2.6

3.4 Results

Snort was evaluated on the basis of network traffic that ranged from 100 Mbps to 1.0 Gbps (divided into five different test scenarios). The other parameters selected for evaluation include network utilization, CPU usage and Snort CPU usage. Snort performance in terms of packets analyzed, packets dropped, alerts/ log and detection status have also been considered for critical evaluation.

3.4.1 Scenario Alpha
Snort was configured to run using performance limiting configuration of a Windows XP SP 2 virtual machine. The results obtained are shown in Fig. 3. It was subjected to heavy background traffic and attack exploits (from a well resourced Linux host). They demonstrate that the performance of Snort deteriorates markedly as network traffic load increases.

3.4.2 Scenario Bravo
Snort was configured to run using the performance-limiting configuration of a Linux virtual machine and the attacker is a well resourced Linux host. The results obtained, as shown in Fig. 4 identify the similar performance limitations as found in scenario alpha. However, an improvement can be observed when Snort runs on the same operating system as that of the attacking host.

3.4.3 Scenario Charlie
Snort was configured to run using a well-resourced Windows platform and the attacker on a Linux host. The results obtained are shown in Fig. 5. Snort performance

declines as a result of being run on a different operating system platform to the attacker. However, an improvement can be observed in comparison to the equivalent virtual scenario.

3.4.4 Scenario Delta

Snort and the attacker both configured using well resourced Linux platform as hosts. The results obtained are shown in Fig. 6. Comparatively an improved performance for Snort can be observed in this scenario as both attacker and Snort are using same OS (Linux).

Attack Platform: Host Linux 2.6 vs Snort Platform: Virtual Windows

Parameter	100 – 200 Mbps	500 – 700 Mbps	800 Mbps – 1.0 Gbps
Network Utilization	12 %	56%	90%
CPU Usage	50 – 70%	90 – 100%	95 – 100%
Snort CPU Usage	40 – 50%	80 – 90%	90-98%
Packets Analysed	72.5%	66%	38 %
Packets Dropped	27.5%	34%	62 %
Alerts & Log	100%	75%	20%

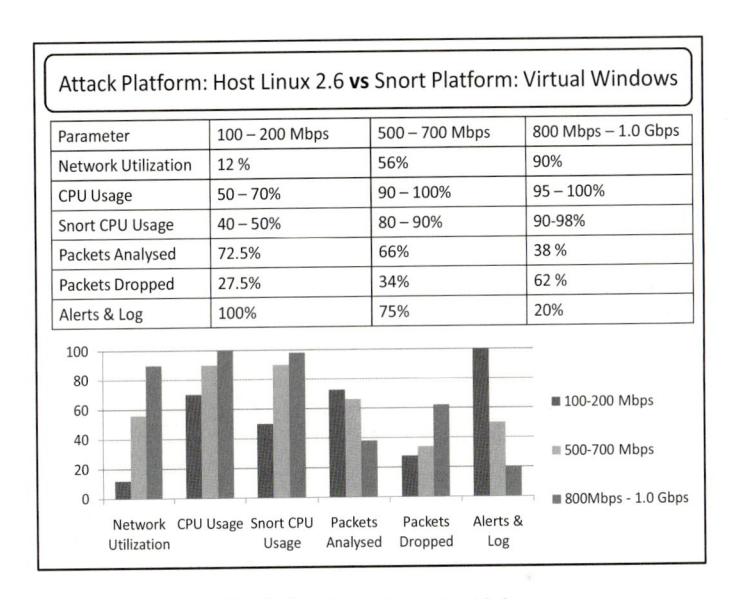

Fig. 3. Results – Scenario Alpha

Attack Platform: Host Linux 2.6 vs Snort Platform: Virtual Linux 2.6

Parameter	100 – 200 Mbps	500 – 700 Mbps	800 Mbps – 1.0 Gbps
Network Utilization	12 %	54 %	90%
CPU Usage	50 – 70%	88 - 95%	90 – 100%
Snort CPU Usage	40 – 50%	75 - 85%	90-95%
Packets Analysed	75 %	62 %	45%
Packets Dropped	25 %	38 %	55 %
Alerts & Log	100%	50 %	30 %

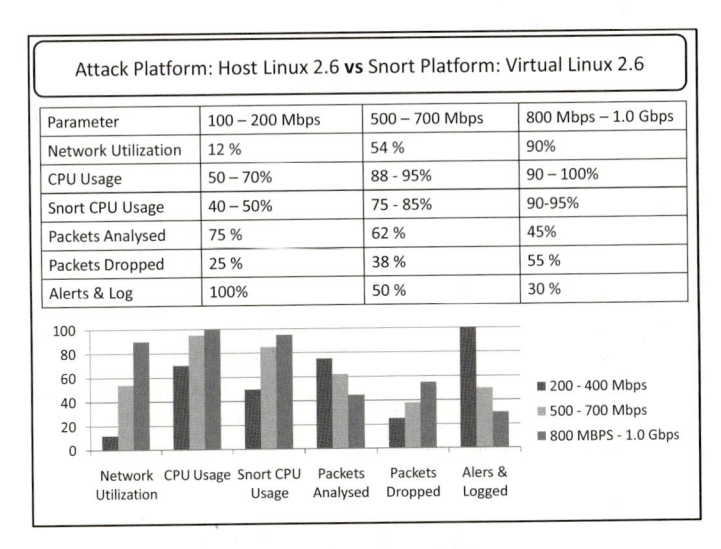

Fig. 4. Results – Scenario Bravo

Attack Platform: Host Linux 2.6 vs Snort Platform: Host Windows

Parameter	100 – 200 Mbps	500 – 700 Mbps	800 Mbps – 1.0 Gbps
Network Utilization	13%	53%	90%
CPU Usage	20 – 30%	30 - 35%	35 – 40%
Snort CPU Usage	15 – 20%	20 - 25%	25-30%
Packets Analysed	98.2 %	38 %	27 %
Packets Dropped	1.8 %	62 %	73 %
Alerts & Log	100%	50 %	20 %

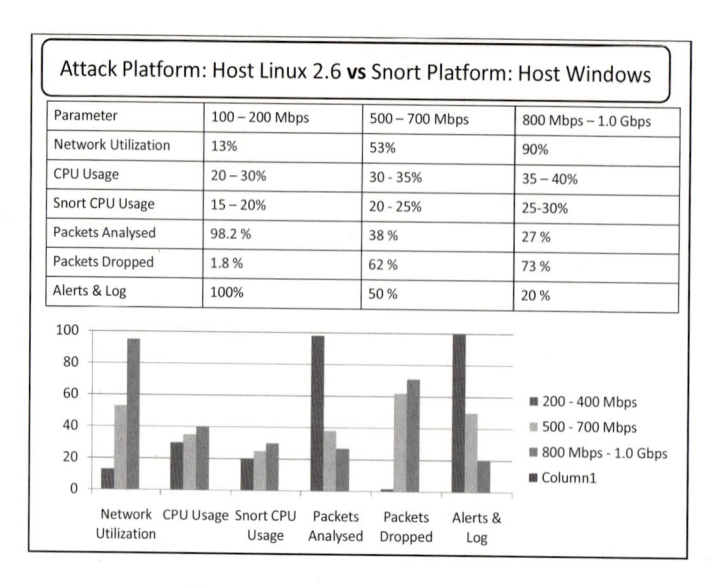

Fig. 5. Results – Scenario Charlie

Attack Platform: Host Linux 2.6 vs Snort Platform: Host Linux 2.6

Parameter	100 – 200 Mbps	500 – 700 Mbps	800 Mbps – 1.0 Gbps
Network Utilization	21%	55%	95%
CPU Usage	18 – 25%	29 - 36%	38 – 43%
Snort CPU Usage	15 – 20%	22 - 27%	29-36%
Packets Analysed	98.5%	47 %	32 %
Packets Dropped	1.5%	53 %	68 %
Alerts & Log	100%	50 %	30 %

Fig. 6. Results – Scenario Delta

3.4.5 Scenario Echo

Snort is configured to run on a well-resourced Linux platform and the attacker on a Windows host as shown in Fig. 7. Similar results to those in scenario Charlie obtained, where the operating system platform used Snort and attacker are reversed.

Attack Platform: Host Windows **vs** Snort Platform: Host Linux 2.6			
Parameter	100 – 200 Mbps	500 – 700 Mbps	800 Mbps – 1.0 Gbps
Network Utilization	15%	54 %	96%
CPU Usage	25 – 30%	32 - 35%	38 – 45 %
Snort CPU Usage	18 – 22%	22 – 26%	27-35%
Packets Analysed	99 %	42 %	35 %
Packets Dropped	1 %	58 %	65 %
Alerts & Log	100%	60 %	30 %

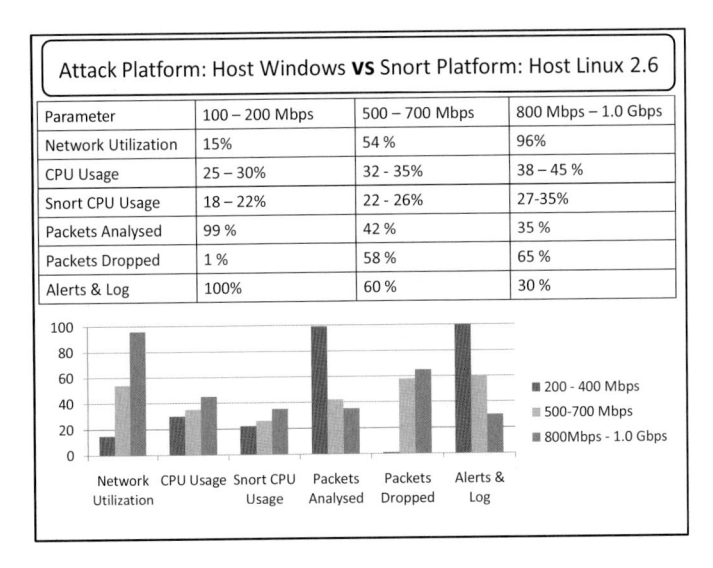

Fig. 7. Results – Scenario Echo

4 Test-Bench II

Fig. 8 describes the Test-Bench II that Snort has been implemented on a fully resourceful host machine built on dual quad core processor using 4.0 Gb RAM, the configuration of the network machines are shown in Table 4. Snort been respectively evaluated on the fully resourceful platforms built on Windows Server 2008 [22], Linux Server 2.6 and Free Berkley Software Distribution (BSD) 7.0 [23] respectively.

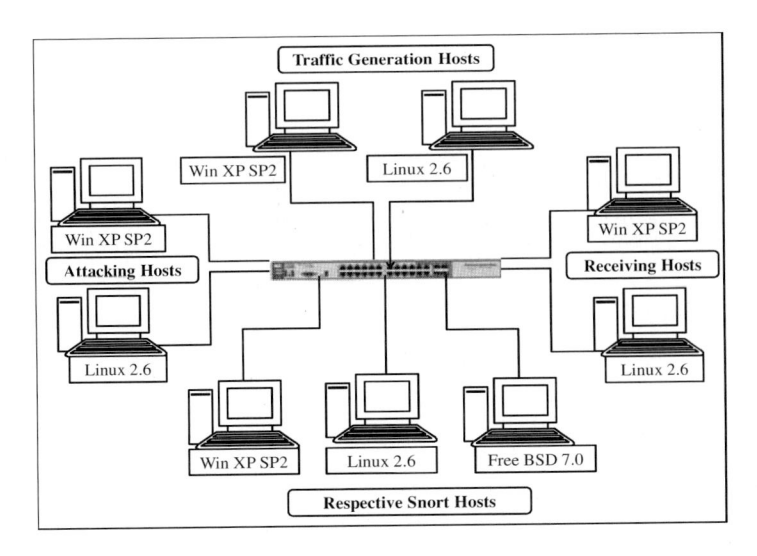

Fig. 8. Test-Bench II

Table 4. Network Description- Test-Benches II and III

Machine Type	Hardware Description	Tools Used
Network traffic/ back ground traffic generator (Win XP SP2)	Dell Precision T3400, Intel Quad-Core, Q6600 2.40 GHz. 2 GB RAM, PCIe, IGbps RJ45, Network Card (Broadcom NetXtremo Gigabit Ethernet), L2 cache 2 x 4.0 MB, FSB 1066 MHz.	Traffic Generators: NetCPS [10], Tfgen [11], Http Traffic Gen [12], LAN Trrafic Version 2 [13] and D-ITG Version 2.6 [14]
Network traffic/ back ground traffic generator (Linux 2.6)	Dell Precision T3400, Intel Quad-Core, Q6600 2.40 GHz. 2 GB RAM, PCIe, IGbps RJ45, Network Card (Broadcom NetXtremo Gigabit Ethernet), L2 Cache 2 x 4.0 MB, FSB 1066 MHz.	Traffic Generators: D-ITG Version 2.6 [14] and hping Version 2 [15]
IDS Machine	Dell Precision T5400, Intel Xeon **Dual Quad-Core 2.0 GHz**, 4 GB RAM, L2 cache 2x6MB, FSB 1066 MHz, PCIe, Network Interface Card, 10 Gbps Chelsio, HD: 1000 GB, Buffer 32 MB, SATA.	IDS: Snort [4]
Receiving Hosts • Win XP SP2 • Linux 2.6	Dell Precision T3400, Intel Quad-Core, Q6600 2.40 GHz. 2 GB RAM, PCIe, IGb/s RJ45, NIC 10 Gbps Chelsio on Win XP SP2 host and Linux 2.6 host has Broadcom NetXtremo Gigabit Ethernet.	• Win XP SP2 – LAN Traffic Generator • Linux 2.6 – D-ITG Traffic Generator
Switch	ProCurve Series 2900 [9], 10Gbps switch with 24x1 Gbps ports and 2x10 Gbps 3CR17762-91-UK ports.	

The system performance is gauged in terms of its packet handling capacity of the application built on respective platforms for different types of network traffic.

4.1 Evaluation Methodology

4.1.1 UDP Traffic

The evaluation methodology is based on the following specifications:

- Different packet sizes (128, 256, 512, 1024 and 1514 bytes) were generated and Snort's performance at the following traffic-load was evaluated: 750 Mbps, 1.0 Gbps, 1.5 Gbps and 2.0 Gbps respectively.
- Varying traffic payload: UDP and Mixed TCP, UDP and ICMP Traffic.
- Snort's performance characteristics were evaluated - packets received, packets analysed, packets dropped and CPU usage at various packet sizes and band widths levels.
- Duration of test: 1, 5 and 10 minutes, we have taken the average value of the results obtained.

The response of the IDS System (Snort) on UDP traffic injected in various packet sizes and bandwidths is shown in Table 5; each scenario is discussed in following paragraphs:

4.1.2 Mixed Traffic

The mixture of TCP (70%), UDP (20%) and ICMP (10%) traffic was generated replicating realistic network flow as follows:

- Generating random packet sizes and observing the response of system – packet handling capacity.
- Traffic bandwidth limited to 1.0 Gbps – supporting commodity hard-ware on account of system implementation as a Test-Bench.
- Recording packet drop statistics in all three Snort platforms built on Free BSD, Linux and windows respectively.

Table 5. Host-based Configuration Results – UDP Traffic (Packet loss %)

Traffic (Bandwidth)	OS	128 Bytes	256 Bytes	512 Bytes	1024 Bytes	1514 Bytes
750 Mbps	Free BSD	15.40	9.450	3.29	6.64	6.26
	Linux	56.91	52.67	27.83	6.72	6.40
	Windows	51.76	50.62	25.32	6.83	6.35
1 Gbps	Free BSD	52.60	32.15	28.40	25.04	24.89
	Linux	72.70	69.04	65.88	55.26	53.35
	Windows	68.05	66.82	61.97	53.60	52.90
1.5 Gbps	Free BSD	66.70	62.03	46.22	41.60	40.80
	Linux	77.60	71.50	67.32	57.10	55.50
	Windows	80.60	74.70	70.23	68.31	64.60
2 Gbps	Free BSD	74.07	69.80	65.30	50.54	49.40
	Linux	78.04	75.80	69.60	59.30	57.30

4.2 Results

4.2.1 UDP Traffic – 750 Mbps

Performance of all operating systems linearly improved from smaller packet size (128 Bytes) to larger one (1514 Bytes); however Free BSD shows a significant edge over the others in all ranges of packet sizes as shown in Fig. 9.

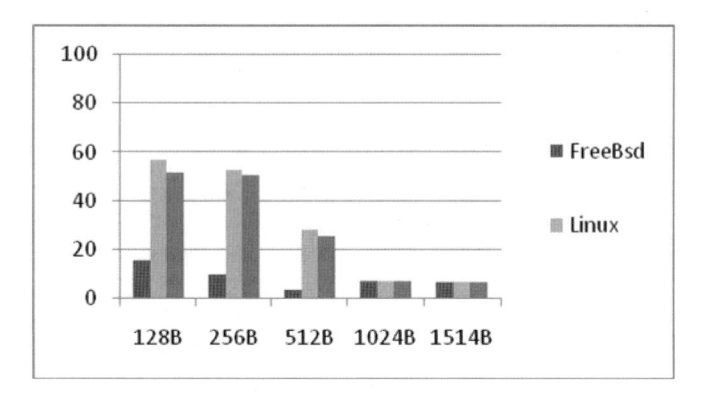

Fig. 9. Results: Packet Dropped (%), UDP Traffic – 750 Mbps

4.2.2 UDP Traffic – 1.0 Gbps

Increase in the bandwidth shows decline in the performance of system result into more packet loss as shown in Fig. 10. Considerably uniform response has been

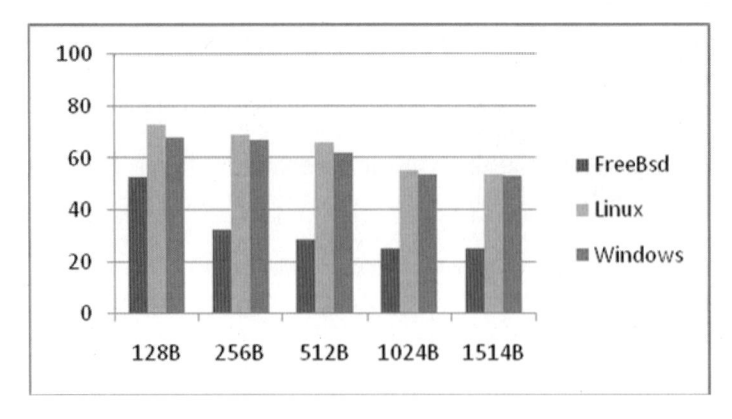

Fig. 10. Results: Packets Dropped (%), UDP Traffic – 1.0 Gbps

observed in all categories of packet sizes from all platforms tested respectively. This scenario also showed a comparatively improved performance from Free BSD however not an ideal one.

4.2.3 UDP Traffic – 1.5 Gbps

A further increase in the traffic bandwidth result into more packet loss by the system. Approximately similar performance was observed in all packet sizes, the response indicates Free BSD performed better followed by Linux and Windows at last as shown in Fig. 11.

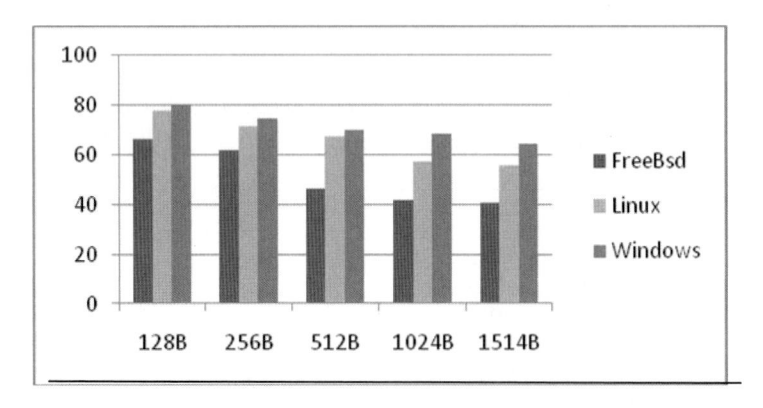

Fig. 11. Results: Packets Dropped (%), UDP Traffic – 1.5 Gbps

4.2.4 UDP Traffic – 2.0 Gbps

At 2.0 Gbps of traffic input, performance of Windows seemed totally compromised at 128 Bytes of packet sizes, the platform lost approximately all the input traffic and performed no evaluation. It gradually increased for higher packet sizes, in a similar pattern as observed for the lower traffic bandwidths. This, however; displayed a highly compromised performance from all platforms identifying the strong limitations

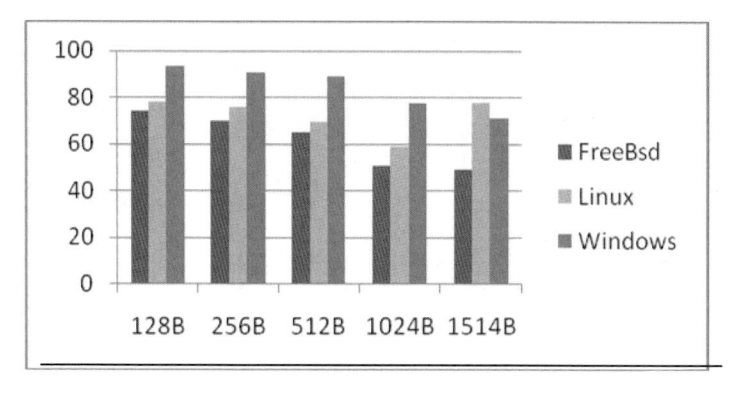

Fig. 12. Results: Packets Dropped (%), UDP Traffic – 2.0 Gbps

to handle the input traffic reaching 2.0 Gbps as shown in Fig. 12. Practically, the system build on Free BSD, Linux and Windows platforms once subjected to 2.0 Gbps of input traffic suffer heavy packet loss.

4.2.5 Mixed Traffic

The main reason to conduct this test is to ascertain the performance of system in realistic network conditions. The results here also followed quite similar pattern of system response. Table 6 describes the results obtained, Free BSD showed quite good performance in terms of handling mixed traffic for the bandwidth of 1.0 Gbps on a multi-core implementation.

Table 6. Host-based Configuration Results – Mixed Traffic

Operating System	Dropped Packets
FreeBSD	21.7 %
Linux	27.2 %
Windows	26.3 %

5 Test-Bench III

Virtualization is a framework for abstracting the resources of a PC into multiple execution platforms by creating multiple machines on a single computer. Each machine operates on the allocated hardware and can afford multiple instances of applications [24]. This concept has been successfully incepted within the industry/ business community. The mechanics of system virtualization to implement network security tools has been considered as an appropriate choice for academia dealing with information security [25, 26].

The concept has been developed to address the issues related to reliability, security, cost and complexity of the network/systems. It has successfully been used for the processing of legacy applications, ensuring load balancing requirements, resource sharing and tasking among virtual machines by using autonomic computing techniques. The technique has also shown merits in the situation where an application failure on one

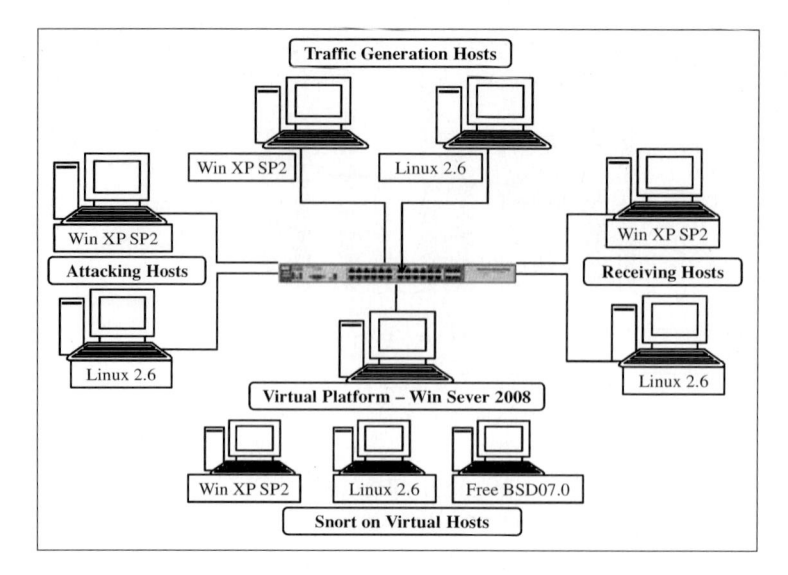

Fig. 13. Test-Bench III

machine does not affect the other. In addition, ease of isolation allows multiple OS platforms to be built on one machine running variable instances of applications. This has made the concept quite fascinating for the research community.

As discussed in [10], the Test-bench III is distributed into three parts and configured around a ProCurve series 2900 switch [9] as shown in Fig. 13.

The basic idea of the evaluation process revolves around packet capturing and evaluation by virtual platforms and Snort. We have selected two machines for traffic generation: Linux 2.6 and Windows XP SP2 platforms respectively. Similarly, the traffic reception machines were also deployed to fulfill network requirements. Details of the traffic generation tools are shown in Table 2 and 4.

The virtual platform running Snort has been configured on a dual quad-core processor. The machine hardware details are listed in Table 4. The system is built on the Windows 2008 Server platform and three separate virtual platforms have been created-Windows XP SP2, Linux 2.6 & Free BSD 7.1. Snort is running simultaneously on all the virtual machines and similar traffic-loads and type are injected onto all platforms.

5.1 Evaluation Methodology

In order to ascertain the capability of Snort to handle high-speed network traffic on virtual platforms we proceed as follows:

- Parallel Snort sessions were run on all virtual machines.
- The machines were injected with similar traffic-load characteristics (UDP and TCP Traffic) for 10 minutes.
- Different packet sizes (128, 256, 512, 1024 and 1460 bytes) were generated and Snort's performance at the following traffic-load was evaluated: 100 Mbps, 250 Mbps, 500 Mbps, 750 Mbps, 1.0 Gbps and 2.0 Gbps respectively.

- Snort's performance characteristics were evaluated - packets received, packets analysed, packets dropped and CPU usage at various packet sizes and band widths levels.
- Packets received were compared at both the host OS and the virtual platforms running the Snort applications.
- During the course of the tests, no changes were made in OS implementation specifically Linux using New Application Program Interface (NAPI) [27] and[2] MMP and Free BSD using Berkley Packet Filter (BPF) [28].

5.2 Results

The results are distributed over UDP and TCP traffic types respectively. It was observed that the total packets transmitted from the traffic-generating PCs was equivalent to the number of packets received at the host machine/ OS running virtual platforms as shown in Table 7; however, this is not the case once the system found[3] non responsive.

5.2.1 UDP Traffic

The results below are described in relation to packet size, bandwidth (i.e. traffic-load), and the virtual OS platform running the Snort application:

5.2.1.1 Snort Response for Packet Sizes of 128 and 256 Bytes

- Linux shows quite good performance for these packet-sizes upto 250 Mbps traffic-load; its performance declined at higher bandwidth levels as shown in Fig. 14. The system found non responsive at the traffic-loads of 750 Mbps and above.

Table 7. Packets Received at Host Operating Systems

Total Packets Received at OS (Millions) – UDP					
Bandwidth	128 Bytes	256 Bytes	512 Bytes	1024 Bytes	1460 Bytes
100 MB	60	35.82	17.77	10.56	6.96
250 MB	178.1	94.14	48.00	18.34	20.22
500 MB	358.3	148.29	92.56	46.2	39.00
750 MB	System Non Responsive		144.72	91.56	45.23
1.0 GB	System Non Responsive			167.40	78.00
2.0 GB	System Non Responsive				

Total Packets Received at OS (Millions) – TCP			
Bandwidth	50 Connections	100 Connections	200 Connections
100 MB	10	26.7	21.60
250 MB	31.86	39.763	48.69
500 MB	67.90	108.56	84.098
750 MB	80.29	113.72	124.58
1.0 GB	102.51	118.144	148.982
2.0 GB	147.54	170.994	221.28

---

[2] Modified Device Drivers Packet Handling Procedures.
[3] In non responsive situation we consider 100% packet loss.

- Windows shows good performance for 128 Bytes packet sizes at 100 Mbps loading only. Its performance is compromised at higher loading levels as shown in Fig. 14. The system also found non responsive at traffic-loads of 750 Mbps and above.
- Free BSD performs slightly better than Windows as shown in Fig. 14. The system also found non responsive at traffic-loads of 750 Mbps and above.

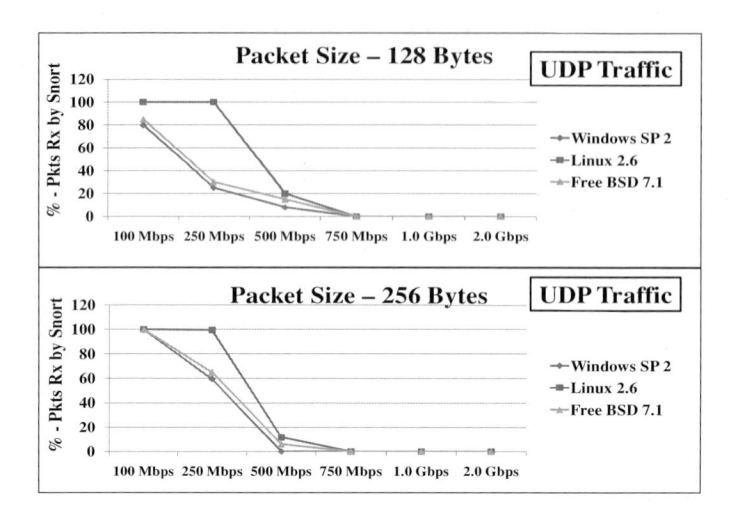

Fig. 14. Snort Packets Received (%) - UDP Traffic (128 Bytes & 256 Bytes)

5.2.1.2 Snort Response for Packet Sizes of 512 and 1024 Bytes

- Linux shows quite good performance for traffic-load up to 500 Mbps for all packet sizes as shown in Fig. 15. The Linux however system found non responsive at traffic-loads of 1.0 Gbps and above for 512 Bytes packet sizes and at 2.0 Gbps for packet sizes of 1024 Bytes.
- Windows also performed satisfactorily at traffic-loads of 250 Mbps and 500 Mbps for packet sizes of 512 Bytes and 1024 Bytes respectively as shown in Fig. 15. The system found non responsive at traffic-loads of 1.0 Gbps and above for packet size of 512 Bytes and 2.0 Gbps for packet sizes of 1024 Bytes.
- Free BSD responds a bit better than Windows as shown in Fig. 15. The system found non responsive at traffic-loads greater than 1.0 Gbps for packet sizes of 512 Bytes and 2.0 Gbps for packet sizes of 1024 Bytes.

5.2.1.3 Snort Response for Packet Size of 1460 Bytes

- Linux shows significantly better performance for packet sizes of 1460 Bytes of packet for traffic-loads up to 1.0 Gbps however, the system found non responsive at 2.0 Gbps loading as shown in Fig. 16.
- Windows also shows good performance upto750 Mbps loading. The system found non responsive at 2.0 Gbps traffic-load as shown in Fig. 16.
- Free BSD responds a bit better than Windows as shown in Fig.16. The system found non responsive at 2.0 GB traffic-load as shown in Fig. 16.

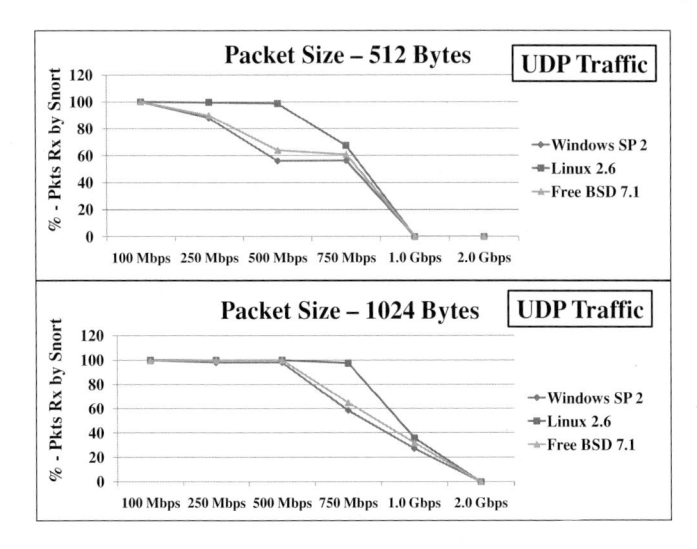

Fig. 15. Snort Packets Received (%) - UDP Traffic (512 Bytes & 1024 Bytes)

5.2.2 TCP Traffic

We have included the results of 512 Bytes packet sizes in this section due to paucity of space. The results have been accumulated on the basis of successful connections (50, 100 and 200 respectively).

5.2.2.1 Snort Response for 50 Connections of 512 Bytes

- Linux exhibits quite good performance up to 750 Mbps loading however, its performance declined at higher traffic-loads as shown in Fig. 16.

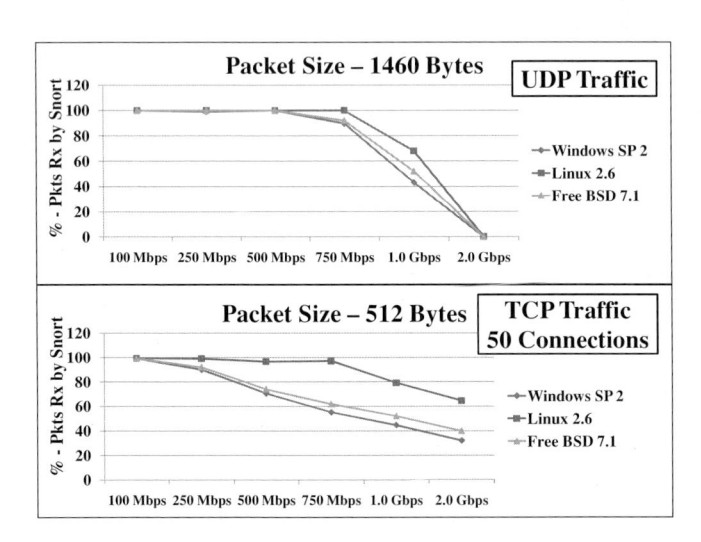

Fig. 16. Snort Packets Rx (%) - UDP (1460 Bytes) & TCP (50 Connections)

- Windows was acceptable up to 250 Mbps loading and its performance reduced for higher traffic-loads as shown in Fig. 16.
- Free BSD performed a bit better than Windows as shown in Fig. 16.

5.2.2.2 Snort Response for 100/ 200 Connections of 512 Bytes

- Linux exhibits quite good performance up to 250 Mbps loading with minimum packet loss, however, its response linearly declined for higher traffic-loads. Windows also exhibits a similar performance level up to 250 Mbps loading levels and its performance declined for higher traffic-loads as shown in Fig. 17.
- Free BSD performs a bit better than Windows as shown in Fig. 17.
- Overall the performance of both categories (100 and 200 Connections is quite similar for packet size 512 Bytes.

6 Analysis

6.1 Test-Bench 1

- As expected, the performance of Snort was found to be dependent on its supporting hard-ware components (CPU, memory, NIC etc). In the virtual scenarios, Snort was found to be less accurate for all categories of background traffic. Conversely, the performance of Snort improved when run natively on its host machine by utilizing all of the available hardware resources.

  The statistics for percentages of dropped packets are shown in Fig. 18. Resource constraints in the virtual machine have affected the overall performance of Snort resulting in a high number of packets dropped and a reduction in alerts logged.

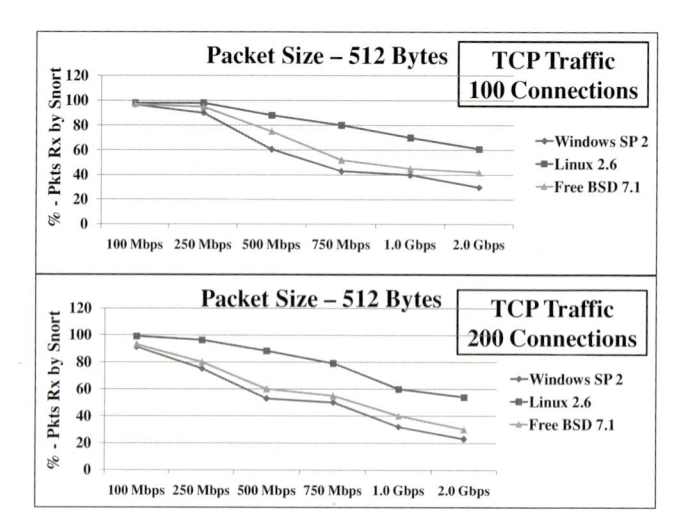

Fig. 17. Snort Packets Received (%) - TCP Traffic (100 & 200 Connections)

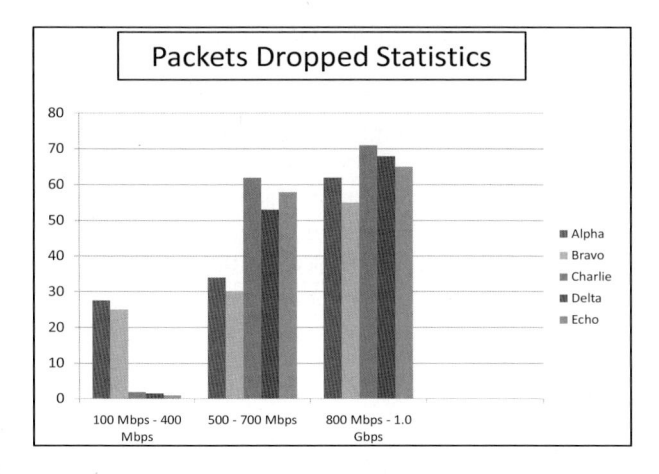

Fig. 18. Packets Dropped

- Background traffic plays a significant role in the performance of Snort. The higher the traffic, the less is the performance of Snort. The impact of background traffic can be ascertained by analyzing the statistics of alerts generated in different categories as shown in Fig. 19.
- Traffic within the range of 100 – 400 Mbps has no significant impact on the performance of Snort when run natively on host machines. However, its performance declines in a virtual setup. Snort was found to be accurate in all scenarios.
- A slight increase in background traffic, in the range of 500 – 700 Mbps causes deterioration in the performance of Snort. This degradation is approximately the same in all scenarios.
- With high background traffic levels, ranging from 800 Mbps – 1.0 Gbps Snort start bleeding. The number of alerts and log entries suffer significant reduction, thus identifying an evident limitation in Snort's detection capability.
- Snort was found to be more effective in the configuration where both attacker and host are on the same OS.
- Snort performance is significantly reduced in the 1.0 Gbps scenarios.
- System performance in relation to packet capture capabilities also found dependent on CPU usage. Higher the CPU usage lesser would be the packet captured for analysis by Snort application. Packets received at virtual platform for evaluation by Snort are significantly less than the packets captured at host platform. However, lesser amount of packets received by virtual platforms result in improved packets analysis statistics by Snort. For example in Windows virtual platform, Snort analyzed 38% of the total packets received at system level whereas in host Windows configuration, this value reduced to 27%. Better packets analysis percentage of virtual platform is due to the fact that the Snort analyzed considerably lesser amount of packets whereas the packets captured for analysis at host level are significantly more thus by no means virtual platform performed better than the fully resourceful host.
- The performance of Snort on a Linux platform was observed to be comparatively better than that of Windows. The results shown in Fig. 20 are based on the

scenarios in which the Snort and attacker are on well resourced host machines (Scenario Charlie and Delta).
- In general, Snort was found to be inaccurate when handling traffic levels above 500 Mbps. There was also a significant performance decline when the traffic load exceeded 500 Mbps.

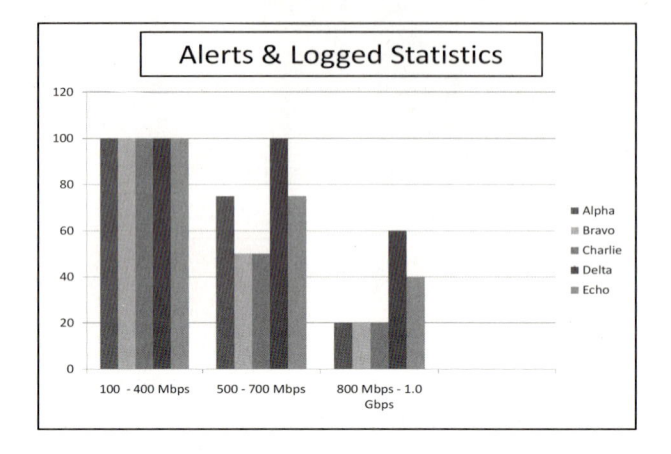

Fig. 19. Alerts & Log (Success Rate)

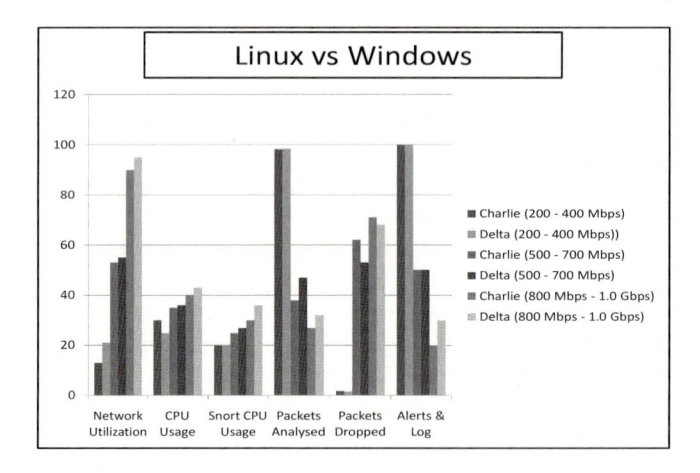

Fig. 20. Comparison - Snort on Linux & Win

6.2 Test-Bench II

The shaded cells in Table 5 indicate the case of I/O disk bottleneck, when the queue for I/O reading and writing exceeds an acceptable limit and the hosting machine is no longer able to process all the traffic (detailed discussion ahead). The overall assessment of system performance indicates following:

- Snort running on Free BSD has achieved the greatest performance in comparison to other OS for all traffic volumes and packet sizes.
- Windows and Linux showed quite similar performance in all scenarios.
- Small sizes of UDP packets are computationally expensive and the performance of Snort declines for increase in traffic bandwidth.
- Considering the 1024 Bytes as an average packet size for normal real life traffic, the raw processing rate of Snort application showed acceptable performance till bandwidth of 750 Mbps for all OS and 1.0 Gbps for Free BSD.
- We have also recorded the CPU and memory usage of the system for packet size 1024 Bytes (UDP Traffic) as shown in Fig. 21. It has been observed that more than 60 % of the hardware strength is available for the traffic ranging from 100 Mbps to 2.0 Gbps.
- The non highlighted results in Table 5 also indicates the Snort design to support multi-core architecture, this is also under the evaluation phase in our test-lab.

6.3 Test-Bench III

- The dynamics of virtualization requires the Host OS and the Virtual Machine Software (VMware Server [29]) to be stored in the physical memory (RAM) of the host machine. The virtual machines (Windows XP SP 2, Linux 2.6 and Free BSD 7.0) running on VMware Server have been respectively allocated virtual RAM and disk space on the physical hard drive of the host machine. The processes/ applications running on the virtual machines use these simulated virtual RAMs and hard disks for the various operations shown in Fig. 22 [10].

The Test-Bench has multiple instances of Snort and packet-capture libraries running on different virtual platforms each with a different OS. The packets captured by each virtual machine are less than the packets received by the Network Interface Card (NIC), thus identifying packet loss somewhere in between. The basic cause of packet loss at each OS apart from the losses incurred by Snort during evaluation is the bottleneck caused by a low disk data transfer rate.

The memory and storage for each virtual machine has actually been allocated on the physical storage resources (i.e. hard disk) of the host machine. Packets received by the NIC without any loss are transferred to the hard disk buffer at the Peripheral Component Interconnect (PCI) rate (4/8 Gbps). From this buffer, these packets are required to be written to the disk at the buffer- to-host transfer rate of 300 MB/sec (SATA Hard Drive [30]); thus a huge gap between the disk-transfer rate and the incoming traffic load exists. In addition, traffic is fed to all virtual machines simultaneously (in parallel mode), the disk is physically only able to write to one location at a time. Thus any disk-write instance to a virtual machine will cause packet drops on another. There are also some additional packet losses due to context switching within the hard disk.

In order to augment our analytical stance of showing hardware as one of the major bottlenecks for the efficacy of the virtualization concept for NIDS in high-speed networks we have utilized the disk queue length counter as shown in Fig. 10. In normal circumstances, the average disk queue length should be three or less (its ideal value) [31]. However; in our test network, it is observed to be always greater than the ideal value for the traffic ranges measured at 2.0 Gbps [10].

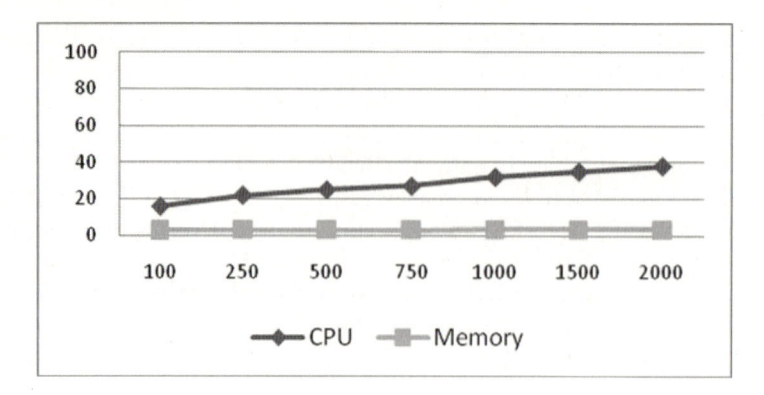

Fig. 21. CPU & Memory usage

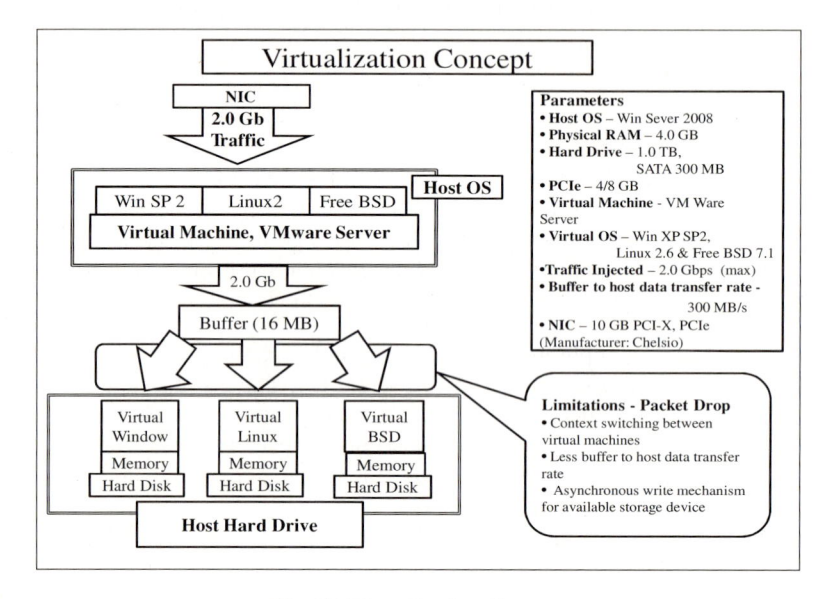

Fig. 22. Virtualization Concept

7 Recommendations

7.1 Processor Speed

Snort has shown a dependency on the hardware strength of host system. Our results identify a decline in the performance of Snort once CPU usage touches 50%. The increase in CPU usage is due to the processing overheads incurred when presented with high background traffic levels. Snort system hardware therefore needs to be compatible with the network traffic load it needs to process. We propose a simple mathematical formula to select a processor for the host system running Snort depending on network traffic loads as follows:

$$Ct = Nt \times K; \quad K = 4.8 \tag{1}$$

where Ct stands for processor speed in GHz, Nt is the maximum value of anticipated network traffic in Gbps and K is constant. The value of K is based on the statistics from the results obtained in this research work. We have verified our proposed formula in different scenarios and found it satisfactory.

7.2 Protocol Oriented Analysis

During our testing work, we also explored a technique which incorporates the analysis of traffic based on protocols. This was undertaken by creating a number of mirror ports on the network switch. The host Snort machines are configured to respond to a specific protocol, thus considerably reducing the processing required to be performed. These protocol-oriented engines reduce the numbers for packet dropped, increase efficiency and reduce response time. The proposed architecture is shown in Fig. 23. The concept also addresses the limitation of Snort being a non multi-threaded application [32].

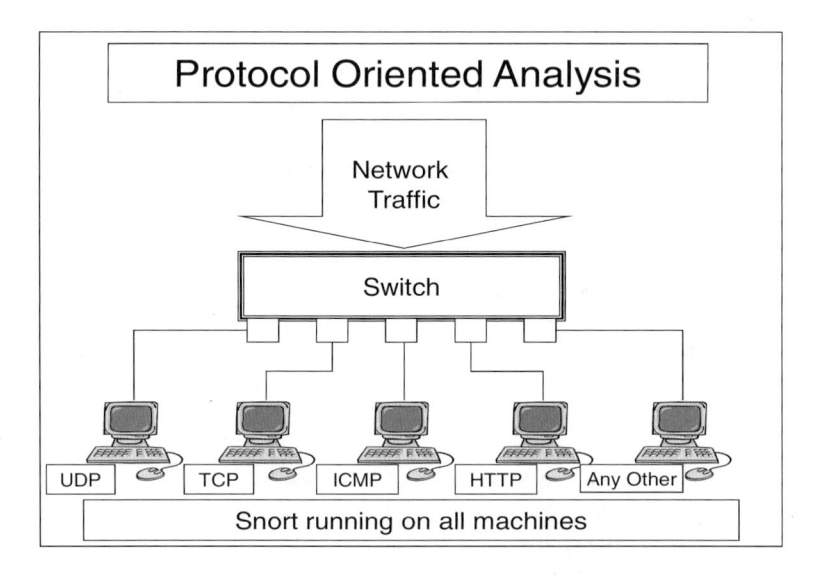

Fig. 23. Protocol Oriented Analysis

7.3 Serialization Concept

Another technique enables the re-evaluation of dropped packets by the Snort engine. Packets dropped by the Snort engine on account of processing overheads and overflow are analyzed by a parallel logic designed to handle packets based on certain attributes. The flow diagram as shown in Fig. 24 can be a good supplement to the Snort packet-dropping weakness under high traffic conditions. We have discussed this concept in our work based on Comparator Logic in [33].

The dropped packets belonging to a complete session can be reassessed and packets showing incomplete sessions get deleted. Similarly, packets showing protocol violations get deleted and others get a chance to be re-evaluated. Other suitable parameters could also be incorporated into the architecture.

7.4 Dynamic Configuration

Site specific NIDS configuration that maximizes the analysis under predefined resource constraints as presented in [34] can also be implemented for Snort.

8 Conclusions

This chapter has focused on ways of determining the efficacy of the widely deployed open-source NIDS, namely Snort, in high-speed network environments. The current development in hardware technologies has opened broad prospects for legacy applications particularly software-based deployed at network edges. Multi-core systems are available and widely used to offer intensive computational opportunities.

The test scenarios employed, involved the evaluation of application under different traffic conditions and observing the response of system to known attack signatures.

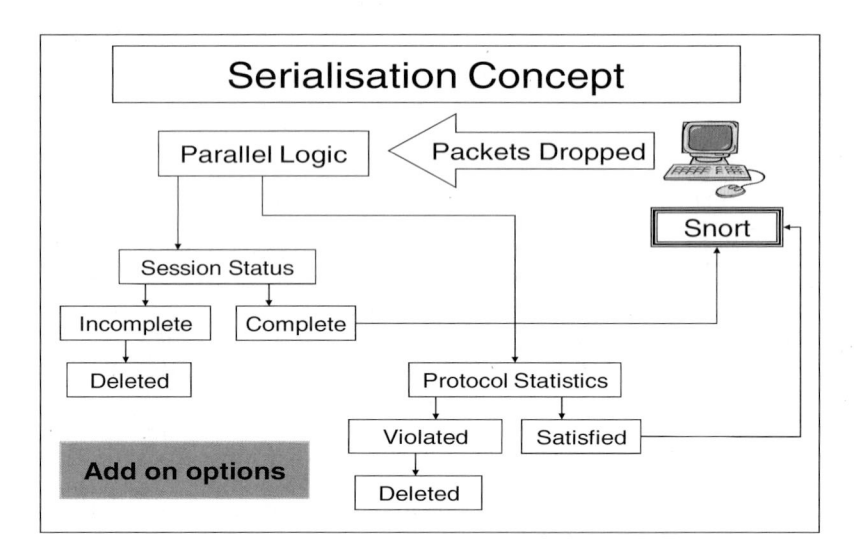

Fig. 24. Serialization Concept

The results obtained have shown a number of significant limitations in Snort at commodity level. It was confirmed that the underlying host hardware plays a prominent role in determining overall system performance. It was also shown that performance is further degraded as the number of virtual instances of NIDS is increased, irrespective of the virtual OS used.

This hardware dependency is exacerbated when running Snort as a virtual machine and it is to be anticipated that running a large number of Snort instances would lead to major degradations in performance and detection levels. In general, any limitations in system configuration would result in poor performance of the IDS and thus, it could be concluded that commodity hardware is not the ideal platform for IDS implementation in high speed environment. The results obtained have shown a number of significant limitations in the use of virtual NIDS, where both packet-handling and processing capabilities at different traffic loads were used as the primary criteria for defining system performance. Furthermore, it was demonstrated a number of significant differences in the performance characteristics of the three different virtual OS environments in which Snort was run.

The performance of Snort was analyzed under realistic network conditions. The results obtained identify a strong dependency of Snort on the host machine configuration. It can be ascertained that Snort is not suitable for all network implementations with high volume of traffic above 750 Mbps. During the course of this study, the suitability of some techniques was identified towards the enhancement of the performance of Snort and they can form the basis of the enhancement of the recommended solutions in future research works.

Acknowledgement. The authors are extremely grateful to Professor Demetres Kouvatsos for his advice and assistance. His in-depth reviews has definitely improved the quality of this work.

References

1. John, D., Tessel, S., Young, F.L.: The Hackers Handbook. Auerbach Publications, New York (2004)
2. Intrusion Detection Systems (IDS) Part 2, Classification; methods; techniques, http://www.windowsecurity.com/articles/IDS-Part2-Classification-methods-techniques.html04
3. Infrastructure Security Report- ARBOR Networks, http://www.arbornetworks.com/report
4. Snort, http://www.Snort.org
5. Paulauskas, N., Sjudutis, J.: Investigation of the Intrusion Detection System "Snort" Performance. Journal of Electrical and Electronics Engineering (2008)
6. Andrew, R.B., Joel, E.: Snort IDS and IPS Toolkit, Syngress, Canada (2007)
7. Case, J., Fedor, M., Schoffstall, M., Davin, J.: Simple Network Management Protocol (SNMP), Network Security Research Group, RFC 1157 (1990)
8. Wu, S., Manber, U.: AGREP, A Fast Approximate Pattern-Matching Tool. In: Proc. of USENIX Winter 1992 Technical Conference, San Francisco, CA, pp. 153–162 (1992)
9. ProCurve Series 2900 Switch, http://www.hp.com/rnd/products/switches/HP_ProCurveSwitch2900Seriesoverview.htm
10. Akhlaq, M., Alserhani, F., Awan, I.U., Mellor, J., Cullen, A.J., Mirchandani, P.: Virtualization Efficacy for Network Intrusion Detection Systems in High-speed Networks. In: Weerasinghe, D. (ed.) IS&DF, vol. 41, pp. 26–41. Springer, Heidelberg (2010)
11. NetCPS, http://www.netchain.com/NetCPS
12. Tfgen, http://www.st.rim.or.jp/~yumo/pub/tfgen
13. Http Traffic Generator, http://www.nsauditor.com/

14. LAN Traffic V2, http://www.topshareware.com/lan-traffic-v2/downloads/1.htm
15. D-ITG V2.6, http://www.grid.unina.it/Traffic/index.php
16. Hping V 2, http://www.hping.org/download.html
17. Metasploit Framework, http://www.metasploit.com/
18. Bandwidth Monitor, http://www.sourceforge.net/projects
19. Nload, http://www.sourceforge.net/projects/nload/
20. Windows XP SP2, http://www.softwarepatch.com/windows/xpsp2.html
21. Linux 2.6, http://www.kernel.org/
22. Windows Server 2008, http://www.microsoft.com/windows_server2008/en/us/default.aspx
23. Free BSD 7.1, http://www.freebsd.org/where.html
24. An Introduction to Virtualization, http://www.kernelthread.com/publications/virtualization
25. Business value of virtualization: Realizing the benefits of integrated solutions, http://h18000.www1.hp.com/products/servers/management/vse/Biz_Virtualization_WhitePaper.pdf
26. Virtualization, http://www.windowsecurity.com/whitepapers/Virtualization.html
27. Linux NAPI, http://www.linuxfoundation.org/collaborate/workgroups/networking/napi
28. Berkley Packet Filter, http://www.freebsd.org/releases/7.1R/relnotes.htm
29. VM Ware Server, http://www.vmware.com/products/server/
30. SATA Technology, http://www.serialata.org/
31. Disk Queue Length Counter, http://www.windowsnetworking.com/articles_tutorials/
32. Alserhani, F., Akhlaq, M., Awan, I.U., Cullen, A.J., Mellor, J., Mirchandani, P.: Evaluating Intrusion Detection Systems in High Speed Networks. In: Proceeding of Fifth IEEE Conf. of Information Assurance and Security, IAS (2009)
33. Subhan, A., Akhlaq, M., Alserhani, F., Awan, I., Cullen, A., Mellor, J., Mirchandani, P.: Smart Logic - Preventing Packet Drop in High Speed Network Intrusion Detection Systems. In: Weerasinghe, D. (ed.) IS&DF, vol. 41, pp. 57–65. Springer, Heidelberg (2010)
34. Akhlaq, M., Alserhani, F., Subhan, A., Awan, I.U., Mellor, J., Mirchandani, P.: High Speed NIDS using Dynamic Cluster and Comparator Logic. In: Proc. of 2010 IEEE 10th International Conference on Computer and Information Technology (CIT), pp. 575–581 (2010)

Appendix I: List of Acronyms

BSD - Berkley Software Distribution
BPF - Berkley Packet Filter
CPU - Central Processing Unit
DDoS - Distributed Denial of Service Attack
D-ITG - Distributed Internet Traffic Generator
FSB - Front Side Bus
Gbps - Giga bits per second
GB - Giga Bytes
HTTP - Hypertext Transfer Protocol
ICMP - Internet Control Message Protocol
IDS - Intrusion Detection Systems
I/O - Input and Output
IPS - Intrusion Prevention Systems
MB - Mega Bytes
MMP - Modified Device Drivers Packet Handling Procedures
NIDS - Network Intrusion Detection Systems
NIC - Network Interface Card
NAPI - New Application Program Interface
OS - Operating System
Pkts - Packets
PC - Personal Computer
PCI X - Peripheral Component Interconnect Extended
PCIe - Peripheral Component Interconnect Express
RX - Received
RPC - Remote Procedure Call
RJ45 - Registered Jack Serial 45
SNMP - Simple Network Management Protocol
TCP - Transmission Control Protocol
UDP - User Datagram Protocol
VRT - Virtual Resource Tree
VM Ware - Virtual Machine Software
Win XP SP2 - Windows XP Service Pack Serial 2

Appendix II

Aho-Corasick Algorithm. The algorithm is multi-pattern exact matching algorithms, based on FSA (finite state automata). In pre-processing stage, Aho-Corasick constructs a state machine called Trie from the strings to be matched. Aho-Corasick state machine implementation can be based on either of two – Non Deterministic Finite Automata (NFA) or Deterministic Finite Automata (DFA).

We have considered NFA which mainly comprise Full, Sparse and Standard (Std) bands. The **Full** matrix representation is one in which all the states have a next state entry for each of the possible input values. However, this kind of a data structure is too memory consuming, and hence not a suitable data structure for storage purposes. In the **Sparse** storage format, the data structure is a bit different. The elements of the new storage matrix will be the number of valid state transitions for the given state, the valid transitions and the corresponding next state. This kind of a storage helps us reduce memory, but speed may be compromised because the random access into the matrix is now lost in this modified structure. Finally the **Standard** mode is the high memory and high performance type.

Aho, Alfred V.; Margaret J. Corasick (June 1975). "Efficient string matching: An aid to bibliographic search". *Communications of the ACM* **18** (6): 333–340

Wu-Manber Algorithm. The algorithm is a high performance multi-pattern matching algorithm, it has two broad types Wu Manber and Wu Manber (Low). The regular **Wu Manber** algorithm has two stages. It uses the bad-character shift and considers the characters from the text in blocks of size B instead of one by one and, consequently, expands the effect of bad-character shift. Wu-Manber algorithm uses a hashing table to index the patterns in the actual matching phase. The best performance of Wu-Manber algorithm is $O(Bn/m)$, where n is the text size and m is minimum length of the pattern. The running time of the Wu-Manber algorithm does not increase in proportion to the size of the pattern set. The performance of the Wu-Manber algorithm is dependent on the minimum length of the patterns. In pre-processing stage, the Wu-Manber algorithm builds three tables, a SHIFT table, a HASH table and a PREFIX table. SHIFT table is used to determine how many characters in the text can be shifted (skipped) when the text is scanned. The HASH and PREFIX tables are used when the shift value is 0 to determine which pattern is a candidate for the match and to verify the match.

Compared with the regular Wu-Manber algorithm, the improved **Wu-Manber (low)** algorithm has three differences, which make it more efficient: 1) a rarest substring with fixed length is chosen for each original pattern as representative to better the quality of the Boyer-moore like SHIFT table; 2) a second Boyer-moore SHIFT table is computed to improve the likelihood of shifting text sliding window continuously; 3) a simple hash function with good randomness property is crafted to build a balanced hash table to accelerate the searching speed of possible matching patterns.

Reference, Boyer Moore: Boyer R. S., Moore J. S: A fast string searching algorithm. In Communications of the ACM. 1977,20 (10), pp. 762-772 (1977).

Unicast QoS Routing in Overlay Networks

Dragos Ilie[1] and Adrian Popescu[2]

[1] dragos.ilie@gmail.com
[2] Blekinge Institute of Technology, Sweden
adrian.popescu@bth.se

Abstract. The goal of quality of service (QoS) routing in overlay networks is to address deficiencies in today's Internet Protocol (IP) routing. This is achieved by application-layer protocols executed on end-nodes, which search for alternate paths that can provide better QoS for the overlay hosts. In the first part of this paper we introduce fundamental concepts of QoS routing and the current state-of-the-art in overlay networks for QoS. In the remaining part of the paper we report performance results for the Overlay Routing Protocol (ORP) framework developed at Blekinge Institute of Technology (BTH) in Karlskrona, Sweden. The results show that QoS paths can be established and maintained as long as one is willing to accept a protocol overhead of maximum 1.5 % of the network capacity.

1 Introduction

One of Internet's characterizing features is its pervasive nature. Online banking, online shopping, voice over IP (VoIP), IP Television (IPTV), video on demand (VoD) and social networking (*e. g.*, Facebook and Twitter) are just a few examples of Internet services that have become an integral part of our lives. Some of these services, in particular those making heavy use of video and audio streams, have strict requirements on how the media streams must be handled during transit in the network. The requirements are typically expressed in the form of constraints on bandwidth[1], packet delay, delay jitter and packet loss. A network that meets these requirements is said to provide QoS [1]. A key issue in providing QoS is that of selecting paths for network traffic such that the stream requirements are satisfied. This can be done by QoS routing, which is a mechanism for optimizing network performance by constrained-path selection and traffic flow allocation.

To be more specific, QoS routing solves the following problem. A *source node* must transfer data over a network to a set of *destination nodes*. The two sets can be overlapping in the sense that some nodes are both senders and receivers. The data is in the form of packets. These packets are grouped into *flows*, where a flow consists of packets sharing the same source and destination address as well as a set of common QoS requirements called the *flow demand*. The problem is how

---
[1] In the field of computer networking, the term *bandwidth* is used to denote data rate or capacity, unless specified otherwise.

to satisfy all flow demands simultaneously. Two sub-problems must be solved in order to meet the flow demands. First, one must find at least one path that connects each source node to the corresponding destination node. Second, the flows must be allocated to these paths without violating their QoS requirements. A more far reaching objective is to solve these two problems while maximizing network throughput and minimizing congestion.

The network architect relies on a number of low-level building blocks in order to design a network with support for QoS. Examples of QoS low-level building blocks routing algorithms and protocols, resource reservation protocols, traffic shapers scheduling algorithms, admission control mechanisms, congestion avoidance and congestion control techniques. A *QoS architecture* defines how building blocks are combined in order to provide QoS.

Integrated Services (IntServ) is the first proposed QoS architecture for IP-based networks [2]. In IntServ, resources are allocated along the path by using the Resource Reservation Protocol (RSVP) [3]. IntServ performs *per-flow* resource management. This has led to skepticism towards IntServ's ability to scale, since core routers in the Internet must handle several hundred thousands flows simultaneously [4].

A second QoS architecture called Differentiated Services (DiffServ) [5] was developed, due to concerns about IntServ's scalability. DiffServ attempts to solve the scalability problem by dividing the traffic into separate forwarding classes. Each forwarding class is allocated resources as stipulated in the service level agreement (SLA) between provider and customer. Packets are classified and mapped to a specific forwarding class at the edge of the network. Inside the core, routers handle the packets according to their forwarding class. Since routers do not have to store state information for every flow, but only have to inspect certain fields in the packet header, it is expected that DiffServ scales much better than IntServ. A major problem with the DiffServ architecture has to do with end-to-end QoS provisioning over multiple DiffServ domains. Premium services cannot be offered unless bilateral SLAs exist between peering domains over the entire end-to-end path. Currently, technical difficulties coupled with the providers' lack of incentive to engage in bilateral SLAs has prevented wide-spread deployment of DiffServ [6,7].

Both architectures are of benefit to services and users located in the same network (*e. g.*, the corporate network), but fail to address a more heterogeneous scenario, where the service provider and users are scattered across the Internet. The main reason for this situation is because of the lack of interaction between network providers or difficulties to align premium services to a common denominator among the providers [8].

Furthermore, another issue to consider in the context of QoS is that of inter-domain routing. From a hierarchical point of view, Internet consists of a large number of autonomous systems (ASs). Interconnected ASs exchange routing information using the Border Gateway Protocol (BGP) [9]. An AS connects to other ASs through peering agreements. A peering agreement is typically a business contract stipulating the cost of routing traffic across an AS along with other policies to be maintained. When there are several routes to a destination the peering agreements force an AS to prefer certain routes over others. For example, given two paths to a destination where the first one is shorter (in terms of hops) and the second one is cheaper, the AS will tend to select the cheaper path. This is called *policy routing* and is one of the reasons for suboptimal routing. With the commercialization of the Internet it is unlikely that problems related to policy routing will disappear in the near future.

In summary, the Internet landscape today consists of a number of "islands" where QoS is provided by IntServ, DiffServ or other customized architectures. As long as peering agreements exist QoS services can be extended beyond the "island" borders. Otherwise, QoS is limited to IP's best-effort service. Consequently, some researchers are investigating the possibility to deploy QoS in overlay networks on top of IP. Overlay networks are more flexible to change architecturally, less dependent on network providers and cheaper to deploy. On the other hand, they face many challenges in the form of fluctuating QoS resources, coordination and scalability.

The remainder of this article is structured as follows. An overview of the current state-of-the-art in overlay networks for QoS is presented in Section 2. That is followed by a brief presentation of essential principals of QoS routing in Section 3. Section 4 introduces basic models for path selection and flow allocation. Simulation results for overlay routing protocols developed at BTH in Sweden are described in Section 5. We conclude with a brief summary and suggestions for future work. This article is based on the work presented in [10].

2 Overlay Networks for QoS

An overlay network utilizes the services of an existing network in an attempt to implement new or better services. Each overlay node establishes *virtual links* with a subset of its peers according to rules specific to the overlay. A virtual link consists of a sequence of one or more physical links (*i. e.*, a physical path). The physical path is selected by inter-domain (*e. g.*, BGP) and intra-domain routing protocols such as the Routing Information Protocol (RIP) and the Open Shortest Path First (OSPF). *Virtual paths* are selected by overlay routing protocols.

An example of an overlay network is shown in Figure 1. The physical interconnections of three ASs are depicted at the bottom of the figure. The gray circles denote nodes that use the physical interconnections to construct virtual paths used by the overlay network at the top of the figure.

The nodes participating in the overlay network perform active measurements to discover the QoS metrics associated with the virtual paths. As an example, assume that an overlay node in AS1 wishes to communicate with another overlay node in AS2. Assume further that AS1 always routes packets to AS2 by using the direct link between them, due to some policy or performance metric. The overlay node in AS1 may discover through active measurements that the path crossing AS3 can actually provide better QoS (*e. g.*, smaller delay), than the direct link to AS2. In this specific case, the AS1 node forwards its traffic to the AS3 node,

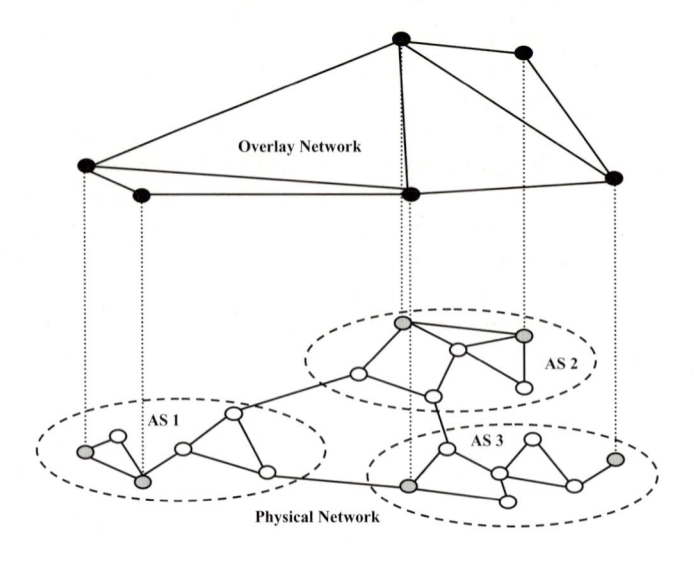

Fig. 1. Overlay network

which in turn forwards the traffic to the destination node (or to the next node on the path if multiple hops are necessary). It is worth to emphasis that the overlay network can choose its routes automatically, without involving the autonomous systems into its decisions. This is the basic argument to motivate routing in overlay networks. Examples of such overlays are the Resilient Overlay Network (RON), OverQoS, the QoS-aware routing protocol for overlay networks (QRON) and the QoS overlay network (QSON).

In RONs [11], strategically placed nodes in the Internet are organized in an application-layer overlay. Nodes belonging to the overlay aid each other in routing packets in such a way as to avoid path failures in the Internet. Each RON node carefully monitors the quality of Internet paths to his neighbors through active measurements. In order to discover the RON topology, RON-nodes exchange routing tables and various quality metrics (*e.g.*, latency, packet loss rate, throughput) using a link-state routing protocol. The path selection is done at the source, which signals to nodes downstream the chosen path. Nodes along the path signal to the source nodes information about link failures pertaining to the selected path. Results involving thirteen sites scattered widely over Internet showed the feasibility of this solution. RON's routing mechanism was able to detect and route around all 32 outages that occurred during the time frame for the experiment, 1 % of the transfers doubled their Transmission Control Protocol (TCP) throughput and 5 % had their loss rate reduced with 5 % .

Following the success of RONs, the authors of [12] propose OverQoS, an overlay-based QoS architecture for enhancing Internet QoS. The key part of the architecture is the controlled-loss virtual link (CLVL) abstraction, which provides statistical loss guarantees to a traffic aggregate between two overlay nodes in the presence of changing traffic dynamics. They demonstrate that their architecture can supply the following QoS enhancements with as Little as 5 % bandwidth overhead: smoothing losses, packet prioritization, as well as statistical bandwidth and loss guarantees.

Another approach involving strategically placed nodes in the Internet is presented in [13]. The authors propose an architecture where each AS has one or more overlay brokers. The overlay brokers are organized into clusters that interconnect with each other to form an overlay service network that runs a QRON. The purpose of QRON is to find an overlay path satisfying a bandwidth constraint. QRON nodes use source routing and a number of backup paths to cope with bandwidth fluctuations. The authors were able to show that the QRON algorithms perform well under a variety of traffic loads while balancing the load among overlay brokers.

In a similar spirit, the QSON architecture[14] advocates a backbone overlay network for QoS routing. This architecture relies on well-established business relationships of two kinds. The first type of business relationships is defined by end-users who purchase QoS services from the QSON provider. The QSON provider is able to supply these services by engaging in SLAs with several Internet service providers (ISPs). This is the second kind of business relationships. The QSON overlay is spanned by QSON proxies located between ISP domains. Each proxy stores a list of paths to the other proxies. The proxies use probes to reserve bandwidth and to inform each other about changes in available bandwidth. Simulation results have shown that QSON is able to provide bandwidth reservation with low control overhead.

3 Principles of QoS Routing

In QoS networks every link and every node has a state described by specific QoS metrics. The *link state* can consist of available bandwidth, delay and cost whereas the *node state* can be a combination of available memory, central processing unit (CPU) utilization and harddisk storage. The link state and the node state may be considered separately or they may be combined. The focus here is on link state.

Routing is the process of finding a path between two hosts in a network. In QoS routing, the path must be selected such that QoS metrics of interest stay within specific bounds. The routing process relies on a *routing algorithm* for computing constrained paths and on a *routing protocol* for distributing state information. In general, the algorithm is independent from the protocol. The coupling between them is decided when the QoS routing architecture is specified. For example, the QoS-enabled OSPF (QOSPF) protocol suggests that a modified Bellman-Ford algorithm should be used for pre-computed paths and Dijkstra's algorithm for on-demand paths [15].

There are three basic forms of storing state information: *local state*, *global state* and *aggregated (partial) state* [16].

When a node keeps local state, it maintains information about its outgoing links only. No information about the rest of the network is available.

A global state is the combination of local states for all nodes in a graph. Global states are imprecise (i. e., they are merely approximations of the global state) due to non-negligible delay in propagating information about local states[2]. When the network size grows, the imprecision grows as well. This makes it hard to maintain an accurate picture about resource availability in the network and in turn has a severe impact on QoS routing.

Aggregated state aims to solve scalability issues in large networks. The basic idea is to group together adjacent nodes into a single *logical node*. The local state of a logical node is the aggregation of local states for physical nodes that are part of the logical node. Similar to the case of global state, this leads to imprecision that grows with the amount of state information aggregated in the logical node.

Imprecision, also called uncertainty, is not generated by aggregation only. Other sources of uncertainty are network dynamics (churn), information kept secret by ISPs due to business reasons, as well as approximate state information due to systematic or random errors in the measurement process [17]. An interesting solution suggested for mitigating these problems is to replace the deterministic state information metrics with random variables. In this case, the routing algorithm must be changed such as to select feasible paths on a probabilistic basis, with the result that the selected paths are those most likely to satisfy the QoS constraints [18]. However, a non-trivial problem with this approach lies in the estimation of the probability distributions for state variables [19].

There are three different classes of routing strategies, each corresponding roughly to one form of maintaining state information: *source routing*, *(flat) distributed routing* and *hierarchical routing* [20].

In source routing the nodes are required to keep global state and the feasible path is computed at the source node. The main advantage in this case is that route computation is performed in a centralized fashion, avoiding some of the problems associated with distributed computation. The centralized computation can guarantee loop-free routes. One disadvantage of source routing is the requirement to maintain global state. In a network where the QoS metrics often change, this requires large communication overhead in order to keep the state information updated. Additionally, due to the propagation delay, the state information may become stale before reaching the destination. This leads to imprecise state information, as explained above. Furthermore, depending on the network size and the number of paths to compute, the source routing algorithm can result in very high computational overhead [20].

Distributed routing, typically, also relies on nodes maintaining global states, but the path computation is performed in a distributed fashion. This diminishes computational overhead and also allows concurrent computation of multiple routes in search for a feasible path. Distributed computation suffers from problems related to distributed state snapshot, deadlock and loop occurrence [16].

---

[2] Link state information is said to become stale when network latency prevents timely updates.

Additionally, when global state is maintained, distributed routing shares with source routing the problems related to imprecise state information.

Some suggestions on using flooding-based algorithms, require nodes to maintain local state only [21,22]. This mitigates problems related to imprecise state information. However, flooding-based algorithms tend to generate large volumes of traffic compared to the other forms of routing.

In hierarchical routing, the network is divided into groups of nodes and the state information is aggregated for the nodes participating in a group. With this form of aggregation, a group appears as a logical node. One node in the group is designated *leader* or *border node* and acts as a gateway for the communication with other logical nodes. Each group can in turn be divided into smaller groups. Using this form of recursion, several hierarchical levels can be created. Nodes maintain global state information for peers within a group and aggregated state information about the other groups. The major advantage of hierarchical routing is scalability[20,23]. In particular, since nodes maintain aggregated state information there is less state information to be transmitted to other nodes, hence less communication overhead. For the same reason, there is also less computational overhead. However, each level of hierarchy induces additional uncertainty in the state information. This problem becomes more difficult when several QoS metrics must be aggregated, since for some topologies there can be no meaningful way to combine the metrics [16]. Some solutions for topology aggregation are presented in [24,25].

In large overlay networks such as Skype, Vuze[3] DHT and Gnutella, the nodes are computers controlled by regular users instead of a central authority. This type of decentralized network is appealing to service providers because the cost of bandwidth and computation required to deliver the service is shifted towards the end-users. Unfortunately, such an overlay network tends to be an unreliable infrastructure. By this, we mean that end-nodes are at owner's whim, and users can turned them off at any time. When a node is turned off a set of links is removed from the overlay network. If the links belong to established QoS paths, their removal will trigger path re-computations. Additionally, traffic flows may need to be reallocated to new paths. This type of node churn is similar to the topology dynamics occurring in mobile ad-hoc networks, when stations move out of radio range. Routing protocols that handle topology dynamics can be classified as *proactive* or *reactive* protocols.

Proactive protocols, such as destination sequence distance vector (DSDV) periodically update the routing tables [26]. In contrast, reactive protocols (*e. g.*, dynamic source routing (DSR) and ad-hoc on-demand distance vector (AODV)) update the routing tables only when routes need to be created or adjusted due to changes to topology [26]. Proactive protocols are in general better at providing QoS guarantees for real-time traffic such as multimedia. Their disadvantage lies in the traffic volume overhead generated by the protocol itself. Reactive protocols scale better than proactive protocols, but will experience higher latency when setting up a new route [26].

---

[3] Formerly known as Azureus.

The focus of this section has been so far placed on single path routing. *Multipath routing* exploits path diversity in the network to find several paths that can satisfy a flow demand. In single path routing, a flow demand is dropped if no suitable path is found. With multipath routing, there is still a chance to satisfy the flow demand, particularly in the case of bandwidth, by spreading the flow over several paths. The ability to spread demands over multiple paths offers additional advantages for load balancing, congestion control, reliability and security [27]. However, multipath routing introduces additional overhead in the form of bandwidth usage and processing requirements. The bandwidth overhead is due to extra link state information that nodes must share. A multipath router must compute and store in memory more than one path for each known source-destination pair. This is the reason for higher processing requirements than in the case of single-path routing.

QoS routing in overlay networks faces additional challenges such as those related to estimation of link state parameters and selfish behavior. We describe them briefly below.

A typical overlay networks consist of nodes, which are computer processes executing at the application layer, and of virtual links[4] implemented on top of TCP or the User Datagram Protocol (UDP). In this scenario, the link state parameters are not readily available but must be estimated through active measurements.

Although active measurement methods exist for most metrics of interest [28,29] there are still open questions to be answered before they can used in practice. One of these issues is that of finding an adequate measurement frequency. In other words, we are asking how long should we wait after obtaining data from one measurement until we start the next measurement. Ideally, one wants to capture all significant changes in the measured metric while keeping the volume of injected traffic at a minimum. Another issue is that of estimating the validity of the measurements when all nodes in the overlay execute them concurrently or when active measurements for different metric types are performed simultaneously.

In fact, the issues related to active measurements are related to a bigger problem: the likely mismatch in the interaction between overlay routing and traffic engineering [30,31]. The goal of overlay routing is to satisfy in an optimal way the flow demands within the overlay network. On the other hand, the goal of traffic engineering is to optimize the performance of the *whole* network, including any existing overlays. Whereas traffic engineering addresses the needs of all nodes under a common administrative domain (*e. g.*, an AS), overlay routing caters for a subset of the Internet, with nodes from many different networks.

The *traffic matrix* concept is central to traffic engineering. A traffic matrix contains estimates of the traffic volumes exchanged by all source-destination pairs in the network. These estimates include both overlay and non-overlay traffic. The traffic matrix is used as input for the inter- and intra-domain routing algorithms.

---

[4] Also called logical links.

If the traffic matrix changes, flows within the network can be re-routed as a result. The overlay routing algorithm reacts to changes in the network layer routes by readjusting the overlay traffic to repair flow demands that are no longer satisfied. Thus, two closed-loop control mechanisms modify each other's input. The selfish behavior of the overlay network can lead to traffic oscillations. Results presented in [30,31] indicate that the price for optimizing the traffic in the overlay can be severe performance degradation in the rest of the network. Gaining a better understanding of the interaction between traffic engineering and selfish routing is an important research topic.

4 Optimization Models

In Section 1 it was stated that QoS routing is a mechanism for optimizing network performance by constrained-path selection and traffic flow allocation. Given one flow demand and information about the network, the goal of constrained-path selection is to find a path such that the flow demand is satisfied. For multiple flow demands, the constrained-path selection algorithm must be run several times, once for each flow demand. In contrast, the goal of flow allocation is to satisfy multiple demands simultaneously. This is the same goal as that of traffic engineering. In fact, flow allocation is one of the components required by traffic engineering. Other required components are topology and state discovery, traffic demand estimation and configuration of network elements [7].

Flow allocation algorithms tend to utilize resources more efficiently since all routes are recomputed when the demands change. When used in combination with multipath routing the efficiency can be increased by exploiting path redundancy. However, a major disadvantage of flow allocation is that ongoing flows can be temporarily disrupted when routes change [7]. Constrained-path selection algorithms handle each changing flow demand by itself, without disrupting other flows. Although different, these two types of algorithms can be combined as it is shown in Section 5.

The algorithms assume that information about network topology is available in the form of a weighted digraph $\mathcal{G}(\mathcal{V}, \mathcal{E})$, where $\mathcal{V}$ is the set of nodes (vertices) in the network and the set $\mathcal{E}$ contains the links (edges). The weight of each link represents a set of metrics of interest, such as bandwidth, delay, jitter, packet loss and cost. In addition to the graph and link weights, information about the flow demands is available as well. A flow demand is expressed as a set of path constraints for the path $P(s, d)$, where $s \in \mathcal{V}$ is the source node and $d \in \mathcal{V}$ is the destination (sink) node. In its simplest form, the flow demand contains only the bandwidth required to transfer data from s to d. It is assumed here that flow demands are tied to the direction of the path.

In the case of a multi-constrained path (MCP) problem we attempt to find one constrained path at a time. This is a feasibility problem. Each link weight in $\mathcal{G}(\mathcal{V}, \mathcal{E})$ is a vector of QoS metrics, where each metric belongs to one of the following types:

additive: delay, jitter, cost
multiplicative: packet loss
min-max: bandwidth, policy flags

Multiplicative weights can be turned into additive weights by taking the logarithm of their product. The constraints on min-max metrics can be dealt with by pruning the links of the graph that do not satisfy the constraints [32]. Therefore, in the remainder of this section we focus on additive link weights only.

For $i = 1, \ldots, m$ we denote by $w_i(u, v)$ the ith additive metric for the link (u, v) between nodes u and v such that $(u, v) \in \mathcal{E}$. The MCP optimization problem for m constraint values L_i on the requested path is shown in Table 1.

Table 1. Multi-constrained path selection problem (MCP)

$$\text{find} \quad \text{path } P$$
$$\text{subject to } w_i(P) = \sum_{(u,v) \in P} w_i(u, v) \leq L_i \text{ for } i = 1, \ldots, m \text{ and } (u, v) \in \mathcal{E}$$

The MCP selection problem problem can be converted to a multi-constrained optimal path (MCOP) selection problem by minimizing or maximizing over one of the metrics w_i. It is also possible to define a path-weight function f over all metrics and to optimize over the path-weight function itself, as shown in Table 2.

Table 2. Multi-constrained optimal path selection problem (MCOP)

$$\text{minimize} \quad f(\boldsymbol{w}(P))$$
$$\text{subject to } w_i(P) = \sum_{(u,v) \in P} w_i(u, v) \leq L_i \text{ for } i = 1, \ldots, m \text{ and } (u, v) \in \mathcal{E}$$

Wang and Crowcroft proved in [33] that MCP problems with two or more constraints are $\mathcal{NP}$-complete. By extension, MCOP problems with two or more constraints are $\mathcal{NP}$-complete as well. The apparent intractability of these problems suggests abandoning the search for exact solutions in the favor of heuristics that have a better chance of running in polynomial time. Chen and Nahrstedt suggest a $O(2L)$ heuristic [21] for the MCP problem, where L is the length of the feasible path.

The results of a study [34] on the $\mathcal{NP}$-complexity of QoS routing found four conditions leading to its appearance:

- graphs with long paths (large hop-count),
- link weights with infinite granularity, or excessively large or small link weights,
- strong negative correlation among link weights,
- "critically constrained" problems, which are problems with constraint values close to the center of the feasible region.

The authors of the study consider that these conditions are unlikely to occur in typical networks. If they are right, the consequence is that the exponential run time behavior of exact algorithms occurs very seldom.

In the flow allocation problem, it is assumed that we know about one or more directed paths connecting a source node s and a destination node d. These paths can be discovered automatically, for example with a K shortest paths (KSP) algorithm [10,35]. Due to space constraints we restrict our focus to bandwidth allocation problems only. We consider the following type of optimization problems: given a digraph $\mathcal{G}(\mathcal{V}, \mathcal{E})$, a set $\mathcal{P}$ of directed paths and a set $\mathcal{D}$ of flow demands for bandwidth, we would like to allocate bandwidth on the paths in $\mathcal{P}$ such as to simultaneously satisfy all demands.

If the traffic volume pertaining to a specific flow is allowed to be distributed over several paths to the destination, this is said to be a feasibility problem for *bifurcated flows*. On the other hand, if the problem includes the requirement that the entire traffic flow between two nodes must be transmitted on a single path, we have a feasibility problem for *non-bifurcated flows*. This problem is known to be computationally intractable [36] for large networks (it is in fact $\mathcal{NP}$-complete). The remainder of this article considers only feasibility problems for bifurcated flows.

We adopt a notation called *link-path formulation* [36] to formalize our problem statement. Using this notation, we let the variable x_{dp} denote bandwidth allocated to demand d on path p. Recall that a demand is a request for a specific amount of bandwidth, h_d, from a source node to a destination node. The source node and the destination node can be connected by more than one path, which explains the use of the index variable p. We use the variables D and E to denote the number of demands in the demand set $\mathcal{D}$ and the number of edges (links) in the set $\mathcal{E}$, respectively. Further, the capacity of a link e is denoted by c_e. The indicator variable δ_{edp} is defined as

$$\delta_{edp} = \begin{cases} 1 & \text{if link } e \text{ is used by demand } d \text{ on path } p, \\ 0 & \text{otherwise.} \end{cases} \quad (1)$$

Our problem statement can now be written as shown in Table 3.

In [36], it is suggested that the pure allocation problem (PAP) described in Table 3 can be reformulated in the form of the linear optimization problem shown in Table 4. The new problem, PAP with modified link-path formulation (PAP-MLPF), has an additional variable z to be modified. Unlike the PAP, this problem *always* has a feasible solution in the sense that a minimum value for z

Table 3. Pure allocation problem (PAP)

$$\begin{aligned} &\text{find} \quad x_{dp} && \text{for all } d \in \mathcal{D}, p \in \mathcal{P} \\ &\text{subject to} \sum_p x_{dp} = h_d, && d = 1, 2, \ldots, D \\ &\qquad\qquad \sum_d \sum_p \delta_{edp} x_{dp} \leq c_e, && e = 1, 2, \ldots, E \end{aligned}$$

Table 4. PAP with modified link-path formulation (PAP-MLPF)

$$\begin{aligned}&\text{minimize}\quad z & &\text{for all } d \in \mathcal{D}, p \in \mathcal{P}\\&\text{subject to } \sum_p x_{dp} = h_d, & & d = 1, 2, \ldots, D\\&\quad\sum_d \sum_p \delta_{edp} x_{dp} \leq z + c_e, & & e = 1, 2, \ldots, E\end{aligned}$$

can be found. If $z < 0$ in the solution, we have a successful bandwidth allocation. Otherwise the value of z indicates how much additional bandwidth is required to obtain feasibility.

5 A Framework for Overlay Routing Protocols

In this section we present performance results for a couple of routing protocols that combine theoretical concepts presented earlier in this paper.

The Overlay Routing Protocol (ORP) framework was developed at BTH in Karlskrona, Sweden as part of the Routing in Overlay Networks (ROVER) project. The framework consists of two protocols: the Route Discovery Protocol (RDP) and the Route Management Protocol (RMP).

RDP is used to find network paths subject to various QoS constraints [10,37]. To achieve this goal, RDP uses a form of selective diffusion based on ideas presented in [21,38].

The purpose of RMP is to alleviate changes in the path QoS metrics, due to node and traffic dynamics. This is done through a combination of path restoration and optimization algorithms for traffic flow allocation on bifurcated paths. The purpose of the flow allocation is to spread the demand on multiple paths towards the destination. The design of RMP is influenced by ideas presented in [23,39].

5.1 Route Discovery Protocol

RDP is a distributed routing protocol relying on local state information. It's distributed path computation is achieved by selective diffusion. RDP uses two different kinds of packets: control packets (CPs), which are used to explore available paths from a source to a destination, and acknowledgement packets (APs) that transport data about available paths back to the source node.

When a node in the overlay wants to open a route to another overlay node it assembles a CP with the desired QoS constraints. The CP is sent to all adjacent nodes connected by links satisfying the QoS constraints. If at least one feasible link is found, the CP is added to the sender's list of active CPs and a timer is started accordingly. If no information is received before the timer expires, the CP is considered lost and it is removed from the active CP list.

If no feasible link exists, the CP is dropped and no further actions are taken. The receive and forward process is repeated at several nodes until one or more CPs reach the destination node, or all CPs are dropped by intermediate nodes.

If all CPs are lost, the nodes on the feasible path eventually experience timeouts and thus are able to free any reserved resources.

The feasible path defined by the first CP that arrives at the destination is copied into an AP. The AP is sent back to the source node over the reverse feasible path[5]. All subsequent CPs that arrive to the destination node are dropped.

Chen and Nahrstedt [21,16] provide worst-case complexity results for the time and communication overhead required to establish a constrained path with this algorithm. For a path length L, the time complexity is $O(2L)$. In the case of RDP, the path length is bounded by the time-to-live (TTL) value carried by CPs and APs. Hence, RDP's time complexity for one QoS request is $O(2\,\text{TTL})$. The communication complexity for one QoS request is on the order $O(E + \text{TTL})$ according to [10].

5.2 RDP Simulation Results

The focus of the simulations is entirely on bandwidth reservations. In what follows, the term *QoS session* is used to denote a request for a directed path with a constraint on minimum available bandwidth. Each session has an associated session duration, which specifies the life length of the path. If a path is successfully established, the amount of bandwidth specified by the path constraint is reserved for the entire session duration. The links in these experiments are error-free and no churn occurs.

We present here results for the following metrics:

call blocking ratio: ratio between the number of infeasible QoS sessions and the total number of QoS sessions arrived at the network,
low-TTL blocking ratio: ratio between infeasible sessions due to low TTL value and the total number of infeasible sessions[6],
bandwidth overhead: ratio between the average number of RDP bytes per second and network capacity (*i. e.*, the aggregated volume of every link in the network).

Results involving additional metrics are available in [10].

The results shown in Figure 2 and Figure 3 are plotted against increasing values of network utilization. The network utilization, ρ, is defined as [13,40]

$$\rho = \frac{\lambda T Q H}{\sum_{e \in \mathcal{E}} b_e} \qquad (2)$$

where T is the average session duration, Q is the average amount of QoS (bandwidth) requested, H is the average path length across all node pairs, and b_e is the available bandwidth on link e. The simulation parameters are summarized in Table 5.

---

[5] Traveling on the reverse feasible path between node v_1 and node v_N means traveling in the opposite direction on the feasible path (*i. e.*, over hops $v_N, v_{N-1}, \ldots, v_1$).
[6] Low-TTL blocking occurs when an AP is dropped because it has traveled the maximum number of hops allowed. The metric does not take into account the possibility that a feasible path may exist for higher TTL values.

Table 5. Simulation parameters

Parameter	Assigned value
Network utilization, ρ	$0.1, 0.25, 0.50, 0.75, 1.00$
Session duration	Generalized Pareto with mean 180 s and 600 s
Requested bandwidth ranges	Uniform, 16–64 Kbps, 128–512 Kbps and 1–2 Mbps
Network size	100 nodes
Topology	Barabási-Albert[41,10]
Link bandwidth	Uniform, 10–10000 Kbps
TTL	8 hops

In the plots the blue color is used for sessions with mean duration of 180 seconds, while red color denotes sessions with mean duration of 600 seconds. Plots for 16-64 Kbps sessions are drawn with solid lines, those for 128–512 Kbps are drawn with dashed lines, and 1-2 Mbps session plots use alternating dots and dashes. Each hollow circle indicates the simulated utilization factor pertaining to the value on the y-axis.

The call blocking ratio and the low-TTL blocking ratio are shown side by side in Figure 2. It can be noticed by observing Figure 2(a) that 180 seconds sessions consistently experience higher call blocking ratio than 600 seconds sessions. The explanation is found in Equation 2. This equation is used to compute the utilization factor, ρ, by adjusting the arrival rate, λ, while the other parameters are kept fixed. For a given ρ value, the arrival rate must be higher for short sessions than for long sessions. A higher arrival rate implies that more feasible paths must be found. This leads to higher bandwidth overhead since more CPs are in the network, as it can be observed in Figure 3. Since less network capacity is available in this case than in the case of low arrival rate, the call blocking ratio is higher. Further support for this assertion is found in Figure 2(b). There it can

(a) Call blocking ratio.

(b) Low-TTL blocking ratio.

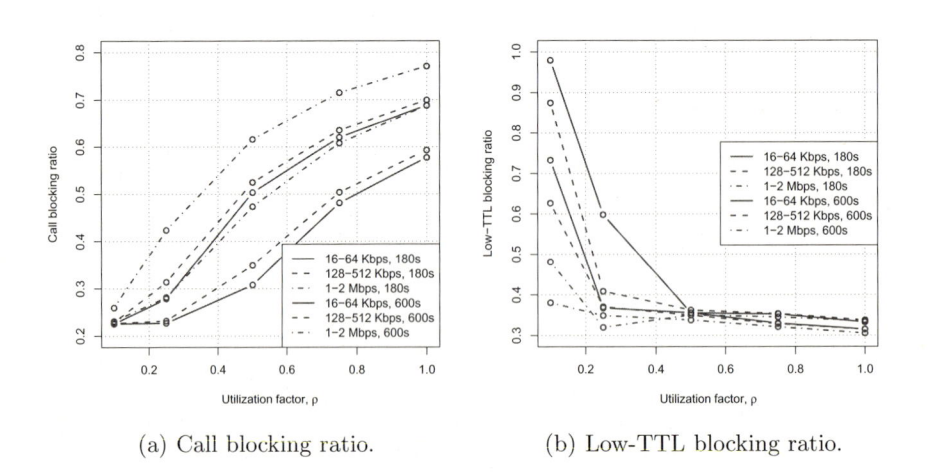

Fig. 2. RDP call blocking

(a) 16–64 Kbps. **(b) 128–512 Kbps and 1–2 Mbps.**

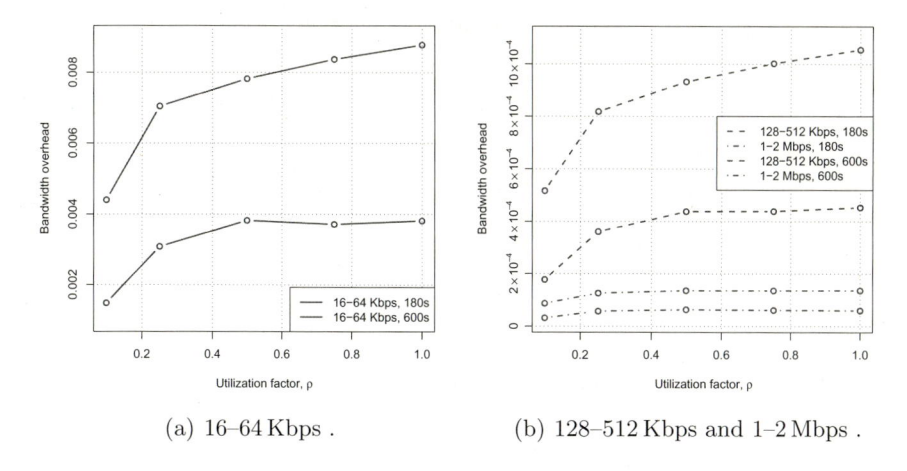

Fig. 3. RDP bandwidth overhead

be observed that, when the utilization factor exceeds 0.25, most call blocking is due to failure to satisfy the constraints of the QoS session, and not due to low-TTL blocking.

Two different graphs are used in Figure 3 to show the RDP bandwidth overhead. This is done because in the case of a single graph, the high bandwidth overhead of 16-64 Kbps sessions would make it hard to distinguish the bandwidth overhead of the remaining sessions. Indeed, for 16-64 Kbps sessions the bandwidth overhead is 40–80 times higher than that of 1–2 Mbps sessions.

The results suggest that one needs to carefully consider the interaction between the TTL value and the bandwidth overhead. This is in fact a trade-off between success in finding feasible paths and efficiency in keeping the overhead low. In our simulations the worst-case cost in the case of RDP, namely 0.9 % bandwidth overhead, occurs for 16–64 Kbps flow demands with 180 seconds mean session duration when the utilization factor is one.

5.3 Route Maintenance Protocol

RMP combines path restoration and flow allocation to handle the type of resource fluctuations described in Section 3. In this paper the focus is on bandwidth constraints, but the protocol as such has support for different types of QoS constraints. RMP relies on two main components: an algorithm for distributing link-state information and an optimization algorithm for flow allocation.

Link-state information is distributed using the *link vector* algorithm proposed by Behrens and Garcia-Luna-Aceves [23]. The difference between a link-state algorithm and a link vector algorithm is that the link-state algorithm is require to broadcast complete topology information. When a link vector algorithm is used, a node uses selective diffusion to disseminate link-state information pertaining only to its preferred paths. This reduces the communication overhead associated with traditional link-state algorithms.

In the case of RMP, the preferred paths are setup using RDP. The QoS information provided by RDP ensures that a node has link-state information about each link in each of its preferred paths. The set of links belonging to a node's preferred paths is called the *source graph* of that node. Nodes exchange source graph information with their neighbors. Additionally, nodes have link-state information about their own outgoing links. The topology information known to a node consists of its own links, its own source graph and the source graphs reported by its neighbors [23].

Nodes report incremental source graph information to their neighbors. Obviously, when a node joins the overlay it receives complete source graphs from its neighbors. Beyond that, information is transmitted only if the link-state changes (*i. e.*, triggered updates). In other words, RMP is a reactive protocol.

A node enters *restoration mode* when a path is broken. A path is considered broken when one or several links are deleted from the source graph or when their updated state information makes it impossible to satisfy the path QoS constraints. In restoration mode the following actions are taken:

i) a broken path error message is sent to the source node of each affected flow,
ii) Yen's KSP algorithm [35] is executed to find the K backup paths to the destinations affected by the topology updates,
iii) the corresponding flow demands and the backup paths are used to construct a PAP-MLPF optimization problem as described in Section 4,
iv) the simplex method [42] is used to solve the PAP-MLPF problem,
v) if the simplex method is successful, the links on the new paths are added to the source graph; otherwise the affected flows are dropped (*i. e.*, packets belonging to them are not forwarded further).

The time complexity of the link vector algorithm after a single link change is $O(n)$, where n is the number of nodes affected by the change. The upper bound for n is given by the length of the longest path in the network. The communication complexity $O(E)$ is asymptotic in the number of links in the network [23].

5.4 RMP Simulation Results

The purpose of the experiments is to evaluate RMP's performance for different levels of churn. In the experiments we focus entirely on bandwidth reservations. A network topology with 100 nodes and 780 links is used for all experiments. The links in these experiments are error-free.

We denote by p_t the total number of preferred paths, by p_r the number of restored paths, and finally by p_f the number of path failures (*i. e.*, paths that could not be restored). The relation $0 \leq p_r + p_f \leq p_t$ always holds.

We focus on the following simulation metrics:

path failure ratio: ratio between the number of path failures and the total number of preferred paths in the network, p_f/p_t,

restored paths ratio: ratio between the number of restored paths and the number of broken paths, $p_r/(p_r + p_f)$ for $p_r + p_f > 0$[7],

bandwidth overhead: ratio between the average number of RDP bytes per second and network capacity (*i.e.*, the aggregated volume of every link in the network),

Additional metrics are available in [10].

We simulate 50 random flow demands (*i.e.*, $p_t = 50$). This value provides an acceptable trade-off between link utilization and the time required to run the simulations. The flow demand bandwidth is uniformly distributed over three different ranges: 16–64 Kbps , 128–512 Kbps and 1–2 Mbps , as in the case of RDP. The source and destination node of each flow demand is selected randomly.

Link bandwidth is interpreted in our simulations as residual capacity after bandwidth is reserved on preferred paths. The residual capacity determines the amount of path diversity within the network. Here, the residual capacity is exponentially distributed with mean value equal to the maximum bandwidth demand multiplied by an integer scaling factor. For example, for the bandwidth range 1–2 Mbps and a scaling factor of 2, the link bandwidth is exponentially distributed with mean value 4 Mbps . Intuitively, a scaling factor of 1 means that 63 % of the links have *less* capacity than the maximum value of the demand range. For a scaling factor of 5 only 18 % of the links have *less* bandwidth than the maximum value of the demand range.

The following scaling factors are used: 1, 2, 3, 4, and 5. The use of exponential distribution with mean value based on the maximum bandwidth demand results in a good mix of links with very little bandwidth as well as links with lots of residual capacity. Using the upper bound of the bandwidth range is a matter of preference. In fact, any value within the bandwidth range can be selected and the mean link bandwidth is scaled accordingly. The residual network capacity increases proportionally with the integer multiple value.

We present simulation results for two different levels of churn: one based on the Gnutella session durations with mean duration of 130 seconds [10] and another based on exponential session durations with mean duration of 30 seconds. The two types of churn correspond roughly to the following scenarios:

Gnutella churn: general purpose peer-to-peer (P2P) overlay network,
Exp(30 s): wireless network with moving stations.

RMP is configured to use 3, 5, and 7 backup paths, respectively. They are abbreviated as 3SPs, 5SPs and 7SPs in the text and figures. A higher number of backup paths increases the chances for successful flow allocation in situations with low residual network capacity.

Figure 4 shows the performance of path restoration for each type of churn. The solid black line at the top of each sub-figure indicates the ratio of path failures to the total number of paths ($p_t = 50$) when no path restoration is in use. Lower path failure ratio values indicate higher RMP success in restoring

---
[7] Instances where $p_r + p_f = 0$ are not used in computing the average.

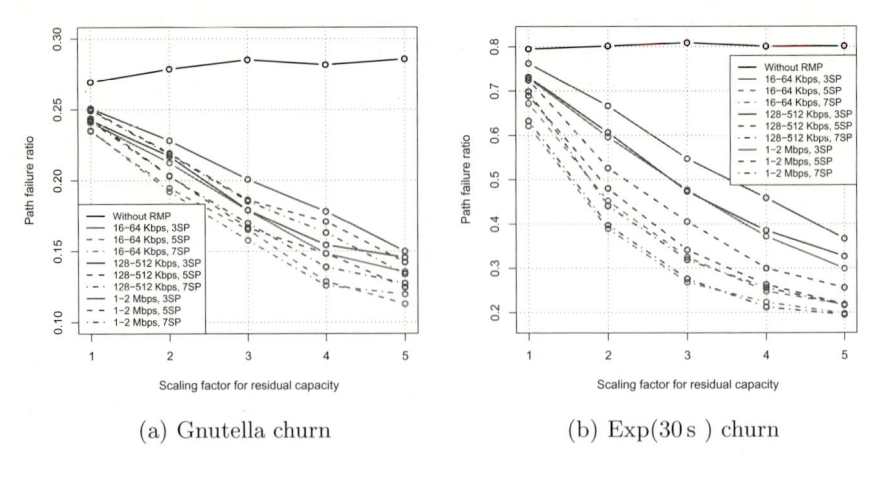

Fig. 4. RMP path restoration

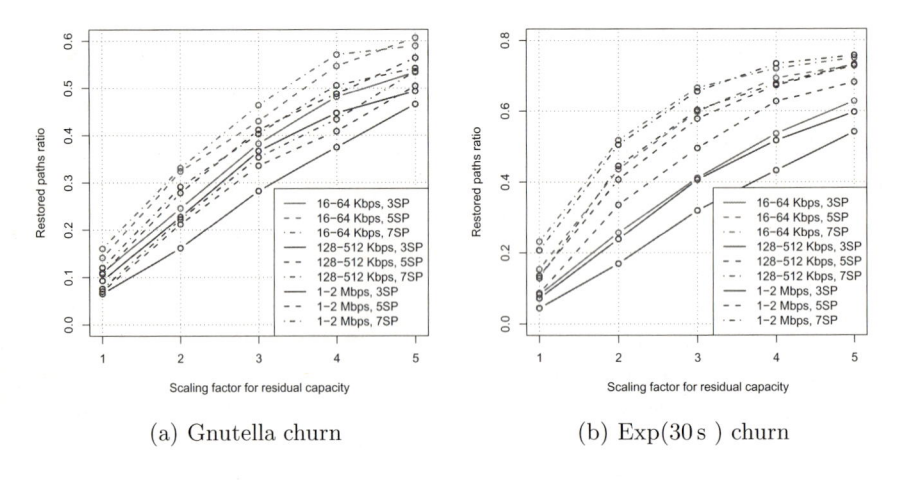

Fig. 5. RMP restored paths ratio

paths. The path failure ratio is lower in the case of Gnutella churn because the session durations are longer on the average, which translates in fewer link failures per time unit.

In both cases of churn, the path failure ratio decreases when RMP is used. Clearly, RMP's success is directly proportional to the amount of residual capacity. In terms of reduced path failure ratio, the largest gains are registered for Exp(30 s) scenarios. In this scenarios, the path failure ratio is very high in the absence of RMP, which means that there is a lot of room for improvement.

Using more backup paths (*i.e.*, 5SPs or 7SPs) shows most gain in the case of Exp(30 s) scenarios for bandwidth range 1–2 Mbps. With less aggressive churn, the usefulness of additional backup paths decreases.

(a) Gnutella churn

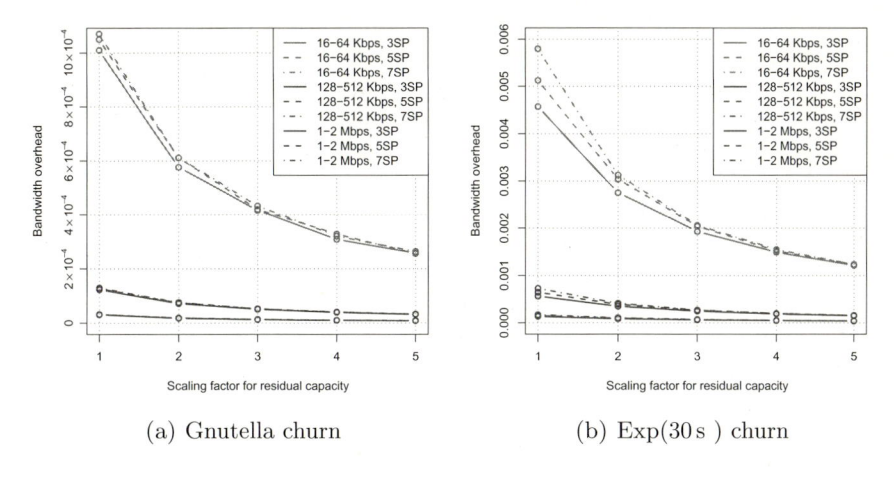

(b) Exp(30 s) churn

Fig. 6. RMP bandwidth overhead

A complementary view of RMP's path restoration success is shown in Figure 5 in terms of the ratio of number of restored paths to the number of broken paths. RMP's largest success in restoring broken paths is experienced in the case of 16-64 Kbps scenarios. In Figure 5(b) for scaling factor 4 and 5, the restored paths ratio for 128-512 Kbps scenarios with 7SPs is slightly higher than the restored path ratio for 16-64 Kbps scenarios with 7SPs. These minimal differences are due to the random selection of source and destination node for each flow demand.

The bandwidth overhead shown in Figure 6 is computed by dividing the bandwidth utilization with the network capacity (i. e., the sum of residual capacity and used capacity). In contrast to RDP's bandwidth overhead, the RMP bandwidth overhead is biased. For example, the 16–64 Kbps scenarios dominate in each of the churn scenarios. This happens because the amount of average residual capacity per link is determined by the product between the scaling factor in use and the upper bound of the bandwidth range. Hence, the network capacity for 16–64 Kbps scenarios is always lower than for the other two bandwidth ranges. Comparison of bandwidth overhead is fair only for graphs within the same bandwidth range.

For each bandwidth range, the largest bandwidth overhead occurs in 7SPs scenarios with Exp(30 s) churn when the scaling factor is equal to one: 0.6 % in the case of 16–64 Kbps scenarios, 0.07 % in the case of 128–512 Kbps scenarios, and 0.02 % in the case of 1–2 Mbps scenarios.

6 Summary

This paper has introduced fundamental concepts related to QoS routing in overlay networks. Furthermore, the paper has reported on performance results for RDP and RMP, which are two protocols under the ORP framework. The focus of the performance results has been on the cost in the form of bandwidth overhead incurred from running these protocols.

The worst-case cost in the case of RDP, namely 0.9% bandwidth overhead, occurs for 16–64 Kbps flow demands with 180 seconds mean session duration when the utilization factor is one. For RMP, the worst-case cost, 0.6%, occurs in Exp(30 s) scenarios with 16-64 Kbps flow demands, 7SPs and scaling factor one for residual capacity. Assuming an additive cost when RDP and RMP are being used together, the worst-case cost is estimated to be as high as 1.5% protocol overhead in terms of bandwidth.

RMP can be quite efficient in restoring broken paths provided enough residual capacity is available. For example in Exp(30 s) scenarios, in spite of aggressive churn, RMP is able to restore up to 40% of broken paths used for transporting 1–2 Mbps flows, with approximately 0.02% bandwidth overhead.

We plan to run similar tests in a more realistic environment, such as PlanetLab. However, a realistic PlanetLab implementation requires that ORP is extended to include two important elements: an overlay network that organizes nodes and transports ORP messages and the ability to measure link-state variables (e. g., available bandwidth, delay and loss).

All experiments presented here focus on a single QoS metric: bandwidth. Additional experiments should be performed to evaluate ORP's performance when several QoS metrics are combined. For RMP, this requires that Yen's KSP algorithm is replaced by the Self-Adaptive Multiple Constraints Routing Algorithm (SAMCRA) [32] or another similar algorithm.

References

1. Crawley, E.S., Nair, R., Rajagopalan, B., Sandick, H.: RFC 2386: A Framework for QoS-based Routing in the Internet, IETF, category: Informational (August 1998), http://www.ietf.org/ietf/rfc2386.txt
2. Braden, R., Clark, D.D., Shenker, S.: RFC 1633: Integrated Services in the Internet Architecture: an Overview, IETF, category: Informational (June 1994), http://www.ietf.org/ietf/rfc1633.txt
3. Wroclawski, J.: RFC 2210: The Use of RSVP with IETF Integrated Services, IETF, category: Standards Track (Septmeber 1997), http://www.ietf.org/ietf/rfc2210.txt
4. Thompson, K., Miller, G.J., Wilder, R.: Wide-area Internet traffic patterns and characteristics. IEEE Network 11(6), 10–23 (1997)
5. Blake, S., Black, D.L., Carlson, M.A., Davies, E., Wang, Z., Weiss, W.: RFC 2475: An Architecture for Differentiated Services (December 1998), http://www.ietf.org/ietf/rfc2475.txt
6. Bouras, C., Sevasti, A.: Service level agreements for DiffServ-based services' provisioning. Journal of Computer Networks 28(4), 285–302 (2005)
7. Wang, Z.: Internet QoS: Architectures and Mechanisms for Quality of Service. Morgan Kaufman Publishers, San Francisco (2000), ISBN: 1-55860-608-4
8. Burgsthaler, L., Dolzer, K., Hauser, C., Jähnert, J., Junghans, S., Macián, C., Payer, W.: Beyond technology: The missing pieces for QoS success. In: Proceedings ot the ACM SIGCOMM Workshops, Karlsruhe, Germany, August 2003, pp. 121–130 (2003)
9. Rekhter, Y., Li, T., Hares, S.: RFC 4271: A Border Gateway Protocol 4 (BGP-4), IETF (January 2006), http://www.ietf.org/ietf/rfc4271.txt

10. Ilie, D.: On unicast QoS routing in overlay networks. Ph.D. dissertation, Blekinge Institute of Technology (BTH), Karlskrona, Sweden (October 2008)
11. Andersen, D.G.: Resilient overlay networks. Master's thesis, Dept. of Electrical Engineering and Computer Science, Massachusetts Institute of Technology (May 2001)
12. Subramanian, L., Stoica, I., Balakrishnan, H., Katz, R.: OverQoS: An overlay based architecture for enhancing Internet QoS. In: Proceedings of NSDI, San Francisco, CA, USA (March 2004)
13. Li, Z., Mohapatra, P.: QRON: QoS-aware routing in overlay networks. IEEE Journal on Selected Areas in Communications 22(1), 29–40 (2004)
14. Lao, L., Gokhale, S.S., Cui, J.-H.: Distributed QoS routing for backbone overlay networks. In: Proceedings of IFIP Networking, Coimbra, Portugal (May 2006)
15. Apostolopoulos, G., Williams, D., Kamat, S., Guerin, R., Orda, A., Przygienda, T.: RFC 2676: QoS Routing Mechanisms and OSPF Extensions, IETF, category: Experimental (August 1999), http://www.ietf.org/rfc2676.txt
16. Chen, S.: Routing support for providing guaranteed end-to-end quality-of-service. Ph.D. dissertation, Engineering College of the University of Illinois, Urbana, IL, USA (1999)
17. Lorenz, D.H.: QoS routing and partitioning in networks with per-link performance-dependent costs. Ph.D. dissertation, Israel Institute of Technology, Haifa, Israel (2004)
18. Lorenz, D.H., Orda, A.: QoS routing in networks with uncertain parameters. IEEE/ACM Transactions on Networking 6(6), 768–778 (1998)
19. Chen, S., Nahrstedt, K.: Distributed QoS routing with imprecise state information. In: Proceedings of ICCCN, Lafayette, LA, USA (October 1998)
20. Shigang, C., Klara, N.: An overview of quality of service routing for the next generation high-speed networks: Problems and solutions. IEEE Network 12(6), 64–79 (1998)
21. Shigang, C., Nahrstedt, K.: Distributed quality-of-service routing in high-speed networks based on selective probing. In: Proceedings of LCN, Lowell, MA, USA, October 1998, pp. 80–89 (1998)
22. Gelenbe, E., Lent, R., Nunez, A.: Self-aware networks and QoS. Proceedings of the IEEE 92, 1478–1489 (2004)
23. Behrens, J., Garcia-Luna-Aceves, J.J.: Distributed, scalable routing based on link-state vectors. In: Proceedings of SIGCOMM, London, UK, August 1994, pp. 136–147 (1994)
24. Lee, W.C.: Topology aggregation for hierarchical routing in ATM networks. ACM SIGCOMM Computer Communications Review 25(2), 82–92 (1995)
25. Lui, K.-S., Nahrstedt, K., Chen, S.: Routing with topology aggregation in delay-bandwith sensitive networks. IEEE/ACM Transactions on Networking 12(1), 17–29 (2004)
26. Schiller, J.: Mobile Communications, 2nd edn. Addison Wesley, Boston (2003) ISBN: 0-321-12381-6
27. He, J., Rexford, J.: Towards internet-wide multipath routing. IEEE Network 22(2), 16–21 (2008)
28. Gummadi, K.P., Saroiu, S., Gribble, S.D.: King: Estimating latency between arbitrary internet end hosts. In: Proceedings of IMW, Marseille, France (November 2002)
29. Prasad, R., Dovrolis, C., Murray, M., Claffy, K.C.: Bandwidth estimation: Metrics, measurement techniques, and tools. IEEE Network 17(6), 27–35 (2003)

30. Liu, Y., Zhang, H., Gong, W., Towsley, D.: On the interaction between overlay routing and traffic engineering. In: Proceedings of IEEE Infocom, Miami, FL, USA (March 2005)
31. Qiu, L., Yang, R., Shenker, S.: On selfish routing in internet-like environments. IEEE/ACM Transactions on Networking 14(4), 725–738 (2006)
32. Van Mieghem, P., Kuipers, F.A.: Concepts of exact QoS routing algorithms. IEEE/ACM Transactions on Networking 12(5), 851–864 (2004)
33. Wang, Z., Crowfort, J.: Quality-of-service routing for supporting multimedia applications. IEEE Journal on Selected Areas in Communications 14(7), 1228–1234 (1996)
34. Kuipers, F.A., Van Mieghem, P.F.A.: Conditions that impact the complexity of QoS routing. IEEE/ACM Transactions on Networking 13(4), 717–730 (2005)
35. Yen, J.Y.: Finding the k shortest loopless paths in a network. Management Science 17(11), 712–716 (1971)
36. Pióro, M., Medhi, D.: Routing, Flow, and Capacity Design in Communication and Computer Networks. Morgan Kaufman Publishers, San Francisco (2004), ISBN: 0-12-557189-5
37. De Vogeleer, K., Ilie, D., Popescu, A.: Constrained-path discovery by selective diffusion. In: Proceedings of HET-NETs, Karlskrona, Sweden (February 2008)
38. Gelenbe, E., Lent, R., Montuori, A., Xu, Z.: Cognitive packet networks: QoS and performance. In: Proceedings of IEEE MASCOTS, Ft. Worth, TX, USA, October 2002, pp. 3–12 (2002)
39. Garcia-Luna-Aceves, J.J.: Loop-free routing using diffusing computations. IEEE/ACM Transactions on Networking 1(1), 130–141 (1993)
40. Shaikh, A.A.: Efficient dynamic routing in wide-area networks. Ph.D. dissertation, University of Michigan, Ann Arbor, MI, USA (1999)
41. Barabási, A.-L., Albert, R.: Emergence of scaling in random networks. Science 286, 509–512 (1999)
42. Luenberger, D.G.: Linear and Nonlinear Programming. Kluwer Academic Publishers, Dordrecht (2004), ISBN: 1-4020-7593-6

Overlay Networks and Graph Theoretic Concepts*

Is-Haka M. Mkwawa[1], Demetres D. Kouvatsos[1], and Adrian Popescu[2]

[1] NetPEn - Networks and Performance Engineering Research Unit,
Informatics Research Institute (IRI),
University of Bradford, Bradford, BD7 1DP, UK
{I.M.Mkwawa1,D.Kouvatsos}@bradford.ac.uk
[2] Dept. of Telecommunication Systems,
School of Engineering Blekinge Institute of Technology,
371 79 Karlskrona, Sweden,
apo@bth.se

Abstract. Overlay networks have shown to be very effective towards the support and enhancement of network performance and the availability of new applications and protocols without interfering with the design of the underlying networks. One of the most challenging open issues in overlay networks, however, is paths overlapping, where overlay paths may share the same physical link and thus, the ability of overlay networks to quickly recover from congestion and path failures is severely affected. This chapter undertakes a review of some graph theoretic based methods for the selection of a set of topologically diverse routers towards the provision of independent paths for better availability, performance and reliability in overlay networks. Moreover, it proposes a graph decomposition-based approach for the maximization of path diversity without degrading network performance of in terms of latency. Some remarks on future developments and challenges in the field of overlay networks are included.

Keywords: Overlay networks, internet, protocols, graph theory, network decomposition, network applications.

1 Introduction

An overlay network is a structured virtual network that is implemented on top of an actual physical network and it consists of nodes connected via logical links, each of which is associated with a path, that includes physical links of the underlay network [1]. The overlay network aims to support new network and application services without affecting the design of the underlying network (e.g., Internet). The applications of overlay networks range from routing overlay to security overlay.

---

* This work was supported in part by the EC NoE Euro-FGI (NoE 028022) and in part by the EC IST project VITAL (IST-034284 STREP).

Routing overlay aims to enhance or replace the existing Internet routing towards improved functionalities, services and network performance. Multicast overlays, one of the earliest mechanisms to employ routing overlays, have proven to be most suitable in deploying wide area multicast systems, such as the one used to broadcast the annual ACM SIGCOMM Conference (c.f., Chu et al [3]). Even though IP multicast has been in place for over 2 decades, it is still not widely used as an Internet service. On the other hand, security overlay, such as that adopted in Virtual Private Networks (VPNs), is based on techniques to control the power of large and distributed collections of servers and provide better network layer authentication and authorization, storage and lookup overlay (c.f., Stoica et al [2]).

In graph theoretic terms, an underlay network may be considered as a graph $G = (S, E)$, where S is a set of nodes (routers) and E is a set of links. A node $\eta_i \in S$, $i \in [1, N]$, where N is the cardinality of S, represents a router, and a link $(\eta_i, \eta_j) \in E$ denotes a bidirectional physical link of the underlying network G. Moreover, an overlay network may be viewed as an embedded tree $\tau = (s, D, S_o, E_o)$ on G, where s is the source node, $s \in S_0$, D is the set of destination nodes, $S_o \subseteq S$ is a set of nodes in G traversed by overlay links and E_o is the set of overlay paths. Moreover, an overlay path $e_o = (s, \eta_0 \ldots, \eta_l, d) \in E_o$ is made up of source node s followed by a sequence of routers $\eta_i \in S_o$ and ending by the destination node $d \in D$. The number of hops in the router sequence is given by l and hence the number of routers is $R = l+1$. Two overlay and underlay networks are depicted in Fig. 1, where all nodes are presented as white circles except those of the underlay nodes connected directly to the overlay nodes, which are displayed by black circles.

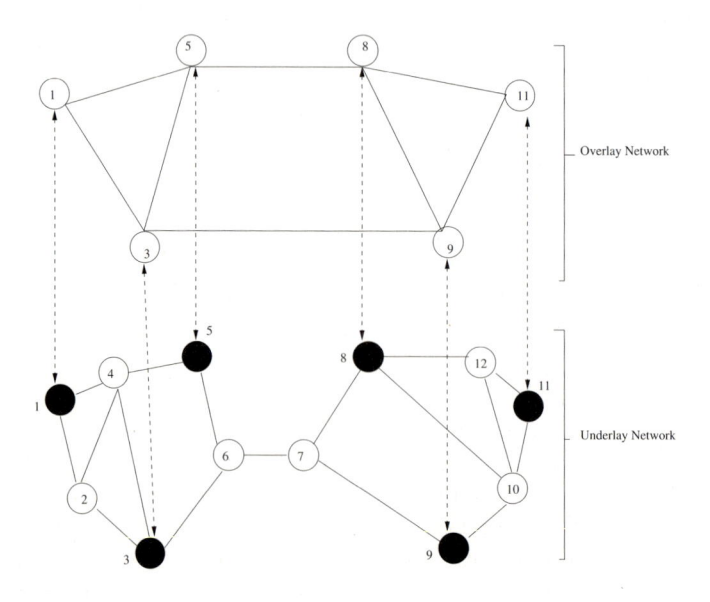

Fig. 1. Overlay and underlay networks

The robustness of the overlay networks depends on the diversity of the overlay paths between the source and destination nodes, usually expressed in terms of administration, linkage and geographical distribution. Apart from the benefits from and particular interests in overlay networks, a major challenge that has so far been largely overlooked is that of path diversity. As overlay paths may share the same physical link, the ability of overlay networks to quickly recover from congestion and path failures is severely affected. Thus, the underlying network topology should be taken properly into consideration during the design, development and deployment of overlay networks.

This tutorial has its roots in Mkwawa et al [4, 5] and addresses some graph theoretic based methods as applied to the optimal selection of a set of topologically diverse routers towards the provision of largely independent paths towards better availability and performance in overlay networks.

The rest of this tutorial is organized as follows: A review on resilient overlay networks is presented in Section 2. The state of the art of underlay aware overlay networks that try to assess the impact of overlay network on the performance of the underlying network is explored in Section 3. The effect of topology on overlay networks is discussed in Section 4. A decomposition criterion towards the improvement in the design and implementation of overlay networks and services is proposed in Section 5. Conclusions follow in Section 6.

2 Resilient Overlay Networks

As it has been reported by several authors in the literature, current Internet routing protocols are very slow to detect and recover failures in links and routers (c.f., [6,7,8,9,10,11,12,13,14,15,16]). Consequently, the performance degradation (i.e. path failures and network congestion) is visible to end users. This shows that, although the Internet routing infrastructure is highly redundant, nevertheless it fails to fully utilise alternative redundant paths. To rectify this problem, several researchers have proposed various types of resilient overlay networks (RONs) (e.g., Rexford et al [16] and Andersen et al [17]). A complete graph (i.e., each node of an overlay network is connected to each other) representing a RON can be seen in Fig. 2.

The main goal of a RON (c.f., Andersen et al [17]) is to enable a group of nodes to communicate with each other under a failure of the underlying paths connecting them. The detection is done by frequently sending probes among overlay nodes at a very short intervals.

The networks trade the overhead of short time probes (at a magnitude of $O(n^2)$), where n is the number of nodes of a RON, for prompting path outages (i.e., failures) or congestion and recovery. However, these networks have not completely avoided path outages and congestion. The research reported in [16, 18] showed that $40 - 50\%$ of the path outages were still unavoidable and the studies explained that all alternative paths through overlay nodes experienced the outages at the same time. This suggests that there is interdependency amongst path outages that is attributed to the fact that no appropriate consideration of

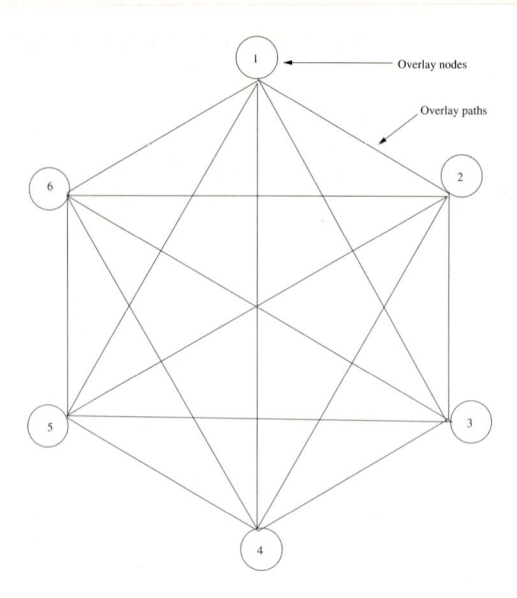

Fig. 2. A RON is deployed as a complete graph

the underlying network topology was taken into account prior to the selection of the overlay nodes and the design of the overlay network.

Thus, a thorough consideration of the underlying network structure can reduce the interdependency of path failures. There are several factors that can lead to path failures such as single point of failures if paths are administered by the same domain and overlay paths that share the same underlying physical links or routers.

3 Topology Aware Overlay Networks

The problem of path diversity that leads to unavoidable outages (c.f., Section 2) generated a lot of interest in the overlay networks community. As a consequence, Junghee et al [19] proposed a novel framework for topology aware overlay networks that enhances the availability and performance of end-to-end communication. The main goal was to maximize path independence towards better availability and performance of overlay services. To achieve the goal, Junghee et al [19] measured the diversity between different Internet Service Providers (ISPs) and also between different overlay nodes inside each ISP. Based on these measurements, a topology aware node placement heuristics were developed to ensure diversity of paths. This allowed the avoidance path failures, which were present in earlier works (c.f., [17]).

More specifically, a selected set of routers that were topologically diverse was selected in [19] by setting a threshold value α. If the correlation coefficient between two overlay nodes i and j, ρ_{ij} was higher than α, then it was judged that these two nodes have similar patterns of path diversity. Note that ρ_{ij} is defined by

$$\rho_{ij} = \frac{E(ij) - E(i)E(j)}{\sigma(i)\sigma(j)} \qquad (1)$$

where the metrics $E(k)$ and $\sigma(k)$, $k = i, j$ are, respectively, the expectation and standard deviation associated with an overlay node k. These metrics can be defined by (c.f., [19])

$$E(i) = \frac{1}{N} \sum_{s \in S, d \in D} I_{s,d} \qquad (2)$$

$$\sigma(i) = \frac{1}{N} \sum_{s \in S, d \in D} (I_{s,d} - E(i))^2 \qquad (3)$$

where N is the number of nodes of the underlying network, $I_{s,d}$ is the number of overlapping routers between an overlay path via overlay node i and underlay path from source node ($s \in S$) and destination node ($d \in S$). Moreover, $E(ij)$ is the joint expectation value related with the overlay nodes i and j given by

$$E(ij) = \frac{1}{N} \sum_{s \in S, d \in D} I_{s,d} J_{s,d} \qquad (4)$$

After identifying the overlay nodes with similar pattern of path diversity, the next task is to cluster them. The main goal at this stage is to maximize path diversity by clustering together the nodes sharing the same routers without degrading the performance of the overlay network. Then a node is selected to be an overlay node at each cluster. The selection criterion here is to minimize network performance degradation in terms of latency (how much time it takes for a packet to get from one designated point to another) and, hence, select a node that has minimum latency to be an overlay node. To this end, path diversity has been maximized without degrading the performance (i.e., lowest latency) of the overlay network.

Junghee et al [19] showed that about 87% of path outages were avoided by employing their proposed topology aware overlay network, which was applied in the evaluation of real-world networks. Nevertheless, further problems (such as path outages) about the credibility of this approach still remain open after maximising path diversity and obtaining lowest latency . For instance, are there any other topologies, rather than those based on RON, that can provide better network performance?

To answer this question, the impact of underlying network topology is considered in the next section.

4 The Impact of Underlay Topology on Overlay Networks

One of the problems of a complete graph is the lack of scalability, therefore RON is not scalable due to its topological (complete graph) nature and frequency of short interval probing [20]. An overlay topology has significant impact on the overlay services such as, routing in terms performance and associated overheads. Moreover, overlay networks require more header packets additional and processing.

The characteristics of overlay networks are based on the degree of each node that determines

- The overlay network neighbour information that each overlay node needs to maintain;
- The maximal number of disjoint paths M that the underlying network could provide for each pair of overlay nodes;
- The resilience, which defines the average number of underlying network disjoint paths that the overlay network can provide for each pair of overlay nodes;
- The distortion, which is defined as the overlay path length penalty (OPLP) in the case two nodes that cannot be directly connected via the underlying network path (c.f., [21, 22]).

The following sections review some topologies proposed in [21] that may be suitable to be adopted as overlay network service topologies.

4.1 The Full Mesh Topology

In the full mesh (FM) topology, the overlay network is a complete graph (c.f., Fig. 3). Since overlay nodes have no control of the underlay links, it is difficult for these nodes to obtain underlay link state information. Therefore, in order to evaluate the performance of the overlay links (expressed by metrics such as latency and bandwidth), the overlay nodes have to continuously send probing packets to neighboring overlay nodes. These probes cause significant performance degradation not only to the overlay links but also to the underlay network. As it can be seen in Fig. 3, many overlay links use the same physical underlay link e.g., both links e_4 and e_5 use the physical underlay link (η_8, η_9). Due to the nature of the full mesh topology, it was demonstrated in Andersen et al [17] that for an overlay network made up of 50 nodes, each node will have at least 33Kbps routing overhead.

4.2 The K-Minimum Spanning Tree Topology

Let $G' = (S', E')$ be a graph representing an overlay network, where S' is the set of nodes and E' is the set of links. Moreover, consider all the spanning trees of G' (i.e., subgraphs of G', which are trees connecting together all the nodes of S'), each of which has an overlay link cost that is defined by the number of physical underlay hops the overlay link passes through. A minimum spanning tree of graph G' can be formed such that all nodes in S' are connected via a subset of the links in E', whose total cost is minimal.

The K-Minimum spanning tree (KMST) topology for an overlay network is based on an integer number $K(K > 0)$ of minimum spanning trees connecting all

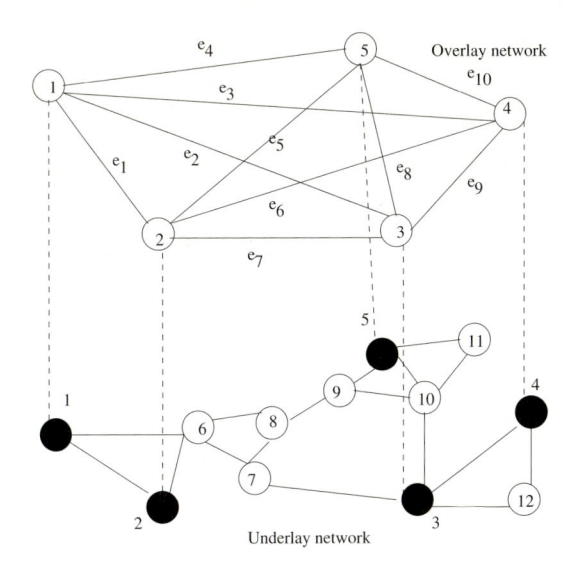

Fig. 3. An overlay network with a Full Mesh (FM) topology

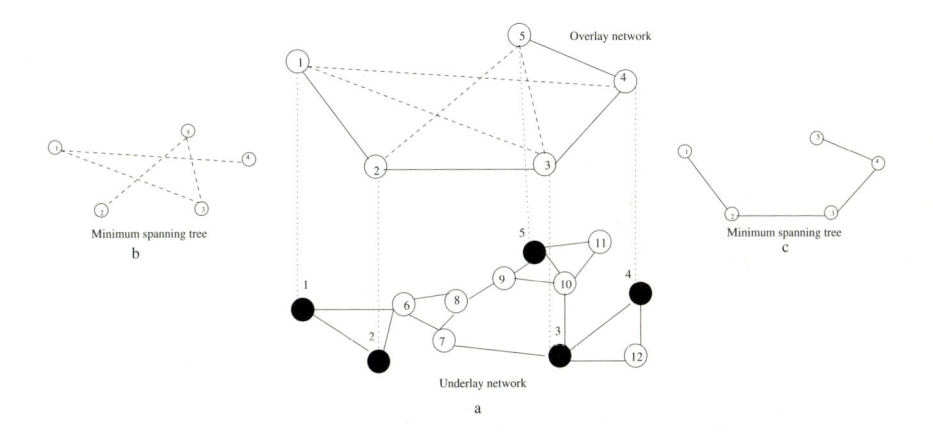

Fig. 4. An overlay network with a 2-minimum spanning tree (2-MST) topology

the overlay network nodes with the lowest cost amongst all the candidate trees. To minimize the state maintenance overhead, Li and Mohapatra [21] proposed an overlay network topology consisting of K minimal disjoint minimum spanning trees in an FM overlay network topology i.e., the K trees have a minimal overlapping of overlay links. However, this cost can take different values depending on the size of K and the trade-off between cost and performance. Fig. 4a shows an overlay network, which is formed by a 2-Minimum spanning tree topology, whose dashed and solid lines belong to two disjoint minimum spanning trees (c.f., Fig. 4b and Fig. 4c).

4.3 The Mesh Tree Topology

The Mesh Tree (MT) overlay network topology (i.e., messages sent on a mesh tree network can take any of several possible paths from source to destination) was suggested by Li and Mohapatra [21] in order to enhance the resilience of overlay multicasting. The MT topology is constructed by first setting up a minimum spanning tree connecting all the overlay nodes of the network. In this context, if two overlay nodes have grandchild-grandparent or uncle-nephew relationship in the minimum spanning tree, there will also be an overlay link connecting these two overlay nodes. Fig. 5a depicts an example of an MT topology. Solid lines constitute a minimum spanning tree (c.f., Fig. 5c) and the dash lines represent mesh links forming a 'mesh' linked-based minimum spanning tree (c.f., Fig. 5b).

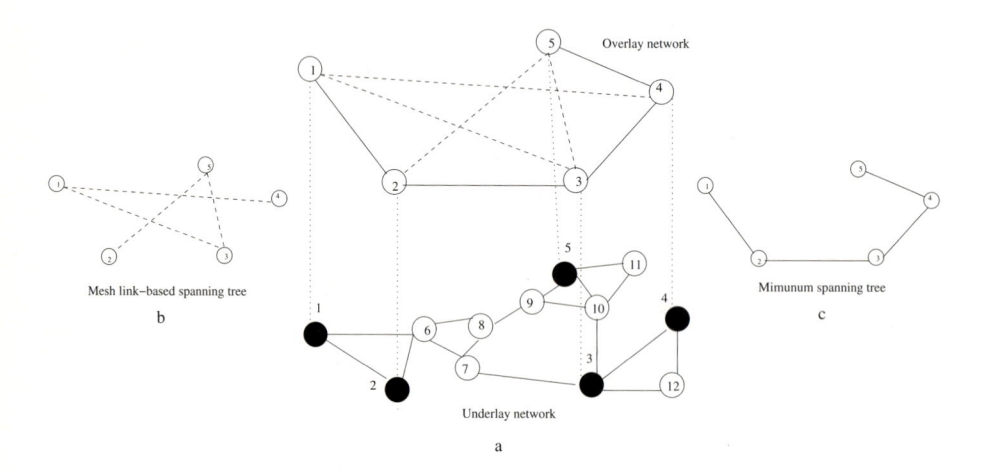

Fig. 5. An overlay network with a Mesh Tree (MT) topology

4.4 The Topology Aware K-Minimum Spanning Tree

The construction of Topology Aware K-Minimum Spanning Tree (TA-KMST) (c.f., [21]) is constrained by the underlying topology and it is illustrated in Fig. 6. This topology is similar to the KMST described in Section 4.2, but the disjoint property of two overlay paths has to be considered in both the overlay and the underlying layers. Therefore, if two overlay paths pass through the same underlaying link, then they are deemed as overlapping paths. For example, in Fig. 6a the overlay links e_2 and e_5 are overlapping as they pass through physical links (η_8, η_9). Thus, the resulting TA-KMSTs is composed of the least number of overlapping paths. The TA-KMST will, therefore, provide each source and destination node with diverse disjoint overlay paths. Fig. 6a shows TA-KMST in which dash and solid lines illustrate two least disjoint minimum spanning trees (c.f., Fig. 6b and Fig. 6c).

4.5 The Adjacent Connection Topology

The adjacent connection (AC) topology makes use of the knowledge gained from the underlay network in order to construct the overlay network. In this method, Li and Mohapatra [21] assumed that the underlying network information connecting all the overlay nodes is known before hand. As the Internet normally uses the shortest path routing strategy [23], constructing the overlay network could be done by the following rule. If there is no other overlay node that is connected to a node of the underlying network's shortest path between a pair of overlay nodes (c.f., Fig. 7), then there must be an overlay link between these two overlay nodes (c.f., [21]).

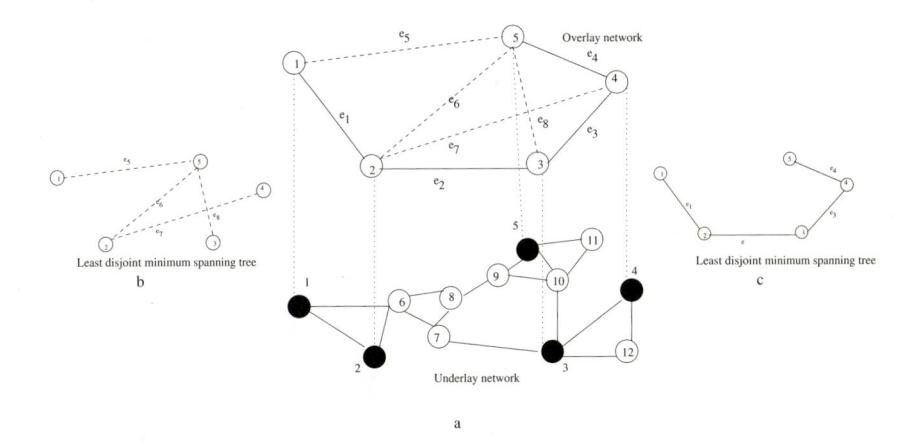

Fig. 6. An overlay network with a Topology Aware K-Minimum Spanning Tree (TA-KMST)

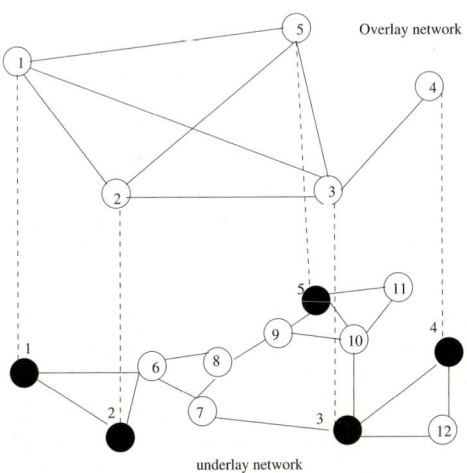

Fig. 7. An overlay network with an Adjacent Connection (AC) topology

4.6 Comparisons of Overlay Topologies

A simulation study focusing on the comparative impact of overlay topologies on the performance services such as messages routing was carried out by Li and Mohapatra [21], based on the estimation of the OPLP and it is displayed in Table 1. Note that the maximum degree of a graph $G'(S', E')$ is the maximum degree of its vertices denoted by $\Delta G'(S', E')$. Moreover, 'Resilience' and 'Distortion' are defined in the beginning of Section 4, N' is the number of nodes of the overlay network, K is the number of the minimal disjoint minimum spanning trees of a graph $G'(S', E')$ and M is the maximal number of disjoint underlay paths for each pair of overlay nodes.

As it can be observed in Table 1, a routing protocol's performance significantly varies on different overlay network topologies.

Table 1. Comparisons of overlay network topologies [21]

	$\Delta G'(S', E')$	Number of Links	Resilience	Distortion
FM	N'-1	n(N'-1)/2	M	Low
KMST	K	$\leq K(N' - 1)$	$< min(M, K)$	High
MT	> 2 and $< N' - 1$	$< 3(N' - 1)$	> 1	High
TA-KMST	K	$\leq K(N' - 1)$	$< min(M, K)$	High
AC	$< N' - 1$	$< N'(N' - 1)/2$	M	Low

5 A Network Decomposition-Based Topology

In Sections 3 and 4, separate considerations as given to the concepts of path diversity and latency. The assessment of the combined impact of latency and path diversity is a difficult optimisation problem as a router might give good path diversity but poor latency and vice versa and which, consequently, will degrade the performance.

In this section, an information theoretic decomposition approach is presented aiming to answer some of the relevant questions such as 'how many routers are needed to gain optimal/suboptimal path diversity?' and 'which routers within the same ISP should be chosen?' (c.f., [4, 5]). More specifically, a network decomposition criterion, which was originally devised by Kouvatsos [24, 25] in the context of the optimal hierarchical design of complex systems represented as undirected graphs $G(S, E)$, can be adopted to facilitate both the maximisation of path diversity and minimisation of latency before selecting any overlay service networks. This combined optimisation process may enhance the performance of overlay service in terms of failure recovery ratio, routing overhead and OPLP.

The decomposition criterion aims to ensure path diversity within a single ISP and between different ISPs. Even though the assurance of paths diversity is guaranteed, nevertheless further analysis and assessment of the performance of each node is needed in order to evaluate the latency overhead. Note that a realistic model proposed by Lastovetsky et al [26] for the design of an Ethernet switched network, can be used to obtain the latency overhead for each source and destination pair. More specifically, the model allows the representation of heterogeneity aspects for both links and nodes and it provides a more intuitive and accurate expression of the execution time of the message passing interface collective operations. To this end, to select a set of routers with the best latency, the decomposition criterion in [24, 25] is also employed focusing on the latency overhead instead of path diversity. For each sub-network (or, cluster), it is sufficient to select one router at random as an overlay node.

Let $N(N > 0)$ be the number of nodes of the underlay network. A theoretical framework may be established by associating dichotomous random variables $\{z_i : i = 1, 2, ..., N\}$ with each underlay node d_i ($i = 1, 2, ..., N$), such that each z_i cuts the domain of all different design solutions (i.e., forms) for the underlay network, say D, into two sets such that z_i takes the value 1 with probability p_i, for the set of forms that the node i fits and the value 0 with probability $q_i = 1 - p_i$ for the set of forms that it doesn't fit (c.f., Alexander [27]). Similarly, all the joint probabilities associated with the subsets of the nodes may be defined. For example, the probability that both variables z_i and z_j fit a domain is given by $P(z_i = 0, z_j = 0)$ (c.f., [24]).

The overall aim of the decomposition process is to partition the set of nodes of the underlay network, $S = \{\eta_1, \eta_2, \ldots, \eta_N\}$ into a number of subnetworks $\{S_1, S_2, \ldots, S_\mu\}, \mu < N$ such that the information transfer between the subnetworks is a minimum i.e., the diverse paths amongst nodes within the subsystems is strong whilst the common paths amongst nodes between subsystems is weak. In this context, the amount of information carried out by the underlay network is given by the entropy function

$$H(S) = -\sum_{\sigma} P(\sigma) log P(\sigma), \quad (5)$$

proposed by Shannon [28], where σ is the joint state $\{s_1, s_2, ..., s_m\}$, s_i being the binary value taken by random variable z_i, $i = 1, ..., m$ and $P(\sigma)$ is the joint state probability.

To this end, let π be an arbitrary partition of S into non-empty subsets $S_1, S_2, \ldots, S_\mu$, such that $S_i \cap S_j = \emptyset$ and $\bigcup_{i=1}^{\mu} S_i = S$. This partition belongs to a certain partition-type Π, which is the set of all partitions, say $\{\pi_1, \pi_2, \ldots, \pi_\mu\}$, $1 < \mu \leq N$ each of which has the same cardinality in each subset $S_1, S_2, \ldots, S_\mu$. The measure of the information transfer (i.e., information flow) between the subnetworks formed when the underlay network follows the Gibbs Theorem (c.f., Watanabe [29]) and it can be given by

$$C^\pi = \{\sum_{\pi \in \Pi} H(S_i) - H(S)\} \geq 0, \quad (6)$$

where $H(S)$ is the amount of information contained in the underlay network whilst $\sum_{\pi \in \Pi} H(S_i)$ is the total information contained in the subnetworks $\{S_i\}$ created by partition π.

By minimizing the redundancy (6), the optimum partition π within a specific partition-type $\Pi = \{\pi_1, \pi_2, \ldots, \pi_\mu\}$ can be determined via a network decomposition criterion (c.f., Kouvatsos [24]), namely

$$\min_\pi \left\{ \sum_{\pi \in \Pi} \rho_{ij}^2 \right\} \qquad (7)$$

where $\sum_{\pi \in \Pi} \rho_{ij}^2$ is the sum of the squares of the weights on the links cut by the partition π which expresses the degree of interconnection between d_i and d_j given by their correlation coefficient ρ_{ij} ($|\rho_{ij}| \leq 1$). Thus, in graph theoretic terms, the decomposition criterion expressed by redundancy (7) can be interpreted as the minimization of the strength of connections of the nodes of the underlay network given by squares of the weights of the links cut by the partition π.

In the context of overlay networks, two overlay nodes should be in the same subsystem when their path diversity patterns are similar to each other over a set of source and destination pairs. Thus, the decomposition criterion expressed by the redundancy (7) clearly implies the decomposition of the underlay network such that nodes with similar pattern are grouped together into subnetworks. The associated correlation coefficient can be defined (c.f., Kouvatsos [25]) by,

$$\rho_{ij} = \frac{E(z_i, z_j) - E(z_i)E(z_j)}{\sigma(z_i)\sigma(z_j)} \qquad (8)$$

where, (c.f., [19])

$$E(z_i) = \frac{1}{K} \sum_{d_s \in S, d_r \in D} L^i_{d_s, d_r} \qquad (9)$$

$$\sigma(z_i)^2 = \frac{1}{K} \sum_{d_s \in S, d_r \in D} (L^i_{d_s, d_r} - E(z_i))^2 \qquad (10)$$

$$E(z_i z_j) = \frac{1}{K} \sum_{d_s \in S, d_r \in D} L^i_{d_s, d_r} L^j_{d_s, d_r} \qquad (11)$$

Moreover, $E(z_i)$ and $E(z_i z_j)$ the marginal and joint expectations of random variables z_i and $z_i z_j$, respectively, $\sigma(z_i)^2$ is the squared of the standard deviation of z_i (i.e., variance) whilst $L^i_{d_s, d_r}$ is a set of source nodes and the number of overlapping routers between the overlay path through overlay node i and the direct link from the source node d_s to the destination node d_r.

Each partition-type belongs to a set $\boldsymbol{\Pi} = \{\Pi_1, \Pi_2, \ldots, \Pi_T\}$ whose cardinality T is determined by $(1/4m\sqrt{3}) \exp(3.14\sqrt{2m/3})$ [24]. Within each partition-type, say Π, a partition with the smallest (minimum) $\sum_{\pi \in \Pi} \rho_{ij}^2$ can be chosen to suggest a possible decomposition of the underlay network. However, as it stands, the redundancy (7) does not provide an unbiased reference for the comparison of the different partitions for each partition-type Π_i ($i = 1, 2, \ldots, T$), the value of $\sum_{\pi \in \Pi} \rho_{ij}^2$ may tend to be lower for some partition-types than others [27]. Therefore, the optimum partition of the network can only be achieved by taking into account the bias amongst asymmetrical partition-types and thus, eliminating their discrepancies with respect to the number of severed links. In this context, the overall optimum partition $\pi \in \boldsymbol{\Pi}$ can be determined by applying a normalisation on $\sum_{\pi \in \Pi} \rho_{ij}^2$ (c.f., [25]), namely,

$$\min_{\pi \in \Pi} \left\{ N(\pi) = \frac{\min_{\pi} \sum \rho_{ij}^2 - E(\sum_{\pi \in \Pi} \rho_{ij}^2)}{(Var(\sum_{\pi \in \Pi} \rho_{ij}^2))^{\frac{1}{2}}} \right\} \quad (12)$$

where, $E(\sum_{\pi \in \Pi} \rho_{ij}^2)$ and $Var(\sum_{\pi \in \Pi} \rho_{ij}^2)$ are the mean and variance of $\sum_{\pi \in \Pi} \rho_{ij}^2$, respectively.

As $N(\pi)$ has the same mean and variance for all partition-types, it may be used to evaluate the relative number of links severed by each partition and thus, meaningfully compare the optimal underlay network decomposition effect of partitions of all types with each other. Note that the normalized function $N(\pi)$ is a generalization to one devised by Alexander [27] in the field of Architecture, where $p_i = 0, 1$ and $|\rho_{ij}| = 1$. Moreover, in the context of the Specific EU Network of Excellence (NoE) Euro-NGI Research Project 'ROVER-NETs: Routing in Overlay Networks' [30], the information theoretic decomposition criterion (7) was initially proposed to maximize path diversity in overlay networks without degrading the network performance. The decomposition criterion (7) was also utilised into the context of the modular design and development process of an operating teaching system on a DEC PDP8/1 computer at UMIST, University of Manchester Institute of Science and Technology [31].

6 Conclusions and Future Work

This tutorial addressed the problem of paths overlapping on overlay networks and highlighted the concept and importance of topology awareness prior to designing and implementing overlay network services and applications. In this context, a review of graph theoretic based methods was undertaken towards the selection of a set of topological diverse routers and, thus, the provision of independent paths for better network availability, performance and reliability. In particular, a decomposition criterion for the optimal design of robust overlay networks was explored, based on the information measure of entropy function and graph theoretic concepts. It was argued that this criterion enhances further the overlay network services towards the maximization of path diversity without degrading overlay network performance in terms of latency.

Repeated applications of the normalised decomposition process for each subsystem $\{S_i, i = 1, 2, \ldots, \mu\}$ will identify a tree structure for the overlay network with one root and m leaves. The root component represents the entire design of the overlay network; the leaves are the individual nodes whilst the intermediate components (i.e., subsystems) are the subsets of nodes representing the most independent subproblems. Thus, the logical structure of the overlay network can have a tree-like representation, which can be constructed in a top-down manner. In this sense, the path diversity can be maximised whilst the latency can be minimised at an early stage of the design process before any irreversible decisions have been taken towards the selection of any overlay service networks. Moreover, any network failure can be rectified within a subnetwork, causing the least possible disturbance to the rest of the network. This kind of optimisation process may enhance the performance of overlay service in terms of failure recovery ratio, routing overhead and OPLP.

Overlay networks are becoming increasingly popular because of their ability to provide reliable and effective services. However, it is possible that it could be problematic to have routing controls in both the overlay and underlay network layers in the future as each of the two networks are unaware of what is happening to the other network. In particular, even though overlay networks generate heavy probing traffic amongst the associated nodes, this traffic is not known to underlying network and vice versa. These problems have to be resolved before the overlay networks have adverse implications on the overall Internet performance.

Experimental performance evaluation studies are required in order to explore further the utility of the network decomposition-based criterion towards the maximisation of path diversity without degrading overlay network performance in terms of latency.

References

1. Wikipedia. Overlay network. Website (2010),
   http://en.wikipedia.org/wiki/Overlay_network
2. Stoica, I., Morris, R., Liben-Nowell, D., Karger, D.R., Kaashoek, M.F., Dabek, F., Balakrishnan, H.: Chord: A Scalable Peer-to-Peer Lookup Protocol for Internet applications. IEEE/ACM Transactions on Networking 11(1), 17–32 (2003)
3. Chu, Y., Rao, G.S., Zhang, H.: A Case for End System Multicast (keynote address). In: SIGMETRICS 2000: Proceedings of the 2000 ACM SIGMETRICS International Conference on Measurement and Modeling of Computer Systems, pp. 1–12. ACM, New York (2000)
4. Mkwawa, I.M., Kouvatsos, D.D., Popescu, A.: On the Decomposition of Topology Aware Overlay Networks: An Information Theoretic Approach. Technical Note RS-OCT06-02, NetPEn Research Unit, University of Bradford, Bradford, UK (2006)
5. Mkwawa, I.M., Kouvatsos, D.D.: Overlay Networks and Graph Theoretic Concepts. In: Proc. of the 5th Int. Working Conference on the Performance Modelling and Evaluation of Heterogeneous Networks, pp. T02/1 – T0/30. Blekinge Int. of Technology, Karlskrona (2008)
6. Labovitz, C., Malan, G.R., Jahanian, F.: Internet Routing Instability. IEEE/ACM Transactions on Networking 6(5), 515–528 (1998)

7. Labovitz, C., Ahuja, A., Jahanian, F.: Experimental Study of Internet Stability and Wide-Area Backbone Failures. Technical Report CSE-TR-382-98, University of Michigan (1998)
8. Labovitz, C., Ahuja, A., Bose, A., Jahanian, F.: Delayed Internet Routing Convergence. In: SIGCOMM, pp. 175–187 (2000)
9. Paxson, V.: End-to-End Routing Behavior in the Internet. In: Proceedings of the ACM SIGCOMM Conference on Applications, Technologies, Architectures and Protocols for Computer Communications. ACM SIGCOMM Computer Communication Review, vol. 26(4), pp. 25–38. ACM Press, New York (1996)
10. Chandra, B., Dahlin, M., Gao, L., Nayate, A.: End-to-End WAN Service Availability. In: Proc. 3rd USITS, San Francisco, CA, pp. 97–108 (2001)
11. Mao, Z., Govindan, R., Varghese, G., Katz, R.: Route Flap Damping Exacerbates Internet Routing Convergence. Tech. Rep. UCB//CSD-02-1184, U.C. Berkeley (June 2002)
12. Basu, A., Riecke, J.G.: Stability Issues in OSPF Routing, 2001. In: Proc. ACM SIGMCOMM (August 2001)
13. Alaettinoglu, C., Jacobson, V., Yu, H.: Toward Millisecond IGP Convergence. In: NANOG 20, Washington D.C (October 2000)
14. Feldmann, A., Maennel, O., Mao, Z., Berger, A., Maggs, B.: Locating Internet Routing Instabilities. In: Proc. ACM SIGCOMM (2004)
15. Feamster, N., Andersen, D.G., Balakrishnan, H., Kaashoek, M.F.: Measuring the Effects of Internet Path Faults on Reactive Routing. In: Proc. of ACM SIGMETRICS 2003, San Diego, CA (June 2003)
16. Rexford, J., Wang, J., Xiao, Z., Zhang, Y.: BGP Routing Stability of Popular Destinations. In: ACM SIGCOMM IMW (Internet Measurement Workshop) 2002 (2002)
17. Andersen, D.G., Balakrishnan, H., Kaashoek, M.F., Morris, R.: Resilient Overlay Networks. In: Symposium on Operating Systems Principles, pp. 131–145 (2001)
18. Andersen, D., Snoeren, A., Balakrishnan, H.: Best-Path vs. Multi-Path Overlay Routing. In: Internet Measurement Conference, Miami, Florida (October 2003)
19. Han, J., Watson, D., Jahanian, F.: Topology Aware Overlay Networks. In: IEEE Infocom, Miami, FL (March 2005)
20. Keralapura, R., Chuah, C., Taft, N., Qiannacco, G.: Can Coexisting Overlays Inadvertently Step on Each Other? In: IEEE International Conference on Network Protocols, pp. 201–214 (2005)
21. Li, Z., Mohapatra, P.: The Impact of Topology on Overlay Routing Service. In: Proceedings of IEEE INFOCOM, 2004 (March 2004)
22. Li, Z., Mohapatra, P.: On Investigating Overlay Service Topologies. Computer Networks 51(1), 54–68 (2007)
23. Hong, X., Xu, K., Gerla, M.: Scalable Routing Protocols for Mobile Ad Hoc Networks. IEEE Network 16(4), 11–21 (2002)
24. Kouvatsos, D.D.: Decomposition Criteria for the Design of Complex Systems. Int. J. of Systems Science 7(10), 1081–1088 (1976)
25. Kouvatsos, D.D.: Mathematical Methods for Modular Design of Complex Systems. Computers and Operations Research 4(1), 55–63 (1977)
26. Lastovetsky, A., Mkwawa, I., O'Flynn, M.: An Accurate Communication Model of a Heterogeneous Cluster Based on a Switch-Enabled Ethernet Network. In: Parallel and Distributed Systems, ICPADS 2006, pp. 15–20 (2006)

27. Alexander, C.: Notes on the Synthesis of the Form. Harvard University Press, Cambridge (1967)
28. Shannon, C.E.: A Mathematical Theory of Communication. The Bell System Technical Journal 27(1), 379–623 (1948)
29. Watanabe, S.: Information Theoretical Analysis of Multivariate. The Bell System Technical Journal 27(1), 379–623 (1948)
30. Popescu, A., Kouvatsos, D.D., Remondo, D., Georgano, S.: Routing in Overlay Networks, Final Report JRA.S.26-ROVER, Specific NoE Euro-NGI Research Project ROVER-NETs (February 2007)
31. Murphy, J.W.: Towards the Modular Design of Computer Systems. Technical Report, UMIST, Manchester (1973)

Author Index

Akhlaq, Monis 988
Al-Dhelaan, Abdullah 988
Alserhani, Faeiz 988
Anisimov, Vladimir V. 258
Arlos, Patrik 14
Assi, Salam A. 357
Atmaca, Tülin 808
Avi-Itzhak, Benjamin 284
Awan, Irfan 988

Balsamo, Simonetta 233
Becker, Monique 477
Belzarena, Pablo 891
Beylot, André-Luc 477
Bradley, Jeremy 331
Brandstätter, Wolfgang 594
Bruneel, Herwig 203, 921

Casares-Giner, Vicente 716
Castel-Taleb, Hind 835
Chaitou, Mohamed 835
Chakka, Ram 612, 642
Chen, Yang 191
Constantinescu, Doru 1
Cullen, Andrea J. 988
Czachórski, Tadeusz 447

de Vogeleer, Karel 784, 795
De Vuyst, Stijn 921
Dhaou, Riadh 477
Do, Tien Van 612, 642

Erman, David 784, 795
Escalle-García, Pablo 716

Ferreira, Fátima 393
Fiedler, Markus 37, 784, 795
Fiems, Dieter 203
Fretwell, Rod J. 141

Gauthier, Vincent 477
Geier, Alfons 594
Giordano, Stefano 979

Harrison, Peter G. 343
Hébuterne, Gerard 835
Horak, Gerhard 594

Ilie, Dragos 795, 1017
Iovanna, Paola 125
Isaksson, Lennart 37

Kavadias, Christoforos 594
Kouvatsos, Demetres D. 141, 357, 594, 665, 682, 767, 859, 979, 1039
Krieger, Udo R. 70, 548

Larijani, Hadi 174
Levy, Hanoch 284
Liang, Lei 191
Lindberg, Peter 37

Markovich, Natalia M. 70
Martin, Jim 682
Mellor, John 988
Min, Geyong 665
Mitrani, Isi 423
Mkwawa, Is-Haka M. 594, 767, 1039
Moser, Scott 682
Mouchos, Harry 859

Naldi, Maurizio 125
Nguyen, Viet-Hung 808
Nogueira, António 98

Pacheco, António 98, 393
Pagano, Michele 301, 571
Pekergin, Ferhan 447
Pla, Vicent 716
Popescu, Adrian 1, 784, 795, 979, 1017, 1039
Popescu, Alexandru 784, 795

Raz, David 284
Remondo, David 746, 979

Sabella, Roberto   125
Salvador, Paulo   98
Secchi, Raffaello   571
Shah, Neelkamal P.   682
Simon, María   891
Sun, Zhili   191, 951

Thomas, Nigel   331, 343
Tiado, M. Issoufou   477
Tsokanos, Athanasios   859

Valadas, Rui   98
Villén-Altamirano, José   509
Villén-Altamirano, Manuel   509

Walraevens, Joris   203
Wang, Lan   665
Wittevrongel, Sabine   921

Zema, Cristiano   125
Zuo, Xiangxiang   665

Printing: AZ Druck und Datentechnik GmbH, Berlin
Binding: Stein+Lehmann, Berlin